The Genera of Fishes

and

A Classification of Fishes

The Genera of Fishes
and
A Classification of Fishes

DAVID STARR JORDAN

Reprinted with a new Foreword by
George S. Myers, Stanford University,
and the Comprehensive Index by
Hugh M. Smith and Leonard P. Schultz,
U.S. National Museum

STANFORD UNIVERSITY PRESS

STANFORD, CALIFORNIA 1963

597
√ 82g

The five parts reprinted herein were originally published in the Stanford University Publications, University Series, as follows: No. 27, *Genera,* pt. I, 1917; No. 36, *Genera,* pt. II, 1919; No. 39, *Genera,* pt. III, 1919; No. 43, *Genera,* pt. IV, 1920; Biological Sciences, Vol. III, No. 2, *Classification,* 1923. (For full citations, see pp. x and xiii of the Foreword.)

Stanford University Press
Stanford, California

© 1963 by the Board of Trustees of the
Leland Stanford Junior University

Library of Congress Catalog Card No. 63-22441

Printed in the United States of America

Contents

Foreword

Jordan's *Genera of Fishes* (four parts, 1917–1920), together with its taxonomic index, *A Classification of Fishes* (1923), form an exceedingly important contribution to the stability of zoological nomenclature. Subsequent to Albert Günther's *Catalogue of the Fishes in the British Museum* (London, 1859–1870), they are the only attempt to bring together references to all generic names applied to fishes and to apportion them to their respective taxonomic groups. Even more important, Jordan attempted to stabilize these generic names and their application in accordance with the *International Rules of Zoological Nomenclature,* first adopted in preliminary form at the Second International Congress of Zoology (Moscow, 1892) and, after emendation, published in a definitive form in Paris in 1905.[1] Now, forty years after the last of the five parts appeared and despite all imperfections, they still stand among a select group of library adjuncts indispensable to any research involving the zoological nomenclature and bibliography of fishes.[2]

Like almost any extensive list or catalog, Jordan's *Genera* and *Classification* contain many errors and omissions; large numbers of new generic names have been proposed and many changes in fish classification and nomenclature have been made during the forty years of later research in fish taxonomy. Almost every working taxonomic ichthyologist who has used these works has annotated them with marginal additions and corrections. It is unfortunate that a completely revised, up-to-date edition of Jordan's work, embodying the corrections of a group of ichthyologists, could not be issued. However, no one has been found with both the ability and the time to carry

---

[1] *Règles internationales de la Nomenclature zoologique adoptées par les Congrès Internationaux de Zoologie,* Paris, 1905: 64 pp. (Including equivalent texts in French, English and German.)

[2] The others are: Günther's *Catalogue,* already mentioned; the *International Rules,* already mentioned (now *Code,* latest revised edition, London, November 1961); the *Catalogue of the Library of the British Museum (Natural History)* (8 vols., London, 1903–1940); the fish sections of the annual volumes of the *Zoological Record* (London, 1865–present); Bashford Dean's *Bibliography of Fishes* (3 vols., New York, 1916–1923); C. D. Sherborn's *Index Animalium* (listing all zoological names from 1758 to 1800, 1 vol., Cambridge, 1902, and from 1800 to 1850, 33 parts, British Museum, London, 1922–1933); and S. A. Neave's *Nomenclator Zoologicus* (5 vols., London, 1939–1950).

through such an extensive revision. Nevertheless, the demand for Jordan's
Genera and *Classification* has continued, although both works have been out
of print and unavailable for the last 25 to 30 years. Under these conditions,
a facsimile reprint of the original five parts seems to be the best that can be
done, especially if it is prefaced by an evaluation of the work, some notes on
its use, and a warning against its abuse.

HISTORY

When Senator and Mrs. Leland Stanford founded Leland Stanford Junior University as a memorial to their son, they chose as its first president
David Starr Jordan, then President of Indiana University. Jordan was a
man of high attainment in biological research, who had also earned an endur-
ing reputation as an educator, having broken the rigid classical curriculum
and having introduced the major-subject system in American higher edu-
cation.

The new university opened in 1891, with a faculty selected by Jordan
principally from Indiana University and from Jordan's own *alma mater*,
Cornell University. To head the Department of Zoology, Jordan appointed
his Indiana colleague and ichthyological collaborator, Charles Henry Gilbert, while the Indiana physiologist Oliver Peebles Jenkins, who had also
worked on the taxonomy of Hawaiian fishes, became Professor of Physiology.
With Jordan, Gilbert, and Jenkins at Stanford, it was certain that Stanford
would become a center of ichthyological research.

Jordan had received his first introduction to ichthyology while a pupil
working briefly under Louis Agassiz at Penikese Island in 1873, and he soon
became a major figure in the ichthyological world. A contemporary, colleague, and friend of the older ichthyologists in Washington (Spencer Fullerton Baird, Theodore Nicholas Gill, George Brown Goode, and Tarleton
Hoffman Bean), Jordan became the teacher of the great "Indiana School"
of ichthyologists (Charles Henry Gilbert, Herbert Edson Copeland, Carl H.
Eigenmann, William Alembert Brayton, Barton Warren Evermann, Seth
Eugene Meek, Oliver Peebles Jenkins, and many others). At Stanford he
continued to attract and inspire new pupils in the field of ichthyology. Among
them were John Otterbein Snyder, Edwin Chapin Starks, Henry Weed
Fowler, Alvin Seale, Albert W. C. T. Herre, William Francis Thompson,
and Carl Leavitt Hubbs.

Few major areas of the world have been untouched by the ichthyological
researches of Jordan and his pupils. Moreover, almost all North American
conservational fishery biology stems from the pioneer work of Jordan and
a group of other North American zoologists on the Bering Sea fur seals,
Gilbert's methodology in studying the life history of the sockeye salmon of
the Fraser River, and W. F. Thompson's early work on the California sar-
dine. Almost no North American taxonomic ichthyologist or fishery biologist
of the present day is not in some way a lineal intellectual descendant of David

Starr Jordan at Indiana University or Stanford University, through the largest group of ichthyological disciples ever known.[3] In 1913, Jordan retired from ichthyological research and from the presidency of Stanford University to devote his entire energy to the furthering of international peace, a movement in which he had become a recognized leader. However, the advent of World War I cut short this new career, and Jordan sadly returned to a quiet life lightened of the active duties and ichthyological projects that had previously filled his days. It was at this time that he took up serious preparation of the *Genera of Fishes*.

Jordan had been forced to deal with innumerable nomenclatural problems during preparation of the monumental *Fishes of North and Middle America*[4] by Jordan and Evermann, and he was well aware of the confusion in taxonomic and other biological research caused by the absence of succinct and universally accepted regulations for the application of generic and specific names. In those earlier days, British zoologists usually paid at least lip service to the so-called "Stricklandian Code."[5] Continental zoologists often followed no rules at all, and after 1886, American zoologists tended to follow the American Ornithologists' Union *Code*.[6] When the *International Rules* had finally been revised sufficiently to embody much of the clarity and strength of the A.O.U. *Code*, Jordan and other American zoologists accepted them. For many years Jordan served as a member of the International Commission on Zoological Nomenclature, a judicial body set up by the International Congress of Zoology to interpret and clarify the *Rules*. Jordan was therefore well qualified to undertake the considerable task of attempting to bring the nomenclature of fish genera, as a whole, into agreement with the *Rules*.

As a general model, Jordan had the *Index Generum Mammalium*[7] of Theodore Sherman Palmer, a meticulously prepared alphabetical list of the generic names applied to living and fossil mammals, together with subfamilial and familial names, a taxonomic index to the whole, and an erudite and ex-

---

[3] Those interested in Jordan's ichthyological history may refer to my own brief summary of 1951, "David Starr Jordan, Ichthyologist, 1851–1931" (Stanford Ichthyological Bulletin, vol. 4, no. 1, pp. 2–6) or to Jordan's own extensive autobiography, *The Days of a Man* (2 vols., Yonkers, New York, 1922). Jordan's exceedingly broad interests are especially well illustrated in his complete bibliography: David Starr Jordan: A bibliography of his writings, 1871–1931, compiled by Alice N. Hays (Stanford University Publications, University Series, Library Studies, vol. I: i–xiii + 195 pp., 1952).

[4] Bulletin of the U.S. National Museum, no. 47, 4 vols., 1896–1900.

[5] Formulated by a group of British zoologists led by the ornithologist Hugh Edwin Strickland and including Charles Darwin, Leonard Jenyns, John Richardson, Richard Owen, William Yarrell, J. O. Westwood, G. R. Waterhouse, and others, with the advice of such continental zoologists as Louis Agassiz and Prince Lucien Bonaparte. The Stricklandian Code was adopted by the British Association for the Advancement of Science in 1842 (see *Report of the British Association*, vol. 11, 1842).

[6] *Code of Nomenclature adopted by the American Ornithologists' Union*, New York, 1886. (Reissued with an index, 1892.)

[7] U.S. Department of Agriculture, Division of Biological Survey, North American Fauna, no. 23, 984 pp., Washington, 1904.

ceptionally informative general introduction. However, Jordan had to deal with more names than Palmer's work included, a somewhat more scattered literature, and a much larger and less well-known group of animals. Moreover, the library facilities in California, and especially at Stanford, were greatly inferior to those in Washington. Finally, although Jordan, like Palmer and Gill, possessed a vast memory for zoological names and references, he had never had either inclination or time for the meticulous type of editorial and bibliographical work characteristic of Palmer's volume. Jordan was well aware of these matters and requested the aid of his old editorial collaborator, B. W. Evermann, in preparing the first part of the *Genera*.

Unlike Palmer, Jordan omitted such subjects as the etymology of each name, the principles of Latinized name formation, sources for finding exact publication dates, and a list of the names of subfamilies, families, and higher categories used for the group being treated. As a result, Palmer's work, especially its first 67 pages, were and still are of interest and use even to a non-mammalian taxonomist.

These omissions in Jordan's work emphasize that the main purposes of the *Genera* were subtly but clearly different from those of the *Index*. Palmer's *Index* appeared before the definitive Paris *Rules* were published, and the author abstained from designating type species himself when others had not done so. Jordan's *Genera* had one primary, immediate purpose: to bring the acceptance and application of the generic names of fishes into accordance with the Paris *Rules* by selecting a nomenclatural type species for all fish genera in the manner laid down by those *Rules*.

The *Genera of Fishes*

The four parts of the *Genera* were published in the paper-bound monograph series of the University. The complete citations are as follows:

JORDAN, DAVID STARR (assisted by Barton Warren Evermann). 1917. The genera of fishes, from Linnaeus to Cuvier, 1758–1833, seventy-five years, with the accepted type of each. A contribution to the stability of scientific nomenclature. Leland Stanford Junior University Publications, University Series: 161 pp. [Continuously paged in Arabic numerals throughout; published August 1917.]

JORDAN, DAVID STARR. 1919. The genera of fishes, part II, from Agassiz to Bleeker, 1833–1858, twenty-six years, with the accepted type of each. A contribution to the stability of scientific nomenclature. Leland Stanford Junior University Publications, University Series: pp. i–ix + 163–284 + i–xiii. [Introductory matter and index each separately paged in Roman numerals; published July 1919.]

JORDAN, DAVID STARR. 1919. The genera of fishes, part III, from Guenther to Gill, 1859–1880, twenty-two years, with the accepted type of each. A contribution to the stability of scientific nomenclature. Leland Stanford Junior University Publications, University Series: pp. 285–410 + i–xv. [Index separately paged in Roman numerals; published October 1919.]

JORDAN, DAVID STARR. 1920. The genera of fishes, part IV, from 1881 to 1920, thirty-nine years, with the accepted type of each. A contribution to the stability of scientific nomenclature. Leland Stanford Junior University Publications, University Series: pp. 411–576 + i–xviii. [Index separately paged in Roman numerals; published August 1920.]

The *Genera* as a whole may be cited much more briefly as follows:

JORDAN, DAVID STARR. 1917–1920. The genera of fishes. . . . In four parts: Stanford University, University Series: 576 pp. (+ total of 55 pp. variously paged in Roman numerals). [Part I, 1917, pp. 1–161; part II, 1919, pp. 163–284; part III, 1919, pp. 285–410; part IV, 1920, pp. 411–576.]

Each part is separately indexed, but the Arabic and Roman pagination of introductory material and indexes vary from part to part, as noted in the citations. There is no general index to the four parts, this having been left for the *Classification,* which is in large part a taxonomic index. The generic names are listed chronologically. Approximately 10,000 names are dealt with.

Nothing in the first part of the *Genera* indicates that other parts would follow, and it seems possible that Jordan at first intended to do no more. This is the only part in which assistance by Evermann is mentioned, and it differs somewhat from the succeeding parts in methodology. As Jordan informed me later, in 1928, Evermann was largely a literary and editorial helper in their collaborative ichthyological work, as indeed he had been in the production of Jordan's *magnum opus, The Fishes of North and Middle America.* There cannot be much doubt that Evermann's help in Part I of the *Genera* was largely editorial. Quite possibly, Evermann devised the format and arrangement of the work, which differ considerably from those of Palmer's *Index.*

The International Commission on Zoological Nomenclature was relatively inactive during and for several years after the end of World War I in 1918. The Commission's parent body, the International Congress of Zoology, which had been meeting at three-year intervals before the war, did not hold its first post-war meeting until 1927 (Budapest). It is evident that Jordan intended to place the *Genera* before the Commission for a decision on the type species of all controversial generic names, and he did distribute copies to the Commissioners. It is likewise evident that the Commission felt that each doubtful case should be taken up separately, with no blanket approval of Jordan's findings.

In hoping for such approval (*Genera,* I, p. 7) Jordan was guilty of the naïveté that characterized much nomenclatural procedure of the day. The Paris *Rules* had been published in 1905, but individual zoologists, each accustomed to handling problems of nomenclature in his own way, were slow to learn of the existence of the *Rules,* to become aware of the way they worked, or to adopt them when the *Rules* indicated that some familiar names needed changing. Nine years after the *Rules* were published, World War I began, and few had time for zoological nomenclature for five or six long years.

Indeed, not until 1930 or later did the majority of continental European and British zoologists begin to seriously follow the *Rules.*

Nor were the easiest ways of complying with the *Rules* always evident even to those presumably most familiar with them. In fact, the methods of applying the *Rules* were not generally well understood until the mid-1920's when a series of *Opinions,* written by the Commission's extremely able secretary, Charles Wardell Stiles, demonstrated a methodology which became standard. Moreover, the *Opinions* interpreted the *Rules,* as courts interpret the law, and the Commission's *Opinions* thus became a body of court law which clarified and supplemented the *Rules.* The earlier *Opinions* themselves show a developing sense of the true legal meaning of the *Rules,* but it is also evident that some of the Commissioners themselves were not too sure how to interpret certain sections. The *Genera* was published before a number of important points had been clarified, and many of Jordan's decisions would undoubtedly have been different if he had had the benefit of nomenclatural clarifications made during the decade of the 1920's.

Beginning with the second part, Jordan used certain terms to indicate the different methods by which a species name was designated or recognized as the nomenclatural type of a generic name. He defined the terms *orthotype, logotype, haplotype,* and *tautotype* on page 165 of Part II. Unfortunately, as Jordan used them, these terms were neither as mutually exclusive as they could have been, nor as exact as rigorous interpretation of the past and future *Rules* required. They were never generally adopted and are no longer used. Moreover, they were not always used by Jordan quite in accordance with his own definitions.[8] The reader of the *Genera* can rarely be sure into which of the modern categories of type designation a particular case falls solely from Jordan's usage of one of his four special terms.

Those who attempt to use Jordan's *Genera* of fish should be fully conversant with the present *International Code,*[9] especially Chapter XV, on generic types. Despite the fact that Chapter XV could bear some revision,[10] it is obvious from it that Jordan's system of annotation is now antiquated and that the user must himself determine the category of type designation for any individual generic name.

For the benefit of zoologists preparing formal taxonomic papers in which generic synonyms are listed, it seems useful to present an example of such a synonymy with the generic type species listed in the approved manner

[8] Compare Jordan's account of Basilewsky's genera (*Genera,* II, p. 262) with my own (Copeia, 1940, p. 201).

[9] *International Code of Zoological Nomenclature Adopted by the XV International Congress of Zoology.* International Commission on Zoological Nomenclature [care of British Museum (Natural History), London], 1961; xviii + 176 pp.

[10] E.g., in Article 68, the "use of typicus or typus," like the formula "gen. n., sp. n." is merely a method of original designation. Moreover, a generic name falling into Article 69(a) ii(i) may be a monotype, as stated, despite formal restriction of monotypy to the preceding Article (68) of the *Code.*

according to the modern *Code*. Most of the commoner methods of type recognition are included. The synonymy, zoological names, and references to the literature (the latter given only as dates) are all imaginary.

Genus OXUS Fitzinger, 1845

Crotaphus Cuvier, 1814, p. 198 (Type species: *Bulus virens* Schmidt, 1798, by monotypy; preoccupied by *Crotaphus* Wendt, 1801, an insect).

Crotaphora Rafinesque, 1816, p. 20 (Substitute name for *Crotaphus* Cuvier, 1814, and therefore taking the same type species: *Bulus virens* Schmidt, 1798; preoccupied by *Crotaphora* Higgins, 1806, a mollusk).

Crotapha Bellini, 1817, p. 201 (Emendation of [substitute for] *Crotaphus* Cuvier, 1814, and therefore taking the same type species: *Bulus virens* Schmidt, 1798; preoccupied by *Crotapha* Johnson, 1805, which is an emendation of *Crotaphus* Wendt, 1801, an insect).

Oxus Fitzinger, 1845, p. 59 (Type species: *Crotaphus niger* Cuvier, 1824, under suspension of the Code by the International Commission on Zoological Nomenclature, Opinion 835, 1960, p. 6).

Holus Peters, 1849, p. 167 (Description but with no species mentioned; type species *Holus freethi* Peters, 1851, by subsequent designation and by monotypy; see Peters, 1851, p. 18, where *Holus freethi* is the first referral of a described species to the genus *Holus*).

Labus Günther, 1865, p. 5 (Type species: *Labus bensonii* Günther, 1865 [equals *Crotaphus labus* Benson, 1841] by absolute tautonymy).

Hololabus Bleeker, 1867, p. 296 (Type species: *Hololabus typus* Bleeker, 1867, by original designation [use of specific name *typus*]).

Syrictes Cope, 1884, p. 171 (Type species: *Labus dyspelor* Cope, 1878, by original designation).

Protolabus Vaillant, 1891, p. 62 (Type species: *Labus albus* Vaillant, 1891, by subsequent designation of Nelson, 1916, p. 42).

THE *Classification*

In the *Genera,* generic names were listed chronologically, with no general index to all four parts, each having been separately indexed. Palmer's alphabetical *Index Generum Mammalium* had a systematic or taxonomic index, and Jordan evidently saw the need for such an index. Moreover, the time seemed ripe for a new outline of the classification of fishes and the *Classification* filled both needs, by listing the generic names under appropriate family, with a page reference to the *Genera* after each genus. Addenda and corrections to the *Genera* are listed in place, under the families, but the only alphabetical indexes in the *Classification* are solely to generic names omitted in the *Genera* (Index, pp. i–iii) and to families and higher groups mentioned in the *Classification* (Index, pp. iv–x).

The *Classification* may be cited as follows:

JORDAN, DAVID STARR. 1923. A classification of fishes including families and genera as far as known. Stanford University Publications, University Series, Biological Sciences, vol. III, no. 2: pp. 77–243 + i–x. [Published January 1923.]

The *Classification* is unique in that it is, to the present day, the only publication subsequent to Günther's *Catalogue* (1859–1870) in which the attempt was made to list and place in its proper family every generic name ever applied to fishes since the beginning of binomial zoological nomenclature in 1758. But as a new arrangement and assessment of the natural groups of fishes, the *Classification* was somewhat less successful.

Prior to World War II the classification of the major groups of fishes according to natural similarities and relationships was a research field in which few men were competent to work. First, taxonomic research has usually been regional, for purely practical reasons. Few men are expert in dealing with the fauna outside their own area or continent. In the past especially, great world-wide museum research collections, where one could study and compare the anatomy of fishes from all parts of the globe, existed only in a few metropolitan and university centers, and the number of such collections has not greatly increased during the past half-century. Moreover, it has been discovered that the kinds of fishes are so numerous, and their relationships so difficult to assess, that most ichthyologists have their hands full with less general problems than group relationships.

As a result, the major classification of fishes has been largely the work of a select group, most notably Francis Willoughby (1635–1672), Peter Artedi (1705–1735), Georges Cuvier (1769–1832), Louis Agassiz (1807–1873), Johannes Müller (1801–1858), Albert Günther (1830–1914, Edward Cope (1840–1896), Theodore Gill (1837–1914), George Boulenger (1858–1937), Tate Regan (1878–1943), and Leo Berg (1876–1950).

Willoughby first recognized some groupings still accepted today. Artedi carried the work much farther and codified existing knowledge of fishes. Cuvier was the first to place the classification of animals on a firm basis of comparative internal anatomy, and specialized on the fishes. Agassiz expanded the system to include fossil fishes. Johannes Müller first recognized the anatomical peculiarities of the more ancient living types, reviewed the sharks and rays, which had been neglected by Cuvier, and pointed out some major groupings of bony fishes. Günther reviewed and described all known living fishes of his day, and made many important improvements. Cope reviewed all the major groupings of bony fishes in a comprehensive study of their skeletons. Gill used Cope's results and carried on a long-continued study of the osteology and relationships of individual groups. Boulenger began a modern reclassification of bony fishes. Regan carried through a brilliant series of studies of the anatomy and classification of nearly every order of living fishes. Berg compiled what was known to form the latest general classification of fishes.

In contrast to the work of these men, Jordan's *Classification* contained no definitions and almost no discussion. Like Berg, Jordan had many non-ichthyological interests, and had neither time for extensive anatomical research nor the inclination to work with higher groups. Jordan's *Classification,* like Berg's later work (1940), was largely a reshuffling of the latest results of others, most notably Regan. In addition, from the beginning Jordan had left matters of the higher classification of fishes to his exceedingly acute older colleague, Gill, who did not die until three years before the publication of Part I of the *Genera.*

Another factor was Jordan's later tendency toward splitting genera and higher categories into finer and finer units. This tendency stemmed both from his departure from active ichthyological research in 1913, and from his inability to work personally with preserved fish specimens for a large part of his life. Many later ichthyologists have wondered why after the 1870's Jordan rarely published ichthyological papers except in collaboration with others. Some have even criticized some of his later work for occasional lack of care with details. Part of this was due to Jordan's extremely busy life, and the necessity of leaving scale and fin-ray counts to a collaborator. However, there was a still more compelling reason. Early in his career, in Indiana, Jordan developed what are today called allergies to both formaldehyde and alcohol and was forced to refrain from handling preserved fishes or breathing the fumes from them. The effect of this disability upon a man as devoted to fish study as Jordan can be appreciated only by a person who saw him trying to work on alcohol-preserved fishes with a collaborator! It can be said that Jordan personally made few or none of the counts and measurements in his collaborative papers after about 1885.

In any event, Jordan was seventy-two when the *Classification* appeared and his tendency to split groups is evident throughout the work. Some of the splitting was well founded. Most was not, and it has not been followed by others. In some ways, Berg's *Classification* of 1940 resembles Jordan's in the splitting up of groups, and he too has not been followed by others. But the character of Berg's book required that dismembered groups be redefined, and he was somewhat more circumspect than Jordan had been.

A Note on Use of the *Genera* and *Classification*

Enough has been said above to make it clear to users that the *Genera* and *Classification* have in part become obsolete, as do all printed works dealing with any scientific discipline. They are of use principally as general works of reference—as starting points for continued systematic research involving the vexatious but necessary problems of zoological nomenclature. However, in no other place can the user find so much information on the generic names of fishes. Only in the *Classification* are the majority of the generic names used for fishes placed in familial categories.[11]

[11] Even here there are a few errors, but I have corrected the familial position of a number of such names in the Index; they are marked [b. in **000**].

In many ichthyological laboratories, the *Classification* is used as a general index to enable relatively untrained assistants to place identified and labeled bottles of fish specimens in the proper categories among the stored research collections. Such usage has one disadvantage—the absence of any single general alphabetical index to generic names in both the *Genera* and the *Classification*. A manuscript alphabetical index to the *Classification* was compiled by the late Hugh McCormick Smith and Leonard Peter Schultz, and typewritten copies of it have long been in use at the United States National Museum. Through the great courtesy of Dr. Schultz and the Library of the Smithsonian Institution, the Smith-Schultz index has been made available to Stanford University Press for use in compiling the new alphabetical index appended to the present reprint.

As I have already mentioned, those who deal with the generic names of fishes must familiarize themselves with the new *International Code*. They must also be familiar with the *Official Lists* of accepted and rejected names and the *Opinions* of the International Commission on Zoological Nomenclature.

In this facsimile reprint, *the four separate indexes to the four parts of the* Genera *and the generic index (A) in the* Classification *have been omitted* but have been merged with the Smith-Schultz index to both the *Genera* and the *Classification*. The familial index (B) to the *Classification* has been retained, with only minor corrections. This solution of a publication difficulty therefore omits the ten Arabic-numbered pages of Part I (pages 153–162) that were devoted to the index for that part, but no other pages originally numbered with Arabic numerals have been omitted. No other pages in any of the four parts have been omitted, whether paged in Arabic or Roman.

GEORGE SPRAGUE MYERS

Stanford University
August 12, 1963

The Genera of Fishes

[PART I]

FROM LINNAEUS TO CUVIER, 1758–1833

SEVENTY-FIVE YEARS

WITH THE ACCEPTED TYPE OF EACH

The Genera of Fishes

CONTENTS

CONTENTS

INTRODUCTION

The leading purpose of the Commission * of the International Congress of Zoology is to give stability to nomenclature. To that end numerous cases of doubt have been resolved by the Commission. But the work done thus far has been largely piecemeal, and by this method it cannot be always made consistent with itself. Recently the Commission has undertaken to fix generic names on a larger scale, covering, for example, all those in use in a given group. In some small classes of animals this has been possible. In larger ones, it demands a study of the literature more detailed than any one has yet attempted.

Recently efforts have been made to secure stability by fiat, using names more or less current, without serious regard to the law of priority. To accept this plan would merely accentuate the confusion already existing and which has arisen through just such disregard of fundamental rules.

It seems to us that the attempts thus far have mostly begun at the wrong end. The need is not to confirm modern errors but to give nomenclature a solid basis for the long future. Stability must rest on a thorough study of the foundations of biological nomenclature, after which the Commission's authority can be used to confirm the results of such studies.

There is no middle ground between using the oldest eligible names in any given group and using whatever names we please. With the latter

---

* The International Commission of Zoological Nomenclature consists (May 1917) of the following persons:

Carl Apstein, Berlin;
F. A. Bather, London;
Joel A. Allen, New York
Raphael Blanchard, Paris, (President);
Ph. Dautzenberg, Paris;
Ernst Hartert, Tring, England;
Dr. Horvath, Budapest;
W. Evans Hoyle, Cardiff;
Dr. Handlirsch, Vienna;
Karl Jordan, Tring;

David Starr Jordan, Stanford University;
J. H. Kolbe, Berlin;
F. S. Monticelli, Naples;
Dr. Roule, Paris;
S. Simon, Paris;
Henry Skinner, Philadelphia;
Leonhard Stejneger, Washington;
Charles Wardell Stiles, Washington, (Secretary).

alternative, Systematic Zoology and Botany would come to a condition of hopeless despair. There can be no finality when the question of acceptance or rejection of names is left as a matter of personal preference.

Binomial nomenclature has its recognized beginning with the Tenth Edition of the *Systema Naturæ of* Linnæus, published in 1758 with the assumed date of January 1. These names of Linnæus constitute the original framework of Zoological Taxonomy. In the nomenclature of this early period, there are two main elements of doubt: the first relates to the eligibility of authors who have for one reason or another not accepted the Linnæan Code; next of authors whose works, published before the Linnæan Code, have been revised more or less and reprinted after the date of the Tenth Edition of the *Systema Naturæ*. In all of these the species are designated by a descriptive phrase, as was the custom before Linnæus began the practice of "scientific bookkeeping" in Systematic Zoology. The writings of Gronow, Schæfer, and Valmont de Bomare come under the first of these heads; those of Klein (*Gesellschaft Schauplatz*), Browne, Catesby, and Osbeck, under the other. Commerson and Plumier, whose manuscript names were published by an author who did not accept them, come under a third head. It is the judgment of the present writer, that the best interests of Ichthyology would have been served by adopting the rule followed by Jordan & Evermann (*Fishes of North and Middle America,* 1898). In this work all writers who use polynomial phrases for the designation of species are disregarded as factors in nomenclature, however regular their practice may be as to genera. It is not a question of justice to able naturalists who, like Gronow and Browne, failed to adopt the Linnæan Code, solely because they had never heard of it. It is the convenience of future naturalists which is now concerned. This would apparently be best served by the exclusion of all these.

The arguments against such exclusion are mainly two: Brisson in 1760, polynomial as to species, had a stronger grasp on the significance of genera than any other ornithologist of his time. He has been called "the Father of Ornithology." Most students of birds wish to retain Brisson's genera as foundation-stones in nomenclature. There are good reasons for accepting Brisson as an exception. Similar exceptions may be demanded in other groups. It is desirable, but not vitally necessary, that all accepted rules be general, without exception; but as a member of the International Commission of Nomenclature, the present senior author has made no objection to the recognition of Brisson. The other argument is this: these writers have published generic names which appear in systematic lists, like that of Sherborne (*Index Animalium*). The Commission has already approved the names of Gronow and Commerson. If the present Commission should decide to reject the whole series of "irregulars," some future Commission may reverse the decision, placing the element of priority above that of regularity. This possibility we cannot forestall, while if once accepted there would be no successful movement for their rejection. Previous decisions of the Commission, as to Gronow and Commerson, point in the direction of general acceptance.

It is especially important to have the status of questioned authors determined as soon as may be, not only for the convenience of ichthyologists but for workers in other fields which may be affected by questions of preoccupation.

More important than the question of acceptance or rejection of some or all of the questioned genera, is the securing of a final decision. This the Commission will be asked to make as soon as practicable. Pending this decision it is perhaps wise for systematic workers to refrain from acceptance of the names questioned.

The other problem is the assignment of generic types to the genera of authors who had no conception of types. In doing this we have followed as closely as may be the rules adopted by the Congress of Zoology, having especial regard to the "first reviser." In some cases we have been in doubt on account of conflicting usages or even rules. But in such cases the weight of authority of the Commission when exercised should serve to turn the scale. With the authors subsequent to Cuvier, 1829, this matter rarely offers any embarrassment. The later authors mostly look upon a genus, not as a pigeon-hole with arbitrary boundaries, but rather as a group of species, with certain definite structural marks clustered around some definite species, the type of the genus.

The writer asks from his colleagues the fullest criticism both as to matters of fact and of opinion, before placing the contents of this paper formally before the International Commission. We would especially request information concerning omissions. There are no doubt numerous generic names overlooked in dictionaries and in obscure publications.

We have arranged in chronological order the generic names of fishes published in the first seventy-five years of the history of ichthyological taxonomy. The determination of the validity of genera is treated only incidentally. Our main problem is the fixation of the type.

Stanford University, California, June 1, 1917.

THE GENERA OF FISHES

PART I

I. LINNÆUS, *Systema Naturæ*, Ed. X, Vol. I, 1758.

KARL LINNÉ (CAROLUS LINNÆUS).

The generic names of Linnæus represent, with one change and two or three additions, the system of Ichthyology as developed in 1738 by his friend and fellow-student, Peter Artedi, a naturalist whose knowledge of fishes was far greater than that of Linnæus. The types of most of the Linnæan genera have been accepted by common consent. As a rule we have not questioned the current application unless compelled to do so by the insistence of established rules.

Petromyzon Linnæus, 230, after Artedi; type PETROMYZON MARINUS L.
Unquestioned.

Raja Linnæus, 231, after Artedi; type RAJA BATIS L.
Spelled also RAIA, by authors.

Squalus Linnæus, 231, after Artedi; type SQUALUS ACANTHIAS L.
Restriction to S. ACANTHIAS and relatives, Rafinesque, *Indice d'Ittiologia Siciliana*, 1810, 45, and by Gill, *Proc. Acad. Nat. Sci. Phila.*, 1862, 497. Gray, *Cat. Chond.*, 1851, following Bonaparte, 1838, uses the name SQUALUS for the allies of CARCHARHINUS COMMERSONIANUS. But this use of the name is no more a specification of a type than is that of Rafinesque, and the more formal choice of type by Gill reinforces the former. Swainson, 1838, chooses CARCHARODON CARCHARIAS L.

Chimæra Linnæus, 236; type CHIMÆRA MONSTROSA L.
Unquestioned.

Lophius Linnæus, 236, after Artedi; type LOPHIUS PISCATORIUS L.
Unquestioned.

Acipenser Linnæus, 237, after Artedi; type ACIPENSER STURIO L.
Unquestioned.

Muræna Linnæus, 244, after Artedi; type MURÆNA HELENA L.
Unquestioned, except by Bleeker, who takes as type MURÆNA ANGUILLA L., the first species named by Artedi.

Gymnotus Linnæus, 246, after Artedi; type GYMNOTUS CARAPO L.
Unquestioned, except by certain authors who take as type GYMNOTUS ELEC-TRICUS L. which appears first in the Twelfth Edition of the *Systema Naturæ*.

Trichiurus Linnæus, 246, after LEPTURUS Artedi; type TRICHIURUS LEPTURUS L.
Monotypic.

Anarhichas Linnæus, 247, after Artedi; type ANARHICHAS LUPUS L.
Monotypic. Usually and more correctly spelled ANARRHICHAS.

Ammodytes Linnæus, 247, after Artedi; type AMMODYTES TOBIANUS L.
Monotypic.

Stromateus Linnæus, 248, after Artedi; type STROMATEUS FIATOLA L.
Unquestioned.

Xiphias Linnæus, 248, after Artedi; type XIPHIAS GLADIUS L.
Monotypic.

Callionymus Linnæus, 249, after Artedi; type CALLIONYMUS LYRA L.
Unquestioned.

Uranoscopus Linnæus, 250, after Artedi; type URANOSCOPUS SCABER L.
Monotypic.

Trachinus Linnæus, 250, after Artedi; type TRACHINUS DRACO L.
Monotypic.

Gadus Linnæus, 251, after Artedi; type GADUS MORHUA L.
By common consent.

Blennius Linnæus, 256, after Artedi; type BLENNIUS OCELLARIS L. as restricted by Jordan & Gilbert, 1883.

Ophidion Linnæus, 259, after Artedi; type OPHIDION BARBATUM L.
Unquestioned.

Cyclopterus Linnæus, 260, after Artedi; type CYCLOPTERUS LUMPUS L.
Unquestioned.

Echeneis Linnæus, 260, after Artedi; type ECHENEIS NAUCRATES L. (misprinted NEUCRATES).
First restriction by Gill, *Proc. Acad. Nat. Sci. Phila.*, 1862, 239. In 1864, *loc. cit.* 60, Gill proposed to adopt as type ECHENEIS REMORA, this being the only species noted by Artedi, and in Linnæus's earlier writings. But as Linnæus referred both species to ECHENEIS, this change seems not warranted.

Coryphæna Linnæus, 261, after Artedi; type CORYPHÆNA HIPPURUS L.
Unquestioned.

Gobius Linnæus, 263, after Artedi; type GOBIUS NIGER L.
By common consent.

Cottus Linnæus, 264, after Artedi; type COTTUS GOBIO L.
First restriction by Cuvier & Valenciennes, *Hist. Nat. Poiss.*, IV, 1829, 142. "Ce genre avait pour type primitive, un petit acanthopterygien de nos rivières, à

tête large," etc. Later restricted to Cottus scorpius L. (Myoxocephalus Steller =Acanthocottus Girard), by Putnam, *Bull. Mus. Comp. Zool.*, I, No. I, 3, 1863.

Scorpæna Linnæus, 266, after Artedi; type Scorpæna porcus L.
Unquestioned.

Zeus Linnæus, 267, after Artedi; type Zeus faber L.
Unquestioned.

Pleuronectes Linnæus, 268, after Artedi; type Pleuronectes platessa L.
First restriction by Fleming, *Philos. Zool.*, 1822. Some writers have taken Pleuronectes maximus L. as type, following a quasi-designation by Fleming in 1828.

Chætodon Linnæus, 272, after Artedi; type Chætodon capistratus L.
First restriction by Cuvier, *Règne Animal*, 1817, 33, and by Jordan & Gilbert, *Synopsis Fish N. A.*, 1883. All authors, except Bleeker, have used the name for the same group. Artedi places first in his list under Chætodon the species called Chætodon arcuatus, a Pomacanthus. Bleeker takes this species as type, replacing Pomacanthus with Chætodon, and Chætodon by Tetragonoptrus.

Sparus Linnæus, 277, after Artedi; type Sparus aurata L.
First restriction by Fleming, *Philos. Zool.*, 1822. This decision has been generally, but not universally accepted.

Labrus Linnæus, 282, after Artedi; type Labrus bimaculatus L., the earliest type designated by Bonaparte (1839), under the name of L. vetulus Bloch.
By common consent restricted to Labrus viridis and closely allied forms.

Sciæna Linnæus, 288; type Sciæna umbra L.
Complex species, to be considered as identical with Cheilodipterus aquila Lacepède. First exact restriction by Cuvier, 1815. Sciæna umbra, based on Artedi, is a mixture of two species, Sciæna aquila (Lac). and Corvina nigra (Bloch). It is the proper type of the genus Sciæna, but its component parts are not congeneric. The two species were confused until Cuvier (*Mém. du Museum*, 1815, and later in the *Règne Animal*, Edition II, 1829, made clear the difference and definitely chose aquila as the type of Sciæna. Jordan & Evermann have adopted Corvina nigra, under the name of Sciæna umbra, as type of Sciæna. An argument can be made for either arrangement, but convenience is best served and probably justice also by accepting the name umbra for the species called aquila and recognizing this as type of Sciæna. The two species concerned should then stand as Sciæna umbra L. and Corvina nigra (Bloch). Bleeker has chosen as type Sciæna cirrosa, the species placed first by Artedi, the type of Umbrina Cuvier.

Perca Linnæus, 289, after Artedi; type Perca fluviatilis L.
By common consent.

Gasterosteus Linnæus, 295, after Artedi; type Gasterosteus aculeatus L.
By common consent.

Scomber Linnæus, 297, after Artedi; type SCOMBER SCOMBRUS L.
By common consent.

Mullus Linnæus, 299; type MULLUS BARBATUS L.
By common consent.

Trigla Linnæus, 300, after Artedi; type TRIGLA LYRA L.
By common consent.

Cobitis Linnæus, 300, after Artedi; type COBITIS TÆNIA L.
By common consent.

Silurus Linnæus, 304, after Artedi; type SILURUS GLANIS L.
By common consent.

Loricaria Linnæus, 307, type LORICARIA CATAPHRACTA L.
Monotypic.

Salmo Linnæus, 308, after Artedi; type SALMO SALAR L.
By common consent.

Trutta Linnæus, as "TRUTTÆ," 308; type SALMO TRUTTA L.
Type by tautonomy.

Osmerus Linnæus, as "OSMERI," 310, after Artedi; type SALMO EPER-
LANUS L.
By common consent.

Coregonus Linnæus, as "COREGONI," 310, after Artedi; type SALMO
LAVARETUS L.
By common consent.

Characinus Linnæus, as "CHARACINI," 311, after Gronow; type SALMO
GIBBOSUS L.
First restriction by Gill, *Proc. U. S. Nat. Mus.,* 1895, 215. Replaces CHARAX
Gronow, EPICYRTUS Müller & Troschel, ANACYRTUS Günther. Its use hinges on
its eligibility, as Linnæus used only the plural form CHARACINI as a section
SALMO. If not accepted, CHARAX Gronow (1763), Scopoli (1777), would replace
it, with the same type. The same slight doubt applies to TRUTTA, OSMERUS and
COREGONUS, all used in the plural form only by Linnæus. In our judgment all are
eligible.

Fistularia Linnæus, 312; type FISTULARIA TABACARIA L.
Monotypic.

Esox Linnæus, 313; type ESOX LUCIUS L.
This type was indicated, somewhat arbitrarily, in Opinion 58 of the International Commission, accepting the view of Cuvier, and current usage both before
and after Linnæus. Klein chose the name LUCIUS for the Pike, this name being
adopted in the *Gesellschaft Schauplatz.* Rafinesque first separated the marine gar-
fishes from the pike, calling the former Esox, the latter LUCIUS. But neither ever
stated formally that ESOX BELONE L. was the type of Esox. In view of the arguments available on either side, we may "let sleeping dogs lie," and follow common
custom, strengthened by the authority of the Commission.

Argentina Linnæus, 315, after Gronow, (*Museum Ichthyologicium,* a pre-Linnæan work) ; type ARGENTINA SPHYRÆNA L.
Monotypic.

Atherina Linnæus, 315, after Artedi ; type ATHERINA HEPSETUS L.
Monotypic.

Mugil Linnæus, 316, after Artedi ; type MUGIL CEPHALUS L.
Monotypic.

Exocœtus Linnæus, 316 ; type EXOCŒTUS VOLITANS L. = E. EVOLANS L. (= HALOCYPSELUS Weinland).
Monotypic.

Polynemus Linnæus, 317, after Gronow and Artedi ; type POLYNEMUS PARADISEUS L.
The first real restriction seems to be that of Günther, *Cat. Fishes,* II, 1860, 319. No type is specified, but the non-congeneric species, P. QUINQUARIUS L., is removed to form the genus PENTANEMUS, a name originally employed by Artedi, but changed to POLYNEMUS by Gronow. As this species, QUINQUARIUS, was the only one known to Artedi or to Gronow, Dr. Gill, with numerous writers, ourselves included, has regarded it as the type of POLYNEMUS. But common usage with the formal selection of P. PARADISEUS L. as type by Jordan & Gilbert, *Synopsis Fishes,* 1883, should prevail.

Clupea Linnæus, 317, after Artedi ; type CLUPEA HARENGUS L.
Unquestioned.

Cyprinus Linnæus, 320, after Artedi ; type CYPRINUS CARPIO L.
Unquestioned.

Mormyrus Linnæus, 327 ; type MORMYRUS CYPRINOIDES L.
Unquestioned.

Balistes Linnæus, 327, after Artedi ; type BALISTES VETULA L.
Unquestioned.

Ostracion Linnæus, 330 ; type OSTRACION CUBICUS L.
As restricted by Swainson 1839, by Bleeker 1865, and in recent usage.
Unquestioned, except by Kaup, who takes as type O. TRIQUETER L., a species referred to LACTOPHRYS Swainson. O. CUBICUS L. is type of CIBOTION Kaup.

Tetraodon Linnæus, 332, after Artedi ; type TETRAODON TESTUDINEUS L.
This genus has been variously treated by authors, but justice and convenience are best served by the choice of T. TESTUDINEUS as type. Bleeker, *Atlas Ichth.,* 1865, appears to be the first reviser. He observes: "C'est une de ces espèces qui est devenue le type du genre Linnéen . . . en effet le TETRAODON TESTUDINEUS, qui est la première des espèces de TETRAODON de la Dixième Edition du *Systema Naturæ.*" Several writers have since indicated as type T. LINEATUS L. a species of OVOIDES.

Diodon Linnæus, 334, after Artedi ; type DIODON HYSTRIX L.
Unquestioned, except by Bleeker, who takes the first species named, DIODON ATINGA L.

Centriscus Linnæus, 336, after Gronow; type CENTRISCUS SCUTATUS L.
Monotypic, a fact overlooked by various authors who choose CENTRISCUS SCOLOPAX L., 1766.

Syngnathus Linnæus, 336, after Artedi; type SYNGNATHUS ACUS L.
By common usage. The earliest restriction as approved by the International Commission is that of Jordan, Opinion, 45, 103, 1912.

Pegasus Linnæus, 358, after Gronow; type PEGASUS VOLITANS L.
Monotypic.

II. GUNNER, *Nachricht von Berglachs, welche* CORYPHÆNOIDES RUPES-TRIS *genannt werden kann*: Throndhjemske Selskab, Schriften III, 1761.

JOHAN ERNST GUNNER.

Coryphænoides Gunner, 43, 50; type CORYPHÆNOIDES RUPESTRIS Gun-ner.

III. SCHÆFER, *Piscium Bavarico Ratisbonensium*, 1761.

JACOB CHRISTIAN SCHÆFER.

The descriptions in this paper are exact and very elaborate ("in universum describiendibus"). The nomenclature is eccentric—in part mononomial, and the names are perhaps not exactly used in the sense of genera. The perch is called PERCA VULGARIS and again PERCA FLUVIA-TILIS.

Names perhaps not eligible as mononomial:

Cernua Schæfer, 37; type CERNUA SEU PERCA FLUVIATILIS MINOR Schæ-fer = PERCA CERNUA L.
Equivalent, if accepted, to GYMNOCEPHALUS Bloch, ACERINA Güldenstadt, CERNUA Fleming.

Schraitzer Schæfer, 38; type SCHRAITZER Ratisbonensium = PERCA SCHRÆTZER L.
Equivalent to LEPTOPERCA Gill.

Asperulus Schæfer 59; type ZINDEL RATISBONENSIS Schæfer = PERCA ZINGEL L.
Equivalent to ZINGEL Oken.

Asper Schæfer 59; type ASPER VERUS Schæfer, "Streber ratisbonensis" = PERCA ASPER L.
Equivalent to ASPRO Cuvier, not of Commerson.

IV. OSBECK, *Reise durch China*, 1762.

PER OSBECK.

Osbeck, a pupil of Linnæus, published in 1757 the record of his travels in China. This work is wholly binomial, but being earlier than 1758, the Latin edition, *"Iter Chinensis,"* 1757, cannot be used in nomenclature. The German edition, *Reise durch China*, bears the date of 1762, and is here considered. A Swedish version dates from 1765, an English edition from 1771. The case is exactly parallel with that of Hasselquist's *"Iter Palestinum,"* published in 1757, and reprinted in German in 1762. Hasselquist, like Osbeck, was a pupil of Linnæus, and adopted the Linnæan Code. In Opinion 57, the Commission of Nomenclature rejected Hasselquist's work and its translation. "The German translation by Gadebusch, published in 1762, does not give validity to the names published in the original edition in 1757." The present writers question the wisdom of this decision.

The names of Osbeck are questioned as translations of work prior to 1758.

Apocryptes Osbeck, 130, 1762; type APOCRYPTES CHINENSIS Osbeck, GOBIUS PECTINIROSTRIS Gmelin.

APOCRYPTES is close to BOLEOPHTHALMUS but distinct. APOCRYPTES Cuvier is a different genus.

Albula Osbeck, 309, 1762; type ALBULA CHINENSIS Osbeck.

Same as SALANX Cuvier, not ALBULA Gronow 1763. If accepted, the genus commonly called ALBULA must receive a new name, BUTYRINUS Lacepède.

V. GRONOW, *Zoophylaceum;* ZOOPHYLACII *Gronoviana . . . Animalia quæ in museo suo adservat, etc.*, 1763.

LORENZ THEODOR GRONOW (LAURENTIUS THEODORUS GRONOVIUS).

In this work of Gronow, printed in 1763, before its author had become acquainted with the *Systema Naturæ*, the genera of fishes are well defined, in a system which runs closely parallel with the system of Artedi (1738), but the species, as with Artedi, have polynomial designations only. In addition to the genera earlier named by Artedi and Linnæus, Gronow has a number of new names. Two of these, AMIA and HEPATUS, conflict with Linnæan genera of 1766. The completed manuscript work of Gronow, written in 1780, called *Systema Ichthyologicum,* in which binomial names were attached to species, after the Linnæan fashion, became the property of the British Museum, by which institution it was published in 1854. This work, as edited by John Edward Gray, does great credit to the scientific discrimination of Gronow; but at that late date, exactly a hundred years after Gronow's first paper, *"Museum Ichthyologicum,"* nearly all of his new names became synonyms.

Fortunately for the interests of nomenclature, most of Gronow's names were adopted in 1777 by Scopoli, *Introd. Hist. Nat.* Several other names have been used by subsequent authors, as Gmelin, Bloch and Cuvier; so that the adoption of the names of Gronow works less confusion in the system than might be expected. The suppression of AMIA, LIPARIS, ZOARCES, CONGER, and SCARUS, with the transposition of AMIA, are the results most to be regretted. As the few new names of Gronow have lain unnoticed for a century and a half, it seems a pity to revive them. The present writer believes, as already stated, that it would be a wise rule to exclude from the system all post-Linnæan writers who failed to adopt the binary designation of species. As however Brisson, 1760, "the Father of Ornithology," was a writer of this type, it may be possible to make an exception in his case, preserving his genera of Birds. But Ichthyology gains little to atone for the confusion resulting from the introduction at this late day of the names of Gronow, Klein and other polynomial writers not hitherto absorbed into the system. The generic names of Gronow have been, however, formally accepted by the International Commission of Zoological Nomenclature (Opinion 20, Smithsonian Miscellaneous Contributions, No. 1938): "Gronow, 1763, is binary, though not consistently binomial. Article 25 demands that an author be binary and Article 2 demands that generic names shall be uninomial. Under these articles, Gronow's names are to be accepted as complying with the conditions prescribed by the Code to render a name available under the Code."

The eligibility of the generic names of Gronow is questioned as not conforming to the Linnæan code in the terminology of species.

Callorhynchus Gronow, 31; type CHIMÆRA CALLORHYNCHUS L. (CALLORHYNCHUS PINNA DORSI etc. Gronow).
    Unquestioned: accepted by later writers.

Cyclogaster Gronow, 55; type CYCLOPTERUS LIPARIS L. (CYCLOGASTER BELGIS KRINGBURGK (Gronow).
    The name LIPARIS used for this genus by Artedi in 1738 was changed to CY-

CLOGASTER by Gronow. LIPARIS was restored by Scopoli in 1777, and has been used by nearly all subsequent authors.

Gonorhynchus Gronow, 56; type CYPRINUS GONORHYNCHUS Gmelin.

Monotypic. The name was adopted by Schlegel, 1846, replacing RHYNCHÆUS Richardson. Also by Scopoli, 1777.

Uranoscopus Gronow, 57; type COTTUS GOBIO L.

Synonym of COTTUS L.

Cynædus Gronow, 60; type SPARUS AURATA L. (CYNÆDUS CAUDA LU-
NULATA etc. Gronow).

This genus is an assemblage of Sparoid fishes, essentially equivalent to SPARUS of Linnæus, and includes species of several modern genera. Jordan (Smithson. Publ. No. 1938: *Opinions on zoological nomenclature*) has proposed to treat CYNÆDUS as a synonym of SPARUS, having SPARUS AURATA as type.

Holocentrus Gronow, 65; type HOLOCENTRUS SOGO Bloch (HOLOCEN-
TRUS MAXILLA SUPERIORE LONGIORE Gronow).

This name, taken from Artedi, was revived by Scopoli, 1777, and by Bloch, 1790, who changed the spelling to HOLOCENTRUM, the form used by Artedi. This neuter form has been used by most authors from Cuvier to Günther.

Coracinus Gronow, 66; type DIPTERODON CAPENSIS Cuv. & Val. (CORA-
CINUS CAUDA-LUNATA Gronow).

Not DIPTERODON Lacepède. If accepted, replaces DICHISTIUS Gill.

Scarus Gronow, 67; type LABRUS VIRIDIS L. (SCARUS VIRIDIS Gronow).

This genus of Gronow unfortunately contains no species of the group later called SCARUS, the two other species enumerated being CICHLIDS from Surinam.

Callyodon Gronow, 72; type SCARUS CROICENSIS Bloch (CALLIODON CAP-
ITE SUBACUTO Gronow).

CALLYODON, if eligible, must replace SCARUS Forskål, 1775, as the name of this large and wide-spread genus. The name SCARUS of Gronow, applied to a species of LABRUS, antedates its use for a parrot-fish. The genus called CALLIODON by Cuvier (C. SPINIDENS), is distinct from CALLYODON of Gronow (CALLIODON of Bloch & Schneider), and must stand as CRYPTOTOMUS Cope. Revived by Scopoli, 1777.

Enchelyopus Gronow, 77, after Klein; type BLENNIUS VIVIPARUS L.
(ENCHELYOPUS CORPORE LITURIS etc. Gronow).

This genus as proposed by Klein in 1744 contained a variety of fishes, eel-shaped but with ventral fins. As used by Gronow it is practically equivalent to ZOARCES Cuvier, and to this type it was restricted by Gill, 1863. ENCHELYOPUS Bloch & Schneider (type GADUS CIMBRIUS L., a species unknown to Gronow), should stand as RHINONEMUS Gill. ENCHELYOPUS Gronow must unfortunately supersede ZOARCES Cuvier.

Pholis Gronow, 78; type BLENNIUS GUNNELLUS L. (PHOLIS MACULIS
ANNULATUS etc. Gronow).

Revived by Scopoli, 1777. Replaces MURÆNOIDES Lacepède and GUNNELLUS Cuvier, (not PHOLIS Cuv. & Val.). Monotypic.

Amia Gronow, 80; type Apogon moluccensis Cuv. & Val. (Amia caput catheoplateum Gronow).

Equivalent to Apogon Lacepède, 1802.

Eleotris Gronow, 83; type Gobius pisonis Gmelin == Eleotris gyrinus Cuvier & Valenciennes (Eleotris capite plageoplateo etc. Gronow).

The name Eleotris was used for the same group by Bloch & Schneider in 1801, and by all later authors.

Clarias Gronow, 100; type Clarias orontis Günther (Clarias Gronow).

Accepted by Scopoli, 1777, and by other writers. Monotypic.

Albula Gronow, 102; type Esox vulpes L. == Albula conorhynchus Bloch & Schneider.

Revived by Bloch & Schneider, 1801. Antedated by Albula Osbeck. Equivalent to Butyrinus Lacepède. Also used by Scopoli, 1777.

Aspredo Gronow, 102, after Linnæus, 1754; type Silurus aspredo L. (Aspredo cirris octo Gronow).

Anableps Gronow; type Cobitis anableps L.

Anostomus Gronow, 112; type Salmo anostomus L. (Anostomus Gronow).

Adopted by Müller & Troschel, 1845. Equivalent to Schizodon Agassiz, 1829. Monotypic. Revived by Scopoli, 1777.

Synodus Gronow, 112; type Salmo synodus L.

This genus, equivalent to Saurus Cuvier, was revived by Scopoli, 1777. Monotypic.

Hepatus Gronow, 113; type Teuthis hepatus L. == Chætodon chirurgus Bloch, not Acanthurus hepatus Cuv. & Val. (Hepatus mucrone reflexo Gronow).

Hepatus Gronow is based on two species, on which two Linnæus based his genus Teuthis. The latter is a substitute for the former, and the two were accepted as identical by Scopoli. Of the two species mentioned by Gronow, hepatus L. and javus L., Gronow had, according to Dr. Günther, a specimen of the first. This we may take as type of both Hepatus and Teuthis hepatus. This example belonged to the species called later Chætodon chirurgus by Bloch. This species must be regarded as the type of the genus Hepatus. Cuvier & Valenciennes have used the name Hepatus L. for an East Indian species, Colocopus lambdurus Gill. They were mistaken in supposing that this species was the one examined by Gronow. The restriction here made is that of the first reviser, Jordan, Tanaka & Snyder, *Cat. Fish Japan*, 1913, 214. If accepted, Hepatus replaces Teuthis and Acanthurus.

Umbra (Krämer) Gronow, 114; type Cyprinodon krämeri Walbaum. (Umbra krameri Gronow).

Revived by Scopoli, 1777, and Müller, 1842, as Umbra crameri Müller.

Erythrinus Gronow, 114; type SALMO ERYTHRINUS Bloch & Schneider, 1801 (= ERYTHRINUS Gronow).

The name was revived by Müller & Troschel, 1846, and is in general use. Monotypic. Used also by Scopoli, 1777.

Cataphractus Gronow, 115; type PEGASUS DRACONIS L. (CATAPHRACTUS CORPORE TETRAGONE Gronow).

This genus is a synonym of PEGASUS L.

Solenostomus Gronow, 119; type FISTULARIA TABACARIA L. (SOLENOSTOMUS CAUDA BIFURCA etc. Gronow).

SOLENOSTOMUS is equivalent to FISTULARIA. The genus SOLENOSTOMUS Lacepède, 1803, typified by FISTULARIA PARADOXA Pallas, must receive a new name, if the names of Gronow are accepted.

Charax Gronow, 123; type SALMO GIBBOSUS L. (CHARAX MAXILLA INFERIORE etc. Gronow).

This name is equivalent to CHARACINI L.=EPICYRTUS Müller & Troschel, 1846, said to be preoccupied, and ANACYRTUS Günther, 1864, presented as a substitute for EPICYRTUS, all with the same type. CHARAX Risso (C. PUNTAZZO) belongs to the very different family SPARIDÆ. Revived by Scopoli, 1777.

Mystus Gronow, 124; type BAGRUS HALEPENSIS Cuv. & Val. (MYSTUS CIRRIS OCTO CAPITE LONGIORIBUS Gronow).

The genus MYSTUS as defined by Gronow contains five species, not congeneric. The name is taken from Russell's *History of Aleppo,* who describes the species called BAGRUS or HYPSELOBAGRUS HALEPENSIS as MYSTUS. Of the generic names included, that of HYPSELOBAGRUS Günther, 1864, is the most recent. The species named "MYSTUS" by Russell may be taken as the type of "MYSTUS" which will, if accepted, supersede HYPSELOBAGRUS. MYSTUS Lacepède, 1805, (MYSTUS CLUPEOIDES Lacepède) is a species of COILIA Gray, 1831, an Engraulid fish. Also Scopoli.

Plecostomus Gronow, 127, after Artedi; type LORICARIA PLECOSTOMUS L. (PLECOSTOMUS DORSO DIPTERYGIO Gronow).

If accepted, replaces HYPOSTOMUS Lacepède; it is now in general use.

Callichthys Gronow, 127; type SILURUS CALLICHTHYS L. (CALLICHTHYS CIRRIS Gronow).

Unquestioned, being accepted by later writers.

Mastacembelus Gronow, 132; type OPHIDIUM SIMACK Walbaum, 1792 = RHYNCHOBDELLA HALEPPENSIS Bloch & Schneider, 1801 (MASTACEMBELUS MAXILLIS SUBACUTIS Gronow).

Revived by Cuvier & Valenciennes. The generic name RHYNCHOBDELLA Bloch & Schneider included both species assigned by Gronow to MASTACEMBELUS. In dividing the group, Cuvier & Valenciennes assigned the former name to OPHIDIUM ACULEATUM Bloch, Gronow's second species, leaving MASTACEMBELUS for his first. Before Gronow, Klein, 1744, had used the name MASTACEMBELUS for an entirely different group. In this he had been followed by Bleeker, who recognized, at first, generic names of earlier date than 1758.

Channna Gronow, 135; type CHANNA ORIENTALIS Bloch & Schneider, 1801 (CHANNA Gronow).

This monotypic genus was accepted by Bloch & Schneider, 1801, and Scopoli.

Gasteropelecus Gronow, 135; type CLUPEA STERNICLA L. (GASTERO-PELECUS Gronow).

Monotypic. The name was revived by Pallas in 1769.

Leptocephalus Gronow, 135; type LEPTOCEPHALUS MORRISI Gmelin, the larva of MURÆNA CONGER L. (LEPTOCEPHALUS Gronow).

This name, based on a larval conger, and revived by Scopoli in 1777, is held to replace CONGER Houttuyn, 1764, and of Cuvier, 1817, given to the adult of the same species. As LEPTOCEPHALUS has been in use more than a century as the collective name of the peculiar translucent expanded larvæ of the Conger and other eels, it would be well to restrict its use to those forms, reserving the genus to which the type belongs, the earliest name given to the adult fish, CONGER Houttuyn.

Gymnogaster Gronow, 136; type TRICHIURUS LEPTURUS L. (GYMNO-GASTER Gronow).

Equivalent to TRICHIURUS L. Monotypic.

Pteraclis Gronow, 136; type CORYPHÆNA VELIFERA Pallas, 1770 (PTER-ACLIS PINNATA Gronow) 1777.

The name PTERACLIS was used by Gronow, 1772. It is monotypic and unquestioned.

VI. LINNÆUS, *Museum Adolphi-Frederici*, II, 1764.

CAROLUS LINNÆUS.

Cepola Linnæus, 63; type CEPOLA RUBESCENS L. = C. TÆNIA L. (also in *Syst. Nat.*, 1766.)

VII. HOUTTUYN, *Natuurlike Historie volgens den Heer Linnæus*, 1764.

MARTIN HOUTTUYN.

This work we have not seen. Mr. Garman (in lit.) quotes:

Conger Houttuyn, VII, 103; type MURÆNA CONGER L.

This name might well be retained for the Conger eel, leaving LEPTOCEPHALUS to its time-honored special use as a designation for the larval forms of Conger and similar eels.

Torpedo Houttuyn, VII, 453; type (not named) RAJA TORPEDO L.

Apparently this fixes the name TORPEDO on the Electric Ray.

VIII. LINNÆUS, *Systema Naturæ*, Ed. XII, 1766.

CAROLUS LINNÆUS.

This edition contains four genera in addition to those given in the Tenth Edition.

Cepola Linnæus, 445; type C. TÆNIA L. = C. RUBESCENS L.
Unquestioned.

Amia Linnæus, 500; type A. CALVA L.

The generic name AMIA appears in the Twelfth Edition of the *Systema Naturæ*, in 1766. It had been used earlier by Gronow, in 1763, for a percoid genus, later called APOGON by Lacepède.

In the opinions already rendered by the Commission, it was decided that the generic names in Gronow's *Zoophylaceum*, published in 1763, between the tenth and twelfth editions of the *Systema Naturæ*, should be adopted, although his names for species were polynomial. Gronow was an excellent ichthyologist, with broader knowledge than Linnæus, and later adopted the Linnæan nomenclature. In view of the fact that his names are not in current usage, and that he had not then accepted binomial nomenclature, most recent authors have rejected them, unless revived by some binomial writer. The transfer of AMIA from the ganoid to a percoid genus is, however, peculiarly undesirable, and it may be urged that general convenience justifies a special exception in this case. If AMIA be used for APOGON, AMIA L. is replaced by AMIATUS Rafinesque, 1815.

Teuthis, 507, after Browne; type CHÆTODON CÆRULEUS Bloch = TEU-
THIS HEPATUS L. in part.

The name TEUTHIS was applied by Linnæus in the twelfth edition of the *Systema Naturæ*, to the two species which formed the genus HEPATUS of Gronow, in 1765. These were named TEUTHIS HEPATUS and TEUTHIS JAVUS. The name TEUTHIS was borrowed from Browne, a non-binomial author. The two Linnæan species belong to different families. The species first named HEPATUS may be regarded as the type of HEPATUS, as already indicated. The name TEUTHIS should properly go with T. HEPATUS, as the name is borrowed from Browne, who applied it to a single species, confused with TEUTHIS HEPATUS by Gronow and Linnæus, the CHÆTODON CÆRULEUS of Bloch. This species is a near relative of the type of HEPATUS. This decision follows the arguments of Dr. Gill. It is reasonable, but not above question. Meanwhile several authors, notably Cantor and Günther, use TEUTHIS for T. JAVUS (SIGANUS Forskål), while others suppress it altogether. Still others misspell it, as THEUTYS, THEUTIS, etc.

Elops Linnæus, 508; type ELOPS SAURUS L.
Monotypic.

IX. MÜLLER, *Delineationes Naturæ*, II, 1767, 141 (fide Sherborne, *Index Animalium*).

Philip Ludwig Statius Müller.
Not seen by us.

Acus Müller, 141 ; type probably Syngnathus acus L.
Equivalent to Syngnathus L.

Orbis Müller, 141 ; type probably Diodon hystrix L.
Equivalent to Diodon L.

We copy these references from Sherborne. In a memoir by Professor Müller, 1774, vol. III, p. 341, Acus appears as a specific name under Syngnathus acus, and again, 1774, IV, 341, as a vernacular under Esox belone. Under Diodon hystrix (III, 327) he quotes "Orbis maximus spinosus." Mr. Garman, who gives us these references, remarks: "I would say that neither of these is available as a generic name." Neither is used in Müller's supplement to the *Systema Naturæ*, 1776.

X. GEOFFROY, *Descriptions de 719 Plantes* etc., 1767.

Étienne Louis Geoffroy.

This paper we have not seen. We copy from Sherborne.

Ichthyocolla Geoffroy, 399; type presumably Acipenser huso L.
The name, meaning fish-glue, was early applied to the fish producing it. Prior to Huso Brandt, as a name for a subgenus of sturgeons.

Harengus Geoffroy, 405; type presumably Clupea harengus L., in which case it is a synonym of Clupea.

Lucius Geoffroy, 407; type presumably Esox lucius L.

Trutta Geoffroy, 719; type presumably Salmo trutta L.

XI. VALMONT DE BOMARE, *Dictionnaire Raisonné Universel d'Histoire Naturelle*, 1764, 1768, 1774, 1791.

Jean Christophe Valmont de Bomare.

The eligibility of Valmont's names is questioned as binomial only by accident, and not accepted as genera by the author himself in 1791.

Of this work we have examined four editions, the first bearing date of 1764; the "new edition," considerably enlarged, of 1768; the second edition, apparently mostly identical with the "new edition," 1775; and the fourth, still larger, in 1791.

The dictionary received no consideration in ichthyological nomenclature until the appearance of the elaborate treatise on the sharks of the world, "PLAGIOSTOMIA" by Samuel Garman, (*Memoirs of the Museum of Comparative Zoology at Harvard College*, vol. XXXVI, 1913).

In this work Mr. Garman makes brief reference to Valmont de Bomare as a worthy author hitherto ignored in taxonomy. "The selection of one authority because he favored binomials more than another, has led to much uncertainty among names and to many changes. It has led authors to belittle and to ignore excellent works which at their time of publication and much later ranked in accuracy and influence among the first of the scientific publications of this period." Mr. Stejneger informs me that a Danish edition of this Dictionary exists.

An examination of four editions of the *Dictionnaire* of Valmont de Bomare shows it to be a compilation pure and simple, that he did not intend to give any new scientific names to animals or plants, but that, in a few cases, he copied binomial appellations from earlier authors which might be construed as revived in a scientific sense.

It is therefore worth while to examine these cases in detail.

FIRST EDITION, 1764.

In the first edition, bearing date of 1764, there is no case of the use of anything resembling scientific nomenclature, although Valmont often gives a Latin equivalent to his French names. It is evident, however, that the work of Artedi, Klein, Gronow, and Linnæus in which genera and species are formally recognized, is unknown to him.

Thus, accompanied by fair descriptions compiled from other authors, he gives the following:

"AIOL, en Latin, SCARUS, un des plus beaux," etc. (vol. I, p. 95).

"ALOSE, ALOSA, poisson de mer qui remonte" (vol. I, p. 105).

"ANCHOIS, en Latin, APUA, petit poisson," etc. (I, 13, 3).

"ANGUILLE, ANGUILLA, poisson allongé," etc. (vol. I, p. 133).

"CONGRE, CONGER, excellent poisson," etc. (vol. II, p. 58).

"DAURADE, AURATA VULGARIS, Espèce de poisson," etc. (vol. II, p. 225).

"ÉPINOCHE, PISCIS ACULEATUS (vol. II, p. 306).

"GLAUCUS, bien des ichthyologues donnent ce nom à trois sortes de poissons, 1 au Derbio, 2 au Liche, 3 au véritable GLAUCUS," etc. (vol. II, p. 512).

"Hareng, Halec, Les harengs sont des poissons de passage," etc. (vol. II, p. 610).

"Torpille, Torpedo ou Tremble, Torpedo poisson," etc. (p. 458). "Grappe marine, Uva marina."

These are plainly not scientific names. While the writer evidently grasps more or less clearly the meaning of genus and species, he has no conception of binomial nomenclature, as distinct from Latin equivalents of the vernacular names in French. Thus no one would take "Piscis aculeatus" as a generic and specific name for the stickleback or Uva marina for the Alga (Sargassum) known as Sea-Grape. Further along (Edition II) occurs "Poisson pétrifiée" indicated as "Ichthyolithus," which certainly is not the name of any genus. No scientific names, generic or specific, can be held to bear date from this first edition, 1764, of the dictionary of Valmont de Bomare.

"Nouvelle Édition," 1768, and "Édition II," 1775.

The "Nouvelle édition," dated 1768, is the one examined by Mr. Garman. Except for the paging it seems substantially identical with the edition of 1775, formally called the "second." It has a few binomial terms, mostly among the sharks. The use of Latin equivalents for the French vernacular is still continued, but these assume more frequently a binomial form, especially in the rather elaborate index. The first edition (octavo) contains no index.

In the editions of 1768 and 1775 the only new names to be considered are Galeus, Mustelus, Vulpecula, and Catulus.

On page 116 (Edition II) we read:

"Cagnot bleu, galeus glaucus, Grand poisson cartilagineux de la famille des chiens de mer." Then follows a fair account of the Great Blue Shark (Squalus glaucus L.; Prionace glauca of recent writers), taken from the description of the "Chien de mer bleu" of Rondelet (de Piscibus, 1558, p. 296). Rondelet begins "Galeus glaucus, en Languedoc, Cagnot bleu, poisson cartilagineux," etc. He uses the name "Chien de mer," or Galeus in a general sense, including the "Aiguillat" (Acanthias), the Emissole, ("Galeus lævis"), the "Chien de Mer Étoilé, Galeus asterias," the "Mélandre, Galeus canis," and the "Chien de Mer bleu, Galeus glaucus."

The other sharks are treated under different heads by Valmont. The entire arrangement appears in the index to the same volume.

"Cagnot bleu, Galeus glaucus," does not appear in the Fourth Edition (1791).

On page cxxxvii "MUSTELLUS" is defined:

"Espèce de Chien de mer, c'est le GALEUS STELLATUS des auteurs."

The name SQUALUS MUSTELUS L. was mostly based on Artedi's references to the "Émissole" or unspotted dog-fish, the GALEUS LÆVIS of Rondelet and Valmont. The specific name MUSTELUS should remain with that species. On the "LÆVIS" the name MUSTELUS of Linck in 1790 was clearly based. The same species, the "Émissole Commune," is clearly the type of MUSTELUS Cuvier (1817). But if the name MUSTELUS Valmont be accepted, its type must be the dog-fish with round spots, SQUALUS STELLATUS of Risso, MUSTELUS ASTERIAS Valmont.

On page ccxxii of the Index occurs the name "Renard Marin, VULPECULA MARINA." This is apparently borrowed from Willughby and it refers to ALOPIAS VULPINUS, the "Sea Fox" of modern authors. The name is not a "scientific" term, but merely a Latın rendering of the vernacular.

CATULUS rests on CATULUS VULGARIS, which is SCYLLIORHINUS CANICULA L. But the name is not available in any case, being preoccupied by CATULUS Kniphof, a genus of insects.

In the fourth edition (1791) Valmont gives a list of the genera of fishes. All those of Linnæus (1766) are enumerated, but no others and none of his own names are included in the list. Evidently he did not regard himself as having made additions to scientific nomenclature.

The eligibility of Valmont's names is questioned as binomial only by accident, and not accepted as genera by the author himself in 1791.

Galeus Valmont de Bomare, I, 371, 1768; type SQUALUS GLAUCUS L.
"CAGNOT BLEU, GALEUS GLAUCUS," with description.
If regarded as eligible GALEUS will replace PRIONACE and CYNOCEPHALUS.

Vulpecula Valmont de Bomare, III, 740, 1768; type VULPECULA MARINA Valmont = SQUALUS VULPINUS Bonnaterre = SQUALUS VULPES Gmelin.
"VULPECULA MARINA; RENARD MARIN," with description.
If eligible, VULPECULA will replace ALOPIAS Rafinesque, 1810.

Catulus Valmont de Bomare, IV, 51, 1768; type SQUALUS CANICULUS L. (CATULUS MAJOR VULGARIS Ray).
According to Sherborne, *Index Animalium*, it is preoccupied by CATULUS Kniphof, *De. Pedic.*, p. 16, 1759, a genus of insects. Equivalent to SCYLLIORHINUS Blainville.

Mustelus Valmont de Bomare, Ed. II, 746, 1768, and Ed. III, 1775, lxxxi; type GALEUS ASTERIAS Valmont = MUSTELUS CANIS Mitchill, 1815 = MUSTELUS STELLATUS Risso, 1826.
"Galeus asterias aut Mustelus stellaris. Chien de mer à taches rondes."

XIII. GOUAN, *Historia Piscium,* 1770.

Antoine Gouan.

Trachipterus Gouan, 104; type Trachipterus gouani = Cepola trachyptera Gmelin.

Lepadogaster Gouan, 105; type Lepadogaster gouani Gouan.

Lepidopus Gouan, 107; type Lepidopus gouani Gouan (Trichiurus caudatus Euphrasen).

XIV. KŒLREUTER, *Piscium Rarorum;* Novi Comm. Act. Petropolit. VIII, 1770.

Joseph Gottlieb Kœlreuter.

Mola Kœlreuter, 337; type Mola aculeata Kœlreuter.
Antedates Mola Cuvier.

XV. FORSTER, *Catalogue of Animals of North America,* 1771.

John Reinhold Forster.

Remora Forster, 20; type Echeneis remora L.
Equivalent to Remora of Catesby and of Gill.

XVI. BRÜNNICH, *Collectio Nova Scriptorum Societatis Scientiarum Hafnensis,* 1771.

M. T. Brünnich.

Regalecus Brünnich, III, 418; type Regalecus remipes Brünnich (Ophidium glesne Ascanius).
Also described in 1788.

XVII. CATESBY AND EDWARDS, *Natural History of Carolina, Florida and the Bahama Islands,* 1731-1750, by Mark Catesby; Edition Second, 1771, by George Edwards.

The large folio volume in which Mark Catesby published the record of his visit to the Bahamas and other parts of America has had an important place in the history of American Ichthyology. Numerous editions of this work with the same plates have been published in German, French and English.

Two of these, the Edwards' Edition of 1771, and the edition quoted as "Catesby, Pisc. Imag., Etc., in 1777," are subsequent to Linnæus and may perhaps deserve consideration in nomenclature, although apparently not eligible in view of Opinion 57, which regards the post-Linnæan translation of Hasselquist as ineligible. These editions contain tables showing the Linnæan names of Catesby's species. Except as an evidence of "revision," these have no bearing on Catesby's "genera." If the generic names with polynomial specific names, of Gronow, Klein, and others are accepted, we can hardly refuse notice to the Latin nouns used by Catesby as republished by Edwards. These nouns have the force of genera, and being built about actual specimens they are mostly monotypic; while those of Gronow and Klein are subdivisions of a system, each covering as a rule many species. The names of Catesby are listed as genera in Sherborne's *Index Animalium* by an author who is rather critical of Latin vernaculars. But Catesby wrote before Artedi and Linnæus had framed the idea of a genus. He was not therefore consciously engaged in the differentiation of generic groups. He was not, to borrow a phrase from Mr. Stejneger, "playing the game." For this reason it seems to us that his names should not be admitted to the system. It is, however, very important to have a decision once for all in this matter.

The names in Edwards' Catesby are of doubtful eligibility as being Latin vernacular nouns rather than genera, and as a reprint virtually unchanged of a pre-Linnæan work.

Umbla Catesby, 1; type Esox BARRACUDA Shaw (SPHYRÆNA PICUDA Bloch & Schneider). "UMBLA MINOR, MAXIMA MAXILLIS LONGIORIBUS, the Barracuda" Catesby.
Equivalent to SPHYRÆNA and prior to it, if accepted in the system.

Mormyrus Catesby, 2; type ULÆMA LEFROYI (Goode). "MORMYRUS EX CINEREO NIGRICANS, the Bone-fish" Catesby.
Identification somewhat uncertain. The name in any event is subsequent to MORMYRUS L.

Saurus Catesby, 2; type SALMO FŒTENS L. "SAURUS EX CINEREO NIGRICANS, the Sea Sparrow Hawk" Catesby.
Identical with SYNODUS Gronow, 1763.

Albula Catesby, 6; type MUGIL CUREMA Cuv. & Val. ("ALBULA BAHA-
MENSIS Catesby).

A synonym of MUGIL L., subsequent to ALBULA Gronow, 1763, and ALBULA
Osbeck, 1761.

Hirundo Catesby, 8; type CYPSELURUS sp., "HIRUNDO" Catesby.

Not HIRUNDO L., 1758, a genus of Swallows.

Turdus Catesby, 9; type LUTIANUS GRISEUS (L.) "TURDUS PINNIS
BRANCHIALIBUS CARENS, the Mangrove Snapper" Catesby.

Not TURDUS, 1758, a genus of Thrushes.

Alburnus Catesby, 12; type CYPRINUS AMERICANUS L. = MENTICIRRUS
AMERICANUS. "ALBURNUS AMERICANUS, the Carolina Whiting"
Catesby, type of CYPRINUS AMERICANUS L., *Syst. Nat.*, X, 321.

If accepted, the genus ALBURNUS must replace MENTICIRRUS Gill, and the genus
of CYPRINIDÆ called ALBURNUS by Rafinesque and Agassiz must receive another
name.

Cugupuguacu Catesby, 14; refers to EPINEPHELUS MACULOSUS (Cuv.
& Val.). ("CUGUPUGUACU BRAZIL, the Hind" Catesby).

Catesby's fish is not that called CUGUPUGUACU by Marcgrave. If this barbarous
name be allowed it will replace EPINEPHELUS. But it is evident that CUGUPUGUACU,
PETIMBUABO and ACARAUNA are not in any sense generic names, but attempts on
the part of Catesby to identify his species with those called in Brazil by these
vernacular names. In the case of ACARAUNA, Catesby is himself doubtful. Even
in case names used in an actual generic sense, as UMBLA, AURATA, UNICORNIS were
allowed, CUGUPUGUACU and ACARAUNA should be rejected.

Saltatrix Catesby, 14; type GASTEROSTEUS SALTATRIX L., which is based
on Catesby's figure, ("SALTATRIX, the Skipjack").

If allowed, will replace POMATOMUS Lacepède, 1802.

Suillus Catesby, 15; type LACHNOLAIMUS SUILLUS Cuvier, "SUILLUS"
Catesby, based on Catesby's figure.

If allowed, SUILLUS will replace LACHNOLAIMUS Cuvier.

Aurata Catesby, 16; type CALAMUS CALAMUS (Cuv. & Val.). "AURATA
BAHAMENSIS, the Porgy," Catesby.

Not AURATA Fleming, 1828, which is SPARUS L. If allowed, AURATA will replace CALAMUS Swainson.

Salpa Catesby, 17; type SPARUS SYNAGRIS L. (LUTIANUS SYNAGRIS,
based on "SALPA PURPURASCENS VARIEGATA, the Lane Snapper,"
Catesby.

Prior to SALPA Forskål, 1775. If allowed, SALPA would replace NEOMÆNIS or
LUTIANUS and perhaps add further to the confusion among the Salpoid Tunicates,
although it is claimed that an earlier name, DAGYSA, must replace SALPA Forskål.

Novacula Catesby, 18; type SCARUS CÆRULEUS Bloch. "NOVACULA CÆ-
RULEA, the Blue-fish," Catesby.

Name a synonym of CALLYODON Gronow, but prior to NOVACULA Cuvier, which
is a synonym of XYRICHTHYS Cuvier.

Petimbuabo Catesby & Edwards, 18; refers to FISTULARIA TABACARIA
L. "PETIMBUABO BRAZIL, the Tobacco Pipe-fish," Catesby.

A synonym of FISTULARIA L., but obviously not intended as a generic name.

Unicornis Catesby, 19; type OSBECKIA SCRIPTA (Osbeck). "UNICORNIS
PISCIS BAHAMENSIS, the Bahama Unicorn-fish," Catesby.

If accepted, would replace OSBECKIA Jordan & Evermann.

Bagre Catesby, 23; type SILURUS CATUS L. (AMEIURUS CATUS L.)
"BAGRE SECUNDÆ SPECIEI MARCGRAVEI AFFINIS, the Cat-fish,"
Catesby.

If accepted, BAGRE must replace AMEIURUS.

Harengus Catesby, 24; type CLUPEA SARDINA Poey. "HARENGUS
MINOR BAHAMENSIS, the Pilchard," Catesby.

Equivalent to HARENGULA Cuv. & Val., and, if allowed, would replace the latter.

Anthea Catesby, 25; refers to MESOPRION ANALIS Cuv. & Val. A species of NEOMÆNIS, Girard. "ANTHEA QUARTUS RONDELETI, the
Mutton-fish," Catesby.

Not intended as a generic name, being wrongly identified with the fourth
ANTHIA of Rondelet.

Remora Catesby, 26; type ECHENEIS REMORA L. "REMORA, the Suck-
ing-fish," Catesby.

Equivalent to REMORA Forster, and of Gill.

Solea Catesby, 27; type PLEURONECTES LUNATUS L., based on Catesby's
figure. "SOLEA LUNATA ET PUNCTATA, the Sole," Catesby.

Not SOLEA of Klein nor of subsequent writers. If allowed, SOLEA would replace
PLATOPHRYS; and SOLEA of Klein, Quensel, Rafinesque and Cuvier would require
a new name.

Orbis Catesby & Edwards, 28; type TETRAODON TESTUDINEUS L. "OR-
BIS LÆVIS VARIEGATUS, the Globe-fish," Catesby.

A synonym of TETRAODON L.

Psittacus Catesby, 29; type LABRUS CATESBÆI Lacepède. "PSITTACUS
PISCIS VIRIDIS BAHAMENSIS, the Parrot-fish," Catesby.

Not PSITTACUS L., 1758, a genus of parrots.

Acus Catesby, 30; type ESOX OSSEUS L. "ACUS MAXIMA SQUAMOSA
VIRIDIS, the Green Gar-fish," Catesby.

Equivalent to PSALISOSTOMUS Klein and LEPISOSTEUS Lacepède.

Acarauna Catesby, 31 ; refers to HOLACANTHUS CILARIS L. "AN ACA-
RAUNA MAJOR PINNIS CORNUTIS AN PARU BRASILIENSIBUS?, the
Angel-fish," Catesby.

This name cannot be used to replace HOLACANTHUS as it represents a very
doubtful identification on the part of Catesby; not at all a generic division.

Vipera Catesby, 9, Appendix; refers to VIPERA MARINA Catesby
(CHAULIODUS SLOANI Bloch & Schneider).

As it is a fish, it cannot belong to the genus VIPERA L. VIPERA MARINA is
apparently merely a vernacular name, sea-viper or viper-fish being intended.

Cataphractus Catesby, 9, Appendix; type SILURUS CATAPHRACTUS L.
= DORAS CATAPHRACTUS of authors = CATAPHRACTUS AMERI-
CANUS Catesby.

This name, being preoccupied, cannot replace DORAS Lacepède.

XVIII. GÜLDENSTADT, *Acerina piscis ad Percæ genus pertinens:*
Nov. Comm. Acad. Petropol., 1774, XIX.

ANTON JOHANN VON GÜLDENSTADT.

Not seen by us.

Acerina Güldenstadt, 455; type PERCA CERNUA L. (ACERINA KABIR
Güldenstadt).
Equivalent to CERNUA Schæfer.

XIX. *Descriptiones Animalium quæ in Itinere Orientali Observavit,*
by PETRUS FORSKÅL (edited after the death of the author
by CARSTEN NIEBUHR). 1775.

Siganus Forskål, X; type SCARUS RIVULATUS Forskål.

No definition. This genus has been of late years generally called TEUTHIS, but
apparently this Linnæan name should remain with the group for which Browne
first used it.

Torpedo Forskål, 1775, 16; type RAJA TORPEDO (not of Linnæus) =
SILURUS ELECTRICUS Gmelin, MALAPTERURUS ELECTRICUS Lacepède.

Forskål describes the Electric Cat-fish of the Nile under the erroneous name
of RAJA TORPEDO L. He questions whether it might be allied to MORMYRUS or
whether it might find a place among the torpedoes of Rondelet, or might it be type
of a new genus. "Aut potius novum constituere genus. Certe determinatur tor-
pedinis CHARACTER GENERICUS: Piscis branchiostegus: apertura lineari, obliqua
supra pinnæ pectorales; corpore nudo: pinnis ventralibus seu abdominalibus;
dentibus numerosissimis densis, subulatis." This statement leaves no question as

to the species in mind, but TORPEDO Houttuyn, 1764, if available, is of still earlier date.

Salaria Forskål, X and 22; type (without specific name) = BLENNIUS BASILISCUS L.

The genus is equivalent to BLENNIUS L. Not SALARIAS Cuvier.

Scarus Forskål, 25; type SCARUS PSITTACUS Forskål.

The type no doubt intended was LABRUS SCARUS L., of the Mediterranean "antiquo nomine σκάρος"; but that species was not mentioned by Forskål, and another must be taken as the type. The name SCARUS was earlier used by Gronow as a synonym of LABRUS. SCARUS of Forskål must give way to CALLYODON of Gronow if the names of Gronow are to be adopted. This is unfortunate, as CALLYODON has been used by most authors as the name of another genus in the same family.

Abu-defduf Forskål, 59; type CHÆTODON SORDIDUS Forskål.

Equivalent to the later GLYPHISODON of Lacepède, 1803. The definition of this genus admits of no question. It occurs in the same paragraph with the equally accurate definition of ACANTHURUS. It may receive objection as a barbarous name. It was probably a "stop-gap" word for which Forskål intended to supply a Latin equivalent. This his editor after his death failed to do and we must apparently take it as it stands: "A generic name is a name without necessary meaning." (*Baird.*)

Acanthurus Forskål, 59; type CHÆTODON SOHAL Forskål; to be replaced by HEPATUS Gronow, if Gronow's names are adopted; otherwise by TEUTHIS L.

Later restricted by authors to the first species named. CH. UNICORNIS.

Besides these names, clearly eligible, Forskål lists a number of subordinate groups or subgenera, under PERCA, SCARUS and SCIÆNA. Some of these are properly and fully defined, and would be accepted without question if in Latin. But all are in Arabic, and may perhaps be taken as vernacular words, as one might divide a genus into "Groupers," "Snappers" and "Porgies." We may perhaps reject them on the same ground as that on which we reject "les sphéroides."

In addition to these more or less formal names are two sections, one under SCARUS called "dentibus Abudjubbe," equivalent to CHEILINUS Lac., and one "dentibus Harid," equivalent to SCARUS. LOUTI and DABA are above reproach, except as to their Arabic origin. ABUHAMRUR is defined by reference to its type species. NAQUA is based on a species referred with doubt to SCIÆNA. GHANAN, SCHOUR and TAHHMEL are names only, identifiable by the correspondence with the Arabic names of their type species.

These names of Forskål doubtfully eligible, being vernacular and not meant as subgeneric.

Naqua Forskål, xvii; type SCIÆNA GIBBA Forskål.

With definition as "Piscis marinis rubri obscuris, an Sciæna 48?" Equivalent to GENYOROGE Cantor, 1850.

Louti Forskål, 44; type PERCA LOUTI Forskål.

With full definition. Equivalent to VARIOLA Swainson, 1839, which apparently it should replace.

Daba Forskål, 44; type PERCA AREOLATA Forskål.

With full definition. EPINEPHELUS Bloch, 1798, which apparently it should replace.

Abuhamrur Forskål, 44; type SCIÆNA HAMRUR Forskål.

With definition by reference to the type species. PRIACANTHUS Cuvier, 1817.

Hobar Forskål, 44; type SCIÆNA BOHAR Forskål.

With short definition. LUTIANUS Bloch, 1790.

Farer Forskål, 44; type SCIÆNA SAMMARA Forskål.

With definition. HOLOCENTRUS Gronow, 1763.

Ghanan Forskål, 44; type SCIÆNA GHANAM Forskål.

Without definition. SCOLOPSIS Cuvier, 1817.

Djabub Forskål, 44; type SCIÆNA JARBUA Forskål.

With definition. THERAPON Cuvier, 1817.

Gaterin Forskål, 44; type SCIÆNA GATERINA Forskål.

With scanty definition. PLECTORHINCHUS Lacepède, 1800.

Schour Forskål, 44; type SCIÆNA NEBULOSA Forskål.

Without definition. LETHRINUS Cuvier, 1817.

Tahhmel Forskål, 44; type SCIÆNA TAHHMEL Forskål.

Without definition. KYPHOSUS Lacepède, 1800, subgenus OPISTHISTIUS Gill, 1862.

XX. KLEIN, *Neuer Schauplatz der Natur, nach den Richtigsten Beobachtungen und Versuchen, in Alphabetischer Ordnung.*

Durch eine Gesellschaft der Gelehrten. Weidmann, Leipzig. (Quoted as "Gesellschaft Schauplatz.")

No author named, the account of the fishes compiled from *Historia Piscium Naturalis* Klein, perhaps by Philip Ludwig Statius Müller, professor at Erlangen. Vol. I, 1775; vol. II, 1776; vol. III, 1776; vol. IV, 1777; vol. V, 1777; vol. VI, 1778; vol. VII, 1779; vol. VIII, 1779; vol. X, 1781.

JAKOB THEODOR KLEIN.

In a recent monograph of the sharks and rays (PLAGIOSTOMIA: Mem. Mus. Comp. Zool., Harvard College, XXXVI, 1916) Mr. Samuel Garman calls attention to the availability in nomenclature of names of genera accepted from Klein, a pre-Linnæan writer, in a post-Linnæan dictionary called "Neuer Schauplatz," or for convenience "Gesellschaft Schauplatz." This publication began, according to Mr. Garman, as a translation of Valmont de Bomare, but later it was extended and improved.

We find no copy of this work in the libraries of Washington and New York. It is probable that the copy in Mr. Garman's possession, which its owner has kindly placed at our disposal, is the only one now in the United States.

Mr. Garman remarks: "The Schauplatz referred to above is anonymous, it is true, but it gives the authorities for its generic and specific names, and thus its citations amount to republication after 1758, by the original authors, previous as the first publication may have been."

All the generic names used by Jacob Theodor Klein in his *Historia Piscium Naturalis,* 1740 to 1744, are here reproduced and accepted, thus bringing them for consideration into eligibility in scientific nomenclature. If accepted they therefore replace nearly all competing names except those of Artedi (1738) accepted by Linnæus (1758), and those of Gronow (1763).

Toward the middle of the eighteenth century the idea of genus among animals as a basis of classification became common property among naturalists. The name of the genus was recognized as consisting of a single word, but, until 1758, the species was indicated by a descriptive phrase attached to the name of the genus. By the device of binomial nomenclature, Linnæus made the system coherent, allotting to genus and species each a single word, the first a noun, the second of the nature of an adjective or genitive. In zoology, scientific nomenclature therefore dates from January 1, 1758, the time of the development of this system by Linnæus in the Tenth Edition of his *Systema Naturæ.*

Prior to Linnæus, on the basis of definite genera with polynomial species, three distinguished ichthyologists had separately developed, without knowledge of each other's work, a system of classification of fishes. These were Peter Artedi, "the Father of Ichthyology," in Upsala, in 1738; Jacob Theodor Klein, in Jena, 1740 to 1744; and Lorenz Theodor Gronow, in Leyden, 1754 to 1780. Of these authors the work of Artedi was the most compact and accurate, that of Klein the most elaborate, and that of Gronow based on the most material. Artedi's work was the basis of Linnæus's classification of the fishes. The principal part of the work of Gronow was published in 1763. His names have been accepted as eligible by the International Commission of Zoological Nomenclature.

In this paper we give a list of these genera of Klein together with the Linnæan type, each as understood or as indicated by the present writer. In deciding on the type, the writer has been materially aided by the possession of a copy of Klein's *Historia Piscium Naturalis*, which was once the property of his commentator, Dr. Johann Julius Walbaum of Greifswald, and which contains profuse annotations in Walbaum's own handwriting.

Dr. Walbaum himself published in 1792 (*Petri Artedi sueci Genera Piscium*, 589-587) a classified record of the genera of Klein, but without actually accepting these as part of the System. Later Walbaum printed an Index to all these early names of various authors, as *Ichthyologia Enodata sive Index Rerum* (Leipzig, 1793). In this Index the generic and other names used by previous writers are arranged in alphabetical order with indication of the Linnæan synonymy. This again does not, in the view of the Zoological Commission (Opinion 21), involve the acceptance of any of these names. A reprint of Klein's *Historiæ Piscium Naturalis* was published, with notes by Walbaum, in 1802.

In the *Gesellschaft Schauplatz* Klein's names are frankly adopted, and the chief serious objection to be raised is that the species are named polynomially. We here omit those names in the *Schauplatz*, as ACIPENSER, CYPRINUS, XIPHIAS, etc., which had been used by Linnæus and were already accepted in the System with the same significance. We are indebted to Mr. Garman for the verification of the list and for the insertion of page references.

Illustrations of the method of the elliptical and disorderly *Schauplatz* are the following. In vol. I, p. 918, 1777, we read:

"Botte. RHOMBUS, ein Kleinisches Fischgeschlecht, welches Linné, PLEURONECTES G. 163, Müller Seitenschwimmer, Richter Butten nennt," etc. Description follows.

In III, p. 512, we find: "GREETE, Kleinische, eine Art platteise, Butten, Richter: s(iehe) Botte, RHOMBUS, des Kleins und unsern Artikel Th. I, s. 918; desgleichen Flunder, PASSER 4 des Kleins und unsern Artikel Th. III, s. 151" etc.

"PASSER 4" is the species called "PASSER ASPER S(EU), SQUAMOSUS Rondelet. This species is PLEURONECTES LIMANDA L. With (eight) other species named polynomially, it is set forth in the *Schauplatz*, after Klein. The quotations from Müller probably refer to the Nürnberg Edition of the *Systema Naturæ*, published in German by Professor Philip

Ludwig Statius Müller of Erlangen, in 1776. This edition in its Supplement contains a few new specific names but no new genera, so far as noticed.*

The reference to "Richter," refers apparently to a work published at Leipzig, in 1754, by J. G. O. Richter, entitled *"Ichthyotheologie, oder Versuch die Menschen aus Betrachtung der Fische zu Bewunderung des Schöpfers zu führen."* This we have not seen, but quote the title from Bosgoed. The same work appears in Dutch, in 1768, as *"Godleerende Vischkunde."*

The names in the "Gesellschaft Schauplatz" are all of doubtful eligibility, because not adopting the Linnæan Code as to species, and possibly because published in an anonymous dictionary.

Conger Klein, I, 22, 1775; type MURÆNA CONGER L. "CONGER PINNA MEMBRANACEA" Klein.

A synonym of CONGER Houttuyn.

Enchelyopus Klein, I, 32, 1775; type TRICHIURUS LEPTURUS L. "ENCHELYOPUS CAPITE PRODUCTO SERPENTINO" Klein.

A synonym of TRICHIURUS L., not ENCHELYOPUS Gronow.

Brama Klein, I, 61, 932, 1775; type CYPRINUS BRAMA L. "BRAMA PRIMI RADIO PINNÆ DORSALIS SIMPLICI" Klein.

BRAMA, if accepted, replaces ABRAMIS Cuvier, and the marine genus now called BRAMA would take the name of LEPODUS Rafinesque.

Galeus Klein, I, 70, 1775; type SQUALUS GALEUS L. "GALEUS ROSTRI EXTREMA PARTE PELLUCIDA" Klein.

Equivalent to GALEORHINUS Blainville, 1816 = GALEUS Cuvier, 1817, not GALEUS Valmont, 1768, nor of Rafinesque, 1810.

Trutta Klein, I, 115, 1775; type SALMO TRUTTA L. "TRUTTA TOTA ARGENTEA" Klein.

A synonym of TRUTTÆ L.

Leuciscus Klein, I, 172, 1775; type CYPRINUS LEUCISCUS L. "LEUCISCUS SUPRA LINEA LATERALIS" Klein.

LEUCISCUS Klein would replace LEUCISCUS Cuvier, Rafinesque, Agassiz, for the same group.

Pelamys Klein, I, 176, 1775; type SCOMBER SCOMBRUS L. "PELAMYS CORPORE CASTIGATO" Klein, the "makarel" of Willughby.

The genus, based on the tunnies and the mackerels, is exactly coterminous with SCOMBER L. If regarded as eligible, the genus of snakes, PELAMYS Daudin, must receive a new name.

*See "On the Fishes Described in Müller's Supplemental Volume to the *Systema Naturæ* of Linnæus," D. S. Jordan, Proc. Acad. Nat. Sci., 1890, p. 48.

Harengus Klein, I, 209, 1775; type CLUPEA HARENGUS L. "HARENGUS VULGARIS" Klein.

Synonym of CLUPEA L.

Amphisilen Klein, I, 280, 1775; type CENTRISCUS SCUTATUS L. "AMPHISILEN" Klein.

Monotypic. A synonym of CENTRISCUS L., and AMPHISILE Cuvier.

Latargus Klein (misprint for LATHARGUS), I, 298, 1775; type ANARHICAS LUPUS L. "LATARGUS ROSTRO RETUSO DENTIS HORRIDIS" Klein.

Synonym of ANARHICHAS L. Monotypic.

Leiobatus Klein, I, 316, 1775; type RAJA OXYRHYNCHUS L. "LEIOBATUS ROSTRO OMNIUM LONGISSIME PRODUCTO" Klein.

Synonym of RAJA L. Includes all smooth rays.

Synagris Klein, I, 442, 1775; type SPARUS AURATA L. "SYNAGRIS DORSO OBSCURE VIRIDE" Klein.

A synonym of SPARUS L. Includes many sparoid fishes.

Mystus Klein, I, 535, 1775; type "MYSTUS FLUVIATILIS" Klein = CYPRINUS BARBUS L., not MYSTUS Gronow, 1763.

Equivalent to BARBUS Cuvier.

Passer Klein, I, 816, 1775; type PLEURONECTES FLESUS L. "PASSER CUTE DENSIS TUBERCULIS."

Not of Brisson, 1760, a genus of birds. A synonym of FLESUS Moreau.

Glaucus Klein, I, 829, 1775; type SCOMBER GLAUCUS L. "GLAUCUS ACULEATUS MACULIS IN UTROQUE LATERE" Klein.

GLAUCUS, if allowable, replaces CÆSIOMORUS Lacepède, a genus distinct from TRACHINOTUS Lacepède, 1803. HYPODIS Rafinesque, 1810, is the same.

Rhombus Klein, I, 918, 1775, VIII, 88, 1779; type PLEURONECTES RHOMBUS L. "RHOMBUS OMNIUM MINIMUS PALMÆ LONGITUDINE."

This name antedates RHOMBUS Da Costa, 1776, a genus of Snails. If eligible, it replaces BOTHUS Rafinesque, and is equivalent to RHOMBUS Cuvier, 1817. The other nominal species of RHOMBUS are synonyms of the Turbot, PSETTA MAXIMA (L.).

Rhombotides Klein, I, 922, 1775; type CHÆTODON CÆRULEUS Bloch. "RHOMBOTIDES OBSCURE CÆRULEUS" Klein, after Catesby.

Synonym of ACANTHURUS Forskål, and of TEUTHIS of Browne and Linnæus, Klein suggests as substitute names, if RHOMBOTIDES is not acceptable, EUROPUS and PSETTA.

Sargus Klein, I, 966, 1775; type SPARUS SARGUS L. "SARGUS PINNIS VENTRALIBUS" Klein.

SARGUS, if allowed, replaces DIPLODUS Rafinesque, and SARGUS Cuvier. The genus of Insects called SARGUS would in this case require a new name.

Dasybatus Klein, I, 991, 1775; type RAJA PASTINACA L. "DASYBATUS CAUDA SQUAMEIS OSSEIS" Klein.

DASYBATUS, if allowed, replaces DASYATIS Rafinesque and TRYGON Adanson. Intended to include all the rough-skinned rays.

Prochilus Klein, I, 1043, 1775; type, as here restricted, "PROCHILUS LÆVIS LATERIBUS CARINATUS" Klein (BLENNIUS PHOLIS L.).

Equivalent to PHOLIS Cuv. & Val., a name used earlier for another genus by Gronow and Scopoli. In this sense PROCHILUS is a subgenus under BLENNIUS. ..

On plates in Bleeker's Atlas, PROCHILUS is used instead of AMPHIPRION, the species named first by Klein being AMPHIPRION EPHIPPIUM (Bloch). We find, however, no formal restriction of PROCHILUS by Bleeker.

Labrax Klein, II, 32, 1775, and VIII, 164, 1779; type PERCA LABRAX L. "LABRAX SIVE LUPUS" Klein.

LABRAX, if allowed, replaces DICENTRARCHUS Gill = LABRAX Cuvier, 1817. Not LABRAX of Pallas, 1810, which is HEXAGRAMMOS.

Percis Klein, II, 45, Ed. 2, 1776; type PERCA CERNUA L., as restricted by Bleeker, *Systema Percarum Revisum*. "PERCIS PINNIS SEX ANTERIORE PARTE DORSALIS 14" Klein.

Equivalent to CERNUA Schæfer 1761, GYMNOCEPHALUS Bloch & Schneider 1801, to ACERINA Cuvier 1817, and CERNUA Fleming 1828; not PERCIS Scopoli 1777, nor of Cuvier.

Narcacion Klein, II, 237, 1776; IV, 726, 1777; type RAJA TORPEDO L. "NARCACION DEMTA CAUDA SINUOSA CIRCULARIS" Klein.

NARCACION, if accepted, replaces TORPEDO Duméril, not of Forskål. If rejected, NARCOBATUS Blainville is next in date. Monotypic.

Mænas Klein, II, 360, 1776; type SPARUS MÆNA L. "MÆNAS DILUTE VIRIDIS" Klein.

MÆNAS would replace MÆNA Cuvier.

Cicla Klein, II, 412, 1776; type LABRUS VIRIDIS L. "CICLA VIRIDIS OPERCULORUM" Klein.

Synonym of LABRUS L. Comprises all the species of LABRUS L., called TURDUS or MERULA by authors, these names also signifying "thrush." If accepted, CICHLA Bloch & Schneider, 1801, would require a new name.

Rhina Klein, II, 587, 1776; type RHINA SQUATINA L. "RHINA SIVE SQUATINA" Klein.

RHINA, if allowed, replaces SQUATINA Duméril.

Rhinobatus Klein, II, 593, 1776; type RAJA RHINOBATUS L. "RHINO-
BATUS SEU SQUATINO RAJA" Klein.

RHINOBATUS would replace RHINOBATUS Bloch & Schneider, 1801, and
SYRRHINA Müller & Henle, based on the same species.

Solea Klein, III, 115, 1776; type PLEURONECTES SOLEA L. "SOLEA
SQUAMIS MINUTIS" Klein.

SOLEA Klein would replace SOLEA Rafinesque, SOLEA Quensel and SOLEA Cuvier.

Pseudopterus Klein, III, 139, 1776; type GASTEROSTEUS VOLITANS L.
"PSEUDOPTERUS QUI PERCA AMBOINENSIS" Klein.

Replaces, if allowed, PTEROIS Cuvier.

Tetragonoptrus Klein, III, 153, 1776; type CHÆTODON CAPISTRATUS L.
"TETRAGONOPTRUS LÆVIS AD CAUDAM BRUNNEA" Klein.

Synonym of CHÆTODON L. Bleeker uses this name in place of CHÆTODON, as
Artedi's "type" of CHÆTODON—that is, the species first named—is a POMACANTHUS.
But Artedi included also true species of CHÆTODON in his list, and the name was
adopted by Linnæus from his own use of it, in the *Amœnitates Academiciœ*.

Batrachus Klein, III, 202, 1776; type LOPHIUS PISCATORIUS L. "BA-
TRACHUS RICTUQUE RANÆ" Klein.

Synonym of LOPHIUS L., not BATRACHUS Bloch & Schneider, 1801.

Mastaccembelus Klein, III, 271, 1776; type ESOX BELONE L. "MAS-
TACCEMBELUS MANDIBULIS LONGISSIMIS" Klein.

Not MASTACEMBELUS Gronow, 1763; stands as BELONE Cuvier.

Platiglossus Klein, III, 300, 1776; type HALICHŒRES MARGINATUS
Rüppell. "PLATIGLOSSUS SUBRUFUS SQUAMULIS LÆVIBUS" Klein.

Accepted by Bleeker as PLATYGLOSSUS. Klein's figure is crude, but recog-
nizable.

Lucius Klein, III, 506, 1776; type ESOX LUCIUS L. "LUCIUS . . . ROS-
TRO QUASI ANSERINO" Klein.

Synonym of ESOX L. = LUCIUS Rafinesque.

Cestracion Klein, III, 523, 1776; type SQUALUS ZYGÆNA L. "CESTRA-
CION FRONTE ACUS" Klein.

CESTRACION, if accepted, replaces SPHYRNA Rafinesque and ZYGÆNA Cuvier.
CESTRACION Cuvier is a different genus, HETERODONTUS Blainville, CENTRACION Gray.

Trichidion Klein, III, 592, 1776; type POLYNEMUS VIRGINICUS L. "TRI-
CHIDION CORPORE OBLONGO" Klein.

A synonym of POLYNEMUS L. as restricted by authors generally. Monotypic.

Asperulus Klein, III, 686, 1776; X, 236, 1781; type PERCA ZINGEL L.
"ASPERULUS VEL ASPREDO DORSO ACUTO" Klein.

Identical with ZINGEL Oken. Monotypic.

Corystion Klein, III, 762, 1776; type TRIGLA LAPPONICA L. "CORYS-
TION CORPORE GRANULATO" Klein.
A synonym of TRIGLA L.

In volume III of the *Gesellschaft Schauplatz*, 1776, pp. 61-73, is
given a list of the genera of fishes. This includes so far as we note
all those proposed by Klein and those of Linnæus also, those of Klein
appearing first. The names are not in alphabetical order, but follow the
sequence of Klein's *Historia Piscum Naturalis*. This seems to imply a
post-Linnæan acceptance of all of Klein's genera, even were their use in
the text rejected. One new name appears:

Pristis Klein, III, 61, 1776; type "PRISTIS, der Sägeschnautz."
Equivalent to PRISTIS Linck.

XXI. SCOPOLI, *Introductio ad Historiam Naturalem*, Prague, 1777.

JOHANN ANTON SCOPOLI.

A descriptive catalogue of the genera of animals and plants. The
genera of fishes are mostly those of Linnæus and Gronow. No types are
named save in the two new, PERCIS and PTERIDIUM.

The following names given by Gronow are not accepted by Scopoli:

ENCHELYOPUS (regarded as a synonym of BLENNIUS), ELEOTRIS (=
GOBIUS), PLECOSTOMUS (= LORICARIA), CORACINUS (= SCIÆNA), CAL-
LORHYNCHUS (= CHIMÆRA), CATAPHRACTUS (= PEGASUS), GYMNO-
GASTER (= TRICHIURUS), CYCLOGASTER (= LIPARIS), HEPATUS, (=
TEUTHIS); AMIA is used only in the Linnæan sense.

Liparis (Artedi) Scopoli, 453; type (not named) CYCLOPTERUS LIPARIS
L.

Percis Scopoli, 454; type COTTUS JAPONICUS Pallas.
Equivalent to HIPPOCEPHALUS Swainson, 1839, not PERCIS Klein.

Pteridium Scopoli, 454; type CORYPHÆNA VELIFERA Pallas (PTERACLIS
Gronow).

The following names of Gronow are introduced into Linnæan nomenclature by Scopoli, without mention of type. These date from Scopoli,
1777, if the *Zoophylaceum* of Gronow, 1763, be not accepted.

Erythrinus, 449

Synodus, 449

Callyodon, 449

Holocentrus, 449
(misprinted Holocenthrus)

Gonorhynchus, 450

Albula, 450

Umbra (Krämer), 450

Anableps, 450

Anostomus, 451

Mystus, 451

Callichthys, 451

Aspredo, 453

Charax, 455

Clarias, 455

Cynædus, 455

Pholis, 456

Mastacembelus, 458

Channa, 459

XXII. FORSTER, *Icones Ineditæ;* Bibliotheca Banksiæ, 1777.

John Reinhold Forster.

Echidna Forster, 181 ; type Echidna variegata Forster = Muræna
echidna Gmelin.

This generic name, according to Kaup, first appears in 1777, in the *Icones
Ineditæ.* It must be retained for the genus of fishes, and is not available for the
genus of mammals named Echidna by Cuvier, = Tachyglossus.

XXIII. KLEIN, *Gesellschaft Schauplatz,* vols. IV, V, 1777.

Jakob Theodor Klein.

Oncotion Klein, IV, 46, 1777 ; type Cyclopterus lumpus L. "Oncotion colore nigricante" Klein.

A synonym of Cyclopterus L.

Cynocephalus Klein, IV, 161, 1777 ; type Squalus glaucus L.
"Cynocephalus glaucus" Klein, as restricted by Gill.

Cynocephalus Boddært is of later date. Equivalent to Prionace Cantor.

Callarias Klein, IV, 327, 1777 ; type Gadus morrhua L. "Callarias
sordide olivaceus" Klein.

Synonym of Gadus L.

Crayracion Klein, IV, 788, 1777 ; type Tetraodon spengleri Bloch.
"Crayracion lævissimus ex terreo rufescens" Klein.

A synonym of Tetraodon L.

The type of this genus was first fixed by Bleeker, *Atlas Ichth.,* 1865, 65.
Bleeker observes: "La première espèce du genre compliqué que Klein en 1742 déjà
nomma Crayracion étant le Tetraodon spengleri des auteurs ou au moins une
espèce extrêmement voisine, je propose d' indiquer sous ce nom générique toutes

les espèces à tentacule nasale non perforée. . . . Le nom de Crayracion est antérieur de plusieurs années à celui de Tetraodon et devrait être substitué à ce dernier si le type du genre n'était pas reconnu à un genre distincte du Crayracion lævissimus de Klein."

This fixes the type of Klein on the species figured by him, but Bleeker is in error in supposing Tetraodon spengleri to be a species with closed nostrils. The proper name for that group—(Arothron Müller: Tetraodon of several authors) seems to be Ovoides Cuvier.

Cataphractus Klein, IV, 828, 1777; type Cottus cataphractus L. "Cataphractus rostro resimo" Klein.

Not Cataphractus Gronow. Stands as Agonus Lacepède.

Capriscus Klein, V, 427, 1777; type Balistes capriscus Gmelin. "Capriscus tribus aculeis" Klein.

Synonym of Balistes L.

Cestreus Klein, V, 460, 1777; type Mugil cephalus L. "Cestreus dorso repando" Klein.

A synonym of Mugil L.

XXIV. FORSTER, *Enchiridion,* 1778.

John Reinhold Forster.

Harpurus Forster, 84; type Harpurus fasciatus Forster.
Equivalent to Hepatus, Teuthis and Acanthurus.

XXV. KLEIN, *Gesellschaft Schauplatz,* vols. VI to X, 1778 to 1781.

Jakob Theodor Klein.

Solenostomus Klein, VI, 32, 1778; type Fistularia tabaccaria L. "Solenostomus cute glabra" Klein.

Synonym of Fistularia. If Solenostomus is accepted, a new generic name is required for Solenostsmus Lacepède.

Sphyræna Klein, VI, 464, 1778; type Esox sphyræna L.

Would replace Sphyræna Röse, unless Umbla Catesby is available.

Gobio Klein, VII, 178, 1779; type Gobius niger L. "Gobio branchiarum operculis et ventre flavicantibus" Klein.

If allowed, Gobio Klein becomes a synonym of Gobius L. and a new name would be required in place of Gobio Cuvier.

Hippurus Klein, VII, 788, 1779; type Coryphæna hippurus L. "Hippurus pinnis branchialibus" Klein.

Synonym of Coryphæna L.

Blennus Klein, VIII, 589, 1779; type BLENNIUS OCELLARIS L. "BLENNUS PINNICEPS COLORIS" Klein.
Synonym of BLENNIUS L.

Psalisostomus Klein, X, 154, 1781; type ESOX OSSEUS L. "PSALISOSTOMUS OMNIUM MAXIMUS" Klein.
If allowed, replaces LEPISOSTEUS Lacepède.

XXVI. HERRMANN, *Schreiben ueber eine neues Americanisches Fischgeschlecht,* STERNOPTIX: Der Naturforscher, 2 Stück, vol. XVI, 1781.

JOHANN HERRMANN.

Sternoptix Herrmann, 8, 36; type STERNOPTIX DIAPHANA Herrmann.
Monotypic.

XXVII. HOUTTUYN, *Beskrivning van Eenige Japanske Visschen*: Actæ Harlemensis, XX, pt. 2, 1782.

MARTIN HOUTTUYN.

Centrogaster Houttuyn, 333; type CENTROGASTER FUSCESCENS Houttuyn.
A synonym of SIGANUS Forskål.

XXVIII. BLOCH, *Naturgeschichte der Ausländischen Fische.* Nine parts, 1785 to 1795. Part 2, 1786.

MARK ELIESER BLOCH.

Kurtus Bloch, II, 122, 1786; type KURTUS INDICUS Bloch.
Also written KYRTUS and CYRTUS. Monotypic.

Macrourus Bloch, II, 150, 1786; type CORYPHÆNA RUPESTRIS Fabricius, the "Ingmingoak" = MACROURUS BERGLAX Lacepède, which is not the same as CORYPHÆNOIDES RUPESTRIS Gunner.

XXIX. BLOCH, *Ueber Zwey Merkwürdige Fisch-Arten:* Abhandlungen Böhmischer Gesellschaft, I, 1787.

Notacanthus Bloch, 278; type NOTACANTHUS CHEMNITZI Bloch.
This name was changed by Bloch in 1797 to ACANTHONOTUS and the species to ACANTHONOTUS NASUS. Monotypic.

XXX. AHL, *De Muræna et Ophichtho, 1787.*

JONAS NICHOLAS AHL.

Ophichthus Ahl, 5; type MURÆNA OPHIS L., as restricted.
Unquestioned.

XXXI. ASCANIUS, *Beretning um Silde-Tusten*: Dansk. Selsk. 1788.

P. ASCANIUS

Regalecus Ascanius, III, 419; type OPHIDIUM GLESNE Ascanius.
Other dates have been quoted, for example "Icones," 1772, a series of plates of
objects in nature. This may not be the oldest.

XXXII. BLOCH, *Charactere und Beschreibung des Geschlechts der
Papageyfische,* CALLYODON: Abhandlungen Böhmischer Gesellschaft, IV, 1788.

MARK ELIESER BLOCH.

Callyodon Bloch, 242; type (presumably) SCARUS CROICENSIS Bloch.
Not seen by us.

XXXIII. BLOCH, *Naturgeschichte der Ausländischen Fische,* III, 1788.

Gymnetrus Bloch, III, 1; type GYMNETRUS HAWKENI Bloch.
A synonym of REGALECUS.

XXXIV. BROWNE, *Civil and Natural History of Jamaica,* Second
Edition, 1789.
PATRICK BROWNE, M. D.

Originally published in 1756, reprinted in 1789, with a table showing
the Linnæan equivalents of species named, and perhaps other revisions.
No changes have been noted in the systematic part. In the original edition of 1756, as in later editions, the author refers to the works of Artedi
and Gronow, with both of whom he was familiar. His recognition of
genus as a technical term and not as a mere Latin noun is unmistakable.
Browne adopts various genera of Artedi (later accepted by Linnæus),
and to these he adds twelve new genera of his own. The species are

indicated by polynomial terms, to which are added the English vernacular names current in Jamaica.

The *History of Jamaica* is the ablest of all the works describing a local fish-fauna, prior to Linnæus, in quality comparable to the work of Osbeck, Hasselquist, and even of Forskål.

The revised republication of the work in 1789 may perhaps make the names of Browne eligible in nomenclature, if other authors rigidly correct as to generic names but polynomial as to species are to be considered. We have examined the original edition and the edition of 1789.

The names of Browne are perhaps not eligible as not binomial and as occurring in a slightly revised reprint of a pre-Linnæan work.

Following the precedent of Opinion 57, rejecting Hasselquist (*Iter Palestinum*), Browne's work would not be regarded as available.

Solenostomus Browne, 441; type "SOLENOSTOMUS CORPORE TERETE SUB-ROTUNDO etc., the Trumpeter" = FISTULARIA TABACCARIA L.

This name, if eligible, antedates SOLENOSTOMUS Klein and SOLENOSTOMUS Lacepède. Monotypic.

Menidia Browne, 441; type "MENIDIA CORPORE PELLUCIDO, LINEA LATERALI LATIORI ARGENTEA" = ATHERINA BROWNI Gmelin, 1789.

This name, if accepted, replaces ANCHOVIELLA Fowler, besides rendering MENIDIA Bonaparte ineligible. Monotypic.

Amia Browne, 442; type "AMIA SUBARGENTEA LABRIS ÆQUALIBUS OSSICULUS BRANCHIOSTEGUS VIGINTIDUOBUS, the Tarpon" = MEGALOPS ATLANTICUS Cuv. & Val.

This name is equivalent to TARPON Jordan & Evermann. It is however preoccupied by AMIA Gronow and by AMIA L. Browne refers also to AMIA, a second species, "the Ten-pounder," which is ELOPS SAURUS L., type of the genus ELOPS.

Mormyra Browne, 445; type "MORMYRA MAJOR CÆRULEA ET AUREO VARIA, the larger Painted Parrot-fish" = SCARUS VETULA Bloch & Schneider.

The name MORMYRA, too close to MORMYRUS L., is subsequent to CALLYODON of Gronow, and to SCARUS Forskål. Three other species of Parrot-fish are placed by Browne in MORMYRA. Synonym of CALLYODON.

Plagusia Browne, 445; type "PLAGUSIA SUBCINEREA CAUDA ATTENUATA, the little Brown Sole with a pointed tail" = PLEURONECTES PLAGUSIA Bloch & Schneider, 1801 = SYMPHURUS PLAGUSIA Jordan & Evermann.

The name PLAGUSIA was adopted for this genus by Cuvier in 1829. It had however been earlier (1806) used by Latreille for a genus of Crustaceans.

Browne's name PLAGUSIA, if available, has priority over SYMPHURUS Rafinesque, 1820, as well as over the numerous later names applied to this group, the best known of which are APHORISTIA Kaup and AMMOPLEUROPS Günther. As there is a still earlier species, PLAGUSIA PLAGIUSA L., in this genus, the present species may perhaps stand as PLAGUSIA ORNATA Lacepède, "PLAGIUSA" being only a variant spelling of the same word.

Helops Browne, 445; type "HELOPS NIGRESCENS VARIE NEBULATIM, the Hog-fish of Catesby" = LABRUS RUFUS L.

If HELOPS, with this type, is accepted, it will eliminate the much confused generic name BODIANUS Bloch, based on the same type (BODIANUS BODIANUS Bloch = LABRUS RUFUS L.). HARPE Lacepède and COSSYPHUS Cuvier are later names for the same genus. The first species indicated by Browne under HELOPS is LACHNOLAIMUS MAXIMUS (Walbaum), the "SUILLUS or Great Hog-fish" of Catesby. This is "HELOPS RUFESCENS IRIDE PARTIM RUBRA, PARTIM ALBIDA, MACULA NIGRA POST PINNIM DORSALEM, the Hog-fish" of Browne. In the interest of nomenclature, it would be better to suppress BODIANUS rather than LACHNOLAIMUS, if HELOPS is found eligible.

Cromis Browne, 449; type "CROMIS SUBARGENTEO OBLONGUS RADIIS ANTERIORIBUS DORSALIS ÆGRE PUNGENTIBUS, the Drummer" = LABRUS CROMIS L. = POGONIAS CHROMIS of authors.

Browne cites four species of his genus CROMIS, the "Silver Shad" (GERRES CINEREUS), the Red-mouth Grunt (HÆMULON PLUMIERI), the Stone Bass (DIAPTERUS BRASILIANUS), and the Drummer (POGONIAS CROMIS). As the last named became LABRUS CROMIS L., we may take it as the type of the genus CROMIS of Browne, if the latter is eligible. CROMIS would then, if accepted, replace POGONIAS Lacepède. CHROMIS Cuvier, 1815, a more correct spelling of the same word, would then give place to HELIASES Cuvier.

Macrocephalus Browne, 449; type "MACROCEPHALUS ARGENTEA MAJOR LINEA LATERALIS RECTA NIGRA, the Snook" = SCIÆNA UNDECIMALIS Bloch.

Monotypic. Equivalent to CENTROPOMUS Lacepède. Preoccupied by MACROCEPHALUS Swederus, 1787, a genus of Insects.

Pelmatia Browne, 449; type "PELMATIA MAJOR SQUAMIS VIX PERSPICUUS, the Mud-fish" = GOBIOMORUS DORMITOR Lacepède.

The name PELMATIA, if accepted, replaces GOBIOMORUS, which in turn has replaced the excellent name PHILYPNUS of Cuvier.

The first species named under PELMATIA by Browne is "PELMATIA MINOR SQUAMIS MAJUSCULUS, the Bullhead." This is DORMITATOR MACULATUS (Bloch). Browne gives a long and correct account of PELMATIA, and notes that his "second sort," the mud-fish, is "most esteemed and grows frequently to the length of 17 to 20 inches. It is the most delicate fish I have yet known, when in full perfection."

Thynnus Browne, 451; type "Thynnus bontii corpore crassiori et breviore etc., The Boneeto" = Scomber pelamys L.

This use of the name Thynnus is later than that of Fabricius, 1775, for a genus of Insects. It antedates Thynnus of Cuvier and it is equivalent to Thynnus of Lütken, a name later changed to Euthynnus by its author. Scomber pelamys L., the Oceanic Bonito, is the Bonito of Cuba and Jamaica, where the Northern Bonito, Sarda sarda, is unknown. Of the "Boneeto," Browne observes that it is "a dry coarse fish, not much esteemed, though a hearty wholesome food." Should stand as Euthynnus Lütken. Monotypic.

Saurus Browne, 452; type "Saurus argenteus cute longitudinalis etc., the Leather Coat" = Scomber saurus Bloch & Schneider = Oligoplites saurus of authors.

Not Saurus Cuv. & Val., which is Synodus Gronow. If Browne's names are accepted, Saurus must replace Oligoplites Gill, its type species standing as Saurus saurus (Bloch & Schneider). Two other species of Saurus are enumerated by Browne, the "Red-tailed Jack" and the "White-fish."

Teuthis Browne, 454; type "Teuthis fusca cæruleo nitens, the Doctor (Turdus rhomboides Catesby)" = Chætodon cæruleus Bloch.

The first application of the name Teuthis (τεύθις, a squid) to a fish is that of Browne in 1756. Linnæus in 1766 accepted Browne's name Teuthis, and substituted it for Gronow's name Hepatus, 1763. The republication of Teuthis in 1789 would help to fix the generic name Teuthis with Browne's original species. Hepatus Gronow was primarily based on a specimen of the West Indian species called Chætodon chirurgus by Bloch, with which Chætodon cæruleus Bloch, the type of Teuthis, is strictly congeneric. The other species concerned, Teuthis javus L., wrongly referred by Gronow to Hepatus, should stand as Siganus Forskål. Cuvier and Valenciennes, according to Günther, were in error in referring Gronow's specimen, the type of Teuthis hepatus L., to an East Indian species, since called Colocopus lambdurus Gill.

The four species entangled in the confusion superimposed upon Browne, should, as Gill has shown, stand as follows:

Hepatus or Teuthis cæruleus (Bloch & Schneider).
Hepatus or Teuthis hepatus L.
Colocopus lambdurus Gill.
Siganus javus (L.).

Rhomboida Browne, 455; type "Rhomboida alepidota argentea pinnis omnis brevibus, the Silver-fish" = Vomer browni Cuv. & Val., 1833.

The generic name Rhomboida antedates that of Vomer Cuvier, and, if accepted, the type species would stand as Rhomboida browni (Cuv. & Val.) = Platysomus spixi Swainson, a species distinct from the northern Vomer setipinnis (Mitchill).

Under Rhomboida, Browne mentions also "Rhomboida major alepidota, the Larger Silver-fish, with long fins (Zeus cauda bifurca Artedi)." This must be

Selene vomer L. A third species is "Rhomboida squamosa ex argentea, the Portugise." This seems to be Pomacanchus arcuatus L., the "Portugais" of the French Antillan fishermen.

XXXV. LINCK, *Magazin Neuestes aus der Physik und Naturgeschichte,* Gotha, 1790.

H. F. Linck.

This work we have not seen; we quote from Gill.

Mustelus Linck, 31; type Squalus mustelus L. = Mustelus lævis of authors.
Equivalent to Pleuracromylon Gill, not quite the same as Mustelus Valmont.

Pristis Linck, 31; type Squalus pristis L.
Unquestioned.

Rhinobatos Linck, 32; type not specified: Raja rhinobatos L.
Equivalent to Rhinobatus Klein.

Callichthys Linck, 32; type not specified: Silurus callichthys L.
Equivalent to Callichthys Gronow.

Alosa Linck, 35 (scarcely defined); type not specified: Clupea alosa L. = Alosa Cuvier.

Thymallus Linck, 35; type not specified: Salmo thymallus L. = Thymallus Cuvier.

Mola Linck, 37; type Diodon mola L.
Equivalent to Mola Koelreuter and Cuvier.

Soarus Linck (misprint for Saurus), 37; type not specified and unidentifiable.

Barbatula Linck, 38; type Cobitis barbatula L.
Replaces Oreias Sauvage; Orthrias Jordan & Fowler.

XXXVI. BLOCH, *Naturgeschichte der Ausländischen Fische,* IV, 1790.

Mark Elieser Bloch.

Bodianus Bloch, IV, 48; type Bodianus bodianus Bloch, by tautonomy.
Replaces Harpe Lacepède. By first restriction, Cuvier and Gill, the type would be Bodianus guttatus Bloch, a species of Enneacentrus Gill. Tautonomy has precedence in this case.

Lutianus Bloch, IV, 105; type Lutianus lutianus Bloch.
Unquestioned. Also spelled Lutjanus.

50 THE GENERA OF FISHES

XXXVII. WHITE, *Journal of a Voyage to New South Wales,* 1790.

J. WHITE.

Enoplosus White, plate 39; type CHÆTODON ARMATUS White.
Lacepède quotes the name ENOPLOSUS from White. We have not seen this paper, and the name may have been first printed by Lacepède.

XXXVIII. VALMONT DE BOMARE, *Dictionnaire* etc., Edition IV, vol. VIII, 1791.

JEAN CHRISTOPHE VALMONT DE BOMARE.

Acus Valmont; type ACUS ARISTOTELIS Valmont = SYNGNATHUS ACUS L.
After Willughby. Same as SYNGNATHUS L.

XXXIX. SHAW, *Description of* STYLEPHORUS CHORDATUS, *a new fish*: Transactions of the Linnæan Society of London, I, 1791.

GEORGE SHAW.

Stylephorus Shaw, I, 90; type STYLEPHORUS CHORDATUS Shaw.
Monotypic.

XL. WALBAUM, *Artedi Piscium,* 1792.

JOHANN JULIUS WALBAUM.

Curimata Walbaum, 80; type SALMO MARCGRAVII Walbaum, based on "CHARAX MAXILLA SUPERIORE LONGIORE" Gronow, which is SALMO CYPRINOIDES L.
CURIMATA should replace CURIMATUS Cuvier for this genus.

XLI. BLOCH, *Naturgeschichte der Ansländischen Fische,* VI, VII, 1792, 1793.

MARK ELIESER BLOCH.

Anthias Bloch, VI, 97, 1792; type LABRUS ANTHIAS L. = ANTHIAS SACER Bloch.
Unquestioned. AYLOPON Rafinesque is a substitute name, ANTHIAS being said to be preoccupied. We do not find it so. ANTHIA, a genus of Beetles, dates from 1801.

Epinephelus Bloch, VII, 11, 1793; type EPINEPHELUS MARGINALIS Bloch = PERCA FASCIATA Forskål, by general consent.

The genus EPINEPHELUS was based on E. AFER, E. MARGINALIS, E. MERRA, and E. RUBER. MARGINALIS and MERRA are congeneric, and belong to the great group called EPINEPHELUS by Gill, Bleeker, and nearly all recent authors. Of these, MARGINALIS is typical. The species named first, AFER, has been on that account chosen as type by Fowler, 1907. This species was separated as the type of ALPHESTES by Bloch & Schneider, 1801. RUBER was named as type by Jordan & Gilbert, 1883. This species under another name (ACUTIROSTRIS Cuv. & Val.) became the type of PAREPINEPHELUS Bleeker, 1875. Justice and convenience are best served by retaining the name EPINEPHELUS for its chief components, as understood by nearly all authors. Otherwise the genus would stand as CERNA Bonaparte, 1837, unless, with Fowler, we recognize EPINEPHELUS GIGAS (PERCA GIGAS) L. as the type of SERRANUS Cuvier, 1817, a change we think unnecessary. If the subgenera of Forskål, with Arabic names, are recognized, EPINEPHELUS must give place to DABA.

Gymnocephalus Bloch, VII, 24, 1793; type PERCA SCHRÆTZER L., equivalent to CERNUA Schæfer, ACERINA Cuvier, and CERNUA Fleming.

Johnius Bloch, VII, 132, 1793; type JOHNIUS CARUTTA Bloch, as restricted by Gill.

Lonchiurus Bloch, VII, 143, 1793; type LONCHIURUS BARBATUS Bloch, 1793 = PERCA LANCEOLATA Bloch, 1788.

Monotypic. Corrected by later writers to LONCHURUS.

Cataphractus Bloch, VII, 80, 1793; type SILURUS CALLICHTHYS L. as here restricted.

A synonym of CALLICHTHYS.

XLII. RÖSE, *Petri Artedi Angermannia-Sueci Synonymia Nominum Piscium* etc. Greifswald. Edition II, 1793.

ANTON FERDINAND RÖSE.

This article, published as a supplement to Walbaum's *Artedi Piscium,* enumerates the generic names of Artedi and others, the species in their original polynomial form. In an Appendix are given a few new generic names, mostly taken from Aristotle. These genera are not described, nor are their species named, but the synonymy is fully given. Some of them had been already used by other authors. In our judgment, these names are eligible, and a few maintain priority of date.

Phycis Röse, 111; type φυκίς Aristotle, PHYCIS TINCA Bloch & Schneider. GADUS BLENNIOIDES Brünnich.

This is identical with PHYCIS Bloch & Schneider, 1801; the latter, but not the former, antedated by PHYCIS Fabricius, 1798, a genus of Insects. PHYCIS Röse replaces EMPHYCUS Jordan & Evermann, 1898.

Cicla Röse, 112; type "CICLA VIX PALMARIS" Röse, κίχλη Aristotle, a species of LABRUS, perhaps L. VIRIDIS L.

CICLA Klein is identical with that of Röse. Schneider, more correctly, wrote the word CICHLA, but applied it to a different group.

Sphyræna Röse, 112; type σφύραινα Aristotle = ESOX SPHYRÆNA L.

This antedates SPHYRÆNA Bloch & Schneider, 1801, for the same genus.

Hepatus Röse, 113; type ἥπατος Aristotle, which is probably LABRUS HEPATUS L., the type of the genus PARACENTROPRISTIS Bleeker.

HEPATUS Gronow is wholly different.

Capriscus Röse, 114; type καπρίσκος Diphili, κάπρος Aristotle, which is probably BALISTES CAPRISCUS L.

A synonym of BALISTES L.

Tænia Röse, 114; type ταινία Aristotle = CEPOLA TÆNIA L.

The genus is equivalent to CEPOLA L., and is preoccupied in Worms by TÆNIA L.

Pholis Röse, 116; type φολὶς Aristotle = BLENNIUS PHOLIS L.

Equivalent to PHOLIS Cuv. & Val., not of Scopoli, 1777.

Citharus Röse, 116; type κιθάρος Aristotle = PLEURONECTUS LINGUA-TULA L.

Equivalent to CITHARUS Bleeker, 1862, and EUCITHARUS Gill, 1888; not CITHARUS Reinhardt, 1838, which is HIPPOGLOSSOIDES.

Liparis Röse, 117; type "LIPARIS NOSTRAS Johnson" = CYCLOPTERUS LIPARIS L.

Equivalent to CYCLOGASTER Gronow and LIPARIS Scopoli.

Chelon Röse, 118; type χάλλων or χέλων Aristotle, CHELON of Gesner, which is probably MUGIL CHELO Cuv. & Val.

Probably equivalent to CHÆNOMUGIL Gill, but the European and American types need further comparison.

XLIII. LATHAM, *Essay on the various species of Saw-Fish:* Transactions of the Linnæan Society of London, 1794.

JOHN F. LATHAM.

Pristis Latham, II, 276; type SQUALUS PRISTIS L.
Same as PRISTIS Linck.

XLIV. VAHL, *Beskrivelse af en nye Fiskeslaegt:* Shrivt. Naturh. Selsk., Kjöbenhavn, III, 1794.

M. VAHL.

Cæcula Vahl, III, 2, 149; type CÆCULA PTERYGERA Vahl.
Monotypic.

XLV. BLOCH, *Naturgeschichte der Ausländischen Fische*, VIII, 1794; IX, 1795.

MARK ELIESER BLOCH.

Platystacus Bloch, VIII, 52, 1794; type PLATYSTACUS COTYLEPHORUS Bloch, as usually restricted.

Ophicephalus Bloch, VIII, 137, 1794; type OPHICEPHALUS PUNCTATUS Bloch.
Unquestioned. Corrected by later writers to OPHIOCEPHALUS.

Sphagebranchus Bloch, IX, 88, 1795; type SPHAGEBRANCHUS ROSTRATUS Bloch.
Monotypic.

Gymnothorax Bloch, IX, 83, 1795; type GYMNOTHORAX MURÆNA Bloch = MURÆNA HELENA L.
A synonym of MURÆNA L. Günther, *Cat. Fish*, VIII, p. 100, 1870, restricts the names to allies of MURÆNA AFRA = LYCODONTIS McClelland, an arrangement apparently not defensible, as GYMNOTHORAX was plainly a substitute name for MURÆNA, and must retain the same type, MURÆNA HELENA L.

Synbranchus Bloch, IX, 86, 1795; type SYMBRANCHUS MARMORATUS Bloch.
Name corrected by later writers to SYMBRANCHUS.

Platycephalus Bloch, IX, 96, 1795; type PLATYCEPHALUS SPATHULA Bloch = COTTUS INSIDIATOR Forskål = CALLIONYMUS INDICUS L.
Unquestioned.

Gastrobranchus Bloch, XII, 51, 1797; type GASTROBRANCHUS CÆCUS Bloch = MYXINE GLUTINOSA L.
A synonym of MYXINE L.

Acanthonotus Bloch, XII, 113, 1797; type ACANTHONOTUS NASUS Bloch, 1797 = NOTACANTHUS CHEMNITZI Bloch, 1787.
A needless substitute name for NOTACANTHUS.

XLVI. FABRICIUS, *Beskrivelse over to sieldne Grönländske Fiske:* Skrivt. Naturhist. Selskab. Kjöbenhavn, 1793, II.

OTTO FABRICIUS.

Campylodon Fabricius, 12; type CAMPYLODON (FABRICII Reinhardt), 1838.
A synonym of NOTACANTHUS Bloch. Mononomial and monotypic.

XLVII. VOLTA, *Ichthyolithologia Veronensis,* 1796.

SERAFINO VOLTA.

Blochius Volta, 53; type BLOCHIUS LONGIROSTRIS Volta (Family BLOCHIIDÆ, fossil).

XLVIII. LACEPÈDE, *sur le Polyodon feuille:* Bull. Sci. Soc. Philom., 1797.

BERNARD GERMAIN ÉTIENNE DE LA VILLE-SUR-ILLON COMTE DE LACEPÈDE.

The name of this author should be written Lacepède. Sherborne, *Index Animalium,* says: "A letter dated 1831 is signed 'b. g. é cte. de Lacepède.' This spelling and accentuation should be adhered to."

Polyodon Lacepède, 49; type POLYODON FOLIUM Lacepède.
Monotypic.

XLIX. CUVIER, *Tableau Élémentaire,* 1798.

GEORGES JEAN LEOPOLD NICOLAS FRÉDÉRIC CUVIER.

Mola Cuvier, 323; type DIODON MOLA L.
Equivalent to MOLA Kœlreuter. Prior to ORTHAGORISCUS Bloch & Schneider, 1801. Monotypic.

Murænophis Cuvier, 329; type MURÆNA HELENA L.
A synonym of MURÆNA L.

L. GEOFFROY SAINT HILAIRE, *Description d'un nouveau Genre de Poisson:* Bull. Soc. Sci. Philom., III, "An X de la République," 1798.

ETIENNE GEOFFROY SAINT HILAIRE.

Polypterus St. Hilaire, 97; type POLYPTERUS BICHIR St. Hilaire.
Monotypic.

LI. RETZIUS, *Lampris, En ny Fiskslagt Beskriven:* Vet. Acad. Nya Handl., Stockholm, XX, 1799.

A. J. RETZIUS.

Lampris Retzius, 91; type ZEUS GUTTATUS Brünnich = ZEUS- REGIUS Bonnaterre.

LII. SHAW, *Naturalist's Miscellany,* 1799.

GEORGE SHAW.

Trachichthys Shaw, 378; type TRACHICHTHYS AUSTRALIS Shaw.
Monotypic.

LIII. CUVIER, *Leçons d'Anatomie Comparée,* 1800.

GEORGES CUVIER.

Ovoides Cuvier, 1, tab. 1; type OVOIDES FASCIATUS Lacepède = TETRO-
DON STELLATUS.
After "LES OVOIDES Lacepède," 1798. Replaces AROTHRON Müller.

LIV. LACEPÈDE, *Histoire Naturelle des Poissons,* *vol. I, 1798; II, 1800.

BERNARD GERMAIN ÉTIENNE DE LA VILLE-SUR-ILLON, COMTE DE
LACEPÈDE (here called "CITOYEN LA CEPÈDE").

The generic names in Volume I, 1798, and Volume II, 1800, to page 160, are nearly all given in French vernacular and are therefore ineligible. Scientific forms for most of these were supplied by Duméril, *Zoologie Analytique,* 1806, from which work the generic names in question must be dated.

Aodon Lacepède, I, 297, 1798; type SQUALUS MASSASA Forskål, as restricted by Jordan & Evermann.
A shark, perhaps imaginary, with no teeth and with long pectorals. The Latin name AODON is used, as well as the French "LES AODONS."

Ovoides (Lacepède), 521, "LES OVOIDES"; type "L'OVOIDE FASCÉ."
A front view of TETRAODON STELLATUS L. Equivalent to OVOIDES Cuvier.

Spheroides (Lacepède) Duméril (LES SPHÉROIDES), II, 22, 1800; type
LE SPHÉROIDE TUBERCULÉ Lacepède.
A front view of TETRAODON SPENGLERI Bloch. A synonym of TETRAODON as here understood.

---

* Of this work we have before us two reprints, neither with the original pagination. This, however, we are able to give, through the kindness of Mr. Henry W. Fowler of the Academy of Natural Sciences of Philadelphia, who has supplied us with a complete list of the genera and the pages on which they occur.
In the original edition the author styles himself "Citoyen La Cepède."

Macrorhynchus (Lacepède) Duméril (LES MACRORHYNQUES) ; type LE MACRORHINQUE ARGENTÉ Lacepède, SYNGNATHUS ARGENTEUS Osbeck.

Probably replaces DICROTUS Günther.

Cæcilia Lacepède), II, 134, 1800 (LES CÆCILIES) ; type CÆCILIA BRANDERIANA Lacepède = MURÆNA CÆCA L.

Not CÆCILIA L., a genus of Amphibians. A synonym of SPHAGEBRANCHUS Bloch.

Monopterus (Lacepède), Duméril, II, 138, 1800 (LES MONOPTÈRES) ; type MONOPTERUS JAVANENSIS Lacepède.

Notopterus Lacepède, II, 189, 1800 ; type GYMNOTUS KAPERAT Bonnaterre (GYMNOTUS NOTOPTERUS Pallas).

Ophisurus Lacepède, II, 195, 1800 ; type MURÆNA SERPENS L. as restricted by Risso, 1826.

Triurus Lacepède, II, 200, 1800 ; type TRIURUS BOUGAINVILLEI Lacepède, later called POMATIAS by Bloch & Schneider.

Unidentified. A deep-sea fish, apparently allied to AULOSTOMATOMORPHA Alcock, but without ventral fins.

Apteronotus Lacepède, II, 208, 1800 ; type APTERONOTUS PASSAN Lacepède = GYMNOTUS ALBIFRONS L.

Must replace STERNARCHUS Bloch & Schneider.

Odontognathus Lacepède, 220, 1800 ; type ODONTOGNATHUS MURICATUS. Lacepède.

Macrognathus Lacepède, II, 283, 1800 ; type OPHIDIUM ACULEATUM L.

Replaces RHYNCHOBDELLA Bloch & Schneider, with the same type. The name MACROGNATHUS was used by Gronow in 1754 (but not in 1763) to designate the group called BELONE Cuvier.

Comephorus Lacepède, II, 312, 1800 ; type CALLIONYMUS BAIKALENSIS. Bonnaterre.

Monotypic.

Rhombus Lacepède, II, 312, 1800 ; type STROMATEUS ALEPIDOTUS L.

Preoccupied, replaced by PEPRILUS Cuvier.

Murænoides Lacepède, II, 324, 1800 ; type BLENNIUS MURÆNOIDES. Sujef (BLENNIUS GUNNELLUS L.).

Equivalent to PHOLIS Scopoli, 1777.

Calliomorus Lacepède, II, 343, 1800 ; type CALLIONYMUS INDICUS L.

Synonym of PLATYCEPHALUS Bloch.

Batrachoides Lacepède, II, 351, 1800; type BATRACHOIDES TAU Lacepède.

Not GADUS TAU L.

Because the body is said to be covered with scales, "molles, petites, minces, rondes, brunes, bordees de blanc et arrosées par une mucosité très abondantes," we cannot identify this genus with the naked GADUS TAU L., which is identical with the type of OPSANUS Rafinesque.

Oligopodus Lacepède, II, 511, 1800; type CORYPHÆNA VELIFERA Pallas.

Equivalent to PTERACLIS Gronow.

Hiatula Lacepède, II, 522, 1800; type HIATULA GARDENIANA Lacepède = LABRUS ONITIS L. = LABRUS HIATULA L.

Not HIATULA Modeer, 1793, a genus of Mollusks. Gives way to TAUTOGA Cuvier.

Tænioides Lacepède, II, 532, 1800; type TÆNIOIDES HERMANNIANUS Lacepède.

Gobioides Lacepède, II, 576, 1800; type GOBIUS BROUSSONETI Lacepède, as restricted by Jordan & Evermann, 1898.

Gobiomorus Lacepède, II, 583, 1800; type GOBIOMORUS DORMITOR Lacepède, as restricted by Jordan.

Replaces PHILYPNUS Bloch & Schneider, 1801.

Gobiomoroides Lacepède, II, 592, 1800; type GOBIOMOROIDES PISON Lacepède = GOBIUS PISONIS L., a species of ELEOTRIS Gronow.

Monotypic.

Gobiesox Lacepède, II, 595, 1800; type GOBIESOX CEPHALUS Lacepède.

Monotypic.

LV. BLOCH & SCHNEIDER, *Systema Ichthyologia,* 1801.

MARK ELIESER BLOCH.

Edited and extended by JOHANN GOTTLOB SCHNEIDER.

Batrachus Bloch & Schneider, 42; type BATRACHUS SURINAMENSIS Schneider, as restricted by Jordan & Evermann.

The name has been usually applied to the congeners of GADUS TAU L., not of Bloch, the group called OPSANUS by Rafinesque. But no ally of the scaleless GADUS TAU is placed in BATRACHOIDES by Lacepède or in BATRACHUS by Schneider, only scaly species being known to either.

Enchelyopus Bloch & Schneider, 50; type, as first restricted, GADUS CIMBRIUS L. (RHINONEMUS CIMBRIUS Gill).

Not ENCHELYOPUS of Gronow, which is ZOARCES Cuvier, nor of Klein, which is TRICHIURUS.

Phycis Bloch & Schneider, 56; type BLENNIUS PHYCIS L. (PHYCIS TINCA Bloch & Schneider).

Equivalent to PHYCIS Röse, 1793; not PHYCIS Fabricius, 1798, a genus of Butterflies.

Periophthalmus Bloch & Schneider, 63; type PERIOPHTHALMUS PAPILIO Bloch & Schneider.

Eleotris Bloch & Schneider, 65; type ELEOTRIS Gronow (GOBIUS PISONIS Gmelin).

The name is borrowed from Gronow, whose ELEOTRIS is the ELEOTRIS of Cuvier and subsequent authors. But ELEOTRIS PISONIS is indicated by Schneider as "a species of doubtful relations, perhaps a PERIOPHTHALMUS." We let the current arrangement stand, though on shaky foundation, unless the names of Gronow are finally accepted. The genus ELEOTRIS of Bloch & Schneider represents an utter confusion of species, many of them not gobies at all.

Brama Bloch & Schneider, 98; type SPARUS RAJI Bloch.

Not BRAMA Klein, which is ABRAMIS Cuvier.

Monocentris Bloch & Schneider, 100; type GASTEROSTEUS JAPONICUS Houttuyn.

Monotypic.

Sphyræna Bloch & Schneider, 109; type ESOX SPHYRÆNA L.

Equivalent to UMBLA Catesby. Subsequent to SPHYRÆNA Röse.

Trichogaster Bloch & Schneider, 164; type TRICHOGASTER FASCIATUS Bloch & Schneider.

Centronotus Bloch & Schneider, 165; type CENTRONOTUS FASCIATUS Schneider.

A synonym of PHOLIS Scopoli.

Percis Bloch & Schneider, 179; type PERCIS MACULATA Schneider.

Not PERCIS Klein nor of Scopoli. Replaced by PARAPERCIS Bleeker. Monotypic.

Trichonotus Bloch & Schneider, 179; type TRICHONOTUS SETIGER Bloch & Schneider.

Monotypic.

Monoceros Bloch & Schneider, 180; type MONOCEROS BIACULEATUS Bloch & Schneider (CHÆTODON UNICORNIS Forskål).

Name preoccupied in Mollusks. Equivalent to NASO Lacepède.

Grammistes Bloch & Schneider, 182; type PERCA SEXLINEATA Thunberg.

A jumble of striped fishes, restricted by Cuvier to the GRAMMISTES of Seba, G. SEXLINEATUS.

Synanceja Bloch & Schneider, 194; type SCORPÆNA HORRIDA L.

By common consent. Commonly written SYNANCEIA.

Amphiprion Bloch & Schneider, 200; type LUTJANUS EPHIPPIUM Bloch.

Amphacanthus Bloch & Schneider, 206; type CHÆTODON GUTTATUS Bloch.

Equivalent to TEUTHIS L. as restricted = SIGANUS Forskål.

Alphestes Bloch & Schneider, 236; type ALPHESTES AFER Bloch & Schneider.

Cephalopholis Bloch & Schneider, 311; type CEPHALOPHOLIS ARGUS Bloch & Schneider.

Monotypic. Replaces ENNEACENTRUS Gill.

Calliodon Bloch & Schneider, 312; type CALLIODON LINEATUS Bloch & Schneider (SCARUS CROICENSIS Bloch).

Equivalent to CALLYODON Gronow.

Cichla Bloch & Schneider, 336; type CICHLA OCELLARIS Bloch & Schneider as restricted by Heckel, 1840.

Name to be changed if CICLA Klein is eligible.

Rhina Bloch & Schneider, 353; type RHINA ANCYLOSTOMUS Bloch & Schneider.

Not of Klein; equivalent to RHAMPHOBATIS Gill.

Rhinobatus Bloch & Schneider, 353; type RAJA RHINOBATUS Forskål.

Equivalent to RHINOBATOS Linck.

Anableps Bloch & Schneider, 389; type ANABLEPS TETROPHTHALMUS Bloch & Schneider (COBITIS ANABLEPS L.).

Equivalent to ANABLEPS Gronow and of Scopoli.

Synodus Bloch & Schneider, 396; type ESOX SYNODUS L.

Replaces SAURUS Cuvier.

Chauliodus Bloch & Schneider, 430; type CHAULIODUS SLOANI Schneider.

Albula Bloch & Schneider, 433; type ALBULA CONORHYNCHUS Schneider (ESOX VULPES L.). ALBULA Gronow.

Not ALBULA Osbeck. To stand as BUTYRINUS Lacepède, if Osbeck's names are accepted.

Pœcilia Bloch & Schneider, 453; type PŒCILIA VIVIPARA Bloch & Schneider.

By common consent.

Polyodon (Lacepède) Bloch & Schneider, 457; type "LE POLYDON FEUILLE" Lacepède, 1798 = SQUALUS SPATHULA Walbaum.

Rhynchobdella Bloch & Schneider, 479; type RHYNCHOBDELLA ORIENTALIS Bloch & Schneider = OPHIDIUM ACULEATUM Bloch.

As restricted by Cuvier. A synonym of MACROGNATHUS Lacepède, 1800.

Sternarchus Bloch & Schneider, 497; type GYMNOTUS ALBIFRONS L.

A synonym of APTERONOTUS Lacepède, 1800.

Orthagoriscus Bloch & Schneider, (misprinted ORTHRAGORISCUS), 510; type DIODON MOLA (Bloch).

Equivalent to MOLA Kœlreuter.

Bogmarus Bloch & Schneider, 518; type BOGMARUS ISLANDICUS Bloch & Schneider (GYMNOGASTER ARCTICUS Brünnich).

A synonym of TRACHIPTERUS Gouan. Monotypic.

Gymnonotus Bloch & Schneider, 521; type GYMNOTUS CARAPO L. (corrected spelling for GYMNOTUS).

Ovum Bloch & Schneider, 530; type OVUM COMMERSONI Bloch & Schneider = "L'OVOIDE FASCÉ, Lacepède, 1798.

Monotypic. Equivalent to OVOIDES Cuvier. OVUM Martin, 1764, is recorded by Sherborne as a "vernacular name only."

Typhlobranchus Bloch & Schneider, 537; type TYPHLOBRANCHUS SPURIUS Bloch & Schneider.

Monotypic, not identified. A river eel of Tropical America, without gill-openings. Doubtless a synonym of SYMBRANCHUS.

Gnathobolus Bloch & Schneider, 556; type ODONTOGNATHUS MUCRONATUS Lacepède.

Equivalent to ODONTOGNATHUS.

Pomatias Bloch & Schneider, 559; type TRIURUS BOUGAINVILLEI Lacepède.

Equivalent to TRIURUS.

Fluta Bloch & Schneider, 565; type MONOPTERUS JAVANENSIS Lacepède.

Equivalent to MONOPTERUS. Like the two preceding, a needless substitute name.

LVI. LACEPÈDE, *Histoire Naturelle des Poissons,* vol. III, 1802; IV and V, 1803.

BERNARD GERMAIN LACEPÈDE.

Scomberoides Lacepède, III, 50, 1802; type SCOMBEROIDES COMMERSONIANUS Lacepède (SCOMBER LYSAN Forskål).

Caranx Lacepède, III, 57, 1802; type SCOMBER CARANGUS Bloch = CARANX HIPPOS L., as restricted by Bleeker, the first reviser.

The generic name CARANX was taken by Lacepède from the manuscripts of Commerson, who first applied the name to SCOMBER SPECIOSUS Forskål. In revising the genus, Bleeker referred SPECIOSUS to a new genus, GNATHANODON. This view Dr. Gill accepted at first, but later selected SPECIOSUS as the type of CARANX, leaving some older name as TRICROPTERUS Rafinesque or CARANGUS Griffith for the bulk of the species of this extensive group. Later, by the process of elimination, CARANX RUBER was selected by Jordan as type of CARANX. On the whole it seems most just, as it is certainly most convenient, to recognize the right of Bleeker as "first reviser." CARANX CARANGUS is congeneric with C. RUBER.

Trachinotus Lacepède, III, 78, 1802; type SCOMBER FALCATUS Forskål.

Caranxomorus Lacepède, III, 82, 1802; type SCOMBER PELAGICUS L.
Synonym of CORYPHÆNA L.

Cæsio Lacepède, III, 85, 1802; type CÆSIO CÆRULEOAUREUS Lacepède.

Cæsiomorus Lacepède, III, 92, 1802; type CÆSIOMORUS BAILLONI Lacepède.
Genus probably valid; equivalent to GLAUCUS Klein.

Coris Lacepède, III, 96, 1802; type CORIS AYGULA Lacepède.

Gomphosus Lacepède, III, 100, 1802; type GOMPHOSUS CÆRULEUS Lacepède.

Naso Lacepède, III, 105, 1802; type CHÆTODON FRONTICORNIS L.
Adopted from Commerson ms. Commerson writes NASEUS, but the form NASO occurs first. Identical with MONOCEROS Bloch & Schneider, the latter preoccupied.

Kyphosus Lacepède, III, 114, 1802; type KYPHOSUS BIGIBBUS Lacepède.

Osphronemus Lacepède, III, 116, 1802; type OSPHRONEMUS GOURAMY Lacepède.
Written OSPHROMENUS by Günther.

Trichopodus Lacepède, III, 125, 1802; type TRICHOPODUS MENTUM Lacepède (OSPHRONEMUS GOURAMY Lacepède).

Monodactylus Lacepède, III, 131, 1802; type MONODACTYLUS FALCIFORMIS Lacepède.
Called PSETTUS by Commerson in ms. as quoted by Lacepède. Monotypic.

Plectorhinchus Lacepède, III, 134, 1802; type PLECTORHINCHUS CHÆTODONOIDES Lacepède.
Monotypic. Replaces DIAGRAMMA Cuvier.

Pogonias Lacepède, III, 137; type POGONIAS FASCIATUS Lacepède = LABRUS CROMIS Lacepède.

Bostrychus Lacepède, III, 144, 1802; type BOSTRYCHUS SINENSIS Lacepède.
Not BOSTRICHUS Geoffroy 1762, a genus of Insects. Replaced by BOSTRICHTHYS Duméril, 1806, by PSILUS Fischer, 1813, and by ICTIOPOGON Rafinesque, 1815.

Bostrychoides Lacepède, III, 144, 1802; type BOSTRYCHOIDES OCULATUS Lacepède.
Monotypic.

Hemipteronotus Lacepède, III, 214, 1802; type HEMIPTERONOTUS QUINQUEMACULATUS Lacepède (CORYPHÆNA PENTADACTYLA Gmelin).

Coryphænoides Lacepède, III, 219, 1802; type CORYPHÆNOIDES HOUT-TUYNI Lacepède (CORYPHÆNA JAPONICA Houttuyn).

Not CORYPHÆNOIDES Gunner. Replaced by BRANCHIOSTEGUS Rafinesque, 1815, which is earlier than LATILUS Cuvier & Val., 1830.

Aspidophorus Lacepède, III, 221, 1802; type ASPIDOPHORUS ARMATUS Lacepède (COTTUS CATAPHRACTUS Lacepède).

Equivalent to AGONUS Bloch & Schneider.

Aspidophoroides Lacepède, III, 227, 1802; type ASPIDOPHOROIDES TRANQUEBAR Lacepède (AGONUS MONOPTERYGIUS Bloch & Schneider).

Scomberomorus Lacepède, III, 292, 1802; type SCOMBEROMORUS PLUMIERI Lacepède (SCOMBER REGALIS Bloch).

Centropodus Lacepède, III, 303, 1802; type SCOMBER RHOMBEUS For-skål.

A synonym of MONODACTYLUS Lacepède.

Centronotus Lacepède, III, 309, 1802; type CENTRONOTUS CONDUCTOR Lacepède (GASTEROSTEUS DUCTOR Lacepède).

Not CENTRONOTUS Bloch & Schneider, 1801 (PHOLIS Scopoli). Replaced by NAUCRATES Rafinesque.

Lepisacanthus Lacepède, III, 320, 1802; type LEPISACANTHUS JAPON-ICUS Lacepède (GASTEROSTEUS JAPONICUS L.).

Equivalent to MONOCENTRIS Bloch & Schneider, 1801. Monotypic.

Cephalacanthus Lacepède, III, 323, 1802; type GASTEROSTEUS SPINA-RELLA L.

The young of some species of DACTYLOPTERUS. Monotypic.

Dactylopterus Lacepède, III, 325, 1802; type DACTYLOPTERUS PIRAPEDA Lacepède (TRIGLA VOLITANS L.).

Equivalent to CEPHALACANTHUS Lacepède.

Prionotus Lacepède, III, 336, 1802; type PRIONOTUS EVOLANS Lacepède.

Monotypic.

Peristedion Lacepède, III, 368, 1802; type PERISTEDION MALARMAT Lacepède (TRIGLA CATAPHRACTA L.).

Monotypic.

Istiophorus Lacepède, III, 374, 1802; type SCOMBER GLADIUS Brous-sonet.

Monotypic. Spelled HISTIOPHORUS by Cuvier.

Apogon Lacepède, III, 411, 1802; type APOGON RUBER Lacepède (MUL-LUS IMBERBIS L.).

Monotypic. Equivalent to AMIA Gronow.

Macropodus Lacepède, III, 416, 1802; type MACROPODUS VIRIDIAURATUS Lacepède.
Monotypic.

Cheilinus Lacepède, III, 529, 1802; type CHEILINUS TRILOBATUS Lacepède.
Monotypic.

Cheilodipterus Lacepède, III, 539, 1802; type CHEILODIPTERUS LINEATUS Lacepède = CHEILODIPTERUS OCTOVITTATUS Cuv. & Val., as restricted by Cuv. & Val.

Hologymnosus Lacepède, III, 556, 1802; type HOLOGYMNOSUS FASCIATUS Lacepède (CORIS ANNULATUS Lacepède, 1802).
Equivalent to CORIS.

Ostorhinchus Lacepède, IV, 23, 1803; type OSTORHINCHUS FLEURIEU Lacepède.
Monotypic; subgenus of APOGON.

Dipterodon Lacepède, IV, 165, 1803; type DIPTERODON HEXACANTHUS Lacepède.
A species of APOGON, as restricted by Jordan & Evermann, *Fishes North Mid. Amer.*, 1106. Not DIPTERODON Cuvier.

Centropomus Lacepède, IV, 248, 1803; type SCIÆNA UNDECIMALIS Bloch, by common consent, as restricted by Cuvier.
Originally a peculiarly confused jumble of species.

Tænianotus Lacepède, IV, 303, 1803; type TÆNIONOTUS TRIACANTHUS Lacepède, restricted by Cuv. & Val., *Hist. Poiss.*, IV, 371.

Micropterus Lacepède, IV, 324, 1803; type MICROPTERUS DOLOMIEU Lacepède.
Monotypic.

Harpe Lacepède, IV, 426, 1803; type HARPE CÆRULEOAUREUS Lacepède (LABRUS RUFUS L.).
A synonym of BODIANUS Bloch and of HELOPS Browne.

Pimelepterus Lacepède, IV, 429, 1803; type PIMELEPTERUS BOSCI Lacepède.
Monotypic. A synonym of KYPHOSUS Lacepède.

Cheilio Lacepède, IV, 432, 1803; type CHEILIO AURATUS Lacepède (LABRUS INERMIS Forskål).

Pomatomus Lacepède, IV, 435, 1803; type POMATOMUS SKIB Lacepède (PERCA SALTATRIX L.).
Monotypic. The name afterwards improperly transferred by Cuvier to a genus of APOGONIDÆ, EPIGONUS Rafinesque.

Leiostomus Lacepède, IV, 438, 1803; type LEIOSTOMUS XANTHURUS Lacepède.
Monotypic.

Centrolophus Lacepède, IV, 441; type PERCA NIGRA L.

Leiognathus Lacepède, IV, 448; type LEIOGNATHUS ARGENTEUS Lacepède (SCOMBER EDENTULUS Bloch).
Monotypic. Prior to EQUULA Cuvier.

Acanthinion Lacepède, IV, 499, 1803; type CHÆTODON RHOMBOIDES L.
Equivalent to TRACHINOTUS.

Chætodipterus Lacepède, IV, 503; type CHÆTODIPTERUS PLUMIERI Lacepède (ZEUS FABER Gmelin).
Monotypic.

Pomacentrus Lacepède, IV, 505, 1803; type CHÆTODON PAVO Bloch.
By general consent.

Pomadasys Lacepède, IV, 515, 1803; type SCIÆNA ARGENTEA Forskål.
Monotypic. Prior to PRISTIPOMA Cuvier.

Pomacanthus Lacepède, IV, 517, 1803; type CHÆTODON ARCUATUS L., as restricted by Cuvier.

Holacanthus Lacepède, IV, 525, 1803; type CHÆTODON TRICOLOR L., as restricted by Cuvier.

Enoplosus (White) Lacepède, IV, 540, 1803; type CHÆTODON ARMATUS White.
Monotypic.

Glyphisodon Lacepède, IV, 542, 1803; type GLYPHISODON MOUCHARRA Lacepède (CHÆTODON SAXATILIS L.).
Equivalent to ABUDEFDUF Forskål. Usually written GLYPHIDODON.

Aspisurus Lacepède, IV, 556, 1803; type CHÆTODON SOHAR Forskål.
Monotypic. Equivalent to ACANTHURUS Forskål, TEUTHIS L.

Acanthopodus Lacepède, IV, 558, 1803; type ACANTHOPODUS ARGENTEUS Lacepède (MONODACTYLUS FALCIFORMIS Lacepède).
A synonym of MONODACTYLUS.

Selene Lacepède, IV, 560, 1803; type SELENE ARGENTEA Lacepède (ZEUS VOMER Cuvier).

Argyreiosus Lacepède, IV, 566, 1803; type ZEUS VOMER L.
Monotypic. A synonym of SELENE.

Gallus Lacepède, IV, 583, 1803; type GALLUS VIRESCENS Lacepède (ZEUS GALLUS L.).
Monotypic. Not GALLUS L., a genus of hens. Replaced by ALECTIS Rafinesque and GALLICHTHYS Cuvier.

Chrysotosus Lacepède, IV, 586, 1803; type ZEUS LUNA Gmelin.
A synonym of LAMPRIS. Monotypic.

Capros Lacepède, IV, 590, 1803; type ZEUS APER L.
Monotypic.

Achirus Lacepède, IV, 658, 1803; type PLEURONECTES ACHIRUS L.
As fixed by Jordan & Gilbert, 1883. ACHIRUS FASCIATUS Lacepède was wrongly supposed to be identical with PLEURONECTES ACHIRUS L.

Makaira Lacepède, IV, 688, 1803; type MAKAIRA NIGRICANS Lacepède.
Monotypic. A synonym of ISTIOPHORUS.

Cirrhitus Lacepède, V, 2, 1803; type CIRRHITUS MACULATUS Lacepède.
Monotypic. Written CIRRHITES by Cuvier.

Cheilodactylus Lacepède, V, 5, 1803; type CHEILODACTYLUS FASCIATUS Lacepède.
Monotypic.

Misgurnus Lacepède, V, 16, 1803; type COBITIS FOSSILIS L.
Monotypic.

Fundulus Lacepède, V, 37, 1803; type FUNDULUS MUDFISH Lacepède
(COBITIS HETEROCLITA Gmelin).

Colubrina Lacepède, V, 40, 1803; type COLUBRINA CHINENSIS Lacepède.
An unidentified Chinese painting, probably fictitious.

Butyrinus Lacepède, V, 45, 1803; type BUTYRINUS BANANA Lacepède
(ESOX VULPES L.).
Equivalent to ALBULA of Gronow and of Schneider, not ALBULA Osbeck. Should apparently replace ALBULA Gronow.

Tripteronotus Lacepède, V, 47, 1803; type TRIPTERONOTUS HAUTIN
Lacepède (SALMO LAVARETUS L.).
Mutilated example. Equivalent to COREGONUS L.

Ompok Lacepède, V, 49, 1803; type OMPOK SILUROIDES Lacepède.
Replaces CALLICHROUS Hamilton.

Macropteronotus Lacepède, V, 84, 1803; type MACROPTERONOTUS
CHARMUTH Lacepède (SILURUS ANGUILLARIS L.).
Identical with CLARIAS Gronow.

Malapterurus Lacepède, V, 90, 1803; type SILURUS ELECTRICUS L.
Monotypic. A synonym of TORPEDO Forskål.

Pimelodus Lacepède, V, 93, 1803; type PIMELODUS MACULATUS Lacepède.
As restricted by Cuvier, Gill and authors.

Doras Lacepède, V, 116, 1803; type SILURUS CARINATUS L.

Pogonathus Lacepède, V, 120, 1803; type POGONATHUS COURBINA Lacepède.
Identical with POGONIAS Lacepède, 1802.

Plotosus Lacepède, V, 129, 1803; type PLATYSTACUS ANGUILLARIS Bloch (SILURUS ANGUILLARIS Forskål).

Ageneiosus Lacepède, V, 132, 1803; type AGENEIOSUS ARMATUS Lacepède (SILURUS MILITARIS L.).

Macrorhamphosus Lacepède, V, 136, 1803; type SILURUS CORNUTUS L.
Equivalent to CENTRISCUS Cuvier, not of Linnæus.

Centranodon Lacepède, V, 138, 1803; type CENTRANODON JAPONICUS Lacepède (SILURUS IMBERBIS Houttuyn) = CALLIONYMUS INDICUS L.
Monotypic and unidentifiable, perhaps based on rough notes of PLATYCEPHALUS.

Hypostomus Lacepède, V, 144, 1803; type HYPOSTOMUS GUACARI Lacepède (LORICARIA PLECOSTOMUS L.).
Monotypic. Identical with PLECOSTOMUS Gronow.

Corydoras Lacepède, V, 147, 1803; type CORYDORAS GEOFFROY Lacepède (CATAPHRACTUS PUNCTATUS Bloch).

Tachysurus Lacepède, V, 150, 1803; type TACHYSURUS SINENSIS Lacepède.
A Chinese picture of some species of ARIUS Cuvier, which name it replaces, if identifiable.

Osmerus Lacepède, V, 229, 1803; type SALMO EPERLANUS L.
Equivalent to OSMERI L.

Coregonus Lacepède, V, 239, 1803; type SALMO LAVARETUS L.
Equivalent to COREGONI L.

Characinus Lacepède, V, 269, 1803; type SALMO GIBBOSUS L.
Equivalent to CHARACINI L. and to CHARAX Gronow, Scopoli.

Serrasalmus Lacepède, V, 283, 1803; type SALMO RHOMBEUS L.
Monotypic. Written SERRASALMO by Cuvier.

Megalops Lacepède, V, 289, 1803; type MEGALOPS FILAMENTOSUS Lacepède.
Monotypic.

Lepisosteus Lacepède, V, 331, 1803; type LEPISOSTEUS GAVIALIS Lacepède (ESOX OSSEUS L.).
Usually written LEPIDOSTEUS.

Scomberesox Lacepède, V, 344, 1803; type SCOMBERESOX CAMPERI Lacepède (ESOX SAURUS Walbaum).
Monotypic.

Aulostomus Lacepède, V, 356, 1803; type AULOSTOMUS CHINENSIS Lacepède (FISTULARIA CHINENSIS L.).
Monotypic. Usually written AULOSTOMA.

Solenostomus Lacepède, V, 360, 1803; type FISTULARIA PARADOXA Pallas.

Monotypic. Not SOLENOSTOMUS Gronow nor of Klein.

Hydrargira Lacepède, V, 378, 1803; type HYDRARGIRA SWAMPINA Lacepède (FUNDULUS MAJALIS Walbaum).

Usually written HYDRARGYRA. Monotypic. Not separable from FUNDULUS.

Stolephorus Lacepède, V, 381, 1803; type ATHERINA JAPONICA Houttuyn (SPRATELLOIDES ARGYROTÆNIA Bleeker).

Restriction of Jordan & Evermann, 1896.

The first species named by Lacepède, ATHERINA JAPONICA Houttuyn, was unknown to him, and very scantily and incorrectly described by its discoverer. Several authors—Bleeker, Jordan & Evermann—have assumed this species to be the type, and that it was congeneric with the second species named, from which Lacepède drew up his generic description and of which he gave a figure.

Study of the fauna of Nagasaki shows that no species of the modern genus STOLEPHORUS has been yet found there and that the ATHERINA JAPONICA of Houttuyn is most likely a description from rough notes or from memory of SPRATELLOIDES ARGYROTÆNIA Bleeker. ATHERINA JAPONICA was the first species to be formally named as type of STOLEPHORUS. It would have been more convenient to assume the one species known to Lacepède to be his type, reverting to the system of Bleeker, who first acted on this supposition but who named no type formally. In that case STOLEPHORUS would become the equivalent of ANCHOVIELLA Fowler, 1911. MENIDIA Browne, if available, would replace both. But it is necessary here as elsewhere to recognize the first formally stated type. If this is regarded as identifiable STOLEPHORUS replaces SPRATELLOIDES.

Mugiloides Lacepède, V, 393, 1803; type MUGILOIDES CHILENSIS Lacepède (ESOX CHILENSIS Molina).

Replaces PINGUIPES Cuv. & Val.

Chanos Lacepède, V, 395, 1803; type CHANOS ARABICUS Lacepède (MUGIL CHANOS L.).

Monotypic.

Mugilomorus Lacepède, V, 397, 1803; type MUGILOMORUS ANNA-CAROLINA Lacepède (ELOPS SAURUS L.).

Synonym of ELOPS L.

Polydactylus Lacepède, V, 419, 1803; type POLYDACTYLUS PLUMIERI Lacepède (POLYNEMUS VIRGINICUS L.).

As understood by us at present, a synonym of POLYNEMUS L.

Buro (Commerson) Lacepède, V, 421, 1803; type BURO BRUNNEUS Lacepède (TEUTHIS HEXAGONATUS Bleeker).

Monotypic. Same as SIGANUS.

Mystus Lacepède, V, 466, 1803; type MYSTUS CLUPEOIDES Lacepède (CLUPEA MYSTUS L.), a species of COILIA Gray.

The name MYSTUS is preoccupied.

Clupanodon Lacepède, V, 468, 1803; type CLUPEA THRISSA L.

As restricted by Rafinesque, 1815, through substitution for CLUPANODON the more euphonious name of THRISSA. The name CLUPANODON has been variously treated. It should properly replace KONOSIRUS Jordan & Snyder, although Lacepède's account of "CLUPANODON THRISSA" is taken mainly from the West Indian "Caillieu-Tassart," OPHISTHONEMA OGLINUM (Le Sueur), the CLUPEA THRISSA of Broussonet, but not of Linnæus.

Mene Lacepède, V, 479, 1803; type MENE ANNA-CAROLINA Lacepède (ZEUS MACULATUS Bloch), MENE MACULATA of authors.

Monotypic.

Dorsuarius Lacepède, V, 482, 1803; type DORSUARIUS NIGRESCENS Lacepède.

A species of KYPHOSUS Lacepède. Monotypic.

Xyster Lacepède, V, 484, 1803; type XYSTER FUSCUS Lacepède.

Equivalent to KYPHOSUS. Monotypic.

Cyprinodon Lacepède, V, 486, 1803; type CYPRINODON VARIEGATUS Lacepède.

Monotypic.

Murænophis Lacepède, V, 627, 1803; type MURÆNA HELENA Lacepède.

Equivalent to MURÆNA L.

Gymnomuræna Lacepède, V, 648, 1803; type GYMNOMURÆNA DOLIATA Lacepède.

As first restricted by Kaup, 1856. Bleeker and Günther have used G. MARMORATA Lacepède as type, thus replacing UROPTERYGIUS Rüppell. A synonym of ECHIDNA.

Murænoblenna Lacepède, V, 652, 1803; type MURÆNOBLENNA OLIVACEA Lacepède.

A synonym of MYXINE L. Monotypic.

Unibranchapertura Lacepède, V, 656, 1803; type SYMBRANCHUS MARMORATUS Bloch.

A synonym of SYMBRANCHUS.

LVII. SEWASTIANOFF, *Acarauna, Piscium Thoracices,* 1796 (1802).

Acarauna Sewastianoff, 357; type ACARAUNA LONGIROSTRIS Sewastianoff (GOMPHOSUS CÆRULEUS Lacepède).

Equivalent to GOMPHOSUS Lacepède, and apparently of later date. Not seen by us.

LVIII. COMMERSON, *Lacepède, Histoire Naturelle des Poissons,* II, 1798; III, 1800; IV, V, 1803.

PHILIBERT COMMERSON.

Commerson was an active and accurate naturalist-explorer who collected mainly in the South Seas. His specimens were accompanied by manuscript names, polynomial in form. The author had a full grasp of the meaning of "genus," as he spoke sometimes of a "genus novissimum," XYSTER, for example. Many of his generic names were adopted by Lacepède and have found their way into the system. The others are here enumerated. These have been formally accepted as eligible by the International Commission (Opinions 23 and 24 as to **Antennarius** and **Aspro**). We think that this decision might well be reconsidered as Commerson was not "binomial" and his names were not adopted by the author who first printed them as synonyms.

The names of Commerson may be ineligible as not binomial as to species and not accepted by the author who published them. Accepted provisionally by the International Commission.

Antennarius Commerson, I, 327, 1798, footnote; type ANTENNARIUS BIVERTEX TOTUS ATER PUNCTO MEDIORUM, LATERUM ALBO Commerson (in foot-note) = LOPHIUS COMMERSONIANUS Lacepède.

Equivalent to CHIRONECTES Cuvier and ANTENNARIUS Cuvier. Regarded as eligible, in Opinion 24, International Commission.

Alticus Commerson, II, 458, 1800, footnote under BLENNIUS; type ALTICUS SALTATORIUS PINNA SPURIA IN CAPITE VERTICE, BLENNIUS SALIENS Lacepède.

Accepted by Jordan & Seale, *Fishes Samoa*, 421, 1806. Equivalent to RUPIS-CARTES Swainson. Cuv. & Val., XI, 337, 1836, adopted the name SALARIAS ALTICUS for the BLENNIUS SALIENS, "un petit SALARIAS qui nous paraît être celui-là même pour lequel Commerson a établi son genre, ALTICUS."

Sciænus Commerson, III, footnote under CARANXOMORUS; type SCIÆNUS EX FUSCO CÆRULESCENS: CARANXOMORUS SACRESTINUS Lacepède, LABRUS FURCATUS Lacepède.

Identical with APHAREUS Cuv. & Val., the name apparently a variant of SCI-ÆNA. Apparently ineligible.

Naseus Commerson, III, 105, under NASO; type NASEUS FRONTICORNIS FUSCUS Commerson (CHÆTODON FRONTICORNIS L.).

Equivalent to NASO Lacepède, NASEUS Cuv. & Val.

Coryphus Commerson, III, 1802, footnote under Coryphæna; type Coryphus chrysurus undique deauratus etc. Commerson, Coryphæna chrysurus Lacepède, Coryphæna hippurus L.

Same as Coryphæna.

Elops Commerson, III, 100, 1802, footnote under Gomphosus; type Gomphosus tricolor Lacepède.

Equivalent to Acarauna Sewastianoff or to Gomphosus Lacepède.

Psettus Commerson, III, 1802, footnote under Monodactylus; type Psettus spinis pinnarum ventralium loco duabus Commerson (Monodactylus falciformis Lacepède).

Revived by Cuvier & Valenciennes. A synonym of Monodactylus Lacepède. Not Psettus Klein.

Odax Commerson, III, 1802, footnote under Scarus; type Odax odon Commerson = Scarus chadri Lacepède = Scarus niger Forskål.

Monotypic. Equivalent to Callyodon Gronow. If the names of Commerson are accepted, Odax Cuvier requires a new name.

Mylio Commerson, III, 131, 1802, footnote under Sparus; type Mylio lineis longitudinalibus pluribus Commerson Ms. = Sparus mylio Lacepède = Chætodon bifasciatus Bloch.

Equivalent to Sparus L.

Aspro Commerson, IV, 273, 1803, footnote; type Aspro dorso dipterygio dentibus raris, et longis et exsertis etc. Commerson Ms. = Cheilodipterus macrodon Lacepède, as restricted by Jordan, Opinion 23, *Zoolgical Nomenclature*, 56, 1910.

A synonym of Cheilodipterus as now restricted. Not Aspro Cuvier.

Opisotomus Commerson, IV, 1803, footnote.

We have been unable to find this name.

Zanclus Commerson, V, 1803, footnote under Chætodon (misprinted Zanchus); type Zanclus transverse fasciatus Commerson Ms. = Chætodon cornutus Bloch.

Monotypic. Revived by Cuvier & Valenciennes, VII, 102, 1831.

Oculeus Commerson, V, 289, 1803, footnote under Megalops; type Oculeus seu megalops postremo pinnæ etc. Commerson (Megalops cyprinoides Lacepède).

Equivalent to Megalops.

Aulus Commerson, V, 1803, footnote under Fistularia; type Aulus urognomon etc., Fistularia tabacaria L.

Equivalent to Fistularia.

Encrasicholus Commerson, V, 382, 1803, footnote under CLUPEA; type
ENCRASICHOLUS MANDIBULA INFERIORE BREVIORE, TÆNIA LATERALI
ARGENTEA Commerson (CLUPEA VITTARGENTEA Lacepède).
Equivalent to ANCHOVIELLA Fowler (MENIDIA Browne).

Pterichthus Commerson, V, 401, 1803, footnote under EXOCŒTUS; type
"PTERICHTHUS PINNIS PECTORALIBUS RADIORUM SEXDECIM VEN-
TRALIBUS INTRA CORPORIS ÆQUILIBRIUM NEQUIDEM AD ANUM
APICE PERTINGENTIBUS."
As indicated by Lacepède, this species, the one placed first by Commerson,
seems to be EXOCŒTUS VOLITANS L. A synonym of EXOCŒTUS.

Halex Commerson, V, 462, 1803, footnote under CLUPEA; type HALEX
CORPORE LATE CATHETEPLATEO etc. Commerson, CLUPEA FASCIATA
Lacepède, a species of LEIOGNATHUS Lacepède.

LIX. PLUMIER, *Lacepède, Histoire Naturelle des Poissons,* II, 1798;
III, 1800; IV, V, 1803.

CHARLES PLUMIER.

Le Père Plumier, a missionary in Martinique, sent to Lacepède numerous paintings of fishes with descriptions and manuscript names. The
generic names used rest on exactly the same basis as those of Commerson. As they were polynomial and as they were not accepted by the author
who published them in synonymy, we think that they should not be considered eligible in scientific nomenclature, a matter which awaits final
decision.

*The names probably not acceptable, as polynomial and as not adopted
by the author who printed them.*

Orbis Plumier, II, 504, 1800, in footnote under TETRAODON; type ORBIS
MINIMUS Plumier = TETRAODON PLUMIERI Lacepède = TETRAODON
SPENGLERI Bloch.
A synonym of TETRAODON.

Monoceros Plumier, II, 1800, in footnote under BALISTES; type MONO-
CEROS PISCIS CLUSII Plumier.
Probably same as MONACANTHUS CILIATUS L. Equivalent to MONOCANTHUS
Cuvier. Not MONOCEROS Zimmermann, 1780, a genus of Gasteropods; not MONO-
CEROS Bloch & Schneider, 1801.

Asellus Plumier, III, 1802, footnote under GOBIOMORUS; type ASELLUS
PALUSTRIS Plumier (PLATYCEPHALUS DORMITATOR Bloch &
Schneider.
Identical with GOBIOMORUS Lacepède, as restricted. PHILYPNUS Cuvier.

Pelamis Plumier, III, 1802, footnote under SCOMBEROIDES; type PEL-
AMIS MINIMA, VULGO SAUTEUR Plumier (SCOMBER SALIENS Bloch).
Identical with OLIGOPLITES Gill. Not PELAMYS Klein.

Trachurus Plumier, III, 1802, footnote under CARANXOMORUS; type
TRACHURUS MAXIMUS SQUAMIS MINUTISSIMIS Plumier = CAR-
ANXOMORUS PLUMIERIANUS Lacepède = SCOMBER TRACHURUS L.
As here restricted, the same as TRACHURUS Rafinesque, 1810.

Scorpius Plumier, III, 1802, footnote under SCORPÆNA; type SCORPIUS
NIGER CORNUTUS Plumier (SCORPÆNA PLUMIERI Lacepède).
The name may be available for the subgenus of SCORPÆNA, having the breast
scaly (PARASCORPÆNA Bleeker). In the European type of SCORPÆNA, SCOR-
PÆNA PORCUS L., the breast is scaleless.

Sarda Plumier, III, 141, 1802, footnote under SPARUS; type SARDA
CAUDA AUREA ET LUNATA Plumier (SPARUS CHRYSURUS Bloch).
Equivalent to OCYURUS Gill. Not SARDA Cuvier, which requires a new name
if the generic names of Plumier are accepted. Monotypic.

Erythrinus (ERITRINUS) Plumier, IV, 347; type ERYTHRINUS POLY-
GRAMMOS, MARIGNAN APUD CARAIBAS Plumier (HONOCENTRUS
SOGO Bloch).
A synonym of HOLOCENTRUS; name preoccupied. Elsewhere written ERITRINUS.

Chrysomelanus Plumier, IV, 160, 1803, footnote under SPARUS; type
CHRYSOMELANUS PISCIS Plumier, SPARUS CHRYSOMELANUS Lace-
pède, ANTHIAS STRIATUS Bloch.
A synonym of EPINEPHELUS.

Aper Plumier, IV, 1803, footnote under SPARUS; type APER SEU TURDUS
ERYTHRINUS, SQUAMIS AMPLIS Plumier (SPARUS ABILDGAARDI
Lacepède).
A synonym of CALLYODON Gronow. Name preoccupied.

Guaperva Plumier, IV, footnote under SELENE; type GUAPERVA MARC-
GRAVII, VULGO LA LUNE Plumier (SELENE ARGENTEA Lacepède).
Identical with SELENE.

Sargus Plumier, IV, 166, footnote under DIPTERODON; type SARGUS
EX AURO VIRGATUS Plumier (DIPTERODON PLUMIERI Lacepède =
SPARUS SYNAGRIS L.
Same as NEOMÆNIS Girard. Not SARGUS Klein nor Cuvier. Not SARGUS
Fabricius, about 1798, a genus of flies.

Pagrus Plumier, IV, 1803, footnote under BODIANUS; type PAGRUS
LEUCOPHÆUS VULGO VIVANET GRIS APUD MARTINICAM Plumier =
BODIANUS VIVANET Lacepède = LUTIANUS GRISEUS (L.).
Not PAGRUS Cuvier, 1817. Equivalent to NEOMÆNIS Girard; same as SALPA
Catesby.

Chromis Plumier, III, 546, 1803, footnote under CHEILODIPTERUS; type CHROMIS SEU TEMBRA AUREO-CÆRULEA LITTURIS FUSCA VARIEGATA Plumier (CHEILODIPTERUS CYANOPTERUS Lacepède), the "grygry" or "grogro" of Martinique, UMBRINA COROIDES Cuv. & Val.

The name is older than UMBRINA, but later than CROMIS Browne. The specific name UMBRINA CYANOPTERA (Lacepède), based on Plumier's figure, must replace COROIDES and BROUSSONETI for this species.

Cheloniger Plumier, IV, 542, 1803, footnote under CHEILODIPTERUS; type CHELONIGER EX AURO ET ARGENTEO VIRGATUS Plumier (CHEILODIPTERUS CHRYSOPTERUS Lac. (PERCA NOBILIS L.).

The name CHELONIGER, if eligible, has priority over CONODON Cuvier.

Cephalus Plumier, V, 1803; type CEPHALUS AMERICANUS VULGO ATOULRI Plumier (MUGIL CEPHALUS L.).

A synonym of MUGIL L.

Trichis Plumier, V, 1803, footnote under CLUPEA ALOSA; type "TRICHIS BELLONII LA PUCELLE" Plumier.

Doubtful, but wrongly identified by Lacepède with CLUPEA ALOSA L.

Acus Plumier, V, 1803, footnote under SPHYRÆNA; type ACUS AMERICANA, ROSTRI LONGIORI Plumier (SPHYRÆNA ACUS Lacepède).

Same as SPHYRÆNA Röse, 1793.

LX. SHAW, *General Zoology, or Systematic Natural History,* vol. IV, 1803; vol. V, 1804.

GEORGE SHAW.

Anguilla Shaw, IV, 15, 1803; type ANGUILLA VULGARIS Shaw (MURÆNA ANGUILLA L.).

The name ANGUILLA has been ascribed to Thunberg, but we find no notice of its use as a generic term prior to Shaw.

Vandellius Shaw, IV, 199, 1803.

Needless substitute for LEPIDOPUS Gouan.

Trichopus Shaw, IV, 392, 1803; type TRICHOPUS PALLASI Shaw (LABRUS TRICHOPTERUS Pallas).

Equivalent to OSPHRONEMUS Lacepède.

Cephalus Shaw, V, 432, 1804; type DIODON MOLA L.

Equivalent to MOLA Cuvier. Not CEPHALUS Plumier.

Trachichthys Shaw, IV, 630, 1803; type TRACHICHTHYS AUSTRALIS Shaw.

Spatularia Shaw, V, 362, 1804; type SPATULARIA RETICULATA (SQUALUS SPATHULA Walbaum).
Synonym of POLYODON Lacepède.

LXI. HERRMANN, *Observationes Zoologicæ*, 1804.

JOHANN HERRMANN.

Notistium Herrmann, 305, 1804; type NOTISTIUM GLADIUS Herrmann.
Equivalent to ISTIOPHORUS Lacepède.

LXII. GIORNA, *Mémoire Sur des Poissons d'Espèces Nouvelles et de genres nouveaux*: Mémoires de l'Académie Impériale de Torino, XVI, 1803-1808.

MICHEL ESPRIT GIORNA.

Lophotes Giorna, 19, 1805; type LOPHOTES CEPEDIANUS Giorna.

Cœlorhynchus Giorna, 18, 1805; type CŒLORHYNCHUS LA VILLE Giorna.

Trachyrhynchus Giorna, 18, 1805; type (not named) LEPIDOLEPRUS TRACHYRINCUS Risso, 1810.

An earlier paper of Giorna is quoted by Dean, *Mémoire sur cinq poissons dont deux sont d'espèces nouvelles . . . et les trois autres sont de nouveaux genres:* Mém. Acad. Imp. Torino. The date of this paper is variously given. The three genera above noted were not named. Giorna would not give the nomenclature until he heard from Lacepède, to whom he had sent descriptions and drawings. The two species indicated were RAIA GIORNA Lacepède and BALISTES BUNIVA Lacepède.

LXIII. QUENSEL, *Forsäk at Narmäre Bestämma och naturligare Uppställa Svensk Arterna af Flunderslagte:* Kong. Vet. Akad. Nya Handlung, XXVII, 1806.

C. QUENSEL.

Solea Quensel, XXVII, 44, 203, 1806; type PLEURONECTES SOLEA L.

LXIV. DUMÉRIL, *Zoologie Analytique*, 1806.

ANDRÉ MARIE CONSTANT DUMÉRIL.

This work furnishes in the Index, pp. 342, 343, Latin equivalents for French vernacular names used by Lacepède, *Hist. Nat. Poiss.*, I, II, 1798, 1800. (Partly examined by us, some of the pages not verified).

Squatina Duméril, 102, 342; type (not named) SQUATINA ANGELUS Duméril (SQUALUS SQUATINA L.).
Equivalent to RHINA Klein.

Torpedo Duméril, 102, 343; type (not named) RAJA TORPEDO L.
Not TORPEDO Forskål. Same as NARCACION Klein. NARCOBATIS Blainville.

Ovoides (Lacepède) Duméril, 108, 342; type L'OVOIDE FASCÉ.
Based on a front view of TETRAODON STELLATUS L.
Equivalent to OVOIDES Cuvier and OVUM Bloch & Schneider.

Apterichthys Duméril, 112, 331; type MURÆNA CÆCA L.
Substitute for CÆCILIA Lacepède, preoccupied. A synonym of SPHAGEBRANCHUS Bloch.

Bostrichthys Duméril, 120, 332; type BOSTRYCHUS SINENSIS Lacepède.
Name a substitute for BOSTRYCHUS, preoccupied.

Spheroides (Lacepède) Duméril, 342; type "LE SPHÉROIDE TUBERCULÉ" Lacepède, 1798.
A front view of TETRAODON SPENGLERI Bloch. A synonym of TETRAODON, as restricted by Bleeker.

Macrorhynchus (Lacepède) Dumèril, 342; type "MACRORHYNQUE ARGENTÉE" Lacepède (SYNGNATHUS ARGENTEUS Osbeck).
A species near DICROTUS PROMETHEOIDES Bleeker. Probably replaces DICROTUS Günther.

LXV. DUMÉRIL, *Dictionnaire des Sciences Naturelles*, 1806.

ANDRÉ MARIE CONSTANT DUMÉRIL.

Torpedo Duméril, pl. 21; type RAJA TORPEDO L.
Not TORPEDO Forskål, 1775; replaced by NARCACION Klein or by NARCOBATIS Blainville.

LXVI. HUMBOLDT, *Ueber den Eremophilus und den Astroblepus, zwei neue Fisch-Gattungen:* Observationes Zoologicæ, and in Philos. Mag., XXIV, 1806.

ALEXANDER VON HUMBOLDT.

Eremophilus Humboldt, 17, 329; type EREMOPHILUS MUTISII Humboldt.

Astroblepus Humboldt, 19, 331; type ASTROBLEPUS GRIXALVII Humboldt.

LXVII. DUMÉRIL, *Dissertation sur les Poissons* CYCLOSTOMES, 1808.

ANDRÉ MARIE CONSTANT DUMÉRIL.

Ammocœtus Duméril; type PETROMYZON PLANERI L.

The larva of LAMPETRA FLUVIATILIS (L.) and of PETROMYZON MARINUS L. Preferably retained, under the usual spelling, AMMOCŒTES, as a designation for larval lampreys. Otherwise replaces LAMPETRA Gray, 1854; type PETROMYZON FLUVIATILIS L.

LXVIII. TILESIUS, *Descriptions de quelques Poissons*: Krusenstern's Reise um die Welt., Mem. Soc. Nat. Moscow, II, 1809.

W. G. VON TILESIUS.

Not seen by us.

Ericius Tilesius, 213, 1809; type ERICIUS (JAPONICUS Houttuyn).

A synonym of MONOCENTRIS Bloch & Schneider.

Hexagrammos (Steller) Tilesius, Actæ Academ. Petropol., II, 335, 1809; type HEXAGRAMMOS STELLERI Tilesius (HEXAGRAMMOS ASPER Tilesius).

Often written HEXAGRAMMUS.

LXIX. GEOFFROY SAINT HILAIRE, *Poissons du Nil, de la Mer Rouge et de la Méditerranée:* in Description de l'Egypte, publée par Napoléon le Grand, Histoire Naturelle, I, 1809-1827.

ÉTIENNE FRANÇOIS GEOFFROY SAINT HILAIRE.

In this huge folio a few species of fishes are described in great detail, with steel engravings. The work is in three parts, published in 1809,

1818, and 1825, respectively, the second and third parts being prepared by Isidore Geoffroy Saint Hilaire.

Polypterus Geoffroy St. Hilaire, I, 1809; type POLYPTERUS BICHIR St. Hilaire.

LXX. PALLAS, *Labraces novum piscium genus Oceani Orientalis*: Mem. Acad. Sci. Petersb., II, 1810.

PETER SIMON PALLAS.

Labrax Pallas, II, 382, 1810; type LABRAX LAGOCEPHALUS Pallas.
Not of Klein nor of Cuvier. A synonym of HEXAGRAMMOS Steller.

LXXI. RISSO, *Ichthyologie de Nice*, 1810.

ANASTASE RISSO.

Cephalopterus Risso, 14; type RAJA GIORNA Lacepède.
Not CEPHALOPTERUS of Geoffroy St. Hilaire, 1809, a genus of birds. Equivalent to MOBULA Rafinesque.

Lepidoleprus Risso, 197; type LEPIDOLEPRUS TRACHYRHYNCHUS Risso.
A synonym of TRACHYRHYNCHUS Giorna.

Tetragonurus Risso, 347; type TETRAGONURUS CUVIERI Risso.

LXXII. RAFINESQUE, *Caratteri di Alcuni Nuovi Generi e Nuove Specie di Animale e Piante della Sicilia*, April 1, 1810.

CONSTANTINE SAMUEL RAFINESQUE-SCHMALTZ
(later written simply "RAFINESQUE").

Carcharias Rafinesque, 10; type CARCHARIAS TAURUS Rafinesque.
The intended type was SQUALUS CARCHARIAS L., but C. TAURUS is the only species actually mentioned. Monotypic. Replaces ODONTASPIS Agassiz. Not CARCHARIAS Cuvier, 1817.

Dalatias Rafinesque, 10; type DALATIAS NOCTURNUS Rafinesque.
Gray in 1851 restricted DALATIAS to D. SPAROPHAGUS Rafinesque. Swainson, 1838, formally restricted DALATIAS to DALATIAS NOCTURNUS, which seems to be a species of CENTROPHORUS Müller & Henle, probably C. GRANULOSUS.

Tetroras Rafinesque, 11; type TETRORAS ANGIOVA Rafinesque.
A second-hand and erroneous description, possibly referring to CETORHINUS MAXIMUS, but unrecognizable. Monotypic.

Isurus Rafinesque, 11; type Isurus oxyrhynchus Rafinesque.
Monotypic. Replaces Oxyrhina Agassiz.

Cerictius Rafinesque, 12; type Cerictius macrourus Rafinesque.
Apparently imaginary.

Alopias Rafinesque, 12; type Alopias macrourus Rafinesque (Squalus vulpinus Bonnaterre).
Monotypic.

Heptranchias Rafinesque, 13; type Squalus cinereus Gmelin.
Monotypic.

Galeus Rafinesque, 13; type Galeus mustelus L.
As restricted by Jordan & Evermann, 1896, after Leach, 1812. Later restricted to Pristiurus melastomus by Garman. Not Galeus Valmont, nor Galeus Cuvier.

Rafinesque describes one new species of Galeus, G. melastomus, but refers in the text to Galeus catulus and Galeus mustelus. Unless Galeus Valmont holds, in place of Prionace, Galeus becomes a synonym of Mustelus Linck.

Hexanchus Rafinesque, 14; type Squalus griseus L.
Monotypic. Replaces Notidanus Cuvier.

Etmopterus Rafinesque, 14; type Etmopterus aculeatus Rafinesque.
Monotypic. Replaces Spinax Cuvier.

Rhina Rafinesque, 14; type Squalus squatina L.
Monotypic. Identical with Rhina Klein, and Squatina Duméril.

Leiobatus Rafinesque, 16; type Leiobatus panduratus Rafinesque.
Monotypic. A synonym of Rhinobatus Linck. Not Leiobatus Klein.

Dipturus Rafinesque, 16; type Raja batis L.
Monotypic. Identical with Raja L.

Dasyatis Rafinesque, 16; type Dasyatis ujo Rafinesque (Raja pastinaca L.).
Monotypic. Identical with Dasybatus Klein and Trygon Adanson.

Orthragus Rafinesque, 17; type Tetraodon mola L.
Monotypic. Equivalent to Mola Kœlreuter.

Diplanchias Rafinesque, 17; type Diplanchias nasus Rafinesque.
Monotypic. Equivalent to Mola.

Typhle Rafinesque, 18; type Typhle hexagonus Rafinesque (Syngnathus typhle L.).
Misspelled Tiphle. Preoccupied by Typhle * Lacepède, 1800, a genus of mammals. Replaced by Typhlinus Rafinesque, 1815, and by Siphonostomus Kaup. Monotypic.

---

* Under the head of Cæcilia, Lacepède refers to its alleged blindness, a character almost unknown among vertebrates. "Parmi lesquels on ne connoît encore qu'un mammifère, nommé Typhle, et le genre des cartilagineux nommés Gastrobranches qui aient paru complètement aveugles."

Siphostoma Rafinesque, 18; type SYNGNATHUS PELAGICUS L.

Monotypic. A synonym of SYNGNATHUS L. as the latter is now restricted.

Hippocampus Rafinesque, 18;   type SYNGNATHUS HIPPOCAMPUS L. (HIPPOCAMPUS HEPTAGONUS Rafinesque).

Oxyurus Rafinesque, 19; type OXYURUS VERMIFORMIS Lacepède.

A larva, probably of CONGER.

Scarcina Rafinesque, 20; type SCARCINA ARGYREA Rafinesque.

A synonym of LEPIDOPUS Gouan.

Luvarus Rafinesque, 22; type LUVARUS IMPERIALIS Rafinesque.

Monotypic.

Bothus Rafinesque, 23; type, as restricted, BOTHUS RUMOLO Rafinesque (PLEURONECTES RHOMBUS L.).

Equivalent to RHOMBUS Klein, of earlier date.

Corystion Rafinesque, 24; type CORYSTION MUSTAZOLA Rafinesque.

Some species of TRACHINUS L. Monotypic.

Merluccius Rafinesque, 25; type GADUS MERLUCIUS L. (MERLUCCIUS SMIRIDUS Rafinesque).

Monotypic.

Phycis Rafinesque, 26; type PHYCIS PUNCTATUS Rafinesque.

Not PHYCIS Fabricius, 1798, a genus of Butterflies. Equivalent to PHYCIS Röse, 1793, which is not preoccupied.

Oxycephas Rafinesque, 31; type OXYCEPHAS SCABRUS Rafinesque.

A species of TRACHYRHYNCHUS Giorna, 1805. Monotypic.

Lepimphis Rafinesque, 33; type LEPIMPHIS HIPPUROIDES Rafinesque (CORYPHÆNA HIPPURUS L.).

Identical with CORYPHÆNA.

Symphodus Rafinesque, 41; type SYMPHODUS FULVESCENS Rafinesque = LUTIANUS ROSTRATUS Bloch = LABRUS SCINA Forskål.

Replaces CORICUS Cuvier.

Trachurus Rafinesque, 41; type SCOMBER TRACHURUS L. (TRACHURUS SAURUS Rafinesque).

By general usage and by tautonomy.

Tricropterus Rafinesque, 41; type by definition, SCOMBER CARANGUS Bloch (SCOMBER HIPPOS L.).

Not separable from CARANX; no species named.

Hypodis Rafinesque, 41; type SCOMBER GLAUCUS L.

Equivalent to GLAUCUS Klein; not separable from CÆSIOMORUS Lacepède.

Centracanthus Rafinesque, 42 (misprinted CENTRACANTUS); type CEN-TRACANTUS CIRRUS Rafinesque (SMARIS INSIDIATOR Cuv. & Val.).

Monotypic; probably a specimen with the dorsal fin torn. This seems to be generically different from SPICARA Rafinesque, SMARIS Cuvier. Afterwards spelled CENTRACANTHA.

Hypacanthus Rafinesque, 43 (misprinted HYPACANTUS) ; type "SCOMBER ACULEATUS L."
But there is no such species of Linnæus. Rafinesque elsewhere identifies it with CENTRONOTUS VADIGO Lacepède, which is the type of CAMPTOGRAMMA Regan, 1903. This name must be replaced by HYPACANTHUS.

Naucrates Rafinesque, 43; type NAUCRATES FANFARUS Rafinesque (GASTEROSTEUS DUCTOR L.).
Replaces CENTRONOTUS Lacepède, preoccupied.

Notognidion Rafinesque, 46; type NOTOGNIDION SCIRENGA Rafinesque.
Unidentified. Monotypic.

Spicara Rafinesque, 51; type SPICARA FLEXUOSA Rafinesque (SPARUS SMARIS L.).
Equivalent to SMARIS Cuvier and having priority. Monotypic.

Aylopon Rafinesque, 52; type LABRUS ANTHIAS L.
A substitute for ANTHIAS Bloch, said to be preoccupied. We do not find it so.

Lopharis Rafinesque, 52; type PERCA LOPHAR Forskål.
Identical with POMATOMUS Lacepède. Monotypic.

Lepterus Rafinesque, 52; type LEPTERUS FETULA Rafinesque (STROMATEUS FIATOLA L.).
Identical with STROMATEUS L. Monotypic.

Gonenion Rafinesque, 53; type GONENION SERRA Rafinesque (PERCA LOPHAR Forskål).
Identical with POMATOMUS Lacepède. Monotypic.

Lepodus Rafinesque, 53; type LEPODUS SARAGUS Rafinesque (SPARUS RAII Bloch).
Monotypic. Equivalent to BRAMA Bloch & Schneider, not of Klein. Replaces BRAMA Cuvier, if Klein's names are eligible.

Tetrapturus Rafinesque, 54; type TETRAPTURUS BELONE Rafinesque.

Argyctius Rafinesque, 55; type ARGYCTIUS QUADRIMACULATUS Rafinesque.
Equivalent to TRACHYPTERUS Gouan.

Tirus Rafinesque, 56; type TIRUS MARMORATUS Rafinesque (ESOX SYNODUS L.).
Equivalent to SYNODUS.

Lucius Rafinesque, 59; type ESOX LUCIUS L. (LUCIUS VORAX Rafinesque).
The author attempts to limit the name Esox to ESOX BELONE L., and its allies, an arrangement not accepted by the International Commission, (Opinion 58).

Sudis Rafinesque, 60; type SUDIS HYALINA Rafinesque.
Monotypic.

Sayris Rafinesque, 60; type SAYRIS RECURVIROSTRA Rafinesque (ESOX SAURUS Walbaum).

A needless substitute for SCOMBERESOX.

Cogrus Rafinesque, 62; type COGRUS MACULATUS Rafinesque.

COGRUS is a tenable subgenus of OPHICHTHUS.

Piescephalus Rafinesque, 63; type PIESCEPHALUS ADHERENS Rafinesque (LEPADOGASTER GOUANI Gouan).

A synonym of LEPADOGASTER Gouan.

Echelus Rafinesque, 63; type by first restriction (Bleeker, *Atlas Ichth., Murœn.*, p. 30) ECHELUS PUNCTATUS Rafinesque (MURÆNA MYRUS L.).

Identical with MYRUS Kaup, 1856. Restricted later by Jordan & Evermann to species of CONGER, but the earliest arrangement must hold.

Nettastoma Rafinesque, 66; type NETTASTOMA MELANURA Rafinesque.

Monotypic.

Dalophis Rafinesque, 68; type DALOPHIS SERPA Rafinesque (SPHAGE-BRANCHUS IMBERBIS De la Roche, 1809).

Equivalent to CÆCULA Vahl.

LXXIII. RAFINESQUE, *Indice d'Ittiologia Siciliana*, May, 1810.

CONSTANTINE SAMUEL RAFINESQUE-SCHMALTZ.

Onus Rafinesque, 12; type ONUS RIALI Rafinesque (GADUS MERLUC-CIUS L.).

A needless substitute for MERLUCCIUS.

Merolepis Rafinesque, 25; type SPARUS MASSILIENSIS Lacepède (SPARUS ZEBRA Brünnich).

Probably a genus distinct from MÆNA Cuvier.

Gaidropsarus Rafinesque, 11, 51; type GAIDROPSARUS MUSTELLARIS Raf-inesque (GADUS MUSTELA L.).

Antedates MOTELLA Cuvier. Monotypic.

Strinsia Rafinesque, 12, 51; type STRINSIA TINCA Rafinesque.

Monotypic.

Symphurus Rafinesque, 13, 52; type SYMPHURUS NIGRESCENS Rafin-esque.

Antedates APHORISTIA Kaup, but not PLAGUSIA Browne.

Solea Rafinesque, 14, 52; type SOLEA BUGLOSSA Rafinesque (PLEURO-NECTES SOLEA L.).

Equivalent to SOLEA Quensel.

Scophthalmus Rafinesque, 14, 53; type PLEURONECTES RHOMBUS L.
As later restricted, identical with BOTHUS Rafinesque and RHOMBUS Klein. SCOPHTHALMUS is based on descriptions, BOTHUS on specimens of the same two species.

Diplodus Rafinesque, 26, 54; type SPARUS ANNULARIS L.
Equivalent to SARGUS Cuvier and prior, but not prior to SARGUS Klein.

Octonus Rafinesque, 29, 54; type OCTONUS OLOSTEON Rafinesque (TRIGLA CATAPHRACTA L.).
Monotypic. Equivalent to PERISTEDION Lacepède.

Cephalepis Rafinesque, 31, 54; type CEPHALEPIS OCTOMACULATUS Rafinesque.
A synonym of REGALECUS Ascanius. Monotypic.

Myctophum Rafinesque, 35, 56; type MYCTOPHUM PUNCTATUM Rafinesque.

Nerophis Rafinesque, 37, 57; type SYNGNATHUS OPHIDION L.
Monotypic.

Carapus Rafinesque, 37, 57; type GYMNOTUS ACUS L.
Prior to FIERASFER Cuvier. Not CARAPUS Cuvier = GITON Kaup.

Sturio Rafinesque, 41, 58; type STURIO VULGARIS Rafinesque (ACIPENSER STURIO L.).

Capriscus Rafinesque, 41, 58; type CAPRISCUS PORCUS Rafinesque (BALISTES CAPRISCUS Gmelin, "the third division of Lacepède").
Equivalent to BALISTES L.

Chlopsis Rafinesque, 42, 58; type CHLOPSIS BICOLOR Rafinesque.
Monotypic.

Xypterus Rafinesque, 43, 59; type XYPTERUS IMPERATI Rafinesque.
Apparently a synonym of REGALECUS Ascanius. Monotypic.

Pterurus Rafinesque, 43, 59; type PTERURUS FLEXUOSUS Rafinesque (SPHAGEBRANCHUS IMBERBIS De la Roche).
Name preoccupied: a synonym of CÆCULA Vahl.

Oxynotus Rafinesque, 45, 60; type SQUALUS CENTRINA L.
Prior to CENTRINA Cuvier, 1817.

Sphyrna Rafinesque, 46, 60; type SQUALUS ZYGÆNA L.
Prior to ZYGÆNA Cuvier.

Torpedo Rafinesque, 48, 60; type TORPEDO OCELLATA Rafinesque (RAJA TORPEDO L.).
Not TORPEDO Forskål.

Mobula Rafinesque, 48, 61; type MOBULA AURICULATA Rafinesque (RAJA MOBULAR Lacepède).
Same as CEPHALOPTERUS Risso, preoccupied.

Cephaleutherus Rafinesque, 48, 61; type CEPHALEUTHERUS MACULATUS Rafinesque.
Apparently a deformed RAJA. Monotypic.

Uroxis Rafinesque, 48, 61; type UROXIS UJUS Rafinesque (RAJA PASTINACA L.).
Monotypic. Equivalent to DASYATIS Rafinesque.

Apterurus Rafinesque, 48, 62; type APTERURUS FABRONI Rafinesque.
Equivalent to MOBULA Rafinesque.

Oxystomus Rafinesque, 49, 62; type OXYSTOMUS HYALINUS Rafinesque, larva of OPHISURUS SERPENS (L.).
A synonym of OPHISURUS Lacepède.

Helmictis Rafinesque, 49, 62; type HELMICTIS PUNCTATUS Rafinesque, 1810.
Probably a species of CÆCULA. Monotypic.

Epigonus Rafinesque, 64; type EPIGONUS MACROPHTHALMUS Rafinesque.
Equivalent to POMATOMUS Cuvier, not of Lacepède. Monotypic.

Gonostoma Rafinesque, 64; type GONOSTOMA DENUDATUM Rafinesque.
Monotypic.

Merlangus Rafinesque, 67; type GADUS MERLUCIUS L.
A needless substitute for ONUS and MERLUCCIUS.

LXXIV. STELLER in *Tilesius*: Mem. Acad. Sci., Petersburg, 1811.

GEORG WILHELM STELLER.

Myoxocephalus Steller, in *Tilesius,* IV, 273; type MYOXOCEPHALUS STELLERI Tilesius.

LXXV. PALLAS, *Zoographia Rosso-Asiatica,* III, 1811.

PETER SIMON PALLAS.

This important work was printed and partly distributed in 1811, the bulk of the edition being withheld until 1831.

Phalangistes Pallas, 113; type COTTUS CATAPHRACTUS L.
As restricted by Jordan & Evermann. Also written PHALANGISTA. Not PHALANGISTA Cuvier, 1800, a genus of mammals. Equivalent to AGONUS Lacepède.

Elæorhoüs Pallas, 122; type CALLIONYMUS BAICALENSIS Pallas.
A synonym of COMEPHORUS Lacepède.

Gasteracanthus Pallas, 228; type Gasteracanthus cataphractus Pallas.

Equivalent to Gasterosteus L.

Coracinus Pallas, 256; type Coracinus chalcis Pallas (Sciæna nigra Bloch).

Equivalent to Corvina Cuvier, 1829. Not Coracinus Gronow, which is Dipterodon Cuvier.

Lebius (Steller Ms.) Pallas, 279; type Labrax superciliosus Pallas, "Lebius, Chirus vel Labrax Steller, Mss. Obs. Ichthyol."

These three names are indicated in Steller's unprinted manuscripts as words from which choice could be made.

Chirus (Steller Ms.) Pallas, 279; type Labrax superciliosus Pallas.

Like the preceding synonym of Labrax or Hexagrammos.

Plagyodus (Steller Ms.) Pallas, 383; type Plagyodus Steller (Alepisaurus æsculapius Bean).

Equivalent to Alepisaurus Lowe, 1833, the name used by Steller only in an oblique case, "Plagyodontem" and without specific name.

LXXVI. LEACH, *Observations on the genus* Squalus, 1812.

William E. Leach.

Mustelus Leach, 62; type Squalus mustelus L.

LXXVII. MONTAGU, *Wernerian Museum*, I, 1812.

George Montagu.

Xipotheca Montagu, I, 82; type Xipotheca tetradens Montagu = Lepidopus caudatus (Euphrasen).

A synonym of Lepidopus.

LXXVIII. FISCHER, *Zoognosia, Tabulis Synopticus Illustrata*, Edition III, vol. I, 1813.

Gotthelf Fischer.

A series of Analytical Keys, leading to genera only.

Histrio Fischer, 70, 78; type (not named) Lophius histrio L., by tautonomy.

Diagnosis erroneous, by misprint "corpus depressum" instead of "corpus compressum." Has precedence over Pterophryne Gill and Pterophrynoides Gill.

Ogcocephalus Fischer, 70, 78; type not named; evidently by definition, Lophius vespertilio L.

Replaces Malthe Cuvier.

Orbis (Lacepède) Fischer, 70; no type named.

Said to be like Diodon but "ad minus 4 dentes in maxilla superiore." Evidently a misprint for "2 dentes." It is probably a synonym of Ovoides Cuvier, the presumable type being Tetraodon lineatus L., the "Orbis" of Salviani and Rondelet,

Odontolepis Fischer, 71, 78; no type named.

A flounder with "pinnæ pectorales tenuissimæ aut nullæ, reliquis conjunctis." The type may be assumed as Symphurus nigrescens. A synonym of Symphurus Rafinesque, 1810. Perhaps species of Monochirus or Microchirus were included.

Psilus Fischer, 74.

A substitute for Bostrychus Lacepède, which is preoccupied. Equivalent to Bostrichthys Duméril.

Psiloides Fischer, 74.

A needless substitute for Bostrychoides Lacepède.

Typhlotes Fischer, 75.

A substitute for Cæcilia Lacepède, preoccupied. A synonym of Sphagebranchus Bloch.

Mustellus Fischer, 78; type (not named) Squalus mustelus L.

A synonym of Mustelus Linck, not of Valmont.

LXXIX. FISCHER, *Recherches Zoologiques:* Mémoires de la Société Impériale des Naturalistes de Moscou, IV, second edition, 1813.

Eleginus Fischer, 252; type Gadus navaga Kœlreuter.

Replaces Tilesia Swainson, 1839 (preoccupied) and Pleurogadus Bean, 1885. Not Eleginus Cuv. & Val., 1830, which becomes Eleginops Gill.

LXXX. LEACH, *Zoological Miscellany,* 1814.

William E. Leach.

Hippocampus Leach, 103; type Syngnathus hippocampus L. (Hippocampus antiquorum Leach).

Same as Hippocampus Rafinesque, 1810.

LXXXI. MITCHILL, *Report in Part on the Fishes of New York,* 1814.

SAMUEL LATHAM MITCHILL.

Stomodon Mitchill, 7; type STOMODON BILINEARIS Mitchill.
Equivalent to MERLUCCIUS Rafinesque.

Morone Mitchill, 18; type MORONE PALLIDA Mitchill (PERCA AMERICANA Gmelin).
As restricted by Gill, 1860.

Tautoga Mitchill, 23; type TAUTOGA NIGER Mitchill (LABRUS ONITIS L.).

Roccus Mitchill, 25; type ROCCUS STRIATUS Mitchill (SCIÆNA LINEATA Bloch).
As restricted by Gill.

LXXXII. RAFINESQUE, *Descrizione di un Nuovo Genere di Pesce*: Specchio delle Scienze, Palermo, 1814.

CONSTANTINE SAMUEL RAFINESQUE.

Leptopus Rafinesque, I, 16; type LEPTOPUS PEREGRINUS Rafinesque.
Monotypic. Name preoccupied and later changed to PODOLEPTUS Rafinesque, 1815. Apparently a synonym of LOPHOTES Giorna.

LXXXIII. RAFINESQUE, *Descrizione di un Nuovo Genere di Pesce Siciliano:* Specchio delle Scienze, II, 1815.

Nemochirus Rafinesque, II, 100, 105; type NEMOCHIRUS ERYTHROPTERUS Rafinesque.
Unrecognized.
The following is Rafinesque's account of this fish:
NEMOCHIRUS. Corpo lanceolato compressissimo ensiforme, fronte diagonale, bocca dentata, un' ala dorsale longitudinale senza raggi sciolti, ala caudale sciolta, nessun' ala anale, le due ale pettorali filiformi avvicinate e situate sotto la gola al posto delle ventrali.
NEMOCHIRUS ERYTHROPTERUS. Corpo argentato, ale rosse, coda lunulata, con un raggio intermedio mucronato sciolto lunghissimo e filiforme, tre macchie fosche da ogni lato del dorso.
Descrizione. Lunghezza totale due palmi, muso ottuso, mascella inferiore più corta con denti acuti, occhi piccoli neri, iride grande argentina con un cerchio rosso esteriore, opercolo doppio: corpo d'un bel colore argenteo, con tre macchie irregolari fosche di ogni lato del dorso, linea laterale dritta, ventre un poco reti-

colato. Ale pettorali rosse lineari-filiformi acute e con un solo raggio, ala dorsale rossa principiando sopra gli occhi e giungendo sino alla coda, della quale è però staccata e con circa 200 raggi molli: coda rossa un poco trifida o quasi lunulata con però il raggio intermedio semplice, sporgente lunghissimo, mentre i laterali sono ramosi con i rami opposti.

We are indebted to Mr. H. M. Scudder for a copy of this rare and forgotten description. If this account is correct it is still unknown to science, though having much in common with STYLEPHORUS CORDATUS Shaw.

LXXXIV. RAFINESQUE, *Précis des Découvertes Somiologiques*, 1814.

Trisopterus Rafinesque, 16; type GADUS CAPELANUS Lacepède (GADUS MINUTUS L.).

According to Risso the name is synonymous with MORUA Risso, and must be older. Equivalent to BRACHYGADUS Gill.

Monochirus Rafinesque; type MONOCHIRUS HISPIDUS Raf.

LXXXV. RAFINESQUE, *Analyse de la Nature, ou Tableau de l'Univers et des Corps Organisés "La Nature est mon Guide et Linneus mon maître."* Palermo, 1815.

In this work the entire animal and plant kingdoms are classified, with definitions of the families and higher groups and lists of the known genera. Three hundred and seventy-seven genera of fishes are enumerated. Many of these are bare names, without explanation of any sort. Others represent changes of names of genera for reasons not expressed, but apparently because current names were too long, too short, or involved a termination (OIDES, OMORUS) expressing resemblance. The fishes are divided into two subclasses, HOLOBRANCHIA with opercles and gill membranes complete, and ATELOSIA, lacking either opercles or gill membranes.

The HOLOBRANCHIA are divided into DERIPIA (Jugulares), GASTRIPIA (Abdominales), THORAXIPIA (Thoracices), APODIA (Apodes).

The ATELEOSIA are divided into ELTROPOMIA (Sturgeons: Sternoptyx, Pegasus) with one opercle and no gill membrane; CHISMOPNEA, with gill-membranes and one opercle (Chimæra, Balistes, Conger, etc.); and TREMAPNEA, with neither opercles nor gill membranes (eels and sharks).

This classification is singularly inept.

Dactyleptus Rafinesque, 82.

Substitute for MURÆNOIDES Lacepède.

Pholidus Rafinesque, 82.

A synonym of PHOLIS Scopoli. Substitute for ENCHELYOPUS Gronow.

Pteraclidus Rafinesque, 82.

Substitute for OLIGOPODUS Lacepède. A synonym of PTERACLIS Gronow.

Pacamus, 82; nomen nudum.

Ictius, 82, "sp. do."

That is, based on species of the preceding, *i. c.,* PACAMUS.

Dropsarus, 82.

Evidently an emendation of GAIDROPSARUS Rafinesque.

Trisopterus, 82.

Brosme, 82.

Batrichtius, 82.

A substitute for BATRACHOIDES Lacepède.

Ceracantha, 82.

Taunis, 82.

Plagiusa, 83, "sp. do."

That is, species of PLEURONECTES; apparently based on PLEURONECTES PLA-GIUSA L., thus equivalent to SYMPHURUS Rafinesque, 1810.

Holacantha, 83.

Substitute for HOLACANTHUS Lacepède.

Pomacantha, 83.

Nasonus, 83.

Substitute for NASO Lacepède.

Alectis, 84.

Substitute for GALLUS Lacepède, which is preoccupied, leaving ALECTIS as valid.

Bostrictis, 84.

Substitute for BOSTRYCHUS Lacepède: a synonym of BOSTRICHTHYS Dumèril, 1806.

Pterops, 84, 90.

Substitute for BOSTRYCHOIDES Lacepède.

Tasica, 84.

Nemipus, 84.

Cephalepis, 84.

Gymnurus, 84.

Substitute for TÆNIOIDES Lacepède.

Polipturus, 84.
Substitute for SCOMBEROMORUS Lacepède.

Orcynus, 84.
Substitute for SCOMBEROIDES Lacepède.

Baillonus, 85.
Substitute for CÆSIOMORUS Lacepède.

Lepicantha, 85.
Apparently shortened from LEPISACANTHUS Lacepède.

Gastrogonus, 85.

Cephimmus, 85.
Substitute for GYMNOCEPHALUS Bloch.

Lepipterus, 85.

Panotus, 85.
Substitute for TÆNIONOTUS Lacepède.

Aylopon, 85.
Substitute for ANTHIAS Bloch.

Lopharis, 85.

Cephacandia, 85.
Substitute for CEPHALACANTHUS Lacepède.

"Gonurus Lac.," 85.
Unexplained; no such name appears in Lacepède.

Lepomus, 86.

Pomagonus, 86.

Mesopodus, 86.

Acaramus, 86.

Clodipterus, 86, also on 88.
Substitute for CHEILODIPTERUS Lacepède.

Macrolepis, 86, "sp. do."
That is, of APOGON Lacepède.

Guebucus, 86.

Micropodus, 86.
Substitute for CHEILIO Lacepède.

Megaphalus, 86.
Substitute for GOBIESOX Lacepède.

Pomacanthis, 86.

Oxima, 86.

Equetus, 86.
Substitute for EQUES Bloch.

Branchiostegus, 86.

Substitute for CORYPHÆNOIDES Lacepède, not of Gunner.　Replaces LATILUS Cuv. & Val., 1830.

"Eleotris Gr., GOBIOMORUS Lacepède," 86.

Epiphthalmus, 86.

Substitute for GOBIOMOROIDES Lacepède.

Lepimphis, 86.

Plecopodus, 87.

Substitute for GOBIOIDES Lacepède.

Piescephalus, 87.

Lumpus, 87, "sp. do."

That is, of CYCLOPTERUS.

Liparius, 87, "sp. do."

That is, of CYCLOPTERUS.

Percis Scopoli, 87.

Said to be same as ASPIDOPHOROIDES Lacepède.

Aygula, 87.

Substitute for CORIS Lacepède.

Octonus, 88.

Gasterodon, 88.

Xysterus, 88.

For XYSTER Lacepède.

Meneus, 88.

For MENE Lacepède.

Buronus, 88.

For BURO Lacepède.

Thrissa, 88.

Substitute for CLUPANODON Lacepède.

Megalops, 88.

Prinodon, 88.

Substitute for CYPRINODON Lacepède.

Maturacus, 88.

Edomus, 88.

Gonipus, 88.

Myxonum, 88.

Substitute for MUGILOIDES Lacepède.

Trichonotus, 88.

Substitute for MUGILOMORUS Lacepède.

Soranus, 88.

Cordorinus, 89.

Substitute for CORYDORAS Lacepède.

Amiatus, 89.

Substitute for AMIA L. AMIATUS becomes eligible if AMIA Gronow be accepted instead of APOGON.

Sayris, 89.

Substitute for SCOMBRESOX Lacepède.

Ramphistoma, 89; "Raf. BELONE Gronow."

The word BELONE was not used in a generic sense by Gronow, but first by Cuvier in 1817. Rafinesque apparently refers to synonymy of the species as quoted from Gronow by Lacepède.* This reference does not seem to justify the substitution of RAMPHISTOMA for BELONE.

Odumphus, 89.

Onopionus, 89.

Guaris, 90.

Typhlinus, 90, "sp. do."

That is, of SYNGNATHUS, earlier called TYPHLE by Rafinesque, which name is used by Lacepède for a "blind" genus of mammals, but without mention of type.

Phyllophorus, 90.

Homolenus, 90.

This and the two preceding are placed in the same family as SYNGNATHUS.

Goniodermus, 90, "sp. do."

Of OSTRACION.

Cephalopsis, 90, "sp. do."

Of DIODON.

Orbidus, 90.

For "SPHEROIDE" of Lacepède.

Oonidus, 90.

For "OVOIDE" of Lacepède.

Tangus, 91.

Said to be the same as "HEPTACA" Rafinesque.

Piratia, 91.

Opictus, 91.

---

*In the synonymy of Esox BELONE Lacepède has this quotation:

"*Belone* et *raphis,* id est *acus.* Petri Artedi Synonymia Piscium etc., auctore J. G. Schneider, etc.

"Gronov. Mus. 1, n. 39. Zooph., p. 117, n. 362."

In a hasty reading, Rafinesque must have ascribed both sentences to Gronow who apparently did not use the name BELONE.

Ictiopogon, 91.
Substitute for Bostrychus Lacepède, which is preoccupied.

Dameus, 91.

Neleus, 91.

Nemochirus, 91.

Dipinotus, 91.

Symphocles, 91.
This and the preceding are placed near Trichiurus.

Melanictis, 92.

Epimonus, 92, "sp. do."
That is, of Balistes, doubtless intended for Monacanthus.

Lophidius, 92.
Variant of Lophius.

Chironectes, "R", 92, "sp. do."
Of Lophius.

Conomus, 92, "sp. do."
Of Lophius.

Branderius, 93.
Substitute for Cæcilia Lacepède, a synonym of Sphagebranchus Bloch.

Anopsus, 93.
Substitute for Muraenoblenna Lacepède.

Gymnopsis, 93.
Substitute for Gymnomuraena Lacepède.

Helmictis, 93.

Rincoxis, 93.

Zebricium, 93.

Pterurus, 93.

Sphyrnias, 93.
Variant of Sphyrna Rafinesque.

Platopterus, 93.
Substitute for Raja L.

Epinotus, 93.

Lymnea, 93.

Podoleptus, 93.
Substitute for Leptopus Rafinesque, 1814. A synonym of Lophotes Giorna.

Megaderus, 93.
Substitute for Echidna Forster.

Ictætus, 93.

Sephenia, 93.

Megabatus, 93.

Apturus, 93.

Lampreda, 94, "sp. do."
Of PETROMYZON.

Pricus, 94, "sp. do."
Of PETROMYZON. This and the preceding doubtless equivalent to AMMO-CŒTES Duméril and LAMPETRA Gray, the former based on the larval state of lampreys.

LXXXVI. CUVIER, *Observations et Recherches Critiques sur differens Poissons de la Méditerrannée, et à leur Occasion sur des Poissons d'autres mers plus ou moins Liés avec Eux:* Mémoires du Museum d'Histoire Naturelle, Paris, I, 1815.

GEORGES LÉOPOLD CHRÉTIEN FRÉDÉRIC DAGOBERT CUVIER.

Glossodus Cuvier, I, 1815; type ARGENTINA GLOSSODONTA Forskål.
Same as ALBULA Gronow, BUTYRINUS Lacepède.

Fierasfer Cuvier, I, 119, 312, 359, 1815; type OPHIDION IMBERBE L.
Same as CARAPUS Rafinesque. (Name in French only; dates from Oken, 1817.)

Xyrichthys Cuvier, I, 317, 329, 355, 1815; type XYRICHTHYS CULTRATUS Cuvier (CORYPHÆNA NOVACULA L.).

Epibulus Cuvier, I, 111; type SPARUS INSIDIATOR Pallas.

Smaris Cuvier, I, 111; type SPARUS SMARIS L.

Myletes Cuvier, I, 115; type MYLETES RHOMBOIDALIS Cuv.

Saurus Cuvier, I, 115; type SALMO SAURUS L.

Chromis Cuvier, 393; type SPARUS CHROMIS L.
Monotypic. Same as HELIASES Cuvier, 1817.

Crenilabrus Cuvier, 357; type LABRUS LAPINA Forskål (Les Crénilabres).
Bonaparte, in 1839, named as type C. PAVO, which is the same as C. LAPINA. Name in French only, dates from Oken, 1817.

Corycus Cuvier, 359; type (not otherwise named) LUTIANUS ROSTRATUS Bloch.
Based on "les deux derniers Lutjans de M. Risso." Called CORICUS by Cuvier in 1817. A synonym of SYMPHODUS Rafinesque.

Tetragonopterus Cuvier, I, 114; type T. ARGENTEUS Cuvier.

Les Serrans Cuvier.

This is not a scientific name. Reference is made to ANTHIAS SACER Bloch.

Diacope Cuvier, 360; type HOLOCENTRUS BENGALENSIS Bloch.

Name preoccupied, replaced by GENYOROGE Cantor, 1850.

Diagramma Cuvier, 360; type ANTHIAS DIAGRAMMA Bloch.

A synonym of PLECTORHINCHUS Lacepède.

Scolopsis Cuvier, 361; type the CURITE of Russell (SCOLOPSIS CURITE Cuvier).

Les Priacanthes Cuvier, 361; type ANTHIAS MACROPHTHALMUS Bloch.

First called PRIACANTHUS by Cuvier in 1817.

Les Pristipomes Cuvier, 361; type LUTIANUS HASTA Bloch.

Called PRISTIPOMUS by Oken in 1817. A synonym of POMADASYS Lacepède.

Julis Cuvier, 362; type LABRUS JULIS L.

By tautonomy and by first restriction.

Boops Cuvier, 453; type SPARUS BOOPS L.

By tautonomy. Later called Box Cuvier.

The genus SPARUS L. is divided into Sargues, SPARUS SARGUS L.; Daurades, SPARUS AURATA L.; and Pagres, SPARUS PAGRUS L. No Latin names are indicated and no type assigned to SPARUS L. "Les Canthères" is a name assigned (p. 485) to SPARUS CANTHARUS L.

Dentex Cuvier, 486; type SPARUS DENTEX L.

By tautonomy,

Les Melettes, 457; type CLUPEA BRÜNNICHI Gmelin.

This group is equivalent to ENGRAULIS Cuvier, 1817. No Latin name is assigned.

On page 14 of this volume the name SCIÆNA UMBRA L., originally based on two species (the "Maigre" and the "Corb," confounded by Linnæus), is definitely restricted to the first of these, the CHEILODIPTERUS AQUILA of Lacepède. This species then becomes the type of SCIÆNA, replacing ARGYROSOMUS DE LA PYLAIE and PSEUDOSCIÆNA Bleeker. The "Corb" remains as CORVINA NIGRA (Bloch).

LXXXVII. BLAINVILLE, *Prodrome d'Une Nouvelle Distribution Systématique du Règne Animal:* Bullétin de la Société Philomatique, 1816.

HENRI MARIE DUCROTAY DE BLAINVILLE.

Descriptions of the genera indicated appeared also in the *Fauna Française,* 1820-1830.

The pagination as here given is taken from Garman, *Plagiostomia,* 1913.

Trygonobatus Blainville, 112; type RAIA PASTINACA L.

Equivalent to DASYATIS Rafinesque.

Aetöbatus Blainville, 112; type RAJA NARINARI Euphrasen.

As restricted by Müller & Henle, 1838; not as restricted by Cantor, 1850. The latter arrangement would accord better with Blainville's obvious purpose of making RAJA AQUILA his "AETOBATIS VULGARIS," the best known of these "raies aigles," his type. The first restriction, however, has the sanction of Agassiz, Günther, and Gill.

Dicerobatus Blainville, 116; type RAJA MOBULAR Lacepède.

A synonym of MOBULA Rafinesque. Monotypic.

Leiobatus Blainville, 121, "Raies lisses"; type LEIOBATUS SLOANI Blainville.

A species of UROTRYGON. Not LEIOBATUS Rafinesque, which is RHINOBATUS; nor LEIOBATUS Klein, a synonym of RAJA.

Narcobatus Blainville, 121; type RAJA TORPEDO L.

Scylliorhinus Blainville, 121, "SQUALES ROUSSETTES"; type SQUALUS CANICULA Lacepède.

As restricted. Prior to SCYLLIUM Cuvier.

Cestrorhinus Blainville, 121; type SQUALUS ZYGÆNA L.

A synonym of CESTRACION Klein and SPHYRNA Rafinesque.

Monopterhinus Blainville, 121, "squales à une seule pinnule dorsale"; type SQUALUS GRISEUS Gmelin.

A synonym of HEXANCHUS Rafinesque. Monotypic.

Acanthorhinus Blainville, 121, "squales épineux"; type SQUALUS ACANTHIAS Lacepède.

Equivalent to SQUALUS L. as restricted by Rafinesque and by Gill.

Heterodontus Blainville, 121; type SQUALUS PHILIPPI Lacepède.

The name seems sufficiently different from HETERODON, a genus of snakes of prior date, although from identical Greek roots. Otherwise would stand as CENTRACION Gray. Monotypic.

Cetorhinus Blainville, 121; type SQUALUS MAXIMUS Gunner.

Monotypic.

Galeorhinus Blainville, 121, "Squales demi-Requins"; type SQUALUS CANIS L.

As restricted by Gill, 1864; SQUALUS MUSTELUS as restricted by Garman, 1913. The former arrangement must hold, replacing EUGALEUS Gill.

Carcharhinus Blainville, 121; type SQUALUS COMMERSONIANUS Blainville.

As restricted by authors.

Echinorhinus Blainville, 121, "Squales bouclés"; type SQUALUS SPINOSUS Gmelin.

Monotypic.

LXXXVIII. CUVIER, *Sur le genre* Chironectes Cuv. (Antennarius Commerson): Mém. du Mus., III, 1817.

GEORGES CUVIER.

Chironectes Cuvier, 418; type Lophius commersonianus Lacepède.

LXXXIX. CUVIER, Hydrocyon, Chalceus etc.: Mémoires du Museum, IV, 1817; V, 1819.

Hydrocyon Cuvier, V, 353, 1819; type Hydrocyon forskali Cuvier ("Les Hydrocyns," I, 115).

Myletes Cuvier, IV, 449, 1817; type Myletes rhomboidalis Cuvier.

Tetragonopterus (Artedi) Cuvier, IV, 455, 1817; type Tetragonopterus argenteus Cuvier.

Chalceus Cuvier, IV, pl. XXI, 1817, V, 351; type Chalceus macrolepidotus Cuvier.

XC. LE SUEUR, *A new genus of Fishes proposed under the name of* Catostomus: Journ. Acad. Nat. Sci. Phila., I, 1817.

CHARLES A. LE SUEUR.

Catostomus Le Sueur, 88; type Cyprinus catostomus Forster (Catostomus hudsonius Le Sueur).

XCI. CUVIER, *Règne Animal,* Ed. 1, 1817. *Le Règne Animal Distribué d'Après son Organisation.* Tome II, 1817 (Reptiles, Fishes, etc.).

GEORGES CUVIER.

This epoch-making work marks the advent of the modern epoch in zoological taxonomy. The entire animal kingdom is considered, its families and genera are all defined, and the entire system is placed on the sound basis of comparative anatomy.

In a number of cases the genera in this work receive French names only. But to all these, Latin forms were immediately supplied in the same year, 1817, by Oken, in the *Isis.*

Ammocœtes (Duméril) Cuvier, 119; type PETROMYZON BRANCHIALIS L.
   The larval form of PETROMYZON and LAMPETRA.

Scyllium Cuvier, 124; type SQUALUS CANICULA L.
   As restricted. A synonym of SCYLLIORHINUS Blainville, 1816.

Carcharias Cuvier, 125; type SQUALUS CARCHARIAS.
   As shown in the figure of Bélon, 60. "Cette figure de Bélon est la seule
bonne. La plupart des autres sont fausses." Equivalent to CARCHARHINUS Blainville. Not CARCHARIAS Rafinesque, 1810.

Lamna Cuvier, 126; type SQUALUS CORNUBICUS Bloch & Schneider
   (SQUALUS NASUS Bonnaterre).

Zygæna Cuvier, 127; type SQUALUS ZYGÆNA L.
   Equivalent to CESTRACION Klein, SPHYRNA Rafinesque.

Galeus Cuvier, 127; type SQUALUS GALEUS L.
   Not GALEUS Valmont, nor of Klein, nor of Rafinesque. Equivalent to GALE-
ORHINUS Blainville. Monotypic.

Mustelus Cuvier, 128; type SQUALUS MUSTELUS L.
   Equivalent to MUSTELUS Linck.

Notidanus Cuvier, 128; type SQUALUS GRISEUS L.
   Monotypic. Equivalent to HEXANCHUS Rafinesque, 1810.

Selache Cuvier, 129; type SQUALUS MAXIMUS L.
   Equivalent to CETORHINUS Blainville, 1816. Monotypic.

Cestracion Cuvier in Oken, 129, ("Les Cestracions" Cuvier); Oken,*
   Isis, 1183; type SQUALUS PHILIPPI Bloch & Schneider.
   Equivalent to HETERODONTUS Blainville, 1816. Not CESTRACION Klein.

Spinax Cuvier, 129; type SQUALUS SPINAX L.
   Equivalent to ETMOPTERUS Rafinesque, 1810.

Centrina Cuvier, 130; type SQUALUS CENTRINA L.
   Equivalent to OXYNOTUS Rafinesque, 1810.

Scymnus Cuvier, 130; type SQUALUS AMERICANUS Gmelin.
   Monotypic. Preoccupied by SCYMNUS Kugelmann, 1814, a genus of Beetles;
replaced by SCYMNORHINUS Bonaparte, 1846.
   DALATIAS Rafinesque may be the same.

---

*Oken, *Isis*, 1817. In *Isis*, 1817, immediately following the publication of the
Règne Animal, Professor L. Oken recapitulated the genera of Cuvier, giving Latin
forms to those left by the author with French names only. These occur on pages
1181, 1182, and 1183 (misprinted in the text as 1781, 1782, and 1783). Several of
these were also Latinized by Cloquet, *Dictionnaire d'Histoire Naturelle,* in 1817,
but the date of Oken is earlier. Still others were Latinized by John Stark, *Elements of Natural History, Edinburgh,* 1828, vol. I.
   As the question of priority is not affected, we leave the names of **Cuvier** latinized by Oken in their proper sequence.

Trygon Adanson in Cuvier, 136; type RAJA PASTINACA L.

This name, quoted from Adanson's Manuscript *Cours d'Histoire Naturelle,* in 1772, has been adopted by several authors. TRYGON appears also as a new genus (equivalent to TÆNIURA) of Geoffroy St. Hilaire in the *Histoire d'Egypte,* with the date of 1825. The name is certainly later than DASYBATUS of Klein, with the same type, and also later than DASYATIS Rafinesque.

Callorhynchus (Gronow) Cuvier, 140; type CHIMÆRA CALLORHYNCHUS L.

Monacanthus (Cuvier) Oken, 152, ("Les Monacanthes" Cuvier); type BALISTES CHINENSIS Bloch.

Alutera (Cuvier) Oken, 153, ("Les Alutères" Cuvier); type BALISTES MONOCEROS L.

Triacanthus (Cuvier) Oken, 153, ("Les Triacanthes" Cuvier); type BALISTES BIACULEATUS Bloch.

Monotypic.

Hippocampus Cuvier, 157; type SYNGNATHUS HIPPOCAMPUS L.

Equivalent to HIPPOCAMPUS Rafinesque.

Characinus Cuvier, 164; no type named.

Equivalent to CHARACINI L., CHARAX Gronow.

Curimatus (Cuvier) Oken, 165, ("Les Curimates" Cuvier); type SALMO EDENTULUS Bloch (SALMO CYPRINOIDES L.).

Name latinized as CURIMATUS Oken, 1182, as CURIMATA by Cloquet. Equivalent to CURIMATA Walbaum, 1792.

Anostomus (Gronow) Cuvier, 165; type SALMO ANOSTOMUS L.

Called LEPORINUS by Cuvier & Valenciennes.

Piabucus (Cuvier) Oken, 166, ("Les Piabuques" Cuvier); type SALMO ARGENTINUS Bloch.

Latinized as PIABUCUS by Oken, as PIABUCA by Müller & Tröschel.

Citharinus Cuvier, 168; type "LE SERRASALME CITHARINE" Geoffroy.

CYTHARINUS Oken, 1182, ("Les Citharines" Cuvier, 1815, I. 115).

Saurus Cuvier, 169; type SALMO SAURUS L.

Scopelus Cuvier, 169; type GASTEROPELECUS HUMBOLDTI Risso.

Equivalent to MYCTOPHUM Rafinesque, 1810.

Aulopus Cuvier, 170; type SALMO FILAMENTOSUS Bloch.

Engraulis Cuvier, 174; type CLUPEA ENCRASICHOLUS L.

Thrissa Cuvier, 176; type CLUPEA SETIROSTRIS Broussonet.

Equivalent to MYSTUS Lacepède, preoccupied. Not THRISSA Rafinesque. Later spelled THRYSSA by Cuvier.

This genus or subgenus may receive a new name, THRISSOCLES Jordan & Evermann, type CLUPEA SETIROSTRIS Broussonet. It is distinguished from MENIDIA Browne or ANCHOVIELLA Fowler, by the greatly prolonged maxillary.

Pristigaster Cuvier, 176; type not named.

Later fixed as PRISTIGASTER CAYANUS Cuvier.

Chirocentrus Cuvier, 178; type Esox CHIROCENTRUS L. (CLUPEA DORAB Gmelin).

Sudis Cuvier, 180; type (not named) SUDIS GIGAS Cuvier.

Equivalent to ARAPAIMA Müller and Troschel, 1846 (VASTRES Valenciennes, 1846).

Not SUDIS Rafinesque, 1810.

Galaxias Cuvier, 183; type Esox TRUTTACEUS Cuvier.

Monotypic.

Microstoma Cuvier, 184; type GASTEROPELECUS MICROSTOMUS Risso.

Monotypic.

Stomias Cuvier, 184; type Esox BOA Risso.

Monotypic.

Salanx Cuvier, 185; type (not named) SALANX REEVESI Cuv. & Val. (ALBULA CHINENSIS Osbeck).

Monotypic. Synonym of ALBULA Osbeck, if the latter name is regarded as eligible.

Belone Cuvier, 185; type Esox BELONE L.

Hemi-Ramphus Cuvier, 186; type Esox BRASILIENSIS L.

Barbus Cuvier, 192; type CYPRINUS BARBUS L.

Gobio Cuvier, 193; type CYPRINUS GOBIO L.

Not GOBIO of Klein, which is identical with GOBIUS L. If Klein's names are accepted, GOBIO Cuvier, based on the common gudgeon of Europe, CYPRINUS GOBIO L., must receive a new name.

Tinca Cuvier, 193; type CYPRINUS TINCA L.

Cirrhinus (Cuvier) Oken, 193, ("Les Cirrhines" Cuvier); type CYPRINUS CIRRHOSUS Bloch.

The name is spelled CIRRHINA by Valenciennes.

Abramis Cuvier, 194; type CYPRINUS BRAMA L.

Equivalent to BRAMA Klein, not of Cuvier.

Labeo Cuvier, 194; type CYPRINUS NILOTICUS (Forskål) Geoffroy.

Not LABEO Bowdich, 1825.

Leuciscus (Klein) Cuvier, 194; type CYPRINUS LEUCISCUS (L.) Bloch.

Identical with LEUCISCUS Klein.

Lebia (Cuvier) Oken, 199, ("Les Lebias" Cuvier), Oken, *Isis*, 1183; type CYPRINODON VARIEGATUS Lacepède.

Monotypic. Identical with CYPRINODON Lacepède, 1803. Written LEBIAS by Cuvier.

Schilbe (Cuvier) Oken, 202, ("Les Schilbe" Cuvier), Oken, *Isis,* 1182; type SILURUS MYSTUS Hasselquist.

Synodontis Cuvier, 203; type PIMELODUS SYNODONTIS Geoffroy.

Bagre (Cuvier) Oken, 204, ("Les Bagres" Cuvier), Oken, *Isis,* 1182; type SILURUS BAGRE L.

By tautonomy. Latinized as BAGRUS by Valenciennes. This generic name must replace FELICHTHYS Swainson and AILURICHTHYS Baird, for the Gaff-top-sail Cat-fish FELICHTHYS MARINUS (L), of American waters, unless the earlier name BAGRE of Catesby, applied to another genus, be deemed eligible. The name BAGRUS, transferred by Cuvier & Valenciennes to another part of the same group of "Bagres," may be replaced by PORCUS Geoffroy, 1817, as BAGRUS is only a variant spelling of BAGRE.

Morrhua (Cuvier) Oken, ("Les Morues"), 212, 1182; type GADUS MORRHUA L.

Same as GADUS L.

Merlangus (Cuvier) Oken, ("Les Merlans"), 213, 1182; type GADUS MERLANGUS L.

Not MERLANGUS Rafinesque, 1810.

Lota (Cuvier), ("Les Lottes") Oken, 215, 1182; type GADUS LOTA Bloch.

By tautonomy.

Mustela (Cuvier), ("Les Mustèles") Oken, 215, 1182; type GADUS MUSTELA L.

Preoccupied in Mammals: replaced by MOTELLA Cuvier, 1829. Synonym of GAIDROPSARUS Raf.

Brosme (Cuvier), ("Les Brosmes") Oken, 216, 1182; type GADUS BROSME Gmelin.

Called BROSMIUS by Cuvier, 1829.

Raniceps (Cuvier) Oken, 217, ("Les Raniceps" Cuvier), Oken, 1182; type GADUS RANINUS Müller.

Platessa Cuvier, 220; type PLEURONECTES PLATESSA L.

Equivalent to PLEURONECTES as restricted by Fleming.

Hippoglossus Cuvier, 221; type PLEURONECTES HIPPOGLOSSUS L.

Rhombus Cuvier, 222; type PLEURONECTES RHOMBUS Cuvier.

Equivalent to RHOMBUS Klein, not of Lacepède. Name used in Mollusks, RHOMBUS Da Costa, 1776. Same as BOTHUS Rafinesque.

Solea Cuvier, 223; type PLEURONECTES SOLEA L.

Equivalent to SOLEA Quensel and SOLEA Rafinesque.

Monochirus (Cuvier) Oken, 223, ("Les Monochires" Cuvier), MONOCHIRUS Oken, 1182; type PLEURONECTES MICROCHIRUS De la Roche.

Monotypic. Equivalent to Rafinesque.

Plagusia (Browne) Cuvier, 224; type PLEURONECTES PLAGUSIA L.

Equivalent to PLAGUSIA Browne, not of Latreille, 1806, a genus of CRUSTA-CEANS.

Lumpus (Cuvier) Oken, 226, ("Les Lumps" Cuvier) Oken, 1182; type
CYCLOPTERUS LUMPUS L.

Equivalent to CYCLOPTERUS L.

Conger (Cuvier) Oken, 231, ("Les Congres" Cuvier), Oken, 1182; type
MURÆNA CONGER L.

Equivalent to CONGER Houttuyn.

Alabes (Cuvier) Oken, 235, ("Les Alabes" Cuvier), Oken, 1182; type
(not named) CHEILOBRANCHUS DORSALIS Richardson.

Said to be identical with CHEILOBRANCHUS Richardson, 1848, which it re-places.

Carapus Cuvier, 237; type GYMNOTUS MACROURUS Bloch.

Not CARAPUS Rafinesque, 1810. The name is written CARAPO by Oken. Re-placed by GITON Kaup.

Fierasfer (Cuvier) Oken, 239, ("Les Fierasfers" Cuvier) Oken, 1182;
type OPHIDIUM IMBERBE L.

Equivalent to CARAPUS Rafinesque, 1810. Monotypic.

Lophotes (Giorna) Cuvier, 243, ("Les Lophotes" Cuvier), the name
written LOPHOTUS by Oken, 1182; type "le Lophote Lacepède"
Giorna (LOPHOTES CEPEDIANUS).

Monotypic.

Clinus Cuvier, 251; type BLENNIUS SUPERCILIOSUS L.

Pholis (Artedi) Cuvier, 251; type BLENNIUS PHOLIS Bloch.

Not PHOLIS Gronow.

Salarias Cuvier, 251; type SALARIAS QUADRIPENNIS Cuvier.

Opistognathus (Cuvier) Oken, 252, ("Les Opisthognathes" Cuvier),
Oken, 1182; type OPISTHOGNATHUS SONNERATI Cuvier.

Monotypic.

Sillago Cuvier, 258; type SILLAGO ACUTA Cuvier.

Julis Cuvier, 261; type LABRUS JULIS L.

This name has been unfortunately transferred by later writers to a great group of the tropical seas, THALASSOMA Swainson.

Crenilabrus Cuvier, 262, ("Les Crenilabres" Cuvier), Oken, 1182; type
LABRUS LAPINA Forskål.

Valenciennes makes the type LABRUS PAVO Risso; but that species is not men-tioned by Cuvier, although identical with L. LAPINA.

Bonaparte, in 1839, named as type C. PAVO, which is the same as C. LAPINA. In 1839, just previous, Swainson observes: "M. Cuvier having expressly stated that the type of his genus CRENILABRUS is the LUTJANUS VERRES of Bloch, I have so retained it, placing all the others under the subgenus CYNÆDUS."

L. VERRES is identical with the type of BODIANUS Bloch. We find no such statement in Cuvier's writings, and L. VERRES is ninth of his original species. CYNÆDUS Swainson, as restricted by Bonaparte, is identical with CTENOLABRUS Cuv. & Val.

In the *Fauna Italica*, 156, 1839, Bonaparte assigned types to certain genera of Labroid fishes. LABRUS GUTTATUS Bloch was indicated as type of HEMIULIS Swainson; LABRUS VETULUS Bloch is considered as type of LABRUS L. This name, VETULUS, was not used by Linnæus, but it is a synonym of LABRUS BIMACULATUS L. and LABRUS MIXTUS L. LABRUS BIMACULATUS may thus be regarded as the type of LABRUS. LABRUS RUPESTRIS Bloch is indicated as type of CYNÆDUS Swainson. This becomes a synonym of CTENOLABRUS Cuv. & Val., of a little earlier date in 1839. LABRUS PAVO of authors, not of Linnæus, is recognized by Bonaparte as type of CRENILABRUS. This species is not named by Cuvier, but it is identical with LABRUS LAPINA Forskål, a species included by Cuvier. In Swain's excellent review of Swainson (1882) the type of each genus is indicated. The type of HEMIULIS (AURATUS) assigned by Swain is a species of CHEILIO Lacepède. The type of CYNÆDUS is C. TINCA = C. DODERLEINI Jordan, not LABRUS TINCA L., a species of CRENILABRUS. Bonaparte's selection of types has, however, priority over Swain's. Bonaparte and Swainson recognize LABRUS JULIS L. as type of JULIS. The same type, as LABRUS PAVO Hasselquist, is assigned to CHLORICHTHYS Swainson by Bonaparte. Swain makes the latter a synonym of THALASSOMA Swainson. Bonaparte, 1846, selected L. JULIS L. as type of ICHTHYCALLUS Swainson, reducing both CHLORICHTHYS and ICHTHYCALLUS to the synonymy of JULIS.

Coricus Cuvier, 263; type LUTIANUS VIRESCENS Risso (L. ROSTRATUS Bloch).

Equivalent to SYMPHODUS Rafinesque.

Epibulus Cuvier, 264; type SPARUS INSIDIATOR Pallas.

Monotypic.

Novacula Cuvier, 265; type CORYPHÆNA NOVACULA L.

By tautonomy. Afterwards restricted by Cuvier & Valenciennes to CORYPHÆNA PENTADACTYLA Lacepède, the genus HEMIPTERONOTUS Lacepède.

A synonym of XYRICHTHYS Cuvier, 1815.

Chromis (Cuvier) Oken, 266, ("Les Chromis" Cuvier), Oken, 1182; type SPARUS CHROMIS Lacepède.

Not CROMIS Browne, 1770, nor CHROMIS Cuvier, 1815. Called HELIASES by Cuvier & Valenciennes.

Plesiops (Cuvier) Oken, 266; type not named.

Afterwards described as PLESIOPS NIGRICANS Rüppell. Equivalent to PHAROPTERYX Rüppell, 1828.

Smaris Cuvier, 269; type SPARUS SMARIS L.

A synonym of SPICARA Rafinesque.

Boops Cuvier, 270; type SPARUS BOOPS L.

Later called Box Cuv. & Val.

Sargus Cuvier, 272; type SPARUS SARGUS L.

Equivalent to SARGUS Klein and DIPLODUS Rafinesque.

Aurata (Cuvier) Oken, 272, ("Les Daurades" Cuvier), Oken, 1183; type SPARUS AURATA L.

Equivalent to SPARUS L., as restricted by Fleming. Later called CHRYSO-PHRYS by Cuvier.

Pagrus Cuvier, 272; type SPARUS ARGENTEUS Bloch & Schneider (SPARUS PAGRUS L.).

Not PAGRUS Plumier.

Dentex Cuvier, 273; type SPARUS DENTEX L.

Diacope Cuvier, 275; type DIACOPE SEBÆ Cuvier.

Name preoccupied in butterflies (DIACOPE Hübner, 1816); replaced by· GEN-YOROGE Cantor, 1850. Equivalent to NAQUA Forskål, 1775, a name doubtfully eligible.

Serranus Cuvier, 276, ("Perche de mer ou serran"); type PERCA CABRILLA L. ("le Serran proprement dit").

SERRANUS CABRILLA is usually assumed as the type of the genus SERRANUS, and therefore of the family SERRANIDÆ. But no species of the genus as thus re-stricted is mentioned by Cuvier by scientific name. Cuvier, however, remarks: "La Méditerranée en produit beaucoup dont les plus communs s'y confondent sous les noms vulgaires de PERCHE DE MER, de SERRAN, etc." This refers to SER-RANUS CABRILLA and SERRANUS SCRIBA. The first of these may be retained as type, following general custom, although its scientific name is not mentioned. Other-wise HOLOCENTRUS GIGAS Bloch & Schneider, which is explicitly mentioned, must be taken, in which case SERRANUS disappears, as a synonym of EPINEPHELUS Bloch, and the name SERRANELLUS Jordan & Eigenmann, a sub-generic term for SERRANUS SCRIBA, would stand for the genus, as in Fowler's arrangement (Proc. Ac. Nat. Sci. Phila., 1907, 266).

Plectropomus (Cuvier) Oken, 277, ("Les Plectropomes" Cuvier), PLECTROPOMUS Oken, 1182, PLECTROPOMA of Cuv. & Val.; type as restricted BODIANUS MACULATUS Bloch.

Cantharus Cuvier, 278; type SPARUS CANTHARUS L. ("Les Canthares," 1815.)

Name preoccupied; replaced by SPONDYLIOSOMA Cantor, 1850.

Pristipomus (Cuvier) Oken, 279, ("Les Pristipomes" Cuvier), Oken, 1182, (PRISTIPOMA of Cuvier & Valenciennes); type LUTJANUS HASTA Bloch.

Equivalent to POMADASYS Lacepède.

Scolopsis Cuvier, 280; type "le Kurite, Russell" (SCOLOPSIDES KURITA Cuv. & Val.).

Called SCOLOPSIDES by Cuvier.

Diagramma (Cuvier) Oken, 280, ("Les Diagrammes" Cuvier), Oken, 1183; type ANTHIAS DIAGRAMMA Bloch.

Equivalent to PLECTORHYNCHUS Lacepède.

Priacanthus (Cuvier) Oken, 281, ("Les Priacanthes" Cuvier), Oken, 1183; type ANTHIAS MACROPHTHALMUS Bloch.

Equivalent to ABUHAMRUR Forskål, a name doubtfully eligible.

Polyprion (Cuvier) Oken, 282, ("Les Polyprions" Cuvier), Oken, 1183; type AMPHIPRION AMERICANUS Bloch & Schneider.

Acerina Cuvier, 283; type PERCA ACERINA Cuvier (PERCA CERNUA L.).

Called CERNUA by Schæfer and by Fleming, ACERINA by Güldenstadt, PERCIS by Klein, and GYMNOCEPHALUS by Bloch.

Stellifer (Cuvier) Oken, 283, ("Les Stellifères" Cuvier), STELLIFER Oken, 1182; type BODIANUS STELLIFER Bloch.

Monotypic.

Pterois (Cuvier) Oken, 286, ("Les Pterois" Cuvier), Oken, 1183; type SCORPÆNA VOLITANS Bloch.

Equivalent to PSEUDOPTERUS Klein, which may not be eligible.

Paralepis Cuvier, 289; type COREGONUS PARALEPIS Risso Ms., PARALEPIS COREGONOIDES Risso.

Monotypic.

Prochilus Cuvier, 294; type SCIÆNA MACROLEPIDOTA Bloch.

Name preoccupied in mammals by PROCHILUS Illiger, 1811. Replaced by DORMITATOR Gill.

Sander (Cuvier) Oken, 294, ("Les Sandres" Cuvier), SANDER Oken, 1182; type PERCA LUCIOPERCA L.

Terapon Cuvier, 295; type HOLOCENTRUS SERVUS Bloch.

Error for THERAPON. Equivalent to DJABUB Forskål, which is doubtfully eligible.

Zingel (Cuvier) Oken, 296, ("Les Cingles" Cuvier), ZINGEL Oken, 1182; type PERCA ZINGEL L., ASPERULUS Schæfer.

Umbrina Cuvier, 297; type SCIÆNA CIRRHOSA L.

Otolithes (Cuvier) Oken, 299, ("Les Otolithes" Cuvier), OTOLITHES Oken, 1182; called OTOLITHUS by Cuvier & Valenciennes; type JOHNIUS RUBER Bloch & Schneider.

Anclyodon (Cuvier) Oken, 299, ("Les Ancylodons" Cuvier), Oken, 1182; type LONCHURUS ANCYLODON Bloch & Schneider.

Monotypic. Preoccupied; replaced by MACRODON Schinz; SAGENICHTHYS Berg, 1895.

Antennarius (Commerson) Cuvier, 310, "Les Chironectes (ANTENNARIUS Commerson)" Cuvier; type LOPHIUS COMMERSONIANUS Lacepède, ANTENNARIUS ANTENNA TRICORNE Commerson, in Lacepède.

Afterwards called CHIRONECTES by Cuvier, the name preoccupied.

Malthe Cuvier, 311; type LOPHIUS VESPERTILIO L.
Equivalent to OGCOCEPHALUS Fischer, 1813.

Thynnus Cuvier, 313; type SCOMBER THYNNUS L.
Not THYNNUS Browne. The name THYNNUS Cuvier is preoccupied, and has been replaced by THUNNUS South and later by ALBACORA Jordan.

Orcynus Cuvier, 314; type SCOMBER GERMON Lacepède (SCOMBER ALALONGA Gmelin).
Name preoccupied by ORCYNUS Rafinesque, 1810, and replaced by GERMO Jordan.

Citula Cuvier, 315; type ("l'espèce est nouvelle") SCIÆNA ARMATA Forskål.

Seriola Cuvier, 315; type CARANX DUMERILI Risso.

Nomeus Cuvier, 315; type GOBIUS GRONOVII Gmelin.
Monotypic.

Vomer Cuvier, 316; type VOMER BROWNI Cuvier.
Equivalent to RHOMBOIDA Browne, 1789, which is probably not eligible.

Spinachia Cuvier, 320; type GASTEROSTEUS SPINACHIA L.
Monotypic.

Lichia Cuvier, 321; type SCOMBER AMIA L.

Blepharis Cuvier, 322; type ZEUS CILIARIS Bloch.
Monotypic; name preoccupied, replaced by BLEPHARICHTHYS Gill. A synonym of ALECTIS Rafinesque.

Equula Cuvier, 323; type CENTROGASTER EQUULA Gmelin.
A synonym of LEIOGNATHUS Lacepède.

Atropus (Cuvier) Oken, 324, ("Les Atropus" Cuvier), Oken, 1182; type BRAMA ATROPUS Bloch & Schneider.
Monotypic. Not separable from CITULA.

Oligopodes (Risso) Cuvier, 328, ("Les Leptopodes Cuvier, OLIGOPODES Risso"); type "l'Oligopode noir, Risso."
Described by Risso, 1826, as OLIGOPUS NIGER, apparently a species of PTERACLIS. Cuvier distinguishes PTERACLIS as "Les Oligopodes" and OLIGOPODES as "Les Leptopodes."

Chelmon Cuvier, 334; type CHÆTODON ROSTRATUS Bloch.

Platax Cuvier, 334; type CHÆTODON TEIRA Bloch.

Heniochus Cuvier, 335; type CHÆTODON MACROLEPIDOTUS Bloch.

Ephippus Cuvier, 335; type CHÆTODON ORBIS Bloch, as restricted by Cuv. & Val., 1831.

Toxotes Cuvier, 338; type LABRUS JACULATOR Bloch & Schneider.
Monotypic.

Anabas Cuvier, 339; type PERCA SCANDENS Daldorf.
Monotypic.

Fiatola Cuvier, 342; type STROMATEUS FIATOLA L.
Equivalent to STROMATEUS L.

Seserinus (Cuvier) Oken, 342, ("Les Seserinus" Cuvier); type "SES-
ERINUS Rondelet" (SESERINUS RONDELETI Cuvier).
This is apparently a young STROMATEUS, MICROCHIRUS (Bonelli).

Premnas Cuvier, 345; type CHÆTODON BIACULEATUS Bloch.
Monotypic.

Temnodon Cuvier, 346; type CHEILODIPTERUS HEPTACANTHUS Lacepède.
Monotypic.   Equivalent to POMATOMUS Lacepède.   The name POMATOMUS
was arbitrarily transferred to a different genus (EPIGONUS Raf.) by Cuvier.

Amphisile (Klein) Cuvier, 350; type CENTRISCUS SCUTATUS L.
Equivalent to CENTRISCUS L. and AMPHISILEN Klein.

XCII. OKEN, *Isis*, 1817.

L. OKEN.

On pages 1182-83 (misprinted 1782-83), Professor Oken gives Latin
equivalents to all the French names in the first edition of the *Règne
Animal* of Cuvier, as indicated above.

XCIII. CLOQUET, *Dictionnaire des Sciences Naturelles de Levrault*
(Articles on Fishes), 1816 to 1830.

HIPPOLYTE CLOQUET.

(Latinizes several of Cuvier's names, as previously done by Oken.)

Zingel Cloquet, 1817; "les cingles" Cuvier, called ZINGEL by Oken;
type PERCA ZINGEL L.

Eptatretus (Duméril) Cloquet, XV, 134, 1819; type GASTROBRANCHUS
DOMBEY Lacepède.
Replaces HEPTATREMA Duméril and BDELLOSTOMA Müller.

XCIV. GEOFFROY ST. HILAIRE, *Suite de l'Histoire des Poissons du Nil*, Plates dated 1817, 1818.

ISIDORE GEOFFROY ST. HILAIRE.

Porcus Geoffroy St. Hilaire, 303, 1818; type SILURUS BAJAD Forskål.
Equivalent to BAGRUS Cuv. & Val. but not to "les Bagres" Cuvier, 1817, BAGRE Oken. PORCUS should probably replace BAGRUS, which is a latinization of BAGRE, the vernacular Spanish name for the larger cat-fishes.

Heterobranchus Geoffroy St. Hilaire, 305, 1818; type HETEROBRANCHUS BIDORSALIS Geoffroy St. Hilaire.

XCV. LE SUEUR, *Description of Several New Species of North American Fishes*: Jour. Ac. Nat. Sci. Phila., 1818.

CHARLES A. LE SUEUR.

Somniosus Le Sueur, 222; type SOMNIOSUS BREVIPINNA Le Sueur, SQUALUS MICROCEPHALUS Bloch & Schneider.

Platirostra Le Sueur, 223; type PLATIROSTRA EDENTULA Le Sueur.
Same as POLYODON.

Hiodon Le Sueur, 366; type HIODON TERGISUS Le Sueur.
Often spelled HYODON.

XCVI. RANZANI, *Descrizione di un Pesce, un Nuovo Genere dei Tænioidei:* Opusculo Sci. Bologna, II, 1818.

CAMILLO RANZANI.

Epidesmus Ranzani, 133; type EPIDESMUS MACULATUS Ranzani.
Synonym of REGALECUS.

XCVII. BLAINVILLE, *Poissons Fossiles*: Nouveau Dictionnaire, XXVII, 1818.

HENRI MARIE DUCROTAY DE BLAINVILLE.

Not seen by us.

Anenchelum Blainville; type ANENCHELUM GLARISIANUM Blainville (fossil: LEPIDOPIDÆ).

Chirurgus Blainville; type (perhaps CHÆTODON CHIRURGUS Bloch).
Not seen by us: probably a synonym of HEPATUS and TEUTHIS.

Palæoniscum Blainville, 320; type PALÆONISCUM FRIESLEBENENSE Blainville (fossil).
Written PALÆONISCUS by Agassiz.

Palæothrissum Blainville, 320; type PALÆOTHRISSUM MACROCEPHALUM Blainville (fossil).
Synonym of PALÆONISCUM.

Palæorhynchum Blainville, XXVII, 314, 1818; type PALÆORHYNCHUM GLARISIANUM Blainville.

Palæobalistum Blainville, 338; type DIODON ORBICULARIS Volta (fossil).

XCVIII. RAFINESQUE, *Description of two new Genera of North American Fishes,* OPSANUS *and* NOTROPIS: American Monthly Magazine and Critical Review, January 1818.

CONSTANTINE SAMUEL RAFINESQUE.

Opsanus Rafinesque, 203; type OPSANUS CERAPALUS Raf. (GADUS TAU L.).
This genus has been usually and wrongly called BATRACHUS.

Notropis Rafinesque, 204; type NOTROPIS ATHERINOIDES Raf. (ALBURNUS RUBELLUS Agassiz).

XCIX. RAFINESQUE, *Discoveries in Natural History*: American Monthly Magazine, September 1818.

Glossodon Rafinesque, 354; type GLOSSODON HARENGOIDES Raf. (HIODON TERGISUS Le Sueur).
A synonym of HIODON, regarded by Rafinesque as too "similar to DIODON in sound."

C. RAFINESQUE, *Further Discoveries in Natural History*: Amer. Monthly Magazine, October 1818.

Pogostoma Rafinesque, 445; type POGOSTOMA LEUCOPS Rafinesque.
A myth drawn by Audubon.

Dinectus Rafinesque, 445; type DINECTUS TRUNCATUS Raf.
A mythical sturgeon, drawn by Audubon.

Litholepis Rafinesque, 445; type LITHOLEPIS ADAMANTINUS Raf.

The "Devil-jack Diamond Fish," drawn by Audubon, a mythical gar-pike of which the scales will "turn a rifle ball."

This remarkable paper is based on paintings of fishes "seen down the river" by Audubon. Rafinesque was, in 1818, a guest at Audubon's house at Hendersonville, Kentucky. One night, bats entered the window. Rafinesque was convinced that they were of a new species, and used Audubon's costly violin to beat them down. Audubon in return showed him paintings of remarkable fishes. Rough copies of these are found in Rafinesque's note-books, preserved in the Smithsonian Institution. Their introduction into Science was a practical joke on the part of the great ornithologist.

Rafinesque also mentions as new genera, by name only, without explanation: CHEILOBUS, MINICULUS, OPLICTIS, LEPTOSOMA and GLANIS. These can have no place in the system.

CI. RAFINESQUE, *Description of Three New Genera of Fluviatile Fish,* POMOXIS, SARCHIRUS *and* EXOGLOSSUM: Jour. Ac. Nat. Sci. Phila., November 1818.

Pomoxis Rafinesque, 417; type POMOXIS ANNULARIS Raf.

Sarchirus Rafinesque, 419; type SARCHIRUS VITTATUS Raf.
Young of LEPISOSTEUS OSSEUS.

Exoglossum Rafinesque, 421; type EXOGLOSSUM ANNULATUM Raf. (CYPRINUS MAXILLINGUA Le Sueur).

Maxillingua Rafinesque, 421; type CYPRINUS MAXILLINGUA Le Sueur. Same as EXOGLOSSUM.

Hypentelium Rafinesque, 421; type EXOGLOSSUM MACROPTERUM Raf. (CATOSTOMUS NIGRICANS Le Sueur).
Replaces HYLOMYZON Agassiz.

CII. RAFINESQUE, *Further account of Discoveries etc.*: Amer. Monthly Magazine, November 1818.

Noturus Rafinesque, 41; type NOTURUS FLAVUS Raf.

CIII. RAFINESQUE, *Prodrome de 70 Nouveaux Genres et d'Animaux Découverts dans l'intérieur des États Unis d'Amérique durant l'Année 1818:* Journal de Physique, de Chymie et d'Histoire Naturelle, Paris, June 1819.

Aplodinotus Rafinesque, 419; type APLODINOTUS GRUNNIENS Raf.
    Spelling corrected by Gill to HAPLOIDONOTUS.

Etheostoma Rafinesque, 419; type ETHEOSTOMA BLENNIOIDES Raf.
    As determined by Agassiz. Later fixed on ETHEOSTOMA FLABELLARIS Rafinesque by Jordan & Evermann. Replaces DIPLESION Rafinesque and HYOSTOMA Agassiz.

Leucops Rafinesque, 419; type POGOSTOMA LEUCOPS Raf.
    A myth of Audubon.

Aplocentrus Rafinesque, 420; type APLOCENTRUS CALLIOPS Raf.
    Mythical; after Audubon.

Calliurus Rafinesque, 420; type CALLIURUS PUNCTULATUS Raf.
    A synonym of MICROPTERUS Lacepède.

Lepomis Rafinesque, 420; type, as stated by the author, LABRUS AURITUS L.

Pomotis Rafinesque, 420; type also stated to be LABRUS AURITUS L.

Apomotis Rafinesque, 420; type LEPOMIS CYANELLUS Raf.

Notemigonus Rafinesque, 421; type NOTEMIGONUS AURATUS Raf. (CYPRINUS CRYSOLEUCAS Mitchill).

Amphiodon Rafinesque, 421; AMPHIODON ALVEOIDES Raf. (Misprint for ALOSOIDES.)
    Replaces ELATTONISTIUS Gill & Jordan.

Amblodon Rafinesque, 421; type AMBLODON BUBALUS Raf.
    The "Buffalo-fish," to which the large blunt pharyngeal teeth of APLODINOTUS were wrongly ascribed. A complex; the name AMBLODON later restricted by Rafinesque to APLODINOTUS GRUNNIENS, while the Buffalo-fish became type of ICTIOBUS Rafinesque.

Cycleptus Rafinesque, 421; type CYCLEPTUS NIGRESCENS (CATOSTOMUS ELONGATUS Le Sueur).

Pilodictis Rafinesque, 422; type PILODICTIS LIMOSUS Raf.
    A myth.

CIV. RAFINESQUE, *Annals of Nature,* I, 1820.

Hemiplus Rafinesque, 6; type HEMIPLUS LACUSTRIS Raf. (CYPRINUS CRYSOLEUCAS Mitchill).
    Same as NOTEMIGONUS Rafinesque, 1819.

CV. RAFINESQUE, *Ichthyologia Ohiensis*, 1820.

It was the fortune of Professor Rafinesque to be one of the first to explore two of the richest fish faunas of the world, that of Sicily and that of the Ohio River. His various papers show his peculiar traits, intense activity, keen philosophical insight, and hopeless slovenliness in method.

Stizostedion Rafinesque, 23; type PERCA SALMONEA Raf. (PERCA VITREA Mitchill).

Lepibema Rafinesque, 23; type PERCA CHRYSOPS Raf.

Pomacampsis Rafinesque, 23; type PERCA NIGROPUNCTATA Raf.
   Mythical, being one of Audubon's practical jokes.

Ichthelis Rafinesque, 27; type LABRUS AURITUS L.
   A needless substitute for LEPOMIS, which is transferred (p. 30) to species of MICROPTERUS Lacepède.

Telipomis Rafinesque, 27; type LEPOMIS CYANELLUS Raf.
   A needless substitute for APOMOTIS.

Aplites Rafinesque, 30; type LEPOMIS PALLIDA Raf.
   A species of MICROPTERUS.

Nemocampsis Rafinesque, 30; type LEPOMIS FLEXUOLARIS Raf.
   A species of MICROPTERUS.

Dioplites Rafinesque, 32; type LEPOMIS NOTATA Raf.
   Also a synonym of MICROPTERUS.

Ambloplites Rafinesque, 33; type LEPOMIS ICHTHELOIDES Raf. (BODI-ANUS RUPESTRIS Raf., 1817).

Aplesion Rafinesque, 36; type ETHEOSTOMA CALLIURA Raf.
   The young of MICROPTERUS.

Diplesion Rafinesque, 37; type ETHEOSTOMA BLENNIOIDES Raf.
   A synonym of ETHEOSTOMA Raf. See opinion 14, Comm. Zool. Nomenc.

Pomolobus Rafinesque, 38; type POMOLOBUS CHRYSOCHLORIS Raf.

Dorosoma Rafinesque, 39; type DOROSOMA HETERURA Raf. (MEGALOPS CEPEDIANA Le Sueur).

Clodalus Rafinesque, 43; type HIODON CLODALUS Le Sueur.
   Same as HIODON.

Minnilus Rafinesque, 45; type MINNILUS DINEMUS Raf.
   This is apparently the same as NOTROPIS Raf., 1818.

Dobula Rafinesque, 45; type not named (CYPRINUS DOBULA L.).
Equivalent to LEUCISCUS Cuvier.

Phoxinus Rafinesque, 45; type not named (CYPRINUS PHOXINUS L.).

Alburnus Rafinesque, 45; type not named (CYPRINUS ALBURNUS L.).

Luxilus Rafinesque, 47; type LUXILUS CHRYSOCEPHALUS Raf. (CYPRINUS CORNUTUS Mitchill).

Chrosomus Rafinesque, 47; type CHROSOMUS ERYTHROGASTER Raf.

Semotilus Rafinesque, 49; type SEMOTILUS DORSALIS Raf. (CYPRINUS ATROMACULATUS Mitchill).

Rutilus Rafinesque, 50; type LEUCISCUS RUTILUS L.

Plargyrus Rafinesque, 50; type RUTILUS PLARGYRUS Raf. (CYPRINUS CORNUTUS Mitchill).

Pimephales Rafinesque, 52; type PIMEPHALES PROMELAS Raf.

Moxostoma Rafinesque, 54; type CATOSTOMUS ANISURUS Raf.

Ictiobus Rafinesque, 55; type CATOSTOMUS BUBALUS Raf.
Spelled ICHTHYOBUS by Agassiz.
The name AMBLODON was based, in part, on the same species, and may be held as tenable for this genus.

Leptops Rafinesque, 64; type PIMELODUS VISCOSUS Raf. (SILURUS OLIVARIS Raf.).

Opladelus Rafinesque, 64; type PIMELODUS NEBULOSUS Raf. (SILURUS OLIVARIS Raf.).
Written HOPLADELUS by Gill. A synonym of LEPTOPS.

Ameiurus Rafinesque, 65; type PIMELODUS CUPREUS Raf. (PIMELODUS NATALIS Le Sueur).

Ilictis Rafinesque, 66; type PIMELODUS LIMOSUS Raf. (SILURUS OLIVARIS Raf.).
A synonym of LEPTOPS.

Picorellr s Rafinesque, 70; type ESOX VITTATUS Raf.
A drawing in Rafinesque's note-book shows this to be a mythical species of "pickerel."

Cylindrosteus Rafinesque, 72; type LEPISOSTEUS PLATOSTOMUS Raf.

Atractosteus Rafinesque, 75; type LEPISOSTEUS FEROX Raf.

Sturio Rafinesque, 79; type ACIPENSER MACULOSUS Le Sueur.

Carpiodes Rafinesque, 56; type CATOSTOMUS CYPRINUS Le Sueur.
As fixed by Jordan & Gilbert, 1883.

Teretulus Rafinesque, 57; type CATOSTOMUS AUREOLUS Le Sueur.
As fixed by Jordan, 1877.

Decactylus Rafinesque, 60; type CATOSTOMUS BOSTONIENSIS Le Sueur
(CYPRINUS TERES Mitchill) as restricted by Jordan.
CATOSTOMUS TERES is probably the tenable name, as the earlier name COM-
MERSONI Lacepède is very doubtful.

Ictalurus Rafinesque, 61; type SILURUS PUNCTATUS Raf.

Elliops Rafinesque, 62; type PIMELODUS MACULATUS Raf.
A synonym of ICTALURUS.

Eurystomus Rafinesque, 65; type CATOSTOMUS MEGASTOMUS Raf.
An Audubonian myth.

Sterletus Rafinesque, 80; type ACIPENSER SEROTINUS Raf. (ACIPENSER
RUBICUNDUS Le Sueur).

Dinectus Rafinesque, 80; type DINECTUS TRUNCATUS Raf.
Another of Audubon's mythical paintings, representing, as suggested by Rafin
esque, "only a sturgeon incorrectly drawn."

Pegedictis Rafinesque, 85; type PEGEDICTIS ICTALOPS Raf.
Apparently a confusion of notes on CATONOTUS FLABELLARIS and COTTUS RICH
ARDSONI. The species is therefore unidentifiable. Best regarded as a synonym of
COTTUS.

Proceros Rafinesque, 87; type PROCEROS MACULATUS Raf.
A myth, not of Audubon but of "Mr. M. of St. Genevieve."

CVI.   TILESIUS, *De Piscium Australium Novo Genere*:   Mémoires
Académie Sciences, Petersburg, 1820.

W. G. VON TILESIUS.

Balistapus Tilesius, 310; type (BALISTES ACULEATUS L.).

CVII.   RISSO, *Alepocephalus, nouveau Genre de Poissons:*   Mémoires
de l'Académie Royale de Turin, XXV, 1820.

ANASTASE RISSO.

Alepocephalus Risso, 262; type ALEPOCEPHALUS ROSTRATUS Risso.

CVIII. GOLDFUSS, *Handbuch Zoologie*, II, 1820.*

Georg August Goldfuss.

Batrachops Goldfuss; type Lophius commersonianus Lacepède.
Substitute for Chironectes, preoccupied. Equivalent to Antennarius.

CIX. LE SUEUR, *Description of a new Genus and Several New Species of Fresh-Water Fish Indigenous to the United States:* Jour. Ac. Nat. Sci. Phila., II, 1821.

Charles A. Le Sueur.

Mollinesia Le Sueur, 2; type Mollinesia latipinna Le Sueur.

CX. HAMILTON, *An Account of the Fishes found in the River Ganges and its Branches:* Edinburgh, 1822.†

Francis Hamilton [formerly Buchanan].
(Often quoted as Francis Hamilton-Buchanan.)

Callichrous Hamilton, 149; type Silurus bimaculatus Bloch.
Synonym of Ompok Lacepède.

Cynoglossus Hamilton, 365; type Cynoglossus lingua Hamilton.
Monotypic.

Bola Hamilton, 368; type Bola coitor Hamilton.
As restricted by Jordan.

Coius Hamilton, 369; type Coius cobojus Hamilton (Perca scandens Daldorf).
As here restricted by us.
The first species named, Coius vacti Hamilton, is Lates colonorum. The genus Coius is grossly unnatural and is best served by relegation to synonymy.
Hamilton observes: "The Cobojus by the natives is considered as the prototype of their genus 'Coi' from which the name Coius is derived."
Equivalent to Anabas Cuvier.

Chanda Hamilton, 370; type Chanda ruconius Hamilton.
The first unquestioned species named. It is a species of Leiognathus Lacepède. The others named belong to Ambassis Cuvier.

* "Acanthiotus Goldfuss" quoted by Agassiz (Nomenclator) is merely a misprint for Acanthonotus.

† We are indebted to Mr. John Smallwood for the record of the genera of Hamilton.

Sisor Hamilton, 379; type SISOR RABDOPHORUS Hamilton.
Monotypic.

Corica Hamilton, 383; type CORICA COBORNA Hamilton.
Monotypic. Apparently an ally of CLUPEOIDES Bleeker. Not CORICUS Cuvier.

Chela Hamilton, 383; type CYPRINUS CACHIUS Hamilton.
The first species named; as restricted by Bleeker, 1862. Günther later restricts it to C. GORA Hamilton. Replaces CACHIA Günther.

Barilius Hamilton, 384; type by tautonomy CYPRINUS BARILA Hamilton.
The first species named.

Bangana Hamilton, 385; type CYPRINUS DERO Hamilton.
The first species named: probably replaces TYLOGNATHUS Heckel and LOBO-CHEILUS van Hasselt.

Puntius Hamilton, 388; type CYPRINUS PUNTIO Hamilton.
By tautonomy. Restricted to CYPRINUS SOPHORE Hamilton by Bleeker.

Danio Hamilton, 390; type CYPRINUS DANGILA Hamilton.
The first species named; as restricted by Bleeker.

Morulius Hamilton, 391; type CYPRINUS MORALA Hamilton.
The first species named; restriction by Bleeker.

Cabdio Hamilton, 392; type CYPRINUS JAYA Hamilton.
The first species named.

Garra Hamilton, 393; type CYPRINUS LAMTA Hamilton.
The first species named; as restricted by Bleeker. Replaces DISCOGNATHUS Heckel.

CXI (A). LEACH & DE LA BÈCHE, *Trans. Zool. Soc. London*, (2) 1822.

WILLIAM E. LEACH.

Dapedium Leach & De La Bèche, 45; type DAPEDIUM POLITUM Leach (fossil).
Written DAPEDIUS by Agassiz.

CXI (B). SCHINZ, *Das Thierreich*, II, 1822.

H. R. SCHINZ.

Macrodon Schinz, 482; type LONCHURUS ANCYLODON Bloch & Schneider.
Replaces ANCYLODON Cuvier, 1817, preoccupied by ANCYLODON Illiger, 1811, a genus of mammals. MACRODON is prior to SAGENICHTHYS Berg, 1895, also a substitute for ANCYLODON. Not MACRODON Müller & Troschel, which is replaced by HOPLIAS Gill, 1903.

CXII. VAN HASSELT, (*Poissons de Java*) : Allgemeine Konst. en Letterbok, II, 1823.

JAN COENRAD VAN HASSELT.

Homaloptera Van Hasselt, 130; type HOMALOPTERA FASCIATA Van Hasselt.

Crossocheilus Van Hasselt, 132; type CROSSOCHEILUS OBLONGUS Van Hasselt.

Hampala Van Hasselt, 132; type HAMPALA MACROLEPIDOTA Van Hasselt.

Lobocheilus Van Hasselt, 133; type LOBOCHEILUS FALCIFER Van Hasselt.
A synonym of BANGANA.

Nemacheilus Van Hasselt, 133; type COBITIS FASCIATUS Valenciennes.

Acanthopsis Van Hasselt, 133; type ACANTHOPSIS DIALYZONA Van Hasselt.

Acanthophthalmus Van Hasselt, 133; type ACANTHOPHTHALMUS FASCIATUS Van Hasselt.

Homaloptera Van Hasselt, 133; type HOMALOPTERA FASCIATA Van Hasselt.

Oxygaster Van Hasselt, 133; type OXYGASTER ANOMALURUS Van Hasselt.

CXIII. DESMAREST, *Première Décade Ichthyologique*, 1823.

ANSELME GAËTAN DESMAREST.

Diabasis Desmarest, 34; type DIABASIS PARRA Desmarest.
Not DIABASIS Hoffmansegg, a genus of beetles, 1819. Replaced by HÆMULON Cuvier.

CXIV. NARDO, *Osservazione Aggiunte all' Adriatica Ittiologia*, 1824.

GIAN DOMENICO NARDO.

Squatinoraja Nardo; type SQUATINORAJA COLONNA (RHINOBATUS COLUMNÆ Müller & Henle).
A synonym of RHINOBATUS.

CXV. MITCHILL, *Description of an Extraordinary Fish*: Ann. Lyc. Nat. Hist. N. Y., I, 1824.

SAMUEL LATHAM MITCHILL.

Saccopharynx Mitchill, 82; type SACCOPHARYNX FLAGELLUM Mitchill.

CXVI. VALENCIENNES, *Description du Cernié*: Mém. du Museum, Paris, XI, 1824.

ACHILLE VALENCIENNES.

Polyprion Valenciennes, 265; type POLYPRION CERNIUM Valenciennes.

CXVII. OTTO, *Propterygia hyposticta,* Nova Acta Acad., 1824.

B. C. OTTO.

Propterygia Otto, 111; type PROPTERYGIA HYPOSTICTA Otto.
A deformed RAJA, with the pectoral fins free from the head, similar to CEPH-ALEUTHERUS Rafinesque.

CXVIII. QUOY & GAIMARD, *Voyage Autour du Monde . . . Exécuté sur les Corvettes l'Uranie et la Physicienne.* Under Captain Louis de Freycinet, 1824.

JEAN RÉNÉ CONSTANTINE QUOY; JOSEPH PAUL GAIMARD.

This important work, which closely preceded the second edition of the *Règne Animal,* adopted a few unpublished names from Cuvier's manuscripts.

Anampses (Cuvier) Quoy & Gaimard, 276; type ANAMPSES CUVIER Q. & G.
Monotypic.

Gerres (Cuvier) Quoy & Gaimard, 293; type GERRES VAIGIENSIS Q. & G., which is SCIÆNA ARGYREA Forskål.

In this paper the genus GERRES is mentioned for the first time. But two species are named, G. VAIGIENSIS and G. GULA. The former belongs to the modern genus XYSTÆMA Jordan & Evermann, the latter to EUCINOSTOMUS Baird. When GERRES was later defined by Cuvier in the *Règne Animal*, SCIÆNA ARGYREA, which is identical with G. VAIGIENSIS, is included, and G. GULA is not mentioned. It would appear that G. VAIGIENSIS must be taken as type, thus replacing XYSTÆMA.

Jordan & Evermann have taken an American species, GERRES LINEATUS (Humboldt) as type. This species is named by Cuvier, but not by Quoy & Gaimard. It cannot therefore serve as the type of GERRES. The genus called GERRES by Jordan & Evermann must apparently stand as DIAPTERUS Ranzani. If GERRES is regarded as preoccupied by GERRIS Fabricius, a genus of insects of earlier date, the substitute name of CATOCHÆNUM Cantor, 1850, should be used.

Percophis Quoy & Gaimard, 351; type PERCOPHIS BRASILIANUS Q. & G.
Monotypic.

Priodon (Cuvier) Quoy & Gaimard, 377; type PRIODON ANNULATUS Q. & G.
A species of NASO Lacepède.

CXVIII (A). HARLAN, *Saurocephalus:* Journ. Ac. Nat. Sci. Phila., III (fossil fish), 1824.

RICHARD HARLAN.

Saurocephalus Harlan, 339; type SAUROCEPHALUS LANCIFORMIS Harlan.

CXIX. KÖNIG, *Icones Fossiles Sectione*, 1825.

C. KÖNIG.

Bucklandium König, 4; type BUCKLANDIUM DILUVII König.
Monotypic.

Ampheristus König, pl. XV, fig. 190; type AMPHERISTUS TOLIACUS König.
Monotypic.

CXX. BOWDICH, *Fishes of Madeira*, 1825.

T. EDWARD BOWDICH.

Labeo Bowdich, 122; type LABEO SPAROIDES Bowdich.
A sparoid fish; not LABEO Cuvier, 1817.

Anomalodon Bowdich, 237; type ANOMALODON INCISUS Bowdich.
Apparently a synonym of POMADASYS Lacepède.

Diastodon Bowdich, 238; type DIASTODON SPECIOSUS Bowdich (LABRUS
   SCROFA Cuv. & Val.).
This genus appears valid.   It is near LEPIDAPLOIS Gill, but with much smaller
scales.

Seleima Bowdich, 238; type SELEIMA AURATA Bowdich.
A synonym of KYPHOSUS Lacepède (K. INCISOR).

Amorphocephalus Bowdich, 238; type AMORPHOCEPHALUS GRANULATUS
   Bowdich.
A synonym of XYRICHTHYS Cuvier.

CXXI.   RISSO, *Histoire Naturelle des Principales Productions de
   l'Europe Méridionale,* vol. III, Paris, 1826.

ANASTASE RISSO.

Lamia Risso, 123; type SQUALUS CORNUBICUS L.
Monotypic. A synonym of LAMNA Cuvier, 1817.

Acanthias Risso, 131; type SQUALUS ACANTHIAS L.
Equivalent to SQUALUS L. as restricted.

Scyphius Risso, 185; type SCYPHIUS FASCIATUS Risso.
A synonym of NEROPHIS Rafinesque, 1810.

Onos Risso, 214; type GADUS MUSTELA Bloch.
A synonym of GAIDROPSARUS Rafinesque.

Lotta Risso, 217; type GADUS ELONGATUS Otto.
A variant of LOTA Cuvier, 1817 ("Les Lottes"); apparently not eligible as a
substitute for MOLVA Fleming.

Mora Risso, 224; type MORA MEDITERRANEA Risso.

Morua Risso, 225; type GADUS CAPELAN Lacepède.
According to Risso, a synonym of TRISOPTERUS Rafinesque, 1814; prior to
BRACHYGADUS Gill (GADUS MINUTUS L.).   Monotypic.

Tripterygion Risso, 241; type TRIPTERYGION NASUS Risso.
Monotypic.

Diana Risso, 267; type DIANA SEMILUNATA Risso.
Monotypic.

Aphia Risso, 287; type APHIA MERIDIONALIS Risso.
Apparently a young ATHERINA. Monotypic.

Fiatola Risso, 289; type FIATOLA FASCIATA Risso.
Monotypic. Apparently a young STROMATEUS.

Ausonia Risso, 341; type Ausonia cuvieri Risso.
A synonym of Luvarus Rafinesque.

Aurata Risso, 355; type Aurata semilunata Risso.
A synonym of Sparus L. as restricted by Fleming.

Macrostoma Risso, 447; type Macrostoma angustidens Risso.
Not Macrostomus Wied, a genus of insects; replaced by Notoscopelus Günther.

Alpismaris Risso, 458; type Alpismaris risso Risso.
The young of Synodus.

Paralepis Risso, 472; type Paralepis coregonoides Risso (Coregonus macrænula Risso, 1810).

CXXI (A). GERMAR, *Keferstein's Deutschland's Geognosie Dargestellt*, 1826.

E. F. Germar.

Ichthyolithus Germar, IV, 96; type Ichthyolithus esociformis Germar.
Perhaps identical with Leptolepis. Ichthyolithus, a term for fossil fishes used by early authors, is perhaps not intended as a scientific name.

CXXII. GEOFFROY ST. HILAIRE, *Histoire Naturelle des Poissons de la Mer Rouge et de la Méditerranée*, 1826.

Isidore Geoffroy St. Hilaire.

Trygon Geoffroy St. Hilaire; type Trygon grabatus Geoffroy St. Hilaire.
This species is not noticed by recent writers. It is apparently a species of Tæniura Müller & Henle, 1842, allied to Tæniura lymma (Forskål). This is generically distinct from the type of Trygon (Adanson) Cuvier, 1817. St. Hilaire makes no reference to Adanson, from whose manuscripts Cuvier took the name. It is plain that the date of the article which contains the name Trygon is subsequent to 1825, as a footnote relates to a paper read in October 1825, and printed in November. Hence we assume the date of the present paper as 1826. In any event Trygon St. Hilaire is preceded by Trygon Cuvier, and both are later than Dasyatis Rafinesque and Dasybatis Klein.

CXXIII. KAUP, *Beiträge zur Amphibiologie und Ichthyologie:* Isis, XIX, 1826.

JOHANN JACOB KAUP.

Not seen by us.

Rachycentron Kaup, 89; type RACHYCENTRON TYPUS.
Name amended in 1827 (*Isis*, p. 624) to RACHYCENTRUM. Replaces ELACATE Cuvier.

Narke Kaup (Ueber NARKE, 65); type RAJA CAPENSIS Gmelin.
Replaces ASTRAPE Gray.

CXXIV. NARDO, *De Proctostego Novum Piscium:* in Diario Chem. et Hist. Nat. Ticino, vol. I, 1827.

GIAN DOMENICO NARDO.

Proctostegus Nardo, I, 18, 42; type PROCTOSTEGUS PROCTOSTEGUS Nardo.
A synonym of DIANA Risso.

CXXV. NARDO, *Prodromus Observationem et Disquisitionem, Adriaticæ Ichthyologiæ:* Giorn, Fisica de Pavia, I, 1827.

Aphanius Nardo, 17, 23; type APHANIUS NANUS Nardo (CYPRINODON CALARITANUS Cuv. & Val.).
A valid genus replacing LEBIAS of authors (not of Cuvier) and MICROMUGIL Gulia, 1861. Local name "Nani vel noni."

Acentrolophus Nardo, 11; type ACENTROLOPHUS MACULATUS Nardo (PERCA NIGRA Gmelin).
Substitute for CENTROLOPHUS Lacepède, regarded as inapplicable.

Leptosoma Nardo, 15, 22; type LEPTOSOMA ATRUM Nardo.
Name preoccupied in Crustacea, Leach, 1819. "Stogio bastardo"; synonym probably of MONOCHIRUS.

Pastinacæ Nardo, 11; type RAJA PASTINACA L.

Lævirajæ Nardo, 11; type RAJA MIRALETUS L.

CXXVI. HARWOOD, *On a Newly Discovered Genus of Serpentiform Fishes:* Philos. Trans., 1827.

J. HARWOOD.

Ophiognathus Harwood, 49; type OPHIOGNATHUS AMPULLACEUS Harwood.
A synonym of SACCOPHARYNX Mitchill.

CXXVII.   CLOQUET, *Nouveau Dictionnaire d'Histoire Naturelle:* Nouvelle Edition, XXX, 1827.

HIPPOLYTE CLOQUET.

Sandat Cloquet, XXX, 126, 129; type PERCA LUCIOPERCA L.
Based on "Les Sandres" Cuvier.   A synonym of SANDER (Cuvier) Oken.

CXXVIII.   RÜPPELL, *Atlas zu der Reise in Nördlichen Afrika:* Fische des Rothen Meeres, 1828.

EDUARD RÜPPELL.

Pharopteryx Rüppell, 15; type PHAROPTERYX NIGRICANS Rüppell.
A synonym of PLESIOPS Cuvier as later shown by Rüppell.

Lutodeira (Van Hasselt) Rüppell, 17; type LUTODEIRA INDICA (MUGIL CHANOS L.).
A synonym of CHANOS.   Monotypic.

Scoliostomus Rüppell, 17; same type.
A synonym of CHANOS.

Haliophis Rüppell, 49; type MURÆNA GUTTATA Forskål.
Monotypic.

Pastinachus Rüppell, 82; type RAJA SEPHEN Forskål.
Same as HYPOLOPHUS Müller & Henle.   Not PASTINACA Nardo.

Asterropteryx Rüppell, 138; type ASTERROPTERYX SEMIPUNCTATUS Rüppell.

CXXIX.   FLEMING, *History of British Animals,* 1828.

JOHN FLEMING.

Selanonius Fleming, 169; type SELANONIUS WALKERI Fleming = SQUALUS NASUS (Bonnaterre).
A synonym of LAMNA Cuvier.

Encrasicholus Fleming, 183; type CLUPEA ENCRASICHOLUS L.
Synonym of ENGRAULIS Cuvier, 1817.   Monotypic.

Gobitis Fleming, 189; type COBITIS TÆNIA L.
A perversion of COBITIS.

Morhua Fleming, 190; type GADUS MORRHUA L.
A synonym of GADUS L.

Molva Fleming, 197; type GADUS MOLVA L.

Cernua Fleming, 212; type PERCA CERNUA L.
Same as CERNUA Schæfer or ACERINA Güldenstadt.

Cataphractus Fleming, 216; type COTTUS CATAPHRACTUS L.
A synonym of AGONUS Bloch & Schneider.

Spinachia Fleming, 219; type GASTEROSTEUS SPINACHIA L.

CXXX.   STARK, *Elements of Natural History,* Edinburgh, 2 vols.,
1828.

JOHN STARK.

Latinization of several of Cuvier's names, some of them differing
from the form chosen by Oken.   These are the following:

Brosmus Stark, 1, 425; type GADUS BROSME Müller.
Same as BROSME Oken, 1817.

Sandrus Stark, 1, 452; type PERCA LUCIOPERCA L.
Same as SANDER Oken.

Stelliferus Stark, 1, 459; type BODIANUS STELLIFER Bloch.
Same as STELLIFER Oken.

Cingla Stark, 1, 465; type PERCA ZINGEL L.
Same as ZINGEL Oken.

Daurada Stark, 1, 465; type SPARUS AURATA L.
A synonym of SPARUS L.

CXXXI.   LESSON, *Description du Nouveau Genre Ichthyophis:*   Mém.
Soc. Nat. Hist. Paris, IV, 397, 1828.

R. P. LESSON.

Ichthyophis Lesson, 397; type ICHTHYOPHIS PANTHERINUS Lesson.
Name preoccupied; replaced by UROPTERYGIUS Rüppell.

CXXXII.   CUVIER & VALENCIENNES, *Histoire Naturelle des
Poissons,* vol. II, 1828.

GEORGES CUVIER and ACHILLE VALENCIENNES.

The first two volumes (the first one being anatomical and descriptive) of this great work preceded the second edition of the *Règne Animal.*   The third and fourth volumes, also dated 1829, appeared in the last
half of the year.   The second edition of the *Règne Animal,* Vol. II, con-

taining the fishes, appeared in the first half of the year 1829, according to Mr. Fowler.

Labrax Cuvier & Valenciennes, II, 55; type PERCA LABRAX L. (LARBAX LUPUS C. & V.).

This is LABRAX of Klein, but not of Pallas, 1811.

Lates Cuvier & Valenciennes, 88; type PERCA NILOTICA L.

Lucioperca Cuvier & Valenciennes, II, 110; type PERCA LUCIOPERCA L. (LUCIOPERCA SANDRA Cuv. & Val.).

A synonym of SANDER Oken.

Huro Cuvier & Valenciennes, 125; type HURO NIGRICÀNS (LABRUS SALMOIDES Lacepède).

The genus is identical with GRYSTES. Both are synonyms of MICROPTERUS Lacepède.

Etelis Cuvier & Valenciennes, II, 127; type ETELIS CARBUNCULUS Cuv. & Val.

Monotypic.

Diploprion (Kuhl & Van Hasselt) Cuvier & Valenciennes, 137; type DIPLOPRION BIFASCIATUM Kuhl & Van Hasselt.

Monotypic.

Pomatomus (Risso) Cuvier & Valenciennes, II, 171; type POMATOMUS TELESCOPIUM Risso.

Not POMATOMUS Lacepède. Stands as EPIGONUS Rafinesque.

Ambassis (Commerson) Cuvier & Valenciennes, II, 175; type CENTROPOMUS AMBASSIS Lacepède (AMBASSIS COMMERSONI Cuv. & Val.), SCIÆNA SAFGHA Forskål.

Aspro Cuvier & Valenciennes, II, 188; type PERCA ASPER L.

Equivalent to ASPER Schæfer and ASPERULUS Klein. Not ASPRO Commerson. PERCA ASPER is probably generically distinct from ZINGEL (Cuvier) Oken. The genus may stand as ASPER Schæfer or ASPERULUS Klein, both names of questionable eligibility.

Mesoprion Cuvier (*Règne Animal*, 143), Cuv. & Val., II, 441; type LUTIANUS LUTIANUS Bloch.

Name a substitute for LUTJANUS, regarded as barbarous.

CXXXIII. BORY DE SAINT VINCENT, *Dictionnaire Classique d'Histoire Naturelle*, XIII, 201, 1828.

JEAN BAPTISTE GEORGE MARIE BORY DE ST. VINCENT.

Sandat Bory de St. Vincent, XIII, 204, 1828; type PERCA LUCIOPERCA L.

A synonym of SANDER Oken.

CXXXIII (A). SEDGWICK & MURCHISON, *Fossil Fishes:* Trans. Geol. Soc. (2), III, 1828.

ADAM SEDGWICK and RODERIC IMPEY MURCHISON.

Dipterus Sedgwick & Murchison, 143; type DIPTERUS VALENCIENNESI Sedgwick & Murchison.

CXXXIV. CUVIER, *Le Règne Animal, distribué d'Après son Organı sation*: Edition II, Vol. II (Fishes etc.), 1829.

GEORGES CUVIER.

A work of great importance, constituting with the first edition in 1817 the foundation of modern Ichthyology. The classification of fishes by Cuvier was for the first time solidly based on the true basis of Comparative Anatomy.

According to Henry Weed Fowler (*Proc. Ac. Nat. Sci. Phila.*, 1907, 264) the first, second, fourth and fifth fascicles of the second volume (*Fishes* etc.) of the Second Edition of the *Règne Animal* appeared in January, February and March 1829. The third was delayed until about July. The third and fourth volumes of the *Histoire Naturelle des Poissons* appeared later in the same year, 1829. These facts do not affect nomenclature, so far as we observe. We are indebted to Mr. Fowler for the pagination of the *Règne Animal*, our two copies being reprints.

Acerina Cuvier, 144 (Cuv. & Val., III, 3); type PERCA CERNUA L.

Identical with CERNUA Schæfer, ACERINA Güldenstadt, PERCIS Klein, and GYMNOCEPHALUS Bloch. The eligibility of these earlier names awaits decision.

Rypticus Cuvier, II, 144 (Cuv. & Val., III, 60); type ANTHIAS SAPONACEUS Bloch.

More correctly written RHYPTICUS.

Pentaceros Cuvier, 145 (Cuv. & Val., III, 30); type PENTACEROS CAPENSIS C. & V.

This name has been regarded as preoccupied by PENTACEROS Schulte, 1760, a star-fish.* It has been replaced by QUINQUARIUS Jordan. This change may not be necessary. According to Prof. Walter K. Fisher PENTACEROS of Schulte is not properly a generic name but a quasi-vernacular.

PENTACEROS Schröter, 1782, a name for the same animal is doubtfully tenable.

---

*Dr. Fisher (*Smithson. Misc. Coll.*, no. 1799, 1908) explains that Schulte ("*Versteinerte Seesterne*") following the pre-Linnæan work of Linck (*De Stel-

Centropristes Cuvier, 145 (Cuv. & Val., III, 56) ; type CENTROPRISTES NIGRICANS Cuv. & Val. (LABRUS STRIATUS L.).

Later written CENTROPRISTIS.

Gristes Cuvier (Cuv. & Val., III, 54) ; type LABRUS SALMOIDES Lacepède.

A synonym of MICROPTERUS Lacepède. Later written GRYSTES.

Chironemus Cuvier (C. & V., 78), 146; type CHIRONEMUS GEORGIANUS Cuv. & Val.

Monotypic.

Pomotis Cuvier, 147 (C. & V., III, 99) ; type "POMOTIS VULGARIS" (PERCA GIBBOSA L.).

No reference is made to Rafinesque's genus POMOTIS, proposed in 1819, with "LABRUS AURITUS L." as type. The description of Cuv. & Val. refers to the species called PERCA GIBBOSA by Linnæus (EUPOMOTIS GIBBOSUS of Gill & Jordan). To this same species Rafinesque later transferred his generic name POMOTIS. Stands as EUPOMOTIS Gill & Jordan.

Centrarchus Cuvier 147 (C. & V., III, 84) ; type CYCHLA ÆNEA Le Sueur.

A synonym of AMBLOPLITES Rafinesque, 1820. LABRUS IRIDEUS Lacepède, usually accepted as the type of CENTRARCHUS, is noted by Cuvier as an uncertain species which the author had not seen. The genus typified by IRIDEUS Lacepède (= MACROPTERUS Lacepède) must stand as EUCENTRARCHUS Gill, and its family as MICROPTERIDÆ, as Fowler has already indicated.

Dules Cuvier, 147 (C. & V., III, 111) ; type DULES AURIGA Cuvier.

Fowler (1907) regards DULES as preoccupied by DULUS Vieillot, 1816, a genus of birds. He gives a new name, EUDULUS, to DULES AURIGA.

Datnia Cuvier, 147 (C. & V., III, 138) ; type COIUS DATNIA Hamilton (DATNIA ARGENTEA Cuv. & Val.).

Pelates Cuvier, 148 (C. & V., III, 145) ; type PELATES QUADRILINEATUS Cuv. & Val.

---

lis Marinis, 1793) divides the star-fishes into three "genera" ("Geschlechte"), to which he gives group names, the five-rayed forms being under the "genus" "quinquifidæ." Under these are several kinds or species ("Arten"), one of them, "Der fünfhörnichte, PENTACEROS, hat fünf tiefe ausgeschweifte Seiten und lange kolbichte oder zugespitzte Strahlen. Die hierher gehörigen Arten sind entweder platt, PLANÆ, oder ab hockericht und bauchericht, GIBBÆ."

Obviously this is not scientific nomenclature.

"Schröter in 1782 (*Musei Gottwaldiani Testaceorum, Stellarum marinum* etc., Nürnberg, 58) used PENTACEROS, but he is not a consistent binomialist, and his 'generic' names are not tenable."

It is evident that PENTACEROS has no standing in nomenclature prior to its use by Cuvier & Valenciennes, unless given it by Schröter in 1782, a matter which awaits decision.

Helotes Cuvier, 148 (C. & V., III, 149); type THERAPON SEXLINEATUS Quoy & Gaimard.
Monotypic.

Trichodon (Steller) Cuvier, about 150 (C. & V., III, 153); type TRACHINUS TRICHODON Steller (TRICHODON STELLERI Cuv. & Val.).
Monotypic.

Myripristis Cuvier (about 150), (C. & V., III, 160); type MYRIPRISTIS JACOBUS Cuv. & Val.

Beryx Cuvier (about 150), (C. & V., III, 222); type BERYX DECADACTYLUS Cuv. & Val.

Pinguipes Cuvier, 153 (C. & V., III, 277); type PINGUIPES BRASILIANUS Cuv. & Val.
Monotypic. A synonym of MUGILOIDES Lacepède.

Upeneus Cuvier (about 160), (C. & V., III, 446); type MULLUS BIFASCIATUS Lacepède.
As first restricted by Bleeker. Later Bleeker transferred the name UPENEUS to the first species named by Cuvier & Valenciennes, MULLUS VITTATUS Forskål, already the type of his own genus UPENEOIDES. MULLUS BIFASCIATUS became then PARUPENEUS. This reversal seems to us unjustified. The name has been spelled HYPENEUS. UPENEUS should stand.

Hemitripterus Cuvier, 164; type COTTUS TRIPTERYGIUS Bloch & Schneider.

Hemilepidotus Cuvier, 165; type COTTUS HEMILEPIDOTUS Tilesius.
Monotypic.

Sebastes Cuvier, 166; type PERCA NORVEGICA Müller.

Blepsias Cuvier, 167; type BLENNIUS VILLOSUS Pallas.
Monotypic.

Apistes Cuvier, 167; type APISTES ALATUS Cuvier.

Agriopus Cuvier ("Les Agriopes"), 168; type BLENNIUS TORVUS Walbaum.

Pelor Cuvier, 168; type SCORPÆNA DIDACTYLA Pallas (PELOR OBSCURUM Cuvier).
Monotypic. Name preoccupied for a genus of beetles, 1813. Replaced by INIMICUS Jordan & Starks (type PELOR JAPONICUM Cuv. & Val.). P. FILAMENTOSUM Cuv. & Val. cannot be taken as the type of PELOR, as Cuvier names but one species, the others being then undescribed. This species, with two of the pectoral rays tipped with filaments, may belong to a different genus.

Oreosoma Cuvier (about 170); type (not named) OREOSOMA ATLANTICUM Cuvier.

Corvina Cuvier, 173; type Sciæna nigra Gmelin; part of Sciæna
umbra L.

The species Sciæna umbra L., the type of the genus Sciæna, was two-fold,
a complex about evenly divided of the species called Sciæna nigra Gmelin and
Cheilodipterus aquila Lacepède. The synonymy was disentangled by Cuvier
(*Mem. du Museum*, 1815) and the name Sciæna umbra retained in place of S.
aquila. In accepting this decision we restore the generic name Corvina, for the
"Corb," which becomes Corvina nigra, while Sciæna replaces Argyrosomus de
la Pylaie and Pseudosciæna Bleeker for the "Maigre," which remains Sciæna
umbra L., the type of Sciæna L.

Hæmulon Cuvier, 175; type Hæmulon elegans Cuvier (Sparus
sciurus Shaw).

Lobotes Cuvier, 177; type Holocentrus surinamensis Bloch.
Monotypic.

Scolopsides Cuvier, 178; type Scolopsides kurita Cuvier.
Called Scolopsis in Edition I.

Dascyllus Cuvier, 179; type Chætodon aruanus L.
Monotypic.

Heliases Cuvier; type (not indicated) Heliastes insolatus Cuv. &
Val.

Identical with Chromis Cuvier, 1815, not Cromis Browne, 1789, nor Chromis
Plumier.

Chrysophris Cuvier, 181; type Sparus aurata L.
A synonym of Sparus L.

Pagellus Cuvier (about 182); type Sparus erythrinus L.

Pentapus Cuvier ("Les Pentapodes") (about 182); type Sparus vittatus Bloch.

Lethrinus Cuvier (about 182); type Sparus chœrohynchus Bloch &
Schneider.

Oblada Cuvier, 185; type Sparus melanurus L.
Monotypic.

Mæna Cuvier; type Sparus mæna L.

Taurichthys Cuvier, 192; type Taurichthys varius Cuvier.
A subgenus of Heniochus Cuvier.

Psettus (Commerson) Cuvier (about 193); type Chætodon rhombeus
Bloch & Schneider.
Synonym of Monodactylus Lacepède.

Dipterodon Cuvier (about 195); type Dipterodon capensis Cuvier.
Not Dipterodon Lacepède, 1803. Monotypic. Replaced by Dichistius Gill.

Pempheris Cuvier, 195; type Kurtus argenteus Bloch & Schneider
(Pempheris touea Cuvier).

Auxis Cuvier, 199; type SCOMBER ROCHEI Risso (SCOMBER THAZARD Lacepède).

Sarda Cuvier, 199; type SCOMBER SARDA Bloch.
Later (1831) called PELAMYS by Cuvier & Valenciennes. Monotypic. Not SARDA Plumier.

Cybium Cuvier, 199; type SCOMBER COMMERSONI Lacepède.
A synonym of SCOMBEROMORUS Lacepède, 1803.

Thyrsites Cuvier (about 200); type SCOMBER ATUN Euphrasen.

Gempylus Cuvier, 200; type GEMPYLUS SERPENS Cuvier.
This is identical with LEMNISOMA of Lesson, but the date of LEMNISOMA is 1830.

Elacate Cuvier, II, 203, as "LES ELACATES," the Latin name dating from C. & V., 1831; type ELACATE MOTTA C. & V.
Synonym of RACHYCENTRON Kaup.

Olistus Cuvier, II, 209; type (not named) OLISTUS MALABARICUS Cuv. & Val.

Scyris Cuvier, II, 209; type "LE GAL D'ALEXANDRIE," SCYRIS INDICA Cuv. & Val.
Not separable from ALECTIS Raf.

Blepharis Cuvier (about 210); type ZEUS CILIARIS Bloch.
Name preoccupied, replaced by BLEPHARICHTHYS Gill. A synonym of ALECTIS.

Peprilus Cuvier, 213; type PEPRILUS CRENULATUS.
Probably the young of PEPRILUS ALEPIDOTUS Cuvier. Monotypic. Same as RHOMBUS Lacepède, not of Klein.

Astrodermus (Bonelli) Cuvier (about 215); type ASTRODERMUS GUTTATUS Bonelli Ms. (DIANA SEMILUNATA Risso).

Axinurus Cuvier, 225; type AXINURUS THYNNOIDES Cuvier.
Monotypic.

Priodon Cuvier, 225; type PRIODON ANNULARIS Cuvier.

Polyacanthus (Kuhl) Cuvier, 227; type TRICHOPODUS COLISA Hamilton.

Helostoma (Kuhl) Cuvier, 228; type HELOSTOMA TEMMINCKI Cuvier.
Monotypic.

Spirobranchus Cuvier, 229; type SPIROBRANCHUS CAPENSIS Cuvier.
Monotypic.

Myxodes Cuvier, 238; type (not named) MYXODES VIRIDIS Cuv. & Val.

Cirrhibarbus Cuvier ("LES CIRRHIBARBES") 238; type (not named) CIRRHIBARBIS CAPENSIS Cuv. & Val.
Monotypic.

Zoarces Cuvier (about 240); type BLENNIUS VIVIPARUS L.
Equivalent to ENCHELYOPUS Gronow, 1763.

Platyptera (Kuhl & van Hasselt) Cuvier, 248; type PLATYPTERA MELANOCEPHALA K. & v. H.

Name preoccupied, replaced by RHYACICHTHYS Boulenger.

Chirus (Steller) Cuvier, about 250; type LABRAX LAGOCEPHALUS Pallas.
Substitute for LABRAX Klein.

Antennarius (Commerson) Cuvier (about 251); type LOPHIUS COMMERSONIANUS Lacepède.

Accepted as a substitute for CHIRONECTES Cuvier, 1817, preoccupied.

Malthe Cuvier, 252; type LOPHIUS VESPERTILIO L.
Synonym of OGCOCEPHALUS Fischer. Monotypic.

Lachnolaimus Cuvier, 257; type LACHNOLAIMUS SUILLUS Cuvier.
SUILLUS of Catesby. Monotypic.

Anampses Cuvier, 259; type ANAMPSES CUVIERI Quoy & Gaimard.

Clepticus Cuvier, 261; type CLEPTICUS GENIZARA Cuvier.
Monotypic.

Malacanthus Cuvier, 264; type CORYPHÆNA PLUMIERI Bloch.

Calliodon Cuvier ("LES CALLIODON"), (about 265); type SCARUS SPINIDENS Q. & G.

Not identical with CALLYODON Gronow nor of Bloch & Schneider. Replaced by CRYPTOTOMUS Cope, 1871.

Odax Cuvier (about 265); type SCARUS PULLUS Forster.
Monotypic. Not ODAX Commerson.

Mallotus Cuvier, II, 305; type SALMO GRŒNLANDICUS Bloch (CLUPEA VILLOSA Gmelin).

Thymallus Cuvier, 306; type SALMO THYMALLUS L.

Regarded by some as preoccupied by THYMALUS Latreille, 1802, a genus of beetles; replaced by CHOREGON Minding, 1832. But the scantily noted THYMALLUS of Linck, 1791, has priority over THYMALUS, if acceptable. The root-words of THYMALLUS and THYMALUS are not identical, THYMALLUS being a Latin name of the Grayling.

Coregonus Cuvier; type SALMO OXYRHYNCHUS L.

Alosa Cuvier, 319; type CLUPEA ALOSA L.
Same as ALOSA Linck.

Chatoëssus Cuvier, 320; type MEGALOPS CEPEDIANUS Le Sûeur.
As restricted by Cuv. & Val. Equivalent to DOROSOMA Rafinesque.

Thryssa Cuvier; type CLUPEA SETIROSTRIS Broussonet (THRISSA Cuv. & Val., 1817).

Not THRISSA Rafinesque, 1815. To be replaced by THRISSOCLES Jordan, a new generic name.

Motella Cuvier, 334; type GADUS MUSTELA L.

Substitute for MUSTELA Cuvier, 1817, preoccupied in mammals. Identical with GAIDROPSARUS Rafinesque, 1810.

Monochir Cuvier, 336; type PLEURONECTES MICROCHIRUS De la Roche.

Substitute for MONOCHIRUS, preoccupied. Not sufficiently differer. .

Osteoglossum (Vandelli) Cuvier (about 390); type OSTEOGLOSSUM VANDELLI Cuvier.

Gymnarchus Cuvier (about 390); type GYMNARCHUS NILOTICUS Cuvier.

Triodon Cuvier (about 390); type TRIODON BURSARIUS Reinwardt.

Anacanthus (Ehrenberg) Cuvier (about 390); type (not named) RAJA UARNAK Forskål.

Monotypic. Replaces HIMANTURA Müller, 1837.

Heterotis Ehrenberg; type SUDIS NILOTICUS Cuv. & Val.

Quoted from the *Règne Animal* by Günther, but we do not find it there. Apparently it appears first in Cuv. & Val., XIX, 465, 1846.

CXXXV. AGASSIZ (AND SPIX), *Selecta Genera et Species Piscium quos in Itinere per Brasiliam, 1817, 1820, Collejit Dr. J. B. de Spix, 1829.*

LOUIS AGASSIZ and JEAN BAPTISTE SPIX.

The date of this work is apparently a little later than that of the Second Edition of the *Règne Animal*.

Acanthicus Spix, 2; type ACANTHICUS HYSTRIX Spix.

Rhinelepis Spix, 2; type RHINELEPIS ASPERA Spix.

Cetopsis Agassiz, 8; type CETOPSIS CÆCUTIENS Agassiz.

Hypophthalmus Spix, 9; type HYPOPHTHALMUS EDENTULUS Spix.

Phractocephalus Agassiz, 10; type PHRACTOCEPHALUS BICOLOR Agassiz.

Platystoma Agassiz, 10; type SILURUS LIMA Bloch & Schneider.

Name preoccupied by a genus of flies, Meigen, 1803.

Glanis Agassiz, 10; type not named.

Name given in dative case only *"In Glanide Agass."* Substitute for BAGRE Cuvier, rejected because of barbarous origin. Afterwards (1856) GLANIS was revived as the generic name for the Greek cat-fish, GLANIS ARISTOTELIS Agassiz, which is perhaps a PARASILURUS.

Ceratorhynchus Agassiz, 10; type SILURUS MILITARIS Bloch & Schneider.

Not L. Name given in genitive only as "CERATORHYNCHI." Genus close to AGENEIOSUS Lacepède but with the two short barbels replaced by erectile bony weapons.

Centrochir Agassiz, 14; type DORAS CROCODILI Humboldt. Based on a species of DORAS "non modo pinna pectorali uniradiata sed etiam appendice quadriradiata primæ caudalis maxime."

Sorubim Spix, 24; type SILURUS LIMA Bloch & Schneider.

Replaces PLATYSTOMA.

Xiphorhynchus Agassiz, 18; type SALMO FALCATUS Bloch.

Salminus Agassiz, 18; type HYDROCYON BREVIDENS Cuvier.

Osteoglossum (Vandelli) Agassiz, 46; type OSTEOGLOSSUM VANDELLI Cuv. & Val.

Ischnosoma Spix, 47; type ISCHNOSOMA BICIRRHOSUM Spix.

A deformed OSTEOGLOSSUM.

Glossodus (Cuvier) Agassiz, 48; type GLOSSODUS FORSKALI Agassiz (ESOX VULPES L.).

Same as ALBULA Gronow.

Anodus Spix, 57; type ANODUS ELONGATUS Spix.

Prochilodus Agassiz, 57; type PROCHILODUS ARGENTEUS Spix.

Leporinus Spix, 58; type LEPORINUS NOVEMFASCIATUS Spix.

Schizodon Agassiz, 58; type SCHIZODON FASCIATUS Agassiz.

Same as ANOSTOMUS Gronow.

Rhaphiodon Agassiz, 59; type RHAPHIODON VULPINUS Agassiz.

Cynodon Spix, 59, 76; type CYNODON GIBBUS Spix.

As restricted by Eigenmann. Not CYNODONTA Schuhmacher, 1817, a genus of mollusks.

Name changed to RHAPHIODON because earlier used in botany.

Xiphostoma Spix, 60, 78; type XIPHOSTOMA CUVIERI Spix.

Micropteryx Agassiz, 102; type (not named) SERIOLA DUMERILI Cuvier.

Substitute for SERIOLA, preoccupied in botany. A synonym of SERIOLA Cuvier.

Corniger Agassiz, 119; type CORNIGER SPINOSUS Agassiz.

Pachyurus Agassiz, 125; type PACHYURUS SQUAMIPINNIS Agassiz.

CXXXVI. CUVIER & VALENCIENNES, *Histoire Naturelle des Poissons*, IV, 1829.

GEORGES CUVIER and ACHILLE VALENCIENNES.

Hoplichthys Cuv. & Val., IV, XIX, 1829; type HOPLICHTHYS LANGS-DORFII Cuv. & Val.

Called HOPLICHTHYS in table of contents, OPLICHTHYS in the text.

Oplichthys Cuv. & Val., IV, 264, 1829; type OPLICHTHYS LANGSDORFII Cuv. & Val.

A variant in spelling.

Bembras Cuv. & Val., IV, 282; type BEMBRAS JAPONICUS Cuv. & Val.

Minous Cuv. & Val., IV, 420, 1829; type MINOUS WOORA Cuv. & Val.

Name changed to CORYTHOBATUS by Cantor, 1850, on account of MINOIS Hübner, 1816, a genus of butterflies.

Hoplostethus Cuv. & Val., IV, 469; type HOPLOSTETHUS MEDITERRANEUS Cuv. & Val.

CXXXVII. COCCO, *Su Alcuni Nuovi Pesce del Mar di Messina*: Archivio della R. Academia Peloritano, 1829.

ANASTASIO COCCO.

Argyropelecus Cocco, 146; type ARGYROPELECUS HEMIGYMNUS Cocco.

CXXXVIII. COCCO, *Su Alcuni Nuovi Pesci del Mar di Messina*: Giorni Sci. Lett. Sicilia, XXVI, no. 77, 1829.

Nyctophus Cocco, 44; type NYCTOPHUS RAFINESQUEI Cocco.

Substitute for MYCTOPHUM.

CXXXIX. COCCO, *Sullo Schedophilus Medusophagus*: Giorn, Cabin de Messina, I, 30 to 32; also quoted as Innom. Messina Ann., III, 1829, p. 57.

Schedophilus Cocco, 30, 57; type SCHEDOPHILUS MEDUSOPHAGUS Cocco.

We have seen none of these papers of Cocco.

CXL. SERVILLE, *Faune Française*, 1820 to 1830.

"Livraison 24 par M. Serville" contains the fishes. This bears no date, but it was probably issued about 1829. It is not quoted by Cuvier, nor does it quote Cuvier. The date is not important, as no question of priority is concerned. This work is commonly ascribed to Blainville, who with Vieillot, Desmarest, Serville and others edited the series. In any event, the work is based on Blainville's *Prodrome*. All of Blainville's generic names ending in BATUS are here changed to BATIS.

AUDINET-SERVILLE.

Dasybatis Serville, 12, 1829 ("RAIES EPINEUSES") ; type RAJA BATIS L.
Same as RAJA.

Narcobatis (Blainville) Serville, 45 ("RAIES TORPILLES") ; type RAIA
TORPEDO L.
Same as NARCACION Klein, TORPEDO Duméril, not of Forskål. Called NARCO-BATUS by Blainville, 1816.

Pristibatis Serville, 49 ("RAIES SCIES") ; type SQUALUS PRISTIS L.
(PRISTIBATIS ANTIQUORUM Blainville).

CXLI. VALENCIENNES, *Poissons Fossiles:* Trans. Geol. Soc., III, 1829.

ACHILLE VALENCIENNES.

Osteolepis Valenciennes, 144; type OSTEOLEPIS MACROLEPIDOTUS Valen-ciennes.

CXLII. CUVIER & VALENCIENNES, *Histoire Naturelle des Poissons*, Vols. V (January) and VI (July) 1830.

GEORGES CUVIER and ACHILLE VALENCIENNES.

Larimus Cuvier & Valenciennes, V, 146; type LARIMUS BREVICEPS Cuv.
& Val.
Monotypic.

Nebris Cuvier & Valenciennes, V, 149; type NEBRIS MICROPS Cuv. &
Val.
Not NEBRIA Latreille, 1802, a genus of beetles. Monotypic.

Lepipterus Cuvier & Valenciennes, V, 151; type LEPIPTERUS FRANCISCI Cuv. & Val.
Monotypic.

Boridia Cuvier & Valenciennes, V, 154; type BORIDIA GROSSIDENS Cuv. & Val.
Monotypic.

Conodon Cuvier & Valenciennes, V, 156; type CONODON ANTILLANUS Cuv. & Val. (PERCA NOBILIS L.).
Monotypic. Same as CHELONIGER Plumier.

Eleginus Cuvier & Valenciennes, 158; type ELEGINUS MACLOVINUS Cuv. & Val.
Monotypic. Not ELEGINUS Fischer, 1813. Replaced by ELEGINOPS Gill.

Micropogon Cuvier & Valenciennes, V, 213; type MICROPOGON LINEATUS Cuv. & Val. (PERCA UNDULATA L.).
The original types of LINEATUS were from New York.

Latilus Cuvier & Valenciennes, V, 368; type LATILUS ARGENTATUS Cuv. & Val. (CORYPHÆNA JAPONICA Houttuyn).
A synonym of BRANCHIOSTEGUS Rafinesque, 1815.

Macquaria Cuvier & Valenciennes, V, 377; type MACQUARIA AUSTRA- LASICA Cuv. & Val.
Monotypic.

Etroplus Cuvier & Valenciennes, V, 486; type ETROPLUS MELEAGRIS Cuv. & Val. (CHÆTODON SURATENSIS Bloch).
As restricted by Bleeker.

Box Cuvier & Valenciennes, VI, 346; type SPARUS BOOPS L. (BOX VUL- GARIS Cuv. & Val.).
A synonym of BOOPS Cuvier.

Scatharus Cuvier & Valenciennes, VI, 375; type SCATHARUS GRÆCUS Cuv. & Val.
Monotypic.

Crenidens Cuvier & Valenciennes, VI, 377; type CRENIDENS FORSKALII Cuv. & Val.
Monotypic.

Aphareus Cuvier & Valenciennes, VI, 485; type LABRUS FURCATUS Lacepède.
Monotypic.

Aprion Cuvier & Valenciennes, VI, 544; type APRION VIRESCENS Cuv. & Val.
Monotypic.

Apsilus Cuvier & Valenciennes, VI, 548; type APSILUS FUSCUS Cuv. & Val.
Monotypic.

CXLIII. BENNETT, *Catalogue of the Fishes of Sumatra*: in Life and
Public Services of Sir Stamford Raffles, 1830.

EDWARD TURNER BENNETT.

Monotaxis Bennett, 688; type MONOTAXIS INDICA Bennett (SCIÆNA
GRANDOCULIS Forskål).
Replaces SPHÆRODON Rüppell.

CXLIV. BRONN, *Ueber Zwei Fossile Fischarten*: Neues Jahrbuch
Mineralogie, 1830.

HEINRICH GEORG BRONN.

Tetragonolepis Bronn, 30; type TETRAGONOLEPIS SEMICINCTUS Bronn.
Monotypic.

CXLV. HAYS, *Saurodon*: Trans. Amer. Philos. Soc., III, 1830.

I. HAYS.

Saurodon Hays, 475; type SAURODON LEANUS Hays (fossil).

CXLVI. CUVIER & VALENCIENNES, *Histoire Naturelle des Poissons*, VII (January) and VIII (October), 1831.

GEORGES CUVIER and ACHILLE VALENCIENNES.

Chelmon Cuvier & Valenciennes, VII, 86; type CHÆTODON ROSTRATUS L.

Zanclus (Commerson) Cuv. & Val., VII, 92, 1831; type CHÆTODON
CORNUTUS L.

Drepane Cuvier & Valenciennes, VII, 132; type CHÆTODON PUNCTA-
TUS L.

Scatophagus Cuvier & Valenciennes, VII, 136; type CHÆTODON ARGUS L.

Psettus (Commerson) Cuvier & Valenciennes, VII, 240; type PSETTUS
COMMERSONI Cuv. & Val. (MONODACTYLUS FALCIFORMIS Lacepède).

Pempheris Cuvier & Valenciennes, VII, 296; type PEMPHERIS OUALEN-
SIS Cuv. & Val.

Toxotes Cuvier & Valenciennes, VII, 310; type SCIÆNA JACULATRIX Pallas.
Monotypic.

Colisa Cuvier & Valenciennes, VII, 359; type COLISA VULGARIS Cuv. & Val. (TRICHOPODUS COLISA Hamilton).
Same as TRICHOGASTER Bloch & Schneider as restricted by Cuv. & Val.

Spirobranchus Cuvier & Valenciennes, VII, 392; type SPIROBRANCHUS CAPENSIS Cuv. & Val.
Monotypic.

Bryttus Cuvier & Valenciennes, VII, 454, 461; type BRYTTUS PUNCTATUS Cuv. & Val.
Same as APOMOTIS Rafinesque.

Nandus Cuvier & Valenciennes, VII, 481; type NANDUS MARMORATUS Cuv. & Val. (COIUS NANDUS Hamilton).

Rhynchichthys Cuvier & Valenciennes, VII, 504; type RHYNCHICHTHYS PELAMIDIS Cuv. & Val.
The young of HOLOCENTRUS.

Lichia Cuvier & Valenciennes, VIII, 340, 1831; type SCOMBER AMIA L.

Chorinemus Cuvier & Valenciennes, VIII, 367; type SCOMBEROIDES COMMERSONIANUS Lacepède.
A synonym of SCOMBEROIDES Lacepède.

Aplodactylus Cuvier & Valenciennes, VIII, 476; type APLODACTYLUS PUNCTATUS Cuv. & Val.
Usually written HAPLODACTYLUS.

Aphritis Cuvier & Valenciennes, VIII, 483; type APHRITIS URVILLI Cuv & Val.
Monotypic.

Bovichtus Cuvier & Valenciennes, VIII, 487; type CALLIONYMUS DIACANTHUS Carmichæl.

CXLVII. JARDINE, *Acestra*, 1831.

SIR WILLIAM JARDINE.

Acestra Jardine; type SYNGNATHUS ÆQUOREUS Rafinesque.
A synonym of NEROPHIS Rafinesque. We take this incomplete reference from Bonaparte, *Catalogo Metodico*, 1846, 91.

CXLVIII. GRAY, *Description of Twelve New Genera of Fish found by General Hardwicke in India*: Zool. Misc., 1831, 7-10.

JOHN EDWARD GRAY.

Centracion Gray, 5; type CENTRACION ZEBRA Gray.
Equivalent to HETERODONTUS Blainville, regarded as preoccupied by HETERODON Latreille, a genus of serpents.

Temera Gray, 7, 152; type TEMERA HARDWICKEI Gray.
Monotypic. Also in Ill. Ind. Zool.

Botia Gray, 8; type BOTIA ALMORHÆ Gray.

Nandina Gray, 8; type CYPRINUS NANDINA Hamilton.

Chaca Gray, 8; type PLATYSTACUS CHACA Hamilton.

Coilia Gray, 9; type COILIA HAMILTONI Gray.

Raconda Gray, 9; type RACONDA RUSSELLIANA Gray.

Moringua Gray, 9; type MORINGUA LATERALIS Gray.
Also in Ill. Ind. Zool., 95.

CXLIX. GRAY, *Illustrations of Indian Zoology, chiefly selected from the collection of General Hardwicke.*

Two volumes of excellent plates, but without text; the plates not paged, but with manuscript numbers. Vol. I, 1830-1832, Vol. II, 1833-1834. The exact dates are uncertain. The first five of the following occur in CXLVIII; also, pp. 8, 9.

Ailia Gray, 85, 1831; type AILIA BENGALENSIS Gray.

Acanthonotus Gray, 85, 1831; type ACANTHONOTUS CUVIERI Gray (AILIA BENGALENSIS Gray).
Name preoccupied. Same as AILIA. Based on injured specimen.

Anacanthus Gray, 85, 1831; type ANACANTHUS BARBATUS Gray.
Name preoccupied by ANACANTHUS Ehrenberg, 1829. Replaced by PSILOCEPHALUS Swainson.

Diplopterus Gray, 87, 1831; type DIPLOPTERUS PULCHER Gray.
Not identified; said to have two anal fins. Name preoccupied in birds, Boie.

Rataboura Gray, 95, 1831; type MURÆNA RATABOURA Hamilton (RATABOURA HARDWICKEI Gray).
Same as MORINGUA Gray, but with line priority.

Bedula Gray, 88, 1833; type BEDULA NEBULOSA Gray.
A synonym of NANDUS Cuv. & Val., 1831.

Pterapon Gray, II, 88, 1833; type PTERAPON TRIVITTATUS Gray (SCIÆNA
JARBUA Forskål).
A synonym of THERAPON Cuvier.

Amora Gray, II, 90, 1833, corrected in manuscript to ANAORA; type
AMORA TENTACULATA Gray.
Not identified. A platycephalus-like fish with tentacles over eye and spinules
on sides. Perhaps a species of THYSANOPHRYS Ogilby.

Apterygia Gray, II, 92, 1833; type APTERYGIA RAMCARATE Gray.
Same as RACONDA Gray.

Tor Gray, II, 96, 1833; type CYPRINUS TOR Hamilton (TOR HAMILTONI
Gray).
Replaces LABEOBARBUS Bleeker.

Bengala Gray, 96, 1833; type CYPRINUS ELANGA Hamilton.
Replaces MEGARASBORA Günther, 1868.

Amanses Gray, II, 98, 1833; type AMANSES HYSTRIX Gray (MONACAN-
THUS SCOPAS Cuvier).

Acarana Gray, II, 98, 1833; type OSTRACION AURITUS Shaw.
Also in Annals Nat. Hist., 1, 110, 1838.

Girella Gray, II, 98, 1833; type GIRELLA PUNCTATA Gray.

CL. GRAY, *Description of three new species of fish including two undescribed genera discovered by John Reeves in China:* Zoological
Miscellany, 1831.

Leucosoma Gray, 4; type LEUCOSOMA REEVESI Gray (ALBULA CHINEN-
SIS Osbeck).
Monotypic. A synonym of SALANX Cuvier and of ALBULA Osbeck.

Samaris Gray, 4; type SAMARIS CRISTATUS Gray.
Monotypic.

CLI. GRAY, *Description of a new genus of Percoid fish, discovered by
Samuel Stutchbury in the Pacific Seas:* Zoological Miscellany,
1831.

Micropus Gray, 20; type MICROPUS MACULATUS Gray.
Name preoccupied by MICROPUS Wolf, 1810, a genus of birds. Replaced by
CARACANTHUS Kröyer, 1844 (C. TYPICUS).

CLI (A).   LESSON, *Voyage Autour du Monde sur la corvette La Coquille, under Captain L. I. Duperrey*, 1830.

R. P. LESSON.

Lemnisoma Lesson, 160; type LEMNISOMA THYRSITOIDES Lesson.
A synonym of GEMPYLUS Cuvier.

CLII.   MINDING, *Lehrbuch, Naturgeschichte der Fische*, 1832.

JULIUS MINDING.

Not seen by us.

Pompilus Minding, 108; type GASTEROSTEUS DUCTOR L.
Same as NAUCRATES.

Choregon Minding, 119; type SALMO THYMALLUS L.
Substitute for THYMALLUS Cuvier, regarded as preoccupied by THYMALUS Latreille, 1802, a genus of beetles. But the two names seen to be from different roots. THYMALLUS Linck, with a word or two of definition, is older than THYMALUS Latreille.

CLIII.   NILSSON, *Prodromus Ichthyologiæ Scandinavicæ*, 1832.

S. NILSSON.

Salvelini Nilsson, 7; type SALMO SALVELINUS L. (SALMO ALPINUS L.).
As group name. The normal form SALVELINUS used by Richardson, *Fauna Boreali Americana*, III, 169, 1836.

CLIV.   AGASSIZ, *Fossile Fischreste:*   Neues Jahrbuch Mineralogie, 1832.

LOUIS (JEAN RODOLPHE) AGASSIZ.

Ptycholepis Agassiz, 142; type PTYCHOLEPIS BOLLENSIS Agassiz.

Uræus Agassiz, 142; type URÆUS FURCATUS Agassiz.
Name preoccupied by Wagler, 1830. Replaced by CATURUS Agassiz, 1834.

Sauropsis Agassiz, 142; type SAUROPSIS LATUS Agassiz.

Pholidophorus Agassiz, 145; type PHOLIDOPHORUS MACROCEPHALUS Agassiz.

Semionotus Agassiz, 144; type SEMIONOTUS BERGERI Agassiz.

Lepidotes Agassiz, 145; type LEPIDOTES GIGAS Agassiz (CYPRINUS ELVENSIS Blainville).
Later written LEPIDOTUS.

Leptolepis Agassiz, 146; type LEPTOLEPIS BRONNI Agassiz.

Acanthoëssus Agassiz, 149; type ACANTHOESSUS BRONNI Agassiz.
Later called ACANTHODES.

CLV. VON MEYER, *Palæologica*, 1832.

H. VON MEYER.

Lepidosaurus von Meyer, 208; type LEPIDOTUS UNGUICULATUS Agassiz.
A synonym of LEPIDOTES.

CLVI. BONAPARTE, *Iconografia della Fauna Italica*: III, 1832-1841.

CARLO LUCIANO PRINCIPE BONAPARTE
(otherwise CHARLES LUCIEN BONAPARTE, Prince of Canino).

An elaborate and finely illustrated work, issued in fascicles, these being numbered but not paged. The number of the fascicle is noted below, with the approximate date of each.

Cerna Bonaparte, 18, 1832; type PERCA GIGAS Brünnich (LABRUS GUAZA L.).
A synonym of EPINEPHELUS Bloch as now restricted.

Microchirus Bonaparte, 28, 1832; type PLEURONECTES MICROCHIRUS De la Roche.
MONOCHIRUS Cuvier, not of Rafinesque.

Monochirus Bonaparte,* 28, 1832; type SOLEA MONOCHIR Bonaparte.
Same as MONOCHIRUS Rafinesque.

Squalius Bonaparte, 96, 1834; type LEUCISCUS SQUALUS Bonaparte (CYPRINUS CEPHALUS L.).
Same as LEUCISCUS.

Telestes Bonaparte, 103, 1834; type TELESTES MUTICELLUS Bonaparte.

Ichthyococcus Bonaparte, 138, 1834; type ICHTHYOCOCCUS OVATUS Bonaparte.
Name later altered to COCCIA by Günther because of its objectionable form.

---

*For sake of completeness we add the remaining new genera of the later fascicles of the *Fauna Italica*.

Lampanyctus Bonaparte, 139, 1834; type LAMPANYCTUS BONAPARTEI Bonaparte.

Chlorophthalmus Bonaparte, 144, 1834; type CHLOROPHTHALMUS AGASSIZI Bonaparte.

Scardinius Bonaparte, 146, 1834; type LEUCISCUS SCARDAFA Bonaparte.

Membras Bonaparte, 1836; type (an "exotic species" not named except by reference to Valenciennes) ; ATHERINA MARTINICA Cuv. & Val.
Replaces KIRTLANDIA Jordan & Evermann.

Menidia Bonaparte, 1836; type (an "exotic species" not directly named) ATHERINA MENIDIA L.

Hepsetia Bonaparte, 1836; type ATHERINA BOYERI L.
Not separable from ATHERINA L.

Ichthyocoris Bonaparte, 1836; type BLENNIUS VARUS Risso.
A synonym of BLENNIUS L.

CLVI (A). RAFINESQUE, *Atlantic Journal and Friend of Knowledge*, I, 1832.

CONSTANTINE SAMUEL RAFINESQUE.

Trinectes Rafinesque; type TRINECTES SCABRA Rafinesque (PLEURONECTES MOLLIS Mitchill).
A subgenus under ACHIRUS.

CLVII. CUVIER & VALENCIENNES, *Histoire Naturelle des Poissons*, IX, 1833.

GEORGES CUVIER and ACHILLE VALENCIENNES.

Hynnis Cuvier & Valenciennes, IX, 95; type HYNNIS GOREENSIS Cuv. & Val.

Nauclerus Cuvier & Valenciennes, IX, 247; type NAUCLERUS COMPRESSUS Cuv. & Val.
Young of NAUCRATES.

Porthmeus Cuvier & Valenciennes, IX, 255; type PORTHMEUS ARGENTEUS Cuv. & Val.

Psenes Cuvier & Valenciennes, IX, 259; type PSENES CYANOPHRYS Cuv. & Val.

Lampugus Cuvier & Valenciennes, IX, 317; type SCOMBER PELAGICUS L.
Same as CORYPHÆNA L.

Aphredoderus (Le Sueur) Cuv. & Val., IX, 445; type APHREDODERUS GIBBOSUS Le Sueur (SCOLOPSIS SAYANUS Gilliams).
Monotypic.

CLVIII. LOWE, *Description of a New Genus of Acanthopterygian Fishes*: Proc. Zool. Soc. London, 1833.

RICHARD THOMAS LOWE.

Alepisaurus Lowe, 104; type ALEPISAURUS FEROX Lowe.
Monotypic. Equivalent to PLAGYODUS Steller, 1811.

CLIX. LOWE, *Characters of a New Genus and of Several New Species of Fishes from Madeira:* Proc. Zool. Soc. London, I, 1833. (Also repeated in other papers.)

Leirus Lowe, 142; type LEIRUS BENNETTI Lowe = CENTROPHORUS OVALIS Cuv. & Val., 1833 = MUPUS IMPERIALIS Cocco, 1833.
Same as MUPUS Cocco of the same date. LEIRUS may be given precedence.

CLX. COCCO, *Lettere al Signor Risso su alcuni Pesci Novelli*: Giorn. Sci. Lett. Sicilia, XLII, no. 124, 1833.

ANASTASIO COCCO.

Tylosurus Cocco; type TYLOSURUS CANTRAINI Cocco (ESOX IMPERIALIS Rafinesque).

CLXI. COCCO, *Su Alcuni Pesci dei Mare di Messina:* Giorn. Sci. Lett. Sicilia, XLII, 1833.

Mupus Cocco, 20; type MUPUS IMPERIALIS Cocco (CENTROLOPHUS OVALIS Cuv. & Val.).
Same as LEIRUS Lowe of the same date.

CLXII. COCCO, *Osservationes Peloritani*, XIII, April 1833.

Ruvettus Cocco, 1833, 18; type RUVETTUS PRETIOSUS Cocco.

CLXIII. NARDO, *De Skeponopodo novo piscium genere et de Guebucu Margravii; Species illi cognata:* Mem. Assem. Nat. Vienna, Isis, 1833, fasc. 416.

GIAN DOMENICO NARDO.

Skeponopodus Nardo, 416; type SKEPONOPODUS TYPUS Nardo (XIPHIAS IMPERATOR Bloch & Schneider).
A synonym of TETRAPTURUS.

CLXIV. AGASSIZ, *Poissons Fossiles*, 1833.

LOUIS AGASSIZ.

Catopterus Agassiz, II, 3; type CATOPTERUS ANALIS Agassiz.
A synonym of DIPTERUS.

Amblypterus Agassiz, II, 3, 28; type AMBLYPTERUS LATUS Agassiz.

Notagogus Agassiz, 10; type NOTAGOGUS PENTLANDI Agassiz.

Microps Agassiz, II, 10; type MICROPS FURCATUS Agassiz.
Perhaps not distinct from PHOLIDOPHORUS; name preoccupied in beetles, Meigen, 1823.

Pachycormus Agassiz, II, 11; type ELOPS MACROPTERUS Blainville.

Thrissops Agassiz, II, 12; type THRISSOPS FORMOSUS Agassiz.

Megalurus Agassiz, II, 13; type MEGALURUS LEPIDOTUS Agassiz.

Aspidorhynchus Agassiz, II, 14; type ESOX ACUTIROSTRIS Blainville.

Saurostomus Agassiz, II, 14; type SAUROSTOMUS ESOCINUS Agassiz.
Near to PACHYCORMUS; perhaps not distinguishable.

Gyrodus Agassiz, II, 16; type GYRODUS MACROPHTHALMUS Agassiz (MICRODON ABDOMINALIS Agassiz).

Microdon Agassiz, II, 16; type MICRODON ELEGANS Agassiz.
Name preoccupied by Meigen, 1803, a genus of DIPTERA.

Pycnodus Agassiz, II, 16; type ZEUS PLATESSUS Blainville (CORYPHÆNA APODA Volta).

Cyclopoma Agassiz, IV, 17, 1833; type CYCLOPOMA GIGAS Agassiz.

Calamostoma Agassiz, II, 18; type SYNGNATHUS BREVICULUS Blainville. Monotypic.

Acanthodes Agassiz, II, 19; type ACANTHOËSSUS BRONNI Agassiz. A needless substitute for ACANTHOËSSUS.

Smerdis Agassiz, IV, 32, 1833; type PERCA MINUTA Blainville.

RECAPITULATION.

SUGGESTED CHANGES IN GENERIC NOMENCLATURE OF FISHES.

A. Changes resting in Priority.

The following changes from current Nomenclature, apparently justified by the law of priority and the accepted rules, result from the present survey.

Acanthoëssus Agassiz: in place of ACANTHODES Agassiz (fossil).
Alabes Cuvier: CHEILOBRANCHUS Richardson.
Alosa Linck: ALOSA Cuvier.
Anacanthus Ehrenberg: HIMANTURA Müller.
Anchoviella Fowler: STOLEPHORUS Bleeker (not Lacepède).
Apteronotus Lacepède: STERNARCHUS Bloch & Schneider.
Bagre Cuvier: FELICHTHYS Swainson, AILURICHTHYS Baird.
Barbatula Linck: OREIAS Sauvage, ORTHRIAS Jordan & Fowler.
Bengala Gray (1833): MEGARASBORA Günther, 1868.
Bodianus Bloch: HARPE Lacepède.
Branchiostegus Rafinesque: LATILUS Cuv. & Val.
Caranx Lacepède: CARANGUS Griffith.
Catonotus Agassiz: ETHESTOMA Jordan, not Raf.
Cephalopholis Bloch & Schneider: BODIANUS Cuv. & Val. (not Bloch). ENNEACENTRUS Gill.
Chelon Röse: CHÆNOMUGIL Gill.
Citharus Röse: EUCITHARUS Gill.
Clupanodon Lacepède: KONOSIRUS Jordan & Snyder.
Conger Houttuyn: CONGER Cuvier.
Corvina Cuvier: SCIÆNA Jordan & Evermann (not of Cuvier), CORACINUS Pallas.
Curimata Walbaum: CURIMATUS Cuvier.
Diapterus Ranzani: GERRES Jordan & Evermann (not of Quoy & Gaimard).
Echeneis L.: LEPTECHENEIS Gill.
Echelus Rafinesque: MYRUS Cuvier.
Etheostoma Rafinesque: DIPLESION Raf.

Eucentrarchus Gill: Centrarchus Jordan & Evermann (not of Cuv. & Val.).

Gaidropsarus Rafinesque: Motella Cuvier.

Gerres Cuvier: Xystæma Jordan & Evermann.

Gnathanodon Bleeker: (Caranx speciosus).

Hemiulis Swainson (as restricted by Bonaparte, 1839): Halichœres Rüppell, Chœrojulis Gill.

Histrio Fischer: Pteryphryne Gill, Pterophrynoides Gill.

Hoplias Gill: Macrodon Müller & Troschel.

Hypacanthus Rafinesque: Camptogramma Regan.

Inimicus Jordan & Starks: Pelor Cuv. & Val.

Leucichthys Dybowski: Argyrosomus Agassiz.

Lycodontis McClelland: Gymnothorax Günther (not of Bloch).

Macrodon Schinz: Ancylodon Cuvier, Sagenichthys Berg.

Macrognathus Lacepède: Rhynchobdella Bloch & Schneider.

Macrorhynchus Lacepède: Dicrotus Günther.

Membras Bonaparte: Kirtlandia Jordan & Evermann.

Merolepis Rafinesque: (Smaris zebra).

Micrometrus Gibbons: Abeona Girard.

Mola Kœlreuter: Mola Cuvier.

Mugiloides Lacepède: Pinguipes Cuvier.

Narke Kaup: Astrape Gray.

Naso Lacepède: Monoceros Bloch & Schneider.

Ompok Lacepède: Callichrous Hamilton.

Ovoides Cuvier: Ovoides Lacepède.

Pastinachus Rüppell: Hypolophus Müller & Henle.

Pentanemus Günther: Polynemus Gill (not of L. as restricted).

Peprilus Cuvier: Rhombus Lacepède.

Phycis Röse: Phycis Bloch & Schneider.

Plagyodus Steller: Alepisaurus Lowe.

Pœcilichthys Agassiz: Etheostoma Jordan & Evermann (not of Rafinesque as restricted).

Polynemus L.: Polydactylus Lacepède.

Porcus St. Hilaire: Bagrus Cuv. & Val.

Quinquarius Jordan: Pentaceros Cuvier.

Remora Forster: Remora Gill.

Sardina Antipa (S. pilchardus).

Sardinella Val., Amblygaster Bleeker, Sardinia Poey.

Sciæna L.: Argyrosomus De la Pylaie, Pseudosciæna Bleeker.

Scylliorhinus Blainville: Catulus Smith.

Sphyræna Röse: SPHYRÆNA Bloch & Schneider.
Spicara Rafinesque: SMARIS Cuv.
Stolephorus Lacepède: SPRATELLOIDES Bleeker.
Syngnathus L.: SIPHOSTOMA Raf.
Thrissocles Jordan*: THRYSSA Cuvier.
Thymallus Linck: THYMALLUS Cuvier.
Tor Gray (1833): LABEOBARBUS Bleeker.
Torpedo Houttuyn: TORPEDO Duméril.
Trisopterus Rafinesque: MORUA Risso, BRACHYGADUS Gill.
Typhlinus Rafinesque: TYPHLE Rafinesque, SIPHONOSTOMUS Kaup.
Xystramia † Jordan: GLOSSAMIA Goode & Bean (not of Gill).
Zoramia ‡ Jordan: MIONORUS Jordan & Seale (not of Krefft).

B. Changes resulting from the operations of Opinions 20, 37.

Taking the precedent of Opinion 20, which admits as eligible the generic names of Gronow, and that of 37, admitting those of Brisson, the following changes seem necessary, the status of Klein being almost identical with that of Gronow.

Amia Gronow: instead of APOGON Lacepède.
Amiatus Rafinesque: AMIA L.
Brama Klein: ABRAMIS Cuvier.
Callyodon Gronow: SCARUS Forskål.
Cestracion Klein: SPHYRNA Rafinesque.
Coracinus Gronow: DIPTERODON Cuvier.
Cyclogaster Gronow: LIPARIS Scopoli.
Dasybatus Klein: DASYATIS Rafinesque.
Enchelyopus Gronow: ZOARCES Cuvier.
Glaucus Klein: HYPODIS Rafinesque, CÆSIOMORUS Lacepède.
Hepatus Gronow: TEUTHIS L., ACANTHURUS Forskål.
Labrax Klein: DICENTRARCHUS Gill.
Lepodus Rafinesque: BRAMA Bloch & Schneider.
Leuciscus Klein: LEUCISCUS Cuvier.
Mænas Klein: MÆNA Cuvier.
Mystus Gronow: HYPSELOBAGRUS Günther.
Narcacion Klein: NARCOBATUS Blainville.

---

* Type CLUPEA SETIROSTRIS Broussonet.
† Type GLOSSAMIA PANDIONIS Goode & Bean: See *Copea,* 1917, p. 46.
‡ Type APOGON GRÆFFEI Günther: See Copea, 1917, p. 46.

Pristis Klein: PRISTIS Linck.

Prochilus Klein: PHOLIS Cuv. & Val. (preoccupied).

Pseudopterus Klein: PTEROIS Cuvier.

Rhina Klein: SQUATINA Duméril.

Rhinobatus Klein: RHINOBATUS Linck.

Rhombus Klein: BOTHUS Rafinesque.

Sargus Klein: DIPLODUS Rafinesque, SARGUS Cuvier.

(new name) CICHLA Bloch & Schneider.

(new name) SOLENOSTOMUS Lacepède.

C. *Changes resulting from the operations of Opinion 24, which legalizes the names of Commerson in Lacepède. Those of Plumier in Lacepède are precisely similar.*

Alticus Commerson: in place of RUPISCARTES Swainson.

Cheloniger Plumier: CONODON Cuvier.

Chromis Plumier: UMBRINA Cuvier. Unless CROMIS Browne is accepted.

Encrasicholus Commerson: ANCHOVIELLA Fowler.

Pagrus Plumier: NEOMÆNIS Girard.

Sarda Plumier: OCYURUS Gill.

(new name) PAGRUS Cuvier.

(new name) SARDA Cuvier.

(new name) ODAX Cuvier.

D. *Hypothetical changes in Nomenclature according to the law of priority but doubtfully eligible, being revised reprints or translations of pre-Linnæan authors; apparently to be rejected under Opinion 57*

Albula Osbeck: SALANX Cuvier.

Apocryptes Osbeck: A valid genus, near BOLEOPHTHALMUS Bloch & Schneider.

Butyrinus Lacepède: ALBULA Gronow.

Cromis Browne: POGONIAS Lacepède.

Heliases Cuvier: CHROMIS Cuvier.

Helops Browne: BODIANUS Bloch.

Menidia Browne: ANCHOVIELLA Fowler, ENCRASICHOLUS Commerson.

Pelmatia Browne: GOBIOMORUS Lacepède, PHILYPNUS Cuvier.

Plagusia Browne: SYMPHURUS Rafinesque.

Rhomboida Browne: VOMER Cuvier.

Saurus Browne: OLIGOPLITES Gill.

E. Changes as under D but the alleged generic names, perhaps to be regarded as Latin vernaculars.

Acus Catesby: PSALISOSTOMUS Klein, LEPISOSTEUS Lacepède.
Alburnus Catesby: MENTICIRRUS Gill.
Bagre Catesby: AMEIURUS Rafinesque.
Harengus Catesby: HARENGULA Cuv. & Val.
Salpa Catesby: NEOMÆNIS Girard.
Saltatrix Catesby: POMATOMUS Lacepède.
Solea Catesby: PLATOPHRYS Swainson.
Suillus Catesby: LACHNOLAIMUS Cuvier.
Unicornis Catesby: OSBECKIA Jordan & Evermann.
(new name) ALBURNUS Rafinesque.
(new name) SOLEA Quensel.

F. Changes in accord with the law of priority, but questionable on account of irregularities not yet passed upon by the Commission of Nomenclature.

Abuhamrur Forskål: PRIACANTHUS Cuvier.
Asper Schæfer: ASPRO Cuvier.
Asperulus Schæfer: ZINGEL Oken.
Cernua Schæfer: ACERINA Güldenstadt, GYMNOCEPHALUS Bloch.
Daba Forskål: EPINEPHELUS Bloch.
Djabub Forskål: THERAPON Cuvier.
Galeus Valmont: PRIONACE Cantor.
Gaterin Forskål: PLECTORHYNCHUS Lacepède.
Ghanan Forskål: SCOLOPSIS Cuvier.
Hobar Forskål: LUTIANUS Bloch.
Louti Forskål: VARIOLA Swainson.
Mustelus Valmont: CYNIAS Gill.
Pleuracromylon Gill: MUSTELUS Leach.
Quinquarius Jordan: PENTACEROS Cuvier.
Ramphistoma Rafinesque: BELONE Cuvier.
Schour Forskål: LETHRINUS Cuvier.
Schraitzer Schæfer: LEPTOPERCA Gill.
Tahmel Forskål: KYPHOSUS Lacepède.
Vulpecula Valmont: ALOPIAS Rafinesque.

G. Changes due to virtual preoccupation of names now in use.

If we follow the current practice of regarding generic names in words otherwise similar and from identical roots, the same word if differing only in gender, in the presence or absence of the initial *h* in words of Greek origin, or by the use of *i* or *o* as a connective, the following changes are necessary in the current nomenclature, as given by Jordan & Evermann, *Fishes N. M. America,* 1898.

Cenisophius Bonaparte: in place of Leucos Heckel, on account of the earlier Leucus.

Centridermichthys Richardson: Trachidermus Heckel; Trachyderma.

Cheilonemus * Baird: Leucosomus Heckel; Leucosoma.

Cremnobates Günther: Auchenipterus Günther; Auchenopterus.

Cynicoglossus Bonaparte: Microstomus Gottsche; Microstoma.

Eudulus Fowler: Dules Cuvier; Dulus.

Evermannellus Fowler: Odontostomus Cocco; Odontostoma.

Haloporphyrus Günther: Lepidion Swainson; Lepidia.

Hemiulis Swainson (as restricted by Bonaparte, 1839); Halichœres Rüppell; Halichœrus.

Hypolophus Müller: Pastinachus Rüppell; Pastinaca.

Nebrodes Garman: Nebrius Rüppell; Nebria.

Notoscopelus Günther: Macrostoma Risso; Macrostomus.

Pisciregia Abbott: Gastropterus Cope; Gastropteron.

Podateles Boulenger: Ateleopus Schlegel; Atelopus.

Quassilabia Jordan & Brayton: Lagochila J. & B.; Lagocheilus.

Scaphirhynchops Gill: Scaphirhynchus Heckel; Scaphorhynchus.

Scytaliscus Jordan & Gilbert: Scytalinus J. & G.; Scytalina.

Stenesthes † Jordan: new name for Stenotomus Gill; Stenotoma.

Thrissocles Jordan: new name for Thryssa Cuvier; Thrissa.

Xiphister Jordan & Gilbert: Xiphidion Girard; Xiphidium.

H. Questionable cases, similar to G.

Catochænum Cantor: Gerres Cuvier, on account of the prior Gerris.

Centracion Gray: Heterodontus Blainville; Heterodon.

Choregon Minding: Thymallus Cuvier; Thymalus.

Corythobatus Cantor: Minous Cuv. & Val.; Minois.

(new name) Chloëa Jordan & Snyder; Chloëia.

(new name) Gramma Poey; Grammia.

(new name) Nebris Cuvier; Nebria; Nebrius.

---

* Type Cyprinus corporalis Mitchill, Cyprinus bullaris Rafinesque.

† Type Sparus chrysops L.

(Pages 153–162, comprising the Index of Part I, have been omitted; Part II Title and Contents occupy the same numbers in the pagination sequence.)

The Genera of Fishes

[PART II]
FROM AGASSIZ TO BLEEKER, 1833–1858
TWENTY-SIX YEARS
WITH THE ACCEPTED TYPE OF EACH

G
II

CONTENTS

Abbreviations of author names, where used, following words in small capitals are as follows:

Ag. Agassiz
B. & C. Bocage & Capello
Baird & Grd. Baird & Girard.
Barn. Barneville
Beyr. Beyrich
Bl. & Schn. Bloch & Schneider.
Blainv. Blainville
Blkr. Bleeker
Bon. Bonaparte
Cuv. Cuvier
Cuv. & Val. Cuvier & Valenciennes.
Dum. Duméril
Egert. Egerton
Ehrenb. Ehrenberg
Eichw. Eichwald
G. & B. Goode & Bean.
Grd. Girard
Gthr. Günther
Guich. Guichenot
Ham. Hamilton

J. & G. Jordan & Gilbert
J. & E. Jordan & Evermann
K. & S. Kner & Steindachner
Kölr. Kölreuter
L. Linnæus
Lac. Lacépède
Lank. Lankester
M. & H. Müller & Henle
M. & T. Müller & Troschel
McCl. M'Clelland
Q. &. G. Quoy & Gaimard
Raf. Rafinesque
Ranz. Ranzani
Reinh. Reinhardt
Rich. Richardson
Steind. Steindachner
Sw. Swainson
T. & S. Temminck & Schlegel
Val. Valenciennes
Woodw. Woodward

INTRODUCTORY NOTE

This present memoir is a continuation of a paper entitled, *The Genera of Fishes, from Linnæus to Cuvier, 1758-1833, seventy-five years, with the accepted type of each, a contribution to the stability of scientific nomenclature,* published by the present writer in 1917.

It is likewise a contribution to the stability of scientific nomenclature. It covers what may be termed the mediæval period of systematic ichthyology, the time in which nomenclature is fairly established, but in which, for want of a general zoological record, many papers have been overlooked or forgotten.

This period is especially marked by the studies of Professor Agassiz on the fossil fishes, by the exhaustive record of the fishes of the East Indies by Dr. Bleeker, and by the unification of ichthyological knowledge by Valenciennes and by Dr. Günther. Günther's *Catalogue of the Fishes of the British Museum,* 1859-1870, must always remain the solid foundation on which systematic zoology is built. This period is also marked by the studies in comparative zoology by Dr. Johannes Müller and by the opening to science of the rich fresh-water fauna of North and South America. In this period wrote also the keenest of taxonomic critics, Dr. Gill, whose conceptions of genera and of families are likely in a large degree to constitute the last word in regard to the status of groups of this grade. Other notable writers whose work has been of a high order are Richardson, Schlegel, Lowe, Egerton, Reinhardt, Poey, Heckel, Kner, Pictet, Holbrook, Baird, Troschel, Owen, McClelland, Rüppell, Henle, Storer, DeKay, Ayres, Cooper, Girard and Cope. Of the many writers in the period preceding the year 1870, Steindachner only, one of the most industrious and accurate, is still living. When the writer visited him in Vienna in 1913 he was still, at the age of 79, hard at work, and one important paper of his bearing date of 1915 has found its way through the vicissitudes of censorship to my library.

In accordance with the philosophy of evolution, the writers at the end of this period have inverted the fish series, the simplest and most ancient types being placed first in taxonomic records, rather than the perch and its relatives, so long regarded as typifying the perfect or completed fish.

Compilations of this sort constitute an exercise in modesty. They can never be made absolutely complete nor free from errors. In the present case, the author cannot be certain that the genotype he indicates is actually the one chosen by the first reviser. The compilation itself deals with the work of great minds, but also with a degree of igno- rance, carelessness and perversity by which names were multiplied with no corresponding increase of ideas. Synonymy in general is the index of lack of actual knowledge, or else, of the failure to deal justly with the work of others, "a burden and a disgrace to science". No system of naming can be beyond the knowledge on which it rests. Ignorance of fact produces confusion in nomenclature.

The science of ichthyology, to which the present series of papers offers a kind of key, is the work of many scholars, each in his own field, each contributing a series of facts, a series of tests of the work of others, or else some improvement in the method of arrangement. As in other branches of science, this work has been done by sincere, devoted men, impelled by a love for this kind of labor, and having in view, as "the only reward they asked, a grateful remembrance of their work". And as future generations may be lenient towards our own shortcomings, we should be generous towards those of our predecessors.

In so far as the present paper assumes to decide questions of nomen- clature, it follows strictly the rules agreed on by the International Com- mission. Of these, the one of most vital importance is concerned with the rigid enforcement of the law of priority.

The writer expresses his special obligation to Mr. Barton A. Bean of the United States National Museum, and to Henry Weed Fowler of the Academy of Natural Sciences of Philadelphia for services in examining rare or forgotten books, and to Mr. Masamitsu Oshima, for valuable aid in compilation. He is also under special obligations to Mr. R. A. Nelgner of the Stanford University Press for his care in ensuring accuracy in this difficult piece of work. In arranging the genera of fossil fishes, he has depended almost entirely on the *Catalogue of the Fossil Fishes in the British Museum,* by Arthur Smith Woodward. The cataloguing of the scattered and multifarious genera of Dr. Bleeker has been made possible by the *Index of the Ichthyological Papers of P. Bleeker,* by Dr. Max Weber and Dr. L. F. de Beaufort. In matters of citation of titles Dean's *Bibliography of Fishes* has been absolutely indispensable. The writer hopes to complete this work by a final paper covering the period from 1859 to 1920. The need of such compilation grows progressively less, as the *Zoological Record* of London, beginning with 1864, contains references to nearly all the generic names of subsequent date, although having occasional surprising omissions.

The writer asks for additions and corrections, that the work may be as helpful as possible to students of fishes.

It will be noted that the genera have been arranged as far as may be in chronological order, publications of the same year being arranged in alphabetical order in accord with the names of authors.

The term *orthotype* is applied to the type of a genus as indicated or distinctly implied by the original author. A *logotype* is one selected by the "first reviser". A *haplotype* is the sole species named under a genus, therefore of necessity an *orthotype* as well. A *tautotype* is a name of a genus identical with the specific name of one of its components.

STANFORD UNIVERSITY, CALIFORNIA,
November 11, 1918: Issued July, 1919.

ADDITIONS AND CORRECTIONS TO PART I

1758

1. LINNÆUS (1758). *Systema Naturæ.* Ed. X, Vol. 1.
   CARL LINNÉ (CAROLUS LINNÆUS). (1707-1778.)

Trigla Linnæus, 300; logotype T. GURNARDUS L., as restricted by Kaup, 1873, Apparently this should replace the later restriction to T. LYRA L. of Jordan & Gilbert, 1883, this species being type of LYRICHTHYS Kaup.

Mormyrus Linnæus, 327; logotype M. CASCHIVE Hasselquist, as restricted by Gill. (Not M. CYPRINOIDES L., as stated in part 1, page 15; this species is the type of MARCUSENIUS Gill.)

Tetraodon Linnæus, 332; logotype T. LINEATUS L. I am compelled to revise my judgment, as given in part I of the *Genera of Fishes,* page 15, in which I present the reasons for regarding T. TESTUDINEUS L. as the type of TETRAODON. This judgment is based on Bleeker's restriction of the genus in 1865. I find that an earlier restriction was made by Bonaparte in the *Fauna Italica,* about 1841, "TETRAODON LINEATUS Bloch" being taken as type following a suggestion of Swainson. Bloch's account includes the synonymy and description of T. LINEATUS L., although the figure he offers may have been based on an example of T. STELLATUS.

In this view of the case, the type of TETRAODON should be T. LINEATUS, a fresh-water species of the Nile, with which the numerous marine species called OVOIDES (AROTHRON) may not be congeneric. According to Dr. Gill, the type of TETRAODON is distinguished from the Pacific species known as OVOIDES by the form of the frontal bones, a matter which needs further elucidation.

In this view the name SPHEROIDES should stand, as used by Jordan and Evermann, for the allies of T. TESTUDINEUS L.

1764

7. HOUTTUYN (1764). *Naturlike historie volgens den Heer Linnæus.*
   MARTIN HOUTTUYN.

Conger Houttyn, VII, 103, and TORPEDO Houttyn, VII, 453, as quoted *in Genera of Fishes,* page 22, are not presented by Houttuyn as generic names, and the reference to them should be canceled.

In a personal letter dated August 23, 1918, Mr. C. Davies Sherborn says: "TORPEDO is the trivial Linnæan name of the species RAJA TORPEDO and CONGER is the trivial Linnæan part of the species name MURÆNA CONGER. Neither name is used scientifically nor is there any intention on Houttuyn's part to do more than quote the Linnæan term for the fish he describes as 'KRAMPFISCH' and 'CONGER'."

The record of this work of Houttuyn should therefore be erased as having no relation to scientific nomenclature.

1767

9. MÜLLER (1767). *Delineationes Naturæ.* II, 141.
   PHILIP LUDWIG STATIUS MÜLLER (1725-1776).

Mr. Sherborn informs me that the generic names ACUS, page 141, and

ORBIS, page 141; "are perfectly valid," that is, properly presented as generic names. ACUS is a synonym of SYNGNATHUS L., ORBIS of DIODON L.

10. GEOFFROY (1767). *Descriptions de 719 plantes,* etc.

ÉTIENNE LOUIS GEOFFROY (1725-1810).

Ichthyocolla Geoffroy, 399; haplotype ACIPENSER HUSO L. This replaces HUSO Brandt, and STERLETUS Raf. Mr. Sherborn informs me that the generic names used by Geoffroy are "perfectly valid" for purposes of nomenclature. These are:

ICHTHYOCOLLA, 399, (HUSO Brandt).
HARENGUS, 405, (CLUPEA L.).
LUCIUS, 467, (ESOX L.).
TRUTTA, 719, (TRUTTA L.).

1772

17A. GRONOW (1772). *(Pteraclis.)* Actæ Helveticæ. VII.

LORENZ THEODOR GRONOW (LAURENTIUS THEODORUS GRONOVIUS). (1730-1777).

Pteraclis Gronow, 44; orthotype P. PINNATA GRONOW = CORYPHÆNA VELIFERA Pallas. As here used, unquestionably binomial.

17B. GÜLDENSTÄDT (1772). *(Sterleta.)* Nov. Comm. Sci. Petropol, XVI.

ANTON JOHANN GÜLDENSTÄDT.

Sterleta Güldenstädt, 533; type ACIPENSER RUTHENUS L. under polynomial designation. If eligible, will replace STERLETUS Brandt and STERLEDUS Bon.

1775

20. KLEIN (1775). *Neuer Schauplatz,* etc.

JAKOB THEODOR KLEIN (1685-1759).

Prochilus Klein, I, 1043; AMPHIPRION EPHIPPIUM Bloch, as restricted by Bleeker, 1877 (P. OVATÆ FIGURÆ MACULA FUSCA MAGNA Klein). This name, if deemed eligible from the date of the *Schauplatz* would replace AMPHIPRION Bl. & Schn., instead of PHOLIS Cuv. & Val. (LIPOPHRYS Gill), as suggested in part I of this work.

1779

31A. KLEIN (1779). *Neuer Schauplatz,* etc.

JACOB THEODOR KLEIN.

Platystacus Klein, VIII, 52, 1779; type (P. COTYLEPHORUS Bloch); a synonym of ASPREDO (Gronow) Scopoli.

1788

31B. BRÜNNICH (1788). *Om den islandske fisk Vaagmaeren (Gymnogaster arcticus).* Kong. Dansk. Selsk. Skr., III, 408-413.

MORTEN THRANE BRÜNNICH (1737-1827)

Gymnogaster Brünnich, III, 408; orthotype G. ARCTICUS Brünnich. A synonym of TRACHIPTERUS Gouan. Not TRICHIURUS L.

1795

45. BLOCH (1795). *Naturgeschichte der ausländischen Fische.* IX.

MARK ELIESER BLOCH (1729-1799).

Gymnothorax Bloch IX, 83, 1795; type as first restricted (Bleeker, 1864) G. RETICULARIS Bloch. Although originally virtually a substitute name for MURÆNA L., the first fixation of types separates the two. GYMNOTHORAX thus replaces LYCODONTIS McCl.

1796
47. VOLTA (1796). *Ichthyolithologia Veronese.*
GIOVANNI SERAFINO VOLTA.

Monopteros Volta, CXCL; orthotype M. GIGAS Volta (fossil). Replaces PLA-
TINX Ag. MONOPTERUS Lac., 1798, is thus preoccupied and must give way
to FLUTA Bl. & Schn. (1801).

1801
55. BLOCH & SCHNEIDER (1801). *Systema Ichthyologiæ.*
MARK ELIESER BLOCH; JOHANN GOTTLOB SCHNEIDER (1750-1822).

Periophthalmus Bloch & Schneider, 63; logotype, as first restricted (by Gill, 1863)
GOBIUS SCHLOSSERI Pallas. The name therefore replaces PERIOPHTHALMODON
Blkr., while the common species known as PERIOPHTHALMUS BARBARUS or
P. KOELREUTERI is the type of EUCHORISTOPUS Gill (1863).

Synanceja Bloch & Schneider, 194; logotype S. VERRUCOSA Bl. & Schn., as first
restricted by Müller, 1843. This limitation to species without vomerine teeth
holds as against Bleeker's restriction to the first species named, S. HORRIDA.
SYNANCEJA thus replaces SPURCO Commerson, SYNANCEICHTHYS Blkr. and
EMMYDRICHTHYS Jordan & Rutter. S. HORRIDA is type of SYNANCIDIUM
Müller. I am not convinced by Bleeker's statement that SYNANCIDIUM is not
a valid genus.

Fluta Bloch & Schneider, 565; orthotype MONOPTERUS JAVANENSIS Lac. Name a
substitute for MONOPTERUS Lac. (1798). The latter is preoccupied by MO-
NOPTEROS Volta (1796). The type species should stand as FLUTA ALBA
(Zuieuw).

1803
56. LACÉPÈDE (1803). *Histoire naturelle des poissons.* IV, V.
BERNARD GERMAIN LACÉPÈDE (1756-1825).

Enoplosus Lacécepede, IV, 540; orthotype CHÆTODON ARMATUS Shaw. This genus
dates from 1803. The name ENOPLOSUS was not used by White in 1790.

Stolephorus Lacépède, V, 381; logotype S. COMMERSONIANUS Lac. In view of the
fact that the descriptions of STOLEPHORUS of Lacépède, Bleeker and Jordan
were drawn from this species, the second-named by Lacépède, it may well be
regarded as the type of STOLEPHORUS. S. JAPONICUS, the first-named species
and the one first formally indicated as type, was very scantily described, and
was unknown to authors until very recently. In this view STOLEPHORUS will
replace or include ANCHOVIELLA Fowler (1911). This reverses the decision
in part I, page 67.

1804
56A. LACÉPÈDE (1804). *Mémoires sur plusieurs animaux de la Nouvelle*
Hollande, etc. Ann. Mus. Nat. Hist., IV, 1804.
BERNARD GERMAIN LACÉPÈDE.

Prionurus Lacépède, IV, 211; orthotype P. MICROLEPIDOTUS Lac.

Platypodus Lacépède, IV, 211; orthotype P. FURCA Lac.
This genus, overlooked by subsequent writers, seems to belong to the
OSPHRONEMIDÆ, an ally of MACROPODUS. It is said to have large fan-like
ventrals of at least eight rays, the dorsal extending from the nape to the
base of the caudal, which is narrow, forked, its lobes half as long as rest
of body; pectorals small; anal very low. This may be, as suggested by
Dr. Günther for MACROPODUS, a domesticated form of POLYACANTHUS. Both
names, if correctly written (MACROPUS; PLATYPUS), are preoccupied.

56B   BORY DE SAINT VINCENT (1804). *Voyage dans les quatre principales l'îles des Mers d'Afrique.* 1801 et 1802.

JEAN BAPTISTE BORY DE ST. VINCENT (1780-1846).

Acinacea Bory de St. Vincent, I, 93; orthotype A. NOTHA Bory ("l'acinacée bâtarde"). An ally of THYRSITES Cuv., overlooked by authors. It differs from THYRSITES (ATUN), if correctly described and figured, in the presence of twenty-nine spines in the dorsal fin, instead of twenty. The type specimen was from the open sea, off Africa. If identical with THYRSITES, ACINACEA has priority. The Latin form as well as the French is given by the author.

1808

67.   DUMÉRIL (1808). *Dissertation sur les poissons Cyclostomes.*

ANDRÉ MARIE CONSTANT DUMÉRIL (1774-1860).

Ammocœtus Duméril; orthotype PETROMYZON BRANCHIALIS L. (larval form of PETRIMYZON MARINUS or of LAMPTERA PLANERI, or both). Best treated as a synonym of PETROMYZON.

1810

72.   RAFINESQUE (1810). *Caratteri di alcuni nuovi generi e nuove specie di animale e piante della Sicillia.*

CONSTANTINE SAMUEL RAFINESQUE (1783-1840).

Typhle Rafinesque, 18; orthotype SYNGNATHUS TYPHLE L. We are unable to find any further notice of the blind genus of mammals "TYPHLE", as noted by Lacépède. TYPHLE should therefore apparently stand, and not be replaced by TYPHLINUS Raf. (1814).

1811

75A.   PERRY (1811). *Arcana or Museum of Natural History.*

GEORGE PERRY.

Congiopus Perry; orthotype C. PERCATUS Perry = BLENNIUS TORVUS Walbaum. Replaces AGRIOPUS Cuv., 1829.

75B.   HUMBOLDT & VALENCIENNES (1811). *Recherches sur les poissons fluviatiles; Humboldt & Bonpland; voyage aux régions équinoxiales du nouveau continent.* II (later edition; the original in 1806).

FRIEDRICH HEINRICH ALEXANDER VON HUMBOLDT (1769-1859) ; ACHILLE VALENCIENNES (1794-1865).

Thrycomycterus Humboldt & Valenciennes; orthotype T. NIGRICANS Humboldt & Val; a synonym of EREMOPHILUS Humboldt, according to Eigenmann. Name spelled TRICHOMYCTERUS by Valenciennes in 1846, and the genus made to include PYGIDIUM Meyer.

1815

85.   RAFINESQUE (1815). *Analyse de la nature,* etc.

CONSTANTINE SAMUEL RAFINESQUE.

Raphistoma Rafinesque, 89; no type named and no diagnosis, except an erroneous reference to Lacépède. Intended to represent ESOX BELONE L. Spelled RAMPHISTOMA by Swainson, who adopts the name. Is apparently not eligible in place of BELONE Cuv., 1817, though an argument might be made for its acceptance, as we know the type intended and there are a few words of inaccurate diagnosis.

Eleuthurus Rafinesque; no diagnosis and no type given. Omitted by inadventure in *Genera of Fishes*, part I. According to Bleeker, Rafinesque's names (undefined and without type), DAMEUS, ELEUTHURUS, RINCOXIS and ZEBRI-SCIUM, were intended for eels). In the account of Rafinesque's *Analyse*, given in *Genera of Fishes*, part I, are a few misprints arising from misreading a manuscript list. Read:

> ICTIAS, p. 82.
> BATRICTIUS, 82.
> CEPHIMNUS, 85.
> EDONIUS, 88.
> PLATISTUS, 89 (for PLATYSTACUS).
> RAPHISTOMA, 89.
> ODAMPHUS, 89.
> GONODERMUS, 91.
> DISPINOTUS, 92.

1817

91. CUVIER (1817). *Le Règne Animal, etc.* Tome II.
GEORGES CUVIER (1769-1832).

Hydrocynus Cuvier, II, 167; logotype H. FORSKALI Cuvier; several species named, the genus restricted by Cuvier (*Mem. Mus.*, 1819) to H. FORSKALI, and the name spelled HYDROCYON. Also in *Cuv. & Val.*, XXII, 509, 1849. The name was afterwards restricted to H. LUCIUS Cuv., a species of XIPHOSTOMA, Ag., but the first adjustment must stand. The name is HYDROCYNUS and the genus contains no American species.

Bodianus Cuvier, *Règne Animal*, II, 276; logotype BODIANUS GUTTATUS Bloch; (not BODIANUS of Bloch) which by tautonomy goes with BODIANUS BODIANUS Bloch, and replaces HARPE Lac. Gill restricts BODIANUS (*Proc. Acad. Nat. Sci. Phil.*, 1862, 237) to B. GUTTATUS Bloch, a species of CEPHALOPHOLIS Bl. & Schn. By the rule of tautonomy, under present rules BODIANUS must stand for HARPE RUFA.

92. OKEN (1817). *Isis.*
LORENZ OKEN (1777-1851).

In *Isis*, 1817, Oken furnishes Latin equivalents to the French generic names used by Cuvier.

Lumpus Oken 1782, orthotype CYCLOPTERUS LUMPUS L. = LUMPUS ANGLORUM of authors. Equivalent for "les LUMPS" of Cuvier, a synonym of CYCLOPTERUS L. Omitted by accident in part I.

1818

97. BLAINVILLE (1818). *(Poissons fossiles.)* Nouveau Dictionnaire d'Histoire Naturelle.
HENRI MARIE DUCROTAY DE BLAINVILLE (1777-1850).

Anormurus Blainville, XXVII, 374; orthotype A. MACROLEPIDOTUS Blainv. (fossil). Perhaps the same as NOTOGONEUS Cope, 1885.

1820

108. GOLDFUSS (1820). *Handbuch der Zoologie.* II.
GEORGE AUGUST GOLDFUSS (1782-1848).

Rhomboides Goldfuss; orthotype PLEURONECTES RHOMBUS L. Substitute for RHOMBUS Cuv.; a synonym of BOTHUS Raf.

Batrachopus Goldfuss (not BATRACHOPS). Orthotype LOPHIUS COMMERSONIANUS Lac. A synonym of ANTENNARIUS.

1822

110. HAMILTON (1822). *An Account of the Fishes Found in the River Ganges and Its Branches.*

FRANCIS HAMILTON (formerly BUCHANAN) (1762-1829).

The reprint of Hamilton's *Fishes of the Ganges* has the pagination changed from that given in the text of the *Genera of Fishes*, page 114. Thus, in the reprint, CHANDA appears on page 109, not 370.

Callichrous Hamilton, 149; logotype SILURUS PABDA Ham. (as restricted by Bleeker, 1862). A synonym of OMPOK Lac.

Coius Hamilton, 369 (95); logotype COIUS POLOTA Ham., as restricted by Fowler. 1905. This designation of type prior to that of Jordan, 1917. Replaces DATNIOIDES Blkr., 1853.

Chanda Hamilton, 370; logotype CHANDA NAMA Ham. (fig. 37) (HAMILTONIA OVATA Sw.). The type of the substitute name HAMILTONIA Sw. (1839) as indicated by Swainson and by Swain (1882) carries the complex genus CHANDA with it. This antedates Fowler's choice (1902) of CHANDA LALA as type. C. NAMA is type of BOGODA Blkr. CHANDA should therefore replace BOGODA.

Corica Hamilton, 253 or 383; orthotype CYPRINUS SOBORNA Ham. The name SOBORNA (misspelled in the *Genera of Fishes*, page 115) is given as GUBORNI by Gray.

1823

112. VAN HASSELT (1823). *Poissons de Java.* Allgemeine Konst en Letterbok. II

JAN CŒNRAD VAN HASSELT (*d.* 1821).

Dermogenys Van Hasselt, 131; orthotype D. PUSILLUS Van Hasselt (omitted in part I). Spelled DERMATOGENYS by Peters.

Diplocheilus Van Hasselt, 329; orthotype D. ERYTHOPTERUS Van Hasselt.

Gonostoma Van Hasselt, 329; orthotype G. JAVANICUM Van Hasselt. Name preoccupied; replaced by ANODONTOSTOMA Blkr.

1825

119. KÖNIG (1825). *Icones fossilium sectiles.*

CLEMENS KÖNIG.

Teratichthys König, 4; orthotype T. ANTIQUITATIS König (fossil). An unknown ally of CARANX.

119A. LE SUEUR (1825). *Description of a New Fish of the Genus Salmo (S. microps).* Journ. Acad. Nat. Sci. Phila.

CHARLES ALEXANDRE LE SUEUR (1778-1846).

Harpodon Le Sueur, V, 50; orthotype SALMO MICROPS Le Sueur = OSMERUS NEHEREUS Ham.

119B. ADANSON (1825). *Cours d'histoire naturelle.*

MICHAEL ADANSON (1727-1806).

Trygon Adanson, II, 170; orthotype RAJA PASTINACA L. A synonym of DASYATIS Raf. (According to Auguste Duméril, Adanson's paper, written in 1772, was printed by Payer in 1825).

1826

121. RISSO (1826). *Histoire naturelle des principales productions de l'Europe méridionale.*

ANTOINE RISSO (1770-1845).

Charax Risso, III, 353; orthotype C. ACUTIROSTRIS Risso = SPARUS PUNTAZZO L. Preoccupied; replaced by PUNTAZZO Blkr.

121A. RILEY (1826). *On a Fossil in the British Museum (Squaloraia dolichognathos) and Discovered in the Lias at Lyme Regis.* Proc. Geol. Soc. I (also in Trans. Geol. Soc., 1833).

HENRY RILEY.

Squaloraia Riley, I, 484; orthotype S. DOLICHOGNATHOS Riley (fossil).

1828

128. RÜPPELL (1828). *Atlas zu der Reise im nördlichen Africa; Fische des Rothen Meeres.*

EDUARD RÜPPELL (1794-1884).

Petroscirtes Rüppell, 110; orthotype P. MITRATUS Rüppell. Omitted by accident from part I; sometimes spelled PETROSKIRTES.

128A. THIENEMANN (1828). *Lehrbuch der Zoologie.*

FRIEDRICH AUGUST THIENEMANN (1793-1858).

Sphyrichthys Thienemann, III, 408; orthotype SQUALUS ZYGÆNA L. A needless substitute for SPHYRNA Raf.

132. CUVIER & VALENCIENNES (1828). *Histoire naturelle des poissons.* II.

GEORGES CUVIER; ACHILLE VALENCIENNES.

Pomatomus Cuvier & Valenciennes, II, 171; type P. TELESCOPIUM RISSO (EPIGONUS MACROPHTHALMUS Raf.). Not POMATOMUS of Lacépède; a synonym of EPIGONUS Raf., 1820, as is also TELESCOPS Blkr.

1829

134. AGASSIZ (and SPIX) (1829). *Selecta genera et species piscium quos in itinere per Brasiliam (1817-1820) ... collejit ... Dr. J. B. de Spix....*

LOUIS AGASSIZ (1807-1873); JEAN BAPTISTE DE SPIX (1807-1873).

Xiphorhynchus Agassiz, 18; orthotype SALMO FALCATUS Bloch. Name preoccupied; replaced by XIPHORHAMPHUS M. & T., 1845; also preoccupied, again replaced by ACESTRORHYNCHUS Eigenmann, 1910.

Pirarara Spix, 23; orthotype PHRACTOCEPHALUS BICOLOR Ag. A synonym of PHRACTOCEPHALUS Ag.

Erythrinus (Gronow) Agassiz, 41; haplotype E. SALVUS Ag., the only species named, the genus credited to Gronow. Same as ERYTHRINUS Gronow.

135. CUVIER (1829). *Le règne animal distribue d'après son organization.* Edition, II, Vol. I.

GEORGES CUVIER.

Polyacanthus (Kuhl) Cuvier, 227; logotype CHÆTODON CHINENSIS Bloch, as restricted by Cuv. & Val. (not TRICHOPODUS COLISA as stated in part I, *Genera of Fishes*).

Rhinoptera (Kuhl) Cuvier, 401; orthotype MYLIOBATIS MARGINATUS Geoffroy St. Hilaire.

136. CUVIER & VALENCIENNES (1829). *Histoire naturelle des poissons.* III.

GEORGES CUVIER; ACHILLE VALENCIENNES.

Ostichthys (Langsdorff) Cuvier & Valenciennes, III, 160; orthotype MYRIPRISTIS JAPONICUS Cuv. & Val. No generic definition; reëstablished by Jordan and Evermann, 1896.

Apistus Cuvier & Valenciennes, IV, 391; logotype APISTUS ALATUS Cuv. & Val. called "les APISTES" by Cuvier. Regarded as preoccupied by APISTA Hübner, a genus of insects, and by ASPISTES Meigen, a genus of flies. If rejected, it may be replaced by PTERICHTHYS Sw., 1839, unless that is regarded as invalid, from the non-binomial PTERICHTHUS of Commerson (1803), which is EXOCŒTUS. HYPODYTES Gistel (1848) is also a substitute name, but the indicated type (LONGISPINIS) is a PARACENTROPOGON of Bleeker, also not mentioned by Cuvier. PROSOPODASYS Cantor, 1849, is also given as a substitute name for APISTUS and must retain the same type. If APISTUS is preoccupied, I think that PTERICHTHYS should replace it. . . . APISTES TRACHINOIDES Cuv. & Val. used by Günther as type of PROSOPODASYS is not eligible for that purpose, the species not being one of those mentioned by Cuvier.

Spurco (Commerson) Cuvier & Valenciennes, IV, 448, 1829; (reference in passing) haplotype SCORPÆNA BRACHIO Cuv. = SYNANCEJA VERRUCOSA Bl. & Schn., same as SYNANCEICHTHYS Blkr.; a synonym of SYNANCEJA as restricted by Müller.

136A. SMITH (1829). *Contributions to the Natural History of South Africa.* Zoological Journal.
ANDREW SMITH (1797-1872).

Rhineodon Alexander Smith, IV, 443; orthotype R. TYPUS Smith. (RHINCODON by misprint.) Also spelled RHINODON.

136B. BANCROFT (1829). *On the Fish Known in Jamaica as the Sea Devil (Cephalopterus manta).* Zool. Journal, IV.
E. N. BANCROFT.

Manta Bancroft, IV, 444; orthotype CEPHALOPTERUS MANTA Bancroft = RAIA BIROSTRIS Walbaum, 1792.

1830

142. CUVIER & VALENCIENNES (1830). *Histoire naturelle des poissons.* V.
GEORGES CUVIER; ACHILLE VALENCIENNES.

Lycogenis (Kuhl & Van Hasselt) Cuvier & Valenciennes, V, 346; orthotype HOLOCENTRUS CILIATUS Lac. A section under SCOLOPSIS Cuv.

Priopis (Kuhl & Van Hasselt) Cuvier & Valenciennes, VI, 503; haplotype PRIOPIS ARGYROZONA Kuhl & Van Hasselt.

143. BENNETT (1830). *Catalogue of the Fishes of Sumatra.* Life and Public Services of Sir Stamford Raffles.
EDWARD TURNER BENNETT (1797-1836).

Leiopsis Bennett, 688; orthotype SPARUS AUROLINEATUS Lac. A synonym of PENTAPUS Cuv. & Val. (Omitted in part I.)

143A. BENNETT (1830). *Obervations on a Collection of Fishes from the Mauritius with Characters of New Genera and Species.* Proc. Zool. Soc. London, 1831, part I.
EDWARD TURNER BENNETT.

Psettodes Bennett, I, 147; orthotype P. BELCHERI Bennett.

Agonostomus Bennett, I, 166; logotype A. TELFARI Bennett. Replaces NESTIS Cuv. & Val.

Apolectus Bennett, I, 169; orthotype A. IMMUNIS Bennett. A synonym of SCOMBER-OMORUS Lac.

1831

146A. CUVIER & VALENCIENNES (1831). *Histoire naturelle des poissons.* VII (January, 1831), VIII, October, 1831.

GEORGES CUVIER; ACHILLE VALENCIENNES.

Drepane Cuvier & Valenciennes, VII, 129; orthotype CHÆTODON PUNCTATUS L.; name preoccupied by DREPANA or DREPANE Schranck, Lepidoptera, 1802; replaced by ENIXE Gistel, 1848, and by HARPOCHIRUS Cantor, 1849.

Spinax (Commerson) Cuvier & Valenciennes, VIII, 333; orthotype S. EDAX Commerson MS. = GASTEROSTEUS CANADUS L., note in passing. Name preoccupied; a synonym of RACHYCENTRON Kaup (ELACATE Cuv.).

146B. BONAPARTE (1831). *Saggio di una distribuzione metodica degli animali vertebrati.* Prospetto del Systema d'Ittiologia Générale; also in Giornale Arcadico, 167-190, 1832.

CHARLES LUCIEN JULES LAURENT BONAPARTE, PRINCE DE CANINO (1803-1857). (CARLO LUCIANO BONAPARTE).

Merou ("Cuvier") Bonaparte, 101; type presumably SERRANUS GIGAS; not defined. Same as EPINEPHELUS Bloch.

Dulichthys Bonaparte, 101; haplotype DULES AURIGA Cuv. Name a substitute for DULES Cuv., regarded as preoccupied by DULUS, a genus of birds. If needed, DULICHTHYS replaces EUDULUS Fowler.

Pogonathus (Lacépède) Bonaparte, 104. This name is revived as a substitute for POGONIAS Lac., 1803, "long since used in ornithology". But POGONIAS Illiger dates from 1811, while POGONIA and POGONIAS, as a genus of birds, are still later in date.

Micropogonias Bonaparte, 104; type PERCA UNDULATA L. Subtitute for MICRO-POGON, which is said to be, but apparently is not, preoccupied.

Sarpa Bonaparte 105; a bare word of reference. The intended type is obviously SPARUS SALPA L. SARPA may stand as a section under BOOPS, the body being deeper and more compressed than in B. BOOPS.

Palamita Bonaparte, 107; (type SCOMBER SARDA Bloch). Name a substitute for PELAMYS Cuv. & Val., 1801, preoccupied in reptiles. But SARDA Cuv., 1829, is older and stands unless the non-binomial SARDA Plumier, 1802 (= OCYURUS Gill) is deemed eligible, in which case PALAMITA would hold.

Priodontichthys Bonaparte, 109; haplotype PRIODON ANNULARIS Cuv. & Val., the young of NASO ANNULATUS Q. & G. Substitute for PRIODON Cuv., preoccupied. A synonym of NASO Lac.

Scopas Bonaparte, 109; type presumably ACANTHURUS SCOPAS Cuv. & Val., being placed next to ACANTHURUS, but neither diagnosis nor type is given. Presumably the same as ZEBRASOMA Sw., but not eligible.

Ctenodon Bonaparte, 109. No description nor type mentioned; presumably intended for ACANTHURUS CTENODON Cuv. & Val., preoccupied; replaced by CTENOCHÆTUS Gill, 1884.

Elops (Commerson) Bonaparte, 112; (type GOMPHOSUS TRICOLOR Lac.). Suggested as a substitute for GOMPHOSUS Lac. (Not ELOPS L., 1758.)

Erythrichthys Bonaparte, 116; type (SALMO ERYTHRINUS Bl. & Schn.). A needless substitute for ERYTHRINUS (Gronow) Scopoli. It renders ERYTHRICHTHYS T. & S. preoccupied, to be replaced by ERYTHROCLES Jordan, 1919.

147A. EICHWALD (1831). *Zoologia specialis potisimæ R. Rossiæ et Poloniæ*
Wilna, 1831.
CARL EDUARD VON EICHWALD (1795-1876).

Nematosoma Eichwald, III, 60; orthotype N. OPHIDIUM Eichw. (not SYNGNATHUS
OPHIDION L.) = SCYPHIUS TERES Nordmann, 1840. Name regarded as preoccupied by NEMOSOMA, a genus of beetles. A synonnym of NEROPHIS Raf.
1810.
Benthophilus Eichwald, III, 77; orthotype GOBIUS MACROCEPHALUS Pallas.

150A. GRAY (1831). *The Zoological Miscellany.* I.
JOHN EDWARD GRAY (1800-1873).

Centracion Gray, I, 5; orthotype C. ZEBRA Gray, Same as CESTRACION Cuv., 1817,
not of Klein. Replaces HETERODONTUS, if that name is regarded as preoccupied
by the genus of snakes, HETERODON, 1802.

1832
151B. MÜNSTER (1832). *Beiträge zur Kunde der Petrefakten.* I.
GEORG (GRAF ZU) MÜNSTER (1776-1844).

Janassa Münster, I, 67; logotype J. ANGULATA Münster (fossil) = TRILOBITES
BITUMINOSUS Schlotheim.

151C. COUCH (1832). *Fishes New to the British Fauna.* Mag. Nat. Hist., V.
JONATHAN COUCH (1788-1870).

Ciliata Couch, V, 15; orthotype C. GLAUCA Couch, a larva of GADUS MUSTELA (L.).
Name may be retained for the five-bearded Rocklings.

151D. NARDO (1832). *Annotazioni di quattro pesci nuovi,* etc.
GIOVANNI DOMENICO NARDO (1802-1877).
(Not seen by us; the title uncertain. The genus GOUANIA is quoted from
Canestrini.)
Gouania Nardo; orthotype G. PROTOTYPUS Nardo = LEPADOGASTER WILLDENOWI
(Risso). Antedates RUPISUGA Sw.

151E. KRYNICKI (1832). *Schilus pallasii descriptus et icone illustratus.* Nouv.
Mém. Soc. Nat. Moscou, II.
J. KRYNICKI.
Schilus Krynicki, II, 441; orthotype S. PALLASI Krynicki = PERCA VOLGENSIS Pallas.
Replaces MIMOPERCA Gill & Jordan, 1877.

152. MINDING (1832). *Lehrbuch: Naturgeschichte der Fische.*
JULIUS MINDING.
Hydronus Minding, 83; orthotype GADUS MERLUCIUS L. A synonym of MERLUC-
CIUS Raf.

153. NILSSON (1832). *Prodromus ichthyologiæ Scandinaviæ.*
SVEN NILSSON (1787-1883).
Salvelini Nilsson, 7; orthotype SALMO SALVELINUS L. = S. ALPINUS L. Group
name, in the plural only; written as SALVELINUS by Richardson (1836).
Carassius Nilsson, IV, 290; orthotype CYPRINUS CARASSIUS L.

153A. REINHARDT (1832). *Bidrag til vor kundskab om Grönlands fiske.*
Dansk. Vid. Selsk. Afh. Kjöbenhavn, V.
JOHANNES REINHARDT (1776-1845).
Triglops Reinhardt, V, 52; orthotype T. PINGELI Reinhardt.

THE GENERA OF FISHES
PART II

1833

164. AGASSIZ (1833). *Recherches sur les poissons fossiles.* II; IV (first part; second part of Vol. II issued in 1844; part of Vol IV in 1833).
Louis Agassiz.

Catopterus Agassiz, II, 3; orthotype C. ovalis Ag. (fossil). A synonym of Dipterus.

Amblypterus Agassiz, II, 3, 28; orthotype A. latus Ag. (fossil).

Gyrolepis Agassiz, II, 6, 172; logotype G. alberti Ag. (fossil).

Tetragonolepis Agassiz, II, 6, 181; logotype T. confluens Ag. (fossil). Name preoccupied; a synonym of Dapedium Leach.

Microps Agassiz, II, 10; orthotype M. furcatus Ag. (fossil). Name preoccupied; replaced by Periurgus Gistel, 1848.

Pygopterus Agassiz, II, 10; orthotype P. humboldti Ag. (fossil).

Notagogus Agassiz, II, 10; orthotype N. pentlandi Ag. (fossil).

Acrolepis Agassiz, II, 11; orthotype A. sedgwicki Ag. (fossil).

Pachycormus Agassiz, II, 11; orthotype Elops macropterus Blainv. (fossil).

Thrissops Agassiz, II, 12; orthotype T. formosus Ag. (fossil).

Megalurus Agassiz, II, 13; orthotype M. lepidotus Ag. (fossil). Name preoccupied in birds (Horsfield, 1821), replaced by Synergus Gistel (1848), also preoccupied; suggested as substitute for both Urocles Jordan, 1919.

Aspidorhynchus Agassiz, II, 14; orthotype Esox acutirostris Blainv. (fossil).

Saurostomus Agassiz, II, 14; orthotype S. esocinus Ag. (fossil).

Sphœrodus Agassiz, II, 15; logotype S. gigas Ag. (fossil). A synonym of Lepidotes Ag.; not Sphærodon Rüppell.

Pycnodus Agassiz, II, 16; orthotype Zeus platessus Blainv. (fossil).

Gyrodus Agassiz, II, 16; orthotype G. macrophthalmus Ag. (fossil).

Microdon Agassiz, II, 16; orthotype M. elegans Ag. (fossil). Name preoccupied in flies (Meigen, 1803), replaced by Proscinetes Gistel, 1848, and, later, by Polypsephis Hay (1899).

Calamostoma Agassiz, II, 18; orthotype Syngnathus breviculus Blainv. (fossil).

Acanthodes Agassiz, II, 19; orthotype Acanthoëssus bronni Ag. (fossil). A substitute, apparently unnecessary, for Acanthoëssus Ag. (1832).

Cyclopoma Agassiz, IV, 17; orthotype C. gigas Ag. (fossil).

Smerdis Agassiz, IV, 32; logotype Perca minuta Blainv. (fossil). Name preoccupied in insects; replaced by Dapalis Gistel (1848).

Gasteronemus Agassiz, V, 17; logotype G. rhombeus Ag. (fossil). An extinct homologue of Mene.

165. BRANDT & RATZEBURG (1833). *Medizinische Zoologie.* II.
Johann Friedrich Brandt (1802-1879); J. T. C. Ratzeburg (1801-1871).

Huso Brandt & Ratzeburg, II, 3; orthotype Acipenser huso L. A synonym of Ichthyocolla Geoffroy, 1767, and of Sterletus Raf., 1820.

Helops Brandt & Ratzeburg II, 3; orthotype ACIPENSER STELLATUS Pallas, 1768, = A. HELOPS Pallas, 1811.

Sterletus Brandt & Ratzeburg, II, 3, 349; orthotype ACIPENSER RUTHENUS L. Name spelled also STERLETA Müller) and STERLEDUS (Bon.). Name preoccupied; replaced by STERLETA Güldenstädt, if binomial. STERLETUS Raf. (1820) is based on an American species, ACIPENSER SEROTINUS Raf., which is the same as ACIPENSER RUBICUNDUS Le Sueur, still earlier called ACIPENSER FULVESCENS by Rafinesque. This species belongs to the group or genus called ICHTHYOCOLLA by Geoffroy in 1767, and HUSO by Brandt & Ratzeburg in 1833.

166. COCCO (1833). *Osservationes Peloritani.*
ANASTASIO COCCO (1799-1854).

Ruvettus Cocco, XIII, 18 (April, 1833); orthotype R. PRETIOSUS Cocco.

167. COCCO (1833). *Lettere al Signor Risso su alcuni pesci novelli.* Giorn. Sci. Lett. Sicilia, XLII.
ANASTASIO COCCO.

Tylosurus Cocco, XLII, No. 124; orthotype T. CANTRAINI Cocco = ESOX IMPERIALIS Raf. = SPHYRÆNA ACUS Lac.

168. COCCO (1833). *Su alcuni pesci de Mare di Messina.* Giorn. Sci. Lett. Sicilia, XLII.
ANASTASIO COCCO.

Mupus Cocco, 20; orthotype M. IMPERIALIS Cocco. The name LEIRUS Lowe, also of 1833, is preoccupied in beetles (1823) and must give place to MUPUS.

169. CUVIER & VALENCIENNES (1833). *Histoire naturelle des poissons.* IX.
GEORGES CUVIER; ACHILLE VALENCIENNES.

Trachurus Cuvier & Valenciennes, IX, 6; haplotype SCOMBER TRACHURUS L. Same as TRACHURUS Raf., 1810.

Hynnis Cuvier & Valenciennes, IX, 95; orthotype H. GOREENSIS Cuv. & Val.

Lactarius Cuvier & Valenciennes, IX, 237; haplotype SCOMBER LACTARIUS Bloch = LACTARIUS DELICATULUS Cuv. & Val.

Nauclerus Cuvier & Valenciennes, IX, 247; orthotype N. COMPRESSUS Cuv. & Val. Young of NAUCRATES Raf.

Porthmeus Cuvier & Valenciennes, IX, 255; orthotype P. ARGENTEUS Cuv. & Val.

Psenes Cuvier & Valenciennes IX, 259; orthotype P. CYANOPHRYS Cuv. & Val.

Gallichthys Cuvier & Valenciennes, IX, 264; orthotype GALLICHTHYS MAJOR Cuv. & Val. Substitute for GALLUS, preoccupied in birds; a synonym of ALECTIS Raf.

Lampugus Cuvier & Valenciennes, IX, 317; orthotype SCOMBER PELAGICUS L. A synonym of CORYPHÆNA L.

Aphredoderus (Le Sueur) Cuvier & Valenciennes, IX, 445; orthotype A. GIBBOSUS Le Sueur = SCOLOPSIS SAYANUS Gilliams. Also spelled APHRODEDIRUS and APHODODERUS by authors.

170. GRAY (1833). *The Illustrations of Indian Zoology, Chiefly Selected from the Collection of General Hardwicke.*
JOHN EDWARD GRAY.

This paper contains plates only, no diagnosis of the new genera being given.

Balitora Gray, II, pl. 5; haplotype B. BRUCEI Gray. A synonym of HOMALOPTERA Van Hasselt.

Bedula Gray, II, 88; logotype B. NEBULOSA Gray. A synonym of NANDUS Cuv. & Val., 1831.

Pterapon Gray, II, 88; haplotype P. TRIVITTATUS Gray = SCIÆNA JARBUA Forskål. A synonym of THERAPON Cuv.

Anaora Gray, II, 90; haplotype A. TENTACULATA Gray. (Name misprinted AMORA, but corrected in manuscript on the plate.) Species not identified; resembles THYSANOPHRYS Ogilby.

Apterygia Gray, II, 92; haplotype A. RAMCARATE Gray. A synonym of RACONDA Gray.

Tor Gray, II, 96; haplotype CYPRINUS TOR Ham. = TOR HAMILTONI Gray. Replaces LABEOBARBUS Rüppell.

Bengala Gray, II 96; haplotype CYPRINUS ELANGA Ham. Replaces MEGARASBORA Gthr.

Amanses Gray, II, 98; haplotype A. HYSTRIX Gray = MONACANTHUS SCOPAS Cuv.

Acarana Gray, II, 98; haplotype OSTRACION AURITUS Shaw. Also in *Ann Nat. Hist.*, I, 110, 1838.

Girella Gray, II, 98; haplotype G. PUNCTATA Gray.

171, 172. LOWE (1833). *Character of a New Genus and of Several New Species of* Proc. Zool. Soc. London, I.
RICHARD THOMAS LOWE (1802-1874).

Alepisaurus Lowe, 104; orthotype A. FEROX Lowe. A synonym of PLAGYODUS Steller, 1811, if the latter be deemed tenable, being given in the dative case only ("Plygyodontem") Often written ALEPIDOSAURUS by authors.

173. LOWE (1833). *Character of a New Genus and of Several New Species of Fishes from Madeira.* Proc. Zool. Soc. London, I.
RICHARD THOMAS LOWE

Leirus Lowe, 142; orthotype L. BENNETTI Lowe = CENTROLOPHUS OVALIS Cuv. & Val. = MUPUS IMPERIALIS Cocco, the three names of the same date. LEIRUS is preoccupied in beetles (Meigen, 1823) and gives place to MUPUS Cocco.

174. NARDO (1833). *De Skeponopodo novo piscium genere et de Guebucu Margravii: species illi cogniti: Isis.* Mem. Assemb. Nat. Vienna.
GIAN DOMENICO NARDO.

Skeponopodus Nardo, fasc. 416; orthotype S. TYPUS Nardo = XIPHIAS IMPERATOR Bl. & Schn. A synonym of TETRAPTURUS Raf.

175. REINHARDT (1833). *Bemärkningen til den Skandinaviske ichthyologic.*
JOHANNES REINHARDT.

Silus Reinhardt, 11; orthotype SALMO SILUS Ascanius. Apparently a synonym of ARGENTINA L.

1834

176. AGASSIZ (1834). *Abgerissene Bemerkungen über fossile Fische.* Neues Jahrb. Mineralogie.
LOUIS AGASSIZ.

Ophiopsis Agassiz, 385; orthotype O. PROCERUS Ag. (fossil).

Saurichthys Agassiz, 386; orthotype S. APICALIS Ag. (fossil).

Propterus Agassiz, 386; orthotype P. MICROSTOMUS Ag. (fossil).

Caturus Agassiz, 387; orthotype URÆUS NUCHALIS Ag. (fossil). Substitute for URÆUS Ag., preoccupied.

Macrosemius Agassiz, 387; orthotype M. ROSTRATUS Ag. (fossil).

Belonostomus Agassiz, 388; orthotype B. TENUIROSTRIS Ag. (fossil).
Dercetis Agassiz, 389; orthotype D. SCUTATUS Ag. (fossil).

177. AGASSIZ (1834). *Verhandlungen der Gesellschaft des vaterländischen Museums.* Böhmen.
LOUIS AGASSIZ.
Halec Agassiz, 67; orthotype H. STERNBERGI Ag. (fossil).

178. AGASSIZ (1834). *Recherches sur les poissons fossiles.* V (part), 1834.
LOUIS AGASSIZ.
Acanthonemus Agassiz, V, 27; orthotype A. FILAMENTOSUS Ag. (fossil).

179. COSTA (1834). *Cenni zoologici, ossia descripzione sommaria delle specie nuovi di animali discoperti in diverse contrade del regno, nell' anno 1834.* (Naples.)
ORONZIO GABRIELE COSTA (1787-1867).
Branchiostoma Costa, 49; orthotype LIMAX LANCEOLATUS Pallas; replaces AMPHIOXUS (Gray) Yarrell.
Nemotherus (Risso) Costa, fig. 9; orthotype N. ERYTHROPTERUS Risso. A synonym of TRACHYPTERUS (not verified).

180. GRIFFITH (1834). *The Class Pisces, Arranged by the Baron Cuvier, with Supplementary Additions by Edward Griffith and Charles Hamilton Smith.*
Carangus Griffith, 335; tautotype SCOMBER CARANGUS Bloch = SCOMBER HIPPOS L. A synonym of CARANX Lac. as restricted; same as CARANGUS Grd.

181. HENLE (1834). *Sur le Narcine, nouveau genre de raies electriques,* etc. Ann. Sci. Nat., II. (Also in German as *Ueber Narcine.*)
FRIEDRICH GUSTAV JACOB HENLE (1807-1855).
Narcine Henle, 31; orthotype TORPEDO BRASILIENSIS Ölfers.

182. MÜNSTER (1834). *Knochenhöhle bei Rabenstein,* etc. Neues Jahrb. Miner.
GEORG MÜNSTER.
Undina Münster, 539; orthotype U. PENICILLATA Münster (fossil).

183. QUOY & GAIMARD (1834). *Voyage de découvertes de "l'Astrolabe,"* etc. (1826-29), *sons le commandement de M. J. Dumont d'Urville.*
JEAN RÉNÉ CONSTANTIN QUOY (1707-1778); PAUL GAIMARD (1790-1858).
Aspidontus (Cuvier) Quoy & Gaimard, 719; orthotype A. TÆNIATUS Cuv. & Val. An ally of PETROSCIRTES Rüppell, distinguished by the lack of elevation of the anterior dorsal spines.

1835

184. AGASSIZ (1835). *Ueber die Familie der Karpfen.* Mem. Soc. Nat. Neufchâtel, I.
LOUIS AGASSIZ.
Acanthopsis Agassiz, I, 36; orthotype COBITIS TÆNIA L. Name preoccupied. A synonym of COBITIS L.
Phoxinus Agassiz, 37; orthotype LEUCISCUS PHOXINUS L. Same as PHOXINUS Raf. 1820.
Rhodeus Agassiz, I, 37; orthotype CYPRINUS AMARUS Bloch.
Chondrostoma Agassiz, I, 38; orthotype CYPRINUS NASUS L.
Aspius Agassiz, I, 38; tautotype CYPRINUS ASPIUS L.
Pelecus Agassiz, I, 39; orthotype CYPRINUS CULTRATUS L.

185. AGASSIZ (1835). *Recherches sur les poissons fossiles.* II.
Louis Agassiz.

Plectrolepis (Agassiz) II, 44; orthotype P. rugosus Ag. (fossil). Name only;
validated by Egerton, 1850. A synonym of Eurynotus Ag. = Notacmon
Gistel, 1848.

Megalodon Agassiz, II, 55; logotype M. lewesiensis Ag. (fossil). Name preoccupied; a synonym of Pachyrhizodus.

Enchodus Agassiz, II, 55 (feuilleton); orthotype Esox lewesiensis Mantell (fossil).

Macropoma Agassiz, II, 55 (euilleton); orthotype M. mantelli Ag. (fossil).

Diplopterus Agassiz, II, 113; orthotype D. macrocephalus Ag. (fossil). Name
preoccupied; replaced by Diplopterax McCoy (1885).

Pleiopterus Agassiz, II, 113; orthotype Osteolepis macrolepidotus Val. (fossil).
A synonym of Astrolepis.

Cheiracanthus Agassiz, II, 125; orthotype C. murchisoni Ag. (fossil).

Cheirolepis Agassiz, II, 128; orthotype C. trailli Ag. (fossil).

Cephalaspis Agassiz, II, 135; orthotype C. lyelli Ag. (fossil).

Eurynotus Agassiz, II, 153; logotype E. crenatus Ag. (fossil). Name preoccupied;
replaced by Notacmon Gistel (1848).

Platysomus Agassiz, II, 6, 161; logotype Stromateus gibbosus Blainv. (fossil).
Name held to be preoccuped by Platysoma Leach, 1817; replaced by Scrotes
Gistel (1848); Uropteryx (Ag.) Woodw. (1907) may be also regarded as
a substitute.

186. AGASSIZ (1835). *Ueber das der Glarner-Schiefer-Formation nach ihren
Fischresten.* Neues Jahrb. Mineral.
Louis Agassiz.

Carangopsis Agassiz, 293; orthotype C. latior Ag. (fossil). Names only; validated
in 1844.

Urosphen Agassiz, 293; orthotype U. fistularis Ag. (fossil). Names only; validated in 1844.

Amphistium Agassiz, 294; orthotype A. paradoxum Ag. (fossil). Names only;
validated in 1844.

Xiphopterus Agassiz, 295; orthotype Esox falcatus Volta (fossil).

Pristigenys Agassiz, 299; orthotype P. macrophthalmus Ag. (fossil). Name only;
described in 1839 (*Poiss. fossiles*, IV, 139).

Cœlogaster Agassiz, 304; orthotype C. analis Ag. (fossil) (name only), and hence
dating only from later usage.

Platinx Agassiz, 304; orthotype P. elongatus Ag. (fossil). A synonym of Monopteros Volta (1794).

Enchelyopus Agassiz, 307; orthotype E. tigrinus Ag. (fossil); name only; validated by Agassiz in 1844. Preoccupied; replaced by Paranguilla Blkr.
(1864).

187. CUVIER & VALENCIENNES (1835). *Histoire naturelle des poissons.* X.
(The series continued after the death of Cuvier, in 1832, until 1848, by
Achille Valenciennes.)

Keris Cuvier & Valenciennes, 304, 1835; orthotype Keris anguinosus Cuv. & Val.
Often written Ceris: the young of Naso Lac.

188. GOTTSCHE (1835). *Die seeländischen Pleuronectes-Arten.* Wiegmann's
Archiv. der Naturgeschichte, II.
Carl Moritz Gottsche (1808-1892).

Limanda Gottsche, 100; orthotype Pleuronectes limanda L.

Microstomus Gottsche, 150; orthotype M. latidens Gottsche. Regarded as preoccupied by Microstoma Risso, 1826; replaced by Cynicoglossus Bon.

Glyptocephalus Gottsche, 156; orthotype Pleuronectes saxicola Faber = P. cynoglossus L.

Hippoglossoides Gottsche, 168; orthotype H. limanda Gottsche = Pleuronectes limandoides Bloch.

Zeugopterus Gottsche, 178; orthotype Pleuronectes hirtus Abildgard.

189. JOANNIS (1835). *Observations sur les poissons du Nil et descriptions de plusieurs espèces nouvelles,* etc. Mag. Zool., 1835.

L. de Joannis.

Mochokus Joannis, pl. 8; orthotype M. niloticus Joannis.

190. MEYEN (1835). *Reise in Peru.* I.

J. Meyen.

Pygidium Meyen, I, 475; orthotype P. fuscum Meyen.

191. MÜLLER (1835). *Vergleichende Anatomie der Myxinoiden, der Cyclostomen mit durchbohrten Gaumen.* (Reprinted in *Arch. Naturg.,* Berlin, for 1846, 247.)

Johannés Müller (1801-1858).

Sturio Müller, 77. (Sturgeons without opercle.) Type presumably Acipenser sturio L. A synonym of Acipenser Ag.

Scaphirhynchus (Heckel) Müller, 77; footnote only; defined as lacking the quadrate bone; logotype Acipenser platorhynchus Raf.

Bdellostoma Müller, 79 (248, reprint); orthotype B. hexatrema Müller; a substitute for Heptatrema Dum. as not correct in fact. Both are synonyms of Eptatretus Cloquet, 1818.

192. DE LA PYLAIE (1835). *Recherches en France, sur les poissons de l'océan, pendant les années 1832 et 1833.* Congrès Scientifique de France tenue a Poitiers, 1834.

M. de la Pylaie (de Fougères).

This singularly. useless paper has been totally overlooked by all natural ists except Dr. Gill, who has noted Pylaie's use of the name Argyrosomus (1835), at a date prior to that of Agassiz (1850). Besides the "new genera" enumerated below, the following new species are indicated by De la Pylaie:

Spinax vitulinus p. 527 ("tête qui rappelle un muffle dé veau").

Scyllium atlanticum (527) ("trop différente," etc.).

Scymnus aquitanensis (527) (la Sénille ou Chénille de l'Île Dieu).

Squalraja cervicata (527) ("pour le distinguer de l'ordinaire, que est acephale").

Squalraia acephala (527) (= Squatina squatina).

Torpedo elliptica (527) ("d'après la structure de ses deux pectorales").

Rajabatis macrophalla (528) (Differing in length of clasper).

Rajabatis microphalla (528). "Neuf autres espèces complètent ce genre; ce sont les R. triptera, polyacantha, variegata, tigrina, monolifera, florigera, melumpseca, mosaica, undulata.

Hippocampus atrichus (529) ("espèce de l'Ocean").

Hippocampus jubatus (529) ("des filamens qui composent, le long de son cou, une espèce de crinière *peu fournie*").

Aptocyclus ostraciodes (529). "Coffret." "Nouvelle espèce encore très bien caractérisée."

ANGUILLA VULGARIS, "mes variétés MACROCEPHALA, ORNITHORHINCHA, PLATYURA,
  également nouvelles pour la science" (529).

MURÆNA HIPPOCREPIS (529) "commune dans certain reservoirs des Salines, à l'Île de
  Noirmontier".

AMMODYTES PICTAVUS (529a) "qui s'en distingue par l'absence de dents, par le
  nombre des rayons et par la couleur de quelques parties de son corps".

MORRHUA VULGARIS CALLARINA (530), a variety "rare à ile Dieu," etc.

GADUS ZONATUS (530), a "variety" of "GADUS BARBATUS".

MUSTELA RUBENS (530), "La Coche Rouge".

MUSTELA QUINQUECIRRHA (530), "a cinq barbillons".

BLENNIUS GATTORUGINE BITENTACULARIS (530).

BLENNIUS PICTAVUS (530), "nouvelles espèces".

BLENNIUS PIETI (530).

PTEROZYGUS BIEVRII (530).

GOBIUS NIGRICANAS (531), "habite les roches de la côte".

THYNNUS ATLANTICUS (531a), the albacore.

THYNNUS OCEANICUS (531a), the tunny.

TRIGLA MEGALOPHTHALMA (531a), "d'après la grosseur remarquable des yeux".

CRENILABRUS PHENODONTUS (532), "a grand dent".

CRENILABRUS MARMORATUS (532).

CRENILABRUS OXYCEPHALUS (532), "n'a pas de tache noir sur la queue; son museau
  est plus pointée".

SARGUS PULCHELLUS (532), "Ruscain, d'après sa chair rouge".

EXOCALLUS INSIGNIS (532a), new name for Boops BOOPS (L).

PERCA OLONNENSIS (532a) ("nouvelle espèce").

PERCA INERMIS (532a). Name only.

ARGYROSOMUS PROCERUS (532a), "Baie de Bourgneuf".

CLUPEA ELONGATA (532a) "une variété N. du CLUPEA ALOSA, le corps beaucoup plus
  allongé que dans l'alose ordinaire".

APTEROGASTERUS ROSTELLATUS (532a), "Jacquine" "D'après la forme de son museau".

LEUCISUS PICTAVA (533). "Chévénau" "le corps est moins large" (que dans le
  LEUCISCUS CHUB de Pennant).

LEUCISCUS OBTUSUS (533) "able ou petit verdon sans taches".

GOBIO PHOXINOIDES (533), "Goujon véronette".

COBITIS PARISIENSIS (534), "Loche Franche".

COBITIS PICTAVA (534), "repandu dans nos rivières".

SOLEA CUNEATA (534), "Sole sétan": "espèce mediocre, mais parfaitement carac-
  terisée".

RHOMBUS GALLINULA (534) ,"poulette de mer".

Squalraia De la Pylaie, 526; orthotype SQUALRAIA ACEPHALA Pylaie = RAJA SQUA-
  TINA L. A synonym of SQUATINA; not SQUALORAIAIA Riley (1826).

Rajabatis De la Pylaie, 528; logotype RAJA MOSAICA Lac. (most of the other
  species named being left undescribed). A synonym of RAJA, possibly not
  intended as a generic name.

Aptocylus De la Pylaie, 529; orthotype "CYCLOGASTERUS VENTRICOSUS" Pallas (error
  for "CYCLOPTERUS"). Replaces CYCLOPTERICHTHYS Steind. (1861).

Pterozygus De la Pylaie, 530; orthotype P. BIEVRII De la Pylaie. "Une loche qui
  presente le singulier caractère d'avoir ses nageoires pectorales et thoracins
  réunies par paires." Probably a monstrous form of COBITIS L.

Exocallus De la Pylaie, 532 (a); orthotype E. INSIGNIS Pylaie (SPARUS BOOPS L.).
  Name a substitute for BOOPS, of which it is a needless synonym.

Apterogasterus De la Pylaie, 532a; orthotype A. ROSTELLATUS Pylaie, "la Jacquine". (Not identified.)

Argyrosomus De la Pylaie, 534; orthotype A. PROCERUS De la Pylaie. A synonym of SCLÆNA L. as restricted by Cuvier; same as PSEUDOSCIÆNA Blkr.

193. RÜPPELL (1835). *Neue Wirbelthiere zu der Fauna von Abyssinien gehörig.*
2 vols.
EDUARD RÜPPELL.

Enneapterygius Rüppell, 2; orthotype E. PUSILLUS Rüppell.

Gazza Rüppell, 4; orthotype G. EQUULÆFORMIS Rüppell.

Pseudochromis Rüppell, 8; logotype P. OLIVACEUS Rüppell.

Halichœres Rüppell, 14; logotype H. BIMACULATUS Rüppell (as restricted by Jordan & Snyder). Regarded as preoccupied by HALICHŒRUS, a genus of seals; replaced by HEMIULIS Sw. (1839) and by CHŒROJULIS Gill (1864).

Lutodeira (Kuhl) Rüppell, 18; orthotype MUGIL CHANOS Forskål. A synonym of CHANOS Lac.

Xenodon Rüppell, 53; orthotype BALISTES NIGER Lac. Preoccupied in reptiles; replaced by ZENODON Sw. (1839), by ODONUS Gistel (1848), by ERYTHRODON Rüppell (1852) and by PYRODON Kaup (1853). The validity of the oldest of these substitute names, ZENODON, may be questioned. See record of Swainson (1839).

Nebrius Rüppell, 62; orthotype NEBRIUS CONCOLOR Rüppell. Name regarded as preoccupied by NEBRIA Latreille and NEBRIS Cuv.; replaced by NEBRODES Garman. But the root is different: νεβρϊοι of Aristotle, referring to some shark.

Uropterygius Rüppell, 83; orthotype N. CONCOLOR Rüppell. This would be a synonym of GYMNOMURÆNA Lac., as restricted by Bleeker, 1864, but not as limited by Kaup, 1856.

Sphærodon Rüppell, 113; orthotype SCIÆNA GRANDOCULIS Forskål. A synonym of MONOTAXIS Bennett.

Pristotis Rüppell, 128; orthotype P. CYANOSTIGMA Rüppell. A synonym of POMACENTRUS Lac.

194. SCHULTZE (1835). *Versammlung von Naturforschers in Bonn.*
CARL AUGUST SIGMUND SCHULTZE (1795-1877).

Hexabranchus Schultze; orthotype presumably BDELLOSTOMA HEXATREMA Müller. If so, a synonym of EPTATRETUS Cloquet. According to Agassiz, a genus of lampreys, presumably one with six gills; no further details available.

1836

195. AGASSIZ (1836). *Recherches sur les poissons fossiles.* III.
LOUIS AGASSIZ.

Zygobatis Agassiz, III, 79; orthotype Z. JUSSIEUI Ag. A synonym of RHINOPTERA Kuhl.

Odontaspis Agassiz, III, 87; orthotype O. FEROX Ag. A synonym of CARCHARIAS Raf. as now limited.

Spinacorhinus Agassiz, III, 94; orthotype S. POLYSPONDYLUS Ag. (fossil). A synonym of SQUALORAIA Riley.

Goniodus Agassiz, III, 183; orthotype SQUALUS SPINOSUS Gmelin = SQUALUS BRUCUS Bonnaterre. A synonym of ECHINORHINUS Blainv.

Amblyurus Agassiz, III, 220; orthotype A. MACROSTOMUS Ag. (fossil). A synonym of DAPEDIUM Leach.

196. AGASSIZ & HIBBERT (1836). *On the Fresh-water Limestones of Burdie-house,* etc. Trans. Roy. Soc. Edinburg.
LOUIS AGASSIZ; SAMUEL HIBBERT (1782-1848).
Megalichthys Agassiz, VIII, 202; orthotype M. HIBBERTI Ag. (fossil).

197. CUVIER & VALENCIENNES (1836). *Histoire naturelle des poissons.* XI.
ACHILLE VALENCIENNES.
Omobranchus (Ehrenberg) Cuvier & Valenciennes, XI, 87; orthotype BLENNE-CHIS FASCIOLATUS Cuv. & Val. A synonym of ASPIDONTUS Cuv.
Cestræus Cuvier & Valenciennes, XI 157; type CESTRÆUS PLICATILIS Cuv. & Val. Near AGONOSTOMUS Bennett, 1830. Not CESTREUS Klein, nor of McClelland, nor Gronow.
Dajaus Cuvier & Valenciennes, XI, 164; haplotype MUGIL MONTICOLA Bancroft; near AGONOSTOMUS Bennett, 1800.
Nestis Cuvier & Valenciennes, XI, 167; logotype NESTIS CYPRINOIDES Cuv. & Val. = AGNOSTOMUS TELFAIRI Bennett, 1830. A synonym of AGONOSTOMUS Bennett.
Blennechis Cuvier & Valenciennes, XI, 279; logotype B. FILAMENTOSUS Cuv. & Val. A synonym of PETROSCIRTES Rüppell.
Chasmodes Cuvier & Valenciennes, XI, 295; logotype BLENNIUS BOSQUIANUS Lac.
Cristiceps Cuvier & Valenciennes, XI, 402; haplotype C. AUSTRALIS Cuv. & Val.
Gunnellus Cuvier & Valenciennes, XI, 419; tautotype BLENNIUS GUNNELLUS L. A synonym of PHOLIS (Gronow) Scopoli (1777), and of MURÆNOIDES Lac. (1803). Often spelled GUNELLUS.

198. HECKEL (1836). *Scaphirhynchus, eine Fischgattung aus der Ordnung der Chondropterygier mit freien Kiemen.* Ann. Wien. Mus., I.
JOHANN JAKOB HECKEL (1790-1857).
Scaphirhynchus Heckel, I, 71; orthotype S. RAFINESQUEI = ACIPENSER PLATORYN-CHUS Raf. Regarded preoccupied by SCAPHORHYNCHUS Maximilian, 1831, a genus of birds; replaced by SCAPHIRHYNCHOPS Gill (see Müller 1835).

199. HECKEL & FITZINGER (1836). *Monographische Darstellung der Gattung Acipenser.* Ann. Wien. Mus., I.
JOHANN JAKOB HECKEL; LEOPOLD J. FITZINGER (1802-1884).
Antaceus Heckel & Fitzinger, 293; orthotype ACIPENSER SCHYPA Eichw.
Lioniscus Heckel & Fitzinger, 370; orthotype ACIPENSER GLABER Fitzinger.

200. LIÉNARD (1836). *Sur quelques poissons de l'île Maurice.* L'Institut, IV.
ÉLIZÉ LIÉNARD.
Platysoma Liénard. Name preoccupied; said to be near DULES (KUHLIA). (Very doubtful; here quoted from Scudder. I fail to find the name in question elsewhere.)

201. MÜNSTER (1836). *Capitosaurus in Keupersandstein; Gyrodus multidens in Jurakalk; achte Belonostomus-Art; Lebias und Cyprinus in Braunköhle im Fichtelgebirge,* etc Neues Jahrb. Mineral.
GEORG MÜNSTER.
Aellopos Münster, 581; orthotype A. ELONGATUS Münster (fossil). An extinct ally of RHINOBATUS.

202. RICHARDSON (1836). *Back's Narrative of an Arctic Land Expedition.*
JOHN RICHARDSON (1787-1865)
Stenodus Richardson (appendix), 521; orthotype SALMO MACKENZIEI Rich. Replaces LUCIOTRUTTA Gthr. (1866).

203. RICHARDSON (1836). *Fauna Borcali-Americana.* III.
JOHN RICHARDSON.

Temnistia Richardson, III, 59; orthotype BLEPSIAS VENTRICOSUS Eschscholtz. A synonym of HEMILEPIDOTUS Cuv.

Salvelinus (Nilsson) Richardson, III, 169; tautotype SALMO SALVELINUS L. First use in the singular of Nilsson's group name "SALVELINI". Replaces BAIONE Dekay (1842) and UMBLA Rapp (1854).

204. RÜPPELL (1836). *Neuer Nachtrag zu Beschreibungen und Abbildungen neuer Fische, im Nil entdeckt.* Museum Senckenberg, II.
WILHELM PETER EDUARD SIMON RÜPPELL.

Labeobarbus Rüppell, II, 14; haplotype LABEOBARBUS NEDGIA Rüppell. Probably not distinct from TOR Gray, 1833.

Varicorhinus Rüppell, II, 14; orthotype V. BESO Rüppell = LABEO VARICORHINUS Cuv. & Val. VARICORHINUS Rüppell is said by Berg to include CAPOËTA, SCAPHIODON, DILLONIA, GYMNOSTOMUS, ONYCHOSTOMA, PTEROCAPOËTA and ACAPOËTA. Some of these have doubtless standing as subgenera.

205. YARRELL (1836). *A History of British Fishes,* etc. 2 vols.
WILLIAM YARRELL (1784-1856).

Amphioxus (Gray) Yarrell, 468; orthotype LIMAX LANCEOLATUS Pallas. A synonym of BRANCHIOSTOMA Costa.

1837

206. AGASSIZ (1837). *Recherches sur les poissons fossiles.* III (part), V (part).
LOUIS AGASSIZ.

Onchus Agassiz, III, 6; orthotype O. MURCHISONI Ag. (fossil).

Ctenacanthus Agassiz, III, 10; logotype C. MAJOR Ag. (fossil).

Oracanthus Agassiz, III, 13; orthotype O. MULLERI Ag. (fossil).

Gyracanthus Agassiz, III, 17; orthotype G. FORMOSUS Ag. (fossil).

Tristychius Agassiz, III, 21; orthotype T. ARCUATUS Ag. (fossil).

Ptychacanthus Agassiz, III, 22; orthotype P. SUBLÆVIS Ag. (fossil). A synonym of TRISTYCHIUS Ag.

Sphenacanthus Agassiz, III, 23; orthotype S. SERRULATUS Ag. (fossil).

Nemacanthus Agassiz, III, 25; logotype N. MONILIFER Ag. (fossil).

Leptacanthus Agassiz, III, 27; logotype L. SEMISTRIATUS Ag. (fossil). Fossil spines of a CHIMÆROID, perhaps synonymous with GANODUS.

Asteracanthus Agassiz, III, 31; orthotype A. ORNATISSIMUS Ag. (fossil).

Pristeacanthus Agassiz, III, 35; orthotype P. SECURIS Ag. (fossil).

Myriacanthus Agassiz, III, 37; logotype M. PARADOXUS Ag. (fossil).

Hybodus Agassiz III, 41; orthotype H. PLICATILIS Ag. (fossil).

Leiacanthus Agassiz, III, 55; orthotype L. FALCATUS Ag. (fossil).

Pleuracanthus Agassiz, III, 66; orthotype P. LÆVISSIMUS Ag. (fossil). Name preoccupied; probably replaced by ORTHACANTHUS Ag.; perhaps the spine of the forms known from the teeth, as DIPLODUS, DICRANODUS, DITTODUS, etc.

Hypsodon Agassiz, V, 100; logotype MEGALODON LEWESIENSIS Mantell (fossil). Specimens not all of one genus, some belonging to PORTHEUS Cope (1872).

207. BONAPARTE (1834-1841). *Iconografia della fauna italica, per le quattro classi degli animali vertebrati.* Tome, III. *Pesce.*
CARLO LUCIANO BONAPARTE (Principe de Canino e de Musignano).

This sumptuous and finely illustrated work was issued in separate fascicles from 1832 to 1842. The fishes seem to begin with fascicle VI in 1837. The last are in fascicle 160, probably issued in 1841. There is no paging, and exact dates, except for a few fascicles, are inaccessible to me. I therefore divide the work into three parts, 1837, 1840 and 1841. Fascicles 156 to 160 are subsequent to Swainson's *Natural History*, 1839, from which they quote. So far as I know, no question of nomenclature hinges on these uncertain dates. Troschel (*Archiv. Naturg.*) gives the date 1884 for fascicles "Lieferungen" VI to XI, 1837 for XVI to XX, 1838 for XXVI to XXIII, 1839 for XXIV to XXVI, and 1840 for XXVII; but these "deliveries" seem not to correspond to the fascicles.

Orectolobus Bonaparte, fasc. 7; orthotype SQUALUS BARBATUS Gmelin (date probably 1837). Replaces CROSSORHINUS M. & H.

Cerna Bonaparte fasc. 10*; orthotype PERCA GIGAS Brünnich = LABRUS GUAZA L. (1837). A synonym of EPINEPHELUS Bloch, as now resticted.

Cynicoglossus Bonaparte, fasc. 19; orthotype PLEURONECTES CYNOGLOSSUS Nilsson (not of L.) = PLEURONECTES KITT Walbaum. Equivalent to MICROSTOMUS Gottsche, which it replaces if MICROSTOMUS is regarded as preoccupied by MICROSTOMA Cuv. (1817).

Batis Bonaparte, fasc 21*; orthotype RAJA RADULA De la Roche (called DASYBATUS in fasc. 68). A synonym of RAJA L.

Læviraja Bonaparte, fasc. 21*; orthotype RAJA OXYRHYNCHUS L.

Microchirus Bonaparte, fasc. 26*; orthotype PLEURONECTES MICROCHIRUS De la Roche. Same as MONOCHIRUS and MONOCHIR Cuv., not of Rafinesque; replaces BUGLOSSUS Gthr.

Monochirus Bonaparte, fasc., 26*; orthotype SOLEA MONOCHIR Bon. Same as MONOCHIRUS Raf.

Pampus Bonaparte, 48; orthotype (not named) but plainly STROMATEUS CANDIDUS Cuv. & Val. Replaces STROMATEIODES Blkr., 1851.

"Diamo il nome di PAMPUS al secundo sottogenere in cui accogliamo quelle specie che non hanno pinne ventralī e portano innanzo ai raggi dorsale e dell' anale parecchie spine terminale superiormente da una lamina tagliente." (Bonaparte.) These peculiar trenchant spines are found in all the species called STROMATEOIDES, though buried in the flesh in the adult.

Dasybatus Bonaparte, fasc. 68; orthotype RAJA RADULA De la Roche. Called BATIS on fascicle 21; also written DASYBATIS. A synonym of RAJA L.; not DASYATIS Raf. (1837).

Membras Bonaparte, fasc. 91; orthotype an "exotic species", evidently ATHERINA MARTINICA Cuv. & Val.

Menidia Bonaparte, fasc. 91; orthotype an "exotic species," evidently ATHERINA MENIDIA L.

Hepsetia Bonaparte, fasc. 91; orthotype ATHERINA BOYERI Risso.

Cyprinopsis (Fitzinger) Bonaparte, fasc. 92; orthotype CYPRINUS CARASSIUS L. (1838). A synonym of CARASSIUS Nilsson.

Squalius Bonaparte, fasc. 96; logotype SQUALIUS TYBERINUS Bon. = LEUCISCUS SQUALUS Bon. A synonym of LEUCISCUS Cuv.

Scardinius Bonaparte, fasc. 96; orthotype LEUCISCUS SCARDAFA Bon.

Telestes Bonaparte, fasc. 103; orthotype TELESTES MUTICELLUS Bon. (1839).

Plagiusa Bonaparte, fasc. 120; orthotype (by implication) PLEURONECTES PLAGIUSA L. A substitute for PLAGUSISA Cuv.; preoccupied ("non PLAGUSIA, nome di crustaceo"). A synonym of SYMPHURUS Raf. (1810).

208. CANTRAINE (1837). *Memoire sur un poisson nouveau (Acanthoderma temmincki) trouvé dans le canal de Messine.* Journ. de Bruxelles.
FRANÇOIS JOSEPH CANTRAINE (1801-1863).

Acanthoderma Cantraine, X, 1-19; orthotype A. TEMMINCKI Cantraine. A synonym of RUVETTUS Cocco.

209. CUVIER & VALENCIENNES (1837). *Histoire naturelle des poissons.* XII.
ACHILLE VALENCIENNES.

Oplopomus (Ehrenberg) Cuvier & Valenciennes, XII, 66 (reference in synonymy);
orthotype O. PULCHER Ehrenb. = GOBIUS OPLOPOMUS Cuv. & Val. Replaces
CENTROGOBIUS Blkr., 1874.

Cryptocentrus (Ehrenberg) Cuvier & Valenciennes, XII, 111; orthotype C.
MELEAGRIS Ehrenb. = GOBIUS CRYPTOCENTRUS Cuv. & Val. Replaces PARA-
GOBIUS Blkr. Ehrenberg's manuscript, *Zoologia: Pisces,* quoted by Cuvier and
Valenciennes, seems never to have been published.

Amblyopus Cuvier & Valenciennes, XII, 139; logotype TÆNOIDES HERRMANNIANUS
Lac. A synonym of TÆNIOIDES Lac. 1803

Apocryptes (Osbeck) Cuvier & Valenciennes, XII, 143; orthotype GOBIUS PECTINI-
ROSTIS Gmelin = APOCRYPTES CHINENSIS Osbeck, 1757, and in later translations (after 1758). APOCRYPTES Cuv. & Val. is plainly a revival of Osbeck's
genus, and the name ought not legitimately to be transferred to another group
as has been done by Günther and Bleeker. Bleeker makes GOBIUS BATO Ham.
the type of APOCRYPTES Cuv. & Val., as this is the first species named by these
authors under APOCRYPTES. But it would seem that Osbeck's haplotype (A.
CHINENSIS) must be accepted in the revival of his genus, in which case
APOCRYPTES (Osbeck) Cuv. & Val. replaces BOLEOPHPHALMUS Cuv. & Val.
(p. 198). As against the views here expressed, it may be urged that the pre-
Linnæan name APOCRYPTES Osbeck has no standing and APOCRYPTES Cuv. &
Val. is really based on GOBIUS BATO, a species actually in hand, as the description shows. APOCRYPTES CHINENSIS (PECTINIROSTRIS) was included by error,
it being a member in fact of the genus BOLEOPHTHALMUS Cuv. & Val., as
later writers have recognized. It may be urged further that the first designation of type of APOCRYPTES is that of Bleeker, who places it on APOCRYPTES
BATO.

A further point is that accepting APOCRYPTES in the Osbeckian sense leaves
the genus of Cuvier and Valenciennes without a name, unless we accept
GOBILEPTES Sw. But the scales in GOBILEPTES are described as large, while
those of APOCRYPTES BATO are "exceedingly small". Swainson obviously drew
his GOBILEPTES from the description of GOBIUS ACUTIPINNIS Cuv. & Val., later
abandoning the genus, which he does not again notice. Among the numerous
genera of sharp-tailed gobies, I have not found the proper place for (GOBI-
LEPTES ACUTIPINNIS. This generic name, GOBILEPRES Sw., with *no species
mentioned,* is interpreted by Bleeker as referring to GOBIUS BATO and hence
is regarded by him as a synonym of APOCRYPTES.

Boleophthalmus Cuvier & Valenciennes, XII, 198; orthotype GOBIUS BODDÆRTI Pallas. Perhaps a synonym of APOCRYPTES (Osbeck) Cuv. & Val., which see
above.

Philypnus Cuvier & Valenciennes, XII, 235; haplotype GOBIOMORUS DORMITOR Lac.
Same as GOBIOMORUS Lac., as first restricted by Jordan. Gill has later limited the name to VALENCIENNEA Blkr.

Hemerocœtes Cuvier & Valenciennes, XII, 311; haplotype CALLIONYMUS ACAN-THORHYNCHUS Forster.

Platyptera (Kuhl & Van Hasselt) Cuvier & Valencenniences, XII, 320; orthotype P. ASPRO Van Hasselt. Name preoccupied, replaced by RHYACICHTHYS Boulenger, 1901.

Halieutæa Cuvier & Valenciennes, XII, 455; haplotype HALIEUTÆA STELLATA Cuv. & Val.

210. EGERTON (1837). *(Catalogue of Fossil Fishes.)*
PHILIP DE MALPAS GREY EGERTON (1806-1881).

Isurus (Agassiz) Egerton; orthotype I. MACROURUS Ag. (fossil). Name only; validated by Agassiz, 1844. Name preoccupied; replaced by SAURICHTHYS Woodw., 1901.

211. FITZINGER (1837). *Vorläufiger Bericht über eine höchst interessante Ent-deckung Dr. Natterers in Brasil. Oken's Isis.*
LEOPOLD JOSEPH FRANZ FITZINGER.

Lepidosiren Fitzinger, 379; orthotype L. PARADOXA Fitzinger.

212. FRIES (1837). *Pterycombus, ett nytt fiskslägte från Ishafvet.* Handl. K. Vetensk. Akad. Stockholm.
BENGT FREDRIK FRIES (1799-1839).

Pterycombus Fries, 14; orthotype P. BRAMA Fries.

213. GEINITZ (1837). *Thüringischen Muschelkalk.*
HANS BRUNO GEINITZ (1814-1900).

Acrodus (Agassiz) Geinitz, 21; orthotype A. GAILLARDOTI Ag. (fossil). (Also in *Poissons fossiles*, III, 139, 1838.)

214. HECKEL (1837). *Ichthyologische Beiträge.* Ann. Wiener Mus., II.
JOHANN JAKOB HECKEL.

Scorpænopsis Heckel, II, 158; orthotype SCORPÆNA NESOGALLICA Cuv. & Val.

Trachydermus Heckel, 11, 159; orthotype T. EFASCIATUS Heckel. Name three times preoccupied as TRACHYDERMA; replaced by CENTRIDERMICHTHYS Rich.

215. KUTORGA (1837). *Beitrag zur Geognosie und Paleontologie Dorpats und seiner nächsten Umgebungen.* St. Petersbourg, IV, for. 1835. Zweiter Bei-trag, 1837.
STEPAN SEMENOVICH KUTORGA (1812-1861).

Ichthyosauroides Kutorga, II, 35; orthotype TRIONYX SPINOSUS Kutorga (fossil). Name rejected as inappropriate, the genus being supposed to be reptilian; an inadequate reason for change of name to HETEROSTIUS Asmuss, 1856.

216. MÜLLER & HENLE (1837). *Gattungen der Haifische und Rochen nach ihrer Arbeit, "Ueber die Naturgeschichte der Knorpelfische".* Ber. Akad. Wiss. Berlin.
JOHANNES MÜLLER; FRIEDRICH GUSTAV JAKOB HENLE.

Tæniura Müller & Henle, 117; logotype TRYGON ORNATA Gray = RAJA LYMMA Forskål.

217. MÜLLER & HENLE (1837). *On the Generic Characters of Cartilaginous Fishes, with Descriptions of New Genera.* Mag. Nat. Hist., II.
JOHANNES MÜLLER; FRIEDRICH GUSTAV JAKOB HENLE.

Hemitrygon Müller & Henle, 90; orthotype TRYGON BENNETTI M. & H.

Ginglymostoma Müller & Henle, 113; orthotype SQUALUS CIRRATUS Gmelin.

Scoliodon Müller & Henle, 114; orthotype CARCHARIAS LATICAUDUS M.. & H.

Gymnura Müller & Henle, 117; orthotype RAJA ASPERRIMA Bl. & Schn. Name preoccupied; replaced by UROGYMNUS M. & H.

Himantura Müller & Henle, 400; orthotype RAJA UARNAK Forskål.

Urogymnus Müller & Henle, 434; logotype RAJA ASPERRIMA Bl. & Schn. Replaces ANACANTHUS Ehrenb. and GYMNURA M. & H., both preoccupied.

218. REDFIELD (1837). *Fossil Fishes of Connecticut and Massachusetts with a Notice of an Undetermined Genus.* Ann. Lyc. Nat. Hist. N. Y.

JOHN HOWARD REDFIELD (1825-1895).

Catopterus Redfield, IV, 39; orthotype C. GRACILIS Redfield (fossil). Preoccupied by CATOPTERUS Ag. (1833); replaced by REDFIELDIUS Hay (1899).

219. REINHARDT (1837). *Bythites, novum genus piscium.* Danske Vid. Selsk. Afh., VI.

JOHANNES CHRISTOPHER HAGEMANN REINHARDT.

Bythites Reinhardt, VI, 77; orthotype B. FUSCUS Reinh.

220. SMITH (1837). *On the Necessity for a Revision of the Groups Included in the Linnæan Genus Squalus.* Proc. Zool. Soc. London.

ANDREW SMITH.

Poroderma Smith, 85; logotype SQUALUS AFRICANUS Gmelin.

Catulus Smith, 85; orthotype SCYLLIUM CAPENSE Smith. Name preoccupied in insects, partly equivalent to SCYLLIORHINUS Blainv., 1816.

Hemiscyllium Smith, 86; logotype SQUALUS OCELLATUS Bonneterre.

Carcharodon Smith, 86; orthotype CARCHARODON CAPENSIS Smith = SQUALUS CARCHARIAS L.

Leptocarias Smith; *Ill. S. Afr. Zool.;* orthotype TRIÆNODON SMITHI Gray; quoted by Agassiz, *Nom. Zool.,* 1846; hence prior to LEPTOCARCHARIAS Gthr., 1870. Replaces LEPTOCARCHARIAS.

221. THOMPSON (1837). *Notes Relating to the Natural History of Ireland, with a Description of a New Genus of Fishes (Echiodon).* Proc. Zool. Soc. London, V.

WILLIAM THOMPSON.

Echiodon Thompson, 55; orthotype E. DRUMMONDI Thompson. A synonym of CARAPUS Raf. (FIERASFER Cuv.).

1838

222. AGASSIZ (1838). *Recherches sur les poissons fossiles.* III, IV (parts).

LOUIS AGASSIZ.

Ptychopleurus Agassiz, III, 67; orthotype PTYCHACANTHUS FAUGASI Ag. A fossil ally of MYLIOBATIS.

Oxyrhina Agassiz, III, 86; logotype O. MANTELLI Ag. (fossil). Name preoccupied, the genus not separable from the living genus ISURUS Raf. (1810).

Orodus Agassiz, III, 96; logotype O. CINCTUS Ag. (fossil.)

Ctenoptychius Agassiz, III, 99; orthotype C. APICALIS Ag. (fossil).

Helodus Agassiz, III, 104; orthotype H. SIMPLEX Ag. (fossil).

Chomatodus Agassiz, III, 107; logotype C. CINCTUS Ag. (fossil). An ally of COCHLIODUS, of undetermined value.

Cochliodus Agassiz, III, 113; orthotype C. CONTORTUS Ag. (fossil).

Strophodus Agassiz, III, 116; logotype S. RETICULATUS Ag. (fossil). A synonym of ASTERACANTHUS.

Ceratodus Agassiz, III, 129; orthotype C. LATISSIMUS Ag. (fossil). Considered by Gistel as preoccupied by the earlier CERATODON; SCROPHA Gistel (1848) has been proposed as a substitute.

Ctenodus Agassiz, III, 137; orthotype C. CRISTATUS Ag. (fossil). Considered as preoccupied by the earlier CTENODON. RHADAMISTA Gistel (1848) has been proposed as a substitute.

Acrodus Agassiz, III, 139; orthotype A. GAILLARDOTI Ag. (fossil). Name first published by Geinitz, 1837; regarded as preoccupied by ACRODON Zimm. (1832), a genus of beetles; ADIAPNEUSTES Gistel (1848) being suggested as substitute. But THECTODUS Meyer & Plieninger (1844) is earlier than ADIAPNEUSTES.

Hoplopteryx Agassiz, IV, 4; logotype H. ANTIQUUS Ag. (fossil).

Sphenocephalus Agassiz, IV, 4; orthotype S. FISSICAUDUS Ag. (fossil).

Acanus Agassiz, IV, 4; logotype A. OVALIS Ag. (fossil). (Printed first without explanation in *Neues Jahrb.*, 1834, 305).

Acrogaster Agassiz, IV, 5; orthotype A. PARVUS Ag. (fossil).

Podocys Agassiz, IV, 5, 135; orthotype P. MINUTUS Ag. (fossil).

Sparnodus Agassiz, IV, 10; orthotype SPARUS MACROPHTHALMUS Volta (fossil). (Printed first without definition, *Neues Jahrb.*, 1835, 300.)

Callipteryx Agassiz, IV, 12; logotype C. SPECIOSUS Ag. (fossil). (Also printed as name only in *Neues Jahrb.*, 1835, 293.)

Semiophorus Agassiz, IV, 14; orthotype KURTUS VELIFER Volta (fossil).

Pygæus Agassiz, IV, 16, 256; logotype P. COLEANUS Ag. (fossil).

Macrostoma Agassiz, IV, 15, 260; orthotype M. ALTUM Ag. (fossil). (Name twice preoccupied.)

223. BUCKLAND (1838). *On the Discovery of Fossil Fishes in the Bagshot Sands at Goldworth Hill, Four Miles North of Guildford.* Proc. Geol. Soc. London, II.

WILLIAM BUCKLAND (1784-1856).

Edaphodon Buckland, II, 687; logotype CHIMÆRA MANTELLI Buckland (fossil). According to Dr. L. Hussakof, the following genera based on Chimæroid teeth are all synonyms of EDAPHODON, namely, ISCHYODUS, EUMYLODUS, PASSALODON, PSITTACODON, DIPRISTIS, DIPHRISSA and BRYACTINUS.

Passalodon Buckland, II, 687; orthotype CHIMÆRA AGASSIZI Buckland (fossil). A synonym of EDAPHODON.

Ameibodon Buckland. This name is quoted by Agassiz from the *Proc. Zool. Soc. London,* 1839. I do not find it there.

224. COCCO (1838). *Su di alcuni salmonidi del mare di Messina; lettera al Ch. D. Carlo Luciano Bonaparte.* Nuovi Ann. Sci. Nat., II.

ANASTASIO COCCO.

Gymnocephalus Cocco. 26; orthotype G. MESSINENSIS Cocco. Name preoccupied; a synonym of CENTROLOPHUS Lac.

Maurolicus Cocco, 32; orthotype M. AMETHYSTINOPUNCTATUS Cocco.

Odontostomus Cocco, 32; orthotype O. HYALINUS Cocco. Regarded as preoccupied by ODONTOSTOMUS (1887), a genus of mollusks; replaced by EVERMANNELLA Fowler.

225. GRAY (1838). *(Aracana.)* Ann. Hist., I.

JOHN EDWARD GRAY.

Aracana Gray, I, 110; logotype A. ORNATA Gray. (Printed without description in 1833.)

226. HECKEL (1838). *Fische aus Caschmir gesammelt und herausgegeben von Carl Freiherrn von Hügel, etc.* Wien.

JOHANN JAKOB HECKEL.

Schizothorax Heckel, 11; logotype S. PLAGIOSTOMUS Heckel as fixed by Bleeker, 1863; limited by Günther in 1868 to S. CURVIFRONS, etc. Same as OREINUS McCl., which it replaces if valid

227. LOWE (1838). *Piscium Maderensium species quædam novæ vel minus rite cognitæ breviter descriptæ,* etc. Trans. Cambridge Phil. Soc., VI.

RICHARD THOMAS LOWE.

Polymixia Lowe, VI, 198; orthotype P. NOBILIS Lowe.

228. MÜLLER & HENLE (1838). *Systematische Beschreibung der Plagiostomen.* Berlin (1838 to 1841).

JOHANNES MÜLLER; FREDERICK GUSTAV JAKOB HENLE.

Chiloscyllium Müller & Henle, 17; logotype SCYLLIUM PLAGIOSUM Bennett.

Crossorhinus Müller & Henle, 21; orthotype SQUALUS BARBATUS Gmelin. A synonym of ORECTOLOBUS Bon.

Stegostoma Müller & Henle 24; orthotype SQUALUS FASCIATUS Bloch.

Aprion Müller & Henle, 32; orthotype CARCHARIAS ISODON M. & H. Name preoccupied in fishes; replaced by APRIONODON Gill, 1861.

Hypoprion Müller & Henle, 34; orthotype CARCHARIAS MACLOTI M. & H.

Prionodon Müller & Henle, 36; logotype SQUALUS GLAUCUS L. Preoccupied; replaced by PRIONACE Cantor, 1849.

Triakis Müller & Henle, II, 36; orthotype T. SCYLLIUM. Spelled TRIACIS by authors.

Triænodon Müller & Henle, 55; orthotype CARCHARIAS OBESUS M. & H.

Galeocerdo Müller & Henle, 59; orthotype G. TIGRINUS M. & H.

Loxodon Müller & Henle, 61; orthotype L. MACRORHINUS M. & H.

Thalassorhinus Müller & Henle, 62; orthotype T. VULPECULA M. & H.

Carcharodon (Smith) Müller & Henle, 70; orthotype C. RONDELETI M. & H. = SQUALUS CARCHARIAS L.

Alopecias Müller & Henle, 74; orthotype SQUALUS VULPES Gmelin = S. VULPINUS Bonnaterre. A revised spelling of ALOPIAS Raf.

Triglochis Müller & Henle, II, 88; orthotype SQUALUS FEROX Risso. A synonym of ODONTASPIS Ag. and of CARCHARIAS Raf.

Centrophorus Müller & Henle, 89; logotype C. GRANULATUS M. & H. Perhaps a synonym of DALATIAS Raf.

Læmargus Müller & Henle, 93; orthotype SQUALUS BOREALIS Scoresby. A synonym of SOMNIOSUS Le Sueur.

Pristiophorus Müller & Henle, 98; orthotype PRISTIS CIRRATUS Latham.

Rhynchobatis Müller & Henle, 111; orthotype RHINOBATUS LÆVIS Bl. & Schn.

Syrrhina Müller & Henle, 113; logotype S. COLUMNÆ M. & H. as restricted by Jordan & Evermann, 1896, the earliest definite restriction I have found. Garman, 1912, indicates SYRRHINA BREVIROSTRIS M. & H. as type. This is a species referred to ZAPTERYX J. & G. As here understood, SYRRHINA is a synonym of RHINOBATUS.

Trygonorhina Müller & Henle, 124; orthotype T. FASCIATA M. & H.

Platyrhina Müller & Henle, 125; logotype RAJA SINENSIS Lac. Name regarded as preoccupied by PLATYRHINUS; replaced by ANALITHIS Gistel (1848) and later by DISCOBATUS Garman (1881).

Astrape Müller & Henle, 130; logotype RAJA CAPENSIS Gmelin. A synonym of NARKE Kaup, 1826.

Uraptera Müller & Henle, 155; logotype U. AGASSIZI M. & H.

Sympterygia Müller & Henle, 155; orthotype S. BONAPARTII M. & H.

Pteroplatea Müller & Henle, 168; orthotype RAJA ALTAVELA L.

Hypolophus Müller & Henle, 170; orthotype RAJA SEPHEN Forskål. A synonym of PASTINACHUS Rüppell, if this name is accepted (not PASTINACA Nardo).

Tæniura Müller Henle, 171; logotype RAJA LYMMA Forskål.

Urolophus Müller & Henle, 173; logotype RAJA CRUCIATA Lac.

Trygonoptera Müller & Henle, 174; orthotype T. TESTACEA M. & H.

Aëtoplatea Müller & Henle, 175; orthotype A. TENTACULATA M. & H.

Ceratoptera Müller & Henle, 186; orthotype CEPHALOPTERA VAMPYRUS Mitchill. A synonym of MANTA Bancroft

Centroscyllium Müller & Henle, 191; orthotype SPINAX FABRICII Reinh.

229. MÜLLER & HENLE (1838). *(Plagiostomen.)* Charlesworth's Magazine of Natural History, 1838.

JOHANNES MÜLLER; FRIEDRICH HENLE.

Leptocharias Müller & Henle, II, 36; orthotype TRIÆNODON SMITHI Gray. Same as LEPTOCARIAS Smith, 1837.

230. NORDMANN (1838). *Bericht an die kaiserliche Akademie der Wissenschaften, über eine neue Fischgattung (Hexacanthus) aus der Familie der Gobioiden.* Bull. Acad. Sci. St. Petersb., III.

ALEXANDER VON NORDMANN.

Hexacanthus Nordmann; orthotype GOBIUS MACROCEPHALUS Pallas. Name regarded as preoccupied by HEXACANTHA; replaced by BENTHOPHILUS Eichw.

231. REINHARDT (1838). *Ichthyologiske bidrag til den Grönlandske fauna,* etc. Danske Vid. Selsk. Afh. Kjöbenhavn, VII.

JOHANNES CHRISTOPHER HAGEMANN REINHARDT.

Himantolophus Reinhardt, 74; orthotype H. GRŒNLANDICUS Reinh.

Citharus Reinhardt, 116; orthotype PLEURONECTES PLATESSOIDES Fabricius, not of Röse, 1793, nor of Bleeker, 1862. A synonym of HIPPOGLOSSOIDES Gottsche.

Gymnelis Reinhardt, VII, 130; orthotype OPHIDIUM VIRIDE Fabricius.

Lycodes Reinhardt, VII, 153; logotype L. VAHLI Reinh.

232. SWAINSON (1838). *The Natural History of Fishes, Amphibians and Reptiles, or Monocardian Animals.* I.

WILLIAM SWAINSON (1789-1855).

Pedalion (Guilding) Swainson, I, 199; II, 195, 329; orthotype (TETRODON MOLA L.). A synonym of MOLA Kölr.

Ariosoma Swainson, I, 220; II, 195; (no type named); logotype, as restricted by Bleeker and by Swain, OPHISOMA ACUTA Sw. = MURÆNA BALEARICA De la Roche; replaced by OPHISOMA Sw., on II, 334; a name preoccupied; replaces CONGERMURÆNA Kaup and CONGRELLUS Ogilby. Several other generic names defined by Swainson in Vol. II (1839) are casually mentioned with a few words of comparison. Tius BREVICEPS (238, 343, 345), CYPSILURUS (I, 299). But when no question of priority arises, these need not be mentioned further.

Breviceps Swainson, I, 328; II, 189; orthotype B. FILAMENTOSUS Sw. = ARIUS BAHIENSIS Castelnau. Name preoccupied; replaced by FELICHTHYS Sw. II,

305. A synonym of BAGRE (Cuv.) Oken, as restricted to its tautotype.

1839
233. AGASSIZ (1839).   *Künstliche Steinkerne von Konchylien und Fische.*
Neues Jahrb. Mineral.
LOUIS AGASSIZ.

Eugnathus Agassiz, 118; logotype E. MINOR Ag. (fossil); regarded as preoccupied by EUGNATHA; replaced by FURO Gistel (1848).

234. AGASSIZ (1839).   *Fishes of the Old Red Sandstone.*   In Murchison's
Silurian System.
LOUIS AGASSIZ.

Holoptychus Agassiz, 68; orthotype H. NOBILISSIMUS Ag. (fossil). Usually written HOLOPTYCHIUS.

Sclerodus Agassiz, 606; orthotype S. PUSTULIFERUS Ag.

Thelodus Agassiz, 606; orthotype T. PARVIDENS Ag. (fossil).

Plectrodus Agassiz, 606, 704; logotype P. MIRABILIS Ag. (fossil).   A synonym of SCLERODUS Ag.

Sphagodus Agassiz (not quoted by Woodward). Name of an Ichthydorulite or fossil fin-spine.

235. AGASSIZ (1839).   *Recherches sur les poissons fossiles.*   II, IV (part).
LOUIS AGASSIZ.

Sphærodus Agassiz, II, 215; logotype S. CONICUS Ag. (fossil).   Not SPHÆRODON Rüppell.

Pristigenys Agassiz, IV, 136 orthotype P. MACROPHTHALMUS Ag. (fossil). (Printed as name only in *Neues Jahrb.*, 1835, 299.)

Ptychodus Agassiz, III, 150; orthotype P. MAMMILARIS Ag. (fossil).

Odonteus Agassiz, IV, 178; orthotype O. SPAROIDES Ag. (fossil).

Pterygocephalus Agassiz, IV, 190; orthotype P. PARADOXUS Ag. (fossil).   (Also as name only, *Neues Jahrb.*, 1835, 295.)

236. BONAPARTE (1839).   *Systema Ichthyologicum.*   Mem. Sci. Nat. Neuf châtel, II, 1839.
CHARLES LUCIEN BONAPARTE.

Pristidurus Bonaparte, II, 11; orthotype GALEUS MELASTOMUS Raf. Commonly written PRISTIURUS, but the form here given is apparently earlier.

237. CUVIER & VALENCIENNES (1839).   *Histoire naturelle des poissons.*
XIII, XIV.
ACHILLE VALENCIENNES.

Cossyphus Cuvier & Valenciennes, XIII, 102; logotype BODIANUS BODIANUS Bloch. Identical with BODIANUS Bloch, as restricted, and HARPE Lac. The name COSSYPHUS is also preoccupied.

Acantholabrus Cuvier & Valenciennes, XIII, 242; logotype LUTJANUS PALLONI Risso.

Ctenolabrus Cuvier & Valenciennes, XIII, 223; logotype LABRUS RUPESTRIS L.

Bagrus Cuvier & Valenciennes, XIV, 388; logotype SILURUS BAJAD Forskål, not BAGRE (Cuv.) Oken, which is FELICHTHYS Sw.   BAGRUS is regarded as preoccupied by BAGRE, both being forms of "LES BAGRES" Cuv. If so, it is replaced by PORCUS Geoffroy St. Hilaire, 1818.

238. LAY & BENNETT (1839). *Fishes.* (In the *Zoology of Captain Beechey's Voyage.*)

G. T. LAY; EDWARD TURNER BENNETT.

Peropus Lay & Bennett, 59; orthotype BLEPSIAS BILOBUS Cuv. & Val. Name preoccupied; replaced by HISTIOCOTTUS Gill, 1888.

239. LOWE (1839). *A Supplement to a Synopsis of the Fishes of Madeira.* Proc. Zool. Soc. London, VII.

RICHARD THOMAS LOWE.

Callanthias Lowe, 76; orthotype C. PARADISÆUS = BODIANUS PELORITANUS Cocco.

Aphanopus Lowe, 79; orthotype A. CARBO Lowe.

Pompilus Lowe, 81; orthotype CORYPHÆNA POMPILUS L. = POMPILUS RONDELETI Lowe. A synonym of CENTROLOPHUS Lac.

Alysia Lowe, 87; orthotype A. LORICATA Lowe = SCOPELUS COCCOI Cocco. Name preoccupied; replaced by RHINSOCOPELUS Lütken.

Acanthidium Lowe, 91; logotype A. PUSILLUM Lowe, as restricted by Jordan and Evermann, 1896. Garman, 1913, chooses the second species named by Lowe, A. CALCEUS, a species of DEANIA, as the type. As first restricted, ACANTHIDIUM is a synonym of ETMOPTERUS Raf.

240. McCLELLAND (1839). *Indian Cyprinidæ.* Asiatic Researches, XIX.

JOHN McCLELLAND (1805-1875).

Oreinus McClelland, 273; logotype O. GUTTATUS McCl. Regarded by Gistel as preoccupied by OREINA, the name ENGLOTTOGASTER substituted for it. Probably a synonym of SCHIZOTHORAX Heckel.

Systomus McClelland, 284; logotype S. IMMACULATUS McCl.

Perilampus McClelland, 288; logotype CYPRINUS DEVARIO Ham. A synonym of BARILIUS.

Opsarius McClelland, 295; logotype O. MACULATUS McCl. Apparently a synonym of BARILIUS Ham.

Psilorhynchus McClelland, 300; logotype CYPRINUS SUCATIO Ham. A synonym of HOMALOPTERA Van Hasselt.

Aplocheilus McClelland, 301; logotype A. CHRYSOSTIGMUS McCl. = ESOX PANCHAX Ham. Name corrected to HAPLOCHILUS Günther.

Schistura McClelland, 306; logotype COBITIS RUPECULA McCl. A synonym of BARBATULA Linck.

Gonorhynchus McClelland, 373; logotype G. BREVIS McCl. Not GONORHYNCHUS of Gronow and Scopoli (1777). A synonym of CROSSOCHEILUS Van Hasselt.

Hymenphysa McClelland, 443; logotype COBITIS DARIO Ham. Corrected to HYMENOPHYSA by Günther; a synonym of BOTIA Gray.

241. McCLELLAND (1839). *Observations on Six New Species of Cyprinidæ with an Outline of a New Classification of the Family.* Journ. Asiatic Soc. Benal, VIII, 941-947.

JOHN McCLELLAND.

Platycara McClelland, VIII, 947; orthotype P. NASUTA McCl. A synonym of GARRA Ham., and DISCOGNATHUS Heckel.

242. MITCHELL (1839). *Three Expeditions into the Interior of Eastern Australia.*
Ed. II.

JAMES MITCHELL.

Tandanus Mitchell, I, 95; orthotype PLOTOSUS TANDANUS Mitchell; replaces COPIDO-
GLANIS Gthr., 1864.

243. MÜLLER (1839). *Vergleichende Anatomie der Myxinoiden, der Cyclostomen
mit durchbohrten Gaumen. IV. Ueber das Gefäs System.* Abh. Akad.
Wiss. Berlin, 1839.

JOHANNES MÜLLER.

Heteropneustes Müller, 243; orthotype SILURUS FOSSILIS Bloch. Said to be the
same as SILURUS SINGIO Ham., 243 (also in *Wiegm. Arch.*, 1839, 115). Replaces SACCOBRANCHUS Cuv. & Val., 1848.

Amphipnous Müller, 244; orthotype UNIBRANCHAPERTURA CUCHIA Ham.

Physogaster Müller, 252; orthotype TETRAODON LUNARIS Bl. & Schn. Name preoccupied; replaced by GASTROPHYSUS Müller. A synonym of LAGOCEPHALUS
Swain.

Arothron Müller, 252; orthotype A. TESTUDINARIUS Müller = TETRAODON RETICU-
LARIS Bl. & Schn. A synonym of OVOIDES Dum.

Cheilichthys Müller, 252; orthotype TETRAODON TESTUDINEUS L. A synonym of
SPHEROIDES Dum.

Chelonodon Müller, 252; orthotype TETRODON PATOCA Ham.

244. MÜNSTER (1839). *Ueber einige merkwürdige Fische aus dem Kupferschiefer und dem Muschelkalk.* Beiträge zur Petrefactenkunde.

GEORG MÜNSTER.

Ascalabos Münster, I, 112; orthotype A. VOITHI Münster (fossil). A synonym of
LEPTOLEPIS.

245. NARDO (1839). *Considerazioni sulla famiglia dei pesci Mola, e sui caratteri
che li distinguono.* Ann. Sci. Lombardo-Veneto, X.

GIOVANNI DOMENICO NARDO.

Pallasia Nardo, V, 10, 112; orthotype P. PALLASI Nardo. A larval form of MOLA
Kölr.

Ranzania Nardo, V, 10, 105; orthotype TETRAODON TRUNCATUS Retzius.

246. OWEN (1839). *On a New Species of the Genus Lepidosiren of Fitzinger and
Natterer, L. Annectens.* Proc. Linn. Soc. London, 1839, I, 27-32.

RICHARD OWEN (1810-1890).

Protopterus Owen, 27; orthotype LEPIDOSIREN ANNECTENS Owen.

247. OWEN (1839). *(Fossil Fishes.)* Report British Association for 1838 (1839).

RICHARD OWEN.

Dictyodus Owen, 142; orthotype D. DESTRUCTOR Owen (fossil) (1854).

248. RANZANI (1839). *Dispositio familiæ Molarum in genera et in species.*
Novi Comment. Acad. Sci. Inst. Bonon., III.

CAMILLO RANZANI.

Ozodura Ranzani, 81; orthotype O. ORSINI Ranzani = MOLA MOLA (L.). A synonym
of MOLA Kölr.

Tympanomium Ranzani, 81; orthotype T. PLANCI Ranzani. A synonym of MOLA.

Trematopsis Ranzani, 81; orthotype T. WILLUGHBEII Ranzani. A synonym of MOLA.

249. RICHARDSON (1839). *Description of Fishes Collected at Port Arthur in Van Diemen's Land.* Proc. Zool. Soc. London, VII.

JOHN RICHARDSON.

Latris Richardson, 98; orthotype L. HECATEIA Rich.

Nemadactylus Richardson, 98; orthotype N. CONCINNUS Rich. Spelled NEMATO-DACTYLUS by Gill.

250. STORER (1839). *A Report on the Fishes of Massachusetts.* Boston Journ. Nat. Hist., II.

DAVID HUMPHREYS STORER (1804-1891).

Cryptacanthodes Storer, 28; orthotype C. MACULATUS Storer.

251. SWAINSON (1839). *The Natural History of Fishes, Amphibians and Reptiles or Monocardian Animals.* II.

WILLIAM SWAINSON.

In a review of Swainson's *Genera of Fishes,* Proc. Acad. Nat. Sci. Phila., 1882, page 272, Dr. Joseph Swain has made identifications of all of the series Swainson's genera, indicating a logotype for each. This designation of type must stand in every case, except when a prior indication has been made, as was done in a few cases by Bonaparte and by Bleeker. Swainson's work is perhaps the last and certainly the most unwelcome of an unfortunate series to which belong Scopoli's *Introductio,* Fischer's *Zoognosia,* Rafinesque's *Analyse de la Nature,* works which treat of the whole scope of Ichthyology, adding new genera here and there, but without special investigation of the animals concerned, leaving their generic names imperfectly defined as a burden to the real students who must give to these names their final meaning. Of the 153 new names introduced by Swainson, about 30 must be retained on the score of priority of date, but there is not a dozen of the whole number based on any clear understanding of the relations of the species indicated. In Chapter I of Vol. 2, on pages 167 to 197, Swainson gives an analytical key to the genera he recognizes. In this part no species are mentioned and in several cases the genera bear other names when they appear later in the regular text in Chapter II of Vol. 2 (197-339).

Erosa Swainson, II, 61; tautotype SYNANCEIA EROSA Langsdorff; type not indicated but understood from context, and from the generic name chosen. The genus is valid and was revived by Jordan & Starks (1904).

Trichosomus Swainson, II, 61, 265; logotype APISTUS TRACHINOIDES Cuv. & Val. Accidentally spelled also TRICHOPHASIA on page 61. Regarded as preoccupied by TRICHOSOMA, a genus of worms, and by TRICHOSOMUS, a genus of bettles; replaced by VESPICULA Jordan & Rich. PROSOPODASYS Cantor, originally a substitute name for APISTUS, is transferred by Günther to A. TRACHINOIDES, an arrangement apparently not allowable, as Cuvier does not mention that species among "les APISTES". TRICHOPHASIA, an error, not a new name, seems also ineligible.

Trichophasia Swainson, II, 61; (used evidently by error of proof-reader for TRICHOSOMUS) which form appears before and after on page 61 and formally on page 279.

Clinitrachus Swainson, II, 75; II, 276; orthotype (on p. 75) BLENNIUS VARIABILIS Raf., a species not mentioned on page 276, hence B. SUPERCILIOSUS L., a species of CLINUS, was indicated as type by Swain. A synonym of CRISTICEPS Cuv. & Val. (1836).

Blennitrachus Swainson, II, 78; II, 274; orthotype PHOLIS QUADRIFASCIATUS Wood. A synonym of CHASMODES Cuv.

Cichlaurus Swainson, II, 173; haplotype (230) LABRUS PUNCTATUS Bloch. Later called CICHLASOMA, which name it should replace.

Dendrochirus Swainson, II, 180; logotype PTEROIS ZEBRA Cuv. & Val.

Pteropterus Swainson, II, 180; logotype (p. 264) PTEROIS RADIATUS Cuv. & Val. A synonym of PTEROIS Cuv.

Psilosomus Swainson, II, 183; logotype TÆNIOIDES HERRMANNIANUS Lac. A synonym of TÆNIOIDES Lac.

Gobileptes Swainson, II, 183; no type named; orthotype intended (as is evident from Swainson's description) GOBIUS ACUTIPINNIS Cuv. & Val. "like GOBIUS, but the caudal lanceolate and the scales large". Bleeker (1876) has, however, fixed the type, none having been named before, on GOBIUS BATO Ham. This raises a question in nomenclature already noted under APOCRYPTES (1837).

Salmostoma Swainson, II, 184; logotype CYPRINUS BACAILA Ham. on page 284, where the genus is called SALMOPHASIA. Replaces SECURICULA Gthr., 1868.

Scrophicephalus Swainson, II, 187; no type named, MORMYRUS CASCHIVE Hasselquist intended ("dorsal fin single, very long; snout produced"). A synonym of MORMYRUS L., as restricted by Gill (1862).

Ariosoma Swainson, II, 194; logotype as restricted by Bleeker, 1864, OPHISOMA ACUTA Sw. = MURÆNA BALEARICA De la Roche. Replaces CONGERMURÆNA Kaup (1856) and CONGRELLUS Ogilby 1898.

Leiodon Swainson, II, 194; logotype LEIOSOMUS MARMORATUS Sw. = TETRAODON CUTCUTIA Ham., as defined by Bleeker (1865). Swain restricts the name LEIOSOMUS as later substituted by Swainson for LEIODON, to LEIOSOMUS LÆVISSIMUS Sw., which is TETRAODON SPENGLERI Bloch, a species of SPHEROIDES. LEIODON Sw. has precedence over LEIODON Owen, 1841, and replaces MONOTRETA Bibron.

Canthigaster Swainson, II, 194; logotype TETRAODON ROSTRATUS Bloch. Called PSILONOTUS later in the text; replaces ANOSMIUS Peters.

Cromileptes Swainson, II, 201; logotype SERRANUS ALTIVELIS Cuv. & Val. Restricted by Bleeker (1875) to S. ALTIVELIS, thus replacing SERRANICHTHYS Blkr. Later restricted by Swain to EPINEPHELUS GIGAS.

Cynichthys Swainson, II, 201; haplotype PERCA FLAVOPURPUREA Bennett = HOLOCENTRUS FLAVOCÆRULEUS Lac. A synonym of EPINEPHELUS (not PLECTORHYNCHUS as stated by Swain).

Variola Swainson, II, 202; haplotype V. LONGIPINNA Sw. = PERCA LOUTI Forskål.

Uriphæton Swainson, II, 202; orthotype SERRANUS PHAËTON Cuv. & Val. = U. MICROLEPTES Sw. This genus is based on a made-up fish, a CEPHALOPHOLIS, with the tail of a FISTULARIA attached. A synonym of CEPHALOPHOLIS Bl. & Schn.

Elastoma Swainson, II, 202; type ETELIS OCULATUS (Cuv. & Val.). A synonym of ETELIS Cuv.

Rabdophorus Swainson, II, 211; haplotype CHÆTODON EPHIPPIUM Bloch. A subdivision of CHÆTODON.

Genicanthus Swainson, II, 212; logotype HOLACANTHUS LAMARCKI (Cuv.) A synonym of HOLACANTHUS Lac.

Microcanthus Swainson, II, 215; haplotype CHÆTODON STRIGATUS Cuv., a wilful misprint for MICRACANTHUS.

Chrysiptera Swainson, II, 216; logotype GLYPHIDODON AZUREUS Q. & G. A subgenus under ABUDEFDUF.

Microgaster Swainson, II, 216; haplotype ETROPLUS CORUCHI Cuv. & Val. = CHÆTODON MACULATUS Bloch. Equivalent to PARETROPLUS Blkr. A synonym of ETROPLUS Cuv. & Val. as restricted.

Chætolabrus Swainson, II, 216; logotype CHÆTODON SURATENSIS Bloch. A synonym of ETROPLUS Cuv.

Chrysoblephus Swainson, II, 221; haplotype SPARUS GIBBICEPS Cuv. A section under SPARUS L.

Argyrops Swainson, II, 221; haplotype SPARUS SPINIFER Forskål. A section under SPARUS L.

Calamus Swainson, II, 221; haplotype PAGELLUS CALAMUS Cuv. & Val. = CALAMUS MEGACEPHALUS Sw.

Lithognathus Swainson, II, 222; haplotype SPARUS CAPENSIS Cuv.

Nemipterus Swainson, II, 223; haplotype DENTEX FILAMENTOSUS Cuv. Replaces SYNAGRIS Gthr., 1859.

Thalassoma Swainson, II, 224; haplotype LABRUS PURPUREUS Rüppell. Stands in place of JULIS Gthr., not of Cuvier.

Urichthys Swainson, II, 224; logotype LABRUS LUNULATUS Rüppell. A synonym of CHEILINUS Lac.

Crassilabrus Swainson, II, 225; haplotype CHEILINUS UNDULATUS Rüppell. A section under CHEILINUS Lac.

Leptoscarus Swainson, II, 226; haplotype SCARUS VAIGIENSIS Q. & G. = SCARUS SPINIDENS Cuv. = CALLIODON VAIGIENSIS Cuv. & Val. Replaces CALOTOMUS Gilbert, and CALLIODON Cuv.; not CALLYODON of Gronow nor CALLIODON of Bl. & Schn.

Hemistoma Swainson, II, 226; haplotype SCARUS PEPO Bennett = HEMISTOMA RETICULATA Sw. A synonym of CALLYODON Gronow or SCARUS Forskål.

Erychthys Swainson, II, 226; logotype LABRUS CROICENSIS Bloch. A synonym of CALLYODON Gronow or SCARUS Forskål.

Petronason Swainson, II, 226; logotype CORYPHÆNA PSITTACUS L. as restricted by Bleeker and by Swain. A synonym of CALLYODON or SCARUS.

Pachynathus Swainson, II, 226 (misprint for PACHYGNATHUS); haplotype P. TRIANGULARIS Sw., based on RAHTEE YELLAKAH Russell = BALISTES CAPISTRATUS Shaw. Name preoccupied as PACHYGNATHUS; replaced by SUFFLAMEN Jordan, 1917. The International Commission of Nomenclature has decided that PACHYNATHUS is to be regarded as a misprint, hence identical with PACHYGNATHUS.

Chlorurus Swainson, II, 227; haplotype SCARUS GIBBUS Rüppell. A synonym of CALLYODON or SCARUS.

Amphiscarus Swainson, II, 227; SCARUS FUSCUS Griffith. A synonym of SIGANUS Forskål.

Sparisoma Swainson, II, 227; haplotype SCARUS ABILDGAARDI Bloch. Equivalent to SCARUS Gthr., which it replaces; not SCARUS of Forskål, which should stand as CALLYODON Gronow, if Gronow's names are allowed.

Hemiulis Swainson, II, 228; logotype LABRUS GUTTATUS Bloch, as restricted by Bonaparte (*Fauna Italica*, 1841). Swain (1882) chose CHEILIO AURATUS Q. & G. as type regarding HEMIULIS as a synonym of CHEILIO. If HALICHŒRES Rüppell is regarded as preoccupied by HALICHŒRUS, HEMIULIS must replace it instead of CHŒROJULIS Gill.

Cynædus Swainson, II, 229; logotype Labrus rupestris Bloch, as restricted by
Bonaparte, Labrus tinca, as restricted by Swain. Swainson offers this name
as a substitute for Crenilabrus Cuv., of which he wrongly assigns Lutianus verres, a Bodianus, as type. Cynædus Sw. (not of Gronow or Scopoli)
is a synonym of Ctenolabrus Cuv. & Val.

Astronotus Swainson, II, 229; haplotype Lobotes ocellatus Spix. Replaces Acara
Heckel and Hygrogonus Gthr., 1862.

Thalliurus Swainson, II, 230; haplotype Labrus blochi = Labrus fuligi nosus
Lac. Replaces Hemigymnus Gthr., 1864.

Labristoma Swainson, II, 230; logotype Pseudochromis olivaceus Rüppell. A
needless substitute for Pseudochromis Rüppell, 1835.

Cichlasoma Swainson, II, 230; haplotype Labrus punctatus Bloch. Called Cichlaurus on page 173; equivalent to Heros Heckel (1840) as restricted; apparently Cichlaurus should stand.

Ichthycallus Swainson, II, 232; logotype Labrus julis L., as restricted by Bonaparte (1839). Swain has chosen (1882) Labrus semipunctatus as type, a
species of Coris Lac. Jordan later took the first species named, dimidiatus,
as type, a species of Iridio. Ichthycallus is a synonym of Julis Cuv. & Val.
"Chlorichthys and Ichthycallus, confused jumble of species, may well be
disposed of as synonyms of Thalassoma and Coris, respectively, although
several other genera are represented in each." (Swain.)

Eupemis Swainson, II, 232; haplotype Labrus fusiformis Rüppell. A synonym of
Cheilio Lac.

Chlorichthys Swainson, II, 232; logotype Labrus bifasciatus Bloch, as restricted
by Swain. Bonaparte selects as type Labrus pavo Hasselquist, a species not
mentioned by Swainson. A synonym of Thalassoma Sw.

Zanclurus Swainson, II, 239; type Histiophorus indicus Cuv. & Val. A synonym
of Istiophorus Lac.

Zyphothyca Swainson, II, 239; type Gempylus coluber Cuv. & Val.; for Xiphotheca Montague. A synonym of Gempylus Cuv.

Leiurus Swainson, II, 242; logotype Gasterosteus aculeatus L. A synonym of
Gasterosteus L.

Polycanthus Swainson, II, 242; type Gasterosteus spinachia L. A synonym of
Spinachia Fleming, 1825. Polycanthus is also preoccupied.

Chirostoma Swainson, II, 242; haplotype Atherina humboldtiana Cuv. & Val.

Meladerma Swainson, II, 243; haplotype M. nigerrima Sw. = Elacate pondi-
cerriana Cuv. & Val. A synonym of Rachycentron Kaup (1826), and of
Elacate Cuv. (1831).

Platylepes Swainson, II, 247; haplotype Scomber lactarius Bl. & Schn. A synonym of Lactarius Cuv. & Val.

Argyrlepes Swainson, II, 247; haplotype Argyrlepes indica Sw., based on Mitta
parah of Russell, pl. 156, a species of Leiognathus. A synonym of
Leiognathus Lac.

Alepes Swainson, II, 248; haplotype Alepes melanoptera Sw., based on Wori
parah Russell, pl. 155, apparently the same as Caranx nigripinnis Day,
a species of Hemicaranx; apparently replaces Hemicaranx Blkr.

Zonichthys Swainson, II, 248; type Scomber fasciatus Bloch. A section or subgenus under Seriola.

Hamiltonia Swainson, II, 250; logotype H. ovata Sw. = Chanda nama Ham.
(fig. 37) as fixed by Swain (1882). Hamiltonia is given as a substitute for

CHANDA Ham., and the fixation of type for the substitute decides that of the original name. CHANDA NAMA is the orthotype of BOGODA Blkr., a genus well defined, AMBASSIS BOGODA being the type species. Swainson's second species, CHANDA LALA Ham., which he spells LATA, belongs to Bleeker's genus PSEUDAMBASSIS. This was chosen as type of CHANDA by Fowler, and still later C. RUCONIUS by Jordan, but Swain's choice of type must hold. HAMILTONIA and BOGODA are therefore both synonyms of CHANDA Ham.

Platysomus Swainson, II, 250; logotype VOMER BROWNI Cuv. & Val. A synonym of VOMER; name preoccupied.

Ctenodon Swainson, II, 255; logotype CTENODON RUPPELLI Sw. = CHÆTODON SOHAL Forskål, not CTENODON Bon. nor of Ehrenberg. A synonym of HEPATUS Gronow, TEUTHIS L. and ACANTHURUS Forskål.

Callicanthus Swainson, II, 256; haplotype ASPISURUS ELEGANS Rüppell = HARPURUS LITURATUS Forster. A section or genus near NASO.

Zebrasoma Swainson, II, 256; haplotype ACANTHURUS VELIFER Bloch.

Xiphichthys Swainson, II, 259; haplotype GYMNETRUS RUSSELLI Shaw. A synonym of REGALECUS Brünnich. (Also, II, 46.)

Xiphasia Swainson, II, 259; type "ZIPHASIA SETIFER" Sw. Replaces NEMOPHIS Kaup (1858) and XIPHOGADUS Gthr. (1862).

Nemotherus (Rafinesque) Swainson (II, 260); orthotype N. ERYTHROPTERUS Raf. Quoted from Rafinesque's *Specchio delle Scienze*, I, 101, 1815. This is evidently a slip for NEMOCHIRUS Raf. NEMOTHERUS (Risso) Costa is a synonym of TRACHYPTERUS. NEMOCHIRUS seems to be an ally of STYLEPHORUS, as yet unrecognized.

Ornichthys Swainson, II, 262; logotype TRIGLA PUNCTATA Bloch. A synonym of PRIONOTUS Lac.

Pteroleptus Swainson, II, 264; haplotype PTEROLEPTUS LONGICAUDA Sw., (KODIPUNGI Russell). A synonym of PTEROIS Cuv.

Macrochyrus Swainson, II, 264; haplotype SCORPÆNA MILES Bennett. A synonym of PTEROIS Cuv.

Brachyrus Swainson, II, 264; logotype PTEROIS ZEBRA Cuv. & Val. A synonym of DENDROCHIRUS Sw., not BRACHYOCHIRUS Nardo. Name spelled BRACHIRUS when first used, page 71.

Pteropterus Swainson, II, 264; haplotype PTEROIS RADIATUS Cuv. & Val.

Platypterus Swainson, II, 265; logotype TÆNIONOTUS LATOVITTATUS Lac. Name preoccupied; replaced by TETRAROGE Gthr., 1860.

Pterichthys Swainson, II, 265; logotype SCORPÆNA CARINATA Bl. & Schn. A synonym of APISTUS Cuv., to be used if that name is regarded as preoccupied by APISTES, etc.; not PTERICHTHUS Commerson (1803), nor PTERICHTHYS Ag.

Gymnapistes Swainson, II, 265; logotype APISTUS MARMORATUS Cuv. & Val. Replaces PENTAROGE Gthr.

Synanchia ("Cuvier") Swainson, 268; a variant of SYNANCEJA: (S. EROSA).

Bufichthys Swainson, II, 268; logotype SCORPÆNA HORRIDA L. which is also type of SYNANCIDIUM Müller, 1844. Where BUFICHTHYS is first mentioned on page 180 it has the definition still earlier assigned to EROSA and here given to "SYNANCHIA". Gill therefore regards it as a synonym of EROSA Sw., rather than as equivalent of SYNANCIDIUM Müller.

Trachicephalus Swainson, II, 268; haplotype SYNANCEIA ELONGATA Cuv. & Val. Preoccupied by TRACHYCEPHALUS, 1838, a genus of reptiles. Replaced by POLYCAULUS Gthr.

Ichthyscopus Swainson, II, 269; logotype URANOSCOPUS INERMIS Cuv. & Val. Replaces ANEMA Gthr.

Enophrys Swainson, II, 271; haplotype COTTUS CLAVIGER Cuv. & Val. .

Gymnocanthus Swainson, II, 271; haplotype COTTUS VENTRALIS Cuv. & Val. Replaces PHOBETOR Kröyer, 1844, and ELAPHOCOTTUS Sauvage. This name should read GYMNACANTHUS. There seems to be no classical warrant for Swainson's persistent use of "CANTHUS" (χάνφος, a donkey) to mean spine (ἄχανϑα).

Canthirhynchus Swainson, II, 272; haplotype COTTUS MONOPTERYGIUS Bloch. A synonym of ASPIDOPHOROIDES Lac. (1802).

Hippocephalus Swainson, II, 272; logotype AGONUS SUPERCILIOSUS Pallas, as restricted by Gill, 1861. A synonym of PERCIS Scopoli (1777), not PERCIS of Klein, nor of Cuvier.

Chirolophis Swainson, II, 275; haplotype BLENNIUS YARRELLI Cuv. & Val. = BLENNIUS ASCANII Walbaum. Replaces BLENNIOPS Nilsson (1855) and CARELOPHUS LOPHUS Kröyer.

Erpichthys Swainson, II, 275; logotype SALARIAS QUADRIPINNIS Rüppell. A synonym of SALARIAS Cuv.

Rupiscartes Swainson, II, 275; haplotype SALARIAS ALTICUS Cuv. & Val. = BLENNIUS TRIDACTYLUS Bl. & Schn. Equivalent to ALTICUS Commerson, but the latter may not be eligible

Cirripectes Swainson, II, 275; haplotype SALARIAS VARIOLOSUS Cuv. & Val. A synonym of ALTICUS or RUPISCARTES.

Blennophis Swainson, II, 276; logotype CLINUS ANGUILLARIS Cuv. & Val., not BLENNOPHIS Val. (1844), which is OPHIOBLENNIUS Gill.

Labrisomus Swainson, II, 277; logotype CLINUS PECTINIFER Cuv. & Val. Often written LABROSOMUS; replaces LEPISOMA Dekay, 1842.

Ophisomus Swainson, II, 277; haplotype BLENNIUS GUNNELLUS L. Substitute for GUNNELLUS Cuv. & Val. A synonym of PHOLIS Scopoli (1777) and MURÆNOIDES Lac. (1803).

Ognichodes Swainson, II, 278; type GOBIOIDES BROUSSONNETI Lac., a needless substitute for GOBIOIDES Lac. Same as AMBLYOPUS Cuv. & Val.

Scartelaos Swainson, II, 279; logotype GOBIUS VIRIDIS Ham. Replaces BOLEOPS Gill, 1863.

Ruppellia (misprinted RUPELLIA) Swainson, II, 281; orthotype GOBIUS ECHINOCEPHALUS Rüppell. Preoccupied, replaced by PARAGOBIODON Blkr. (1879).

Amphichthys Swainson, II, 282; haplotype BATRACHUS RUBIGENES Sw., from Pernambuco. This fish is said to have but one dorsal fin and no teeth on the palate. It is undoubtedly based on BATRACHUS CRYPTOCENTRUS Cuv. & Val. It therefore replaces MARCGRAVIA Jordan (1886) as also the substitute name, MARCGRAVICHTHYS Ribeiro.

Salmophasia Swainson, II, 284; logotype CYPRINUS BACALLA Ham. (S. OBLONGA Sw.). Called SALMOSTOMA on page 184. A synonym of SALMOSTOMA, which replaces SECURICULA Gthr. 1868).

Chedrus Swainson, II, 285; haplotype CHEDRUS GRAYI Sw. = CYPRINUS CHEDRA Ham. A synonym of BARILIUS Ham.

Esomus Swainson, II, 285; haplotype ESOMUS VITTATUS Sw. = CYPRINUS DAURICA Ham. Replaces NURIA Cuv. & Val. (1842).

Clupisudis Swainson, II, 286; haplotype SUDIS NILOTICA Ehrenb. Replaces HETEROTIS Ehrenb. (1843).

Laurida Swainson, II, 287; logotype SALMO FŒTENS L. A synonym of SYNODUS Bl. & Schn.

Triurus Swainson, II, 288; haplotype TRIURUS MICROCEPHALUS Sw. = OSMERUS NEHEREUS Ham. A synonym of HARPODON Le Sueur. Preoccupied by TRIURUS Lac. (1800).

Mormyrhynchus Swainson, II, 291; haplotype M. GRONOVII Sw. = ANOSTOMUS Gronow. Same as ANOSTOMUS (Gronow), Scopoli (1777) = SCHIZODON Ag., 1829).

Trichosoma Swainson, II, 292; haplotype.TRICHOSOMA HAMILTONI Sw.; preoccupied by TRICHOSOMUS Sw., 265; and by TRICHOSOMA Rudolphi (1819), a genus of worms. Equivalent to THRISSA Cuv. (1817), not of Rafinesque (1815); replaced by THRISSOCLES Jordan (1917).

Setipinna Swainson, II, 292; logotype CLUPEA PHASA Ham. (SETIPINNA MEGALURA Sw.). Replaces TELARA Gthr. (1868).

Platygaster Swainson, II, 294; logotype CLUPEA AFRICANA Bloch. Preoccupied in bees, 1809. A synonym of ILISHA Gray, 1846.

Cypselurus Swainson, II, 296 (misprinted CYPSILURUS); logotype EXOCŒTUS NUTTALLI Le Sueur = EXOCŒTUS FURCATUS Mitchill. Based on a young example, with barbel at the chin. In view of Swainson's general carelessness and of the irregularity of his proof-reading, the form CYPSILURUS has been officially deemed a misprint by the International Communion of Nomenclature. Same as PTENICHTHYS Müller, 1843.

Ramphistoma (Rafinesque) Swainson, 296 orthotype R. VULGARIS Sw., same as BELONE Cuv.

Leptodes Swainson, II, 298; logotype CHAULIODUS SLOANI Bl. & Schn. A synonym of CHAULIODUS Bl. & Schn., (1801).

Tilesia Swainson, II, 300; orthotype GADUS GRACILIS Tilesius; synonym of ELEGINUS Fischer.

Cephus Swainson, II, 300; haplotype GADUS MACROCEPHALUS Tilesius. A synonym of GADUS L.

Lepidion Swainson, II, 300; haplotype GADUS LEPIDION Risso = LEPIDION RUBESCENS Sw., regarded as preoccupied by LEPIDEA (1817), a genus of worms; replaced by HALOPORPHYRUS Gthr. (1862).

Psetta (Aristotle) Swainson, II 302; haplotype PLEURONECTES MAXIMUS L. Sufficiently distinct from PSETTUS Cuv.

Platophrys Swaison, II, 302; haplotype PLEURONECTES OCELLATUS Spix & Ag. Replaces RHOMBOIDICHTHYS Blkr. (1856).

Brachirus Swainson, II, 303; logotype PLEURONECTES ORIENTALIS Bl. & Schn. Preoccupied by BRACHIRUS and BRACHYRUS Sw.; replaced by EURYGLOSSA Kaup (1858).

Plagusia (Browne) Swainson, 302; orthotype P. BILINEATA Bloch; name preoccupied, same as PARAPLAGUSIA Blkr.

Sturisoma Swainson, II, 304; haplotype LORICARIA ROSTRATA Spix.

Hoplisoma Swainson, II, 304; haplotype CATAPHRACTUS PUNCTATUS Bloch = CORYDORAS GEOFFROYI Lac. A synonym of CORYDORAS Lac., 1803.

Silonia Swainson, II, 305; logotype PIMELODUS SILONDIA Ham. = SILONIA LURIDA Sw. Replaces SILUNDIA Cuv. & Val. (1830).

Felichthys Swainson, II, 305; logotype, as restricted by Swainson and Bleeker, F. FILAMENTOSUS Sw. = ARIUS BAHIENSIS Castelnau. Name a substitute for BREVICEPS Sw., preoccupied; a synonym of BAGRE (Cuv.) Oken (1817), as restricted through tautonymy.

Cyclopium Swainson, II, 305; haplotype PIMELODUS CYCLOPUM Humboldt = CY-CLOPIUM HUMBOLDTI Sw. Replaces STYGOGENES Gthr. (1864).

Clupisoma Swainson, II, 306; haplotype SILURUS GARUA Ham., CLUPISOMA AR-GENTATA Sw. Replaces SCHILBEICHTHYS Blkr., 1858.

Pachypterus Swainson, II, 306; logotype SILURUS ATHERINOIDES Bloch. Preoccupied in Coleoptera (1833); replaced by PSEUDEUTROPIUS Blkr. (1863).

Pusichthys Swainson, II, 307; haplotype SCHILBE URANOSCOPUS Rüppell. A synonym of SCHILBE Cuv. (1817).

Cotylephorus Swainson, II, 308; haplotype PLATYSTACUS COTYLEPHORUS Bloch = C. BLOCHI Sw. A synonym of PLATYSTACUS Bloch, and of ASPREDO (Gronow) (Scopoli).

Pteronotus Swainson, II, 309; haplotype HETEROBRANCHUS SEXTENTACULATUS Ag. Preoccupied in mollusks; gives way to RHAMDIA Blkr. (1858).

Diacantha Swainson, II, 310; logotype DIACANTHA ZEBRA Sw. = COBITIS GETO Ham. (pl. 11). A synonym of BOTIA Gray, 1831.

Canthophrys Swainson, II, 310; logotype CANTHOPHRYS ALBESCENS Sw., based on COBITIS No. 3, Hamilton. Said by Swain to be a species of BOTIA Gray (1831).

Acoura Swainson, II, 310; logotype COBITIS SAVONA Ham. (ACOURA OBSCURA Sw.). A synonym of NEMACHEILUS Van Hasselt; spelled ACOURUS on an earlier page.

Somileptes Swainson II, 311; logotype S. BISPINOSA Sw. = COBITIS GONGOTA Ham. A synonym of COBITIS L., 1758.

Platysqualus Swainson, II, 311; logotype SQUALUS TIBURO Sw. (not of L.), based on a figure by Russell, representing SPHYRNA TUDES Cuv. A synonym of SPHYRNA Raf.

Pterocephala Swainson, II, 321; haplotype RAJA GIORNA Lac. A synonym of MOBULA Raf.

Tetrosomus Swainson, II, 323; haplotype OSTRACION TURRITUS Forskål. A section of OSTRACION L.

Rhinesomus Swainson, II, 324; logotype OSTRACION TRIQUETER Bloch. A synonym of LACTOPHRYS Sw.

Lactophrys Swainson, II, 324; logotype OSTRACION TRIGONUM L.

Platycanthus Swainson, II, 324; haplotype OSTRACION AURITUS Shaw. A synonym of ARACANA Gray.

Rhinecanthus Swainson, II, 325; logotype BALISTES ORNATISSIMUS Lesson = BAL-ISTES ACULEATUS L. A synonym of BALISTAPUS Tilesius.

Zenodon (Rüppell) Swainson, II, 194, 325; orthotype BALISTES NIGER Lac. This name ZENODON, being ascribed to Rüppell, is probably a careless misprint of ˋXENODON Rüppell. Preoccupied in reptiles by XENODON Boie, 1827. The name is spelled ZENODON on page 194 also. ZENODON is, however, accepted by Fowler (1904). If adopted, it replaces ODONUS Gistel (1848), ERYTHRODON Rüppell (1852) and PYRODON Kaup (1855)—all substitutes offered for XENODON.

Chalisoma Swainson, II, 325; logotype BALISTES PULCHERRIMUS Lesson = BALISTES VETULA L. A synonym of BALISTES L.

Canthidermis Swainson, II, 325; logotype BALISTES ANGULOSUS Q. & G.

Melichthys Swainson, II, 325; logotype BALISTES RINGENS Osbeck (not of L.). Name changed to MELANICHTHYS by Gthr. (1870).

Capriscus (Willughby) Swainson, II, 326; logotype BALISTES ERYTHROPTERON Lesson; a synonym of BALISTAPUS Tilesius.

Leiurus Swainson, II, 326; logotype LEIURUS MACROPHTHALMUS Sw., based on
LAMA YELLAKAH Russell, pl. 22 = BALISTES STELLARIS Bl. & Schn. Name
preoccupied by LEIURUS Sw. (p. 242); replaced by ABALISTES Jordan &
Seale (1911).

Chætodermis Swainson, II, 327; logotype MONOCANTHUS SPINOSISSMUS Q. & G.

Psilocephalus Swainson, II, 327; haplotype BALISTES BARBATUS Gray. Substitute
for ANACANTHUS Gray (1831), which is preoccupied by Ehrenberg, 1829.

Cantherines Swainson, II, 327; haplotype MONACANTHUS NASUTUS Q. & G.
Replaces LIOMONACANTHUS Blkr. (1866).

Cirrhisomus Swainson, II, 328; haplotype TETRAODON SPENGLERI Bloch. Same as
CHEILICHTHYS Müller, a synonym of SPHEROIDES Dum.

Leisomus Swainson, II, 328; logotype L. MARMORATUS Sw. = TETRAODON CUTCUTIA
Ham., as restricted by Blekeer, 1865. The genus is called LEIODON on page
194, without mention of type. LEISOMUS is preoccupied. LEIODON replaces
MONOTRETA Bibron. Swain restricts LEISOMUS and therefore LEIODON, to the
first species named, L. LÆVISSIMUS Sw., a species of SPHEROIDES (TETRAODON
SPENGLERI Bloch).

Lagocephalus Swainson, II, 328; logotype TETRAODON STELLATUS Donovan. Replaces GASTROPHYSUS Müller.

Trichoderma Swainson, II, 328; logotype BALISTES SCAPUS Lac. A synonym of
AMANSES Gray (1833).

Psilonotus Swainson, II, 328; logotype TETRAODON ROSTRATUS Bloch. Called CAN-
THIGASTER on page 194, and therefore a synonym of the latter.

Molacanthus Swainson, II, 329; logotype MOLACANTHUS PALLASI Sw. = DIODON
MOLA Pallas. Young of MOLA Kölr. = MOLA Cuv.

Astrocanthus Swainson, II, 331; haplotype LOPHIUS STELLATUS Vahl. A synonym
of HALIEUTÆA Cuv. & Val., 1837.

Phyllopteryx Swainson, II, 332; haplotype SYNGNATHUS FOLIATUS Shaw.

Acus (Willughby) Swainson, II, 333; logotype SYNGNATHUS ÆQUOREUS "Yarrell".
A synonym of NEROPHIS Raf.

Solegnathus Swainson, II, 333; haplotype SYNGNATHUS HARDWICKI Gray. Spelled
SOLENOGNATHUS by Günther.

Ophisoma Swainson, II, 334; logotype O. ACUTA Sw., as restricted by Bleeker
(1864). Swain, in 1882, chose O. OBTUSA Sw. as type. The latter species is a
CONGER, the former belongs to CONGERMURÆNA Kaup, as restricted. The
name is preoccupied by OPHISOMUS Sw., page 277. It is called ARIOSOMA on
page 194. ARIOSOMA thus replaces CONGERMURÆNA Kaup, and also CON-
GRELLUS Ogilby, all based on the same type, earlier described as MURÆNA
BALEARICA De la Roche.

Pterurus Swainson, II, 334, logotype PTERURUS MACULATUS Sw. = MURÆNA
RAITOBORUA Ham. A synonym of RAITABOURA Gray (1831) or MORINGUA
Gray (1831).

Leptognathus Swainson, I, 221, and II, 334; haplotype LEPTOGNATHUS OXYRHYN-
CHUS Sw. = MURÆNA SERPENS L. A synonym of OPHISURUS Lac., as
restricted by Risso.

Pachyurus Swainson, II, 335; haplotype MORINGUA LINEARIS Gray; preoccupied by
PACHYUBUS Ag. (1829). A synonym of PSEUDOMORINGUA Blkr. (1864).

Ophichthys Swainson, II, 336; haplotype UNIBRANCHAPERTURA CUCHIA Ham.
Name preoccupied by OPHICHTHUS Ahl. (1789). A synonym of AMPHIP-
NOUS Müller (1839).

Rupisuga Swainson, II, 339; haplotype R. NICENSIS Sw. = LEPODOGASTER BALBIS Risso. A synonym of GOUANIA Nardo (1833).

1840

252. BENNETT (1840). *Narrative of a Whaling Voyage Around the Globe.*
F. D. BENNETT.

Elagatis F. D. Bennett, II, 283; orthotype SERIOLA BIPINNULATA Q .& G.

253. BONAPARTE (1840). *Iconografia della fauna italica.* Fascicles 68 to 138, part 27. According to Troschel, fascicle 91 dates from 1837, 92 to 96 from 1838, 103 from 1839, and fascicle 138 dates from 1840.
CARLO LUCIANO BONAPARTE.

Stomocatus Bonaparte, 129*; logotype (not named) CYPRINUS CATOSTOMUS Forster. A synonym of CATOSTOMUS Le Sueur; species with the dorsal fin short, "especie . . . che han breve la pinna dorsale".

Ichthyocoris Bonaparte, fasc. 134; orthotype BLENNIUS VARUS Risso. A synonym of BLENNIUS L.

Ichthyococcus Bonaparte, fasc. 138; orthotype I. OVATUS Bon. Name later altered by Günther to COCCIA on account of its crude form (1840).

Lampanyctus Bonaparte, fasc. 138; orthotype L. BONAPARTEI Bon.

Chlorophthalmus Bonaparte, fasc. 138; orthotype C. AGASSIZI Bon.

254. CUVIER & VALENCIENNES (1840) *Histoire naturelle des poissons.* XV.
ACHILLE VALENCIENNES.

Galeichthys Cuvier & Valenciennes, XV, 28; logotype GALEICHTHYS FELICEPS Cuv. & Val., as restricted by Bleeker.

Pangasius Cuvier & Valenciennes, XV, 45; tautotype PIMELODUS PANGASIUS Ham.

Silundia Cuvier & Valenciennes, XV, 49; tautotype PIMELODUS SILONDIA Ham. A synonym of SILONIA Sw .

Arius Cuvier & Valenciennes, XV, 53; tautotype PIMELODUS ARIUS Ham. As so understood, a synonym of TACHYSURUS Lac., if the latter is correctly identified. Bleeker takes the first species named, ARIUS STRICTICASSIS Cuv. & Val., as type, thus replacing NOTARIUS Gill.

Trachelyopterus Cuvier & Valenciennes, XV, 220; orthotype T. CORIACEUS Cuv. & Val.

Arges Cuvier & Valenciennes, XV, 333; logotype ARGES SABALO Cuv. & Val.

Brontes Cuvier & Valenciennes, XV, 341; haplotype BRONTES PRENADILLA Cuv. & Val. Name two or three times preoccupied; replaced by STREPHON Gistel (1848).

Saccobranchus Cuvier & Valenciennes, XV, 399 haplotype SILURUS SINGIO Ham. = HETEROPNEUSTES Müller, 1839.

Chaca Cuvier & Valenciennes, XV, 444; haplotype PLATYSTACUS CHACA Ham.

255. EICHWALD (1840). *(Fossil Fishes.)* Bull. Acad. Sci. St. Petersburg.
CARL EDUARD VON EICHWALD.

Astrolepis Eichwald, VIII, 79; orthotype A. ORNATUS Eichw. (fossil). Later spelled ASTEROLEPIS by Eichwald and by most writers.

Bothriolepis Eichwald, VII, 79; orthotype B. PRISCA Eichw. (fossil).

256. HECKEL (1840). *Johann Natterer's neue Flussfische Brasilien's nach den Beobachtungen und Mittheilungen des Entdeckers beschrieben.* Abth. 1, *Die Labroiden.* Ann. Wien. Mus., II.

JOHANN JAKOB HECKEL.

Uaru Heckel, II, 330; orthotype U. AMPHACANTHOIDES Heckel.

Symphysodon Heckel, II, 332; orthotype S DISCUS Heckel.

Pterophyllum Heckel, II, 334; orthotype PLATAX SCALARIS Cuv. & Val.

Acara Heckel, II, 338; logotype A. CRASSISPINIS Heckel, as restricted by Gill (1858). A synonym of ASTRONOTUS Sw. (HYGROGONUS Gthr.).

Heros Heckel, II, 362; logotype H. SEVERUS Heckel, as restricted by J. & G., 1883. A synonym of CICHLAURUS or CICHLASOMA Sw.

Geophagus Heckel, II, 383; logotype G. MEGASEMA Heckel.

Chætobranchus Heckel, II, 401; logotype C. FLAVESCENS Heckel.

Crenicichla Heckel, II, 416; logotype C. VITTATA Heckel.

Batrachops Heckel, II, 432; logotype B. RETICULATUS Heckel.

Monocirrhus Heckel II, 439; orthotype M. POLYACANTHUS Heckel.

257. LOWE (1840). *Certain New Species of Madeiran Fishes.* Proc. Zool. Soc. London, 36-39.

RICHARD THOMAS LOWE.

Cheilopogon Lowe, VII, 38; orthotype CYPSELURUS PULCHELLUS Lowe. A manuscript name, withdrawn because of the earlier name CYPSELURUS Sw. Applied to young flying fish with barbels; equivalent also to PTENICHTHYS Müller.

258. MÜNSTER (1840). *Ueber einige Placoiden im Kupferschiefer zu Richselsdorf; Beiträge zur Petrefactenkunde.*

GEORG MÜNSTER.

Dictea Münster, III, 124; otrhotype D. STRIATA Münster (fossil). A synonym of JANASSA Münster.

259-261. MÜNSTER (1840). *(Braun, Petrefakten von Bayreuth.)*

GEORG MÜNSTER.

Saurorhynchus (Münster) Braun, 73 (not defined); orthotype SAURICHTHYS TENUIROSTRIS (Münster) (fossil). Name apparently eligible, as being applied to a species already described.

262. NORDMANN (1840). *Observations sur la fauna pontique; im Demidoff, voyage dans la Russie meridionale et la Crimée.* III.

ALEXANDER VON NORDMANN.

Percarina Nordmann, 357; orthotype P. DEMIDOFFI Nordmann.

263. OWEN (1840). *Odontography; or a Treatise on the Comparative Anatomy of the Teeth.*

RICHARD OWEN.

Petalodus Owen, 61; orthotype P. HASTINGSI Owen (fossil).

Rhizodus Owen, 75; orthotype MEGALICHTHYS HIBBERTI Ag. (fossil). A synonym of MEGALICHTHYS Ag.

264. RANZANI (1840). *De novis speciebus piscium; dissertationes quotuor.* Novi Comment. Acad. Sci. Inst. Bonon., IV.

CAMILLO RANZANI.

Syacium Ranzani, 20; orthotype S. MICRURUM Ranzani. Replaces HEMIRHOMBUS Blkr. (1862).

Diapterus Ranzani, IV, 340; orthotype D. AURATUS Ranzani.   Replaces GERRES J. & E.; not of Cuvier as properly restricted.

265.   RICHARDSON (1840).   *On Some New Species of Fishes from Australia.* Proc. Zool. Soc. London, VIII.

JOHN RICHARDSON.

Oplegnathus Richardson, 27; orthotype O. CONVAII Rich.   Name later corrected by Richardson to HOPLEGNATHUS CONWAYI; altered by Günther to HOPLOGNATHUS, in which form it is preoccupied.   If necessary, may be replaced by SCARODON T. & S.

266.   THOMPSON (1840).   *On a New Genus of Fishes from India (Bregmaceros macclellandi* Cantor, *MS.).*   Ann. Mag. Nat. Hist., IV.

WILLIAM THOMPON.

Bregmaceros (Cantor) Thompson, IV, 184; orthotype B. MACCLELLANDI (Cantor) Thompson.

1841

267.   AGASSIZ (1841).   *On the Fossil Fishes Found by Mr. Gardner in the Province of Ceará, in the North of Brazil.*   Edinb. New Philos. Journ., XXX.

LOUIS AGASSIZ.

Cladocyclus Agassiz, XXX, 83; orthotype C. GARDNERI Ag. (fossil).

Rhacolepis Agassiz, XXX, 83; logotype R. BUCCALIS Ag. (fossil).   (Misprinted PHACOLEPIS.)

Calamopleurus Agassiz, XXX, 84; orthotype C. CYLINDRICUS Ag. (fossil).   This genus belongs to the ELOPIDÆ, near to ELOPS, RHACOLEPIS and NOTELOPS.   See Jordan & Branner, *Smithsonian Misc. Coll.,* 1793-1908 ("Cretaceous Fishes of Ceará, Brazil").

268.   BONAPARTE (1841).   *Iconografia della fauna italica.*   3d part (fascicles 156-160 dating probably 1841).

CARLO LUCIANO BONAPARTE.

Pristiurus Bonaparte (1841); orthotype GALEUS MELASTOMUS Raf. (Written PRISTIURUS in *Saggio,* p. 121, 1831, and in *Fauna Italica.*   Called PRISTIDURUS in *Mem Soc. Sci. Neufchâtel,* II, p. 11, 1839), PRISTIURUS is the form in general use, but PRISTIDURUS may have priority of definition.

Syrraxis (Jourdan) Bonaparte, fasc. 160; orthotype NARCINE (SYRRAXIS)INDICA Henle.   Manuscript name based on specimens in the Museum of Lyons "ben ordinato di celebri Prof. Jourdan," identified by Bonaparte as NARCINE INDICA Henle.   A synonym of NARCINE Henle.

Agassiz quotes also a genus "POWERIA Bonaparte".   This is certainly an error.   Troschel writes, "GON. POWERLE" for GONOSTOMA.   This was doubtless mistaken for "Gen."

269.   EICHWALD (1841).   *Fauna Caspio-Caucasica non nullis observationibus novis illustrata.*   St. Petersburg.

CARL EDUARD VON EICHWALD.

Sypterus Eichwald; orthotype (PERCA LOPHAR Forskål).   A synonym of POMATO-MUS Lac. (not verified).

270.   FLEMING (1841).   *Description of a New Species of Skate (Hieroptera abredonensis) New to the British Fauna.*   Edinb. New Phil. Journ., XXXI, 1841.

JOHN FLEMING (1785-1857).

Hieroptera Fleming, XXXI, 236; orthotype H. ABREDONENSIS Fleming (a monstrous form of RAJA CLAVATA *fide* Gill). A synonym of RAJA L.

271. HECKEL (1841). *Ueber eine neue Gattung (Aulopyge hugeli) von Süsswasserfische in Europa; aus einen Schreiben an den Akademiker Dr. Brandt.* Bull. Sci. Acad. St. Petersb., VIII.

JOHANN JAKOB HECKEL.

Aulopyge (Brandt) Heckel, I, 1021; orthotype A. HUGELI Heckel (also in *Russegger's Reisen,* p. 1021).

272. HOGG (1841). *(Protomelus and Amphibichthys.)* Am. Mag. Nat. Hist., VII.

JOHN HOGG.

Protomelus Hogg, VII, 359; orthotype LEPIDOSIREN ANNECTENS Owen. A synonym of PROTOPTERUS Owen.

Amphibichthys Hogg, 362; orthotype LEPIDOSIREN PARADOXA Fitzinger. A synonym of LEPIDOSIREN Fitzinger.

273. LOWE (1841). *Synopsis of the Fishes of Madeira,* etc. Trans. Zool. Soc. London, II.

RICHARD THOMAS LOWE.

Aplurus Lowe, II, 180; orthotype A. SIMPLEX Lowe. A synonym of RUVETTUS Cocco.

Prometheus Lowe, II, 181; orthotype GEMPYLUS PROMETHEUS Cuv. & Val. Name preoccupied; replaced by PROMETHICHTHYS Gill (1893).

274. MILLER (1841). *The Old Red Sandstone; or New Walks in an Old Field.* Edinburgh.

HUGH MILLER (1802-1856).

Pterichthys (Agassiz) Miller, XXII; orthotype P. MILLERI Ag. (fossil). Name preoccupied—Swainson (1839); replaced by PTERICHTHYODES Blkr. (1839 and by MILLERICHTHYS S. A. Miller, 1892.

275. MÜNSTER (1841). *(Asterodon.)* Beiträge zur Petrefaktenkunde, IV.

GEORG MÜNSTER.

Asterodon Münster, IV, 140; orthotype A. BRONNI Münster (fossil). Probably identical with COLOBODUS Ag. and of earlier date.

276. NARDO (1841). *Proposizione per la formazioni di un nuovo generi di pesci, intitolato Brachyochirus.* Ann. Sci. Reale Lombardo-Veneto, Padova.

GIOVANNI DOMENICO NARDO.

Brachyochirus Nardo I; orthotype GOBIUS APHYA Risso = GOBIUS MINUTUS L. A section of GOBIUS; same as POMATOSCHISTUS Gill (1862). Perhaps to be regarded as preoccupied by BRACHYRUS and BRACHIRUS of Swainson (1839) of similar etymology.

277 OWEN (1841). *On the Structure of Fossil Teeth from the Central of Cornstone Division of the Old Red Sandstone, Indicative of a New Genus of Fishes, or Fish-like Batrachia,* etc. Micr. Journ., I.

RICHARD OWEN.

Dendrodus Owen, I, 4; logotype D. BIPORCATUS Owen (fossil).

Sphyrænodus (Agassiz) Owen, 129; orthotype S. PRISCUS Ag. (fossil). Names only; validated by Agassiz in 1844.

278. SYKES (1840). *The Fishes of Dukhun.* Ann. Mag. Nat. Hist., IV.
WILLIAM HENRY SYKES.

Rohtee Sykes, II, 364; orthotype R. VIGORSI Sykes. Replaces OSTEOBRAMA Heckel.

1842

279. AGASSIZ (1942). *(Diplacanthus.)* In Duff's *Geology of Moray.*
LOUIS AGASSIZ.

Diplacanthus (Agassiz) Duff, 71; orthotype D. CRASSISSIMUS Duff = D. STRIATUS
Ag. (fossil); more fully defined in *Poissons fossiles Vieux Grès Rouge.*
Name misspelled DIPLOCANTHUS by Duff.

280. CUVIER & VALENCIENNES (1842). *Histoire naturelle des poissons.* XVI.
ACHILLE VALENCIENNES.

Dangila Cuvier & Valenciennes, XVI, 229; logotype DANGILA CUVIERI Val.

Nuria Cuvier & Valenciennes, XVI, 238; logotype NURIA THERMOICOS Cuv. & Val.
A synonym of ESOMUS Sw.

Rohita Cuvier & Valenciennes, XVI, 242; tautotype CYPRINUS ROHITA Ham., ROHITA
RUCHANANI Val. Apparently identical with LABEO Cuv.

Capoëta Cuvier & Valenciennes, XVI, 278; tautotype CAPOËTA FUNDULUS Val. =
CYPRINUS CAPOËTA Güldenstadt (logotype C. AMPHIBIA Cuv. & Val.) as
restricted by Bleeker.

281. DEKAY (1842). *Zoology of New York; or, The New York Fauna. Comprising Detailed Descriptions of All the Animals Hitherto Observed Within
the State Borders,* etc. Natural History of New York State Geological Survey,
Albany.
JAMES ELLSWORTH DEKAY (1792-1851).

Lepisoma Dekay, 11; orthotype L. CIRRHOSUM Dekay = CLINUS PECTINIFER Cuv. &
Val. Regarded as preoccupied in beetles by LEPISOMUS Kirby. A synonym
of LABROSOMUS Sw.

Pileoma Dekay, 16; haplotype PILEOMA SEMIFASCIATA Dekay = ETHEOSTOMA CAPRODES
Raf. A synonym of PERCINA Haldeman, of earlier date in the same year.

Boleosoma Dekay, 20; haplotype B. TESSELLATUM Dekay = ETHEOSTOMA OLMSTEDI
Storer.

Uranidea Dekay, 61; haplotype URANIDEA QUIESCENS Dekay.

Apeltes Dekay, 68; orthotype GASTEROSTEUS QUADRACUS Mitchill.

Argyrea Dekay, 141; haplotype ATHERINA NOTATA Mitchill. Name preoccupied; a
synonym of MENIDIA Bon.

Amblyopsis Dekay, 187; haplotype AMBLYOPSIS SPELÆUS Dekay.

Stilbe Dekay, 204; haplotype CYPRINUS CRYSOLEUCAS Mitchill. A synonym of NO-
TEMIGONUS Raf.

Baione Dekay, 244; haplotype B. FONTINALIS Dekay (the young of SALMO FONTINALIS
Mitchill). A synonym of SALVELINUS Rich., 1836.

Acanthosoma Dekay, 350; orthotype A. CARINATUM Dekay. Same as MOLACANTHUS
Sw., a larval form of MOLA Cuv.

Pastinaca (Belon) Dekay, 373; orthotype RAJA PASTINACA L. Name used for the
same genus by several authors. A synonym of DASYATIS Raf.

282. GRAY (1842). *Description of Two Fresh-water Fish from New Zealand.*
(In Dieffenbach, Ernst: *Travels in New Zealand.* II.)
JOHN EDWARD GRAY.

Ptycholepis Gray, II, 218; orthotype MUGIL SALMONEUS Forster. A synonym of CHANOS Lac.

283. HALDEMAN (1842). *Description of Two New Species of the Genus Perca, from the Susquehannah River.* Journ. Acad. Nat. Hist.

SAMUEL STEHMAN HALDEMAN (1712-1890).

Percina Haldeman, VIII, 330; logotype PERCA NEBULOSA Haldeman = ETHEOSTOMA CAPRODES Raf. Same as PILEOMA Dekay and a little earlier in date.

284. HECKEL (1842). *Ichthyologie (von Syrien).* In Russegger (Joseph von) : *Reisen in Europa, Asien und Africa,* etc. Part I.

JOHANN JAKOB HECKEL.

Gibelion Heckel, I, 1014; orthotype CYPRINUS CATLA Ham. Replaces CATLA Cuv. & Val.

Cyprinion Heckel, I, 1015; logotype C. MACROSTOMUS Heckel.

Dillonia Heckel, I, 1020; orthotype CHONDROSTOMA DILLONI Cuv. & Val.

Scaphiodon Heckel, I, 1020; logotype S. TINCA Heckel.

Gymnostomus Heckel, I, 1030; logotype G. ARIZA Heckel, as restricted by Bleeker (1860) ; logotype CAPOËTA SYRIACA Cuv. & Val., as limited by Günther (1868). Apparently replaces TYLOGNATHUS Gthr., but not of Heckel.

Chondrochilus Heckel, I, 1031; orthotype C. NASICUS. A synonym of CHONDROSTOMA Ag.

Chondrorhynchus Heckel, I, 1031; logotype CH. SOETTA Heckel. A synonym of CHONDROSTOMA Ag.

Ballerus Heckel, I, 1032; tautotype CYPRINUS BALLERUS L.

Acanthobrama Heckel, I, 1033; logotype A. MARMID Heckel, as restricted by Bleeker.

Osteobrama Heckel, I, 1033; logotype CYPRINUS COTIO Ham. A synonym of ROHTEE Sykes (1841).

Pachystomus Heckel, I, 1038; orthotype P. SCHACRA Heckel. Name preoccupied. Same as SHACRA Blkr.

Argyreus Heckel, I, 1040; haplotype CYPRINUS ATRONASUS Mitchill. Name preoccupied. Equivalent to RHINICHTHYS Ag., 1850.

Leucosomus Heckel, I, 1042; haplotype L. CHRYSOLEUCUS Heckel, not of Mitchill = CYPRINUS CORPORALIS Mitchill. Considered as preoccupied by LEUCOSOMA Gray; replaced by CHEILONEMUS Baird (1855).

Glossodon Heckel, I, 1003; haplotype CYPRINUS SMITHI Rich.; identical with the type of GLOSSODON Raf. (1818). A synonym of HIODON Le Sueur (1818).

Luciobarbus Heckel, I, 1054; orthotype L. ESOCINUS Heckel.

Pseudophoxinus Heckel, I, 1063; orthotype PHOXINELLUS ZEREGI Heckel.

Discognathus Heckel, I, 1071; logotype D. VARIABILIS Heckel, as restricted by Bleeker, 1863. A synonym of GARRA Ham., as restricted by Bleeker.

Tylognathus Heckel, I, 1073; logotype T. DIPLOSTOMUS Heckel, as restricted by Bleeker, 1863, not TYLOGNATHUS, as restricted by Günther, 1868.

285. JENYNS (1842). *The Zoology of the Voyage of H.M.S. "Beagle", During the Years 1832-1836; edited by Charles Darwin: Fishes.* London.

LEONARD JENYNS (1800-1893) (afterwards LEONARD JENYNS BLOMEFIELD).

Arripis Jenyns, 14; orthotype CENTROPRISTES GEORGIANUS Cuv. & Val.

Prionodes Jenyns, 46; orthotype P. FASCIATUS Jenyns.

Stegastes Jenyns, 63; orthotype S. IMBRICATUS Jenyns = CHÆTODON LURIDUS Broussonet. This is probably identical with MICROSPATHODON Gthr., 1862, and of earlier date.

Acanthoclinus Jenyns, 92; orthotype A. FUSCUS Jenyns.

Mesites Jenyns, 118; orthotype M. ATTENUATUS Jenyns. A synonym of GALAXIAS Cuv.

Aplochiton Jenyns, 130; orthotype A. ZEBRA Jenyns. Spelled HAPLOCHITON by Günther.

Iluocœtes Jenyns, 166; orthotype I. FIMBRIATUS Jenyns.

Phucocœtes Jenyns, 168; orthotype P. LATITANS Jenyns.

286. McCLELLAND (1842). *Remarks on a New Genus of Thoracic Percoid Fishes, Cestreus.* Journ. Nat. Hist. Calcutta, IV.
JOHN McCLELLAND.

Cestreus McClelland, II, 150, 151; orthotype C. MINIMUS McCl. = ATHERINA DANIUS Ham. An ally of ELEOTRIS, near HYPSELEOTRIS, CARASSOPS, etc., small species with the habit of GAMBUSIA. D. V-9:A. 12. vertebræ 26, outer teeth minute, hooked (not "forked," as stated by Day).

287. McCLELLAND (1842). *On the Fresh-water Fishes Collected by William Griffin in His Travels, from 1835 to 1842.* Journ. Nat. Hist. Calcutta, II, 560-589.
JOHN McCLELLAND.

Racoma McClelland, II, 576; logotype R. LABIATA McCl., according to Bleeker.

Glyptosternon McCllelland, II, 584; logotype G. RETICULATUS McCl. Spelled GLYP-TOSTERNUM by Günther.

Olyra McClelland, II, 588; logotype O. LONGICAUDATA McCl.

288. MÜNSTER (1842). *Beschreibungen einiger neuer Fische aus den lithographi-schen Schiefern von Bayern.* Beiträge zur Petrefaktenkunde.
GEORG MÜNSTER.

Dorypterus (Germar) Münster, V. 35; orthotype D. HOFFMANNI Germar (fossil).

Globulodus Münster, V, 47; orthotype T. ELEGANS Münster (fossil).

Thaumas Münster, V. 62; orthotype T. ALIFER Münster (fossil). A fossil homo-logue of SQUATINA Dum.

Capitodus Münster, V, 68; orthotpe C. SUBTRUNCATUS Münster (fossil).

Soricidens Münster, V. 68; orthotype S. HAUERI Münster (fossil).

289. MÜNSTER (1842) *Neues Jahrb. für Mineralogie,* etc.
GEORG MÜNSTER.

Æthalion Münster, 41; logotype A. INFLATUS Münster (fossil).

Libys Münster, 45; orthotype L. POLYJTERUS Münster (fossil).

290. OWEN (1842). *(Raphiosaurus.)* Trans. Geol. Soc., London.
RICHARD OWEN.

Raphiosaurus Owen, VI, 413; orthotype R. SUBULIDENS Owen (fossil).

291. RICHARDSON (1842). *Catalogue of Fish Found at King George's Sound.* (In Eyre, E. J.: *Journals of Expeditions of Discovery into Central Aus-tralia.* I.)
JOHN RICHARDSON.

Centrogenys Richardson, IX, 120; orthotype SCORPÆNA VAIGIENSIS Blkr.

292. SCHOMBURGK (1842). *The Natural History of the Fishes of (British) Guiana.* Edinburgh.
ROBERT HERMANN SCHOMBURGK.

Ellipesurus Schomburgk, II, 184; orthotype E. SPINICAUDA Schomburgk (misprinted ELIPESURUS).

293. TEMMINCK & SCHLEGEL (1842). *Fauna Japonica, sive descriptio animalium quæ in itinere per Japoniam suscepto annis 1823-30 collegit, notis observationibus et adumbrationibus illustravit P. F. de Siebold. Conjunctis studiis C. J. Temminck et H. Schlegel pro vertebratis, atque W. de Haan pro invertebratis elaborata.* (Part 1R, *Pisces.*) Edited by Carl Theodor Ernst von Siebold, the text of fishes mostly by Hermann Schlegel.

> CŒNRAAD JACOB TEMMINCK (1770-1858); HERMANN SCHLEGEL (1804-1884).

> Temminck and Schlegel gave no specific names when but one species occurred in the genus. The names of type species have been given by subsequent writers.

Perca-Labrax Temminck & Schlegel, 2; orthotype LABRAX JAPONICUS Cuv. & Val. Name rejected by Bleeker and by Boulenger because no new genus was intended, LABRAX being merely entered as a subgenus of PERCA. Replaced by LATEOLABRAX Blkr. But an argument can be made for the retention of PERCALABRAX.

Aulacocephalus Temminck & Schlegel, 15; orthotype A. TEMMINCKI Blkr.

Anoplus Temminck & Schlegel, 17; orthotype ANOPOLUS BANJOS Rich. Name preoccupied; replaced by BANJOS Blkr. (1879).

294. VALENCIENNES (1842). *Determination des poissons: Voyage autour du monde . . . sur . . . la "Bonite,"* etc. Zoologie, I, Paris.

> ACHILLE VALENCIENNES.

Oxuderces Valenciennes, 181; orthotype O. DENTATUS Val.

1843.

295. AGASSIZ (1843). *Recherches sur les poissons fossiles.* III, in part.

> LOUIS AGASSIZ.

Orthacanthus Agassiz, III, pl. 45; orthotype O. CYLINDRICUS Ag. (fossil). Regarded by Woodward as a synonym of PLEURACANTHUS Ag., which preoccupied name it may replace.

Pleurodus Agassiz, III, 174; orthotype P. RANKINEI Ag. (fossil). No definition until accepted by Hancock and Althey, 1872. Regarded as preoccupied by PLEURODON (1840), replaced by PLEUROPLAX Woodw., 1889.

Cladacanthus Agassiz, III, 176; orthotype C. PARADOXUS Ag. (fossil); name only, defined first by Davis. *Trans. Roy. Dublin Soc.,* I, 1883, 365. A synonym of ERISMACANTHUS.

Cricacanthus Agassiz, III, 176; orthotype C. JONISI Ag. Name only, hence having no place in the system.

Gyropristis Agassiz, III, 177; orthotype G. OBLIQUUS Ag. Name only, hence without place in the system.

Cladodus Agassiz, III, 196; logotype C. MIRABILIS Ag. (fossil).

Sphenonchus Agassiz, III, 201; logotype S. ELONGATUS Ag. (fossil). A synonym, as restricted, of ACRODUS Ag.

Diplodus Agassiz, III, 204; logotype D. GIBBOSUS Ag. (fossil). Regarded by Woodward as a synonym of PLEURACANTHUS Ag. Both names are preoccupied. DIPLODUS has been replaced by DIDYMODUS Cope (1883), which is also preoccupied (as DIDYMODON Blake, 1863, a mammal) and again replaced by DICRANODUS Garman (1885). ORTHACANTHUS Ag. may replace PLEURACANTHUS, and some of the names given by Owen, as DITTODUS, have priority over DICRANODUS. It is not yet certain that the teeth called DIPLODUS and the spines called ORTHACANTHUS belong to the same fish.

Otodus Agassiz, III, 206; logotype O. OBLIQUUS Ag., not O. APPENDICULATUS Ag. (fossil). Apparently not separable from the living genus, LAMNA Cuv. (1817).

Corax Agassiz, III, 244; logotype GALEUS PRISTODONTUS Morton (fossil).

Hemipristis Agassiz, III, 237; orthotype H. SERRA Ag. (fossil). A genus represented by a living homologue, DIRRHIZODON ELONGATUS Klunzinger in the Red Sea.

Meristodon Agassiz, III, 286; orthotype OXYRHINA PARADOXA Ag. (fossil). Probably a synonym of HYBODUS.

Sphenodus Agassiz, III, 298; orthotype LAMNA LONGIDENS Ag. (fossil). Name regarded as preoccupied as SPHENODON Gray (1831); replaced by ORTHACODUS Woodw. (1889).

Carcharopsis Agassiz, III, 313; orthotype C. PROTOTYPUS Ag. (fossil). Name only; same as PRISTICLADODUS DENTATUS McCoy. A synonym of DICRENODUS Romanowsky(1853), which name Woodward adopts, as being first defined and with a described type.

Zygobatis Agassiz, III, 328; logotype RHINOPTERA STUDERI Ag. (fossil). Apparently a synonym of RHINOPTERA Kuhl.

Ganodus Agassiz, III, 339; orthotype CHIMÆRA OWENI Ag. (fossil).

Psittacodon Agassiz, III, 340; logotype ISCHYODUS PSITTACINUS Egert. (fossil). A synony of GANODUS.

Actinobatis Agassiz, III, 372; orthotype RAJA ORNATA Ag. (fossil). Regarded as a synonym of RAJA L. If so, the living species, RAJA ORNATA Garman, must receive a new name.

Scylliodus Agassiz, III, 377; orthotype S. ANTIQUUS Ag. (fossil). An ally of THYELLINA and SCYLLIORHINUS.

Arthropterus Agassiz, III, 379; orthotype A. RILEYI Ag. (fossil). A bit of some unrecognized ray.

Asterodermus Agassiz, III, 381; orthotype A. PLATYPTERUS (fossil). Name preoccupied by ASTRODERMUS Bonelli, 1829.

Thyellina Agassiz, III, 378; orthotype T. ANGUSTA Ag. (fossil). A fossil ally of SCYLLIORHINUS Blainv.

Narcopterus Agassiz, III, 382; orthotype N. BOLCANUS Ag. (fossil). Name only, hence not part of the system. An ally of TORPEDO Dum.

Cyclarthrus Agassiz, III, 382; orthotype C. MACROPTENES Ag. (fossil). A fin of some ray, imperfectly known.

Euryarthra Agassiz, III, 382; orthotype E. MUNSTERI Ag. (fossil). An ally of RHINOBATUS Linck and a synonym of ÆLLOPOS Münster.

296. EGERTON (1843). *On Some New Species of Fossil Chimæroid Fishes, with Remarks on Their Affinities* Proc. Geol. Soc., IV.

PHILIP DE MALPAS GREY EGERTON.

Ischyodus Egerton, IV, 155; orthotype CHIMÆRA COLEI (Buckland) Ag. (fossil).

Elasmodus Egerton, IV, 156; orthotype E. HUNTERI Egert. (fossil).

297. HALL (1843). *Natural History of New York.* (Palæontology.)

JAMES HALL.

Sauripteris Hall, IV. 282; orthotype S. TAYLORI Hall (fossil). Usually written SAURIPTERUS.

298. HECKEL (1843). *Abbildungen und Beschreibungen der Fische Syriens, nebst einer neuen Classification und Characteristik sämmtlicher Gattungen der Cyprinen.* Stuttgart.

Leucos (Bonaparte) Heckel, 48; logotype L. RUBELLUS Bon. Name regarded as preoccupied as LEUCUS Kaup, a genus of birds; replaced by CENISOPHIUS Bon.

Phoxinellus Heckel, 50; orthotype P. ALEPIDOTUS Heckel, as restricted by Bleeker, who later separated P. ZEREGI as PSEUDOPHOXINUS, 1859. Same as PARAPHOXINUS Blkr., 1863.

Mola Heckel, appendix, 257; orthotype CYPRINUS MOLA Ham. Preoccupied in fishes; replaced by AMBLYPHARYNGODON Blkr.

299. HECKEL (1843). *Ichthyologie (von Syrien)*. In Russegger (Joseph von): *Reisen in Europa, Asien und Africa, mit besonderer Rücksicht auf naturwissenschaftlichen Verhältnisse der betreffenden Länder, unternommen in den Jahren 1835 bis 1841.* Part 2, 1843.
JOHANN JAKOB HECKEL.

Cyrene Heckel, II, 284; logotype C. OCELLATA Heckel. A synonym of DANGILIA Cuv. & Val. (1842).

Schizopyge Heckel, II, 285; logotype SCHIZOTHORAX CURVIFRONS Heckel, as restricted by Bleeker, 1863.

Aspidoparia Heckel, II, 288; orthotype A. SARDINA Heckel.

300. LOWE (1843). *Notice of Fishes Newly Observed or Discovered in Madeira During the Years 1840, 1841 and 1842.* Proc. Zool. Soc. London.
RICHARD THOMAS LOWE.

Taractes Lowe, 82; orthotype T. ASPER Lowe.

Antigonia Lowe, 85; orthotype A. CAPROS Lowe.

Hesperanthias Lowe, 14; orthotype SERRANYS OCULATUS Cuv. & Val. A synonym of ETELIS Cuv. & Val.

Cubiceps Lowe, 82; orthotype PSENES GRACILIS Lowe. A synonym of PSENES Cuv. & Val.

Echiostoma Lowe, 87; orthotype E. BARBATUM Lowe.

Metopias Lowe, 90; orthotype M. TYPHLOPS Lowe. Preoccupied; replaced by MELAMPHAËS Gthr., 1864.

Gadella Lowe, 91; orthotype G. GRACILIS Lowe = GADUS MARALDI Risso. Replaces URALEPTUS Costa (1858).

Diaphasia Lowe, 92; orthotype GYMNOTUS ACUS Brünnich. A synonym of CARAPUS Raf.; a substitute for the "barbarous" name, FIERASFER.

301. LOWE (1843). *A History of the Fishes of Madeira,* etc.
RICHARD THOMAS LOWE.

Pleurothyris Lowe, 64; orthotype STERNOPTYX OLFERSI Cuv. Misprinted PLEUROTHYSIS by Günther. A synonym of ARGYROPELECUS Cocco.

302. MÜLLER (1843). *Beiträge zur Kenntniss der natürlichen Familien der Knochenfische.* Sitzb. Akad. Wiss. Berlin.
JOHANNES MÜLLER.

Encheliophis Müller, 153; orthotype E. VERMICULARIS Müller.

303. MÜLLER (1843). *Beiträge zur Kenntniss der natürlichen Familien der Fische.* Wiegmann's Arch. Naturgesch.
JOHANNES MÜLLER.

Cotylis Müller & Troschel, 297. haplotype LEPADOGASTER NUDUS Bl. & Schn. = GOBIESOX GYRINUS J. & E., not CYCLOPTERUS NUDUS L., which is not identified. Later restricted by Günther, 1861, COTYLIS FIMBRATA M. & T., a species described later and not at first included by Müller. (*Horæ Ichthyologiæ*, III, 297. For CoTYLIS Günther, not Müller, the name CoTYLICHTHYS Jordan, 1919, has been substituted. A synonym of GOBIESOX Lac.

Sicyases Müller & Troschel, 298; orthotype S. SANGUINEUS M. & T.

Synancidium Müller, 302; orthotype SCORPÆNA HORRIDA L., "SYNANCEIA mit Vomer-zähnen". A synonym of SYNANCEJA Bl. & Schn., as restricted.

Ptenichthys Müller, 312; logotype (EXOCŒTUS FURCATUS Mitchell; "EXOCŒTUS mit Barbfaden"). A synonym of CYPSELURUS Sw., 1839.

Hemiodus Müller, 316; orthotype H. CRENIDENS Müller.

Macrodon Müller, 163; orthotype M. TRAHIRA Müller = SYNODUS MALABARICUS Bl. & Schn. Name preoccupied by MACRODON Schinz, 1822; replaced by HOPLIAS Gill.

Euanemus Müller, 318; orthotype E. COLUMBETES Müller = HYPOPHTHALMUS NUCHALIS Spix. A synonym of AUCHENIPTERUS Cuv. & Val., as restricted by Bleeker and by Eigenmann.

Calophysus Müller, 318; orthotype PIMELODUS MACROPTERUS Lichtenstein.

Mormyrops Müller, 324; logotype MORMYRUS ANGUILLOIDES Geoffroy St. Hiraire.

Gastrophysus Müller, 330; orthotype TETRAODON LUNARIS L. Substitute for PHY-SOGASTER Müller, preoccupied. A synonym of LAGOCEPHALUS Sw.

Arapaima Müller, 326; orthotype SUDIS GIGAS Cuv. A sybstitute for SUDIS Cuv., preoccupied by SUDIS Raf. (1810).

Heterotis (Ehrenberg) Müller, 326; orthotype SUDIS NILOTICUS Cuv. A synonym of CLUPISUDIS Sw.

304. MÜNSTER (1843). *Beiträge zur Petrefaktenkunde.* VI.
GEORG MÜNSTER.

Wodnika Münster, VI, 48; orthotype ACRODUS ALTHAUSI Münster (fossil).

Byzenos Münster, VI, 50; orthotype B. LATIPINNATUS Münster (fossil). A synonym of JANASSA.

305. PORTLOCK (1843). *(Geological Survey of Londonderry.)*
JOSEPH ELLISON PORTLOCK.

Tristychius Portlock, 464; orthotype T. MINOR Portlock (fossil).

306. RICHARDSON (1843). *Description of the Lurking Macherie (Machærium subducens) from the Northern Coast of New Holland.* Ann. Mag. Nat. Hist. VII.

JOHN RICHARDSON.

Machærium Richardson, XII, 175; orthotype M. SUBDUCENS Rich. Name pre-occupied; replaced by CONGROGADUS Gthr., 1862.

307, 308. TEMMINCK & SCHLEGEL (1843). *Fauna Japonica* (part).
COENRAAD JACOB TEMMINCK; HERMANN SCHLEGEL.

Acropoma Temminck & Schlegel, 41; orthotype ACROPOMA (JAPONICUM Gthr.).

Aploactis Temminck & Schlegel, 51; orthotype A. ASPERA Rich.

Glaucosoma Temminck & Schlegel, 1843, 62; orthotype G. BURGERI Rich.

Caprodon Temminck & Schlegel, 64; orthotype CAPRODON = ANTHIAS SCHLEGELI Gthr.

1844

309. AGASSIZ (1844). *Recherches sur les poissons fossiles.* I, II, III, IV, V (parts completed).
LOUIS AGASSIZ.

Placosteus Agassiz, I, XXIII; orthotype P. MÆANDRINUS Ag. (fossil); name only, replaced in 1845 by PSAMMOSTEUS Ag.

Chelonichthys Agassiz, I, XXXIII; logotype C. ASMUSSI Ag. (fossil); name only. A synonym of ICHTHYOSAUROIDES Kutorga.

Glyptosteus Agassiz, I, XXXIV; orthotype G. RETICULATUS Ag. (fossil); name only. A synonym of BOTHRIOLEPIS.

Psammolepis Agassiz, I, XXXIV; orthotype P. PARADOXUS Ag. (fossil); name only, replaced in 1845 by PSAMMOSTEUS Ag.

Scaphodus Agassiz, I, XLIII; orthotype GYRONCHUS OBLONGUS Ag. (fossil). A synonym of GYRONCHUS.

Pododus Agassiz, II, 83; orthotype P. CAPILLATUS Ag. (fossil); name only. Identified by Woodward, 1901, as a synonym of MESOLEPIS Young.

Megalichthys Agassiz, II, 89, 154; orthotype M. HIBBERTI Ag. (fossil). Not MEGALICHTHYS Ag. 1836. Should apparently stand as CENTRODUS McCoy.

Conodus Agassiz, II, 105; orthotype C. FEROX Ag. (fossil); name only; preoccupied as CONODON Cuv. & Val., 1830. A synonym of CATURUS Ag.

Thrissonotus Agassiz, II, 128; orthotype T. COLEI Ag. (fossil); name only. Same as OXYGNATHUS Egert.; defined by Egerton, 1858.

Microspondylos Agassiz, II, 139; orthotype M. ESCHERI Ag. (no definition: shales of Glarus).

Cricodus Agassiz, II, 156; orthotype DENDRODUS INCURVUS Duff. (fossil).

Amblysemus Agassiz, II, 165; orthotype A. GRACILIS Ag. (fossil).

Macropoma Agassiz, II, 174; orthotype M. EGERTONI Ag. (fossil); name only. Defined by Egerton, *Mem. Geol. Surv.*, IX, No. 10, 1858, from which the name should date. A synonym of EURYCORMUS Wagner, 1863, and of earlier date.

Uronemus Agassiz, II, 178; orthotype U. LOBATUS Ag. (fossil).

Hoplopygus Agassiz, II, 178; orthotype H. BINNEYI Ag. (fossil). A synonym of CŒLACANTHUS.

Gyrosteus Agassiz, II, 179; orthotype G. MIRABILIS Ag. (fossil); name only; validated by Egerton, *Phil. Trans.*, 1858, 883.

Ctenolepis Agassiz, II, 180; orthotype C. CYCLUS Ag. (fossil). Undetermined scales; defined by Woodward, 1895.

Periodus Agassiz, II, 201; orthotype P. KOENIGI Ag. A synonym of PYCNODUS Ag.

Gyronchus Agassiz, II, 202; orthotype G. OBLONGUS Ag. (fossil).

Acrotemnus Agassiz, II, 203; orthotype A. FABA Ag. (fossil).

Colobodus Agassiz, II, 237; orthotype C. HOGARDI Ag. (fossil). Apparently a synonym of ASTERODON Münster.

Phyllodus Agassiz, II, 238; orthotype P. TOLIAPICUS Ag. (fossil).

Acanthoderma Agassiz, II, 251; logotype A. SPINOSUM Ag. (fossil). Name preocupied. See AGOREION Gistel, 1847.

Acanthopleurus Agassiz, II, 253; orthotype PLEURACANTHUS SERRATUS Ag. (fossil). This name may be regarded as preoccupied by ACANTHOPLEURA, 1829.

Rhinellus Agassiz, II, 260; orthotype R. FURCATUS Ag. (fossil), regarded as preoccupied by RHINELLA Fitzinger, 1826; replaced by ICHTHYOTRINGA Cope, 1878.

Glyptocephalus Agassiz, II, 264; orthotype G. RADIATUS Ag. (fossil). Name preoccupied, Gottsche, 1835. A synonym of BUCKLANDIUM König.

Chondrosteus Agassiz, II, 280; orthotype C. ACIPENSEROIDES Ag. (fossil); name only; validated by Egerton, *Phil. Trans.*, 1858, 871.

Nothosomus Agassiz, II, 292; orthotype N. OCTOSTYCHIUS Ag. (fossil); name only; described by Egerton, *Mem. Geol. Surv.*, 1858, Dec. IX, No. 6. An ally of PHOLIDOPHORUS.

Coccolepis Agassiz, II, 300; orthotype C. BUCKLANDI Ag. (fossil).

Plectrolepis Agassiz, II, 306; orthotype P. RUGOSUS Ag. (fossil) ; name only; defined by Egerton, 1850. A synonym of NOTAĆMON Gistel, which replaces EURYNOTUS Ag.

Oxyrhina Agassiz, III, 276; orthotype ISURUS SPALLANZANI Raf. Preoccupied; replaced by ·PLECTROSTOMA Gistel (1848). A synonym of ISURUS Raf., 1810.

Cœlacanthus Agassiz, IV, 83; orthotype C. GRANULATUS Ag. (fossil).

Rhamphosus Agassiz, IV, 270; orthotype CENTRISCUS ACULEATUS Blainv. (fossil).

Urosphen Agassiz, IV, 284; orthotype U. FISTULARIS Ag. (fossil). (Printed as name only in *Neues Jahrb.*, 1835, 293.)

Notæus Agassiz, V, 12; orthotype N. LONGICAUDUS Ag. (fossil). Name regarded as preoccupied, as NOTEUS Ehrenb., 1830; replaced by AMIOPSIS Kner. A synonym of CYCLURUS Ag.

Cyclurus Agassiz, V, 12; orthotype V. VALENCIENNESI Ag. (fossil). Regarded as preoccupied by CYCLURA Harlan, 1825, a reptile, replaceable by AMIOPSIS Kner. A fossil homologue of AMIA L. (AMIATUS Raf.).

Sphenolepis Agassiz, V, 13; orthotype CYPRINUS SQUAMOSSEUS Blainv. (fossil). Name preoccupied, same as NOTOGONEUS Cope, 1885; perhaps a synonym of ANORMURUS Blainv., 1818.

Autolepis Agassiz, V, 14; orthotype A. TYPUS Ag.

Enchelyopus Agassiz, V, 16; orthotype E. TIGRINUS Ag. (fossil). (As name only in *Neues Jahrb.*, 1835, 307.) Name preoccupied; replaced by PARANGUILLA Blkr., 1864.

Pamphractus Agassiz, V, 20; orthotype P. HYDROPHILUS Ag. (fossil). A synonym of BOTHROLEPIS.

Coccosteus Agassiz, 22; orthotype C. DECIPIENS Ag. (fossil).

Polyphractus Agassiz, V, 29; orthotype P. PLATYCEPHALUS Ag. (fossil). A synonym of DIPTERUS Sedgwick & Murchison.

Carangopsis Agassiz, V, 39; logotype C. LATIOR Ag. (fossil). Published as name only, *Neues Jahrb.*, 1835, 293.

Amphistium Agassiz, V, 44; orthotype A. PARADOXUM Ag. (fossil). (Noted as name only, *Neues Jahrb.*, 1835, 294.)

Halecopsis Agassiz, V, 44; orthotype H. LÆVIS Ag. (fossil) ; names only; given definition by Woodward, 1901.

Palimphyes Agassiz, V, 46; logotype P. LONGUS Ag. (fossil).

Archæus Agassis, V, 49; orthotype A. GLARISIANUS Ag. (fossil).

Isurus Agassiz, V, 51; orthotype ISURUS MACRURUS Egert. (fossil). (Printed as name only by Egerton, *Cat Foss. Fish,* 1837.) Name preoccupied; replaced by ISURICHTHYS Woodward, 1901.

Pleionemus Agassiz, V, 52; orthotype P. MACROSPONDYLUS Ag. (fossil). Name only. A fragment of an ARCHÆUS.

Ductor Agassiz, V, 53; orthotype D. LEPTOSOMUS Ag .(fossil). Printed in 1834 as name only in *Verhandl. Mus. Böhmen,* 66.

Glyptopomus Agassiz, 57; orthotype G. MINOR Ag. (fossil).

Placothorax Agassiz, 134; orthotype P. PARADOXUS Ag. (fossil). A synonym BOTHRIOLEPIS.

Platygnathus Agassiz, V, 61; orthotype P. JAMESONI Ag. (fossil). A synonym of HOLOPTYCHUS.

Glyptolepis Agassiz, V, 62; logotye G. LEPTOPTERUS Ag. (fossil).

Goniognathus Agassiz, V, 63; logotype G. MAXILLARIS Ag. (fossil) ; name only. A synonym of AMPHERISTUS König.

Phyllolepis Agassiz, V, 67; orthotye P. CONCENTRICA Ag. (fossil).

Nemopteryx Agassiz, V, 75; orthotype CYCLURUS CRASSUS Egert. (fossil).

Holosteus Agassiz, V, 85; orthotye H. ESOCINUS Ag. (fossil).

Hemirhynchus Agassiz, V, 87; orthotype HISTIOPHORUS DESHAYESI Ag. (fossil). Name preoccupied; perhaps the same as HOMORHYNCHUS Van Beneden, 1873.

Istieus Agassiz, V, 91; logotype I. GRANDIS Ag. (fossil). An extinct ally of PTEROTH-RISSUS Hilgendorf.

Sphyrænodus Agassiz, V, 98; orthotype SPHYRÆNODUS PRISCUS Ag. (fossil). Name printed by Owen, 1841.

Cœlorhynchus Agassiz, V, 92; orthotype C. RECTUS Ag. (fossil); name preoccupied.

Hypsodon Agassiz, V, 101; orthotype H. TOLIAPICUS Ag., name only (fossil). Apparently a synonym of ally of MEGALOPS Lac.

Osmeroides Agassiz, V, 103; orthotype O. MONASTERI Ag. (fossil). The name has been transferred by authors to SALMO LEWESIENSIS Mantell, the type of HOLCOLEPIS von der Marck. OSMEROIDES should replace SARDINIOIDES von der Marck.

Rhamphognathus Agassiz, V, 9, 104; orthotye R. PARALEPOIDES Ag. (fossil).

Mesogaster Agassiz, V, 9, 105; orthotype M. SPHYRÆNOIDES Ag. (fossil)

Spinacanthus Agassiz, V, 107; orthotype S. BLENNIOIDES Ag. (fossil).

Acrognathus Agassiz, V, 108; orthotype A. BOOPS Ag. (fossil).

Rhynchorinus Agassiz, V, 139; orthotype R. BRANCHIALIS Ag. (fossil); name only; validated by Woodward, 1901.

Cœlocephalus Agassiz, V, 139; orthotype C. SALMONEUS Ag. (fossil); name only. Preoccupied before defined by CŒLOCEPHALUS Clark, 1860, and by CŒLO-CEPHALUS Gilbert & Cramer, 1897; replaced by EOTHYNNUS Woodw., 1901.

Halecopsis Agassiz, V, 139; orthotype H. LÆIS Ag. (fossil) = OSMEROIDES INSIGNIS Delvaux & Ortlieb (fossil); name only; validated by Woodward, 1901.

Elopides Agassiz, V, 139; orthotype E. COULONI Ag. (fossil); name only; validated by Woodward, 1901.

310. COCCO (1844). *Intorno ad alcuni nuovi pesci del mare di Messina.* Lettera del Prof. Anastasio Cocco al Signor Augusto Krohn.

ANASTASIO COCCO.

Krohnius Cocco, 1; orthotype K. FILAMENTOSUS Cocco. Young of CŒLORHYNCHUS Giorna.

Bibronia Cocco, 15; orthotype B. LIGULATA Cocco. Larva of SYMPHURUS Raf.

Peloria Cocco, 21; orthotype P. HECKELI Cocco. A larva of PLATOPHRYS Sw.

Coccolus (Bonaparte) Cocco, 21; orthotype C. ANNECTENS Cocco. A larva of PLA-TOPHRYS Sw.

311. CUVIER & VALENCIENNES (1844). *Histoire naturelle des poissons.* XVII.

ACHILLE VALENCIENNES.

Catla Cuvier & Valenciennes, XVII, 140; tautotype CYPRINUS CATLA Ham. A synonym of GIBELION Heckel.

Sclerognathus Cuvier & Valenciennes, XVII, 472; logotype, restriction of Günther, 1868, SCLEROGNATHUS CYPRINELLA Cuv. & Val.; the original diagnosis drawn up from CATOSTOMUS CYPRINUS Le Sueur, a species of CARPIODES Raf. A synonym of ICTIOBUS Raf., as here accepted. This view is not taken by Fowler, who regards CARPIODES CYPRINUS as type of SCLEROGNATHUS and gives to I. CYPRINELLA the new name of MEGASTOMATOBUS.

312. EGERTON (1845). *Description of a Fossil Ray from Mount Lebanon (Cyclobatis oligodactylus).* Quart. Journ. Geol. Soc., I.

PHILIP DE MALPAS GREY EGERTON.

Cyclobatis Egerton, IV, 442; orthotype C. OLIGODACTYLUS Egert. (fossil).

313. EICHWALD (1844). *Ueber die Fische des devonischen Systems in der Gegend von Pawlowsk.* Bull. Soc. Nat. Moscow, XVII.

CARL EDWARD VON EICHWALD.

Sclerolepis Eichwald, XVII, 828; orthotype S. DECORATA Eichw. (fossil). A synonym of GLYPTOLEPIS.

Microlepis Eichwald VXII, 830; logotype M. EXILIS Eichw. (fossil). A doubtful genus, the name preoccupied in reptiles.

Chiastolepis Eichwald, XVIII, 831; orthotype C. CLATHRATA Eichw. (fossil).

[GOLDFUSS (1844). SCLEROCEPHALUS HAEUSERI Goldfuss, 1844, described as a ganoid fish, is a reptile.]

314. KONINCK (1844). *Description des animaux fossiles qui se trouvent dans le terrain carbonifère de Belgique.*

LAURENT (GUILLAUME) DE KONINCK (1809-1887).

Campodus Koninck, 617; orthotype C. AGASSIZIANUS Koninck (fossil).

315. McCLELLAND (1844). *Description of a Collection of Fishes Made at Chasun and Ningpo in China by G. R. Playfair.* Journ. Nat. Hist. Calcutta, IV.

JOHN McCLELLAND.

Cossyphus McClelland, IV, 403; orthotype C. ATER McCl. Preoccupied; replaced by PHAGORUS McCl.

Phagorus McClelland, IV, 403; orthotype COSSYPHUS ATER McCl. = MACROPTERONOTUS JAGUR Ham. Replaces COSSYPHUS McCl., preoccupied.

Chætomus McClelland, IV, 405; orthotype C. PLAYFAIRI McCl.

Murænesox McClelland, V, 408; logotype M. TRICUSPIDATA McCl.

316. McCLELLAND (1844). *Apodal Fishes of Bengal.* Journ. Nat. Hist. Calcutta, V.

JOHN McCLELLAND.

Lycodontis McClelland, V, 173; logotype L. LITERATA McCl., as restricted by Jordan & Evermann, 1896. A synonym of GYMNOTHORAX Bloch, as here understood.

Thærodontis McClelland, V, 174; orthotype T. RETICULATA McCl. A synonym of GYMNOTHORAX.

Strophidon McClelland, V, 187; orthotype LYCODONTIS LONGICAUDATA McCl. A synonym of GYMNOTHORAX, the name wrongly transferred by Bleeker and Günther to the genus PSEUDOCHIDNA Blkr.

Ophicardia McClelland, V, 191; orthotype O. PHAYRIANA McCl. A synonym of FLUTA Bl. & Schn. = MONOPTERUS Lac.

Pneumabranchus McClelland, V, 192; logotype P. STRIATUS McCl. A synonym of AMPHIPNOUS Müller.

Ophisternon McClelland, V, 197; logotype O. BENGALENSIS McCl. A synonym o: SYMBRANCHUS Bloch.

Ptyobranchus McClelland, V, 200; logotype P. ARUNDINACEUS McCl. A synonym of RAITOBOURA Gray = MORINGUA Gray.

Ophithorax McClelland, V, 212; logotype "OPHISURUS OPHIS Lac." (MURÆNA MACULOSA Cuv.). The species called MURÆNA OPHIS by Linnæus is the type of the genus OPHICHTHUS of Ahl. MURÆNA OPHIS Bloch is probably the same and M. OPHIS Lac. seems to be drawn from the same material. The figure of Lacépède is the sole basis of Cuvier's MURÆNA MACULOSA. All these accounts seem to refer to the fish called OPHICHTHUS HAVANNENSIS (Bl. & Schn.). The MACULOSUS of Kaup and Günther is a different fish, a species of MYRICHTHYS. McClelland places "OPHISURUS OPHIS Lac." first among his species of OPHITHORAX. We therefore take it as his type, although his description is rather dubious. "Dorsal fin broad, anal narrow; body and dorsal covered with large round spots. Pectoral fin very minute. Seas of Europe." The genus is defined: "Pectoral fins very small so as to be scarcely perceptible." It is possible that this description applies to the species called OPHICHTHYS MACULOSUS by Günther (not by Cuvier). This is a species of MYRICHTHYS, thus far without tenable name. Using Günther's description of OPHICHTHYS MACULOSUS as the type, the species may be named MYRICH-THYS GUNTHERI Jordan. It seems best to regard OPHITHORAX as a synonym of OPHICHTHYS Ahl, rather than to substitute it on doubtful grounds for MYRICHTHYS or for CHLEVASTES (COLUBRINUS).

317. MEYER & PLIENINGER (1844). *Paläontologie von Württemberg.* CHRISTIAN ERICH VON MEYER (1801-1894); WILHELM HEINRICH THEODOR VON PLIENINGER (1795-1879).

Thectodus Meyer & Plieninger, 116; logotype T. INFLATUS Müller & Plieninger (fossil). A synonym of ACRODUS.

318. MÜLLER & TROSCHEL (1844). *Synopsis generum et specierum familiæ Characinarum.* Wiegmann's *Archiv. Naturg.* (Also in *Horæ Ichthyologicæ,* III, 1845.)
JOHANNES MÜLLER; FRANZ HERMANN TROSCHEL (1810-1882).

Chilodus Müller & Troschel, 85; haplotype CHILODUS PUNCTATUS M. & T. Pre-occupied; replaced by MICRODUS Kner, 1859; also preoccupied, and by CÆNO-TROPUS Gthr., 1864.

Alestes Müller & Troschel, 88; orthotype SALMO NILOTICUS Geoffroy St. Hilaire.

Brycon Müller & Troschel, 90; logotype CHARACINUS MICROLEPIDOTUS Cuv.

Exodon Müller & Troschel, 91; haplotype EXODON PARADOXUS M. & T. Said to be preoccupied; replaced by HYSTRICODON Gthr. 1864

Epicyrtus Müller & Troschel, 92; haplotype SALMO GIBBOSUS Gronow. Preoccupied; replaced by ANACYRTUS Gthr., 1864. A synonym of CHARAX (Gronow) Scopoli.

Xiphorhamphus Müller & Troschel, 92; orthotype SALMO FALCATUS Bloch. Sub-stitute for XIPHORHYNCHUS Ag., preoccupied. This name is also preoccupied and changed to ACESTRORHYNCHUS Eigenmann (1903). "Nomen XIPHORHYN-CHUS usurpatum in Ornithologiæ" (M. & T.).

Hydrolycus Müller & Troschel, 93; orthotype HYDROCYON SCOMBEROIDES Cuv.

Pygocentrus Müller & Troschel, 94; logotype SERRASALMO PIRAYA Cuv.

Pygopristis Müller & Troschell, 95 logotype SERRASALMO DENTICULATUS Cuv.

Catoprion Müller & Troschel, 96; haplotype SERRASALMO MENTO Cuv.

Myleus Müller & Troschel, 98; orthotype MYLEUS SETIGER Müller.

319. NARDO (1844). *Proposizione per la formazione di un nuovo genere di Selachi chiamato Caninoa o Caninotus, che costiturebbe una nuova sottofamiglia prossima ai Notidani.* Ann. Soc. Lombardo-Veneto, III.
GIOVANNI DOMENICO NARDO.

Caninoa Nardo, 8; orthotype SQUALUS BARBARUS (Chierigini MSS.) = C. CHIERE-GHINI Nardo. An anomalous shark, not since recognized, if existing.

Caninotus Nardo, 8; orthotype CANINOA CHIEREGHINI Nardo. (SQUALUS BAR-
BARUS Chierigini MSS.) Name a provisional substitute for CANINOA.

320. PETERS (1844). *Ueber einem den Lepidosirens annectens verwandten,
mit Lungen und Kiemen zugleich versehenen Fisch aus den Sumpfen von
Quellimane (Rhinocryptis amphibia).* Mon. Akad. Berlin.
WILHELM CARL HARTWIG PETERS (1815-1883).

Rhinocryptis Peters, 414; orthotype R. AMPHIBIA Peters = LEPIDOSIREN ANNECTENS
Owen. A synonym of PROTOPTERUS Owen.

320A. REUSS (1844). *Geognostische Skizzen aus Böhmen.*
AUGUST EMANUEL VON REUSS (1811-1873).

Thaumaturus Reuss, II, 264; orthotype T. FURCATUS Reuss (fossil). An extinct
form of trout.

321. RICHARDSON (1844). *Description of a Genus of Chinese Fish (Hapalogenys).* Ann. Mag. Nat. Hist., XIII.
JOHN RICHARDSON.

Hapalogenys Richardson, XIII, 463; logotype H. NITENS Rich. = POGONIAS NIGRI-
PINNIS T. & S.

322. RICHARDSON (1844). *Description of a New Genus of Gobioid Fish
(Channichthys).* Ann. Mag. Nat. Hist. XIII.
JOHN RICHARDSON.

Channichthys Richardson; orthotype C. RHINOCERATUS Rich. Name afterward
spelled CHÆNICHTHYS, as dictated by a purist, an unwarranted change.

,323. RICHARDSON (1844). *Generic Characters of an Undescribed Australian
Fish (Patæcus fronto).* Ann. Mag. Nat. Hist., XIV.
JOHN RICHARDSON.

Patæcus Richardson, XIV, 280; orthotype P. FRONTO Rich.

324. RICHARDSON (1844). *Ichthyology.* (In *The Zoology of the Voyage of
H.M.S. "Sulphur", Under the Command of Captain Sir Edward Belcher,
During the Years 1836-42.* Edited by R. B. Hinds).
JOHN RICHARDSON.

Chæturichthys Richardson, 54; orthotype C. STIGMATIAS Rich.

Centridermichthys Richardson, 74; orthotype C. ANSATUS Rich. = TRACHIDERMUS
FASCIATUS Heckel. Replaces TRACHIDERMUS Heckel, regarded as preoccupied
by TRACHYDERMA.

Calloptilum Richardson, 94; orthotype C. MIRUM Rich. A synonym of BREGMA-
CEROS Cantor.

Astronesthes Richardson, 97; orthotype A. NIGRA Richardson.

325. RICHARDSON (1844). *The Zoology of the Voyage of H.M.S. "Erebus
and Terror", Under the Command of Captain Sir J. C. Ross . . . During
1839-43.* Edited by J. Richardson . . . and J. E. Gray. London.
JOHN RICHARDSON.

Notothenia Richardson, 5; logotype N. CORIICEPS Rich.

Harpagifer Richardson, 11; orthotype CALLIONYMUS BISPINIS Forster.

Chænichthys Richardson, 12; orthotype C. RHINOCERATUS Rich.; a substitute for
CHANNICHTHYS.

Pagetodes Richardson, 15; orthotype (not given a specific name, a cat having run
away with the type specimen).

Emmelichthys Richardson, 47; orthotype E. NITIDUS Rich.
Cheilobranchus Richardson, 50; logotype C DORSALIS Rich. A synonym of ALABES Cuv.
Prymnothonus Richardson, 51; orthotype P. HOOKERI Rich. A larval form.
Xystophorus Richardson, 52; very young of NAUCRATES.
Aleuterius (Cuvier) Richardson, 66; after ALUTERA (Cuv.) Oken. One of many varied spellings of ALUTERA, derived from "les ALUTERES", as ALUTARIUS, ALUTERUS, etc. ALUTERA has priority.
Oxybeles Richardson, 74; orthotype O. HOMEI Rich. A synonym of CARAPUS (FIERASFER).
Molarii (Molarius) Richardson, 79; logotype "MURÆNA OPHIS" = ECHIDNA NEBULOSA (plural form only). A synonym of ECHIDNA Forster.
Channo-Muræna Richardson, 96 orthotype ICHTHYOPHIS VITTATUS Rich.
Congrus Richardson, 107; orhtotype MURÆNA CONGER L. A variant spelling of CONGER.
Psammoperca Richardson, 115; orthotype Ps. DATNIOIDES Rich. = LABRAX WAIGIENSIS Cuv. & Val.; replaces CNIDON M. & T. (1849.
Gadopsis Richardson, 122; orthotype G. MARMORATUS Rich.

326. TEMMINCK & SCHLEGEL (1844). *Fauna Japonica* (part).
COENRAAD JACOB TEMMINCK; HERMANN SCHLEGEL.

Chætopterus Temminck & Schlegel, 78; orthotype C. SIEBOLDI Blkr. Name preoccupied; replaced by PRISTIPOMOIDES Blkr.
Melanichthys Temminck & Schlegel, 75; orthotype CRENIDENS MELANICHTHYS Rich. A synonym of GIRELLA Gray.
Ditrema Temminck & Schlegel, 77; orthotype D. TEMMINCKI Blkr.
Hypsinotus Temminck & Schlegel, 84; orthotype H. RUBESCENS Gthr. A synonym of ANTIGONIA Lowe.
Scarodon Temminck & Schlegel, 89; logotype S. FASCIATUS T. & S. A synonym of OPLEGNATHUS Rich., 1840; available if the latter is regarded as preoccupied.

327. VALENCIENNES (1844). *Ichthyologie des îles Canaries, ou Histoire naturelle des poissons, rapportés par MM. Webb & Berthelot.* In Webb, *Histoire naturelle des îles Canaries.* II, 1835-1850.
ACHILLE VALENCIENNES.

Nemobrama Valenciennes, 40; orthotype N. WEBBI Val. = POLYMIXIA NOBILIS Lowe. A synonym of POLYMIXIA Lowe.
Crius Valenciennes, 43; orthotype C. BERTHELOTI Val. = SCHEDOPHILUS MEDUSOPHAGUS. A synonym of SCHEDOPHILUS Cocco.
Asellus Valenciennes, 76; orthotype ASELLUS CANARIENSIS Val. A synonym of MORA Risso.

1845

328. AGASSIZ (1845). *Rapport sur les poissons fossiles de l'argile de Londres.* Rept. British Assoc. Adv. Sci.
LOUIS AGASSIZ.

Rhinocephalus Agassiz, 294; orthotype R. PLANICEPS Ag. (fossil). Name only, noted by Woodward, 1901.
Sciænurus Agassiz, 295; orthotype S. BOWERBANKI Ag. (fossil) A synonym of SPARNODUS.

Pomaphractus Agassiz, 307; orthotype P. EGERTONI Ag. (fossil); name only; identification by Woodward, 1901. A synonym of BRYCHÆTUS (Ag.) Woodw.

Cœlopoma Agassiz, 307; orthotpe C. COLEI Ag. (fossil); name only; preoccupied (Adams, 1867) and replaced by EOCŒLOPOMA Woodw., 1901.

Eurygnathus Agassiz, 307; orthotype E. CAVIFRONS Ag. (fossil). Name only; validated by Woodward, 1901; preoccupied and replaced by ESOCELOPS Woodw.

Percostoma Agassiz, 307; orthotype P. ANGUSTA Ag. Name only; noted by Woodward, *Cat.*, 519, 1901.

Cœloperca Agassiz, 307; orthotype CŒLOPERCA LATIFRONS Ag. (fossil). Name only; noted by Woodward, *Cat.*, IV, 519.

Bothrosteus Agassiz, 307; orthotype B. MINOR Ag. (fossil). Name only, noted by Woodward, *Cat.*, IV, 611, 1901.

Calopomus Agassiz, 307; orthotype C. POROSUS Ag. (fossil), name only.

Brachygnathus Agassiz, 307; orthotype B. TENUICEPS Ag. (fossil). Name only, noted by Woodward *Cat.*, 1901, 519 ;name preoccupied.

Podocephalus Agassiz, 307 orthotype P. NITIDUS Ag. (fossil). Name only; noted by Woodward, *Cat.*, 519, 1901.

Brychætus Agassiz, 308; orthotype B. MULLERI Ag. (undefined) (fossil); first defined by Woodward, 1901.

Merlinus Agassiz, 308; orthotype M. CRISTATUS Ag. (fossil), noted by Woodward, 1901.

Scombrinus Agassiz, 308; orthotype S. NUCHALIS Ag. (fossil); name only; validated by Woodward, 1901.

Phalacrus Agassiz, 308; orthotpe P. CYBIOIDES Ag. (fossil); name only; referred by Woodward to synonymy of EOTHYNNUS Woodw.

Rhonchus Agassiz, 308; orthotype R. CARANGOIDES Ag. (fossil). Name only; explained by Woodward, *Cat.*, IV, 458. A synonym of EOTHYNNUS Woodw.

Naupygus Agassiz, 308; orthotype N. BUCKLANDI Ag. (fossil). Name only; noted by Woodward, *Cat.*, 611 (1901).

Acestrus Agassiz, 308; orthotype A. ORNATUS Ag. (fossil). Name only; validated by Woodward, 1901, perhaps to be regarded as preoccupied by ACESTRA Jardine, 1831.

Laparus Agassiz, 308; orthotype L. ALTICEPS Ag. (fossil: name only; discussed by Woodward, 1901).

Labrophagus Agassiz, 308; orthotype L. ESOCINUS Ag. (fossil). Name only; noticed by Woodward, *Cat.*, IV, 611, 1901.

Loxostomus Agassiz, 308; orthotype L. MANCUS Ag. (fossil). Name only; noted by Woodward, *Cat.*, IV, 612, 1901.

Pachycephlus Agassiz, 308; orthotype P. CRISTATUS Ag. (fossil). Name only; noted by Woodward, *Cat.*, IV, 612, 1901.

Most of the foregoing names from Agassiz's list of the Egerton collection have been identified from labeled specimens and several have justly come into use as dating, not from 1845, but from their first definition. Woodward gives the following additional generic names, of which nothing is known, and which, until defined, have no place in the system.

Echenus (POLITUS).

Gadopsis (BREVICEPS).

Phasganus (DECLIVIS).

Rhipidolepis (ELEGANS).

Synophrys (HOPEI).

329. AGASSIZ (1845). *Monographie des poissons fossiles du Vieux Grèx Rouge, ou Systèma Dévonien* (Old Red Sandstone) *des îles Britanniques et de Russie*. Part 2.

LOUIS AGASSIZ.

Diplacanthus Agassiz, 34, 40; orthotype D. STRIATUS Ag.

Lamnodus Agassiz, 83; orthotype DENDRODUS BIPORCATUS Owen (fossil). A syno nym of DENDRODUS.

Psammosteus Agassiz, 103; logotype P. MÆANDRINUS Ag. (fossil).

Homacanthus, 113; orthotype H. ARCUATUS Ag. (fossil).

Haplacanthus Agassiz, 114; orthotype H. MARGINALIS Ag. (fossil).

Odontacanthus Agassiz, 114; logotype O. CRENATUS Ag. (fossil). A synonym of ASTROLEPIS.

Narcodes Agassiz, 115; orthotype N. PUSTULIFER Ag. (fossil). A supposed synonym of ASTROLEPIS.

Byssacanthus Agassiz, 116; orthotype B. CRENULATUS Ag. (fossil).

Naulas Agassiz, 116; orthotype N. SULCATUS Ag. (fossil).

Climatius Agassiz, 119; orthotype C. RETICULATUS (Ag. fossil).

Cosmacanthus Agassiz, 120; orthotype C. MALCOLMSONI Ag. (fossil).

Parexus Agassiz, 120; orthotype P. INCURVUS Ag. (fossil).

Homothorax Agassiz, 134; orthotype H. FLEMINGI Ag. (fossil). A synonym of BOTHRIOLEPIS.

Chelyophorus Agassiz, 135; orthotype C. VERNEUILI Ag. (fossil).

Actinolepis Agassiz, 141; orthotype A. TUBERCULATUS Ag. (fossil). A supposed synonym of ASTROLEPIS.

Stagonolepis Agassiz (not quoted by Woodward; referred to the DIPTERINI).

330. BLEEKER (1845). *Bijdragen tot geneeskunde topographie van Batavia. Generisch oversicht der fauna*. II, 1845. Arch. Nederl. Indie.

PIETER BLEEKER (1819-1878).

Pinjalo Bleeker, II, 251 (1845); orthotype CÆSIO PINJALO Blkr. = P. TYPUS Blkr.

Heterodon Bleeker, 323; orthotype H. ZONATUS Blkr. Name preoccupied in ser pents; replaced by HETEROGNATHODON Blkr., 1848.

Mesopristes Bleeker, 523; orthotype M. MACRACANTHUS Blkr.; a synonym of DATNIA Cuv. & Val.

331. BONAPARTE (1845). *Catalogo metodico dei pesci Europei*. Atti Sci. Ital. (Reprint as a separate volume, Naples, 1846.)

CARLO L. PRINCIPE BONAPARTE.

Some of the names here used may have occurred in other papers of 1844 and 1845.

Glaucostegus Bonaparte, 14; orthotype G. CEMICULUS Bon. ("RHINA RHINOBATUS SHAW = RH. CEMICULUS Geoffroy.") A synonym of RHINOBATUS.

Acanthias Bonaparte, 15; orthotype SQUALUS ACANTHIAS L. A. VULGARIS Bon. same as ACANTHIAS Risso. A synonym of SQUALUS L., as restricted.

Lepidorhinus Bonaparte, 16; orthotpe SQUALUS SQUAMOSUS Broussonnet.

Scymnorhinus Bonaparte, 16; orthotype S. LICHIA. Replaces SCYMNUS Cuv., pre- occupied.

Spinax Bonaparte, 16; tautotype SQUALUS SPINAX L. Same as SPINAX Cuv. A synonym of ETMOPTERUS Raf.

Sterledus Bonaparte, 21; orthotype ACIPENSER RUTHENUS Bon. Same as STERLETA Güldenstädt.

Cenisophius Bonaparte, 29; orthotype LEUCISCUS PAUPERUM Bon. Replaces LEUCOS Heckel, preoccupied in birds, as LEUCUS.

Gardonus Bonaparte, 29; orthotype LEUCISCUS DECIPIENS Ag. A synor.ym of CENISOPHIUS.

Pigus Bonaparte, 29; tautotype CYPRINUS PIGUS Cloquet. Apparently a synonym of CENISOPHIUS.

Cephalus Bonaparte, 30; orthotype CYPRINUS CEPHALUS L. Name preoccupied; a synonym of LEUCISCUS Cuv.

Microlepis Bonaparte, 30; no diagnosis; orthotype M. TURSKYI Bon. A synonym of LEUCISCUS Cuv.

Hegerius Bonaparte, 31; orthotype SCARDINIUS HEGERI Bon., HEGERIUS TYPUS Bon. An ally of TELESTES; the name later spelled HEEGERIUS.

Scardinius Bonaparte, 31; logotype LEUCISCUS SCARDAFA Bon.

Pollachius (Nilsson) Bonaparte 45; orthotype GADUS POLLACHUS L. = GADUS VIRENS L.

Coccolus Bonaparte, 47; orthotype C. ANNECTENS Bon., a substitute for KROHNIUS Cocca (K. FLAMENTOSUS, which is the larva of the MACRURID, CŒLORHYNCHUS, Giorna).

Cynoglossa Bonaparte, 48; orthotype PLEURONECTES CYNOGLOSSUS L. A synonym of CYNICOGLOSSUS Bon., not CYNGLOSSUS Ham.

Plagusia Bonaparte, 51; orthotype PLAGUSIA LACTEA Bon. A substitute for PLAGUSIA Cuv., preoccupied. A synonym of SYMPHURUS Raf.

Scorpichthes Bonaparte, 62; logotype COTTUS SCORPIUS L., in plural form only. A synonym of MYOXOCEPHALUS Stellar.

Gastronemus (Cocco) Bonaparte, 68; orthotype PHAROPTERYX BENOITI Rüppell. Name preoccupied. A synonym of MORA Risso.

Tinimogaster (Cocco) Bonaparte, 68; no definition; quoted by Bonaparte in connection with GASTRONEMUS Cocco. A synonym of MORA Risso, apparently a substitute for GASTRONEMUS, preoccupied.

Selenia Bonaparte, 75; orthotype CARANX LUNA Geoffroy St. Hilaire = SCOMBER GUARA *Bonnaterre*. Apparently replaces URASPIS Blkr.

332. BRODIE (1845). *(Fossil Insects.)*
PETER BELLINGER BRODIE.

Oxygonius (Agassiz) Brodie, 16; orthotype O. TENUIS Ag. (fossil). Young fishes, indeterminable.

Ceramurus (Egerton) Brodie, 17; orthotype C. MACROCEPHALUS Egert. (fossil).

333. COSTA (1845). *(Fossil Fishes.)* Atti Acad. Pontan. Napoli.
ORONZIO GABRIELE COSTA.

Megastoma Costa, V, 287; orthotype M. APENNINUM Costa (fossil). A synonym of LEPTOLEPIS.

334. KRÖYER (1845). *Icelus Hamatus.* Naturhist. Tidsskr., I.
HENRIK NICOLAJ KRÖYER (1799-1870).

Icelus Kröyer, I, 253; orthotype I. HAMATUS Kröyer.

Phobetor Kröyer, I, 263; orthotype COTTUS TRICUSPIS Reinh. A synonym of GYMNOCANTHUS Sw.

335. KRÖYER (1845). *Carelophus stromii.* Naturhist. Tidsskr., I.
HENRIK NICOLAJ KRÖYER.

Carelophus Kröyer, I; orthotype C. STRŒMI Kröyer. A synonym of CHIROLOPHIS Sw.

336. KRÖYER (1845). *Caracanthus typicus.* Naturhist. Tidsskr., I.
HENRIK NICOLAJ KRÖYER.

Caracanthus Kröyer, I, 267; orthotype C. TYPICUS Kröyer = MICROPUS MACULATUS Gray. Replaces MICROPUS Gray, preoccupied.

337. KRÖYER (1845). *Ceratis hollbolli.* Naturhist. Tidsskr., I.
HENRIK NICOLAJ KRÖYER.

Ceratias Kröyer, I, 639; orthotype C. HOLLBOLLI Kröyer.

338. McCLELLAND (1845). *Description of Four Species of Fishes from the Rivers at the Foot of the Boutan Mountains.* Journ. Nat. Hist. Calcutta, V, 1840, 272, 282.
JOHN McCLELLAND.

Ctenops McClelland, V, 281; orthotype C. NOBILIS McCl. A synonym of OSPHRO-MENUS Lac.

339. MÜLLER & TROSCHEL (1845). *Horæ Ichthyologicæ. Beschreibung und Abbildung neuer Fische; die Familie Characinen.* I. Berlin.
JOHANNES MÜLLER: FRANZ HERMANN TROSCHEL.

Distichodus Müller & Troschel, I, 12; orthotype SALMO NILOTICUS Hasselquist = CHARACINUS NEFASCH Lac.
Agoniates Müller & Troschel, I, 19; haplotype AGONIATES HALECINUS M & T.
Cnidon Müller & Troschel, I, 21; orthotype CNIDON CHINENSIS M. & T.

340. OWEN (1845). *Odontography.*
RICHARD OWEN.

Pisodus (Agassiz) Owen, 138; orthotype P. OWENI Ag. (fossil). An ally of AL-BULA Gronow.

341. RICHARDSON (1845). *Generic Characters of Gasteroschisma melampus, a Fish which Inhabits Port Nicholson, New Zealand.* Ann. Mag. Nat. Hist,, XV.
JOHN RICHARDSON.
Gasteroschisma Richardson, 346; orthotype G. MELAMPUS Rich.

342. RICHARDSON (1845). *Report on the Ichthyology of the Seas of China and Japan.* Rept. Brit. Assoc. Adv. Sci.
JOHN RICHARDSON.

Ilisha (Gray) Richardson, 306; haplotype ILISHA ABNORMIS Gray = ALOSA ELONGATA Bennett. Replaces PELLONA Cuv. & Val., and ZUNASIA Jordan & Thompson.

343. SOUTH (1845). *Encyclopdia Metropolitana.* (Edition for 1845.)
JOHN FLINT SOUTH.
Thunnus South, V, 620; orthotype SCOMBER THYNNUS L. A substitute for THYN-NUS Cuv., preoccupied in butterfles.

344. TEMMINCK & SCHLEGEL (1845). *Fauna Japonica* (issued in parts).
CŒNRAAD JACOB TEMMINCK; HERMANN SCHLEGEL.
Erythrichthys Temminck & Schlegel, 1845, 117; orthotype E. SCHLEGELI Gthr. Name preoccupied; replaced by ERYTHROCLES Jordan, 1919.
Scombrops Temminck & Schlegel, 118; orthotype S. CHEILODIPTEROIDES Blkr. = SCOMBER BOOPS Houttuyn.
Dictyosoma Temminck & Schlegel, 139; orthotype D. BURGERI Van der Hoeven.
Cirrhilabrus Temminck & Schlegel, 167; orthotype C. TEMMINCKI Blkr.

1846

345. AGASSIZ (1846). *Nomenclator Zoologicus. Pisces.*

Louis Agassiz.

Hypeneus Agassiz, 190; logotype Mullus bifasciatus Lac. A corrected spelling of Upeneus.

346. AGASSIZ (1846). *Nomen manuscriptum in diversis museo usurpatum. Nomina systematica generum piscium, tam viventium quam fossilium.* Addenda to *Nomenclator Zoologicus. Pisces.*

Louis Agassiz.

The three names which follow occur in the *Nomenclator Zoologicus* as manuscript names used by Agassiz on labels in different museums. They have no place in the system.

Glaphyrodus Agassiz, Addenda, 3; species not named (same as Oxyrhina). A synonym of Isurus Raf.

Holodus Agassiz, Addenda, 3; species not named. A synonym of Heptranchias (Notidanus).

Oroptychus Agassiz, Addenda, 4 species not named ("Cestraciontes").

Toxophorus. This name is given by Agassiz, page 63, with no indication as to its author.

347. BIBRON (1846). *Revue Zoologique.*

Gabriel Bibron (1806-1848).

Chilomycterus Bibron, 40; logotype Diodon reticulatus L.

348. BRISOUT DE BARNEVILLE (1846). *Note sur le groupe des Gobiesoces.* Rev. Zool.

Charles N. F. Brisout de Barneville (1823-1893).

Tomicodon Brisout de Barneville, 144; orthotype T. chilensis Barn. A synonym of Sicyases M. & T.

Sicyogaster Brisout de Barneville, 144; orthotype Gobiesox marmoratus Jenyns. Name preoccupied; replaced by Gill.

Chorisochismus Brisout de Barneville, 209; orthotype C. nudus Barn. = Cyclopterus dentex Pallas.

349. BRISOUT DE BARNEVILLE (1846). *Note sur le nouveau genre de la famille des Discoboles (Trachelochismus).* Rev. Zool.

Charles N. F. Brisout de Barneville.

Trachelochismus Brisout de Barneville, 214; orthotype Cyclopterus pinnulatus Forster.

350. COSTA (1846). *Fauna del regno di Napoli, ossia enumerazione di tutti gli animali che abitano le diverse regioni di questo regno e le acque che le bagnano,* etc. Vol. III.

Oronzio Gabriele Costa.

Cynoponticus Costa, pl. 28; orthotype C. ferox Costa = Conger savanna Cuv.

351. CUVIER & VALENCIENNES (1846). *Histoire naturelle des poissons.* XVIII, 1835.

Achille Valenciennes.

Grundulus Cuvier & Valenciennes, XVIII, 216; haplotype Fundulus bogotensis Humboldt. Eigenmann identifies this with Characin, Ctenocharax bogotensis Regan, 1907.

Orestias Cuvier & Valenciennes, XVIII, 221; logotype Orestias cuvieri Val.

Vastres Cuvier & Valenciennes, XIX, 433; orthotype VASTRES CUVIERI Val. =
SUDIS GIGAS Cuv. Same as SUDIS Cuv., preoccupied by SUDIS Raf., 1810;
replaced by ARAPAIMA M. & T., 1846.

Heterotis (Ehrenberg MS.) Cuvier & Valenciennes, XIX, 465; orthotype HETEROTIS
EHRENBERGI Val. I do not find that Ehrenberg had published this name elsewhere. A synonym of CLUPISUDIS Sw.

Lebiasina Cuvier & Valenciennes, XIX, 531; haplotype LEBIASINA BIMACULATA Val.

Pyrrhulina Cuvier & Valenciennes, XIX, 534; haplotype P. FILAMENTOSA Val.

352. EICHWALD (1846). *Geognosy of Russia* (in Russian). Reproduced in
1860 in *Lethæa Rossica.*
CARL EDUARD VON EICHWALD.

Aulacosteus Eichwald; logotype (in 1860) A. COCHLEARIFORMIS Eichw. Same as
PTYCTODUS Pander and, if eligible, of prior date.

353. KEYSERLING (1846). *Reise in das Petchoraland im Jahre 1843.*
ALEXANDER VON KEYSERLING.

Dimeracanthus Keyserling, 292b; orthotype D. CONCENTRICUS Keyserling (fossil).

354. KRÖYER (1846). *(Danmarks Fiske.)*
HENRIK NIKOLAJ KRÖYER.

Acantholepis Kröyer, III, 98; orthotype ARGENTINA SILUS. A synonym of SILUS
Reinhardt and probably of ARGENTINA L.

355. LOWE (1846). *On a New Genus of the Family Lophiidæ (les Pectorales
pediculées, Cuv.) Discovered in Madeira.* Proc. Zool. Soc. London, XIV.
RICHARD THOMAS LOWE.

Chaunax Lowe, III, 339; orthotype C. PICTUS Lowe.

356. MEYER (1846). *Geologie von Jena.* Schmidt und Schleiden.
HERMANN VON MEYER.

Tholodus Meyer, I, 199; orthotype T. SCHMIDTI Meyer (fossil). Undetermined;
perhaps an ASTERODON, perhaps a reptile.

357. MÜLLER (1846). *(Arapaima.)* Abhandl. Akad. Wiss. Berlin.
JOHANNES MÜLLER.

Arapaima Müller; haplotype SUDIS GIGAS Cuv. Replaces SUDIS Cuv., preoccupied,
and VASTRES Cuv. & Val. of the same year but of later date, as Valenciennes
quotes ARAPAIMA.

358. MÜNSTER (1846). *Ueber die im Korallenkalk des Lindner Berges bei
Hannover vorcommenden Ueberreste von Fischen, etc.* Beiträge zur Petre-
facten-Kunde.
GEORG MÜNSTER.

Radamas Münster, VII, 11, 28; orthotype ASIMA JUGLERI Giebel (fossil). Same as
ASIMA Giebel and of prior date.

359. NORWOOD & OWEN (1846). *Description of a New Fossil Fish from the
Palæozoic Rocks of Indiana.* Amer. Journ. Sci., I.
JOSEPH GRANVILLE NORWOOD; DAVID DALE OWEN.

Macropetalichthys Norwood & Owen, I, 371; orthotype M. RAPHEIDOLABIS Norwood & Owen (fossil).

359A. RICHARDSON (1846). *Stokes: Discoveries in Australia.*
JOHN RICHARDSON.

Assiculus Richardson, 492; orthotype A. PUNCTATUS Rich. A synonym of PSEUDO-CHROMIS Rüppell.

360. PETERS (1846). *Ueber eine neue Gattung von Labyrinthfischen . . . aus Quellimane.* Arch. Anat.
WILHELM CARL HARTWIG PETERS.

Ctenopoma Peters, 481; orthotype C. MULTISPINIS Peters.

361. TEMMINCK & SCHLEGEL (1846). *Fauna Japonica* (part).
COENRAAD JACOB TEMMINCK; HERMANN SCHLEGEL.

Plecoglossus Temminck & Schlegel, 229; orthotype P. ALTIVELIS T. & S.

Ateleopus Temminck & Schlegel, 255; orthotype A. JAPONICUS Blkr. Name pre-occupied as ATELOPUS; replaced by PODATELES Boulenger.

362. TSCHUDI (1846). *Untersuchungen über die Fauna Peruana.*
JOHANN JACOB TSCHUDI (1818-1882).

Cheilotrema Tschudi, 13; orthotype C. FASCIATUM Tschudi.

Chætostoma (Heckel) Tschudi, 25; orthotype C. LOBORHYNCHUS Heckel. Said to be thus originally written by Heckel; spelled CHÆTOSTOMUS by Kner (1854) and by subsequent writers.

Discopyge Tschudi, 32; orthotype D. TSCHUDII Tschudi.

363. VALENCIENNES (1846). *Nouvelles observations sur l'Eremophilus.*
Recueil d'Observ. Zool., II.
ACHILLE VALENCIENNES.

Trichomycterus Valenciennes, 347; orthotype T. NIGRICANS Humboldt, 1811. Same as THRYCOMYCTERUS.

1847

364. BLEEKER (1847). *Siluroideorum bataviensium conspectus diagnosticus.*
Verh. Batav. Genootsch., XXI.
PIETER BLEEKER.

Ketengus Bleeker, 9; orthotype K. TYPUS Blkr.

Osteogeneiosus Bleeker, XXI, 49; orthotype O. MACROCEPHALUS Bleeker

Batrachocephalus Bleeker, XXI, 158, 375; orthotype B. MICROPOGON Blkr.

365. BLEEKER (1847). *Pharyngognathorum Siluroideorumque species novæ Javanenses.* Nat & Geneesk. Arch. Neder.-Indie, IV.
PIETER BLEEKER.

Chœrodon Bleeker, 10; orthotype LABRUS MACRODONTUS Lac. Replaces CHŒROPS Rüppell, 1852; later written CHOIRODON.

366. BRISOUT DE BARNEVILLE (1847). *Note sur le genre Centropristis Cuvier.*
Rev. Zool.
CHARLES N. F. BRISOUT DE BARNEVILLE.

Homodon Brisout de Barneville, 133; logotype CENTROPRISTIS GEORGIANUS Cuv. & Val. A synonym of ARRIPIS Jenyns.

Myriodon Brisout de Barneville, 133; orthotype CENTROPRISTIS SCORPÆNOIDES Cuv. & Val. = SCORPÆNA WAIGIENSIS Q. & G. A synonym of CENTROGENYS.

367. BRISOUT DE BARNEVILLE (1847). *Note- sur un nouveau genre d'Anguilliformes*. Rev. Zool.

CHARLES N. F. BRISOUT DE BARNEVILLE.

Ichthyapus Brisout de Barneville, 219; orthotype I. ACUTIROSTRIS Barn. A synonym of SPHAGEBRANCHUS Bloch.

368. CUVIER & VALENCIENNES (1847). *Histoire naturelle des poissons*. XX.

ACHILLE VALENCIENNES.

Sardinella Cuvier & Valenciennes, XX 261; logotype SARDINELLA AURITA Val.

Harengula Cuvier & Valenciennes, XX, 277; logotype HARENGULA LATULUS Val. (said to be from France, but apparently an American species, not the French "Blanquette", which is reputed to be the sprat (SPRATELLA SPRATTUS) *fide* Cligny).

Pellona Cuvier & Valenciennes, XX, 300; logotype PELLONA ORBIGNYANA Val. = PRISTIGASTER FLAVIPINNIS Val. A synonym of ILISHA Gray (1846).

Rogenia Cuvier & Valenciennes, XX, 340; haplotype ROGENIA ALBA Val. Young of CLUPEA L. ("Whitebait").

Clupeonia Cuvier & Valenciennes, XX, 345; logotype CLUPEONIA JUSSIEUI Val. Very near HARENGULA Cuv. & Val.

Spratella Cuvier & Valenciennes, XX, 356; logotype SPRATELLA PUMILA Val. Young of CLUPEA SPRATTUS L.

Kowala Cuvier & Valenciennes, XX, 362; logotype KOWALA ALBELLA Cuv. & Val. Very close to CLUPEONIA, if not identical with it.

Meletta Cuvier & Valenciennes, XX, 366; type MELETTA VULGARIS Val. (young of SPRATELLA SPRATTUS). A synonym of SPRATELLA Cuv. & Val.

Alausa Cuvier & Valenciennes, XX, 389; type ALAUSA VULGARIS = CLUPEA ALOSA L. A variant spelling of ALOSA Cuv.

Dussumieria Cuvier & Valenciennes, XX, 467; haplotype DUSSUMIERIA ACUTA Val.

369. EGERTON (1847). *On the Nomenclature of the Fossil Chimæroid Fishes*. Quart. Journ. Geol. Soc., III.

PHILIP DE MALPAS GREY EGERTON.

Dictyopyge Egerton, III, 276; orthotype CATOPTERUS MACRURUS Egert. (fossil).

Ischypterus Egerton, III, 277; orthotype F. LATUS Egert. (fossil). A synonym of SEMIONOTUS.

370. GIEBEL (1847). *Vier neue Fische aus dem dunkeln Kreide-Schiefer von Glarus*. Neues Jahrb. Mineral.

CHRISTOPH GOTTFRIED ANDREAS GIEBEL (1820-1881).

Palymphyes Giebel, 666; logotype P. CRASSUS Giebel.

Pachygaster Giebel, 667; orthotype P. SPINOSUS Giebel (fossil).

371. GIEBEL (1847). *Fauna der Vorwelt . . . monographisch dargestellt*.

CHRISTOPH GOTTFRIED ANDREAS GIEBEL.

Centrodus Giebel, 344; orthotype C. ACUTUS Giebel (fossil).

Chilodus Giebel, I, 352; orthotype (not named in Woodward). A tooth of dubious nature.

Oxytes Giebel, 364; orthotype O. OBLIQUA Giebel (fossil). A fossil homologue of CARCHARIAS Raf., ODONTASPIS Ag.

372. GUICHENOT (1847). *Notice sur un nouveau genre de Percoides (Pomanotis).* Rev. Zool., X.

ALPHONSE GUICHENOT.

Pomanotis Guichenot, 390; orthotype P. RUBESCENS Guich. A genus not since recognized. The type P. RUBESCENS was brought by Dussumier from Alipey,* a locality I do not find on the maps, but evidently in India or eastward. The genus is not related to EUCENTRARCHIDÆ, as its author suggests, but belongs probably to the NANDIDÆ (D. XIV, 14. A III, 9) near PRISTOLEPIS (CATOPRA).

373. KNER (1847). *Ueber die beiden Arten Cephalaspis lloydii und C, lewisii Agassiz, und einige diesen zunächst stehenden Schalenreste.* Naturwiss. Abh., I.

RUDOLF KNER.

Pteraspis Kner, I, 165; orthotype CEPHALASPIS LEWISI Ag. (fossil).

374. MEYER (1847). *Palæontographica.*

HERMANN VON MEYER.

Tylodus Meyer, I, 102; orthotype T. GLABER Meyer (fossil).

375. MEYER (1847). *Vorläufige Uebersicht der in dem Muschelkalk Ober-Schlesiens vorkommenden Saurier, Fische, Crustaceen und Echinodermen, nach der Sammlung des Herrn Mentzel in Königshütte.* Uebersicht Schles. Ges. Breslau.

HERMANN VON MEYER.

Omphalodus Meyer, 574; orthotype O. CHORZOWENSIS Meyer (fossil). A synonym of ASTERODON.

376. PLIENINGER (1847). *Microlestes antiquus und Sargodon tomicus in der Grenzenbreccie des Keupers von Degerloch.* Jahresschr. Ver. Vaterl. Naturk. Württemberg.

WILHELM HEINRICH THEODOR VON PLIENINGER.

Sargodon Plieninger, III, 165; orthotype S. TOMICUS Plieninger (fossil).

377. WOOD (1847). *Description of a Species of Shark (Leiodon echinatum).* Proc. Boston Nat. Hist. Soc., II.

WILLIAM WOOD.

Leiodon Wood, 174; orthotype L. ECHINATUM Wood. Name preoccupied; a synonym of SOMNIOSUS Le Sueur.

1848

378. AGASSIZ (1848). *Two New Fishes from Lake Superior, Percopsis and Rhinichthys.* Proc. Boston Soc. Nat. Hist., III.

LOUIS AGASSIZ.

Percopsis Agassiz, 80; haplotype PERCOPSIS GUTTATUS Ag. = SALMO OMISCOMAYCUS ,Walbaum (1792). Also in *Lake Superior,* 1850, 284.

Rhinichthys Agassiz, 80; haplotype CYPRINUS ATRONASUS Mitchill. Also in *Lake Superior,* 1850, 353.

---

* Alipey is probably the same as Alleppi or Allapalli, a seaport of Travancore in southern India, 33 miles south of Cochin.

379. BEYRICH (1848). *Ueber Xenacanthus decheni und Holacanthodes gra-cilis, zwei Fische aus der Formation des Rothliegenden in Nord-Deutschland.* Archiv. f. Mineral., XXII.

HEINRICH ERNST BEYRICH (1815-1896).

Xenacanthus Beyrich, 24; orthotype ORTHACANTHUS DECHENI Goldfuss (fossil). Regarded as a synonym of ORTHACANTHUS Ag. (PLEURACANTHUS).

Holacanthodes Beyrich, 24; orthotype H. GRACILIS Beyr. (fossil). A synonym of ACANTHOESSUS and of ACANTHODES.

380. BLEEKER (1848). *A Contribution to the Ichthyology of Sumbawa.* Journ. Ind. Arch. IX, 1848.

PIETER BLEEKER.

Heterognathodon Bleeker, I, 101; orthotype H. BIFASCIATUS Blkr., a substitute for HETERODON, preoccupied. A synonym of PENTAPUS Cuv. & Val.

381. CUVIER & VALENCIENNES (1848). *Histoire naturelle des poissons.* XXI.

ACHILLE VALENCIENNES.

Fario Cuvier & Valenciennes, XXI, 277; tautotype SALMO FARIO L. = SALMO ARGEN-TEUS Val. Same as TRUTTA L., a subgenus of SALMO.

Salar Cuvier & Valenciennes, XXI, 314; type SALAR AUSONII Val., a form or ally of SALMO FARIO L. A synonym of FARIO Val. and TRUTTA L.

382. DANA (1848). *Fossils of the Exploring Expedition Under the Command of Charles Wilkes, U. S. N.; a Fossil Fish from Australia, and a Belemnite from Tierra del Fuego.* Amer. Journ. Sci., V.

JAMES DWIGHT DANA (1813-1895).

Urosthenes Dana, V, 433; orthotype U. AUSTRALIS Dana (fossil).

383. EWALD (1848). *Ueber Menaspis, eine neue fossile Fischgattung.* Ber. Akad Wiss. Berlin.

AUGUST EWALD.

Menaspis Ewald, 83; orthotype M. ARMATA Ewald (fossil).

384. GIEBEL (1848). *Die Fische der Vorwelt in seiner Berücksichtigung der lebenden Fische.*

CHRISTOPH GOTTFRIED ANDREAS GIEBEL.

Tharsis Giebel, 145; orthotype T. GERMARI Giebel (fossil). A synonym of LEPTO-LEPIS.

Asima Giebel, 184; orthotype A. JUGLERI Giebel (fossil). A synonym of RADAMAS Münster.

Elonichthys Giebel, 249; logotype E. GERMARI Giebel (fossil).

385. GISTEL (1848). *Naturgeschichte des Thierreichs für höhere Schulen.*

JOHANNES GISTEL.

A remarkable dull and confused publication, not hitherto noticed by ichthyologists. Gistel speaks feelingly of his "höchst mühsamer Ver-gleichung, zeiträubenden Nachschlagens, in vielen hundert Werken, etc., aus meinen Manuscripten aufzuführen". It is not often that so fatiguing a piece of work presents itself in such a crabbed form. Gistel insists that personal names of genera are allowable only in Botany and in the lowest classes of animals ("den niedrigsten Thierklassen").

The following may replace names actually preoccupied:

ABRON instead of PLATYSTOMA.

DAPALIS instead of SMERDIS.

ENIXE instead of DREPANE.

HYPODYTES instead of PARACENTROPOGON.

NOTACMON instead of EURYNOTUS.

ODONUS instead of XENODON (ERYTHRODON).

PERIURGUS instead of MICROPS.

PROSCINETES instead of MICRODON.

SECUTOR instead of DEVEXIMENTUM.

STREPHON instead of BRONTES.

STROTES instead of PLATYSOMUS.

The following are also admissible if a name differing in gender or slightly in form but from the same roots as an earlier name is to be rejected as preoccupied:

ANALITHIS instead of PLATYRHINA.

ALAZON instead of PŒCILIA.

ENGLOTTOGASTER instead of OREINUS.

FURO instead of EUGNATHUS.

MYRMILLO instead of MUSTELUS.

PRENES instead of SCATOPHAGUS.

RHADAMISTA instead of CTENODUS.

SCROPHA instead of CERATODUS.

PEDITES instead of MACROPODUS.

The twice preoccupied name of UNDINA Gistel, IX, is offered as a substitute for LINKIA Nardo, said to be a fish, LINKIA TYPUS Nardo, preoccupied by LINCKIA Ag., a genus of ECHINODERMS, for which he offers the substitute name of CATANTES. But both LINKIAS are starfishes and neither a fish. Another generic name, PHYGAS Gistel, X, is proposed in place of POLIA (Della Chiaje, *Isis*, 1832, 647, SCIÆNA), said to be preoccupied in worms, by POLIA Grube, which genus Gistel renames CYRYX. But I am informed by Mr. Bean that POLIA SIPUNCULUS Della Chiaje is itself a worm also, not a fish.

Carcharias Gistel, VIII; type SQUALUS ACANTHIAS L.; substitute for ACANTHIAS Risso, regarded as preoccupied by ACANTHIA. CARCHARIAS was already twice preoccupied. A synonym of SQUALUS L., as restricted.

Pelonectes Gistel, VIII; type (ACANTHOËSSUS BRONNI Ag.). Substitute for ACANTHODES Ag. (1833), wrongly said to be preoccupied in crustacea (Serville, 1835). A synonym of ACANTHOËSSUS Ag. (1832).

Gremilla Gistel, VIII; type (PERCA CERNUA L.). A substitute for ACERINA Cuv.; but CERNUA Schäfer and ACERINA Güldenstädt, as well as GYMNOCEPHALUS Bloch & Schneider are still older than GREMILLA.

Adiapneustes Gistel, VIII; type (ACRODUS GAILLARDOTI Ag.). A substitute for ACRODUS Ag. (1838), wrongly regarded as preoccupied by ACRODON Zimmermann (1840).

Histiodromus Gistel, VIII; type (SALMO ANOSTOMUS L.). Name a substitute for ANASTOMUS Gronow, Scopoli, Cuvier. All these writers have priority over ANASTOMUS Illiger (1835), a genus of birds. The name has often been written ANOSTOMUS.

Nomalus Gistel, VIII; type LONCHURUS ANCYLODON Bl. & Schn. (A JACULIDENS Cuv. & Val.) Substitute for ANCYLODON, preoccupied. A synonym of MACRODON Schinz (1822).

Hypodytes Gistel, VIII; orthotype "Aspistes Q. & G.", which is Apistus longispinis Cuv. & Val. The name is a substitute for Apistes ("Aspistes")
regarded as preoccupied.

As the species of Quoy & Gaimard (1834) (longispinis) is the type of
Paracentropogon Blkr., Hypodytes may supersede that name, rather than
replace it by Prosopodasys Cantor (1849). Gistel offers Hypodytes as a
substitute for "Aspistes Quoy, *Astrolabe Fische*," although Apistes occurs
in the *Règne Animal*, 1829 (Index), and Apistus in *Cuv. & Val.*, IV, also
in .1829, in the *latter* part of the same year. As Hypodytes is expressly
used to replace Aspistus (not Aspistes) of Quoy & Gaimard and as the
chief species figured by those authors is A. longispinis, the type of the
later genus Paracentropogon, Hypodytes may replace the latter, if Apistes
or Apistus is deemed occupied.

Scrofaria Giestel, VIII; orthotype Ausonia cuvieri Risso. Name an unexplained
substitute for Ausonia; a synonym of Luvarus Raf.

Gliscus Gistel, VIII; type (Blochius longirostris Volta) (fossil). A substitute
for Blochius, a personal name.

Strephon Gistel, VIII; orthotype (Brontes prenadilla (Cuv. & Val.). A substitute for Brontes Cuv. & Val. several times preoccupied.

Thalassoklephtes Gistel, VIII; orthotype (Caninoa chiereghini Nardo). A
substitute for Caninoa Nardo, as a genus unworthy of a prince: "Ist denn
der Prinz von Canino mit einem Hayen vergleichbar?"

Scropha Gistel, VIII; type (Ceratodus latissimus Ag.) (fossil). A substitute
for Ceratodus, regarded as preoccupied by Ceratodon Brisson.

Capellaria Gistel, VIII; type (Lophius commersonianus Lac.). Substitute for
Chironectes Cuv., preoccupied. A synonym of Antennarius Cuv.

Cæso Gistel, VIII; type (Sparus berda Forskål). A substitute for "Chrysophrys
Rüppell" (berda). A synonym of Sparus L. as understood by me.

Cerdo Gistel,· VIII; type (Gadus callarias L.) as here restricted. Cerdo is
offered as a substitute for "Asellus Klein." But Klein does not accept
Asellus as a generic name, but discusses its application to various cod-like
fishes. He refers to "Cerdones" in a discussion of dried codfish. We may
therefore regard Cerdo Gistel as a synonym of Gadus L.

Echemythes Gistel, VIII; type (Zeus luna Gmelin). A substitute for Chrysotosus Lac. A synonym of Lampris.

Ronchifex Gistel, VIII; type (Harpe rufa Lac.). substitute for Cossyphus
Cuv., preoccupied; Bodianus Bloch and Harpe Lac. are names still older
for the same genus.

Rhadamista Gistel, VIII; type (Ctenodus cristatus Ag. (fossil). A substitute
for Ctenodus Ag. (1838) regarded as preoccupied by Ctenodon Wagner
(1830).

Pirene Gistel, IX; type (Chætodon aruanus L.). Substitute for Dascyllus;
preoccupied as Dascillus; Pirene is itself preoccupied and gives way to
Tetradrachmum Cantor, 1849.

Enixe Gistel, IX; type Chætodon punctatus L.). A substitute for Drepane
Cuv., regarded as preoccupied by Drepane or Drepana Schranck. It has
priority over Harpochirus Cantor (1849).

Secutor Gistel, IX; orthotype Sparus insidiator (Bloch). A substitute for
Equula Cuv., but with a different type. Has priority over Deveximentum
Fowler, if this subgenus is distinct from Leiognathus Lac.

Furo Gistel, IX; type (Eugnathus orthostomus Ag. fossil). Substitute for Eugnathus Ag. regarded as preoccupied by Eugnatha, a genus of spiders.

Notacmon Gistel, IX; type (Eurynotus crenatus Ag. fossil). Substitute for Eurynotus Ag., preoccupied by a genus of beetles (Kirby, 1817).

Podager Gistel, IX; logotype Gerres waigiensis Q..& G. Name a substitute for Gerres, regarded as preoccupied by Gerris Fabricius, a genus of insects. But Podager is itself preoccupied in birds, and if a new name is needed, it must give place to Catochænum Cantor (1849). But the root words, Gerris and Gerres, are different.

Synergus Gistel, IX; orthotype (Megalurus lepidotus Ag. fossil). A substitute for Megalurus Ag. preoccupied in mammals, but Synergus is preoccupied also by Synergus Hartig, 1840, a genus of Hymenoptera, and Megalurus may be replaced by Urocles Jordan as already indicated.

Ellops Gistel, IX, type (Acipenser helops Pallas). A substitute for Helops, ἔλλοψ being the original Greek form.

Apeches Gistel, IX; logotype Johnius, carutta Bl. & Schn. A substitute for Johnius, a name regarded as objectionable. In general Gistel offers a substitute for all personal names everywhere except among plants and the lowest animals.

Neanis Gistel, IX; type (Labrus julis L.). Name a substitute for Julis Cuv., regarded as preoccupied by Julus L., a genus of myriapods. But the two names are of entirely distinct origin.

Exoles Gistel, IX; type (Squalus nasus Bonnaterre). A substitute for Lamia Bon., itself a synonym of Lamna Cuv.

Apepton Gistel, IX; orthotype Lepadogaster piger Nardo. A needless synonym of Gouania Nardo, a personal name.

Anepistomon Gistel, IX; type (Leptorynchus capensis Smith). A substitute for Leptorhynchus Smith, preoccupied. A synonym of Ophisurus, as restricted.

Spanius Gistel, IX; type (Leptosoma atra Nardo). Name a substitute for Leptosoma, preoccupied; probably a synonym of Monochirus Raf. "Soglio bastardo".

Encrotes Gistel, IX; type Cyprinodon variegatus Lac. A substitute for Lebia or Lebias Cuv. on account of the name Lebia Latreille, 1802, an earlier name wrongly attributed to Bonaparte. But Lebia Cuv. (later written Lebias) is a synonym of Cyprinodon Lac.

Maina Gitsel, IX; type (Sparus chœrorhynchus Bl. & Schn.). Name a substitute for Lethrinus Cuv. unreasonably regarded as preoccupied by Lethrus Fabricius, a genus of beetles.

Massaria Gistel, IX; type Cyclopterus liparis L. A needless substitute for Liparis, used in Botany.

Pedites Gistel, IX (Macropodus viridiauratus Lac.). Name a substitute for Macropodus Lac. corrected by Günther as Macropus, in that form preoccupied by Macropus Thunberg, a genus of beetles. Perhaps to be adopted as the roots are the same and Macropodus an incorrect form.

Paschalestes Gistel, IX; type (Macquaria australasica Cuv. & Val. Substitute for Macquaria Cuv. & Val., rejected, as a personal name.

Englottogaster Gistel, IX; type Oreinus guttatus McCl. Substitute for Oreinus McCl., regarded as preoccupied in beetles as Oreina. But the generic value of Oreinus is doubtful.

Proscinetes Gistel, X; type MICRODON ELEGANS Ag. (fossil). Substitute for MICRODON Ag., preoccupied; replaces POLYPSEPHIS Hay, 1899, which is also a substitute for MICRODON.

Periergus Gistel, X; type MICROPS FURCATUS Ag. (fossil). A substitute for MICROPS Ag., preoccupied in beetles.

Arnion Gistel, X; type (MUGIL CEPHALUS L.). A substitute for MUGIL L. for no assigned reason.

Myrmillo Gistel, X; orthotype MUSTELUS VULGARIS. A substitute for MUSTELUS Cuv., regarded as preoccupied by MUSTELA L. Same as PLEURACROMYLON Gill.

Analithis Gistel, X; orthotype PLATYRHINA SINENSIS (Bl. & Schn.). Substitute for PLATYRHINA M. & H., regarded as preoccupied by PLATYRHINUS, replaced by DISCOBATUS Garman (1881).

Strotes Gistel, X; type (STROMATEUS GIBBOSUS Blainv.). Name a substitute for PLATYSOMUS Ag. preoccupied by a genus of beetles.

Abron Gistel, X; orthotype SILURUS LIMA Ag. A substitute for PLATYSTOMA Ag., 1829, preoccupied in insects.

Plectrostoma Gistel, X; type OXYRHINA MANTELLI Ag. (fossil). Substitute for OXYRHINA Ag., preoccupied. A synonym of ISURUS Raf.

Euporista Gistel, X; type PLEURONECTES PLAGIUSA L. A substitute for PLAGUSIA Bon., preoccupied; a synonym of SYMPHURUS Raf.

Deportator Gistel, X; type (SILURUS ANGUILLARIS Forskål). Name a substitute for PLOTOSUS Lac., needlessly regarded as preoccupied by PLOTUS L.

Alazon Gistel, X; orthotype (PŒCILIA VIVIPARA). A substitute for PŒCILIA Cuv. & Val., regarded as preoccupied by PŒCILUS Bon., a beetle.

Apodometes Gistel, X; type (PSAMMODUS RUGOSUS Ag., fossil). A substitute for PSAMMODUS Ag. regarded as preoccupied by PSAMMODIUS, a genus of beetles, but the words are not from the same roots.

Demiurga Gistel, X; type (RHINA ANCYLOSTOMA Bl. & Schn.). A substitute for RHINA Bon., etc., regarded as preoccupied by RHINA Olivier (1807), a genus of beetles. But this genus of rays was called RHINA in 1801, by Bloch & Schneider. The same name was used by non-Linnæan writers in place of SQUATINA. For this purpose it was first binomially used in 1810. If RHINA is used for SQUATINA, DEMIURGA could replace it for this genus, instead of RHAMPHOBATIS Gill (1861).

Prenes Gistel, X; type (CHÆTODON ARGUS Bloch). Name a substitute for SCATOPHAGUS Cuv., regarded as preoccupied by SCATOPHAGA. Has priority over CACODOXUS Cantor, 1849.

Agoreion Gistel, X; type ACANTHODERMA SPINOSUM Ag. (fossil). Name a substitute for "SCLERODERMA Ag.", used in fungi, also as SCLERODERMUS in bees, but I find no SCLERODERMA in Agassiz. It was probably the preoccupied name, ACANTHODERMA Ag., which Gistel had in mind; for this AGOREION may perhaps be substituted.

Creotroctes Gistel, X; type SCOMBER SARDA L. Name an unexplained substitute for SARDA Cuv. It is a synonym of PALAMITA Bon., offered in 1832 as a substitute for PELAMYS Cuv. & Val., 1831, which is the same as SARDA Cuv., 1829.

Denius Gistel, X; type (SPARUS SARGUS L.). Name a substitute for SARGUS Cuv., preoccupied. A synonym of DIPLODUS Raf.

Borborodes Gistel, X; type (SQUALUS LICHIA Cuv., SQUALUS LICHIA Bonnaterre). Substitute for SCYMNUS Cuv., preoccupied in beetles. A synonym of SCYMNORHINUS Bon., 1846.

Dapalis Gistel, XI; type (PERCA MINUTA Blainv.) (fossil). Name a substitute for SMERDIS Ag., 1833, preoccupied by SMERDIS Leach, a genus of beetles.

Peroptera Gistel, XI; said to be a fish and substituted for PERIOPTERA Gistel, "deleatur". I find no reference elsewhere to either name.

Orthocolus Gistel XI; type SALMO THYMALLUS L. Substitute for THYMALLUS Cuv., regarded as preoccupied by THYMALLUS Latr., 1803, a genus of beetles. But the root word is different. In any event, CHOREGON Minding has priority

Trompe Gistel, XI, 109; type (SCIÆNA JACULATRIX Pallas). Name an unexplained substitute for TOXOTES Cuv.

Aphobus Gistel, XI; type (TRACHIDERMIS FASCIATUS Heckel). Name a substitute for TRACHIDERMIS Heckel preoccupied as TRACHYDERMA, etc., a synonym of CENTRIDERMICHTHYS Rich., 1844.

Chætichthys Gistel, XI; type (TRACHINUS TRICHODON Steller). Name a substitute for TRICHODON Cuv., deemed preoccupied by TRICHODA and TRICHODES, but the root meaning is different.

Lithulcus Gistel, XI; type (TRICHOPODUS MENTUM Lac.). Substitute for TRICHOPODUS (TRICHOPUS) (Lac.) Cuv. & Val.; used also in botany. The oldest use of the name in Zoology is that of Lacépède, 1802 (III, 125), a genus of fishes already called (116) OSPHRONEMUS by the same author.

Odonus Gistel, XI; type (BALISTES NIGER Lac.). A substitute for XENODON Rüppell, preoccupied. ODONUS is prior to ERYTHRODON Rüppell, 1852, and to PYRODON Kaup. If ZENODON Sw. (1839) is regarded as eligible, not as a mere misprint, it will be preferred to ODONUS.

Eudynama Gistel, XIII, 108; orthotype (SPARUS AURATA L.). A substitute for CHRYSOPHRYS Cuv. A synonym of SPARUS L. as restricted.

Eupnœa Gistel, 105; orthotype PLAGIUSA LACTEA Bon. (SYMPHURUS NIGRESCENS Raf.). A synonym of SYMPHURUS Raf.

Melantha Gistel, 109; orthotype SCIÆNA NIGRA Gmelin, "die Seekrähe". A needless synonym of CORVINA Cuv.

Attilus Gistel, 109; orthotype SCIÆNA (CIRRHOSA L.), "der Schattenfisch". A needless synonym of UMBRINA Cuv.

386. GUICHENOT (1848). *Notice sur l'établissement d'un nouveau genre de Chétodons (Megaprotodon).* Rev. Zool., XI.

ALPHONSE GUICHENOT.

Megaprotodon Guichenot 12; orthotype CHÆTODON BIFASCIALIS Cuv. & Val.

387. GUICHENOT (1848). *Peces de Chile.* (In Claudio Gay: *Historia fisica y politica de Chile.*)

ALPHONSE GUICHENOT.

Boxaodon (Guichenot) Gay, II, 208; orthotype B. CYANESCENS Guich. An ally of EMMELICHTHYS Rich.

Mendosoma (Guichenot) Gay, II, 212; logotype M. LINEATUM Guich.

Seriolella (Guichenot) Gay, II, 238; orthotype S. POROSA Guich.

Merlus (Guichenot) Gay, II, 238; orthotype M. GAYI Guich. Apparently a variant of MERLUCCIUS, of which it is a synonym.

388. HECKEL (1848). *Eine neue Gattung von Pœcilien mit rochenartigem Anklammerungs-Organe (Xiphophorus).* Sitzb. Akad. Wiss. Wein.

JOHANN JAKOB HECKEL.

Xiphophorus Heckel, I, 163; orthotype X. HELLERI Heckel.

389. JERDON (1848). *On the Fresh-water Fishes of Southern India.* Madras Journ. Lit. Sci., XV.

THOMAS CLAVERHILL JERDON (1811-1873).

Pristolepis Jerdon, XV, 141; orthotype P. MARGINATUS Jerdon. Replaces CATO-PRA Blkr.

390. M'COY (1848). *On Some New Fossil Fish of the Carboniferous Period.* Ann. Mag. Nat. Hist., II.

FREDERICK M'COY.

Isodus M'Coy, 3; orthotype I. LEPTOGNATHUS M'Coy (fossil).

Colonodus M'Coy, II, 5; orthotype C. LONGIDENS M'Coy (fossil).

Osteoplax M'Coy, II, 6; orthotype O. EROSA M'Coy (fossil).

391 M'COY (1848). *On Some New Ichthyolites from the Scotch Old Red Sandstone.* Ann. Mag. Nat. Hist., II.

FREDERICK M'COY.

Physonemus M'Coy, II, 117; orthotype P. ARCUATUS M'Coy (fossil).

Erismacanthus M'Coy, II, 118; orthotype E. JONESI M'Coy (fossil).

Asteroptychius (Agassiz) M'Coy, II, 118; orthotype A. ORNATUS (Ag.) M'Coy (fossil). Name only, in *Poissons fossiles,* III, 176, 1843; made valid by M'Coy, 1848.

Platyacanthus M'Coy, II, 120; orthotype P. ISOSCELES M'Coy (fossil). Name preoccupied. A synonym of ORACANTHUS Ag.

Polyrhizodus M'Coy, II, 125; orthotype P. MAGNUS M'Coy (fossil).

Glossodus M'Coy, II, 127; logotype G. LINGUA-BOVIS M'Coy (fossil). Not GLOSSODON Raf.

Climaxodus M'Coy, II, 128; orthotype C. IMBRICATUS M'Coy (fossil). A synonym of JANASSA.

Cheirodus M'Coy, II, 130; orthotype C. PES-RANÆ M'Coy (fossil).

Petrodus M'Coy, II, 132; logotype P. PATELLIFORMIS M'Coy (fossil).

Tripterus M'Coy, II, 306; orthotype T. POLLEXFENI M'Coy (fossil). A synonym of OSTEOLEPIS Val. Name preocucpied; replaced by TRIPLOPTERUS M'Coy (1855).

Gyroptychius M'Coy, II, 307; orthotype G. ANGUSTUS M'Coy (fossil).

Conchodus M'Coy, II, 312; orthotype C. OSTREÆFORMIS M'Coy (fossil).

392. MÜLLER (1848). *(Fossile Fische.) Reise in den äussersten Norden und Osten Siberiens während der Jahr 1843-44.* Herausgegeben von A. Th. v. Middendorf, I.

JOHANNES MÜLLER.

Lycoptera Müller, I, 262; orthotype L. MIDDENDORFI Müller (fossil).

393. MÜLLER & TROSCHEL (1848). *Reisen in British-Guiana in den Jahren 1840-44.* (Im Auftrag Sr. Majestät des Königs von Preussen ausgeführt von Richard Schomburgk: *Versuch einer Fauna und Flora von British-Guiana. (Fische.)* III.)

JOHANNES MÜLLER; FRANZ HERMANN TROSCHEL.

Acharnes Müller & Troschel, III, 622; orthotype A. SPECIOSA M. & T.

Polycentrus Müller & Troschel, III, 622; orthotype P. SCHOMBURGKI M. & T.

394. RICHARDSON (1848). *Fishes: The Zoology of the Voyage of H.M.S. "Samarang"; Under the Command of Captain Sir Edward Belcher, During the years 1843-1846.*

JOHN RICHARDSON.

Choridactylus Richardson, 8; orthotype C. MULTIBARBIS Rich. Later spelled CHO-
RISMODACTYLUS.

Sthenopus Richardson, 10; orthotype S. MOLLIS Rich. Name preoccupied; replaced
by TRICHOPLEURA Kaup.

Podabrus Richardson, 11; logotype P. CENTROPOMUS Rich. Preoccupied in beetles;
replaced by VELLITOR Jordan & Starks (1894).

Nemichthys Richardson, 25; orthotype N. SCOLOPACEUS Rich.

Aperioptus Richardson, 27; orthotype A. PICTORIUS Rich.

1849

395. AYRES (1849). *On a Very Curious Fish, for which the Name Malacosteus
niger Is Proposed.* Journ. Bost. Soc. Nat. Hist.

WILLIAM O. AYRES.

Malacosteus Ayres, 53; orthotype M. NIGER Ayres.

396. BLEEKER (1849). *A Contribution to the Knowledge of the Ichthyological
Fauna of Celebes.* Journ. Ind. Arch., III.

PIETER BLEEKER.

Apogonoides Bleeker, III, 70; orthotype A. MACASSARIENSIS Blkr. This forgotten
genus is based on a species without teeth on the plate; with the preopercle
entire, no opercular spines: D .VI-1, 9; caudal bilobed. APOGONOIDES seems
to differ from the later APOGONICHTHYS only in the presence of 6, not 7,
dorsal spines and in the bilobed caudal.

Amblygaster Bleeker, 70; orthotype A. CLUPEOIDES Blkr. A synonym of SARDINELLA
Cuv. & Val. The true sardines (SARDINIA Poey) are well distinguished from
SARDINELLA externally by the strong concentric grooves on the opercles.

Dipterygonotus Bleeker, 70; orthotype D. LEUCOGRAMMICUS Blkr. An ally or
synonym of EMMELICHTHYS Rich.

Pogonognathus Bleeker, III, 73; orthotype P. BARBATUS Blkr. = ANACANTHUS
BARBATUS Gray. A synonym of PSILOCEPHALUS Sw.

397. BLEEKER (1849). *Overzigt der te Batavia voorkomende Gladschubbige
Labroïden, met deschrivning van 11 nieuwe soorten.* Verh. Bat. Gen. XXII,
1849, 1-64.

PIETER BLEEKER.

Scarus (Forskål) Bleeker XII, 3; orthotype LABRUS CRETENSIS L. (not SCARUS
Forskål). A synonym of SPARISOMA Sw.

398. BLEEKER (1849). *Bijdrage tot de kennis der Percoïden van den Malayo-
Molukschen Archipel, mit beschrijvning van 22 nieuwen soorten.* Verh. Batav.
Genootsch., XXII.

PIETER BLEEKER.

Mulloides Bleeker, 12; orthotype MULLUS FLAVOLINEATUS Lac.

Upeneoides Bleeker, 63; orthotype MULLUS VITTATUS Forskål

399. BLEEKER (1849). *Bijdrage tot de kennis der Blennioïden en Gobioïden van
den Soenda-Molukschen Archipel, met beschrijvning van 42 nieuwe soorten.*
Verh. Batav. Genootsch., XX.

PIETER BLEEKER.

Philypnoides Bleeker, XXIII, 19; orthotype P. SURAKARTENSIS Blkr. (Young of OPHICEPHALUS GACHUA Ham.) A synonym of OPHICEPHALUS Bloch.

400. BLEEKER (1849). *Bijdrage tot de kennis der ichthyologische fauna van het eiland Madura, met beschrijving van eenige nieuwe species.* Verh. Batav. Genootsch., XXII.

PIETER BLEEKER.

Leptonurus Bleeker, XXII, 14; orthotype L. CHRYSOSTIGMA Blkr. A synonym of COILIA Gray.

Hippichthys Bleeker, XXII, 15; orthotype H. HEPTAGONUS Blkr. (not HIPPO-CAMPUS HEPTAGONUS Raf.). Probably the same as CORYTHROICHTHYS Kaup and of earlier date.

Anodontostoma Bleeker, XXI, 15; orthotype A. HASSELTI Blkr. = CHATOËSSUS CHACUNDA Cuv. & Val.; replaces GONOSTOMA Van Hasselt, 1823, preoccupied.

401. BLEEKER (1849). *Bijdrage tot de kennis der ichthyologische fauna van het eiland Bali, met beschrijving van eenige nieuwe species.* Verh. Batav. Genootsch., XXII.

PIETER BLEEKER.

Stomianodon Bleeker, XXII, 10; orthotype S. CHRYSOPHEKADION Blkr. A synonym of ASTRONESTHES Rich.

402. CANTOR (1849). *Catalogue of Malayan Fishes.* Journ. Roy. Asiat. Soc. Bengal, XVIII (1849). Reprinted as a separate volume, 1850.

THEODORE EDWARD CANTOR.

Genyoroge Cantor, 12; loogtype DIACOPE SEBÆ. Name a substitute for DIACOPE Cuv. & Val., preoccupied.

Prosopodasys Cantor, 44; orthotype APISTUS ALATUS Cuv. A substitute for APISTUS Cuv., regarded as preoccupied by APISTA, hence must retain the same type. PTERICHTHYS Sw., 1839, has priority over PROSOPADASYS.

Corythobatus Cantor, 45; orthotype APISTUS MINOUS Cuv. & Val. = MINOUS MOORA Cuv. & Val. Substitute for MINOUS, regarded as preoccupied by MINOËS.

Spondyliosoma Cantor, 50; orthotype SPARUS CANTHARUS L. Substitute for CAN-THARUS Cuv., preoccupied.

Catochænum Cantor, 55; orthotype GERRES WAIGENSIS Cuv. & Val. Substitute for GERRES, unnecessarily regarded as preoccupied by GERRIS, a genus of insects. PODAGER Gistel (1848), also a substitute for GERRES, is preoccupied. I think that GERRES should hold.

Ilarches Cantor, 100; orthotype CHÆTODON ORBIS Bloch. Substitute for EPHIPPUS Cuv., regarded as preoccupied by EPHIPPIUM.

Diphreutes Cantor, 159; orthotype CHÆTODON MACROLEPIDOTUS L., replacing HENIO-CHUS Cuv. (1817), regarded as preoccupied by HENIOCHE, 1816.

Harpochirus Cantor, 162; orthotype CHÆTODON PUNCTATUS Gmelin. Same as ENIXE Gistel, 1848, an earlier substitute for DREPANE Cuv., regarded as pre-occupied by DREPANA, DREPANIA, DREPANIS and DREPANIUS.

Cacodoxus Cantor, 163; orthotype CHÆTODON ARGUS Gmelin. Same as PRENES Gistel, 1848. Substitute for SCATOPHAGUS Cuv. & Val. regarded as pre-occupied by SCATOPHAGA.

Cannorhynchus Cantor, 211; orthotype FISTULARIA TABACARIA L. A synonym of FISTULARIA.

Synaptura Cantor, 222; logotype PLEURONECTES COMMERSONIANUS Lac., as restricted by Bleeker.

Tetradrachmum Cantor, 240; logotype CHÆTODON ARUANUS L. Same as PIRENE Gistel (1848). A substitute for DASCYLLUS Cuv., regarded as preoccupied by DASCILLUS and often spelled DASCILLUS.

Prionace Cantor, 399; logotype SQUALUS GLAUCUS L. Substitute for PRIONODON M. & H., preoccupied.

Rhachinotus Cantor, 422; orthotype RAJA AFRICANA Bl. & Schn. A substitute for ANACANTHUS Ehrenb.. 1837, preoccupied. A synonym of UROGYMNUS M. & H.

Stoasodon Cantor, 434; orthotype RAJA NARINARI Euphrasen. Substitute for AËTOBATIS as restricted by Müller & Henle, the proper type of AËTOBATIS Blainv. being regarded by tautonomy as RAJA AQUILA L.

403. CUVIER & VALENCIENNES (1849). *Histoire naturelle des poissons.* XXII.
ACHILLE VALENCIENNES.

Parodon Cuvier & Valenciennes, XXII, 50; haplotype PARODON SUBORBITALE Val.

Brycinus Cuvier & Valenciennes, XXII, 157; haplotype BRYCINUS MACROLEPIDOTUS Val.

Piabucina Cuvier & Valenciennes, XXII, 161; haplotype PIABUCIANA ERYTHRINOIDES Val.

Distichodus Cuvier & Valenciennes, XXI, 172; haplotype SALMO NEFASCH Lac. = SALMO NILOTICUS Hasselquist.

Tometes Cuvier & Valenciennes, XXII, 225; logotype TOMETES TRILOBATUS Val. A synonym of MYLEUS M. & T

Mylesinus Cuvier & Valenciennes, XXII, 234; haplotype MYLESINUS SCHOMBURGKI Val.

Chalcinus Cuvier & Valenciennes, XXII, 258; logotype CHALCINUS BRACHYPOMUS Val.

Cynopotamus Cuver & Valenciennes, XXII, 317; logotype HYDROCYON ARGENTEUS Val.

Saurida Cuvier & Valenciennes, XXII, 499; logotype SALMO TUMBIL Bloch.

Farionella Cuvier & Valenciennes, XXII, 507; haplotype FARIONELLA GAYI Val. A synonym of APLOCHITON Jenyns (1842).

404. GERMAR (1849). *Steinkohle Wittenbergs.*
ERNST FRIEDRICH GERMAR (1786-1853).

Styracodus Germar, 70; orthotype CENTRODUS ACUTUS Giebel (fossil). A synonym of CENTRODUS.

405, 406. GIEBEL (1849). *Fauna der Vorvelt.*
CHRISTIAN GOTTFRIED ANDREAS GIEBEL.

Asima Giebel, 184; orthotype RADAMAS JUGLERI Münster. A synonym of RADAMAS.

407. HECKEL (1849). *Abhandlung über eine neue fossile Fischgattung, Chiro-centrites, und die ersten Ueberreste eines Siluroiden aus der Vorwelt.* Sitzb. Akad. Wiss. Wien, II.
JOHANN JAKOB HECKEL.

Chirocentrites Heckel, I, 18; logotype C. MICRODON Heckel (fossil). A synonym of THRISSOPS.

408. HECKEL (1849). *Ichthyologie (von Syrien).* In Russegger, Joseph von, *Reisen in Europa, Asien und Africa, mit besonderer Rücksicht auf die natur-wissenschaftlichen Verhältnisse der betreffenden Länder unternommen in den Jahren 1835 bis 1841,* etc. II. part 3. (To this part Dean assigns the date 1843.)

Isodus Heckel, II, 342; orthotype I. SULCATUS Heckel (fossil). A synonym of ENCHODUS Ag.

Pycnosterina Heckel, II, 337; orthotype P. RUSSEGGERI Heckel (fossil).

409. HOEVEN, VAN DER (1849). *Handboek der dierkunde; tweede verbeterde uitgave; met bijvoegsels en aanmerkingen door Leuckart.* Amsterdam.

JANUS VAN DER HOEVEN (1802-1868).

Pteronemus Van der Hoeven, II, 177; orthotype CHEILODACTYLUS FASCIATUS Lac. A synonym of CHEILODACTYLUS.

Sphyra Van der Hoeven, II; type SQUALUS ZYGÆNA L. Emendation of SPHYRNA Raf.

410. JORDAN (1849). *(Triodus.)* Neues Jahrb. von Mineralogie.

AUGUST JORDAN.

Triodus Jordan, 843; orthotype T. SESSILIS Jordan = PLEURACANTHUS DECHENI (fossil). A synonym of ORTHACANTHUS (PLEURACANTHUS).

411. M'COY (1849). *On Some New Fossil Fish of the Carboniferous Period.* Ann. Mag. Nat. Hist., ser. II.

FREDERICK M'COY.

Centrodus M'Coy, II, 3; orthotype C. STRIATULUS M'Coy (fossil). Same as MEGA-LICHTHYS Ag., 1844, not of 1838.

412. MEYER (1849). *Fische,* etc., *aus dem Muschelkalk Ober-Schlesiens.* Palæontographica, I.

HERRMANN VON MEYER.

Palæobates Meyer, I, 234; orthotype PSAMMODUS ANGUSTISSIMUS Ag. (fossil).

413. MÜLLER & TROSCHEL (1849). *Horæ Zoologicæ. Beschreibung und Abbildung neuer Fische.* III.

JOHANNES MÜLLER; FRANZ HERMANN TROSCHEL.

Sciades Müller & Troschel, III, 6; logotype SCIADES PICTUS M. & T.

Eutropius Müller & Troschel, III, 6; logotype HYPOPHTHALMUS NILOTICUS Rüppell.

Citharinus Müller & Troschel, III, 8; tautotype SERRASALMO CITHARINUS Cuv. = SALMO UNIMACULATUS Bloch.

Hemiodus Müller & Troschel, III, 9; logotype ANODUS CRENIDENS M. & T. = SALMO UNIMACULATUS Bloch.

Ariodes Müller & Troschel, III, 9; logotype ARIODES ARENARIUS M. & T.

Erethistes Müller & Troschel, III, 12; orthotype E. PUSILLUS M. & T.

Carapus (Cuvier) Müller & Troschel, III, 13; logotype, restricted by Eigenmann, to FASCIATUS Pallas = GYMNOTUS CARAPO L. (CARAPUS Cuv., but not of Rafinesque). A synonym of GYMNOTUS L.

Sternopygus Müller & Troschel, III, 13; orthotype GYMNOTUS MACRURUS Bl. & Schn.

Rhamphichthys Müller & Troschel, III, 15; orthotype GYMNOTUS ROSTRATUS L.

Cnidon Müller & Troschel, III, 21; orthotype C. CHINENSIS M. & T. = LABRAX WAIGIENSIS Cuv. & Val. A synonym of PSAMMOPERCA Rich.

Cichlops Müller & Troschel, III, 24; orthotype C. MICROPHTHALMUS M. & T. Name preoccupied; replaced by DAMPIERIA Castelnau, 1875, or by LABRACINUS Schlegel, 1875.

Caprophonus Müller & Troschel, III, 28; orthotype C. AURORA M. & T. = ANTI-GONIA CAPROS Lowe. A synonym of ANTIGONIA Lowe.

413A. NEWMAN (1849). *(Devil Fish.)* The Zoologist.
EDWARD NEWMAN.

Brachioptilon Newman, 74; orthotype B. HAMILTONI Newman. A synonym of
MANTA Bancroft.

414. REID (1849). *An Account of a Specimen of the Vaagmær . . . Thrown
Ashore in the Firth of Forth.* Ann. Mag. Nat. Hist., III.
JOHN REID.

Vogmàrus Reid, III, 456; orthotype BOGMARUS ISLANDICUS Bl. & Schn. A substitute for BOGMARUS, both names synonyms of TRACHIPTERUS Gouan.

415. REINHARDT (1849). *Nye sydamerikanske ferskvandsfiske.* Vidensk.
Meddel-Naturh. Foren. Kjöbenhavn.
JOHANN THEODOR REINHARDT (1816-1882).

Hydropardus Reinhardt, 46; orthotype H. RAPAX Reinh. == RAPHIODON VULPINUS
Ag. A synonym of RAPHIODON.

416. SISMONDA (1849). *Descrizione dei pesci e dei crustacei fossili nel Piemonte,*
Mem. Reale Accad. Sci. Torino, ser. X.
EUGENIO SISMONDA.

Trigonodon Sismonda, 25; orthotype T. OWENI Sismonda (fossil).

417. SMITH (1849). *Illustrations from the Zoology of South Africa; Consisting
Chiefly of Figures and Descriptions of Objects of Natural History Collected
During an Expedition into the Interior of South Africa in 1834-36.*
ANDREW SMITH.

Leptorhynchus Smith, pl. 6; orthotype L. CAPENSIS Smith. A synonym of OPHI-
SURUS Lac., as restricted.

Cheilobarbus Smith, pl. 10; logotype BARBUS CAPENSIS Smith.

Pseudobarbus Smith, pl. 11; orthotype BARBUS BURCHELLI Smith.

Abrostomus Smith, pl. 12; logotype A. UMBRATUS Smith.

Tilapia Smith; orthotype CHROMIS NILOTICA Cuv. & Val. Replaces CHROMIS Gthr.,
not of Cuvier.

Atimostoma Smith, XXIV; orthotype A. CAPENSE Smith. A synonym of PSENES
Cuv. & Val.

Rhineodon Smith, pl. 26; orthotype R. TYPICUS Smith. Misprinted RHINCODON;
often written RHINODON.

Xiphiurus Smith; orthotype X. CAPENSIS Smith. Name preoccupied as XIPHIURA,
a synonym of GENYPTERUS Philippi.

418. THIOLLIÈRE (1849). *Un nouveau gisement de poissons fossiles dans le
Jura du département de l'Ain.* Ann. Soc. Agric. Lyons, I.
VICTOR JOSEPH DE L'ISLE THIOLLIÈRE (1801-1859).

Spathobatis Thiollière, I, 63; orthotype S. BUGESIACUS Thiollière (fosil). A synonym of AELLOPOS Münster.

419. WILLIAMSON (1849). *On the Microscopic Structure of the Scales and
Dermal Teeth of Some Ganoid and Placoid Fishes.* Phil. Trans. Roy. Soc.
London.
WILLIAM CRAWFORD WILLIAMSON.

Pholidotus Williamson, 444; orthotype P. LEACHI Williamson (fossil). A synonym
of DAPEDIUM Leach.

1850

420. AGASSIZ (1850). *Lake Superior: Its Physical Character, Vegetation, and Animals, Compared with Those of Other and Similar Regions,* etc.

LOUIS AGASSIZ.

Pœcilosoma Agassiz, 299; haplotype ETHEOSTOMA VARIATEUM Kirtland. Preoccupied; replaced by PŒCILICHTHYS Ag., 1854.

Argyrosomus Agassiz, 339; haplotype SALMO CLUPIFORMIS Ag., not of Mitchill = SALMO HARENGUS Rich. Name preoccupied (Pylaie, 1834); replaced by THRISSOMIMUS Gill (1909). A subgenus under LEUCICHTHYS Dybowski.

Hypsolepis (Baird) Agassiz, 368; orthotype CYPRINUS CORNUTUS Mitchill. Written HYPSILEPIS by Cope.

421. BLEEKER (1850). *Bijdrage tot de kennis der Mænoiden van den Sœnda-Molukschen Archipel.* Verh. Batav. Genootsch., XXIII.

PIETER BLEEKER.

Pentaprion Bleeker, I, 104; orthotype P. GERREOIDES Blkr.

422. BLEEKER (1850). *Bijdrage tot de kennis der visschen met doolhofvormige kieuwen van den Sœnda-Molukschen Archipel.* Verh. Batav. Genootsch, XXIII, 1850.

PIETER BLEEKER.

Betta Bleeker, XXIII, 12; orthotype B. TRIFASCIATA Blkr.

423. COSTA (1850). *(Fossil Fishes: Ittioliti.)* Atti Accad. Pontan, V.

ORONZIO GABRIELE COSTA.

Sarginites Costa, V, 285; orthotype, S. PYGMÆUS Costa (fossil). A synonym of LEPTOLEPIS.

Histiurus Costa, V, 288; orthotype H. ELATUS Costa. Preoccupied as ISTIURUS Cuv., 1829, emended as HISTIURUS Ag.; replaced by KNIGHTIA Jordan, 1907; type K. EOCÆNA Jordan.

Rhynchoncodes Costa, V, 317; orthotype R. SCACCHI Costa (fossil). A synonym of PROPTERUS Ag.

Blenniomœus Costa, V, 319; orthotype B. LONGICAUDA Costa (fossil). A synonym of NOTAGOGUS Ag.

Sauropsidium Costa, V, 322; orthotype S. LÆVISSIMUM Costa.

Megalurites Costa, V, 370; orthotype DIODON SCILLÆ Ag. (fossil). A synonym or fossil ally of DIODON.

424. DIXON (1850). *The Geology and Fossils of the Tertiary and Cretaceous Formations of Sussex.*

A. FRANCIS DIXON.

Platylæmus Dixon, 205; orthotype P. COLEI Dixon (fossil).

Aulodus Dixon, 366; orthotype A. AGASSIZI Dixon (fossil). A synonym of PTYCHODUS Ag.

Plethodus Dixon, 366; logotype P. EXPANSUS Dixon (fossil). (Not PLETHODON Tschudi, a genus of lizards.)

Pomognathus Dixon, 367; orthotype P. EURYPTERYGIUS Dixon (fossil). A synonym of HALEC Ag.

Prionolepis (Egerton) Dixon, 368; orthotype P. ANGUSTUS Egert. (fossil).

Homonotus Dixon, 372; orthotype B. ELEGANS Dixon (fossil).

Berycopsis Dixon, 372; orthotype B. ELEGANS Dixon (fossil).

Stenostoma Dixon, 373; orthotype S. PULCHELLA Dixon (fossil).
Pachyrhizodus (Agassiz) Dixon, 374; orthotype P. BASALIS Dixon (fossil).
Tomognathus Dixon, 376; logotype P. MORDAX Dixon (fossil).

425. EGERTON (1850). *Palichthyologic Notes, Supplemental to the Work of Prof. Agassiz on the Ganoidei heterocerci.* Quart. Journ. Geol. Soc. Lond.

PHILIP DE MALPA GREY EGERTON.

Plectrolepis (Agassiz) Egerton, VI, 4; orthotype P. RUGOSUS (Ag.) Egert. (fossil). A synonym of EURYNOTUS Ag. = NOTACMON Gistel.

426. FISCHER VON WALDHEIM (1850). *Sur un poisson fossile de la Grèce (Platacanthus).* Bull. Soc. Imp. Nat. Moscou., XXIII.

GOTTHELF FISCHER VON WALDHEIM (1771-1853).

Platacanthus Fischer von Waldheim, XXIII, 286; orthotype P. UBINOI Waldheim (fossil). An extinct ally of DICENTRARCHUS Gill = LABRAX Cuv., not Pallas.

427. FISCHER VON WALDHEIM (1850). *Ommatolampes et Trachelacanthus. genera piscium fossilium nova, in literis Eduardo ab Eichwald datis, descripta.* Mosquæ.

GOTTHELF FISCHER VON WALDHEIM.

Ommatolampes Fischer von Waldheim, 4; orthotype O. EICHWALDI Fischer.
Trachelacanthus Fischer von Waldheim, 9; orthotype T. STSCHUROVSKII Waldheim (fossil).

428. GIBBES (1850). *Mososaurus and Allied Genera.* Smithsonian Contr. Knowledge, II.

ROBERT WILSON GIBBES (1809-1866).

Conosaurus Gibbes, 10; orthotype C. BOWMANI Gibbes. Name regarded as preoccupied by CONIOSAURUS; replaced by CONOSAUROPS Leidy (1868).

429. GUICHENOT (1850). *Exploration scientifique de l'Algérie.* Paris.

ALPHONSE GUICHENOT.

Gadiculus Guichenot, 101; orthotype G. ARGENTEUS Guich.

430. HECKEL (1850). *Beiträge zur Kenntniss der fossilen Fische Oesterreichs.* Denkschr. Akad. Wiss. Wien, I.

JOHANN JAKOB HECKEL.

Saurorhamphus Heckel I, 217; orthotype S. FREYERI Heckel (fossil).
Lepidopides Heckel, I, 239; LEPIDOPUS LEPTOSPONDYLUS Heckel (fossil). A synonym of ANENCHELUM Blainv. and probably of the living genus LEPIDOPUS

431. JERDON (1850). *(Pristolepis.)* Madras Journ. Lit. Sci., XVI.

THOMAS CLAVERHILL JERDON.

Pristolepis Jerdon, XVI, 298; orthotype P. MARGINATUS Jerdon. Jerdon (*Ann. Mag. Nat. Hist.*, 1865, 298) claims priority of PRISTOLEPIS over CATOPRA Blkr. CATOPRA was first noted in 1851.

432. LANKESTER (1850). *(Fishes of the Old Red Sandstone.)*

EDWIN RAY LANKESTER (1847-   ).

Zenaspis Lankester, 43; orthotype CEPHALASPIS SALWEYI Egert. (fossil). A synonym of CEPHALASPIS.

433. LOWE (1850). *An Acconut of Fishes Discovered or Observed in Madeira Since the Year 1842.* Proc. Zool. Soc. London, XVIII.

RICHARD THOMAS LOWE.

Phænodon Lowe, 250; orthotype P. RINGENS Lowe. A synonym of ASTRONESTHES Rich.

434. PICTET (1850). *Description de quelques poissons fossiles du Mont Liban.* Mém. Soc. Phys. Hist. Nat. Genève, XII, 1849, published in 1850.
FRANÇOIS JULES PICTET (1809-1872).

Petalopteryx Pictet, 20; orthotype P. SYRIACUS Pictet (fossil).

Eurypholis Pictet, 28 logotype E. BOISSIERI Pictet (fossil). A synonym of SAUROR-HAMPHUS.

Spaniodon Pictet, 33; logotype S. BLONDELI Pictet (fossil).

Coccodus Pictet, 51; orthotye C. ARMATUS Pictet (fossil).

435. RICHARDSON (1850). *Notices of Australian Fish.* Proc. Zool. Soc. London. XVIII.
JOHN RICHARDSON.

Sciænoides (Solander MS.). Richardson, 62; haplotype S. ABDOMINALIS Solander = SCIÆNA MACROPTERA Forster; in synonymy only; a species of DACTYLOSPARUS Gill.

Threpterius Richardson, 68; orthotype T. MACULOSUS Rich.

Olisthops Richardson, 74; orthotype O. CYANOMELAS Rich. (Also in *Ann. Mag. Nat. Hist.,* VIII, 185, 290,) Name spelled OLISTHEROPS by Günther.

436. TEMMINCK & SCHLEGEL (1850). *Fauna Japonica* (last part).
COENRAAD JACOB TEMMINCK; HERMANN SCHLEGEL.

Pseudoblennius Temminck & Schlegel, 313; orthotype P. PERCOIDES Gthr.

Pseudoclinus Temminck & Schlegel, pl. LXXIX; orthotype PSEUDOBLENNIUS PERCOIDES Gthr. A slip of the pen for PSEUDOBLENNIUS.

Velifer Temminck & Schlegel, 312; orthotype V. HYPSELOPTERUS Blkr.

437. THIOLLIÈRE (1850). *Un nouveau gisement de poissons fossiles dans le Jura de département de l'Ain.* Ann. Soc. Agric. Lyon, I.
VICTOR JOSEPH DE L'ISLE THIOLLIÈRE.

Disticholepis Thiollière, III, 136; orthotype M. FOURNETI Thiollière. A synonym of MACROSEMIUS Ag.

Oligopleurus Thiollière, III, 154; orthotype O. ESOCINUS Thiollière (fossil).

1851

438. BLEEKER (1851). *Visschen van Banka.* Nat. Tijdschr. Neder.-Indie, 1851.
PIETER BLEEKER.

Gnathanodon Bleeker, 160; orthotype SCOMBER SPECIOSUS Forskål. Not CARANX of Lacépède, as first restricted (to C. CARANGUS) by Bleeker.

439. BLEEKER (1851). *Nieuwe bijdrage tot de kennis der ichthyologische fauna van Borneo, met beschrijving van eenige nieuwe soorten van zoetwatervisschen.* Nat. Tijdschr. Neder.-Indie, II.
PIETER BLEEKER.

Wallago Bleeker, 202; orthotype WALLAGO DINEMA Blkr.

Achiroides Bleeker, 262; logotype PLAGUSIA MELANORHYNCHUS Blkr.

Clupeoides Bleeker, 275; orthotype C. BORNEENSIS Blkr. A synonym of CORICA Gray.

Luciocephalus Bleeker, 278; orthotype DIPLOPTERUS PULCHER Gray. Replaces DIPLOPTERUS Gray, preoccupied in birds.

440. BLEEKER (1851). *Over einige nieuwe geslachten en soorten van Makreel-agtige Visschen van den Indischen Archipel.* Nat. Tijdschr. Neder.-Indie, I, 1851.

PIETER BLEEKER.

Decapterus Bleeker, II, 358; orthotype CARANX KURRA Cuv. & Val.

Selar Bleeker, 359; logotype CARANX HASSELTI Blkr. (C. BOOPS Cuv. & Val. indicated as type by Jordan & Evermann, 1898, was not named in Bleeker's first account (only S. HASSELTI, S. KUHLI, S. MACRURUS, S. MALAM).

Carangoides Bleeker, 366; orthotype CARANX PRÆUSTUS Bennett, the only other species here mentioned being C. ATROPUS, type of ATROPUS Cuv. Defined to include CITULA, ATROPUS and OLISTUS as well as CARANGOIDES proper, as restricted by Jordan & Evermann.

Leioglossus Bleeker, 367; orothotype L. CARANGOIDES Blkr.

Stromateoides Bleeker, I, 368; orthotype STROMATEUS CINEREUS Bloch; a synonym of PAMPUS Bon.

441. BLEEKER (1851). *Derde bijdrage tot de kennis der ichthyologische fauna van Borneo, etc.* Nat. Tijdschr. Neder.-Indie, II.

PIETER BLEEKER.

Tetrabranchus Bleeker, II, 69; orthotype T. MICROPHTHALMA Blkr. A synonym of SYMBRANCHUS Bloch.

Catopra Bleeker, 172; orthotype C. FASCIATA Blkr. A synonym of PRISTOLEPIS Jerdon, 1850.

442. BLEEKER (1851). *Cheilinoides, een nieuw geslacht van gladschubbige Labroiden van Batavia.* Nat. Tijdschr. Neder.-Indie, II

PIETER BLEEKER.

Cheilinoides Bleeker, II, 71; orthotype C. CYNANOPLEURA Blkr.

443. BLEEKER (1851). *Vierde bijdrage tot den kennis der ichthyologische fauna van Borneo, met beschrijvning van eenige nieuwe soorten van zoetwatervisschen.* · Nat. Tijdschr. Neder.-Indie, II.

PIETER BLEEKER.

Bagroides Bleeker, II, 204; ·orthotype B. MELANOPTERUS Blkr.

444. BLEEKER (1851). *Nieuwe bijdrage tot de kennis der ichthyologische fauna van Celebes.* Nat. Tijdschr. Neder.-Indie, II, 1851.

PIETER BLEEKER.

Megalaspis Bleeker, II, 213; orthotype SCEMBER CORDYLA L. (CARANX ROTTLERI Blkr.).

Selaroides Bleeker, II, 213; orthotype CARANX LEPTOLEPIS Kuhl & Van Hasselt. Later called LEPTASPIS, the reason not stated.

Spratelloides Bleeker, 214; orthotype CLUPEA ARGYROTÆNIA Blkr.

445. BLEEKER (1851). *Bijdrage tot de kennis der ichthyologische fauna van de Banda eilanden.* Nat. Tijdschr. Neder.-Indie, II.

PIETER BLEEKER.

Labroides Bleeker, II, 249; orthotype L. PARADISEUS Bleeker.

Syngnathoides Bleeker, 259; orthotype S. BLOCHI Blkr. Replaces GASTEROTOKEUS Heckel (1856).

446. GAIMARD (1851). *Voyage de la Commission Scientifique du Nord. Voyage en Skandinavie, en Lapponie, à Spitzberg et aux Faroë,* Paris, VII. Zoologie (1851).

JOSEPH PAUL GAIMARD.

Eperlanus Gaimard, VII; orthotype SALMO EPERLANUS L. A synonym of OSMERUS L. (Not seen.)

447. GIRARD (1851). *On the Genus Cottus auct.* Proc. Boston Soc. Nat. Hist., III.

CHARLES FRÉDÉRIC GIRARD (1822-1895).

Acanthocottus Girard, III, 185; logotype COTTUS GRŒNLANDICUS Cuv. & Val. A section under MYOXOCEPHALUS Steller.

448. GIRARD (1851). *Some Additional Observations on the Nomenclature and Classification of the Genus Cottus.* Proc. Boston Soc. Nat. Hist., III (for 1848), 1851.

CHARLES FRÉDÉRIC GIRARD.

Cottopsis Girard, III, 303; orthotype COTTUS ASPER Rich. A section of COTTUS L.

449. GRAY (1851). *Description of a New Form of Lamprey from Australia with a Synopsis of the Family.* Proc. Zool. Soc. London.

JOHN EDWARD GRAY.

Geotria Gray, 143; orthotype GEOTRIA AUSTRALIS Gray.

Velasia Gray, 143; orthotype V. CHILENSIS Gray.

Caragola Gray, 143; orthotype C. LAPICIDA Gray. Same as MORDACIA Gray.

Mordacia Gray, 144; orthotype PETROMYZON MORDAX Rich. A synonym of CARAGOLA Gray.

Lampetra Gray, 235; logotype PETROMYZON FLUVIATILIS L.

450. LÜTKEN (1851). *Nogle bemaerkinger om naeseborenes stiling hos de i gruppe med Ophisurus staaende slaegter af aalefamilien.* Vidensk. Meddel. Naturhist. Foren. Kjöbenjavn.

CHRISTIAN FREDERIK LÜTKEN (1827-1901).

Myrophis Lütken, 1; orthotype M. PUNCTATUS Lütken.

Chilorhinus Lütken, 1; orthotype C. SUENSONI Lütken.

451. MEYER (1851). *Fossile Fische aus dem Muschelkalk von Jena, Querfurt und Esperstadt.* Palæontographica.

HERMANN VON MEYER.

Tholodus (Meyer) Schmid & Schleiden, I, 199; orthotype T. SCHMIDI (fossil). Perhaps a tooth of a reptile.

452, 453. MEYER (1851). *Palæontographica.* I.

HERMANN VON MEYER.

Nephrotus Meyer, I, 243; orthotype OMPHALODUS CHORZOWENSIS Meyer (fossil). A synonym of ASTERODON.

454. M'COY (1851). *(Steganodictyum.)* Mag. Nat. Hist., VIII.

FREDERICK M'COY.

Steganodictyum M'Coy, VIII, 481; orthotype S. CORNUBICUM M'Coy (fossil). (Shields of a PTERASPID mistaken for part of a sponge.)

455. THOMPSON (1851). *On Percopsis pellucida.* Proc. Boston Soc. Nat. Hist., III.

ZADOCK THOMPSON (1796-1856).

Salmoperca Thompson, III, 164; haplotype SALMOPERCA PELLUCIDA Thompson = PERCOPSUS GUTTATIS Ag. = SALMO OMISCOMAYCUS Walbaum. A synonym of PERCOPSIS Ag., 1848.

456. WAGNER (1851). *Beiträge zur Kenntniss der in den lithographischen Schiefern abgelagerten urweltlichen Fische.* Abh. Bayer. Akad. Wiss., VI.
JOHANN ANDREAS WAGNER (1797-1861).

Mesodon Wagner, VI, 56; orthotype GYRODUS MACROPTERUS Ag. (fossil). Name preoccupied in mollusks, Rafinesque, 1819; replaceable by TYPODUS Quenstedt, 1858.

Strobilodus Wagner, VI, 75; orthotype S. GIGANTEUS Wagner (fossil). A synonym of CATURUS Ag.

1852

457. BLEEKER (1852). *Bijdrage tot de kennis der Makreelachtige visschen van den Soenda-Molukschen Archipel.* Verh. Batav. Genootsch., XXIV, 1852.
PIETER BLEEKER.

Leptaspis Bleeker, XXIV, 71; orthotype CARANX LEPTOLEPIS Kuhl & Van Hasselt. Called earlier SELAROIDES, the reason for changing not known to me.

458. BLEEKER (1852). *Bijdrage tot de kennis der Plagiostomen van den Indischen Archipel.* Verh. Batav. Genootsch., XXIV.
PIETER BLEEKER.

Hemigaleus Bleeker, XXIV, 45; orthotype H. MICROSTOMA Blkr. Preoccupied in mammals as HEMIGALEA.

459. BLEEKER (1852). *Diagnostische beschrijvning van nieuwe of weinig bekende visschsoorten van Sumatra.* Nat Tijdschr. Neder.-Indie, III, 1852.
PIETER BLEEKER.

Pristipomoides Bleeker, III, 574; orthotype P. TYPUS Blkr. Replaces CHÆTOPTERUS T. & S., preoccupied

460. BLEEKER (1852). *Derde bijdrage tot de kennis der ichthyologische fauna van Celebes.* Nat. Tijdschr. Neder.-Indie, III.
PIETER BLEEKER.

Carangichthys Bleeker, III, 760; orthotype C. TYPUS Blkr. A synonym of CARANX Lac. (based on young specimens).

461. DUMÉRIL (1852). *Monographie de la famille des torpediniens, ou poissons plagiostomes électriques.* Rev. Mag. Zool., IV.
AUGUSTE HENRI ANDRÉ DUMÉRIL (1812-1870).

Hypnos A. Duméril, 279; orthotype H. SUBNIGRUM Dum. Name preoccupied; replaced by HYPNARCE Waite, 1902.

462. EGERTON (1852). *Fish: Elasmodus, etc.* (In *Figures and Descriptions of British Organic Remains.*) Mem. Geo. Surv. United Kingdom.
PHILIP DE MALPAS GREY EGERTON.

Lophistomus Egerton, Dec. VI, No. 10; orthotype L. DIXONI Egert. (fossil).

463. FISCHER VON WALDHEIM (1852). *Sur quelques poissons fossiles de la Russie.* Bull. Soc. Imp. Nat. Moscou, XXV.
GOTTHELF FISCHER VON WALDHEIM.

Pycnacanthus Fischer von Waldheim, XXV, 174; orthotype (not named by Woodward) (fossil). (An unrecognized fragment.)

Siphonodus Fischer von Waldheim, XXV, 175; orthotype S. PANDERI Waldheim (fossil).

464. GERVAIS (1852). *Zoologie et Paléontologie Françaises.* II.
FRANÇOIS LOUIS PAUL GERVAIS.
Sandroserrus Gervais, II, 8; orthotype PERCA REBOULI Gervais (fossil).

465. LOWE (1852). *An Account of Fishes Discovered or Observed in Madeira Since the Year 1842.* Ann. Mag. Nat. Hist. London, ser. X.
RICHARD THOMAS LOWE.
Leptorhynchus Lowe, X, 54; orthotype L. LEUCHTENBERGI Lowe. Name preoccupied; replaced by BELONOPSIS Brandt. A synonym of NEMICHTHYS Rich.

466. QUENSTEDT (1852). *Handbuch der Petrefaktenkunde.*
FRIEDRICH AUGUST VON QUENSTEDT (1809-1889).
Selachidea Quenstedt, 173; orthotype S. TOROLOSI Quenstedt (fossil). Probably a synonym of HYBODUS.
Serrolepis Quenstedt, 207; orthotype S. SUEVICUS Dames. (fossil).
Pleurolepis Quenstedt, 214; orthotype P. SEMICINCTUS Quenstedt (fossil). A synonym of TETRAGONOLEPIS Bronn.

467. ROEMER (1852). *Die Kreidebildungen von Texas und ihre organischen Einschlüsse. Mit einem die Beschreibung von Versteinerungen aus paläozoischen und tertiären Schichten enthaltenden Anhange und mit 11 von Chohenach der Natur auf Stein gezeichneten Tafeln.* Bonn.
CARL FERDINAND VON ROEMER.
Ancistrodon (Derby) Roemer, 30; orthotype later called A. TEXANUS Dames (fossil). Teeth undetermined. Name preoccupied in snakes as AGKISTRODON; replaced by ANKISTRODUS Koninck, 1870, scarcely better, and by GRYPODON Hay, 1899.

468. RÜPPELL (1852). *Verzeichniss der in dem Museum der senckenbergischen naturforschenden Gesellschaft aufgestellten Sammlungen.* Vierte Abtheilung. *Fische und deren Skelette.*
WILHELM PETER EDUARD SIMON RÜPPELL.
Microichthys Rüppell, 1; orthotype M. COCCOI Rüppell.
Pharopteryx Rüppell, 16; orthotype P. BENOIT Rüppell. Not PHAROPTERX of Rüppell, 1828, which is a synonym of PLESIOPS Cuv. A synonym of MORA Risso.
Chœrops Rüppell, 20; orthotype C. MELEAGRIS Rüppell. A synonym of CHŒRODON Blkr.
Sarotherodon Rüppell, 21; orthotype S. MELANOTHERON Rüppell.
Erythrodon Rüppell, 34; orthotype BALISTES NIGER Lac., BALISTES ERTHRODON Gthr. Substitute for XENODON Rüppell, 1835, preoccupied, but the name ODONUS Gistel (1848) has priority and ZENODON Sw. (1829) is still older, if available.

1853

469. AGASSIZ (1853). *Recent Researches of Professor Agassiz.* (Extract from letter to J. D. Dana dated Cambridge, June 9, 1853). Amer. Journ. Sci., XVI.
LOUIS AGASSIZ.
In this paper four new genera are briefly described. The species concerned were not named until later.
Chologaster Agassiz, 135; (haplotype CHOLOGASTER CORNUTUS Ag.). Short account of the genus, the species named later.
Heterandria Agassiz, 135; (logotype H. FORMOSA Ag.). Stated to contain two species named later, H. FORMOSA and H. HOLBROOKI, the latter being GAM-

BUSIA PATRUELIS Baird & Grd. After consultation with Professor Poey
FORMOSA was chosen by Jordan as the type of HETERANDRIA. This name
replaces GIRARDINUS Poey.

Zygonectes Agassiz, 135; logotype POECILIA OLIVACEA Storer = SEMOTILUS NOTATIUS
Raf. Brief notice of genus; its species indicated in 1854 (*Fish. Tennessee
River*).

Melanura Agassiz, 135; orthotype EXOGLOSSUM ANNULATUM Raf. Note on the
genus, the type indicated on page 217 of the same journal. A synonym of
UMBRA Gronow.

470. AGASSIZ (1853). *Extraordinary Fishes from California Constituting a New
Family (Holconoti or Embiotocidæ).* Amer. Journ. Sci., XVI.
LOUIS AGASSIZ.

Embiotoca Agassiz, 386; orthotype E. JACKSONI Ag.

471. BAIRD & GIRARD (1853). *Descriptions of New Species of Fishes Collected
by Mr. John H. Clark on the U. S. and Mexican Boundary Survey, under
Lt.-Col. Jas. D. Graham.* Proc. Acad. Nat. Sci. Phila.
SPENCER FULLERTON GIRARD (1823-1887); CHARLES GIRARD.

Ceratichthys Baird & Girard, 391; orthotype C. VIGILAX Baird & Girard (not of
Baird & Girard, 1856). No description. Replaces CLIOLA Grd., 1856.

472. BAIRD & GIRARD (1853). *Descriptions of New Fishes from the River Zuñi.*
Proc. Acad. Nat. Sci. Phila.
SPENCER FULLERTON BAIRD; CHARLES GIRARD.

Gila Baird & Girard, VI, 368; logotype GILA ROBUSTA Grd.

473. BLEEKER (1853). *Diagnostische beschrivningen van nieuwe of weinig
bekende visschsoorten van Batavia.* Tintal I-IV, Nat. Tijdschr. Neder.-
Indie, IV, 1853.
PIETER BLEEKER.

Murænichthys Bleeker, 506; orthotype M. GYMNOPTERUS Blkr.

474. BLEEKER (1853). *Nieuwe bijdrage tot de kennis de ichthyologische fauna
van Ternate en Halmaheira (Giolo).* Nat. Tijdschr. Neder.-Indie, IV.
PIETER BLEEKER.

Polypterichthys Bleeker, II, 608; orthotype P. VALENTINI Blkr. A synonym of
AULOSTOMUS.

475. BLEEKER (1853). *Zevende bijdrage tot de kennis der ichthyologische
fauna van Borneo. Zoetwatervisschen van Sambas, Pontianak en Pengaron.*
Nat. Tijdschr. Neder.-Indie, V.
PIETER BLEEKER.

Datnioides Blkr., V, 440; orthotype CIUS POLOTA Ham. A synonym of COIUS Ham.,
as restricted by Fowler, 1905, to COIUS POLOTA. This antedates the restriction by Jordan (*Genera of Fishes*, 1917, 114) to C. COBOJUS Ham.

476. BLEEKER (1853). *Bijdrage to de kennis der Murænoïden en Symbranchoïden
van den Indischen Archipel.* Verh. Batav. Genootsch., XXV.
PIETER BLEEKER.

Stethopterus Bleeker, XXV, 24, 36; orthotype OPHISURUS VIMINEUS Rich. A synonym of LEIURANUS Blkr. and having line priority.

Leiuranus Bleeker, XXV, 36; orthotype L. LACEPEDEI Blkr. (OPHISURUS SEMI-CINCTUS Bennett.). Same as STETHOPTERUS Blkr., the latter having line priority, the name LEIURANUS preferred by the author.

477. BLEEKER (1853). *Nalezingen op de ichthyologie van Japan.* Verh. Batav. Genootsch., XXV.

PIETER BLEEKER.

Atherinichthys Bleeker, 40; orthotype ATHERINA HUMBOLDTIANA Cuv. & Val. A synonym of CHIROSTOMA Sw.

Atherinoides Bleeker, 40; orthotype ATHERINA VOMERINA Cuv. & Val. = ATHERINA HUMBOLDTIANA Cuv. & Val. A synonym of CHIROSTOMA Sw.

Etrumeus Bleeker, XXV, 48; orthotype CLUPEA MICROPUS T. & S.

478. BLEEKER (1853). *Nalezingen op de ichthyologische fauna van Bengalen en Hindostan.* Verh. Batav. Genootsch., XXV.

PIETER BLEEKER.

Bogoda Bleeker, 89; CHANDA BOGODA Ham. = CHANDA NAMA Ham. A synonym of CHANDA as now restricted.

Badis Bleeker, 106; orthotype LABRUS BADIS Ham. = BADIS BUCHANANI Blkr.

Bagarius Bleeker, 121; tautotype PIMELUDUS BAGARIUS Ham.

479. COSTA (1853). *(Fossil Fishes of Italy.)* Atti Soc. Pontan, VII.

ORONZIO GABRIELE COSTA.

Ionoscopus Costa, 2; orthotype I. PETRAROIÆ Costa (fossil). Name changed in 1864 to Œnoscopus Costa for no obvious reasons.

Calignathus Costa, VII, 17; orthotype BLENNIOMŒUS LONGICAUDA Costa (fossil). A synonym of NOTAGOGUS Ag

Glossodus Costa, VII, 26; orthotype PYCNODUS MANTELLI Ag. (fossil). Name preoccupied; to be replaced by COSMODUS Sauvage (1879).

480. GERVAIS (1853). *Remarques sur les poissons fluviatiles de l'Algérie, et description de deux genres nouveaux souls les noms de Coptodon et Tellia.* Ann. Nat. (Zool).), XIX.

FRANÇOIS LOUIS PAUL GERVAIS.

Coptodon Gervais, II, 15; orthotype C. ZILLI Gervais. Also in *Bull. Soc. Agr. Hérault*, 1853, 80.

Tellia Gervais, XIX, 15; orthotype TELLIA APODA Gervais.

481. KAUP (1853). *Uebersicht der Lophobranchier.* Archiv. Naturg.

JOHANN JACOB KAUP (1802-1873).

All these genera, except HEMIMARSUPIUM, are repeated in Kaup's *Catalogue of the Lophobranchiate Fishes in the British Museum*, 1856.

Acentronura Kaup, 230; orthotype HIPPOCAMPUS GRACILLIMUS T. & S.

Gasterotokeus (Heckel) Kaup, 230; orthotype SYNGNATHUS BIACULEATUS Bloch = SYNGNATHOIDES BLOCHI Blkr. A synonym of SYNGNATHOIDES Blkr.

Halicampus Kaup, 231; logotype H. GRAYI Kaup.

Trachyrhamphus Kaup 231; orthotype T. SERRATUS Kaup (not SYNGNATHUS SERRATUS T. & S.).

Corythroichthys Kaup, 231; logotype C. ALBIROSTRIS Heckel. A synonym of HIPPICHTHYS Blkr.

Ichthyocampus Kaup 231; logotype I. BELCHERI Kaup.

Leptonotus Kaup, 232; orthotype SYNGNATHUS BLAINVILLEANUS Eydoux & Gervais.

Leptoichthys Kaup, 232; orthotype L. FISTULARIS (Bibron).

Stigmatophora Kaup, 233; logotype SYNGNATHUS ARGUS Rich.

Doryrhamphus Kaup, 233; orthotype D. EXCISUS Kaup.

Choëroichthys Kaup, 233; orthotype C. VALENCIENNESI Kaup.

Doryichthys Kaup, 233; logotype D. BILINEATUS Heckel.

Hemimarsupium Kaup, 234; orthotype TYPHLUS GONDOTII Bibron. For this name Kaup has substituted in 1856 the name HEMITHYLACUS. Replaces CŒLONOTUS Peters, 1855.

Microphis Kaup, 234; logotype SYNGNATHUS DEOKHATA Ham.

482. KÖLLIKER (1853). *Weitere Bemerkungen über die Helmichthyiden.* Verh. Phys. Med. Ges. Würzburg., IV.
RUDOLPH ALBERT VON KÖLLIKER (1817-1905).

Tilurus Kölliker, IV, 100; orthotype T. GEGENBAURI Kölliker. A larva of some form allied to CONGPR.

Hyoprorus Kölliker, IV, 101; orthotype H. MESSINENSIS Kölliker. A larva of NETASTOMA Raf.

483. OWEN (1853). *(Fossil Fishes.)* Quarterly Journ. of Geology.
RICHARD OWEN.

Parabatrachus Owen, IX, 67; orothotype P. COLEI Owen (fossil). Same as CENTRODUS M'Coy, a preoccupied name.

484. POMEL (1853). *Catalogue méthodique des vertèbres fossiles de la Loire.*
A. POMEL.

Propalæoniscus Pomel, 133; orthotype P. AGASSIZI Pomel (fossil) (undescribed). Probably the same as ELONICHTHYS.

Cobitopsis Pomel, 134; orthotype C. EXILIS Pomel (fossil).

Pœcilops Pomel, 135; orthotype P. BREVICEPS Pomel (fossil).

485. ROMANOVSKII 1833). *Ueber eine neue Gattung versteinerte Fisch-Zähne.*
Bull. Soc. Imp. Nat. Moscou, XXVI.
GHENNADI DANILOVICH ROMANOVSKII.

Dicrenodus Romanovskii, XXVI, 407; orthotype D. OKENSIS Romanovskii (fossil). Same as CARCHAROPSIS Ag.

1854

486. AGASSIZ (1854). *Notice of a Collection of Fishes from the Southern Bend of the Tennessee River in the State of Alabama.* Am. Journ. Sci., XVII.
LOUIS AGASSIZ.

Pœcilichthys Agassiz, 304; orthotye ETHEOSTOMA VARIATUM Kirtland. Substitute for PŒCILOSOMA Ag., preoccupied.

Catonotus Agassiz, 305; haplotype C. LINEOLATUS Ag. This name has been regarded as a synonym of ETHEOSTOMA, but a decision of the International Committee makes ETHEOSTOMA identical with HYOSTOMA.

Hadropterus Agassiz, 305; haplotype H. NIGROFASCIATUS Ag.

Hyostoma Agassiz, 305; haplotype H. NEWMANI Ag. = ETHEOSTOMA BLENNIOIDES Raf. 1818. A synonym of ETHEOSTOMA Raf. and of DIPLESION Raf. (1820).

Hybopsis Agassiz, 358; haplotype H. GRACILIS Ag., identified by me with RUTILUS AMBLOPS Raf., a member of the group called ERINEMUS Jordan. If this identification is correct, Agassiz must have overlooked the characteristic barbel. His description fairly fits another species of the same waters, without a barbel, Hybopsis stramineus Cope (Minnilus microstomus Raf)., a near ally of Alburnops blennius, type of Alburnops. Girard and Jordan identify Hybopsis with Erinemus; Cope with the Alburnops group. Girard's view seems on the basis of evidence available to be correct, and Hybopsis replaces Erinemus.

Hypsolepis (Baird) Agassiz, 359; haplotype Cyprinus cornutus Mitchill. A subgenus under Notropis. A synonym of Luxilus Raf., later written Hypsilepis.

487. AGASSIZ (1854). *Additional Notes on the Holconoti.* Amer. Journ. Sci. XVIII (May, 1854).
Louis Agassiz.

Rhacochilus Agassiz, 367; orthotype R. toxotes Ag.
Amphistichus Agassiz, 367; orthotype A. argenteus Ag.
Holconotus Agassiz, XXVII, 367; orthotype H. rhodoterus Ag.

488. AYRES (1854). *(Description of New Fishes from California.)* Proc. Calif. Acad. Nat. Sci., I (1857).
William O. Ayres.

Clypeocottus Ayres, 12; orthotype C. robustus Ayres = Aspicottus bison Grd. A synonym of Aspicottus Grd. of the same date.

489. AYRES (1854). *(Description of Four Species of Fishes, Camarina nigricans, Poronotus simillimus Ayres, Johnius nobilis, and Seriphus politus.)* Proc. Calif. Acad. Nat. Sci., I (1857).
William O. Ayres.

Seriphus Ayres, 80; orthotype Seriphus politus Ayres.
Camarina Ayres, 81; orthotype Camarina nigricans Ayres. A synonym of Girella Gray.

490. BAIRD & GIRARD (1854). *Descriptions of New Species of Fish Collected in Texas, New Mexico and Sonora, by Mr. John H. Clark on the U. S. and Mexican Boundary Survey, and in Texas by Capt. Stewart Van Vliet, U. S. A.* Proc. Acad. Nat. Sci. Phila.
Spencer Fullerton Baird; Charles Girard.

Herichthys Baird & Girard, 25; haplotype H. cyanoguttatus Baird & Girard. A synonym of Heros Heckel, itself a synonym of Astronotus Sw.
Ailurichthys Baird & Girard, 26; haplotype Silurus marinus Mitchill (name corrected to Ælurichthys by Gill). A synonym of Bagre (Cuvier) Oken, 1817, and Felichthys Sw., 1839.
Astyanax Baird & Girard, 26; haplotype A. argentatus Biard & Girard.

491. BAIRD & GIRARD (1854). *Notice of a New Genus of Cyprinidæ (Cochlognathus).* Proc. Acad. Nat. Sci. Phila.
Spencer Fullerton Baird; Charles Girard.

Cochlognathus Baird & Girard, 150; haplotype C. ornatus Baird & Girard.

492. BLEEKER (1854). *Species piscium bataviensium novæ vel minus cognitæ.* Nat. Tijdschr. Neder.-Indie, VI.
Pieter Bleeker.

Seriolichthys Bleeker, VI, 196; orthotype Seriola bipinnulata Q. & G. A synonym of Elagatis Bennett.

493. BLEEKER (1854). *Bijdrage tot de kennis der ichthyologische fauna van het eiland Flores.* Nat Tijdschr. Neder.-Indie, VI, 1854, 311-338.
PIETER BLEEKER.

Apogonichthys Bleeker, 321; logotype A. PERDIX Blkr.

Labrichthys Bleeker, 415; orthotype L. CYANOTÆNIA Blkr.

494. BLEEKER (1854). *Vijfde bijdrage tot de kennis der ichthyologische fauna van Amboina.* Nat. Tijdschr. Neder.-Indie, VI.
PIETER BLEEKER.

Tropidichthys Bleeker, VI, 501; logotype T. MARGARITATUS Blkr. A synonym of CANTHIGASTER Sw.

494A. BLEEKER (1854). *Bijdrage tot de Kennis van de ichthyologische fauna van Celebes.* Nat. Tidschr. Neder.-Indie, VII.
PIETER BLEEKER.

Chonerhinos Bleeker, 259; orthotype TETRAODON MODESTUS Blkr.; replaces XENOPTERUS Bibron (1855).

495. BRANDT (1854). *(Belonopsis.)* Mém. Acad. Sci. St. Petersbourg.
JOHANN FRIEDRICH BRANDT.

Belonopsis Brandt, 174; orthotype LEPTORHYNCHUS LEUCHTENBERGI Lowe. Substitute for LEPTORHYNCHUS, preoccupied; a synonym of NEMICHTHYS Rich.

496. COSTA (1854). *Fauna del regno di Napoli, ossia enumerazione di tutti gli animali che abitano le diverse regioni di questo regno e le acque che le bagnano; contenente la descrizione di nuovi o poco estattamente conosciuti; con figure ricavata da original viventi e depinte al naturale di O. G. Costa (continuata da A. Costa).*
ORONZIO GABRIELE COSTA.

Helmichthys Costa, tavola 31; orthotype H. DIAPHANUS Costa. A larva of CONGER.

497. EGERTON (1854). *On Some New Genera and Species of Fossil Fishes.* Ann. Mag. Nat. Hist., XIII.
PHILIP DE MALPAS GREY EGERTON.

Æchmodus Egerton, X, 367; logotype TETRAGONOLEPIS CONFLUENS Ag. (fossil). A substitute for TETRAGONOLEPIS Ag., not of Bronn. A synonym of DAPEDIUM Leach.

Histionotus Egerton, XIII, 434; orthotype H. ANGULARIS Egert. (fossil).

Legnonotus Egerton, III, 435; orthotype L. COTHAMIENSIS Egert. (fossil).

498. EGERTON (1854). *Palæichthyologic Notes. ...Supplemental to the Works of Prof. Agassiz.* Quart. Journ. Geol. Soc. On a fossil (DIPTERONOTUS CYPHUS) from the upper beds of the new red sandstone at Bromsgrove, X.
PHILIP DE MALPAS GREY EGERTON.

Dipteronotus Egerton, X, 369; orthotype D. CYPHUS Egert.

499. EICHWALD (1854). *Lethæa Rossica, ou Paléntologie de la Russie décrite et figurée.*
CARL EDUARD VON EICHWALD.

Thyestes Eichwald, I, 108; orthotype T. VERRUCOSUS Eichw. (fossil).

500. GIBBONS (1854). *Description of New Species of Viviparous Marine and Fresh-water fishes, from the Bay of San Francisco and from the River and Lagoons of the Sacramento.* Proc. Acad. Nat. Sci. Phila., VII. (These genera were printed in advance, in the local proceedings of the society, in the *Daily Placer Times and Transcript*, San Francisco, May and June, 1854 (the genus MYTILOPHAGUS excepted.)

WILLIAM P. GIBBONS.

Hysterocarpus Gibbons, May 18; orthotype H. TRASKI Gibbons (later published in *Proc. Acad. Nat. Sci. Phila.*)

Cymatogaster Gibbons, May 18; haplotype C. AGGREGATUS Gibbons (not CYMATO-GASTER of Gibbons, *Proc. Acad. Nat. Sci. Phila.*, 1854, 123, or of *Placer Times*, June 21, which is a synonym of HOLCONOTUS Ag.)

Hyperprosopon Gibbons. May 18; orthotype H. ARGENTEUS Gibbons.

Micrometrus Gibbons, May 30; restricted by Alexander Agassiz, 1861, to logotype M. MINIMUS Gibbons. Replaces ABEONA Grd.

Pachylabrus Gibbons, June 21; orthotype P. VARIEGATUS Gibbons. A synonym of RHACOCHILUS Ag.

501. GIBBONS (1854). *Description of New Species of Viviparous Marine and Fresh-water Fishes, from the Bay of San Francisco, and from the River and Lagoons of the Sacramento.* Proc. Acad. Nat. Sci. Phila., 1854, VII.

WILLIAM P. GIBBONS.

In this paper are repeated the descriptions in the *Daily Placer Times and Transcript*, with the addition of the genus MYTILOPHAGUS, and the transfer of the name CYMATOGASTER to the group called HOLCONOTUS by Agassiz.

Mytilophagus Gibbons, 125; orthotype M. FASCIATUS Gibbons (not in the *Placer Times*). A synonym of AMPHISTICHUS Ag.

502. GIRARD (1854). *On a New Genus of American Cottoids.* Proc. Boston Soc. Nat. Hist., IV (for 1851), 1854.

CHARLES FRÉDÉRIC GIRARD.

Triglopsis Girard, IV, 16; haplotype T. THOMPSONI Grd. Name changed on account of the earlier TRIGLOPS to PTYONOTUS by Gthr., 1860; this change is not necessary.

502A. GIRARD (1854). *Description of New Fishes Collected by A. L. Heermann, Naturalist Attached to the Survey of the Pacific Railroad, Under Lieut. R. S. Williams, U. S. A.* Proc. Acad. Nat. Sci. Phila., VII.

CHARLES FRÉDÉRIC GIRARD.

Aspicottus Girard, 130; orthotype A. BISON Grd.

Leptocottus Girard, 130; orthotype L. ARMATUS Grd.

Scorpænichthys Girard, 131; orthotype HEMITRIPTERUS MARMORATUS Ayres.

Ophiodon Girard, 133; orthotype O. ELONGATUS Grd.

Atherinopsis Girard, 134; orthotype A. CALIFORNIENSIS Grd.

Pogonichthys Girard, 136; logotype POGONICHTHYS INÆQUILOBUS Baird & Girard.

Lavinia Girard, 137; haplotype L. EXILICAUDA Baird & Girard.

Platichthys Girard, 139; orthotype P. RUGOSUS Grd. = PLEURONECTES STELLATUS Pallas.

Pleuronichthys Girard, 139; orthotype P. CŒNOSUS Grd.

Parophrys Girard, 139; orthotype P. VETULUS Grd.

Psettichthys Girard, 140; orthotype P. MELANOSTICTUS Grd.

503. GIRARD (1854). *Enumeration of the Species of Marine Fishes Collected at San Francisco, California, by Dr. C. B. R. Kennerly, Naturalist, Attached to* *the Survey of the Pacific Railroad Route Under Lieut. A. W. Whipple.* Proc. Acad. Nat. Sci., Phila., VIII.

CHARLES FRÉDÉRIC GIRARD.

Porichthys Girard, 141; orthotype P. NOTATUS Grd.

Heterostichthus Girard, 143; orthotype H. ROSTRATUS Grd.

Apodichthys Girard, 150; logotype A. FLAVIDUS Grd.

Phanerodon Girard, 153; orthotype P. FURCATUS Grd.

504. GIRARD (1854). *Abstract of a Report to Lieut. James M. Gilliss, U. S. N., Upon the Fishes Collected During the U. S. Naval Astronomical Expedition to Chili.* Proc. Acad. Nat. Sci. Phila. (Complete report printed in 1855.)

CHARLES FRÉDÉRIC GIRARD.

Percichthys Girard, 197; logotype PERCA TRUCHA Cuv. & Val.

Percilia Girard, 197; orthotype P. GILLISSI Grd.

Basilichthys Girard, 198; orthotype ATHERINA MICROLEPIDOTA Jenyns.

Heterognathus Girard, 198; logotype ATHERINA HUMBOLDTIANA, Cuv. & Val. A synonym of CHIROSTOMA Sw.

Nematogenys Girard, 198; orthotype TRICHOMYCTERUS INERMIS Guich.

Cheirodon Girard, VII, 199; orthotype CHEIRODON PISCICULUS Grd. Name spelled CHIRODON by Günther.

505. GIRARD (1854). *Notice Upon the Viviparous Fishes Inhabiting the Pacific Coast of North America, with an Enumeration of the Species Observed.* Proc. Acad. Nat. Sci. Phila., VII.

CHARLES FRÉDÉRIC GIRARD.

Damalichthys Girard, 321; orthotype D. VACCA Grd. = EMBIOTOCA ARGYROSOMA Grd.

Abeona Girard, 322; orthotype A. TROWBRIDGEI Grd. A synonym of MICROMETRUS Gibbons, as restricted by A. Agassiz (1861).

Ennichthys Girard, 322; logotype HOLCONOTUS MEGALOPS Grd. A synonym of HYPERPROSOPON Gibbons.

506. GRONOW (1854). *Catalogue of Fishes Collected and Described by Laurence Theodore Gronow, Now in the British Museum.* Edited from the manuscript, by Albert Günther (written about 1770, an admirable paper, had it seen the light 80 years earlier).

LORENZ THEODOR GRONOW.
(Laurentius Theodorus Gronovius.)

Pastinaca Gronow, 11; logotype P. LÆVIS Gronow, RAJA PASTINACA Gronow. A synonym of DASYATIS Raf.

Holacanthus Gronow, 23; orthotype H. LEIONOTHOS Gronow. Name preocupied; a synonym of TETRAODON L. as understood.

Cestreus Gronow, 49; orthotype C. CAROLINENSIS Gronow. Name preoccupied by CESTREUS McCl.; equivalent to CYNOSCION Gill.

Boops Gronow, 58; orthotype B. ASPER Gronow = PRIACANTHUS BOOPS Cuv. A synonym of PRIACANTHUS Cuv.

Gonopterus Gronow, 77; orthotype G. MŒRENS Gronow. A synonym of ZANCLUS Cuv.

Adonis Gronow, 93; logotype A. PAVONINUS Gronow = BLENNIUS OCELLARIS L. A synonym of BLENNIUS L.

Gonocephalus Gronow, 105; orthotype G. MACROCEPHALUS Gronow = TRIGLA VOLITANS L. A synonym of DACTYLOPTERUS Lac.

Sarda Gronow, 119; logotype Sarda immaculata Gronow = Scomber pelagicus L., not of Cuvier; a synonym of Coryphæna L.

Thynnus Gronow, 121; logotype T. moluccensis Gronow; name preoccupied; a synonym of Scomberoides Lac.

Trachurus Gronow, 124; logotype T. cordyla L. Name preoccupied; same as Megalaspis Blkr.

Merlucius Gronow, 129; orthotype M. lanatus Gronow. Same as Merluccius Raf.

Glanis Gronow, 135; orthotype G. imberbis Gronow = Mystus chihala Ham. A species of Notopterus. Name preoccupied.

Macrognathus Gronow, 147; logotype Centriscus scolopax L. Name preoccupied; a synonym of Macrorhamphosus Lac.

Chromis Gronow, 149; orthotype C. epicurorum Gronow = Gasterosteus saltatrix L. Name preoccupied; a synonym of Pomatomus Lac.

Cephalinus Gronow, 159; logotype Blennius torvus. A synonym of Congiopus Perry (Agriopus Cuv.).

Trichopterus Gronow, 162; orthotype T. indicus Gronow = Cheilodactylus fasciatus Lac. A synonym of Cheilodactylus Lac.

Cordylus Gronow, 163; orthotype Scomber scombrus L. A synonym of Scomber L.

Mystus Gronow, 165; type M. carolinensis Gray. Name preoccupied; a synonym of Bagre Cuv. = Felichthys Sw.

Lepturus Gronow, 165; orthotype Lepturus brevirostris Gronow = Coryphænoides rupestris Gunner. A synonym of Coryphænoides Gunner.

Dascillus Gronow, 171; type not named; a clupeid fish, part of the description lost; not Dascyllus Cuv.

Amia Gronow, 173; orthotype A. percæformis Gronow = Apogon moluceensis Blkr. This is the first use of the name Amia by a binomial author (after Gronow's non-binomial use in 1763) in place of Apogon Lac. This was followed afterward by Bleeker, Gill and others.

Stethochætus Gronow, 174; orthotype S. biguttatus Gronow; a synonym of Setipinna Sw. (Telara Gthr.).

Acronurus Gronow, 190; logotype Teuthis hepatus L. A synonypm of Hepatus or Teuthis.

507. HECKEL (1854). *Ueber fossile Fische aus Chiavon und das geologische Alter der sie enthaltenden schichten.* Sitzb. Akad. Wiss. Wien, IX, 1854, 322-334.

Johann Jacob Heckel

Galeodes Heckel, XI, 324; orthotype Galeocerdo priscus Zigno (fossil). A fossil ally of Galeocerdo.

508. HECKEL (1854). *Bericht ueber die vom Herrn Cavaliere Achille de Zigno hier angelangte Sammlung fossiler Fische.* Sitzb. Akad. Wiss. Wien, XI, 1854.

Johann Jacob Heckel

Solenorhynchus Heckel, XI, 126; orthotype S. elegans Heckel (fossil).

Enneodon Heckel XI, 127; orthotype E. echinus Heckel (fossil). Name preoccupied, replaced by Heptadiodon; a fossil ally of Diodon.

Vomeropsis Heckel, XI, 134; logotype Vomer longispinus Ag. (fossil).

509. HYRTL (1854). *Beitrag zur Anatomie von Heterotis ehrenbergii C. et V.* Denkschr. Akad. Wiss. Wien, VIII.

Carl Joseph Hyrtl (1811-1894).

Helicobranchus Hyrtl, VIII, 73; orthotype Sudis niloticus Cuv. & Val. A substitute for Heterotis Ehrenb.; preoccupied, already replaced by Clupisudis Sw., 1839.

510. KNER (1854). *Die Panzerwelze des K. K. Hofnaturalien-Cabinetes zu Wien. I. Abtheilung, Loricarinæ.* Denkschr. Akad. Wiss. Wien, 1854. (An abstract in *Sitzb. Akad. Wiss. Wien*, X, 113.)

Rudolph Kner.

Hemiodon Kner, about 65; logotype H. depressus Kner.

Acestra Kner, 93; orthotype A. acus Kner. Name preoccupied; replaced by Farlowella Eigenmann (1890).

511. KNER (1854). *Die Hypostomiden. Zweite Hauptgruppe der Familie der Panzerfische (Loricata vel Goniodontes).* Denkschr. Akad. Wiss. Wien, 1853. This paper appeared in 1854, an abstract in *Sitzb. Akad. Wiss. Wien,* 1853.

Rudolph Kner.

Ancistrus Kner VII, 272; logotype Hypostomus temmincki Val, as restricted by Bleeker, 1863; Eigenmann (1912) indicates as type A. cirrhosus Kner, a congeneric species.

512. MARCUSEN (1854). *Vorläufige Mittheilung aus einer Abhandlung über die Familie der Mormyren.* Bull. Acad. Sci. St. Pétersb., XII.

Johann Marcusen

Petrocephalus Marcusen, XII, 10; orthotype Mormyrus bane Val.

513. OWEN (1854). *Descriptive Catalogue of the Fossil Organic Remains of Reptilia and Pisces Contained in the Museum of the Royal College of Surgeons of England.* London, XIX.

Richard Owen.

Eucynodus Owen, 160; orthotype E. hunteri Owen (fossil).

Acropoma Owen, 164; orthotype A. alta Owen (fossil). Name preoccupied by Acropoma T. & S.

Toxopoma Owen, 164 orthotype T. politum Owen (fossil).

Cariniceps Owen, 165; orthotype C. compressus Owen (fossil); name only, referred by Woodward to synonym of Eothynnus.

Platygenys Owen, 165; orthotype P. rugosus Owen (fossil).

Pygacanthus Owen, 166; orthotype P. altus Owen (fossil).

Planesox Owen, 169; orthotype P. vorax Owen (fossil). Name only, noted by Woodward, *Cat.*, V, 519, 1901. Same as Percostoma Ag.

Platops Owen, 170; orthotype P. subulidens Owen (fossil). Undefined; supposed to be a synonym of Brychætus (Ag.)

Sciænurus Owen, 171; orthotype S. longior Owen (fossil). Name only; noted by Woodward, *Cat.*, 519, 1901.

514. POEY (1854). *Nuevo genero de peces Escombridos, Epinnula magistralis Poey.* Memorias de Cuba, I.

Felipe Poey y Aloy (1799-1891).

Epinnula Poey, I, 369 (April, 1854); orthotype E. magistralis Poey.

515. POEY (1854). *Los Guajacones, peces de agua dulce.* Memorias de Cuba,, I. (June, 1854.)

Felipe Poey y Aloy.

Gambusia Poey, I, 382; orthotype G. punctata Poey.

Girardinus Poey I, 383; orthotype G. METALLICUS Poey. A syonym of HETERAN-
DRIA Ag. as restricted by Poey.

Limia Poey, I, 382; orthotype L. CUBENSIS Poey. A synonym of PŒCILIA Cuv.
(ALAZON Gistel).

516. RAPP (1854). *Die Fische des Bodensees untersucht und beschrieben.*
Jahresschr. Ver. Vaterl. Naturg. Würtemberg, X.

WILHELM LUDWIG VON RAPP.

Umbla Rapp, 32; orthotype SALMO UMBLA = SALVELINUS ALPINUS L. A synonym
of SALVELINUS Rich.

517. RICHARDSON (1854). *Vertebrates, Including Fossil Mammals.* (In Forbes,
Edward (editor), *The Zoology of the Voyage of H. M. S. "Herald," Under
the Command of Captain Henry Kellett, R. N., During the Years 1845-1851.*
London.

JOHN RICHARDSON.

Prilonotus (Müller) Richardson; orthotype P. CAUDACACINCTUS Rich. A synonym
of CANTHIGASTER Sw. or PSILONOTUS Sw.

Anchisomus (Kaup) Richardson, 156; logotype TETRAODON SPENGLERI Bloch. A
synonym of SPHEROIDES L. as restricted.

518. THIOLLIÈRE (1854). *Description des poissons fossiles provenant des gisements coralliens du Jura dans le Bugey.* Lyons.

VICTOR JOSEPH DE L'ISLE THIOLLIÈRE.

Holochondrus Thiollière, 4; orthotype (not named) (name only) (fossil).

Spathobatis Thiollière, 7; orthotype S. BUGESIACUS Thiollière (fossil).

Belemnobatis Thiollière, 8; orthotype B. SISMONDÆ Thiollière (fossil.

Phorcynus Thiollière, 9; orthotype P. CATULINA Thiollière (fossil). An ally of
SQUATINA Dum.

519. TROSCHEL (1854). *Ueber die fossilien Fische aus der Braunkohle des
Siebengebirges.* Verh. Naturh. Ver. Preuss. Rheinl. Westphal.

FRANZ HERMANN TROSCHEL.

Tarsichthys Troschel, XI, 11; orthotype LEUCISCUS TARSIGER Troschel (fossil).
A fossil analogue of TINCA Cuv.

1855

520. AGASSIZ (1855). *Synopsis of the Ichthyological Fauna of the Pacific Slope
of North America, Chiefly from the Collections Made by the Expedition
Under the Command of Captain C. Wilkes,* etc. Amer. Journ. Sci., XIX.

LOUIS AGASSIZ.

Bubalichthys Agassiz, 192; orthotype CARPIODES URUS Ag. Regarded as a synonym
of ICTIOBUS Raf.

Ptychostomus Agassiz, 203; logotype CATOSTOMUS AUREOLUS Le Sueur. A synonym
of MOXOSTOMA Raf.

Hylomyzon Agassiz, 205; haplotype CATOSOMUS NIGRICANS Le Sueur. A synonym
of HYPENTELIUM Raf.

Acrocheilus Agassiz, 211; orthotype LEUCISCUS ALUTACEUS Rich.

Campostoma Agassiz, 218; orthotype RUTILUS ANOMALUS Raf.

Hyborhynchus Agassiz, 223; haplotype MINNILUS NOTATUS Raf.

Hybognathus Agassiz, 223; haplotype H. NUCHALIS Ag.

Ptychocheilus Agassiz, 227; logotype PTYCHOCHEILUS GRACILIS Ag. = CYPRINUS OREGONENSIS Rich.

Mylocheilus Agassiz, 229; haplotype MYLOCHEILUS LATERALIS Ag.

521. AYRES (1855). *Description of New species of California Fishes.* Proc. Cal. Acad. Sci., I, 1855.

WILLIM O. AYRES.

Leptogunnellus Ayres, 26; orthotype L. GRACILIS Ayres. A synonym of LUMPENUS.

Anarrhichthys Ayres, I, 33; orthotype A. OCELLATUS Ayres.

Mylopharodon Ayres, 33; orthotype M. ROBUSTUS Ayres.

Cebedichthys Ayres, I, 59; orthotype C. VIOLACEUS Ayres.

Notorhynchus Ayres, I, 72; orthotype N. MACULATUS Ayres.

Calycilepidotus Ayres, I, 76; orthotype C. SPINOSUS Ayres.

522. BASILEWSKI (1855). *Ichthyographia Chinæ Borealis.* Nouv. Mém. Soc. Nat. Moscou, X.

STEPHANUS BASILEWSKI.

Nasus Basilewski, 234; orthotype CYPRINUS NASUS L. A synonym of CHONDROSTOMA Ag.

Leptocephalus Basilewski, 234; logotype L. MONGOLICUS Basilewski. Name preoccupied in eels; replaced by CHANODICHTHYS Blkr.

Cephalus Basilewski, 235; orthotype C. MANTSCHURICUS Basilewski. Name preoccupied; a synonym of HYPOPHTHALMICHTHYS Blkr.

Culter Basilewski, 236; logotype C. ALBURNUS Basilewski, as restricted by Bleeker and Günther. Under the head of CULTER Basilewski mentions especially (CYPRINUS CULTRATUS L.) the type of PELECUS Ag., although he does not exactly specify this as type. He then proceeds to describe certain Chinese species. For some of these the name CULTER has been kept, although Basilewski plainly intended to make his type CYPRINUS CULTRATUS, in which case CULTER would be a synonym of PELECUS. At present we follow the authority of Bleeker and Günther.

Eperlanus Basilewsky, 242; orthotype E. CHINENSIS Basilewsky = SALANX HYALOCRANIUS Abbott. Name preoccupied; a synonym of SALANX Cuv.

Osteoglossum Basilewski, X, 244; logotype O. PRIONOSTOMA Basilewski. Name preoccupied by OSTEOGLOSSUM Vandelli. A synonym of CHÆTOMUS McCl.

Apterigia Basilewski, 247; logotype A. SACCOGULARIS Basilewski. A synonym of MONOPTERUS Lac. or FLUTA Bl. & Schn.

523. BIANCONI (1855). *Specimina zoologica Mosambicana quibus vel novæ vel minus notæ animalium species illustrantur. Bononiæ.*

GIOVANNI GIUSEPPE BIANCONI (1809-1878).

Megalepis Bianconi, 270; orthotype M. ALESSANDRAI Bianconi (= UPENOIDES VITTATUS, *fide* Günther). A synonym of UPENOIDES Blkr.

524. BIBRON (1855). *Note sur un travail inédit de Bibron relatif aux poissons plectognathes gymnodontes (Diodons et Tetrodons).* Rev. Mag. Zool., VIII.

GABRIEL BIBRON.

This is an important piece of work, but left incomplete at the death of its author, hence becoming in some degree a stumbling block in taxonomy. It was ably continued with the same material by Hollard in 1857. Edited by Auguste Henri Duméril.

Promecocophalus Bibron, 279; orthotype TETRAODON LÆVIGATUS L. A synonym of LAGOCEPHALUS Sw.

Dilobomycter Bibron, 279; orthotype T. RETICULARIS Bl. & Schn.   A synonym of
OVOIDES.

Amblyrhynchotus Bibron, 279; orthotype TETRAODON HONCKENI Bloch. (in French
form only).   Apparently distinct from LAGOCEPHALUS Sw.

Stenometopus Bibron, 279; logotype TETRAODON TESTUDINEUS L.   A synonym of
SPHEROIDES Dum.

Dichotomycter Bibron, 279; orthotype TETRAODON PLUVIATILIS Ham.   A synonym
of TETRAODON L.   Later called BRACHYCEPHALUS by Hollard.

Epipedorhynchus Bibron, 279; orthotype TETRAODON PREYCINETI Bibron.   No diagnosis.   A synonym of OVOIDES Cuv.

Geneion Bibron, 280; orthotype G. MACULATUS Bibron.   Apparently a synonym of
LAGROCEPHALUS Sw.

Catophorhynchus Bibron, 280; orthotype C. LAMPRIS Bibron.   Probably a synonym of LAGOCEPHALUS Sw.

Batrachops Bibron, 280; orthotype TETRAODON PSITTACUS Bl. & Schn.   Name preoccupied; replaced by COLOMESUS Gill.

Monotreta Bibron, 280; orthotype TETRODON CUTCUTIA Ham. (French form only),
the Latin given by Hollard, 1857; also spelled MONOTRETUS.   A synonym of
LEIODON Sw.

Rhynchotus Bibron, 280; orthotype TETRAODON GRONOVII Cuv.   A synonym of
CANTHIGASTER Sw.

Ephippion Bibron, 280; orthotype E. MACULATUM Bibron = TETRODON GUTTIFER
Bennett, 1830.   Regarded as twice preoccupied as EPHIPPIUM, therefore
replaced by HEMICONIATUS Gthr., 1870.

Aphanacanthus Bibron, 280; orthotype T. RETICULATUS Bibron.   No diagnosis.
Regarded by Bleeker as identical with MONOTRETA (= LEIODON).   A synonym of LEIODON Sw.

Xenopterus Bibron, 281; orthotype X. BELLENGERI Bibron = TETRAODON NARITUS
Rich.   A synonym of CHONERHINOS Blkr. (1854).   CHONERHINOS has clear
priority, as claimed by Bleeker.

525. BLEEKER (1855).   *Over eenige visschen van Van Diemensland.*   Verh. Akad.
Amsterdam, II, 1855, 1-30.
PIETER BLEEKER.

Upeneichthys Bleeker, 7; orthotype UPENEUS POROSUS Cuv. & Val.

Brachionichthys Bleeker, II, 12; logotype B. POLITUS Blkr.

Gnathanacanthus Bleeker, II, 21; orthotype G. GOETZI Blkr.   Replaces HOLOXENUS
Gthr.

526. BLEEKER (1855).   *Derde bijdrage tot de kennis der ichthyologische fauna
van de Kokos-eilanden.*   Neder. Tijdschr. Dierk., III.
PIETER BLEEKER.

Amphiprionichthys Bleeker, III, 172; orthotype MICROPUS UNIPINNA Gray.   Replaces MICROPUS Gray, preoccupied.

527. BLEEKER (1855).   *Bijdrage tot de kennis der ichthyologische fauna van de
Batoe-eilanden.*   Nat Tijdschr. Neder.-Indie, VIII.
PIETER BLEEKER.

Dinematichthys Bleeker, 318; orthotype D. ILUOCŒTOIDES Blkr.

528. BLEEKER (1855).   *Visschen van de Duizend-eilanden.*   Nat. Tijdschr.
Neder.-Indie, VIII, 1855, 344.
PIETER BLEEKER.

Serranichthys Bleeker, 344; orthotype SERRANUS ALTIVELIS Cuv. & Val. A synonym of CROMILEPTES Sw. as restricted by Bleeker and by Swain.

529.   BLEEKER (1855). *Zesde bijdrage tot de kennis der ichthyologische fauna van Amboina.* Nat. Tijdschr. Neder-Indie, VIII.

PIETER BLEEKER.

Uraspis Bleeker, VI, 418; orthotypt U. CARANGOIDES Blkr. Near SELENIA Bon.; perhaps a synonym of the latter.

530.   BLEEKER (1855). *Nalezingen op de vischfauna van Sumatra. Visschen van Lahat en Sibogha.* Nat. Tijdschr. Neder.-Indie, IX, 1855, 257-280.

PIETER BLEEKER.

Luciosoma Bleeker, IX, 258; orthotype BARBUS Cuv. & Val.

Schismatorhynchos Bleeker, IX, 258; orthotype LOBOCHEILUS HETERORHYNCHOS Blkr.

Clupeichthys Bleeker, IX, 275; orthotype C. GONIOGNATHUS Blkr. A synonym of CORICA Ham.

Epalzeorhynchos Bleeker, IX, 279, 258; orthotype BARBUS KALOPTERUS Blkr.

531.   BRONN (1855). *Lethæa Geognostica, oder Abbildungen und Beschreigen der für die Gebirgs-Formation bezeichnendsten Versteinerungen.* Stuttgart, Ed. I, 1834; Ed. III, 1855.

HEINRICH GEORG BRONN (1800-1862).

Archæoteuthis (Roemer) Bronn, I, 520, ed. III; orthotype PTERASPIS DUNENSIS Huxley (fossil). A synonym of PTERASPIS Kner.

Heptadiodon Bronn, III, 677; orthotype ENNEODON ECHINUS Heckel (fossil). Name substitute for ENNEODON Heckel, preoccupied.

532.   CASTELNAU (1855). *Expédition dans les parties centrales de l'Amérique du Sud, de Rio de Janeiro à Lima, et de Lima à Para; executée . . . pendant les années 1834 à 1844.* (Tome 3.)

(COMTE) FRANÇOIS L. DE LAPORTE DE CASTELNAU (1812-1880).

Plataxoides Castelnau, 21; orthotype P. DUMERILI Castelnau. A synonym of PTEROPHYLLUM Heckel.

Genidens Castelnau, 33; tautotype BAGRUS GENIDENS Cuv. & Val. = GENIDENS CUVIERI Castelnau.

Sternarchorhynchus Castelnau, 95; orthotype STERNARCHUS OXYRHYNCHUS M. & T. = STERNARCHORHYNCHUS MULLERI Castelnau; replaces RHAMPHOSTERNARCHUS Gthr., 1870.

Acanthodemus Castelnau; logotype HYPOSTOMUS AURANTIACUS Castelnau; name in French only (ACANTHODÈMES), latinized by Marschall *(Nomenclator Zoologicus,* 1873).

533.   COSTA (1855). *Su di un nuovo genere di pesce fossile (Cacus).* Atti Reale Accad. Sci. Napoli, II.

ORONZIO GABRIELE COSTA.

Cacus Costa, II, 234. (Noted by Dean and Eastman, not by Woodward. Probably a misprint for CÆUS Costa (C. LEOPOLDI, 1864).)

534.   EGERTON (1855). *Fish: . . .* (In *Figures and Descriptions of British Organic Remains.*) Mem. Geol. Surv. United Kingdom.

PHILIP DE MALPAS GREY EGERTON.

Oxygnathus Egerton, VIII, No. 9; orthotype O. ORNATUS Egert. (fossil).

535. FILIPPI (1855). *(Gadopsis.)* Zeitschr. Akad. Wiss. Wien, VII.
FILIPPO DE FILIPPI (1814-1867).
Gadopsis Filippi, VII, 170; orthotype OLIGOPUS ATER Risso. Name preoccupied by
GADOPSIS Rich. (1844) It may be replaced by the new name, VERATER Jordan, orthotype OLIGOPUS ATER Risso.

536. HOLBROOK (1855). *Ichthyology of South Carolina.* (Edition of 1855
nearly but not all destroyed by fire; replaced in 1860 by a second edition.
The generic names should date from 1855.
JOHN EDWARDS HOLBROOK (1796-1871).
Diplectrum Holbrook, 32; orthotype D. FASICULARIS Holbrook = PERCA FORMOSA L.
Lagodon Holbrook, 59; orthotype SPARUS RHOMBOIDES L.
Bothrolæmus Holbrook, 80; orthotype TRACHINOTUS PAMPANUS Cuv. & Val.
(TRACHINOTUS CAROLINUS). A synonym of TRACHINOTUS.
Homoprion Holbrook, 168 (second edition, 1860); orthotype H. LANCEOLATUS
Holbrook. A synonym of STELLIFER (Cuvier) Oken.

537. KAUP (1855). *Enchelynassa, neue Gattung aus der Familie der Aale.* Arch.
Naturgesch. XXI.
JOHANN JACOB KAUP.
Enchelynassa Kaup, 213; orthotype E. BLEEKERI Kaup. Perhaps the same as ENCHLYCORE Kaup (1856).

538. KAUP (1855). *Uebersicht über die Species einiger Familien der Sclerodermen.*
Wiegmann's Arch. Naturg., XXI.
JOHANN JACOB KAUP.
Cibotion (Kaup), 215; orthotype OSTRACION CUBICUS L. A synonym of OSTRACION as here restricted.
Capropygia (Gray) Kaup, 220; orthotype C. UNISTRIATA Gray.
Centaurus Kaup, 221; orthotype OSTRACION BOOPS Rich. A larva of MOLA.
Kentrocapros Kaup, 221; orthotype OSTRACION ACULEATUS Houttuyn.
Anoplocapros Kaup, 221; orthotype A. GRAYI Kaup.
Dicotylichthys Kaup, 230; orthotype D. PUNCTULATUS Kaup.
Cyanichthys Kaup, 231; orthotype C. CÆRULEUS Kaup (not of Q. & G.) = DIODON
TIGRINUS Cuv. A synonym of CHILOMYCTERUS Bibron.
Cyclichthys Kaup, 231; logotype DIODON ORBICULARIS Bloch.
Pyrodon Kaup; orthotype BALISTES NIGER Lac. Name a substitute for XENODON,
preoccupied. Later than ZENODON Sw., ODONUS Gistel and ERYTHRODON
Rüppell, substitutes for XENODON.

539. KNER (1855). *Ichthyologische Beiträge.* Abh. Sitzb. Akad. Wiss. Wien, XVII.
RUDOLF KNER.
Bunocephalus Kner, XVII, 95; type B. VERRUCOSUS Bloch.
Oxydoras Kner, XVII, 115; orthotype DORAS NIGER Kner, not of Val. = OXYDORAS
KNERI Blkr.
Clarotes Kner, XVII, 313; orthotype PIMELODUS LATICEPS Rüppell.
Pareiodon Kner, XVII, 611; orthotype P. MICROPS Kner. Spelled PARIODON by
Günther.

540. M'COY (1855). *Descriptions of the British Palæozoic Fossils in the Geological Museum of the University of Cambridge.*
FREDERICK M'COY.
Diplopterax M'Coy, 586; orthotype (D. MACROLEPIDOTUS Ag.) (fossil). Name a
substitute for DIPLOPTERUS, preoccupied.

Triplopterus M'Coy, 589; orthotype TRIPTERUS POLLEXFENI M'Coy (fossil). A synonym of OSTEOLEPIS; name a substitute for TRIPTERUS, preoccupied.

Pœcilodus M'Coy, 638; logotype P. JONESI (Ag.) M'Coy (fossil) as restricted by Woodward.

Pristicladodus M'Coy, 642; orthotype P. DENTATUS M'Coy (fossil). A synonym of DICRENODUS Romanovskii.

541. MEYER (1855). *Physichthys höningshausi aus dem Uebergangskalke der Eifel.* Palæontographica, IV.

HERMANN MEYER.

Physichthys Meyer, IV, 80; orthotype ASTEROLEPIS HŒNINGSHAUSI Ag. (fossil). A synonym of MACROPETALICHTHYS.

542. NILSSON (1855). *Skandinavisk fauna.* Fjerde Delen: *Fiskarna.* Lund.

XXXIV.

SVEN NILSSON.

Blenniops Nilsson, IV, 184; orthotype BLENNIUS GALERITA Pennant, not of Linnæus. A synonym of CHIROLOPHIS Sw.

Ctenodon Nilsson, IV, 190; orthotype CLINUS MACULATUS Fries. Name three times preoccupied; replaced by LEPTOCLINUS Gill.

543. PETERS (1855). *Uebersicht der in Mossambique beobachteten Seefische.* Monatsber. Akad. Wiss. Berlin.

WILHELM CARL HARTWIG PETERS.

Ctenopoma Peters, XXI, 1, 247; orthotype C. MULTISPINIS Peters, not CTENOPOMA of Heckel, 1856.

Pteragogus Peters XXI 261; orthotype COSSYPHUS OPERCULARIS Peters.

Anosmius Peters, XXI, 274; orthotype T. TÆNIATUS Peters. A synonym of CÆNTHIGASTER Sw.

Cœlonotus Peters XXI, 465; orthotype C. ARGULUS Peters.

544. ROEMER (1855). *Palæoteuthis, eine Gattung nakter Cephalopoden in devonischen Schichten der Eifel.* Palæontographica, IV.

CARL FERDINAND VON ROEMER.

Palæoteuthis Roemer, IV, 72; orthotype PTERASPIS DUNENSIS Huxley (fossil)) (wrongly supposed to be a cuttle-fish). A synonym of PTERASPIS Kner.

545. STORER (1855). *A History of the Fishes of Massachusetts.* Mem. Amer. Acad. Arts. Sci. Boston, V.

DAVID HUMPHREYS STORER.

Cheilonemus (Baird) Storer, 285; haplotype LEUCISCUS PULCHELLUS Storer = CYPRINUS CORPORALIS Mitchill. Same as LEUCOSOMUS Heckel, regarded as preoccupied by LEUCOSOMA.

546. VALENCIENNES (1855). *Ichthyologie de la Voyage de la Vénus.*

ACHILLE VALENCIENNES.

Doydixodon Valenciennes; orthotype D. FREMINVILLEI Val.

Passer (Klein) Valenciennes, 344; haplotype P. MARCHIONESSARUM Val. Name preoccupied in birds. A synonym of PLATOPHRYS Sw.

Smecticus Valenciennes, 305; orthotype SMECTICUS BICOLOR Val. A synonym of RYPTICUS.

1856

547. AGASSIZ (1856). *The Habits of the Glanis of Aristotle.* Proc. Amer. Acad.
Sci. Arts. Sci. III.
LOUIS AGASSIZ.

Glanis Agassiz; orthotype GLANIS ARISTOTELIS Ag.; not GLANIS Ag., 1829, a substitute for BAGRE; not GLANIS Gronow, 1854. Same as PARASILURUS Blkr.

548. ASMUSS (1856). *Das vollkommenste Hautskelet der bisher bekannten
Thierreihe. An fossilien Fischen des alten rothen Sandsteins aufgefunden
und aus ihren Resten erläutert.* Abh. Erlangung der Magisterwürde, Dorpat.
HERMANN ASMUSS.

Heterosteus Asmuss, 7; orthotype ASTEROLEPIS ASMUSSI Ag. A synonym of ICH-
THYOSAUROIDES Kutorga.

Homostius Asmuss, 8; orthotype H. FORMOSISSIMUS Asmuss (fossil). (Name corrected to HOMOSTEUS.)

549. BLEEKER (1856). *Beschrijvningen van nieuwe en weinig bekende vischsoorten van Amboina, versameld op eene reis door den Molukschen Archipel,
gedaan in het gevolg van den Gouverneur-Generaal Duymaer van Twist in
September en October, 1855.* Act. Soc. Sci. Indo-Nederl., I.
PIETER BLEEKER.

Pteroidichthys Bleeker, I, 33; orthotype P. AMBOINENSIS Blkr.

Duymæria Bleeker, I, 52; logotype CRENILABRUS AURIGARIA Rich.

550. BLEEKER (1856). *Beschrijvningen van nieuwe en weinig bekende vischsoorten van Menado en Makassar grootendeels verzameld op eene reis naar
den Molukschen Archipel in het gevolg van den Gouverneur-General Duymaer
van Twist.* Act. Soc. Sci. Indo-Nederl., I.
PIETER BLEEKER.

Heterophthalmus Bleeker, 43; orthotype H. KATOPRON Blkr. Name preoccupied
in beetles, Blanchard, 1851 . A synonym of ANOMALOPS Kner, 1858.

Rhomboidichthys Bleeker, I, 67; orthotype R. MYRIASTER Blkr. A synonym of
PLATOPHRYS Sw.

Leptocephalichthys Bleeker, I, 69; orthotype L. HYPSELOSOMA Blkr. A larval
form allied to CONGER.

551. BLEEKER (1856). *Bijdrage tot de kennis der ichthyologische fauna van
het eiland Boeroe.* Nat. Tijdschr. Neder.-Indie. XI.
PIETER BLEEKER.

Scorpænichthys Bleeker, XI, 385, 402; orthotype SCORPÆNA Gibbos. Name preoccupied (Girard, 1854) ; a synonym of SCORPÆNOPSIS Heckel.

Pholidichthys Bleeker, 406; orthotype P. LEUCOTÆNIA Blkr.

Gobiodon (Kuhl & Van Hasselt) Bleeker, 407; logotype GOBIUS HETEROSPILUS
Blkr.

Culius Bleeker, 411; orthotype ELEOTRIS NIGER Blkr. = CHEILODIPTERUS CULIUS Ham.
A synonym of ELEOTRIS as restricted.

Butis Bleeker, 42; orthotype BUTIS GYMNOPOMUS Blkr. (tautotype CHEILODIPTERUS
TERUS. BUTIS Ham.).

552. BLEEKER (1856). *Bijdrage tot de kennis der ichthyologische fauna van Nias.*
Nat. Tijdschr. Neder.-Indie, XII.
PIETER BLEEKER.

Solenognathus Bleeker, 218; orthotype Syngnathoides blochi Blkr.; also written Solegnathus. A synonym of Syngnathoides Blkr., equivalent to Gasterotokeus Heckel.

Xiphochilus Bleeker, 223; orthotype X. typus Blkr.

553. BLEEKER (1856). *Nieuwe bijdrage tot de kennis der ichthyologische fauna van Bali.* Nat. Tijdschr. Neder.-Indie, XIII.

Pieter Bleeker.

Belobranchus Bleeker, 300; orthotype Eleotris belobrancha Cuv. & Val.

554. COSTA (1856). *Ittiologia fossile Italiana. Edizione simile all' opera dell' Agassiz sui pesci fossili alla quale è destinata a sopplimento.* Disp. v. Napoli.

Oronzio Gabriele Costa.

Dichelospondylus Costa, 13; orthotype D. longirostris Costa (fossil). A synonym of Belonostomus Ag.

Ophirachis Costa, 13; orthotype O. deperditus Costa (fossil). Probably a synonym of Belonostomus.

Anomiophthalmus Costa, 30; orthotype A. vetustus Costa (fossil). Probably identical with Cœlodus and Cosmodus and of earlier date than the latter.

555. DUMÉRIL (1856). *Ichthyologie analytique ou classification des poissons, suivant la méthode naturelle, à l'aide de tableaux synoptiques.* Mém. Acad. Sci. Paris, XXVII.

André Marie Constant Duméril.

Heptatrema C. Duméril; orthotype Petromyzon cirrhatus Forster. A synonym of Eptatretus Cloquet.

Pastinaca Duméril; tautotype Raja pastinaca L. A synonym of Dasyatis Raf.

Ptérure Duméril, 169; orthotype Syngnathus æquoreus Gmelin (name in French only). A synonym of Nerophis Raf.

Péromere Duméril, 170; logotype Syngnathus rondelett De la Roche (name in French only). A synonym of Typhle or Typhlinus Raf. In this volume of Duméril a list of genera of eels is given by Kaup in advance of its publication elsewhere. But in the same year, apparently before the appearance of Duméril's work, Kaup published his *Catalogue of Apodal Fishes,* and several of the names given by him to Duméril were dropped in this completed work.

Auchenichthys (Kaup) Duméril), 198; orthotype A. typus Kaup. Name only; an ally of Murænichthys.

Leiurus (Kaup) Duméril, 198; orthotype Muræna colubrina Boddaert. Name twice preoccupied in Swainson; replaced by Chlevastes Jordan & Snyder.

Scytallurus (Kaup) Duméril, 199; orthotype Sphagebranchus imberbis De la Roche. A synonym of Cæcula Vahl, 1794.

Enchelycotte (Kaup) Duméril, 200; orthotype Enchelycore eurhina Kaup. A manuscript name changed to Enchelycore when printed in 1854.

Memarchus (Kaup) Duméril, 201; orthotype Gymnotus albifrons L. No definition; a synonym of Apteronotus Lac. = Sternarchus Cuv.

Altona (Kaup) Duméril, 201; orthotype Gymnotus rostratus L. (no definition), a synonym of Rhamphichthys M. & T.

Unipertura (Kaup) Duméril, 201; orthotype Unibranchapertura lævis Lac. An abridgment of Unibranchapertura Lac.; a synonym of Fluta Bl. & Schn. if regarded as a new name.

Giton (Kaup) Duméril, 201; orthotype Gymnotus fasciatus Pallas. = Gymnotus carapo L. A synonym of Gymnotus L. No definition.

Aptérichthe C. Duméril, 205; orthotype Muræna cæca L.; name in French only; given in Latin form, 1806. A synonym of Sphagebranchus Bloch.

Macrops Duméril, 279; orthotype Serranus aculeatus (oculatus) Cuv. & Val. (name three times preoccupied). A synonym of Etelis Cuv. & Val.

Macrones Duméril, 484; orthotype Bagrus lamarrii Cuv. & Val. Name preoccupied in beetles (Newman, 1841); replaced by Aoria Jordan, type Bagrus lamarrii Cuv. & Val.

Conostome (for Conostomus) Duméril, 484; orthotype Pimelodus conirostris Cuv. & Val. Called Conorhynchus by Bleeker.

Diplomyste (Le) Duméril, 487; haplotype Arius papillosus Cuv. & Val. Latinized as Diplomystes by Bleeker; Diplomystax by Günther.

Hydrocion "ou mieux Hydrocynus Duméril," restricted to "une seule espèce du Nil" (Hydrocyon forskali Cuv.). A synonym of Hydrocynus Cuv., 1817.

556, 557. GIRARD (1856). *Contributions to the Ichthyology of the Western Coast of North America from Specimens in the Smithsonian Institute ...* Proc. Acad. Nat. Sci. Phila.

CHARLES GIRARD.

Paralabrax Girard, 131; orthotype Labrax nebulifer Grd.

Homalopomus Girard, 132; orthotype H. trowbridge Grd. A synonym of Merluccius Raf.

Oligocottus Girard, 133; logotype O. maculatus Grd.

Leiocottus Girard, 133; orthotype L. hirundo Grd.

Artedius Girard, VIII, 134; logotype A. lateralis Grd.

Oplopoma Girard, 135; orthotype O. pantherina Grd. A synonym of Ophiodon Grd., 1854.

558. GIRARD (1856). *Researches Upon the Cyprinoid Fishes Inhabiting the Fresh Waters of the United States West of the Mississippi Valley, from Specimens in the Museum of the Smithsonia Institution.* Proc. Acad. Nat. Sci. Phila., VIII.

CHARLES GIRARD.

Minomus Girard, 173; haplotype Catostomus insignis Baird & Grd. A synonym of Catostomus Le Sueur, section Decactylus Raf., 1818.

Acomus Girard, 173; logotype Catostomus forsterianus Rich. = Cyprinus catostomus Le Sueur. A synonym of Catostomus Le Sueur.

Dionda Girard, 176; logotype D. episcopa Grd.

Algoma Girard, 180; logotype A. amara Grd. Not separable from Dionda, both scarcely distinct from Hybognathus

Orthodon Girard, 182; orthotype Gila microlepidota Ayres.

Algansea Girard, 182; orthotype Leuciscus tincella Cuv & Val.

Agosia Girard, 186; logotype A. chrysogaster Grd.

Nocomis Girard, 190; logotype N. nebrascensis Grd. = Rutilus kentuckiensis Raf. = Leuciscus biguttatus Kirtland.

Meda Girard, 192; haplotype Meda fulgida Grd.

Cliola Girard, 192; logotype Ceratichthys vigilax Baird & Grd. A synonym of Ceratichthys Baird & Grd., as first used; the name later taken for the genus Nocomis.

Alburnellus Girard, 193; orthotype Alburnus dilectus Grd. A synonym of Notropis Raf. (1818).

Codoma Girard, 194; logotype CODOMA ORNATA Grd.

Alburnops Girard, 194; logotype ALBURNOPS BLENNIUS Grd.

Cyprinella Girard, 196; logotype LEUCISCUS BUBALINUS Baird & Grd. Probably a valid genus allied to NOTROPIS and ALBURNOPS.

Richardsonius Girard, 201; haplotype ABRAMIS BALTEATUS Baird & Grd. Probably distinct from LEUCISCUS Cuv. and from TIGOMA Grd.

Tiaroga Girard, 204; haplotype TIAROGA COBITIS Grd.

Moniana Girard, 199; logotype LEUCISCUS LUTRENSIS Baird & Grd. Not separable from CYPRINELLA.

Tigoma Girard, 205; logotype GILA PULCHELLA Baird & Grd. Probably a genus distinct from LEUCISCUS Cuv. and from RICHARDSONIUS Grd.

Cheonda Girard, 207; logotype CHEONDA COOPERI Grd. Probably not distinct from TIGOMA.

Siboma Girard, 209; haplotype SIBOMA CRASSICAUDA Baird & Grd. A section or genus close to TIGOMA.

Hudsonius Girard, 210; orthotype CLUPEA HUDSONIA Clinton. A section or subgenus of NOTROPIS or ALBURNOPS.

Clinostomus Girard, 211; logotype LUXILUS ELONGATUS Kirtland (not CLINOSTOMUM, a genus of worms). A section or subgenus near NOTROPIS.

Ceratichthys (Baird MS.) Girard, 212; haplotype SEMOTILUS BIGUTTATUS Kirtland (not CERATICHTHYS Baird & Grd., 1854). A synonym of NOCOMIS Grd.

559. HOLMES (1856). *Contributions to the Natural History of the American Devilfish, with Descriptions of a New Genus from the Harbor of Charleston, South Carolina.* Proc. Elliott Soc. Nat. Hist., I.

FRANCIS S. HOLMES

Diabolichthys Holmes, 39; orthotype D. ELLIOTTI Holmes. A synonym of MANTA Bancroft.

560. HECKEL (1856). *Neue Beiträge zur Kenntniss der fossilen Fische Oesterreichs.* Denkschr. Akad. Wiss. Wien, XI.

JOHANN JAKOB HECKEL.

Thrissopterus Heckel, XI, 245; orthotype T. CATULLI Heckel (fossil). A synonym of MONOPTEROS Volta (PLATINX Ag.)

Elopopsis Heckel, XI, 251; logotype E. FENZLII Heckel (fossil).

Carangodes Heckel, XI, 262; orthotype C. CEPHALUS Heckel (fossil). Perhaps to be regarded as preoccupied by CARANGOIDES Blkr., 1851.

Ctenopoma Heckel, XI, 272; orthotype PYGÆUS JEMELKA Heckel (fossil); preoccupied by CTENOPOMA Peters, 1855.

561. HECKEL (1856). *Ueber den Bau und die Eintheilung der Pycnodonten,* etc. Sitzb. Akad. Wiss. Wien, VIII, 433-464, 1854 (1856).

JOHANN JAKOB HECKEL.

Stemmatodus Heckel, XI, 202; orthotype PYCNODUS RHOMBUS Ag. (fossil). Preoccupied; replaced by STEMMATIAS Hay, 1899.

Cœlodus Heckel, XI, 202, 455; orthotype CŒLODUS SATURNUS Heckel (fossil). If regarded as preoccupied by CŒLODON Lund, may be replaced by COSMODUS Sauvage, 1879.

562. KAUP (1856). *Uebersicht der Aale.* Arch. Naturg.

JOHANN JACOB KAUP.

In this paper ocurs the first diagnosis of the many genera of eels, published in the same year as the *Catalogue of the Apodal Fish*. The pagination of the volume last mentioned is given in parentheses in the following list:

563. KAUP (1856). *Catalogue of the Apodal Fish in the Collection of the British Museum.* London.
JOHANN JACOB KAUP.

Congerodon Kaup, 37; orthotype C. INDICUS Kaup. (Not in *Cat. Apodal Fishes.*)

Centrurophis Kaup, 42 (2); logotype C. SPADICEUS Kaup. A synonym of OPHICHTHUS Ahl.

Pœcilocephalus Kaup, 43 (5); orthotype P. BONAPARTEI Kaup. A synonym of OPHICHTHUS Ahl.

Microdonophis Kaup, 43 (6) orthotype M. ALTIPINNIS Kaup. A synonym of OPHICHTHUS.

Cœcilophis Kaup, 44 (6); orthotype C. COMPAR Kaup. A synonym of OPHICHTHUS.

Herpetoichthys Kaup, 44 (7); orthotype H. ORNATISSIMUS Kaup. A synonym of OPHICHTHUS.

Brachysomophis Kaup, 45 (9); orthotype B. HORRIDUS Kaup.

Elapsopsis Kaup, 45 (9); orthotype E. VERSICOLOR Kaup (misprinted ELAPSOPIS). A synonym of OPHICHTHUS.

Mystriophis Kaup, 45 (10); orthotype OPHISURUS ROSTELLATUS Rich.

Murænopsis (Le Sueur) Kaup, 46 (11); orthotype MURÆNOPHIS OCELLATUS Le Sueur. This name was not used by Le Sueur, who wrote MURÆNOPHIS after Lacépède.

Echiopsis Kaup, 46 (13); orthotype OPHISURUS INTERTINCTUS Rich. A synonym of MYSTRIOPHIS Kaup.

Scytalophis Kaup, 46 (13); orthotype S. MAGNIOCULIS Kaup.

Leptorhinophis Kaup, 46 (14); orthotype OPHISURUS GOMESI Castelnau. A synonym of SCYTALOPHIS Kaup.

Pisodonophis Kaup, 47 (17); logotype P. CANCRIVORUS Kaup, as restricted by Bleeker. Later written PISOODONOPHIS by Kaup, and since spelled PISODONTOPHIS by purists.

Lamnostoma Kaup, 49 (24); logotype L. PICTUM Kaup.

Anguisurus Kaup, 50 (24); orthotype A. PUNCTULATUS Kaup. A synonym of CŒCULA Vahl.

Callechelys Kaup, 51 (28); orthotype C. GUICHENOTI Kaup.

Cirrhimuræna Kaup, 52 (27); orthotype C. CHINENSIS Kaup.

Ophisurapus Kaup, 52 (29); ortthotype O. GRACILIS Kaup. Later written OPHISURAPHIS A synonym of SPHAGEBRANCHUS Bloch.

Myrus Kaup, 53 (31); orthotype MURÆNA MYRUS L. = MYRUS VULGARIS Kaup. A synonym of ECHELUS Raf., as restricted by Bleeker.

Enchelynassa Kaup, 59 (72); orthotype E. BLEEKERI Kaup.

Eurymyctera Kaup, 59 (72); orthotype E. CRUDELIS Kaup. A synonym of GYMNOTHORAX.

Siderea Kaup, 59 (70); orthotype S. PFEIFFERI Kaup. Later written SIDERA. A synonym of GYMNOTHORAX, as restricted by Bleeker.

Enchelycore Kaup, 60 (72); orthotype MURÆNA NIGRICANS Bonnaterre.

Thyrsoidea Kaup, 60 (74); tautotype MURÆNA THYRSOIDEA Rich., but first restricted to MURÆNA MACRURA Bleeker, by Bleeker. As understood by me, a synonym of GYMNOTHORAX. As used by Bleeker, the same as EVENCHELYS J. & E.

Limamuræna Kaup 65 (95): orthotype L. GUTTATA Kaup. A synonym of MURÆNA L. as restricted by Ahl, 1789.

Polyuranodon Kaup, 65 (96); orthotype Muræna polyuranodon Blkr. A synonym of Gymnothorax.

Pœcilophis Kaup, 66 (98); orthotype Gymnothorax catenatus Bloch. A synonym of Echidna Forster.

Aphthalmichthys Kaup, 68 (105); orthotype A. javanicus Kaup.

Murænoblenna (Lacépède) Kaup, 67; orthotype Ichthyophis tigrinus Lesson, not of Lacépède, whose type M. olivacea is a Myxine. Kaup observes: "Ich kenne M. olivacea nicht."

Congermuræna Kaup, 71 (108); logotype Muræna balearica De la Roche, as restricted by Bleeker. A synonym of Ariosoma Sw., equivalent to Congrellus Ogilby, 1898, who restricts Congermuræna to C. habenata, which probably belongs to Gnathophis Blkr. The name is spelled Congromuræna by authors.

Uroconger Kaup, 71 (110); orthotype U. lepturus Kaup.

Esunculus Kaup, 143; orthotype E. costai Kaup. A larva, allied to Alepocephalus. (Not in *Uebersicht.*)

564. KAUP (1856). *Einiges über die Unterfamilie Ophidinæ.* Arch. Naturg., XXII.
Johann Jacob Kaup.

Cepolophis Kaup, I, 96; orthotype Ophidium viride Fabricius. A synonym of Gymnelis Reinh.

565. KAUP (1856). *Catalogue of the Lophobranchiate Fish in the Collection of the British Museum.* London.
Johann Jacob Kaup.

The various genera of Kaup, 1853, Hemimarsupium excepted, are in this paper with their redescribed species

Hemithylacus Kaup, 61; orthotype Syngnathus leiaspis Kaup. An unexplained substitute for Hemimarsupium Kaup, 1853. Same as Cœlonotus Peters (1855).

566. LEIDY (1856). *Notice of Remains of Extinct Vertebrated Animals Discovered by Professor E. Emmons.* Proc. Acad. Nat. Sci. Phila., 1856.
Joseph Leidy (1825-1891).

Ischyrhiza Leidy, VIII, 256; logotype I. mira Leidy (fossil). Supposed to be a synonym of Pachyrhizodus.

567. LEIDY (1856). *Notices of Extinct Vertebrata Discovered by Dr. F. V. Hayden, During the Expedition to the Sioux Country Under the Command of Lieut. G. K. Warren.* Proc. Acad. Nat. Sci. Phila., VIII.
Joseph Leidy.

Mylognathus Leidy, 312; orthotype M. priscus Leidy (fossil). Probably a synonym of Edaphodon.

568. LEIDY (1856). *Indications of Five Species, with Two New Genera, of Extinct Fishes.* Proc. Acad. Nat. Sci. Phila., VII.
Joseph Leidy.

Edestus Leidy, VII, 414; orthotype E. vorax Leidy (fossil).

Sicarius Leidy, VII, 414; orthotype S. extinctus Leidy (fossil). A synonym of Petalodus.

569. LEIDY (1856). *Description of Two Ichthyodorulites (Stenacanthus nitidus and Cylindracanthus ornatus).* Amer. Journ. Sci., XXI.

JOSEPH LEIDY.

Stenacanthus Leidy, XXI, 8; orthotype S. NITIDUS Leidy (fossil) (also in *Proc. Acad Nat. Sci. Phila.*, 1856, 11.). A synonym of BOTHRIOLEPIS.

Cylindracanthus Leidy, XXI, 8; orthotype C. ORNATUS Leidy. Regarded as a synonym of CŒLORHYNCHUS Ag., which name is preoccupied.

570. LEIDY (1856). *Remarks on Certain Extinct Species of Fishes.* Proc. Acad. Nat. Sci., VIII, 301. (Also in *Trans. Philos. Soc.*, XI, 1857.)

JOSEPH LEIDY.

Protosphyræna Leidy, 56, VIII, 302 (and XI, 95); orthotype P. FEROX Leidy (fossil).

Cimolichthys Leidy, VIII, 302 (also XI, 95); orthotype C. LEWEIENSIS Leidy (fossil).

571. NEWBERRY (1856). *Description of Several New Genera and Species of Fossil Fishes from the Carboniferous Strata of Ohio.* Proc. Acad. Nat. Sci. Phila., VIII.

JOHN STRONG NEWBERRY (1822-1892).

Mecolepis Newberry, VIII, 96; logotype M. GRANULATUS Newberry (fossil). Name preoccupied, replaced by EURYLEPIS Newberry (1857).

Compsacanthus Newberry, 100; orthotype PLEURACANTHUS LÆVIS Newberry (fossil). A synonym of PLEURACANTHUS.

572. PANDER (1856). *Monographie der fossilen Fische des silurischen Systems der russisch-baltischen Gouvernements.* St. Petersburg.

CHRISTIAN HEINRICH PANDER.

Rytidolepis Pander, 48; orthotype R. QUENSTEDTI Pander (fossil). Besides this nominal genus, the following additional generic names have been given in this paper by Pander. All are based on fragments of dermal armor, not identified and some of them, according to Eichwald (1860), likely to belong to Eurypterid crustaceans rather than to fishes. These are COCCOPELTUS, CYPHOMALEPIS, TRACHYLEPIS and PHLEBOLEPIS. The following is the list of these nominal genera and their type species as given by Woodward:

RYTIDOLEPIS (UENSTEDTI), 48.
SCHEDIOSTEUS (MUSTELENSIS), 49.
COCCOPELTUS (ASMUSSI), 50.
CYPHOMALEPIS (EGERTONI), 51.
TRACHYLEPIS (FORMOSA), 52.
STIGMOLEPIS (OWENI), 53.
DASYLEPIS (KEYSERLINGI), 54.
LOPHOLEPIS (SCHMIDTI), 55.
DICTYOLEPIS (BRONNI), 56.
ONISCOLEPIS (MAGNA, etc.), 58.
MELITTOMALEPIS (ELEGANS), 60.
PHLEBOLEPIS (ELEGANS), 60.
PRIONACANTHUS (DUBIA), 70.

Stigmolepis Pander, 53; orthotype S. OWENI Pander (fossil).

Cœlolepis Pander, 56; logotype C. LÆVIS Pander (fossil).

Tolypelepis Pander, 61; orthotype T. UNDULATUS Pander (fossil) (fragment of a PTERASPID).

Lophosteus Pander, 62; orthotype L. superbus Pander (fossil).

Pachylepis Pander, 67; orthotype P. glaber Pander (fossil).   A synonym of Thelodus Ag.

Thelolepis Pander, 67; orthotype Thelodus parvidens Ag. (fossil).   A synonym of Thelodus Ag.

Nostolepis Pander, 68; orthotype N. striata Pander (fossil).

Rabdacanthus Pander 69; orthotype R. truncatus Pander (fossil).

Odontotodus Pander, 75; orthotype O. rootsikulliensis Pander (fossil).   A

Gomphodus Pander 76; orthotype G. sandelensis Pander (fossil).

synonym of Stigmolepis Pander.

573.   RICHARDSON (1856). *Ichthyology.* (In *Encyclopedia Britannica,* I, London.)
JOHN RICHARDSON.

Xanthichthys (Kaup) Richardson, XII, 313; orthotype Balistes curassavicus Gmelin.

574.   THOMPSON (1856).   *The Natural History of Ireland.*   London.
WILLIAM THOMPSON.

Couchia Thompson, IV, 188; orthotype Couchia minor Thompson.   A larval form of Ciliata.

1857

575.   BAIRD (1857). *Report on the Fishes Observed on the Coast of New Jersey and Long Island During the Summer of 1854.*   From the ninth annual report of the Smithsonian Institution for 1854.
SPENCER FULLERTON BAIRD.

Eucinostomus Baird, 345; orthotype E. argenteus Baird = Gerres gula Cuv. & Val.

576.   BELLOTTI (1857).   *Stoppani Studii Geol. e Paleontol.*   Lombardia.
CRISTOFORO BELLOTTI.

Urolepis Bellotti, 431; orthotype U. macroptera Bellotti (fossil).

Heptanema Bellotti, 435; orthotype H. paradoxum Bellotti (fossil).

Ichthyorhynchus Bellotti, 436; orthotype I. curioni Bellotti (fossil).   A synonym of Saurorhynchus.

577.   BLEEKER (1857).   *Achtste bijdrage tot de kennis der vischfauna van Amboina.*   Act. Soc. Sci. Indo-Nederl., II.
PIETER BLEEKER.

Cirrhitichthys Bleeker, II, 39; orthotype Cirrhites graphidopterus Blkr. = Cirrh. aprinus Cuv. & Val.

Oxycirrhites Bleeker, II, 39; orthotype O. typus Blkr.

578.   BLEEKER (1857).   *Nieuwe nalezingen op de ichthyologie van Japan.*   Verh. Bat. Gen.
PIETER BLEEKER.

Lateolabrax Bleeker, XXVI, 53; orthotype Labrax japonicus Cuv & Val.   Same as Perca-Labrax T. & S., but this word may not have been meant as a new name.

579.   BLEEKER (1857).   *Bijdrage tot de kennis der ichthyologische fauna van de Sangi-eilanden.*   Nat. Tijdschr. Neder.-Indie, XIII.
PIETER BLEEKER.

Scorpænodes Bleeker, 371; orthotype Scorpæna polylepis Blkr. Replaces Sebastopsis Gill.

Eleotriodes Bleeker, XIII, 372; orthotype E. sexguttata Blkr. E. heldsingeni Blkr., sometimes named as type, occurs in a later paper.

Bogoda Bleeker, 468; orthotype Chanda nama Ham. A synonym of Chanda as restricted.

580. COSTA (1857). *Descrizione di alcuni pesci fossili del Libano.* Mem. Reale Accad. Napoli, II, 97-112.

Oronzio Gabriele Costa.

Imogaster Costa, II, 102; orthotype I. auratus Costa (fossil).

Omosoma Costa, II, 105; orthotype O. sachelalmæ Costa (fossil).

Rhamphornimia Costa, II, 108; orthotype R. rhinelloides Costa (fossil).

581. EGERTON (1857). *Paleolithologic Notes on Some Fish Remains from the Neighborhood of Ludlow.* Quart. Journ. Geol. Sci., XIII.

Philip de Malpas Grey Egerton.

Auchenaspis Egerton, XIII, 286; orthotype A. salteri Egert. (fossil). A synonym of Thyestes Eichw.

581A. EMMONS (1857). *American Geology,* etc. Ed. I.

Ebenezer Emmons (1799-1863).

Rabdiolepis Emmons, VI, 45; orthotype R. speciosus Emmons (fossil) (fragment of a scale) (not Rhabdolepis Troschel).

Microdus Emmons, VI, 48; orthotype M. lævis Emmons (a shark tooth of uncertain relations).

582. GERVAIS (1857). *Sur un poisson labroide fossile dans les sables marins de Montpellier (Labrodon pavimentatum).* Mém. Acad. Sci. Montpellier, III.

François Louis Paul Gervais.

Labrodon Gervais, III, 515; orthotype L. pavimentatum Gervais (fossil).

583. GIRARD (1857). *Contributions to the Ichthyology of the Western Coast of the United States, from Specimens in the Museum of the Smithsonian Institution.* Proc. Acad. Nat. Sci. Phila., VIII.

Charles Girard.

Chiropsis Girard, IX, 201; orthotype Chirus constellatus Grd. (male of H. decagrammus). A synonym of Hexagrammos Steller.

Zaniolepis Girard, IX, 202; orthotype Z. latipinnis Grd. Misquoted as Zaniodermis by Günther (1860).

584. HOLLARD (1857). *Études sur les gymnodontes et en particulier sur leur ostéologie et sur les indications qu'elle peut fournir pour leur classification.* Ann. Sci. Nat. (Zool.), VIII.

Henri Louis Gabriel Marc Hollard (1801-1866).

Apsicephalus Hollard, 275; orthotype Tetrodon fluviatilis Ham. No type named; said to be equivalent to Dichotomycter Bibron, but including several other of Bibron's genera. Apparently a valid genus to be called Dichotomycter.

Brachycephalus Hollard, VIII, 275; orthotype not named, the genus said to be equivalent to Tetraodon of Bibron. T. lineatus L. may be assumed as type. A synonym of Tetraodon L.

585. NEWBERRY (1857). *Fossil Fishes from the Devonian of Ohio.* Proc. Nat. Inst. Washington, I.

John Strong Newberry

Agassichthys Newberry, 3; orthotype PLACOTHORAX AGASSIZI Meyer (fossil). A synonym of MACROPETALICHTHYS Ag.

Machæracanthus Newberry, 6; orthotype M. PERACUTUS Newberry (fossil).

Onychodus Newberry, 7; orthotype O. SIGMOIDES Newberry (fossil).

586. NEWBERRY (1857). *Letter Changing the Genus Mekolepis to Eurylepis, the Former Being Preoccupied.* Proc. Acad. Nat. Sci. Phila., IX.

JOHN STRONG NEWBERRY.

Eurylepis Newberry, 150; orthotype MECOLEPIS GRANULATUS Newberry (fossil). Substitute for MECOLEPIS Newberry, preoccupied.

587. PETERS (1857). *Neue Chromidengattung (Hemichromis).* Monatsber. Akad. Wiss. Berlin.

WILHELM CARL HARTWIG PETERS.

Hemichromis Peters, 403; orthotype H. FASCIATUS Peters.

588. PHILIPPI (1857). *Ueber einige chilenische Vögel und Fische,* etc. Arch. Naturgesch.

RUDOLPH AMANDUS PHILIPPI (b. 1808).

Genypterus Philippi, 268; orthotype G. NIGRICANS Philippi.

Thysanochilus Philippi, 268; orthotype T. VALDIVIANUS Philippi. A synonym of VELASIA Gray, suppressed (but printed) by Troschel, editor of the *Archiv.* Not THYSANOCHEILUS Kner, 1865.

Rhynchobatis Philippi, 270; orthotype a ray from Juan Fernandez, not given a specific name, only the dried head with long, flat snout preserved. Name preoccupied as RHYNCHOBATUS. This animal is described as a ray with a long, thin, flat snout, rounded at the tip apparently like that of the shark, MITSUKURINA. Its surface below, except at the smooth base, is covered by small stellate prickles, curved backward at tip. Larger spinules cover the head with six still larger ones forming a curve around the eye. As this form, obviously distinct, is without a name, it has been called TARSISTES PHILIPPI Jordan.

589. RICHARDSON (1857). *On Siphonognathus, a New Genus of Fistularidæ.* Proc. Zool. Soc. London.

JOHN RICHARDSON.

Siphonognathus Richardson, 237; orthotype S. ARGYROPHANES Rich.

590. RÜTIMEYER (1857). *(Encheiziphius.)* Verh. Ges. Basel, I.

CARL LUDWIG RÜTIMEYER.

Encheiziphius Rütimeyer, I, 561; orthotype E. TERETIROSTRIS Rütimeyer (fossil). A synonym or fossil homologue of ISTIOPHORUS.

591. TROSCHEL (1857). *Beobachtungen über die Fische in den Eisennieren des Saarbrücker Steinkohlgebirges.* Verh. Naturh. Ver. Preuss. Rheinl. Westphal.

FRANZ HERMANN TROSCHEL.

Rhabdolepis Troschel, 15; logotype PALÆONISCUM MACROPTERUM Bronn (fossil). A synonym of ELONICHTHYS Giebel.

592. WAGNER (1857). *Characteristik neuer Arten von Knorpelfischen aus den lithographischen Schiefern der Umgegend von Solenhofen.* Gelehrte Anz. Bayer. Akad. Wiss., XLIV.

JOHANN ANDREAS WAGNER.

Palæoscyllium Wagner, XLIV, 291; othotype P. FORMOSUM Wagner (fossil).

1858

593. AGASSIZ (1858). *Remarks on a New Species of Skate from the Sandwich Islands.* Proc. Boston Soc. Nat. Hist., 1858, VI.

Louis Agassiz.

Goniobatis Agassiz, VI, 385; orthotype Raja flagellum Bl. & Schn. A synonym of Aëtobatis Blainv., as restricted by Müller & Henle. Same as Stoasodon Cantor, 1850.

594. AGASSIZ (1858). *Remarks on New Fishes from Nicaragua Collected by Julius Fræbel.* Proc. Boston Soc. Nat. Hist., VI.

Louis Agassiz.

Parachromis Agassiz, VI, 401; orthotype P. gulosus Ag. (not identifiable, said to be a Cichlid (Chromid) resembling Dentex, but with the lateral line broken. Probably a synonym of Cichlaurus Sw.

Hypsophrys Agassiz, VI, 401; orthotype H. unimaculatus Ag. No description. A Cichlid said to resemble Chrysophrys (Sparus) with a large, black shoulder spot (as in Astronotus helleri Steind. and other species). Probably a synonym of Astronotus Sw.

Baiodon Agassiz, VI, 401; orthotype B. fasciatus Ag. A slender form, said to resemble Boops Cuv. Possibly the same as Theraps irregularis Gthr., but not identifiable.

Amphilophus Agassiz, VI, 401; orthotype A. frœbeli Ag., apparently Heros labiatus Gthr. or H. lobocheilus Gthr. species in which the male has both lips greatly enlarged, "peculiar in not having the ordinary fleshy lips, but a large, triangular lobe projecting above and below jaw like the fleshy appendages of some bats" (Agassiz). It is possible that the species thus provided should stand as a distinct genus Amphilophus.

595. AYRES (1858). *On New Fishes of the Californian Coast.* Proc. Calif. Acad. Sci., 1858-1861 (1853), II.

William O. Ayres.

Anoplopoma Ayres, II, 27; orthotype A. merlangus Ayres. = Gadus fimbria Pallas.
Stereolepis Ayres, II, 28; orthotype S. gigas Ayres.

596. BLEEKER (1858). *Zesde bijdrage* (etc.) *vischfauna van Sumatra (Visschen van Padang, etc.).* Act. Soc. Neder.-Indie, III.

Pieter Bleeker.

Helicophagus Bleeker, 45; orthotype H. typus Blkr.

597. BLEEKER (1858). *Vierde bijdrage tot de kennis der ichthyologische Fauna van Japan.* Act. Soc. Sci. Indo-Neerl., III.

Pieter Bleeker.

Sirembo Bleeker, 22; orthotype Brotula imberbis T. & S.
Congerodon (Kaup) Bleeker, 29; orthotype C. indicus Kaup MS. = Cogrus lepturus Rich. Same as Uroconger and of the same original date, 1856.
Triacanthodes Bleeker, 37; orthotype Tricanthus anomalus T. & S.

598. BLEEKER (1858). *Tiende bijdrage tot de kennis der vischfauna van Celebes.* Act. Soc. Sci. Indo-Neerl., III.

Pieter Bleeker.

Gunnellichthys Bleeker, III, 9; orthotype G. pleurotænia Blkr.

599. BLEEKER (1858). *Bijdrage tot de kennis der vischfauna van den Goram-Archipel.* Nat Tijdschr. Neder.-Indie, XV.

PIETER BLEEKER.

Pseudoplesiops Bleeker, XV, 217; orthotype P. TYPUS Bleeker.

600. BLEEKER (1858). *Vierde bijdrage* (etc.) *van Bilitou.* Nat. Tijdschr. Neder.-Indie, XV.

PIETER BLEEKER.

Leiocassis Bleeker, 225; orthotype L. MICROPOGON Blkr. Corrected by Günther to LIOCASSIS.

Silurichthys Bleeker, 225; orthotype S. PHAIOSOMA Blkr.

601. BLEEKER (1858). *Twaalfde bijdrage tot de kennis der vischfauna van Borneo, visschen van Sinkawang.* Act. Soc. Sci. Indo-Neerl., V.

PIETER BLEEKER.

Sauridichthys Bleeker, 2; orthotype S. OPHIODON Blkr. (1858). A synonym of HARPODON Le Sueur.

Hexanematichthys Bleeker, V, 2; orthotype BAGRUS SUNDAICUS Blkr.

Sciadeichthys Bleeker, 99; orthotype SCIADES PICTUS M. & T. Not SCIADEICHTHYS Blkr., 1863.

602. BLEEKER (1858). *De visschen van den Indischen Archipel,* etc. I. *Siluri.* Title also in Latin as *Prodromus.*

PIETER BLEEKER.

Most of these genera occur also in later papers, 1862, 1863; the type species often indicated later.

Pseudodoras Bleeker, I, 53; orthotype DORAS NIGER Val.

Hemidoras Bleeker, I, 53; orthotype DORAS STENOPELTIS Kner. A synonym of DORAS Lac. as restricted by Bleeker.

Davalla Bleeker, 58; orthotype HYPOPHTHALMUS DAVALLA Schomburgk.

Pseudobagrus Bleeker, 60; orthotype BAGRUS AURANTIACUS T. & S.

Melanodactylus Bleeker 60; orthotype M. ACUTIVELIS Blkr. = PIMELODUS NIGRODIGITATUS Lac.

Chrysichthys Bleeker, 60; orthotype PIMELODUS AURATUS Geoffroy St. Hilaire.

Octonematichthys Bleeker, 60; orthotype O. NIGRITUS Blkr.

Rita Bleeker, 60; orthotype PIMELODUS RITA Ham. (R. BUCHANANI Blkr.).

Selenaspis Bleeker, I, 62; haplotype SILURUS HERZBERGI Bloch.

Cephalocassis Bleeker, 62; orthotype ARIUS TRUNCATUS Cuv. & Val.

Netuma Bleeker, 62; orthotype N. THALASSINA Blkr. = N. NASUTA Blkr.

Guiritinga Bleeker, 62; orthotype G. COMMERSONI Blkr.

Ostogeneiosus Bleeker, 63; orthotype ARIUS MILITARIS Blkr.

Batrachocephalus Bleeker, 63; orthotype B. MICROPOGON Blkr.

Ketengus Bleeker, 190; orthotype K. TYPUS Blkr.

Conorhynchos Bleeker, 191; orthotype PIMELODUS CONIROSTRIS Cuv. & Val.

Hemipimelodus Bleeker, 195; orthotype H. BORNEENSIS Blkr.

Zungaro Bleeker, 196; orthotype PIMELODUS ZUNGARO Humboldt = ZUNGARO HUMBOLDTI Blkr.

Heptapterus Bleeker, 197; orthotype H. SURINAMENSIS Blkr.

Rhamdia Bleeker, 197; orthotype PIMELODUS QUELEN Q. & G.

Pinirampus Bleeker, 198; orthotype P. TYPUS Blkr. = PIMELODUS PIRINAMPU Spix.

Auchenaspis Bleeker, 198; orthotype PIMELODUS BISCUTATUS Geoffroy. Name preoccupied; replaced by AUCHENOGLANIS Gthr.

Trachcorystes Bleeker, 200; orthotype T. TYPUS Blkr.
Rama Bleeker, 201; orthotype PIMELODUS RAMA Ham. = RAMA BUCHANANI Blkr.
Acrochordonichthys Bleeker, 204; orthotype A. PLATYCEPHALUS Blkr.
Gagata Bleeker, 204; orthotype G. TYPUS Blkr.
Schilbeichthys Bleeker, 253; orthotype SILURUS GARUA Ham. A synonym of CLUPI-
SOMA Sw.
Belodontichthys Bleeker, 253; orthotype B. MACROCHIR Blkr. = WALLAGO DINEMA
Blkr.
Pseudosilurus Bleeker, 253; orthotype P. BIMACULATUS Blkr. A synonym of
OMPOK Lac.
Siluranodon Bleeker, 253; orthotype SILURUS AURITUS Geoffroy.
Kryptopterus Bleeker, 253; tautotype SILURUS CRYPTOPTERUS Blkr. Corrected to
CRYPTOPTERUS by Günther.
Kryptopterichthys Bleeker, 253; orthotype SILURUS PALEMBRANGENSIS Blkr.
Micronema Bleeker, 253; orthotype M. TYPUS Blkr.
Phalacronotus Bleeker, 253; orthotype P. LEPTONEMA Blkr.
Hemisilurus Bleeker, 255; orthotype H. HETERORHYNCHUS Blkr.
Schilbeodes Bleeker, 258; orthotype SILURUS GYRINUS Mitchill.
Aspredinichthys Bleeker, 328; orthotype ASPREDO TIBICEN Temminck.
Bunocephalichthys Bleeker, 329; orthotype BUNOCEPHALUS VERRUCOSUS Kner.
Piratinga Bleeker, 355; orthotype BAGRUS RETICULATUS Kner.
Platynematichthys Bleeker, 357; orthotype BAGRUS PUNCTULATUS Kner.
Piramutana Bleeker, 355; orthotype BARGUS PIRAMUTA Kner.

603. BLEEKER (1858). *Over de geslachten der Cobitinen.* Nat. Tijdschr. Neder-
Indie, XVI, 1858-1859, 302-304.
PIETER BLEEKER.
Lepidocephalus Bleeker, 303; orthotype C. MICROCHIR Blkr.
Cobiitichthys Bleeker, 304; orthotype C. ENALIOS Bleeker. A synonym of MIS-
GURNUS Lac.

604. BLEEKER (1858). *Vischsoorten gevangen bij Japara, versameld door S. A.
Thurkow.* Nat. Tijdschr. Neder.-Indie, XVI, 1858-1859, 406-409.
PIETER BLEEKER.
Oxyurichthys Bleeker, 408; orthotype O. MICROLEPIS Blkr. or GOBIUS BELOSSO Blkr.

605. BLYTH (1858). *(Amblyceps).* Journ. Asiat. Soc. Bengal, XXVII.
EDWARD BLYTH.
Amblyceps Blyth, XXVIII, 281; haplotype A. CÆCUTIENS Blyth.

606. BRONN (1858). *Beiträge zur triasischen Fauna und Flora der bituminöschen
Schiefer von Raibl in Kärnthen.* Neues Jahrb. Min.
HEINRICH GEORG BRONN (1800-1862).
Belonorhynchus Bronn, 12; orthotype B. STRIOLATUS Bronn (fossil). A synonym
of SAURORHYNCHUS.
Pholidopleurus Bronn, 17; orthotype P. TYPUS Bronn (fossil).
Thoracopterus Bronn 21; orthotype T. NIEDERRISTI Bronn (fossil).

607, 612. COSTA (1858). *(Uraleptus.)* Archiv. Naturg. (1858).
ORONZIO GABRIELE COSTA.
Uraleptus Costa, 87; orthotype GADUS MARALDI Risso. A synonym of GADELLA
Lowe.

280 THE GENERA OF FISHES, PART II

613. DUMÉRIL (1858). *Reptiles et poissons de l'Afrique occidentale,* etc.
Comtes Rendus Acad. Sci. Paris, 1858, LXI.
AUGUSTE HENRI ANDRÉ DUMÉRIL.

Chromichthys A. Duméril, X, 257; orthotype C. ELONGATUS Guich.

614. EGERTON (1858). *On Chondrosteus, an Extinct Form of the Sturionidæ,
Found in the Lias Formation at Lyme Regis.* Phil. Trans. London, 1858.
PHILIP DE MALPAS GREY EGERTON.

Chondrosteus (Agassiz) Egerton, 871; orthotype C. ACIPENSEROIDES Ag. (fossil).
Gyrosteus (Agassiz) Egerton, 883; orthotype G. MIRABILIS Ag. (fossil).

615. EGERTON (1858). *Fish: Cosmolepis,* etc. (In *Figures and Descriptions of
British Organic Remains.)* Mem. Geol. Surv. United Kingdom, 1858, 9 dec.
PHILIP DE MALPAS GREY EGERTON.

Cosmolepis Egerton, Dec. IX, No. 1; orthotype C. EGERTONI Ag. (fossil).
Endactis Egerton, Dec. IX, No. 4; orthotype E. AGASSIZI Egert. (fossil). A syno-
nym of CATURUS.
Centrolepis Egerton, Dec. IX, No. 5; orthotype C. ASPER Ag. (fossil).
Nothosoma (Agassiz) Egerton, Dec. IX, No. 6; orthotype N. OCTOSTYCHIUS Ag.
(fossil).
Pleuropholis Egerton, Dec. IX, No. 7; logootype PHOLIDOPHORUS LÆVISSIMUS Ag.
(fossil).
Macropoma (Agassiz) Egerton, Dec. IX, No. 10; orthotype M. EGERTONI Ag.
(fossil). Replaces EURYCORMUS Wagner (1863).

616. GILL (1858). *Prodromus descriptionis subfamiliæ Gobinarum squamis
cycloideis piscium, a cl. W. Stimpsono in mare Pacifico acquisitorum.* Ann.
Lyceum Nat. Hist. N. Y., 1858-1862, VII, 12-16.
THEODORE NICHOLAS GILL (1837-1914).

Chænogobius Gill, 12; orthotype C. ANNULARIS Gill.
Lepidogobius Gill, 14; orthotype GOBIUS GRACILIS Grd. = G. LEPIDUS Grd.

617. GILL (1858). *Description of a New Genus of Pimelodinæ from Canada.*
Ann. Lyceum Nat. Hist. N. Y., 1853-1858, VI, 3.
THEODORE NICHOLAS GILL.

Synechoglanis Gill, 39; orthotype SYNECHOGLANIS BEADLEI Gill = ICTALURUS
PUNCTATUS (Raf.). A synonym of ICTALURUS Raf.

618. GILL (1858). *Prodromus descriptionis familiæ Gobioidarum generum novorum.*
Ann. Lyceum Nat. Hist. N. Y., 1858-1862, VII, 16-19.
THEODORE NICHOLAS GILL.

Tridentiger Gill, 16, Dec.; orthotype SICYDIUM OBSCURUM, T. & S.
Triænophorus Gill, 17; orthotype T. TRIGONOCEPHALUS Gill. Name preoccupied,
replaced by TRIÆNOPHORICHTHYS Gill.

619. GILL (1858). *(Triænophorichthys.)* Proc. Acad. Nat. Sci. Phila., 1859.
THEODORE NICHOLAS GILL.

Triænophorichthys Gill 195; orthotype TRIÆNOPHORUS TRIGONOCEPHALUS Gill.
Substitute for TRIÆNOPHORUS Gill, preoccupied.

620. GILL (1858). *Synopsis of the Fresh-water Fishes of the Western Portion of the Island of Trinidad, West Indies.* Ann. Lyceum Nat. Hist. N. Y., 1853-1858, VI, 363-430.

THEODORE NICHOLAS GILL.

Ctenogobius Gill, 374; orthotype C. FASCIATUS Gill.

Pimelonotus Gill, VI, 387; orthotype P. VILSONI. A synonym of RHAMDIA Blkr. of the same year.

Batrachoglanis Gill, VI, 389; orthotype PIMELODUS RANINUS Cuv. & Val. A synonym of PSEUDOPIMELODUS Blkr. of about the same date.

Hoplosternum Gill, VI, 395; orthotype CALLICHTHYS LÆVIGATUS Val.

Pterygoplichthys Gill, VI, 408; orthotype HYPOSTOMUS DUODECIMALIS Cuv. & Val.

Pœcilurichthys Gill, 414; logotype P. BREVOORTI Gill.

Hemigrammus Gill, 416; haplotype H. UNILINEATUS Gill.

Stevardia Gill, 423; orthotype S. ALBIPINNIS Gill.

Corynopoma Gill, 425; orthotype C. RIISEI Gill. A synonym of STEVARDIA Gill.

Nematopoma Gill, 428; orthotype N. SEARLESI Gill. A synonym of STEVARDIA Gill.

621. HECKEL & KNER (1858). *Die Süsswasserfische der oesterreichischen Monarchie mit Rücksicht auf die angränzenden Länder.*

JOHANN JAKOB HECKEL; RUDOLPH KNER.

Carpio Heckel, 64; orthotype CYPRINUS KOLLARI Heckel. A supposed hybrid of CYPRINUS and CARASSIUS.

Blicca Heckel, 120; orthotype CYPRINUS BLICCA Gmelin.

Leucaspius Heckel & Kner, 145; orthotype L. ABRUPTUS Heckel & Kner.

Idus Heckel, 147; orthotype CYPRINUS IDUS, L.

622. KADE (1858). *Ueber die devonischen Fischreste cines diluvial Blockes.* Meseritz, 1858.

G. KADE.

Gyrolepis Kade, 17, 18; orthotype G. POSNANIENSIS Kade (fossil). Name preoccupied; perhaps a DIPLOPTERAX.

Archæacanthus Kade, 19; orthotype A. QUADRISULCATUS Kade (fossil). A synonym of ONCHUS.

Spirodus Kade, 20; orthotype S. REGULARIS Kade (fossil).

623. KAUP (1858). *Uebersicht der Familie Gadidæ.* Arch. Naturg., 1858, XXIV, pt. 1.

JOHANN JACOB KAUP.

Xenocephalus Kaup, 83; orthotype X. ARMATUS Kaup.

Physiculus Kaup, 88; orthotype P. DALWIGKII Kaup.

Lotella Kaup, 88; orthotype LOTA PHYCIS T. & S.

Molvella Kaup, 90; orthotype M. BOREALIS Kaup = GADUS MUSTELA L. A synonym of CILIATA Couch.

Brotella Kaup, 92; orthotype BROTULA IMBERBIS T. & S. A synonym of SIREMBO Blkr.

Hoplophycis Kaup, 93; orthotype XIPHIURUS CAPENSIS Smith. A section under GENYPTERUS Philippi.

Brotulophis Kaup, 93; orthotype B. ARGENTISTRIATUS Kaup.

624. KAUP (1858). *On Nemophis, a New Genus of Riband-shaped Fishes.* Proc. Zool. Soc. London, 1858, pt. 26.

JOHANN JACOB KAUP.

Nemophis Kaup 168; orthotype N. LESSONI Kaup.

625. KAUP (1858). *Uebersicht der Soleinæ,* etc. Arch. Naturg.

JOHANN JACOB KAUP.

Monochirus Kaup, 94; orthotype M. MACULIPINNIS Kaup. Not of Rafinesque, 1814; a synonym of ACHIRUS Lac.

Grammichthys Kaup 94; orthotype ACHIRUS LINEATUS L. A synonym of ACHIRUS Lac.

Æsopia Kaup, 97; logotype SOLEA CORNUTA Cuv. (as restricted by Günther, 1862).

Euryglossa Kaup, 99; orthotype PLEURONECTES ORIENTALIS Bl. & Schn.

Eurypleura Kaup, 100; logotype E. LEUCORHYNCHA Kaup. A needless substitute for ACHIROIDES Blkr.

Gymnachirus Kaup, 101; orthotype G. NUDUS Kaup.

Aseraggodes Kaup, 103, logotype A. GUTTULATUS Kaup.

Heteromycteris Kaup, 103; orthotype H. CAPENSIS Kaup.

Apionichthys Kaup, 104; orthotype A. DUMERILI Kaup.

626. KAUP (1858). *Uebersicht der Plagusinæ, der fünften Subfamilie der Pleuronectidæ.* Arch. Naturg., 1858, XXIV, pt. 1.

JOHANN JACOB KAUP.

Cantoria Kaup, 106; orthotype PLAGUSIA CANTORI Blkr. = CANTORIA PINANGENSIS Kaup.

Aphoristia Kaup, 106; orthotype ACHIRUS ORNATUS Lac. A snyonym of SYMPHURUS Raf.

Arelia Kaup, 106; logotype PLEURONECTES AREL Bl. & Schn.

Trulla Kaup, 106; orthotype PLAGUSIA TRULLA Cantor.

Icania Kaup, 106; orthotype ACHIRUS CYNOGLOSSUS Buchanan.

627. KAUP (1858). *Einiges über die Acanthopterygiens à joue cuirassée Cuv.* Arch. Naturg., 1858, XXIV, pt. 1.

JOHN JACOB KAUP

Polemius Kaup, 333; orthotype APISTUS ALATUS Cuv. & Val. Kaup transfers APISTUS Cuv. to A. TRACHINOIDES, a species not mentioned by Cuvier, making PROSOPODASYS, Cantor's substitute name (1849), a synonym of APISTUS. POLEMIUS must be regarded as a synonym of PTERICHTHYS Sw. if APISTUS is rejected.

Cocotropus Kaup, 333; orthotype CORYTHOBATUS ECHINATUS Cantor.

Peristethus Kaup, 336; orhotype (TRIGLA CATAPHRACTA L.). A change in spelling of PERISTEDION a name also revised to PERISTETHIDIUM.

Trichopleura Kaup, 338; orthotype STHENOPUS MOLLIS Rich. A synonym of STHENOPUS.

Hoplocottus Kaup, 339; logotype PODABRUS COTTOIDES Rich. (first species named), a species of PSEUDOBLENNIUS T. & S., 1850. PODABRUS is preoccupied and HOPLOCOTTUS was made to contain both PODABRUS and TRACHIDERMIS Heckel. A synonym of PSEUDOBLENNIUS T. & S.

628. KNER (1858). *Ichthyologische Beiträge.* II. Abtheilung. Sitzb. Akad. Wiss. Wien, XXVI.

RUDOLPH KNER.

Asterophysus Kner, XXVI, 402; orthotype A. BATRACHUS Kner.

Centromochlus Kner, XXVI, 430; orthotype C. MEGALOPS Kner.

629. LEIDY (1858). *Descriptions of Some Remains of Fishes from the Carboniferous and Devonian Formations of the United States.* Journ. Acad. Nat. Sci. Phila., III.

JOSEPH LEIDY.

Apedodus Leidy, III, 164; orthotype A. priscus Leidy (fossil). Probably a synonym of Dendrodus.

630. LEIDY (1858). *Notices of Some Remains of Extinct Fishes (Hadrodus, etc.).* Proc. Acad. Nat. Sci. Phila., 1857 (1858), 167-168.

Joseph Leidy.

Hadrodus Leidy, 167; orthotype H. priscus Leidy.

Phasganodus Leidy, 167; orthotype P. dirus Leidy (fossil). A synonym of Cimolichthys Leidy.

Turseodus Leidy, 167; orthotype T. acutus Leidy (fossil).

631. VON DER MARCK (1858). *(Fossil Fishes.)* Zeitschrift der Deutschen Geologie, X.

W. von der Marck (*d.* 1900).

Pelargorhynchus von der Marck, X, 242; logotype P. dercetiformis Marck (fossil).

Sardinius von der Marck, X, 245; orthotype Osmerus cordieri Ag. (fossil). Not Sardinia Poey, 1860, nor Sardina Antipa, 1914, which is the same.

Sardinioides von der Marck, X, 245; orthotype Osmeroides monasteri Ag. (fossil). A synonym of Osmeroides Ag.

Echidnocephalus von der Marck, X, 247; orthotype E. troscheli Marck (fossil). An extinct analogue of Halosaurus.

Ischyrocephalus von der Marck, X, 248; orthotype I. gracilis Marck (fossil). A synonym of Enchodus.

Platycormus von der Marck, X, 351; orthotype Beryx germanus Ag. (fossil).

632. PANDER (1858). *Die Ctenodipterinen des devonischen Systems.* St. Petersburg, 1858, VIII, 65.

Christian Heinrich Pander.

Cheirodus Pander, 33; orthotype C. jerofejewi Pander (fossil). A synonym of Conchodus McCoy; the name preoccupied.

Ptyctodus Pander, 48; orthotype P. obliquus Pander (fossil).

633. PHILIPPI (1858). *Beschreibung neuer Wirbelthiere aus Chile.* Arch. Naturg.

Rudolph Amandus Philippi.

Chilopterus Philippi, 306; no specific name given. A larval lamprey; probably of Caragola Gray.

633A. QUENSTEDT (1858). *Der Jura.* Tübingen, 1856–1858, VI.

Friederich August von Quenstedt.

Desmacanthus Quenstedt, 34; orthotype D. cloacinus Quenstedt (fossil). A synonym of Nemacanthus.

Lycodus Quenstedt, 240; orthotype Saurichthys gigas Quenstedt (fossil).

Pachylepis Quenstedt, 241; orthotype (problematical cranial bones) (fossil).

Chimæracanthus Quenstedt, 347; orthotype. A. aalensis Quenstedt (fossil). A synonym of Ischyodus.

Psilacanthus Quenstedt, 347; orthotype P. aalensis Quenstedt (fossil).

Typodus Quenstedt, 781; logotype T. annulatus Quenstedt (fossil). Apparently a synonym of Mesodon, which, being preoccupied, it replaces.

Kokkoderma Quenstedt, 810; orthotype K. suevicum Quenstedt (fossil). Later spelled Coccoderma.

634. REINHARDT (1858). *Stegophilus insidiosus, en ny mallefisk fra Brasilien og dens levemaade.* Vidensk. Meddel. Naturh. Foren. Kjöbenhavn, 1859, 79-97.
JOHANN THEODOR REINHARDT.

Stegophilus Reinhardt, tab. 2; orthotype S. INSIDIOSUS Reinh.

635. ROUAULT (1858.) *Note sur les vertèbres fossiles des terrains sedimentaired. de l'ouest de la France.* Comtes Rendus Acad. Sci. Paris, 1858, XLVII, 99-103.
MARIE ROUAULT.

Nummopalatus Rouault, XLVII, 101; orthotype N. EDWARDSIUS Rouault (fossil). A synonym of LABRODON Gervais.

Machærius Rouault, XLVII, 102; orthotype M. ARCHIACI Rouault (fossil). Name preoccupied; a synonym of MACHÆRACANTHUS Newberry.

636. THIOLLIÉRE (1858). *Notice sur les poissons fossiles du Bugey, et sur application de la méthode de Cuvier à leur classement.* Bull. Soc. Géol. France, 1858, 2 sér., XV, 782-794.
VICTOR JOSEPH DE L'ISLE THIOLLIÈRE.

Attakeopsis Thiollière, XV, 784; orthotype A. DESORI Thiollière (fossil). Name only; validated in 1873. A synonym of IONOSCOPUS Costa.

Callopterus Thiollière, XV, 784; orthotype C. AGASSIZI Thiollière (fossil). Name occupied as CALOPTERUS in beetles (1838); may be replaced by AINIA Jordan, from the locality Ain in France.

637. WEINLAND (1858). *A New Division of the Five Species of Flying-fish Found Along the Coast of North America, Which Have Hitherto All Been Referred to the Genus Exocetus.* Proc. Boston Soc. Nat. Hist., VI, 385.
DAVID FRIEDRICH WEINLAND.

Halocypselus Weinland, VI, 385; orthotype "EXOCŒTUS MESOGASTER" = EXOCŒTUS VOLITANS L. = E. EVOLANS L., the two being identical. as Dr. Lönnberg has shown by the examination of Linnæus' type of E. VOLITANS.

OMISSIONS IN PART II.

As these pages go through the press the following errors and omissions are noted:

On page 173, Bennett's *Fishes of Mauritius*, No. 143A, is in two parts, both in 1831, not 1830. The first is the collection of Captain Belcher (November), the second that of Charles Telfair (December).

Insert—

Apolectus Bennett, I, 146 (November); orthotype A. IMMUNIS Bennett. Name preoccupied, replaced by APODONTIS Bennett.

Apodontis Bennett, I, 169 (December); orthotype APOLECTUS IMMUNIS Bennett, substitute for APOLECTUS Bennett, preoccupied by APOLECTUS Cuv. & Val. (October) 1830.

Under **Agonostomus** read A. TELFAIRI.

On page 174, to 142A Cuvier & Valenciennes add—

Apolectus Cuvier & Valenciennes, VIII, 438; orthotype A. STROMATEUS Cuv. & Val. = STROMATEUS NIGER Bloch.

On page 219 the name COCCOLUS was given as substitute for KROHNIUS, a larval MACROID, as stated on page 226.

The Genera of Fishes

[PART III]
FROM GUENTHER TO GILL, 1859–1880
TWENTY-TWO YEARS
WITH THE ACCEPTED TYPE OF EACH

G
III

INTRODUCTORY NOTE

The present memoir is the third part of a catalogue of the generic names applied to fishes subsequent to January 1, 1858. With the other parts it is a contribution to the stability of scientific nomenclature, the genera being listed in order of date accompanied by the indication of the type species of each.

The first part (1758 to 1833) was intended to serve as an aid to the International Commission of Zoological Nomenclature by bringing together the generic names of fishes as proposed in the formative period of classification, together with the data by which authors who failed to conform to the Linnæan system could be considered, each on his merits. I had at first no intention of continuing the series beyond 1833, but the data gathered for the first part seemed to make the compilation of the catalogue worth while.

Part II covers what I have termed the mediæval period of taxonomy—from 1833 to 1858, this completing the first century of names in ichthyology.

Part III represents the early modern period, dating from Darwin's "Origin of Species" in 1858. In this epoch most authors began to see that the fundamental basis of classification must be genetic and that the problems involved in a natural grouping are vastly more complex than appeared to the earlier authors. To paraphrase a saying of Dr. Elliott Coues, genera and species are but larger and smaller twigs of a tree which we try to arrange as nearly as possible in accordance with nature's ramifications.

This view applied to taxonomy involved an extensive subdivision of accepted genera, the genus becoming a genetic conception rather than a convenient pigeon-hole into which to throw species.

The effort to reform classification in Ichthyology was undertaken almost simultaneously by three eminent systematists, Dr. Pieter Bleeker of Java, Dr. Albert Günther of the British Museum, and Dr. Theodore Gill of the Smithsonian Institution. Of these writers Dr. Bleeker had by far the greatest experience in field work, but his use of the writings of other authors was not discriminating, and the names he chose for genera, though technically correct, were largely clumsy and tasteless. Dr. Günther had at his disposal a greater range of material than any other contemporary writer. His groupings were, however, of unequal value, and he was more distinctly "conservative" than either Bleeker or Gill. It has long been a sort of tradition in the British Museum that whatever is not found there probably does not exist. Dr. Gill saw more clearly than either of the other two the nature of the classification of the future. His generic names have as a rule a sound basis, the exceptions being mainly among those based on erroneous records of other writers.

In addition to the large expansion of generic names among species already known, the years between 1858 and 1880 mark great accessions of knowledge of the life of the oceanic depths, a series of discoveries due mostly to the voyages of the Challenger and the Albatross, and made known in England by Dr. Günther, and in America by Goode and Bean. To this period belong also the extensive studies of the fresh-water fishes of the United States and the marine fishes of the Eastern Pacific as conducted by Jordan and Gilbert. Most of the excellent papers of Franz Steindachner in Vienna, Léon Vaillant in Paris, Christian F. Lütken in Copenhagen, and Joseph Leidy and Edward D. Cope in Philadelphia, occur in this period.

I express again my obligations to Barton A. Bean of the United States National Museum, to Masamitsu Oshima, now Scientific Expert of the Fisheries Bureau at Taihoku, Formosa, and to Henry Weed Fowler of the Academy of Natural Sciences of Philadelphia for assistance in verifying records and in examining rare publications. Dr. Louis Hussakof of New York, Theodore D. A. Cockerell of Boulder, Colorado, Allan R. McCulloch, Edgar R. Waite, and J. Douglas Ogilby have also furnished valuable notes. I am again indebted to Mr. R. A. Nelgner of the Stanford University Press for his efforts in securing accuracy and in the preparation of the index. My obligations to Weber and de Beaufort's *Index of the Ichthyological Papers of P. Bleeker,* to Bashford Dean's *Bibliography of Fishes,* to Oliver Perry Hay's *Fossil Vertebrata of North America,* and to Arthur Smith Woodward's *Catalogue of the Fossil Fishes in the British Museum* have been great and constant.

DAVID STARR JORDAN.

STANFORD UNIVERSITY, CALIFORNIA,
August 10, 1919; Issued October 25, 1919.

THE GENERA OF FISHES

PART III

1859

638. ANDERSON (1859). *(On the Fossil Fishes and Yellow Sandstone.)* Dura Den.
JOHN ANDERSON.

Glyptolæmus (Huxley) Anderson, 63; orthotype G. KINAIRDI Huxley (fossil).
A synonym of GLYPTOPOMUS Ag.

Phaneropleuron (Huxley) Anderson, 67; orthotype P. ANDERSONI Huxley (fossil).

639. BLEEKER (1859). *Over eenige vischsoorten van de zuidkust-wateren van Java.* Nat. Tijdschr. Neder.-Indie, XIX, 1859, 329-352.
PIETER BLEEKER.

Scarichthys Bleeker, 334; orthotype SCARUS NÆVIUS Blkr.

640. BLEEKER (1859). *Vischsoorten van Siam, . . . door Fr. Castelnau.*
Nat. Tijdschr. Neder.-Indie, XX.
PIETER BLEEKER.

Balantiocheilos Bleeker, 102; orthotype B. MELANOPTERUS Blkr.

Morara Bleeker, 102; orthotype CYPRINUS MORAR Ham. = MORARA SIAMENSIS Blkr.
A synonym of ASPIDOPARIA Heckel.

Acanthobrama, Bleeker, 102; orthotype A. SIMONI Blkr.

Pseudoxiphophorus Bleeker, 102; orthotype XIPHOPHORUS BIMACULATUS Steind.

641. BLEEKER (1859). *Conspectus systematis Cyprinorum.* Nat. Tijdschr.
Neder.-Indie, XX, 421-441. (In this brief synopsis no types are indicated.
These are supplied in the *Atlas*, 1863.)
PIETER BLEEKER.

Lissorhynchus Bleeker, 422; tautotype GOBIO LISSORHYNCHUS Ham. Recorded
later by Bleeker as a synonym of GARRA Ham.

Diplocheilichthys Bleeker, 423; orthotype LOBSCHEILOS PLEUROTÆNIA Blkr.

Discognathichthys Bleeker, 423; orthotype probably DISCOGNATHUS RUFUS Heckel
(= CYPRINUS LAMTA Ham.) (not verified). Said by Bleeker to be a synonym of GARRA Ham. = DISCOGNATHUS Heckel.

Barbichthys Bleeker, 424; orthotype B. LÆVIS Blkr.

Rohitichthys Bleeker, 424; orthotype LABEO SENEGALENSIS Cuv. & Val.

Chrysophekadion Bleeker, 424; orthotype ROHITA CHRYSOPHEKADION Blkr. = C.
POLYPOROS Blkr. A synonym of MORULIUS or of LABEO.

Semiplotus Bleeker, 424; orthotype CYPRINUS SEMIPLOTUS McCl.

Opistocheilus Bleeker, 425; orthotype O. PROPIUS Blkr. = SCHIZOTHORAX PLAGIOSTO-
MUS Heckel. A synonym of SCHIZOTHORAX Heckel.

Pseudogobio Bleeker, 425; orthotype GOBIO ESOCINUS T. & S.

Mrigala Bleeker, 427; orthotype CYPRINUS MRIGALA Ham. = M. NINENSIS Blkr.
A synonym of CIRRHINA Cuv.

Acheilognathus Bleeker, 427; orthotype A. MELANOGASTER Blkr.

Smiliogaster Bleeker, 428; orthotype LEUCISCUS BELANGERI Cuv. & Val. A synonym of ROHTEE Sykes.

Amblyrhynchichthys Bleeker, 430; orthotype BARBUS TRUNCATUS Blkr.

Albulichthys Bleeker, 430; orthotype SYSTOMUS ALBULOIDES Blkr.

Tambra Bleeker, 430; orthotype CYPRINUS ABRAMIOIDES Sykes.

Gonoproktopterus Bleeker, 430; orthotype BARBUS KOLUS Sykes. Same as HYPSELOBARBUS Blkr.

Hypselobarbus Bleeker, 430; orthotype H. TYPUS Blkr.

Hemibarbus Bleeker, 431; orthotype GOBIO BARBUS T. & S.

Rohteichthys Bleeker, 431; orthotype R. MICROLEPIS Blkr.

Pseudophoxinus Bleeker, 431; orthotype PHOXINELLUS ZEREGI Heckel.

Barbodes Bleeker, 431; orthotype SYSTOMUS BELINKA McCl.

Anematichthys Bleeker, 431; orthotype BARBUSA APOGON Kuhl.

Siaja Bleeker, 431; tautotype CAPOËTA SIAJA Blkr.

Cyclocheilichthys Bleeker, 431; orthotype C. ARMATUS Blkr.

Pseudoculter Bleeker, 432; orthotype CULTER PEKINENSIS Basil.

Hemiculter Bleeker, 432; orthotype CULTER LEUCINOCULUS Basil.

Chanodichthys Bleeker, 432; orthotype LEPTOCEPHALUS MONGOLICUS Basilewsky.

Leptobarbus Bleeker, 432; orthotype BARBUS HOEVENI Blkr.

Trinematichthys Bleeker, 433; orthotype LEUCISCUS TRINEMA Blkr. A synonym of LUCIOSOMA Blkr.

Amblypharyngodon Bleeker, 433; orthotype; CYPRINUS MOLA Ham.

Hypophthalmichthys Bleeker 433; orthotype LEUCISCUS MOLITRIX Cuv. & Val. Replaces ABRAMOCEPHALUS Steind.

Devario Bleeker 433; tautotype PERILAMPUS DEVARIO Ham. A synonym of DANIO Ham.

Thynnichthys Bleeker, 433; orthotype LEUCISCUS THYNNOIDES Blkr.

Sarcocheilichthys Bleeker, 435; orthotype LEUCISCUS VARIEGATUS T. & S.

Pseudorasbora Bleeker, 435; orthotype LEUCISCUS PARVUS T. & S.

Rasbora Bleeker, 435; tautotype CYPRINUS RASBORA Ham.

Gnathopogon Bleeker, 435; orthotype CAPOËTA ELONGATA T. & S. Replaces SQUALIDUS Dybowski and LEUCOGOBIO Gthr.

Rasborichthys Bleeker, 435; orthotype LEUCISCUS HELFRICHI Blkr.

Elopichthys Bleeker, 436; orthotype LEUCISCUS BAMBUSA Rich.

Shacra Bleeker, 437; tautotype CYPRINUS SHACRA Ham.

Bendilisis Bleeker, 437; orthotype CYPRINUS BENDELISIS Ham. A synonym of BARILIUS Ham.

Iaus (Heckel) Bleeker 438; a misprint for IDUS.

Laubuca Bleeker, 438, tautotype CYPRINUS LAUBUCA Blkr.

Fundulichthys Bleeker, 439; orthotype FUNDULUS VIRESCENS T. & S.

Macrochirichthys Bleeker, 439; orthotype M. URANOSCOPUS Blkr.

Hemixiphophorus Bleeker, 440; (type not named). A synonym of GAMBUSIA Poey.

642. BLEEKER (1859). *(Palinurichthys.)* Enumeratio piscium archipelago Ind.

PIETER BLEEKER.

Palinurichthys Bleeker, 22 (Nov., 1859); orthotype CORYPHÆNA PERCIFORMIS Mitchill. A substitute for PALINURUS Dekay, preoccupied.

643. BLEEKER (1859). *(Pterichthyodes.)* Verh. Nat. Hist. Indie, VI.
PIETER BLEEKER.
Pterichthyodes Bleeker, xi; orthotype PTERICHTHYS MILLERI Ag. (fossil). Substitute for PTERICHTHYS Ag., preoccupied.

644. BLYTH (1859). *(Andamia.)* Journ. Asiat. Soc. Bengal, 1858.
EDWARD BLYTH.
Andamia Blyth, 270; orthotype A. EXPANSA Blyth.

645. FILIPPI & VERANY (1859). *Sopra alcuni pesci nuovi o poco noti del Mediterraneo.* Mem. Accad. Sci. Torino, 1859, 2. ser., XVIII.
FILIPPO DE FILIPPI; J. B. VERANY.
Navarchus Filippi & Verany, XVIII, 7; orthotype N. SULCATUS. Filippi & Verany. A synonym of PSENES Cuv. & Val.
Pteridium Filippi & Verany, XVIII, 11; orthotype OLIGOPUS ATER Risso. (Not PTERIDIUM of Scopoli). Replaced by VERATER Jordan, 1919.

646. GILL (1859). *Description of a New Generic Form of Gobiinæ from the Amazon River.* Ann. Lyc. Nat. Hist. N. Y., 1858-1862, VII, 45-48.
THEODORE NICHOLAS GILL.
Euctenogobius Gill, 45; orthotype E. BADIUS Gill.

647. GILL (1859). *Description of a New Genus of Salarianæ from the West Indies.* Proc. Acad. Nat. Sci. Phila, 1859 (1860).
THEODORE NICHOLAS GILL.
This and the papers which follow were printed in 1859 when probably extras were issued. The publication as a whole was in 1860.
Entomacrodus Gill, 168; orthotype E. NIGRICANS Gill.

648. GILL (1859). *Description of New South American Type of Siluroids, Allied to Callophysus.* Proc. Acad. Nat. Sci. Phila., 1859 (1860).
THEODORE NICHOLAS GILL.
Pimelotopis Gill, 196; orthotype P. LATERALIS Gill = CALLOPHYSUS MACROPTERUS. A synonym of CALLOPHYSUS M. & T .

649. GILL (1859). *Description of a Third Genus of Hemirhamphinæ.* Proc. Acad. Nat. Sci. Phila. (1859) 1860.
THEODORE NICHOLAS GILL.
Euleptorhamphus Gill, 156; orthotype E. BREVOORTI Gill.

650. GILL (1859). *Description of a Type of Gobioids Intermediate Between Gobinæ and Tridentigerinæ.* Proc. Acad. Nat. Sci. Phila., 1859 (1860).
THEODORE NICHOLAS GILL.
Evorthodus Gill, 195; orthotype E. BREVICEPS Gill.

651. GILL (1859). *Description of Hyporhamphus, a New Genus of Fishes Allied to Hemirhamphus Cuv.* Proc. Acad. Nat. Sci. Phila., 1859 (1860).
THEODORE NICHOLAS GILL.
Hyporhamphus Gill, 131; orthotype H. TRICUSPIDATUS Gill.

652. GILL (1859). *Description of New Generic Types of Cottoids, from the Collection of the North Pacific Exploring Expedition Under Com. John Rodgers.* Proc. Nat. Sci. Phila., 1859 (1860).
THEODORE NICHOLAS GILL.

Boreocottus Gill, 165; orthotype B. AXILLARIS Gill. A synonym of MYOXOCEPHALUS Steller.

Ceratocottus Gill, 165; orthotype COTTUS DICERAUS Pallas.

Porocottus Gill, 166; logotype P. QUADRIFILIS Gill.

653. GILL (1859). *On Dactyloscopus and Leptoscopus, Two New Genera of the Family of Uranoscopidæ.* Proc. Acad. Nat. Sci. Phila., 1859 (1860).
THEODORE NICHOLAS GILL.

Dactyloscopus Gill, 132; orthotype D. TRIDIGITATUS Gill.

Leptoscopus Gill, 133; orthotype URANOSCOPUS MACROPYGOS Rich.

654. GILL (1859). *On the Genus Callionymus of Authors.* Proc. Acad. Nat. Sci. Phila., 1859 (1860).
THEODORE NICHOLAS GILL.

Synchiropus Gill, 129; orthotype CALLIONYMUS LATERALIS Rich.

Dactylopus Gill, 130; orthotype CALLIONYMUS DACTYLOPUS Bennett = DACTYLOPUS BENNETTI Gill, not DACTYLOPUS, 1862, a genus of crustacea. Replaces VULSUS Gthr.

655. GILL (1859). *Notes on a Collection of Japanese Fishes, Made by Dr. J. Morrow.* Proc. Acad. Nat. Sci. Phila., 1859 (1860).
THEODORE NICHOLAS GILL.

Acanthogobius Gill, 145; orthotype GOBIUS FLAVIMANUS T. & S.

Rhinogobius Gill, 145; orthotype R. SIMILIS Gill.

Glossogobius Gill, 146; orthotype GOBIUS PLATYCEPHALUS Rich.

Luciogobius Gill, 146; orthotype L. GUTTATUS Gill.

656. GIRARD (1859). *Ichthyological Notices.* Proc. Acad. Nat. Sci. Phila., 1858 (1859).
CHARLES FRÉDÉRIC GIRARD.

Pterognathus Girard, 57; orthotype NEOCLINUS SATIRICUS Grd., hypothetical name, revived by Jordan & Evermann (1896).

Myrichthys Girard, 58; orthotype M. TIGRINUS Grd.

Typhlichthys Girard, 63; haplotype T. SUBTERRANEUS Grd.

Arlina Girard, 64; haplotype A. EFFULGENS Grd. Same as BOLEOSOMA Dekay, 1842.

Estrella Girard, 64; orthotype E. ATROMACULATA Grd. A synonym of BOLEOSOMA Dekay.

Oligocephalus Girard, 67; haplotype A. LEPIDUS Grd.

Alvordius Girard, 68; haplotype ALVORDIUS MACULATUS Grd. A section of HADROP-TERUS Ag.

Alvarius Girard, 101; haplotype A. LATERALIS Grd.

Boleichthys Girard, 103; logotype B. EXILIS Grd.

Adinia Girard, 117; haplotype A. MULTIFASCIATA Grd.

Lucania Girard, 118; haplotype LUCANIA VENUSTA Grd.

657. GIRARD (1859). *Fishes.* (In *General Report Upon Zoology of the Several Pacific Railroad Routes.* 1857.)
CHARLES GIRARD.

Nautichthys Girard, 74; orthotype BLEPSIAS OCULOFASCIATUS Grd.

Xiphidion Girard, 119; orthotype X. MUCOSUM Grd. Name preoccupied as XIPHI-DIUM, replaced by XIPHISTER J. & G., 1880.

Thaleichthys Girard, 325; orthotype P. STEVENSI Grd. = SALMO PACIFICUS Rich.

Ichthyomyzon Girard, 379; logotype PETROMYZON ARGENTEUS Kirtland = AMMO-CŒTES CONCOLOR Kirtland.

Scolecosoma Girard, 384; haplotype AMMOCŒTES CONCOLOR Kirtland. Larva of ICHTHYOMYZON Grd.

658. GIRARD (1859). *Notes Upon Various New Genera and New Species of Fishes, in the Museum of the Smithsonian Institution, and Collected in Connection with the United States and Mexican Boundary Survey, Major William Emory, Commissioner.* Proc. Acad. Nat. Sci. Phila., 1858 (1859), 167-171.

CHARLES GIRARD.

Neomænis Girard, 167; orthotype LOBOTES EMARGINATUS Baird & Grd.= LABRUS GRISEUS L.

Orthopristis Girard, 167; orthotype O. DUPLEX Grd. = PERCA CHRYSOPTERA L.

Carangus Girard, 168; tautotype CARAX CARANGUS Cuv. & Val. = SCOMBER HIPPOS L. Same as CARANGUS Griffith. A synonym of CARANX Lac., as restricted.

Doliodon Girard, 168; orthotype GASTEROSTEUS CAROLINUS L. A synonym of TRACHINOTUS Lac.

Chloroscombrus Girard, 168; logotype SERIOLA COSMOPOLITA Cuv. & Val. = SCOMBER CHRYSURUS L.

Gobionellus Girard, 168; logotype G. HASTATUS Grd.

Neoconger Girard, 171; orthotype N. MUCRONATUS Grd.

Neomuræna Girard, 171; orthotype N. NIGROMARGINATA Grd. A synonym of MURÆNA L.

658A. GRAY (1859). *Description of a New Genus of Lophobranchiate Fishes from Western Australia.* Proc. Zool. Soc. London.

JOHN EDWARD GRAY.

Haliichthys Gray, 38; orthotype H. TÆNIOPHORUS Gray.

659. GÜNTHER (1859). *Catalogue of the Acanthopterygian Fishes of the British Museum.*

ALBERT GÜNTHER (1830-1914).

Anoplogaster Günther, I, 12; orthotype HOPLOSTETHUS CORNUTUS Cuv. & Val.

Percalabrax (Schlegel) Günther, I, 71; orthotype LABRAX JAPONICUS J. & S.

Trachypoma Günther, I, 167; orthotype T. MACRACANTHUS Gthr.

Pogonoperca Günther, I, 169; orthotype P. OCELLATA Gthr.

Anyperodon Günther, I, 210; orthotype SERRANUS LEUCOGRAMMICUS Cuv. & Val.

Oligorus Günther, I, 251; orthotype GRYSTES MACQUARIENSIS Cuv. & Val.

Odontonectes Günther, I, 265; orthotype CÆSIO ERYTHROGASTER Kuhl & Van Hasselt.

Hyperoglyphe Günther, I, 337; orthotype DIAGRAMMA POROSA Rich.

Synagris (Klein) Günther, I, 373; logotype DENTEX FURCOSUS Cuv. & Val. A section under NEMIPTERUS Sw., not SYNAGRIS Klein as restricted.

Pachymetopon Günther, I, 424; orthotype P. GRANDIS Gthr.

Proteracanthus Günther, I, 426; orthotype CRENIDENS SARISSOPHORUS Cantor.

Tephræops Günther, I, 431; tautotype CRENIDENS TEPHRÆOPS Rich.

Gymnocrotaphus Günther, I, 432; orthotype G. CURVIDENS Gthr.

Lembus Günther, I, 505; orthotype L. MACULATUS Gthr. A synonym of GOBIOMORUS Lac.; PHILYPNUS Cuv.

660. GÜNTHER (1859). *On the Reptiles and Fishes Collected by the Rev. H. B. Tristram in Northern Africa.* Proc. Zool. Soc. London, 1859, 469-474.

ALBERT GÜNTHER.

Haligenes Günther, 471; orthotype H. TRISTRAMI Gthr. A synonym of TILAPIA Smith.

661. HYRTL (1859). *Anatomische Untersuchung des Clarotes (Gonocephalus) heuglini Kner, mit einer Abbildung und einer osteologischen Tabelle der Siluroiden.* Denkschr. Akad. Wiss. Wien, 1859, XVI.

CARL JOSEPH HYRTL.

Notophthalmus Hyrtl, XVI, 17; orthotype HYPOPHTHALMUS MARGINATUS Cuv. & Val. Name preoccupied in Salamanders. A synonym of HYPOPHTHALMUS Spix.

662. KAUP (1859). *Description of a New Species of Fish, Peristethus rieffeli.* Proc. Zool. Soc. London, 1859, pt. 27.

JOHANN JACOB KAUP.

Peristethus Kaup, 103; orthotype P. CATAPHRACTUS. An attempt at amendment of PERISTEDION Lac.

663. KNER (1859). *Zur Familie der Characinen.* III. *Folge der Ichthyologischen Beiträge.* Denkschr. Akad. Wiss. Wien, 1859, XVII.

RUDOLPH KNER (1819-1869).

Microdus Kner, XVIII, 151; orthotype CHILODUS PUNCTATUS M. & T. Preoccupied, replaced by CÆNOTROPUS Gthr., 1864.

Rhytiodus Kner, XVII, 165; logotype R. MICROLEPIS Kner.

Centrophorus Kner, XVII, 167; orthotype PAREIODON MICROPS Kner. Name preoccupied in sharks. Intended as a substitute for PAREIODON, on account of an earlier genus, PARODON, of different etymology.

Bryconops Kner XVII, 179; logotype B. ALBURNOIDES Kner.

664. LEIDY (1859). *Some Remains of Cartilaginous Fishes Discovered by Dr. F. V. Hayden and F. B. Meek in the Carboniferous Formations of Kansas.* Proc. Acad. Nat. Sci. Phila., 1859 (1860), 3.

JOSEPH LEIDY

Xystracanthus Leidy, 3; orthotype X. ARCUATUS Leidy (fossil).

665. PICTET (1859). *Description des fossiles du terrain Neocomien des Voirons.* Genève, 1859, 1-54.

FRANÇOIS JULES PICTET; PERCIVAL DE LORIOL.

Spathodactylus Pictet & Loriol, III, 2; orthotype S. NEOCOMIENSIS Pictet (fossil).

Crossognathus Pictet & Loriol, III, 27; orthotype C. SABAUDIANUS Pictet (fossil).

666. RATH (1859). *Beitrag zur Kenntniss der fossilen Fische des Plattenberges im Canton Glarus.* Zeitschr. Deutsch. Geol. Ges., 1859, II, 108-132.

J. J. GERHARD VOM RATH (1830-1888).

Archæoides Rath, 112; orthotype A. LONGICOSTATUS .Rath (fossil). A synonym of ARCHÆUS.

Thyrsitocephalus Rath, XI, 114; orthotype T. ALPINUS Rath (fossil).

Palæogadus Rath, XI, 126, orthotype P. TROSCHELI Rath (fossil). A synonym of of NEMOPTERYX.

Palæobrosmius Rath, XI, 130; orthotype P. ELONGATUS Rath (fossil). A synonym of NEMOPTERYX Ag.

667. STEINDACHNER (1859). *Beiträge zur Kenntniss der fossilen Fischfauna Oesterreichs.* Sitzb. Akad. Wiss. Wien, math.-nat. Cl., 1859, XXXVII, 673-701.

FRANZ STEINDACHNER (1834-   ).

This is the first of a very long series of papers mostly dealing with individual species and with unusual accuracy. All these papers are illustrated by excellent figures.

Scorpænopterus Steindachner, XXXVII, 694; orthotype S. SILURIDENS Steind. (fossil).

Aipichthys Steindachner, XXVIII, 763; orthotype A. PRETIOSUS Steind. (fossil).

1860

668. AYRES (1860). *On New Fishes of the California Coast.* Proc. Cal. Acad. Sci., II, 1861.

WILLIAM O. AYRES.

Halias Ayres, II, 52; orthotype BROSMIUS MARGINATUS Ayres. Name preoccupied in butterflies; replaced by BROSMOPHYCIS Gill.

Seriphus Ayres, II, 80; orthotype S. POLITUS Ayres.

Camarina Ayres, 81; orthotype C. NIGRICANS Ayres. A synonym of GIRELLA Gray.

669. BLEEKER (1860). *Over eenige vischsoorten van de Kaap de Goede Hoop.* Nat. Tijdschr. Neder.-Indie, XXI, 1860, 49-80.

PIETER BLEEKER.

Pagrichthys Bleeker, XXI, 60; orthotype P. CASTELNAUI Blkr.

670. BLEEKER (1860). *Zesde bijdrage tot de kennis der vischfauna van Japan.* Act. Soc. Sci. Indo-Neerl., VIII.

PIETER BLEEKER.

Pseudobagrus Bleeker, 87; orthotype BAGRUS AURANTIACUS Schlegel.

Salangichthys Bleeker, 101; orthotype SALANX MICRODON Blkr.

671. BLEEKER (1860). *Achtste bijdrage tot de kennis der vischfauna van Sumatra. Visschen van Benkoelen, Priaman, Tandjong, Palembang en Djambi.* Act. Soc. Sci. Indo-Neerl., VIII, 1860, 1-88.

PIETER BLEEKER.

Cossyphodes Bleeker, 37; orthotype LABRUS MACRODON Lac. A synonym of CHŒRODON Blkr.

Lais Bleeker, 46; orthotype L. HEXANEMA Blkr. Name preoccupied by LAIS Gistel, 1848, VIII, a substitute for CYNTHIA, preoccupied as a genus of TUNICATES; replaceable by LAIDES Jordan, new name.

Silurodes Bleeker, 47; orthotype S. HYPOPHTHALMUS Blkr.

Bagrichthys Bleeker, 54; orthotype BAGRUS HYPSELOPTERUS Blkr.

672. BLEEKER (1860). *Negende bijdrage tot de kennis der vischfauna van Sumatra. Visschen uit de Lematang-Enim en van Benkoelen.* Act. Soc. Sci. Indo-Neerl., VIII, 1860, 1-12.

PIETER BLEEKER.

Akysis Bleeker, 234; orthotype PIMELODUS VARIEGATUS Blkr.

673. BLEEKER (1860). *Dertiende bijdrage tot de kennis der vischfauna van Borneo.* Act. Soc. Sci. Indo-Neerl., VIII, 1860, 1-64.

PIETER BLEEKER.

Trypauchenichthys Bleeker, XIII, 63; orthotype T. TYPUS Blkr.

674. BLEEKER (1860). *Elfde bijdrage tot de kennis der vischfauna van Amboina.* Act. Soc. Sci. Indo-Neerl., VIII, 1860, 1-14.

PIETER BLEEKER.

Scopelosaurus Bleeker, XI, 12; orthotype H. HOEDTI Blkr.

675. BLEEKER (1860). *Dertiende bijdrage tot de kennis der vischfauna van Celebes. Visschen van Bonthain, Badjoa, Sindjai, Lagoesi en Pompenoea.* Act. Soc. Sci. Indo-Neerl., VIII, 1860, 1-60.

PIETER BLEEKER.

Soleichthys Bleeker, 14; orthotype S. HETERORHINOS Blkr.

Macolor Bleeker, 25; orthotype DIACOPE MACOLOR Cuv. & Val. = MACOLOR TYPUS Blkr; replaces PROAMBLYS Gill.

Oxyurichthys Bleeker, 44; orthotype GOBIUS BELOSSO Blkr.

676. BLEEKER (1860). *Ordo Cyprini: Karpers.* Nat. Tijdschr. Neder.-Indie, VII, 1-492.

PIETER BLEEKER.

Girardinichthys Bleeker, 481; orthotype G. INNOMINATUS Blkr. Replaces LIMNUR-GUS Gthr.

677. BLYTH (1860). *On Some Fishes from the Sitang and Tributary Streams.* Journ. Asiat. Soc. Bengal, XXIX, 138-174.

EDWARD BLYTH.

Bogota Blyth, 139; orthotype B. INFUSCATA Blyth. Apparently the young of PRIACANTHUS Cuv., not BOGODA Blkr.

Sciænoides Blyth, 139; logotype S. BIAURITUS Blyth (Jan., 1860). Same as COLLICH-THYS Gthr., June, 1860, but preoccupied by the tentative name SCIÆNOIDES Rich., 1843.

Microzeus Blyth, XIX, 142; orthotype M. ARMATUS Blyth.

Batasio Blyth, XXIX, 148; tautotype PIMELODUS BATASIO Ham.; replaces GAGATA Blkr.

Hara Blyth, 152; tautotype PIMELODUS HARA Ham. (H. BUCHANANI Blyth). A synonym of ERETHISTES M. & T.

Pseudecheneis Blyth, XXIX, 154; orthotype GLYPTOSTERNON SULCATUS McCl.

Glyptothorax Blyth, XXIX, 154; orthotype G. TRILINEATUS Blyth. A synonym of GLYPTOSTERNON McCl.

Exostoma Blyth, XXIX, 155; orthotype E. BERMOREI Blyth.

Syncrossus Blyth, 166; logotype S. BERDMOREI Blyth. A synonym of BOTIA Gray.

Prostheacanthus Blyth, 167; orthotype P. SPECTABILIS Blyth. A synonym of ACANTHOPSIS Van Hasselt.

Pangio Blyth, 169; orthotype COBITIS PANGIA Ham. A synonym of ACANTHOPH-THALMUS Van Hasselt, as restricted by Bleeker.

Apua Blyth, 169; orthotype A. FUSCA Blyth.

678. CANESTRINI (1860). *Zur Systematik der Percoiden.* Verh. Zool. Bot. Ver. Wien, X.

GIOVANNI CANESTRINI (1835-   ).

Datnioides (Brisout de Barneville) Canestrini, 305; orthotype not named; teeth on vomer, none on pelatines; not DATINIODES Blkr., 1853. Apparently a synonym of THERAPON Cuv.

Apostata (Heckel) Canestrini, 309; orthotype A. CALCARIFER Heckel = CONODON NOBILIS (L.). A synonym of CONODON Cuv. & Val., 1830.

Asproperca (Heckel) Canestrini, 309; orthotype A. ZEBRA Heckel. A synonym of PERCINA Haldeman.

679, 680. CANESTRINI (1860). *(Fishes.)* Verh. Akad. Wiss. Wien.
GIOVANNI CANESTRINI.

Sphærichthys Canestrini, X, 707; orthotype S. OSPHROMENOIDES Canestrini.

Trichopsis (Kner) Canestrini, X, 708; orthotype T. STRIATUS Blkr. A synonym of CTENOPS McCl.

681. COQUAND (1860). *Description de la géologie du département de Charente.* II.
H. COQUAND.

Orthodon Coquand, II, 126; orthotype O. CONDAMYI (fossil). Name preoccupied in CYPRINIDÆ; an ally of THYELLINA.

682. EGERTON (1860). *(Fossil Fishes.)* Rept. British Association for 1859 (1860).
PHILIP DE MALPAS GREY EGERTON.

Brachyacanthus Egerton, 116; orthotype B. SCUTIGER Egert. (fossil). A synonym of CLIMATIUS Ag.

683. EICHWALD (1860). *Lethæa Rossica ou paléontologie de la Russie décrite et figurée.* Stuttgart. *Pisces.* Vol. I.
KARL EDUARD VON EICHWALD.

Aulacosteus Eichwald, I, 1548; logotype A. COCHLEARIFORMIS Eichw. (fossil); repeated from a Russian paper of 1846. A synonym of PTYCTODUS Pander (1858) and originally of prior date.

Homacanthus Eichwald, I, 1601; orthotype H. TRIANGULARIS Eichw. (fossil). Regarded as a synonym of CALODODUS Ag.; name preoccupied by Agassiz.

684. GILL (1860). *Notes on the Nomenclature of North American Fishes.* Proc. Acad. Nat. Sci. Phila., Vol. XI, for 1860 (printed as a whole in 1861).
THEODORE GILL.

Astroscopus (Brevoort) Gill, 20; orthotype URANOSCOPUS ANOPLOS Cuv. & Val. (young of U. Y-GRÆCUM Cuv. & Val).

Palinurichthys Gill, 20 (Jan., 1860); orthotype CORYPHÆNA PERCIFORMIS Mitchill. Substitute for PALINURUS, preoccupied. A synonym of PALINURICHTHYS Blkr, 1859.

Leptoblennius Gill, 21; orthotype BLENNIUS SERPENTINUS Storer.

685. GILL (1860). *Conspectus piscium in expeditione ad oceanum pacificum Septentrionalem. . . . Stimpson collectorum; Sicydianæ.* Proc. Acad. Nat. Sci. Phila., 1860, 100-102.
THEODORE GILL.

Sicyopterus Gill, 101; orthotype S. STIMPSONI Gill.

Sicyogaster Gill, 102; orthotype S. CONCOLOR Gill. Preoccupied, replaced by LENTIPES Gthr.

686. GILL (1860). *Monograph of the Genus Labrosomus Sw.* Proc. Nat. Sci. Phila., 1860, 102-108.
THEODORE GILL.

Malacoctenus Gill, 103; orthotype CLINUS DELALANDI Cuv. & Val.

Gobioclinus Gill, 103; orthotype CLINUS GOBIO Cuv. & Val.

Blennioclinus Gill, 103; orthotype CLINUS BRACHYCEPHALUS Cuv. & Val.

Auchenionchus Gill, 103; orthotype CLINUS VARIOLOSUS Cuv. & Val. Misprinted ANCHENIONCHUS.

Calliclinus Gill, 103; orthotype CLINUS GENIGUTTATUS.

Ophioblennius Gill, 103; orthotype BLENNOPHIS WEBBI Val. Substitute for BLEN-NOPHIS Val., preoccupied.

Ophthalmolophus Gill, 104; orthotype CLINUS LATIPINNIS Cuv. & Val.

687. GILL (1860). *Monograph of the Genus Labrax of Cuvier.* Proc. Acad. Nat. Sci. Phila., 1860, 108-120.

THEODORE GILL.

Dicentrarchus Gill, 109; orthotype PERCA ELONGATA St. Hilaire = PERCA LABRAX L. Same as LABRAX (Klein) Cuv., 1817, not LABRAX Pallas, 1811.

688. GUICHENOT (1860). *Notice sur un nouveau poisson du genre des Tricho-myctères.* Rev. Mag. Zool., 2d sér., XII.

ALPHONSE GUICHENOT.

Astemomycterus Guichenot, XII, 525; orthotype A. PUSILLUS Guich.

689. GUICHENOT (1860). *Notice sur un nouveau poisson du groupe Ctenolabres (Labrastrum).* Rev. Mag. Zool., 2d sér., XII.

ALPHONSE GUICHENOT.

Labrastrum Guichenot, XII, 152; orthotype (CTENOLABRUS FLAGELLIFER Cuv. & Val. A synonym of DUYMÆRIA Blkr.

690. GÜNTHER (1860). *Catalogue of the Fishes of the British Museum.* II.

ALBERT GÜNTHER.

Atypus Günther, II, 64; orthotype A. STRIGATUS Gthr. Name preoccupied in spiders; replaced by ATYPICHTHYS Gthr., 1862.

Agrammus Günther, II, 94; orthotype LABRAX AGRAMMUS T. & S.

Glyptauchen Günther, II, 121; orthotype APISTUS PANDURATUS Rich.

Centropogon Günther, II, 128; logotype COTTUS AUSTRALIS White.

Pentaroge Günther, II, 132; orthotype APISTUS MARMORATUS Cuv. & Val. A synonym of GYMNAPISTES Sw.

Tetraroge Günther, II, 132; logotype APISTUS BARBATUS Cuv. & Val.

Polycaulus Günther, II, 175; orthotype SYNANCEIA ELONGATA Cuv. & Val.

Ptyonotus Günther, II, 175; orthotype TRIGLOPSIS THOMPSONI Grd. Needlessly proposed as a substitute for TRIGLOPSIS on account of the earlier TRIGLOPS.

Lepidotrigla Günther, II, 196; logotype TRIGLA ASPERA Cuv. & Val. = TRIGLA CAVILLONE Lac.

Agnus Günther, II, 229; orthotype URANOSCOPUS ANOPLOS Cuv. & Val. A synonym of ASTROSCOPUS Gill.

Anema Günther, 229; logotype as restricted by Gill, 1861, URANOSCOPUS ANOPLUS Cuv. & Val. A synonym of ASTROSCOPUS Brevoort.

Kathetostoma Günther, 231; orthotype URANOSCOPUS LÆVIS Bl. & Schn. Some-times spelled CATHETOSTOMA.

Epicopus Günther, 248; orthotype MERLUS GAYI Guich. A synonym of MERLUCCIUS Raf.

Platystethus Günther, II, 301; orthotype SCIÆNA CULTRATA Forster; preoccupied; replaced by BATHYSTETHUS Gill, 1893.

Collichthys Günther, II, 312; orthotype SCIÆNA LUCIDA Rich. Same as SCIÆNOIDES Blyth, not of Richardson.

Pentanemus (Artedi) Günther, 330; orthotype POLYNEMUS QUINQUARIUS L.

Galeoides Günther, II, 332; orthotype POLYNEMUS POLYDACTYLUS Vahl.

Dicrotus Günther, II, 349; orthotype D. ARMATUS Gthr.

Hypsiptera Günther, II, 386; orthotype H. ARGENTEA Gthr. A larva related to GAIDROPSARUS Raf.

Neptomenus Günther, II, 389; orthotype N. BRAMA Gthr.

Pammelas Günther, II, 485 (June); orthotype CORYPHÆNA PERCIFORMIS Mitchill. Substitute for PALINURUS Dekay, preoccupied. A synonym of PALINURICH-THYS Blkr., 1859.

691. HOLBROOK (1860). *Ichthyology of South Carolina.* (Second edition; the first, in 1855, destroyed by fire.)

JOHN EDWARDS HOLBROOK.

(See No. 536, 1855.)

692. KAUP (1860). *Neue aalähnliche Fische des Hamburger Museums.* Abh. Naturw. Verein Hamburg, 1859 (1860), IV, Abth. 2.

JOHANN JACOB KAUP.

Gnathophis Kaup, 7; orthotype G. HETEROGNATHUS Kaup. (An ally of CONGERMU-RÆNA HABENATA). Corresponds to CONGERMURÆNA as restricted by Ogilby, not of Bleeker.

Tæniophis Kaup, 10; orthotype T. WESTPHALI Kaup. A synonym of GYMNO-THORAX.

Cryptopterus Kaup, 11; orthotype C. PUNCTICEPS Kaup. Probably a synonym of OPHICHTHYS. Name preoccupied.

Crotalopis Kaup, 12; orthotype C. PUNCTIFER Kaup.

Echiopsis Kaup, 13; orthotype OPHISURUS INTERTINCTUS Rich. A synonym of CROTALOPSIS Kaup.

Hoplunnis Kaup, 19; orthotype H. SCHMIDTI Kaup.

Priodonophis Kaup, 22; orthotype GYMNOTHORAX OCELLATUS Ag.

693. KAUP (1860). *Hoplarchus, neues Genus der Familie Labridæ.* Arch. Naturg., 1860, XXVI, pt. I.

JOHANN JACOB KAUP.

Hoplarchus Kaup, 128; orthotype H. PENTACANTHUS Kaup. A synonym of CICH-LAURUS Sw. = (HEROS Heckel).

694. KAUP (1860). *On Some New Genera and Species of Fishes Collected by Doctors Kerferstein and Heckel at Messina.* Ann. Mag. Nat. Hist., VI, 270-273.

JOHANN JACOB KAUP.

Stomiasunculus Kaup, V, 270; orthotype S. BARBATUS Kaup. A larva, allied to STOMIAS.

Porobronchus Kaup, V, 272; orthotype P. LINEARIS Kaup. A larva of CARAPUS (FIERASFER).

695. KAUP (1860). *Ueber die Chætodontidæ.* Wiegm. Archiv. Naturg., XXVI, pt. I, 133-156.

JOHANN JACOB KAUP.

Citharœdus Kaup, 136; logotype CHÆTODON ORNATISSIMUS Solander.

Coradion Kaup, 137; orthotype CHÆTODON CHRYSOZONUS Kuhl & Van Hasselt.

Eteira Kaup, XXVI, 137; orthotype CHÆTODON TRIANGULARIS Rüppell. A synonym of MEGAPROTODON Guich.

Linophora Kaup, 137; orthotype CHÆTODON AURIGA Forskål.

Centropyge Kaup, XXVI, 140; orthotype C. TIBICEN Kaup. A synonym of HOLA-CANTHUS Lac.

Therapaina Kaup, 140; orthotype CHÆTODON STRIGATUS (Langsdorff) Cuv. & Val. A synonym of MICRACANTHUS Sw.

696. KNER (1860). *Ueber Belonesox belizanus,* etc. Sitzb. Akad. Wiss. Wien, XL.
RUDOLPH KNER.

Belonesox Kner, XL, 419; orthotype B. BELIZANUS Kner.

697. KNER (1860). *Ueber einige noch unbeschriebene Fische.* Sitzb. Akad. Wiss. Wien, XXXIX.
RUDOLF KNER.

Centropus Kner, XXXIX, 531; orthotype C. STAUROPHORUS Kner. Name pre-occupied. A synonym of AMPHIPRIONICHTHYS Blkr.

Trichopsis attributed to Kner. A synonym of OSPHRONEMUS Lac. Seems to be a slip of the pen for TRICHOPUS.

698. MARCK (1860). (*Fossile Fische.*) Verh. Nat. Ver. Preuss-Rhein. Westph.
W. VON DER MARCK.

Palæoscyllium von der Marck, 47; orthotype P. DECHENI Marck (fossil). Name only, described in *Palæontographica*, XI, 67, 1863. Name preoccupied. An ally of THYELLINA.

699. MOLIN (1860). *Primitivæ Musei Archigymnasii Patavini (Pisces fossiles).* Sitzb. Akad. Wiss. Wien, XL.
RAFFAELE MOLIN.

Protogaleus Molin, XL, 585; orthotype P. CUVIERI Molin (fossil). An ally or synonym of SCOLIODON M. & H.

700. PANDER (1860). *Ueber die Saurodipterinen, Dendrodipterinen, Glyptolepi-den und Cheirolepiden des devonischen Systems.* St. Petersb., IX.
CHRISTIAN HEINRICH PANDER.

Polyplacodus Pander, 28; orthotype P. INCURVUS Pander (not of Duff) = P. WENJUKOWI Rohon (fossil). A synonym of CRICODUS Ag.

701. PLIENINGER (1860). (*Xystrodus.*) Neues Jahrbuch.
WILHELM HEINRICH THEODOR VON PLIENINGER.

Xystrodus Plieninger, 695; orthotype X. FINITIMUS Plieninger.

702. POEY (1860). *Peces Ciegos de la Isla de la Cuba.* Memorias de Cuba, II. July, 1860.
FELIPE POEY Y ALOY.

Lucifuga Poey, II, 95; orthotype L. SUBTERRANEUS Poey.

703. POEY (1860). *Poissons de Cuba: Espèces nouvelles.* In *Memorias sobre la historia natural de la Isla de Cuba,* etc. II, 115–356.
FELIPE POEY Y ALOY.

Dinemus Poey, II, 107, 161; orthotype D. VENUSTUS Poey. A synonym of POLY-MIXIA, Lowe.

Monoprion Poey, II, 123; orthotype M. MACULATUS Poey. A synonym of APOGON Lac., AMIA Gronow.

Verilus Poey, II, 124; orthotype V. SORDIUS Poey.

Latebrus Poey, II, 168; orthotype L. OCULATUS Poey.

Inermia Poey, II, 193; orthotype I. VITTATA Poey. An ally of EMMELICHTHYS Rich.

Furcaria Poey, II, 194; orthotype F. PUNCTA Poey. An ally or subgenus of CHROMIS Cuv.

Sphyrænops Poey, II, 249; orthotype S. BAIRDIANUS Poey.

Joturus Poey, 263; orthotype J. PICHARDI Poey.

Erotelis Poey, II, 273; orthotype E. VALENCIENNESI Poey = ELEOTRIS SMARAGDUS Cuv. & Val.

Chonophorus Poey, II, 274; orthotype C.BUCCULENTUS Poey = GOBIUS BANANA Cuv. & Val., probably = GOBIUS TAIASICA Lichtenstein. Replaces AWAOUS Steind. of the same year but a few days later.

Smaragdus Poey, II, 279; orthotype GOBIUS SMARAGDUS Cuv. & Val. = SMARAGDUS VALENCIENNESI Poey. A synonym of GOBIONELLUS Grd.

Trifarcius Poey, II, 305; orthotype T. RIVERENDI Poey. A synonym of CYPRINODON Lac.

Rivulus Poey, II, 307; orthotype R. CYLINDRACEUS Poey.

Sardinia Poey, II, 311; orthotype S. PSEUDOHISPANICA Poey (not of Jordan & Evermann, which is SARDINELLA ANCHOVIA Cuv. & Val.). Same as SARDINA Antipa, which it replaces unless it be regarded as preoccupied by SARDINIUS von der Marck.

Pontinus Poey, II, 172; logotype P. CASTOR Poey.

704. SUCKLEY (1860). *Report of the Fishes Collected on the Survey (U. S. Pacific R. R.).* Report on the SALMONIDÆ. (Also in several other papers.)
GEORGE SUCKLEY (1830-   ).

Oncorhynchus Suckley, 312; orthotype (male of) SALMO SCOULERI Rich. = SALMO GORBUSCHA Walbaum. (The males of the Pacific salmon taken as a genus distinct from the females.)

705. STEINDACHNER (1860). *Beitrag zur Kenntniss der Gobioiden.* Sitzb. Math. Naturw. Classe Wien., XLII.
FRANZ STEINDACHNER

Cyclogobius Steindachner, XLII, 284; orthotype GOBIUS LEPIDUS Grd. A synonym of LEPIDOGOBIUS Gill.

Awaous Steindachner, XLII, 289; type (after "les AWAOUS" Cuv. & Val.; GOBIUS OCELLARIS Cuv. & Val. from Hawaii) "GOBIUS (AWAOUS Val.) LITERATUS Heckel," from the Philippines, is the only species named except for the reference to Valenciennes. The name is a latinization of "les AWAOUS". AWAOUS is a synonym of CHONOPHORUS Poey. As AWAOUS was first used as a scientific name (as a subgenus of GOBIUS) in an article presented on July 12, 1860, it is probable that CHONOPHORUS Poey, issued in July, 1860, has actual priority.

Oplopomus (Ehrenberg) Steindachner, XLII, 290; orthotype OPLOPOMUS PULCHER Ehrenb. = GOBIUS OPLOPOMUS Cuv. & Val. (XII, 66). Not OPLOPOMA Grd., 1856. Same as CENTROGOBIUS Blkr. (1874).

Gobiopsis Steindachner, XLII, 291; orthotype GOBIUS MACROSTOMUS Heckel. Very close to WAITEA Jordan, 1906.

706. TROSCHEL (1860). *Leptopterygius, neue Gattung der Discoboli.* Arch. Naturg. XXVI, 1.
FRANZ HERMANN TROSCHEL.

Leptopterygius Troschel, XXVI, 205; orthotype L. coccoi Troschel.   A synonym of GOUANIA Nardo = RUPISUGA Sw.

707. WAGNER (1860). *Fossile Fische aus einem neuentdeckten Lager in Süd-bayerischem Tertiärgebirge.*. Südbayer. Akad., I, 1860.
ANDREAS WAGNER.

Alosina Wagner, 54; orthotype A. SALMONEA Wagner (fossil).

708. WAGNER (1860). *Die Griffelzähner (Stylodontes) eine neue aufgestellte Familie . . . der rautenschuppige Ganoiden.* Gelehrte Anz. Bayer. Akad. Wiss., L.

ANDREAS WAGNER.

Homœolepis Wagner, 92; orthotype H. DROSERA Wagner (fossil). A synonym of TETRAGONOLEPIS Bronn.

Heterostichus Wagner, 93; orthotype HETEROSTICHUS LATUS Wagner (fossil). Name preoccupied; replaced by HETEROSTROPHUS Wagner, 1863.

709. WAGNER (1860). *Zur Charakteristik der Gattungen Sauropsis und Pachy-cormus, etc.* Gelehrte Anz. Bayer. Akad. Wiss., L.
ANDREAS WAGNER.

Hypsocormus Wagner, 214; orthotype H. INSIGNIS Wagner (fossil).

Euthyonotus Wagner, 214; orthotype E MICROPODIUS Ag. (fossil) (Esox INCOG-NITUS Blainv.).

1861

710. AGASSIZ, A. (1861). *Notes on the Described Species of Holconoti Found on the Western Coast of North America.* Proc. Boston Soc. Nat. Hist., 1861, VIII, 122-134.
ALEXANDER AGASSIZ (1835-1910).

Sargosomus (Agassiz) Alex. Agassiz, 128; orthotype S. FLUVIATILIS Agassiz = HYSTEROCARPUS TRASKI Gibbons. A synonym of HYSTEROCARPUS Gibbons.

Metrogaster (Agassiz) Alex. Agassiz, VIII, 128; orthotype CYMATOGASTER AGGREGA-SUS Gibbons. A synonym of CYMATOGASTER Gibbons.

Bramopsis (Agassiz) Alex. Agassiz, VIII, 132; orthotype B. MENTO Agassiz = HYPERPROSOPON ARGENTEUS Gibbons. A synonym of HYPERPROSOPON.

Tæniotoca Alexander Agassiz, VIII, 133; orthotype EMBIOTOCA LATERALIS Agassiz.

Hypsurus Alexander Agassiz, VIII, 133; orthotype EMBIOTOCA CARYI Agassiz.

711. BLEEKER (1861). *Iets over de geslachten der Scaroiden en hunne Indisch-archipelagische soorten.* Versl. Akad. Amsterdam, XII, 1861, 228-244.
PIETER BLEEKER.

Pseudodax Bleeker, 229; orthotype ODAX MOLUCCANUS Cuv. & Val.

Callyodontichthys Bleeker, 230; orthotype SCARUS FLAVESCENS Blkr. (not of Bloch & Schneider).

Pseudoscarus Bleeker, 231; logotype P. MICRORHINOS Blkr.

712. BLEEKER (1861). *(Labridæ.)* Proc. Zool. Soc. London, 1861.
PIETER BLEEKER.

Ophthalmolepis Bleeker, 413; orthotype JULIS LINEOLATUS Cuv. & Val.

713. BLEEKER (1861). *Zesde bijdrage tot de kennis der vischfauna van Timor.* Nat. Tijdschr. Neder.-Indie, XXII, 1861, 247-261.
PIETER BLEEKER.

Oxymetopon Bleeker, VI, 12; orthotype O. TYPUS Blkr.

714. BLEEKER (1861). *Notice sur le genre Trachinus et ses espèces.* Ann. Sci. Nat., XVI; reprinted in 1862.

PIETER BLEEKER.

Echiichthys Bleeker, 379; orthotype E. VIPERA Blkr.

Pseudotrachinus Bleeker, 379; orthotype P. RADIATUS Blkr.

715. CASTELNAU (1861) *Mémoire sur les poissons de l'Afrique Australe.* Paris, VII.

FRANÇOIS LAPORTE DE CASTELNAU.

Boopsidea Castelnau, 25; orthotype. B. INORNATA Castelnau.

Ichthyorhamphos Castelnau, 35; orthotype I. PAPPEI Castelnau. A synonym of OPLEGNATHUS.

Sandelia Castelnau, 36; orthotype S. BAINSI Castelnau. A synonym of SPIRO-BRANCHUS.

Stromatoidea Castelnau 44; orthotype S. LAYARDI Castelnau, not STROMATEOIDES Blkr., 1851.

Gnathendalia Castelnau, 57; orthotype G. VULNERATA Castelnau.

Hydrocyonoides Castelnau, 66; orthotype H. CUVIERI Castelnau; a synonym of SARCODACES.

Algoa Castelnau, 69; orthotype A. VIRIDIS Castelnau.

Athæna Castelnau, 72; orthotype A. FASCIATA Castelnau.

Aledon Castelnau, 75; orthotype A. STORERI Castelnau. A synonym of Mola.

716. COSTA (1861). *(Fossil Fishes.): Palæontologia Regno Napoli.* III.

ORONZIO GABRIELE COSTA.

Centropterus Costa, III, 123; orthotype C. LIVIDUS Costa (fossil). An ally of ETMOPTERUS Raf.

717. EGERTON (1861). *Description of Tristichopterus, Acanthodes, Climatius,* etc. (In *Figures and Descriptions of British Organic Remains.*) Mem. Geol. Surv., 1861.

PHILIE DE MALPAS GREY EGERTON.

Holophagus Egerton, X, 19; orthotype H. GULA Egert. (fossil). A synonym of UNDINA Münster.

Tristichopterus Egerton, 51; orthotype T. ALATUS Egert. (fossil).

718. FILIPPI (1861). *Note Zoologische.* Arch. Zool. Anat. Fisiol., I.

FILIPPO DE FILIPPI.

Lebistes Filippi, I, 69; orthotype L. PŒCILOIDES Filippi.

719. GILL (1861). *Catalogue of the Fishes of the Eastern Coast of North America, from Greenland to Georgia.* Issued as a supplement to the *Proc. Acad. Nat. Sci. Phila.,* 1861, and dated February, 1861.

THEODORE GILL.

In this paper numerous new generic names are introduced in connection with the type species, the genera being defined later. Under present usage, these names date from the indication of the type species. No question of priority is here concerned except with REINHARDTIUS Gill, called PLATYSOMATICHTHYS by Bleeker in 1862, before REINHARDTIUS was defined, and HYPSOBLENNIUS Gill, called ISESTHES by Jordan & Gilbert, 1881, before a definition of HYPSO-BLENNIUS was published.

Note. The most of the many new genera, indicated by American authors from 1860 to 1875, occur in the proceedings of scientific societies in Philadelphia, New York and Boston, and especially in those of the Academy of Natural Sciences in Philadelphia. The bound volumes of these proceedings were issued in the year subsequent to the reading of the papers, as for example that of 1861 in 1862. But separate copies of most of these papers were issued in the date year. In this record, the papers are credited to the date year. In the few cases in which questions of priority are involved, I have tried to secure the exact dates concerned.

Uranoblepus 5; orthotype (not named) SYNANCEIA ELONGATA Cuv. & Val. A substitute for TRACHICEPHALUS Sw., preoccupied. A synonym of POLYCAULUS Gthr., 1860.

Ctenolucius Gill, 8; orthotype not named = LUCIOCHARAX INSCULPTUS Steind. It is described as an ally of XIPHOSTOMA, a South American CHARACIN, identical with LUCIOCHARAX Steind., 1876, which name it should replace.

Triloburus Gill, 30; orthotype PERCA TRIFURCA L. = P. PHILADELPHICA L. A section of CENTROPRISTES Cuv.

Promicropterus Gill, 31; orthotype RYPTICUS MACULATUS Holbrook = BODIANUS BISTRISPINUS Mitchill.

Anisotremus Gill, 32; orthotype PRISTIPOMA RODO Cuv. & Val. = SPARUS VIRGINICUS L.

Bairdiella Gill, 33; orthotype BODIANUS ARGYROLEUCUS Mitchill = DIPTERODON CHRYSURUS Lac.

Palinurichthys Gill, 34; orthotype PALINURUS PERCIFORMIS Storer. Substitute for PALINURUS, preoccupied; a synonym of PALINURICHTHYS Blkr., 1860.

Poronotus Gill, 35; orthotype RHOMBUS TRIACANTHUS Storer.

Lepturus (Artedi) Gill, 35; orthotype TRICHIURUS LEPTURUS L. A revival of a pre-Linnæan name. A synonym of TRICHIURUS L.

Blepharicthys Gill, 36; orthotype ZEUS CRINITUS Mitchill. A substitute for BLEPHARIS Cuv., preoccupied. A synonym of ALECTIS Raf.

Solenostomus (Gronow) Gill, 38; orthotype FISTULARIA TABACARIA L. A synonym of FISTULARIA L.

Pygosteus (Brevoort) Gill, 39; orthotype GASTEROSTEUS OCCIDENTALS Cuv. & Val. Val. = G. PUNGITIUS L.

Trichidion (Klein) Gill, 40; logotype POLYNEMUS VIRGINICUS L. A revival of a pre-Linnæan name. A synonym of POLYNEMUS L.

Argyrotænia Gill, 40; orthotype AMMODYTES VITTATUS Storer. A synonym of AMMODYTES L.

Oncocottus Gill, 42; orthotype COTTUS HEXACORNIS Storer = COTTUS QUADRICORNIS Rich. Misprinted ONCHOCOTTUS.

Hypleurochilus Gill, 44; logotype BLENNIUS MULTIFILIS Grd. = BLENNIUS GEMINATUS Wood.

Dormitator Gill, 44; orthotype ELEOTRIS SOMNOLENTUS Grd. = SCIÆNA MACULATA Bloch.

Hypsoblennius Gill, 44; orthotype BLENNIUS HENTZ Le Sueur (no description of genus). Replaces ISESTHES J. & G.

Leptoblennius Gill, 44; orthotype BLENNIUS SERPENTINUS Storer.

Centroblennius Gill, 45; logotype LUMPENUS NUBILUS Rich.

Leptoclinus Gill, 45; orthotype LUMPENUS ACULEATUS Reinh.

Reinhardtius Gill, 50; orthotype PLEURONECTES CYNOGLOSSUS Fabricius, not of L.=

PLEURONECTES HIPPOGLOSSOIDES Walbaum. Replaces PLATYSOMATICHTHYS Blkr., 1862.

Chænopsetta Gill, 50; orthotype PLATESSA OBLONGA Storer. A synonym of PARALICHTHYS Grd.

Drepanopsetta Gill, 50; orthotype PLEURONECTES PLATESSOIDES Fabricius. A synonym of HIPPOGLOSSOIDES Gottsche.

Myzopsetta Gill, 51; orthotype PLATESSA FERRUGINEA Storer. A synonym of LIMANDA Gottsche.

Lophopsetta Gill, 51; orthotype PLEURONECTES MACULATUS Mitchill.

Glossichthys Gill, 51; orthotype PLEURONECTES PLAGUSIA L. A synonym of·SYMPHURUS Raf.

Trachinocephalus Gill, 53; orthotype SAURUS MYOPS Cuv. & Val.

Alausella Gill, 54; orthotype CLUPEA PARVULA Mitchill. A synonym of POMOLOBUS Raf.

Opisthonema Gill, 54; orthotype "O. THRISSA Gill" (not CLUPEA THRISSA Osbeck) = MEGALOPS OGLINA Le Sueur.

Brevoortia Gill, 55; orthotype ALOSA MENHADEN Storer = CLUPEA TYRANNUS Latrobe.

Conorhynchus (Nozemann) Gill, 55; orthotype BUTYRINUS VALPES Storer (ALBUTA VALPES). Revival of a pre-Linnæan name, 1757; a synonym of ALBULA Gronow.

Isognatha (Dekay) Gill, 56; orthotype ANGUILLA OCEANICA Storer = MURÆNA CONGER L. A synonym of LEPTOCEPHALUS Gronow = CONGER Cuv.

Ariopsis Gill, 56; orthotype ARIUS MILBERTI = SILURUS FELIS L. A synonym of HEXANEMATICHTHYS Blkr., 1858.

Ceratacanthus Gill, 57; orthotype MONACANTHUS AURANTIACUS.

Aprionodon Gill, 59; orthotype SQUALUS PUNCTATUS Mitchill. A substitute for APRION; preoccupied in fishes.

Cestracion (Klein) Gill, 59; orthotype SQUALUS ZYGÆNA L. A synonym of SPHYRNA Raf.

Eugomphodus Gill, 60; orthotype CARCHARIAS GRISEUS Storer. A section under ODONTASPIS Agassiz = CARCHARIAS Raf.

Narcacion (Klein) Gill, 61; orthotype (RAJA TORPEDO L.). A revival of a pre-Linnæan name; same as TORPEDO Dum.

720. GILL (1861). *Synopsis of the Subfamily of Clupeinæ,* etc. Proc. Acad. Nat. Sci. Phila., 1861.
THEODORE GILL.

Alausella Gill, 36; orthotype CLUPEA PARVULA Mitchill, the young of some species of POMOLOBUS, perhaps P. MEDIOCRIS (repeated from *Catalogue Fishes East Coast).*

Opisthonema Gill, 37; orthotype CLUPANODON THRISSA Lac. (not of Osbeck) = MEGALOPS OGLINA Le Sueur (repeated from *Catalogue).*

Opisthopterus Gill, 37, 38; orthotype PRISTIGASTER TARTOOR Cuv. & Val.

Brevoortia Gill, 37; orthotype CLUPEA MENHADEN Mitchill = CLUPEA PATRONUS Latrobe. (Repeated from *Catalogue.)*

721. GILL (1861). *Synopsis of the Family of Percinæ.* Proc. Acad. Nat. Sci. Phila.
THEODORE GILL.

Asperulus (Klein) Gill, 46; orthotype PERCA ZINGEL L. Equivalent to ZINGEL Oken = ASPRO Cuv.

Kuhlia Gill, 48; orthotype PERCA CILIATA Cuv. & Val.
Hypopterus Gill, 51; orthotype PSAMMOPERCA MACROPTERA Gthr.
Percosoma Gill, 51; orthotype PERCICHTHYS CHILENSIS Grd.
Deuteropterus Gill, 51; orthotype PERCA MARGINATA Cuv. & Val.
Liopropoma Gill, 52; orthotype PERCA ABERRANS Poey.

722. GILL (1861). *Synopsis generum Rhyptici et Affinum.* Proc. Acad. Nat. Sci. Phila.
THEODORE GILL.
Dermatolepis Gill, 54; orthotype D. PUNCTATUS Gill.

723. GILL (1861). *On Several New Generic Types of Fishes.* Proc. Acad. Nat. Sci. Phila., 1861.
Podothecus Gill, 77, 259; orthotype P. PERISTETHUS Gill = AGONUS ACIPENSIRINUS Pallas.
Brachyopsis Gill, 77; orthotype not named; said to be same as AGONUS Sw.
Hoplopagrus Gill, 78; orthotype H. GUNTHERI Gill.
Rhinoscion Gill, 78; orthotype AMBLODON SATURNUS Grd.
Stephanolepis Gill, 78; orthotype MONACANTHUS SETIFER Bennett.

724. GILL (1861). *Revision of the Genera of North American Sciæninæ.* Proc. Acad. Nat. Sci. Phila.
THEODORE GILL.
Cynoscion Gill, 81; orthotype OTOLITHUS REGALIS Cuv. & Val.
Anomiolepis Gill, 82; orthotype a Chinese species unnamed, allied to CORVINA TRIDENTIFER Rich., but without small teeth in upper jaw.
Plagioscion Gill, 82; no species indicated; logotype SCIÆNA SQUAMOSISSIMA Heckel, of Brazil.
Rhinoscion Gill, 78, 85; orthotype AMBLODON SATURNUS Grd. A synonym of CHEILOTREMA Tschudi.
Menticirrhus Gill, 86; orthotype PERCA ALBURNUS L. = CYPRINUS AMERICANUS L., 1758.
Pachypops Gill, 87; orthotype P. TRIFILIS Gill.
Genyonemus Gill, 87; orthotype LEIOSTOMUS LINEATUS Ayres.

725. GILL (1861). *Two New Species of Marine Fishes.* Proc. Acad. Nat. Sci. Phila.
THEODORE GILL.
Hyporthodus Gill, 98; orthotype H. FLAVICAUDA Gill = SERRANUS NIVEATUS Cuv. & Val. A synonym of EPINEPHALUS Bloch.
Sarothrodus Gill, 99; orthotype CHÆTODON CAPISTRATUS Gill. Name a substitute for CHÆTODON as used generally, the name being tranferred to POMACANTHUS, to the species called CHÆTODON by Artedi (1738, before Linnæus), none of them being true CHÆTODONS. This change seems needless and not allowable.

726. GILL (1861). *Haploidonotinæ.* Proc. Acad. Nat. Sci. Phila.
THEODORE GILL.
Haploidonotus Gill, 101; orthotype APLODINOTUS GRUNNIENS Raf. A revision of the spelling of APLODINOTUS.

727. GILL (1861). *Synopsis of the Uranoscopoids.* Proc. Acad. Nat. Sci. Phila.
THEODORE GILL.

Nematagnus Gill, 113; orthotype URANOSCOPUS FILIBARBIS Cuv. & Val.

Upselonphorus Gill, 113; orthotype URANOSCOPUS Y-GRÆCUM Cuv. & Val. A synonym of ASTROSCOPUS Brevoort.

Cathetostoma Gill, 114; orthotype URANOSCOPUS LÆVIS Bloch. A revised spelling of KATHETOSTOMA.

Genyagnus Gill, 115; orthotype URANOSCOPUS MONOPTERYGIUS Bloch.

Gnathagnus Gill, 115; orthotype URANOSCOPUS ELONGATUS T. & S.

728. GILL (1861). *Notes on Some Genera of Fishes of the Western Coast of North America.* Proc. Acad. Nat. Sci. Phila.
THEODORE GILL.

Atractoperca Gill, 164; orthotype LABRAX CLATHRATUS Grd. A synonym of PARALABRAX Grd.

Archoplites Gill, 165; orthotype AMBLOPLITES INTERRUPTUS Grd.

Parephippus Gill, 165; orthotype EPHIPPUS ZONATUS Grd. A synonym of CHÆTODIPTERUS Lac.

Hypsypops Gill, 165; orthotype GLYPHISODON RUBICUNDUS Girard.

Sebastodes Gill, 165; orthotype SEBASTES PAUCISPINIS Ayres.

Acantholebius Gill, 166; orthotype CHIROPSIS NEBULOSUS Grd. = HEX. STELLERI Tilesius. A synonym of HEXAGRAMMOS Steller.

Pleurogrammus Gill, 166; orthotype LABRAX MONOPTERYGIUS Pallas.

Grammatopleurus Gill, 166; orthotype LABRAX LAGOCEPHALUS Pallas. A synonym of HEXAGRAMMOS Steller.

Megalocottus Gill, 166; orthotype COTTUS PLATYCEPHALUS Pallas.

Clinocottus Gill, 166; orthotype OLIGOCOTTUS ANALIS Grd.

Blennicottus Gill, 166; orthotype OLIGOCOTTUS GLOBICEPS Grd.

Hypsagonus Gill, 167, 259; orthotype ASPIDOPHORUS QUADRICORNIS Cuv. & Val.

Paragonus Gill, XIII, 167, 259; orthotype AGONUS ACIPENSERINUS Tilesius. A synonym of PODOTHECUS Gill.

Agonopsis Gill, XIII, 167, 259; orthotype ASPIDOPHORUS CHILOËNSIS Kröyer.

Leptagonus Gill, XIII, 167, 259; orthotype AGONUS SPINOSISSIMUS Kröyer.

Brachyopsis Gill, 167, 259; logotype AGONUS ROSTRATUS Tilesius.

Anoplagonus Gill, 167; orthotype ASPIDOPHOROIDES INERMIS Gthr.

Brosmophycis Gill, 168; orthotype BROSMIUS MARGINATUS Ayres.

729. GILL (1861). *On a New Type of Aulostomatoids, Found in Washington Territory.* Proc. Acad. Nat. Sci. Phila.
THEODORE GILL.

Aulorhynchus Gill, 169; orthotype A. FLAVIDUS Gill.

730. GILL (1861). *Description of a New Generic Type of Blennioids.* Proc. Acad. Nat. Sci. Phila.
THEODORE GILL.

Anoplarchus Gill, 261; orthotype A. PURPURESCENS Gill.

731. GILL (1861). *Monograph of the Tridigitate Uranoscopoids.* Proc. Acad. Nat. Sci. Phila.
THEODORE GILL.

Myxodagnus Gill, 269; orthotype M. OPERCULARIS Gill.

732. GILL (1861). *Synopsis of the Polynematoids.* Proc. Acad. Nat. Sci. Phila.
THEODORE GILL.

Polistonemus Gill, 277; orthotype POLYNEMUS MULTIFILIS T. & S.

733. GILL (1861). *Synopsis of the Sillaginoids.* Proc. Acad. Nat. Sci. Phila.

THEODORE GILL.

Leptoperca Gill, 502; orthotype PERCA SCHRÆTZER L.   Same as SCHRAITZER Schaefer, if the latter is eligible

Sillaginodes Gill, 504; orthotype SILLAGO PUNCTATA Cuv. & Val.

Sillaginopsis Gill, 505; orthotype SILLAGO DOMINA Cuv. & Val.

734. GILL (1861). *Snyopsis of the Chœnichthyoids.* Proc. Acad. Nat. Sci. Phila.

THEODORE GILL.

Champsocephalus Gill, 509; orthotype CHÆNICHTHYS ESOX Gthr.

735. GILL (1861). *Synopsis of the Notothenioids.* Proc. Acad. Nat. Sci.

THEODORE GILL.

Macronotothen Gill, 520; orthotype NOTOTHENIA ROSSI Rich.

Eleginops Gill, 522; logotype APHRITIS UNDULATUS Jenyns.

736. GILL (1861). *Analytical Synopsis of the Squali and Revision of the Nomenclature of the Genera, Followed by "Squalorum generum novorum descriptiones diagnosticæ,"* 330-422; reprint, 1-47. Ann. Lyc. Nat. Hist. N. Y., VII, Dec., 1861.

THEODORE GILL.

Isuropsis Gill, VII, 32 (397); orthotype OXYRHINA GLAUCA M. & H. (not SQUALUS GLAUCUS L.).

Hypoprionodon Gill, 35 (400); orthotype CARCHARIAS HEMIODON Val.

Platypodon Gill, 35 (400): orthotype CARCHARIAS MENISORRAH M. & H.

Eulamia Gill, 35 (400); orthotype "CARCHARIAS LAMIA Raf." = CARCHARHINUS COMMERSONIANUS Blainv. A synonym of CARCHARHINUS Blainv.

Isogomphodon Gill, 35 (400); orthotype CARCHARIAS OXYRHYNCHUS M. & H.

Lamiopsis Gill, 35 (400); orthotype CARCHARIAS TEMMINCKI M. & H.

Isoplagiodon Gill, VII, 35 (400); orthotype CARCHARIAS SORRAH Val.

Aprionodon Gill, VII, 35 (400); orthotype CARCHARIAS ISODON M. & H. Substitute for APRION, preoccupied.

Cynocephalus (Klein) Gill, VII, 35 (400); logotype SQUALUS GLAUCUS L. Name preoccupied in monkeys, unless dating from Klein *(Schauplatz).* A synonym of PRIONACE Cantor.

Boreogaleus Gill, VII, 36 (401); orthotype SQUALUS ARCTICUS Faber.

Chænogaleus Gill, VII, 36 (401); orthotype HEMIGALEUS MACROSTOMA Gthr. Probably not distinct from HEMIGALEUS, which preoccupied name it should perhaps replace.

Reniceps Gill, VII, 37 (402); orthotype SQUALUS TIBURO L.

Cestracion (Klein) Gill, 37 (402); orthotype SQUALUS ZYGÆNA L. A synonym of SPHYRNA Raf., not CESTRACION Cuv., 1817.

Eusphyra Gill, VII, 37 (402); orthotype ZYGÆNA BLOCHI Cuv.

Halælurus Gill, 41 (406) orthotype SCYLLIUM BURGERI M. & H.

Cephaloscyllium Gill, 42 (403); orthotype SCYLLIUM LATICEPS A. Duméril.

Parascyllium Gill, VII, 42 (412); orthotype HEMISCYLLIUM VARIOLATIUM Dum.

Synchismus Gill, 42 (407); orthotype SQUALUS TUBERCULATUS Bl. & Schn. A synonym of CHILOSCYLLIUM M. & H.

Rhina (Klein) Gill, VII, 42 (407); orthotype RHINA SQUATINA Raf. = SQUALUS SQUATINA L., after Klein, 1745-1775. A synonym of SQUATINA Constant Duméril, 1806.

Rhamphobatis Gill, VII, 42 (407) ; type RHINA ANCYLOSTOMA. A synonym of RHINA Bl. & Schn., not of RHINA Klein.

737. GILL (1861). *(Torpedo.)* Ann. Lyc. N. Y., VIII, 1861.
THEODORE GILL.

Tetranarce Gill, VII, 387; orthotype TORPEDO OCCIDENTALIS *Storer.* A section or subgenus under TORPEDO Dum.; misprinted TETRONARCE.

Cyclonarce Gill, VII, 387; orthotype RAJA TIMLEI Bl. & Schn. A synonym of NARCINE Henle.

Gonionarce Gill, VII, 387; orthotype NARCINE INDICA Henle. A synonym of NARCINE Henle.

738. GILL (1861). *Observations on the Genus Cottus,* etc. Proc. Boston Soc. Nat. Hist., VIII.
THEODORE GILL.

Potamocottus Gill, 40; logotype P. PUNCTULATUS Gill. A synonym of COTTOPSIS Grd.

739. GULIA (1861). *Tentamen ichthyologiæ Melitensis,* etc. (In Latin; also in Maltese as *Catalogu tal hut ta Malta.)*
GAVINO GULIA (*d.* 1889).

Micromugil Gulia, 11; orthotype M. TUMIDUS Gulia = CYPRINODON CALARITANUS Cuv. & Val. A synonym of APHANIUS Nardo.

740. GÜNTHER (1861). *Catalogue of the Fishes of the British Museum.* III.
ALBERT GÜNTHER.

Latrunculus Günther, III, 80; orthotype GOBIUS ALBUS Parnell.

Lentipes Günther, III, 96; orthotype SICYOGASTER CONCOLOR Gill. A substitute for SICYOGASTER Gill, preoccupied.

Vulsus Günther, III, 151; otrhotype CALLIONYMUS DACTYLOPUS Bennett. A synonym of DACTYLOPUS Gill, wrongly stated to be preoccupied.

Thalassophryne Günther, III, 174; orthotype T. MACULOSA Gthr.

Saccarius Günther, III, 183; orthotype S. LINEATUS Gthr.

Auchenopterus Günther, III, 275; orthotype A. MONOPHTHALMUS Gthr. Name re garded as preoccupied as AUCHENIPTERUS Cuv. & Val.; replaced by CREMNOBATES Gthr.

Asternopteryx (Rüppell) Günther, II, 288; orthotype A. GUNNELLIFORMIS Rüppell.

Acronurus (Gronow) Günther, II, 345; logotype ACANTHURUS ARGENTEUS Q. & G. Larval form of HEPATUS Gronow (ACANTHURUS Forskål).

Myxus Günther, III, 466; logotype M. ELONGATUS Gthr.

Diplocrepis Günther, III, 506; orthotype LEPADOGASTER PUNICEUS Rich.

Crepidogaster Günther, III, 507; logotype C. TASMANIENSIS Gthr. Name pre- occupied; replaced by ASPASMAGASTER Waite.

Psychrolutes Günther, III, 516; orthotype P. PARADOXUS Gthr.

741. GÜNTHER (1861). *On a New Genus of Australian Fresh-water Fishes, Nannoperca australis.* Ann. Mag. Nat. Hist., VII. Also in Proc. Zool. Soc. London, 1861, 116.
ALBERT GÜNTHER.

Nannoperca Günther, VII, 410; orthotype N. AUSTRALIS Gthr.

742. GÜNTHER (1861). *On Three New Trachinoid Fishes, Crapatalus,* etc. Ann. Mag. Nat. Hist., 1861, VII.
ALBERT GÜNTHER.

Crapatalus Günther, VII, 86; orthotype C. NOVÆZELANDIÆ Gthr.

743. GÜNTHER (1861). *A Preliminary Synopsis of the Labroid Genera.* Ann. Mag. Nat. Hist., VIII.

Hypsigenys Günther VIII, 383; logotype LABRUS MACRODONTUS Lac. A synonym of CHŒRODON Blkr.

Semicossyphus Günther, VIII, 384; orthotype COSSYPHUS RETICULATUS Cuv. & Val.

Stethojulis Günther, VIII, 386; logotype JULIS STRIGIVENTER Bennett.

Hemigymnus Günther, IV, 386; orthotype MULLUS FASCIATUS Thunberg (1776). A synonynm of THALLIURUS Sw.

743A. GÜNTHER (1861). *On a Collection of Fishes Sent by Capt. Dow from the Pacific Coast of Central America.* Proc. Zool. Soc. London.
ALBERT GÜNTHER.

Cremnobates Günther, 374; orthotype AUCHENOPTERUS MONOPHTHALMUS Gthr. A substitute for AUCHENOPTERUS Gthr., preoccupied as AUCHENIPTERUS.

744. JEITTELES (1861). *Zoologische Mittheilungen. I. Ueber zwei für die Fauna Ungarns neue Fische, Lucioperca volgensis Cuv. und Val. und Alburnus maculatus Kessler.* Verh. Zool. Bot. Ges. Wien, XI.
LUDWIG HEINRICH JEITTELES (1830-1883).

Alburnoides Jeitteles, XI, 325; orthotype ALBURNUS MACULATUS Kessler.

745. KEYSERLING (1861). *Neue Cypriniden aus Persien.* Zeits. Ges. Naturw. Berlin, XVII.
EUGEN VON KEYSERLING (1833-1889).

Bungia Keyserling, XVII, 18; orthotype B. NIGRESCENS Keys.

746. KRÖYER (1861). *Nogle Bidrag tel Nordisk ichthyologii.* Nat. Tidskr. 1861.
HENRIK NICOLAI KRÖYER.

Careproctus Kröyer, I, 257; orthotype C. REINHARDTI Kröyer.

746A. MOLIN (1861). *De Rajibus tribus Bolcanis. XLII. Raffaele Molin.* Sitzb. Akad. Wiss. Wien.

Alexandrinus Molin, 579; orthotype A. MOLINI Zigno (fossil) (*Mem. Reale Inst. Veneto,* VIII, 299, 1874) = RAJA MURICATA Volta, 1796. A synonym of TÆNIURA M. & H.
WILHELM CARL HARTWIG PETERS.

747. PETERS (1861). *Ueber zwei neue Gattungen von Fischen . . . aus dem Ganges,* etc. Mon. Akad. Wiss. Berlin.
WILHELM CARL HARTWIG PETERS.

Pterocryptis Peters, 712; orthotype PTEROCRYPTIS GANGETICA Peters.
Acanthocobitis Peters, 712; ortohtype A. LONGIPINNIS Peters.

748. POEY (1861). *Poissons de Cuba.* Memorias de Cuba, II, June. 1861.
FELIPE POEY Y ALOY.

Hollardia Poey, II, 348; orthotype H. HOLLARDI Poey.
Limia Poey, II, 383; orthotype PŒCILIA VITTATA Guich. A synonym of PŒCILIA Bl. & Schn.
Prospinus Poey, II, 388; logotype PLECTROPOMA CHLOROPTERUM Cuv. & Val. A synonym of ALPHESTES Bl. & Schn.

Decaptus Poey, II, 391; orthotype SERIOLA PINNULATA. Poey. A synonym of ELAGATIS Bennett.

749. SALTER (1861). *(Byssacanthus.)* Mem. Geol. Soc., 1861.
JOHN WILLIAM SALTER (1820-1869).

Byssacanthus Salter, 244; orthotype a finspine allied to SPHENACANTHUS (fossil).

750. SCHMID (1861). *Die Fischfauna der Trias bei Jena.* Nova Acta Acad. Cæs. Leop.-Carol. XXIX, 1862 (separates in 1861).
E. E. SCHMID.

Thelodus Schmid, XXIX, 27; logotype T. INFLATUS Schmid (fossil). A synonym of ASTERODON.

751. SYMONDS (1861). *Old Bones.*
WILLIAM SAMUEL SYMONDS (1818-1887).

Lophodus Symonds, 103; (orthotype ACRODUS KEUPERIANUS M. & S.) (fossil) (not LOPHODUS Romanowsky, 1864).

752. VALENCIENNES (1861). *(Acanthomullus.)* Comptes Rendus Acad. Sci. Paris, LII, 1861.
ACHILLE VALENCIENNES.

Acanthomullus Valenciennes, LII, 300; orthotype A. ISABELLÆ Val. (fossil).

1862

753. BLEEKER (1862). *Notice sur le genre Trachinus et ses espèces.* Versl. Akad. Amsterdam (reprinted from 1861).
PIETER BLEEKER.

Pseudotrachinus Bleeker, 113; orthotype P. RADIATUS Blkr.
Echiichthys Bleeker, 117; orthotype E. VIPERA Blkr.

754. BLEEKER (1862). *Conspectus generum Labroideorum analyticus.* Versl. Akad. Amsterdam, XIII, 1862, 94-109.
PIETER BLEEKER.

Pseudocheilinus Bleeker, 95; orthotype CHEILINUS HEXATÆNIA Blkr.
Pseudocoris Bleeker, 97; orthotype JULIS HETEROPTERUS Blkr.
Pseudojulis Bleeker, 99; orthotype P. GIRARDI Blkr.
Platyglossus (Klein) Bleeker, 99; orthotype HALICHŒRES MARGINATUS Rüppell.
Hemitautoga Bleeker, 101; orthotype TAUTOGA NOTOPHTHALMUS Blkr. A synonym of CORIS Lac.
Macropharyngodon Bleeker, 100; orthotype JULIS GEOFFROYI Blkr.
Guntheria Bleeker, 101; orthotype JULIS TRIMACULATA, Q. & G.
Leptojulis Bleeker, 100; orthotype L. CYANOPLEURA Blkr.
Hemitautoga Bleeker, 101; orthotype TAUTOGA NOTOPHALMUS Blkr. A synonym of GUNTHERIA Blkr.
Pseudolabrus Bleeker, 101; orthotype LABRUS RUBIGINOSUS T .& S. (preoccupied) = L. JAPONICUS Houttuyn.
Novaculichthys Bleeker, 102; orthotype LABRUS MACROLEPIDOTUS Bloch.
Diproctacanthus Bleeker, 104; orthotype LABROIDES XANTHURUS Blkr.

755. BLEEKER (1862). *Sixième mémoire sur la fauna ichthyologique de l'île Batjan.* Versl. Akad Amsterdam, XIV, 1862, 99-112.
PIETER BLEEKER.

310   THE GENERA OF FISHES, PART III

Enchelyopus (Klein) Bleeker, 109; orthotype TRICHIURUS LEPTURUS L., not of Gronow. A synonym of TRICHIURUS L.

Eleutheronema Bleeker, 110; orthotype POLYNEMUS TETRADACTYLUS Shaw.

756. BLEEKER (1862). *Sur quelques genres de la famille des Pleuronectoides,* Versl. Akad. Amsterdam, XIII.
PIETER BLEEKER.

Citharus Bleeker, XIII, 424; tautotype PLEURONECTES CITHARUS Spinola = P. LINGUATULA L. Same as CITHARUS Röse, 1793; the later CITHARUS, being preoccupied, has been replaced by EUCITHARUS Gill.

Hemirhombus Bleeker, XIII, 425; orthotype H. GUINEENSIS Bleeker. A synonym of SYACIUM Ranzani.

Clidoderma Bleeker, XIII, 425; orthotype PLEURONECTES ASPERRIMA, T. & S.

Pseudorhombus Bleeker, XIII, 426; orthotype RHOMBUS POLYSPILOS Blkr. Includes RHOMBISCUS Jordan & Snyder.

Citharichthys Bleeker, XIII, 427; orthotype C. CAYENNENSIS Blkr. = C. SPILOPTERUS Günther.

Arnoglossus Bleeker, XIII, 427; tautotype PLEURONECTES ARNOGLOSSUS Bl. & Schn. = PLEURONECTES LATERNA Walbaum.

Platysomatichthys Bleeker, XIII, 426; orthotype PLEURONECTES PINGUIS Fabricius. A synonym of REINHARDTIUS Gill.

Brachyprosopon Bleeker, XIII, 428; orthotype PLEURONECTES MICROCEPHALUS Donovan = P. KITT Walbaum. A synonym of CYNICOGLOSSUS Bon.

Pseudopleuronectes Bleeker, XIII, 428; orthotype PLATESSA PLANA Storer.

Heteroprosopon Bleeker, XIII, 429; orthotype PLEURONECTES CORNUTUS T. & S. A synonym of PLEURONICHTHYS Girard.

757. BLEEKER (1862). *Notices ichthyologiques* (I-X). Versl. Akad. Amsterdam, XIV, 1862, .123-141.
PIETER BLEEKER.

Pseudupeneus Bleeker, 134; orthotype P. PRAYENSIS Blkr.

758. BLEEKER (1862). *Description de quelques espèces nouvelles de Silures de Suriname.* Versl. Akad. Amsterdam, XIV.
PIETER BLEEKER.

Pseudorhamdia Bleeker, 384; orthotype PIMELODUS ASCITA Blkr. (not SILURUS ASCITA L.) = PIMELODUS MACULATUS Lac. A synonym of PIMELODUS Lac. as restricted by Gill and authors generally. Bleeker (1862) limits PIMELODUS to P. BAGRE, allowing it to replace FELICHTHYS Sw. and AILURICHTHYS Baird (as also BAGRE Cuv., Oken).

759. BLEEKER (1862). *Notice sur les genres Parasilurus, Eutropiichthys, Pseudeutropius et Pseudopangasius.* Versl. Akad. Amsterdam, XIV, 1862, 390-399.
PIETER BLEEKER.

Parasilurus Bleeker, 394; orthotype P. JAPONICUS T. & S. = SILURUS ASOTUS L. Same as GLANIS Ag., 1856, not of Gronow, 1854.

Pseudeutropius Bleeker, 398; orthotype P. BRACHYPTERUS Blkr.

Eutropiichthys Bleeker, 398; orthotype E. VACHA Blkr.

Pseudopangasius Bleeker, 399; orthotype P. POLYURANODON Blkr.

760. BLEEKER (1862). *Notice sur les genres Trachelyopterichthys, Hemicetopsis et Pseudocetopsis.* Versl. Akad. Amsterdam, XIV.
PIETER BLEEKER.

Trachelyopterichthys Bleeker, XIV, 402; orthotype T. TÆNIATUS Kner.
Hemicetopsis Bleeker, 403; orthotype CETOPSIS CANDIRA Ag.
Pseudocetopsis Bleeker, 403; orthotype CETOPSIS GOBIOIDES Kner.

761. BLEEKER (1862). *Atlas ichthyologique des Indes Orientales Néerlandaises,
publié sous les auspices du gouvernement colonial néerlandais.* Tome II,
Siluroides, etc.

PIETER BLEEKER.

This notable paper was soon followed by another in which most of
these genera occur. This is *Systema Silurorum revisum.* Ned. Tijdschr.
Dierk., I, 1863, 77-122.

Pseudancistrus Bleeker, 2; orthotype HYPOSTOMUS BARBATUS Cuv. & Val.
Hemiancistrus Bleeker, 2; orthotype ANCISTRUS MEDIANS Kner.
Parancistrus Bleeker, 2; orthotype HYPOSTOMUS AURANTIACUS Castelnau.
Pseudacanthicus Bleeker, 2; orthotype HYPOSTOMUS SERRATUS Cuv. & Val.
Pseudorinelepis Bleeker, 3; orthotype RINELEPIS GENIBARBIS Cuv. & Val.
Loricariichthys Bleeker, 3; orthotype LORICARIA MACULATA Bloch.
Pseudoloricaria Bleeker, 3; orthotype LORICARIA LÆVIUSCULA Cuv. & Val.
Parahemiodon Bleeker, 3; orthotype P. TYPUS Blkr. A synonym of LORICARIICH-
      THYS Blkr.
Hemiloricaria Bleeker, 3; orthotype H. CARACASENSIS Blkr.
Pseudohemiodon Bleeker, 3; orthotype HEMIODON PLATYCEPHALUS Kner.
Rineloricaria Bleeker, 3; orthotype L. LIMA Kner.
Hemiodontichthys Bleeker, 4; orthotype H. ACIPENSERINUS Kner.
Oxyloricaria Bleeker, 5; orthotype L. BARBATA Kner.
Rhinodoras Bleeker, III, 5; orthotype OXYDORAS D'ORBIGNYI (Kröyer). A synonym
      of OXYDORAS.
Lithodoras Bleeker, .5; orthotype DORAS LITHOGASTER Heckel. A synonym of
      CENTROCHIR Ag.
Platydoras Bleeker, 5; orthotype DORAS COSTATUS L. A synonym of DORAS, as
      restricted by Günther and Eigenmann, not by Bleeker.
Pterodoras Bleeker, 5; orthotype DORAS GRANULOSUS Val. A syonym of CEN-
      TROCHIR Ag.
Acanthodoras Bleeker, 5; orthotype SILURUS CATAPHRACTUS L.
Astrodoras Bleeker, 5; orthotype DORAS ASTERIFRON: Kner. A synonym of CEN-
      TROCHIR Ag.
Amblydoras Bleeker, 6; orthotype DORAS AFFINIS Kner.
Brachysynodontis Bleeker, 6; orthotype SYNODONTIS BATENSODA Rüppell.
Pseudosynodontis Bleeker, 6; orthotype SYNODOTIS SERRATUS Rüppell.
Hemisynodontis Bleeker, 6; orthotype PIMELODUS MEMBRANACEUS Geoffroy.
Leiosynodontis Bleeker, 6; orthotype SYN. MACULOSUS (young of SILURUS SCHAL
      Bl. & Schn.).
Pseudauchenipterus Bleeker, 6; orthotype SILURUS NODOSUS Bloch.
Parauchenipterus Bleeker, 7; orthotype SILURUS GALEATUS L. A synonym of
      TRACHYCORYSTES Blkr.
Auchenipterichthys Bleeker, 7; orthotype AUCH. THORACATUS Kner.
Hemiarius Bleeker, 7; orthotype CEPHALOCASSIS STORMI Blkr.
Pseudarius Bleeker, 8; orthotype ARIUS ARIUS Cuv. & Val. A synonym of TACHY-
      SURUS Lac. = ARIUS Cuv. & Val.
Diplomystes (Duméril) Bleeker, 8; orthotype ARIUS PAPILLOSUS Cuv. & Val.
      (after les DIPLOMYSTES Dum.). Spelled DIPLOMYSTAX by Günther, 1864.

Paradiplomystes Bleeker, 8; orthotype PIMELODUS CORUSCANS Lichtenstein.

Bagrichthys Bleeker, 9; orthotype B. HYPSELOPTERUS Blkr.

Pseudobagrichthys Bleeker, 9; orthotype BAGROIDES MACROPTERUS Blkr.

Hemibagrus Bleeker, 9; orthotype BAGRUS NEMURUS Cuv. & Val.

Aspidobagrus Bleeker, 9; orthotype BAGRUS GULIO Cuv. & Val.

Hypselobagrus Bleeker, 10; orthotype BAGRUS MACRONEMA Blkr. Identical with MYSTUS (Gronow) Scopoli, 1763, as restricted by Jordan.

Hemisorubim Bleeker, 10; orthotype PLATYSTOMA PLATYRHYNCHOS Cuv. & Val.

Brachyplatystoma Bleeker, 10; orthotype PLATYSTOMA VAILLANTI Cuv. & Val.

Pseudoplatystoma Bleeker, 10; orthotype SILURUS FASCIATUS L.

Hemiplatystoma Bleeker, 10; orthotype PLATYSTOMA TIGRINUM Cuv. & Val. A synonym of PSEUDOPLATYSTOMA.

Sorubimichthys Bleeker, 10; orthotype SORUBIM JANDIA Spix.

Platystomatichthys Bleeker, 10; orthotype PLATYSTOMA STURIO Kner.

Leiarius Bleeker, 10; orthotype A. LONGIBARBIS Castelnau.

Pseudariodes Bleeker, 11; orthotype SILURUS CLARIAS Bloch, not of Linnæus, after Hasselquist. A synonym of PIMELODUS Lac.

Malacobagrus Bleeker, 11; orthotype PIMELODUS FILAMENTOSUS Lichtenstein. Perhaps synonym of PIRATINGA.

Pirarara (Spix) Bleeker, 11; orthotype SILURUS HEMILIOPTERUS Bl. & Schn.

Parabagrus Bleeker, 11; orthotype PIMELODUS PUSILLUS Ranzani.

Pseudocalophysus Bleeker, 12; orthotype PIMELODUS CTENODUS Ag. = CALOPHYSUS MACROPTERUS. A synonym of CALOPHYSUS Müller.

Aglyptosternon Bleeker, 12; orthotype SILURUS COUS L.

Tetranematichthys Bleeker, 14; orthotype AGENIOSUS QUADRIFILIS Kner.

Psudageneiosus Bleeker, 14; orthotype AGENEIOSUS BREVIFILIS Cuv. & Val.

Pseudohypophthalmus Bleeker 15; orthotype H. FIMBRIATUS Kner = H. EDENTULUS Spix. A synonym of HYPOPHTHALMUS Spix.

Paracetopsis (Guichenot) Bleeker, 16; orthotype P. BLEEKERI Guich.

Wallago Bleeker, 17; orthotype SILURUS ATTU Bl. & Schn.

Silurodes Bleeker, 17; orthotype S. HYPOPHTHALMUS Blkr.

Catastoma (Kuhl & Van Hasselt) Bleeker, 28; orthotype ARIUS NASUTUS Val. An undefined synonym of NETUMA Blkr.

762. BLEEKER (1862). *Beskrivning en afdeeling van eene nieuwe soort van Brama (Abramis).*

PIETER BLEEKER.

Brama (Klein) Bleeker, 211; type CYPRINUS BRAMA L. ABRAMIS Cuv., not BRAMA Bloch.

763, 764. COSTA (1862). *(Fossil Fishes.)* Atti Reale Inst. Napoli, VI.

GABRIELE ORONZIO COSTA.

Giffonus Costa, VI, App. 26; orthotype G. DEPERDITUS Costa (fossil).

Urocomus Costa, VI, 32; orthotype U. PICENUS Costa (fossil). Doubtful fragments.

765. (COSTA 1862). *Di un novello genere di pesci Esocetidi.* Ann. Mus. Zool. Univ. Napoli, 1862, I.

ACHILLE COSTA (1823-   ).

Grammiconotus Costa, I, 54; orthotype G. BICOLOR Costa = SCOMBEROSOX RONDELETI Cuv. & Val. Differs from SCOMBERESOX in lacking the air bladder; jaws short.

766. COUCH (1862). *The History of the Fishes of the British Islands.* I.
JONATHAN COUCH.

Polyprosopus Couch, I, 67; orthotype P. RASHLEIGHANUS Couch. = SQUALUS
MAXIMUS Gunner. A synonym of CETORHINUS Blainv.

767. DYBOWSKI (1862). *Versuch einer Monographie der Cypriniden Livlands.*
Arch. Naturg. Biol. Dorpat (1864, separates in 1862).
BENEDIKT IVAN DYBOWSKI (1834-   ).

Owsianka Bybowski, 146; orthotype ASPIUS OWSIANKA Cernay = O. CZERNAYI
Dybowski = SQUALIUS BILINEATUS Heckel. A synonym of LEUCASPIUS
Heckel.

768. GILL (1862). *Notice of a New Species of Hemilepidotus and Remarks on
the Group (Temnistiæ) of Which It Is a Member.* Proc. Acad. Nat. Sci.
Phila., XIV.

Oncocottus Gill, 13; orthotype COTTUS QUADRICORNIS L.

769. GILL (1862). *On the Subfamily of Argentininæ.* Proc. Acad. Nat. Sci. Phila.
THEODORE GILL.

Retropinna Gill, 14; orthotype ARGENTINA RETROPINNA Rich. = RETROPINNA
RICHARDSONI Gill.

Mesopus Gill, 14; orthotype ARGENTINA PRETIOSA Grd. An error for HYPOMESUS;
an error at once corrected by Gill. HYPOMESUS should therefore stand.

Hypomesus Gill, 15; orthotype ARGENTINA PRETIOSA Grd. A synonym of MESOPUS
Gill, page 14, one or the other being left through defective proof-reading.

770. GILL (1862). *Appendix to the Synopsis of the Subfamily of Percinæ.* Proc.
Acad. Nat. Sci. Phila.
THEODORE GILL.

Chorististium Gill, 15; orthotype LLOPROPOMA RUBRA Poey.
Siniperca Gill, 16; orthotype PERCA CHUATSI Basilewsky.

771. GILL (1862). *Notes on the Sciænoids of California.* Proc. Acad. Nat. Sci.
Phila., I.

Cirrimens Gill, 17; orthotype UMBRINA OPHIOCEPHALA Jenyns.
Brachydeuterus Gill, 17; orthotype LARIMUS AURITUS Cuv.
Odontoscion Gill, 18; orthotype CORVINA DENTEX Cuv. & Val.
Atractoscion Gill, 18; orthotype OTOLITHUS ÆQUIDENS Cuv. & Val.
Archoscion Gill, 18; orthotype OTOLITHUS ANALIS Jenyns.
Apseudobranchus Gill, 18; orthotype OTOLITHUS TOEROE Cuv. & Val. = CHEILODIP-
TERUS ACOUPA Lac. A synonym of CYNOSCION Gill.
Isopisthus Gill, 18; orthotype ANCYLODON PARVIPINNIS Cuv. & Val.

772. GILL (1862). *Synopsis of the Family of Cirrhitoids.* Proc. Acad. Nat. Sci.
Phila.
THEODORE GILL.

Amblycirrhitus Gill, 105; orthotype CIRRHITES FASCIATUS Cuv. & Val.
Cirrhitopsis Gill, 109; orthotype C. AUREUS T. & S. A synonym of CIRRHITICH-
THYS Blkr.
Dactylosargus Gill, 110; orthotype APLOPODACTYLUS ARCIDENS Rich.
Crinodus Gill, 112; orthotype HAPLODACTYLUS LOPHODON Gthr.
Acantholatris Gill, 114; orthotype CHÆTODON MONODACTYLUS Carmichael.
Chirodactylus Gill, 114; orthotype CHEILODACTYLUS ANTONII Cuv. & Val.
Dactylopagrus Gill, 114; orthotype CHEILODACTYLUS CARPONEMUS Cuv. & Val.
Called DACTYLOSPARUS on page 117.

Latridopsis Gill, 115; orthotype ANTHIAS CILIARIS Bloch.
Dactylosparus Gill, 117; orthotype CHEILONEMUS CARPONEMUS Cuv. & Val. A synonym of DACTYLOPAGRUS Gill; the two names left through defective proof-reading.
Goniistius Gill, 120; orthotype CHEIL. ZONATUS Cuv. & Val.

773. GILL (1862). *On the Limits and Arrangements of the Family of Scombroids.* Proc. Acad. Nat. Sci. Phila.
THEODORE GILL.

Grammatorcynus Gill, 125; orthotype THYNNUS BILINEATUS Rüppell. Misspelled GRAMMATORYCNUS.
Gymnosarda Gill, 125; orthotype THYNNUS UNICOLOR Rüppell.
Orcynopsis Gill, 125; orthotype SCOMBER UNICOLOR Geoffroy St. Hillaire. (Misprinted ORYCNOPSIS.)
Lepidocybium Gill, 125; orthotype CYBIUM FLAVOBRUNNEUM Smith. A synonym of SCOMBEROMORUS Lac.
Acanthocybium Gill, 125; orthotype CYBIUM SARA Bennett.
Thyrsitops Gill, 125; orthotype THYRSITES LEPIDOPOIDES Cuv. & Val.
Eupleurogrammus Gill, 126; orthotype TRICHIURUS MUTICUS Gray.
Zenopsis Gill, 126; orthotype ZEUS NEBULOSUS T. & S.
Cyttopsis Gill, 126; orthotype ZEUS ROSEUS Lowe.
Chondroplites Gill, 126; orthotype STROMATEUS ATOUS Cuv. & Val.
Poronotus Gill, 126; orthotype STROMATEUS TRIACANTHUS Peck.
Hoplocoryphis Gill, 127; orthotype SCHEDOPHILUS MACULATUS Gthr.
Clara Gill, 127; orthotype EQUULA LONGIMANUS Cantor. A synonym of PENTAPRION Blkr.
Psenopsis Gill, 127; orthotype TRACHINOTUS ANOMALUS T. & S.

774. GILL (1862). *Descriptions of New Species of Alepisauroidæ.* Proc. Acad. Nat. Sci. Phila.
THEODORE GILL.

Caulopus Gill, 128; orthotype ALEPIDOSAURUS BOREALIS Gill.

775. GILL (1862). *On the West African Genus Hemichromis,* etc. Proc. Acad. Nat. Sci. Phila.
THEODORE GILL.

Epiplatys Gill, 136; orthotype E. SEXFASCIATUS Gill.
Marcusenius Gill, 139; orthotype MORMYRUS CYPRINOIDES L. Name later, page 444, by some confusion transferred to MORMYRUS ANGUILLOIDES L.
Mormyrodes Gill, 139; orthotype MORMYRUS HASSELQUISTI Geoffroy.
Hyperopisus Gill, 139; orthotype MORMYRUS DORSALIS Geoffroy.

776. GILL (1862). *Catalogue of the Fishes of Lower California . . . Collected by Mr. J. Xantus.* Proc. Acad. Nat. Sci. Phila.
THEODORE GILL.

Lepidaplois Gill, 10; orthotype COSSYPHUS AXILLARIS Cuv. & Val.
Chœrojulis Gill, 142; orthotype JULIS SEMICINCTUS Ayres. A sybstitute for HALICHŒRES Rüppell, preoccupied by HALICHŒRUS; a synonym of HEMIULIS Sw., as restricted by Bonaparte.
Iniistius Gill, 143; orthotype XYRICHTHYS PAVO Cuv. & Val.
Malacocentrus Gill, 143; orthotype XYRICHTHYS TÆNIURUS Cuv. & Val.

Oxycheilinus Gill, 143; orthotype Cheilinus arenatus Cuv. & Val.
Euschistodus Gill, 145; orthotype E. declivifrons Gill. A synonym of Glyphiso-don Lac.

777. GILL (1862). *On a New Genus Allied to Aulorhynchus and on the Affinities of the Family Aulorhynchoidæ to Which It Belongs.* Proc. Acad. Nat. Sci. Phila., XIV.
Aulichthys (Brevoort) Gill, 234; orthotype A. japonicus Brevoort.
Acentrachme Gill, 234; orthotype Amphisile scutata Cuv. = Centriscus scuta-tus L. A synonym of Centriscus L. Amphisile (Klein) Gill is the genus Æoliscus Jordan & Starks. Amphisile Cuv. (scutatus) is Centriscus.
Orthichthys Gill, 234; orthotype Centriscus velitaris Pallas.
Centriscops Gill, 234; orthotype Centriscus humerosus Rich.

778. GILL (1862). *Remarks on the Relations of the Genera and Other Groups of Cuban Fishes.* Proc. Acad. Nat. Sci. Phila.
THEODORE GILL.
Haliperca Gill, 236; logotype Serranus bivittatus Cuv. & Val.
Mentiperca Gill, 236; orthotype Serranus luciopercanus Poey.
Hypoplectrus Gill, 236; orthotype Plectropoma puella Cuv. & Val.
Hypoplectrodes Gill, 236 (no definition); orthotype Plectropoma nigrorubrum Cuv. & Val. Same as Gilbertia Jordan & Eigenmann, 1890.
Acanthistius Gill, 236; orthotype Plectropoma serratum Cuv. & Val.
Ocyurus Gill, 236; orthotype Sparus chrysurus Bloch.
Proamblys Gill, 236; orthotype Diacope nigra Cuv. = Diacope macolor Cuv. & Val. A synonym of Macolor Blkr. (1860).
Gonioplectrus Gill, 236 orthotype Plectropoma hispanum Cuv. & Val.
Plectroplites Gill, 236; orthotype Datnia ambigua Rich.
Rhomboplites Gill, 236; orthotype Centropristes aurorubens Cuv. & Val.
Brachyrhinus Gill, 236; orthotype Serranus creolus Cuv. & Val. = S. furcifer Cuv. & Val. Name preoccupied; replaced by Paranthias Guich. (1868).
Platyinius Gill, 236; orthotype Mesoprion vorax Poey.
Gonioperca Gill, 236; orthotype Serranus albomaculatus Jenyns. A synonym of Paralabrax Grd.
Labroperca Gill, 236; orthotype Serranus labriformis Jenyns. A synonym of Epinephelus Bloch.
Mycteroperca Gill, 236; orthotype Serranus olfax Jenyns.
Hypoplites Gill, 236; orthotype Mesoprion retrospinis Cuv. & Val.
Evoplites Gill, 236; orthotype Mesoprion pomacanthus Blkr.
Schistorus Gill, 237; orthotype Serranus mystacinus Poey.
Lioperca Gill, 237; orthotype Serranus inermis Cuv. & Val. A synonym of Dermatelepis Gill.
Rhinoberyx Gill, 237; orthotype Rhynchichthys brachyrhynchus Blkr. A larval form allied to Holocentrus
Plectrypops Gill, 237; orthotype Holocentrum retrospinis Guich.
Centroberyx Gill, 238; orthotype Beryx lineatus Cuv. & Val.
Synistius Gill, 238; orthotype Gerres longirostris Rüppell.
Nandopsis Gill, 238; orthotype Chromis tetracanthus Poey.
Odontoscion Gill, 238; orthotype Johnius dentex Cuv. & Val.
Orycnus Gill, 238; orthotype Scomber thynnus L. A substitute for Orcynus Cuv., preoccupied, of which name it was originally a misprint (Cooper); a synonym of Thunnus South.

Sarothrodus Gill, 238; orthotype CHÆTODON STRIATUS L. Substitute for CHÆTO-
DON L., the name transferred to POMACANTHUS, the first species named by
Artedi being a POMACANTHUS.

Prognathodes Gill, 238; orthotype CHELMO PELTA Gthr. (not PROGNATHODUS Owen).

Carangops Gill, 238; orthotype CARANX HETEROPYGUS Poey. A synonym of HEMI-
CARANX Blkr. of the same date. Both are apparently synonyms of ALEPES Sw.,
1839.

Trachurops Gill, 238; orthotype SCOMBER PLUMIERI Bloch = SCOMBER CRUMENOPH-
THALMUS Bloch.

Remora Gill, 239; orthotype ECHENEIS REMORA L. This adjustment restricts
ECHENEIS L. to E. NAUCRATES L. It is earlier than any other and must stand.

Remilegia Gill, 239; orthotype ECHENEIS SCUTATA Gthr.

Phtheirichthys Gill, 239; orthotype ECHENEIS LINEATA Menzies.

Caulolatilus Gill, 240; orthotype LATILUS CHRYSOPS Cuv. & Val.

Lophogobius Gill, 240; orthotype GOBIUS CRISTAGALLI Cuv. & Val.

Dormitator Gill, 240; logotype D. GUNDLACHI Poey.

Gnathypops Gill, 241; logotype OPISTHOGNATHUS MAXILLOSUS Poey.

Lonchopisthus Gill, 241; orthotype OPISTHOGNATHUS MICROGNATHUS Poey.

779. GILL (1862). *Catalogue of the Fishes of Lower California . . . Collected by*
Mr. J. Xantus. Part II. Proc. Acad. Nat. Sci. Phila.
THEODORE GILL.

Pomacanthodes Gill, 244; orthotype P. ZONIPECTUS Gill.

Incisidens Gill, 244; orthotype CRENIDENS SIMPLEX Rich.

Opisthistius Gill, 245; orthotype SCIÆNA TAHMEL Forskål.

Orthostæchus Gill, 255; orthotype O. MACULICAUDA Gill.

Microlepidotus Gill, 255; orthotype M. INORNATUS Gill.

Genytremus Gill, 256; orthotype PRISTIPOMA BILINEATUM Cuv. & Val. A section
under ANISOTREMUS Gill.

Genyatremus Gill, 256; orthotype DIAGRAMMA CAVIFRONS Cuv. & Val.

Pristocantharus Gill, 256; orthotype PRISTIPOMA CANTHARINUM Jenyns. A synonym of ORTHOPRISTIS Grd.

Nematistius Gill, 258; orthotype N. PECTORALIS Gill.

Eustomatodus Gill, 261; logotype DECAPTERUS KURROIDES Blkr. A synonym of
DECAPTERUS Blkr.

Gymneipignathus Gill, 261; orthotype DECAPTERUS MACROSOMA Blkr.

Evepigymnus Gill, 261; orthotype DECAPTERUS HYPODUS Gill.

780. GILL (1862). *Notes on a Collection of the Fishes of California. . . . by*
Samuel Hubbard. Proc. Acad. Nat. Sci. Phila.
THEODORE GILL.

Hypocritichthys Gill, 275; orthotype HYPERPROSOPON ANALIS Alex. Agassiz

Brachyistius Gill, 275; orthotype B. FRENATUS Gill.

Oxylebius Gill, 277; orthotype O. PICTUS Gill.

Sebastopsis Gill, 278; orthotype SEBASTES POLYLEPIS Cuv. & Val. A synonym of
SCORPÆNODES Blkr.

Eucyclogobius Gill, 279; orthotype GOBIUS NEWBERRYI Grd.

Melanotænia Gill, 280; orthotype ATHERINA NIGRANS Rich.

Melanogrammus Gill, 280; orthotype GADUS ÆGLIFINUS L.

Brachygadus Gill, 280; orthotype GADUS MINUTUS Yarrell. A synonym of TRISOP-
TERUS Raf., 1814.

781. GILL (1862). *Synopsis of the Species of Lophobranchiate Fishes of Western North America.* Proc. Acad. Nat. Sci. Phila.
THEODORE GILL.

Dermatostethus Gill, 283; orthotype D. PUNCTIPINNIS Gill.

782. GILL (1862). *Note on Some Genera of Fishes of Western North America.* Proc. Acad. Nat. Sci. Phila.
THEODORE GILL.

Sebastichthys Gill, 329; orthotype SEBASTES NIGROCINCTUS Ayres.

Paratractus Gill, 330; orthotype CARANX PISQUETOS Cuv. & Val. = SCOMBER CRYSOS Mitchill. A section under CARANX.

Caularchus Gill, 330; orthotype LEPADOGASTER RETICULATUS Grd. (GOBIESOX MÆANDRICUS Grd.).

Eumicrotremus Gill, 330; orthotype CYCLOPTERUS ORBIS Müller.

Hypsifario Gill, 330; orthotype SALMO KENNERLYI Suckley, a landlocked form of ONCORHYNCHUS NERKA (Walbaum). The genus may be valid, as distinguished by the long and numerous gill-rakers.

Lepidopsetta Gill, 330; orthotype PSETTICHTHYS UMBROSUS Grd.

Hypsopsetta Gill, 330; orthotype PLEURONICHTHYS GUTTULATUS Grd.

Orthopsetta Gill, 330; orthotype PSETTICHTHYS SORDIDUS Grd.

Uropsetta Gill, 330; orthotype HIPPOGLOSSUS CALIFORNICUS Ayres. A synonym of PARALICHTHYS Grd.

Hydrolagus Gill, 331; orthotype CHIMÆRA COLLIEI Lay & Bennett.

Gyropleurodus Gill, 331; orthotype CESTRACION FRANCISCI Grd.

Holorhinus Gill, 331; orthotype RHINOPTERA VESPERTILIO Grd. = MYLIOBATIS CALIFORNICUS Gill. A section under MYLIOBATIS Dum.

Entosphenus Gill, 331; orthotype PETROMYZON TRIDENTATUS Rich.

783. GILL (1862). *Synopsis of the Carangoids of the Eastern Coast of North America.* Proc. Acad. Nat. Sci. Phila.
THEODORE GILL.

Naucratopsis Gill, 441; orthotype SERIOLA GIGAS Gthr.

Halatractus Gill, 442; logotype SERIOLA DUMERILI Cuv. & Val. A substitute for SERIOLA, used in botany; a reason for change no longer recognized.

784. GILL (1862). *Description of a New Generic Type of Mormyroids,* etc. Proc. Acad. Nat. Sci. Phila.
THEODORE GILL.

Isichthys Gill, 443; orthotype I. HENRYI Gill.

Gnathonemus Gill, 444; orthotype MORMYRUS PETERSI Gthr.

785. GILL (1862). *On the Classification of the Families and Genera of the Squali of California.* Proc. Acad. Nat. Sci. Phila.
THEODORE GILL.

Rhinotriacis Gill, 486; orthotype R. HENLEI Gill.

Tropidodus Gill, 489; orthotype CESTRACION PANTHERINUS Val.

Entoxychirus Gill, 496; orthotype SQUALUS UYATO Raf. A synonym of CENTROPHORUS M. & H.

786. GILL (1862). *On the Limits and Affinity of the Family of Leptoscopoids.* Proc. Acad. Nat. Sci. Phila.
THEODORE GILL.

Dactylagnus Gill, 504; orthotype D. MUNDUS Gill.

787. GILL (1862). *Synopsis of the Genera of the Subfamily of Pimelodinæ.* Proc. Boston Soc. Nat. Hist. for 1861 (1862), VIII.

THEODORE GILL.

Branchiosteus Gill, 52; orthotype OLYRA LATICEPS McCl.

788. GÜNTHER (1862). *Catalogue of the Fishes in the British Museum.* IV.

ALBERT GÜNTHER.

Lepidozygus Günther, IV, 15; orthotype POMACENTRUS TAPEINOSOMA Blkr.

Microspathodon Günther, IV, 35; orthotype GLYPHIDODON CHRYSURUS Cuv. & Val. Apparently a synonym of STEGASTES Jenyns, 1842.

Parma Günther, IV, 57; logotype P. MICROLEPIS Gthr.

Tautogolabrus Günther, IV, 89; orthotype CTENOLABRUS BURGALL Gthr. = LABRUS ADSPERSUS Walbaum.

Trochocopus Günther, IV, 100; logotype T. OPERCULARIS Gthr.

Decodon Günther, IV, 101; orthotype COSSYPHUS PUELLARIS Poey.

Doratonotus Günther, IV, 124; orthotype D. MEGALEPIS Gthr.

Cymolutes Günther, IV, 207; logotype JULIS PRÆTEXTATUS Q. &. G.

Coridodax Günther, IV, 243; SPARUS PULLUS Forster.

Pseudetroplus (Bleeker) Günther, IV, 266; orthotype CHÆTODON SURATENSIS Bloch.

Chromis (Cuvier) Günther, IV, 267; logotype CHROMIS NILOTICUS Cuv., not as restricted to the original tautotype SPARUS CHROMIS L. This understanding is that of most modern authors following Gill and Bleeker. C. CHROMIS and C. NIOLOTICUS are both included in the CHROMIS of Cuvier. CHROMIS Gthr. is a synonym of TILAPIA Smith.

Theraps Günther, IV, 284; orthotype T. IRREGULARIS Gthr.

Mesonauta Günther, IV, 300; orthotype HEROS INSIGNIS Heckel.

Petenia Günther, IV, 301; orthotype P. SPLENDIDA Gthr.

Hygrogonus Günther, IV, 303; orthotype LOBOTES OCELLATUS Ag. A synonym of ASTRONOTUS Sw.

Mesops Günther, IV, 311; logotype GEOPHAGUS CUPIDO Heckel. Name preoccupied, replaced by BIOTODOMA Eigenmann & Kennedy, 1903.

Satanoperca Günther, IV, 312; logotype GEOPHAGUS ACUTICEPS Heckel.

Uronectes Günther, IV, 325; orthotype OPHIDIUM PARRII Ross. Name preoccupied in crustacea, replaced by LYCOCARA Gill, 1884.

Boreogadus Günther, IV, 336; logotype GADUS FABRICII Rich.

Halargyreus Günther, IV, 342; orthotype H. JOHNSONI Gthr.

Pseudophycis Günther, IV, 350; orthotype LOTA BREVIUSCULA Rich.

Læmonema Günther, IV, 356; orthotype PHYCIS YARRELLI Lowe.

Haloporphyrus Günther, 358; orthotype GADUS LEPIDION Risso. Substitute for LEPIDION Sw. regarded as preoccupied by LEPIDIA, 1817.

Xiphogadus Günther IV, 374; orthotype XIPHASIA SETIFER Sw. A needless substitute for XIPHASIA Sw. (1839).

Hyperoplus Günther, IV, 384; orthotype AMMODYTES LANCEOLATUS Le Sauvage.

Bleekeria Günther, IV, 387; orthotype B. KALLOLEPIS Gthr.

Congrogadus Günther, IV, 388; orthotype MACHÆRIUM SUBDUCENS, Rich. A substitute for MACHÆRIUM, preoccupied.

Malacocephalus Günther, IV, 396; orthotype M. LÆVIS Gthr.

Tephritis Günther, IV, 406; orthotype PLEURONECTES SINENSIS Lac. Name preoccupied, replaced by TEPHRINECTES Gthr., and later by VELIFRACTA Jordan.

Lepidorhombus Günther, IV, 407; orthotype PLEURONECTES MEGASTOMA Donovan.
Phrynorhombus Günther, IV, 414; orthotype RHOMBUS UNIMACULATUS Risso.
Brachypleura Günther, IV, 419; orthotype B. NOVÆZEELANDIÆ Gthr.
Engyprosopon Günther, IV, 431; RHOMBUS MOGKII Blkr.
Psammodiscus Günther, IV, 457; orthotype P. OCELLATUS Gthr.
Rhombosolea Günther, IV, 458; logotype R. MONOPUS Gthr.
Ammotretis Günther, IV, 458; orthotype A. ROSTRATUS Gthr.
Peltorhamphus Günther, IV, 460; orthotype P. NOVÆZEELANDIÆ Gthr.
Buglossus Günther, IV, 462; orthotype PLEURONECTES VARIEGATUS Donovan. A synonym of MICROCHIRUS Bon.
Microbuglossus Günther, IV, 462; orthotype SOLEA HUMILIS Cantor.
Pegusa Günther, IV, 467; logotype SOLEA AURANTIACA Gthr.
Pardachirus Günther, IV, 478; lgootype ACHIRUS MARMORATUS Lac.
Liachirus Günther, IV, 479; orthotype L. NITIDUS Gthr.
Anisochirus Günther, IV, 480; orthotype SYNAPTURA PANOIDES Blkr.
Soleotalpa Günther, IV, 489; orthotype S. UNICOLOR Gthr. A synonym of APIONICHTHYS Kaup.
Ammopleurops Günther, IV, 490; orthotype PLAGUSIA LACTEA Bon. A synonym of SYMPHURUS Raf.

789. GÜNTHER (1862). *Descriptions of New Species of Reptiles and Fishes,* etc. Proc. Zool. Soc. London.
ALBERT GÜNTHER.
Tyntlastes Günther, 194; orthotype, AMBLYOPUS SAGITTA Gthr.

790. GÜNTHER (1862). *Note on Pleuronectes sinensis Lac.* Ann. Mag. Nat. Hist. X.
ALBERT GÜNTHER.
Tephrinectes Günther, 475; orthotype PLEURONECTES SINENSIS Lac. Substitute for TEPHRITIS Gthr., preoccupied; replaces VELIFRACTA Jordan, a later substitute name.

791. JOHNSON (1862). *Notes on Rare and Little Known Fishes Taken at Madeira.* Ann. Mag. Nat. Hist.
JAMES YATE JOHNSON.
Pseudomuræna Johnson, 167; orthotype P. MADERENSIS Johnson. A synonym of PRIODONOPHIS Kaup.
Synaphobranchus Johnson, 169; orthotype S. KAUPI Johnson = MURÆNA PINNATA Gronow.
Nesiarchus Johnson, 173; orthotype N. NASUTUS Johnson.
Setarches Johnson, 177; orthotype S. GUNTHERI Johnson.

792. JONES (1862). *(Estheria.)*
THOMAS RYMER JONES (1819-1911).
Estheria Jones, 112; orthotype E. MIDDENDORFFI Jones (fossil).

793. KNER (1862). *Kleinere Beiträge zur Kenntniss der fossilen Fische Oesterreichs.* Sitzb. Akad. Wiss Wien, XLV.
RUDOLF KNER.
Palimphemus Kner, XLV, 490; orthotype P. ANCEPS (fossil).

794. MORRIS & ROBERTS (1862). *On the Carboniferous Limestone of Oreton and Farlow Clee Hills, Shropshire,* etc. Quart. Journ. Geol., XVIII.
JOHN MORRIS; GEORGE E. ROBERTS.

Characodus (Agassiz) Morris & Roberts, XVIII, 99; orthotype C. ANGULATUS Ag. (fossil); name only, defined by Davis, 1883. Regarded by Woodward as a synonym of COPODUS; not CHARACODUS Owen, 1867, earlier in definition.

Ctenopetalus (Agassiz MS.) Morris & Roberts, XVIII, 100; orthotype CTENOP-TYCHIUS SERRATUS Ag. (fossil); name only, defined by Davis, 1881. Regarded as a synonym of CTENOPTYCHIUS.

Harpacodus (Agassiz) Morris & Roberts, XVIII, 100; orthotype CTENOPTYCHIUS DENTATUS Ag. (fossil); name only, defined by Davis (1881). Regarded as a synonym of CTENOPTYCHIUS.

Copodus (Agassiz) Morris & Roberts, XVIII, 100; logotype PSAMMODUS CORNUTUS Ag. (fossil); name only, defined by Davis, 1883.

Deltodus (Agassiz) Morris & Roberts, XVIII, 100; orthotype PŒCILODUS SUBLÆVIS (Ag.) McCoy (fossil).

Deltoptychius (Agassiz) Morris & Roberts XVIII, 100; orthotype COCHLIODUS ACUTUS (Ag.) McCoy (fossil).

Xystrodus (Agassiz) Morris & Roberts, XVIII, 101; orthotype COCHLIODUS STRIATUS Ag. (fossil). Name preoccupied (Plieninger, 1860), replaced by PLATYXYSTRODUS Hay (1899).

Lobodus (Agassiz) Morris & Roberts, XVIII, 101; logotype L. PROTOTYPUS Ag. (fossil); name only, defined by Davis, 1883. Regarded as a synonym of COPODUS.

Mesogomphus (Agassiz) Morris & Roberts, XVIII, 101; orthotype M. LINGUA Ag. (fossil); name only, defined by Davis, 1883. Regarded as a synonym of COPODUS.

Pleurogomphus (Agassiz) Morris & Roberts, XVIII, 101; orthotype P. AURICU-LATUS Ag. (fossil); name only, defined by Davis, 1883. Regarded as a synonym of COPODUS.

Mylax (Agassiz) Morris & Roberts, XVIII, 101; orthotype M. BATOIDES Ag. (fossil); name only, defined by Davis, 1883. Regarded as a synonym of COPODUS.

Tomodus (Agassiz) Morris & Roberts, XVIII, 101; orthotype T. CONVEXUS (Ag.) (fossil); name only, defined by Davis, 1883 (not TOMODUS Trautschold, 1879), later called OXYTOMODUS Trautschold, 1880.

Streblodus (Agassiz) Morris & Roberts, XVIII, 101; orthotype COCHLIODUS OBLONGUS (Agassiz) Portlock (fossil).

Psephodus (Agassiz) Morris & Roberts, XVIII, 101; orthotype COCHLIODUS MAGNUS Portlock (fossil).

Petalorhynchus Agassiz) Morris & Roberts, XVIII, 101; orthotype PETALODUS PSITTACINUS (Ag.) McCoy (fossil); name only, defined by Davis, 1881.

795. VALENCIENNES (1862). *Description de quelques espèces de poisson . . . envoyés de Bourbon par M. Morel.* Comptes Rendus, LIV.

ACHILLE VALENCIENNES.

This little paper of Valenciennes has been overlooked by writers. In it are described two species of IREX (ELAGATIS), I. INDICUS from Bourbon Island and I. AMERICANUS from St. Thomas, the former with D. VI-I, 24-2, the latter D. VI-I, 22-2. Günther counts D. VI-I, 25-1 in ELAGATIS BIPIN-NULATUS from Amboyna. Poey counts D. VI-I, 26-2 in ELAGATIS PINNULATUS from Havana. In a specimen from Long Island, Jordan and Evermann count D. VI-I, 27-2, regarding all as belonging to one pelagic species.

Irex Valenciennes, 1204; orthotype IREX INDICUS Val. A synonym of ELAGATIS Bennett (1834), same as SERIOLICHTHYS Blkr. (1854) and DECAPTUS Poey (1860).

796. WAGNER (1862). *Monographie der fossilen Fische aus den Lithographischen Schiefern Bayerns.* Abh. Bayer. Akad. Wiss., IX.

ANDREAS WAGNER.

Mesturus Wagner, IX, 338; orthotype M. VERRUCOSUS Wagner (fossil).

797. WINKLER (1862). *Description de quelques nouvelles espèces de poissons fossiles du calcaires lithographique de Solenhofen d'Oeningen.* Nat. Verh. Holl. Maatsch. Wetensch. Haarlem, XVI.

TIBERIUS CORNELIUS WINKLER (1822-1898).

Brachyichthys Winkler, XIV, 47; orthotype B. TYPICUS Winkler (fossil). Should replace EULEPIPOTUS (1868) and HETEROLEPIDOTUS Egert. (1872).

1863

798. BLEEKER (1863). *Mémoire sur les poissons de la côte de Guinée.* Nat. Verh. Holl. Maatsch. Wetsch., 2 Verz. Deel XVIII, 1863, 1-136.

PIETER BLEEKER.

Glaucus (Klein) Bleeker, 14; orthotype (after Klein) SCOMBER GLAUCUS L. A synonym of CÆSIOMORUS Lac.

Melanogenes Bleeker, 36; orthotype M. MACROCEPHALUS Blkr.

Pseudotolithus Bleeker, 59; orthotype P. TYPUS Blkr.

799. BLEEKER (1863). *Description de quelques espèces de poissons . . . de Chine . . . Musée de Leide, par M. G. Schlegel.* Nat Tijdschr. Dierk. Indie, I.

PIETER BLEEKER.

Hemisciæna Bleeker, 141; orthotype COLLICHTHYS LUCIDUS Günther. A synonym of COLLICHTHYS Gthr.

Pseudosciæna Bleeker, 142; orthotype CORVINA AMBLYCEPS Blkr., species named. Same as ARGYROSOMUS Pylaie, 1835. A synonym of SCIÆNA L., as restricted by Cuvier.

800. BLEEKER (1863). *Systema Cyprinoideorum revisum.* Neder. Tijdschr. Dierk., I., 1863, 179-186.

PIETER BLEEKER.

Aplocheilichthys Bleeker, 116; orthotype A. TYPUS Blkr. A synonym of APLOCHEILUS McCl. The numerous genera here defined are also contained in a second paper written at about the same time.

Gobionichthys Bleeker, 156; orthotype GOBIO MICROCEPHALUS Blkr.

Crossocheilichthys Bleeker, 192; orthotype LOBOCHEILOS COBITIS Blkr.

Isocephalus (Heckel) Bleeker, 194; orthotype CYPRINUS BANGON Ham. Same as BANGANA Ham.

Rohitodes Bleeker, 195; orthotype LABEO CEPHALUS Cuv. & Val.

Acra Bleeker, 197; tautotype CYPRINUS ACRA Ham.; orthotype as indicated by Bleeker, CYPRINUS SYRIACUS.

Leptobarbus Bleeker, 202; orthotype L. HOEVENI Blkr.

Cirrhinichthys Bleeker, 202; orthotype C. DUSSUMIERI.

Opsariichthys Bleeker, 203; orthotype LEUCISCUS UNCIROSTRIS T. & S.; also under
No. 805, p. 263.

Paraphoxinus Bleeker, 209; orthotype PHOXINELLUS ALEPIDOTUS Heckel. A synonym of PHOXINELLUS Heckel.

Trachybrama (Heckel) Bleeker, I, 210; orthotype ACANTHOBRAMA MARMID Heckel.
A synonym of ACANTHOBRAMA Heckel.

Paracheilognathus Bleeker, 213; orthotype CAPOËTA RHOMBEA T. & S.

Pseudoperilampus Bleeker, 214; orthotype P. TYPUS Blkr.

801. BLEEKER (1863). *Onzième notice sur la faune ichthyologique de l'île d'Obi.*
Neder. Tijdschr. Dierk., I, 1863, 228-238.
PIETER BLEEKER.

Paramia Bleeker, 233; orthotype CHEILODIPTERUS LINEATUS Lac. A substitute for
CHEILODIPTERUS (Lac.) Cuv., which name he transfers to the genus called
TEMNODON Cuv. (POMATOMUS Lac.) because the first species named belongs
to POMATOMUS. A synonym of CHEILODIPTERUS as now restricted.

Rhombotides (Klein) Bleeker, 235; orthotype CHÆTODON TRIOSTEGUS L. A synonym of HEPATUS Gronow or TEUTHIS L., as restricted.

Parapercis Bleeker, 236; orthotype SCIÆNA CYLINDRICA Bloch.

Parupeneus Bleeker, 342; orthotype MULLUS BIFASCIATUS Lac. A synonym of
UPENEUS Cuv., as first restricted by Bleeker.

802. BLEEKER (1863). *Septième mémoire sur la faune ichthyologique de l'île
Timor.* Neder. Tijdschr. Dierk., I 1863, 262-276.
PIETER BLEEKER.

Paradules Bleeker, 267; orthotype DULES RUPESTRIS Cuv. & Val. A synonym of
KUHLIA Gill.

Pseudechidna Bleeker, 272; orthotype MURÆNA BRUMMERI Blkr. (STROPHIDON
Gthr., not of McClelland).

803. BLEEKER (1863). *Sur les genres de la famille des Cobitioïdes.* Versl.
Akad. Amsterdam, XV, 1863, 32-44.
PIETER BLEEKER.

Paracobitis Bleeker, 37; orthotype COBITIS MALAPTERURA Cuv. & Val.

Lepidocephalichthys Bleeker, 42; orthotype COBITIS HASSELTI Cuv. & Val.

803A. BLEEKER (1863). *Systema silurorum revisum.* Neder. Tijdschr. Dierk. I.
PIETER BLEEKER.

Pseudopimelodus Bleeker, I, 101; orthotype PIMELODUS RANINUS Cuv. & Val.

804. BLEEKER (1863). *Deuxième notice sur la faune ichthyologique de l'île de
Flores.* Neder. Tijdschr. Dierk., I, 1863, 248-252.
PIETER BLEEKER.

Eurypegasus Bleeker, 250; orthotype PEGASUS DRACONIS L. = P. VOLITANS L. A
synonym of PEGASUS L.

805. BLEEKER (1863). *Notice sur les noms de quelques genres de la famille des
Cyprinoides.* Versl. Akad. Amsterdam, XV, 1863, 261-264.
PIETER BLEEKER.

Carpionichthys Bleeker, 262; orthotype CYPRINUS CARPIO L. A synonym of
CYPRINUS L.

Paraschizothorax Bleeker, 262; orthotype SCHIZOTHORAX HUGELI Heckel.

806. COOPER (1863). *On New Genera and Species of California Fishes.* 4 articles. Proc. Cal. Acad. Sci.

JAMES GRAHAM COOPER (1830-).

Dekaya Cooper, 72; orthotype D. ANOMALA Cooper. A synonym of CAULOLATILUS Gill; the name DEKAYIA already used.

Ayresia Cooper, 73; orthotype A. PUNCTIPINNIS Cooper.

Orycnus Cooper, 77; orthotype SCOMBER THYNNUS L. Originally a slip of the pen for ORCYNUS, offered as a substitute for ORCYNUS, twice preoccupied. A synonym of THUNNUS.

Gibbonsia Cooper, III, 108; orthotype MYXODES ELEGANS Cooper.

Gillichthys Cooper, 109; orthotype G. MIRABILIS Cooper.

807. DOÛMET (1863). *Description d'un nouveau genre de poissons de la Mediterannée.* Rev. Mag. Zool., XV.

P. NAPOLÉON DOUMET-ADANSON.

Trachelocirrus Doûmet, XV, 212; orthotype T. MEDITERRANEUS Doûmet. A synonym of PSENES Cuv. & Val. Doûmet correctly regards this genus as identical with NAVARCHUS Filippi & Verany. The names CUBICEPS Lowe and ATIMOSTOMA Smith have been also bestowed upon these little pelagic fishes as well as PSENES, the name they should bear.

808. GILL (1863). *(Platygobio.)* Trans. Am. Philos. Soc., V.

THEODORE GILL.

Platygobio Gill, V, 12, 178; orthotype POGONICHTHYS COMMUNIS Grd. = CYPRINUS GRACILIS Rich.

809, GILL (1863). *(Sturgeons.)* Trans. Am. Philos. Soc., V.

THEODORE GILL.

Scaphirhynchops Gill, V, 12, 178; orthotype ACIPENSER PLATORHYNCHUS Raf. Substitute for SCAPHIRHYNCHUS regarded as preoccupied.

810. GILL (1863). *On a New Family Type (Chænopsis) of Fishes Related to the Blennioids.* Ann. Lyc. Nat. Hist. N. Y., VIII.

THEODORE GILL.

Chænopsis Gill, 141; orthotype C. OCELLATUS Poey.

811. GILL (1863). *On a Remarkable New Type of Fishes Allied to Nemophis.* Ann. Lyc. Nat. Hist. N. Y., VIII.

THEODORE GILL.

Plagiotremus Gill, 138; orthotype P. SPILISTIUS Gill.

812. GILL (1863). *Catalogue of the North American Sciænoid Fishes.* Proc. Nat. Acad. Sci. Phila.

THEODORE GILL.

Sciænops Gill, 30; orthotype PERCA OCELLATA L.

813. GILL (1863). *Catalogue of the Fishes of Lower California . . . Collected by Mr. J. Xantus,* Proc. Acad. Nat. Sci., Phila., XV.

THEODORE GILL.

Pronotogrammus Gill, 80 orthotype P. MULTIFASCIATUS Gill.

Lepidamia Gill, 81; orthotype APOGON KALOSOMA Blkr.

Archamia Gill, 81; orthotype APOGON BLEEKERI Gthr.

Glossamia Gill, 82; orthotype APOGON APRION Rich.

Xenichthys Gill, 82; orthotype X. xanti Gill.

Moronopsis Gill, 82; orthotype Dules marginatus Cuv. & Val. A synonym of Kuhlia Gill.

Rhamphoberyx Gill, 87; orthotype R. pœcilopus Gill. Young form of Myripristis Cuv.

Rhombochirus Gill, 88; orthotype Echeneis osteochir Cuv. & Val.

Remoropsis Gill, 88; orthotype Echeneis brachypterus Lowe

814. GILL (1863). *Descriptions of Some New Species of Pediculati.* Proc. Acad. Nat. Sci. Phila., XV.

Theodore Gill.

Pterophryne Gill, 90; orthotype Chironectes lævigatus Cuv. Name preoccupied; replaced by Pterophrynoides Gill, 1878. A synonym of Histrio Fischer.

Histiophryne Gill, 90; orthotype Chironectes bougainvillei Cuv.

Halieutichthys (Poey) Gill, 90; orthotype H. reticulatus Poey = Lophius aculeatus Mitchill.

815. GILL (1863). *Descriptive Enumeration of a Collection of Fishes from the Western Coast of Central America . . . by Captain J. M. Dow.* Proc. Acad. Nat. Sci. Phila., XV.

Theodore Gill.

Ophioscion Gill, 165; orthotype O. typicus Gill.

Amblyscion Gill, 165; orthotype A. argenteus Gill. A section under Larimus Cuv.

Oligoplites Gill, 166; orthotype O. occidentalis Gill (not L.) = Scomber saurus Bl. & Schn.

Chænomugil Gill, 169; orthotype Mugil proboscideus Gthr. Apparently a synonym of Chelon Röse, 1793.

Rhinomugil Gill, 169; orthotype Mugil corsula Hamilton-Buchanan.

Halophryne Gill, 170; orthotype Batrachoides diemensis Le Sueur.

Leptarius Gill, 170; orthotype Leptarius dowi Gill. A synonym of Selenaspis Blkr.

Notarius Gill, 171; orthotype Arius grandicassis Cuv. & Val. Equivalent to Arius Blkr., 1863, not of 1858, and not of Cuvier & Valenciennes.

Urotrygon Gill, 173; orthotype U. mundus Gill.

816. GILL (1863). *On an Unnamed Generic Type Allied to Sebastes.* Proc. Acad. Nat. Sci. Phila., XV.

Theodore Gill.

Sebastoplus Gill, 208; orthotype Scorpæna kuhli Bowdich.

817. GILL (1863). *Description of a New Generic Type of Ophidioids.* Proc. Acad. Nat. Sci. Phila., XV.

Theodore Gill.

Leptophidium Gill, 210; orthotype L. profundorum Gill. Name preoccupied in snakes. Replaced by Lepophidium Gill, 1895.

818. GILL (1863). *Synopsis of the Pomacentroids of the Western Coast of North and Central America.* Proc. Acad. Nat. Sci. Phila., XV.

Theodore Gill.

Dischistodus Gill, 213; orthotype Pomacentrus fasciatus Cuv. & Val.

Acanthochromis Gill, 214; orthotype Dascyllus polyacanthus Blkr.

Pomataprion Gill 216; orthotype Hypsypops dorsalis Gill. A synonym of Microspathodon Gthr.

819. GILL (1863). *Notes on the Labroids of the Western Coast of North America.* Proc. Acad. Nat. Sci. Phila., XV.
THEODORE GILL.

Lepidaplois Gill, 222; orthotype COSSYPHUS AXILLARIS Cuv. & Val.

Euphysocara Gill, 222; orthotype COSSYPHUS ANTHIOIDES Gthr.

Gymnopropoma Gill, 222; orthotype COSSYPHUS BILUNULATUS Cuv. & Val. A synonym of LEPIDAPLOIS Gill.

Achœrodus Gill, 222; orthotype COSSYPHUS GOULDI Rich.

Dimalacocentrus Gill, 223; orthotype NOVACULA KALLOSOMA Blkr. A synonym of NOVACULICHTHYS Blkr.

Oxyjulis Gill, 223; orthotype JULIS MODESTUS Grd.; the name MODESTUS preoccupied in JULIS; replaced by HALICHŒRES CALIFORNICUS Gthr.

820. GILL (1863). *Synopsis of the Family of Lepturoids and Description of a Remarkable New General Type.* Proc. Acad. Nat. Sci. Phila., XV.
THEODORE GILL.

Evoxymetopon (Poey) Gill, 228; orthotype E. TÆNIATUS Poey.

821. GILL (1863). *Synopsis of the North American Gadoid Fishes.* Proc. Acad. Nat. Sci. Phila.

Brachygadus Gill, 230; orthotype GADUS MINUTUS L. A synonym of TRISOPTERUS Raf.

Micromesistius Gill, 231; orthotype GADUS POUTASSOU Risso. Called BRACHY-MESISTIUS on page 233.

Brachymesistius Gill, 233; orthotype GADUS POUTASSOU Risso. A lapsus for MICROMESISTIUS.

Urophycis Gill, 240; orthotype BLENNIUS REGIUS Walbaum.

Rhinonemus Gill, 241; orthotype GADUS CIMBRIUS L.

822. GILL (1863). *Descriptions of the Genera of Gadoid and Brotuloid Fishes of North America.* Proc. Acad. Nat. Sci. Phila., XV.
THEODORE GILL.

Leptogadus Gill, 248; orthotype GADUS BLENNIOIDES Pallas. Perhaps a synonym of GADICULUS Guich.

Odontogadus Gill, 248; orthotype GADUS EUXINUS Nordmann.

Nematobrotula Gill, 252; orthotype BROTULA ENSIFORMIS Gthr.

Stygicola Gill, 252; orthotype LUCIFUGA DENTATA Poey.

Hoplobrotula Gill, 253; orthotype BROTULA ARMATA T. & S.

823. GILL (1863). *Synopsis of the Family of Lycodidæ.* Proc. Acad. Nat. Sci. Phila., XV.

Macrozoarces Gill, 258; orthotype BLENNIUS ANGUILLARIS Peck.

824. GILL (1863). *Descriptions of the Gobioid Genera of the Western Coast of Temperate North America.* Proc. Acad. Nat. Sci. Phila., XV.
THEODORE GILL.

Pomatoschistus Gill, 263; orthotype GOBIUS MINUTUS L. Apparently a synonym of BRACHYOCHIRUS Nardo.

Deltentosteus Gill, 263; orthotype GOBIUS QUADRIMACULATUS.

Coryphopterus Gill, 263; orthotype C. GLAUCOFRÆNUM Gill. A synonym of RHINO-GOBIUS Gill.

Coryphogobius Gill, 263; obviously a lapse for CORYPHOPTERUS.

Pterogobius Gill, 266; orthotype GOBIUS VIRGO T. & S.

Synechogobius Gill, 266; orthotype GOBIUS HASTA T. & S. A synonym of ACANTHO-
GOBIUS Gill.

825. GILL (1863). *On the Gobioids of the Eastern Coast of the United States*,
Proc. Acad. Nat. Sci. Phila., XV.
THEODORE GILL.

Gymnogobius Gill, 269; orthotype GOBIUS MACROGNATHOS Blkr. A synonym of
CHÆNOGOBIUS Gill.

Ophiogobius Gill, 269; orthotype GOBIUS OPHIOCEPHALUS Jenyns.

Crystallogobius Gill, 269; orthotype GOBIUS NILSSONI Duben & Koren.

Boreogobius Gill, 269; orthotype GOBIUS STUVITZI Duben & Koren.

Mogurnda Gill, 270; orthotype ELEOTRIS MOGURNDA Rich.

Ophiocara Gill, 270; orthotype ELEOTRIS OPHIOCEPHALUS Cuv. & Val.

Gobiomorphus Gill, 270; orthotype ELEOTRIS GOBIODES Cuv. & Val.

Hypseleotris Gill, 270; orthotype ELEOTRIS CYPRINOIDES Cuv. & Val.

Odonteleotris Gill, 270; orthotype ELEOTRIS MACRODON Blkr.

Calleleotris Gill, 270; orthotype ELEOTRIS STRIGATA Cuv. & Val. A synonym of
VALENCIENNEA.

Ptereleotris Gill, 271; orthotype ELEOTRIS MICROLEPIS Blkr.

826. GILL (1863). *On the Genus Periophthalmus of Schneider*. Proc. Acad.
Nat. Sci. Phila.
THEODORE GILL.

Euchoristopus Gill, 271; orthotype GOBIUS KOELREUTERI Pallas. Replaces PERIOPH-
THALMUS of recent authors, of which the logotype is P. SCHLOSSERI.

Boleops Gill, 271; orthotype BOLEOPHTHALMUS AUCUPATORIUS Rich. A synonym of
SCARTELAOS Sw.

827. GILL (1863). *Note on the Genera of Hemirhamphinæ*. Proc. Acad. Nat. Sci.
Phila., XV.
THEODORE GILL.

Zenarchopterus Gill, 273; orthotype HEMIRAMPHUS DISPAR Cuv. & Val.

Oxyporhampus Gill, 273; orthotype HEMIRAMPHUS CUSPIDATUS Cuv. & Val.

828. GÜNTHER (1863). *On New Species of Fishes from Victoria, South Australia*.
Ann. Mag. Nat. Hist., XI.
ALBERT GÜNTHER.

Melambaphes Günther, XI, 115; orthotype GLYPHIDODON NIGRORIS Gthr., not of
Cuv. & Val. = M. GUENTHERI Gill.

Rhombosolea Günther, XI, 117; orthotype R. FLESOIDES Gthr.

829. GÜNTHER (1863). *On New Species of Fishes from the Essequibo*. Ann.
Mag. Nat. Hist., XII.
ALBERT GÜNTHER.

Helogenes Günther, XII, 443; orthotype H. MARMORATUS Gthr.

Crenuchus Günther, XII, 443; orthotype C. SPILURUS Gthr.

830. JOHNSON (1863). *Description of Three New Genera of Marine Fishes
Obtained at Madeira*. Ann. Mag. Nat. Hist., XIV.
JAMES YATES JOHNSON.

Diretmus Johnson, 403; orthotype D. ARGENTEUS Johnson.

Halosaurus Johnson, 407; orthotype H. OWENI Johnson.
Chiasmodon Johnson, 408; orthotype C. NIGER Johnson. Spelled CHIASMODUS by Günther.

831. KAUP (1863). *Einige japanische Fische.* Nederl. Tijdschr. Dierk., I.
JOHANN JACOB KAUP.
Helotosoma Kaup, I, 162; orthotype H. SERVUS Kaup; said to be a synonym of ATYPICHTHYS Gthr.

832. KNER (1863). *Ueber einige fossile Fische aus den Kreide- und Tertiar- schichten von Comen und Podsused.* Sitzb. Akad. Wiss. Wien, XLVIII.
RUDOLF KNER.
J. Y. JOHNSON.
Amiopsis Kner, XLVIII, 126; orthotype A. PRISCA Kner (fossil). Fossil homologue of AMIA L.
Scombroclupea Kner, XLVIII, 132; orthotype S. PINNULATA Kner (fossil). The earlier name of this species, CLUPEA MACROPHTHALMA Heckel, 1849, is pre-occupied; Ranzani, 1842.

833. KNER & STEINDACHNER (1863). *Neue Beiträge zur Kenntniss der fossilen Fische Oesterreichs.* Denkschr. Akad. Wiss. Wien, XXI.
RUDOLPH KNER; FRANZ STEINDACHNER.
Pseudosyngnathus Kner & Steindachner, XXI, 28; orthotype SYNGNATHUS OPIS-THOPTERUS Ag. (fossil).

834. KNER & STEINDACHNER (1863). *Neue Gattungen und Arten von Fischen aus Central-Amerika gesammelt von Prof. Moritz Wagner.* Abh. Münchner. Akad. Wiss., X.
RUDOLPH KNER; FRANZ STEINDACHNER.
Date given by Dean as 1870; Troschel quotes these genera as of 1867; Günther, *Zoological Record,* notes them as previous to 1864.
Saccodon Kner & Steindachner, X, 31; orthotype S. WAGNERI Kner & Steindachner.
Pseudochalceus Kner & Steindachner, X, 31; orthotype P. LINEATUS Kner & Stein-dachner.
Chalcinopsis Kner & Steindachner, X, 31; logotype C. STRIATULUS Kner & Stein-dachner. A synonym of BRYCON M. & T., not CHALCINOPSIS Holmberg, 1891, which is PSEUDOCORYNOPOMA.

835. MACKIE (1863). *On a New Species of Hybodus from the Lower Chalk.* The Geologist, VI.
SAMUEL JOSEPH MACKIE
Nemacanthus Mackie, VI, 243 (orthotype ACRODUS KEUPERIANUS Murchison & Strickland (fossil). A synonym of ACRODUS; name preoccupied, 1837.

836. VON DER MARCK (1863). *Fossile Fische, Krebse und Pflanzen aus dem Plattenkalk der jüngsten Kreide in Westphalen.* Palæontographica, XI, Cassel.
W. VON DER MARCK.
Macrolepis von der Marck, XI, 26; orthotype M. ELONGATUS Marck (fosil).
Rhabdolepis von der Marck, XI, 26; orthotype R. CRETACEUS Marck (fossil). Name preoccupied; replaced by HOLCOLEPIS von der Marck, 1868.
Palæolycus von der Marck, XI, 31; orthotype P. DREGINENSIS Marck (fossil).
Scrobodus (Münster) von der Marck, XI, 38; orthotype S. SUBOVATUS Münster (fossil). A synonym of LEPIDOTES.

328   THE GENERA OF FISHES, PART III

Microcœlia von der Marck, XI, 48; orthotype M. GRANULATA Marck (fossil).
Leptosomus von der Marck, XI, 49; orthotype L. GUESTPHALICUS Marck (fossil). (Name several times preoccupied.)
Tachynectes von der Marck, XI, 51; logotype T. MACRODACTYLUS Marck (fossil).
Enchelurus von der Marck, XI, 58; orthotype E. VILLOSUS Marck (fossil).
Leptotrachelus von der Marck, XI, 59; orthotype L. ARMATUS Marck (fossil).

837. PUTNAM (1863). *List of the Fishes Sent by the Museum in Exchange,* etc. Bull. Mus. Comp. Zool. Harvard College, I.
FREDERICK WARD PUTNAM (1839-1915).

Nothonotus (Agassiz) Putnam, I, 3; orthotype ETHEOSTOMA MACULATA Kirtland.
Microperca Putnam, I, 4; orthotype M. PUNCTULATA Putnam.
Cottogaster Putnam, I, 4; orthotype BOLEOSOMA TESSELLATUM Thompson, not of Dekay = COTTOGASTER PUTNAMI J. & G.
Pleurolepis (Agassiz) Putnam, I, 5; orthotype ETHEOSTOMA PELLUCIDUM Ag. Name preoccupied; replaced by VIGIL Jordan, 1919.
Anarmostus (Scudder) Putnam, I, 12; logotype DIABASIS FLAVOLINEATUS Desmarest. Name preoccupied; a synonym of HÆMULON Cuv.
Bathystoma (Scudder) Putnam, I, 12; logotype PERCA MELANURA L. = HÆMULON JENIGUANO Poey.

838. SCHAFHÄUTL (1863). *Südbayerns Lethæa Geognostica.*
K. E. SCHAFHÄUTL (1803-1890).

Diaphyodus Schafhäutl, 246; orthotype D. OVALIS Schafhäutl (fossil). A synonym of LABRODON.

839. SIEBOLD (1863). *Die Süsswasserfische von Mitteleuropa.*
CARL THEODOR ERNST VON SIEBOLD (1804-1885).

Abramidopsis Siebold, 117; orthotype A. LEUCKARTI Siebold. Thought to be a hybrid of ABRAMIS and RUTILUS.
Bliccopsis Siebold, 142; orthotype B. ABRAMORUTILUS Siebold. Thought to be a hybrid: BLICCA and RUTILUS.

840. STEINDACHNER (1863). *Ichthyologische Mittheilungen: V. Ueber einige Labroiden.* Verh. Zool. Bot. Ges. Wien, XIII.
FRANZ STEINDACHNER.

Cheiliopsis Steindachner, 1113; orthotype C. BIVITTATUS Steind.

841. STEINDACHNER (1863). *Beiträge zur Kenntniss der fossilen Fischfauna Oesterreichs.* Sitz. Akad. Wiss. Wien, XLVII.
FRANZ STEINDACHNER.

Calamostoma Steindachner, XLVII, 133; logotype ACANTHURUS CANOSSÆ Heckel (fossil). Name preoccupied, replaced by AULORHAMPHUS De Zigno, 1887.

842. STEINDACHNER (1863). *Beiträge zur Kenntniss der Sciænoiden Brasiliens mit der Cyprinodonten Mexicos.* Sitz. Akad. Wiss. Wien, XLVIII.
FRANZ STEINDACHNER.

Diplolepis Steindachner, 2; orthotype SCIÆNA SQUAMOSISSIMA Heckel. Name preoccupied; a synonym of PLAGIOSCION Gill.
Pœcilodes Steindachner, XLVIII, 176; orthotype P. BIMACULATUS Steind. Synonym of PSEUDOXIPHOPHORUS Blkr.

843. WAGNER (1863). *Monographie der fossilen Fische aus den lithographischen Schiefern Bayerns*. IX.

ANDREAS WAGNER.

Heterostrophus Wagner, IX, 614; orthotype H. LATUS Wagner (fossil). Printed HETEROSTICHUS in *Anz. Bayr. Akad. Wiss.*, 1860, p. 63. This name is preoccupied.

Plesiodus Wagner, IX, 632; orthotype P. PUSTULOSUS Wagner (fossil). A synonym of LEPIDOTES.

Eurycormus Wagner, IX, 707; orthotype E. SPECIOSUS Wagner (fossil). A synonym of MACROPOMA Ag.

Liodesmus Wagner, IX, 709; orthotype PHOLIDOPHORUS GRACILIS Ag. (fossil).

Macrorhipis Wagner, IX, 723; orthotype PACHYCORMUS MUENSTERI Wagner (fossil). A synonym of IONOSCOPUS.

1864

844. VAN BENEDEN & DE KONINCK (1864). *Notice sur le Palædaphus insignis*. Bull. Roy. Soc. Belge, XVII.

PIERRE J. VAN BENEDEN (1809-1894); L. J. DE KONINCK.

Palædaphus Van Beneden & De Koninck, XVII, 150; orthotype P. INSIGNIS Beneden & Koninck (fossil).

845. BLEEKER (1864). *Notice sur la faune ichthyologique de Siam*. Versl. Akad. Amsterdam, XVI, 1864, 352-358.

PIETER BLEEKER.

Heterobagrus Bleeker, 355; orthotype H. BOCOURTI Blkr.

846. BOCAGE & CAPELLO (1864). *Sur quelque espèces inédites de Squalidæ . . . que frequentent les côtes du Portugal*. Proc. Zool. Soc. London.

JOSÉ VINCENT BARBOZA DU BOCAGE (1823-1907);

FELIX DE BRITO CAPELLO (1838-   ).

Centroscymnus Bocage & Capello, 263; orthotype C. CŒLOLEPIS Bocage & Capello.

Scymnodon Bocage & Capello, 263; orthotype S. INGENS Bocage & Capello.

847. CANESTRINI (1864). *Studi sui Lepadogaster del Mediterraneo*. Arch. Zool. Anat. Fisiol., III.

GIOVANNI CANESTRINI.

Mirbelia Canestrini, 189; logotype LEPADASTER DECANDOLLII Risso.

848. COCCHI (1864). *Monografia dei Pharyngodopilidæ nuovo famiglia dei pesci Labroidei*. Ann. Mus. Firenze, I.

IGINO COCCHI (1828-   ).

Egertonia Cocchi, I, 121; orthotype E. ISODONTA Cocchi (fossil).

Pharyngodopilus Cocchi, I, 123; logotype P. CANARIENSIS Cocchi (fossil). A synonym of LABRODON.

Taurinichthys Cocchi, I, 152; orthotype SCARUS MIOCENICUS Michelotti (fossil).

849. COPE (1864). *On a Blind Silurid . . . from Pennsylvania*. Proc. Acad. Nat Sci. Phila.

EDWARD DRINKER COPE (1840-1897).

Gronias Cope, 231; orthotype GRONIAS NIGRILABRIS Cope.

850. COSTA (1864). *(Fossil Fishes.)* Atti, Pontan.

ORONZIO GABRIELE COSTA.

Oenoscopus Costa, VIII, 59; orthotype IONOSCOPUS PETRAROIÆ Costa (fossil). Name a (needless?) substitute for IONOSCOPUS.

Hyptius Costa, 80; orthotype H. SEBASTIANI Costa. A synonym of SAUROPSIDIUM Costa, 1850.

Platycerhynchus Costa, VIII, 98; orthotype P. RHOMBEUS Costa (fossil). An indeterminable ally of BELONOSTOMUS.

851. EGERTON (1864). *On Some Ichthyolites . . . from New South Wales.* Quart. Journ. Geol. Soc., XX.

PHILIE DE MALPAS GREY EGERTON.

Myriolepis Egerton, XX, 3; orthotype M. CLARKEI Egert. (fossil).

Cleithrolepis Egerton, XX, 3; orthotype C. GRANULATUS Egert. (fossil).

852. GILL (1864). *Description of a New Labroid Genus Allied to Trochocopus Gthr.* Proc. Acad. Nat. Sci. Phila., XVI.

THEODORE GILL.

Pimelometopon Gill, 58; orthotype LABRUS PULCHER Ayres.

853. GILL (1864). *Notes on the Nomenclature of Genera and Species of the Family Echeneidoidæ.* Proc. Acad. Nat. Sci. Phila., XVI.

THEODORE GILL.

Leptecheneis Gill, 60; orthotype ECHENEIS NAUCRATES L. A synonym of ECHENEIS L. as restricted by Gill, 1860.

854. GILL (1864). *Critical Remarks on the Genera Sebastes and Sebastodes of Ayres.* Proc. Acad. Nat. Sci. Phila., XVI.

THEODORE GILL.

Sebastomus Gill, 147; orthotype SEBASTES ROSACEUS Grd.

Sebastosomus Gill, 147; orthotype SEBASTES MELANOPS Grd.

855. GILL (1864). *Second Contribution to the Selachology of California.* Proc. Acad. Nat. Sci. Phila., XVI.

THEODORE GILL.

Eugaleus Gill, 148; orthotype SQUALUS GALEUS L. Substitute for GALEUS Cuv., preoccupied by GALEUS Raf. A synonym of GALEORHINUS Blaniv., as restricted by Gill.

Pleuracromylon Gill, 148; orthotype MUSTELUS LÆVIS = GALEUS MUSTELUS Gill. A synonym of MUSTELUS Cuv., not of VALMONT, the last probably not eligible.

856. GILL (1864). *Several Points in Ichthyology.* Proc. Acad. Nat. Sci. Phila., XVI.

THEODORE GILL.

Electrophorus Gill, 151; orthotype GYMNOTUS ELECTRICUS L.

Gymnotes Gill, 152; orthotype GYMNOTUS ÆQUILABIATUS Humboldt. A synonym of STERNOPYGUS M. & T.

Hypopomus Gill, 152; orthotype RHAMPHICHTHYS MULLERI Kaup. Replaces BRACHYRHAMPHICHTHYS Gthr., 1870.

857. GILL (1864). *Notes on the Paralepidoids and Microstomatoids and on Some Peculiarities of Arctic Ichthyology.* Proc. Acad. Nat. Sci. Phila., XVI.

Arctozenus Gill, 188; orthotype PARALEPIS BOREALIS Reinh.

858. GILL (1864.) *Note on the Cyclopteroids of Eastern North America.* Proc. Acad. Nat. Sci. Phila., **XVI.**
THEODORE GILL.

Eumicrotremus Gill, 190; orthotype CYCLOPTERUS SPINOSUS Müller.

859. GILL (1864). *Synopsis of the Pleuronectoids of California and Northwestern America.* Proc. Acad. Nat. Sci. Phila., XVI.
THEODORE GILL.

Lepidopsetta Gill, 195; orthotype PLATICHTHYS UMBROSUS Grd.
Hypsopsetta Gill, 195; orthotype PLEURONICHTHYS GUTTULATUS Grd.
Metoponops Gill, 198; orthotype M. COOPERI Gill = PSETTICHTHYS SORDIDUS Grd. A synonym of ORTHOPSETTA Gill.

860. GILL (1864). *On the Affinity of Several Doubtful British Fishes.* Proc. Acad. Nat. Sci. Phila., XVI.
THEODORE GILL.

Helminthodes Gill, 203; orthotype OXYBELES LUMBRICOIDES Blkr.
861. GILL (1864). *Note on the Family of Stichæoids.* Proc. Acad. Nat. Sci. Phila., XVI.

Anisarchus Gill, 210; orthotype CLINUS MEDIUS Reinh.
Eumesogrammus Gill, 210; orthotype CLINUS PRÆCISUS Kröyer.

862. GILL (1864). *Synopsis of the Pleuronectoids of the Eastern Coast of North America.* Proc. Acad. Nat. Sci. Phila., XVI.
THEODORE GILL.

Pomatopsetta Gill, 216; orthotype PLATESSA DENTATA Storer = PLEURONECTES PLATESSOIDES Fabricius. A synonym of HIPPOGLOSSOIDES Gottsche.
Lophopsetta Gill, 216; orthotype PLEURONECTES MACULATUS Mitchill.
Liopsetta Gill, 216; orthotype PLATESSA GLABRA Storer.
Myzopsetta Gill, 217; orthotype PLATESSA FERRUGINEA Storer. A synonym of LIMANDA Gotsche.
Chænopsetta Gill, 218; orthotype PLATESSA OCELLARIS Dekay. A synonym of PARALICHTHYS Grd.
Euchalarodus Gill, 221; orthotype E. PUTNAMI Gill, the male of LIOPSETTA GLABRA (Storer). A synonym of LIOPSETTA Gill.
Ancylopsetta Gill, 224; orthotype A. QUADROCELLATA Gill. A synonym of PSEUDORHOMBUS Blkr.

863. GILL (1864). *Synopsis of the Eastern American Sharks.* Proc. Acad. Nat. Sci. Phila., XVI.
THEODORE GILL.

Eugomphodus Gill, 26; orthotype SQUALUS LITTORALIS Mitchill. A synonym of CARCHARIAS Raf.
Platypodon Gill, 264; orthotype CARCHARIS MENISORRAH M. & H.
Isistius Gill, 264; orthotype SCYMNUS BRASILIENSIS Q. & G.
Euprotomicrus Gill, 264; orthotype SCYMNUS LABORDEI M. & H.
Rhinoscymnus Gill, 264; orthotype SCYMNUS ROSTRATUS Risso.

864. GILL (1864). *Review of Holbrook's Ichthyology of South Carolina.* Am. Journ. Sci. Arts, XXXVII.
THEODORE GILL.

Enneacanthus Gill, 92; orthotype Pomotis obesus Grd.

Acantharchus Gill, 92; orthotype Centrarchus pomotis Baird.

Hyperistius Gill, 92; orthotype Centrarchus hexacanthus Cuv. A section under Pomoxis Raf.

Mesogonistius Gill, 92; orthotype Pomotis chætodon Baird.

Chænobryttus Gill, 92; orthotype Calliurus melanops Grd. = Pomotis gulosus Cuv. & Val. Misprinted Chænobrythus.

Eucentrarchus Gill, 93; orthotype Perca iridea Bosc. A substitute for Centrarchus as currently used—not of Cuvier & Valenciennes, which is a synonym of Ambloplites Raf.

865. GRAY (1864). *Notice of a Portion of a New Form of Animal (Myriosteon higginsi) Probably Indicating a New Group of Echinodermata.* Proc. Zool. Soc. London.

John Edward Gray.

Myriosteon Gray; orthotype M. higginsi Gray; a tube from a rostral cartilage of Pristis, supposed to be an echinoderm. A synonym of Pristis Latham.

866. GUICHENOT (1864). *L. Maillard: Notes sur l'île de la Réunion.* Fauna Ichthiologique.

Alphonse Guichenot.

Glyphodes Guichenot, 3; orthotype G. aprionoides Guich. Name preoccupied in Lepidoptera.

Cotylopus Guichenot (appendix) 10; logotype C. acutipinnis Guich.

867. GÜNTHER (1864). *Catalogue of the Fishes in the British Museum.* V.

Albert Günther.

Copiodoglanis Günther, V, 25; logotype Plotosus tandanus Mitchill. A synonym of Tandanus Mitchill.

Amphilius Günther, V, 115; orthotype Pimelodus platychir Gthr.

Notoglanis Günther, V, 136; haplotype Pimelodus multiradiatus Kner = P. arakaima Schomburgk. A section under Rhamdia.

Euclyptosternum Günther, V, 180; orthotype Silurus cous L. A substitute for Aglyptosternon Blkr., 1863, the thorax having a sucking disk, which Bleeker's name would deny. The change is not admissible.

Diplomystax Günther, V, 180; orthotype Arius papillosus Cuv. & Val. An emendation of Diplomystes Blkr.

Paradiplomystax Günther, V, 180; orthotype Pimelodus coruscans Lichtenstein. An emendation of Paradiplomystes Blkr.

Rhinoglanis Günther, V, 216; orthotype R. typus Gthr.

Cnidoglanis Günther, V, 217; logotype Plotosus megastomus Rich.

Callomystax Günther, V, 218; orthotype Pimelodus gagata Ham. = Gagata aypus Blkr. An unwarranted substitute for Gagata Blkr., incorrectly defined. A synonym of Batasio Blyth.

Stygogenes Günther, V, 223; orthotype S. humboldti Gthr. Perhaps a synonym of Cyclopium Sw.

Scleromystax Günther, V, 225; orthotype Callichthys barbatus Q. & G.

Liposarcus Günther, V, 234; logotype Hypostomus multiradiatus Hancock.

Cochliodon (Heckel) Günther, V, 238; tautotype Hypostomus cochliodon Kner.

Heterythrinus Günther, V, 283; orthotype Erythrinus salmoneus Gronow. A synonym of Erythrinus (Gronow) Scopoli.

Cænotropus Günther, V, 297; orthotype CITHARINUS CHILODUS Cuv. & Val. = CHILODUS PUNCTATUS M. & T. A substitute for CHILODUS M. & T. and MICRODUS Kner, both names preoccupied.

Brachyalestes Günther, V, 314; logotype MYLETES NURSE Rüppell.

Hemibrycon Günther, V, 318; haplotype TETRAGONOPTERUS POLYODON Gthr.

Creatochanes Günther, V, 318; logotype SALMO MELANURUS Bloch.

Scissor Günther, V, 331; orthotype S. MACROCEPHALUS Gthr.

Creagrutus Günther, V, 339; orthotype LEPORINUS MUELLERI Gthr.

Anacyrtus Günther, V, 345; logotype SALMO GIBBOSUS L. Substitute for EPICYRTUS, preoccupied.

Rœstes Günther, V, 345; orthotype CYNOPOTAMUS MOLOSSUS Kner.

Rœboides Günther, V, 345; logotype ANACYRTUS GUATEMALENSIS Gthr.

Hystricodon Günther, V, 349; orthotype EXODON PARADOXUS M. & T. = EPICYRTUS EXODON Cuv. & Val. A substitute for EXODON M. & T., said to be preoccupied.

Sarcodaces Günther, V, 353; orthotype SALMO ODOË Bloch.

Oligosarcus Günther, V, 353; orthotype O. ARGENTEUS Gthr.

Ichthyborus Günther, V, 364; orthotype I. MICROLEPIS Gthr.

Prototroctes Günther, V, 383; orthotype P. MARÆNA Gthr.

Coccia Günther, V, 387; orthotype GONOSTOMUS OVATUS Cocco. A substitute for ICHTHYOCOCCUS Bon., a name given in honor of Cocco. Such names have "always been considered as a nuisance," hence changed by Günther, an action without warrant under accepted rules.

Notoscopelus Günther, V, 405; logotype LAMPANYCTUS RESPLENDENS Rich. Replaces MACROSTOMA Risso, preoccupied as MACROSTOMUS.

Dasyscopelus Günther, V, 405; orthotype MYCTOPHUM ASPERUM Rich.

Ceratoscopelus Günther, V, 405; orthotype SCOPELUS MADERENSIS Lowe.

Melamphaës Günther, V, 433; orthotype METOPIAS TYPHLOPS Lowe, substitute for METOPIAS, preoccupied.

868. GÜNTHER (1864). *Description of a New Fossil Fish from the Lower Chalk (Folkestone).* Geol. Mag., I.
ALBERT GÜNTHER.

Plinthophorus Günther, I, 115; orthotype P. ROBUSTUS Gthr. (fossil).

869. GÜNTHER (1864). *On Some New Species of Central American Fishes.* Proc. Zool. Soc. London.
ALBERT GÜNTHER.

Microdesmus Günther, 26; orthotype M. DIPUS Gthr.

870. GÜNTHER (1864). *On a New Genus of Pediculate Fish (Melanocetus johnsoni) from the Sea of Madeira.* Proc. Zool. Soc. London.
ALBERT GÜNTHER.

Melanocetus Günther, 301; orthotype M. JOHNSONI Gthr.

871. GÜNTHER (1864). *On a New Generic Type of Fishes Discovered by the Late Dr. Leichardt in Queensland.* Ann. Mag. Nat. Hist.
ALBERT GÜNTHER.

Scleropages Günther, XIV, 195; orthotype S. LEICHARDTI Gthr.

872. GÜNTHER (1864). *Report on a Collection of Reptiles and Fishes Made by Dr. Kirk in the Zambesi and Nyassa Regions .* Proc. Zool. Soc. London.
ALBERT GÜNTHER.

Peletrophus Günther, 314; logotype P. MICROLEPIS Günther.

873. GÜNTHER (1864). *Zoological Record for 1864.*
ALBERT GÜNTHER.

Auchenoglanis Günther, 165; orthotype PIMELODUS BISCUTATUS Geoffroy St. Hilaire. A substitute for AUCHENASPIS Blkr. (not Egerton), preoccupied.

Gillia Günther, 157; orthotype GILLICHTHYS MIRABILIS Cooper. A substitute for GILLICHTHYS, deemed "barbarous."

874. JÄCKEL (1864). *Die Fische Bayerns.* Abh. Zool. Min. Ver. Regensburg.
ANDREAS JOHANNES JÄCKEL.

Scardiniopsis Jäckel, 101; orthotype S. ANCEPS Jäckel. Supposed to be a hybrid of RUTILUS and SCARDINIUS.

875. KNER (1864). *Specielles Verzeichniss der Kaiserlichen Fregatte "Novara" gesammelten Fische.* Sitzb. Akad. Wiss. Wien, LIII.
RUDOLF KNER.

Pseudomugil Kner, LIII, 543; orthotype P. SIGNIFER Kner.

Chœroplotosus Kner, LIII, 545; orthotype PLOTOSUS LIMBATUS Cuv. & Val.

Silurodon Kner, LIII, 546; orthotype S. HEXANEMA Kner.

876. MARCUSEN (1864). *Die Familie der Mormyren, etc.* Mem. Acad. Sci. St. Petersb., VII.
JOHANN MARCUSEN.

Petrocephalus Marcusen, 1; logotype MORMYRUS BANE Lac.

Phagrus Marcusen, 1; orthotype MORMYRUS DORSALIS Geoffroy St. Hilaire.

877. PETERS (1864). *Ueber einige neue Säugethiere, Amphibien und Fische.* Sitzb. Akad. Wiss. Berlin.
WILHELM PETERS.

Saurenchelys Peters, 397; orthotype S. CANCRIVORA Peters. A synonym of CHLOPSIS Raf.

Diaphanichthys Peters, 399; orthotype D. BREVICAUDUS Peters. Larva, allied to CONGER.

878. POWRIE (1864). *On the Fossiliferous Rock of Forfarshire and Their Contents.* Quart. Journ. Geol., XX.
JAMES POWRIE.

Ictinocephalus (Page) Powrie, XX, 419; orthotype DIPLACANTHUS GRACILIS Egert. (fossil). Listed name only as I. GRANULATUS by Page, *Rept. Brit. Assn.,* 1858, 104.

Euthacanthus Powrie, XX, 425; orthotype E. MACNICOLI Powrie. A synonym of CLIMATIUS Ag.

879. ROMANOWSKY (1864). *Description de quelques restes de poissons fossiles trouvés dans le calcaire carbonifère du gouvernement de Toula.* Bull Soc. Imp. Nat. Moscou, II.
GHENNADI DANILOVICH ROMANOWSKY (= ROMANOVSKII).

Lophodus Romanowsky, XXXVI, 160; logotype L. IRREGULARIS Romanowsky (fossil). Apparently a synonym of PSEPHODUS.

Plintholepis Romanowsky, XXXVII, 169; orthotype P. RETRORSA Romanowsky (fossil).

Sporolepis Romanowsky, XXXVII, 169; logotype S. PYRIFORMIS Romanowsky (fossil).

1865

880. BLEEKER (1865). *Rhinobagrus et Pelteobagrus, deux genres nouveaux de Siluroides de Chine.* Neder. Tijdschr. Dierk., II, 1865, 710.
PIETER BLEEKER.

Rhinobagrus Bleeker, 9; orthotype R DUMERILI Blkr.

Pelteobagrus Bleeker, 9; orthotype SILURUS CALVARIUS Basilewsky = PIMELODUS FULVIDRACO Rich. Replaces FLUVIDRACO Jordan & Starks.

881. BLEEKER (1865). *Paralaubuca, un genre nouveau de Cyprinoïdes de Siam.* Neder. Tijdschr. Dierk., II, 1865, 15-17.
PIETER BLEEKER.

Paralaubuca Bleeker, II, 16; orthotype P. TYPUS Blkr.

882. BLEEKER (1865). *Notices sur quelques genres et espèces de Cyprinoïdes de Chine.* Neder. Tijdschr. Dierk., II, 1865, 28-29.
PIETER BLEEKER.

Parabramis Bleeker, 22; orthotype ABRAMIS PEKINENSIS Basil.

Acanthobrama Bleeker, 25; orthotype A. SINONI Blkr.

Paracanthobrama Bleeker, 23; orthotype P. GUICHENOTI Blkr.

Pseudolaubuca Bleeker, 28; orthotype P. SINENSIS Blkr.

883. BLEEKER (1865). *Poissons inédits indo-archipélaiques de l'ordre des Murènes.* Neder. Tijdschr. Dierk., II, 1865, 55-62.
PIETER BLEEKER.

Brachyconger Bleeker, II, 236; orthotype MURÆNA SAVANNA Cuv. A synonym of CYNOPONTICUS Costa.

Achirophichthys Bleeker, II, 242; orthotype A. TYPUS Blkr. A pparently the young of BRACHYSOMOPHIS.

884. BLEEKER (1865). *Systema Murænorum revisum.* Neder. Tijdschr. Dierk., II, 1865, 113-122.
PIETER BLEEKER.

Paranguilla Bleeker, 113; orthotype ENCHELYOPUS TIGRINUS Ag. (fossil). Name a substitute for ENCHELYOPUS Ag., preoccupied.

Pseudomoringua Bleeker, 114; orthotype MORINGUA LUMBRICOIDEA Rich.

Oxyconger Bleeker, 116; orthotype CONGER LEPTOGNATHUS Blkr.

885. BLEEKER (1865). *Quatrième notice sur la faune ichthyologique de l'île de Bouro.* Neder. Tijdschr. Dierk., II, 1865.
PIETER BLEEKER.

Tetrades Bleeker, 145; orthotype TETRADRACHMUM XANTHOSOMA Blkr. Apparently a lapsus for TETRADRACHMUM.

886. BLEEKER (1865). *Notice sur le genre Paraploactis et description de son espèce type.* Neder. Tijdschr. Dierk., 1865, 168-170.
PIETER BLEEKER.

Paraploactis Bleeker, II, 169; orthotype P. TRACHYDERMA Blkr.

886A. BLEEKER (1865). *Description de quelques espèces de poissons du Japan, du Cap de Bonne Espérance et de Suriname . . . au Musée de Leide.* Neder. Tijdschr. Dierk., II, 250-269.

Parajulis Bleeker, 251; orthotype JULIS PŒCILOPTERUS T. & S. A synonym of HALICHŒRES Rüppell (HEMIULIS Sw.).

887. BLEEKER (1865). *Sixième notice sur la faune ichthyologique de Siam.*
Neder. Tijdschr. Dierk., II, 1865, 171-176.
PIETER BLEEKER.
Parastromateus Bleeker, 174; orthotype STROMATEUS NIGER Bloch. A synonym of
APOLECTUS Cuv. & Val.

888. BLEEKER (1865). *Énumération des espèces de poissons actuellement connues
de l'île de Ceram.* Neder. Tijdschr. Dierk., II, 1865, 182-183.
Solenichthys Bleeker, 183; logotype FISTULARIA PARADOXUS Pallas. A substitute for
SOLENOSTOMUS Lac., regarded as preoccupied by SOLENOSTOMUS Gronow;
and of Klein, which is a synonym of FISTULARIA L. Later (1873) Bleeker
changed SOLENICHTHYS to SOLENOSTOMATICHTHYS.

889. BLEEKER (1865). *Énumeration des espèces de poissons actuellement connues
de l'île d'Amboine.* Neder. Tijdschr. Dierk., II, 1865, 270-283.
PIETER BLEEKER.
Paradiodon Bleeker, 271; orthotype DIODON HYSTRIX L., the name DIODON being
transferred to DIODON ATINGA, the first species named by Linnæus. A syno-
nym of DIODON L. as now understood.
Pseudalutarius Bleeker, 273; orthotype ALUTERES NASICORNIS T. & S. Name
unnecessarily changed to PSEUDALUTERES Blkr., 1866.
Paraplagusia Bleeker, 274; orthotype PLAGUSIA BILINEATA Blkr.
Pseudamia Bleeker, 284; orthotype APOGON POLYSTIGMA Blkr.

890. BLEEKER (1865). *Atlas ichthyologique,* etc. V. *Baudroies, Ostracions,* etc.
PIETER BLEEKER.
Atopomycterus (Kaup) Bleeker, 49; orthotype DIODON NYCTHEMERUS Cuv..
Uranostoma (Bibron) Bleeker, 63; orthotype not stated, passing reference to a
manuscript name in the Museum of Paris. (A synonym of SPHEROIDES or of
TETRAODON.
Pleuranacanthus (Bibron) Bleeker, 65; orthotype P. ARGENTATUS Bibron =
TETRAODON SCELERATUS Forster. A manuscript name in the Museum of
Paris, noted by Bleeker in *Synonymy.* Same as PROMECOCEPHALUS Bibron;
a synonym of LAGOCEPHALUS Sw.

891. COPE (1865). *Partial Catalogue of the Cold-blooded Vertebrata of Michigan.
Supplemental Note on a Peculiar Genus of Cyprinidæ.* Proc. Acad. Nat. Sci.
Phila.
EDWARD DRINKER COPE.
Ericymba Cope, 88; orthotype E. BUCCATA Cope.

892. COSTA (1865). *Studi sopra i terrene ad ittioliti delle provincie Napolitane,* etc.
Atti Reale Accad. Sci. Napoli.
ORONZIO GABRIELE COSTA.
Heterolepis Costa, II, 4; orthotype (not stated by Woodward) (fossil).
Andreiopleura Costa, II, 27; orthotype A. VETUSTISSIMA Costa (fossil). Called
A. ESIMIA (EXIMIA) Costa on page 10.

893. DAY (1865). *The Fishes of Malabar.*
FRANCIS DAY (1829-1889).
Paranandus Day, 130; orthotype CATOPRA MALABARICA. A synonym of PRISTOLEPIS
Jerdon (CATOPRA).

Platacanthus Day, 296; orthotype P. AGRENSIS Day. A synonym of LEPIDO-CEPHALICHTHYS Blkr.

Brachygramma Day, 304; orthotype B. JERDONI Day. A synonym of AMBLY-PHARYNGODON Blkr.

Paradanio (Bleeker) Day, 319; orthotype PERILAMPUS AUROLINEATUS Day. A synonym of DANIO Ham.

894. DUMÉRIL (AUGUSTE) (1865). *Histoire naturelle des poissons ou ichthyologie générale.* Vol. I.

AUGUSTE HENRI ANDRÉ DUMÉRIL.

Paratrygon Duméril, 594; orthotype a specimen from Brazil, identified as TRYGON AIREBA M. & H. = RAJA ORBICULARIS Bl. & Schn., usually regarded as replacing POTAMOTRYGON Garman, 1877, but Eigenmann considers it as the same as DISCEUS Garman.

895. GILL (1865). *Synopsis of the Fishes of the Gulf of St. Lawrence and Bay of Fundy.* Canadian Naturalist, II.

THEODORE GILL.

Stilbius Gill, 262 (18, reprint); orthotype CYPRINUS AMERICANUS Lac. A substitute for STILBE, preoccupied. A synonym of NOTEMIGONUS Raf.

Micristius Gill, 24 (reprint); logotype FUNDULUS ZONATUS Cuv. & Val. A synonym of ZYGONECTES Ag.

896. GILL (1865). *On the Genus Caulolatilus.* Proc. Acad. Nat. Sci. Phila.

THEODORE GILL.

Prolatilus Gill, 66; orthotype LATILUS JUGULARIS Cuv. & Val.

897. GILL (1865). *On the Cranial Characters of Gadus proximus Grd.* Proc. Acad. Nat. Sci. Phila.

THEODORE GILL.

Microgadus Gill, 69; orthotype GADUS PROXIMUS Grd.

898. GILL (1865). *On a New Genus of Serraninæ.* Proc. Acad. Nat. Sci. Phila.

THEODORE GILL.

Trisotropis Gill, 104; orthotype JOHNIUS GUTTATUS Bl. & Schn. = PERCA VENENOSA L.

Enneacentrus Gill, 105; orthotype PERCE PUNCTATA L. A synonym of CEPHA-LOPHOLIS Bl. & Schn.

Petrometopon Gill, 105; orthotype SPARUS CRUENTATUS Lac.

899. GILL (1865). *On a New Generic Type of Sharks.* Proc. Acad. Nat. Sci. Phila.

THEODORE GILL.

Micristodus Gill, 177; orthotype M. PUNCTATUS Gill. A synonym of RHINEODON Smith.

900. GILL (1865). *Note on the Family of Myliobatoids with Description of a New Species of Aëtobatis.* Ann. Lyc. N. Y., VII.

THEODORE GILL.

Mylorhina Gill, 136; orthotype RHINOPTERA LALANDI M. & H.

Micromesus Gill, 136; orthotype RHINOPTERA ADSPERSA M. & H.

901. GÜNTHER (1865). *Description of a New Characinoid Genus of Fish (Phago loricatus) from West Africa.* Ann. Mag. Nat. Hist., XV.

ALBERT GÜNTHER.

Phago Günther, XV, 3; orthotype P. LORICATUS Gthr.

902, 903. GÜNTHER (1865). *Zoological Record.* II.
ALBERT GÜNTHER.

Herpetoichthys (Smith) Günther, 159; orthotype ERPETOICHTHYS MALABARICUS Smith. Name preoccupied, replaced by CALAMOICHTHYS Smith, 1866.

904. JOHNSON (1865). *Description of a New Genus of Trichuroid Fishes Obtained at Madeira, with Remarks, etc.* Proc. Zool. Soc. London.
JAMES YATE JOHNSON.

Nealotus Johnson, 434; orthotype N. TRIPES Johnson.

905. KNER (1865). *Einige neue Fische,* etc. Anz. Akad. Wiss. Wien.
RUDOLF KNER.

Thysanocheilus Kner, I, 185; orthotype T. ORNATUS Kner, not THYSANOCHILUS Philippi, 1857.

Leius Kner, 186; orthotype L. FEROX Kner. A synonym of ISISTIUS Gill.

906. KNER (1865). *Reise der Oesterreichischen Fregatte "Novara" um die Erde in den Jahren 1857-59, unter . . . B. von Wüllersdorf-Urbain.*
RUDOLPH KNER.

Pseudomugil Kner, 275; orthotype P. SIGNIFER Kner.

Chœroplotosus Kner, 300; orthotype C. DECEMFILIS Kner. A synonym of CNIDOGLANIS Gthr.

Silurodon Kner, 305; orthotype S. HEXANEMA Kner.

Opsarius Kner, 355; orthotype LEUCISCUS PARVUS T. & S. A synonym of PSEUDORASBORA Blkr.

907. KNER (1865). *Psalidostoma, eine neue Characinengattung aus dem Weissen Nil.* Sitzb. Akad. Wiss. Wien. L.
RUDOLPH KNER.

Psalidostoma Kner, L, 99; orthotype P. CAUDIMACULATUM Kner.

908. LANKESTER (1865). *On the British Species of Cephalaspis and the Scotch Pteraspis.* Rept. Brit. Assn. Adv. Sci. for 1864.
EDWIN RAY LANKESTER.

Scaphaspis Lankester, 58; orthotype CEPHALASPIS LLOYDI Ag. (fossil). = CEPHALASPIS ROSTRATUS Ag. A synonym of PTERASPIS Ag.

909. LEIDY (1865). *(Tomodon).* Smithsonian Contributions.
JOSEPH LEIDY.

Tomodon Leidy, XIV, 102; orthotype TOMODON HORRIFICUS Leidy (fossil); originally described as a reptile tooth, hence needlessly replaced by DIPLOTOMODON Leidy, 1868.

910. LIOY (1865). *Sopre alcuni Avanzi di Plagiostomi fossili del Vicentino,* etc. Atti Soc. Ital. Sci. Nat. Milano, VIII.
PAOLO LIOY (1836-   ).

Uropterina Lioy, VIII, 113; orthotype U. PLATYRHACHIS Lioy (fossil).

Alopiopsis Lioy, VIII, 403; orthotype A. PLEJODON Leidy (GALEUS CUVIERI Ag.) (fossil). A synonym of PROTOGALEUS and SCOLIODON.

Ptericephalina Lioy, 414; orthotype CLUPEA LEPTOSTEA Ag. (fossil). An ally of SPRATELLA Val. (the sprat).

911. MARCK & SCHLÜTER (1865).   *Neue Fische und Krebse aus der Kreide*
     von Westphalien. XV. *Palæontographica.*
       W₁ VON DER MARCK; AUGUST JOSEPH SCHLÜTER (1836-   ).
Megapus Schlüter, XV, 275; orthotype M. GUESTPHALICUS Schlüter (fossil). A syno-
    nym of CHEIROTHRIX.
Telepholis von der Marck, XV, 276; orthotype T. ACROCEPHALUS Marck (fossil).
Dactylopogon von der Marck, XV, 278; orthotype D. GRANDIS Marck (fossil).
Holcolepis von der Marck, XV, 278; orthotype RHABDOLEPIS CRETACEUS Marck
    (fossil). Replaces RHABDOLEPIS, preoccupied.
Brachyspondylus von der Marck, XV, 283; orthotype B. CRETACEUS Marck (fossil).
Dermatoptychus von der Marck, XV, 287; orthotype D. MACROPHTHALMUS Marck
    (fossil). A synonym of OSMEROIDES Ag.
Archæogadus von der Marck, XV, 291; orthotype H. GUESTPHALICUS Marck
    (fossil). A synonym of HALEC Ag.

912. OWEN (1865).   *Description of Portions of a Large Extinct Fish (Stereo-*
     dus) . . . from the Middle Beds of the Maltese Miocene.
                          RICHARD OWEN.
Stereodus Owen, II, 147; orthotype S. MELITENSIS Owen (fossil).

913. PETERS (1865).   *Ueber eine neue Percoidengattung Plectroperca aus*
     Japan, etc. Mon. Ber. Akad. Wiss. Berlin.
                          WILHELM PETERS.
Plectroperca Peters, 121; orthotype P. BERENDTI Peters = P. CHUATSI Basilewsky.
    A synonym of SINIPERCA Gill.

914. POEY (1865).   *Repertorio fisico-natural de la isla de Cuba: Peces Ciegos.*
                          FELIPE POEY Y ALOY.
Stygicola (Gill) Poey, 116; orthotype LUCIFUGA DENTATUS Poey.

915. SMITH, JOHN ALEXANDER (1865).   *Description of Erpetoichthys, a*
     New Genus of Ganoid Fish from Old Calabar, etc. Edinburg Roy. Phys. Soc.
                          JOHN ALEXANDER SMITH.
Erpetoichthys Smith, 273; orthotype CALAMOICHTHYS MALABARICUS Smith. Name
    preoccupied; replaced by CALAMOICHTHYS Smith.

916. TICKELL (1866).   *Description of a Proposed New Genus of the Gadidæ.*
                    Journ. Asiat. Soc. Bengal, XXXIV.
                          S. R. TICKELL.
Asthenurus Tickell, XXXIV, 32; orthotype A. ATRIPINNIS Tickell. A synonym
    of BREGMACEROS Thompson.

1866

917. BLANCHARD (1866).   *Les poissons des eaux douces de la France,* etc.
                    CHARLES ÉMILE BLANCHARD (1820-1900).
Cyprinopsis Blanchard, 336; orthotype CYPRINUS CARASSIUS L. A synonym of
    CARASSIUS Nilsson.
918. BLEEKER (1866).   *Systema Balistidorum, Ostracionidorum, Gymnodontido-*
     rumque revisum. Neder. Tijdschr. Dierk., III, 1866, 8-10.
                          PIETER BLEEKER.

Parabalistes Bleeker, 10; orthotype BALISTES CHRYSOSPILUS Blkr.

Pseudobalistes Bleeker, 11; orthotype BALISTES FLAVIMARGINATUS Rüppell.

Pseudomonacanthus Bleeker, 12; orthotype MONACANTHUS MACRURUS Blkr.

Paramonacanthus Bleeker, 12; orthotype MONACANTHUS CURTORHYNCHUS Blkr.

Acanthaluteres Bleeker, 13; orthotype MONACANTHUS PERONI Hollard.

Brachaluteres Bleeker, 13; orthotype ALUTERIUS TROSSULUS Rich.

Liomonacanthus Bleeker, 13; orthotype MONACANTHUS PARDALIS Rüppell. A synonym of CANTHERINES Sw.

Pseudaluteres Bleeker, 14; orthotype ALUTERES NASICORNIS T. & S. Originally called PSEUDALUTARIUS by Bleeker in 1865.

Paraluteres Bleeker, 14; orthotype ALUTARIUS PRIONURUS Blkr.

Oxymonacanthus Bleeker, 16; orthotype O. CHRYSOSPILUS Blkr. = BALISTES LONGIROSTRIS Bl. & Schn.

Trichodiodon Bleeker, 18; orthotype DIODON PILOSUS Mitchill. Probably a larval DIODON.

Crayracion (Klein) Bleeker, 18; orthotype C. LÆVISSIMUS EX TERREO RUFESCENS Klein = TETRAODON SPENGLERI Bloch. Probably a synonym of TETRAODON, but Bleeker regards it as a species of OVOIDES as now understood.

Acanthostracion Bleeker, 27; orthotype OSTRACION QUADRICORNIS L.

919. BLEEKER (1866). *Sur les espèces d'Exocet de l'Inde archipélagique.* Neder. Tijdschr. Dierk., III, 1866, 105-129.
PIETER BLEEKER.

Parexocœtus Bleeker, III, 126; orthotype EXOCŒTUS MENTO Cuv. & Val.

920. BLEEKER (1866). *Révision des Hémirhamphes de l'Inde archipélagique.* Neder. Tijdschr. Dierk., III, 1866, 136-170.
PIETER BLEEKER.

Hemirhamphodon Bleeker, 118; orthotype HEMIRHAMPHUS PHAIOSOMA Blkr.

921. BLEEKER (1866). *Description du Narcacion Polleni, espèce inédite des mers de l'île de la Réunion.* Neder. Tijdschr. Dierk., III, 1866, 171-173.
PIETER BLEEKER.

Narcacion (Klein) Bleeker, III, 171; orthotype RAJA TORPEDO L. Equivalent to TORPEDO Dum.

922. COSTA (1866). *(Ittiolite.)* Ann. Accad. Aspir. Nat., V.
GABRIELE ORONZIO COSTA.

Lobodus Costa, V, 81; orthotype L. PEDEMONTANUS Costa (fossil).

923. GÜNTHER (1866). *Catalogue of Fishes in the British Museum.* VI.
ALBERT GÜNTHER.

Hucho Günther, VI, 125; orthotype SALMO HUCHO L. = HUCHO GERMANORUM Gthr.

Brachymystax Günther, VI, 162; orthotype SALMO COREGONOIDES Pallas.

Luciotrutta Günther, VI, 164; orthotype SALMO MACKENZIEI Rich. A synonym of STENODUS Rich.

Potamorrhaphis Günther, VI, 234; orthotype BELONE TÆNIATA Gthr.

Arrhamphus Günther, VI, 276; orthotype A. SCLEROLEPIS Gthr.

Fitzroyia Günther, VI, 307; orthotype LEBIAS MULTIDENTATA Jenyns.

Characodon Günther, VI, 308; orthotype C. LATERALIS Gthr.

Limnurgus Günther, VI, 309; orthotype L. VARIEGATUS Gthr. A needless substitute for GIRARDINICHTHYS Blkr., called "barbarous".

Jenynsia Günther, VI, 331; orthotype LEBIAS LINEATA Jenyns; a synonym of FITZROYIA.

Platypœcilus Günther, VI, 350; orthotype P. MACULATUS Gthr.

923A. GÜNTHER (1866). *On the Fishes of the States of Central America,* etc. Proc. Zool. Soc. London.

ALBERT GÜNTHER.

Neetroplus Günther, 603; orthotype N. NEMATOPUS Gthr.

924. HUXLEY (1866). *British Fossils: Crossopterygian Ganoids.* Mem. Geol. Surv. United Kingdom.

THOMAS HUXLEY (1825-1895).

Eurypoma Huxley, 32; orthotype MACROPOMA EGERTONI Ag. (fossil). A synonym of EURYCORMUS and of MACROPOMA.

925. GUICHENOT (1866). *Le Trigle polyommate, nouveau genre de poissons,* (etc.). *Ichthyologie.* Ann. Soc. Maine-et-Loire, IX.

ALPHONSE GUICHENOT.

Hoplonotus Guichenot, IX, 3; orthotype TRIGLA POLYOMMATA Rich. Preoccupied in beetles, 1851; replaced by PTERYGOTRIGLA Waite (1899) and later by OTOHIME Jordan & Starks (1905).

926. GUICHENOT (1866). *Le Zancle centrognathe, nouveau genre de Chétodons.* Ann. Soc. Maine-et-Loire, IX.

ALPHONSE GUICHENOT.

Gnathocentrum Guichenot, IX, 4; orthotype G. CENTROGNATHUM Guich. = CHÆTO- DON CORNUTUS L. A synonym of ZANCLUS.

927. GUICHENOT (1866). *L'Argentine leioglosse, nouveau genre de Salmonoides.* Ann. Soc. Maine-et-Loire, IX.

ALPHONSE GUICHENOT.

Glossanodon Guichenot, IX, 9; orthotype ARGENTINA LEIOGLOSSA Cuv. & Val.

928. GUICHENOT (1866). *Notice sur un nouveau genre de la famille des Cottoides,* etc. Mem. Soc. Sci. Nat. Cherbourg.

ALPHONSE GUICHENOT.

Agonomalus Guichenot, 252; orthotype ASPIDOPHALUS PROBOSCIDALIS Val.

929. KNER (1866). *Neue Fische aus dem Museum der Herren J. Cæsar Godeffroy & Sohn in Hamburg.* Sitzb. Akad. Wiss. Wien, LIV

RUDOLPH KNER.

Strabo Kner & Steindachner, LIV, 372; orthotype S. NIGROFASCIATUS K. & S. A synonym of MELANOTÆNIA Gill.

930. KNER (1866). *Die Fische der bituminösen Schiefer von Raibl in Kärnthen.* Sitzb. Akad. Wiss. Wien, LIII.

RUDOLPH KNER.

Graphiurus Kner, LIII, 155; orthotype G. CALLOPTERUS Kner (fossil).

Orthurus Kner, LIII, 163; orthotype O. STURI Kner (fossil).

Megalopterus Kner, LIII, 174; orthotype M. RAIBLIANUS Kner (fossil)

Peltopleurus Kner, LIII, 180; orthotype P. SPLENDENS Kner (fossil).

931. NEWBERRY & WORTHEN (1866). *Descriptions of New Species of Verte-*
tebrates Mainly from the Subcarboniferous Limestones and Coal Measures
of Illinois. Geol. Surv. Ills., II.

JOHN STRONG NEWBERRY; AMOS H. WORTHEN.

Antliodus Newberry & Worthen, II, 33; logotype A. PARVULUS Newberry &
Worthen. A synonym of PETALODUS Owen.

Dactylodus Newberry & Worthen, II, 33; logotype D. PRINCEPS Newberry &
Worthen (fossil).

Aspidodus Newberry & Worthen, II, 92; logotype PSEPHODUS CRENULATUS
St. John & Worthen (fossil). A synonym of PSEPHODUS.

Sandalodus Newberry & Worthen, II, 102; orthotype S. ANGUSTUS Newberry &
Worthen. (fossil).

Rinodus Newberry & Worthen, II, 106; orthotype R. CALCEOLUS Newberry &
Worthen (fossil). A synonym of PTYCTODUS.

Trigonodus Newberry & Worthen, II, 111; logotype T. MAJOR, Newberry &
Worthen (fossil). Apparently a synonym of PSEPHODUS.

Drepanacanthus Newberry & Worthen, II, 120 (fossil); orthotype D. ANCEPS
Newberry & Worthen. A synonym of PHYSONEMUS.

932. OWEN (1866). *On a Genus and Species of Sauroid Fish. (Thlattodus*
suchoides Ow.) from the Kimmeridge Clay of Norfolk. Geol. Mag.

RICHARD OWEN.

Thlattodus Owen, III, 55; orthotype T. SUCHOIDES Owen (fossil). A synonym
of CATURUS.

933. OWEN (1866). *On a Genus and Species of Sauroid Fish (Ditaxiodus impar*
Ow.) from the Kimmeridge Clay of Culham, Oxfordshire. Geol. Mag.

RICHARD OWEN.

Ditaxiodus Owen, 107; orthotype D. IMPAR Owen (fossil). A synonym of
CATURUS.

934. PETERS (1866). *Mittheilung über Fische, etc.* Mon. Akad. Wiss. Berlin.

WILHELM PETERS.

Auliscops Peters, 510; orthotype A. SPINESCENS Peters = AULORHYNCHUS FLAVIDUS.
Gill. A synonym of AULORHYNCHUS Gill.

Labracoglossa Peters, 513; orthotype L. ARGENTEIVENTRIS Peters.

Nematocentris Peters, 516; orthotype N. SPLENDIDA Peters. A synonym of MELA-
NOTŒNIA Gill.

935. PICTET & HUMBERT (1866). *Nouvelles recherches sur les poissons*
fossiles du Mont Liban. Arch. Sci. Génève, XXVI.

FRANÇOIS JULES PICTET; A. HUMBERT.

Cheirothrix Pictet & Humbert, 51; orthotype C. LIBANICUS Pictet & Humbert
(fossil). Spelled CHIROTHRIX by Woodward.

Solenognathus Picetet & Humbert, 54; orthotype S. LINEOLATUS Pictet & Humbert
(fossil). Name preoccupied; same as CHARITOSOMUS von der Marck, 85.

Opisthopteryx Pictet & Humbert, 78; (misprinted OPISTOPTERYX); orthotype
MESOGASTER GRACILIS Pictet.

Pseudoberyx Pictet & Humbert, 32; logotype P. SYRIACUS Pictet & Humbert
(fossil).

Aspidopleurus Pictet & Humbert, 107; orthotype A. CATAPHRACTUS Pictet & Humbert. A synonym of PRIONOLEPIS Egert.

936. PLAYFAIR (1866). *The Fishes of Zanzibar.*
ROBERT PLAYFAIR.

Tripterodon Plaiyfair, 42; orthotype T. ORBIS Playfair.

937. REINHARDT (1866). *Om trende formentligt ubeskrevne Fisk of Chara-cinernes eller Karpelaxenes familie* Overs. Dansk. Forh. Kjöbenhavn.
JOHANN THEODOR REINHARDT (1816-1882).

Piabina Reinhardt, 49; orthotype P. ARGENTEA Reinh.

Characidium Reinhardt, 56; orthotype C. FASCIATUM Reinh.

938. SCHMIDT (1866). *Ueber Thyestes verrucosus . . . silurische Fischreste auf der Insel Oesel.* Verh. Min. Ges. St. Petersburg.
FRIEDRICH SCHMIDT (1832-1908).

Tremataspis Schmidt, I, 233; orthotype CEPHALASPIS SCHRENKI Pander (fossil).
Apparently a synonym of STIGMOLEPIS Pander.

939. SMITH (1866). *Description of Calamoichthys, a New Genus of Ganoid Fish from Old Calabar,* etc. Trans. Roy. Soc. Edinburg, XXIV.
JOHN ALEXANDER SMITH.

Calamoichthys J. A. Smith, V, 654; orthotype HERPETOICHTHYS CALABARICUS Smith.
A substitute for HERPETOICHTHYS or ERPETOICHTHYS, preoccupied in eels.

940. STEINDACHNER (1866). *(Richardsonia.)* Sitzb. Akad. Wiss. Wien.
FRANZ STEINDACHNER.

Richardsonia Steindachner, 53; orthotype ARGENTINA RETROPINNA Rich. A synonym
of RETROPINNA Gill.

941. STEINDACHNER (1866). *Zur Fischfauna von Port Jackson in Australien.*
Sitzb. Akad. Wiss. Wien, LIII.
FRANZ STEINDACHNER.

Schuettea Steindachner, LIII, 449; orthotype S. SCALARIPINNIS Steind.

Heterochœrops Steindachner, LIII, 461; orthotype H. VIRIDIS Steind

942. STEINDACHNER (1866). *Ichthyologischer Bericht ueber eine nach Spanien und Portugal unternomener Reise.* IV. Sitzb. Akad. Wiss. Wien.
FRANZ STEINDACHNER.

Pachychilon Steindachner, LIV, 263; orthotype SQUALIUS PICTUS Heckel & Kner.

943. STEINDACHNER (1866). *Zur Fischfauna Kaschmirs und der benachbarten Landesstreiche.* Verh. Zool. Bot. Ges. Wien, XVI.
FRANZ STEINDACHNER.

Kneria Steindachner, 769; orthotype K. ANGOLENSIS Steind.

Ctenopharyngodon Steindachner, 782; orthotype C. LATICEPS Steind.

Schizopygopsis Steindachner, 786; orthotype S. STOLICZKÆ Steind.

Diptychus Steindachner, 787; orthotype D. MACULATUS Steind.

Ptychobarbus Steindachner, 789; orthotype P. CONIROSTRIS Steind

944. TROSCHEL (1866). *Ein Beitrag zur ichthyologischen Fauna der Insel der Grünen Vorgebirges (Cape Verde Islands).* Arch. Naturg.
FRANZ HERMANN TROSCHEL.

Onychognathus Troschel, 231; orthotype O. CAUTUS Troschel.

945. YOUNG (1866). *On the Affinities of Platysomus and Allied Genera.*
Quart. Journ. Geol., XXII.
JOHN YOUNG.

Amphicentrum Young, XXII, 306; orthotype A. GRANULOSUM Young (fossil). A synonym of CHEIRODUS.

Euryosomus Young, XXII, 311; orthotype E. MACRURUS Young (fossil). A synonym of GLOBULODUS.

Mesolepis Young, XXII, 313; logotype M. WARDI Young (fossil).

Cycloptychius Young, 318; orthotype C. CARBONARIUS (Huxley) Young (fossil).

946. YOUNG (1866). *Notice of New Genera of Carboniferous Glyptodipterines.* Quart. Journ. Geol. Soc., XXII.

JOHN YOUNG.

Rhizodopsis Young, XXII 596; orthotype HOLOPTYCHIUS SAUROIDES Williamson, not of Binney = RHIZODUS GRANULATUS (Ag.), fossil.

Dendroptychius (Huxley) Young, XXII, 601; orthotype HOLOPTYCHIUS SAUROIDES Binney (fossil). Identical with STREPSODUS Young and with page priority.

Strepsodus (Young) XXII, 602; orthotype HOLOPTYCHIUS SAUROIDES Binney (fossil).

Rhomboptychius (Huxley) Young, XXII, 604; orthotype (MEGALICHTHYS INTERMEDIUS Woodw.) (fossil). A synonym of CENTRODUS.

1867

947. COPE (1867). *Description of a New Genus (Phenacobius teretulus) of Cyprinoid Fisches from Virginia.* Proc. Acad. Nat. Sci. Phila.

EDWARD DRINKER COPE.

Phenacobius Cope, 96; orthotype P. TERETULUS Cope.

948. COPE (1867). *On the Distribution of Fresh-water Fishes in the Allegheny Region of Southwestern Virginia.* Journ. Acad. Nat. Sci. Phila.

EDWARD DRINKER COPE.

Hemioplites Cope, 217; orthotype H. SIMULANS Cope. An aberrant specimen of ENNEACANTHUS.

949. COPE (1867). *Synopsis of the Cyprinidæ of Pennsylvania: Supplement on Some New Species of American and African Fishes.* Trans. Amer. Philos. Soc., XIII.

EDWARD DRINKER COPE.

Photogenis Cope, 378; tautotype SQUALIUS PHOTOGENIS Cope = PHOTOGENIS LEUCOPS Cope. P. SPILOPTERUS, a species of the group EROGALA, has been named as type by Jordan, but the tautotype takes precedence.

Cryptosmilia Cope, XIII, 401; orthotype C. LUNA Cope. A synonym of DREPANE Cuv. (ENIXE Gistel).

Enteromius Cope, XIII, 405; orthotype E. POTAMOGALIS Cope.

Sphagomorus Cope, XIII, 407; orthotype HIPPOGLOSSUS ERUMEI Cuv. An ally of PSETTODES Bennett, but with simple, not barbed, teeth.

950. DAY (1867). *On Some New or Imperfectly Known Fishes of Madras.* Proc. Zool. Soc. London.

FRANCIS DAY.

Gogrius Day, 563; orthotype PIMELODUS GOGRA Sykes = GOGRIUS SYKESI Day. A synonym of RITA Blkr.

951. GÜNTHER (1867). *On the Identity of Alepisaurus (Lowe) with Plagyodus (Steller).* Ann. Mag. Nat. Hist., XIX.

ALBERT GÜNTHER.

Plagyodus (Steller, in *Pallas*, 1811) Günther, XIX, 185; orthotype Plagyodus (Steller) Pallas = Alepisaurus æsculapius Bean. Revival of an old name which may be intenable as mononomial and printed only in the dative case ("Plagyodontem") by its author.

952. GÜNTHER (1867). *Desription of Some Little Known Species of Fishes in . . . the British Museum.* Proc. Zool. Soc. London.
Albert Günther.
Champsodon Günther, 102; orthotype C. vorax Gthr.

953. GÜNTHER (1867). *On a New Form of Mudfish from New Zealand.* Ann. Mag. Nat. Hist., XX.
Albert Günther.
Neochanna Günther, X, 305; orthotype N. apoda Gthr.

954. GÜNTHER (1867). *Additions to the Knowledge of Australian Reptiles and Fishes.* Ann. Mag. Nat. Hist., XX.
Albert Günther.
Sticharium Günther, XX, 62, orthotype S. dorsale Gthr.
Notograptus Günther, XX, 63; orthotype N. guttatus Gthr.

955. GÜNTHER (1867). *New Fishes from Gaboon and the Gold Coast.* Ann. Nat. Hist.
Albert Günther.
Gymnallabes Günther, XX, 111; orthotype G. typus Gthr.
Nannocharax Günther, XX, 112; orthotype N. fasciatus Gthr.
Xenocharax Günther, XX, 112; orthotype X. spilurus Gthr.

956. GUICHENOT (1867). *Notice sur le lophiopside, nouveau genre de poisson de la famille de lophioides,* etc. Mem. Soc. Nat. Sci. Cherbourg, XIII.
Alphonse Guichenot.
Lophiopsides Guichenot, 21; orthotype Lophius vomerinus Cuv. & Val. Distinguished from Lophius by the absence of vomerine teeth; perhaps lost with age.

957. GUICHENOT (1867). *Notice sur le Neosébaste, nouveau genre,* etc. Mem. Soc. Nat. Sci. Cherbourg, XIII.
Alphonse Guichenot.
Neosebastes Guichenot, XIII, 83; orthotype N. scorpænoides Guich.

958. GUICHENOT (1867). *Notice sur le Sériolophe, nouveau genre de poissons de la famille des scombéroides,* etc. Mem. Soc. Nat. Sci. Cherbourg, XIII, 90-95.
Alphonse Guichenot.
Leptolepis (Van Hasselt) Guichenot, 68; orthotype L. argenteus Van Hasselt. Name of type only, no definition. Possibly intended for Pampus argenteus; name preoccupied in fossils.
Seriolophus Guichenot, 90; orthotype S. carangoides Guich. A synonym of Nematistius Gill.

959. GUICHENOT (1867). *Notice sur le Salarichthys nouveau genre de poissons,* etc. Mem Soc. Nat. Sci. Cherbourg.
Alphonse Guichenot.

Salarichthys Guichinot, 14; orthotype SALARIAS VOMERINUS Cuv. & Val. = SALARIAS TEXTILIS Q. & G.   (Written SALARIICHTHYS by Jordan, 1890.)

960. HUXLEY (1867).   *(Fossil Fishes.)*   Trans. Royal Irish Society, XXIV.
THOMAS HENRY HUXLEY.

Campylopleuron Huxley & Wright, XXIV, 353 (fossil).   Probably a synonym of CTENODUS.

961. JOHNSON (1867).   *Description of a New Genus of Spinacidæ Founded on a Shark Obtained at Madeira.*   Proc. Zool. Soc. London.
JAMES YATE JOHNSON.

Machephilus Johnson, 713; orthotype M. DUMERILI Johnson = SQUALUS SQUAMOSUS Bonnaterre.   A synonym of LEPIDORHINUS Bon.

962. KNER (1867).   *Neuer Beitrag zur Kenntniss der fossilen Fische von Comen bei Görz.*   LVI.   Sitzb. Akad. Wiss. Wien.
RUDOLF KNER.

Hemisaurida Kner, LVI, 172; orthotype H. NEOCOMIENSIS Kner (fossil).

Hemirhynchus Kner, LVI, 182; logotype H. HECKELI Kner (fossil).   A synonym of BELONOSTOMUS Ag.

963. KNER (1867).   *Nachtrag zu den fossilen Fischen von Raibl.*   Sitzb. Akad. Wiss. Wien, LV.
RUDOLF KNER.

Pterygopterus Kner, LV, 722; orthotype P. APUS Kner (fossil).   Probably a synonym of THORACOPTERUS.

964. KNER (1867).   *(Teleosaurus.)*   Sitzb. Akad. Wiss. Wien, LVI.
RUDOLF KNER.

Teleosaurus Kner, LVI, 905; orthotype T. TENUISTRIATUS Kner (fossil).   A synonym of SAURORHYNCHUS.

965. KNER (1867).   *(Scarostoma.)*   Sitzb. Akad. Wiss. Wien.
RUDOLF KNER.

Scarostoma Kner, LVI, 715; orthotype S. INSIGNE Kner.   An ally of OPLEGNATHUS Rich., perhaps generically distinct.

966. KREFFT (1867).   *Descriptions of Some New Australian Fresh-water Fishes.*   Proc. Zool. Soc. London.
GERARD KREFFT.

Mionorus Krefft, 943; orthotype M. LUNATUS Krefft (scales 30).   An ally of APOGON (AMIA).

967. LANKESTER (1867).   *On Didymaspis, a New Genus of Cephalaspidian Fishes.*   Geol. Mag., IV.
EDWIN RAY LANKESTER.

Didymaspis Lankester, IV, 152; orthotype D. GRINDRODI Lank. (fossil).

968. OWEN (1867).   *(Dittodus, etc.)*   Trans. Odontograph Soc. London, V.
RICHARD OWEN.

Dittodus Owen, V, 325; logotype DITTODUS DIVERGENS Owen (fossil).   Probably founded on teeth of ORTHACANTHUS.

Mitrodus Owen, V, 338; orthotype M. QUADRICORNIS Owen (fossil). A synonym of GYRACANTHUS Ag.

Ganacrodus Owen, V, 340; orthotype G. HASTULA Owen (fossil). Probably identical with ELONICHTHYS.

Ageleodus Owen, V, 340; orthotype A. DIADEMA Owen (fossil). Regarded as identical with CALLOPRISTODUS Traquair, 1888, and of earlier date.

Ochlodus Owen, V, 346; orthotype O. CRASSUS Owen (fossil). Probably a synonym of DICRANODUS and ORTHACANTHUS.

Ganolodus Owen, V, 354; logotype G. CRAGGESI Owen (fossil). A synonym of RHIZODOPSIS.

Aganodus Owen, V, 359; logotype A. APICALIS Owen (fossil). A synonym of ORTHACANTHUS (DICRANODUS).

Pternodus Owen, V, 363; orthotype P. PRODUCTUS Owen (fossil). A synonym of ORTHACANTHUS (DICRANODUS).

Sagenodus Owen, V, 365; orthotype S. INEQUALIS Owen (fossil).

Characodus Owen, V, 366; logotype C. CONFERTUS Owen (fossil). A synonym of RHIZODOPSIS, not CHARACODON Gthr., 1866, not CHARACODUS (Ag.) Davis, 1883.

Gastrodus Owen, 370; orthotype G. PRÆPOSITUS Owen (fossil). A synonym of RHIZODOPSIS.

969. POEY (1867). *Revista de los peces descritos par Poey*. In Repertorio Fisico-Natural de la Isla de Cuba.
FELIPE POEY Y ALOY.

Mullhypeneus Poey, II, 160; orthotype UPENEUS FLAVOVITTATUS Poey. A section under UPENEUS Cuv.

970. POEY (1867). *Monografia de las Morenas Cubanas*. Repertorio Fisico-Natural de la Isla de Cuba, II.
FELIPE POEY Y ALOY.

Macrodonophis Poey, II, 251; orthotype M. MORDAX Poey. A synonym of CROTALOPSIS Kaup.

Uranichthys Poey, II, 256; logotype MURÆNA HAVANNENSIS Bl. & Schn. A synonym of OPHICHTHUS Ahl.

Pythonichthys Poey, II, 265; orthotype P. SANGUINEUS Poey.

971. QUENSTEDT (1867). *Handbuch der Petrefacten*.
FRIEDRICH AUGUST VON QUENSTEDT.

Cyclospondylus Quenstedt, Ed. 2, 260; orthotype THRISSOPS MICROPODIUS Ag. (fossil). A synonym of EUTHYONOTUS.

972. SAUVAGE (1867). *Catalogue des poissons des formations sécondaires du Boulonnais*.
HENRI ÉMILE SAUVAGE (1844-   ).

"Eulepidotæ" Sauvage, 20; a plural term applied to typical species of LEPIDOTES. Not a generic name.

Curtodus Sauvage, II, 53; logotype C. CORALLINUS Sauvage (fossil).

Auluxacanthus Sauvage, II 63; orthotype A. DUTERTREI Sauvage (fossil). A synonym of ISCHYDODUS Egert.

973. STEINDACHNER (1867). *Ueber einige neue und seltene Meeresfische aus China*. Sitzb. Akad. Wiss. Wien, LV.
FRANZ STEINDACHNER.

Paramonacanthus Steindachner, LV, 591; orthotype MONACANTHUS KNERI Steind.
Tæniolabrus Steindachner, LV, 713; orthotype T. FILAMENTOSUS Steind. A synonym of TRICHONOTUS.

974. STEINDACHNER (1867). *Ueber einige Fische aus dem Fitzroy-Flusse. bei Rockhampton in Ost-Australien.* Sitzb. Akad. Wiss. Wien, LV.
FRANZ STEINDACHNER.
Lepidoblennius *Steindachner*, LV, 11; orthotype L. HAPLODACTYLUS Steind.
Neosilurus Steindachner, LV, 14; orthotype N. HYRTLI Steind.

975. STEINDACHNER (1867). *Ichthyologische Notizen: Ueber eine neue Gattung und Art der Gruppe, Trypauchenina.* Sitzb. Akad. Wiss. Wien, LV, 1867.
FRANZ STEINDACHNER.
Ctenotrypauchen, Steindachner, LV, 530; orthotype C. CHINENSIS Steind. A synonym of TRYPAUCHENICHTHYS Blkr.

1868

976. BLEEKER (1868). *Troisème notice sur la faune ichthyologique de l'île d'Obi.* Versl. Akad. Amsterdam (2), II, 1868, 275.
PIETER BLEEKER.
Valenciennea Bleeker, 288; haplotype ELEOTRIS HASSELTI Blkr. (ELEOTRIS STRIGATA Cuv. & Val. named in a later paper.) This is the earlier and therefore the proper form of this name, later written VALENCIENNESIA).

977. BLEEKER (1868). *Cinquième notice sur le faune ichthyologique de Solor.* Versl. Akad. Amsterdam, II.
Valenciennesia Bleeker, 275; orthotype ELEOTRIS STRIGATA Cuv. & Val. Earlier written VALENCIENNEA by Bleeker. (See No. 1115, p. 371.)

978. BLEEKER (1868). *Notice sur la faune ichthyologique de l'île de Waigiou.* Versl. Akad. Amsterdam, II.
PIETER BLEEKER.
Tetragonoptrus (Klein) Bleeker, 300; orthotype (after Klein) CHÆTODON CAPISTRATUS L. A syonym of CHÆTODON.

979. BLEEKER (1868). *Description de trois espèces inédites de Chromidoïdes de Madagascar.* Versl. Akad. Amsterdam (2), II, 1868, 307-314.
PIETER BLEEKER.
Paratilapia Bleeker, 308; orthotype P. POLLENI Blkr.
Paretroplus Bleeker, 313; orthotype P. DAMII Blkr.

980. BLEEKER (1868). *Description de trois espèces inédites des poissons des îles d'Amboine et de Waigiou.* Versl. Akad Amsterdam, II.
PIETER BLEEKER.
Heteroconger Bleeker, II, 332; orthotype H. POLYZONA Blkr., 400.

981. BURMEISTER (1868). *Petromyzon macrostomus, description de una nueva especie de pez.* Anales Mus. Pub. Buenos Ayres, I.
HERMANN CARL CONRAD BURMEISTER (1807-1892).
Exomegas Burmeister, I, 36; orthotype PETROMYZON MACROSTOMUS Burmeister.

982. CAPELLO (1868). *Descripção de dos peixes novos provenientés dos mares de Portugal,* etc. Journ. Sci. Math., etc., Lisboa.

FELIX DE BRITO CAPELLO.

Pseudotriakis Capello, 316; orthotype P. MICRODON Capello. Also written PSEUDO-TRIACIS.

983. CAPELLO (1868). *Descripção de dos peixes novos provenientés dos mares Lias of Lyme Regis.* Quart. Journ. Geol. Soc., XXIV.

FELIX DE BRITO CAPELLO.

Osteorachis Egerton, XXIV, 500; orthotype O. MACROCEPHALUS Egert. (fossil).

Isocolum Egerton, XXIV, 501; orthotype I. GRANULATUM Egert. (fossil). A synonym of OSTEORACHIIS.

Eulepidotus Egerton, XXIV, 505; orthotype DAPEDIUS FIMBRIATUS Ag. (fossil). Unnecessarily "withdrawn" by Egerton and replaced by HETEROLEPIDOTUS, 1872, on account of the name "EULEPIDOTÆ" applied to typical species of LEPIDOTES by Sauvage, 1867.

984. DAY (1868). *On Some New Fishes from Madras.* Proc. Zool. Soc. London.

FRANCIS DAY.

Priacanthichthys Day, 193; orthotype P. MADERASPATENSIS Day = EPINEPHELUS LATIFACIATUS T. & S. A synonym of EPINEPHELUS.

984A. GEINITZ (1868). *Die fossilen Fischschuppen aus der Planerkalke in Strehlen.* Jahresber. Ges. Natur.- u. Heilkunde Dresden.

HANS BRUNO GEINITZ.

(Reviewed by T. D. A. Cockerell, *American Naturalist,* January, 1917.)

In addition to his own generic names, Geinitz quotes from Dr. Steinla's unpublished *Katalog,* in lit. the definitions, a number of generic names proposed for scales of Cretaceous fishes. For most of these no type is assigned, and most of them appear from Geinitz' account to fall into the rank of synonyms.

Perigrammatolepis (Steinla) Geinitz, 39; orthotype SALMO LEWESIENSIS Mantell. "Es ist dies normale Form für OSMEROIDES LEWESIENSIS, welche den Abbildungen bei Agassiz am nächsten entspricht." From this I infer that LEWESIENSIS should be considered as type, but Geinitz seems to place PERIGRAMMATOLEPIS in the synonymy of his CYCLOLEPIS (= AULOLEPIS).

Credemnolepis (Steinla) Geinitz, 39; orthotype not stated; name only. Apparently defined on page 48 as CODONOLEPIS. Said to be a synonym of CYCLOLEPIS Geinitz (AULOLEPIS).

Kymatolepis (Steinla) Geinitz, 39; orthotype named K. GEINITZI by Cockerell (1919).

Cyclolepis Geinitz, 39; orthotype C. AGASSIZI Geinitz (fossil). (Scales, like those of SALMO.)

Aspidolepis Geinitz, 40; orthotype A. STEINLAI Geinitz (fossil). (Scales, similar to those of PORONOTUS.)

Dypterolepis (Steinla) Geinitz, 40; orthotype not named. Said to contain part of the species of LEPTOLEPIS Ag. Apparently a synonym of HOLCOLEPIS.

Kymatopetalolepis (Steinla) Geinitz, 40; orthotype not named. Apparently SALMO LEWESIENSIS Mantell. If so, a synonym of HOLCOLEPIS.

Micropetalolepis (Steinla) Geinitz, 40; orthotype not named. Apparently founded on a scale of SALMO LEWESIENSIS Mantell. If so, a synonym of HOLCOLEPIS.

Leptogrammatolepis (Steinla) Geinitz, 40; orthotype not named. Apparently a scale of SALMO LEWESIENSIS. If so, a synonym of HOLCOLEPIS.

Petalolepis (Steinla) Geinitz, 42; orthotype OSMEROIDES DIVARICATUS Geinitz. Apparently a syr.onym of HOLCOLEPIS.

Mixogrammatolepis (Steinla) Geinitz, 43; orthotype a Cretaceous scale of CLADO-CYCLUS STREHLENSIS Geinitz (BERYX ORNATUS Geinitz), regarded as a syno-nym of CLADOCYCLUS. But this must be an error, as the South American genus CLADOCYCLUS is quite distinct from these forms found in the British Chalk.

Heliolepis (Steinla) Geinitz, 43; orthotype Cretaceous scale of "CLADOCYCLUS."

Oolepis (Steinla) Geinitz 43; orthotype a Cretaceous scale of "CLADOCYCLUS."

Coinolepis (Steinla) Geinitz, 43; orthotype a Cretaceous scale of "CLADOCYCLUS."

Polypterolepis (Steinla) Geinitz, 43; orthotype a Cretaceous scale of "CLADO-CYCLUS."

Ptycholepis (Steinla) Geinitz, 43; orthotype a Cretaceous scale of "CLADOCYCLUS."

Pterogrammatolepis (Steinla) Geinitz, 43; orthotype a Cretaceous scale of "CLADO-CYCLUS."

Agrammatolepis (Steinla) Geinitz, 43; orthotype a Cretaceous scale (misspelled ACRAMMATOLEPIS). Apparently a synonym of HEMICYCLUS Geinitz.

Crommiolepis (Steinla) Geinitz, 43; orthotype Cretaceous scales referred to CLADO-CYCLUS STREHLENSIS Geinitz. "Eine Anzahl kleiner Schuppen von denen ein Theil jedenfalls an CL. STREHLENSIS gehört. On the rest of these scales Geinitz founds his new genus HEMICYCLUS. The type of CLADOCYCLUS (C. GARDNERI Ag.) is a South American OSTEOGLOSSID, and none of these fishes of the German Chalk seems related to it.

Hemicyclus Geinitz, 44; orthotype CLADOCYCLUS STREHLENSIS Geinitz, in part (fossil scales).

Prionolepis (Steinla) Geinitz, 45; orthotype ZEUS LEWESIENSIS Mantell = BERYX ORNATUS Ag. Cretaceous scales. Apparently a synonym of HOPLOPTERYX Ag.

Goniolepis (Steinla) Geinitz, 45; orthotype Cretaceous scales. (HOPLOPTERYX?)

Hemigonolepis (Steinla) Geinitz, 45; orthotype Cretaceous scales. (HOPLOP-TERYX?)

Hemicyclolepis (Steinla) Geinitz, 45; orthotype Cretaceous scales. (HOPLOP-TERYX?)

Lophoprionolepis (Steinla) Geinitz, 47; orthotype AMIA LEWESIENSIS Mantell = MACROPOMA MANTELLI Ag., orthotype Cretaceous scales. A synonym of MACROPOMA Ag.

Acrogrammatolepis Geinitz, 47; orthotype A. STEINLAI Geinitz (fossil).

Hemilampronites Geinitz, 48; orthotype H. STEINLAI Geinitz (fossil). (Scales, similar to those of HYPORHAMPHUS.)

The following additional names for Cretaceous scales are also quoted, but I have at present no further data for them:

Psygmatolepis (Steinla) Geinitz, 40; orthotype Cretaceous scales. Misspelled on plate.

Leptogrammatolepis (Steinla) Geinitz, 40; orthotype Cretaceous scales.

Microgrammatolepis (Steinla) Geinitz, 40; orthotype Cretaceous scales.

Codonolepis (Steinla), 48; orthotype not named. Spelled CREDEMNOLEPIS on page 40. A synonym of CYCLOLEPIS.

985. GÜNTHER (1868). *Catalogue of the Fishes in the British Museum.* VII. ALBERT GÜNTHER.

Osteochilus Günther, VII, 40; logotype ROHITA MELANOPLEURA Blkr.

Barynotus Günther, VII, 61; logotype B. LAGENSIS Gthr.

Gymnocypris Günther, VII, 169; orthotype G. DOBULA Gthr.

Megarasbora Günther, VII, 198; orthotype CYPRINUS ELANGA Ham.

Aphyocypris Günther, VII, 201; orthotype A. CHINENSIS Gthr.

Xenocypris Günther, VII, 205; orthotype X. ARGENTEA Gthr.

Mystacoleucus Günther, VII, 206; orthotype SYSTOMUS PADANGENSIS Blkr.

Pteropsarion Günther, VII, 284; orthotype BARILIUS BAKERI Day.

Opsaridium (Peters) Günther, VII, 286; orthotype O. ZAMBEZENSE Peters (MS.). A synonym of BARILIUS Ham.

Bola Günther, 293; tautotype CYPRINUS BOLA Ham. = BOLA GOHA Gthr. Name preoccupied by BOLA Ham.; replaced by RAIAMAS Jordan, 1919. "RAJAH MAS" is the angler's name for the type-species in India.

Schacra Günther, VII, 294; orthotype CYPRINUS SHACRA Ham. Same as SHACRA Blkr.

Squaliobarbus Günther, VII, 297; orthotype LEUCISCUS CURRICULUS Rich.

Ochetobius Günther, VII, 297; orthotype OPSARIUS ELONGATUS Kner.

Eustira Günther, VII, 331; orthotype E. CEYLONENSIS Gthr.

Securicula Günther, VI, 332; logotype CYPRINUS GORA Ham. A synonym of SALMOSTOMA Sw.

Cachius Günther, VII, 338; orthotype CYPRINUS CACHIUS Ham. A synonym of CHELA Ham. as restricted by Bleeker, 1863.

Oreonectes Günther, VII, 369; orthotype O. PLATYCEPHALUS Gthr.

Cetengraulis Günther, VII, 383; logotype ENGRAULIS EDENTULUS Cuv.

Pterengraulis Günther, VII, 384; orthotype CLUPEA ATHERINOIDES L.

Heterothrissa Günther, VII, 385; orthotype ENGRAULIS BREVICEPS Cantor.

Lycengraulis Günther, VII, 385; logotype ENGRAULIS GROSSIDENS Cuv.

Lycothrissa Günther, VII, 385; orthotype ENGRAULIS CROCODILUS Blkr.

Telara Günther, VII, 385; tautotype CLUPEA TELARA Ham. A synonym of SETIPINNA Sw.

Pellonula Günther, VII, 452; orthotype P. VORAX Gthr.

Chirocentrodon Günther, VII, 463; orthotype C. TÆNIATUS Gthr.

Xenomystus Günther, VII, 481; orthotype NOTOPTERUS NIGRI Gthr.

Graodus Günther, VII, 485; orthotype G. NIGROTAENIATUS Gthr. A synonym of ALBURNOPS Grd. (the teeth lost).

986. GÜNTHER (1868). *Report on a Collection of Fishes Made at St. Helena by J. C. Melliss,* Esq. Proc. Zool. Soc. London.
ALBERT GÜNTHER.

Holanthias Günther, 226; ANTHIAS FRONTICINCTUS Gthr.

987. GÜNTHER (1868). *Descriptions of Fresh-water Fishes from Surinam and Brazil.* Proc. Zool. Soc. London.
ALBERT GÜNTHER.

Hypoptopoma Günther, 234; orthotype H. THORACATUM Gthr.

Aphyocharax Günther, 245; orthotype A. PUSILLUS Gthr.

988. GÜNTHER (1868). *Additions to the Ichthyological Fauna of Zanzibar.* Ann. Mag. Nat. Hist.
ALBERT GÜNTHER.

Tholichthys Günther I, 457; orthotype T. OSSEUS Gthr. Larva of CHÆTODON.

989. GUICHENOT (1868). *Index generum et specierum anthiadidorum,* etc.
Ann. Soc. Maine-et-Loire.
ALPHONSE GUICHENOT.

Paranthias Guichenot, X, 87; orthotype SERRANUS CREOLUS Cuv. & Val. = S.
FURCIFER Cuv. & Val. Replaces BRACHYRHINUS Gill.

990. HANCOCK & ATTHEY (1868). *Notes on the Remains of Some Reptiles
and Fishes from the Shales of the Northumberland Coal-fields.* Trans.
Nat. Hist. Soc. Northumberland.
ALBANY HANCOCK (1806-1873) ; T. ATTHEY.

Acanthodopsis Hancock & Atthey, I, 364; orthotype A. WARDI Hancock & Atthey
(fossil).

991. KNER (1868). *Ueber einige Fische aus dem Museum . . . Godeffroy in
Hamburg.* Sitzb. Akad. Wiss. Wien.
RUDOLPH KNER.

Anomalops Kner, LVIII, 26; orthotype A. GRÆFFEI Kner.
Sparopsis Kner, LVIII, 27; orthotype S. LATIFRONS Kner. A synonym of APRION Cuv.
Micropus Kner, 28; orthotype MICROPTERYX POLYCENTRUS Kner. Name four times
preoccupied; replaced by ORQUETA Jordan.
Bunocottus Kner, LVIII, 28; orthotype B. APUS Kner.
Orthostomus Kner, LVIII, 29; orthotype O. AMBLYOPINUS Kner.
Opisthocentrus Kner, 49; orthotype O. QUINQUEMACULATUS Kner = OPHIDIUM
OCELLATUM Tiles.
Urocentrus Kner, LVIII, 51; orthotype U. PICTUS Kner. A synonym of PHOLIS
Scopoli.

992. KNER (1868). *Ueber Conchipoma gadiforme . . . aus dem Rothliegcndem
(der Untern Dyas) von Lebach bei Saarbrücken in Rheinpreussen.* Sitzb.
Akad. Wiss. Wien.
RUDOLPH KNER.

Conchopoma Kner, LVII, 279; orthotype C. GADIFORME Kner (fossil).

993. LEIDY (1868). *Remarks on Conosaurus of Gibbes.* Proc. Acad. Nat. Sci. Phila.
JOSEPH LEIDY.

Conosaurops Leidy, 202; orthotype CONOSAURUS BOWMANI Gibbes (fossil). Substitute for CONOSAURUS Gibbes; regarded as preoccupied by CONIOSAURUS,
a reptile.
Diplotomodon Leidy, 202; orthotype TOMODON HORRIFICUS Leidy. A substitute for
TOMODON Leidy, preoccupied.

993A. VON DER MARCK (1868). *(Holcolepis.)* Palæontographica.
W. VON DER MARCK.

Holcolepis von der Marck, 278; orthotype RHABDOLEPIS CRETACEUS von der Marck.
Name a substitute for RHABDOLEPIS Marck, 1863, preoccupied.

994. MARTENS (1868). *Ueber einige ostasiatische Süsswasserthiere.* Arch.
Nat. XXIV.
EDUARD VON MARTENS.

Octonema Martens, about 60; orthotype HOMALOPTERA ROTUNDICAUDA Martens.
Name preoccupied, replaced by LEFUA Herzenstein.

995. PETERS (1868). *Naturwissenschaftliche Reise nach Mossambique.* Zoologie.
WILHELM PETERS.

Nothobranchius Peters, 60; orthotype CYPRINODON ORTHONOTUS Peters.
Belonichthys Peters, 108; orthotype B. ZAMBEZENSIS Peters.

996. PETERS (1868). *Ueber eine neue Gattung der Flederthiere (Peronymus) und ueber neue Gattungen und Arten der Fische.* Mon. Akad. Wiss. Berlin.
WILHELM PETERS.

Lycocyprinus Peters, 145; orthotype PŒCILIA SEXFASCIATA Peters = HAPLOCHILUS INFRAFASCIATUS Gthr.

997. PETERS (1868). *Ueber die von Herrn Dr. F. Jagor in dem ostindischen gesammelten Fische.* Mon. Akad. Wiss. Berlin.
WILHELM PETERS.

Enchelyurus Peters, 268; orthotype E. FLAVIPES Peters.

998. PETERS (1868). *Ueber eine von dem Baron Carl von der Decken entdeckte neue Gattungen von Welsen . . . aus Ostafrica.* Mon. Akad. Wiss. Berlin.
WILHELM PETERS.

Chiloglanis Peters, 599; orthotype C. DECKENI Peters.

999. POEY (1868). *Synopsis piscium cubensium,* etc. In Repertorio Fisico-Natural de la Isla de Cuba, II.
FELIPE POEY Y ALOY.

Promicrops Poey, 287; orthotype SERRANUS GUAZA Poey.
Gramma Poey, 296; orthotype G. LORETO Poey. Not GRAMMIA Rambeau, 1866, a genus of Lepidoptera.
Tropidinius (Gill) Poey, 296; orthotype MESOPRION ARNILLO Poey = APSILUS DENTATUS Guich.
Brachygenys (Scudder) Poey, 310; orthotype HÆMULON TÆNIATUM Poey.

1000. SCHIÖDTE (1868). *Om stillings udvikling hos flynder fiskene.* Nat. Tid., V.
JORGEN MATTHIAS CHRISTIAN SCHIÖDTE (1815-1884).

Bascanius Schiödte, V, 269; orthotype B. TÆDIFER Schiödte. A larval flounder, probably young of PLATOPHRYS or SYMPHURUS; name preoccupied as BASCANION.

1869.

1001. BLEEKER (1869). *Description de deux espèces inédites d'Alticus, de Madagascar.* Versl. Akad. Amsterdam.
PIETER BLEEKER.

Alticus (Commerson) Bleeker, 234; orthotype A. SALTATORIUS Commerson. Same as RUPISCARTES Sw., and of earlier date if Commerson's non-binomial names are accepted.

1002. BLEEKER (1869). *Neuvième notice sur la faune ichthyologique du Japon.* Versl. Akad. Amsterdam (2), III, 1869, 237-259.
PIETER BLEEKER.

Pseudopriacanthus Bleeker, 241; orthotype PRIACANTHUS NIPHONIUS Cuv. & Val.
Sarcocheilichthys Bleeker, 252; orthotype LEUCISCUS VARIEGATUS T. & S.

1003. BOCOURT (1869). *Descriptions de quelques reptiles et poissons nouveaux appartenant à la faune tropicale de l'Amerérique. Nouv. Arch. Mus. Hist. Nat.* Paris.
FIRMIN N. BOCOURT (1819-   ).

Paralonchurus Bocourt, V, 21; orthotype P. PETERSI Bocourt.

Polycirrhus Bocourt, V, 22; orthotype P. DUMERILI Bocourt. Name preoccupied in worms; replaced by POLYCLEMUS Berg, 1895.

1004. BRANDT (1869). *Einige Worte ueber die europäisch-asiatisch Störarten.* Mélanges Biologiques, VII. Bull. Acad. Sci. St. Petersburg.

JOHANN FRIEDRICH BRANDT.

Schipa Brandt, VII, 113; orthotype ACIPENSER SCHYPA Eichw. = A. NUDIVENTRIS Lovetzky. A synonym of ANTACEUS Heckel & Fitzinger.

Sterledus Brandt, VII, 113; orthotype ACIPENSER RUTHENUS L. Apparently a variant spelling of STERLETUS Heckel & Fitzinger, not a new name and not STERLETUS Raf. 1820.

1005. COPE (1869). *Description of Some Extinct Fishes Previously Unknown.* Proc. Boston Soc. Nat. Hist.

EDWARD DRINKER COPE.

Phacodus Cope, XII, 311; orthotype P. IRREGULARIS Cope (fossil). Name preoccupied; replaced by CROMMYODUS Cope, 1870.

Leptomylus Cope, XII, 313; orthotype L. DENSUS (fossil).

Plinthicus Cope, XII, 316; orthotype P. STENODON Cope (fossil). An ally of AËTOBATUS.

1006. DAY (1869). *Remarks on Some Fishes in the Calcutta Museum.* II. Proc. Zool. Soc. London.

FRANCIS DAY.

Mayoa Day, 553; orthotype M. MODESTA Day.

1007. DYBOWSKI (1869). *Vorläufige Mittheilungen über die Fischfauna des Onon-Flusses und die Ingoda in Transbaikalien.* Verh. Zool. Bot. Ges. Wien, XIX, 945-948.

BENEDIKT IVAN DYBOWSKI.

Gobiobarbus Dybowski, XIX, 951; orthotype CYPRINUS LABEO Pallas. A synonym of HEMIBARBUS Blkr.

Micraspius Dybowski, XIX, 953; orthotype M. MIANOWSKI Dybowski.

Pseudaspius Dybowski, XIX, 953; orthotype CYPRINUS LEPTOCEPHALUS Pallas.

Ladislavia Dybowski, XIX, 954; orthotype L. TACZANOWSKII Dybowski.

1008. GERVAIS (1869). *Paléontologie Générale.*

FRANÇOIS LOUIS PAUL GERVAIS (1816-1879).

Dipristis Gervais, 240; orthotype D. CHIMÆROIDES Gervais (fossil). Not DIPRISTIS Marsh of the same date, also a CHIMÆROID fragment.

1009. GÜNTHER (1869). *Account of the Fishes of the States of Central America, Based on Collections Made by Capt. J. M. Dow, F. Godman and V. Salvin.* Trans. Zool. Soc. London.

ALBERT GÜNTHER.

Megalobrycon Günther, 423; orthotype M. CEPHALUS Gthr. A synonym of BRYCON M. & T.

1010. GUICHENOT (1869). *Notice sur quelques poissons inédits de Madagascar et de Chine.* Nouv. Arch. Mus. Paris., V.

ALPHONSE GUICHENOT.

Paragonus Guichenot, 202; orthotype P. STURIOIDES Guich. PARAGONUS Gill, 1861, and PARAGONUS Guich., 1869, seem both to be synonyms of PODOTHECUS Gill.

1011. MACDONALD (1869). *On . . . a New Genus of Mugilidæ Inhabiting the Fresh Waters of Viti Levu of the Feejee Group,* etc. Proc. Zool. Soc. London.
JOHN DENIS MACDONALD.

Gonostomyxus Macdonald, 38; orthotype G. LOALOA Macdonald.

1011A. MASON (1869). *(Acrodontosaurus.)* Quart. Journ. Geol., XXV.
J. W. MASON.

Acrodontosaurus Mason, XXV, 444; orthotype A. GARDNERI Mason (fossil). A synonym of PACHYRHIZODUS.

1012. OWEN (1869). *Notes on Two Ichthyodorulites Hitherto Undescribed.*
RICHARD OWEN.

Lepracanthus Owen, VI, 481; orthotype L. COLEI (Egert.) Owen (fossil).

1013. STEINDACHNER (1869). *Ichthyologische Notizen.* VIII. Sitzb. Akad. Wiss. Wien, LX.
FRANZ STEINDACHNER.

Hæmulopsis Steindachner, 60; logotype HÆMULON CORVINÆFORME Steind. A synonym of BRACHYDEUTERUS Gill.

1014. STEINDACHNER (1869). *Ichthyologische Notizen.* IX. *Ueber eine neue Gattung und Art der Cyprinoiden de China.* Sitzb. Akad. Wiss. Wien, LX.
FRANZ STEINDACHNER.

Abramocephalus Steindachner, LX, 302; orthotype A. MICROLEPIS Steind. A synonym of HYPOPHTHALMICHTHYS Blkr.

1870

1015. BLEEKER (1870). *Atlas des Indes Orientales Nederlandaises . . .* VII. *Pleuronectes,* etc.
PIETER BLEEKER.

Rhinoplagusia Bleeker, 27; orthotype PLAGUSIA JAPONICA T. & S. Replaces USINOSITA Jordan & Snyder, 1901.

1016. BLEEKER (1870). *Mededeeling omtrent eenige nieuwe vischsoorten van China.* Versl. Akad. Amsterdam (2), IV, 1870, 251-253.
PIETER BLEEKER.

Acanthorhodeus Bleeker, IV, 253; orthotype A. MACROPTERUS Blkr.

Pseudobrama Bleeker, IV, 253; logotype P. DABRYI Blkr.

Saurogobio Bleeker, IV, 253; orthotype S. DUMERILI Blkr.

Rhinogobio Bleeker, IV, 253; orthotype R. TYPUS Blkr.

Luciobrama Bleeker, IV, 253; orthotype L. TYPUS Blkr.

Leptobotia Bleeker, IV, 256; orthotype L. ELONGATA Blkr.

1017. COPE (1870). *Partial Synopsis of the Fresh-water Fishes of North Carolina.* Proc. Amer. Philos. Soc., 1870, XI, 448-495.
EDWARD DRINKER COPE.

Hypohomus Cope, XI, 449; orthotype H. AURANTIACUS Cope.

Labidesthes Cope, XI, 455; orthotype CHIROSTOMA SICCULUM Cope.

Hemitremia Cope, XI, 462; orthotype H. VITTATA Cope, later called LEUCISCUS FLAMMEUS Jordan, the name VITTATUS being preoccupied in LEUCISCUS, to which genus the species was referred.

Placopharynx Cope, XI, 467; orthotype P. CARINATUS Cope.

1018. COPE (1870). *Observations on the Fishes of the Tertiary Shales of Green River, Wyoming Territory.* Proc. Amer. Philos. Soc., XI.
EDWARD DRINKER COPE.

Asineops Cope, XI, 381; orthotype A. SQUAMIFRONS Cope (fossil).

Erismatopterus Cope, XI, 427; orthotype E. RICKSECKERI Cope (fossil).

1019. COPE (1870). *Contribution to the Ichthyology of the Marañon.* Proc. Amer. Philos. Soc. 1870, 559-701.
EDWARD DRINKER COPE.

Stethaprion Cope, 562; orthotype S. ERYTHROPS Cope.

Plethodectes Cope, 563; orthotype P. ERYTHRURUS Cope. A synonym of CHALCEUS Cuv.

Holotaxis Cope, 563; logotype H. MELANOSTOMUS Cope.

Odontostilbe Cope, 566; orthotype O. FUGITIVA Cope.

1020. COPE (1870). *On the Fishes of a Fresh-water Tertiary Lake in Idaho, Discovered by Capt. Clarence King.* Proc. Amer. Philos. Soc. XI.
EDWARD DRINKER COPE.

Lithichthys Cope, XI, 316; orthotype L. PUSILLUS Cope (fossil).

Diastichus Cope, XI, 539; orthotype D. MACRODON Cope (fossil).

Oligobelus Cope, XI, 541; orthotype O. ARCIFERUS Cope (fossil). Said to be near SEMOTILUS Raf.

Anchybopsis Cope, XI, 543; orthotype A. LATUS Cope (fossil). Referred later to synonymy of "LEUCOS Heckel" (SIPHATELES Cope).

Rhabdofario Cope, XI, 546; orthotype R. LACUSTRIS Cope (fossil). A very large trout from the ancient Lake Idaho.

1021. COPE (1870). *Second Addition to the History of the Fishes of the Cretaceous of the United States.* Proc. Amer. Philos. Soc., XI, 240-244.
EDWARD DRINKER COPE.

Sphagepœa Cope, XI, 241; orthotype S. ACICULATA Cope (fossil). An ally of EDAPHODON Leidy.

Pneumatosteus Cope, XI, 242; orthotype P. NAHUNTICUS Cope (fossil). An ally of LEPISOSTEUS Lac.

Crommyodus Cope, XI, 243; orthotype PHACODUS IRREGULARIS Cope (fossil). A substitute for PHACODUS Cope, preoccupied.

1022. COPE (1870). *On the Saurodontidæ.* Proc. Amer. Philos. Soc., 1871, XI, 529-538.
EDWARD DRINKER COPE.

Ichthyodectes Cope, 536; orthotype I. CTENODON Cope (fossil).

1023. DAY (1870). *On the Fishes of the Andaman Islands.* Proc. Zool. Soc. London.
FRANCIS DAY.

Jerdonia Day, 700; orthotype PLATACANTHUS MACULATUS Day.

1024. DUMÉRIL (1870). *Histoire générale des poissons.* Vol. II.
AUGUSTE DUMÉRIL.

Parapegasus (Bleeker) Duméril, 493; orthotype PEGASUS NATANS L.

Typhlus (Bibron MS.) Duméril, 540; logotype ICHTHYOCAMPUS PONDICERIANUS
Kaup (in synonymy). A synonym of ICHTHYOCAMPUS Kaup.

Atelurus Duméril, 584; orthotype A. GERMANI Dum.

Entelurus Duméril, 605; orthotype SYNGNATHUS ÆQUOREUS L. A synonym of
NEROPHIS Raf.

Hymenolomus Duméril, 607; orthotype SYNGNATHUS HYMENOLOMUS Rich. =
H. RICHARDSONI Dum. A synonym of PROTOCAMPUS Gthr., 1870.

1025. GASCO (1870). *Intorno ad un nuovo genere di pesci,* etc. Bull. Assoc.
Nat. e Med. Napol., April, 1870, 59-61.
FRANCESCO GASCO.

Vexillifer Gasco, 59; orthotype V. DEPHILIPPII Gasco. An immature and very
singular form of some fish, perhaps of CARAPUS ACUS.

1026. GIEBEL (1870). *Trachypoma marmoratum, ein neuer Wels aus dem
Amazonenstrome.* Zeitschr. Ges. Naturw., XXXVII.
CHRISTOPH GOTTFRIED ANDREAS GIEBEL.

Trachypoma Giebel, 97; orthotype TRACHYPOMA MARMORATUM Giebel = EREMO-
PHILUS MUTISI. A synonym of EREMOPHILUS Humboldt, the name preoccupied.

1027. GÜNTHER (1870). *Catalogue of the Fishes in the British Museum.* VIII.
ALBERT GÜNTHER.

Rhamphosternarchus Günther, VIII, 4; logotype STERNARCHUS OXYRHYNCHUS
M. & T. A synonym of STERNARCHORHYNCHUS Castelnau.

Brachyrhamphichthys Günther, VIII, 6; logotype RHAMPHICHTHYS ARTEDII Kaup.
A synonym of HYPOPOMUS Gill.

Paramyrus Günther, VIII, 51; logotype ECHELUS MICROCHIR Blkr.

Myroconger Günther, VIII, 93; orthotype M. COMPRESSUS Gthr.

Helminthostoma (Cocco) Günther, VIII, 145; orthotype HELMINTHOSTOMA DELLA-
CHIAJE Cocco. Larva of CARAPUS (FIERASFER); passing reference to an
unpublished name.

Nannocampus Günther, VIII, 178; orthotype N. SUBOSSEUS Gthr.

Urocampus Günther, VIII, 179; orthotype U. NANUS Gthr.

Protocampus Günther, VIII, 193; orthotype SYNGNATHUS HYMENOLOMUS Rich.
Same as HYMENOLOMUS Dum. of the same year. Günther's work is dated
"May 20, 1870," Duméril's merely "1870". Neither writer quotes the other;
the name of Günther has probably precedence.

Hemiconiatus Günther, VIII, 272; orthotype TETRAODON GUTTIFER Bennett. Same
as EPHIPPION Bibron, 1855, which it replaces if the latter is regarded as preoccupied by EPHIPPIUM.

Liosaccus Günther, VIII, 287; logotype T. CUTANEUS Gthr.

Trichocyclus Günther, VIII, 316; orthotype T. ERINACEUS Gthr. Probably a larval
DIODON; equivalent to TRICHODIODON Blkr.

Leptocarcharias (Smith MS.) Günther, VIII, 384; orthotype TRIÆNODON SMITHI
M. & H. A variant of LEPTOCARIAS Smith.

Psammobatis Günther, VIII, 470; orthotype P. RUDIS Gthr.

1028. HANCOCK & ATTHEY (1870). *Notes on an Undescribed Fossil from the Newsham Coalfields near Newcastle-upon-Tyne.* Ann. Mag. Nat. Hist., V.
ALBANY HANCOCK; T. ATTHEY.

Archichthys Hancock & Atthey, V, 268; orthotype A. SULCIDENS Hancock & Atthey. Apparently a synonym of STREPSODUS.

1029. KLUNZINGER (1870). *Synopsis der Fische des Rothen Meeres.* I. *Percoiden-Mugiloiden.* Verh. K.-K. Zool.-Bot. Ges. Wien, 1870.
CARL BENJAMIN KLUNZINGER (1834-   ).

Dirrhizodon Klunzinger, XX, 664 (Nov. 2); orthotype D. ELONGATUS Klunzinger. A living homologue of HEMIPRISTIS Ag.

Pseudoserranus Klunzinger, XX, 687; logotype PERCA LOUTI Forskål, as restricted by Bleeker (based on PERCA LOUTI, CABRILLA, and SCRIBA). Regarded by me, following Bleeker, as a synonym of VARIOLA Sw. Klunzinger protests against this arrangement in a later paper, PSEUDOSERRANUS (type CABRILLA) being intended as a substitute for SERRANUS, which name of Cuvier he uses for EPINEPHELUS.

Paracentropristes Klunzinger, XX, 687; orthotype LABRUS HEPATUS L.

Pristiapogon Klunzinger, XX, 715; orthotype P. FRENATUS Klunzinger. A synonym of APOGON Lac.

Polysteganus Klunzinger, 762; orthotype P. CÆRULEOPUNCTATUS Klunzinger.

Gymnocranius Klunzinger, XX, 764; orthotype DENTEX RIVULATUS Rüppell. Replaces PARADENTEX Blkr.

1030. DE KONINCK (1870). *Notice sur un nouveau genre de poisson fossile de la craie supérieure.* Bull. Acad. Soc. Roy. Bruxelles, XXIX.
LAURENT GUILLAUME DE KONINCK (1809-1887).

Ankistrodus De Koninck, XXIX, 77; orthotype A. SPLENDENS De Koninck. A substitute for ANCISTRODON Rœmer, regarded as preoccupied. Both these names, being regarded as preoccupied by AGKISTRODON or ANCISTRODON Beauvais, a genus of snakes, are replaced by GRYPODON Hay, 1899.

1031. LANKESTER (1870). *Fishes of the Old Red Sandstone.* Paleont. Soc., I.
EDWIN RAY LANKESTER.

Eucephalaspis Lankester, I, 43; orthotype CEPHALASPIS LYELLI Ag. (fossil). A synonym of CEPHALASPIS.

Hemicyclaspis, Lankester, I, 43; orthotype CEPHALASPIS MURCHISONI Egert. (fossil).

Eukeraspis Lankester, I, 56; orthotype SCLERODUS PUSTULIFERUS Ag. (fossil). A synonym of SCLERODUS, that name and also PLECTRODUS being rejected for the inadequate reason that the structures which gave the name were wrongly supposed to be teeth.

Kallostrakon Lankester, I, 61; orthotype K. PODURA Lank. (fossil). Fragments of a PTERASPID.

1032. LEIDY (1870). *Remarks on Ichthyodorulites and on Certain Fossil Mammalia.* Proc. Acad. Nat. Sci. Phila., X, 12-13.
JOSEPH LEIDY.

Xiphactinus Leidy, 12; orthotype X. AUDAX Leidy. Perhaps the same as PORTHEUS Cope.

1033. LEIDY (1870). *Remarks on Fossil Vertebrates.* Proc. Acad. Nat. Sci. Phila.
JOSEPH LEIDY.

Mylocyprinus Leidy, 70; orthotype M. ROBUSTUS Leidy.

1034. MARSH (1870). *Notice of Some New Tertiary and Cretaceous Fishes.*
Amer. Assoc. Adv. Sci. for 1869 (1870).
OTHNIEL CHARLES MARSH (1831-1899).

Embalorhynchus Marsh, 228; orthotype E. KINNEYI Marsh (fossil). A fossil
analogue of ISTIOPHORUS.

Dipristis Marsh, 230; orthotype D. MIERSI Marsh (fossil). Apparently a synonym
of EDAPHODON.

1035. NEWBERRY (1870). *Fossil Fishes from the Devonian Rocks of Ohio.*
Proc. N. Y. Lyc. Nat. Hist., I, 152-153.
JOHN STRONG NEWBERRY.

Aspidophorus Newberry, I, 153; orthotype A. CLAVATUS Newberry (fossil). Name
preoccupied, replaced by ASPIDICHTHYS Newberry, 1873.

1036. NEWBERRY & WORTHEN (1870). *Descriptions of Fossil Vertebrates.*
Geol. Surv. Ills., IV.
JOHN STRONG NEWBERRY; A. H. WORTHEN.

Cymatodus Newberry & Worthen, IV, 108; orthotype C. OBLONGUS Newberry &
Worthen (fossil). Probably a synonym of JANASSA.

Lophodus Newberry & Worthen, IV, 360; orthotype L. VARIABILIS Newberry &
Worthen (fossil). Name preoccupied, replaced by AGASSIZODUS St. John &
Worthen, 1875. A synonym of CAMPODUS Koninck.

Peltodus Newberry & Worthen, IV, 362; orthotype P. UNGUIFORMIS Newberry &
Worthen (fossil). A synonym of JANASSA.

Listracanthus Newberry & Worthen, IV, 371; orthotype L. HYSTRIX Newberry &
Worthen (fossil).

1037. POWRIE (1870). *On the Earliest Known Vestiges of Vertebrate Life.* . . . .
Fish Remains from the Old Red Sandstone of Forfarshire. Trans. Geol. Soc.
Edinburgh, I.
JAMES POWRIE.

Cephalopterus Powrie, I, 298; orthotype C. PAGEI Powrie (fossil). Name preoccupied; a fragment of a supposed CEPHALASPID.

1038. ST. JOHN (1870). *Descriptions of Fossil Fishes from the Upper Coal
Measures of Nebraska.* Proc. Amer. Philos. Soc., XI.
ORESTES H. ST. JOHN.

Peripristis St. John, XI, 434; orthotype CTENOPTYCHIUS SEMICIRCULARIS Newberry &
Worthen (fossil). A synonym of CTENOPTYCHIUS.

1039. SAUVAGE (1870). *Synopsis des poissons tertiaires de Licata (Sicilia).*
Ann. Sci. Nat. XIV.
HENRI ÉMILE SAUVAGE.

Pseudovomer Sauvage, XIV, 10; orthotype P. MINUTUS Sauvage (fossil).

Acanthonotos Sauvage, XIV, 10; logotype A. ARMATUS Sauvage (fossil). Name
preoccupied by ACANTHONOTUS Bloch; replaced by HEMITHYRSITES Sauvage,
1873.

Tydeus Sauvage, XIV, 23; orthotype T. MEGISTOSOMA Sauvage (fossil). (Name
preoccupied; replaced by ANAPTERUS Sauvage, 1873.)

1040. STEINDACHNER & KNER (1870). *Ueber einige Pleuronectiden, . . . von Decastris Bay und von Viti-Levu.* Sitzb. Akad. Wiss. Wien, LXI, 421-446.

FRANZ STEINDACHNER; RUDOLPH KNER.

Stichæopsis Steindachner & Kner, LXI, 441; orthotype S. NANA Steindachner & Kner.

1041. STEINDACHNER (1870). *Ichthyologische Notizen.* X. Sitzb. Akad. Wiss. Wien, LXI.

FRANZ STEINDACHNER.

Parapriacanthus Steindachner, LXI, 623; orthotype P. RANSONNETI Steind.; replaces PEMPHERICHTHYS Klunzinger.

1871

1042. VAN BENEDEN (1871). *Recherches sur quelques poissons fossiles de la Belgique.* (Later included in *Bull. Acad. Roy. Belg.*, XXXI.)

PIERRE J. VAN BENEDEN.

Brachyrhynchus Van Beneden, XXXI, 495; orthotype ENCHEIZIPHIUS TERETIROSTRIS Rütimeyer (fossil). Name twice preoccupied; same as ENCHEIZIPHIUS.

Xiphiorhynchus Van Beneden, XXXI, 499; orthotype X. ELEGANS Van Beneden (fossil).

Trigloides Van Beneden, XXXI, 501; orthotype T. DEJARDINI Van Beneden (fossil).

Hannovera Van Beneden, XXXI, 504; orthotype H. AURATA Beneden (fossil). An extinct ally of CETORHINUS Blainv.

Scomberodon Van Beneden, XXXI, 504; orthotype S. DUMONTI Van Beneden (fossil). A synonym or extinct analogue of SCOMBEROMORUS Lac.

1043. COPE (1871). *On Two Extinct Forms of Physostomi of the Neotropical Region.* Proc. Amer. Philos. Soc., XII.

EDWARD DRINKER COPE.

Prymnetes Cope, XII, 52; orthotype P. LONGIVENTER Cope (fossil).

Anædopogon Cope, XII, 53; orthotype A. TENUIDENS Cope (fossil). An Elopid fish not identical with CLADOCYCLUS Ag., which seems to belong to the OSTEOGLOSSIDÆ.

1044. COPE (1871). *Note on Some Cretaceous Vertebrates in the State Argicultural College of Kansas,* etc. Proc. Amer. Philos. Soc., XII.

EDWARD DRINKER COPE.

Anogmius Cope, XII, 170-354; orthotype A. CONTRACTUS Cope (fossil); not ANOGMIUS Cope, 1873 (A. ARATUS Cope, 1873). Regarded by Woodward as probably a synonym of PACHYRHIZODUS.

1045. COPE (1871). *Report on Recent Reptiles and Fishes . . . Hayden Expedition . . . Wyoming,* etc. Ann. Rept. U. S. Geol. Surv. for 1870 (1871).

EDWARD DRINKER COPE.

Apsopelix Cope, 424; orthotype A. SAURIFORMIS Cope (fossil).

1046. COPE (1871). *Contribution to the Ichthyology of the Lesser Antilles.* Trans. Amer. Philos. Soc., 1871, XIV, 445-483.

EDWARD DRINKER COPE.

Cryptotomus Cope, XIV, 462; orthotype C. ROSEUS Cope. Description of a valid genus from an injured specimen.

Eleutheractis Cope, XIV, 467; orthotype E. CORIACEUS Cope. A synonym of RYPTICUS Cuv.

Holopterura Cope, XIV, 482; orthotype H. PLUMBEA Cope.

1047. CUNNINGHAM (1871). *Notes on the Reptiles, Amphibia, Fishes, Mollusca, and Crustacea Obtained During the Voyage of H.M.S. Nassau in the Years 1866-69.* Trans. Linn. Soc. London (Zool.), 1871, XXVII, 465-502.

ROBERT OLIVER CUNNINGHAM.

Maynea Cunningham, XXVII, 471; orthotype M. PATAGONICA Cunningham.

1048. DAY (1871). *On the Fresh-water Siluroid Fishes of India and Burmah.* Proc. Zool. Soc. London, 1871, 703-721.

FRANCIS DAY.

Ailiichthys Day, 712; orthotype A. PUNCTATA Day.

1049. DELFORTRIE (1871). *Les broyeurs (Cestraciontes) du tertiaire aquitanien.* Acad. Soc. Linn. Bordeaux, XXVIII.

EUGÈNE DELFORTRIE.

Gymnodus Delfortrie, XXVIII, 232; orthotype G. HETERODON Delfortrie (fossil). An extinct analogue of DIODON.

1050. GÜNTHER (1871). *Description of New Percoid Fish from the Macquarie River.* Proc. Zool. Soc. London, 1871, 320.

ALBERT GÜNTHER.

Ctenolates Günther, 320; orthotype CTENOLATES MACQUARIENSIS Gthr. = DATNIA AMBIGUA Rich.

1051. GÜNTHER (1871). *Report on Several Collections of Fishes Recently Obtained for the British Museum.* Proc. Zool. Soc. London, 1871, 652-675.

ALBERT GÜNTHER.

Blennodesmus Günther, 667; orthotype B. SCAPULARIS Gthr.

Halidesmus Günther, 668; orthotype H. SCAPULARIS Gthr.

Nannæthiops Günther, 669; orthotype N. UNITÆNIATUS Gthr.

Pœciloconger Günther, 673; orthotype P. FASCIATUS Gthr.

1052. KLUNZINGER (1871). *Synopsis der Fische des Rothen Meeres.* II. Theil. Verh. K.-K. Zool.-Bot. Ges. Wien, 1871, XXI, 441-668.

CARL BENJAMIN KLUNZINGER.

Pempherichthys Klunzinger, XXI, 470; orthotype P. GUNTHERI Klunzinger. A synonym of PARAPRIACANTHUS Steind.

Gobiichthys Klunzinger, 479; orthotype G. PETERSI Klunzinger. A synonym of OXYURICHTHYS Blkr.

1053. LE HON (1871). *Préliminaire mémoire des poissons tertiaires de la Belgique.* HENRI LE HON (1809-1872).

Scaldia Le Hon, 7; orthotype SCALDIA BIFORIS Le Hon (fossil). A synonym of THAUMAS or perhaps of the living genus SQUATINA.

Palanarrhichas Le Hon, 10; orthotype P. CRASSUS Le Hon (fossil).

Goniobatis Le Hon, 10; orthotype G. OMALIUSI Le Hon (fossil). Name preoccupied; a synonym of AËTOBATUS.

1054. LEIDY (1871). *Report on the Vertebrate Fossils of the Tertiary Formations of the West.* Rept. Geol. Surv. Terr. for 1870 (1871).
JOSEPH LEIDY.
Oncobatis Leidy, 70; orthotype O. PENTAGONUS Leidy (fossil). An ally of RAJA.

1055. LÜTKEN (1871). *Oneirodes eschrichtii Ltk., en ny grönlandsk Tudsefisk.* Overs. Dansk. Vid. Selsk. Kjöbenhavn, 1871, 56-74.
CHRISTIAN FREDERIK LÜTKEN.
Oneiroides Lütken, 56; orthotype O. ESCHRICHTII Lütken.

1056. POEY (1871). *Genres de poissons de la fauna de Cuba, appartenant à la famille de Percidæ.* Ann. Lyc. Nat. Hist. N. Y., X.
FELIPE POEY Y ALOY.
Menephorus Poey, X, 50; orthotype SERRANUS DUBIUS Poey.

1872

1057. BARRANDE (1872). *Indications sommaires des vestiges de poissons connus dans les dépôts Siluriens,* etc. (Bohème).
JOACHIM BARRANDE.
Gompholepis Barrande, I, 644; orthotype G. PANDERI Barrande (fossil).

1058. BLEEKER (1872). *Atlas ichthyologique,* etc. VI. *Pleuronectes, Clupées,* etc.
PIETER BLEEKER.
Leptogaster Bleeker, 168; orthotype SPRATELLOIDES ARGYROTÆNIA Blkr. A synonym of SPRATELLOIDES (an error probably overlooked in proof-reading).
Paralosa Bleeker, —. This name is quoted from Bleeker, but I fail to find it anywhere. Said to be a synonym of HARENGULA Cuv. & Val.

1059-1060. CASTELNAU (1872). *Contributions to the Ichthyology of Australia.* Proc. Zool. Acclim. Soc. Victoria, 1872, I, 29-247.
*FRANCIS L. CASTELNAU.
Microperca Castelnau, I, 48; orthotype M. YARRÆ Castelnau. Name preoccupied by MICROPERCA Putnam, 1863. A synonym of NANNOPERCA Gthr.
Murrayia Castelnau, I, 61; orthotype M. GUENTHERI Castelnau. A synonym of MACQUARIA Cuv. & Val.
Riverina Castelnau, I, 64; orthotype R. FLUVIATILIS Castelnau. Probably a synonym of MACQUARIA Cuv. & Val.
Neotephræops Castelnau, I, 68; orthotype CRENIDENS ZEBRA Rich. Same as GIRELLICHTHYS Klunzinger. Both are synonyms of MELAMBAPHES Gthr.
Neoplatycephalus Castelnau, I, 87; orthotype N. GRANDIS Castelnau.
Neosphyræna Castelnau, I, 96; orthotype N. MULTIRADIATA Castelnau = DINOLESTES MUELLERI Klunzinger. A synonym of DINOLESTES Klunzinger.
Richardsonia Castelnau I, 112; orthotype R. INSIGNIS Castelnau = HISTIOPTERUS LABIOSUS Rich. Name preoccupied; replaced by PARISTIOPTERUS Blkr., 1876, and later by MACCULLOCHIA Waite.

---

* François Laporte Comte de Castelnau, while in South Africa and Australia, wrote his name in English fashion as Francis L. Castelnau.

Atherinosoma Castelnau, I, 136; orthotype A. vorax Castelnau.

Yarra Castelnau, I, 231; orthotype Y. singularis Castelnau. A synonym of Geotria Gray.

Neomordacia Castelnau, I, 232; orthotype N. howittii Castelnau. A synonym of Geotria Gray.

Neocarassius Castelnau, I, 236; orthotype N. ventricosus Castelnau.

Aploactisoma Castelnau, I, 244; orthotype A. schomburgki Castelnau. A synonym of Aploactis T. & S.

Vincentia Castelnau, I, 245; logotype V. waterhousei Castelnau. A synonym of Apogon Lac.

Heteroscarus Castelnau, I, 245; logotype H. filamentosus Castelnau.

Ophioclinus Castelnau, I, 246; orthotype O. antarcticus Castelnau.

Heteroclinus Castelnau, I, 247; orthotype H. adelaidæ Castelnau.

1061. COPE (1872). *On the Fishes of the Ambyiacu River.* Proc. Acad. Nat. Sci. Phila. for 1871, 250-292 (dated Jan. 16, 1872).

Edward Drinker Cope.

Læmolyta Cope, 258; logotype Anostomus tæniata Kner.

Iguanodectes Cope, 260; orthotype I. tenuis Cope.

Triportheus Cope, 263; logotype T. albus Cope.

Zathorax Cope, 271; orthotype Z. monitor Cope.

Physopyxis Cope, 273; orthotype P. lyra Cope.

Dianema Cope, 276; orthotype D. longibarbis Cope.

Brochis Cope, 277; orthotype B. cæruleus Cope.

Otocinclus Cope, 283; orthotype O. vestitus Cope.

Pariolius Cope, 289; orthotype P. armillatus Cope.

1062. COPE (1872). *Report on the Reptiles and Fishes Obtained by the Naturalists of the Expedition.* Hayden's Geol. Surv. Wyoming, etc.

Edward Drinker Cope.

Coliscus Cope, 437; orthotype Coliscus parietalis Cope = young of Pimephales promelas Raf. A synonym of Pimephales Raf.

Sarcidium Cope, 440; orthotype S. scopiferum Cope. A synonym of Phenacobius Cope, 1867.

1063. COPE (1872). *Report on the Recent Reptiles and Fishes of the Survey, Collected by Campbell Carrington and C. M. Dawes.* Hayden's Rept. Geol. Surv. Montana, etc., for 1871 (1872).

Edward Drinker Cope.

Apocope Cope, 472; logotype A. carringtoni Cope.

Protoporus Cope, 473; orthotype P. domninus Cope (young of Tigoma lineata Grd.). A synonym of Tigoma Grd.

Myloleucus Cope, 475; orthotype M. pulverulentus Cope. Myloleucus is a synonym of Tigoma Grd. In studying these species, Professor Cope by accident replaced teeth of a Siphateles in his type example of Myloleucus pulverulentus (*fide* Snyder and Fowler). The group called Rutilus by Jordan & Evermann (*Fishes of N. and M. Amer.*, 1896. 243) is a composite one. Rutilus, with the teeth 6-5, should stand by itself. The name Leucos Heckel, preoccupied by Leucus Kaup, with teeth 5-5, should probably give place to

CENISOPHIUS Bon., while the American species should be distributed between SIPHATELES Cope, HESPEROLEUCUS Snyder, and LEUCIDIUS Snyder.

1064. COPE (1872). *A New Genus of Sauro'dont Fishes, Erisichthe, from the Kansas Niobrara.* Proc. Acad. Nat. Sci. Phila.

EDWARD DRINKER COPE.

Erisichthe Cope, 280; orthotype E. NITIDA Cope (fossil). A synonym of PROTOSPHYRÆNA Leidy.

1065. COPE (1872). *On the Families of Fishes of the Cretaceous Formation of Kansas.* Proc. Amer. Philos. Soc., 1872, XII, 327-357.

EDWARD DRINKER COPE.

Portheus Cope, XII, 331; orthotype P. MOLOSSUS Cope (fossil). Possibly a synonym of XIPHACTINUS Leidy, 1870.

Daptinus Cope, XII, 339; orthotype SAUROCEPHALUS PHLEBOTOMUS Cope (fossil). A synonym of SAURODON Hays, itself a synonym of SAUROCEPHALUS Harlan.

Empo Cope, XII, 347; orthotype E. NEPAHOLICA Cope (fossil).

Stratodus Cope, XII, 349; orthotype S. APICALIS Cope (fossil).

1066. COPE (1872). *On the Tertiary Coal and Fossils of Osino, Nevada.* Proc. Amer. Philos. Soc., 1872, XII, 478-481.

EDWARD DRINKER COPE.

Trichophanes Cope, XII, 479; orthotype T. HIANS Cope (fossil).

Amyzon Cope, XII, 480; orthotype A. MENTALE Cope (fossil).

1067. COPE (1872). *Notices of New Vertebrata from the Upper Waters of Bitter Creek, Wyoming Territory.* Proc. Amer. Philos. Soc., for 1871, XII, 483-486 (reprints dated Aug. 20, 1872).

EDWARD DRINKER COPE.

Rhineastes Cope, XII, 486; orthotype R. PELTATUS Cope (fossil).

1068. DYBOWSKI (1872). *Zur Kenntniss der Fischfauna des Amurgebietes.* Verh. Bot. Ges. Wien, 1872, 200-222.

BENEDIKT IVAN DYBOWSKI

Actenolepis Dybowski, 210; logotype A. DITMARII Dybowski = PERCA CHUATSI Basilewsky. A synonym of SINIPERCA Gill.

Onychodon Dybowski, 211; orthotype CEPHALUS MANTSCHURICUS Basilewsky. Same as ABRAMOCEPHALUS Steind., both synonyms of HYPOPHTHALMICHTHYS Blkr.

Gobiosoma Dybowski, 211; orthotype G. AMURENSIS Dybowski. Name preoccupied by GOBIOSOMA Grd. A synonym of SAUROGOBIO Blkr.

Megalobrama Dybowski, 213; orthotype M. AKOLKOVII. A synonym of PARABRAMIS Blkr.

Squalidus Dybowski, 215; orthotype S. CHANKAËNSIS Dybowski. A synonym of GNATHOPOGON Blkr. (LEUCOGOBIO Gthr.).

Plagiognathus Dybowski, 216; orthotype P. JELSKII Dybowski. Name preoccupied; replaced by PLAGIOGNATHOPS Berg.

Barbodon Dybowski, 216; orthotype B. LACUSTRIS Dybowski. A synonym of SARCOCHEILICHTHYS Blkr.

1069. EGERTON (1872). *Fish: Heterolepidotus, Isocolum,* etc. Mem. Geol. Surv. United Kingdom, 1872, XIII, decade 7.

PHILIP DE MALPAS GREY EGERTON.

Heterolepidotus Egerton, XIII, No. 2; orthotype H. LATUS Egert. (fossil).

Platysiagum Egerton, XIII, No. 6; orthotype P. SCLEROCEPHALUM Egert. (fossil).

Palæospinax Egerton, XIII, No. 7; orthotype THYELLINA PRISCA (fossil).

Drepanephorus Egerton, XIII, No. 9; orthotype ACRODUS RUGOSUS Ag. (fossil). An extinct analogue of the living HETERODONTUS or CENTRACION.

1070. EGERTON (1872). *On Prognathodus guentheri Egerton, a New Genus of Fossil Fishes from the Lias of Lyme Regis.* Quart. Journ. Geol. Soc., 1872, XXVIII, 233-237.

PHILIP DE MALPAS GREY EGERTON.

Prognathodus Egerton, XXVIII, 233; orthotype P. GUENTHERI Egert. (fossil). Not PROGNATHODES Gill, the root meaning being different.

1071. GÜNTHER (1872). *Description of Thrissopater salmoneus n. sp.* (In *Great Britain and Ireland Geological Survey. Figures and Descriptions Illustrative of British Organic Remains.* Decade 13, London, 1872.)

ALBERT GÜNTHER.

Thrissopater Günther, No. 1; orthotype T. SALMONEUS Gthr. (fossil).

1072. GÜNTHER (1872). *Zoological Record for 1871.*

ALBERT GÜNTHER.

Macruronus Günther, 103; orthotype CORYPHÆNOIDES NOVÆ-ZELANDIÆ Hector.

1073. GÜNTHER (1872). *On a New Genus of Characinoid Fishes from Demerara.* Proc. Zool. Soc. London, 1872, 146.

ALBERT GÜNTHER.

Nannostomus Günther, 146; orthotype N. BECKFORDI Gthr.

1074. GÜNTHER (1872). *Description of Two New Fishes (Lanioperca mordax, Chilodactylus allporti) from Tasmania.* Ann. Mag. Nat. Hist., 1872, ser. 4, X, 183-184.

ALBERT GÜNTHER.

Lanioperca Günther, 183; orthotype L. MORDAX, Gthr. = DINOLESTES MUELLERI Klunzinger.

"The same fish is described as DINOLESTES MUELLERI (g. et. sp. n.) by Klunzinger, *Arch. f. Nat.,* 1872, p. 29, Taf. 3. (this name has the priority); and still later as NEOSPHYRÆNA MULTIRADIATA (g. et. sp. n.) by Castelnau, *Proc. Zool. Soc. Victor,* 1872, p. 96."—(Günther).

The name DINOLESTES stands.

1075. GÜNTHER (1872). *Notice on Two New Fishes (Symphorus tæniolatus, Mugil meyeri) from Celebes.* Ann. Mag. Nat. Hist., 1872, IV, ser. 9, 438-440.

ALBERT GÜNTHER.

Symphorus Günther, 438; orthotype S. TÆNIOLATUS Gthr.

1076. HANCOCK & ATTHEY (1872). *Descriptive Notes on a Nearly Entire Specimen of Pleurodus . . . in the Coal Measures of Newsham.* Trans. Nat. Hist. Soc. Northumberland, IX.

ALBANY HANCOCK; T. ATTHEY.

Pleurodus (Agassiz) Hancock & Atthey, IX, 408; orthotype PLEURODUS RANKINEI Ag. (fossil). Name regarded as preoccupied by PLEURODON Wood, 1840; replaced by PLEUROPLAX Woodw. (1889).

1077. HUTTON (1872). *Contributions to the Ichthyology of New Zealand.* Trans. Proc. New Zealand Instit., 1872, V, 259-272.

FREDERICK WOLLASTON HUTTON (1836-1905).

Calloptilum Hutton, 266; orthotype C. PUNCTATUM Hutton. Name preocupied, replaced by AUCHENOCEROS Gthr.

1078. HUTTON & HECTOR (1872). *Fishes of New Zealand,* etc. Wellington, 1872, 1-133.

FREDERICK WOLLASTON HUTTON; JAMES HECTOR.

Phosichthys Hutton & Hector, 55; orthotype P. ARGENTEUS Hutton & Hector (later corrected to PHOTICHTHYS).

1079. KLUNZINGER (1872). *Zur Fischfauna von Süd-Australien.* Arch. Naturg., 1872, 38 Jahrg., pt. 1, 17-47.

CARL BENJAMIN KLUNZINGER.

Paradules Klunzinger, 20; logotype P. OBSCURUS Klunzinger. Name preoccupied; a synonym of NANNOPERCA Gthr.

Girellichthys Klunzinger, 22; orthotype CRENIDENS ZEBRA Rich. Same as NEOTE-PHRÆOPS Castelnau, and a little earlier; both apparently synonyms of MELAMBAPHES Gthr.

Dinolestes Klunzinger, 29; orthotype D. MUELLERI Klunzinger; replaces LANIOPERCA and NEOSPHYRÆNA, both a little later in the same year.

1080. PETERS (1872). *Ueber eine neue Gattung von Fischen aus der Familie der Cataphracti Cuv., Scombrocottus salmoneus, von der Vancouvers-Insel.* Monatsb. Akad. Wiss. Berlin, 1872, 568-570.

WILHELM CARL HARTWIG PETERS.

Scombrocottus Peters, 568; orthotype S. SALMONEUS Peters = GADUS FIMBRIA Pallas. A synonym of ANOPLOPOMA Ayres.

1081. THIOLLIÈRE (1872). *Description des poissons fossiles des gisements coralliens du Jura dans le Bugey.* (Second edition, the first in 1854.)

VICTOR JOSEPH DE L'ISLE THIOLLIÈRE.

Attakeopsis Thiollière, 23; orthotype A. DESORI Thiollière (fossil). A synonym of IONOSCOPUS Costa; first printed as name only in 1858.

1873

1082. BARKAS (1873). *Illustrated Guide to the Fish* . . . (and other) *Remains of the Northumberland Carboniferous Strata.*

THOMAS PALLISTER BARKAS (1819-1891).

Orthognathus Barkas, 38; orthotype O. RETICULOSUS Barkas (fossil). A synonym of RHIZODOPSIS Young.

Labyrinthodontosaurus Barkas, 75; orthotype L. SIMMI Barkas = HOLOPTYCHIUS SAUROIDES Binney. A synonym of STREPSODUS Huxley.

1083. VAN BENEDEN (1873). *Sur un nouveau poisson du terrain Bruxellien.* Bull. Acad. Roy. Belgique, XXXV.

PIERRE J. VAN BENEDEN.

Homalorhynchus Van Beneden, XXXV, 210; orthotype PALÆORHYNCHUM BRUXEL-LIENSE Le Hon (fossil). A synonym of HEMIRHYNCHUS Ag.

1084. BLEEKER (1873). *Mémoire sur la faune ichthyologique de Chine.* Neder. Tijdschr. Dierk., IV, 1873, 113-154.

PIETER BLEEKER.

Silurodon Bleeker, IV, 125; orthotype S. HEXANEMA Blkr.

Leptopegasus Bleeker, IV, 125; orthotype PEGASUS NATANS L. A synonym of PARAPEGASUS Dum. (1870).

Paragobiodon Bleeker, IV, 129; orthotype GOBIUS ECHINOCEPHALUS Rüppell (not P. MELANOSOMA Blkr., as stated by Jordan & Seale, *Fishes of Samoa*). Replaces RUPPELLIA Sw., 1839, preoccupied.

Tylometopon (Van Beneden) Bleeker, IV, 133; orthotype T. DUSSUMIERI Blkr.

Odontolabrax Bleeker, IV, 149; orthotype O. TYPUS Blkr.

1084A. BLEEKER (1873). *Révision des espèces indo-archipélagiques du groupe des Anthianini.* Neder. Tijdschr. Dierk., IV, 1872, 155-169.

PIETER BLEEKER.

Pseudanthias Bleeker, 158; orthotype ANTHIAS PLEUROTÆNIA Blkr.

Dactylanthias Bleeker, 158; orthotype D. APLODACTYLUS Blkr.

Odontanthias Bleeker, 158; orthotype ANTHIAS RHODOPELUS Gthr.

Plectranthias Bleeker, 158; orthotype PLECTROPOMA ANTHIOIDES Gthr.

1085. BLEEKER (1873). *Mededeelingen omtrent eene herziening der Indisch-Archipelagische soorten van Epinephelus, Lutjanus, Dentex en verwante geslachten.* Versl. Akad. Amsterdam (2) VII, 1873, 40-46.

PIETER BLEEKER.

Gnathodentex Bleeker, VII, 41; orthotype PENTAPUS AURILINEATUS Blkr.

1086. BLEEKER (1873). *Sur le genre Parapristipoma et sur l'identité spécifique des Perca trilineata Thunb., Pristipoma japonicum Cv. et Diagramma japonicum Blkr.* Arch. Neerl. Sci. Nat., VIII, 1873, 19-24.

PIETER BLEEKER.

Parapristipoma Bleeker, VIII, 22; orthotype PERCA TRILINEATA Thunberg.

1087. CASTELNAU (1873). *Contributions to the Ichthyology of Australia.* Proc. Zool. Soc. Victoria, II, 37-158.

FRANCIS LAPORTE CASTELNAU.

Lacepedia Castelnau, II, 42; orthotype L. CATAPHRACTA Castelnau.

Ruppelia Castelnau, II, 51; orthotype R. PROLONGATA Castelnau. A synonym of PARAPLESIOPS Blkr. Name preoccupied by RUPPELLIA Sw., a genus of gobies, and by a still earlier genus of insects.

Zantecla Castelnau, II, 87; orthotype Z. PUSILLA Castelnau. A synonym of MELANOTÆNIA Gill.

Ellerya Castelnau, II, 95; orthotype E. UNICOLOR Castelnau. A synonym of GOBIODON Kuhl & Van Hasselt.

Neomyripristis Castelnau, II, 98; orthotype N. AMŒNUS Castelnau.

Neocirrhites Castelnau, II, 101; orthotype N. ARMATUS Castelnau. A synonym of CIRRHITICHTHYS.

Neosudis Castelnau, II, 118; orthotype N. VORAX Castelnau.

Neotrygon Castelnau, II, 121; orthotype RAJA TRYGONOIDES Castelnau.

Edelia Castelnau, II, 123; logotype E. VITTATA Castelnau. A synonym of NANNOPERCA Gthr.

Bostockia Castelnau, II, 126; orthotype B. POROSA Castelnau.

Neochætodon Castelnau, II, 130; orthotype N. VITTATUS Castelnau. A synonym of MICROCANTHUS Sw.

Hectoria Castelnau, II, 151; orthotype OLIGORUS GIGAS Owen. A synonym of POLYPRION Val.

1088. CASTELNAU (1873). *Edible Fishes of Victoria.* Exhibition Essays.
FRANCIS LAPORTE CASTELNAU.

Bleeckeria Castelnau, —; orthotype B. CATAFRACTA Castelnau = PLESIOPS BLEEKERI Gthr. (*fide* Ogilby). Name preoccupied as BLEEKERIA Gthr. A synonym of PARAPLESIOPS Blkr., 1875.

1089. COPE (1873). *On the Extinct Vertebrata of the Eocene of Wyoming,* etc.
Rept. Geol. Surv. for 1872.
EDWARD DRINKER COPE.

Clastes Cope, 633; orthotype C. ANAX Cope (fossil) = LEPIDOSTEUS ATROX Leidy. An extinct homologue of LEPIDOSTEUS.

Pappichthys Cope, 636; orthotype P. PLICATUS Cope (fossil).

Astephus Cope, 637; orthotype RHINEASTES CALVUS Cope.

1090. COPE (1873). *A Contribution to the Ichthyology of Alaska.* Proc. Amer.
Philos. Soc., 1873, XIII, 24-32.
EDWARD DRINKER COPE.

Bathymaster Cope, XIII, 31; orthotype B. SIGNATUS Cope.

1091. COPE (1873). *On Some New Batrachia and Fishes from the Coal Measures of Linton, Ohio.* Proc. Acad. Nat. Sci. Phila.
EDWARD DRINKER COPE.

Conchiopsis Cope, 341; orthotype C. FILIFERUS Cope (fossil). A synonym of CŒLACANTHUS Ag.

Peplorhina Cope, 343; orthotype P. ANTHRACINA Cope (fossil). Regarded as an amphibian by Dr. Newberry.

1092. COSTA (1873). *Ittioliti fossili Italiani.*
ORONZIO GABRIELE COSTA.

Omalopleurus Costa, 59; orthotype O. SPECIOSUS Costa (fossil). A synonym of DAPEDIUM Leach.

1093. FITZINGER (1873). *Die Gattungen der europäischen Cyprinen nach ihren aüsseren Merkmalen.* Sitzb. Akad. Wiss. Wien, LXVIII
LEOPOLD JOSEPH FRANZ JOHANN FITZINGER.

Zopa Fitzinger, LXVIII, 152; orthotype CYPRINUS SOPA Pallas. Apparently a synonym of BALLERUS Heckel.

Vimba Fitzinger, LXVIII, 152; tautotype CYPRINUS VIMBA L. An ally of ABRAMIS Cuv.

Rubellus Fitzinger, LXVIII, 152, 162; orthotype CYPRINUS RUTILUS L. A synonym of RUTILUS Raf.

Orfus Fitzinger, LXVIII, 152, 163; orthotype LEUCISCUS VIRGO Heckel (Orfus Germanorum Marsilius, 1726, *Pisces Danubienses*) = CYPRINUS PIGUS Lac. A synonym of CENISOPHIUS Bon.

Cephalopsis Fitzinger, LXVIII, 152, 165; logotype LEUCISCUS SVALLIZE Heckel & Kner. A synonym of LEUCISCUS Cuv.

Habrolepis Fitzinger, LXVIII, 152, 167; orthotype LEUCISCUS UKLIVA Heckel. A synonym of TELESTES Bon.

Bathystoma Fitzinger, LXVIII, 152, 168; orthotype LEUCISCUS MICROLEPIS. Name preoccupied; a synonym of TELESTES Bon.

Machærochilus Fitzinger, LXVIII, 152, 170; orthotype CHONDROSTOMA PHOXINUS Heckel. A synonym of CHONDROSTOMA Ag.

1094-1095. GOLDENBURG (1873). *Fauna Saræpontana fossilis.*
F. GOLDENBURG (1873-   ).

Leiolepis Goldenburg, 5; orthotype (not named by Woodward). A synonym of AMBLYPTERUS Ag.

1096. GÜNTHER (1873). *New Fishes from Angola.* Ann. Mag. Nat. Hist., 1873, ser. 4, XII, 142-144.
ALBERT GÜNTHER.

Channalabes Günther, 142; orthotype C. APUS Gthr.

Bryconæthiops Günther, 143; orthotype B. MICROSTOMA Gthr.

1097. GÜNTHER (1873). *Report on a Collection of Fishes from China.* Ann. Mag. Nat. Hist., ser. 4, XII, 239-250.
ALBERT GÜNTHER.

Lophiogobius Günther, 241; orthotype L. OCELLICAUDA Gthr. Not identical with LOPHOGOBIUS Gill.

Myloleucus Günther, 247; orthotype LEUCISCUS ÆTHIOPS Basilewsky. Name preoccupied by MYLOLEUCUS Cope.

Toxabramis Günther, 249; orthotype T. SWINHONIS Gthr.

Psephurus Günther, 250; orthotype POLYODON GLADIUS Martens.

1098. GÜNTHER (1873). *Andrew Garrett's Fische der Südsee.* Band I. Journ. Mus. Godeffroy, 1873, 1-128.
ALBERT GÜNTHER.

Gymnoscopelus Günther, 91; orthotype G. APHYA Gthr.

Holotrachys Günther, 93; orthotype MYRIPRISTIS LIMA Cuv. & Val.

Diplophos Günther, 101; orthotype D. TÆNIA Gthr.

1099. HAAST (1873). *Notes on Some Undescribed Fishes of New Zealand.* Trans. New Zealand Instit., 1873, V, 272-278.
JOHANN FRANZ JULIUS VON HAAST (1822-1887).

Synnema Haast, V, 274; orthotype ANEMA MONOPTERYGIUM Blkr.

Bowenia Haast, V, 276; orthotype B. NOVÆ-ZELANDIÆ Haast.

1100. KAUP (1873). *Ueber die Familie Triglidæ, nebst einigen Worten über die Classification.* Archiv. Naturgeschichte.
JOHANN JACOB KAUP.

Microtrigla Kaup, 86; logotype TRIGLA PAPILIO Cuv. & Val. Differs from LEPIDOTRIGLA in the armed scales of the lateral line.

Chelidonichthys Kaup, 87; logotype TRIGLA HIRUNDO L.

Lyrichthys Kaup, 88; logotype TRIGLA LYRA L.   A valid subgenus, if TRIGLA
   GURNARDUS L. is taken as type of TRIGLA, setting aside the later choice of
   type (TRIGLA LYRA L.) by Jordan & Gilbert, 1883.

Palænichthys Kaup, 90; logotype TRIGLA ASPERA Cuv. & Val.   A synonym of
   LEPIDOTRIGLA Gthr.

   These subgenera of Kaup have been overlooked by authors.   They may
   be thus briefly defined, after Kaup:

   a. Scales moderate, evenly placed; preorbital not produced.
   b. Lateral line armed with strong spines; dorsal short. MICROTRIGLA.
   bb. Lateral line unarmed; head well armed. (LEPIDOTRIGLA) PALÆNICHTHYS.
   aa. Scales very small, crowded.
   c. Lateral line unarmed; preorbital produced. CHELIDONICHTHYS.
   cc. Lateral line with strong spines.
   d. Preorbital expanded in a broad, emarginate "lyre-like projection."
        LYRICHTHYS.
   dd. Preorbital not expanded. TRIGLA.

1101. LANKESTER (1873). *On Holaspis sericeus and on the Relationships of
   the Fish-Genera Pteraspis, Cyathaspis, and Scaphaspis.* Geol. Mag., X.
                     EDWIN RAY LANKESTER.

Holaspis Lankester, X, 242; orthotype H. SERICEUS Lank. (fossil). Name preoccupied.; a synonym of PALÆASPIS Claypole.

1102. LEIDY (1873). *Contributions to the Extinct Vertebrate Fauna of the
   Western Territories.* I. Rept. U. S. Geol. Surv. for 1867 (1873).
                        JOSEPH LEIDY.

Protautoga Leidy, 15; orthotype P. CONIDENS Leidy.

Eumylodus Leidy, 309; orthotype LAQUEATUS Leidy (fossil). A synonym of
   EDAPHODON Buckland.

1103. LEIDY (1873). *Notice of Remains of Fishes in the Bridger Tertiary Formations of Wyoming.* Proc. Acad. Nat. Sci. Phila.
                        JOSEPH LEIDY.

Hypamia Leidy, 98; orthotype H. ELEGANS Leidy (fossil).

Protamia Leidy, 98; orthotype AMIA GRACILIS Leidy (fossil).

Phareodus Leidy, 99; P. ACUTUS Leidy (fossil). Genus undefined, but not the
   species; replaces DAPEDOGLOSSUS Cope.

1104. LÜTKEN (1873). *Nogle nye eller mindre fuldstændigt kjendte Pandser-
   maller, isaer fra det nordlige Sydamerica.* Vid. Meddel., Kjöbenhavn, 1873,
   202-220.
                     CHRISTIAN FREDERIK LÜTKEN.

Xenomystus Lütken, 217; orthotype X. GOBIO Lütken. Name preoccupied; replaced
   by HEMIPSILICHTHYS Eigenmann.

   1105. MARCK (1873). *(Fossile Fische.)* Palæontographia, XXII, 1873.
                     W. VON DER MARCK.

Thrissopteroides von der Marck, XXII, 61; orthotype T. ELONGATUS von der Marck
   (fossil).

1106. NEWBERRY (1873). *Descriptions of Fossil Fishes.* Rept. Geol. Surv. Ohio. I. (*Devonian System*, 290-324.)

JOHN STRONG NEWBERRY.

Cyrtacanthus Newberry, I, 306; orthotype C. DENTATUS Newberry (fossil). Probably a synonym of ERISMACANTHUS.

Liognathus Newberry, I, 306; orthotype L. SPATULATUS Newberry. Name preoccupied by LEIOGNATHUS Lac.; replaced by LISPOGNATHUS Miller, 1892.

Rhynchodus Newberry, I, 307; orthotype R. SECANS Newberry (fossil).

Dinichthys Newberry, I, 313; logotype D. HERTZERI Newberry (fossil).

Aspidichthys Newberry, I, 322; orthotype ASPIDOPHORUS CLAVATUS Newberry (fossil). Replaces ASPIDOPHORUS Newberry, preoccupied.

1107. POEY (1873). *Grammicolepis brachiusculus, tipo de una nueva familia en la clase de los peces.* Anal. Soc. Española Hist. Nat., Madrid, 1873, II, 403-406.

FELIPE POEY Y ALOY.

Grammicolepis Poey, II, 403; orthotype G. BRACHIUSCULUS Poey.

1108. SAUVAGE (1873). *Note sur le Sebastes minutus.* Ann. Sci. Nat. (Zool.), V.

HENRI ÉMILE SAUVAGE.

Sebastopsis Sauvage, V, 1; orthotype SEBASTES MINUTUS Cuv. Identical with the earlier SEBASTOPSIS Gill, both being synonyms of SCORPÆNODES Blkr.

1109. SAUVAGE (1873). *Mémoire sur la faune ichthyologique de la période tertiaire et plus spécialement sur les poissons fossiles d'Oran (Algérie) et sur ceux découverts par M. R. Alby à Licata en Sicile.* Ann. Sci. Géol., 1873, IV, No. 1.

HENRI ÉMILE SAUVAGE.

Trichiurichthys Sauvage, IV, 130; orthotype T. INCERTUS Sauvage (fossil).

Hemithyrsites Sauvage, IV, 136; orthotype ACANTHONOTUS ARMATUS Sauvage (fossil); replaces ACANTHONOTOS Sauvage, preoccupied by ACANTHONOTUS Bloch.

Pseudeleginus Sauvage, IV, 164; orthotype P. ALBYI Sauvage (fossil).

Parascopelus Sauvage, IV, 196; orthotype SCOPELUS LACERTOSUS Sauvage (fossil).

Anapterus Sauvage, IV, 199; orthotype TYDEUS MEGISTOSOMA Sauvage (fossil). Name a substitute for TYDEUS Sauvage, preoccupied.

1110. TRAQUAIR (1873). *On a New Genus (Ganorhynchus) of Fossil Fish of the Order Dipnoi.* Geol. Mag., 1873, X, 552-555.

RAMSAY HEATLEY TRAQUAIR (1840-1912).

Ganorhynchus Traquair, 555; orthotype G. WOODWARDI Traquair (fossil).

1111. VAILLANT (1873). *Recherches sur les poissons des eaux douces de l'Amérique septentrionale designés par M. L. Agassiz sous le nom d'Etheostomatidæ.* Nouv. Archiv. Mus. Hist. Nat. Paris, 1873, IX, 5-154.

LÉON LOUIS VAILLANT.

Plesioperca Vaillant, IX, 36; orthotype P. ANCEPS Vaillant = HADROPTERUS NIGROFASCIATUS Ag. A synonym of HADROPTERUS Ag.

Astatichthys Vaillant, IX, 106; orthotype ETHEOSTOMA CÆRULEA Storer. A synonym of OLIGOCEPHALUS Grd.

1874

1112. BARKAS (1874). *On the Microscopical Structure of Fossil Teeth from the Northumberland Coal Measures.* Monthly Review Dental Surgery.

WILLIAM JAMES BARKAS.

Petalodopsis Barkas, II, 538; orthotype P. MIRABILIS Barkas = SAGENODUS INÆQUALIS Owen (fossil). Probably a synonym of SAGENODUS.

1113. BLEEKER (1874). *Typi nonnulli generici piscium neglecti.* Versl. Akad. Amsterdam (2), VIII, 367-371.

PIETER BLEEKER.

Campylomormyrus Bleeker, VIII, 367; orthotype MORMYRUS TAMANDUA Gthr.

Oxymormyrus Bleeker, VIII, 367; orthotype MORMYRUS ZANCLIROSTRIS Gthr.

Solenomormyrus Bleeker, VIII, 368; orthotype CENTRISCUS NILOTICUS Bl. & Schn.

Pararhychobdella Bleeker, VIII, 368; orthotype RHYNCHOBDELLA MACULATA Reinwardt.

Polyacanthonotus Bleeker, VIII, 368; orthotype NOTACANTHUS RISSOANUS Filippi & Verany.

Gunnellops Bleeker, VIII, 368; orthotype GUNNELLUS ROSEUS Cuv. & Val.

Lycodalepis Bleeker, VIII, 369; orthotype LYCODES MUCOSUS Rich.

Paralycodes Bleeker, VIII, 369; orthotype LYCODES VARIEGATUS Gthr.

Acanthocepola Bleeker,, VIII, 369; orthotype CEPOLA KRUSENSTERNI Schlegel.

Macuroplus Bleeker, VIII, 369; orthotype MACRURUS SERRATUS Lowe.

Paramacrurus Bleeker, VIII, 370; orthotype LEPIDOLEPRUS AUSTRALIS Rich.

Oxymacrurus Bleeker, VIII, 370; orthotype MACRURUS JAPONICUS T. & S.

Parabembras Bleeker, VIII, 370; orthotype BEMBRAS CURTUS T. & S.

Octogrammus Bleeker, VIII, 370; orthotype "LABRAX OCTOGRAMMUS Schlegel" = OCTOGRAMMUS PALLASI Blkr. A synonym of HEXAGRAMMOS Steller.

Parachætodon Bleeker, VIII, 371; orthotype CHÆTODON OLIGACANTHUS Blkr.

1114. BLEEKER (1874). *Notice sur les genres Amblyeleotris, Valenciennesia et Brachyeleotris.* Versl. Akad. Amsterdam (2), VIII, 1874, 372-376.

PIETER BLEEKER.

Amblyeleotris Bleeker, VIII, 372; orthotype ELEOTRIS PERIOPHTHALMUS Blkr.

Eleotrioides Bleeker, VIII, 372; orthotype E. CYANOSTIGMA Blkr. Not ELEOTRIODES Blkr., 1857; a synonym of BRACHYELEOTRIS Blkr.

1115. BLEEKER (1874). *Esquisse d'un système naturel des Gobioides.* Arch. Néerl. Sci. Nat., IX, 1874, 289-331.

PIETER BLEEKER.

Guavina Bleeker, IX, 302; orthotype ELEOTRIS GUAVINA Cuv. & Val.

Oxyeleotris Bleeker, IX, 302; orthotype ELEOTRIS MARMORATA Blkr.

Culius Bleeker, IX, 303; orthotype CULIUS FUSCUS Blkr = ELEOTRIS NIGRA Q. & G. A synonym of ELEOTRIS Gronow.

Belobranchus Bleeker, IX, 304; orthotype B. QUOYI Blkr.

Gymneleotris Bleeker, IX, 304; orthotype ELEOTRIS SEMINUDUS Gthr.

Butis Bleeker, IX, 304; orthotype CHEILODIPTERUS BUTIS Ham.

Gymnobutis Bleeker, IX, 304; orthotype ELEOTRIS GYMNOCEPHALUS Steind.

Prionobutis Bleeker, IX, 305; orthotype ELEOTRIS DASYRHYNCHUS Gthr. A synonym of BUTIS Blkr.

Odontobutis Bleeker, IX, 305; orthotype ELEOTRIS OBSCURA T. & S.

Priolepis (Ehrenberg) Bleeker, IX, 305; orthotype ASTERROPTERYX SEMIPUNCTATUS Rüppell. A synonym of ASTERROPTERYX Rüppell, noted in passing.
Brachyeleotris Bleeker, IX, 306; orthotype ELEOTRIS CYANOSTIGMA Blkr.
Hetereleotris Bleeker, IX, 306; orthotype GOBIUS DIADEMATUS Rüppell.
Valenciennesia Bleeker, IX, 307; orthotype ELEOTRIS STRIGATA Blkr.

The names VALENCIENNEA, VALENCIENNESIA, and ELEOTRIODES have been confused by Bleeker. VALENCIENNEA appeared (1868) without diagnosis, with mention only of ELEOTRIS HASSELTI Blkr. As this was a previously known species, the name may be regarded as established and the species in question as its type. Later in 1868 VALENCIENNESIA appears also without explanation, in connection with ELEOTRIS STRIGATA Blkr. Meanwhile, the name ELEOTRIODES had been given in 1857, in the same way, without diagnosis, to ELEOTRIS SEXGUTTATA Cuv. & Val. In 1874 Bleeker makes ELEOTRIODES a synonym of VALENCIENNESIA, with E. STRIGATA as type, and still later in 1874, in his *Esquisse Général,* ELEOTRIODES appears as a separate genus, its type being specified as ELEOTRIS HELDSDINGENI Blkr. Clearly Bleeker regarded these earlier names as names only, but as each of them refers to a species already known, they must have place in the system. That VALENCIENNEA, 1868 (HASSELTI), is distinct from ELEOTRIODES, 1857 (SEXGUTTATA), is quite probable.
Eleotriodes Bleeker IX, 307; orthotype E. HELSDINGENI Blkr. (not ELEOTRIODES Blkr., 1857).
Orthostomus (Kner) Bleeker, IX, 308; orthotype GOBIUS AMBLYOPINUS Kner.
Oxymetopon Bleeker, IX, 308; orthotype O. TYPUS Blkr.
Pseudogobiodon Bleeker, IX, 309; orthotype GOBIUS CITRINUS Rüppell.
Paragobiodon Bleeker, IX, 309; orthotype GOBIUS MELANOSOMA Blkr. Replaces RUPPELLIA Sw., preoccupied.
Alepidogobius Bleeker, IX, 310; orthotype GOBIOSOMA FASCIATUM Playfair.
Gobiopterus Bleeker, IX, 311; orthotype APOCRYPTES BRACHYPTERUS Blkr.
Leptogobius Bleeker, IX, 311; orthotype GOBIUS OXYPTERUS Blkr.
Triænopogon Bleeker, IX, 312; orthotype TRIÆNOPHORICHTHYS BARBATUS Gthr.
Sicydiops Bleeker, IX, 314; orthotype SICYDIUM XANTHURUS Blkr.
Microsicydium Bleeker, IX, 314; orthotype SICYDIUM GYMNAUCHEN Blkr.
Brachygobius Bleeker, IX, 315; orthotype GOBIUS DORIÆ Gthr.
Callogobius Bleeker, IX, 315; orthotype ELEOTRIS HASSELTI Blkr. A synonym of VALENCIENNEA.
Platygobius Bleeker, IX, 316; orthotype GOBIUS MACRORHYNCHUS Blkr.
Mesogobius Bleeker, IX, 317; orthotype GOBIUS BATRACHOCEPHALUS Pallas.
Stenogobius Bleeker, IX, 317; orthotype GOBIUS GYMNOPOMUS Blkr.
Oligolepis Bleeker, IX, 318; orthotype GOBIUS MELANOSTIGMA Blkr.
Gnatholepis Bleeker, IX, 318; orthotype GOBIUS ANJERENSIS Blkr.
Hypogymnogobius Bleeker, IX, 318; orthotype GOBIUS XANTHOZONA Blkr.
Hemigobius Bleeker, IX, 319; orthotype GOBIUS MELANURUS Blkr.
Actinogobius Bleeker, IX, 319; orthotype GOBIUS OMMATURUS Blkr.
Cephalogobius Bleeker, IX, 320; orthotype GOBIUS SUBLITUS Cantor.
Centrogobius Bleeker, IX, 321; orthotype GOBIUS NOTACANTHUS Blkr. Same as OPLOPOMUS Steind., not OPLOPOMA Grd.
Acentrogobius Bleeker, IX, 321; orthotype GOBIUS CHLOROSTIGMA Blkr.
Porogobius Bleeker, IX, 321; orthotype GOBIUS SCHLEGELI Gthr.

Amblygobius Bleeker, IX, 322; orthotype Gobius sphinx Blkr.

Cryptocentrus (Ehrenberg), Bleeker, IX, 322; orthotype Gobius cryptocentrus Cuv. & Val. Same as Paragobius Blkr.

Zonogobius Bleeker, IX, 323; orthotype Gobius semifasciatus Kner.

Odontogobius Bleeker, IX, 323; orthotype Gobius bynoënsis Blkr.

Stigmatogobius Bleeker, IX, 323; orthotype Gobius pleurostigma Blkr.

Amblychæturichthys Bleeker, IX, 324; orthotype Chæturichthys hexanema Blkr.

Parachæturichthys Bleeker, IX, 325; orthotype Chæturichthys polynema Blkr.

Periophthalmodon Bleeker, IX, 326; orthotype Gobius schlosseri Pallas = Gobius barbarus L. A synonym of Periophthalmus Bl. & Schn., as restricted by Gill, 1863.

Apocryptodon Bleeker, IX, 327; orthotype Apocryptes madurensis Blkr.

Parapocryptes Bleeker, IX, 327; orthotype Apocryptes microlepis Blkr.

Gobileptes (Swainson) Bleeker, IX, 327; type Apocryptes bato Cuv. & Val. The first assignment of type, but evidently not the species Swainson had in mind as Gobileptes.

Pseudapocryptes Bleeker, IX, 328; orthotype Apocryptes lanceolatus Cantor.

Brachyamblyopus Bleeker, IX, 329; orthotype Amblyopus brachysoma Blkr.

Odontamblyopus Bleeker, IX, 330; orthotype Gobioides rubicundus Ham.

Trypauchenichthys Bleeker, IX, 331; orthotype T. typus Blkr.; replaces Cteno-trypauchen Steind.

1116. BLEEKER (1874). *Révision des espèces d'Ambassis et de Parambassis de l'Inde archipélagique.* Nat. Verh. Holl. Maatsch. Wetensch., 3. verz., II, 1874, 83-106.

Pieter Bleeker.

Parambassis Bleeker, 102 (19, reprint); orthotype Ambassis apogonides Blkr.

1117. BLEEKER (1874). *Révision des espèces insul-indiennes de la famille des Synanceoides.* Nat. Verh. Holl. Maatsch. Wetensch., 3. verz., II, 1874, 1-22.

Pieter Bleeker.

Synanchia (Swainson) Bleeker, 11; orthotype Synanceia erosa Cuv. & Val. A synonym of Erosa Sw. Synanchia Sw. is plainly only a "wilful misprint" for Synanceia. In any case, Erosa occurs on an earlier page in the same work and was intended for the same species.

Leptosynanceia Bleeker, 17; orthotype Synancia astroblepa Rich. In this paper Bleeker cancels his earlier genus Synanceichthys (verrucosa), regarding it as intergrading with Synanceja (horrida). But the two seem fairly separable. The earliest subdivision of Synanceja, that of Müller in 1843, leaves verrucosa as the type of Synanceja, S. horrida being set off as Synancidium. This arrangement apparently must stand. Synanceichthys, with Spurco (Commerson) and Emmydrichthys Jordan and Snyder, become synonyms of Synanceja.

1118. COPE (1874). *On Some Batrachians and Nematognathi Brought from the Upper Amazon by Prof. Orton.* Proc. Acad. Nat. Sci. Phila., 1874, 120-137.

Edward Drinker Cope.

Dysichthys Cope, 133; orthotype D. coracoideus Cope.

1119. COPE (1874). *On the Plagopterinæ and the Ichthyology of Utah.* Proc. Amer. Philos. Soc., XIV, 122-139.

Edward Drinker Cope.

Plagopterus Cope, 130; orthotype P. ARGENTISSIMUS Cope.
Lepidomeda Cope, 131; orthotype L. VITTATA Cope.

1120. COPE (1874). *Description of Some Species of Reptiles Obtained by Dr. John F. Bransford, Assistant Surgeon United States Navy, . . . Nicaragua,* etc. Proc. Acad. Nat. Sci. Phila., 1874, 64-72.

EDWARD DRINKER COPE.

Protistius Cope, 66; orthotype P. SEMOTILUS Cope.

1121. COPE (1874). *Review of the Vertebrata of the Cretaceous Period Found West of the Mississippi River.* Bull. U. S. Geol. Surv. Terr., I.

EDWARD DRINKER COPE.

Pelecorapis Cope, 39; orthotype P. VARIUS Cope (fossil).
Tetheodus Cope, 43; orthotype T. PEPHREDO Cope (fossil).
Sporetodus Cope, 47; orthotype S. JANEVAII Cope. A synonym of PTYCHODUS Ag.

1122. DYBOWSKI (1874). *Fische des Baikal-Wassersystemes.* Verh. K.-K. Zool.-Bot. Ges. Wien, XXIV.

BENEDIKT IVAN DYBOWSKI

Leucichthys Dybowski, 390; logotype SALMO OMUL Pallas, nearly or quite identical with ARGYROSOMUS Ag., which is preoccupied and replaced by THRISSOMIMUS Gill.

1123. GILL (1874). *On the Identity of Esox lewini With the Dinolestes mülleri of Klunzinger.* Am. Mag. Nat. Hist., XIV.

THEODORE GILL.

Gobiopus Gill, XIV, 159; orthotype not stated, a goby with the "first dorsal fin atrophied," as in LUCIOGOBIUS and LEUCOPSARION. Apparently a synonym of LUCIOGOBIUS Gill.

1124. GÜNTHER (1874). *Descriptions of New Species of Fishes in the British Museum.* Ann. Mag. Nat. Hist., 1874, XIV, 368-371; 453-475.

ALBERT GÜNTHER.

Rhamphocottus Günther, 369; orthotype R. RICHARDSONI Gthr.
Gastromyzon Günther, 454, 453-475; orthotype G. BORNEENSIS Gth.

1125. HAAST (1874). *On Cheimarrichthys forsteri, a New Genus Belonging to the New Zealand Fresh-water Fishes.* Trans. New Zealand Instit., 1874, VI, 103-104.

JOHANN FRANZ JULIUS VON HAAST.

Cheimarrichthys Haast, 103; orthotype C. FORSTERI Haast.

1126. JONES (1874). *A New Fish (Lefroyia bermudensis).* Zoologist, 1874, II, 3837-3838.

JOHN MATTHEW JONES.

Lefroyia Jones, 3837; orthotype L. BERMUDENSIS Jones. A synonym of CARAPUS Raf. (FIERASFER Cuv.).

1127. KESSLER (1874). *Pisces.* (In *Fedtschensko's Expedition to Turkestan.* Zoogeographical Researches.) Bull. Soc. Sci. Moscou, 1874, XI, 1-63.

KARL THEODOROVICH KESSLER (1815-1881).

Pseudodon Kessler, XI, 40; orthotype P. LONGICAUDA Kessler; name preoccupied. A synonym of PARACOBITIS Blkr., 1863.

Diplophysa Kessler, 57; logotype D. STRAUCHI Kessler.

1128.. LÜTKEN (1874). *Siluridæ novæ Brasiliæ centralis a clarissimo J. Rein-hardt in provincia Minas-Geraës circa oppidulum Lagoa Santa, præcique in flumine rio das Velhas,* etc. Overs Dansk. Vid. Selsk., Kjöbenhavn, 1874, 29-36.

CHRISTIAN FREDERIK LÜTKEN.

Leporellus Lütken, 129; orthotype LEPORINUS PICTUS Kner.

1129. LÜTKEN (1874). *Characinæ novæ Brasiliæ centralis a clarissimo J. Rein-hardt in provincia Minas-Geraës circa oppidulum Lagoa Santa in lacu eiusdem nominis, flumine rio das Velhas,* etc. Oversigt Dansk. Vid. Forh. Kjöben-havn, 1874 (1875), 127-143.

CHRISTIAN FREDERIK LÜTKEN.

Glanidium Lütken 31; orthotype G. ALBESCENS Lütken.

Bagropsis Lütken, 32; orthotype B. REINHARDTI Lütken.

1130. POEY (1874). *Monographie des poissons de Cuba compris dans la sous famille de Sparini.* Ann. Lyc. Nat. Hist. N. Y., X.

FELIPE POEY Y ALOY.

Grammateus Poey, 182; orthotype PAGELLUS MICROPS Guich. = PAGELLUS PENNA Cuv. & Val. A section under CALAMUS Sw.

1131. PROBST (1874). *Beitrag zur Kenntniss der fossilen Fische aus der Molasse von Baltringen.* Jahresb. Ver. Vaterl. Naturk. Württemberg, XXX.

JOSEF PROBST (1823-   ).

Sparoides Probst, XXX, 291; logotype S. UMBONATUS Probst (fossil).

1132. SAUVAGE (1874). *Notice sur les poissons tertiaires de l'Auvergne.* Bull. Soc. Hist. Nat. Toulouse, 1873-74, VIII, 171-198.

HENRI ÉMILE SAUVAGE.

Prolebias Sauvage, 187; logotype P. GREGATUS (Aymard) Sauvage (fossil).

Pachystetus (Aymard) Sauvage, 190; orthotype LEBIAS AYMARDI Sauvage = PACHYSTETUS GREGATUS Aymard MS. (noted in synonymy). A synonym of PROLEBIAS Sauvage.

1133. SAUVAGE (1874). *Notices Ichthyologiques.* Rev. Mag. Zool., 1874, II, 332-340.

HENRI ÉMILE SAUVAGE.

Oreias Sauvage, 334; orthotype O. DABRYI Sauvage. Apparently a synonym of BARBATULA Linck; same as ORTHRIAS Jordan & Fowler of later date.

Sinibarbus Sauvage, 335; orthotype S. VITTATUS Sauvage.

Doliichthys Sauvage, 336; orthotype D. STELLATUS Sauvage.

Lepidoblennius Sauvage, 337; orthotype L. CALEDONICUS Sauvage. Name preoccu-pied (Steindachner, 1867), replaced by SAUVAGEA Jordan and Seale (1906).

Heteroleuciscus Sauvage, 339; orthotype H. JULLIENI Sauvage.

1134. SAUVAGE (1874). *Révision des espèces du groupe des Épinoches.* Nouv. Arch. Mus., X, 5-32.

HENRI ÉMILE SAUVAGE.

Gasterostea Sauvage, 7; orthotype GASTEROSTEUS PUNGITIUS L. A synonym of
PYGOSTEUS Brevoort (1861).

Gastræa Sauvage, 7; orthotype GASTEROSTEUS SPINACHIA L. A synonym of SPI-
NACHIA Cuv.

1135. SAUVAGE (1874). *Étude sur les poissons du Lias supérieur de la Lozère
et de la Bourgogne.* Rev. Sci. Nat., II.

HENRI ÉMILE SAUVAGE.

Cephenoplosus Sauvage, II, 428; orthotype C. TYPUS Sauvage (fossil). A synonym
of PACHYCORMUS Ag.

1136. STEINDACHNER (1874). *Ichthyologische Beiträge.* I. Sitzb. Akad.
Wiss. Wien, 1874, LXX, 1. Abth., 355-390.

FRANZ STEINDACHNER.

Pikea Steindachner, 375; orthotype GRYSTES LUNULATUS Guich.
Hemianthias Steindachner, 380; orthotype ANTHIAS PERUANUS Steind.

1137. TRAUTSCHOLD (1874). *Die Kalkbrüche von Mjatschkoeva,* etc. **Nouv.**
Mem. Soc. Imp. Natur., XIII, Moscou.
HERMANN VON TRAUTSCHOLD (1817-1902).

Solenodus Trautschold, 293; orthotype S. CRENULATUS Trautschold.
Octinaspis Trautschold, pl. 28; orthotype PETRODUS SIMPLICISSIMUS Trautschold
(fossil).

1138. WINKLER (1874). *Mémoire sur les dents de poissons du terrain bruxellien.*
Arch. Mus. Teyler, Haarlem.
TIBERIUS CORNELIUS WINKLER.

Plicodus Winkler, III, 301; orthotype GINGLYMOSTOMA THIELENSE Noetling (fossil).
A fossil homologue of GINGLYMOSTOMA.

1875

1139. BLEEKER (1875). *Notice sur les Elcotriformes et description de trois
espèces nouvelles.* Arch. Néerl. Sci. Nat., X, 1875, 101-112.
PIETER BLEEKER.

Pogoneleotris Bleeker, 107; orthotype ELEOTRIS HETEROLEPIS Gthr.

1140. BLEEKER (1875). *Description du genre Parascorpis et du son espèce-
type.* Arch. Néerl. Sci. Nat., 1875, X, 380-382.
PIETER BLEEKER.

Parascorpis Bleeker, 380; orthotype P. TYPUS Blkr.

1141. BLEEKER (1875). *Révision des Sicydiini et Latrunculi de l'Insulindes.*
Versl. Acad. Amsterdam, IX.
PIETER BLEEKER.

Leptogobius Bleeker, 22; orthotype GOBIUS OXYPTERUS Blkr. (See also No. 1115.)

1142. BLEEKER (1875). *Recherches sur la faune de Madagascar et de ses
dépendances d'après les découvertes de François P. L. Pollen et D. C. van Dam.*
4e Partie. *Poissons de Madagascar et de l'île de la Réunion.* Leiden.
PIETER BLEEKER.

Paracæsio Bleeker, 92; orthotype Cæsio xanthurus Blkr.

Paracirrhites Bleeker, 93; orthotype Cirrhites forsteri Bloch. A synonym of Cirrhitus Lac.

1143. BLEEKER (1875). *Sur la famille des Pseudochromidoides et révision de ses espèces insulindiennes.* Verh. Akad. Amsterdam, XV, 1875, 1-32.

Pieter Bleeker.

Paraplesiops Bleeker, 3; orthotype Plesiops bleekeri Gthr. Replaces Bleeckeria Castelnau, 1873, preoccupied.

Pseudogramma Bleeker, 24; orthotype P. polyacanthus Blkr.

1144. CASTELNAU (1875). .*Researches on the Fishes of Australia.* Philadelphia Centennial Expedition of 1876. Official Record, etc., 1875.

Francis Laporte de Castelnau.

A carelessly prepared paper, as usual with Castelnau. I am under obligation to Mr. Allan R. McCulloch of the Australian Museum for the identification of these and other nominal genera proposed for Australian fishes, by Castelnau, De Vis and others.

Neoniphon Castelnau, 4; orthotype N. armatus Castelnau.

Breviperca Castelnau, 6; orthotype B. lineata Castelnau.

Neomesoprion Castelnau, 8; orthotype N. unicolor Castelnau. Apparently a synonym of Lutianus Bloch.

Aida Castelnau, 10; orthotype A. inornata Castelnau. A synonym of Melanotænia Gill.

Neolethrinus Castelnau, 11; orthotype N. similis Castelnau.

Neoblennius Castelnau, 11; orthotype N. fasciatus Castelnau.

Neosillago Castelnau, 16; orthotype N. marmorata Castelnau.

Pseudobatrachus Castelnau, 24; orthotype Batrachus striatus Castelnau, perhaps identical with Batrachomœus Ogilby of later date.

Stenophus Castelnau, 27; orthotype S. marmoratus Castelnau.

Neogunellus Castelnau, 27; orthotype N. sulcatus Castelnau. A synonym of Ophiclinus Castelnau.

Dampieria Castelnau, 30; orthotype D. lineata Castelnau; replaces Cichlops M. & T., preoccupied; same as Labracinus (Schlegel) Blkr., of slightly later date.

Neoatherina Castelnau, 31; orthotype N. australis Castelnau. A synonym of Melanotænia Gill.

Torresia Castelnau, 36; orthotype T. australis Castelnau. A synonym of Chœrodon Blkr.

Neoodax Castelnau, 37; orthotype N. waterhousei Castelnau. A synonym of Odax Cuv.

Othos Castelnau, 43; orthotype O. cephalotes Castelnau.

Neorhombus Castelnau, 43; orthotype N. unicolor Castelnau.

Neoplotosus Castelnau, 45; orthotype N. waterhousei Castelnau.

Neoscopelus Castelnau, 46; orthotype Scopelus cephalotes Castelnau; name preoccupied.

Blanchardia Castelnau, 47; orthotype B. maculata Castelnau. A synonym of Notograptus Gthr.

1145. COLLETT (1875). *Norges fiske, med bemærkninger om deres Udbredelse.* Christiania.

Robert Collett (1842-1913).

Cottunculus Collett, 20; orthotype C. MICROPS Collett.

Latrunculodes Collett, 60; orthotype GOBIUS NILSSONI Düben & Koren.

1146. COPE (1875). *The Vertebrata of the Cretaceous Formations of the West.* Rept. U. S. Geol. Surv. Terr., 1875, II, 1-303.

EDWARD DRINKER COPE.

Syllæmus Cope, 180; orthotype S. LATIFRONS Cope (fossil).

Isótænia Cope, 222; orthotype I. NEOCESARIENSIS Cope (fossil).

Pelecopterus Cope, II, 244; orthotype ICHTHYODECTES PERNICIOSUS Cope (fossil). A synonym of PROTOSPHYRÆNA Leidy.

Bryactinus Cope, 282; orthotype B. AMORPHUS Cope (fossil). A synonym of EDAPHODON Leidy.

Diphrissa Cope, 283; orthotype ISCHYODUS SOLIDULUS Cope (fossil). A synonym of EDAPHADON.

1147. COPE (1875). *On a New Genus (Osphyolax pellucidus) of Lophobranchiate Fishes.* Proc. Acad. Nat. Sci. Phila., 1875, 450.

EDWARD DRINKER COPE.

Osphyolax Cope, 450; orthotype O. PELLUCIDUS Cope. Apparently a synonym of NEROPHIS Raf.

1148. DAY (1875). *The Fishes of India.*

FRANCIS DAY.

Pseudosynanceia Day, 163; orthotype P. MELANOSTIGMA Day.

1149. FRITSCH (1875). *Ueber die fauna der Gaskohle des Pilsner und Rakonitzer Beckens.* Sitzb. Königl. Böhm. Ges. Wiss. Prag.

ANTON JAN FRITSCH (1832-1914).

Osmerolepis Fritsch, IX, 70; orthotype O. RETICULATA Fritsch (fossil).

1150. NEWBERRY (1875). *Descriptions of Fossil Fishes.* Rept. Geol. Surv. Ohio, 1875, II, pt. 2 (Palæontology), 1-64.

JOHN STRONG NEWBERRY.

Asterosteus Newberry, 35; orthotype A. STENOCEPHALUS Newberry (fossil).

Acanthaspis Newberry, 36; orthotype A. ARMATUS Newberry.

Acantholepis Newberry, 38; orthotype A. PUSTULOSUS Newberry. Name preoccupied, replaced by ECZEMATOLEPIS Miller, 1892.

Platyodus Newberry, 58; orthotype P. LINEATUS Newberry.

Heliodus Newberry, 62; logotype H. LESLEYI Newberry (fossil). A synonym of PALÆDAPHUS Van Beneden.

1151. PETERS (1875). *Ueber eine neue mit Halieutæa verwandte Fischgattung, Dibranchus, aus dem atlantischen Ocean,* etc. Monatsb. Akad. Wiss. Berlin, 1875 (1876), 736-742.

WILHELM CARL HARTWIG PETERS.

Dibranchus Peters, 736; orthotype D. ATLANTICUS Peters.

1152. POEY (1875). *Enumeratio piscium cubensium.* Anal. Soc. Española Hist. Nat. Madrid, 1875, IV, 75-161; V, 131-218 (1876).

FELIPE POEY Y ALOY.

Moharra Poey, 124; orthotype GERRES RHOMBEUS Cuv. A section under DIAPTERUS Ranzani.

Microgobius Poey, V, 168; orthotype M. SIGNATUS Poey.

1153. SAUVAGE (1875). *Note sur les poissons fossiles.* Bull. Soc. Géol. France, 1875, sér. 3, III, 631-642.

HENRI ÉMILE SAUVAGE.

Lepidocottus Sauvage, 637; orthotype COTTUS ARIES Ag. (fossil).

Paraperca Sauvage, 639; orthotype P. PROVINCIALIS Sauvage (fossil).

Trachinopsis Sauvage, 641; orthotype T. IBERICA Sauvage (fossil).

1154. ST. JOHN & WORTHEN (1875). *Description of Fossil Fishes.* Géol. Surv. Ill., 1875, VI, 245-488.

ORESTES H. ST. JOHN; A. H. WORTHEN.

Phœbodus St. John & Worthen, 251; orthotype P. SOPHIÆ St. John & Worthen (fossil).

Bathycheilodus St. John & Worthen, 252; orthotype B. MACISAACSI St. John & Worthen (fossil). A synonym of PHŒBODUS St. John & Worthen.

Lambdodus St. John & Worthen, 280; orthotype L. COSTATUS St. John & Worthen (fossil).

Hybocladodus St. John & Worthen, 284; orthotype H. PLICATILIS St. John & Worthen (fossil).

Thrinacodus St. John & Worthen, 289, orthotype T. NANUS St. John & Worthen (fossil).

Mesodmodus St. John & Worthen, 290 logotype M. EXSCULPTUS St. John & Worthen (fossil).

Agassizodus St. John & Worthen, 311; orthotype LOPHODUS VARIABILIS St. John & Worthen (fossil). Substitute for LOPHODUS Newberry & Worthen, preoccupied. Both are synonyms of CAMPODUS Koninck, 1844.

Periplectrodus St. John & Worthen, 325; logotype P. WARRENI St. John & Worthen (fossil).

Stemmatodus St. John & Worthen, 330; logotype S. CHEIRIFORMIS St. John & Worthen (fossil). Name preoccupied, replaced by STEMMATIAS Hay, 1899.

Leiodus St. John & Worthen, 335; logotype L. CALCARATUS St. John & Worthen (fossil). Not LEIODON Sw.

Desmiodus St. John & Worthen, 341; orthotype D. TUMIDUS St. John & Worthen (fossil).

Venustodus St. John & Worthen, 344; orthotype CHOMATODUS VENUSTUS Leidy = V. LEIDYI St. John & Worthen (fossil).

Harpacodus St. John & Worthen, 354; orthotype H. OCCIDENTALIS St. John & Worthen.

Lisgodus St. John & Worthen, 363; logotype L. CURTUS St. John & Worthen (fossil). Perhaps a synonym of PETALODUS.

Tanaodus St. John & Worthen, 367; logotype T. BELLICINCTUS St. John & Worthen (fossil).

Calopodus St. John & Worthen, 403; orthotype; C. APICALIS St. John & Worthen (fossil).

Fissodus St. John & Worthen, 413; orthotype F. BIFIDUS St. John & Worthen (fossil).

Cholodus St. John & Worthen, 415; orthotype C. INÆQUALIS St. John & Worthen (fossil). Perhaps a synonym of FISSODUS St. John & Worthen.

Acondylacanthus St. John & Worthen, 432; logotype A. GRACILIS St. John & Worthen (fossil).

Geisacanthus St. John & Worthen, 440; orthotype G. STELLATUS St. John & Worthen (fossil). A synonym of COSMACANTHUS Ag.

Bythiacanthus St. John & Worthen, 444; orthotype B. VANHORNEI St. John & Worthen (fossil).

Glymmatacanthus St. John & Worthen, 446; orthotype G. IRISHI St. John & Worthen (fossil).

Amacanthus St. John & Worthen, 464; orthotype HOMACANTHUS GIBBOSUS Newberry & Worthen (fossil).

Marracanthus St. John & Worthen, 465; orthotype HOMACANTHUS RECTUS Newberry & Worthen (fossil).

Batacanthus St. John & Worthen, 468; orthotype B. BACULIFORMIS St. John & Worthen.

Gampsacanthus St. John & Worthen, 471; orthotype G. TYPUS St. John & Worthen (fossil).

Lecracanthus St. John & Worthen, 475; orthotype L. UNGUICULUS St. John & Worthen (fossil).

1155. SAUVAGE (1875). *(Poissons fossiles.)* Bibl. École des Hautes Études, XIII.

HENRI ÉMILE SAUVAGE.

Heterothrissops, Sauvage, 46; orthotype H. INTERMEDIUS Sauvage (fossil). A synonym of EUTHYNOTUS Wagner.

Pseudothrissops Sauvage, 46; orthotype P. MICROPTERUS Sauvage = ESOX INCOGNITUS Blainv. (fossil). A synonym of EUTHYNOTUS.

1156. STEINDACHNER (1875). *Beiträge der Kenntniss der Chromiden des Amazonenstromes.* Sitzb. Akad. Wiss. Wien, 1875, LXXI, 61-137.

FRANZ STEINDACHNER.

Acaropsis Steindachner, 99; orthotype ACARA NASSA Heckel.

Crenicara Steindachner, 99; orthotype C. ELEGANS Steind.

Dicrossus (Agassiz) Steindachner, 102; orthotype D. MACULATUS Steind.

Saraca Steindachner, 125; orthotype S. OPERCULARIS Steind.

1157. STEINDACHNER (1875). *Ueber eine neue Gattung und Art . . . aus der Familie der Pleuronectiden und über eine neue Thymallus-Art.* Sitzb. Akad. Wiss. Wien, 1875, LXX, 363-371.

FRANZ STEINDACHNER.

Oncopterus Steindachner, 365; orthotype O. DARWINI Steind.

1158. STEINDACHNER (1875). *Ichthyologische Beiträge.* II. Sitzb. Akad. Wiss. Wien.

FRANZ STEINDACHNER

Neolabrus Steindachner, 461; orthotype N. FENESTRATUS Steind.

Atherinella Steindachner, 477 (35, reprint); orthotype A. PANAMENSIS Steind.

1159. STEINDACHNER (1875). *Ichthyologische Beiträge.* III. Sitzb. Akad. Wiss. Wien, 1875 (June), LXXII, 29-96.

FRANZ STEINDACHNER.

Parapsettus Steindachner, 68 (50, reprint); orthotype P. PANAMENSIS Steind.

Neoliparis Steindachner, 72 (54, reprint); orthotype LIPARIS MUCOSUS Ayres.

Atherinops Steindachner, 89; (61, reprint); orthotype ATHERINOPSIS AFFINIS Ayres.

Cottoperca Steindachner, 69; orthotype C. ROSENBERGI Steind.

1160. TRAQUAIR (1875). *On Some Fossil Fishes from the Neighborhood of Edinburgh.* Ann. Mag. Nat. Hist., ser. 4, XV, 258-268.

RAMSAY HEATLEY TRAQUAIR.

Nematoptychius Traquair, 259; orthotype PYGOPTERUS GREENOCKI (Ag.) (fossil).

Wardichthys Traquair, 266; orthotype W. CYCLOSOMA Traquair (fossil).

1876.

1161. BLEEKER (1876). *Notice sur les genres Gymnocæsio, Pterocæsio, Paracæsio et Liocæsio.* Versl. Med. Akad. Nat., 1876, 149-154.

PIETER BLEEKER.

Gymnocæsio Bleeker, 152; orthotype CÆSIO GYMNOPTERUS Blkr.

Pterocæsio Bleeker, 153; orthotype CÆSIO MULTIRADIATUS Steind.

Liocæsio Blkr., 153; orthotype CÆSIO CYLINDRICUS Blkr.

1162. BLEEKER (1876). *Sur la famille des Pseudochromidoides et révision de ses espèces insulindiennes.* Verh. Akad. Amsterdam, XV, 1-32.

PIETER BLEEKER.

Leptochromis Bleeker, 21; orthotype PSEUDOCHROMIS CYANOTÆNIA Blkr.

1163. BLEEKER (1876). *Systema Percarum revisum.* Arch. Néerl. Sci. Nat., XI, 1876, Pars I, 247-288; Pars II, 289-340.

PIETER BLEEKER.

Phæthonichthys Bleeker, 256; orthotype SERRANUS PHÆTON Cuv., a fallacious specimen with the tail of a FISTULARIA fastened on the body of a CEPHALOPHOLIS. A synonym of CEPHALOPHOLIS (same as URIPHÆTON Sw.)

Percamia Bleeker, 260; orthotype MICROPERCA YARRÆ Castelnau, substitute for MICROPERCA Castelnau, preoccupied by MICROPERCA Putnam 1863.

Telescops Bleeker, 261; orthotype POMATOMUS TELESCOPIUM Risso. Substitute for POMATOMUS Cuvier, not of Lacépède. A synonym of EPIGONUS Raf.

Oxylabrax Bleeker, 264; orthotype SCIÆNA UNDECIMALIS Bloch. Name offered as a substitute for CENTROPOMUS Lac., as the first species named under CENTROPOMUS belongs to another genus.

Mesopristes Bleeker, 267; orthotype M. MACRACANTHUS Blkr. (Note in passing, a synonym of DATNIA Cuv.)

Paristiopterus Bleeker, 268; orthotype RICHARDSONIA INSIGNIS Castelnau = HISTIOPTERUS LABIOSUS Gthr. A substitute for RICHARDSONIA Castelnau, 1872, preoccupied.

Pseudopentaceros Bleeker, 270; orthotype PENTACEROS RICHARDSONI Smith. Replaces GILCHRISTIA Jordan.

Paradentex Bleeker, 278; orthotype GYMNOCRANIUS RIVULATUS Klunzinger. A synonym of GYMNOCRANIUS Klunzinger.

Paraconodon Bleeker, 272; orthotype CONODON PACIFICI Gthr.

Mænas (Klein) Bleeker, 273; orthotype SPARUS MÆNA L. = MÆNA VULGARIS Cuv. & Val. A synonym of MÆNA Cuv.

Hemilutjanus Bleeker, 277; orthotype PLECTROPOMA MACROPHTHALMUS Tschudi. This name is no doubt badly chosen, but it cannot be replaced by POMODON Boulenger, 1895.

Banjos Bleeker, 277; orthotype ANOPLUS BANJOS Richardson = BANJOS TYPUS Blkr. A substitute for ANOPLUS T. & S., preoccupied.

Synagris (Klein) Bleeker, 278; orthotype DENTEX VULGARIS Cuv. & Val. A synonym of DENTEX Cuv.

Puntazzo Bleeker, 284; orthotype CHARAX PUNTAZZO Cuv. & Val. = P. ANNULARIS Blkr. A substitutue for CHARAX Risso, regarded as preoccupied by CHARAX Gronow (1763).

Mænichthys (Kaup) Bleeker, 291; orthotype DITREMA TEMMINCKI Blkr. (passing reference in synonymy). A synonym of DITREMA T. & S.

Pseudambassis Bleeker, 292; orthotype CHANDA LALA Ham.

Therapaina (Kaup) Bleeker, 298; orthotype CHÆTODON STRIGATUS Langsdorff (passing notice in synonymy). A synonym of MICROCANTHUS Sw.

Helotosoma (Kaup) Bleeker, 298; orthotype H. SERVUS Kaup, said to be ATYPUS STRIGATUS Gthr. A synonym of ATYPICHTHYS Gthr.

Cæsiosoma (Kaup, 1863) Bleeker, 299; orthotype C. SIEBOLDI Kaup = SCORPIS ÆQUIPINNIS Rich., said to be a synonym of SCORPIS.

Tylometopon (Van Beneden) Bleeker, 299; orthotype BRAMA RAJII Bl. & Schn. (passing notice in synonymy). A synonym of BRAMA Cuv.

Hemitaurichthys Bleeker, 304; orthotype CHÆTODON POLYLEPIS Blkr.

Chelmonops Bleeker, 304; orthotype CHÆTODON TRUNCATUS Kner.

Tetragonoptrus (Klein) Bleeker, 305; orthotype CHÆTODON STRIATUS L. A synonym of CHÆTODON L. as understood.

Chætodontops Bleeker, 304; orthotype CHÆTODON COLLARIS Bloch.

Lepidochætodon Bleeker, 306; orthotype CHÆTODON UNIMACULATUS Bloch.

Oxychætodon Bleeker, 306; orthotype CHÆTODON LINEOLATUS Q. & G.

Gonochætodon Bleeker, 306; orthotype CHÆTODON TRIANGULUM Kuhl & Van Hasselt.

Chætodontoplus Bleeker, 307; orthotype HOLACANTHUS SEPTENTRIONALIS T. & S.

Acanthochætodon Bleeker, 308; orthotype HOLACANTHUS ANNULARIS Lac.

Amblytoxotes Bleeker, 311; orthotype TOXOTES SQUAMOSUS Hutton.

Labracinus (Schlegel) Bleeker, 320; orthotype CICHLOPS CYCLOPHTHALMUS M. & T. Same as CICHLOPS M. & T., preoccupied. Passing notice in synonymy. A synonym of DAMPIERIA Castelnau, 1875.

Pteronemus (Van der Hoeven) Bleeker, 315; orthotype CHEILODACTYLUS FASCIATUS Lac.) (notice in synonymy). A synonym of CHEILODACTYLUS Lac.

Cirrhiptera (Kuhl & Van Hasselt) Bleeker, 322; (orthotype PLESIOPS NIGRICANS Rüppell) (note in synonymy). A synonym of PLESIOPS Cuv.

Larimodon (Kaup MS., 1862) Bleeker, 329; orthotype CORVINA DENTEX Cuv. & Val. (misprinted LAMNODON in a later paper). A synonym of ODONTOSCION Gill.

Paraplesiops Bleeker, 332; orthotype PLESIOPS BLEEKERI Gthr.

Brachymullus Bleeker, 333; orthotype UPENEUS TETRASPILUS Gthr.

Paraplesichthys (Kaup, 1862) Bleeker, 335; orthotype ANCYLODON PARVIPINNIS Gill (note in synonymy). A synonym of ISOPISTHUS Gill.

Bogota (Blyth) Bleeker, 336; passing reference. A synonym of PRIACANTHUS Cuv.

Macrocephalus (Browne) Bleeker, 336; orthotype SCIÆNA UNDECIMALIS Bloch. A substitute for CENTROPOMUS Lac. and OXYLABRAX Blkr. on the supposition that the names in the reprint of Browne's *Jamaica*, 1789, would be allowed as eligible.

1164. BLEEKER (1876). *Description de quelques espèces insulindiennes inédites des genres Oxyurichthys, Paroxyurichthys et Cryptocentrus.* Verh. Akad. Amsterdam (2) IX, 1876, 138-148.

PIETER BLEEKER.

Paroxyurichthys Bleeker, 140; orthotype P. TYPUS Blkr.

1165. BLEEKER (1876). *Genera familiæ Scorpænoideorum conspectus analyticus.* Versl. Akad. Amsterdam (2) IX, 1876, 294-300.

PIETER BLEEKER.

Parascorpæna Bleeker, 296; orthotype SCORPÆNA PICTA Kuhl & Van Hasselt.

Parapterois Bleeker, 296; orthotype PTEROIS HETERURUS Blkr.

Paracentropogon Bleeker, 297; orthotype APISTUS LONGISPINIS Cuv. & Val. Probably to be regarded as a synonym of HYPODYTES Gistel.

Pseudopterus (Klein) Bleeker, 296; orthotype (after Klein) GASTEROSTEUS VOLITANS L. A synonym of PTEROIS Cuv.

Amblyapistus Bleeker, 297; orthotype APISTUS TÆNIANOTUS Cuv. & Val.

Pteroidichthys Bleeker, 297; orthotype P. AMBOINENSIS Blkr.

Cottapistus Bleeker, 298; orthotype APISTUS COTTOIDES Cuv. & Val.

Paraploactis Bleeker, 300; orthotype P. TRACHYDERMA Blkr.

1166. BLEEKER (1876). *Notice sur les genres et les espèces des Chætodontoides de la sous-famille des Taurichthyiformes.* Versl. Akad. Amsterdam (2) X, 308-320.

PIETER BLEEKER.

Hemichætodon Bleeker, 313; orthotype CHÆTODON CAPISTRATUS Bloch.

1167. BLEEKER (1876). *Description de quelques espèces inédites de Pomacentroides de l'Inde Archipélagique.* Versl. Akad. Amsterdam.

PIETER BLEEKER.

Paraglyphidodon Bleeker, XII, 38; orthotype P. OXYCEPHALUS Blkr.

1168. CASTELNAU (1876). *Mémoire sur les poissons appelés barramundi par les aborigènes du nord-est de l'Australie.* Journ. Zool. (Gervais), 1876, V, 129-136.

FRANCIS L. CASTELNAU.

Neoceratodus Castelnau, 132; orthotype N. BLANCHARDI Castelnau. Prof. W. Baldwin Spencer of the University of Melbourne regards NEOCERATODUS BLANCHARDI as an unknown or perhaps fictitious fish, not to be identified with the Dipneustan, CERATODUS FORSTERI Krefft, which is the type of the genus EPICERATODUS Teller. Mr. McCulloch, however, thinks that EPICERATODUS is a synonym of NEOCERATODUS, Castelnau's account being an inaccurate description of the same fish. NEOCERATODUS BLANCHARDI is said to come from Fitzroy River, where the FORSTERI does not exist. According to Professor Spencer, the name BARRAMUNDA or BARRAMUNDI, used by Castelnau, belongs to the Serranoid fish SCLEROPAGES LEICHARDTI, a valuable foodfish, which EPICERATODUS is not.

1169. COPE & YARROW (1876). *Report on the Collections of Fishes Made in Portions of Nevada, etc., in 1871-1874, by Prof. E. D. Cope and Dr. H. C. Yarrow.* Wheeler's Survey West of 100th Meridian, VI, Zoölogy.

EDWARD DRINKER COPE; HENRY CRÉCY YARROW.

Eritrema Cope, 645; orthotype APOCOPE HENSHAVII Cope. A synonym of APOCOPE Cope.

Pantosteus Cope, 673; orthotype MINOMUS PLATYRHYNCHUS Cope.

1170. COPE (1876). *On a New Genus of Fossil Fishes (Cyclotomodon vagrans).*
Proc. Acad. Nat. Sci. Phila, 1876, V, 356-357.
EDWARD DRINKER COPE.

Cyclotomodon Cope, 355; orthotype C. VAGRANS Cope (fossil).

1171. COPE (1876). *Description of Some Vertebrate Remains from the Fort
Union Beds of Montana.* Proc. Acad. Nat. Sci. Phila., 1876.
EDWARD DRINKER COPE.

Hedronchus Cope, 259; orthotype H. STERNBERGI Cope. A tooth, perhaps not of a
fish.

Myledaphus Cope, 260; orthotype M. BIPARTITUS Cope.

1172. DAVIS (1876). *On a Bone Bed in the Lower Coal Measures, etc.* Quart.
Journ. Geol. Soc., XXXII, 332-340.
JAMES WILLIAM DAVIS (1846-1893).

Hoplonchus Davis, XXXII, 336; orthotype H. ELEGANS Davis (fossil).

1173. DAY (1876). *The Fishes of India.* (Part 2.)
FRANCIS DAY.

Apocryptichthys Day, 302; orthotype APOCRYPTES CANTORIS Day.

Nangra Day, 494; orthotype N. PUNCTATA Day.

1174. EGERTON (1876). *Notice of Harpactes velox, a Predaceous Ganoid Fish . . .
from the Lias of Lyme Regis.* Geol. Mag.
PHILIP DE MALPAS GREY EGERTON.

Harpactes Egerton, 441; orthotype HARPACTES VELOX Egerton (fossil). Name preoccupied, later replaced by HARPACTIRA Egert.; a synonym of OSTEORACHIS
Egert.

Harpactira Egerton, 576; orthotype H. VELOX Egert. = OSTEORACHIS MACRO-
CEPHALUS Egert. (fossil). Name a substitute for HARPACTES. preoccupied;
a synonym of OSTEORACHIS Egert.

1175. GOODE (1876). *Catalogue of the Fishes of the Bermudas, etc.* Bull. U. S.
Nat. Mus., V.
GEORGE BROWN GOODE (1851-1896).

Pareques (Gill) Goode, 50; orthotype GRAMMISTES ACUMINATUS Bl. & Schn.

1176. GÜNTHER (1876). *Remarks on Fishes, With Descriptions of New Fishes
in the British Museum, Chiefly from Southern Seas.* Ann. Mag. Nat. Hist.,
1876, ser. 4, XVII, 389-472.
ALBERT GÜNTHER.

Holoxenus Günther, 393; orthotype H. CUTANEUS Gthr. A synonym of GNATHANA-
CANTHUS Blkr.

Neophrynichthys Günther, 396; orthotype PSYCHROLUTES LATUS Hutton.

1177. GÜNTHER (1876). *Contributions to Our Knowledge of the Fish-Fauna
of the Tertiary Deposits of the Highlands of Pedang, Sumatra.* Geol. Mag.,
1876, II, dec. 3, 433-440.
ALBERT GÜNTHER.

Hexapsephus Günther, 439; orthotype H. GUENTHERI Woodw. (fossil).

1178. JORDAN (1876). *Concerning the Fishes of the Ichthyologia Ohiensis.*
Bull. Buffalo Soc. Nat. Hist., 1876, 91-97.
DAVID STARR JORDAN (1851-   ).

Erimyzon Jordan, 95; orthotype CYPRINUS OBLONGUS Mitchill.

1179. JORDAN (1876). *Manual of the Vertebrates of the Northern United States, Including the District East of the Mississippi River and North of North Carolina and Tennessee, Exclusive of Marine Species.*
DAVID STARR JORDAN.

Eutychelithus Jordan, 242; orthotype CORVINA RICHARDSONI Cuv. & Val., a deformed example of APLODINOTUS GRUNNIENS (Raf.) A synonym of APLODINOTUS Raf.

Eucalia Jordan, 248; orthotype GASTEROSTEUS INCONSTANS Kirtland. (Also in *Proc. Acad. Nat. Sci. Phila.,* 1877, 65.)

Erinemus Jordan, 279; orthotype CERATICHTHYS HYALINUS Cope = RUTILUS AMBLOPS Raf. A synonym of HYBOPSIS Ag.

Lythrurus Jordan, 272; orthotype "SEMOTILUS DIPLÆMIUS Raf." = ALBURNUS UMBRATILIS Grd. = NOTROPIS LYTHRURUS Jordan. S. DIPLÆMIUS Raf. is something different.

1180. KESSLER (1876). *Description of Fishes Collected by Col. Prejevalsky in Mongolia.* (Text in Russian.) (In *Prejevalsky, N. Mongolia i Strana Tangutov.* II, pt. 4, 1-36.)
KARL THEODOROVICH KESSLER.

Megagobio Kessler, 16; orthotype M. NASUTUS Kessler.

1181. NELSON (1876). *A Partial Catalogue of the Fishes of Illinois.* Bull. Ill. Mus. Nat. Hist., 1876, I, 33-52.
EDWARD WILLIAM NELSON (1855-   ).

Sternotremia Nelson, 39; orthotype S. ISOLEPIS Nelson, the young of APHREDODERUS SAYANUS, the vent farther back. A synonym of APHREDODERUS Le Sueur.

1182. MARCK (1876). *Fossile Fische von Sumatra.* Palæontologia, XXII.
W. VON DER MARCK.

Protosyngnathus von der Marck, XXII, 406; orthotype P. SUMATRENSIS Marck (fossil). A fossil homologue of AULORHYNCHUS Gill.

1183. OWEN (1876). *Catalogue of Fossils.* Report on South Africa.
RICHARD OWEN.

Hypterus Owen, IX, orthotype H. BAINI Owen (fossil). Names only; quoted by Woodward, 1891, as a synonym of ATHERSTONIA Woodw., 1889.

1184. PETERS (1876). *Uebersicht der während der von 1874 bis 1876 unter der Commando des Hrn. Capitän z. S. Freiherrn von Schleinitz ausgeführten Reise S.M.S. Gazelle . . . Fische.* Monatsb. Akad. Wiss. Berlin, 1876, 831-854.
WILHELM CARL HARTWIG PETERS.

Idiacanthus Peters, 846; orthotype I. FASCIOLA Peters.

Stigmatonotus Peters, 838; orthotype S. AUSTRALIS Peters.

1185. STEINDACHNER (1876). *Die Süsswasser Fische der Südöstlichen Brasilien.* III. Sitzb. Akad. Wiss. Wien, LXXIV, November, 1876.
FRANZ STEINDACHNER.

Wertheimeria Steindachner, 660 (101, reprint); orthotype W. MACULATA Steind.

STEINDACHNER, 1876 387

Harttia Steindachner, 668 (110, reprint); orthotype H. LORICARIFORMIS Steind.

1186. STEINDACHNER (1876). *Ichthyologische Beiträge.* V. Sitzb. Akad. Wiss. Wien, 1876, LXXIV, 49-240.

FRANZ STEINDACHNER.

Hippoglossina Steindachner, 61 (13, reprint); orthotype H. MACROPS Steind.

Curimatopsis Steindachner, 81; orthotype CURIMATUS MICROLEPIS Steind.

Lutkenia Steindachner, 85; orthotype L. INSIGNIS Steind. Name preoccupied; replaced by STICHANODON Eigenmann, 1903.

Paragoniates Steindachner, 117 (69, reprint); orthotype P. ALBURNUS Steind.

Pimelodina Steindachner, 149 (101, reprint); orthotype P. FLAVIPINNIS Steind.

Lophiosilurus Steindachner, 155 (106, reprint); orthotype L. ALEXANDRI Steind.

Achiropsis Steindachner, 158 (110, reprint); orthotype SOLEA NATTERERI Steind.

Cynolebias Steindachner, 172 (124, reprint); orthotype C. POROSUS Steind.

Siphagonus Steindachner, 188 (140, reprint); orthotype AGONUS SEGALIENSIS Tilesius. A synonym of BRACHYOPSIS Gill.

Blakea Steindachner, 196 (148, reprint); orthotype MYXODES ELEGANS Cooper. A synonym of GIBBONSIA Cooper.

Bembrops Steindachner, 211 (163, reprint); orthotype B. CAUDIMACULA Steind.

1877

1187-1188. ALLEYNE & MACLEAY (1877). *The Ichthyology of the Chevert Expedition.* Proc. Linn. Soc. New S. Wales, 1877, 261-268; 321-359.

HAYNES GIBBES ALLEYNE; WILLIAM MACLEAY.

Pseudolates Alleyne & Macleay, 262; orthotype P. CAVIFRONS Alleyne & Macleay = HOLOCENTRUS CALCARIFER Bloch. A synonym of LATES Cuv.

Homalogrystes Alleyne & Macleay, 268; orthotype H. GUENTHERI Alleyne & Macleay. A synonym of ACANTHOCHROMIS Gill.

Gerreomorpha Alleyne & Macleay, 274; orthotype G. ROSTRATA Alleyne & Macleay.

Heptadecacanthus Alleyne & Macleay, 343; orthotype H. LONGICAUDUS Alleyne & Macleay. A synonym of ACANTHOCHROMIS Gill.

Cheilolabrus Alleyne & Macleay, 345; orthotype C. MAGNILABRUS Alleyne & Macleay. A synonym of THALLIURUS Sw. = HEMIGYMNUS Gthr.

1189. BLEEKER (1877). *Notice sur les espèces nominales de Pomacentroïdes de l'Inde archipélagique.* Arch. Néerl. Sci. Nat., XII, 1877, 38-41.

PIETER BLEEKER.

Glyphidodontops Bleeker, 41; orthotype GLYPHIDODON ANTJERIUS Blkr. A synonym of CHRYSIPTERA Sw.

1190. BLEEKER (1877). *Mémoire sur les Chromides marins ou Pomacentroïdes de l'Inde archipélagique.* Nat. Verh. Holl. Maatsch. Wetensch., 3 Verz. II, 1877, 1-166.

PIETER BLEEKER.

Parapomacentrus Bleeker, 65; orthotype POMACENTRUS POLYNEMA Blkr.

Amblypomacentrus Bleeker, 68; orthotype POMACENTRUS BREVICEPS Blkr.

Eupomacentrus Bleeker, 73; orthotype POMACENTRUS LIVIDUS Blkr.

Brachypomacentrus Bleeker, 73; orthotype POMACENTRUS ALBIFASCIATUS Blkr.

1191. CASTELNAU (1877). *Australian Fishes. New or Little Known Species.* Proc. Linn. Soc. New S. Wales, 1877, II, 225-248.

FRANCIS LAPORTE CASTELNAU.

Brisbania Castelnau, 241; orthotype B. STAIGERI Castelnau.

Neosilurus Castelnau, 238; orthotype N. AUSTRALIS Castelnau. Name preoccupied, replaced by CAINOSILURUS Macleay.

Neoarius Castelnau, 237; orthotype ARIUS CURTISI Castelnau.

Beridia Castelnau, 229; orthotype B. FLAVA Castelnau. A synonym of GNATHANA-CANTHUS Blkr.

1192. COPE (1877). *On Some New and Little Known Reptiles and Fishes from the Austroriparian Region.* Proc. Amer. Philos. Soc., XVII.

EDWARD DRINKER COPE.

Xystroplites (Jordan) Cope, 67; orthotype X. LONGIMANUS Cope = POMOTIS HOLBROOKI Cuv. & Val. (In this case Cope suppressed his own proposed generic name in favor of one given almost simultaneously by Jordan.) Jordan's type, X. GILLII, seems to be the same as POMOTIS PALLIDUS Ag.

1193. COPE (1877). *Report on the Geology of the Region of the Judith River, Montana,* etc. Bull. U. S. Geol. Surv., III.

EDWARD DRINKER COPE.

Anogmius Cope, 584; orthotype A. ARATUS Cope, not ANOGMIUS Cope, 1871.

Pelycorapis Cope, 587; orthotype P. VARIUS Cope (fossil).

1194. COPE (1877). *Contribution to the Knowledge of the Ichthyological Fauna of the Green River Shales.* Bull. U. S. Geog. Surv. Terr., 1877, III, 807-819.

EDWARD DRINKER COPE.

Dapedoglossus Cope, 807; orthotype D. TESTIS Cope (fossil). A synonym of PHAREODUS Leidy.

Amphiplaga Cope, 812; orthotype A. BRACHYPTERA Cope (fossil).

Mioplosus Cope, 813; orthotype M. LABRACOIDES Cope (fossil).

Priscacara Cope, 816; orthotype P. SERRATA Cope (fossil).

1195. COPE (1877). *Notice of a New Locality of the Green River Shales Containing Fishes, Insects, and Plants in a Good State of Preservation.* Pal. Bull., XXV.

EDWARD DRINKER COPE.

Diplomystus Cope, 139; orthotype D. DENTATUS (fossil). Name regarded as preoccupied by DIPLOMYSTES Gthr., replaced by COPEICHTHYS Dollo.

1196. COPE (1877). *Descriptions of Extinct Vertebrates from the Permian and Triassic Formations of the United States.* Proc. Amer. Philos. Soc., XVII.

EDWARD DRINKER COPE.

Strigilina Cope, 182; orthotype JANASSA GURLEIANA Cope (fossil). A synonym of JANASSA.

1197. COPE (1877). *On Some New or Little Known Reptiles and Fishes of the Cretaceous, No. 3, of Kansas.* Proc. Amer. Philos. Soc., XVII.

EDWARD DRINKER COPE.

Oricardinus Cope, 177; orthotype O. TORTUS Cope (fossil).

Ptyonodus Cope, 192; orthotype SAGENODUS VINSLOVI Cope (fossil). A synonym of SAGENODUS Owen.

1198. CORNUEL (1877). *Description de débris de poissons fossiles . . . du calcaire neocomien . . . Haute Marne.* Bull. Géol. Soc. France.

J. Cornuel.

Ellipsodus Cornuel, 617; orthotype E. incisus Cornuel (fossil).

1199. DYBOWSKI (1877). *(Fishes.)*

Benedikt Ivan Dybowski.

(I have failed to find the work in which these Siberian fishes are described.)

Pristiodon Dybowski, VIII, 26; orthotype P. siemionovi Dybowski = Leuciscus idella Val. A synonym of Ctenopharyngodon Steind.

Perccottus Dybowski, VIII, 28; orthotype P. glenii Dybowski (a goby; allied to Gobiomorus Lac.)

1200. FRITSCH (1877). *Die Reptilien und Fische der Böhmischen Kreideformation.* Prag.

Anton Jan Fritsch.

Sphœrolepis Fritsch, 46; orthotype S. kounoviensis Fritsch (fossil).

1201. GARMAN (1877). *On the Pelvis and External Sexual Organs of Selachians, With Especial Reference to the New Genera Potamotrygon and Disceus.* Proc. Boston Soc. Nat. Hist., 1877, XIX, 197-215.

Samuel Garman (1846-   ).

Malacorhina Garman, 203; orthotype Raja mira Garman.

Disceus Garman, 208; orthotype Trygon strongylopterus Schomburgk.

Potamotrygon Garman, 210; orthotype Trygon hystrix M. & H.

1202. GILL (1877). *Synopsis of the Fishes of Lake Nicaragua.* Proc. Acad. Nat. Sci. Phila., 1877, 175-218.

Theodore N. Gill.

Bramocharax Gill, 189; orthotype B. bransfordi Gill.

1203. GILL (1877). *Annual Record of Industry and Science for 1876.*

Theodore Gill.

Lycichthys Gill, CLXVII; orthotype Anarrhichas latifrons Steenstrup & Hallgrimmson.

1204. GILL & JORDAN (1877). *(The Sun-fish.)* Field & Forest, 1877, 190.

Theodore Gill; David Starr Jordan.

Eupomotis Gill & Jordan, 190; orthotype Sparus aureus Walbaum = Labrus gibbosus L.

1205. GÜNTHER (1877). *Preliminary Notes on New Fishes Collected in Japan During the Expedition of H.M.S. Challenger.* Ann. Mag. Nat. Hist., ser. 4, XX, 433-446.

Albert Günther.

Bathythrissa Günther, 443; orthotype B. dorsalis Gthr. A synonym of Pterothrissus Hilgendorf a little earlier in the same year.

1206. HILGENDORF (1877). *(Pterothrissus.)* Act. Soc. Leopoldina, III, Sept. 3, 1877.

Franz Martin Hilgendorf (1839-189?).

Pterothrissus Hilgendorf, 127 (Sept. 3, 1877); orthotype P. GISSU Hilgendorf; replaces BATHYTHRISSA Gthr. (Nov. 1, 1877).

1207. JORDAN (1877). *On the Fishes of Northern Indiana.* Proc. Acad. Nat. Sci. Phila., 1877, 42-82.

DAVID STARR JORDAN.

Imostoma Jordan, 49; orthotype HADROPTERUS SHUMARDI Grd.

Copelandia Jordan, 56; orthotype C. ERIARCHA Jordan, an abnormal variant of ENNEACANTHUS GLORIOSUS (Holbrook). A synonym of ENNEACANTHUS Gill, based on an aberrant specimen. Also in *Bull. Buffalo Soc. Nat. Hist.,* 1877, 136.

Xenotis Jordan, 76; orthotype POMOTIS FALLAX Baird & Girard. A section under LEPOMIS Raf.

1208. JORDAN (1877). *A Partial Synopsis of the Fishes of Upper Georgia.* Ann. Lyc. N. Y., XI, 307-377.

DAVID STARR JORDAN.

Xenisma Jordan, 322; orthotype HYDRARGYRA CATENATA Ag. A subgenus under FUNDULUS.

Helioperca Jordan, 355; orthotype POMOTIS INCISOR Cuv. & Val. = LABRUS PALLIDUS Mitchill. A section under LEPOMIS.

1209. JORDAN (1877). *Notes on Cottidæ, Etheostomatidæ, Percidæ, Centrachidæ, Aphododeridæ, Dorysomatidæ, and Cyprinidæ, with Revisions of the Genera and Descriptions of New or Little Known Species.* Bull. U. S. Nat. Mus., 1877, X, 1-68.

DAVID STARR JORDAN.

Ammocrypta Jordan, 5; orthotype A. BEANI Jordan.

Nanostoma (Putnam) Jordan, 6; orthotype PŒCILICHTHYS ZONALIS Cope; regarded as preoccupied by NANNOSTOMUS Gthr.; replaced by RHOTHŒCA Jordan.

Ericosma Jordan & Copeland, 8; orthotype ALVORDIUS EVIDES Jordan & Copeland. A section under HADROPTERUS Ag.

Rheocrypta Jordan, 9; orthotype R. COPELANDI Jordan. A section under COTTOGASTER Putnam (1863).

Xystroplites Jordan, 24; orthotype X. GILLI Jordan. A section under EUPOMOTIS Gill & Jordan. (See No. 1192.)

Cynoperca Gill & Jordan, 44; orthotype LUCIOPERCA CANADENSIS C. H. Smith.

Mimoperca (Gill & Jordan) Jordan, 44; orthotype PERCA VOLGENSIS Pallas. A synonym of SCHILUS Krynicki (1832).

Elassoma Jordan, 50; orthotype E. ZONATA Jordan. Name (from ἐλασσώμα, a diminution) wrongly expanded to ELASSOSOMA in the *Zoological Record* for 1878.

Asternotremia (Nelson) Jordan, 51; orthotype STERNOTREMIA ISOLEPIS Nelson. A needless substitute name for STERNOTREMIA Nelson, thought to be anatomically not exact. A synonym of APHREDODERUS Le Sueur.

Episema Cope & Jordan, 64; orthotype PHOTOGENIS SCABRICEPS Cope. Name preoccupied, replaced by PARANOTROPIS Fowler.

Elattonistius Gill & Jordan, 67; orthotype HYODON CHRYSOPSIS Rich. = AMPHIODON ALOSOIDES Raf. A synonym of AMPHIODON Raf.

1210. JORDAN (1877). *Catalogue of Fishes of Ohio.* (In J. H. Klippart's *Report of the Fish Commission of Ohio for 1876.* Appendix (1877).
DAVID STARR JORDAN.
Erichæta Jordan, 147; orthotype POMOTIS INCISOR Cuv. & Val. A slip for HELIOPERCA due to uncorrected proof. A section under LEPOMIS.

1211. JORDAN & BRAYTON (1877). *On Lagochila, a New Genus of Catostomoid Fishes.* Proc. Acad. Nat. Sci. Phila., 1877, 280-283.
DAVID STARR JORDAN; ALEMBERT WINTHROP BRAYTON (1849-   ).
Lagochila Jordan & Brayton, 280; orthotype L. LACERA Jordan & Brayton. Name regarded as preoccupied by LAGOCHEILUS; replaced by QUASSILABIA Jordan & Brayton.

1212. KESSLER (1877). *The Aralo-Caspian Expedition.* IV. *Fishes of the Aralo-Caspio-Pontine Region.* Trans. St. Petersburg Nat. Hist. Soc.
KARL THEODOROVICH KESSLER (1815-1881).
Clupeonella Kessler, 187; orthotype C. GRIMMI Kessler.

1213. LEIDY (1877). *Descriptions of Vertebrate Remains Chiefly from the Phosphate Beds of South Carolina.* Journ. Acad. Nat. Sci. Phila., VIII.
JOSEPH LEIDY.
Mesobatis Leidy, 244; orthotype AËTOBATIS EXIMIUS Leidy (fossil). A synonym or homologue of AËTOBATUS.
Gryphodobatis Leidy, 249; orthotype G. UNCUS Leidy (fossil).
Acrodobatis Leidy, 250; orthotype A. SERRA Leidy (fossil). A homologue of GINGLYMOSTOMA M. & H.
Xiphodolamia Leidy, 252; orthotype X. ENSIS Leidy (fossil).

1214. MALM (1877). *Göteborgs och Bohusläns fauna.* Ryggradsdjuren.
AUGUST WILHELM MALM (1821-1882).
Æglefinus Malm, 481; orthotype GADUS ÆGLEFINUS L. = ÆGLEFINUS LINNÆI Malm. A synonym of MELANOGRAMMUS Gill.
Amblyraja Malm, 607; orthotype RAJA RADIATA Donovan.
Leucoraja Malm, 609; orthotype RAJA FULLONICA L.

1215. MARSH (1877). *Introduction and Succession of Vertebrate Life in America.* Amer. Journ. Sci., XIV.
OTHNIEL CHARLES MARSH.
Heliobatis Marsh, 376; orthotype H. RADIANS Marsh (fossil). Both names undefined, hence becoming valid only when located by Woodward, 1889. A synonym of XIPHOTRYGON Cope, 1879 (X. ACUTIDENS).

1216. PETERS (1877). *Ueber Epigonichthys cultellus, eine neue Gattung und Art der Leptocardii.* Monatsb. Akad. Wiss. Berlin, 1876 (1877), 322-327.
WILHELM (CARL HARTWIG) PETERS.
Epigonichthys Peters, 322; orthotype E. CULTELLUS Peters.

1217. PETERS (1877). *Ueber eine merkwürdige von Dr. Buchholz entdeckte neue Gattung von Süsswasserfischen, Pantodon buchholzi, etc.* Monatsb. Akad. Wiss. Berlin, 1876 (1877), 195-200.
WILHELM (CARL HARTWIG) PETERS.
Pantodon Peters, 196; orthotype P. BUCHHOLZI Peters.

1218. PETERS (1877). *(Fossil Fishes.)* Preuss. Akad.

WILHELM (CARL HARTWIG) PETERS.

Hemitrichas Peters, 682; orthotype H. SCHISTICOLA Peters (fossil).

1219. PROBST (1877). *Beitrag zur Kenntniss der fossile Fische aus der Molasse von Baltringen.* II. Jahr. Verein. Vaterl. Naturk. Württemberg.

JOSEF PROBST.

Bates Probst, XXXIII, 88; logotype B. SPECTABILIS Probst (fossil). Not BATIS Bon.; an ally of MYLIOBATIS.

1220. SAUVAGE (1877). *Sur les Lepidotus palliatus et Sphærodus gigas.* Bull. Soc. Géol. France.

HENRI ÉMILE SAUVAGE.

Eurystethus Sauvage, 629; orthotype E. BRONGNIARTI Sauvage (fossil).
Desmichthys Sauvage, 634; orthotype D. DAUBREI Sauvage (fossil).

1221. TRAQUAIR (1877). *On the Agassizian Genera Amblypterus, Palæoniscus, Gyrolepis and Pygopterus.* Quart. Journ. Geol. Soc., 1877, XXXIII, 548-578.

RAMSAY HEATLEY TRAQUAIR.

Gonatodus Traquair, 555; logotype AMBLYPTERUS PUNCTATUS Ag. (fossil).
Cosmoptychius Traquair, 553; orthotype C. STRIATUS Traquair (fossil). A synonym of ELONICHTHYS Giebel.
Acentrophorus Traquair, 565; orthotype PALÆONISCUS VARIANS Kirby (fossil).
Rhadinichthys Traquair, 559; logotype PALÆONISCUS ORNATISSIMUS L. (fossil).

1222. TRAQUAIR (1877). *Monograph on the Ganoid Fishes of the British Carboniferous Formations.* I.

RAMSAY HEATLEY TRAQUAIR.

Microconodus Traquair, 33; orthotype M. MOLYNEUXI Traquair (fossil), name only. Quoted by Woodward, 1891. A synonym of GONATODUS.

1223. WINTHER (1877). *Om de danske fiske af slægten Gobius.* Nat. Tidskr. Kjöbenhavn, XI.

GEORG WINTHER (1844-1879).

Lebetus Winther, 49; orthotype GOBIUS SCORPIOIDES Collett.

1878.

1224. BARKAS (1878). *On the Microscopical Structure of Fossil Teeth from the Northumberland Coal Measures.* Monthly Review of Dental Surgery, VII.

WILLIAM JAMES BARKAS

Hybodopsis Barkas, 191; orthotype H. WARDI Barkas (fossil).

1225. BLEEKER (1878). *Sur deux espèces inédites de Cichloïdes de Madagascar.* Versl. Akad. Amsterdam (2), XII, 1878, 192-198.

PIETER BLEEKER.

Paracara Bleeker, 193; orthotype P. TYPUS Bleeker. A synonym of PARATILAPIA Blkr.

1226. BLEEKER (1878).   *Quatrième mémoire sur la faune ichthyologique de la Nouvelle-Guinée.* Arch. Néerl. Sci. Nat. XIII, 1878, 35-66.
PIETER BLEEKER.

Bathygobius Bleeker, 54; logotype GOBIUS NEBULOPUNCTATUS Cuv. & Val. Replaces MAPO Smitt (G. SOPORATOR Cuv. & Val.).
Symphysanodon Bleeker, 61; orthotype S. TYPUS Blkr.

1227. BOSNIASKI (1878). *Studii sui pesci fossili del mioceno del Gabbro.* Atti. Soc. Tosc. Sci. Nat. Pisa.
S. DE BOSNIASKI.

Acanthonemopsis Bosniaski, I, 19; orthotype A. CAPELLINII Bosniaski (fossil).

1228. CASTELNAU (1878). *Notes on the Fishes of the Norman River.* Proc. Linn. Soc. New S. Wales, 1878, III, 41-51.
FRANCIS L. CASTELNAU

Pseudoambassis Castelnau, 43; logotype P. MACLEAYI Castelnau. Apparently a synonym of AMBASSIS Cuv. Preoccupied by PSEUDAMBASSIS Blkr.
Acanthoperca Castelnau, 45; orthotype A. GULLIVERI Castelnau. A synonym of AMBASSIS.
Gulliveria Castelnau, 45; logotype G. FUSCA Castelnau = APOGON APRION Rich. A synonym of· GLOSSAMIA Gill.

1229. CASTELNAU (1878). *On Several New Australian (Chiefly) Fresh-water Fishes.* Proc. Linn. Soc. New S. Wales, 1878, 140-144.
FRANCIS L. CASTELNAU.

Aristeus Castelnau, 141; logotype A. FITZROYENSIS Castelnau. Name preoccupied; a synonym of MELANOTÆNIA Gill, 1862.
Eumeda Castelnau, 143; orthotype E. ELONGATA Castelnau. A synonym of NEOSILURUS Steind. (not NEOSILURUS Castelnau).

1230. CASTELNAU (1878). *Essay on the Ichthyology of Port Jackson.* Proc. Linn. Soc. New S. Wales, 1878, III, 347-402.
FRANCIS LAPORTE CASTELNAU.

Neoanthias Castelnau, 366; orthotype N. GUNTHERI Castelnau. A synonym of CAPRODON T. & S.
Agenor Castelnau, 371; orthotype AGENOR MODESTUS Castelnau. A synonym of SCORPIS Cuv. & Val.
Zeodrius Castelnau, 377; orthotype Z. VESTITUS Castelnau. A synonym of GONIISTIUS Gill (1862).

1231. COLLETT (1878). *Fiske fra Nordhavs-expeditionens sidste togt sommeren 1878.* Forh. Vidensk. Selsk. Christiania, 1878, XIV, 1-106.
ROBERT COLLETT

Rhodichthys Collett, 99; orothotype R. REGINA Collett.
Paraliparis Collett, 32; orthotype P. BATHYBII Collett.

1232. COPE (1878). *Synopsis of the Fishes of the Peruvian Amazon, Obtained by Prof. Orton During His Expeditions of 1873 and 1877.* Proc. Amer. Philos. Soc., 1878, XVII, 673-701.
EDWARD DRINKER COPE.

Epapterus Cope, XVII, 677; orthotype E. DISPILURUS Cope.

Chænothorax Cope, 679; orthotype C. BICARINATUS Cope.
Potamorhina Cope, 685; orthotype CURIMATUS PRISTIGASTER Steind.
Metynnis Cope, 693; orthotype M. LUNA Cope.
Gastropterus Cope, 700; orthotype G. ARCHÆUS Cope. A synonym of BASILICH-THYS Girard; name regarded as preoccupied by GASTROPTERON.

1233. COPE (1878). *Descriptions of Fishes from the Cretaceous and Tertiary Deposits West of the Mississippi River.* Bull. U. S. Geol. Surv., IV.
EDWARD DRINKER COPE.

Triænaspis Cope, 67; orthotype LEPTOTRACHELUS VIRGATULUS Cope (fossil).
Ichthyotringa Cope, IV, 69; orthotype I. TENUIROSTRIS Cope (fossil). A synonym of RHINELLUS Ag.

1234. CLARKE (1878). *On Two New Fishes (Argyropelecus intermedius and Aegæonichthys appelti).* Trans. Proc. New Zealand Instit., 1877 (1878), X, 243-246.
F. E. CLARKE.

Aegæonichthys Clarke, 245; orthotype A. APPELTI Clarke.

1235. DUBALEN (1878). *Note sur un poisson mal connu du bassin de l'Adour.* Bull. Soc. Borda. Dax.
—— DUBALEN.

Aturius Dubalen, 157; orthotype A. DUFOURI Dubalen = CYPRINUS LEUCISCUS L. A synonym of LEUCISCUS Cuv.

1236. GILL (1878). *On a Remarkable Generic Type (Elopomorphus jordani).* Ann. Mag. Nat. Hist., 1878, ser. 5, III, 112.
THEODORE NICHOLAS GILL.

Elopomorphus Gill, 112; orthotype E. JORDANI Gill.

1237. GILL (1878). *Note on the Antennariidæ.* Proc. U. S. Nat. Mus., I, 221.
THEODORE GILL.

Sympterichthys Gill, 222; orthotype LOPHIUS LÆVIS Lac.

1238. GILL (1878). *Note on the Ceratiidæ.* Proc. U. S. Nat. Mus., I.
THEODORE GILL.

Mancalias Gill, 227; orthotype CERATIAS URANOSCOPUS Murray.

1239. GILL (1878). *Myxocyprinus Gill.* Johnson's Cyclopædia.
THEODORE GILL

Myxocyprinus Gill, 1574; orthotype CARPIODES ASIATICUS Blkr.

1240. GÜNTHER (1878). *Preliminary Notices of Deep-Sea Fishes Collected During the Voyage of H. M. S. Challenger.* Ann. Mag. Nat. Hist., ser. 5, II, 17-28, 179-187, 248-251.
ALBERT GÜNTHER.

Bathydraco Günther, 18; orthotype B. ANTARCTICUS Gthr.
Melanonus Günther, 19; orthotype M. GRACILIS Gthr.
Bathynectes Günther, 20; logotype B. LATICEPS Gthr.
Typhlonus Günther, 21; orthotype T. NASUS Gthr.
Aphyonus Günther, 22; orthotype A. GELATINOSUS Gthr.
Acanthonus Günther, 23; orthotype A. ARMATUS Gthr.
Bathyophis Günther, 181; orthotype B. FEROX Gthr.

Bathysaurus Günther, 182; logotype B. FEROX Gthr.
Bathypterois Günther, 183; logotype B. LONGIFILIS Gthr.
Ipnops Günther, 186; orthotype I. MURRAYI Gthr.
Bathylagus Günther, 248; logotype B. ANTARCTICUS Gthr.
Bathytroctes Günther, 249; logotype B. MICROLEPIS Gthr.
Platytroctes Günther, 249; orthotype P. APUS Gthr.
Xenodermichthys Günther, 250; orthotype X. NODULOSUS Gthr.
Cyema Günther, 251; orthotype C. ATRUM Gthr.

1241. HILGENDORF (1878). *Ueber das Vorkommen einer Brama-Art und einer neuen Fischgattung Centropholis aus der Nachbarschaft des Genus Brama in den japanischen Meeren.* Sitzb. Ges. Naturf. Freunde Berlin, 1878, 1-2.
FRANZ MARTIN HILGENDORF.
Centropholis Hilgendorf, 1; orthotype C. PETERSI Hilgendorf.

1242. HILGENDORF (1878). *Ueber einige neue japanische Fischgattungen.* Sitzb. Ges. Naturf. Freunde Berlin, 1878, 155-157.
FRANZ MARTIN HILGENDORF
Megaperca Hilgendorf, 155; orthotype M. ISCHINAGI Hilgendorf. A synonym of STEREOLEPIS Ayres.
Liobagrus Hilgendorf, 155; orthotype L. REINI Hilgendorf.

1243. JORDAN (1878). *Manual of the Vertebrates of the Northern United States, Including the District East of the Mississippi River and North of North Carolina and Tennessee, Exclusive of Marine Species.* 2d ed., 1-407.
DAVID STARR JORDAN.
Quassilabia Jordan & Brayton, 406; orthotype LAGOCHILA LACERA Jordan & Brayton. Substitute for LAGOCHILA Jordan & Brayton, regarded as preoccupied by LAGOCHEILUS, a genus of mollusks.
Tauridea Jordan & Rice, 255; orthotype COTTUS RICEI Nelson.
Prosopium (Milner) Jordan, 261; orthotype COREGONUS QUADRILATERALIS Rich. The name PROSOPIUM was proposed by James W. Milner of the U. S. Fish Commission in a *Review of the American Whitefishes,* written in 1878, but left incomplete through the untimely death of the author.
Ulocentra Jordan, 223; orthotype, ARLINA ATRIPINNIS Jordan.
Minytrema Jordan, 318; orthotype CATOSTOMUS MELANOPS Raf.

1244. JORDAN & BRAYTON (1878). *Contributions to North American Ichthyology, III. B. On the Distribution of the Fishes of the Allegany Region of South Carolina, Georgia, and Tennessee, With Descriptions of New or Little Known Species.* Bull. U. S. Nat. Mus., 1878, XII, 1-95.
DAVID STARR JORDAN; ALEMBERT WINTHROP BRAYTON (1849-   ).
Hydrophlox Jordan, 18; orthotype HYBOPSIS RUBRICROCEUS Cope.
Erogala Jordan & Brayton, 20; orthotype PHOTOGENIS STIGMATURUS Jordan. A section under CYPRINELLA Grd.
Cristivomer Gill & Jordan, 69; orthotype SALMO NAMAYCUSH Walbaum.
Ioa Jordan & Brayton, 88; orthotype PŒCILICHTHYS VITREUS Cope.
Vaillantia Jordan, 89; orthotype BOLEOSOMA CAMURUM Forbes.

1245. JORDAN (1878). *Report on the Collection of Fishes Made by Dr. Elliott Coues in Dakota and Montana,* etc. Bull. U. S. Geol. Surv., IV.
DAVID STARR JORDAN.
Couesius Jordan, 785; orthotype NOCOMIS MILNERI Jordan = GOBIO PLUMBEUS Ag.

Zophendum Jordan, 786; orthotype HYBORHYNCHUS SIDERIUS Cope = AGOSIA CHRYSOGASTER Grd. A synonym of AGOSIA Grd.

Chriope Jordan, 787; orthotype HYBOPSIS BIFRENATUS Cope.

Symmetrurus Jordan, 788; orthotype POGONICHTHYS ARGYREIOSUS Grd., the young of P. INÆQUILOBUS Grd. A synonym of POGONICHTHYS.

1246. JORDAN (1878). *Notes on a Collection of Fishes from the Rio Grande at Brownsville, Texas.* Bull. U. S. Geol. Surv. Terr., 1878, IV, 397-406, 663-667.
DAVID STARR JORDAN.

(A few very young EMBIOTOCOID fishes from San Francisco were unfortunately mixed with this collection and wrongly described as new.)

Sema Jordan, 399; orthotype S. SIGNIFER Jordan; a larval form of CYMATOGASTER AGGREGATUS Gibbons. A synonym of CYMATOGASTER Gibbons.

Dacentrus Jordan, 667; orthotype D. LUCENS Jordan; the young of HYSTEROCARPUS TRASKI Gibbons. A synonym of HYSTEROCARPUS Gibbons.

1247. JORDAN (1878). *A Catalogue of Fishes of the Fresh Waters of North America.* Bull. U. S. Geol. Surv. Terr., 1878, IV., 407-442.
DAVID STARR JORDAN.

Chasmistes Jordan, 417; orthotype CHASMISTES LIORUS Jordan, a species wrongly supposed to be the same as CATOSTOMUS FECUNDUS Cope & Yarrow.

1248. JORDAN & GILBERT (1878). *Notes on the Fishes of Beaufort Harbor, North Carolina.* Proc. U. S. Nat. Mus., I.
DAVID STARR JORDAN; CHARLES HENRY GILBERT (1859-   ).

Chriolax Jordan & Gilbert, 374; orthotype TRIGLA EVOLANS L. A synonym of PRIONOTUS Lac.

1249. JORDAN (1878). *(Notes on Fishes of Ohio.)* Second Report Ohio Fish Commission; John H. Klippart, Commissioner.
DAVID STARR JORDAN.

Mascalongus Jordan, 92; orthotype ESOX MASQUINONGY Mitchill = ESOX NOBILIOR Thompson.

1250. KONINCK (1878). *Faune du calcaire carbonifère de la Belgique; première partie.* Ann. Mus. Roy. Hist. Nat. Belgique, 1878, II, 1-152.
LAURENT DE KONINCK.

Benedenius (Traquair) Koninck, 14; orthotype B. DENEENSIS Traquair (fossil). Name preoccupied, replaced by BENEDENICHTHYS Traquair, 1890.

Serratodus Koninck, 53; orthotype S. ELEGANS Koninck (fossil). A synonym of CTENOPTYCHIUS Ag.

Stichæacanthus Koninck, 70; orthotype S. COEMANSI Koninck (fossil).

1251. MACLEAY (1878). *Description of Some New Fishes from Port Jackson and King George's Sound.* Proc. Linn. Soc. New S. Wales, 1878, III, 33-38.
WILLIAM MACLEAY.

Isosillago Macleay, 34; orthotype I. MACULATA Macleay. A synonym of SILLAGINODES Gill.

1252. NEWBERRY (1878). *Descriptions of New Paleozoic Fishes.* Ann. N. Y. Acad. Sci., 1878, I, 188-192.
JOHN STRONG NEWBERRY.

Diplurus Newberry, 127; orthotype D. LONGICAUDATUS Newberry (fossil).

Diplognathus Newberry, 188; orthotype D. MIRABILIS Newberry (fossil).
Archæobatis Newberry, 190; orthotype A. GIGAS Newberry (fossil).

1253. NEWTON (1878). *The Chimæroid Fishes of the British Cretaceous Strata.*
Mem. Geol. Surv. United Kingdom, 1878, Monograph IV, 1-50.
EDWIN TULLEY NEWTON (1840-   ).
Elasmognathus Newton, 43; orthotype E. WILLETTII Newton (fossil). Name preoccupied in HEMIPTERA, replaced by ELASMODECTES Newton, 1888.

1254. SAUVAGE (1878). *Note sur quelques Cyprinidæ et Cobitinæ d'espèces inédites, provenant des eaux douces de la Chine.* Bull. Soc. Philom. Paris, VII, sér. 2, 86-90.
HENRI ÉMILE SAUVAGE.
Agenigobio Sauvage, 87; orthotype A. HALSONETI Sauvage.
Crossostoma Sauvage, 89; orthotype C. DAVIDI Sauvage.
Paramisgurnus Sauvage, 90; orthotype P. DABRYANUS Sauvage.

1255. SAUVAGE (1878). *Note sur quelques poissons d'espèces nouvelles provenant des eaux douces de l'Indo-Chine.* Bull. Soc. Philom. Paris, 1878, VII, sér. 2, 233-242.
HENRI ÉMILE SAUVAGE.
Cosmochilus Sauvage, 240; orthotype C. HARMANDI Sauvage.

1256. SAUVAGE (1878). *Description de poissons nouvelles ou imparfaitement connus,* etc. Nouv. Arch. Mus. Nat. Hist. Paris, II.
HENRI ÉMILE SAUVAGE.
Pseudosebastes Sauvage, 115; orthotype SEBASTES BOUGAINVILLEI Cuv. & Val.
Eusebastes Sauvage, 115; orthotype SEBASTES SEPTENTRIONALIS Gaimard. A synonym of SEBASTES Cuv.
Elaphocottus Sauvage, 142; logotype COTTUS. PISTILLIGER Pallas. A synonym of GYMNOCANTHUS Sw.

1257. STEINDACHNER (1878). *Ichthyologische Beiträge.* VII. Sitzb. Akad. Wiss. Wien, 1878 (1879), LXXVIII, 1. Abth., 377-400.
FRANZ STEINDACHNER.
Neomyxus Steindachner, 384; orthotype MYXUS SCLATERI Steind.
Leptobrama Steindachner, 388; orthotype L. MUELLERI Steind.
Cratinus Steindachner, 395; orthotype C. AGASSIZI Steind.

1258. STREETS (1878). *Contributions to the Natural History of the Hawaiian and Fanning Islands and Lower California.* Bull. U. S. Nat. Mus., 1878, VII, 43-102.
THOMAS HALE STREETS (1847-   ).
Sebastapistes (Gill) Streets, 62; orthotype SEBASTES STRONGIA Cuv. & Val.

1259. VAILLANT & BOCOURT (1878). *Études sur les poissons.* Mission Scientifique au Mexique, etc., IV.
LÉON LOUIS VAILLANT; FIRMIN BOCOURT.
Itaiara Vaillant & Bocourt, 67; orthotype SERRANUS ITAIARA Lichtenstein = PERCA GUTTATA L. A synonym of PROMICROPS Poey.

1260. WINKLER (1878). *Mémoire sur quelques restes de poissons du système heersien.* Arch. Mus. Teyler Haarlem, IV.
TIBERIUS CORNELIUS WINKLER.

Cycloides Winkler, 3; orthotype C. INCISUS Winkler (fossil). A fragment perhaps not of a fish.

Trigonodus Winkler, 14; orthotype SQUATINA PRIMA Noetling (fossil). A synonym of THAUMAS Münster.

1261. WINKLER (1878). *Deuxième Mémoire sur les dents de poissons fossiles du terrain bruxellien.* Arch. Mus. Teyler Haarlem.

TIBERIUS CORNELIUS WINKLER.

Trichiurides Winkler, 31; orthotype T. SAGITTIDENS Winkler (fossil).

1879.

1262. BASSANI (1879). *Vorläufige Mittheilungen über die (fossil) Fischfaune der Insel Lesina.* Verh. Geol. Reichsanst. Wien, 1879, 162-170.

FRANCESCO BASSANI.

Aphanepygus Bassani, 162; Orthotype A. ELEGANS Bassani (fossil). A synonym of PETALOPTERYX Pictet.

Prochanos Bassani, 165; orthotype P. RECTIFRONS Bassani (fossil).

Hemielopopsis Bassani, 166; orthotype H. SUESSI Bassani (fossil). An extinct homologue of HISTIALOSA.

1263. BEAN (1879). *Description of Some Genera and Species of Alaskan Fishes.* Proc. U. S. Nat. Mus., II.

TARLETON HOFFMAN BEAN (1846-1918.)

Melletes Bean, 354; orthotype M. PAPILIO Bean.

Dallia Bean, 358; orthotype D. PECTORALIS Bean.

1264. BLEEKER (1879). *Mémoire sur les poissons à pharyngiens labyrinthiformes de l'Inde archipélagique.* Verh. Akad. Amsterdam, XIX, 1879, 1-56.

PIETER BLEEKER.

Pseudosphromenus Bleeker, 2; orthotype OSPHROMENUS OPERCULARIS Blkr.

Parosphromenus Bleeker, 20; orthotype P. DEISSNERI Blkr.

1265. BLEEKER (1879). *Révision des espèces insulindiennes de la famille des Callionymoïdes.* Versl. Akad. Amsterdam (2), XIV, 1879, 79-107.

PIETER BLEEKER.

Eleutherochir Bleeker 103; orthotype E. OPERCULARIOIDES Blkr.

1266. BLEEKER (1879). *Sur quelques espèces inédites ou peu connues de poissons de Chine appartenant au Muséum de Hambourg.* Verh. Akad. Amsterdam XVIII, 1879, 1-17.

PIETER BLEEKER.

Hemiglyphidodon Bleeker, 8; orthotype H. PLAGIOMETOPON Blkr.

1267. BLEEKER (1879). *Énumération des espèces de poissons actuellement connues du Japon et description de trois espèces inédites.* Verh. Akad. Amsterdam, XVIII, 1879, 1-33.

PIETER BLEEKER.

Lepidorhynchus Bleeker, 21; orthotype L. VILLOSUS Blkr.

1268. CAMPBELL (1879). *On a New Fish (Discus aureus).* Trans. Proc. New Zeal. Instit., XI, 297-298.

W. D. CAMPBELL.

Discus Campbell, 298; orthotype D. AUREUS Campbell.

1269. CASTELNAU (1879). *On a New Ganoid Fish from Queensland.* Proc. Linn. Soc. New S. Wales, 1878, III, 164-165.

FRANCIS L. CASTELNAU.

Ompax Castelnau, 165; orthotype O. SPATULOIDES Castelnau. This species is based on a drawing, probably a rough representation of EPICERATODUS Teller. Macleay doubts the actual existence of the genus or species.

1270. CASTELNAU (1879). *On Several New Australian Fresh-water Fishes.* Proc. Linn. Soc., III.

FRANCIS L. CASTELNAU.

Aristeus Castelnau, 141; orthotype A. FITZROYENSIS Castelnau. Preoccupied in CRUSTACEA; replaced by RHOMBATRACTUS Gill. A synonym of MELANOTÆNIA Gill.

1271. COLLETT (1879). *On a New Fish of the Genus Lycodes from the Pacific.* Proc. Zool. Soc. London.

ROBERT COLLETT.

Lycodopsis Collett, 386; orthotype LYCODES PACIFICUS Collett (issued August 1, 1879). Same as LEURYNNIS Lockington (March 25, 1880).

1272. COPE (1879). *A Sting Ray (Xiphotrygon acutidens) from the Green River Shales of Wyoming.* Amer. Natural., 1879, XIII, 333.

EDWARD DRINKER COPE.

Xiphotrygon Cope, 333; orthotype X. ACUTIDENS Cope (fossil).

1273. DAVIS (1879). *Notes on Pleurodus . . . and . . . Three Spines of Cestracionts from the Lower Coal Measures.* Quart. Journ. Geol. Soc., XXXV.

JAMES WILLIAM DAVIS.

Phricacanthus Davis, 186; orthotype P. BISERIALIS Davis (fossil). Probably a synonym of ORACANTHUS Ag.

1274. DAVIS (1879). *On Ostracacanthus . . . a Fossil Fish from the Coal Measures Southeast of Halifax, Yorkshire.* Reprint Brit. Assn. Adv. Sci., 1879.

JAMES WILLIAM DAVIS.

Ostracacanthus Davis, 343; orthotype O. DILATATUS Davis.

1275. GILL (1879). *(Rhombatractus.)* Amer. Naturl., 1879.

THEODORE GILL.

Rhombatractus Gill, 709; orthotype ARISTEUS FITZROYENSIS Castelnau. Substitute for ARISTEUS Castelnau, preoccupied. A synonym of MELANOTÆNIA Gill, 1862.

1276. GOODE & BEAN (1879). *A Preliminary Catalogue of the Fishes of the St. Johns River and the East Coast of Florida . . .* Proc. U. S. Nat. Mus., II.

GEORGE BROWN GOODE; TARLETON HOFFMAN BEAN.

Jordanella Goode & Bean, 117; orthotype J. FLORIDÆ Goode & Bean.

1277. GOODE & BEAN (1879). *Description of a New Genus and Species of Fish,*
Lopholatilus chamæleonticeps, from the South Coast of New England. Proc.
U. S. Nat. Mus., II, 205-209.

GEORGE BROWN GOODE; TARLETON HOFFMAN BEAN.

Lopholatilus Goode & Bean, 205; orthotype L. CHAMÆLEONTICEPS Goode & Bean.

1278. JORDAN (1879). *Description of New Species of North American Fishes.*
Proc. U. S. Nat. Mus., II, 235-241.

DAVID STARR JORDAN

Xiphister Jordan, 241; orthotype XIPHIDION MUCOSUM Grd. A substitute for
XIPHIDION Grd., regarded as preoccupied by XIPHIDIUM, a genus of grass-
hopper.

1279. JORDAN (1879). *Note on a Collection of Fishes Obtained in the Streams*
of Guanajuato and in Chapala Lake, Mexico, by Prof. A. Dugès. Proc. U. S.
Nat. Mus., II, 299-301.

DAVID STARR JORDAN.

Goodea Jordan, 299; orthotype G. ATRIPINNIS Jordan.

1280. GORJANOVIC-KRAMBERGER (1879). *Beiträge zur Kenntniss der fos-*
silen Fische der Karpathen. Palæontographica, XXVI.

DRAGUTIN GORJANOVIC-KRAMBERGER.

Megalolepis Kramberger, 61; orthotype M. CASHCÆNSIS Kramberger (fossil).

1281. KESSLER (1879). *Beiträge zur Ichthyologie von Central-Asien.* Bull.
Acad. Sci. St. Petersb., 1879, XXV, 282-310.

KARL THEODOROVICH KESSLER.

Aspiorhynchus Kessler, 289; orthotype, A. PREZWALSKII Kessler.

1282. LOCKINGTON (1879). *Description of New Genera and Species of Fishes*
from the Coast of California. Proc. U. S. Nat. Mus., II, 326-332.

WILLIAM NEALE LOCKINGTON (1842-1902).

Leurynnis Lockington, 326; orthotype LEURYNNIS PAUCIDENS Lockington (male of
LYCODOPSIS PACIFICUS Collett) (issued March 25, 1880). A synonym of LYCO-
DOPSIS Collett (Aug. 1, 1879).

Odontopyxis Lockington, 326; orthotype O. TRISPINOSUS Lockington.

1283. LOCKINGTON (1879). *On a New Genus of Scombridæ.* Proc. Acad. Nat.
Sci. Phila., 1879, 133-136.

WILLIAM NEALE LOCKINGTON.

Chriomitra Lockington, 133; orthotype C. CONCOLOR Lockington. A synonym of
SCOMBEROMORUS Lac.

1284. SAUVAGE (1879). *Notice sur la faune ichthyologique de l'Ogôoué.* Bull.
Soc. Philom. Paris, sér. 7, III, 90-103.

HENRI ÉMILE SAUVAGE.

Micracanthus Sauvage, 95; orthotype M. MARCHII Sauvage. Name preoccupied by
MICROCANTHUS (MICRACANTHUS) Sw. (1839). Being apparently a valid
genus, the name has been replaced by OSHIMIA Jordan, 1919, in honor of
Masamitsu Oshima.

Doumea Sauvage, 96; orthotype D. TYPICA Sauvage.
Atopochilus Sauvage, 96; orthotype A. SAVORGNANI Sauvage.

1285. SAUVAGE (1879). *Description de gobioides nouveaux ou peu connus de la collection du Museum d'Histoire Naturelle.* Bull. Sci. Philom. Paris.

HENRI ÉMILE SAUVAGE.

Giuris Sauvage, 15; logotype G. VANICOLENSIS Sauvage (no generic diagnosis), near HYPSELEOTRIS Gill.

1286. SAUVAGE (1879). *(Uranoplosus.)* Bull. Soc. Yonne, XXXIII.

HENRI ÉMILE SAUVAGE.

Uranoplosus Sauvage, 47; orthotype U. COTTEAUI Sauvage (fossil).
Cosmodus Sauvage, 48; orthotype C. GRANDIS Sauvage (specific name preoccupied) = CŒLODUS MAJOR Woodward. May replace CŒLODUS Heckel (1856), regarded as preoccupied by CŒLODON Lund (1839).

1287. STEINDACHNER (1879). *Zur Fischfauna des Magdalenen-Stromes.* Denkschr. Akad. Wiss. Wien, XXXIX, 19-78.

FRANZ STEINDACHNER.

Luciocharax Steindachner, 67; orthotype L. insculptus Steind. A synonym of CTENOLUCIUS Gill (1861), but in the latter case no specific name was given.

1288. STEINDACHNER (1879). *Ueber einige neue und seltene Fisch-Arten aus den k.-k. zoologischen Museum zu Wien, Stuttgart, und Warschau.* Denkschr. Akad. Wiss. Wien, XLI, 1-52.

FRANZ STEINDACHNER.

Parequula Steindachner, 8; orthotype P. BICORNIS Steind.

1289. TRAUTSCHOLD (1879). *Ueber Fischzähne des Moskauer Jura.* Bull. Soc. Imp. Nat. Moskau, XIV.

HERMANN VON TRAUTSCHOLD.

Cymatodus Trautschold, 53; C. PLICATULUS Trautschold (fossil). Name preoccupied; probably a synonym of MESOLOPHODUS Woodw.
Cranodus Trautschold, 54; orthotype C. ZONODUS Trautschold (fossil).
Chiastodus Trautschold, 58; orthotype C. OBVALLATUS Trautschold (fossil).
Arpagodus Trautschold, 59; orthotype A. RECTANGULUS Trautschold (fossil). A synonym of CAMPODUS Koninck.

1290. WAAGEN (1879). *Paleontology of India.*
W. WAAGEN (1841-1900).

Helodopsis Waagen, I, 14; logotype H. ELONGATA Waagen (fossil).
Thaumatacanthus Waagen, I, 78; orthotype T. BLANFORDI Waagen (fossil).

1880.

1291. BLEEKER (1880). *Musei Hamburgensis species piscium novæ minusque cognitas descripsit et depingi curavit.* Abh. Naturw. Ver. Hamburg, 1880, VII, 25-30.

PIETER BLEEKER.

Hemiglyphidodon Bleeker, 8; orthotype H. PLAGIOMETOPON Blkr.

1292. COLLETT (1880). *Den Norske Nordhavs Expedition, 1876-1878. Zoologi, Fiske.* Christiania.

ROBERT COLLETT.

Rhodichthys Collett, 154; orthotype R. REGINA Collett.

1293. COPE (1880). *Second Contribution to the History of the Vertebrata of the Permian Formation of Texas.* Proc. Amer. Philos. Soc., XIX, 38-58.

EDWARD DRINKER COPE.

Ectosteorhachis Cope, 52; orthotype E. NITIDUS Cope (fossil). A synonym of PARABATRACHUS Owen.

1294. FONTANNES (1880). *Nouvelle contribution à la faune et à la flore des marnes pliocènes à Brissopsis d'Eurre (Drôme). Type nouveau de la famille des Clupeoides.* Ann. Soc. Agric. Hist. Nat. Lyon, V, 423-425.

CHARLES FRANCISQUE FONTANNES (1839-1886).

Clupeops (Sauvage), Fontannes, 209; orthotype C. INSIGNIS Sauvage (fossil). Miocene from le Bassin de Crest at Eurre, Drôme (publication needs verification).

1295. GARMAN (1880). *Synopsis and Descriptions of the American Rhinobatidæ.* Proc. U. S. Nat. Mus.

SAMUEL GARMAN.

Platyrhinoidis Garman, 516; orthotype PLATYRHINA TRISERIATA J. & G.

Discobatus Garman, 523; orthotype RAJA SINISIS Lac.; substitute for PLATYRHINA M. & H., regarded as preoccupied by PLATYRHINUS; but ANALITHIS Gistel (1848) is a still earlier substitute name.

1296-1297. GIGLIOLI (1880). *Elenco dei mammiferi, degli uccelli e dei rettili ittiofagi . . . et catalogo . . . dei pesci italiani.* Esposizione Internazionale della Pesca, Berlin.

Pomatomichthys Giglioli, 20; orthotype P. CONSTANCIÆ Giglioli = POMATOMUS TELESCOPIUM Risso. A synonym of EPIGONUS Raf.

Thynnichthys Giglioli, 25; orthotype THYNNUS THUNNINA Cuv. & Val. = SCOMBER ALLETERATUS Raf. Name preoccupied; a synonym of EUTHYNNUS Lütken (1883).

Pelamichthys Giglioli, 25; orthotype SCOMBER UNICOLOR Geoffroy St. Hilaire. A synonym of ORCYNOPSIS Gill, 1862.

1298. GOODE (1880). *Fishes from the Deep Water on the South Coast of New England Obtained by the United States Fish Commission in the Summer of 1880.* Proc. U. S. Nat. Mus., 1880, III, 467-486.

GEORGE BROWN GOODE.

Amitra Goode, 478; orthotype A. LIPARINA Goode. Name regarded as preoccupied by AMITRUS; replaced by MONOMITRA Goode, 1883.

1299. GOODE (1880). *Descriptions of Seven New Species of Fishes from Deep Soundings on the Southern New England Coast, With Diagnosis of Two Undescribed Genera of Flounders and a Genus Related to Merlucius.* Proc. U. S. Nat. Mus. 1880, III, 337-350.

Monolene Goode, 338; orthotype M. SESSILICAUDA Goode.
Thyris Goode, 344; orthotype T. PELLUCIDUS Goode; larva of MONOLENE Goode.
Name preoccupied, replaced by DELOTHYRIS Goode, 1883.
Hypsicometes Goode, 347; orthotype H. GOBIOIDES Goode.

1300. GÜNTHER (1880). *Report on the Shore Fishes.* (In *Zoölogy of the Voyage of H. M. S. Challenger.* I, 1-82.

ALBERT GÜNTHER.

Bathyanthias Günther, 6; orthotype B. ROSEUS Gthr.
Zanclorhynchus Günther, 15; orthotype Z. SPINIFER Gthr.
Murænolepis Günther, 17; orthotype M. MARMORATUS Gthr.
Lepidopsetta Günther, 18; orthotype L. MACULATA Gthr. Name preoccupied among flounders; replaced by MANCOPSETTA Gill.
Thysanopsetta Günther, 22; orthotype T. NARESI Gthr.
Lophonectes Günther, 28; orthotype L. GALLUS Gthr.
Læops Günther, 28; orthotype L. PARVICEPS Gthr.
Propoma Günther, 39; orthotype P. ROSEUM Gthr.
Lioscorpius Günther, 40; orthotype L. LONGICEPS Gthr.
Acanthaphritis Günther, 43; orthotype A. GRANDISQUAMIS Gthr.
Tetrabrachium Günther, 45; orthotype T. OCELLATUM Gthr.
Anticitharus Günther, 47; orthotype A. POLYSPILUS Gthr.
Pœcilopsetta Günther, 48; orthotype P. COLORATA Gthr.
Nematops Günther, 57; orthotype N. MICROSTOMA Gthr.

1301. HAY (1880). *On a Collection of Fishes from Eastern Mississippi.* Proc. U. S. Nat. Mus., 1880, III, 488-515.

OLIVER PERRY HAY (1846-   ).

Opsopœodus Hay, 507; orthotype O. EMILIÆ Hay.

1302. HECTOR (1880). *Notice of a New Fish (Hypolycodes haastii).* Trans. Proc. New Zealand Instit., 1880 (1881), 13, 194-195.

JAMES HECTOR.

Hypolycodes Hector, 194; orthotype H. HAASTII Hector.

1303. HILGENDORF (1880). *Ueber eine neue berkenswerthe Fischgattung, Leucopsarion, von Japan.* Monatsb. Akad. Wiss. Berlin.

FRANZ MARTIN HILGENDORF

Leucopsarion Hilgendorf, 340; orthotype L. PETERSI Hilgendorf.

1304. JORDAN & GILBERT (1880). *Notes on a Collection of Fishes from San Diego, California.* Proc. U. S. Nat. Mus., III, 1880, 23-34.

DAVID STARR JORDAN; CHARLES HENRY GILBERT.

Roncador Jordan & Gilbert, 28; orthotype CORVINA STEARNSI Steind.
Leuresthes Jordan & Gilbert, 29; orthotype ATHERINA TENUIS Ayres.

1305. JORDAN & GILBERT (1880). *Description of a New Flounder (Xystreurys liolepis) from Santa Catalina, California.* Proc. U. S. Nat. Mus., 1880, III, 34-36.

DAVID STARR JORDAN; CHARLES HENRY GILBERT.

Xystreurys Jordan & Gilbert, 34; orthotype X. LIOLEPIS J. & G.

1306. JORDAN & GILBERT (1880). *Description of a New Flounder (Pleuro-nichthys verticalis) from the Coast of California, with Notes on Other Species.* Proc. U. S. Nat. Mus., 1880, III, 49-51.

DAVID STARR JORDAN; CHARLES HENRY GILBERT.

Atheresthes Jordan & Gilbert, 51; orthotype PLATYSOMATICHTHYS STOMIAS J. & G.

1307. JORDAN & GILBERT (1880). *On the Generic Relations of Platyrhina exasperata.* Proc. U. S. Nat. Mus., LIII.

DAVID STARR JORDAN; CHARLES HENRY GILBERT.

Zapteryx Jordan & Gilbert, 53; orthotype PLATYRHINA EXASPERATA J. & G.

1308. JORDAN & GILBERT (1880). *Descriptions of Two New Species of Fishes (Ascelichthys rhodorus and Scytalina cerdale) from Neah Bay, Washington Territory.* Proc. U. S. Nat. Mus., 1880, III, 264-268.

DAVID STARR JORDAN; CHARLES HENRY GILBERT.

Ascelichthys Jordan & Gilbert, 264; orthotype A. RHODORUS J. & G.

Scytalina Jordan & Gilbert, 266; orthotype S. CERDALE J. & G. Name regarded as preoccupied by SCYTALINUS, replaced by SCYTALISCUS J. & G.

1309. JORDAN & GILBERT (1880). *Description of a New Species of Deep-water Fish (Icichthys lockingtoni) from the Coast of California.* Proc. U. S. Nat. Mus., 1880, III, 305-308.

DAVID STARR JORDAN; CHARLES HENRY GILBERT.

Icichthys Jordan & Gilbert, 305; orthotype I. LOCKINGTONI J. & G.

1310. KLUNZINGER (1880). *Die von müllersche Sammlung australischer Fische in Stuttgart.* Sitzb. Akad. Wiss. Wien, LXXX, 325-430.

CARL BENJAMIN KLUNZINGER.

Platychærops Klunzinger, 399; orthotype P. MUELLERI Klunzinger.

1311. KŒNEN (1880). *Ueber Coccosteus-Arten aus dem Devon von Birken.* Zeitschr. Deutsch. Geol. Ges., 1880, XXXII, 673-675.

ADOLF VON KŒNEN (1837-1915).

Brachydirus Kœnen, 675; orthotype COCCOSTEUS BIRKENSIS Kœnen (fossil).

1312. LOCKINGTON (1880). *Description of a New Genus and Some New Species of California Fishes (Icosteus ænigmaticus and Osmerus attenuatus).* Proc. U. S. Nat. Mus., 1880, III, 63-68.

WILLIAM NEALE LOCKINGTON.

Icosteus Lockington, 63; orthotype I. ÆNIGMATICUS Lockington.

1313. LOCKINGTON (1880). *Description of a New Chiroid Fish (Myriolepis zonifer) from Monterey Bay, California.* Proc. U. S. Nat. Mus., 1880, III, 248-251.

WILLIAM NEALE LOCKINGTON.

Myriolepis Lockington, 248; orthotype, M. ZONIFER Lockington. Based on a young specimen; name preoccupied, replaced by ERILEPIS Gill, which supplants EBISUS Jordan & Snyder, 1901, the adult fish.

1314. PETERS (1880). *Ueber die von der chinesischen Regierung zu der internationalen Fischerei-Austellung gesandte Fischsammlung aus Ningpo.* Monatsb. Akad. Wiss. Berlin, 1880, 921-927.

WILHELM (CARL HARTWIG) PETERS.

Distœchodon Peters, 924; orthotype D. TUMIROSTRIS Peters.

Mylopharyngodon Peters, 925; orthotype LEUCISCUS ÆTHIOPS.

1315. PETERS (1880). *Ueber eine Sammlung von Fischen, welche Dr. Gerlach in Hongkong gesandt hat.* Monatsb. Akad. Wiss. Berlin, 1880, 1029-1037.

WILHELM (CARL HARTWIG) PETERS.

Semilabeo Peters, 1032; orthotype S. NOTABILIS Peters.
Cranoglanis Peters, 1030; orthotype C. SINENSIS Peters.

1316. ROCHEBRUNE (1880). *Description de quelques espèces nouvelles de poisson propres à Sénégambie.* Bull. Soc. Philom. Paris, 1880, sér. 7, IV, 159-169.

ALPHONSE TRÉMEAU DE ROCHEBRUNE (1834-   ).

Sparactodon Rochebrune, 162; orthotype S. NALNAL Rochebrune. A synonym of POMATOMUS Lac.

1317. ST. JOHN & WORTHEN (1880). *Descriptions of Fossil Fishes.* Geological Survey of Illinois, VII.

ORESTES H. ST. JOHN; A. H. WORTHEN.

Vaticinodus St. John & Worthen, 80; logotype V. TENERRIMUS St. John & Worthen (fossil). A synonym of SANDALODUS Newberry & Worthen.
Stenopterodus St. John & Worthen, 100; orthotype S. ELONGATUS St. John & Worthen (fossil). A synonym of DELTODUS.
Chitonodus St. John & Worthen, 109; logotype COCHLIODUS LATUS Leidy (fossil).
Deltodopsis St. John & Worthen, 158; logotype D. AFFINIS St. John & Worthen (fossil).
Orthopleurodus St. John & Worthen, 190; logotype O. CONVEXUS St. John & Worthen (fossil). A synonym of SANDALODUS.

1318. SAUVAGE (1880). *Description des gobioides nouveaux ou peu connus de la collection du Muséum d'Histoire Naturelle.* Bull. Soc. Philom. Paris, 1880, sér. 7, IV, 40-58.

HENRI ÉMILE SAUVAGE.

Cayennia Sauvage, 57; orthotype C. GUICHENOTI Sauvage.

1319. SAUVAGE (1880). *Description de quelques poissons d'espèces nouvelles,* etc. Bull. Sci. Philom. Paris, IV.

HENRI ÉMILE SAUVAGE.

Pseudopristipoma Sauvage, 220; orthotype PRISTIPOMA LEUCURUM Cuv. & Val.

1320. SAUVAGE (1880). *Notice sur quelques poissons de l'île Campbell et de l'Indo-Chine.* Bull. Soc. Philom. Paris, 1880, sér. 7, IV, 228-233.

HENRI ÉMILE SAUVAGE.

Paratylognathus Sauvage, 227; orthotype P. DAVIDI Sauvage.
Probarbus Sauvage, 232; orthotype P. JULLIENI Sauvage.

1321. SAUVAGE (1880). *Notice sur les poissons tertiaires de Céreste (Basses Alpes).* Bull. Soc. Géol. France (3), VIII, 1880.

HENRI ÉMILE SAUVAGE.

Enoplophthalmus Sauvage, 449; orthotype E. SCHLUMBERGERI Sauvage (fossil).
Properca Sauvage, 452; logotype PERCA ANGUSTA Ag. (fossil).

1322. SAUVAGE (1880). *Synopsis des poissons et des reptiles des terrains jurassiques de Boulogne-sur-Mer.* Bull. Soc. Géol. Paris, 1880, sér. 3, VIII, 524-547.

HENRI ÉMILE SAUVAGE.

Athrodon Sauvage, 530; orthotype A. DOUVILLII Sauvage (fossil).

1323. SAUVAGE (1880). *Nouvelles recherches sur les poissons fossiles decouverts par M. Alby à Licata en Sicilie.* Ann. Soc. Géol. XI.

HENRI ÉMILE SAUVAGE.

Podopteryx Sauvage, 17; orthotype P. ALBYI Sauvage (fossil).

Parequula Sauvage, 25; orthotype P. ALBYI Sauvage (fossil); name preoccupied, Steindachner, 1879.

Paraleuciscus Sauvage, 38; orthotype P. ECNOMI Sauvage (fossil).

1324. STEINDACHNER (1880). *Ichthyologische Beiträge.* VIII. Sitzb. Akad. Wiss. Wien, LXXX, 119-191.

FRANZ STEINDACHNER.

Typhlogobius Steindachner, 141; orthotype T. CALIFORNIENSIS Steind; replaces OTHONOPS Rosa Smith.

1325. STEINDACHNER (1880). *Ichthyologische Beiträge.* IX. Sitzb. Akad. Wiss. Wien, LXXXII (July 15, 1880), 238-266.

FRANZ STEINDACHNER.

Ptychochromis Steindachner, 248; orthotype TILAPIA OLIGACANTHUS Blkr.

Ancharius Steindachner, 251; orthotype A. FUSCUS Steind.

Hypoptychus Steindachner, 257; orthotype H. DYBOWSKII Steind.

Neozoarces Steindachner, 263; orthotype N. PULCHER Steind.

1326. STOCK (1880). *Note on a Spine . . . from the Coal Measures of Northumberland.* Ann. Mag. Nat. Hist., V.

THOMAS STOCK.

Lophacanthus Stock, V, 217; orthotype L. TAYLORI Stock (fossil). A synonym of ORACANTHUS Ag..

1327. THOMINOT (1880). *Note sur un poisson de genre nouveau appartenant à la famille des scombérodés, voisin des sérioles.* Bull. Soc. Philom. Paris, 1880, sér. 7, IV, 173-174.

ALEXANDRE THOMINOT.

Lepidomegas Thominot, 173; orthotype L. MUELLERI Thominot. A synonym of SERIOLA Cuv.; based on a very old example with the free spines absorbed.

1328. TRAUTSCHOLD (1880). *Ueber Tomodus Agass.* Bull. Soc. Imp. Moscou. LV, 139-140.

HERMANN VON TRAUTSCHOLD.

Oxytomodus Trautschold, 140; orthotype TOMODUS ARGUTUS Trautschold (fossil).

OMISSIONS

1758

1. LINNÆUS (1758). *Systema Naturæ.* Ed. X.

CAROLUS LINNÆUS.

Myxine Linnæus, 650; orthotype M. GLUTINOSA L. Overlooked in Part I, p. 16, being placed by Linnæus among the worms.

1777

23. KLEIN (1777). *Schauplatz,* IV, etc.

JAKOB THEODOR KLEIN.

Cynocephalus Klein; type SQUALUS GLAUCUS L. The name CYNOCEPHALUS was applied to a genus of bats by Boddaert in 1768.

1803
62. GIORNA (1803). *Mémoire*, etc.
MICHEL ESPRIT GIORNA.

Trachyrincus Giorna. This is the original spelling, and the apparent date is 1803.

1815
83. RAFINESQUE (1815). *Descrizione*, etc.
CONSTANTINE SAMUEL RAFINESQUE.

Nemochirus Rafinesque; a synonym of TRACHIPTERUS Gouan, according to Hubbs.

1839
251. SWAINSON (1839). *Natural History*, etc.
WILLIAM SWAINSON.

Lepidosoma ("Risso") Swainson; haplotype "LEPIDOSOMA TRACHYRHYNCHUS Risso." Evidently a slip for LEPIDOLEPRUS Risso.

1843
295. AGASSIZ (1843). *Recherches,* etc., III.
LOUIS AGASSIZ.

Physonemus Agassiz, 176; haplotype P. SUBTERES Ag. (fossil). Name only; with locality, lower Subcarboniferous of Armagh, described later by M'Coy, 1848, with a different type, probably not congeneric with it.

Glyphis Agassiz, III, 243; orthotype G. HASTALIS Ag. (fossil).

1844
318A. MÜLLER (1844). *Ueber den Bau und die Grenzen der Ganoiden und das natürliche System der Fische.* Abh. Akad. Wiss. Berlin, 1844, 117-216, reprinted elsewhere.
JOHANNES MÜLLER.

This epoch-making memoir is the most important of Müller's contributions to Ichthyology. In it he continues his bad habit of inserting new genera in brief footnotes irrelevant to the text.

Tribranchus (Peters) Müller; orthotype T. ANGUILLARIS Peters = ANGUILLA AUSTRALIS Rich. A synonym of ANGUILLA Shaw.

324. RICHARDSON (1844). *Voyage Erebus,* etc.
JOHN RICHARDSON.

Rhynchana Richardson, 44; orthotype R. GREYI Rich. = CYPRINUS GONORHYNCHUS Gmelin. A synonym of GONORHYNCHUS (Gronow) Scopoli.

1846
351. CUVIER & VALENCIENNES (1846). *Histoire naturelle,* etc. XVIII, XIX.
ACHILLE VALENCIENNES.

Panchax Cuvier & Valenciennes, XVIII, 380; tautotype ESOX PANCHAX Ham. = PANCHAX BUCHANANI Cuv. & Val., a synonym of APLOCHEILUS M'Clelland.

Vandellia Cuvier & Valenciennes, XVIII, 386; haplotype V. CIRRHOSA Cuv. & Val.

1847
375. MEYER (1847). *Vorläufige Uebersicht,* etc.
HERMANN VON MEYER.

Cenchrodus Meyer, 574; logotype C. GŒPPERTI Meyer (fossil).

Hemilopas Meyer, 575; orthotype H. MENTZELI Meyer (fossil). (This and the preceding indeterminable fragments of a jaw, probably of COLOLOBUS Ag.)

1848

390. M'COY (1848). 411 M'Coy with CENTRODUS is the same paper as 390, the proper date being 1848.

391. M'COY (1848). *Ichthyodorulites,* etc.
FREDERICK M'COY.

Platacanthus M'Coy, 120; as originally written, later emended to PLATYACANTHUS. **Dipriacanthus** M'Coy, 121; logotype D. STOKESII M'Coy (fossil).

391. M'COY (1848). *Scotch Old Red Sandstone,* etc.
FREDERICK M'COY.

Chirodus M'Coy, 130; orthotype C. PES-RANÆ M'Coy. This name was originally and correctly spelled CHIRODUS.

1849

427A. GEINITZ (1849). *Das Quadersandsteingebirge.*
HANS BRUNO GEINITZ.

Aulolepis Geinitz, 86; orthotype A. REUSSI Geinitz. "CYCLOIDEN-SCHUPPE" Geinitz, 1839 = CYCLOLEPIS Geinitz, 1868. AULOLEPIS apparently replaces CYCLOLEPIS Geinitz, 1868.

1851

439. BLEEKER (1851). *Borneo,* II, etc.
PIETER BLEEKER.

Clupeoides Bleeker, 275; orthotype C. BORNEËNSIS Blkr. is a valid genus distinct from CORICA, according to Weber.

1852

464. GERVAIS (1852). *Zoologie et Paléontologie Françaises.* I.
FRANÇOIS LOUIS PAUL GERVAIS.

Onchosaurus Gervais, 268; orthotype O. RADICALIS Gervais (fossil). Wrongly identified as the tooth of a Mososaur. Replaces ISCHYRHIZA Leidy.

1853

482A. M'COY (1853). *On the Supposed Fish Remains Figured on Plate 4 of the Silurian System.* Quart. Journ. Geol. Soc., IX.
FREDERICK M'COY.

Leptocheles M'Coy, 14; orthotype L. MURCHISONI M'Coy (fossil). A synonym of ONCHUS.

483. OWEN (1853). *Notice of a Batrachoid in British Coal Shale.* Quart. Journ. Geol. Soc., IX, 67-70.
RICHARD OWEN.

Parabatrachus Owen, 67, etc.

1854

503. GIRARD (1854). *Enumeration,* etc.
CHARLES GIRARD.

Heterostichus is misspelled on page 258.

512A. MORRIS (1854). *Catalogue of British Fossils.*

JOHN MORRIS.

Gyrosteus (Agassiz) Morris, 328; orthotype G. MIRABILIS Ag. First validation of GYROSTEUS and perhaps of some other nominal genera of Agassiz. See No. 614.

1855

521. AYRES (1855). *Description,* etc.

WILLIAM O. AYRES.

Cebidichthys Ayres. This is the original and correct spelling, not CEBEDICHTHYS.

530. BLEEKER (1855). *Sumatra,* etc.

PIETER BLEEKER.

Clupeichthys Bleeker, 275; orthotype C. GONIOGNATHUS Blkr. This is a valid genus, according to Weber, distinct from CORICA.

535A. GERVAIS (1855). *Description d'un poisson fossile du terrain crétacé de la Drome,* etc. Ann. Sci. Nat., III.

FRANÇOIS PAUL LOUIS GERVAIS.

Histialosa Gervais, 322; orthotype H. THIOLLERI Gervais (fossil).

543. PETERS (1855). *Uebersicht,* etc.

WILHELM PETERS.

Opsaridium Peters, 269; orthotype O. ZAMBEZENSE Peters. Description given as from manuscript by Günther, 1868.

1856

561. HECKEL (1856). *Pycnodonten,* etc.

JOHANN JAKOB HECKEL.

Stemmatodus Heckel, 202; orthotype PYCNODUS RHOMBUS Ag. This generic name is not preoccupied. STEMMATODUS St. John & Worthen (1875) is preoccupied and replaced by STEMMATIAS Hay, 1899.

572. PANDER (1856). *Monographie.* etc.

CHRISTIAN HEINRICH PANDER.

Besides the nominal genera of Russian Silurian fossils listed by Woodward, the following additional names are given by Marshall. These we are not able to verify.

DREPANODUS, 20.

ACODUS, 21.

MACHAIRODUS, 22.

PALTODUS, 24.

ACONTIODUS, 28.

PRIONIODUS, 29.

BELODUS, 30.

CTENOGNATHUS, 32.

GNATHODUS, 33.

PRIONOGNATHUS, 34

STROSIPHERUS, 73.

MONOPLEURODUS, 78.

1857

582A. GIEBEL (1857). *Dichelodus, ein neuer Fisch im Mansfelder Kupferschiefer.* Zeitschr. Gesammt. Naturw., IX, 121-126.

CHRISTOPH (GOTTFRIED ANDREAS) GIEBEL.

Dichelodus Giebel, 121; orthotype D. ACUTUS Giebel (fossil).

583. GIRARD (1857). *Contributions,* etc.
CHARLES GIRARD.

Chiropsis Girard, 201; haplotype, as indicated in the *Pacific R. R. Survey,* is CHIRUS PICTUS Grd. = LABRUS SUPERCILIOSUS Pallas.

1859

664A. PAGE (1859). *(Fossil Fishes; Old Red Sandstone of Scotland.)* Rept. Brit. Assn. Adv. Sci., 1858 (1859).
D. PAGE.

Ictinocephalus Page, 104; orthotype I. GRANULATUS Page (fossil). Name only, identical with ISCHNACANTHUS Powrie, 1864.

1863

Blepharichthys Gill, misprinted in text on page 302.

1866

834. KNER & STEINDACHNER. *Neue Gattungen,* etc. Abh. Königl. Bayerischen Akad. Wiss. München, X.
RUDOLPH KNER; FRANZ STEINDACHNER.

The correct date of this paper is 1866, the second part issued in 1868, the whole with title page in 1870, according to Hussakof.

1874

1128 and **1129** LÜTKEN (1874). LEPORELLUS Lütken, 129 should come under No. 1129, and GLANIDIUM Lütken, 31, and BAGROPSIS Lütken, 32, should come under 1128, a slip made in printing.

The Genera of Fishes

[PART IV]

FROM 1881 TO 1920

THIRTY-NINE YEARS

WITH THE ACCEPTED TYPE OF EACH

G
IV

INTRODUCTORY NOTE

The present memoir is the fourth and concluding part of a catalogue of generic names applied to fishes subsequent to the establishment of the Binomial or Linnæan system of nomenclature, which dates from January 1, 1758.

It covers the late modern period in Ichthyology, signalized especially by explorations of the deep sea, of fossil beds, and of the darker continents. The unexampled explorations of Eigenmann in South America belong to this period, as also the records of Boulenger of the fishes of South Africa and the varied researches of Regan in the rich material of the British Museum. In this period the fauna of Australia, Japan, and Siberia have been critically and successfully studied. Our knowledge of fossil fishes is assuming definite form through the work of Woodward, Zittel, Hay, and others who have given Palæontology a constructive turn.

At the end of this period is recorded the death (December 10, 1919) of the veteran ichthyologist, Franz Steindachner, of Vienna, whose excellent papers mark the years from 1859 to 1915.

The present writer, as a life-long teacher of Biology, may be pardoned a personal reference. Himself a student of Agassiz and Gill of the earlier periods, he has had the honor of being also a friend, to some extent, a *protégé,* of Baird, Günther, Vaillant, Lütken, Hilgendorf, and Cope. In addition he has numbered among his own students many of those active in systematic ichthyology during the late modern period.

Among these are Gilbert, Eigenmann, Evermann, Bollman, Snyder, Starks, Jenkins, Meek, Fowler, Hubbs, Seale, Richardson (R. E.), Oshima, Abbott, Burke, Brayton, Cramer, Culver, Davis (B. M.), Fesler, Fordice, Gilbert (J. Z.), Goss, Greeley, Heller, Herre, Kirsch, McGregor, Pierson, Metz, Rutter, Snodgrass, Swain, Thoburn (W. W.), Thompson (W. F.), Weymouth, Williams (T. M.), Otaki (K.), and Woolman.

It is a pleasure to repeat my acknowledgments to the naturalists mentioned in Part III, as having given valuable assistance. I am also indebted to Arthur W. Henn of the American Museum, for looking up obscure names. My thanks are especially due to Masamitsu Oshima for the arrangement of the generic names given in the *Zoological Record.*

DAVID STARR JORDAN.

STANFORD UNIVERSITY, CALIFORNIA
JULY 25, 1920; ISSUED AUGUST 15, 1920.

THE GENERA OF FISHES
PART IV

1881

1329. BEAN (1881). *Descriptions of New Fishes from Alaska and Siberia.* Proc. U. S. Nat. Mus., 1881, IV, 144–159.

TARLETON HOFFMAN BEAN.

Ptilichthys Bean, 157; orthotype P. GOODEI Bean.

1330. BEAN (1881). *Notes on a Collection of Fishes Made by Captain Henry E. Nichols, U. S. N., in British Columbia and Southern Alaska.* Proc. U .S. Nat. Mus., 1881, IV, 436-474.

TARLETON HOFFMAN BEAN.

Delolepis Bean, 465; orthotype D. VIRGATUS Bean.

1331. COPE (1881). *A New Genus of Catostomidæ (Lipomyzon).* American Naturalist, XV (January, 1881).

EDWARD DRINKER COPE.

Lipomyzon Cope, 59; orthotype CHASMISTES BREVIROSTRIS Cope.

1332. DAVIS (1881). *On the Zoological Position of the Genus Petalorhynchus Ag., a Fossil Fish from the Mountain Limestone.* Rept. Brit. Assoc. Adv. Sci.

JAMES WILLIAM DAVIS.

Diodontopsodus Davis, 646; orthotype (PRISTODUS FALCATUS Ag.) (fossil). Replaces PRISTODUS Ag. (62), which until 1883 was a name only.

1333. DAVIS (1881). *On the Genera Ctenoptychius,* etc. Ann. Mag. Nat. Hist., VII.

JAMES WILLIAM DAVIS.

Ctenopetalus (Agassiz) Davis, 426; orthotype PETALODUS SERRATUS (Ag.) Owen. A synonym of CTENOPTYCHIUS Ag.

Harpacodus (Agassiz) Davis, 426; orthotype PETALODUS (Ag.) Owen (fossil). A synonym of CTENOPTYCHIUS Ag.

1334. DAVIS (1881). *On Anodontacanthus . . . Fossil Fishes from the Coal Measures.* Quart. Journ. Geol. Soc., XXXVII.

JAMES WILLIAM DAVIS.

Anodontacanthus Davis, 427; orthotype A. ACUTUS Davis (fossil). A synonym of ORACANTHUS Ag.

1335. GARMAN (1881). *New and Little Known Reptiles and Fishes in the Museum Collections.* Bull. Mus. Comp. Zool., VIII.

SAMUEL GARMAN.

Stypodon Garman, 90; orthotype S. SIGNIFER Garman.

1336. GILL (1881). *Record of Recent Scientific Progress: A Deep-sea Rockfish.* Ann. Rept. Smithsonian Inst. for 1880.

THEODORE GILL.

Sebastolobus Gill, 375; orthotype SEBASTES MACROCHIR Gthr.

1337. GÜNTHER (1881). *Account of the Zoological Collection Made During the Survey of H. M. S. Alert in the Straits of Magellan and on the Coast of Patagonia.* Proc. Zool. Soc. London, 2–141.

Melanostigma Günther, 20; orthotype M. GELATINOSUM Gthr.

1338. JORDAN & GILBERT (1881). *Notes on the Fishes of the Pacific Coast of the United States.* Proc. U. S. Nat. Mus., 1881.

DAVID STARR JORDAN; CHARLES HENRY GILBERT.

Polistotrema (Gill) Jordan & Gilbert, 30; orthotype GASTROBRANCHUS DOMBEY Cuv. = BDELLOSTOMA STOUTI Lockington. Also noted on page 18 by Jordan & Jouy.

1339. JORDAN & GILBERT (1881). *Description of Thirty-three New Species of Fishes from Mazatlan, Mexico.* Proc. U. S. Nat. Mus., IV.

DAVID STARR JORDAN; CHARLES HENRY GILBERT.

Etropus Jordan & Gilbert, 364; orthotype ETROPUS CROSSOTUS J. & G.

1340. JORDAN & GILBERT (1881). *Description of Nineteen New Species of Fishes from the Bay of Panama.* Bull. U. S. Fish Commission, I, for 1881.

DAVID STARR JORDAN; CHARLES HENRY GILBERT.

Cerdale Jordan & Gilbert, 332; orthotype C. IONTHAS J. & G.

1341. GORJANOVIĆ-KRAMBERGER (1881). *Studien über die Gattung Saurocephalus Harlan; ein Beitrag zur Neocom-Fischfauna der Insel Lesina.* Jahrb. Geol. Reichsanstalt Wien, XXXI.

DRAGUTIN GORJANOVIĆ-KRAMBERGER.

Solenodon Kramberger, 373; orthotype S. NEOCOMIENSIS Kramberger (fossil) = SAUROCEPHALUS LYCODON Kner. Regarded by Woodward as a synonym of ENCHODUS Ag., the name SOLENODON preoccupied; replaced by HOLCODON Kramberger, 1885.

1342. LOCKINGTON (1881). *Description of a New Genus and Species of Cottidæ.* Proc. U. S. Nat. Mus., 1881, 141–144.

WILLIAM NEALE LOCKINGTON.

Chitonotus Lockington, 141; orthotype C. MEGACEPHALUS Lockington.

1343. LUNEL (1881). *Liste de quelques espèces de poissons, nouvelles pour la faune de l'île Maurice.* Mém. Soc. Phys. Genève, XXVII, 266–387.

GODEFROY LUNEL.

Penetopteryx Lunel, 275; orthotype P. TÆNIOCEPHALUS Lunel.

1344. MACLEAY (1881). *Descriptive Catalogue of the Fishes of Australia.* Proc. Linn. Soc. New S. Wales, 1881. Also issued in two bound volumes.

WILLIAM MACLEAY.

Neopempheris Macleay, 517; orthotype N. RAMSAYI Macleay. A synonym of LEPTOBRAMA.

Teratorhombus Macleay, 126; orthotype T. EXCISICEPS Macleay. Said to be a synonym of PSEUDORHOMBUS Blkr.

Cainosilurus Macleay, 147; orthotype NEOSILURUS AUSTRALIS Castelnau. Substitute for NEOSILURUS Castelnau, preoccupied.

1345. MOREAU (1881). *Histoire naturelle des poissons de France.*

ÉMILE MOREAU (1823-1896).

Flesus Moreau, II, 301; orthotype PLEURONECTES FLESUS L.

1346. POEY (1881). *Peces.* (In *Apuntes para la fauna puerto-riqueña,* by Juan Gundlach. Ann. Soc. Española Hist. Nat. Madrid, X, 317–350.

FELIPE POEY Y ALOY.

Monosira Poey, 326; orthotype M. STAHLI Poey. A synonym of AMIA Gronow or APOGON Lac.

1347. RZEHAK (1881). *Ueber das Vorkommen und die geologische Bedeutung der Clupeidengattung.* Brünn.

A. RZEHAK.

Anaglyphus Rzehak, 79; orthotype A. INSIGNIS Rzehak (fossil).

Eupalæoniscus Rzehak, 79; orthotype PALÆONISCUM FRIESLEBENENSIS Blainv. (fossil). An unexplained substitute for PALÆONISCUM Blainv.

1348. SMITH. R. (1881). *Description of a New Gobioid Fish (Othonopseos) from San Diego, California.* Proc. U. S. Nat. Mus., IV.

ROSA SMITH (1859-   ) (afterwards Mrs. Carl. H. Eigenmann).

Othonops Rosa Smith, 19; orthotype O. EOS Smith = TYPHLOGOBIUS CALIFORNIENSIS Steind. A synonym of TYPHLOGOBIUS Steind.

1349. STEINDACHNER (1881). *Ichthyologische Beiträge.* X. Sitzb. Akad. Wiss. Wien, 1881, LXXXIII, 179–219.

FRANZ STEINDACHNER.

Cyclopterichthys Steindachner, 192; orthotype C. GLABER Steind. A synonym of APTOCYCLUS Pylaie.

Breitensteinia Steindachner, 213; orthotype B. INSIGNIS Steind.

1350. STEINDACHNER (1881). *Ichthyologische Beiträge.* XI. Sitzb. Akad. Wiss. Wien, LXXXIII, 393–408 (May 5, 1881).

FRANZ STEINDACHNER.

Schedophilopsis Steindachner, 100; orthotype S. SPINOSUS Steind. = ICOSTEUS ÆNIGMATICUS Lockington. A synonym of ICOSTEUS Lockington.

Parachela Steindachner, 404 (12); orthotype P. BREITENSTEINI Steind. *(Borneo).*

1351. THOMINOT (1881). *Sur deux genres nouveaux de poissons faisant partie de la famille des squamipinnes et rapportés d'Australie par J. Verreaux.* Bull. Soc. Philom. Paris, 1881, VII, sér. 8, 140-142.

ALEXANDRE THOMINOT.

Tilodon Thominot, 140; orthotype T. AUSTRALIS Thominot.

Diodyxodon Thominot, 142; orthotype D. AUSTRALIS Thominot. Not DOYDIXODON Val.

1352. TRAQUAIR (1881). *Notice of New Fish-Remains from the Blackband Ironstone of Borough Lee, Near Edinburgh.* Geol. Mag., 1881, II, decade VIII, 34-37.

RAMSAY HEATLEY TRAQUAIR.

Cynopodius Traquair, 35; orthotype C. CRENULATUS Traquair (fossil).

Euctenius Traquair, 36-334; logotype CTENOPTYCHIUS UNILATERALIS Barkas (fossil).

Ganopristodus Traquair, 37; orthotype G. SPLENDENS Traquair (fossil). A synonym of URONEMUS Ag.

1353. TRAQUAIR (1881). *Report on Fossil Fishes Collected by the Geological Survey of Scotland in Eskdale and Liddesdale.* Trans. Roy. Soc. Edinburgh, 1881, XXX, 15-71.

RAMSAY HEATLEY TRAQUAIR.

Phanerosteon Traquair, 39; orthotype P. MIRABILE Traquair (fossil).
Holurus Traquair, 43; orthotype H. PARKI Traquair (fossil).
Canobius Traquair, 46; orthotype C. RAMSAYI Traquair (fossil).
Cheirodopsis Traquair, 56; orthotype C. GEIKIEI Traquair (fossil).
Tarrasius Traquair, 61; orthotype T. PROBLEMATICUS Traquair.

1354. TRAQUAIR (1881). *Notice of New Fish Remains from the Blackband Ironstone of Borough Lee, Near Edinburgh.* Geol. Mag., decade VIII, 491-494.
RAMSAY HEATLEY TRAQUAIR.
Cryphiolepis Traquair, 491; orthotype CŒLACANTHUS STRIATUS Traquair (fossil).

1355. VETTER (1881). *Die Fische aus dem lithographischen Schiefer im Dresdner Museum.* Mitt. Mus., IV, Dresden, 1881.
BENJAMIN VETTER (1848-1893).
Eusemius Vetter, 51; orthotype E. BEATÆ Vetter (fossil).
Diplolepis Vetter, 91; orthotype (SAUROPSIS LONGIMANUS Ag.) (fossil). A synonym of SAUROPSIS Ag., the name twice preoccupied.
Agassizia Vetter, 97; orthotype EUGNATHUS TITANIUS Wagner (fossil). Name twice preoccupied, replaced by ASTHENOCORMUS Woodw., 1895.
Lophiurus Vetter, 116; orthotype L. MINUTUS Vetter (fossil). A synonym of LIODESMUS Wagner.

1356. WHITEAVES (1881). *Some Remarkable Fossil Fishes from the Devonian Rocks of Scaumenac Bay, . . . Quebec . . .* Canadian Naturalist, X.
JOSEPH FREDERICK WHITEAVES (1835-1909).
Eusthenopleuron Whiteaves, 30; orthotype E. FOORDI Whiteaves (fossil).

1882

1357. BASSANI (1882). *Descrizioni dei pesci fossili di Lesina . . .*
FRANCESCO BASSANI.
Opsigonus (Kramberger) Bassani, 200; orthotype O. MEGALURIFORMIS Kramberger (fossil).

1358. BOGHDANOV (1882). *Sketch of the Expeditions and Natural History Investigations in the Aral-Caspian Region from 1720 to 1874.* (Not verified.)
MODEST NIKOLAEVICH BOGHDANOV (1841-1888).
Kessleria Boghdanov, II, 3; orthotype SCAPHIRHYNCHUS FEDTSCHENKOI Kessler. This generic name is rejected by L. S. Berg, as a nomen nudum, which received its first definition in Jordan's *Guide to the Study of Fishes*, II, 20, 1905. But Boghdanov's species, S. FEDTSCHENKOI Kessler, was described in 1872 and was a well known species. KESSLERIA should therefore be preferred to PSEUDOSCAPHIRHYNCHUS Nikolski.

1359. DÖDERLEIN (1882). *Ein Stomiatide aus Japan.* Arch. Naturg. (2), XLVIII, 26-31.
LUDWIG DÖDERLEIN (1855-   ).
Lucifer Döderlein, 26; orthotype L. ALBIPINNIS Döderlein.

1360. FACCIOLÀ (1882). *Note sui pesce del stretto di Messina.* X (part 1). 166-168. Naturalista Siciliano, I, for 1882.
LUIGI FACCIOLÀ.
Alepichthys Facciolà, 167; orthotype A. ARGYROGASTER Facciolà.

1361. GIGLIOLI (1882). *New and Very Rare Fish from the Mediterranean.*
Naturalist, XXV.

ENRICO HILLYER GIGLIOLI.

Paradoxichthys Giglioli, 535; orthotype NOTACANTHUS RISSOANUS Giglioli. A
synonym of POLYACANTHONOTUS.

Teratichthys Giglioli, 574; orthotype T. GARIBALDIANUS Giglioli. A synonym of
POLYACANTHONOTUS Blkr.; name preoccupied in fossil fishes.

1362. GILL (1882). *Supplementary Note on the Pediculati.* Proc. U. S. Nat.
Mus., V.

THEODORE GILL.

Lophiomus Gill, 552; orthotype LOPHIUS SETIGERUS Vahl.

1363. GOODE & BEAN (1882). *Reports on the Results of Dredging Under the
Supervision of Alexander Agassiz, on the East Coast of the United States,
During the Summer of 1880, by the U. S. Coast Survey Steamer Blake,* etc.
Bull. Mus. Comp. Zool. Harvard College, 1882, X, 183-226.

GEORGE BROWN GOODE; TARLETON HOFFMAN BEAN.

Notosema Goode & Bean, 193; orthotype N. DILECTA Goode & Bean.
Chalinura Goode & Bean, 199; orthotype C. SIMULA Goode & Bean.
Barathrodemus Goode & Bean, 200; orthotype B. MANATINUS Goode & Bean.
Dicrolene Goode & Bean, 202; orthotype D. INTRONIGER Goode & Bean.
Lycodonus Goode & Bean, 208; orthotype L. MIRABILIS Goode & Bean.
Poromitra Goode & Bean, 214; orthotype P. CAPITO Goode & Bean.
Cyclothone Goode & Bean, 221; orthotype C. LUSCA Goode & Bean.

1364. GOODE & BEAN (1882). *Benthodesmus, a New Genus of Deep-sea Fishes,
allied to Lepidopus.* Proc. U. S. Nat. Mus., 1882, V, 379-383.

GEORGE BROWN GOODE; TARLETON HOFFMAN BEAN.

Benthodesmus Goode & Bean, 380; orthotype LEPIDOPUS ELONGATUS Clarke.

1365. GOODE & BEAN (1882). *Descriptions of Twenty-five New Species of
Fish from the Southern United States, and Three New Genera, Letharchus,
Ioglossus, and Chriodorus.* Proc. U. S. Nat. Mus., 1882, V, 412-437.

GEORGE BROWN GOODE; TARLETON HOFFMAN BEAN.

Baiostoma Bean, 413; orthotype B. BRACHIALIS Bean. A synonym of ACHIRUS Lac.
Ioglossus Bean, 419; orthotype I. CALLIURUS Bean.
Chriodorus Goode & Bean, 431; orthotype C. ATHERINOIDES Goode & Bean.
Letharchus Goode & Bean, 436; orthotype L. VELIFER Goode & Bean.

1366. GUIMARAËS (1882). *Description d'un nouveau poisson de Portugal.* Jorn.
Sci. Lisboa, VIII, for 1881 (1882).

ANTONIO ROBERTO GUIMARAËS.

Pseudohelotes Guimaraës, 222; orthotype P. GUENTHERI Capello (Ms.).

1367. HAY (1882). *On a Collection of Fishes from the Lower Mississippi.*
Bull. U. S. Fish Comm., II, 57–75.

OLIVER PERRY HAY.

Tirodon Hay, 68; orthotype T. AMNIGENUS Hay.

1368. JORDAN (1882). *The Fishes of Ohio.* Rept. Geol. Surv. Ohio, IV.

DAVID STARR JORDAN.

Coccotis Jordan, 852; orthotype HYPSILEPIS COCCOGENIS Cope.
Erimystax Jordan, 858; orthotype LUXILUS DISSIMILIS Kirtland.

420 THE GENERA OF FISHES, PART IV

1369. JORDAN & GILBERT (1882). *Notes on Fishes Observed About Pensacola, Florida, and Galveston, Texas,* etc. Proc. U. S. Nat. Mus., V, 1882.
DAVID STARR JORDAN; CHARLES HENRY GILBERT.
Ioglossus (Bean) Jordan & Gilbert, 297; orthotype I. CALLIURUS Bean (description in detail on page 419).

1370. JORDAN & GILBERT (1882). *Notes on a Collection of Fishes from Charleston, South Carolina, With Descriptions of Three New Species.* Proc. U. S. Nat. Mus., 1882, V, 580-620.
DAVID STARR JORDAN; CHARLES HENRY GILBERT.
Querimana Jordan & Gilbert, 588; orthotype MYXUS HARENGUS Gthr.
Pneumatophorus Jordan & Gilbert, 593; orthotype SCOMBER PNEUMATOPHORUS De la Roche = SCOMBER COLIAS Gmelin.

1371. JORDAN & GILBERT (1882). *List of Fishes Now in the Museum of Yale College Collected by Prof. Frank H. Bradley at Panama,* etc. Proc. U. S. Nat. Mus., 1882.
DAVID STARR JORDAN; CHARLES HENRY GILBERT.
Emblemaria Jordan & Gilbert, 627; orthotype E. NIVIPES J. & G.

1372. JORDAN & GILBERT (1882). *A Review of the Siluroid Fishes Found on the Pacific Coast of Tropical America,* etc. Bull. U. S. Fish Comm., II.
DAVID STARR JORDAN; CHARLES HENRY GILBERT.
Cathorops Jordan & Gilbert, II, 54; orthotype ARIUS HYPOPHTHALMUS Steind.

1373. JORDAN & GILBERT (1882). *Synopsis of the Fishes of North America.* Bull. U. S. Nat. Mus., 1882 (September), XVI, 1–1018.
DAVID STARR JORDAN; CHARLES HENRY GILBERT.
Creolus Jordan & Gilbert, XXXVI; orthotype SERRANUS FURCIFER J. & G. Substitute for BRACHYRHINUS Gill, preoccupied. Withdrawn in the text, being a synonym of PARANTHIAS Guich.
Trycherodon (Forbes) Jordan & Gilbert, 247; orthotype T. MEGALOPS Forbes. A synonym of OPSOPŒDUS Hay.
Euthynnus (Lütken) Jordan & Gilbert, 429; orthotype THYNNUS PELAMYS L.
Bothragonus (Gill) Jordan & Gilbert, 728; orthotype HYPSAGONUS SWANI Steind.
Isesthes Jordan & Gilbert, 757; orthotype BLENNIUS GENTILIS Grd.; a synonym of HYPSOBLENNIUS Gill.
Isopsetta (Lockington) Jordan & Gilbert, 832; orthotype LEPIDOPSETTA ISOLEPIS Lockington.
Xenistius Jordan & Gilbert, 920; orthotype XENICHTHYS CALIFORNIENSIS Steind.

1374. GORJANOVIC-KRAMBERGER (1882). *Die jüngsttertiäre Fischfauna Croatiens.* Parts I-II. Beitr. Palæont. Oesterr.-Ungarn, 1882, II, 86-135 (*ibid.* 1883, II, 65-85).
DRAGUTIN GORJANOVIC-KRAMBERGER.
Metoponichthys Kramberger, 104; orthotype M. LONGIROSTRIS Kramberger (fossil). Same as PROANTIGONIA Kramberger, and having page priority.
Proantigonia Kramberger, 131; orthotype P. RABDOBOJANA Kramberger (fossil). A synonym of METOPONICHTHYS.

1375. MACLEAY (1882). *Notes on the Pleuronectidæ of Port Jackson, With Descriptions of Two Hitherto Unobserved Species.* Proc. Linn. Soc. New S. Wales, 1882, VII, 11-15.
WILLIAM MACLEAY.

Lophorhombus Macleay, 14; orthotype L. CRISTATUS. A synonym of LOPHONECTES Gthr.

1376. QUENSTEDT (1882). *(Fossile Fische.)* Württ. Jahresh. XXXVIII.

FRIEDRICH AUGUST VON QUENSTEDT.

Bdellodus Quenstedt, 137; orthotype B. BOLLENSIS Quenstedt (fossil).

1377. SAUVAGE (1882). *Étude sur les poissons des Faluns de Brétagne.* Mém. Soc. Sci. Nat. Saône-et-Loire, IV.

HENRI ÉMILE SAUVAGE.

Stylodus Sauvage, IV, 77; orthotype S. LEBESCONTEI Sauvage (fossil).

1378. SAUVAGE (1882). *Description de quelques poissons de la collection du museum d'histoire naturelle.* Bull. Soc. Philom. Paris (7), VI, 168-176.

HENRI ÉMILE SAUVAGE.

Saccostoma Sauvage, 171; orthotype S. GULOSUS Sauvage. Name preoccupied, replaced by CHASMIAS Jordan & Snyder; this also preoccupied and replaced by CHASMICHTHYS Jordan & Snyder.

Cosmochilus Sauvage, 180; orthotype C. HARMANDI Sauvage.

1379. STEINDACHNER (1882). *Ichthyologische Beiträge.* XII. Sitzb. Akad. Wiss. Wien, LXXXVI, 61-82.

FRANZ STEINDACHNER.

Pachychilon Steindachner, 71; orthotype SQUALIUS PICTUS (Heckel & Kner).

1380. VAILLANT (1882). *Sur un poisson des grandes profondeurs de l'Atlantique, l'Eurypharynx pelecanoides.* Comptes Rendus Acad. Sci. Paris, 1882, 1226-1228.

LÉON LOUIS VAILLANT.

Eurypharynx Vaillant, 1226; orthotype E. PELECANOIDES Vaillant.

1883

1381. BEAN (1883). *Notes on Collection of Fishes Made in 1882 and 1883 by Capt. Henry E. Nichols, U. S. N., in Alaska and British Columbia,* etc. Proc. U. S. Nat. Mus., VI, 353-362.

TARLETON HOFFMAN BEAN.

Prionistius Bean, 355; orthotype P. MACELLUS Bean.

1382. COPE (1883). *A New Chondostrean (Crossopholis magnicaudatus) from the Eocene.* Amer. Naturalist, 1883, XVII, 1152–1153.

EDWARD DRINKER COPE.

Crossopholis Cope, 1152; orthotype C. MAGNICAUDATUS Cope (fossil).

1383. COPE (1883). *On the Fishes of the Recent and Pliocene Lakes of the Western Part of the Great Basin and of the Idaho Pliocene Lake.* Proc. Acad. Nat. Sci. Phila., 134-166.

EDWARD DRINKER COPE.

Siphateles Cope, 146; orthotype S. VITTATUS Cope. Young individual of a genus apparently valid, confused with MYLOLEUCUS Cope.

1384. COPE (1883). *On Some Vertebrata from the Permian of Illinois.* Proc. Acad. Nat. Sci. Phila., 108–110.

EDWARD DRINKER COPE.

Didymodus Cope, 108; orthotype DIPLODUS GIBBOSUS Ag. (fossil). Substitute for DIPLODUS Ag., preoccupied. DIDYMODUS is regarded as preoccupied by DIDYMODON and again replaced by DICRANODUS Garman. But there are apparently older synonyms. These teeth probably belong to the forms called PLEURACANTHUS or ORACANTHUS, known from fin-spines only. But it is impossible to coördinate the teeth with the spines, hence the occurrence of two parallel sets of generic names.

Thoracodus Cope, 108; orthotype T. EMYDINUS Cope (fossil). A synonym of JANASSA Münster.

1385. COPE (1883). *Fourth Contribution to the History of the Permian Formation of Texas.* Proc. Amer. Phila. Soc., XX, 628-629.
EDWARD DRINKER COPE.

Gnathorhiza Cope, 629; orthotype G. SERRATA Cope (fossil).

1386. COPE (1883). *On a New Extinct Genus of Percidæ from Dakota Territory.* Amer. Journ. Sci., XXV.
EDWARD DRINKER COPE.

Plioplarchus Cope, 416; orthotype P. WHITII Cope (fossil).

1387. COPE (1883). *Permian Fishes and Reptiles.* Proc. Acad. Nat. Sci. Phila.
EDWARD DRINKER COPE.

Ectosteorhachis Cope, 69; orthotype E. CICERONIUS Cope.

1388. DAMES (1883). *Ueber eine tertiäre Wirbelthier-Fauna von der westlichen Insel des Birket el Qurun im Fajum (Aegypten).* Sitzb. Akad. Wiss. Berlin, VI, 129-152.
WILHELM BARNIM DAMES.

Propristis Dames, 136; orthotype P. SCHWEINFÜRTHI Dames (fossil).

Progymnodon Dames, 148; orthotype P. HILGENDORF Dames (fossil). An extinct analogue of DIODON L.

1389. DAVIS (1883). *On the Fossil Fishes of the Carboniferous Limestone Series of Great Britain.* Scient. Trans. Roy. Dublin Soc., 1883, II, ser. 1, 327-600.
JAMES WILLIAM DAVIS.

Compsacanthus Davis, 354; orthotype C. CARINATUS Davis (fossil).

Lispacanthus Davis, 359; logotype L. GRACILIS Davis (fossil).

Gnathacanthus Davis, 363; logotype G. TRIANGULARIS (fossil).

Cladacanthus (Agassiz) Davis, 364; orthotype C. PARADOXUS Ag. (fossil) (printed as name only in *Poissons Fossiles,* III, 176, 1843). A synonym of ERISMACANTHUS M'Coy (1848).

Chalazacanthus Davis, 370; orthotype C. VERRUCOSUS Davis (fossil).

Carcharopsis (Agassiz) Davis, 381; orthotype C. PROTOTYPUS Ag. (fossil) (name only in 1843). A synonym of DICRENODUS Romanowsky.

Glyphanodus Davis, 386; orthotype G. TENUIS Davis (fossil). A synonym of PETALODUS.

Rhamphodus Davis, 402; orthotype R. DISPAR Davis.

Diclitodus Davis, 410; orthotype D. SCITULUS Davis (fossil).

Tomodus (Agassiz) Davis, 446; orthotype T. CONVEXUS Morris & Roberts (fossil).

Copodus (Agassiz) Davis, 464; orthotype C. CORNUTUS Davis (fossil).

Lobodus (Agassiz) Davis, 468; orthotype L. PROTOTYPUS Ag. (fossil). A synonym of COPODUS (Ag.) Davis.

Mesogomphus (Agassiz) Davis, 470; orthotype M. LINGUA Ag. (fossil). A synonym of COPODUS.

Rhymodus (Agassiz) Davis, 473; orthotype R. TRANSVERSUS Ag. (fossil). A synonym of COPODUS.

Pinacodus (Agassiz) Davis, 476; orthotype P. GONOPLAX Ag. (fossil). A synonym of COPODUS.

Characodus (Agassiz) Davis, 474; orthotype C. ANGULATUS Ag. (fossil) (name only, printed by Morris & Roberts, 1862). Name preoccupied; a synonym of COPODUS.

Dimyleus (Agassiz) Davis, 478; orthotype D. WOODI (Ag.) Davis (fossil).

Mylacodus Davis, 480; orthotype M. QUADRATUS Davis (fossil). A synonym of COPODUS.

Homalodus Davis, 484; logotype H. TRAPEZIFORMIS Davis (fossil). A synonym of PSAMMODUS Ag.

Petalodopsis Davis, 499; orthotype P. TRIPARTITUS Davis (fossil). Name preoccupied; a synonym of CTENOPTYCHIUS.

Pristodus (Agassiz) Davis, 519; orthotype P. FALCATUS Davis (fossil).

Phoderacanthus Davis, 534; orthotype P. GRANDIS Davis (fossil).

1390. DE VIS (1883). *Descriptions of Some New Queensland Fishes.* Proc. Linn. Soc. New S. Wales, 1883, VII, 367–371.

CHARLES W. DE VIS.

Cleidopus De Vis, 367; orthotype C. GLORIA-MARIS De Vis. A synonym of MONOCENTRIS Bl. & Schn.

1391. FACCIOLÀ (1883). *Note sui pesci dello stretto di Messina.* Naturalista Siciliano, 1883, II, 145–148.

LUIGI FACCIOLÀ.

Pelopsia Facciolà, 145; orthotype P. CANDIDA Facciolà. Said to be the young of CHLOROPHTHALMUS Ag.

1392. GAUDRY (1883). *Les enchaînements du monde animal dans les temps géologiques. Fossiles primaires.*

JEAN ALBERT GAUDRY (1827-1908).

Megapleuron Gaudry, 239; orthotype M. ROCHEI Gaudry (fossil). A synonym of SAGENODUS.

1393. GIGLIOLI (1883). *Intorno a due nuovi pesci dal golfo di Napoli.* Zool. Anz., 1883, VI, 397–400.

ENRICO HILLYER GIGLIOLI.

Bellottia Giglioli, 399; orthotype B. APODA Giglioli.

1394. GIGLIOLI (1883). *New and Very Rare Fish from the Mediterranean.* (Part 2.) Nature.

ENRICO HILLYER GIGLIOLI.

Bathyophilus Giglioli, 199; orthotype B. NIGERRIMUS Giglioli.

1395. GILL (1883). *(Pediculate Fishes.)* Forest & Stream, November 8.

THEODORE GILL.

Cryptopsaras Gill, November 8, 1883; orthotype C. COUESI Gill.

Typhlopsaras Gill, 284; orthotype T. SHUFELDTI Gill. Apparently a synonym of MANACALIAS Gill.

1396. GILL (1883). *Diagnosis of New Genera and Species of Deep-sea Fish-like Vertebrates.* Proc. U. S. Nat. Mus., VI, 253–260.

THEODORE GILL.

Bathymyzon Gill, 254; orthotype PETROMYZON BAIRDI Gill.

Histiobranchus Gill, 255; orthotype H. INFERNALIS Gill.

Hyperchoristus Gill, 256; orthotype H. TANNERI Gill.

Sigmops Gill, 256; orthotype S. STIGMATICUS Gill.

Stephanoberyx Gill, 258; orthotype S. MONÆ Gill.

Plectromus Gill, 257; orthotype P. SUBORBITALIS Gill.

Caulolepis Gill, 258; orthotype C. LONGIDENS Gill.

Bassozetus Gill, VI, 259; orthotype B. NORMALIS Gill.

1397. GILL & RYDER (1883). *Diagnosis of New Genera of Nemichthyoid Eels.* Proc. U. S. Nat. Mus., 1883, VI, 260–262.

THEODORE GILL; JOHN ADAM RYDER (1852–95).

Labichthys Gill & Ryder, 261; orthotype L. CARINATUS Gill & Ryder.

Spinivomer Gill & Ryder, 261; orthotype S. GOODEI Gill & Ryder.

Serrivomer Gill & Ryder, 260; orthotype S. BEANI Gill & Ryder.

1398. GILL & RYDER (1883). *On the Anatomy and Relations of the Eurypharyngidæ.* Proc. U. S. Nat. Mus., 1883, VI, 262–273.

THEODORE GILL; JOHN ADAM RYDER.

Gastrostomus Gill & Ryder, 271; orthotype G. BAIRDI Gill & Ryder.

1399. GOODE (1883). *The Generic Names Amitra and Thyris Replaced.* Proc. U. S. Nat. Mus., VII.

GEORGE BROWN GOODE.

Monomitra Goode, 109; orthotype AMITRA LIPARINA Goode. Substitute for AMITRA Goode, 1880, regarded as preoccupied by AMITRUS Schönherr, a beetle.

Delothyris Goode, 109; orthotype THYRIS PELLUCIDUS Goode. Substitute for THYRIS Goode, preoccupied; larval form of MONOLENE Goode.

1400. JORDAN & GILBERT (1883). *Notes on the Nomenclature of Certain North American Fishes.* Proc. U. S. Nat. Mus., VI.

DAVID STARR JORDAN; CHARLES HENRY GILBERT.

Scytaliscus Jordan & Gilbert, 111; orthotype SCYTALINA CERDALE J. & G. Substitute for SCYTALINA, regarded as preoccupied by SCYTALINUS.

1401. KAYSER (1883). *Richthofen's Beiträge zur Palæontologie von China.* Berlin, 1883.

FRIEDRICH HEINRICH EMANUEL KAYSER (1845–   ).

Leptodus Kayser, 161; orthotype L. RICHTHOFENI Kayser (fossil).

1402. KOENEN (1883). *Beiträge zur Kenntniss der Placodermen des Norddeutschen Oberdevons.* Abh. Ges. Wiss. Gottingen, XXX.

ADOLF VON KOENEN.

Brachydirus Koenen, 26; orthotype COCCOSTEUS INFLATUS Koenen (fossil).

Anomalichthys Koenen, 38; orthotype A. SCABER Koenen (fossil).

1403. MACLEAY (1883). *.On a New and Remarkable Fish of the Family Mugilidæ.* Proc. Linn. Soc. New S. Wales, VIII, 2–6.

WILLIAM MACLEAY.

Æschrichthys Macleay, 5; orthotype A. GOLDIEI Macleay.

1404. MACLEAY (1883). *Contributions to a Knowledge of the Fishes of New Guinea,* No. 4. Proc. Linn. Soc. New S. Wales, VIII, 252–282.

WILLIAM MACLEAY.

Tetracentrum Macleay, 256; orthotype T. APOGONOIDES Macleay.

1405. NEWBERRY (1883). *Some Interesting Remains of Fossil Fishes Recently Discovered.* Trans. N. Y. Acad. Sci., II, 144–147.

JOHN STRONG NEWBERRY.

Mylostoma Newberry, 145; logotype M. TERRELLI Newberry.

1406. ST. JOHN & WORTHEN (1883). *Descriptions of Fossil Fishes; a Partial Revision of the Cochliodonts and Psammodonts, Including Notices of Miscellaneous Material Acquired from the Carboniferous formations of the United States.* Geol. Surv. Ill., VII, 55–264.

ORESTES H. ST. JOHN; A. H. WORTHEN.

Tæniodus St. John & Worthen, 75; logotype T. REGULARIS St. John & Worthen (fossil). A synonym, as here restricted, of PSEPHODUS Ag.

Stenopterodus St. John & Worthen, 102; logotype S. PLANUS St. John & Worthen (fossil).

Vaticinodus St. John & Worthen, 82; logotype V. VETUSTUS St. John & Worthen (fossil).

Chitonodus St. John & Worthen, 112; logotype C. SPRINGERI St. John & Worthen (fossil).

Deltodopsis St. John & Worthen, 160; logotype D. ANGUSTUS St. John & Worthen (fossil).

Orthopleurodus St. John & Worthen, 192; logotype O. CONVEXUS St. John & Worthen (fossil).

Acondylacanthus St. John & Worthen, 241; logotype A. RECTUS St. John & Worthen (fossil).

Eunemacanthus St. John & Worthen, 246; orthotype CTENACANTHUS COSTATUS Newberry & Worthen (fossil). A synonym of CTENACANTHUS.

Gymnatacanthus St. John & Worthen, 249; logotype G. RUDIS St. John & Worthen (fossil).

Batacanthus St. John & Worthen, 253; orthotype B. NECIS St. John & Worthen (fossil).

Pnigeacanthus St. John & Worthen, 259; orthotype P. TRIGONALIS St. John & Worthen (fossil).

1407. SAUVAGE (1883). *Note sur les poissons du muschelkalk de Pontpierre (Lorraine).* Bull. Soc. Geol. (3), XI, 492–496.

HENRI ÉMILE SAUVAGE

Meristodon Sauvage, 480; orthotype M. JURENSIS Sauvage (fossil).

Paralates Sauvage, 485; orthotype P. BLEICHERI Sauvage (fossil).

Sparosoma Sauvage, 487; orthotype S. OVALIS Sauvage (fossil). Preoccupied by SPARISOMA Sw.

1408. SAUVAGE (1883). *Sur une collection de poissons recuellis dans le lac Biwako (Japon) par M. F. Steenackers.* Bull. Soc. Philom., Paris, 1883, ser. 7, VII, 144–150.

HENRI ÉMILE SAUVAGE.

Tribolodon Sauvage, 149; orthotype T. PUNCTATUM Sauvage.

1409. SAUVAGE (1883). *Descriptions de quelques poissons de la collection du museum d'histoire naturelle.* Bull. Soc. Philom., Paris, 1883, ser. 7, VII, 156–161.

HENRI ÉMILE SAUVAGE.

Petroscirtes Sauvage, 158; orthotype P. GERMANI Sauvage. Name preoccupied by PETROSCIRTES Rüppell.

1410. STEINDACHNER & DÖDERLEIN (1883). *Beiträge zur Kenntniss der Fische Japans.* Part 1. Denkschr. Akad. Wiss. Wien, 1883, XLVII, 1–34.

FRANZ STEINDACHNER; LUDWIG DÖDERLEIN.

Melanostoma Steindachner & Döderlein, 5; orthotype M. JAPONICUM Steindachner & Döderlein. Name preoccupied; replaced by SYNAGROPS Gthr., 1887.

Neoditrema Steindachner & Döderlein, 32; orthotype N. RANSONNETI Steindachner & Döderlein.

Cypselichthys Steindachner & Döderlein, 14; orthotype C. JAPONICUS Steindachner & Döderlein = LABRACOGLOSSA ARGENTIVENTRIS Peters. A synonym of LABRACOGLOSSA Peters.

Paracirrhites Steindachner & Döderlein, 25; orthotype P. JAPONICUS Steindachner & Döderlein. Name preoccupied; replaced by ISOBUNA Jordan, 1907.

Doderleinia Steindachner, 29; orthotype D. ORIENTALIS Steindachner & Döderlein.

Acanthocephalus (Döderlein) Steindachner, 29; orthotype DODERLEINIA ORIENTALIS Steind. A manuscript synonym of DODERLEINIA, noted in passing, being preoccupied.

Labracopsis Döderlein, 235; orthotype L. JAPONICUS Döderlein.

1411. STEINDACHNER & DÖDERLEIN (1883). *Beiträge zur Kenntniss der Fische Japans.* Part 2. Denkschr. Akad. Wiss. Wien, 1883, XLVIII, 1–40.

FRANZ STEINDACHNER; LUDWIG DÖDERLEIN.

Argo Döderlein, 34; orthotype A. STEINDACHNERI Döderlein. A synonym of TARACTES Lowe.

Pentaceropsis Steindachner & Döderlein, 13; logotype HISTIOPTERUS RECURVIROSTRIS Rich.

Malakichthys Döderlein, 32; orthotype M. GRISEUS Döderlein. Replaces SATSUMA Smith & Pope.

1412. THOMINOT (1883). *Notice sur un poisson de genre nouveau appartenant à la famille des Sparidés.* Bull. Soc. Philom., Paris, VII.

ALEXANDRE THOMINOT.

Parhaplodactylus Thominot, 140; orthotype HAPLODACTYLUS LOPHODON Gthr.

1413. THOMINOT (1883). *Note sur le genre Aplodon, poisson de la famille des Sparidés voisin des girelles.* Bull. Sci. Philom., Paris, VII.

ALEXANDRE THOMINOT.

Aplodon (Duméril) Thominot, 141; logotype A. MARGARITIFERUM Duméril. Mss. also written HAPLODON; both forms preoccupied. Said to be a synonym of INCISIDENS Gill.

1414. TRAUTSCHOLD (1883). *Ueber Edestus und einige andere Fischreste des moskauer Bergkalks.* Bull. Soc. Imp. Nat. Moscou, LVII, pt. 2, 160–174.

HERMANN VON TRAUTSCHOLD.

Euacanthus Trautschold, 172; orthotype E. MARGARITATUS Trautschold (fossil).

1884.

1415. CLAYPOLE (1884). *Preliminary Note on Some Fossil Fishes Recently Discovered in the Silurian Rocks of North America.* Amer. Naturalist, XVIII, 1222–1226.

EDWARD WALLER CLAYPOLE (1835–1901).

Palæaspis Claypole, 1224; orthotype P. AMERICANA Claypole (fossil).

1416. DAVIS (1884). *On some Remains of Fossil Fishes from the Yoredale Series at Leyburn in Wenslaydale.* Quart. Journ. Geol. Soc., 1884, XL, 614–634.

JAMES WILLIAM DAVIS.

Gomphacanthus Davis, 618; orthotype G. ACUTUS Davis (fossil).

Hemicladodus Davis, 620; orthotype H. UNICUSPIDATUS Davis (fossil). An unidentified fragment of a tooth.

Astrabodus Davis, 630; orthotype A. EXPANSUS Davis (fossil). A synonym of PSAMMODUS.

Echinodus Davis, 631; orthotype E. PARADOXUS Davis (fossil).

Cyrtonodus Davis, 631; orthotype C. GIBBUS Davis (fossil).

Diplacodus Davis, 633; orthotype D. BULBOIDES Davis (fossil).

1417. DAVIS (1884). *Description of a New Genus of Fossil Fishes from the Lias.* Ann. Mag. Nat. Hist., ser. 5, XIII, 448–453.

JAMES WILLIAM DAVIS.

Lissolepis Davis, 449; orthotype L. SERRATUS Davis (fossil).

1418. DE VIS (1884). *Descriptions of New Genera and Species of Australian Fishes.* Proc. Linn. Soc., New S. Wales, for 1883 (1884), VIII, 283–289.

CHARLES W. DE VIS.

Dactylophora De Vis, 284; orthotype D. SEMIMACULATA De Vis. Same as PSILOCRANIUM Macleay of the same year.

Leme De Vis, 286; orthotype L. MORDAX De Vis.

1419. DE VIS (1884). *Fishes from South Sea Islands.* Proc. Linn. Soc., New S. Wales, VIII, 445–457.

CHARLES W. DE VIS.

Harpage De Vis, 447; orthotype H. ROSEA De Vis.

Nesiotes De Vis, 453; orthotype N. PURPURASCENS De Vis. A synonym of PSEUDOCHROMIS Rüppell.

Trachycephalus De Vis, 455; orthotype T. BANKANIENSIS De Vis. Name preoccupied; a synonym of AMPHIPRIONICHTHYS Blkr., 1855.

1420. FACCIOLÀ (1884). *Note sui pesci dello stretto di Messina.* Naturalista Sicili., III, 111–114.

LUIGI FACCIOLÀ.

Hypsirhynchus Facciolà, 112; orthotype H. HEPATICUS Facciolà.

1421–1422. FILHOL (1884). *Explorations sous-marines: voyage du Talisman.* La Nature, 1884, 182–186.

HENRI FILHOL (1843——).

Neostoma Filhol, 184; orthotype N. BATHYPHILUM Filhol.

Eustomias Filhol, 185; orthotype E. OBSCURUS Filhol.

1423. GARMAN (1884). *New Sharks: Chlamydoselachus anguineus and Heptranchias pectorosus.* Bull. Essex Instit., 1884, XVI, 3–15.

SAMUEL GARMAN.

Chlamydoselachus Garman, 8; orthotype C. ANGUINEUS Garman.

1424. GILBERT (1884). *A List of Fishes Collected in the East Fork of the White River, Indiana, with Descriptions of Two New Species.* Proc. U. S. Nat. Mus., VII, 199–205.

CHARLES HENRY GILBERT.

Hypargyrus (Forbes) Gilbert, 200; orthotype HYBOPSIS TUDITANUS Cope. A synonym of CERATICHTHYS Baird (CLIOLA Grd.).

Serraria Gilbert, 205; orthotype HADROPTERUS SCIERUS Swain.

1425. GILL (1884). *On the Anacanthine fishes.* Proc. Acad. Nat. Sci., Phila., 1884.

THEODORE GILL.

Lycenchelys Gill, 180; orthotype LYCODES MURÆNA Collett.

Lycocara Gill, 180; orthotype OPHIDIUM PARRII Ross. A substitute for URONECTES Gthr., preoccupied.

1426. GILL (1884). *Synopsis of the genera of the superfamily Teuthidoidea* (families Teuthididæ and Siganidæ). Proc. U. S. Nat. Mus., VII, 1884.

THEODORE GILL.

Ctenochætus Gill, 279; orthotype ACANTHURUS STRIGOSUS Bennett. Replaces CTENODON (Bon.) Klunzinger, preoccupied.

Colocopus Gill, 279; orthotype C. LAMBDURUS Gill = ACANTHURUS HEPATUS Bl. & Schn., not TEUTHIS HEPATUS L.

1427. GILL (1884). *Synopsis of the Plectognath Fishes.* Proc. U. S. Nat. Mus., VII.

THEODORE GILL.

Masturus Gill, 425; orthotype ORTHAGORISCUS OXYUROPTERUS Blkr.

1428. GILL (1884). *(Derichthys,* etc.) American Naturalist, XVIII.

THEODORE GILL.

Derichthys Gill, 433; orthotype D. SERPENTINUS Gill.

Acanthochænus Gill, 433; orthotype A. LUTKENI Gill. Perhaps a synonym of STEPHANOBERYX Gill.

1429. GÜRICH (1884). *(Eupleurodus)* Zeitschr. Deutsch. Geol. Ges., XXXVI.

GEORG GÜRICH (1859–   ).

Eupleurodus Gürich, 142; orthotype E. SULCATUS Gürich (fossil) = AMPHALODUS CHORZOWIENSIS Meyen. A synonym of ASTERODON Münster.

1430. JORDAN & SWAIN (1884). *A Review of the American Species of Marine Mugilidæ.* Proc. U. S. Nat. Mus., VII, 261–275.

DAVID STARR JORDAN; JOSEPH SWAIN (1857–   ).

Liza Jordan & Swain, 261; orthotype MUGIL CAPITO Cuv.

1431. JORDAN & SWAIN (1884). *A Review of the Species of the Genus Hæmulon.* Proc. U. S. Nat. Mus., VII.

DAVID STARR JORDAN; JOSEPH SWAIN

Lythrulon Jordan & Swain, 284; orthotype HÆMULON FLAVIGUTTATUS Gill.

1432. KLUNZINGER (1884). *Die Fische des Rothen Meeres. Eine kritische Revision mit Bestimmungs-Tabellen. I. Theil. Acanthopteri veri Owen,* Stuttgart, 133.

CARL BENJAMIN KLUNZINGER.

Hyposerranus Klunzinger, 3; logotype SERRANUS MORRHUA Cuv. & Val. A substitute for EPINEPHELUS Bloch. because not originally clearly defined. A synonym of EPINEPHELUS Bloch, as restricted. "Bleeker nent die SERRANUS jetzt EPINEPHELUS. Diese Sucht, alte Namen hervorzusuchen, halte ich für verfehlt, wenigstens in fällen wie hier . . . Bloch hat allerdings verschiedene Arten dieser Gattung unter EPINEPHELUS aufgeführt, die Gattungsdiagnose ist aber nichtssagend" (Klunzinger). But relatively few original diagnoses are adequate until amended, and to give a new name with each new definition would lead to endless confusion. In fact, much of the chaos which rules are intended to remove is due to the method advocated by Klunzinger and justified by rule 12 of the British Association (1865, 1869), now superseded, whereby a name not fully defined should be changed if its meaning is not clearly expressed at first.

Anisochætodon Klunzinger, 54; logotype CHÆTODON AURIGA. A substitute for LINOPHORA Kaup and OXYCHÆTODON Blkr., the definition being changed. A synonym of LINOPHORA Kaup.

Ctenodon (Bonaparte) Klunzinger, 85; orthotype ACANTHURUS STRIGOSUS Bennett. Name preoccupied; replaced by CTENOCHÆTUS Gill, 1884.

Hypocaranx Klunzinger, 92; logotype SCOMBER SPECIOSUS Bloch. A synonym of GNATHANODON Blkr. Under the new name of HYPOCARANX, Klunzinger includes Bleeker's genera, HEMICARANX, GNATHANODON, SELAROIDES, and URASPIS. His conception of a genus is that of a compartment of species, its name to be changed whenever the walls are altered. But in modern nomenclature a genus is a group of species clustered around a type which may be arbitrarily chosen. The oldest type-name included (unless preoccupied) gives its name to the whole. HYPOCARANX is inadmissible, being merely a synonym of GNATHANODON Blkr., its first species mentioned (SPECIOSUS) being the type of GNATHANODON. The same remarks apply to PSEUDOSERRANUS Klunzinger (1870), a substitute for the earlier names SERRANUS and VARIOLA.

1433. LANKESTER (1884). *Report on Fragments of Fossil Fishes from the Paleozoic Strata of Spitzbergen.* Svensk. Vet. Akad. Handl., 1884, XX, 1–7.

EDWIN RAY LANKESTER.

Lophostracon Lankester, 5; orthotype, L. SPITZBERGENSE Lankester (fossil).

1434. MACLEAY (1884) *On a New Genus of Fishes from Port Jackson.* Proc. Linn. Soc., New S. Wales, VIII, 439–443.

WILLIAM MACLEAY.

Psilocranium Macleay, 439; orthotype P. COXII Macleay. Same as DACTYLOPHORA De Vis, of the same year. McCulloch gives the latter name precedence.

1435. SAUVAGE (1884). *Contribution a la faune ichthyologique du Tonkin.* Bull. Soc. Zool. France, 1884, VII, ser. 7, 150–155.

HENRI ÉMILE SAUVAGE.

Gymnognathus Sauvage, 214; orthotype G. HARMANDI Sauvage. Name preoccupied; gives place to ASPIDOLUCIUS Berg.

1436. SAUVAGE (1884). *Sur un siluroide de l'île de la Réunion.* Bull. Soc. Philom., Paris, 1884, VII, sér. 8, 147–148.

HENRI ÉMILE SAUVAGE.

Laimumena Sauvage, 147; orthotype L. BORBONICA Sauvage.

1437. STEINDACHNER (1884). *Ichthyologische Beiträge,* XIII. Sizgb. Akad. Wiss. Wien, 1883 (1884), 1 Abth., LXXXVIII, 1065–1114.

FRANZ STEINDACHNER.

Parapercis Steindachner, 1071; orthotype P. RAMSAYI Steind. Name preoccupied; replaced by NEOPERCIS Steind.

Peronedys Steindachner, 1083; orthotype P. ANGUILLARIS Steind.

Neopatæcus Steindachner, 1087; orthotype PATÆCUS MACULATUS Gthr.

Austrolabrus Steindachner, 1100; orthotype LABRICHTHYS MACULATA Macleay.

1438. THOMINOT (1884). *Note sur un poisson de la famille des Cyprinodontidæ.* Bull. Soc. Philom., Paris, VII, sér. 8, 149–158.

ALEXANDRE THOMINOT.

Rhodeoides Thominot, 150; orthotype R. VAILLANTI Thominot.

1439. TRAQUAIR (1884). *Notice of New Fish remains from the Blackband Ironstone of Borough Lee, near Edinburgh,* No. 5. Geol. Mag., 1884, III, dec. 1, 64–65.

RAMSAY HEATLEY TRAQUAIR.

Aganacanthus Traquair, 64; orthotype A. STRIATUS Traquair (fossil).

1885.

1440. BEAN (1885). *On Stathmonotus, a New Genus of Fishes Related to Murænoides, from Florida.* Proc. U. S. Nat. Mus., 1885, VIII, 191–192.

TARLETON HOFFMAN BEAN.

Stathmonotus Bean, 191; orthotype S. HEMPHILLI Bean.

1441. CLAYPOLE (1885). *On the Recent Discovery of Pteraspidian Fish in the Upper Silurian Rocks of North America.* Quart. Journ. Geol. Soc., XLI.

EDWARD WALLER CLAYPOLE.

Palæaspis Claypole, 62; orthotype P. AMERICANA Claypole (fossil) : (detailed account).

1442. COCCO (1885). *Indice ittiologico del mare di Messina.* Naturalista Siciliana, V.

ANASTASIO COCCO.

Symproptopterus Cocco, 294; orthotype not stated by Hilgendorf. A Paralepis, with dorsal and anal joined to caudal; perhaps a monstrosity; mentioned by Bonaparte, 1846.

1443. COPE (1885). *Eocene Paddle-Fish and Gonorhynchidæ.* Amer. Nat., XIX, 1090–91.

EDWARD DRINKER COPE.

Notogoneus Cope, 1091; orthotype N. OSCULUS Cope (fossil).

1444. DAVIS (1885). *The Fossil Fishes of the Chalk of Mount Lebanon.* Royal Dublin Society, 1885, ser. 2., III, 457–636.

JAMES WILLIAM DAVIS.

Centrophoroides Davis, 478; orthotype C. LATIDENS Davis (fossil).

Rhinognathus Davis, 480; orthotype R. LEWISI Davis. Name preoccupied, replaced by SCAPANORHYNCHUS Woodw., 1889.

Spathiurus Davis, 503; orthotype S. DORSALIS Davis (fossil).

Amphilaphurus Davis, 504; orthotype A. MAJOR Davis (fossil) = SPATHIURUS DORSALIS Davis. A synonym of SPATHIURUS.

Xenopholis Davis, 548; orthotype X. CARINATUS Davis (fossil).

Exoccetoides Davis, 551; orthotype E. MINOR Davis (fossil).

Lewisia Davis, 593; orthotype L. OVALIS Davis = CLUPEA LATA Ag. (fossil). A synonym of SPANIODON Pictet.

Pantopholis Davis, 599; orthotype P. DORSALIS Davis (fossil).

Eurygnathus Davis, 601; logotype E. MAJOR Davis (fossil). A synonym of ENCHODUS Ag.

Phylactocephalus Davis, 605; orthotype P. MICROLEPIS Davis (fossil). A synonym of HALEC Ag.

1445. DE VIS (1885). *New Australian Fishes in the Queensland Museum.* Proc. Linn. Soc. New S. Wales, IX, 389–400, 537–547, etc.

CHARLES W. DE VIS.

Herops De Vis, 392; orthotype H. MUNDA De Vis. A synonym of KUHLIA Gill.

Homodemus De Vis, 395; orthotype H. CAVIFRONS De Vis. A synonym of OLIGORUS Gthr.

Autisthes De Vis, 398; orthotype A. ARGENTEUS De Vis. A synonym of THERAPON Cuv.

Hephæstus De Vis, 399; orthotype H. TULLIENSIS De Vis, THERAPON FULIGINOSUS Macleay. A synonym of THERAPON Cuv.

Neoniphon De Vis, 537; orthotype N. HASTA De Vis. Preoccupied by Castelnau, 1875.

1446. DE VIS (1885). *New Fishes in the Queensland Museum.* Proc. Linn. Soc. New S. Wales, 1885, IX, 869–887.

CHARLES W. DE VIS.

Onar De Vis, 875; orthotype O. NEBULOSUM De Vis. A synonym of PSEUDOCHROMIS Rüppell.

Torresia De Vis, 881; orthotype T. LINEATA De Vis.

Julichthys De Vis, 884; orthotype J. INORNATA De Vis.

1447. FACCIOLÀ (1885). *Su di alcuni rari Pleuronettidi del Mar di Messina. Nota preliminare.* Nat. Sicil., IV, 261–266.

LUIGI FACCIOLÀ.

Charybdia Facciolà, 265; orthotpye PELORIA RUPPELII Cocco. A larval flounder, perhaps young of PLATOPHRYS Sw.

1448. FISCHER (1885). *Ichthyologische und herpetologische Bemerkungen.* Jahrb. Hamburg. Wiss. Anstalten, 1885, II, 49–119.

JOHANN GUSTAV FISCHER (1819–1889).

Sclerocottus Fischer, 58; orthotype S. SCHRADERI Fischer.

Gymnelichthys Fischer, 61; orthotype G. ANTARCTICUS Fischer.

1449. FORBES (1885). *Description of New Illinois Fishes.* Bull. Illinois Lab. Nat. Hist., II.

STEPHEN ALFRED FORBES (1848–).

Oxygeneum Forbes, 136; orthotype O. PULVERULENTUM Forbes.

1450. GARMAN (1885). *The Generic Name of the Pastinacas or Sting Rays.*
Proc. U. S. Nat. Mus., VIII.

SAMUEL GARMAN.

Dasybatus (Klein) Garman, 221; orthotype RAJA PASTINACA L. A revival of a
non-binomial name of Klein as reprinted by Walbaum in place of DASYATIS
Raf. and TRYGON (Adanson) Geoffroy.

1451. GARMAN (1885). *Chlamydoselachus anguineus Garm., a Living Representative of Cladodont Sharks.* Bull. Mus. Comp. Zool. Harvard College, XII.

SAMUEL GARMAN.

Pternodus Garman, 6; orthotype PHŒBODUS SPRINGERI St. John & Worthen (fossil).
Name preoccupied by PTERNODUS Owen, 1867; probably a synonym of
PHŒBODUS.

Dicranodus Garman, 30; orthotype DIDYMODUS TEXENSIS Cope (fossil). Substitute
for DIDYMODUS Cope, regarded as preoccupied by DIDYMODON Blake, 1863, a
mammal; perhaps same as DITTODUS Owen and ORTHACANTHUS Ag.

1452. GILL (1885). *Riverside Natural History.* Edited by John Sterling Kingsley.
III. *Lower Vertebrates,* Chapter ACANTHOPTERYGIANS.

THEODORE GILL.

Phrynotitan Gill, 255; orthotype BATRACHUS GIGAS, the author not stated, the name
not found by me.

1453. GOODE & BEAN (1885). *On American Fishes in the Linnæan Collection.*
Proc. U. S. Nat. Mus., VIII.

GEORGE BROWN GOODE; TARLETON HOFFMAN BEAN.

Borborys (Broussonet Ms.) Goode & Bean. 205; orthotype COBITIS HETEROCLITUS L.
A synonym of FUNDULUS Lac.

1454. GOODE & BEAN (1885). *Descriptions of New Fishes Obtained by the
United States Fish Commission, Mainly from Deep Water Off the Atlantic
and Gulf Coasts.* Proc. U. S. Nat. Mus., VIII, 589–605.

GEORGE BROWN GOODE; TARLETON HOFFMAN BEAN.

Neobythites Goode & Bean, 601; orthotype N. GILLI G. & B.
Porogadus Goode & Bean, 602; orthotype P. MILES G. & B.
Bathyonus Goode & Bean, 603; logotype B. CATENA G. & B.

1455. GOODE & BEAN (1885). *Description of a New Genus and Species of Pediculate Fishes (Halieutella lappa).* Proc. Biol. Soc. Wash., II, 88.

GEORGE BROWN GOODE; TARLETON HOFFMAN BEAN.

Halieutella Goode & Bean, 88; orthotype H. LAPPA G. & B.

1456. JORDAN (1885). *Identification of the Species of Cyprinidæ and Catostomidæ Described by Dr. Charles Girard, etc.* (1856). Proc. U. S. Nat. Mus., VIII.

DAVID STARR JORDAN.

Luxilinus Jordan, 126; orthotype LUXILUS OCCIDENTALIS Baird & Girard = young of
LAVINIA EXILICAUDA Baird & Grd. A synonym of LAVINIA Baird & Grd.

1457. JORDAN (1885). *A Catalogue of the Fishes Known to Inhabit the Waters
of North America North of the Tropic of Cancer.* Rept. U. S. Fish Comm.
for 1885 (1887). (Reprint issued 1885.)

DAVID STARR JORDAN.

Crystallaria Jordan & Gilbert, 78; orthotype PLEUROLEPIS ASPRELLUS Jordan.
Rhothœca Jordan, 80; orthotype PŒCILICHTHYS ZONALIS Cope. Substitute for
NANOSTOMA Putnam, regarded as preoccupied by NANNOSTOMUS Gthr.

Artediellus Jordan, 110; orthotype Cottus uncinatus Reinh.

Icelinus Jordan, 110; orthotype Artedius quadriseriatus Lockington.

Otophidium (Gill MS.) Jordan, 126; orthotype Genypterus omostigma J. & G.

Pleurogadus (Bean) Jordan, 130; orthotype Gadus navaga Kölr. A substitute for Tilesia, preoccupied. A synonym of Eleginus Fischer, 1813, not of Cuvier & Valenciennes, 1830.

Aramaca (Jordan & Goss) Jordan, 133; orthotype Citharichthys pætulus J. & G. = Pleuronectes aramaca Dönndorf = Pleuronectes papillosus L. A synonym of Syacium Ranz.

Eopsetta Jordan & Goss, 135; orthotype Hippoglossoides jordani Lockington.

Lyopsetta Jordan & Goss, 135; orthotype Hippoglossoides exilis J. & G.

Inopsetta Jordan and Goss, 136; orthotype Parophrys ischyrus J. & G.

1458. JORDAN & MEEK (1885). *A Review of the American Species of Flying-fish.* (Exocœtus.) Proc. U. S. Nat. Mus., VIII.

David Starr-Jordan; Seth Eugene Meek (1859–1914).

Fodiator Jordan & Meek, 45; orthotype Exocœtus acutus Cuv. & Val.

1459. KRAMBERGER (1885). *Palæontologische beiträge.* Soc. Hist. Nat. Croatia Rad. Jugoslav. Akad., LXXII.

Dragutin Gorjanovic Kramberger.

Holcodon Kramberger, 19; orthotype Solenodon neocomiensis Kramberger = Saurocephalus lycodon Kner (fossil). Name a substitute for Solenodon Kramberger, preoccupied; thought to be a synonym of Enchodus.

Hypsospondylus Kramberger, 31; orthotype H. bassanii Kramberger (fossil).

Mesiteia Kramberger, 54; orthotype M. emiliæ Kramberger.

1460. KUNISCH (1885). *Dactylolepis gogoliensis.* Zool. Geol. Ges., XXXVII, 588–594.

Hermann Kunisch (1853–1893).

Dactylolepis Kunisch, 588; orthotype D. gogoliensis Kunisch (fossil). A synonym of Cololobus Ag.

1461. LAUBE (1885). *Ein Beiträge zur Kenntniss der Fische des böhmischen Turons.* Denk. Akad. Wiss. Wien., 1. Abth., 285–298.

Gustav Carl Laube (1839–   ).

Protelops Laube, 286; orthotype P. geinitzii Laube (fossil).

1462. MARCK (1885). *Fische der oberen Kreide Westfalens.* Paleontographica. 1885, XXXI, 233–267.

W. von der Marck.

Charitosomus Marck, 257; orthotype C. formosus Hosius & von der Marck (fossil).

1463. MOCQUARD (1885). *Sur un nouveau genre de Blenniidæ voisin des Clinus.* Bull. Soc. Philom. Paris, 1885, 18–20.

M. Mocquard.

Acanthoclinus Mocquard, 18; orthotype A. chaperi Mocquard. Name preoccupied; replaced by Paraclinus Mocquard (1886).

1464. NEWBERRY (1885). *Description of Some Gigantic Placoderm Fishes Recently Discovered in the Devonian of Ohio.* Trans. N. Y. Acad. Sci., 1885, V, 25–28.

John Strong Newberry.

Titanichthys Newberry, 27; orthotype T. agassizi Newberry.

1465. NEWBERRY (1885). *Description of Some Peculiar Screw-like Fossils from the Chemung Rocks.* Ann. N. Y. Acad. Sci., XVIII, 217–220.

JOHN STRONG NEWBERRY.

Spiraxis Newberry, 219; orthotype S. MAJOR Newberry. Name preoccupied; replaced by PROTOSPIRAXIS Williams.

1466. NOETLING (1885). *Die Fauna des sammländischen Tertiäres. (Pisces.)* Abh. Geol. Spec. Preuss., 1885, VI, 3–106.

FRITZ NOETLING (1857–   ).

Pseudosphærodon Noetling, 104; orthotype P. HILGENDORFI Noetling (fossil).

1467. OGILBY (1885). *Notes and Descriptions of Some Rare Port Jackson Fishes.* Proc. Linn. Soc. New S. Wales, X.

JAMES DOUGLAS OGILBY (1848–   ).

Petraites Ogilby, 225; orthotype P. HEPTACOLUS Ogilby.

1468. ROCHEBRUNE (1885). *Poissons de la Rivière Casamence, Afrique occidentale.* Bull. Sci. Philom. Paris, IX.

ALPHONSE TRÉMEAU DE ROCHEBRUNE.

Gyrinostomus Rochebrune, 96; orthotype G. MARCHEI Rochebrune.

1469. STEINDACHNER & DÖDERLEIN (1885). *Beiträge zur Kenntniss der Fische Japans.* (3) Denkshr. Akad. Wiss. Wien, XLIX.

FRANZ STEINDACHNER; LUDWIG DÖDERLEIN.

Neopercis Steindachner, 212; orthotype PERCIS RAMSAYI Steind.   Substitute for PARAPERCIS Steind., preoccupied.

1886.

1470. BASSANI (1886). *Sui fossili e sull' età degli scisti bituminosi triasici di Besano in Lombardia.* Atté. Soc. Ital. Sci. Nat., XXIX.

FRANCESCO BASSANI.

Nothosomus Bassani, 37; orthotype N. BELLOTTI Bassani (fossil).

1471. COLLETT (1886). *On a New Pediculate Fish from the Sea off Madeira (Linophryne lucifer).* Proc. Zool. Soc. London, 138–142.

ROBERT COLLETT.

Linophryne Collett, 138; orthotype L. LUCIFER Collett.

1472. COPE (1886). *Fossil Fishes from Brazil.* Proc. Amer. Phil. Soc., XXIII.

EDWARD DRINKER COPE.

Apocopodon Cope, 2; orthotype A. SERICEUS Cope (fossil).

Chiromystus Cope, 4; orthotype C. MAWSONI Cope (fossil).

1473. COPE (1886). *On Two New Forms of Polyodont and Gonorhynchid Fishes from the Rocky Mountains.* Mem. Nat. Acad. Sci., Washington, III.

EDWARD DRINKER COPE.

Crossophilus Cope, 162; orthotype C. MAGNICAUDATUS Cope.

1474. COPE (1886). *An Interesting Connecting Genus of Chordata.* Amer. Nat., XX.

EDWARD DRINKER COPE.

Mycterops Cope, 1027; orthotype M. ORDINATUS Cope. (Said by Woodward to be an EURYPTERID CRUSTACEAN.)

1475. FRITSCH (1886). *Ergebnisse der Vergleichungen an den elektrischen Organen der Torpedinen.* Arch. Anat. Physiol., 358–370.

GUSTAV THEODOR FRITSCH.

Gymnotorpedo Fritsch, 365; logotype TORPEDO OCCIDENTALIS Storer.

Fimbriotorpedo Fritsch, 365; logotype TORPEDO MARMORATA Risso = RAJA TORPEDO L. A synonym of TORPEDO Duméril, and of authors.

1476. GOODE & BEAN (1886). *Reports on the Results of Dredging, under the supervision of Alexander Agassiz, in the Gulf of Mexico (1877–78) and in the Caribbean Sea (1879–80), by the U. S. Coast Survey steamer "Blake,"* etc. Bull. Mus. Comp. Zool., Harvard Coll., 1886, XII, no. 5, 153–170.

GEORGE BROWN GOODE; TARLETON HOFFMAN BEAN.

Barathronus Goode & Bean, 164; orthotype, B. BICOLOR G. & B.

Benthosaurus Goode & Bean, 168; orthotype B. GRALLATOR G. & B.

1477. JORDAN (1886). *List of Fishes Collected at Havana, Cuba, in December, 1883, with notes and descriptions.* Proc. U. S. Nat. Mus. IX.

DAVID STARR JORDAN.

Scartella Jordan, 50; orthotype BLENNIUS MICROSTOMUS Poey.

1478. JORDAN & EVERMANN (1886). *Description of Six New Species of Fishes from the Gulf of Mexico, with Notes on Other Species.* Proc. U. S. Nat. Mus., 1886, IX, 466–476. (September 17.)

DAVID STARR JORDAN; BARTON WARREN EVERMANN.

Steinegeria Jordan & Evermann, 467; orthotype S. RUBESCENS J. & E.

1479. JORDAN (1886). *Notes on Typical Specimens of Fishes Described by Cuvier and Valenciennes and preserved in the Musée d'Histoire Naturelle in Paris.* Proc. U. S. Nat. Mus., 1886, IX, 525–546.

DAVID STARR JORDAN.

Marcgravia Jordan, 546; orthotype BATRACHUS CRYPTOCENTRUS Cuv. & Val. A synonym of AMPHICHTHYS Sw., the name later changed to MARCGRAVICHTHYS by Ribeiro, because MARCGRAVIA is used in Botany.

1480. JORDAN (1886). *A Preliminary Catalogue of the Fishes of the West Indies.* Proc. U. S. Nat. Mus., IX.

DAVID STARR JORDAN.

Amiichthys (Poey), Jordan, 586; orthotype ———— DIAPTERUS Poey (1861). Only the specific name of this fish was published by Poey.

1481. JORDAN & FORDICE (1886). *A Review of the American Species of Belonidæ.* Proc. U. S. Nat. Mus., IX.

DAVID STARR JORDAN; MORTON W. FORDICE.

Athlennes Jordan & Fordice, 342; orthotype BELONE HIANS Cuv. & Val. See ABLENNES, of which word ATHLENNES was an accidental slip in writing.

Ablennes Jordan & Fordice, 342; orthotype BELONE HIANS Cuv. & Val. Misprinted ATHLENNES, an orthography accepted by Jordan & Evermann, but altered to the correct Greek form from ἀβλέννης (sans mucosité, an old name for the needlefish), by a decision of the International Commission on Nomenclature.

1482. LARRAZET (1886). *Des Pièces de la peau de quelques sélaciens fossiles.* Bull. Soc. Geol. France, 1886, III, ser. 14, 255–277.

AUGUSTIN LARRAZET (1855–   ).

Dynatobatis Larrazet, 258; logotype D. PARANENSIS Larrazet (fossil).
Acanthobatis Larrazet, 258; orthotype A. EXIMIA Larrazet (fossil).

1483. MACLEAY (1886). *A Remarkable Fish from Lord Howe Island.* Proc. Linn. Soc. New South Wales, 1886, X, 718–720.

WILLIAM MACLEAY.

Ctenodax Macleay, 718; orthotype C. WILKINSONI Macleay. A synonym of TETRA-GONURUS Risso.

1484. MIKLUKHO-MACLAY & MACLEAY (1886). *Plagiostomata of the Pacific.* III. Proc. Linn. Soc. N. South Wales, 1886, X, 674–684.

NICOLAI NICOLAEVICH MIKLÚKHO-MACLAY; WILLIAM MACLEAY.

Discobatis Miklukho-Maclay & Macleay, 676; orthotype D. MARGINIPINNIS Miklukho-Maclay & Macleay = RAJA LYMMA Forskål. A synonym of TÆNIURA M. & H.; not DISCOBATUS Garman.

1485. ROMANOVSKY (1886). *(Lyrolepis.)* Geol. Soc. St. Petersburg, XXII.

GHENNADI DANILOVICH ROMANOVSKY.

Lyrolepis Romanovsky, 304; orthotype L. CAUCASICUS Romanovsky (fossil).

1486. STORMS (1886). *Note sur un nouveau genre de poisson fossile de l'argile rupélienne.* Ann. Soc. Geol. Belgique. XIII. 261–266.

RAYMOND STORMS.

Amphodon Storms, 265; orthotype A. BENEDENI Storms (fossil). Regarded as pre-occupied by AMPHODUS Peters, 1872; replaced by SCOMBRAMPHODON Woodw., 1901.

1487. SZAJNOCHA (1886). *Ueber fossile Fische von Monte Bolca bei Verona.* Pam. Akad. Krakau, XII, (text in Polish).

WLADYSLAW SZAJNOCHA (1857–   ).

Hacquetia Szajnocha, 3; orthotype H. BOLCENSIS Szajnocha (fossil).

1488. TRAQUAIR (1886). *On Harpacanthus, a New Genus of Carboniferous selachian spines.* Ann. Mag. Nat. Hist., 1886, 5 ser., XVIII, 493–496.

RAMSAY HEATLEY TRAQUAIR.

Harpacanthus Traquair, 493; orthotype TRISTYCHIUS FIMBRIATUS Stock. (fossil).

1489. VAILLANT (1886). *Considerations sur les poissons des grandes profondeurs et en particulier sur . . . les Abdominales.* Comptes Rendus, Ac. Sci. Paris, CIII.

LÉON LOUIS VAILLANT.

Anomalopterus Vaillant, 1239; orthotype A. PINGUIS Vaillant.
Leptoderma Vaillant, 1239; orthotype L. MACROPS Vaillant.

1490. WETTSTEIN (1886). *Ueber die Fischfauna des tertiaren Glarnerschiefers.* Abh. Schweiz. Palæont. Ges., XIII, 1–103.

ALEXANDER WETTSTEIN (1861–1887).

Scopeloides Wettstein, 55; orthotype OSMERUS GLARISIANUS Ag. (fossil).
Archæteuthis Wettstein, 67; orthotype A. GLARONESIS Wettstein (fossil).
Cyttoides Wettstein, 91; orthotype C. GLARONESIS Wettstein (fossil).

1491. YATES (1886). *Catalogue of Fossils in Lorenzo G. Yates Collection, Santa Barbara, Cal.*

LORENZO G. YATES.

Platychodus Yates, 20; orthotype PTYCHODUS MORTONI Mantell (fossil). A synonym of PTYCHODUS Ag. (perhaps a slip of the pen for the latter name).

1887.

1492. BEAN (1887). *Description of a New Genus and Species of Fish, Acrotus willoughbyi from Washington Territory.* Proc. U. S. Nat. Mus., X, 631–632.

TARLETON HOFFMAN BEAN.

Acrotus Bean, 631; orthotype A. WILLOUGHBYI Bean.

1493. BOULENGER (1887). *An Account of Fishes Collected by Mr. C. Buckley in eastern Ecuador.* Proc. Zool. Soc. London, 1887, II, 274–283.

GEORGE ALBERT BOULENGER (1858–   ).

Nannoglanis Boulenger, 278; orthotype N. FASCIATUS Boulenger.

Leptogoniates Boulenger, 281; orthotype L. STEINDACHNERI Boulenger.

1494. DAMES (1887). *Ueber Titanichthys pharao . . . aus der Kreideformation Aegyptens.* Sitzber. Ges. Nat. Freunde Berlin, 69–72.

WILHELM BARNIM DAMES (1843–1898).

Titanichthys Dames, 69; orthotype T. PHARAO Dames. Name preoccupied; replaced by GIGANTICHTHYS Dames, p. 137, 1888.

1495. FORIR (1887.) *Contributions a l'étude du système Cretacé de la Belgique.* Ann. Mem. Soc. Geol. Belgique, XIV, 25–56.

H. FORIR.

Anomœodus Forir, 25; orthotype PYCNODUS SUBCLAVATUS Ag. (fossil).

1496. GÜNTHER (1887). *Report on the Deep-Sea Fishes Collected by H. M. S. "Challenger" during the years 1873–76. In Report on the Scientific Results of the Voyage of H. M. S. "Challenger" during the years 1873–76,* XXII, pt. 57, 1–268.

ALBERT GÜNTHER.

Synagrops Günther, 16; orthotype MELANOSTOMA JAPONICUM Döderlein. Substitute for MELANOSTOMA Döderlein, preoccupied.

Malacosarcus Günther, 30; orthotype M. MACROSTOMUS Gthr.

Diceratias Günther, 53; orthotype CERATIAS BISPINOSUS Gthr.

Lyocetus Günther, 57; orthotype L. MURRAYI Gthr.

Salilota Günther, 95; orthotype HALOPORPHYRUS AUSTRALIS Gthr.

Catætyx Günther, 104; orthotype SIREMBO MESSIERI Gthr.

Pteroidonus Günther, 106; orthotype P. QUINQUARIUS Gthr.

Mixonus Günther, 108; orthotype BATHYNECTES LATICEPS Gthr.

Nematonus Günther, 114; orthotype BATHYONUS PECTORALIS G. & B.

Diplacanthopoma Günther, 115; orthotype D. BRACHYSOMA Gthr.

Nematonurus Günther, 124; orthotype N. ARMATUS (Hector).

Cetonurus Günther, 124; orthotype C. CRASSICEPS Gthr.

Trachonurus Günther, 124; orthotype MACRURUS VILLOSUS Gthr.

Lionurus Günther, 124; orthotype CORYPHÆNOIDES FILICAUDA Gthr.

Mystaconurus Günther, 124; orthotype M. LONGIBARBIS Gthr.; a synonym of HYMENOCEPHALUS Giglioli.

Optonurus Günther, 147; orthotype O. DENTICULATUS Günther.

Lyconus Günther, 158; orthotype L. PINNATUS Gthr.

Polyipnus Günther, 170; orthotype P. spinosus Gthr.
Nannobrachium Günther, 199; orthotype N. nigrum Gthr.
Omosudis Günther, 201; orthotype O. lowi Gthr.
Opostomias Günther, 208; orthotype Echiostoma macripnus Gthr.
Pachystomias Günther, 210; orthotype Echiostoma microdon Gthr.
Photonectes Günther, 212; orthotype Lucifer albipennis Döderlein.   Substitute
for Lucifer Döderlein, preoccupied.

1497. GÜNTHER (1887). *Descriptions of Two New Species of Fishes from Mauritius,* etc. Proc. Zool. Soc. London.

ALBERT GÜNTHER.

Hoplolatilus Günther, 550; orthotype Latilus fronticinctus Gthr.

1498. HERZENSTEIN & VARPAKHOVSKII (1887). *Notizen ueber die Fischfauna des Amur-Beckens und der angrenzenden Gebiete* (Russian text). Trudui St. Petersb. Nat., 1887, XVIII, 1–58.

SALOMON MARKOVICH HERZENSTEIN (1854–1894) ;
NIKOLAI ARKADEVICH VARPAKHOVSKII (1862–   ).

Octonema Herzenstein, 47; orthotype O. pleskei Herzenstein.   Name preoccupied;
replaced by Lefua Herzenstein.

1499. MATTHEW (1887). *Preliminary Notice of a New Genus of Silurian Fishes (Diplaspis acadicus).* Bull. Nat. Hist. Soc. New Brunswick, VI.

GEORGE FREDERICK MATTHEW (1837–   ).

Diplaspis Matthew, 69; orthotype Pteraspis acadica Matthew (fossil.).   A synonym
of Cyathaspis Lankester.

1500. MORTON (1887). *(Eurumetopos.)*   Proc. Roy. Soc. Tasmania for 1887.
(1888.)

ALLPORT MORTON.

Eurumetopos Morton, 77; orthotype E. johnstoni Morton.   A synonym of Hyperoglyphe Gthr.

1501. OGILBY (1887). *On a New Genus of Percidæ.* Proc. Zool. Soc. London,
1887, 616–18.

JAMES DOUGLAS OGILBY.

Chthamalopteryx Ogilby, 616; orthotype C. melbournensis Ogilby.   A synonym
of Parequula Steind.

1502. OGILBY (1887). *On a New Genus and Species of Australian Mugilidæ.*
Proc. Zool. Soc. London, 614–616.

JAMES DOUGLAS OGILBY.

Trachystoma Ogilby, 614; orthotype T. multidens Ogilby.

1503. PHILIPPI (1887). *Historia natural. Sobre los tiburones y algunos otros peces de Chile.* Anal. Univ. Chile, 1887, LXXI, 1–42.

RUDOLPHO AMANDUS PHILIPPI.

Graus Philippi, 40; orthotype G. nigra Philippi.

1504. RAMSAY & OGILBY (1887). *On a New Genus and Species of Labroid Fish from Port Jackson.* Proc. Linn. Soc. New South Wales, 1887 (2), II,
631–634.

EDWARD PIERSON RAMSAY; JAMES DOUGLAS OGILBY.

Eupetrichthys Ramsay & Ogilby, 631; orthotype E. ANGUSTIPES Ramsay & Ogilby.

1505. RAMSAY & OGILBY (1887). *Notes on the Genera of Australian Fishes.* Proc. Linn. Soc. New South Wales, 1887, 2 ser., II, 181–184.

EDWARD PIERSON RAMSAY; JAMES DOUGLAS OGILBY.

Percalates Ramsay & Ogilby, 182; orthotype LATES COLONORUM Gthr.

1506. SCHLÜTER (1887). *Panzerfische, etc.* Niederrhein. Ges., Bonn, 1887, 120. (Also in Nat. Ver. Preuss. West., XLIV, 1888, 126.)

CLEMENS (AUGUST JOSEPH) SCHLÜTER (1836–   ).

Ceraspis Schlüter, 120; orthotype C. CARINATUS Schlüter (fossil). Name preoccupied, according to Hilgendorf.

Drepanaspis Schlüter, 120; orthotype D. GEMUENDENSIS Schlüter (fossil).

1507. STEINDACHNER & DÖDERLEIN (1887). *Beiträge zur Kenntniss der Fische Japans, IV.* Denksch. Kais. Akad. Wiss. Wien., 1887, LIII, 258–296.

FRANZ STEINDACHNER; LUDWIG DÖDERLEIN.

Myxocephalus Steindachner & Döderlein, 281; orthotype M. JAPONICUS Steindachner & Döderlein.

1508. STORMS (1887). *Notes sur les poissons fossiles du terrain rupélien.* Bull. Soc. Belg. Geol. Paléont., I, 98–112.

RAYMOND STORMS.

Platylates Storms, 98; orthotype P. RUPELIENSIS Storms (fossil).

1509. VARPAKHOVSKII (1887). *Ueber die Gattung Hemiculter Bleeker und uber eine neue Gattung, Hemiculterella.* Bull. Acad. Sci. St. Petersb., 14–23.

NIKOLAI ARKADEVICH VARPAKHOVSKII.

Hemiculterella Varpakhovskii, 23; orthotype H. SAUVAGEI Varpakhovskii.

1510. WILLIAMS (1887). *On the Fossil Faunas of the Upper Devonian, the Genesee Section, New York.* Bull. U. S. Geol. Surv., XLI, 1–104.

HENRY SHALER WILLIAMS (1847–1916).

Prospiraxis Williams, 86; orthotype SPIRAXIS MAJOR Newberry (fossil). Substitute for SPIRAXIS Newberry, preoccupied.

1511. ZIGNO (1887). *Due novi pesci fossili della famiglia dei Balistini, scoperto nel terreno Eoceno del Veronese.* Mem. Soc. Ital., VI.

ACHILLE DE ZIGNO (1813–1892).

Probalistum (Massalongo) Zigno, 3; orthotype OSTRACION IMPERIALIS Massalongo (fossil). A synonym of SPINACANTHUS Ag.

1512. ZITTEL (1887). *Handbuch der Paleontologie. I. Abtheilung. Palæozoologie, Band III. Vertebrata,* 1887–90, pp. 1–900.

KARL ALFRED VON ZITTEL (1839–1904).

Chalcodus Zittel, 72; orthotype C. PERMIANUS Zittel (fossil).

Metopacanthus Zittel, 110; orthotype ISCHYODUS ORTHORHINUS Egerton (fossil). A synonym of MYRIACANTHUS Ag.

Chimæropsis Zittel, 113; orthotype C. PARADOXA Zittel (fossil).

Isopholis Zittel, 216; logotype OPHIOPSIS MUNSTERI Ag. A synonym of EUGNATHUS Ag.; regarded as preoccupied. But FURO Gistel, 1848, is a still earlier substitute.

1888.

1513. BASSANI (1888). *Sopra un nuovo genere di fisostomi scoperto nell' Eocene medio del Friuli, in provincia di Udine.* Atti. Acad. Sci. Napoli, 1889, 2 ser., III, 1-4.

FRANCESCO BASSANI.

Omiodon Bassani, 2; orthotype O. CABASSI Bassani (fossil).

1514. DAMES (1888). *Amblypristis cheops, aus dem Eocän Aegyptens.* Sitzber. Ges. Naturf. Berlin, 1888, 106–109.

WILHELM BARNIM DAMES.

Amblypristis Dames, 106; orthotype A. CHEOPS Dames (fossil).

1515. DAMES (1888). *Die Ganoiden der deutschen Muschelkalk.* Paleont. Abh., IV, Heft 2, 131–180 (1–50 reprint).

WILHELM BARNIM DAMES.

Gigantichthys Dames, 137 (7 reprint); orthotype TITANICHTHYS PHARAO Dames (fossil). Substitute for TITANICHTHYS Dames, preoccupied.

Crenilepis Dames, 171 (40 reprint); orthotype C. SANDBERGERI Dames (fossil; a fragment of a scale from the Muschelkalk of Krainberg, Germany.)

1516. DAY (1888). *Supplement to the Fishes of India.*

FRANCIS DAY.

Acanthonotus (Tickell) Day, 807; orthotype A. ARGENTEUS Day. Name preoccupied, replaced by MATSYA Day, 1889.

1517. EIGENMANN & EIGENMANN (1888). *Preliminary Notes on South American Nematognathi.* Proc. Cal. Acad. Sci., 119–172.

CARL H. EIGENMANN (1863–); ROSA SMITH EIGENMANN (1860–).

Luciopimelodus Eigenmann & Eigenmann, 122; logotype PIMELODUS PATI Val.

Pimelodella Eigenmann & Eigenmann, 136; orthotype PIMELODUS CRISTATUS M. & T.

Duopalatinus Eigenmann & Eigenmann, 136; orthotype PLATYSTOMA EMARGINATUM Cuv. & Val.

Sciadeoides Eigenmann & Eigenmann, 136; orthotype SCIADES MARMORATUS Gill.

Steindachneria Eigenmann & Eigenmann, 137; orthotype PLATYSOMA PARAHYBÆ Steind. Name preoccupied; replaced by STEINDACHNERIDION Eigenmann.

Hassar Eigenmann & Eigenmann, 158; orthotype DORAS ORESTES Steindachner.

Decapogon Eigenmann & Eigenmann, 165; orthotype CALLICHTHYS ADSPERSUS Steindachner.

Neoplecostomus Eigenmann & Eigenmann, 171; orthotype PLECOSTOMUS MICROPS Steindachner.

Rhamdella Eigenmann & Eigenmann, 172; logotype R. ERIARCHA Eigenmann & Eigenmann.

1518. EIGENMANN & EIGENMANN (1888). *A List of American Species of Gobiidæ and Callionymidæ, with notes on the Specimens Contained in the Museum of Comparative Zoology, at Cambridge, Massachusetts.* Proc. Cal. Acad. Sci., 1888, 2 ser., 51–78.

CARL H. EIGENMANN; ROSA SMITH EIGENMANN.

Barbulifer Eigenmann & Eigenmann, 70; orthotype B. PAPILLOSUS Eigenmann & Eigenmann = GOBIOSOMA CEUTHŒCUM J. & G.

Clevelandia Eigenmann & Eigenmann, 73; orthotype GOBIOSOMA LONGIPINNE Steind.

1519. FRITSCH (1888). *Fauna der Gaskohle und der Kalksteine der Permforma-tion Böhmens, II.*

ANTON JAN FRITSCH.

Dipnoites Fritsch, 86; orthotype D. PERNERI Fritsch (fossil). A cranial bone.

1520. FACCIOLÀ (1888). *Annunzio ittiologico.* Natur. Sicil., VII.

LUIGI FACCIOLÀ.

Sympodoichthys Facciolà, 167; orthotype S. FASCIATUS Facciolà.

1521. GARMAN (1888). *On an Eel from the Marshall Islands.* Bull. Essex In-stitute, 1889, XX, 114–116.

SAMUEL GARMAN.

Rhinomuræna Garman, 114; orthotype R. QUÆSITA Garman.

1522. GILL (1888). *On the Classification of the Mail-Checked Fishes.* Proc. U. S. Nat. Mus., XI.

THEODORE GILL.

Histiocottus Gill, 573; orthotype PEROPUS BILOBUS Lay & Bennett. Substitute for PEROPUS Lay & Bennett, preoccupied.

1523. GILL (1888). *Gleanings Among the Pleuronectids, and Observations on the Name Pleuronectes.* Proc. U. S. Nat. Mus., XI.

THEODORE GILL.

Eucitharus Gill, 600; orthotype PLEURONECTES LINGUATULA L. Substitute for CITHARUS Blkr., preoccupied by CITHARUS Reinh., 1838; itself a synonym of CITHARUS Röse, 1793.

Cyclopsetta Gill, 601; orthotype HEMIRHOMBUS FIMBRIATUS G. & B.

Trichopsetta Gill, 601; orthotype ARNOGLOSSUS VENTRALIS G. & B.

1524. GILL (1888). *Some Extinct Scleroderms.* Amer. Naturalist, 1888, XXII, 446–448.

THEODORE NICHOLAS GILL.

Protacanthodes Gill, 448; orthotype PROTOBALISTUM OMBONI Zigno (fossil).

1525. GILL (1888). *Note on the Genus Dipterodon.* Proc. U. S. Nat. Mus., XI.

THEODORE GILL.

Dichistius Gill, 68; orthotype DIPTERODON CAPENSIS Cuv. Substitute for DIPTERODON Cuv. (1829), preoccupied by DIPTERODON Lac. (1803).

1526. GOODE & BEAN (1888). *Three Cruises of the "Blake," by Alexander Agassiz, II.*

GEORGE BROWN GOODE; TARLETON HOFFMAN BEAN.

Steindachneria Goode & Bean, 26; orthotype (S. ARGENTEA G. & B.) A short diagnosis but no type named; fully defined in Oceanic Ichthyology, 1896, 419. Meanwhile a genus of South American Silurids had been named STEINDACHNERIA by Eigenmann, who gave the present genus of MACRURIDS the name of STEINDACHNERELLA. But the first diagnosis is adequate to hold the name, and a new name, STEINDACHNERIDION, has been assigned by Eigen-mann to his genus of SILURIDS.

1527. HERZENSTEIN (1888). *Wissenschaftliche Resultate der von N. M. Prze-walski nach Central-Asien unternommenen Reisen.* Zoologischer Theil, Band III, 2 Abth., 1 Lief., 91 p.

SALOMON MARKOVICH HERZENSTEIN.

Lefua Herzenstein, 3; orthotype OCTONEMA PLESKEI Herzenstein, substitute for OCTONEMA Herzenstein & Varpachovski, preoccupied. Replaces ELXIS Jordan & Fowler.

1528. HERZENSTEIN (1888). *Wissenschaftliche Resultate der von N. M. Prze-walski nach Central-Asien unternommenen Reisen.* Zoologischer Theil, III, 2 Abth., 2 Lief., Fische, 181–262.

SALOMON MARKOVICH HERZENSTEIN.

Chuanchia Herzenstein, 224; orthotype C. LABIOSA Herzenstein.

Platypharodon Herzenstein, 226; orthotype P. EXTREMUS Herzenstein.

1529. HILGENDORF (1888). *Fische aus dem Victoria Nyanza (Ukerewe-see)* . . . *Dr. G. A. Fischer.* Sitzber. Ges. Naturforscheude Freunde, Berlin.

FRANZ MARTIN HILGENDORF.

Haplochromis Hilgendorf, 76; orthotype H. OBLIQUIDENS Hilgendorf.

1530. JENKINS & EVERMANN (1888). *Description of Eighteen New Species of Fishes from the Gulf of California.* Proc. U. S. Nat. Mus., X, 137–158.

OLIVER PEEBLES JENKINS (1850–   ); BARTON WARREN EVERMANN.

Hermosilla Jenkins & Evermann, 144; orthotype H. AZUREA Jenkins & Evermann.

Psednoblennius Jenkins & Evermann, 156; orthotype P. HYPACANTHUS Jenkins & Evermann.

1531. JORDAN (1888). *On the Generic Name of the Tunny.* Proc. Acad. Nat. Sci. Phila.

DAVID STARR JORDAN.

Albacora Jordan, 180; orthotype SCOMBER THYNNUS L. A substitute for THYNNUS Cuv., preoccupied, but THUNNUS South is older.

Germo Jordan, 180; orthotype SCOMBER ALALUNGA Gmelin.

1532. JORDAN (1888). *Manual of the Vertebrates, etc.* Edition V.

DAVID STARR JORDAN.

Miniellus Jordan, 56; orthotype HYBOPSIS PROCNE Cope. A synonym of ALBUR-NOPS Grd.

1533. JORDAN & BOLLMAN (1888). *List of Fishes Collected at Green Turtle Cay, in the Bahamas, by Charles L. Edwards; with Descriptions of Three New Species.* Proc. U. S. Nat. Mus., XI, 549–553.

DAVID STARR JORDAN; CHARLES HARVEY BOLLMAN (1868–1889).

Stilbiscus Jordan & Bollman, 549; orthotype S. EDWARDSI Jordan & Bollman. An ally of MORINGUA Gray.

1534. JORDAN & GOSS (1888). *A Review of the Flounders and Soles (Pleuro-nectidæ) of America and Europe.* Ann. Rept. U. S. Comm. Fish and Fisheries for 1886 (1888).

DAVID STARR JORDAN; DAVID KOP GOSS (1863–1900).

Azevia Jordan & Goss, 271; orthotype CITHARICHTHYS PANAMENSIS Steindachner.

Quenselia Jordan, 306; orthotype PLEURONECTES OCELLATUS L.

Acedia Jordan, 321; orthotype APHORISTIA NEBULOSA G. & B.

1535. KIRSCH (1888). *Note on a Collection of Fishes Obtained in the Gila River at Fort Thomas, Arizona, by Lieut. W. L. Carpenter, U. S. Army.* Proc. U. S. Nat. Mus., XI, 555–558.

PHILIP HENRY KIRSCH (1851–1900).

Xyrauchen Kirsch, 556; orthotype CATOSTOMUS CYPHO Lockington = CATOSTOMUS TEXANUS C. C. Abbott.

1536. MOCQUARD (1888). *Revision des Clinus de la Collection du Museum, etc.* Bull. Sci. Philom. Paris, I.

F. MOCQUARD.

Paraclinus Mocquard, 41; orthotype ACANTHOCLINUS CHAPERI Mocquard. Substitute for ACANTHOCLINUS Mocquard, preoccupied.

1537. REIS (1888). *Die Cælacanthinen mit besonder Berücksichten der im Weissen Jura Bayerns vorkommenden Gattungen.* Paleontographica, XXV.

OTTO M. REIS.

Rhabdoderma Reis, 71; orthotype CŒLACANTHUS ELEGANS Ag. (fossil). A synonym of CŒLACANTHUS Ag.

1538. STEINDACHNER (1888). *Ichthyologische Beiträge, XIV.* S. B. Ak. Wiss. Wien., XCIX.

FRANZ STEINDACHNER.

Hoplotilapia Steindachner, 60; orthotype PARATILAPIA RETRODENS Steindachner.

1539. TRAQUAIR (1888). *Further Notes on Carboniferous Selachii.* Geol. Mag. 3, dec. V.

RAMSAY HEATLEY TRAQUAIR.

Callopristodus Traquair, 85; orthotype CTENOPTYCHIUS PECTINATUS Ag. (fossil).
Dicentrodus Traquair, 86; orthotype CLADODUS BICUSPIDATUS Traquair (fossil).
Chondrenchelys Traquair, 103; orthotype C. PROBLEMATICA Traquair (fossil).

1540. TRAQUAIR (1888). *Notes on the Nomenclature of the Fishes of the Old Red Sandstone of Great Britain.* Geol. Mag., 1888, III. dec. 5, 507–517.

RAMSAY HEATLEY TRAQUAIR.

Microbrachium Traquair, 510; orthotype M. DICKII Traquair (fossil).
Mesacanthus Traquair, 511; orthotype ACANTHODES PUSILLUS Ag. (fossil).
Rhadinacanthus Traquair, 511; orthotype DIPLACANTHUS LONGISPINUS Ag. (fossil). A synonym of DIPLACANTHUS Ag.
Thursius Traquair, 516; orthotype T. PHOLIDOTUS Traquair (fossil).

1541. TRAUTSCHOLD (1888). *Ueber Edestus protopirata Trd.* Zeitschr. Deutsch. Geol. Ges., 1888, XL, 750–753.

HERMANN VON TRAUTSCHOLD.

Protopirata Trautschold, 750; orthotype EDESTUS PROTOPIRATA Trautschold (fossil). = P. CENTRODON Trautschold. A synonym of EDESTUS Leidy.

1542. VAILLANT (1888). *(Poissons.) Mission scientifique du Cap Horn, 1882–83.* VI. Zoologie, Paris, 1888, 1–35.

LÉON (LOUIS) VAILLANT.

Enantioliparis Vaillant, 22; orthotype E. PALLIDUS Vaillant.

1543. VAILLANT (1888). *Expéditions scientifiques du "Travailleur" et du "Talisman" pendant les années 1880, 1881, 1882, 1883. Poissons.* Paris, 1–406.

LÉON VAILLANT.

Gyrionemus Vaillant, 18, 45; orthotype G. NUMMULARIS Vaillant = DIRETMUS ARGENTEUS G. & B. A synonym of DIRETMUS Gthr.
Opisthoproctus Vaillant, 105; orthotype O. SOLEATUS Vaillant.

Eustomias Vaillant, 112; orthotype E. obscurus Vaillant.

Scopelogadus Vaillant, 141, 385; orthotype S. cocles Vaillant. A synonym of Melamphaes Gthr.

Anomalopterus Vaillant, 160; orthotype A. pinguis Vaillant. This genus and the next noticed in 1886.

Leptoderma Vaillant, 165; orthotype L. macrops Vaillant.

Alexeterion Vaillant, 182; orthotype L. parfaiti Vaillant.

Brosmiculus Vaillant, 292; orthotype B. imberbis Vaillant.

Lycodophis Vaillant, 311; orthotype L. albus Vaillant. A synonym of Lycenchelys Gill.

Gymnolycodes Vaillant, 312; orthotype G. edwardsi Vaillant.

Neostoma Vaillant, 385; logotype N. bathyphilus Vaillant.

1544. WOODWARD (1888). *A Synopsis of the Vertebrate Fossils of the English Chalk.* Proc. Geol. Assoc. London, 1888, X, 273–338.

Arthur Smith Woodward (1864–   ).

Synechodus Woodward, 288; orthotype Hybodus dubrisiensis Mackie (fossil).

Elasmodectes (Newton) Woodward, 301; orthotype Elasmognathus willetti Newton (fossil). Substitute for Elasmognathus Newton, preoccupied.

Neorhombolepis Woodward, 304; orthotype N. excelsus Woodw. (fossil).

————

1889.

1545. ALCOCK (1889) *On the Bathybial Fishes of the Bay of Bengal and Neighboring Waters, Obtained During the Season of 1888-9.* Ann. Mag. Nat. Hist., 1889, IV, 376–399, 450–461.

Alfred William Alcock.

Paracentroscyllium Alcock, 379; orthotype P. ornatum Alcock.

Brephostoma Alcock, 383; orthotype B. carpenteri Alcock.

Pycnocraspedum Alcock, 386; orthotype P. squamipinne Alcock.

Paradicrolene Alcock, 387; orthotype P. multifilis Alcock.

Saccogaster Alcock, 389; orthotype S. maculatus Alcock.

Glyptophidium Alcock, 390; orthotype G. argenteum Alcock.

Halosaurichthys Alcock, 454; orthotype H. carinicauda Alcock.

Coloconger Alcock, 456; orthotype C. raniceps Conger.

Sauromuraenesox Alcock, 457; orthotype S. vorax Alcock.

Dysomma Alcock, 459; orthotype D. bucephalus Alcock.

Gavialiceps Alcock, 460; orthotype G. tæniola Alcock.

1546. BEAN (1889). *Description of a New Cottoid Fish Collected by the U. S. Fish Com. (Synchirus gilli).* Proc. U. S. Nat. Mus., XII, 641–642.

Tarleton Hoffman Bean.

Synchirus Bean, 642; orthotype S. gilli Bean.

1547. COLLETT (1889). *Diagnoses de poissons nouveaux provenant des campagnes de "L'Hirondelle."* Bull. Soc. Zool. France, 1889, XIV, 291–293, 122–132.

Robert Collett.

Conchognathus Collett, 123; orthotype C. grimaldii Collett. A synonym of Simenchelys Gill.

Photostomias Collett, 291; orthotype P. guerni Collett.

1548. COPE (1889). *Storms on the Adhesive Disk of Echeneis.* Amer. Naturalist, 1889, XXIII, 254–255.

EDWARD DRINKER COPE.

Opisthomyzon Cope, 255; orthotype ECHENEIS GLARONENSIS Wettstein (fossil).

1549. DAY (1889). *W. T. Blanford, the Fauna of British India, including Ceylon and Burmah; (Fishes).*

FRANCIS DAY.

Matsya Day, 292; orthotype ACANTHONOTUS ARGENTEUS Day. A substitute for ACANTHONOTUS (Tickell) Day, preoccupied.

1550. DEECKE (1889). *Ueber Fische aus verschiedenen Horizonten der Trias.* Palæontographica, 1888–89, XXV., 97–138.

JOHANNES ERNST WILHELM DEECKE.

Allolepidotus Deecke, 114; logotype PHOLIDOPHORUS RUPPELLI Bellotti (fossil).

Archæosemionotus Deecke, 121; orthotype A. CONNECTENS Deecke (fossil). Probably a synonym of SEMIONOTUS.

Prohalecites Deecke, 125; orthotype PHOLIDOPHORUS PORRO Bellotti (fossil).

1551. EIGENMANN & EIGENMANN (1889). *Notes from the San Diego Biological Laboratory. The Fishes of Cortez Banks, etc.* West. Amer. Sci., 1889, VI, 123–132; 147–150.

CARL H. EIGENMANN; ROSA SMITH EIGENMANN.

Paricelinus Eigenmann & Eigenmann, 131; orthotype P. HOPLITICUS Eigenmann & Eigenmann.

1552. EIGENMANN & EIGENMANN (1889). *Description of New Nematognathoid Fishes from Brazil.* West. Amer. Sci., VI, 8–10.

CARL H. EIGENMANN; ROSA SMITH EIGENMANN.

Microlepidogaster Eigenmann & Eigenmann, 8; orthotype M. PERFORATUS Eigenmann & Eigenmann.

1553. EIGENMANN & EIGENMANN (1889). *Preliminary Description of New Species and Genera of Characinidæ.* West. Amer. Sci., 1889, VI, 7–8.

CARL H. EIGENMANN; ROSA SMITH EIGENMANN.

Psectrogaster Eigenmann & Eigenmann, 7; orthotype P. RHOMBOIDES Eigenmann & Eigenmann.

1554. EIGENMANN & EIGENMANN (1889). *Preliminary Notes on South American Nematognathi,* II. Proc. Cal. Acad. Sci. (2), II, 28–56.

CARL H. EIGENMANN; ROSA SMITH EIGENMANN.

Acentronichthys Eigenmann & Eigenmann, 29; orthotype A. LEPTOS Eigenmann & Eigenmann.

Nemuroglanis Eigenmann & Eigenmann, 29; orthotype N. LANCEOLATUS Eigenmann & Eigenmann.

Farlowella Eigenmann & Eigenmann, 32; orthotype ACESTRA ACUS Kner. Substitute for ACESTRA Kner, preoccupied in Mollusks.

Oxyropsis Eigenmann & Eigenmann, 39; orthotype O. WRIGHTIANA Eigenmann & Eigenmann.

Hisonotus Eigenmann & Eigenmann, 41; orthotype H. NOTATUS Eigenmann & Eigenmann.

Parotocinclus Eigenmann & Eigenmann, 41; orthotype OTOCINCLUS MACULICAUDA Steind.

Panaque Eigenmann & Eigenmann, 44; orthotype CHÆTOSTOMUS NIGROLINEATUS Peters.

Delturus Eigenmann & Eigenmann, 45; orthotype D. PARAHYBÆ Eigenmann & Eigenmann.

Hemipsilichthys Eigenmann & Eigenmann, 46; orthotype XENOMYSTUS GOBIO Lütken. Substitute for XENOMYSTUS Lütken, preoccupied.

Tridens Eigenmann & Eigenmann, 53; orthotype T. MELANOPS Eigenmann & Eigenmann.

Pseudostegophilus Eigenmann & Eigenmann, 54; orthotype STEGOPHILUS NEMURUS Gthr.

Miuroglanis Eigenmann & Eigenmann, 56; orthotype M. PLATYCEPHALUS Eigenman & Eigenmann.

1555. EIGENMANN & EIGENMANN (1889). *A Revision of the Edentulous Genera of Curimatinæ.* Ann. N. Y. Acad. Sci., IV, 409–440.
CARL H. EIGENMANN; ROSA SMITH EIGENMANN.

Curimatella Eigenmann & Eigenmann, 415; logotype C. LEPIDURUS Eigenmann & Eigenmann.

Semitapicis Eigenmann & Eigenmann, 415; logotype CURIMATUS SPILURUS Gthr.

1556. FRAAS (1889). *Kopfstacheln von Hybodus und Acrodus sogenannte Ceratodus heteromorphus Ag.* Jahrb. Ver. Vaterl. Natur. Württenberg, XLV.
EBERHARD FRAAS (1862–1915).

Hybodonchus Fraas, 235; orthotype CERATODUS HETEROMORPHUS Ag. (part) (fossil).

Acrodonchus Fraas, 235; orthotype confused with CERATODUS HETEROMORPHUS Ag. (fossil).

1557. FRITSCH (1889). *Fauna der Gasköhle und der Kalksteine der Permformation Böhmens,* Vol. II.
ANTON JAN FRITSCH.

Tubulacanthus Fritsch, 113; orthotype T. SULCATUS Fritsch (fossil).

Brachiacanthus Fritsch, 113; orthotype B. SEMIPLANUS Fritsch (fossil). (Not BRACHYACANTHUS Egert.)

Platyacanthus Fritsch, 113; orthotype P. VENTRICOSUS Fritsch (fossil).

1558. GIGLIOLI (1889). *On a Supposed New Genus and Species of Pelagic Gadoid Fishes from the Mediterranean.* Proc. Zool. Soc. London, 1889. 328–332.
ENRICO HILLYER GIGLIOLI.

Eretmophorus Giglioli, 328; orthotype E. KLEINENBERGI Giglioli.

1559. GÜNTHER (1889). *On Some Fishes from Kilima-Njaro District.* Proc. Zool. Soc. London. 1889, 70–72.
ALBERT GÜNTHER.

Oreochromis Günther, 70; orthotype O. HUNTERI Gthr.

1560. GÜNTHER (1889). *Third Contribution to Our Knowledge of Reptiles and Fishes from the Upper Yang-tze-Kiang.* Ann. Mag. Nat. Hist., 1889; 6 ser. IV, 218–229.
ALBERT GÜNTHER.

Rhynchocypris Günther, 225; orthotype R. VARIEGATA Gthr.

Scombrocypris Günther. 226; orthotype S. STYANI Gthr.

Scomberopsis Günther, 226; orthotype S. STYANI Gthr.
Parapelecus Günther, 227; orthotype P. ARGENTEUS Gthr.

1561. GÜNTHER (1889). *Report on the Pelagic Fishes; in Report on the Scientific Results of the Voyage of H. M. S. "Challenger" during the years 1873-76,* Vol. XXXI, pt. 78.

ALBERT GÜNTHER.

Lepidothynnus Günther, 15; orthotype L. HUTTONI Gthr.
Auchenoceros Günther, 24; orthotype CALLOPTILUM PUNCTATUM Hutton. Substitute for CALLOPTILUM Hutton, preoccupied.
Halaphya Günther, 38; orthotype H. ELONGATA Gthr.

1562. HILGENDORF (1889). *Ueber eine Fischsammlung von Haiti, welche zwei neue Arten, Pœcilia tridens und Eleotris maltzani, enthält.* Sitzber. Ges. Naturf. Freunde Berlin, 1889. 51–55.

FRANZ (MARTIN) HILGENDORF.

Acropœcilia Hilgendorf, 52; orthotype PŒCILIA TRIDENS Hilgendorf.

1563. JAEKEL (1889). *Die Selachier aus dem oberen Muschelkalk Lothringens.* Abh. Geol. Spec. Elsass-Lothringen. III, 275–332.

OTTO (MAX JOHANNES) JAEKEL. (1863–).

Polyacrodus Jaekel, 321; orthotype HYBODUS POLYCYPHUS Jaekel (fossil).

1564. JORDAN & BOLLMAN (1889). *Descriptions of New Species of Fishes Collected at the Galapagos Islands and Along the Coast of the United States of Colombia, 1887–88.* Proc. U. S. Nat. Mus., 1889 (1890), XII, 149–183.

DAVID STARR JORDAN; CHARLES HARVEY BOLLMAN.

Xenocys Jordan & Bollman, 160; orthotype X. JESSIÆ Jordan & Bollman.
Bollmannia Jordan, 164; orthotype B. CHLAMYDES Jordan.
Runula Jordan & Bollman, 171; orthotype R. AZALEA Jordan & Bollman.
Engyophrys Jordan & Bollman, 176; orthotype E. SANCTI-LAURENTII Jordan & Bollman.

1565. JORDAN & EIGENMANN (1889). *A Review of the Sciænidæ of America and Europe.* Report U. S. Fish. Comm., 1886 (1889), XIV, 343–446.

DAVID STARR JORDAN; CARL H. EIGENMANN.

Corvula Jordan & Eigenmann. 379; orthotype JOHNIUS BATABANUS Poey.
Callaus Jordan, 401; orthotype CORVINA DELICIOSA Tschudi.

1566. NEWBERRY (1889). *The Paleozoic Fishes of North America.* Monogr. U. S. Geol. Surv., 1889 (issued August, 1890), XVI, 1–340.

JOHN STRONG NEWBERRY.

Heteracanthus Newberry, 65; orthotype H. POLITUS Newberry (fossil). Name preoccupied; replaced by GAMPHACANTHUS Miller.
Ganiodus Newberry, 67; orthotype G. HERTZERI Newberry (fossil).
Callognathus Newberry, 69; orthotype C. REGULARIS Newberry (fossil).
Sphenophorus Newberry, 91; orthotype S. LILLEYI Newberry (fossil).
Holonema Newberry, 92; orthotype PTERICHTHYS RUGOSUS Claypole (fossil).
Glyptaspis Newberry, 157; orthotype G. VERRUCOSUS Newberry (fossil).
Actinophorus Newberry, 174; orthotype A. CLARKII Newberry (fossil).
Mazodus Newberry, 178; orthotype M. KEPLERI Newberry (fossil).
Trachosteus Newberry, 188; orthotype DINICHTHYS MIRABILIS Newberry (fossil).
Stethacanthus Newberry, 198; orthotype S. TUMIDUS Newberry (fossil).

1567. OGILBY (1889). *Description of a New Genus and Species of Deep-Sea Fish from Lord Howe Island.* Proc. Linn. Soc. New S. Wales, 1888, 1313–1316 (1889).

JAMES DOUGLAS OGILBY.

Sternoptychides Ogilby, 1313; orthotype S. AMABILIS Ogilby.

1568. PFEFFER (1889). *Uebersicht der von Herrn Dr. Franz Stuhlmann in Aegypten, auf Zanzibar und dem gegenüberliegenden Festlande gesammelten Reptilien, Amphibien, Fische,* etc. Jahrb. Wiss. Anstalt Hamburg, VI, Pt. 2, 1–36.

GEORG JOHANN PFEFFER (1854–).

Anoplopterus Pfeffer, 15; orthotype A. URANOSCOPUS Pfeffer. A synonym of AMPHILIUS Gthr.

1569. SCHULZE (1889). *Fauna piscium Germaniæ. Verzeichniss der Fische der Stromgebiete der Donau, des Rheines, der Ems, Weser, Elbe, Oder, Weichsel, des Pregels und der Memel.* Jahresber. Naturw. Ver. Magdeburg, 1889, 137–213.

ERWIN SCHULZE (1861–).

Liparus Schulze, 61; orthotype CYPRINUS RUTILUS L. A synonym of RUTILUS Raf. if this unverified citation is correct.

Epitomynis Schulze, 174; logotype SALMO HUCHO L. A needless substitute for HUCHO and SALVELINUS.

Metallites Schulze, 184; orthotype CYPRINUS RUTILUS L. A synonym of RUTILIUS Raf.

Epitrachys Schulze, 209; orthotype PERCA FLUVIATILIS L. An absurdly needless synonym of PERCA L.

1570. VAILLANT (1889). *Sur les poissons des eaux douces de Borneo.* Congr. Intern. Zool. Paris, 1889, 81–82, Comptes Rendus.

LÉON VAILLANT.

Lepidoglanis Vaillant, 82; orthotype L. MONTICOLA Vaillant.

1571. ROHON (1889). *Ueber unter-Silurische Fische.* Mélange, Geol. Pal.; Acad. Sci. St. Petersb., I, 7–15.

JOSEPH VICTOR ROHON.

Gyrolepidotus Rohon, 8; orthotype G. SCHMIDI Rohon (fossil).

1572. VINCIGUERRA (1889). *Viaggio di Leonardo Fea in Burmania e regioni vicine: Pesci.* Ann. Mus. Civ. Storia Nat. Genova, II, ser. 9, 129–362.

DECIO VINCIGUERRA.

Scaphiodonichthys Vinciguerra, 285; orthotype S. BURMANICUS Vinciguerra.

Helgia Vinciguerra, 330; logotype HOMALOPTERA BILINEATA Blyth.

1574. WOODWARD (1889). *Preliminary Notes on Some New and Little Known British Jurassic Fishes.* Geol. Mag., VI, 448–455.

ARTHUR SMITH WOODWARD.

Leedsia Woodward, 451; orthotype L. PROBLEMATICA Woodw. (fossil). Name pre-occupied; replaced by LEEDSICHTHYS Woodw. (1889).

1575. WOODWARD (1889). *Notes on Some New and Little Known British Jurassic Fishes.* Ann. Mag. Nat. Hist., 6 ser., IV, 405–407.

ARTHUR SMITH WOODWARD.

Leedsichthys Woodward, 406; orthotype LEEDSIA PROBLEMATICUS Woodw. (fossil). Substitute for LEEDSIA Woodw., preoccupied.

Browneichthys Woodward, 407; orthotype B. ORNATUS Woodw. (fossil).

1576. WOODWARD (1889). *On Atherstonia, a New Genus of Palæoniscid Fishes from the Karoo Formation of South Africa, and a Tooth of Ceratodus from the Stormberg Beds of the Orange Free State.* Ann. Mag. Nat. Hist., 1899, 6 ser., IV, 239–243.

ARTHUR SMITH WOODWARD.

Atherstonia Woodward, 239; orthotype A. SCUTATA Woodw. (fossil).

1577. WOODWARD (1889). *Catalogue of the Fossil Fishes in the British Museum of Natural History, I, 1889.*

ARTHUR SMITH WOODWARD.

Mesolophodus Woodw., 61; orthotype M. PROBLEMATICUS Woodw. (fossil).

Sclerorhynchus Woodward, 76; orthotype S. ATAVUS Woodw. (fossil).

Pleuroplax Woodward, 173; orthotype PLEURODUS RANKINEI Ag. (fossil). Substitute for PLEURODUS Hancock & Atthey, preoccupied.

Cantioscyllium Woodward, 347; orthotype C. DECIPIENS Woodw. (fossil).

Orthacodus Woodward, 349; orthotype LAMNA LONGIDENS Ag. (fossil). Substitute for SPHENODUS Ag., regarded as preoccupied by SPHENODQN Gray.

Scapanorhynchus Woodward, 351; logotype RHYNCHOGNATHUS LEWISI Davis (fossil). A substitute for RHYNCHOGNATHUS Davis, preoccupied. The fossil homologue of the living shark, MITSUKURINA Jordan.

1578. WOODWARD (1889). *On the Palæontology of Sturgeons.* Proc. Geol. Assoc. London, 1889, XI, 24–44.

ARTHUR SMITH WOODWARD.

Pholidurus Woodward, 44; orthotype P. DISJECTUS Woodw. (fossil).

1890.

1579. ALCOCK (1890). *On the Bathybial Fishes Collected in the Bay of Bengal During the Season 1888–1890.* Ann. Mag. Nat. Hist., 1890, VI, ser. 6, 197–222.

ALFRED WILLIAM ALCOCK.

Bathyseriola Alcock, 202; orthotype B. CYANEA Alcock.

Ponerodon Alcock, 203; orthotype P. VASTATOR Alcock.

Paroneirodes Alcock, 206; orthotype P. GLOMEROSUS Alcock.

Tauredophidium Alcock, 212; orthotype T. HEXTII Alcock.

Thaumastomias Alcock, 220; orthotype T. ATROX Alcock.

1580. ALCOCK (1890). *On the Bathybial Fishes of the Arabian Sea, Obtained During the Season 1889–1890.* Ann. Mag. Nat. Hist., 1890, VI, ser. 6, 295–311.

ALFRED WILLIAM ALCOCK.

Paradicrolene Alcock, 297; orthotype P. VAILLANTI Alcock.

Monomitopus Alcock, 297; orthotype SIREMBO NIGRIFINNIS Alcock.

Dermatorus Alcock, 298; orthotype D. TRICHIURUS Alcock.

Scopelogenys Alcock, 302; orthotype S. TRISTIS Alcock.

Narcetes Alcock, 305; orthotype N. ERIMELAS Alcock.

Aulastomatomorpha Alcock, 307; orthotype A. PHOSPHEROPS Alcock.

Promyllantor Alcock, 310; orthotype P. PURPUREUS Alcock.

1581. ALCOCK (1890). *Descriptions of Some New and Rare Species of Fishes from the Bay of Bengal, Obtained During the Season of 1888–1889.* Journ. Asiat. Soc. Bengal, 1890, LVIII, 296–305.

ALFRED WILLIAM ALCOCK.

Parascombrops Alcock, 296; orthotype P. PELLUCIDUS Alcock.
Bathymyrus Alcock, 305; orthotype B. ECHINORHYNCHUS Alcock.

1582. ALCOCK (1890). *List of Pleuronectidæ Obtained in the Bay of Bengal in 1888 and 1889, with Descriptions of New and Rare Species.* Journ. Asiat. Soc. Bengal, 1889, LVII, 279–295, 293–305.

ALFRED WILLIAM ALCOCK.

Scianectes Alcock, 284; orthotype S. MACROPHTHALMUS Alcock.

1583. ALCOCK (1890). *On Some Undescribed Shore-Fishes from the Bay of Bengal.* Ann. Mag. Nat. Hist., 1890, VI, ser. 6, 425–443.

ALFRED WILLIAM ALCOCK.

Psettylis Alcock, 437; logotype P. PELLUCIDA Alcock.

1584. *BEAN (1890). *New Fishes Collected off the Coast of Alaska and the Adjacent Region Southward.* Proc. U. S. Nat. Mus., 1890, XIII, 37–45.

TARLETON HOFFMAN BEAN.

Bothrocara Bean, 39; orthotype B. MOLLIS Bean.
Poroclinus Bean, 40; orthotype P. ROTHROCKI Bean.
Dasycottus Bean, 42; orthotype D. SETIGER Bean.
Malacocottus Bean, 43; orthotype M. ZONURUS Bean.

1585. DAVIS (1890). *On the Fossil Fishes of the Cretaceous Formations of Scandinavia.* Trans. Roy. Dublin Soc., IV, 363–434.

JAMES WILLIAM DAVIS.

Bathysoma Davis, 424; orthotype B. LUTKENI Davis (fossil).

1586. EIGENMANN & EIGENMANN (1890). *Additions to the Fauna of San Diego.* Proc. Cal. Ac. Sci., III, 1–24.

CARL H. EIGENMANN; ROSA SMITH EIGENMANN.

Stenobrachius Eigenmann & Eigenmann, 4; orthotype S. LEUCOPSARUM Eigenmann & Eigenmann. A synonym of NANNOBRACHIUM Gthr.
Catablemella Eigenmann & Eigenmann, 3, 24; orthotype C. BRACHYCHIR Eigenmann & Eigenmann. A synonym of NOTOSCOPELUS Gthr. (MACROSTOMA Risso).

1587. FATIO (1890). *Faune des Vertebrés en Suisse.* V. Histoire Naturelle des Poissons.

VICTOR FATIO.

Spirlingus Fatio, 389; orthotype CYPRINUS BIPUNCTATUS Bloch. An ally or synonym of BLICCA.

1588. FORBES (1890). *A New Genus of Percidæ from New Zealand.* Trans. N. Z. Inst., XXII, for 1890.

H. O. FORBES.

Plagiogeneion Forbes, 273; orthotype THERAPON RUBIGINOSUS Hutton.

1589. GARMAN (1890). *On the Species of the Genus Anostomus.* Bulletin Essex Inst., XXII, 15–23.

SAMUEL GARMAN.

Schizodontopsis Garman, 19; logotype S. PROXIMUS Garman.

1590. GARMAN (1890). *On a Genus and Species of the Characines (Henochilus wheatlandi, etc.).* Bull. Essex Instit., 1890, XXII, 49–52.

SAMUEL GARMAN.

Henochilus Garman, 49; orthotype H. WHEATLANDI Garman.

1591. GILBERT (1890). *A Preliminary Report on the Fishes Collected by the Steamer "Albatross" on the Pacific Coast of North America during the Year 1889, etc.* Proc. U. S. Nat. Mus., 1890, XIII, 49–126.

CHARLES HENRY GILBERT.

Leuroglossus Gilbert, 57; orthotype L. STILBIUS Gilbert. Closely allied to GLOSSA-NODON (LIOGLOSSUS) Guichenot; perhaps the same.

Calotomus Gilbert, 70; orthotype C. XENODON Gilbert. A synonym of LEPTOSCARUS Sw.

Radulinus Gilbert, 88; orthotype R. ASPRELLUS Gilbert.

Bathyagonus Gilbert, 89; orthotype B. NIGRIPINNIS Gilbert.

Xenochirus Gilbert, 90; orthotype X. TRIACANTHUS Gilbert. Name preoccupied; replaced by XENERETMUS Gilbert.

Gillellus Gilbert, 98; orthotype G. SEMICINCTUS Gilbert.

Cryptotrema Gilbert, 101; orthotype C. CORALLINUM Gilbert.

Plectobranchus Gilbert, 102; orthotype P. EVIDES Gilbert.

Lucioblennius Gilbert, 103; orthotype L. ALEPIDOTUS Gilbert.

Aprodon Gilbert, 106; orthotype A. CORTEZIANA Gilbert.

Lycodapus Gilbert, 107; orthotype L. FIERASFER Gilbert.

Lioglossina Gilbert, 122; orthotype L. TETROPHTHALMA Gilbert.

1592. GILBERT (1890). *Supplementary List of Fishes Collected at Galapagos Islands and Panama, with Description of One New Genus and Three New Species.* Proc. U. S. Nat. Mus., 1890, XIII, 449–455.

CHARLES HENRY GILBERT.

Dialommus Gilbert, 452; orthotype D. FUSCUS Gilbert.

1593. GILL (1890). *Osteological Characteristics of the Family Murænosocidæ.* Proc. U. S. Nat. Mus., XIII.

THEODORE GILL.

Congresox Gill, 231; orthotype CONGER TALABON Cuv.

1594. HERZENSTEIN (1890). *Ichthyologische Bemerkungen . . . Mélanges Biologiques.* Bull. Ac. Sci. St. Petsb., XIII.

SALOMON MARKOVICH HERZENSTEIN.

Cheiragonus Herzenstein, 116; orthotype HYPSAGONUS GRADIENS Herzenstein = AS-PIDOPHORUS QUADRICORNIS Cuv. & Val. A synonym of HYPSAGONUS Gill.

Dinogunellus Herzenstein, 119; orthotype STICHÆUS GRIGORJEWI Herzenstein.

1595. HUTTON (1890). *List of New Zealand Fishes.* Proc. N. Z. Inst., XXII.

FREDERICK WOLLASTON HUTTON.

Neptotichthys Hutton, 278; orthotype DITREMA VIOLACEA Hutton.

1596. JAEKEL (1890). *Ueber Phaneropleuron und Hemictenodus n. g.* Sitzber. Ges. Naturf. Freunde Berlin, 1890, 1–8.

OTTO (MAX JOHANNES) JAEKEL.

Hemictenodus Jaekel, 7; logotype CTENODUS OBLIQUUS Hancock & Atthey (fossil). As restricted by Traquair, 1890, a synonym of SAGENODUS.

1597. JORDAN (1890). *List of Fishes Obtained (by the "Albatross") in the Harbor of Bahia and in Adjacent Waters.* Proc. U. S. Nat. Mus., XIII.

DAVID STARR JORDAN.

Verecundum Jordan, 330; orthotype V. RASILE Jordan.

1598. JORDAN (1890). *A Review of the Labroid Fishes of America and Europe.* Rept. U. S. Fish Comm., XV, for 1887 (1890).

DAVID STARR JORDAN.

Lappanella Jordan, 622; orthotype CTENOLABRIS IRIS Cuv. & Val.

Xyrula Jordan, 656; orthotype XYRICHTHYS JESSIÆ Jordan.

1599. JORDAN & EIGENMANN. *A Review of the Genera and Species of Serranidæ Found in the Waters of America and Europe.* Bull. U. S. Fish Comm. for 1888 (1890), VIII.

DAVID STARR JORDAN; CARL H. EIGENMANN.

Gilbertia Jordan & Eigenmann, 333, 346; orthotype PLECTROPOMA SEMICINCTUM Cuv. & Val. A synonym of HYPOPLECTRODES Gill, an undescribed genus, known by its indicated type.

Garrupa Jordan & Eigenmann, 353; orthotype SERRANUS NIGRITUS Holbrook.

Serranellus Jordan & Eigenmann, 399; orthotype PERCA SCRIBA L.

1600. LOHEST (1890). *Recherches sur les Poissons des Terrains Paléoziques de Belgique. Poissons des Psammites du Condroz, Famennien Supérieur.* Ann. Soc. Geol. Belg., XV, 112–203.

MAXIMIN LOHEST.

Pentagonolepis Lohest, 159; orthotype P. KONINCKI Lohest (fossil).

1601. RENAULT & ZEILLER (1890). *La Flore Fossile de Commentry.* St. Etienne, 1890.

B. RENAULT; R. ZEILLER.

Fayolia Renault & Zeiller, orthotype not named (fossil). Egg: the spiral egg-case of a HETERODONTID Shark, mistaken for a plant. (Quoted from Meunier).

1602. ROHON (1890). *Die Jura-Fische von Ust-Balei in Ost-Sibirien.* Mem. Ac. Petersb., XXXVIII, no. 1, 1–15.

JOSEF VICTOR ROHON.

Palæoniscinotus Rohon, 8; logotype P. CZEKANOWSKII Rohon (fossil).

Baleiichthys Rohon, 13; logotype B. GRACIOSA Rohon (fossil). A synonym of PHOLIDOPHORUS Ag.

1603. SCHILTHUIS (1890). *On a Collection of Fishes from the Congo; with Descriptions of Some New Species.* Tijdschr. Nederl. Dierk. Ver., 1890–92, 2 ser., III, 83–92.

LOUISE SCHILTHUIS.

Lamprologus Schilthuis, 85; orthotype L. CONGOENSIS Schilthuis.

1604. TRAQUAIR (1890). *On Phlyctænius, a New Genus of Coccosteidæ.* Geol. Mag., 1890, III, dec. 7, 55–60.

RAMSAY HEATLEY TRAQUAIR.

Phlyctænius Traquair, 55; orthotype COCCOSTEUS ACADIUS Whiteaves (fossil). Name regarded as preoccupied; replaced by PHLYCTÆNASPIS Traquair.

1605. TRAQUAIR (1890). *Note on Phlyctænius, a New Genus of Coccosteidæ.* Geol. Mag., VII, 144.
RAMSAY HEATLEY TRAQUAIR.
Phlyctænaspis Traquair, 144; orthotype COCCOSTEUS ACADICUS Whiteaves. Substitute for PHLYCTÆNIUS Traquair, regarded as preoccupied by PHLYCTÆNIUM Zittel.

1606. TRAQUAIR (1890). *List of the Fossil Dipnoi and Ganoidei of Fife and the Lothian.* Proc. Roy. Phys. Soc. Edinburgh, 1890, XVII, 385–400.
RAMSAY HEATLEY TRAQUAIR.
Drydenius Traquair, 399; orthotype D. INSIGNIS Traquair (fossil).

1507. TRAQUAIR (1890). *Observations on Some Fossil Fishes from the Lower Carboniferous Rocks of the Eskdale, Dumfriesshire.* Ann. Mag. Nat. Hist., 1890, 6 ser., VI, 491–494.
RAMSAY HEATLEY TRAQUAIR.
Benedenichthys Traquair, 492; orthotype PALÆONISCUS DENEENSIS Van Beneden (fossil). A substitute for BENEDENIUS Traquair, 1878, preoccupied.
Mesopoma Traquair, 493; logotype CANOBIUS PULCHELLUS Traquair (fossil).

1608. TRAQUAIR (1890). *On the Fossil Fishes Found at Achanarras Quarry, Caithness.* Ann. Mag. Nat. Hist., 1890, VI, ser. 6, 479–486.
RAMSAY HEATLEY TRAQUAIR.
Palæospondylus Traquair, 485; orthotype P. GUNNII Traquair (fossil).

1609. WHITEFIELD (1890). *Observations on a Fossil Fish from the Eocene Beds of Wyoming.* Bull. Amer. Mus. Nat. Hist., 1890, III, 117–120.
ROBERT PARR WHITEFIELD (1828–1910).
Protocatostomus Whitefield, 117; orthotype P. CONSTABELI Whitefield (fossil). A synonym of NOTOGONEUS Cope.

1610. WOODWARD (1890). *The Fossil Fishes of the Hawkesbury Series at Gosford.* Mem. Geol. Surv. New South Wales, Palæont., 1890, no. 4, 1–55.
ARTHUR SMITH WOODWARD.
Gosfordia Woodward, 4; orthotype G. TRUNCATA Woodw. (fossil).
Apateolepis Woodward, 12; orthotype A. AUSTRALIS Woodw. (fossil).
Pristisomus Woodward, 32; orthotype P. LATUS Woodw. (fossil).

1611. WOODWARD (1890). *On Some Upper Cretaceous Fishes of the Family Aspidorhynchidæ.* Proc. Zool. Soc. London, 1890, 629–636.
ARTHUR SMITH WOODWARD.
Apateopholis Woodward, 634; orthotype RHINELLUS LANIATUS Davis (fossil).

1612. WOODWARD (1890). *A Synopsis of the Fossil Fishes from the English Lower Oölites.* Proc. Geol. Assoc., XI, 289–306.
ARTHUR SMITH WOODWARD.
Scaphodus Woodward, 301; orthotype S. HETEROMORPHUS Woodw. (fossil).

1613. ZIGNO (1890). *Nuove aggiunte all' ittiofauna dell' epoca eocena.* Mem. Instit. Veneto, 1890, XXIII, 9–33.
ACHILLE DE ZIGNO.
Aulorhamphus Zigno, 19; orthotype CALAMOSTOMA BOLCENSIS Steind. (fossil). Replaces CALAMOSTOMA Steind., preoccupied.
Histiocephalus Zigno, 29; orthotype H. BASSANI Zigno (fossil).

1891.

1614. ALCOCK (1891). *On the Deep-Sea Fishes Collected by the "Investigator"* *in 1890–91.* Ann. Mag. Nat. Hist., 1891, 6 ser., VIII, 16–34, 119–138.
ALFRED WILLIAM ALCOCK.

Malthopsis Alcock, 26; orthotype M. LUTEUS Alcock.
Halicmetus Alcock, 27; orthotype H. RUBER Alcock.
Lamprogrammus Alcock, 32; orthotype L. NIGER Alcock.
Bathyclupea Alcock, 130; orthotype B. HOSKYNII Alcock.
Dysommopsis Alcock, 137; orthotype D. MUCIPARUS Alcock.

1615. COMPTER (1891). *A Fossil Fish from the Muschelkalk of Jena.* Zeitsch.
f. Naturw., LXIV.
G. COMPTER.

Dolichopterus Compter, 41; orthotype D. VOLITANS Compter (fossil).

1616. COPE (1891). *On Some New Fishes from South Dakota.* Amer. Naturalist, XXV.
EDWARD DRINKER COPE.

Gephyrura Cope, 654; orthotype G. CONCENTRICA Cope (fossil).
Proballostomus Cope, 655; orthotype P. LONGULUS Cope (fossil).
Oligoplarchus Cope, 656; orthotype O. SQUAMIPINNIS Cope (fossil).

1617. COPE (1891). *On the Characters of Some Paleozoic Fishes.* Proc. U. S.
Nat. Mus., 1891, XIV, 447–463.
EDWARD DRINKER COPE.

Styptobasis Cope, 447; orthotype S. KNIGHTIANA Cope (fossil).

1618. EIGENMANN, R. S. (1891). *New California Fishes.* Amer. Naturalist.
XXV.
ROSA SMITH EIGENMANN.

Perkinsia Rosa Smith Eigenmann, 153; orthotype P. OTHONOPS Eigenmann. Near
ETRUMEUS Blkr.

1619. EIGENMANN & EIGENMANN (1891). *Additions to the Founa of San* *Diego.* Proc. Calif. Acad. Sci. (2), III, 1–24.
CARL H. EIGENMANN; ROSA SMITH EIGENMANN.

Diaphus Eigenmann & Eigenmann, 4; orthotype SCOPELUS ENGRAULIS Gthr.
Tarletonbeania Eigenmann & Eigenmann, 6; orthotype T. TENUA Eigenmann &
Eigenmann.

1620. FELIX (1891). *(Otomitla.)* Palæontographica, XXXVII.
JOHANNES PAUL FELIX (1859–   ).
Otomitla Felix, 189; orthotype O. SPECIOSA Felix (fossil).

1621. GILBERT (1891). *Descriptions of Apodal Fishes from the Tropical Pacific.*
Proc. U. S. Nat. Mus., 1891, XIV, 347–354.
CHARLES HENRY GILBERT.

Xenomystax Gilbert. 348; orthotype X. ATRARIUS Gilbert.
Ilyophis Gilbert, 351; orthotype I. BRUNNEUS Gilbert.

1622. GILBERT (1891). *Descriptions of Thirty-four New Species of Fishes Col-* *lected in 1888 and 1889, Principally among the Santa Barbara Islands and in* *the Gulf of California.* Proc. U. S. Nat. Mus., 1891, XIV, 539–566.
CHARLES HENRY GILBERT.

Chriolepis Gilbert, 557; orthotype C. MINUTILLUS Gilbert.

1623. GILL (1891). *On Eleginus of Fischer, Otherwise Called Tilesia or Pleuro-gadus.* Proc. U. S. Nat. Mus., XIV.

THEODORE GILL.

Eleginops Gill, 305; orthotype ELEGINUS MACLOVINUS Cuv. & Val. A substitute for ELEGINUS Cuv. & Val., preoccupied by ELEGINUS Fischer, 1813.

1624. GILL (1891). *On the Genera Labrichthys and Pseudolabrus.* Proc. U. S. Nat. Mus., XIV.

THEODORE GILL.

Pictilabrus Gill, 403; orthotype LABRUS LATICLAVIUS Rich.

1625. HOLMBERG (1891). *Sobre algunos peces nuevos o poco conocidos de la Republica Argentina.* Rev. Argent., I, 1891, 180–193.

EDUARDO LADISLAO HOLMBERG.

Chalcinopelecus Holmberg, 190; orthotype, C. ARGENTINUS Holmberg. A synonym of PSEUDOCORYNOPOMA Perugia, of the same year.

1626. HOLT (1891). *Survey of Fishing Grounds, West Coast of Ireland. Preliminary Note on the Fish Obtained During the Cruise of the "Fingal," 1890.* Sci. Dublin Soc. (2), VII, 121–123.

ERNEST W. L. HOLT.

Nettophichthys Holt, 122; orthotype N. RETROPINNATUS Holt.

1627. KRAMBERGER (1891). *Palæichthyolozki prilozi (Collectæ palæoichthyo-logicæ).* Dis. II, Rad. Jugoslav. Akad., 1891, 106, 58–129.

DRAGUTIN GORJANOVIC KRAMBERGER.

Apostasis Kramberger, 104; orthotype ACANUS GAUDRYI Kramberger (fossil).

1628. LÜTKEN (1891). *Om en med stegophiler og trichomycterer beslægtet syd-amerikansk mallefisk (Acanthopoma annectens Ltk., n. g. et sp.).* Vid. Meddel. Kjobenhavn, 1891, 53–60.

CHRISTIAN FREDERIK LÜTKEN.

Acanthopoma Lütken, 53; orthotype A. ANNECTENS Lütken.

1629. MEUNIER (1891). *Note rectificative sur un fossile corallier recemment decrit.* Comptes Rendus Acad. Sci. Paris, CXII, 1891, 1154–5.

STANISLAS MEUNIER.

Vaillantoonia Meunier, 1155; orthotype CYCADOSPADIX VIREI Meunier (fossil). Egg-case of a CHIMÆROID from Coralline Jurassic deposits at Verdun. It was first described in the Comptes Rendus 1891, p. 356, as a species of fossil CYCAD, but it was shown later by Dr. Vaillant to be a CHIMÆROID egg-case.

1630. PERUGIA (1891). *Appunti sopra alcuni pesci sud-americani conservati nel Museo Civico di Storia Naturale di Genova.* Ann. Mus. Civ. Storia Nat. Genova, 1891, 2 ser., X, 605–657 (April).

ALBERTO PERUGIA.

Pseudocorynopoma Perugia, 646; orthotype P. DORIÆ Perugia (April). Replaces BERGIA Steind. of July, 1891.

1631. SAUVAGE (1891). *(Poissons fossiles). Études des gîtes minéraux de la France. Bassin houiller et permien d'Autun et d'Epinac.* Fasc. III. 31 p.

HENRI ÉMILE SAUVAGE.

Ædua Sauvage, 16; orthotype Æ. GAUDRYI Sauvage (fossil).
Archæoniscus Sauvage, 19; orthotype A. ROCHEI Sauvage (fossil).

1632. SAUVAGE (1891). *Note sur quelques poissons du Lias supérieur de l'Yonne, XLV.*

HENRI EMILE SAUVAGE.

Parathrissops Sauvage, 37; orthotype P. MILLOTI Sauvage (fossil). A synonym of EUTHYNOTUS.

1633. STEINDACHNER (1891). *Ichthyologische Beiträge, XV.* S. B. Ak. Wiss. Wien, C.

FRANZ STEINDACHNER.

Bergia Steindachner, 173 (July 1891); orthotype B. ALTIPINNIS Steind. = PSEUDOCORYNOPOMA DORIÆ Perugia. A synonym of PSEUDOCORYNOPOMA Perugia. April, 1891. BERGIA is preoccupied in Insects (Scott, 1881).

1634. TELLER (1891). *Ueber den Schadel eines fossilen Dipnoers, Ceratodus sturii, n. sp. aus den Schlichten den oberen Trias der Nordalpen.* Abh. Geol. Reichsanstalt, Wien, 1891, XV, pt. 3, 1–39.

FRIEDRICH JOSEPH TELLER (1852–1913).

Epiceratodus Teller, 37; orthotype CERATODUS FORSTERI Krefft (Australian species). It is not certain whether this living species from Australia is identical or not with the fish intended by the loosely written account of NEOCERATODUS BLANCHARDI Castelnau, 1876. EPICERATODUS is probably a synonym of NEOCERATODUS.

1635. VAILLANT (1891). *Note sur un nouveau genre de Siluroïdes (Diastatomycter) de Borneo.* Bull. Soc. Philom. Paris, 1891, VIII, ser. 3, 181–182.

LÉON LOUIS VAILLANT.

Diastatomycter Vaillant, 182; orthotype D. CHAPERI Vaillant. A synonym of HEMISILURUS Blkr.

1636. WALCOTT (1891). *Supposed Trenton Fossil Fish.* Amer. Geol., VIII, 178–180. (Editorial comment on Walcott's discoveries at Canyon City, Colo.)

Palæchimæra Walcott, 178; orthotype P. PRISMA Walcott (fossil). Name only; called DICTYORHABDUS in 1892.

1637. WOODWARD (1891). *The Devonian Fish-Fauna of Spitzbergen.* Ann. Mag. Nat. Hist., 1891, VI, ser. 8, 1–15.

ARTHUR SMITH WOODWARD.

Porolepis Woodward, 8; orthotype GYROPTYCHIUS POSNANIENSIS Kade (fossil).
Asteroplax Woodward, 11; orthotype A. SCABRA Woodw. (fossil).

1638. WOODWARD (1891). *Catalogue of the Fossil Fishes in the British Museum Nat. Hist., II.*

ARTHUR SMITH WOODWARD.

Palæomylus Woodw., 39; logotype RHYNCHODUS FRANGENS Newberry (fossil).
Elasmodectes (Newton) Woodward, 88; orthotype ELASMOGNATHUS WILLETTII Newton. A substitute for ELASMOGNATHUS Newton, preoccupied.
Apateacanthus Woodward, 118; orthotype PRISTACANTHUS VETUSTUS Clarke.
Glyptognathus Woodward, 390; orthotype GLYPTOPOMUS MINOR Ag.

Uropteryx (Agassiz) Woodward, 541; orthotype PLATYSOMUS STRIATUS Ag. = STROMATEUS GIBBOSUS Blainv. (fossil). Name only, noted in synonymy from Agassiz ms. A synonym of STORTES Gistel, which replaces PLATYSOMUS Ag., preoccupied.

1892.

1639. ALCOCK (1892). *Natural History Notes from H. M. Indian Marine Survey Steamer "Investigator," Lieut. G. S. Gunn, R. N., Commanding. Series II, no. 5. On the Bathybial Fishes Collected During the Season of 1891-2.* Ann. Mag. Nat. Hist., 1892 (6), IX, 345–365.

ALFRED WILLIAM ALCOCK.

Hephthocara Alcock, 349; orthotype H. SIMUM Alcock.

1640. BELLOTTI (1892). *Un nuovo Siluroide giapponese.* Atti. Soc. Ital., XXXIV, 99–101.

CRISTOFORO BELLOTTI.

Neobagrus Bellotti, 100; orthotype N. FUSCUS Bellotti = LIOBAGRUS REINI Hilgendorf. A synonym of LIOBAGRUS Hilgendorf.

1641. BOULENGER (1892). *On Some New or Little Known Fishes Obtained by Dr. J. W. Evans and Mr. Spencer Moore, during Their Recent Expedition to the Province of Matto Grosso, Brazil.* Ann. Mag. Nat. Hist., (6), X, 1892, 9–12.

GEORGE ALBERT BOULENGER.

Brachychalcinus Boulenger, 11; orthotype B. RETROSPINA Boulenger.

1642. BOULENGER (1892). *Description of a New Blennioid Fish from Kamtschatka.* Proc. Zool. Soc. London, 1892, 583–585.

GEORGE ALBERT BOULENGER.

Blenniophidium Boulenger, 583; orthotype B. PETROPAULI Boulenger = OPHIDIUM OCELLATUM Tilesius. A synonym of OPISTHOCENTRUS Kner (1868).

1643. CLAYPOLE (1892). *A New Gigantic Placoderm from Ohio.* Am. Geol., X, 1892, 1–4.

EDWARD WALLER CLAYPOLE.

Gorgonichthys Claypole, 2; orthotype G. CLARKI Claypole. A synonym of DINICHTHYS Newberry.

1644. EIGENMANN (1892). *The Percopsidæ of the Pacific Slope.* Science, XX, 1892, 233–234.

CARL H. EIGENMANN.

Columbia Eigenmann, 234; orthotype C. TRANSMONTANA Eigenmann.

1645. GARMAN (1892). *The Discoboli. Cyclopteridæ, Liparopsidæ, and Liparididæ.* Mem. Mus., C. Z., XIV, no. 2, 1892, 1–96.

SAMUEL GARMAN.

Cyclopteroides Garman, 37; orthotype C. GYRINOPS Garman.
Liparops Garman, 42; orthotype CYCLOPTERUS STELLERI Pallas.

1646-1647. HERZENSTEIN (1892). *Ichthyologische Bemerkungen aus dem Zoologischen Museum der Kaiserlichen Akademie der Wissenschaften.* Mel. Biol., XIII, 1892, 219–235.

SALOMON MARKOVICH HERZENSTEIN.

Argyrocottus Herzenstein, 219; orthotype A. ZANDERI Herzenstein.

Gymnodiptychus Herzenstein, 225; orthotype DIPTYCHUS DYBOWSKII Kessler.

Acanthogobio Herzenstein, 228; orthotype A. GUENTHERI Herzenstein. A synonym of HEMIBARBUS Blkr.

Pungtungia Herzenstein, 231; orthotype P. HERZI Herzenstein.

1648. JORDAN & DAVIS (1892). *A Preliminary Review of the Apodal Fishes or Eels Inhabiting the Waters of America and Europe.* Rept. U. S. Fish Comm., 1888, 581–677 (1892).

DAVID STARR JORDAN; BRADLEY MOORE DAVIS (1871–　).

Rabula Jordan & Davis, 590; orthotype MURÆNA AQUÆ-DULCIS Cope.

Bascanichthys Jordan & Davis, 621; orthotype CŒCULA BASCANIUM Jordan.

Quassiremus Jordan & Davis, 622; orthotype OPHICHTHYS EVIONTHAS Jordan & Bollman.

Scytalichthys Jordan & Davis, 635; orthotype OPHICHTHYS MIURUS J. & G.

Ahlia Jordan & Davis, 639; orthotype MYROPHIS EGMONTIS Jordan.

Venefica Jordan & Davis, 651; orthotype NETTASTOMA PROCERUM. G. & B.

Gordiichthys Jordan & Davis, 644; orthotype G. IRRETITUS Jordan & Davis.

Avocettina Jordan & Davis, 655; orthotype NEMICHTHYS INFANS Gthr.

1649. LÜTKEN (1892). *Spolia Atlantica. Scopelini Musei Zoologici Universitatis Hauniensis. Bidrag til Kundskab om det aabne Havs Laxesild eller Scopeliner.* Vid. Selsk. Skr. (6), VII, 1892, 223–297.

CHRISTIAN FREDERIK LÜTKEN.

Rhinoscopelus Lütken, 242; orthotype SCOPELUS COCCOI Cocco.

Pseudoscopelus Lütken, 285; orthotype P. SCRIPTUS Lütken.

1650. MILLER (1892). *North American Geology and Palæontology.* (First Appendix, 665–768.)

SAMUEL A. MILLER (1836–1897).

Eczematolepis Miller, 715; orthotype ACANTHOLEPIS PUSTULOSUS Newberry (fossil). Substitute for ACANTHOLEPIS Newberry, preoccupied.

Gamphacanthus Miller, 715; orthotype HETERACANTHUS POLITUS Newberry (fossil). Substitute for HETERACANTHUS Newberry, preoccupied.

Haplolepis Miller, 715; orthotype MECOLEPIS CORRUGATA Newberry (fossil). Substitute for MECOLEPIS and EURYLEPIS Newberry, both preoccupied.

Millerichthys Miller, 716; orthotype PTERICHTHYS MILLERI Ag. (fossil). A synonym of PTERICHTHYODES Blkr., being also a substitute for PTERICHTHYS Ag., 1843, preoccupied.

Lispognathus Miller, 716; orthotype LIOGNATHUS SPATULATUS Newberry (fossil). Substitute for LIOGNATHUS Newberry, preoccupied by LEIOGNATHUS Lac., 1803.

Icanodus Miller, 716; orthotype TOMODUS CONVEXUS St. John & Worthen. Substitute for TOMODUS Davis, preoccupied.

Œstophorus Miller, 716; orthotype SPHENOPHORUS LILLEYI Newberry (fossil). Substitute for SPHENOPHORUS Newberry, preoccupied.

Tegeolepis Miller, 717; orthotype ACTINOPHORUS CLARKII Newberry (fossil). Substitute for ACTINOPHORUS Newberry, 1888, preoccupied.

Ponerichthys Miller, 717; orthotype DINICHTHYS TERRELLI Newberry (fossil). A synonym of DINICHTHYS Newberry.

Xenodus Miller, 718; orthotype GONIODUS HERTZERI Newberry.

1651. OGILBY (1892). *On Some Undescribed Reptiles and Fishes from Australia.* Rec. Austral. Mus. II, 1892, 23–26.

JAMES DOUGLAS OGILBY.

Hyperlophus Ogilby, 24; orthotype CLUPEA SPRATELLOIDES Ogilby.

1652. PERUGIA (1892). *Descripzione di due nuove specie di pesci raccolti in Sarawak dai Sign. G. Doria ed O. Beccari.* Ann. Mus. Genov. (2), X, 1892, 1009–1010.

ALBERTO PERUGIA.

Eucirrichthys Perugia, 1009; orthotype E. DORIÆ Perugia.

1653. PERUGIA (1892). *Intorno ad alcuni pesci raccolti al Congo dal Capìtano Giacomo Bove.* Ann. Mus. Genova (2), X, 1892, 967–977.

ALBERTO PERUGIA.

Peltura Perugia, 972; orthotype P. BOVEI Perugia. Name regarded as preoccupied by PELTOURA Milne-Edwards, 1840, replaced by PHRACTURA Boulenger.

1654. POHLIG (1892). *Altpermische Saurierhäkiten, Fische und Medusen der Gegend von Friedrichroda.* 1 Thür. Leuckart Festschrift, 1892, 59–64.

HANS POHLIG.

Lepidopterus Pohlig, 63; orthotype L. CRASSUS Pohlig (fossil).

1655. REIS (1892). *Zur Osteologie und Systematik der Belonorhynchiden und Tetragonolepiden.* Geogn. JB, IV, 1892, 143–166.

OTTO M. REIS

Saurorhynchus Reis, 145; orthotype BELONOSTOMUS ACTUS Ag. (fossil).

1656. TRAQUAIR (1892). *Notes on the Devonian Fishes of Campbelltown and Scaumenac Bay in Canada.* No. 2. Geol. Mag. (3) X.

RAMSAY HEATLEY TRAQUAIR.

Doliodus Traquair, 145; orthotype DIPLODUS PROBLEMATICUS Woodw. (fossil). Not DOLIODON Grd.

1657. VAILLANT (1892). *Sur le genre Megapleuron.* Comptes Rendus Acad. Sci. Paris, CXIV.

LÉON VAILLANT.

Megapleuron Vaillant, 1083; orthotype CERATODUS ROCHEI Gaudry (fossil).

1658. WALCOTT (1892). *Preliminary Notes on the Discovery of a Vertebrate Fauna in Silurian (Ordovician) Strata.* Bull. Geol. Soc. Amer. III, 153–172.

CHARLES DOOLITTLE WALCOTT (1850–   ).

Dictyorhabdus Walcott, 165; orthotype D. PRISCUS Walcott; called, without description, PALÆCHIMÆRA in 1891.

Astraspis Walcott, 166; orthotype A. DESIDERATA Walcott (fossil). An OSTRACO-PHORE, supposed to be allied to THYESTES or, according to Eastman, to PSAM-MOSTEUS; the earliest recorded vertebrate fossil, from Trenton rocks, Canyon City, Colo.

Eriptychius Walcott, 167; orthotype E. AMERICANUS Walcott.

1659. WOODWARD (1892). *Further Contributions to Knowledge of the De-vonian Fish-Fauna of Canada.* Geol. Mag. (3) IX, 1892, 481–485.

ARTHUR WILLIAM WOODWARD.

Protodus Woodward, 1; orthotype P. JEXI (Traquair) Woodw. (fossil).

1660. WOODWARD (1892). *On Some Teeth of New Chimæroid Fishes from the Oxford and Kimmeridge Clays of England.* Ann. Mag. Nat. Hist. (6) X, 1892, 13–16.

ARTHUR SMITH WOODWARD.

Pachymylus Woodward, 14; orthotype P. LEEDSI Woodw. (fossil).
Brachymylus Woodward, 14; orthotype B. ALTIDENS Woodw. (fossil).

1893.

1661. ALCOCK (1893). *Natural History Notes from H. M. Indian Marine Survey Steamer "Investigator," Commander C. F. Oldham, R. N., Commanding. Series II, No. 9, An Account of the Deep Sea Collection Made During the Season of 1892–93.* Journ. Asiat. Soc. Bengal, LXII, 1893, 177–184.

ALFRED WILLIAM ALCOCK.

Bathypercis Alcock, 178; orthotype B. PLATYRHYNCHUS Alcock.

1662. BEAN (1893). *Description of a New Blennioid Fish from California.* Proc. U. S. Nat. Mus. XVI.

TARLETON HOFFMAN BEAN.

Plagiogrammus Bean, 699; orthotype P. HOPKINSI Bean.

1663. CLAYPOLE (1893). *The Cladodont Sharks of the Cleveland Shale.* Amer. Geol. XI, 325–331.

EDMUND WALLER CLAYPOLE.

Monocladodus Claypole, 329; orthotype M. CLARKI Claypole (fossil).

1664. COPE (1893). *On Symmorium, and the Position of the Cladodont Sharks.* Amer. Naturalist, X, VII, 1893, 999–1001.

EDWARD DRINKER COPE.

Symmorium Cope, 999; orthotype S. RENIFORME Cope (fossil).

1665. COPE (1893). *A New Extinct Species of Cyprinidæ.* P. Ac. Phila., 1893, 19–20.

EDWARD DRINKER COPE.

Aphelichthys Cope, 19; orthotype A. LINDAHLII Cope (fossil).

1666. EIGENMANN & BEESON (1893). *Preliminary Note on the Relationship of the Species Usually United Under the Generic Name Sebastodes.* Amer. Natural. XXVII, 1893, 668–671.

CARL H. EIGENMANN; CHARLES H. BEESON.

Primospina Eigenmann & Beeson, 669; orthotype SEBASTODES MYSTINUS J. & G.
Acutomentum Eigenmann & Beeson, 669; orthotype SEBASTODES OVALIS Ayres.
Pteropodus Eigenmann & Beeson, 670; orthotype SEBASTODES MALIGER J. & G.
Auctospina Eigenmann & Beeson, 670; orthotype SEBASTES AURICULATUS Grd.

1667. FRITSCH (1893). *Fauna der Gaskohle und der Kalkstein der Permformation Böhmens.* III. Heft. 2, 1893, 49–80.

ANTON JAN FRITSCH.

Traquairia Fritsch, 50; orthotype ACANTHODES PYGMÆUS Fritsch (fossil).
Protacanthus Fritsch, 55; orthotype P. PINNATUS Fritsch (fossil)
Trissolepis Fritsch, 76; orthotype SPHÆROLEPIS KOUNOVIENSIS Fritsch (fossil).

1668. FRITSCH (1893). *(Electrolepis, etc.)*   Archiv. Naturw. Landeselwrchf. Böhmen, IX.

ANTON JAN FRITSCH.

Dipnolepis Fritsch, 66; orthotype D. JAHNI Fritsch (fossil). Dermal scutes similar to those of ENCHODUS.

Electrolepis Fritsch, 72; orthotype E. HORRIDA. Fritsch (fossil). Berycoid scales.

Spinacites Fritsch, 72; orthotype S. RADIATUS Fritsch (fossil). Berycoid fin spine.

1669. GILBERT (1893). *Report on the Fishes of the Death Valley Expedition, Collected in Southern California and Nevada in 1891, with Descriptions of New Species.* North Amer. Fauna, VII, 1893, 229–234.

CHARLES HENRY GILBERT.

Empetrichthys Gilbert, 233; orthotype E. MERRIAMI Gilbert.

1670. GILL (1893). *A Comparison of Antipodal Faunas.* Mem. Ac. Washington, VI, 1893, 91–124.

THEODORE GILL.

Promethichthys Gill, 123; orthotype GEMPYLUS PROMETHEUS Cuv. & Val. Substitute for PROMETHEUS Lowe, preoccupied.

Pagrosomus Gill, 97, 123; orthotype CHRYSOPHRYS UNICOLOR Q. & G. = SPARUS AURATUS Forster. Called SPAROSOMUS, on pages 116 and 123, by a slip in proof-reading.

Sparosomus Gill, 123; orthotype CHRYSOPHRYS UNICOLOR Q. & G. = SPARUS AURATUS Forster. Lapsus for PAGROSOMUS Gill, not SPARISOMA Sw.

Capromimus Gill, 123; orthotype PLATYSTETHUS ABBREVIATUS Hector.

Rhombocyttus Gill, 123; orthotype CYTTUS TRAVERSI Hutton.

Bathystethus Gill, 123; orthotype SCIÆNA CULTRATA Forster. Substitute for PLATYSTETHUS Gthr., preoccupied.

Evistius Gill, 123; orthotype PLATYSTETHUS HUTTONI Gthr.

Ericentrus Gill, 123; orthotype CLINUS RUBRUS Hutton.

Notoclinus Gill, 124; orthotype BLENNIUS FENESTRATUS Forster.

Cologrammus Gill, 124; orthotype CLINUS FLAVESCENS Hutton.

Caulopsetta Gill, 124; orthotype PLEURONECTES SCAPHUS Forster.

1671. GILL (1893). *(Erilepis.)*   Science XXII.

THEODORE GILL.

Erilepis Gill, 52; orthotype MYRIOLEPIS ZONIFER Lockington. A substitute for MYRIOLEPIS Lockington, preoccupied in fossil fishes (MYRIOLEPIS Egerton).

1672. GÜNTHER (1893). *Second Report on the Reptiles, Batrachians, and Fishes Transmitted by Mr. H. H. Johnston from British Central Africa.* Proc. Zool. Soc. London, 1893, 616–628.

ALBERT GÜNTHER.

Engraulicypris Günther, 626; orthotype E. PINGUIS Gthr.

1673. HÆCKEL (1893). *Zur Phylogenie der Australischen Fauna, in Richard Semon's Forschungsreisen in Australien und dem Malayischen Archipel.* Jena.

ERNST (HEINRICH PHILIPP AUGUST) HÆCKEL (1834–1918).

Paramphioxus Hæckel; orthotype EPIGONICHTHYS CULTELLUS Peters. A synonym of EPIGONICHTHYS Peters.

1674. HOLMBERG (1893). *Dos peces Argentinos.* Revista del Jardin Zoologique de Buenos Aires, I.

EDUARDO LADISLAO HOLMBERG.

Aristommata Holmberg, 96, 354; orthotype A. INEXPECTATE Holmberg. Said to be a synonym of HYPOPTOPOMA Gthr.

1675. JORDAN & FESLER (1893). *A Review of the Sparoid Fishes of America and Europe.* Rep. U. S. Fish Comm. for 1889–91, 421–554 (1893).

DAVID STARR JORDAN ; BERT FESLER (1868–   ).

Raizero Jordan & Fesler, 438; orthotype MESOPRION ARATUS Gthr.

Rabirubia Jordan & Fesler, 438; orthotype MESOPRION INERMIS Peters.

Isaciella Jordan & Fesler, 500; orthotype PRISTIPOMA BREVIPINNE Steind.

Isacia Jordan & Fesler, 501; orthotype PRISTIPOMA CONCEPTIONE Cuv. & Val.

Medialuna Jordan & Fesler, 536; orthotype SCORPIS CALIFORNIENSIS Steind.

Sectator Jordan & Fesler, 536; orthotype PIMELEPTERUS OCYURUS Jordan.

1676. MICHAEL (1893). *Ueber eine Neue Lepidosteiden-Gattung aus dem oberen Keuper Oberschlesiens.* Zeits. Deutsch. Ges., XLIV, 710–729.

RICHARD MICHAEL (1869–   ).

Prolepidotus Michael, 71; orthotype P. GALLINEKI Michael (fossil). A synonym of LEPIDOTES Ag.

1677. PERUGIA (1893). *Di alcuni pesci raccolti in Sumatra dal Dott. Elio Modigliani.* Ann. Mus. Stor. Nat. Geneva, XIII.

ALBERT PERUGIA.

Modigliania Perugia, 246; orthotype M. PAPILLOSA Perugia.

1678. PFEFFER (1893). *Ostafrikanische Fische gesammelt von Herrn Dr. F. Stuhlmann in Jahre 1888 und 1889.* Jahrb. Hamburg Anst., X, 1893, 131–177.

GEORG JOHANN PFEFFER (1854–   ).

Salarigobius Pfeffer, 141; orthotype S. STUHLMANNI Pfeffer.

1679. ROHON (1893). *Die Obersilurischen Fische von Oesel. II. Theil. Selachii, Dipnoi, Ganoidei, Pteraspidæ, and Cephalaspidæ.* Mem. Ac. St. Petersb., XLI, 1893, No. 5, 1–124.

JOSEF VICTOR ROHON.

Ancistrodus Rohon, 51; orthotype A. GRACILIS Rohon (fossil). Name regarded as preoccupied by ANCISTRODON, AGKISTRODON, etc.; replaced by GRYPODON Hay (1899).

Rhabdiodus Rohon, 53; orthotype R. PARVIDENS Rohon (fossil).

Tylodus Rohon, 57; logotype T. DELTOIDES Rohon (fossil).

Chelomodus Rohon, 60; orthotype C. DIGITIFERUS Rohon.

Palæosteus Rohon, 63; orthotype P. SCHMIDTI Rohon (fossil).

Gyropeltus Rohon, 67; orthotype G. LAHUSENI Rohon (fossil).

1680. SAUVAGE (1893). *Recherches sur les poissons du terrain permien d' Autun. Études des gîtes minéraux de la France, 1893.*

HENRI ÉMILE SAUVAGE.

Prosauropsis Sauvage, 4; orthotype PACHYCORMUS ELONGATUS Sauvage (fossil).

1681. SCHMIDT (1893). *Ueber neue silurische Fischfunde auf Oesel.* Jahrb. Mineral., 1893, I, 99–101.

FRIEDRICH SCHMIDT (1832–1908).

Tolypaspis Schmidt, 100; orthotype TOLYPELEPIS UNDULATUS Pander; substitute for TOLYPELEPIS Pander, preoccupied.

1682. SMITT (1893). *A History of Scandinavian Fishes* (by B. Fries, C. U. Ekström, and C. Sundevall). Edited by
FREDRIK ADAM SMITT (1839–1904).

Molua Smitt, 526; orthotype GADUS MOLVA L. A variant of MOLVA Fleming, reverting to the original spelling of Rondelet.

1683. TRAQUAIR (1893). *Notes on the Devonian Fishes of Campbelltown and Scaumenac Bay in Canada,* No. 3. Geol. Mag. (3). X, 1893, 262–267.
RAMSAY HEATLEY TRAQUAIR.

Scaumenacia Traquair, 262; orthotype PHANEROPLEURON CURTUM Whiteaves (fossil).

1684. VAILLANT (1893). *Sur une collection de poissons recueillie par M. Chaper à Borneo.* Bull. Soc. Zool. France, XVIII, 1893, 55–62.
LÉON VAILLANT.

Oxybarbus Vaillant, 57; orthotype BARBUS HETERONEMA Blkr.

1685. VAILLANT (1893). *Sur un poisson provenant du voyage de M. Bonvalotet du Prince Henri d'Orleans.* Bull. Soc. Philom. Paris (8), 1893, V, 197–204.
LÉON VAILLANT.

Anopleutropius Vaillant, 198; orthotype A. HENRICI Vaillant.

1686. VAILLANT (1893). *Sur un nouveau genre de poissons voisin de Fierasfer.* Comptes Rendus Acad. Paris, CXVII.
LÉON VAILLANT.

Rhizoiketicus Vaillant, 745; orthotype R. CAROLINENSIS Vaillant, from the Caroline Islands.

1687. WOODWARD (1893). *Palæichthyological Notes.* Ann. Mag. Nat. Hist. (6), XI, 1893, 281–287.
ARTHUR SMITH WOODWARD.

Ganolepis Woodward, 286; orthotype G. GRACILIS Woodw. (fossil).

1894.

1688. ALCOCK (1894). *Natural History Notes from H. M. Indian Marine Survey Steamer "Investigator." Series II, no. 11. An Account of a Recent Collection of Bathybial Fishes from the Bay of Bengal and from the Laccadive Sea.* Journ. Asiat. Soc. Bengal, LXIII, 1894, 115–137.
ALFRED WILLIAM ALCOCK.

Chascanopsetta Alcock, 128; orthotype C. LUGUBRIS Alcock.

1689. BEAN, B. A. (1894). *Description of Two New Flounders, Gastropsetta frontalis, and Cyclopsetta chittendeni. ("Albatross" Collections.)* Proc. U. S. Nat. Mus., XVII.
BARTON APPLER BEAN (1860–   ).

Gastropsetta Bean, 633; orthotype G. FRONTALIS Bean.
Cyclopsetta Bean, 634; orthotype C. CHITTENDENI Bean.

1690. BOULENGER (1894). *Descriptions of New Fresh Water Fishes from Borneo.* Ann. Mag. Nat. Hist. (6), XIII, 1894, 245–251.
GEORGE ALBERT BOULENGER.

Nematabramis Boulenger, 249; orthotype N. EVERETTI Boulenger.

1691. COPE (1894). *On the Fishes Obtained by the Naturalist Expedition in Rio Grande do Sul.* Proc. Amer. Phil. Soc., XXXIII, 1894, 84–108.
EDWARD DRINKER COPE.
Osteogaster Cope, 102; orthotype CORYDORAS EQUES Steind.

1692. COPE (1894). *On Three New Genera of Characinidæ.* Amer. Naturalist, XLVIII, 1894, 67.
EDWARD DRINKER COPE.
Diapoma Cope, 67; orthotype D. SPECULIFERUM Cope.
Chorimycterus Cope, 67; orthotype C. TENUIS Cope.
Asiphonichthys Cope, 67; orthotype A. STENOPTERUS Cope.

1693. COPE (1894). *New and Little Known Palæozoic and Mesozoic Fishes.* Journ. Ac. Nat. Sci. Phila., IX.
EDWARD DRINKER COPE.
Macrepistius Cope, 441; orthotype M. ARENATUS Cope (fossil).

1694. CLAYPOLE (1894). *On a New Placoderm, Brontichthys clarki, from the Cleveland Shale.* Amer. Geol., XIV, 1894, 379–80.
EDWARD WALLER CLAYPOLE (1835–1901).
Brontichthys Claypole, 379; orthotype B. CLARKI Claypole (fossil).

1695. EIGENMANN (1894). *Notes on Some South American Fishes.* Ann. N. York Ac., VII, 1894, 625–627.
CARL H. EIGENMANN.
Cryptops Eigenmann, 626; orthotype STERNOPYGUS HUMBOLDTI Steind. Name preoccupied; replaced by EIGENMANNIA J. & E., 1896.

1696. EIGENMANN & BRAY (1894). *A Revision of the American Cichlidæ.* Ann. N. York Acad., VII, 1894, 607–624.
CARL H. EIGENMANN; WILLIAM L. BRAY.
Retroculus Eigenmann & Bray, 614; orthotype R. BOULENGERI Eigenmann & Bray.
Æquidens Eigenmann & Bray, 616; orthotype HEROS TETRAMERUS Heckel.

1697. FRITSCH (1894). *Fauna der Gäskohle und der Kalksteine der Permformation Böhmens, III, e, Palæoniscidæ, I.* Prag., 1894, 81–104.
ANTON JAN FRITSCH.
Pyritocephalus Fritsch, 86; orthotype PALÆONISCUS SCULPTUS Fritsch (fossil).
Sceleptophorus Fritsch, 88; orthotype S. BISERIALIS Fritsch (fossil).

1697A. GILL (1894). *An Australasian Subfamily of Fresh-Water Atherinoid Fishes.* Am. Nat. XXVIII, 708–709.
THEODORE GILL.
Rhombatractus Gill, 709; orthotype ARISTEUS FRITZROYENSIS Castelnau. Substitute for ARISTEUS Castelnau, preoccupied; a synonym of MELANOTÆNIA Gill.

1698. GOODE & BEAN (1894). *On Harriotta, a New Type of Chimæroid Fish from the Deeper Waters of the Northwestern Atlantic.* Proc. U. S. Nat. Mus., 1894, XVII, 471–473.
GEORGE BROWN GOODE; TARLETON HOFFMAN BEAN.
Harriotta Goode & Bean, 471; orthotype H. RALEIGHANA G. & B.

1699. GOODE & BEAN (1894). *A Revision of. the Order Heteromi, Deep-Sea Fishes, with a Description of the New Generic Types Macdonaldia and Lipogenys.* Proc. U. S. Nat. Mus.. XVII, 1894, 455–470.

GEORGE BROWN GOODE; TARLETON HOFFMAN BEAN.

Cetomimus Goode & Bean, 453; orthotype C. GILLII G. & B.

Rondeletia Goode & Bean, 454; orthotype R. BICOLOR G. & B.

Gigliolia Goode & Bean, 465; orthotype G. MOSELEYI G. & B.

Macdonaldia Goode & Bean, 467; orthotype NOTACANTHUS ROSTRATUS Collett.

Lipogenys Goode & Bean, 469; orthotype L. GILLII G. & B.

1700. HILGENDORF (1894). *Neue Characinidengattung, Petersius, aus dem Kingani-Flusse in Deutsch-Ostafrika.* Sitzber. Ges. Naturf. Berlin, 1894, 172–173.

FRANZ MARTIN HILGENDORF.

Petersius Hilgendorf, 172; orthotype P. CONSERIALIS Hilgendorf.

1701. JAEKEL (1894). *Die eocänen Selachier vom Monte Bolca. Ein Beitrag zur Morphogenie der Wirbeltiere.* Berlin, 1894.

OTTO (MAX JOHANNES) JAEKEL.

Hemiptychodus Jaekel, 137; orthotype PTYCHODUS MORTONI Mantell.

Promyliobatis Jaekel, 152; orthotype MYLIOBATIS GAZOLÆ Zigno (fossil).

Pseudogaleus Jaekel, 170; orthotype P. VOLTAI Jaekel (fossil).

1702. LANDOIS (1894). *Die Familie Megistopodus, Riesenbauchflosser.* N. Jahrb. Min., 1894, II, 228–235.

HERMANN LANDOIS (1835–   ).

Megistopus Landois, 228; orthotype MEGAPUS GUESTFALICUS Schlüter (fossil). Substitute for MEGAPUS Schlüter, 1868, preoccupied; a synonym of CHIROTHRIX Pictet & Humbert.

1702A. OGILBY (1894). *Description of Five New Species from the Australasian Region.* Proc. Linn. Soc. N. S. W., IX, June 27, 1894.

JAMES DOUGLAS OGILBY.

Scleropteryx (De Vis) Ogilby, 372; orthotype S. BICOLOR De Vis (MS.), OPHIOCLINUS DEVISI Ogilby. Name only, noted in synonymy by Ogilby; perhaps a synonym of OPHIOCLINUS.

1703. PERUGIA (1894). *Viaggio di Lamberto Loria nella Papuasia orientale. Pesci d'aqua dolce.* Ann. Mus. Genova (2), XIV, 1894, 546–553.

ALBERTO PERUGIA.

Lambertia Perugia, 550; orthotype L. ATRA Perugia.

1704. SAUVAGE (1894). *(Poissons fossiles.)* Bull. Sci. Yonne, XLVIII.

HENRI ÉMILE SAUVAGE.

Protosauropsis Sauvage, 4; orthotype PACHYCORMUS ELONGATUS Sauvage (fossil). Written PROSAUROPSIS by Woodward.

1705. STEINDACHNER (1894). *Die Fische Liberias.* Notes Leyden Mus., XVI, 1894, 1–96.

FRANZ STEINDACHNER.

Pelmatochromis Steindachner, 40; orthotype PARATILAPIA BUTTIKOFERI Steind.

Neolebias Steindachner, 78; orthotype N. UNIFASCIATUS Steind.

1706. STORMS (1894). *Troisième note sur les poissons du terrain Rupélien.* Mém. Soc. Belge. Geol., VIII, 1894, 67–82.

RAYMOND STORMS.

Amylodon Storms, 71; orthotype A. DELHEIDI Storms (fossil).

1707. TRAQUAIR (1894). *Notes on Paleozoic Fishes.* Ann. Mag. Nat. Hist. (6), XIV, 1894, 368–374.

RAMSAY HEATLEY TRAQUAIR.

Euphyacanthus Traquair, 371; orthotype E. SEMISTRIATUS Traquair (fossil).

1708. VAILLANT (1894). *Sur une collection de poissons recuellie en Basse-California et dans le Golf par M. Léon Diguet.* Bull. Soc. Philom. (8), VI, 1894, 69–75.

LÉON VAILLANT.

Neomugil Vaillant, 73; orthotype N. DIGUETI Vaillant.
Atopoclinus Vaillant, 73; orthotype A. RINGENS Vaillant.

1895.

1709. BERG (1895). *Enumeracion sistematica y sinonimica de los Peces de las Costas Argentina y Uruguaya.* An. Mus. Buenos Aires, IV, 1–120.

CARLOS BERG.

Parona Berg, 39; orthotype PAROPSIS SIGNATA Jenyns. Substitute for PAROPSIS Jenyns, 1842, preoccupied.

Sagenichthys Berg, 52; orthotype LONCHURUS ANCYLODON Bl. & Schn. Substitute for ANCYLODON Cuv., preoccupied, but MACRODON Schinz, 1817, is a still earlier substitute name.

Polyclemus Berg, 54; orthotype POLYCIRRHUS DUMERILI Bocourt. Substitute for POLYCIRRHUS Bocourt, preoccupied.

Phricus Berg, 65; orthotype APHRITIS URVILLI Cuv. Substitute for APHRITIS Cuv., preoccupied in DIPTERA (Latreille, 1804).

Thalassothia Berg, 67; orthotype THALASSOPHYNE MONTEVIDENSIS Berg.

1710. BOULENGER (1895). *Viaggio del dottor Alfredo Borelli nella Republica Argentina e nel Paraguay.* XII. Poissons. Boll. Mus. Torino, X, 1895, no. 196.

GEORGE ALBERT BOULENGER.

Nanognathus Boulenger, 3; orthotype N. BORELLII Boulenger.

1711. BOULENGER (1895). *Catalogue of the Fishes in the British Museum. Second Edition, Catalogue of the Perciform Fishes.* Vol. I, 1895.

GEORGE ALBERT BOULENGER.

Pseudalphestes Boulenger, 139; orthotype PLECTROPOMA PICTUM Tschudi.

Pomodon Boulenger, 144; orthotype PLECTROPOMA MACROPHTHALMUM Tschudi. An unwarranted substitute for the tasteless and inappropriate name HEUILUTJA-NUS Blkr.

Dinoperca Boulenger, 153; orthotype HAPALOGENYS PETERSII Day.

Chelidoperca Boulenger, 304; orthotype CENTROPRISTIS HIRUNDINACEUS Cuv. & Val.

1712. GARMAN (1895). *The Cyprinodonts.* Mem. Mus. Harvard, XIX, no. 1, 1–179.

SAMUEL GARMAN.

Glaridodon Garman, 40; orthotype G. LATIDENS Garman. Name preoccupied; later replaced by GLARIDICHTHYS Garman.

Cnestrodon Garman, 43; orthotype PŒCILIA DECEMMACULATUS Jenyns.

Pterolebias Garman, 141; orthotype P. LONGIPINNIS Garman.

1713. GILBERT (1895). *The Ichthyological Collections of the Steamer "Albatross" During the Years 1890 and 1891.* Rept. U. S. Fish Comm. for 1893, IX (1895). CHARLES HENRY GILBERT.

Gyrinichthys Gilbert, 444; orthotype G. MINYTREMUS Gilbert.

Rhinoliparis Gilbert, 445; orthotype R. BARBULIFER Gilbert.

Lyconectes Gilbert, 446; orthotype L. ALEUTICUS Gilbert.

Bathyphasma Gilbert, 448; orthotype B. OVIGERUM Gilbert.

Derepodichthys Gilbert, 456; orthotype D. ALEPIDOTUS Gilbert.

Lyconema Gilbert, 471; orthotype L. BARBATUM Gilbert.

1714. GILL (1895). *The Differential Characters of Characinoid and Erythrinoid Fishes.* Proc. U. S. Nat. Mus., 1895, XVIII, 205–209. THEODORE GILL.

Hoplerythrinus Gill, 208; orthotype ERYTHRINUS UNITÆNIATUS Spix.

1715. GILL (1895). *Note on the Fishes of the Genus Characinus.* Proc. U. S. Nat. Mus., XVIII. THEODORE GILL.

Myloplus Gill, 214; orthotype (by implication) MYLETES ASTERIAS M. & T. Substitute for MYLETES M. & T., not of Cuvier, whose original haplotype was SALMO NILOTICUS L., not MYLETES RHOMBOIDALIS Cuv., as erroneously stated in Part I, p. 93, of *Genera of Fishes.*

1716. GILL (1895). *The Genera of Branchiostomidæ.* Amer. Naturalist, XXIX, 457–459. THEODORE GILL.

Amphioxides Gill, 457; orthotype BRANCHIOSTOMA PELAGICUM Gthr.

1717. GILL (1895). *The Genus Leptophidium.* Amer. Nat., XXIX, 1895, 167–168. THEODORE GILL.

Lepophidium Gill, 167; orthotype LEPTOPHIDIUM PROFUNDORUM Gill. Substitute for LEPTOPHIDIUM Gill, preoccupied in Snakes.

1718. GOODE & BEAN (1895). *Oceanic Ichthyology. A Treatise on the Deep-Sea and Pelagic Fishes of the World, Based Chiefly upon the Collections made by the Steamers "Blake," "Albatross," and "Fish Hawk" in the Northwestern Atlantic.* Special Bull. U. S. Nat. Mus., 553 pp., 1895. GEORGE BROWN GOODE; TARLETON HOFFMAN BEAN.

Conocara Goode & Bean, 39; orthotype C. MACDONALDI G. & B.

Talismania Goode & Bean, 43; logotype BATHYTROCTES HOMOPTERUS Vaillant.

Bathylaco Goode & Bean, 57; orthotype B. NIGRICANS G. & B.

Benthosema Goode & Bean, 75; orthotype SCOPELUS MUELLERI Gmelin.

Collettia Goode & Bean, 83; orthotype MYCTOPHUM RAFINESQUEI Cocco.

Lampadena Goode & Bean, 85; orthotype L. SPECULIGERA G. & B.

Æthoprora Goode & Bean, 86; orthotype SCOPELUS METOPOCLAMPUS Cocco.

Bonapartia Goode and Bean, 102; orthotype B. PEDALIOTA G. & B. Name preoccupied; later replaced by ZAPHOTIAS G. & B.

Yarrella Goode & Bean, 103; orthotype Y. BLACKFORDI G. & B.

Grammatostomias Goode & Bean, 110; orthotype G. DENTATUS G. & B.

Aldrovandia Goode & Bean, 132; orthotype HALOSAURUS ROSTRATUS Gthr.

Hypoclydonia Goode & Bean, 236; orthotype H. BELLA G. & B.

Helicolenus Goode & Bean, 248; orthotype SCORPÆNA DACTYLOPTERA De la Roche.

Hilgendorfia Goode & Bean, 280; orthotype PARALIPARIS MEMBRANACEUS Gthr.

Dicromita Goode & Bean, 319; orthotype D. AGASSIZII G. & B.

Benthocometes Goode & Bean, 327; orthotype NEOBYTHITES ROBUSTUS G. & B.

Celema Goode & Bean, 329; logotype POROGADUS NUDUS Vaillant.

Alcockia Goode & Bean, 329; orthotype POROGADUS ROSTRATUS Gthr.

Mœbia Goode & Bean, 331; orthotype BATHYNECTES GRACILIS Gthr.

Penopus Goode & Bean, 335; orthotype P. MACDONALDI G. & B.

Hypsirhynchus "Facciolà, 111, pl. 11" (quoted in Goode & Bean, Oceanic Ichth., 379, 1895); orthotype H. HEPATICUS Facciolà.

Moseleya Goode & Bean, 417; orthotype MACRURUS LONGIFILIS Gthr. Name preoccupied; later replaced by DOLLOA Jordan.

Abyssicola Goode & Bean, 417; orthotype MACRURUS MACROCHIR Gthr.

Caulophryne Goode & Bean, 496; orthotype C. JORDANI G. & B.

Valenciennellus (Jordan & Evermann) in Goode & Bean. 513; orthotype MAUROLICUS TRIPUNCTULATUS Esmark. (Afterwards in Jordan & Evermann Fish, N. M. A. I., 1896, 577.)

Vinciguerria (Jordan & Evermann), in Goode & Bean, 513; orthotype MAUROLICUS ATTENUATUS Cocco. Advance notice, detailed account in Jordan & Evermann Fish, N. M. A. I., 1896, 577.

Manducus Goode & Bean, 514; orthotype GONOSTOMA MADERENSE Johnson.

Escolar (Jordan & Evermann) in Goode & Bean, August 23, 1896, 518; orthotype THYRSITOPS VIOLACEUS Bean. Later (1896) called BIPINNULA by Jordan & Evermann, October 3, 1896, through a slip in proof-reading.

Lophiodes Goode & Bean, 537; orthotype LOPHIUS MUTILUS Alcock.

1719. JORDAN (1895). *Description of Evermannia, a New Genus of Gobioid Fishes.* Proc. Calif. Acad. Sci. (2), IV, 592.

DAVID STARR JORDAN.

Evermannia Jordan, 592; orthotype GOBIOSOMA ZOSTERURUM J. & G.

1720. JORDAN (1895). *The Fishes of Sinaloa.* Proc. Calif. Acad. Sci. (2), V, 377–514.

DAVID/STARR JORDAN (Assisted by EDWIN CHAPIN STARKS, GEORGE BLISS CULVER (1873–) & THOMAS MARION WILLIAMS (1871–).

Anchovia Jordan & Evermann, 411; orthotype ENGRAULIS MACROLEPIDOTUS K. & S. (Described fully in Fish, N. M. Amer., Jordan & Evermann. 1896, 449).

Thyrina Jordan & Culver, 419; orthotype T. EVERMANNI Jordan & Culver.

Xystæma Jordan & Evermann, 471; orthotype MUGIL CINEREUS Walbaum. A synonym of GERRES, as at first used by Quoy & Gaimard; not GERRES of most writers.

Ulæma Jordan & Evermann, 471; orthotype DIAPTERUS LEFROYI Goode.

Aboma Jordan & Starks, 497; orthotype A. ETHEOSTOMA Jordan & Starks.

Enneanectes Jordan & Gilbert, 501; orthotype TRIPTERYGIUM CARMINALE J. & G.

Alexurus Jordan, 511; orthotype A. ARMIGER Jordan.

1721. JORDAN & STARKS (1895). *The Fishes of Puget Sound.* Proc. Calif. Acad. Sci. (2), V, 1895, 785–852.

DAVID STARR JORDAN; EDWIN CHAPIN STARKS.

Zalarges Jordan & Williams, 793; orthotype Z. NIMBARIUS Jordan & Williams. Said to be a synonym of VINCIGUERRIA Jordan.

Ruscarius Jordan & Starks, 805; orthotype R. MEANYI Jordan & Starks.

Astrolytes Jordan & Starks, 807; orthotype ARTEDIUS FENESTRALIS J. & G.

Gilbertina Jordan & Starks, 811; orthotype G. SIGALUTES Jordan & Starks. Name preoccupied; later replaced by GILBERTIDIA Berg.

Pallasina (Cramer) Jordan & Starks, 815; orthotype SIPHAGONUS BARBATUS Steind.

Stelgis (Cramer) Jordan & Starks, 821; orthotype AGONUS VULSUS J. & G.

Averruncus Jordan & Starks, 824; orthotype A. EMMELANE Jordan & Starks.

Xystes Jordan & Starks, 824; orthotype X. AXINOPHRYS Jordan & Starks.

Lethotremus (Gilbert) Jordan & Starks, 827; orthotype L. VINOLENTUS Jordan & Starks.

Ronquilus Jordan & Starks, 838; orthotype BATHYMASTER JORDANI Gilbert.

Quietula Jordan & Evermann, 839; orthotype GILLICHTHYS Y-CAUDA Jenkins & Evermann.

Bryostemma Jordan & Starks, 841; orthotype BLENNIUS POLYACTOCEPHALUS Pallas.

Xererpes Jordan & Gilbert, 846; orthotype APODICHTHYS FUCORUM J. & G.

Xiphistes Jordan & Starks, 847; orthotype XIPHISTER CHIRUS J. & G. Name preoccupied in HEMIPTERA; the genus itself, however, is of doubtful value.

1722. KIRKALDY (1895). *A Revision of the Genera and Species of the Branchiostomidæ.* Quart. Journ. Mur. Sci., XXXVII.

J. W. KIRKALDY.

Heteropleuron Kirkaldy, 314; orthotype BRANCHIOSTOMA BASSANUM Gthr.

1723. KOENEN (1895). *Ueber einige Fischreste des norddeutschen und Böhmischen Devons.* Abh. Ges. Götting., XL, no. 2, 1–37.

ADOLF VON KOENEN.

Platyaspis Koenen, 21; orthotype P. TENUIS Koenen (fossil).

Holopetalichthys Koenen, 25; orthotype H. NOVAKI Koenen (fossil).

1724. KRAMBERGER (1895). *De Piscibus fossilibus Comeni, Mrzleci, Lesinæ et M. Libanonis, et appendix de Piscibus oligocænicis ad Tüffer, Sagor et Trifail.* Agram, 1895.

DRAGUTIN GORJANOVIC KRAMBERGER.

Opsigonus Kramberger, 9; orthotype O. SQUAMOSUS Kramberger (fossil).

Ancylostylos Kramberger, 42; orthotype A. GIBBUS Kramberger (fossil).

Lobopterus Kramberger, 44; orthotype L. PECTINATUS Kramberger (fossil).

Acanthophoria Kramberger, 45; orthotype PAGELLUS LIBIANICUS Pictet (fossil). A synonym of ACROGASTER Ag.

1725. OGILBY (1895). *On Two New Genera and Species of Fishes from Australia.* Proc. Linn. Soc. New South Wales (2), X, 320–324.

JAMES DOUGLAS OGILBY.

Centropercis Ogilby, 320; orthotype C. NUDIVITTATUS Ogilby.

Tropidostethus Ogilby, 323; orthotype T. RHOTHOPHILUS Ogilby. Name preoccupied; replaced by Iso Jordan & Starks.

1726. STARKS (1895). *Description of a New Genus and Species of Cottoid from Puget Sound.* Proc. Acad. Nat. Sci. Phila, 1895, 410–412.

EDWIN CHAPIN STARKS.

Jordania Starks, 410; orthotype J. ZONOPE Starks.

1727. STRÖMMAN (1895). *Leptocephalids in the University Zoological Museum at Upsala.* S. B. Bohmisch. Ges., 1895, No. XXXIII, 42 pp.

H. STRÖMMAN.

Euleptocephalus Strömman, 51; logotype LEPTOCEPHALUS SICANUS Facciolà.

1728. VINCIGUERRA (1895). *Esplorazione del Giuba a dei suoi affluenti compiuta dal Cap. V. Bottego durante gli anni 1892–93 sotto gli auspicii della Società geographia Italiana.* Pesci Ann. Mus. Genova (2), XV, 1895, 21–60.

DECIO VINCIGUERRA.

Neobola Vinciguerra, 56; orthotype N. BOTTEGOI Vinciguerra.

1729. WEBER (1895). *Fische von Ambon, Java, Thursday Island, dem Burnett-Fluss und von der Südküste von New Guinea.* Semon's Zoolog. Forschungs-reisen, V, 259–276.

MAX (CARL WILHELM) WEBER (1852–   ).

Stiphodon Weber, 269; orthotype S. SEMONI Weber.

1730. WOODWARD (1895). *The Fossil Fishes of the Tabragar Beds (Jurassic ?).* Mem. Geol. Surv. New S. Wales, Pal., No. 9, 1–27.

ARTHUR SMITH WOODWARD.

Aphnelepis Woodward, 9; orthotype A. AUSTRALIS Woodw. (fossil).
Ætheolepis Woodward, 12; orthotype A. MIRABILIS Woodw. (fossil).
Archæomæne Woodward, 15; orthotype A. TENUIS Woodw. (fossil).

1731. WOODWARD (1895). *Catalogue of Fossil Fishes in the British Museum, etc., III.*

ARTHUR SMITH WOODWARD.

Asthenocormus Woodward, 380; orthotype EUGNATHUS TITANIUS Wagner (fossil). Substitute for AGASSIZIA Vetter, preoccupied.
Microconodus (Traquair) Woodward, 437; orthotype M. MOLYNEUXI Traquair (fossil). In synonymy: a synonym of GONATODUS Traquair.
Ctenolepis (Agassiz) Woodward, 530; orthotype C. CYCLUS Ag. (fossil). Name only in Agassiz, Poiss. Foss., II, 1844, 180.

1732. WOOLMAN (1895). *Report on a Collection of Fishes from the Rivers of Central and Northern Mexico.* Bull. U. S. Fish Comm. for 1894 (1895).

ALBERT JEFFERSON WOOLMAN (1868–1918).

Evarra Woolman, 64; orthotype E. EIGENMANNI Woolman.

1896.

1733. ALCOCK (1896). *Natural History Notes from H. M. Indian Marine Survey Steamer "Investigator." Series II, no. 23. A Supplementary List of the Marine Fishes of India, with Descriptions of Two New Genera and Eight New Species.* Journ. Asiat. Soc. Bengal, LXV, 301–338.

ALFRED WILLIAM ALCOCK.

Boopsetta Alcock, 305; orthotype B. UMBRARUM Alcock.
Scopelarchus Alcock, 307; orthotype S. GUENTHERI Alcock.

1734. ALESSANDRI (1896). *Ricerche sui pesci fossili di Paranà (Repubblica Argentina).* Atti. Acc. Sci. Torino, XXXI.

GULIO DE ALESSANDRI.

Protautoga Alessandri, 729; orthotype P. LONGIDENS Alessandri (fossil).

1735. BEAN & BEAN (1896). *Notes on Fishes Collected in Kamtschatka and Japan by Leonhard Stejneger and Nikolai A. Grebnitzki, with a Description of a New Blenny.* (Dec. 30, 1896.)
TARLETON HOFFMAN BEAN; BARTON APPLER BEAN.

Phoilidapus Bean & Bean, 389; orthotype P. GREBNITZKII Bean & Bean.

1736. BOULENGER (1896). *Descriptions of New Fishes from the Upper Shiré River, British Central Africa, Collected by Dr. Percy Rendall, and Presented to the British Museum by Sir Harry H. Johnston.* Proc. Zool. Soc. London, 1896, 915–920.
GEORGE ALBERT BOULENGER.

Docimodus Boulenger, 917; orthotype D. JOHNSTONI Boulenger.

Corematodes Boulenger, 919; orthotype C. SHIRANUS Boulenger.

1737. COLLETT (1896). *Poissons provenant des campagnes du Yacht l'Hirondelle. Résultats des campagnes scientifiques accomplies sur son Yacht par Albert I, Prince Souverain de Monaco.* Fascicule X, Monaco, 1896, 198 pp.
ROBERT COLLETT.

Halosauropsis Collett, 143; orthotype HALOSAURUS MACROCHIR Gthr.

1738. GARMAN (1896). *Cross Fertilization and Sexual Rights and Lefts Among Vertebrates.* Amer. Naturalist, XXX, 232.
SAMUEL GARMAN.

Glaridichthys Garman, 232; orthotype GIRARDINUS UNINOTATUS Poey. Substitute for GLARIDODON Garman, preoccupied.

1739. GILL (1896). *The Differential Characters of the Syngnathid and Hippocampid Fishes.* Proc. U. S. Nat. Mus., XVI. (April 23, 1896.)
THEODORE GILL.

Phycodurus Gill, 159; orthotype PHYLLOPTERYX EQUES Gthr.

1740. GILL (1896). *Lipophrys a substitute for Pholis.* Amer. Naturalist, XXX, 498.
THEODORE GILL.

Lipophrys Gill, 498; orthotype BLENNIS PHOLIS L. Substitute for PHOLIS Cuv. & Val., preoccupied by PHOLIS (Gronow) Scopoli.

1741. GRASSI & CALANDRUCCIO (1896). *Sullo svilluppo dei murenoidi.* Atti. Acc. Lincei Roma, V.
GIOVANNI BATTISTA GRASSI; SALVATORE CALANDRUCCIO.

Todarus Grassi & Calandruccio, 349; orthotype NETTASTOMA BREVIROSTRE Facciolà.

1742. GÜNTHER (1896). *Report on the Collection of Reptiles and Fishes Made by Messrs. Potanin & Berezowski in the Chinese Provinces Kansu and Sze-Chuen.* Ann. Ac. St. Petersb., 1896, 199–219.
ALBERT GÜNTHER.

Onychostoma Günther, 211; orthotype O. LATICEPS Gthr.

Leucogobio Günther, 212; orthotype L. HERZENSTEINI Gthr. A synonym of GNATHOPOGON Blkr.

1743. HERZENSTEIN (1896). *Ueber einige neue und seltene Fische des Zoologischen Museums der Kaiserlichen Akademie der Wissenschaften.* Ann. Ac. St. Petersb., 1896, 1–14.
SALOMON MARKOVICH HERZENSTEIN.

Coreoperca Herzenstein, 11; orthotype C. HERZI Herzenstein.
Nemalycodes Herzenstein, 14; orthotype N. GRIGORJEWI Herzenstein.

1744. JORDAN (1896). *Notes on Fishes Little Known or New to Science. (Con-- tributions to Biology from the Hopkins Seaside Laboratory.)* Proc. Cal. Ac. Sci., VI, 1896 (June 19).

DAVID STARR JORDAN.

Zaprora Jordan, 202; orthotype Z. SILENUS Jordan.

Emmydrichthys Jordan & Rutter, 221; orthotype E. VULCANUS Jordan & Rutter, based on an example with a deformed dorsal fin, the last two spines bitten off and the wound healed, the coloration jet black, otherwise agreeing fairly with SYNANCEICHTHYS VERRUCOSUS (Bl. & Schn.). A synonym of SPURCO (Commerson) Cuv. & Val. or SYNANCEICHTHYS Blkr.

Tarandichthys Jordan & Evermann, 225; orthotype ICELINUS FILAMENTOSUS Gilbert.

Ulca Jordan & Evermann, 227; orthotype HEMITRIPTERUS MARMORATUS Gilbert.

Bryssetæres Jordan & Evermann, 230; orthotype GOBIESOX PINNIGER Gilbert.

Arbaciosa Jordan & Evermann, 230; orthotype GOBIESOX HUMERALIS Gilbert.

Rimicola Jordan & Evermann, 231; orthotype GOBIESOX MUSCARUM Meek & Pierson.*

Starksia Jordan & Evermann, 231; orthotype LABRISOMUS CREMNOBATES Gilbert.

Exerpes Jordan & Evermann, 232; orthotype AUCHENOPTERUS ASPER Jenkins & Evermann.

1745. JORDAN & EVERMANN (1896). *A Check List of the, Fishes and Fish-like Vertebrates of Northern and Middle America.* Rept. U. S. Fish Comm. for 1895 (1896) (issued December, 1896). This Catalogue was issued in 1896, soon after the publication of Part 1 of the Fishes of North and Middle America by the same authors, and in advance of parts 2 and 3 (1898) and part 4 (1900). Most of the genera described in detail in the later parts are first characterized in this Check List, from which several names later defined should date. In this and other papers by these authors several names are first presented as appellations for subgenera, a matter of no importance so far as their use in nomenclature is concerned.)

DAVID STARR JORDAN; BARTON WARREN EVERMANN.

Rhonciscus Jordan & Evermann, 387; orthotype PRISTIPOMA CROCRO Cuv. & Val.

Rhencus Jordan & Evermann, 387; orthotype PRISTIPOMA PANAMENSE Steind.

Exonautes Jordan & Evermann, 332; orthotype EXOCŒTUS EXSILIENS Müller.

Eslopsarum Jordan & Evermann, 330; orthotype CHIROSTOMA JORDANI Woolman. A section under CHIROSTOMA Sw.

Evapristis Jordan & Evermann, 388; orthotype ORTHOPRISTIS LETHOPRISTIS Jordan & Fesler.

Otrynter Jordan & Evermann, 388; orthotype STENOTOMUS CAPRINUS Bean.

Salema Jordan & Evermann, 390; orthotype PERCA UNIMACULATA Bloch. (Not SALEIMA Bowdich, a Portuguese variant of the Spanish name.)

Buccone Jordan & Evermann, 394; orthotype CYNOSCION PRÆDATORIUS J. & G.

Elattarchus Jordan & Evermann, 397; orthotype ODONTOSCION ARCHIDIUM J. & G.

Zonoscion Jordan & Evermann, 401; orthotype POLYCIRRHUS RATHBUNI Jordan & Bollman.

---

* Charles John Pierson.

Zaclemus (Gilbert) Jordan & Evermann, 401; orthotype PARALONCHURUS GOODEI Gilbert.

Zalembius Jordan & Evermann, 403; orthotype CYMATOGASTER ROSACEUS J. & G.

Iridio Jordan & Evermann, 412; orthotype LABRUS RADIATUS L. (1758).

Emmeekia Jordan & Evermann, 413; orthotype PSEUDOJULIS VENUSTUS Jenkins & Evermann.

Julidio Jordan & Evermann, 413; orthotype PSEUDOJULIS ADUSTUS Gilbert.

Euscarus Jordan & Evermann, 416; orthotype LABRUS CRETENSIS L.

Loro Jordan & Evermann, 418; orthotype SCARUS GUACAMAIA Cuv.

Zenion Jordan & Evermann, 418; orthotype CYTTUS HOLOLEPIS G. & B.

Angelichthys Jordan & Evermann, 420; orthotype CHÆTODON CILIARIS L.

Xesurus Jordan & Evermann, 421; orthotype PRIONURUS PUNCTATUS Gill.

1746. JORDAN & EVERMANN (1896). *The Fishes of North and Middle America; a Descriptive Catalogue of the Species of Fish-like Vertebrates Found in the Waters of North America, North of the Isthmus of Panama.* Part. I, Bull. U. S. Nat. Mus., No. XLVII, 1896, 1340 pp. (March 18, 1896.)
DAVID STARR JORDAN; BARTON WARREN EVERMANN.

Haustor Jordan & Evermann, 137; orthotype GADUS LACUSTRIS Walbaum = PIMELODUS NIGRICANS Le Sueur.

Rabidus Jordan & Evermann, 146; orthotype NOTURUS FURIOSUS Jordan & Meek.

Iotichthys Jordan & Evermann, 243; orthotype LEUCISCUS PHLEGETHONTIS Cope.

Opsopœa Jordan & Evermann, 249; orthotype OPSOPŒODUS BOLLMANI Gilbert..

Azteca Jordan & Evermann, 258; orthotype NOTROPIS AZTECUS Woolman. Name preoccupied; later replaced by AZTECULA Jordan & Evermann.

Orcella Jordan & Evermann, 289; orthotype NOTROPIS ORCA Woolman.

Yuriria Jordan & Evermann, 321; orthotype HYBOPSIS ALTUS Jordan.

Eigenmannia Jordan & Evermann, 341; orthotype STERNOPYGUS HUMBOLDTI Steind. Substitute for CRYPTOPS Eigenmann, preoccupied.

Verma Jordan & Evermann, 374; orthotype SPHAGEBRANCHUS KENDALLI Gilbert.

Scutica Jordan & Evermann, 404; orthotype GYMNOMURÆNA NECTURA J. & G. A synonym of UROPTERYGIUS Rüppell.

Tarpon Jordan & Evermann, 409; orthotype MEGALOPS ATLANTICUS Cuv. & Val.

Jenkinsia Jordan & Evermann, 418; orthotype DUSSUMIERIA STOLIFERA J. & G.

Lile Jordan & Evermann, 431; orthotype SARDINELLA STOLIFERA J. & G.

Anchovia Jordan & Evermann, 449; orthotype ENGRAULIS MACROLEPIDOTUS K. & S.

Mitchillina Jordan & Evermann, 453; orthotype ALEPOCEPHALUS BAIRDII G. & B.

Spirinchus Jordan & Evermann, 522; orthotype OSMERUS THALEICHTHYS Ayres.

Nansenia Jordan & Evermann, 528; orthotype MICROSTOMUS GRŒNLANDICUS Reinh.

Kenoza Jordan & Evermann, 626; orthotype ESOX AMERICANUS Gmelin.

Fontinus Jordan & Evermann, 645; orthotype FUNDULUS SEMINOLIS Grd.

Gambusinus Jordan & Evermann, 649; orthotype FUNDULUS RATHBUNI Jordan & Meek.

Cololabis (Gill) Jordan & Evermann, 726; orthotype SCOMBRESOX BREVIROSTRIS Peters.

Lethostole Jordan & Evermann, 792; orthotype CHIROSTOMA ESTOR Jordan. A section under CHIROSTOMA Sw.

Kirtlandia Jordan & Evermann, 794; orthotype CHIROSTOMA VAGRANS G. & B. A synonym apparently of MEMBRAS Bonaparte.

Eurystole Jordan & Evermann, 802; orthotype ATHERINELLA ERIARCHA J. & G.

Ostichthys (Langsdorff) Jordan & Evermann, 846; orthotype MYRIPRISTIS JAPONI-CUS C. & V.

Bipinnula Jordan & Evermann, 878; orthotype THYRSITOPS VIOLACEUS Bean (Oct. 3, 1896). A synonym of ESCOLAR J. & E. (Aug. 23, 1896), the two names both left standing through a slip in proof-reading.

Palometa Jordan & Evermann, 966; orthotype RHOMBUS PALOMETA Jordan & Bollman.

Swainia Jordan & Evermann, 1040; orthotype ETHESTOMA SQUAMATUM Gilbert & Swain.

Torrentaria Jordan & Evermann, 1080; orthotype ETHEOSTOMA SAGITTA Jordan & Swain.

Rafinesquiellus Jordan & Evermann, 1082; orthotype ETHEOSTOMA POTTSI Grd.

Nivicola Jordan, 1082; orthotype ETHEOSTOMA BOREALE Jordan.

Claricola Jordan & Evermann, 1093; orthotype ETHEOSTOMA JULIÆ Meek.

Psychromaster Jordan & Evermann, 1099; orthotype ETHEOSTOMA TUSCUMBIA Gilbert & Swain.

Copelandellus Jordan & Evermann, 1100; orthotype ETHEOSTOMA QUIESCENS Jordan.

Enneistus Jordan & Evermann, 1147; orthotype BODIANUS ACANTHISTIUS Gilbert.

Archoperca Jordan & Evermann, 1171; orthotype MYCTEROPERCA BOULENGERI Jordan & Starks.

Xystroperca Jordan & Evermann, 1181; orthotype MYCTEROPERCA PARDALIS Gilbert.

Ocyanthias Jordan & Evermann, 1227; orthotype HOLANTHIAS MARTINICENSIS Guich.

Osbeckia Jordan & Evermann, 424; orthotype BALISTES SCRIPTUS Osbeck.

Chapinus Jordan & Evermann, 424; orthotype OSTRACION BICAUDALIS L.

Rosicola Jordan & Evermann, 429; orthotype SEBASTOSOMUS PINNIGER Gill.

Eosebastes Jordan & Evermann, 430; orthotype SEBASTICHTHYS AURORA Gilbert.

Hispanicus Jordan & Evermann, 431; orthotype SEBASTICHTHYS RUBRIVINCTUS J. & G.

Rastrinus Jordan & Evermann, 437; orthotype ICELUS SCUTIGER Bean.

Zesticelus Jordan & Evermann, 443; orthotype ACANTHOCOTTUS PROFUNDORUM Gilbert.

Stellerina .(Cramer) Jordan & Evermann, 447; orthotype BRACHYOPSIS XYOSTERNUS J. & G.

Sarritor (Cramer) Jordan & Evermann, 448; orthotype ODONTOPYXIS FRENATUS Gill.

Lyoliparis Jordan & Evermann, 451; orthotype LIPARIS PULCHELLUS Ayres.

Caremitra Jordan & Evermann, 451; orthotype CAREPROCTUS SIMUS Gilbert.

Allochir Jordan & Evermann, 452; orthotype CAREPROCTUS MELANURUS Gilbert.

Allurus Jordan & Evermann, 452; orthotype CAREPROCTUS ECTENES Gilbert. Name preoccupied; replaced in 1900 by ALLINECTES J. & E.

Amitrichthys Jordan & Evermann, 453; orthotype PARALIPARIS CEPHALUS Gilbert.

Sicya Jordan & Evermann, 456; orthotype SICYDIUM GYMNOGASTER Ogilby-Grant.

Lythrypnus Jordan & Evermann, 458; orthotype GOBIUS DALLI Gilbert.

Zalypnus Jordan & Evermann, 459; orthotype GOBIUS EMBLEMATICUS J. & G.

Ilypnus Jordan & Evermann, 460; orthotype LEPIDOGOBIUS GILBERTI Eigenmann & Eigenmann.

Rathbunella Jordan & Evermann, 463; orthotype BATHYMASTER HYPOPLECTUS Gilbert.

Arctoscopus Jordan & Evermann, 464; orthotype TRICHODON JAPONICUS Steind.

Esloscopus Jordan & Evermann, 465; orthotype DACTYLOSCOPUS ZELOTES J. & E.

Mnierpes Jordan & Evermann, 468; orthotype CLINUS MACROCEPHALUS Gthr.

Scartes Jordan & Evermann, 471; orthotype SALARIAS RUBROPUNCTATUS Cuv. & Val. Name preoccupied; replaced in 1898 by SCARTICHTHYS J. & E.

Rhodymenichthys Jordan & Evermann, 474; orthotype GUNNELLUS RUBERRIMUS Cuv. & Val. = BLENNIUS DOLICHOGASTER Pallas.

Ulvaria Jordan & Evermann, 475; orthotype PHOLIS SUBBIFURCATUS Storer.

Furcella Jordan & Evermann, 480; orthotype LYCODES DIAPTERUS Gilbert. Name preoccupied; replaced by FURCIMANUS J. & E. (1898).

Chilara Jordan & Evermann, 482; orthotype OPHIDIUM TAYLORI Grd.

Rissola Jordan & Evermann, 483; orthotype OPHIDIUM MARGINATUM De Kay.

Bellator Jordan & Evermann, 488; orthotype PRIONOTUS MILITARIS G. & B.

Vulsiculus Jordan & Evermann, 489; orthotype PERISTEDION IMBERBE Poey.

Remorina Jordan & Evermann, 490; orthotype ECHENEIS ALBESCENS T. & S.

Caulistius Jordan & Evermann, 491; orthotype GOBIESOX PAPILLIFER Gilbert.

Embassichthys Jordan & Evermann, 506; orthotype CYNICOGLOSSUS BATHYBIUS Gilbert.

Nexilarius (Gilbert) Jordan & Evermann, 512; orthotype EUSCHISTODUS CONCOLOR Gill. Misprinted NEXILARIS.

1747. OGILBY (1896). *On a New Genus and Species of Fishes from Maroubra Bay.* Proc. Linn. Soc. New S. Wales, XXI, 23–25.

JAMES DOUGLAS OGILBY.

Apogonops Ogilby, 23; orthotype A. ANOMALUS Ogilby. This genus also occurs in an article entitled *On a New Genus (Apogonops) and Species of Australian Fishes.* Zool. Anzeiger, XIX, 256, 1896.

1748. OGILBY (1896). *Descriptions of Two New Genera and Species of Australian Fishes.* Proc. Linn. Soc. New S. Wales, XXI, 136–142.

JAMES DOUGLAS OGILBY.

Macrurrhynchus Ogilby, 136; orthotype M. MAROUBRÆ Ogilby. A synonym of ASPIDONTUS Ehrenberg.

Dermatopsis Ogilby, 138; orthotype D. MACRODON Ogilby.

1749. OGILBY (1896). *Rough-Backed Herrings.* Proc. Linn. Soc. N. S. W., XXI. 504.

JAMES DOUGLAS OGILBY.

Potamalosa Ogilby, XXI, 504; orthotype CLUPEA NOVÆ-HOLLANDIÆ Cuv. & Val.

1750. OSTROUMOFF (1896). *Zwei neue Relicten-Gattungen im Amazonen Meere.* Zool. Anz.. XIX, 30.

ALEKSEYEI ALEKSANDROVICH OSTROUMOFF (1858–   ).

Asperina Ostroumoff, 30; orthotype A. IMPROVISA Ostroumoff. a deep-sea form of UMBRINA CIRRHOSA (L.). A synonym of UMBRINA Cuv.

1751. PHILIPPI (1896). *Peces nuevos de Chile.* Ann. Univ. Santiago de Chile. LXLIII.

RUDOLPH AMANDUS PHILIPPI.

Acmonotus Philippi, 375; orthotype A. CHILENSIS Philippi. Probably a synonym of LEPTONOTUS Kaup.

1752. RUTTER (1896). *Notes on Fresh-Water Fishes of the Pacific Slope of North America. II. The Fishes of Rio Yaqui, Sonora, with the Description of a New Genus of Siluridæ.* Proc. Cal. Ac. Sci., VI. (June 23, 1896.)

CLOUDSLEY M. RUTTER (1875–1905, about).

Villarius Rutter, 258; orthotype V. PRICEI Rutter. This genus is supposed to differ from HAUSTOR J. & E. (March 18, 1896) (fork-tailed AMEIURUS) in the presence of small scattered villi on the sides of the body. It is claimed that such villi exist also in several other species of HAUSTOR, at least in the young. VILLARIUS is therefore perhaps, but not certainly, a synonym of HAUSTOR, the latter having three months' priority.

1753. STARKS (1896). *List of Fishes Collected at Port Ludlow, Washington* Proc. Calif. Acad. Sci. (2), VI, 549–562.
EDWIN CHAPIN STARKS.

Axyrias Starks, 554; orthotype A. HARRINGTONI Starks. A near ally of ARTEDIUS Grd., perhaps a synonym.

1754. SEALE (1896). *Note on Deltistes, a New Genus of Catostomoid Fishes* Proc. Cal. Ac. Sci., V., 1896.
ALVIN SEALE (1870–   ).

Deltistes Seale, 269; orthotype CHASMISTES LUXATUS Cope.

1755. STORMS (1896). *Première note sur les poissons wemméliens (Éocène supérieur de la Belgique)*. Bull. Soc. Belg. Geol. Pal., X.
RAYMOND STORMS.

Ctenodentex Storms, 199; orthotype DENTEX LÆKENIENSIS Van Beneden (fossil).

1897.

1756. BOULENGER (1897). *An Account of the Fresh-Water Fishes Collected in Celebes by Drs. P. & F. Sarasin.* Proc. Zool. Soc. London, 1897, 426–429.
GEORGE ALBERT BOULENGER

Telmatherina Boulenger, 428; orthotype T. CELEBENSIS Boulenger.

1757. BOULENGER (1897). *Description of a New Ceratopterine Eagle-Ray from Jamaica.* Ann. Nat. Hist. (6), XX, 227–228.
GEORGE ALBERT BOULENGER.

Ceratobatis Boulenger, 227; orthotype C. ROBERTSI Boulenger.

1758. EASTMAN (1897). *Tamiobatis Vetustus, a New Form of Fossil Skate.* Amer. J. Sci. (4), 85–90.
CHARLES ROCHESTER EASTMAN (1868–1918).

Tamiobatis Eastman, 85; orthotype T. VETUSTUS Eastman. (Fossil cranium from Devonic or carboniferous, Kentucky.)

1759. EASTMAN (1897). *On the Occurrence of Fossil Fishes in the Devonian of Iowa.* Iowa Geol. Surv., VII, Ann. Rep. for 1896, 108–116.
CHARLES ROCHESTER EASTMAN.

Synthetodus Eastman, 111; orthotype S. TRISULCATUS Eastman (fossil, dental elements, perhaps DIPNEUSTAN, Devonic, Iowa).

1760. EIGENMANN (1897). *Steindachneria.* Amer. Nat., 1897, XXXI, 158–159.
CARL H. EIGENMANN.

Steindachnerella Eigenmann, 159; orthotype STEINDACHNERIA ARGENTEA G. & B., 1888. Substitute for STEINDACHNERIA G. & B., a name regarded as preoccupied by STEINDACHNERIA Eigenmann, 1897, as no type had been mentioned with the earlier STEINDACHNERIA of Goode & Bean. STEINDACHNERIA of Goode & Bean apparently holds, and a new name, STEINDACHNERIDION Eigenmann, 1919, has been given to Eigenmann's genus.

1761. GILBERT & CRAMER (1897). *Report on the Fishes Dredged in Deep Water Near the Hawaiian Islands, with descriptions and figures of Twenty-three New Species.* Proc. U. S. Nat. Mus., 1896, XIX, 403–435. (Feb. 5, 1897.)

CHARLES HENRY GILBERT; FRANK CRAMER.

Argyripnus Gilbert & Cramer, 414; orthotype A. EPHIPPIATUS Gilbert & Cramer.

Cœlocephalus Gilbert & Cramer, 422; orthotype C. ACIPENSERINUS Gilbert & Cramer. Name preoccupied; replaced by MATÆOCEPHALUS Berg.

Pelecanichthys Gilbert & Cramer, 432; orthotype P. CRUMENALIS Gilbert & Cramer.

1762. GILL & TOWNSEND (1897). *Diagnosis of New Species of Fishes Found in Bering Sea.* Proc. Soc. Washington, XI, 231–234.

THEODORE GILL; CHARLES HASKINS TOWNSEND.

Ericara Gill & Townsend, 232; orthotype E. SALMONEA Gill & Townsend.

1763. GILBERT (1897). *Descriptions of Twenty-two New Species of Fishes Collected by the Steamer "Albatross," of the U. S. Fish Comm.* Proc. U. S. Nat. Mus., 1896, XIX, 437–457. (Feb. 5, 1897.)

CHARLES HENRY GILBERT.

Emmnion (Jordan) Gilbert, 454; orthotype E. BRISTOLÆ Jordan.

Ulvicola Gilbert & Starks, 455; orthotype U. SANCTÆ-ROSÆ Gilbert & Starks.

1764. NIKOLSKII (1897). *Les Reptiles, Amphibiens, et Poissons recueillis par Mr. N. Zaroudny dans la Perse orientale* (Russian text). Ann. Mus. St. Petersb., 1897, 306–348.

ALEKSANDR MIKHAILOVICH NIKOLSKII (1858–   ).

Apiostoma Nikolskii, 345; orthotype A. ZARUDNYI Nikolskii. A synonym of SCHIZOTHORAX Heckel.

Apiorhynchus Nikolskii, 345; orthotype APIOSTOMA ZARUDNYI Nikolskii. Apparently a synonym of APIOSTOMA Nikolskii.

1765. OGILBY (1897). *New Genera and Species of Australian Fishes.* Proc. Linn. Soc. New S. Wales, 1897, XXII, 62–95.

JAMES DOUGLAS OGILBY.

Omochætus Ogilby, 72; orthotype HYPERLOPHUS COPEI Ogilby = MELETTA VITTATA Castelnau. Perhaps a synonym of HYPERLOPHUS Ogilby.

Monothrix Ogilby, 87; orthotype M. POLYLEPIS Ogilby.

Austrophycis Ogilby. 91; orthotype A. MEGALOPS Ogilby.

1766. OGILBY (1897). *Some New Genera and Species of Fishes.* Proc. Linn. Soc. New S. Wales, XXII, 245–257.

JAMES DOUGLAS OGILBY.

Scolenchelys Ogilby, 246; orthotype MURÆNICHTHYS AUSTRALIS Macleay. A synonym of MURÆNICHTHYS Blkr.

Myropterura Ogilby, 247; orthotype M. LATICAUDATA Ogilby.

Goodella Ogilby, 249; orthotype G. HYPOZONA Ogilby. A synonym of TRACHINO-CEPHALUS Gill.

1767. OGILBY (1897). *On Some Australian Eleotrinæ.* Proc. Linn. Soc. New S. Wales, XII. 725–757.

JAMES DOUGLAS OGILBY.

Carassiops Ogilby, 732; orthotype ELEOTRIS COMPRESSUS Krefft.

Krefftius Ogilby, 736; orthotype ELEOTRIS AUSTRALIS Krefft. A synonym of Mo-
GURNDA Gill.

Mulgoa Ogilby, 740; orthotype ELEOTRIS COXII Krefft.

Ophiorrhinus Ogilby, 745; logotype ELEOTRIS GRANDICEPS Krefft. A synonym of
PHILYPNODON Blkr.

1768. PERUGIA (1897). *Di alcuni Pesci raccolti nell' Alto Paraguay dal Cav.
Guido Boggiani.* Ann. Mus. Genova (2), XVIII, 147–150.

ALBERTO PERUGIA.

Boggiania Perugia, 148; orthotype B. OCELLATA Perugia.

1769. PLATE (1897). *Ein neuer Cyclostom mit grossen, normal entwickelten Augen, Macrophthalmia chilensis, n. g. n. sp.* Sitzber. Ges. Naturf. Berlin, 1897,
137–141.

LUDWIG HERMANN PLATE (1862–   ).

Macrophthalmia Plate, 137; orthotype M. CHILENSIS Plate.

1770. PRIEM (1897). *Sur les dents d'élasmobranches de divers gisements séno-
niens. (Villedieu, Meudon, et Foix-les-Caves.)* Bull. Soc. Geol. France, XXV,
40–56.

FERNAND PRIEM (–1919).

Pseudocorax Priem, 47; orthotype CORAX AFFINIS Ag.

1771. VAILLANT (1897). *Siluroide nouveau de. l'Afrique orientale (Chimarrhoglanis leroyi).* Bull. Mus. Paris, 1897, 81–84.

LÉON VAILLANT.

Chimarrhoglanis Vaillant, 81; orthotype C. LEROYI Vaillant. A synonym of
AMPHILIUS Gthr.

1898.

1772. ALCOCK (1898). *Natural History Notes from H. M. Indian Marine Survey
Ship "Investigator," Commander T. H. Heming, R. N., commanding. Series
II, No. 25. A Note on the Deep-Sea Fishes, Including Another Probably
Viviparous Ophidioid.* Ann. Mag. Nat. Hist. (7), II, 136–156.

ALFRED WILLIAM ALCOCK.

Benthobatis Alcock, 144; orthotype B. MORESBYI Alcock.

1773. BASSANI (1898). *Aggiunti all' ittiofauna eocenica di Monte Bolca e
Postate.* Palont. Italica, III.

FRANCESCO BASSANI.

Oncolepis Bassani, 79; orthotype O. ISSELI Bassani (fossil).

1774. BOULENGER (1898). *Description of a New Genus of Cyprinoid Fishes
from Siam.* Ann. Mag. Nat. Hist (7), I, 450–451.

GEORGE ALBERT BOULENGER.

Catlocarpio Boulenger, 450; orthotype C. SIAMENSIS Boulenger.

1775. BOULENGER (1898). *Poissons nouveaux du Congo.* Ann. Mus. Congo,
Zool. I, 1–20, 21–38.

GEORGE ALBERT BOULENGER.

Stomatorhinus Boulenger, 9; orthotype MORMYRUS WALKERI Gthr.

Myomyrus Boulenger, 10; orthotype M. MACRODON Boulenger.
Genyomyrus Boulenger, 17; orthotype G. DONNYI Boulenger.
Eugnathichthys Boulenger, 26; orthotype E. EETVELDII Boulenger.

1776. BOULENGER (1898). *On a Collection of Fishes from the Rio Jura, Brazil.*
Trans. Zool. Soc. London, XIV, 421–428.
GEORGE ALBERT BOULENGER.
Steatogenys Boulenger, 428; orthotype RHAMPHICHTHYS ELEGANS Steind.

1777. BOULENGER (1898). *On a New Genus of Salmonoid Fishes from Altai
Mountains.* Ann. Mag. Nat. Hist. (7), I, 329–331.
GEORGE ALBERT BOULENGER.
Phyllogephyra Boulenger, 330; orthotype P. ALTAICA Boulenger.

1778. BOULENGER (1898). *Descriptions of Two New Siluroid Fishes from Bra-
zil.* Ann. Mag. Nat. Hist. (7), II, 477–478.
GEORGE ALBERT BOULENGER.
Leptodoras Boulenger, 477; orthotype OXYDORAS ACIPENSERINUS Gthr.

1779. BOULENGER (1898). *A Revision of the African and Syrian Fishes of the
Family Cichlidæ, Part I.* Proc. Zool. Soc. London, 1898, 132–152.
GEORGE ALBERT BOULENGER.
Chromidotilapia Boulenger, 151; orthotype C. KINGSLEYÆ Boulenger.

1780. BOULENGER (1898). *Report on the Collection of Fishes Made by Mr.
J. E. S. Moore in Lake Tanganyika . . . 1895–1896.* Trans. Zool. Soc.,
XV, 1–30.
GEORGE ALBERT BOULENGER.
Telmatochromis Boulenger, 10; orthotype T. VITTATUS Boulenger.
Julidochromis Boulenger, 11; orthotype J. ORNATUS Boulenger.
Bathybates Boulenger, 15; orthotype B. FEROX Boulenger.
Eretmodus Boulenger, 16; orthotype E. CYANOCINCTUS Boulenger.
Tropheus Boulenger, 17; orthotype T. MOORII Boulenger.
Simochromis Boulenger, 19; orthotype CHROMIS DIAGRAMMA Gthr.
Petrochromis Boulenger, 20; orthotype P. POLYODON Boulenger.
Perissodus Boulenger, 20; orthotype P. MICROLEPIS Boulenger.
Plecodus Boulenger, 22; orthotype P. PARADOXUS Boulenger.

1781. DEAN (1898). *New Species and a New Genus of American Palæozoic
Fishes, . . . from a Nearly Completed MS. by John Strong Newberry.*
Trans. N. Y. Acad. Sci., XV. Edited by Bashford Dean.
BASHFORD DEAN (1867–   ).
Stenognathus (Newberry) Dean, about 300; orthotype DINICHTHYS CORRUGATUS
Terrell.

1782. EASTMAN (1898). *Dentition of Devonian Pyctodontidæ.* Amer. Natural-
ist, XXXII, 473–488; 545–560.
CHARLES ROCHESTER EASTMAN.
Phlyctænacanthus Eastman, 552; orthotype P. TELLERI Eastman (a fossil ICHTHYO-
DORULITE from Devonic, Wisconsin). A synonym of ECZEMATOLEPIS Miller
= ACANTHOLEPIS Newberry.
Belemnacanthus Eastman, 552; orthotype B. GIGANTEUS Eastman (fossil spine,
perhaps of an OSTRACODERM, from the Devonic of Eifel).

1783. EVERMANN & KENDALL (1898). *Descriptions of New or Little Known Genera and Species of Fishes from the United States.* Bull. U. S. Fish Commission, XVII, 125–133.

BARTON WARREN EVERMANN; WILLIAM CONVERSE KENDALL.

Signalosa Evermann & Kendall, 127; orthotype S. ATCHAFALAYÆ Evermann & Kendall.

Lyosphæra Evermann & Kendall, 131; orthotype L. GLOBOSA Evermann & Kendall.

Ogilbia (Jordan & Evermann) Evermann & Kendall, 132; orthotype O. CAYORUM Evermann & Kendall,

1784. GOELDI (1898). *Primeira Contribução para o conhecimento dos peixes do valle dos Amazonas e dos Guyanas, etc.* Bol. Mus. Paraense. Hist. Nat., Para II.

EMIL AUGUST GOELDI.

Phrenatobius Goeldi, orthotype P. CISTERNARUM Goeldi. A little known subterranean fish.

1785. HAY (1898). *Notes on Species of Ichthyodectes, including the New Species I. cruentus, and on the Related and Herein Established Genus Gillicus.* Amer. J. Sci. (4), VI, 225–232.

OLIVER PERRY HAY.

Gillicus Hay, 230; orthotype PORTHEUS ARCUATUS Cope. Said to be a synonym of ICHTHYODECTES Cope.

1786. IHERING (1898). *Description of a New Fish from Saõ Paulo.* Proc. Acad. Nat. Sci. Phila,, 1898, 108–109.

HERMANN VON IHERING (1850–   ).

Paulicea Ihering, 109; orthotype P. JAHU Ihering.

1787. JAEKEL (1898). *Ueber Hybodus Agassiz.* Sitzber. Naturf. Berlin, 1898, 135–146.

OTTO (MAX JOHANNES) JAEKEL.

Orthybodus Jaekel, 138; orthotype HYBODUS GROSSICONUS Ag. (fossil).

Parhybodus Jaekel, 143; orthotype HYBODUS LONGICONUS Ag. (fossil).

1788. JAEKEL (1898). *Der verschiedenen Rochentypen.* Sitzber. Naturf. Berlin, 1898, 44–53.

OTTO (MAX JOHANNES) JAEKEL.

Lagarodus Jaekel, 50; orthotype PSAMMODUS SPECULARIS Romanowskii (fossil).

1789. JORDAN (1898). *Description of a Species of Fish (Mitsukurina owstoni) from Japan, the Type of a Distinct Family of Lamnoid Sharks.* Proc. Calif. Acad. Sci. (3), I, 199–201.

DAVID STARR JORDAN.

Mitsukurina Jordan, 199; orthotype M. OWSTONI Jordan, a living homologue of the extinct genus SCAPANORHYNCHUS Woodw.

1790. JORDAN & EVERMANN (1898). *The Fishes of North and Middle America; a Descriptive Catalogue of the Species of Fish-like Vertebrates found in the Waters of North America, North of the Isthmus of Panama.* Pt. II. Bull. U. S. Nat. Mus. No. 47, 1241–2182. (Oct. 3, 1898.)

DAVID STARR JORDAN; BARTON WARREN EVERMANN.

Nector Jordan & Evermann, 1433; orthotype BAIRDIELLA ARMATA Gill.

Zestis Jordan & Evermann, 1439; orthotype STELLIFER OSCITANS J. & G.

Zestidium Jordan & Evermann, 1442; orthotype STELLIFER ILLECEBROSUS Gilbert.

Stellicarens Jordan & Evermann, 1445; orthotype STELLIFER ZESTOCARUS Gilbert.

Sigmurus (Gilbert) Jordan & Evermann, 1447; orthotype CORVINA VERMICULARIS Gthr.

Azurina (Jordan & McGregor) Jordan & Evermann, 1544; orthotype A. HIRUNDO Jordan & McGregor.

Forcipiger (Jordan & McGregor) Jordan & Evermann, 1671; orthotype CHELMON LONGIROSTRIS Cuv. & Val.

Emmelas Jordan & Evermann, 1773; orthotype SEBASTODES GLAUCUS Hilgendorf.

Alcidea Jordan & Evermann, 1886; orthotype PARICELINUS THOBURNI Gilbert.

Archistes (Jordan & Gilbert) Jordan & Evermann, 1900; orthotype A. PLUMARIUS J. & G.

Steligistrum (Jordan & Gilbert) Jordan & Evermann, 1921; orthotype S. STEJNE-GERI J. & G.

Oxycottus Jordan & Evermann, 2015; orthotype OLIGOCOTTUS ACUTICEPS Gilbert.

Nautiscus Jordan & Evermann, 2019; orthotype NAUTICHTHYS PRIBILOVIUS J. & G.

Occa Jordan & Evermann, 2043; orthotype BRACHYOPSIS VERRUCOSUS Lockington.

Gurnardus Jordan & Evermann, 2148; orthotype PRIONOTUS GYMNOSTETHUS Gilbert.

Merulinus Jordan & Evermann, 2148; orthotype TRIGLA CAROLINA L.

1791. JORDAN & EVERMANN (1898). *The Fishes of North and Middle America; a Descriptive Catalogue of the Species of Fish-like Vertebrates Found in the Waters of North America North of the Isthmus of Panama.* Part III. Bull. U. S. Nat. Mus., No. 47, pp. 2183–3136. (Nov., 1898.)
DAVID STARR JORDAN; BARTON WARREN EVERMANN.

Enypnias Jordan & Evermann, 2232; orthotype GOBIUS SEMINUDUS Gthr.

Dæctor Jordan & Evermann, 2325; orthotype THALASSOPHRYNE DOWI J. & G.

Bryssophilus Jordan & Evermann, 2329; orthotype GOBIESOX PAPILLIFER Gilbert.

Pterognathus (Girard) Jordan & Evermann, 2355; orthotype NEOCLINUS SATIRICUS Grd.

Corallicola Jordan & Evermann, 2369; orthotype AUCHENOPTERUS NIGRIPINNIS Steind.

Homesthes (Gilbert) Jordan & Evermann, 2394; orthotype H. CAULOPUS Gilbert.

Blenniolus Jordan & Evermann, 2386; orthotype BLENNIUS BREVIPINNIS Gthr.

Scartichthys Jordan & Evermann, 2395; orthotype SALARIAS RUBROPUNCTATUS Cuv. & Val. A substitute for SCARTES J. & E., 1896, preoccupied.

Enedrias Jordan & Evermann, 2414; orthotype GUNNELLUS NEBULOSUS T. & S.

Ernogrammus Jordan & Everman, 2441; orthotype STICHÆUS ENNEAGRAMMUS Kner.

Ozorthe Jordan & Evermann, 2441; orthotype STICHÆUS HEXAGRAMMUS T. & S.

Embryx Jordan & Evermann, 2456; orthotype LYCODOPSIS CRASSILABRIS Gilbert.

Lycias Jordan & Evermann, 2463; orthotype LYCODES NEBULOSUS Kner.

Theragra (Lucas) Jordan & Evermann, 2535; orthotype GADUS CHALCOGRAMMUS Pallas.

Emphycus Jordan & Evermann, 2552; orthotype PHYCIS TENUIS Mitchill. A substitute for PHYCIS Raf., which is preoccupied. but PHYCIS Röse (1793), its equivalent, is apparently valid.

Verasper (Jordan & Gilbert) Jordan & Evermann, 2618; orthotype V. MOSERI J. & G.

Ramularia Jordan & Evermann. 2663; orthotype ANCYLOPSETTA DENDRITICA Gilbert.

Perissias Jordan & Evermann, 2667 orthotype PLATOPHYRYS TÆNIOPTERUS Gilbert.

Albatrossia Jordan & Evermann, 2573; orthotype Macrourus pectoralis Gilbert.

Bogoslovius Jordan & Evermann, 2574; orthotype B. clarki J. & G.

Aztecula Jordan & Evermann, 2799; orthotype Notropis aztecus Woolman. A substitute for Azteca J. & E., preoccupied.

Zaphotias (Goode & Bean) Jordan & Evermann, 2826; orthotype Bonapartia pedaliota G. & B. Substitute for Bonapartia G. & B., preoccupied.

Rhynchias (Gill) Jordan & Evermann, 2841; orthotype Ammodytes septipinnis Pallas, a species still unrecognized.

Zalocys (Jordan & McGregor) Jordan & Evermann, 2848; orthotype Z. stilbe Jordan & McGregor.

Sigmistes (Rutter) Jordan & Evermann, 2863; orthotype S. caulias Rutter.

Aspistor Jordan & Evermann, 2863; orthotype Arius luniscutis Cuv. & Val.

Crystallichthys (Jordan & Gilbert) Jordan & Evermann, 2864; orthotype C. mirabilis J. & G.

Allinectes Jordan & Evermann, 2866; orthotype Careproctus ectenes Gilbert. Substitute for Allurus J. & E., preoccupied by Allurus Forster (1862), a genus of Hymenoptera; also by Allurus Eisen (1874), a genus of worms.

Prognurus (Jordan & Gilbert) Jordan & Evermann, 2866; orthotype P. cypselurus J. & G.

Sicyosus Jordan & Evermann, 2867; orthotype Sicydium gymnogaster Ogilvie-Grant. Substitute for Sicya Jordan & Evermann, preoccupied.

Alectrias Jordan & Evermann, 2869; orthotype Blennius alectrolophus Pallas.

Furcimanus Jordan & Evermann, 2869; orthotype Lycodes diapterus Gilbert. Substitute for Furcella J. & E., preoccupied.

Flammeo Jordan & Evermann, 2871; orthotype Holocentrum marianum Cuv. & Val.

1792. JORDAN & McGREGOR (1898). *List of Fishes Collected at the Revillagi-gedo Archipelago and Neighboring Islands.* Rept. U. S. Fish Comm., for 1898, 273–284.

        David Starr Jordan; Richard Crittenden McGregor.

    The contents of this paper were included in the appendix to Part III of the *Fishes of North and Middle America,* of the same date.

Zalocys Jordan & McGregor, 276; orthotype Z. stilbe Jordan & McGregor.

Forcipiger Jordan & McGregor, 279; orthotype Chelmo longirostris Cuv. & Val.

1793. JORDAN & GILBERT (1898). *The Fishes of Bering Sea. In Jordan, Fur Seals, and Fur-Seal Islands of the North Pacific Ocean.* Part III.

        David Starr Jordan; Charles Henry Gilbert.

    The genera and species described in this paper were mostly included in Jordan & Evermann's *Fishes of North America,* Part III, of the same date.

Therobromus (Lucas) Jordan & Gilbert, 440; orthotype Therobromus callorhini Lucas.*

Archistes Jordan & Gilbert, 454; orthotype A. plumarius J. & G.

Stelgistrum Jordan & Gilbert, 456; orthotype S. stejnegeri J. & G.

Crystallichthys Jordan & Gilbert, 476; orthotype C. mirabilis J. & G.

Prognurus Jordan & Gilbert, 478; orthotype P. cypselurus J. & G.

Albatrossia Jordan & Gilbert, 487; orthotype Macrourus pectoralis Gilbert.

Verasper Jordan & Gilbert, 490; orthotype V. moseri J. & G.

---

* Frederick Augustus Lucas (1852–   ).

1794. NEWBERRY (1898). *New Species and New Genus of American Palæozoic Fishes, Together with Notes on the Genera Oracanthus, Dactylodus, Polyrhizodus, Sandalodus, Deltodus.* Trans. N. Y. Acad., XVI, 1898, 282–304.

JOHN STRONG NEWBERRY.

Stenognathus Newberry, 303; orthotype DINICHTHYS CORRUGATUS Terrell (fossil).

1795. OGILBY (1898). *A Contribution to the Zoology of New Caledonia.* Proc. Linn. Soc. New S. Wales, XXII, 762–770.

JAMES DOUGLAS OGILBY.

Trichopharynx Ogilby, 769; orthotype GOBIUS CRASSILABRIS Gthr. A synonym of CHONOPHORUS Poey (AWAOUS Steind.).

1796. OGILBY (1898). *New Genera and Species of Fishes.* Proc. Linn. Soc. New S. Wales, XXIII, 32–41, 280–299.

JAMES DOUGLAS OGILBY.

Cinetodus Ogilby, 32; orthotype ARIUS FROGGATI Ramsay.

Nedystoma Ogilby, 32; orthotype HEMIPIMELODUS DAYI Ramsay & Ogilby.

Pachyula Ogilby, 33; orthotype HEMIPIMELODUS CRASSILABRIS Ramsay. A synonym of HEMIPIMELODUS.

Thysanophrys Ogilby, 40; orthotype PLATYCEPHALUS CIRRONASUS Rich.

Tæniomembras Ogilby, 41; orthotype ATHERINA MICROSTOMA Gthr. Apparently a synonym of HEPSETIA Bon.

Endorrhis Ogilby, 283; orthotype COPIDOGLANIS LONGIFILIS Macleay. A synonym of PARAPLOTOSUS.

Congrellus Ogilby, 288; orthotype MURÆNA BALEARICA Delaroche. A synonym of ARIOSOMA Sw., as restricted ; first noted by Jordan & Evermann.

Bathycongrus Ogilby, 292; orthotype CONGROMURÆNA NASICA Alcock.

Eucentronotus Ogilby, 294; orthotype E. ZIETZI Ogilby. A synonym of PERONEDYS Steind.

Creedia Ogilby, 298; orthotype C. CLATHRISQUAMIS Ogilby.

1797. OGILBY (1898). *Additions to the Fauna of Lord Howe Island.* Proc. Linn. Soc. New S. Wales, 1898, XXIV, 730–745.

JAMES DOUGLAS OGILBY.

Howella Ogilby, 733; orthotype H. BRODIEI Ogilby.

Machærope Ogilby, 736; orthotype M. LATISPINIS Ogilby.

Diancistrus Ogilby, 743; orthotype D. LONGIFILIS Ogilby.

1798. OGILBY (1898). *On Some Australian Eleotrinæ.* Proc. Linn. Soc. New S. Wales, XXII, 783–893.

JAMES DOUGLAS OGILBY.

Caulichthys Ogilby, 784; orthotype ELEOTRIS GUENTHERI Blkr. A synonym of HYPSELEOTRIS Gill.

Austrogobio Ogilby, 788; orthotype CARASSIOPS GALII Ogilby.

1799. PRIEM (1898). *Sur la faune ichthyologique des Assises Montiennes du Bassin de Paris et en particulier sur Pseudolates heberti, Gervais.* Bull. Soc. Geol. France (3), XXVI, 399–412.

FERNAND PRIEM.

Pseudolates Priem, 408; orthotype LATES HERBERTI Gervais (fossil). Name preoccupied by Macleay, 1877; replaced by PROLATES Priem.

1800. SMITT (1898). *Poissons de l'Expédition scientifique à la Terre de Feu.* Bih. Svenska Ak., XXIII–IV, no. 5, 1–80.

FREDERIK ADAM SMITT (1839–1904).

Dissostichus Smitt, 3; orthotype D. ELEGINOIDES Smitt.

1801. STEINDACHNER (1898). *Die Fische der Sammlung Plate. Fauna Chilensis . . . nach der Sammlung von Dr. L. Plate.* Zool. Jahrb. Suppl., IV, 281–337.

FRANZ STEINDACHNER.

Platea Steindachner, 323; orthotype P. INSIGNIS Steind.

1802. STORMS (1898). *Première note sur les Poissons Wemméliens (Éocène Supérieur) de la Belgique.* Mém. Soc. Belge. Geol., X, 1898, 198–240.

RAYMOND STORMS.

Ctenodentex Storms, 199; orthotype DENTEX LÆKENIENSIS Van Beneden (fossil).

Eomyrus Storms, 225; orthotype E. DOLLOI Storms (fossil).

1803. TRAQUAIR (1898). *Notes on Paleozoic Fishes, No. 2.* Ann. Mag. Nat. Hist. (7), II, 67–70.

RAMSAY HEATLEY TRAQUAIR.

Farnellia Traquair, 69; orthotype F. TUBERCULATA Traquair (fossil).

1804. VINCIGUERRA (1898). *I pesci dell' ultima Spedizione del Cap. Bottego.* Ann. Mus. Genova (2), XIX, 240–261.

DECIO VINCIGUERRA.

Oxyglanis Vinciguerra, 250; orthotype O. SACCHII Vinciguerra.

1805. WOODWARD (1898). *Notes on Some Type Specimens of Cretaceous Fishes from Mount Lebanon in the Geneva Museum.* Ann. Mag. Nat. Hist., VII, series 2.

ARTHUR SMITH WOODWARD.

Thrissopteroides Woodward, 408; orthotype T. TENUICEPS Woodw. = CLUPEA ELONGATA Davis (fossil).

1899.

1806. ABBOTT (1899). *The Marine Fishes of Peru.* Proc. Acad. Nat. Sci. Phila., 1899, 324–364.

JAMES FRANCIS ABBOTT (1875–   ).

Pisciregia Abbott, 342; orthotype P. BEARDSLEEI Abbott. A synonym of BASILICHTHYS Grd.

1807. ALCOCK (1899). *New Species of Fish.* Proc. Asiat. Soc. Bengal, 1899, 781.

ALFRED WILLIAM ALCOCK.

Halimochirus Alcock. 78; orthotype H. CENTRISCOIDES Alcock.

1808. AMEGHINO (1899). *Sinopsis geologico-paleontologica. Supplemento (Adiciones y correcciones).* La Plata, 1899, 1–13.

FLORENTINO AMEGHINO (1854–1911).

Paraikichthys Ameghino, 10; orthotype P. ORNATISSIMUS Ameghino (fossil).

1809. BERG (1899). *Substitucion de nombres genericos.* II Comun. Mus. Buenos Aires, I, 37–80.

CARLOS BERG.

Gilbertidia Berg, 42; orthotype GILBERTINA SIGOLUTES Jordan & Starks. Substitute for GILBERTINA Jordan & Starks, preoccupied.

Matæocephalus Berg, 43; orthotype CŒLOCEPHALUS ACIPENSERINUS Gilbert & Cramer. Substitute for CŒLOCEPHALUS Gilbert & Cramer, preoccupied.

1810. BOULENGER (1899). *Matériaux pour la faune du Congo. Poissons nouveaux du Congo. Troisième Partie: Silures, Acanthopterygiens, Mastacembles, Plectognathes,* Ann. Mus. Congo, Zool. I, 39–58. *Quatrième Partie: Polyptéres, Clupes, Mormyres, Characins,* 59–96. *Cinquieme Partie: Cyprins, Silures, Cyprinodontes, Acanthopterygiens,* 97–128.

GEORGE ALBERT BOULENGER.

Chrysobagrus Boulenger, 40; orthotype C. BREVIBARBIS Boulenger.

Gephyroglanis Boulenger, 42; logotype G. CONGICUS Boulenger.

Steatocranus Boulenger, 52; orthotype S. GIBBICEPS Boulenger.

Teleogramma Boulenger, 53; orthotype T. GRACILE Boulenger.

Odaxothrissa Boulenger, 64; orthotype O. LOSERA Boulenger.

Paraphago Boulenger, 76; orthotype P. ROSTRATUS Boulenger.

Neoborus Boulenger, 78; orthotype N. ORNATUS Boulenger.

Micralestes Boulenger, 86; orthotype M. HUMILIS Boulenger.

Pseudoplesiops Boulenger, 90; orthotype P. NUDICEPS Boulenger. (Name preoccupied: Bleeker, 1858); replaced by NANOCHROMIS Pellégrin.

Chelæthiops Boulenger, 101; orthotype C. ÉLONGATUS Boulenger.

Parailia Boulenger, 105; orthotype P. CONGICA Boulenger.

Xenochromis Boulenger, 125; orthotype X. HECQUI Boulenger.

1811. BOULENGER (1899). *Second Contribution to the Ichthyology of Lake Tanganyika. On the Fishes Obtained by the Congo Free State Expedition under Lieut. Lemaire in 1898.* Trans. Zool. Soc. London, XV, 87–96.

GEORGE ALBERT BOULENGER.

Trematocara Boulenger, 89; orthotype T. MARGINATUM Boulenger.

Grammatotria Boulenger, 90; orthotype G. LEMAIRII Boulenger.

Xenotilapia Boulenger, 92; orthotype X. SIMA Boulenger.

1812. BOULENGER (1899). *Description of a New Genus of Perciform Fishes from the Cape of Good Hope.* Ann. S. African Mus., I, 379–380.

GEORGE ALBERT BOULENGER.

Atyposoma Boulenger, 379; orthotype A. GUERNEYI Boulenger.

1813. BOULENGER (1899). *Description of a New Genus of Gobioid Fishes from the Andes of Ecuador.* Ann. Mag. Nat. Hist. (7), IV, 125–126.

GEORGE ALBERT BOULENGER.

Oreogobius Boulenger, 125; orthotype O. ROSENBERGI Boulenger.

1814. BOULENGER (1899). *Descriptions of Two New Homalopterine Fishes from Borneo.* Ann. Nat. Hist. (7), IV, p. 228–229.

GEORGE ALBERT BOULENGER.

Glaniopsis Boulenger, 228; orthotype G. HANITSCHI Boulenger.

1815. EIGENMANN (1899). *A case of Convergence.* Science IX.

CARL H. EIGENMANN.

Troglichthys Eigenmann, 282; orthotype TYPHLICHTHYS ROSÆ Eigenmann (a blind fish from a cave in Missouri).

1816. EVERMANN & MARSH (1899). *Descriptions of New Genera and Species of Fishes from Puerto Rico.* Rept. U. S. Fish Comm. for 1899.

BARTON WARREN EVERMANN; MILLARD CALEB MARSH.

Gillias Evermann & Marsh, 357; orthotype G. JORDANI Evermann & Marsh.

Auchenistius Evermann & Marsh, 359; orhotype A. STAHLI Evermann & Marsh.

Coralliozetus Evermann & Marsh, 362; orthotype C. CARDONÆ Evermann & Marsh.

1817. GARMAN (1899). *Reports on an Exploration off the West Coasts of Mexico, Central and South America, and off the Galapagos Islands, in Charge of Alexander Agassiz, by the U. S. Fish Commission Steamer "Albatross" during 1891, Lieut. Commander Z. L. Tanner, U. S. N., Commanding. The Fishes.* Mem. Mus. Harvard, XXIV, 1–431.

SAMUEL GARMAN.

Centristhmus Garman, 47; orthotype C. SIGNIFER Garman.

Ectreposebastes Garman, 53; orthotype E. IMUS Garman.

Dolopichthys Garman, 81; orthotype D. ALLECTOR Garman.

Halieutopsis Garman, 89; orthotype H. TUMIFRONS Garman.

Dibranchopsis Garman, 96; orthotype HALIEUTÆA SPONGIOSA Gilbert.

Dibranchichthys Garman, 99; orthotype D. NUDIVOMER Garman.

Bothrocaropsis Garman, 127; orthotype B. ALALONGA Garman.

Leucicorus Garman, 146; orthotype L. LUSCIOSUS Garman.

Holcomycteronus Garman, 163; orthotype H. DIGITATUS Garman.

Eretmichthys Garman, 164; orthotype E. PINNATUS Garman.

Pseudonus Garman, 169; orthotype P. ACUTUS Garman.

Sciadonus Garman, 171; orthotype S. PEDICELLARIS Garman.

Microlepidium Garman, 180; orthotype LEPIDION VERECUNDUM Gilbert.

Leptophycis Garman, 182; orthotype L. FILIFER Garman.

Lychnopoles Garman, 224; orthotype L. ARGENTEOLUS Garman.

Dactylostomias Garman, 279; orthotype D. FILIFER Garman.

Congrosoma Garman, 308; orthotype C. EVERMANNI Garman.

Atopichthys Garman, 327; logotype A. ESUNCULUS Garman. A catchall for unplaced larval eels not known to be young CONGERS.

1818–1819. HAY (1899). *On Some Changes in the Names, Generic and Specific, of Certain Fossil Fishes.* Amer. Naturalist, XXXIII, 783–792.

OLIVER PERRY HAY.

Stemmatias Hay, 784; logotype STEMMATODUS CHEIRIFORMIS St. John & Worthen (fossil). Substitute for STEMMATODUS St. John & Worthen, preoccupied by STEMMATODUS Heckel.

Hybocladodus Hay, 784; orthotype HELODUS COMPRESSUS St. John & Worthen (fossil).

Platyxystrodus Hay, 785; orthotype COCHLIODUS STRIATUS Ag. (fossil). Substitute for XYSTRODUS (Ag.) Morris & Roberts, preoccupied by XYSTRODUS Plieninger (1860).

Polypsephis Hay, 788; orthotype MICRODON ELEGANS Ag. Substitute for MICRODON Ag., preoccupied in DIPTERA Meigen, 1803; but PROSCINETES Gistel (1848) is a still earlier substitute name.

Redfieldius Hay, 789; orthotype CATOPTERUS GRACILIS Redfield. Substitute for CATOPTERUS Redfield, preoccupied by CATOPTERUS Ag.

Grypodon Hay, 790; orthotype ANCISTRODON TEXANUS Dames (fossil). Substitute for ANCISTRODON Rœmer, and AGKISTRODUS de Koninck, regarded as preoccupied by AGKISTRODON Beauvais (ANCISTRODON of authors).

1820. JORDAN & SNYDER (1899). *Notes on a Collection of Fishes from the Rivers of Mexico, with Descriptions of Twenty New Species.* Bull. U. S. Fish Comm. for 1899, 116–147.

DAVID STARR JORDAN; JOHN OTTERBEIN SNYDER (1867–   ).

Istlarius Jordan & Snyder, 118; orthotype I. BALSANUS Jordan & Snyder.

Xystrosus Jordan & Snyder, 123; orthotype X. POPOCHE Jordan & Snyder.
Falcula Jordan & Snyder, 124; orthotype F. CHAPALÆ Jordan & Snyder.
Xenendum Jordan & Snyder, 127; orthotype X. CALIENTE Jordan & Snyder.

1821. KARPINSKII (1899). *Ueber die Reste von Edestiden und die neue Gattung von Helicoprion.* Mem. Acad. Sci. St. Petersb., VIII (and in other papers).
ALEKSANDR PETROVICH KARPINSKII (1846–).
Helicoprion Karpinskii, 307; orthotype H. BESSENOWI Karpinskii (fossil).

1822. OGILBY (1899). *Contributions to Australian Ichthyology.* Proc. Linn. Soc. New S. Wales, 1899, XXIV, 154–186.
JAMES DOUGLAS OGILBY.
Euristhmus Ogilby, 154; logotype PLOTOSUS ELONGATUS Castelnau.
Ostophycephalus Ogilby, 155; orthotype O. DURICEPS Ogilby.
Austrocobitis Ogilby, 158; orthotype MESITES ATTENUATUS Jenyns. A substitute for MESITES Jenyns, preoccupied; a synonym of GALAXIAS Cuv.
Epinephelides Ogilby, 170; orthotype E. LEAI Ogilby.
Anagramma Ogilby, 175; orthotype CALLANTHIAS ALLPORTI Gthr. A synonym of CALLANTHIAS Gthr.

1823–1824. PRIEM (1899). *Sur des poissons fossiles éocènes d'Egypte et de Roumanie, et rectification rélative à Pseudolates heberti (Gervais).* Bull. Soc. Geol. France, 1899, 3 ser. XXVII, 241–252.
FERNAND PRIEM.
Scorpænoides Priem, 248; orthotype S. POPOVICII Priem. (fossil). (Not SCORPÆNODES Blkr.)
Prolates Priem, 252; orthotype LATES HEBERTI Gervais (fossil). A substitute for PSEUDOLATES Priem, preoccupied.

1825. SMITT (1899). *Preliminary Notes on the Arrangement of the Genus Gobius, with an Enumeration of Its European Species.* Ofv. Vet. Ak. Forh., 1899, 543–555.
FREDERIK ADAM SMITT (1839–1904).
Caffrogobius Smitt, 540; orthotype GOBIUS NUDICEPS Cuv. & Val.
Lebistes Smitt, 543; orthotype L. SCORPIOIDES Smitt. Not LEBISTES Filippi, 1862.
Mugilogobius Smitt, 543; for "Indian & Japanese Species," no type named, but in a personal letter in 1903 Dr. Smitt stated to me that his type species was the one named CTENOGOBIUS ABEI Jordan & Snyder. Accepting this version, MUGILOGOBIUS is substantially the same as VAIMOSA Jordan & Seale, which name it may replace.
Mapo Smitt, 543; orthotype GOBIUS SOPORATOR Cuv. & Val. A synonym of BATHYGOBIUS Blkr.
Proterorhinus Smitt, 544; orthotype GOBIUS MARMORATUS Pallas.
Eichwaldia Smitt, 545; orthotype GOBIUS CASPIUS Eichwald.

1826. STEWART (1899). *Leptichthys, a New Genus of Fishes from the Cretaceous of Kansas.* Amer. Geologist, XXIV.
ALBAN F. STEWART.
Leptichthys Stewart, 78; orthotype L. AGILIS Stewart (fossil).

1827. TRAQUAIR (1899). *Report on Fossil Fishes Collected by the Geological Survey of Scotland in the Silurian Rocks of the South of Scotland.* Trans. Roy. Soc. Edinburgh, XXXIX, 827–864.
RAMSAY HEATLEY TRAQUAIR.

Lanarkia Traquair, 832; logotype L. HORRIDA Traquair (fossil).
Ateleaspis Traquair, 834; orthotype A. TESSELLATA Traquair (fossil).
Birkenia Traquair, 837; orthotype B. ELEGANS Traquair (fossil).
Lasanius Traquair, 840; logotype L. PROBLEMATICUS Traquair (fossil).

1828. VARPACHOVSKII (1899). *Sur la faune ichthyologique de la fleuve Obi.* Ann. Mus. Acad. Imp. St. Petersb.

NICOLAI ARCADEVICH VARPACHOVSKII.

Oreoleuciscus Varpachovskii, 263; orthotype CHONDROSTOMA POTANINI Kessler.

1829. WAITE (1899). *Scientific Results of the Trawling Expedition of H. M. C. C. S. "Thetis." Fishes.* Mem. Austr. Mus., IV, 1–132.

EDGAR RAVENSWOOD WAITE (1866–   ).

Paratrachichthys Waite, 64; orthotype TRACHICHTHYS TRAILLI Hutton.
Pterygotrigla Waite, 108; orthotype TRIGLA POLYOMMATA Rich. Replaces HOPLO-NOTUS Guichenot, preoccupied, apparently same as OTOHIME Jordan & Starks of later date.

1830. WILLISTON (1899). *A New Genus of Fishes from the Niobara Cretaceous.* Kans. Univ. Quarterly, 1899, VIII, 113–115.

SAMUEL WENDALL WILLISTON (1852–1918).

Leptecodon Williston, 113; orthotype L. RECTUS Williston (fossil).

1831. WOODWARD (1899). *Additional Notes on Some Type Specimens of Cretaceous Fishes from Mount Lebanon in the Edinburgh Museum of Science and Art.* Ann. Mag. Nat. Hist. (7), IV, 317–321.

ARTHUR SMITH WOODWARD.

Nematonotus Woodward, 318; orthotype CLUPEA BOTTÆ Pictet & Humbert (fossil).

1832. WOODWARD (1899). *Note on Some Cretaceous Clupeoid Fishes with Pectinated Scales. (Ctenothrissa and Pseudoberyx.)* Ann. Mag. Nat. Hist. 1899, (7), III, 489–492.

ARTHUR SMITH WOODWARD.

Ctenothrissa Woodward, 490; orthotype BERYX VEXILLIFER Pictet (fossil).

1900.

1833. BOULENGER (1900). *On Some Little Known African Siluroid Fishes of the Subfamily Doradinæ.* Ann. Mag. Nat. Hist. (7), VI, 520–529.

GEORGE ALBERT BOULENGER.

Euchilichthys Boulenger, 523; logotype ATOPOCHILUS GUENTHERI Schilthuis.
Phractura Boulenger, 527; orthotype PELTURA BOVEI Perugia. Substitute for PELTURA Perugia, regarded as preoccupied by PELTOURA, 1840.
Andersonia Boulenger, 528; orthotype A. LEPTURA Boulenger.

1834. BOULENGER (1900). *Matériaux pour la faune du Congo. Poissons nouveaux du Congo. Sixième Partie.* Ann. Mus. Congo. Zool., I, 129–164.

GEORGE ALBERT BOULENGER.

Leptocypris Boulenger, 133; orthotype L. MODESTUS Boulenger.
Spathodus Boulenger, 152; orthotype S. ERYTHRODON Boulenger.

1835. BRAUER (1900). *Aus der Tiefen des Weltmeeres* (by Prof. Carl Chun). *Schilderung von der deutschen Tiefsee-Expedition.* Jena, 1900, 549 pp.

AUGUST BRAUER.

Megalopharynx Brauer, 521; orthotype M. LONGICAUDATUS Brauer.

1836. DELFIN (1900). *Nota del Ictiolojia. El nuevo jenero Cilus.* Act. Soc. Chili, X, 53–60.

FREDERICO TEOBALDO DELFIN.

Cilus Delfin, 56; orthotype C. MONTII Delfin.

1837. DOLLO (1900). *Cryodraco antarcticus, poisson abyssal nouveau, recueilli par l'expédition antarctique belge,* etc. Bull. Ac. Belgique, 1900, 128–137.

LOUIS DOLLO (1857–   ).

Cryodraco Dollo, 129; orthotype C. ANTARCTICUS Dollo.

1838. DOLLO (1900). *Gerlachea australis, poisson abyssal nouveau recueilli par l'expédition antarctique belge,* etc. Bull. Ac. Belgique, 1900, 194–206.

LOUIS DOLLO.

Gerlachea Dollo, 195; orthotype G. AUSTRALIS Dollo.

1839. DOLLO (1900). *Racovitzia glacialis, poisson abyssal nouveau recueilli par l'expédition antarctique belge,* etc. Bull. Ac. Belgique, 1900, 316–327.

LOUIS DOLLO.

Racovitzia Dollo, 317; orthotype R. GLACIALIS Dollo.

1840. EIGENMANN & NORRIS (1900). *Sobre algunos peixes de S. Paulo, Brazil.* Revista Mus. Paulista, IV, 1900, 349–362.

CARL H. EIGENMANN; ALLEN A. NORRIS.

Imparfinis Eigenmann & Norris, 351; orthotype I. PIPERATUS Eigenmann & Norris.
Goeldiella Eigenmann & Norris, 353; orthotype PIMELODUS EQUES M. & T.
Iheringichthys Eigenmann & Norris, 354; orthotype PIMELODUS LABROSUS Kröyer.
Perugia Eigenmann & Norris, 355; orthotype PIRINAMPUS AGASSIZI Steind.
Bergiella Eigenmann & Norris, 355; orthotype PIMELODUS WESTERMANNI Reinh.
Catabasis Eigenmann & Norris, 358; orthotype C. ACUMINATUS Eigenmann & Norris.

1841. GILBERT (1900). *Results of the Branner-Agassiz Expedition to Brazil. III. The Fishes.* Proc. Wash. Ac., II, 161–184.

CHARLES HENRY GILBERT.

Brannerella Gilbert, 180; orthotype B. BRASILIENSIS Gilbert.

1842. JORDAN (1900). *Notes on Recent Fish Literature.* Amer. Naturalist, XXXIV, 897–899.

DAVID STARR JORDAN.

Dolloa Jordan, 897; orthotype CORYPHÆNOIDES LONGIFILIS Gthr. Substitute for MOSELEYA G. & B., preoccupied.

1843. JORDAN & EVERMANN (1900). *The Fishes of North & Middle America; a Descriptive Catalogue of the Species of Fish-like Vertebrates Found in the Waters of North America, North of the Isthmus of Panama.* Part IV, Bull. U. S. Nat. Mus., No. 47, 3137–3313.

DAVID STARR JORDAN; BARTON WARREN EVERMANN.

Orcula Jordan & Evermann, 3140; orthotype NOTROPIS ORCA Woolman. A substitute for ORCELLA J. & E., preoccupied in CETACEA.

Rhegma (Gilbert) Jordan & Evermann, 3169; orthotype R. THAUMASIUM Gilbert.

1844. JORDAN & SNYDER (1900). *A List of Fishes Collected in Japan by Keinosuke Otaki, and by the United States Steamer "Albatross," with Descriptions of Fourteen New Species.* Proc. U. S. Nat. Mus., XXIII, 335–380. (Dec. 20, 1890.)

DAVID STARR JORDAN; JOHN OTTERBEIN SNYDER.

Otakia Jordan & Snyder, 345; orthotype O. RASBORINA Jordan & Snyder.

Ischikauia Jordan & Snyder, 346; orthotype OPSARIICHTHYS STEENACKERI Sauvage.

Konosirus Jordan & Snyder, 349; orthotype CHATOËSSUS NASUS Bloch. A synonym of CLUPANODON Lac., as now properly restricted.

Bryttosus Jordan & Snyder, 354; orthotype SERRANUS KAWAMEBARI T. & S.

Eteliscus Jordan & Snyder, 355; orthotype ETELIS BERYCOIDES Hilgendorf. Equivalent to CORUSCULUS Jordan & Snyder. Both are synonyms of DODERLEINIA Steind.

Insidiator Jordan & Snyder, 368; orthotype PLATYCEPHALUS RUDIS Gthr. = PLATYCEPHALUS MEERDERVOORTI Blkr.

Trifissus Jordan & Snyder, 373; orthotype T. IOTURUS Jordan & Snyder. A synonym of TRIDENTIGER Gill.

Rhombiscus Jordan & Snyder, 379; orthotype RHOMBUS CINNAMOMEUS T. & S. A synonym of PSEUDORHOMBUS Blkr.

Kareius Jordan & Snyder, 379; orthotype PLEURONECTES SCUTIFER Steind.

Usinosita Jordan & Snyder, 380; orthotype PLAGUSIA JAPONICA T. & S., misprinted USINOSTIA. A synonym of RHINOPLAGUSIA Blkr.

Zebrias Jordan & Snyder, 380; orthotype SOLEA ZEBRINA T. & S.

Areliscus Jordan & Snyder, 380; orthotype SOLEA JOYNERI Gthr.

1845. KYLE (1900). *On a New Genus of Flat-Fishes from New Zealand.* Proc. Zool. Soc. London, 1900, 986–992.

HARRY M. KYLE.

Apsetta Kyle, 986; orthotype A. THOMPSONI Kyle.

1846. LOOMIS (1900). *Die Anatomie und die Verwandschaft der Ganoid-und Knochen-Fische aus der Kreide-Formation von Kansas, U. S. A.* Palæontgr., XLVI, 213–284.

FREDERICK BREWSTER LOOMIS (1873–   ).

Pseudothryptodus Loomis, 225; orthotype P. INTERMEDIUS Loomis (fossil).

Thryptodus Loomis, 229; logotype PLETHODUS ZITTELI Loomis (fossil). Apparently replaces ANOGMIUS Cope, 1877 (A. ARATUS), not of 1871 (A. CONTRACTUS).

Syntegomodus Loomis, 252; orthotype S. ALTUS Loomis (fossil).

1847. NIKOLSKII (1900). *Pseudoscaphirhynchus rossikowi n. gen. et spec.* Ann. Mus. St. Petersb., IV, 257–259.

ALEKSANDR MIKHAILOVICH NIKOLSKII.

Pseudoscaphirhynchus Nikolskii, 257; orthotype P. ROSSIKOWI Nikolskii = SCAPHIRHYNCHUS HERMANNI (Sewertov) Kessler. A synonym of KESSLERIA Bogdanow.

1848. PELLÉGRIN (1900). *Poissons nouveaux ou rares du Congo Français.* Bull. Mus. Paris, 1900, 175–176, 348–354.

JACQUES PELLÉGRIN.

Mesoborus Pellégrin, 179; orthotype M. CROCODILUS Pellégrin.

Hemistichodus Pellégrin, 352; orthotype H. VAILLANTI Pellégrin. Replaces MONO-STICHODUS Vaillant, said to be "nomen nudum."

Monostichodus (Vaillant) Pellégrin, 352; orthotype (HEMISTICHODUS VAILLANTI Pellégrin). A "nomen nudum" quoted by Pellégrin, said to be synonymous with HEMISTICHODUS.

1849. PELLÉGRIN (1900). *Poisson nouveau du lac Baikal.* Bull. Mus. Paris, 1900, 354–356.

JACQUES PELLÉGRIN.

Cottocomephorus Pellégrin, 354; orthotype C. MEGALOPS Pellégrin.

1850. WAITE (1900). *Additions to the Fish-Fauna of Lord Howe Island.* Rec. Austr. Mus., III, 193–209.

EDGAR RAVENSWOOD WAITE.

Acanthocaulus Waite, 206; orthotype PRIONURUS MICROLEPIDOTUS Lac. A substitute for PRIONURUS Lac., 1804, wrongly supposed to be preoccupied, the date of the latter being assumed as 1830, the time of a reprint.

Euchilomycterus Waite, 208; orthotype E. QUADRATICATUS Waite.

1851. WILLISTON (1900). *Some Fossil Teeth from the Kansas Cretaceous.* Kansas Quart., VIII, 27–42.

SAMUEL WENDALL WILLISTON.

Leptostyrax Williston, 42; orthotype L. BISCUSPIDATUS Williston (fossil).

1852. WOODWARD (1900). *On a New Ostracoderm (Euphanerops longævus) from the Upper Devonian of Scaumenac Bay, Province of Quebec, Canada.* Ann. Mag. Nat. Hist. (7), V, 416–419.

ARTHUR SMITH WOODWARD.

Euphanerops Woodward, 416; orthotype E. LONGÆVUS Woodw. (fossil).

1853. WOODWARD (1900). *Evidence of an Extinct Eel (Urenchelys anglicus) from the Upper Devonian of Scaumenac Bay, Province of Quebec, Canada.* Ann. Mag. Nat. Hist. (7), V, 321–323.

ARTHUR SMITH WOODWARD.

Urenchelys Woodward, 322; logotype ANGUILLA HAKELENSIS Davis (fossil).

Pronotacanthus Woodward, 322; orthotype ANGUILLA SAHEL-ALMÆ Davis (fossil).

1854. ZITTEL (1900). *Handbuch der Palæontologie, III.* Pt. XII, 1890.

KARL A. ZITTEL.

Phlyctænium Zittel, 159; orthotype COCCOSTEUS ACADICUS Whiteaves (fossil). A substitute for PHLYCTÆNIUS Traquair, preoccupied, but PHLYCTÆNASPIS Traquair, 1890, has priority.

1901.

1855. ABBOTT (1901). *List of Fishes Collected in the River Pei-Ho at Tien-Tsin, China, by Noah Fields Drake, with Descriptions of Seven New Species.* Proc. U. S. Nat. Mus., XXIII, 483–491. (February 25, 1901.)

JAMES FRANCIS ABBOTT (1876–).

Culticula Abbott, 485; orthotype C. EMMELAS Abbott.

1856. AMMON (1901). *Ueber eine Tiefbohrung durch den Buntsandstein und die Zechssteinschichten bei Meelrichstadt an der Rhön.* Geogn. Jahresb. München, XIII.

LUDWIG VON AMMON.

Ephippites Von Ammon, 59; orthotype E. PEISSENBERGIENSIS Ammon (fossil).

1857. BOULENGER (1901). *Third Contribution to the Ichthyology of Lake Tanganyika. Report on the Collection Made by Mr. J. E. S. Moore in Lake Tanganyika and Kivu during His Second Expedition, 1899–1900.* Trans. Zool. Soc. London, XVI, 137–178.

GEORGE ALBERT BOULENGER.

Gephyrochromis Boulenger, 156; orthotype G. MOORII Boulenger.
Asprotilapia Boulenger, 159; orthotype A. LEPTURA Boulenger.

1858. BOULENGER (1901). *Diagnosis of New Fishes Discovered by Mr. W. L. S. Loat in the Nile.* Ann. Mag. Nat. Hist. (7), VII, 444–446.

GEORGE ALBERT BOULENGER.

Cromeria Boulenger, 445; orthotype C. NILOTICA Boulenger.
Physailia Boulenger, 445; orthotype P. PELLUCIDA Boulenger.

1859. BOULENGER (1901). *Notes on the Classification of Teleostean Fishes. I. On the Trachinidæ and Their Allies.* Ann. Mag. Nat. Hist. (7), VII, 261–271.

GEORGE ALBERT BOULENGER.

Rhyacichthys Boulenger, 267; orthotype P. ASPRO Van Hasselt. A substitute for PLATYPTERA (Kuhl & Van Hasselt) Cuv. & Val.; preoccupied in files, Meigen, 1803.

1860. BOULENGER (1901). *On the Fishes Collected by Dr. W. J. Ansorge in the Niger Delta.* Proc. Zool. Soc. London, 1901, 4–10.

GEORGE ALBERT BOULENGER.

Polycentropsis Boulenger, 8; orthotype P. ABBREVIATA Boulenger.

1861. BOULENGER (1901). *On Some Deep-Sea Fishes Collected by Mr. F. W. Townsend in the Sea of Oman.* Ann. Mag. Nat. Hist. (7), VII, 261–263.

GEORGE ALBERT BOULENGER.

Parascolopsis Boulenger, 262; orthotype P. TOWNSENDI Boulenger.

1862. BRAUER (1901). *Ueber einige von der Valdivia-Expedition gesemmelte Tiefseefische und ihre Augen.* Sitzber. Ges. Marburg, 1901, 115–130.

AUGUST BRAUER.

Winteria Brauer, 126; orthotype W. TELESCOPA Brauer.
Dolichopteryx Brauer, 127; orthotype D. LONGIPES Brauer.
Gigantura Brauer, 128; orthotype G. CHUNI Brauer.

1863. COLLETT (1901). *Om fem for Norges fauna nye fiske.* Arch. Math. Naturv. Christiania, XXIII, 125.

ROBERT COLLETT.

Bathyalopex Collett, 5; orthotype CHIMÆRA MIRABILIS Collett.

1863A. DEAN (1901). *Further Notes on the Relationships of the Arthrognathi. Palæontological Notes.* Mem. N. Y. Acad. Sci., II, 110 to 123.

BASHFORD DEAN (1867–   ).

Selenosteus Dean, 110; orthotype S. KEPLERI Dean (fossil).
Stenosteus Dean, 110; orthotype S. GLABERI Dean (fossil).

1864. EIGENMANN & NORRIS (1901). *Bergiaria.* Comun. Mus. Buenos Aires, 272.

CARL H. EIGENMANN; ALLEN A. NORRIS.

Bergiaria Eigenmann & Norris, 272; orthotype PIMELODUS WESTERMANNI Reinh. Substitute for BERGIELLA Eigenmann & Norris, preoccupied.

1865. FOWLER (1901). *Description of a New Hemiramphid.* Proc. Acad. Nat. Sci. Phila., 1901, 293–294.

HENRY WEED FOWLER.

Hemiexocœtus Fowler, 293; orthotype H. CAUDIMACULATUS Fowler.

1866. FOWLER (1901). *Note on the Odontostomidæ.* Proc. Acad. Nat. Sci. Phila., 1901, 211–212.

HENRY WEED FOWLER.

Evermanella Fowler, 211; orthotype ODONTOSTOMUS HYALINUS Cocco. Substitute for ODONTOSTOMUS Cocco, preoccupied in Mollusks, Beck, 1837.

1867. GARMAN (1901). *Genera and Families of Chimæroids.* Proc. New England Zool. Club, II, 75–77.

SAMUEL GARMAN.

Rhinochimæra Garman, 75; orthotype HARRIOTTA PACIFICA Mitsukuri.

1868. GILL & TOWNSEND (1901). *The Largest Deep-Sea Fish.* Science (2), XIV, 937–938.

THEODORE NICHOLAS GILL; CHARLES HASKINS TOWNSEND (1859–   ).

Macrias Gill & Townsend, 937; orthotype M. AMISSUS Gill & Townsend.

1869. GREELEY (1901). *Notes on the Tide-Pool Fishes of California, with a Description of Four New Species.* Bull. U. S. Fish Comm., XIX, 7–20.

ARTHUR WHITE GREELEY (1875–1904).

Rusciculus Greeley, 13; orthotype R. RIMENSIS Greeley.

Dialarchus Greeley, 14; orthotype D. SNYDERI Greeley.

Eximia Greeley, 18; orthotype E. RUBELLIO Greeley. Name preoccupied: Karl Jordan, 1894; it may be replaced by the new name GREELEYA Jordan, in honor of the very able young naturalist, discoverer of the species.

1870. JENKINS (1901). *Descriptions of Fifteen New Species of Fishes from the Hawaiian Islands.* Bull. U. S. Fish Comm., XIX, 387–407.

OLIVER PEEBLES JENKINS (1850–   ).

Eumycterias Jenkins, 399; orthotype E. BITÆNIATUS Jenkins. A synonym of CANTHIGASTER Sw.

Scaridea Jenkins, 468; orthotype S. ZONARCHA Jenkins.

1871. JORDAN & SNYDER (1901). *Descriptions of Three New Species of Fishes from Japan.* Proc. Calif. Acad. Sci. (3), II, 381–386.

DAVID STARR JORDAN; JOHN OTTERBEIN SNYDER.

Ereunias Jordan & Snyder, 377; orthotype E. GRALLATOR Jordan & Snyder.

Draciscus Jordan & Snyder, 379; orthotype D. SACHI Jordan & Snyder.

1872. JORDAN & SNYDER (1901). *List of Fishes Collected in 1883 and 1885 by Pierre Louis Jouy and preserved in the United States National Museum, with Descriptions of Six New Species.* Proc. U. S. Nat. Mus., XXIII, 740–769. (July 21, 1901.)

DAVID STARR JORDAN; JOHN OTTERBEIN SNYDER.

494 THE GENERA OF FISHES, PART IV

Chasmias Jordan & Snyder, 761; orthotype C. MISAKIUS Jordan & Snyder. Name preoccupied, replaced by CHASMICHTHYS Jordan & Snyder.

Watasea Jordan & Snyder, 765; orthotype W. SIVICOLA Jordan & Snyder.

1873. JORDAN & SNYDER (1901). *A Preliminary Check-List of the Fishes of Japan.* Annot. Zool. Jap., III, 31–159.

DAVID STARR JORDAN; JOHN OTTERBEIN SNYDER.

Corusculus Jordan & Snyder, 75; orthotype ANTHIAS BERYCOIDES Hilgendorf. A synonym of DODERLEINIA Steind., the definition being that of ETELISCUS duplicated by accident.

1874. JORDAN & SNYDER (1901). *Descriptions of Nine New Species of Fishes Contained in Museums of Japan.* Journ. Coll. Sci. Tokyo, XV, 301–311.

DAVID STARR JORDAN; JOHN OTTERBEIN SNYDER.

Bentenia Jordan & Snyder, 306; orthotype B. ÆSTICOLA Jordan & Snyder.

Ebisus Jordan & Snyder, 308; orthotype E. SAGAMIUS Jordan & Snyder. A synonym of ERILEPIS Gill.

1875. JORDAN & SNYDER (1901). *Fishes of Japan.* Am. Nat., XXXV.

DAVID STARR JORDAN; JOHN OTTERBEIN SNYDER.

Chasmichthys Jordan & Snyder, 941; orthotype CHASMIAS MISAKIUS Jordan & Snyder. Substitute for CHASMIAS Jordan & Snyder, preoccupied.

1876. JORDAN & SNYDER (1901). *A Review of the Apodal Fishes of Japan, with Descriptions of Nineteen New Species.* Proc. U. S. Nat. Mus., XXIII, 837–890. (Aug. 28, 1901.)

DAVID STARR JORDAN; JOHN OTTERBEIN SNYDER.

Chlevastes Jordan & Snyder, 867; orthotype OPHICHTHYS COLUBRINUS Boddaert.

Xyrias Jordan & Snyder, 868; orthotype X. REVULSUS Jordan & Snyder.

Æmasia Jordan & Snyder, 883; orthotype Æ. LICHENOSA Jordan & Snyder.

Scuticaria Jordan & Snyder, 886; orthotype ICHTHYOPHIS TIGRINUS Lesson.

1877. JORDAN & SNYDER (1901). *A Review of the Gobioid Fishes of Japan, with Descriptions of Twenty-one New Species.* Proc. U. S. Nat. Mus., XXIV, 33–132. (Sept. 25, 1901.)

DAVID STARR JORDAN; JOHN OTTERBEIN SNYDER.

Vireosa Jordan & Snyder, 38; orthotype V. HANÆ Jordan & Snyder.

Hazeus Jordan & Snyder, 51; orthotype H. OTAKII Jordan & Snyder. A synonym of GNATHOLEPIS Blkr.

Chloëa Jordan & Snyder, 78; orthotype GOBIUS CASTANEA O'Shaughnessy. Not CHLOEIA Savigny, 1817, a genus of worms. CHLOËA is a personal name.

Suruga Jordan & Snyder, 96; orthotype S. FUNDICOLA Jordan & Snyder.

Sagamia Jordan & Snyder, 100; orthotype S. RUSSULA Jordan & Snyder.

Ainosus Jordan & Snyder, 109; orthotype GOBIUS GENEIONEMA Hilgendorf.

Astrabe Jordan & Snyder, 119; orthotype A. LACTISELLA Jordan & Snyder.

Clariger Jordan & Snyder, 120; orthotype C. COSMURUS Jordan & Snyder.

Eutæniichthys Jordan & Snyder, 122; orthotype E. GILLI Jordan & Snyder.

1878. JORDAN & SNYDER (1901). *A Review of the Hypostomide and Lophobranchiate Fishes of Japan.* Proc. U. S. Nat. Mus., XXIV, 1–20. (Oct. 1. 1901.)

DAVID STARR JORDAN; JOHN OTTERBEIN SNYDER.

Zalises Jordan & Snyder, 2; orthotype Z. UMITENGU Jordan & Snyder. A synonym of PEGASUS L.

Yozia Jordan & Snyder, 8; orthotype Y. WAKANOURÆ Jordan & Snyder.

1879. JORDAN & SNYDER (1901). *A Review of the Cardinal Fishes of Japan.*
Proc. U. S. Nat. Mus., XXIII, 891–913. (Oct. 12, 1901.)

DAVID STARR JORDAN; JOHN OTTERBEIN SNYDER.

Telescopias Jordan & Snyder, 909; orthotype T. GILBERTI Jordan & Snyder.

1880. JORDAN & STARKS (1901). *Descriptions of Three New Species of Fishes
from Japan.* Proc. Calif. Acad. Sci. (3), II, 381–386.

DAVID STARR JORDAN; EDWIN CHAPIN STARKS (1867–   ).

Snyderina Jordan & Starks, 381; orthotype S. YAMANOKAMI Jordan & Starks.

1881. JORDAN & STARKS (1901). *A Review of the Atherine Fishes of Japan.*
Proc. U. S. Nat. Mus., XXIV, 198–206. (Oct. 4, 1901.)

DAVID STARR JORDAN; EDWIN CHAPIN STARKS.

Atherion Jordan & Starks, 203; orthotype A. ELYMUS Jordan & Starks.

Iso Jordan & Starks, 204; orthotype I. FLOS-MARIS Jordan & Starks.   Replaces
TROPIDOSTHETHUS Ogilby, preoccupied.

1882. LAUBE. (1901). *Synopsis der Wirbelthierfauna der Böhmischen Braun-
koklenformation und Beschreibung neuer, oder bisher unvollständig bekannter
Arten.* Prag., 1901, 80 pp.

GUSTAVE CARL LAUBE (1839–   ).

Protothymallus Laube, 23; orthotype THAUMATURUS LUSATUS Laube (fossil).

1883. TRAQUAIR (1901). *Notes on the Lower Carboniferous Fishes of Eastern
Fifeshire.* Geol. Mag. (2), Dec., IV, VIII, 110–114.

RAMSAY HEATLEY TRAQUAIR.

Eucentrurus Traquair, 114; orthotype E. PARADOXUS Traquair (fossil).

1884. WELLBURN (1901). *On the Fish Fauna of the Millstone Grits of Great
Britain.* Geol. Mag. (2), Dec. IV, 216–222.

EDGAR D. WELLBURN.

Euctenodopsis Wellburn, 220; orthotype E. TENUIS Wellburn (fossil).

1885. WILLEY (1901). *Dolichorhynchus indicus, a New Acraniate.* Quart. Journ.
Micr. Sci. (2), XLIV, 269–271.

ARTHUR WILLEY (1867–   ).

Dolichorhynchus Willey, 269; orthotype D. INDICUS Willey.

1886. WOODWARD (1901). *Catalogue of Fossil Fishes in the British Museum
Natural Hist.* Part IV, etc., 636 pp.

ARTHUR SMITH WOODWARD.

Elopides (Agassiz) Woodward, 23; orthotype E. COULONI Ag. (fossil).   Printed as
name only in *Poissons Fossiles,* V, 1844, 139. A fossil homologue of ELOPS L.

Notelops Woodward, 27; orthotype RHACOLEPIS BRAMA Ag. = CALAMOPLEURUS
CYLINDRICUS Ag. (fossil).   A synonym of CALAMOPLEURUS Ag.

Esocelops Woodward, 46; orthotype EURYGNATHUS CAVIFRONS Ag. (fossil).   A substitute for EURYGNATHUS Ag., preoccupied.

Chanoides Woodward, 63; orthotype CLUPEA MACROPOMA Ag. (fossil).

Brychætus Woodward, 76; orthotype B. MUELLERI Woodw. (fossil).   Name only in
1845; validated in 1901.

Histiothrissa Woodward, 131; orthotype SARDINIUS MACRODACTYLUS Von der
Marck (fossil).

Halecopsis Woodward, 133; orthotype OSMEROIDES INSIGNIS Delvaux & Ortlieb (fossil).

Apateodus Woodward, 258; orthotype PACHYRHIZODUS GLYPHODUS Blake (fossil).

Pachylebias Woodward, 294; orthotype LEBIAS CRASSICAUDATUS Ag. (fossil).

Rhynchorhinus (Agassiz) Woodward, 342; orthotype R. BRANCHIALIS Ag. (fossil). Name printed without explanation in 1844.

Protaulopsis Woodward, 371; orthotype P. BOLCENSIS Woodward (fossil).

Dinopteryx Woodward, 406; orthotype HOPLOPTERYX SPINOSUS Davis (fossil).

Isurichthys Woodward, 453; logotype ISURUS MACRURUS Ag. (fossil). Name a substitute for ISURUS Ag., preoccupied.

Eothynnus Woodward, 457; orthotype CŒLOCEPHALUS SALMONEUS Ag. (fossil). Name a substitute for CŒLOCEPHALUS Ag., undefined and preoccupied by Clark, 1860, in COLEOPTERA.

Phalacrus (Agassiz) Woodward, 458; orthotype P. CYBIOIDES Ag. (fossil). Note on type specimen. A synonym of EOTHYNNUS Woodward.

Cariniceps (Owen) Woodward, 458; orthotype C. COMPRESSUS Owen (fossil). Note on type specimen. A synonym of EOTHYNNUS Woodw.

Rhonchus (Agassiz) Woodward, 458; orthotype R. CARANGOIDES Ag. (fossil). Note on type specimen. A synonym of EOTHYNNUS Woodw.

Scombrinus (Agassiz) Woodward, 461; orthotype S. NUCHALIS Ag. (fossil). Validation of a nomen nudum printed in 1845.

Eocœlopoma Woodward, 470; orthotype E. COLEI Woodw. (fossil).

Scombramphodon Woodward, 474; orthotype AMPHODON BENEDENI Storms (fossil). Replaces AMPHODON Storms, 1887, regarded as preoccupied by AMPHODUS Peters, 1872.

Acestrus (Agassiz) Woodward, 494; orthotype A. ORNATUS Woodw. (fossil). Nomen nudum as printed in 1845, validated in 1901; perhaps to be regarded as preoccupied by ACESTRA Jardine, 1831.

Auchenilabrus (Agassiz) Woodward, 552; orthotype A. FRONTALIS Ag. Name only, reprinted from Agassiz, 1845, 308, when it is also nomen nudum.

Eocottus Woodward, 580; orthotype GOBIUS VERONENSIS Volta (fossil).

1902.

1887. BOULENGER (1902). *Report on the Collection of Natural History Made in the Antarctic Regions during the Voyage of the "Southern Cross."* Pisces, London, 174–189.

GEORGE ALBERT BOULENGER.

Pleuragramma Boulenger, 187; orthotype P. ANTARCTICUM Boulenger. Not PLEUROGRAMMUS Gill.

1888. BOULENGER (1902). *Additions à la faune ichthyologique du Bassin du Congo.* Ann. Mus. Cong. Zool., II, 19–57.

GEORGE ALBERT BOULENGER.

Microthrissa Boulenger, 26; orthotype M. ROYAUXI Boulenger.

Leptoglanis Boulenger, 42; orthotype L. XENOGNATHUS Boulenger.

Paraphractura Boulenger, 47; orthotype P. TENUICAUDA Boulenger.

Trachyglanis Boulenger, 48; orthotype T. MINUTUS Boulenger.

Belonoglanis Boulenger, 50; orthotype B. TENUIS Boulenger.

1889. BOULENGER (1902). *Descriptions of New Characinid Fish Discovered by Dr. W. J. Ansorge in Southern Nigeria.* Ann. Mag. Nat. Hist. (7), IX, 144–145.

GEORGE ALBERT BOULENGER.

Citharidium Boulenger, 144; orthotype C. ANSORGII Boulenger.

1890. BOULENGER (1902). *Description of a New Deep-Sea Gadid Fish from South Africa.* Ann. Mag. Nat. Hist. (7), IX, 335–336.

GEORGE ALBERT BOULENGER.

Tripterophycis Boulenger, 335; orthotype T. GILCHRISTI Boulenger.

1891. BOULENGER (1902). *Notes on the Classification of Teleostean Fishes.* Ann. Mag. Nat. Hist. (7), IX, 197–204.

GEORGE ALBERT BOULENGER.

Geophyroberyx Boulenger, 203; orthotype TRACHICHTHYS DARWINI Johnston.

1892. BOULENGER (1902). *Description of a New South-African Galeid Selachian.* Ann. Mus. Nat. Hist (7), X, 52–63.

GEORGE ALBERT BOULENGER.

Scylliogaleus Boulenger, 51; orthotype S. QUECKETTI Boulenger.

1893. BOULENGER (1902). *Diagnoses of New Cichlid Fishes Discovered by Mr. J. E. S. Moore in Lake Njassa.* Ann. Mag. Nat. Hist. (7), 69–71.

GEORGE ALBERT BOULENGER.

Hemitilapia Boulenger, 70; orthotype H. OXYRHYNCHUS Boulenger.
Cyrtocara Boulenger, 70; orthotype C. MOORI Boulenger.

1894. BOULENGER (1902). *On the Genus Ateleopus of Schlegel.* Ann. Mag. Nat. Hist. (7), X, 402–403.

GEORGE ALBERT BOULENGER.

Podateles Boulenger, 403; orthotype ATELEOPUS JAPONICUS Blkr. Substitute for ATELEOPUS T. & S., regarded as preoccupied by ATELOPUS Duméril & Bibron, a genus of reptiles.

1895. BOULENGER (1902). *Contributions to the ichthyology of the Congo.* Proc. Zool. Soc. London, 1902, 234–237.

GEORGE ALBERT BOULENGER.

Allabenchelys Boulenger, 234; orthotype A. LONGICAUDA Boulenger.
Chilochromis Boulenger, 236; orthotype C. DUPONTI Boulenger.

1896. BRAUER (1902). *Diagnosen von neuen Tiefseefischen welche von der Valdivia-Expedition gesammelt sind.* Zool. Anz., XXV, 277–298.

AUGUST BRAUER.

Dissomma Brauer, 278; orthotype D. ANALE Brauer.
Triplophos Brauer, 282; orthotype T. ELONGATUS Brauer.
Macrostomias Brauer, 283; orthotype M. LONGIBARBATUS Brauer.
Melanostomias Brauer, 284; orthotype M. VALDIVIÆ Brauer.
Bathylynchus Brauer, 289; orthotype B. CYNÆUS Brauer.
Macropharynx Brauer, 290; orthotype M. LONGICAUDATUS Brauer. Called MEGALOPHARYNX by Brauer in 1900.
Cœlophrys Brauer, 291; orthotype C. BREVICAUDATA Brauer.
Gigantactis Brauer, 295; orthotype G. VANHOEFFENI Brauer.
Aceratias Brauer, 296; logotype A. MACRORHYNCHUS Brauer.
Stylophthalmus Brauer, 298; orthotype S. PARADOXUS Brauer.

1897. EASTMAN (1902). *On Campyloprion, a New Form of Edestus-like Denti-tion.* Geol. Mag. (8), IX, 148–152.

CHARLES ROCHESTER EASTMAN.

Campyloprion Eastman, 151; orthotype C. ANNECTENS Eastman (fossil). Shark, locality unknown.

1898. FRITSCH & BAYER (1902). *Nové Ryby Ceskeho útvarn Krédového.* Ceska Ak. Preze, II, 1–18.

ANTON FRITSCH; EDWIN BAYER.

Schizospondylus Fritsch & Bayer, 8; orthotype S. DUBIUS Fritsch & Bayer.
Parelops Fritsch & Bayer, 13; orthotype P. PARAZAKII Fritsch & Bayer (fossil).

1899. GILCHRIST (1902). *South African Fishes.* Marine Investigations in South Africa, II, 101–113.

J. D. F. GILCHRIST.

Choridactylodes Gilchrist, 101; orthotype C. NATALENSIS Gilchrist.
Melanonosoma Gilchrist, 106; orthotype M. ACUTICAUDATUM Gilchrist.
Paralichthodes Gilchrist, 108; orthotype P. ALGOENSIS Gilchrist.

1900. GRACIANOV (1902). *Die Ichthyo-fauna des Baikalsees; Dnevn.* Zool. Otd. Obsc. Liub. Jest. Moska, III, 18–61. (In Russian.)

VALERIAN GRACIANOV.

Procottus Gracianov (Gratzianow), 41; orthotype COTTUS JEITTELESI Dybowski.

1901. GÜNTHER (1902). *Third Notice of New Species of Fishes from Morocco.* Novitat. Zool. (Tring.), IX, 446–448.

ALBERT GÜNTHER.

Pterocapoeta Günther, 446; orthotype P. MAROCCANA Gthr.

1902. GÜNTHER (1902). *Last Account of Fishes Collected by Mr. R. B. N. Walker, on the Gold Coast.* Proc. Zool. Soc. London, 1902, II, 330–339.

ALBERT GÜNTHER.

Notoglanidium Günther, 336; orthotype N. WALKERI Gthr.

1903. JORDAN & EVERMANN (1902). *Notes on a Collection of Fishes from the Island of Formosa.* Proc. U. S. Nat. Mus., XXV, 315–368. (Sept. 24, 1902.)

DAVID STARR JORDAN; BARTON WARREN EVERMANN.

Zacco Jordan & Evermann, 322; orthotype LEUCISCUS PLATYPUS T. & S.
Evenchelys Jordan & Evermann, 327; orthotype GYMNOTHORAX MACRURUS Bloch. Same as THYRSOIDEA Kaup, as restricted by Bleeker, but by tautonomy Kaup's genus THYRSOIDEA must have MURÆNA THYRSOIDEA Rich., as type.

1904. JORDAN & FOWLER (1902). *A Review of the Trigger-Fishes, File-Fishes, and Trunk-Fishes of Japan.* Proc. U. S. Nat. Mus., XXV, 251–286. (Sept. 17, 1902.)

DAVID STARR JORDAN; HENRY WEED FOWLER.

Rudarius Jordan & Fowler, 270; orthotype R. ERCODES Jordan & Fowler.
Lactoria Jordan & Fowler, 278; orthotype OSTRACION CORNUTUS L.

1905. JORDAN & FOWLER (1902). *A Review of the Cling-Fishes (Gobiesocidæ) of the Waters of Japan.* Proc. U. S. Nat. Mus., XXV. (Sept. 19, 1902.)

DAVID STARR JORDAN; HENRY WEED FOWLER.

Aspasma Jordan & Fowler, 414; orthotype LEPADOGASTER MINIMUS Döderlein.

1906. JORDAN & FOWLER (1902). *A Review of the Ophidioid Fishes of Japan.* Proc. U. S. Nat. Mus., XXV, 743–786. (Dec. 2, 1902.)

DAVID STARR JORDAN; HENRY WEED FOWLER.

Heirichthys Jordan & Fowler, 744; orthotype H. ENCRYPTES Jordan & Fowler. A synonym of CONGROGADUS Gthr.

1907. JORDAN & SNYDER (1903). *A Review of the Discobolous Fishes of Japan.* Proc. U. S. Nat. Mus., XXIV, 343–51. (Feb. 10, 1902.)

DAVID STARR JORDAN; JOHN OTTERBEIN SNYDER.

Crystallias Jordan & Snyder, 349; orthotype C. MATSUSHIMÆ Jordan & Snyder.

1908. JORDAN & SNYDER (1902). *A Review of the Trachinoid Fishes and Their Supposed Allies found in the Waters of Japan.* Proc. U. S. Nat. Mus., XXIV, 461–497. (Mar. 28, 1902.)

DAVID STARR JORDAN; JOHN OTTERBEIN SNYDER.

Pteropsaron Jordan & Snyder, 470; orthotype P. EVOLANS Jordan & Snyder.
Ariscopus Jordan & Snyder, 479; orthotype A. IBURIUS Jordan & Snyder.
Stalix Jordan & Snyder, 495; orthotype S. HISTRIO Jordan & Snyder.

1909. JORDAN & SNYDER (1902). *A Review of the Labroid Fishes and Related Forms found in the Waters of Japan.* Proc. U. S. Nat. Mus., XXV, 595–662. (May 2, 1902.)

DAVID STARR JORDAN; JOHN OTTERBEIN SNYDER.

Verreo Jordan & Snyder, 619; orthotype COSSYPHUS OXYCEPHALUS Blkr.
Ampheces Jordan & Snyder, 628; orthotype ANAMPSES GEOGRAPHICUS Cuv. & Val.

1910. JORDAN & SNYDER (1902). *Description of Two New Species of Squaloid Sharks from Japan.* Proc. U. S. Nat. Mus., XXV, 79–81. (Sept. 2, 1902.)

DAVID STARR JORDAN; JOHN OTTERBEIN SNYDER.

Deania Jordan & Snyder, 80; orthotype D. EGLANTINA Jordan & Snyder, regarded by Garman as a synonym of ACANTHIDIUM Lowe, but the logotype of that genus, as first restricted is an ETMOPTERUS. Not DEANEA, 1875, a genus of PROTOZOA, the name of the fish genus of different origin, being dedicated to Bashford Dean.

1911. JORDAN & SNYDER (1902). *A Review of the Blennioid Fishes of Japan.* Proc. U. S. Nat. Mus., XXV, 441–504. (Sept. 26, 1902.)

DAVID STARR JORDAN; JOHN OTTERBEIN SNYDER.

Zacalles Jordan & Snyder, 448; orthotype Z. BRYOPE Jordan & Snyder; name preoccupied, replaced by CALLIBLENNIUS Barbour.
Azuma Jordan & Snyder, 463; orthotype A. EMMNION Jordan & Snyder.
Zoarchias Jordan & Snyder, 480; orthotype Z. VENEFICUS Jordan & Snyder.
Abryois Jordan & Snyder, 486; orthotype A. AZUMÆ Jordan & Snyder.

1912. JORDAN & SNYDER (1902). *On Certain Species of Fishes Confused with Bryostemma polyactocephalum.* Proc. Nat. Mus., XXV. (Nov. 5, 1902.)

DAVID STARR JORDAN; JOHN OTTERBEIN SNYDER.

Bryolophus Jordan & Snyder, 618; orthotype B. LYSIMUS Jordan & Snyder.

1913. JORDAN & STARKS (1902). *A Review of the Hemibranchiate Fishes of Japan.* Proc. U. S. Nat. Mus., XXVI, 57–73. (Dec. 2, 1902.)

DAVID STARR JORDAN; EDWIN CHAPIN STARKS.

Æoliscus Jordan & Starks, 71; orthotype AMPHISILE STRIGATA Gthr.

1914. LAMBE (1902). *New Genera and Species from the Belly River Series (Mid-Cretaceous).* Contr. Canad. Pal., III, 25–81.

LAWRENCE M. LAMBE.

Diphyodus Lambe, 30; orthotype D. LONGIROSTRIS Lambe (fossil).

1915. MEEK (1902). *A Contribution to the Ichthyology of Mexico.* Field Mus. Zool., III, 63–128.

SETH EUGENE MEEK.

Zoogoneticus Meek, 91; logotype Z. DIAZI Meek.

Chapalichthys Meek, 97; orthotype CHARACODON ENCAUSTUS Jordan & Snyder.

Skiffia Meek, 102; orthotype S. LERMÆ Meek.

Melaniris Meek, 117; orthotype M. BALSANUS Meek. A synonym of THYRINA Jordan & Culver.

1916. POPTA (1902). *Notes from the Leyden Museum.*

CANNA M. L. POPTA.

Paracrossochilus Popta, 200; orthotype P. BICORNIS Popta = CROSSOCHILUS VITTATUS Boulenger.

1917. RIBEIRO (1902). *Oito especies de peixes do Rio Pomba.* Bol. Soc. Nac. Agric., Rio de Janeiro.

ALIPIO DE MIRANDA RIBEIRO.

Platycephalus Ribeiro, 7; type SCIÆNA UNDECIMALIS Bloch. A preoccupied name substituted for CENTROPOMUS Lac. "in accordance with the laws of priority."

1918. ROHON (1902). *Beiträge zur Anatomie und Histologie der Psammosteiden.* Sitzber. Böhmisch. Ges., 1901, No. XVI, 1–31.

JOSEF VICTOR ROHON.

Ganosteus Rohon, 36; logotype G. TUBERCULATUS Rohon.

1919. SMITH (1902). *The Smallest Known Vertebrate.* Science (2), XV, 30–31.

HUGH McCORMICK SMITH (1865–   ).

Mistichthys Smith, 30; orthotype M. LUZONENSIS Smith.

1920. SMITH (1902). *Description of a New Species of Blenny from Japan.* Bull. U. S. Fish Comm., XXI, 93–94.

HUGH McCORMICK SMITH.

Eulophias Smith, 93; orthotype E. TANNERI Smith.

1921. STEINDACHNER (1902). *Ueber zwei neue Fischarten aus dem Rothen Meere.* Anz. Ak. Wiss., XXXIX, 336–338.

FRANZ STEINDACHNER.

Beanea Steindachner, 337; orthotype B. TRIVITTATA Steind.

1922. VAILLANT (1902). *Résultats zoologiques de l'expédition scientifique Néerlandaise au Borneo central.* Notes Leyden Mus., XXIV, 1–166.

LÉON VAILLANT.

Pseudolais Vaillant, 51; orthotype P. TETRANEMA Vaillant.

Sosia Vaillant, 81; orthotype S. CHAMÆLEON Vaillant.

Gyrinochilus Vaillant, 107; orthotype G. PUSTULOSUS Vaillant.

Parhomaloptera Vaillant, 129; orthotype P. OBSCURA Vaillant.

Ellopostoma Vaillant, 145; orthotype APERIOPTUS MEGALOMYCTER Vaillant.

1923. WAITE (1902). *Notes on Fishes from Western Australia.* Rec. Austral. Mus., IV, 179–194.

EDGAR RAVENSWOOD WAITE (1866–   ).

Hypnarce Waite, 180; orthotype HYPNOS SUBNIGRUM Duméril. Substitute for HYPNOS Duméril, preoccupied.

Gilbertella Waite, 182; orthotype SERRANUS ARMATUS Castelnau. A synonym of EPINEPHELIDES Ogilby.

1924. WEBER (1902). *Siboga Expeditie. Introduction et description de l'Expédition.* Leiden, 1902, 159 pp.

MAX (CARL WILHELM) WEBER (1852–   ).

Photoblepharon Weber, 108; orthotype SPARUS PALPEBRATUS Boddaërt.

1925. WELLBURN (1902). *On the Fish Fauna of the Pendleside Limestones.* Proc. Yorkshire Geol. and Polytechnic Soc. New Series, XIV, 465–473.

EDGAR D. WELLBURN.

Marsdenius Wellburn, 466; orthotype M. SUMMITI Wellburn (fossil DIPLACANTHID: Carboniferous).

1926–1927. WOODWARD (1902). *The Fossil Fishes of the English Chalk.* I. Palæontogr. Soc., 1902, 1–56.

ARTHUR SMITH WOODWARD.

Trachichthyoides Woodward, 29; orthotype T. ORNATUS Woodw. (fossil).

1903.

1928. BERG (1903). *On the Taxonomy of the Cottidæ from Lake Baikal* (Russian text). Ann. Mus. Zool. St. Petersb., VIII, 99–114.

LEW SEMENOWITCH BERG.

Baicalocottus Berg, 100; orthotype COTTUS GREWINGKI Dybowski. A synonym of COTTOCOMEPHORUS Pellégrin.

Batrachocottus Berg, 109; orthotype COTTUS BAIKALENSIS Dybowski.

1929. BOULENGER (1903). *On the Fishes Collected by Mr. G. L. Bates in Southern Cameroon.* Proc. Zool. Soc. London, 1903, I, 21–29.

GEORGE ALBERT BOULENGER.

Microsynodontis Boulenger, 26; orthotype M. BATESII Boulenger.

1930. BOULENGER (1903). *Descriptions of Two New Deep-Sea Fishes from South Africa.* Rept. Govt. Biologist Cape of Good Hope, 66–68.

GEORGE ALBERT BOULENGER.

Tripterophycis Boulenger, 66; orthotype T. GILCHRISTI Boulenger.

1931. EIGENMANN (1903). *The Fresh-Water Fishes of Western Cuba.* Bull. U. S. Fish Comm. for 1902, 211–236 (1903).

CARL H. EIGENMANN.

Toxus Eigenmann, 226; orthotype T. RIDDLEI Eigenmann.

1932. EIGENMANN (1903). *New Genera of South American Fresh-Water Fishes and New Names for Old Genera.* Smithson. Collec. XLV, 114–148. (Dec. 19, 1903.)

CARL H. EIGENMANN.

Lahiliella Eigenmann, 144; orthotype ANASTOMUS NASUTUS Kner.

Anisistia Eigenmann, 144; orthotype ANODUS NOTATUS Schomburgk.

Holoshesthes Eigenmann, 144; orthotype CHIRODON PEQUIRA Steind. Later corrected to HOLESTHES, but the first spelling should apparently hold.

Moenkhausia Eigenmann, 145; orthotype TETRAGONOPTERUS XINGUENSIS Steind.

Markiana Eigenmann, 145; orthotype TETRAGONOPTERUS NIGRIPINNIS Perugia.

Othonophanes Eigenmann, 145; orthotype BRYCON LABIATUS Eigenmann. Misprinted ORTHOPHANES in the Zoological Record; also misprinted as ORTHONOPHANES.

Holoprion Eigenmann,´145; orthotype CHIRODON AGASSIZI Steind. A synonym of APHYOCHARAX Gthr.

Holopristis Eigenmann, 145; orthotype TETRAGONOPTERUS OCELLIFER Steind.

Stichonodon Eigenmann, 145; orthotype LUETKENIA INSIGNIS Steind. Substitute for LUETKENIA Steind., preoccupied.

Bryconodon Eigenmann, 146; orthotype BRYCON ORTHOTÆNIA Gthr.

Acestrorhynchus Eigenmann, 146; orthotype SALMO FALCATUS Bloch. A substitute for XIPHORHYNCHUS Ag. and XIPHORHAMPHUS M. & T., both names preoccupied.

Acestrorhamphus Eigenmann, 146; orthotype HYDROCYON HEPSETUS Cuv.

Acestrocephalus Eigenmann, 146; orthotype XIPHORHAMPHUS ANOMALUS Steind.

Evermannella Eigenmann, 146; orthotype CYNOPOTAMUS BISERIALIS Garman. Name preoccupied; replaced by EUCYNOPOTAMUS Fowler, 1904.

Gilbertella Eigenmann, 147; orthotype ANACYRTUS ALATUS Steind,, preoccupied by Waite, 1902; replaced by GILBERTOLUS Eigenmann & Ogle.

Myleocollops Eigenmann, 147; orthotype METYNNIS GOELDII Eigenmann.

Acnodon Eigenmann, 147; orthotype MYLEUS OLIGACANTHUS M. & T.

Boulengerella Eigenmann, 147; orthotype XIPHOSTOMA LATERISTRIGA Boulenger.

Mylossoma Eigenmann, 148; orthotype MYLETES ALBISCOPUS Cope.

Colossoma Eigenmann, 148; orthotype MYLETES OCULUS Cope.

Orthomyleus Eigenmann, 148; orthotype MYLETES ELLIPTICUS Gthr.

Piaractus Eigenmann, 148; orthotype MYLETES BRACHYPOMUS Cuv.

1933. EIGENMANN & KENNEDY (1903). *On a Collection of Fishes from Paraguay, with a Synopsis of the American Genera of Cichlids.* Proc. Acad. Nat. Sci. Phila., 1903, 497–537. (Sept. 4, 1903.)

CARL H. EIGENMANN; CLARENCE HAMILTON KENNEDY.

Biotæcus Eigenmann & Kennedy, 533; orthotype SARACA OPERCULARIS Steind. A substitute for SARACA Steind., said to be preoccupied.

Biotodoma Eigenmann & Kennedy, 533; orthotype B. TRIFASCIATUS Eigenmann & Kennedy.

1934. FOWLER (1903). *Descriptions of Several Fishes from Zanzibar Island, Two of Which Are New.* Proc. Acad. Sci. Phila., LV, 161–176. (June 4, 1903.)

HENRY WEED FOWLER.

Graviceps Fowler, 170; orthotype PETROSCIRTES ELEGANS Steind.

1935. FOWLER (1903). *New and Little Known Mugilidæ and Sphyrænidæ.* Proc. Acad. Nat. Sci. Phila., 1903, 743–752. (Dec. 16, 1903.)

HENRY WEED FOWLER.

Œdalechilus Fowler, 748; orthotype AGONOSTOMA MONTICOLA (Griffith).

Agriosphyræna Fowler, 749; orthotype SPHYRÆNA BARRACUDA Walbaum.

FOWLER, 1903   503

1936. FOWLER (1903). *Descriptions of New, Little Known, and Typical Atheri-nidæ.* Proc. Acad. Nat. Sci. Phila., 1903, 727–742. (Dec. 16, 1903.)
HENRY WEED FOWLER.

Atherinomorus Fowler, 730; orthotype ATHERINA LATICEPS Poey. A synonym of HEPSETIA Bon.

Ischnomembras Fowler, 730; orthotype I. GABUNENSIS Fowler. A synonym of MENIDIA Bon.

Phoxargyrea Fowler, 732; orthotype P. DAYI Fowler. Apparently a synonym of MENIDIA Bon.

1937. GILL (1903). *Note on the Fish Genera Named Macrodon.* Proc. U. S. Mus., XXVI, 1015–1016
THEODORE NICHOLAS GILL.

Hoplias Gill, 1016; orthotype MACRODON TAREIRA Müller. A substitute for MACRO-DON Müller, preoccupied.

1938. GILCHRIST (1903). *Descriptions of New South African Fishes.* Marine Invest. S. Africa, II, 203–211.
J. D. F. GILCHRIST.

Cyttosoma Gilchrist, 113; orthotype C. BOOPS Gilchrist.

Trachichthodes Gilchrist, 203; orthotype BERYX AFFINIS Gthr. Replaces AUSTRO-BERYX McCulloch.

Læmonemodus Gilchrist, 208; orthotype L. COMPRESSICAUDA Gilchrist.

Selachophidium Gilchrist, 209; orthotype S. GUENTHERI Gilchrist.

1939. HAY (1903). *On a Collection of Upper Cretaceous Fishes from the Mount Lebanon, Syria, with Descriptions of Four New Genera and Nineteen New Species.* Bull. Amer. Mus., XIX, 395–452.
OLIVER PERRY HAY.

Stenopropoma Hay, 407; orthotype S. HAMATA Hay (fossil) (BELONORHYNCHIDÆ).

Eubiodectes Hay, 415; orthotype CHIROCENTRITES LIBANICUS Pictet & Humbert (fossil).

Anguillavus Hay, 436; orthotype A. QUADRIPINNIS Hay (fossil eel).

Enchelion Hay, 441; orthotype E. MONTIUM Hay (fossil).

1940. HELLER & SNODGRASS (1903). *Papers from the Hopkins-Stanford Galapagos Expedition, 1898–1899. XV. New Fishes.* Proc. Washington Acad., V, 189–229.
EDMUND HELLER (1875–   ); ROBERT EVANS SNODGRASS (1875–   ).

Evolantia Heller & Snodgrass, 189; orthotype EXOCŒTUS MICROPTERUS Cuv. & Val.

Galeagra Heller & Snodgrass, 193; orthotype G. PAMMELAS Heller & Snodgrass.

Nexilosus Heller & Snodgrass, 204; orthotype N. ALBEMARLEUS Heller & Snodgras.

Petrotyx Heller & Snodgrass, 222; orthotype P. HOPKINSI Heller & Snodgrass.

Eutyx Heller & Snodgrass, 224; orthotype E. DIAGRAMMUS Heller & Snodgrass.

Allector Heller & Snodgrass, 228; orthotype A. CHELONIÆ Heller & Snodgrass.

1941. JAEKEL (1903). *Ueber die Epiphyse und Hypophyse.* SB. Ges. Naturf. Berlin, 1903, 27–58.
OTTO JAEKEL.

Pachyosteus Jaekel, 39; orthotype P. BULLA Jaekel (fossil).

1942. JAEKEL (1903). *Ueber Ramphodus . . . einen neuen devonischen Holocephalen von Wildungen.* Sitzb. Ges. Naturf. Freunde. Berlin, 383–393.
OTTO JAEKEL.

Ramphodus Jaekel, 383; orthotype .... Name preoccupied
as RHAMPHODUS. A synonym of RHYNCHODUS.

1943. JENKINS (1903). *Report on the Collection of Fishes Made in the Hawaiian Islands, with Descriptions of New Species.* Bull. U. S. Fish Comm., XXII, for 1902, 415–516. (Sept. 23, 1903.)

OLIVER PEEBLES JENKINS.

Cirrhitoidea Jenkins, 489; orthotype C. BIMACULA Jenkins.
Chlamydes Jenkins, 503; orthotype C. LATICEPS Jenkins.
Eviota Jenkins, 542; orthotype E. EPIPHANES Jenkins.

1944. JORDAN (1903). *Supplementary Note on Bleekeria mitsukurii and on Certain Japanese Fishes.* Proc. U. S. Nat. Mus., XXVI, 693–696. (Apr. 9, 1903.)

DAVID STARR JORDAN.

Embolichthys Jordan, 693; orthotype BLEEKERIA MITSUKURII J. & E.
Zen Jordan, 694; orthotype CYTTOPSIS ITEA Jordan & Fowler.
Chasmichthys Jordan, 696; orthotype CHASMIAS MISAKIUS Jordan & Snyder. Substitute for SACCOSTOMA Sauvage and for CHASMIAS Jordan & Snyder, both names preoccupied.

1945. JORDAN (1903). *Generic Names of Fishes* (correspondence). Amer. Nat., XXXVII, 360.

DAVID STARR JORDAN.

Falcularius Jordan & Snyder, 360; orthotype FALCULA CHAPALÆ Jordan & Snyder. Name a substitute for FALCULA Jordan & Snyder, preoccupied.
Xeneretmus (Gilbert) Jordan, 360; orthotype XENOCHIRUS TRIACANTHUS Gilbert. Name a substitute for XENOCHIRUS Gilbert, preoccupied.

1946. JORDAN & EVERMANN (1903). *Descriptions of New Genera and Species of Fishes from the Hawaiian Islands.* Bull. U. S. Fish Comm. for 1902, XXII, 161–208. (Apr. 11, 1903.)

DAVID STARR JORDAN; BARTON WARREN EVERMANN.

Fowleria Jordan & Evermann, 180; orthotype APOGON AURITUS Blkr.
Bowersia Jordan & Evermann, 182; orthotype B. VIOLESCENS J. & E.
Verriculus Jordan & Evermann, 191; orthotype V. SANGUINEUS J. & E.
Quisquilius Jordan & Evermann, 203; orthotype Q. EUGENIUS J. & E.
Vitraria Jordan & Evermann, 205; orthotype V. CLARESCENS J. & E.
Osurus Jordan & Evermann, 206; orthotype PARAPERCIS SCHAUINSLANDI Steind.

1947. JORDAN & EVERMANN (1903). *Description of a New Genus and Two New Species of Fishes from the Hawaiian Islands.* Bull. U. S. Fish Comm., XXII, for 1902, 209–210. (April 11, 1903.)

DAVID STARR JORDAN; BARTON WARREN EVERMANN.

Iracundus Jordan & Evermann, 209; orthotype I. SIGNIFER J. & E.

1948. JORDAN & FOWLER (1903). *A Review of the Elasmobranchiate Fishes of Japan.* Proc. U. S. Nat. Mus., XXVI, 593–674. (March 30, 1903.)

DAVID STARR JORDAN; HENRY WEED FOWLER.

Zameus Jordan & Fowler, 632; orthotype CENTROPHORUS SQUAMULOSUS Gthr. A synonym of SCYMNODON Bocage & Capello.

1949. JORDAN & FOWLER (1903). *A Review of the Cobitidæ, or Loaches, of the Rivers of Japan.* Proc. U. S. Nat. Mus., XXVI, 765–774. (April 3, 1903.)

DAVID STARR JORDAN; HENRY WEED FOWLER.

Elxis Jordan & Fowler, 768; orthotype E. NIKKONIS Jordan & Fowler. A synonym of LEFUA Herzenstein.

Orthrias Jordan & Fowler, 769; orthotype O. OREAS Jordan & Fowler. A synonym of OREIAS Sauvage and probably also of BARBATULA Linck.

1950. JORDAN & FOWLER (1903). *A Review of the Dragonets (Callionymidæ) and Related Fishes of the Waters of Japan.* Proc. U. S. Nat. Mus., XXV, 939–959. (May 3, 1903.)

DAVID STARR JORDAN; HENRY WEED FOWLER.

Draconetta Jordan & Fowler, 939; orthotype D. XENICA Jordan & Fowler.

Calliurichthys Jordan & Fowler, 941; orthotype CALLIONYMUS JAPONICUS Houttuyn.

1951. JORDAN & FOWLER (1903). *A Review of the Cyprinoid Fishes of Japan.* Proc. U. S. Nat. Mus., XXVI, 811–862. (July 6, 1903.)

DAVID STARR JORDAN; HENRY WEED FOWLER.

Abbottina Jordan & Fowler, 835; orthotype A. PSEGMA Jordan & Fowler.

Zezera Jordan & Fowler, 837; orthotype SARCOCHILICHTHYS HILGENDORFI Ishikawa.

Biwia Jordan & Fowler, 838; orthotype PSEUDOGOBIO ZEZERA Ishikawa.

1952. JORDAN & FOWLER (1903). *A Review of the Siluroid Fishes, or Catfishes of Japan.* Proc. U. S. Nat. Mus., XXVI, 897–911. (July 7, 1903.)

DAVID STARR JORDAN; HENRY WEED FOWLER.

Fluvidraco Jordan & Fowler, 904; orthotype PSEUDOBAGRUS RANSONNETI Steind.

1953. LAHILLE (1903). *Nota sobre un genera nuevo de Escombrido.* An. Mus. Buenos Aires (3), II, 1903, 375–376.

FERNANDO LAHILLE.

Chenogaster Lahille, 375; orthotype C. HOLMBERGI Lahille.

1954. REGAN (1903). *On a Collection of Fishes Made by Dr. Goeldi at Rio Janeiro.* Proc. Zool. Soc. London, 1903, 59–68.

CHARLES TATE REGAN.

Mylacrodon Regan, 62; orthotype M. GOELDII Regan.

1955. REGAN (1903). *A Revision of the Fishes of the Family Lophiidæ.* Ann. Mag. Nat. Hist., XI.

CHARLES TATE REGAN.

Chirolophius Regan, 279; orthotype LOPHIUS NARESI Gthr. A synonym of LOPHIODES G. & B. (1895).

1956. REGAN (1903). *On the Systematic Position and Classification of the Gadoid or Anacanthine Fishes.* Ann. Mag. Nat. Hist. (7), XI, 459–466.

CHARLES TATE REGAN.

Gadomus Regan, 459; orthotype BATHYGADUS LONGIFILIS G. & B.

Melanobranchus Regan, 459; orthotype BATHYGADUS MELANOBRANCHUS Vaillant.

1957. REGAN (1903). *On the Genus Lichia of Cuvier.* Ann. Mag. Nat. Hist. (7), XII, 348–350.

CHARLES TATE REGAN.

Campogramma Regan, 350; orthotype LICHIA VADIGO Risso. A synonym of HYPACANTHUS Raf.

1958. REGAN (1903). *On a Collection of Fishes from the Azores.* Ann. Mag. Nat. Hist., XII.

CHARLES TATE REGAN.

Macristium Regan. 345; orthotype M. CHAVESI Regan.

1959. REGAN (1903). *Descriptions de poissons nouveaux . . . Musée d'Histoire Naturelle de Génève.* Revue Suisse Zool. Génève.
CHARLES TATE REGAN.
Bedotia Regan, 416; orthotype B. MADAGASCARIENSIS Regan.

1960. RIBEIRO (1903). *Pescados do "Annie."* Bol. Soc. Nat. Agric. Rio de Janeiro.
ALIPIO DE MIRANDA RIBEIRO.
Pseudopercis Ribeiro, 4; orthotype P. NUMIDA Ribeiro.

1961. SAUVAGE (1903). *Noticia sobre los Peces de la Caliza litografica de la Provincia de Lerida (Cataluña).* Mem. Ac. Barcel. (3), IV, No. 35, 1–32.
HENRI EMILE SAUVAGE.
Vidalia Sauvage, 15; orthotype V. CATALUNICA Sauvage (fossil).

1962. SCHMIDT (1903). *Faune de la Mer Ochotsk et du Japon.*
PETR YULIEVICH SCHMIDT.
Cynopsetta Schmidt, 19; orthotype C. DUBIA Schmidt, name only; defined by Jordan & Starks, 1906.

1963. STEINDACHNER (1903). *Ueber einige neue Reptilien und Fischarten des Hofmuseums in Wien.* Sitzb. Ak. Wien, CXII, 15–21.
FRANZ STEINDACHNER.
Gymnocharacinus Steindachner, 20; orthotype G. BERGII Steind.

1964. TRAQUAIR (1903). *The Lower Devonian Fishes of Gemünden.* Trans. Roy. Soc. Edinb., XL, 723–739.
RAMSAY HEATLEY TRAQUAIR.
Gemundina Traquair, 734; orthotype G. STURTZI Traquair (fossil).
Hunsruckia Traquair, 736; orthotype H. PROBLEMATICA Traquair (fossil).

1965. VOLZ (1903). *Fische von Sumatra.* Zool. Anz., XXVI, 553–559.
WALTER VOLZ.
Trypauchenopsis Volz, 555; orthotype T. INTERMEDIUS Volz.

1966. WAITE (1903). *New Records of Recurrences of Rare Fishes from Eastern Australia.* Rec. Austr. Mus., V. 56–61.
EDGAR RAVENSWOOD WAITE.
Prosoplismus Waite, 59; orthotype HISTIOPTERUS RECURVIROSTRIS Rich. A synonym of PENTACEROPSIS Steindachner & Döderlein, noted by Waite, Ann. Mag. Nat. Hist., XII, 288.

1904.

1967. BOULENGER (1904). *Description of New West African Fresh-Water Fishes.* Ann. Mag. Nat. Hist. (7), XIV, 16–20.
GEORGE ALBERT BOULENGER.
Procatopus Boulenger, 20; orthotype P. NOTOTÆNIA Boulenger.

1968. DEAN (1904). *Notes on Japanese Myxinoids. A New Genus Paramyxine and a New Species Homea okinoseana.* Journ. Coll. Sci. Tokyo, XIX, No. 2, 1–23.
BASHFORD DEAN (1867–   ).
Paramyxine Dean, 14; orthotype P. ATAMI Dean.

1969. DOLLO (1904). *Résultats du voyage du S. Y. Belgica. Expédition Antarctique Belge.* Anvers.

Louis Dollo.

Copeichthys Dollo, 159; orthotype DIPLOMYSTUS DENTATUS Cope (fossil). Substitute for DIPLOMYSTUS Cope (1877), regarded as preoccupied by DIPLOMYSTES Gthr. (1864).

1970. EASTMAN (1904). *Description of Bolca Fishes.* Bull. Mus. Comp. Zool. Harvard, XLVI, 1–36.

Charles Rochester Eastman.

Histionotopterus Eastman, 32; orthotype HISTIOCEPHALUS BASSANI Zigno (fossil, Monte Bolca). A substitute for HISTIOCEPHALUS Zigno, preoccupied in worms.

1971. FOWLER (1904). *Description of a New Lantern Fish.* Proc. Acad. Nat. Sci. Phila. for 1903, 754–755. (Jan. 15, 1904.)

Henry Weed Fowler.

Centrobranchus Fowler, 754; orthotype C. CHŒROCEPHALUS Fowler.

1972. FOWLER (1904). *Note on the Characinidæ.* Proc. Ac. Nat. Sci. Phila., LIII. (Jan., 1904.)

Henry Weed Fowler.

Eucynopotamus Fowler, 119; orthotype CYNOPOTAMUS BISERIALIS Garman. A substitute for EVERMANNELLA Eigenmann, preoccupied by EVERMANELLA Fowler, 1901.

1973. FOWLER (1904). *New, Little Known, and Typical Berycoid Fishes.* Proc. Acad. Nat. Sci. Phila., LV, 222–238. (Apr. 7, 1904.)

Henry Weed Fowler.

Sargocentron Fowler, 235; orthotype HOLOCENTRUM LEO Cuv.

1974. FOWLER (1904). *Notes on Fishes from Arkansas, Indian Territory, and Texas.* Proc. Acad. Nat. Sci. Phila., LV, 242–249.

Henry Weed Fowler.

Paranotropis Fowler, 245; orthotype PHOTOGENIS SCABRICEPS Cope. A substitute for EPISEMA Cope & Jordan, preoccupied.

1975. FOWLER (1904). *A Collection of Fishes from Sumatra.* Journ. Ac. Nat. Sci. Phila.

Henry Weed Fowler.

Rastrum Fowler, 509; orthotype ALEPES SCITULA Fowler.
Boulengerina Fowler, 512; orthotype DULES TÆNIURUS Cuv. & Val. Name preoccupied in snakes, Dollo, 1886; replaced by SAFOLE Jordan, 1912.
Equulites Fowler, 513; orthotype LEIOGNATHUS VERMICULATUS Fowler.
Eubleekeria Fowler, 516; orthotype EQUULA SPLENDENS Cuv.
Deveximentum Fowler, 517; orthotype ZEUS INSIDIATOR Bloch. A synonym of SECUTOR Gistel.
Æthaloperca Fowler, 522; orthotype PERCA ROGAA Forskål.
Bennettia Fowler, 524; orthotype ANTHIAS JOHNI Bloch.
Parkia Fowler, 525; orthotype LUTJANUS FURVICAUDATUS Fowler.
Anemura Fowler, 527; orthotype DENTEX NOTATUS Day.
Odontoglyphis Fowler, 527; orthotype DENTEX TOLU Fowler.
Euthyopteroma Fowler, 527; orthotype DENTEX BLOCHI Blkr. (= SPARUS JAPONICUS Bloch).

Eutherapon Fowler, 527; orthotype THERAPON THERAPS Cuv.

Euelatichthys Fowler, 527; orthotype DIAGRAMMA AFFINE Gthr.

Spilotichthys Fowler, 528; orthotype HOLOCENTRUS RADJABAU Lac.

Lethrinella Fowler, 529; orthotype SPARUS MINIATUS Fowler.

Pertica Fowler, 530; orthotype GERRES FILAMENTOSUS Cuv. & Val. A synonym of GERRES Cuv. (XYSTÆMA J. & E.), as restricted.

Actinicola Fowler, 533; orthotype LUTJANUS PERCULA Lac.

Octocynodon Fowler, 535; orthotype JULIUS MINIATUS Cuv. & Val.

Grammoplites Fowler, 550; orthotype COTTUS SCABER L.

1976. GILBERT & STARKS (1904). *The Fishes of Panama Bay.* Mem. Calif. Ac., IV, 1–304.

CHARLES HENRY GILBERT; EDWIN CHAPIN STARKS.

Guentheridia Gilbert & Starks, 158; orthotype TETRAODON FORMOSUS Gthr.

1977. GILCHRIST (1904). *Descriptions of New South African Fishes.* Mar. Invest. S. Africa, III, 1–16.

J. D. F. GILCHRIST.

Cyttosoma Gilchrist, 6; orthotype C. BOOPS Gilchrist.

1978. GILL (1904). *Extinct Pediculate and Other Fishes.* Science (2), XX, 845–846.

THEODORE NICHOLAS GILL.

Bradyurus Gill, 846; orthotype SYMPHODUS SZAJNOCHÆ Zigno (fossil).

1979. HILGENDORF (1904). *Ein neuer Scyllium artiger Haifisch. Proscyllium habereri, . . . von Formosa.* Sitzber. Ges. Naturf. Berlin, 1904, 39–41.

FRANZ MARTIN HILGENDORF.

Proscyllium Hilgendorf, 39; orthotype P. HABERERI Hilgendorf.

1980. JORDAN (1904). *Notes on Fishes Collected in the Tortugas Archipelago.* Bull. U. S. Fish. Comm., XXII, 539–544.

DAVID STARR JORDAN.

Elacatinus Jordan, 542; orthotype E. OCEANOPS Jordan.

Acteis Jordan, 543; orthotype MALOCOCTENUS MOOREI Evermann & Marsh.

Ericteis Jordan, 543; orthotype E. KALISHERÆ Jordan.

1981. JORDAN & SNYDER (1904). *Notes on the Collection of Fishes from Oahu Island and Laysan Island, Hawaii, with Descriptions of Four New Species.* Proc. U. S. Nat. Mus., XXVII, 939–948. (June 2, 1904.)

DAVID STARR JORDAN; JOHN OTTERBEIN SNYDER.

Ariomma Jordan & Snyder, 942; orthotype A. LURIDA Jordan & Snyder.

1982. JORDAN & SNYDER (1904). *On a Collection of Fishes Made by Mr. Alan Owston in the Deep Waters of Japan.* Smithson. Collect., I, 230–240.

DAVID STARR JORDAN; JOHN OTTERBEIN SNYDER.

Trismegistus Jordan & Snyder, 238; orthotype T. OWSTONI Jordan & Snyder. A giant LIPARIS.

1983. JORDAN & STARKS (1904). *List of Fishes Dredged by the Steamer "Albatross" off the Coast of Japan in the Summer of 1900, with Descriptions of New Species and a Review of the Japanese Macrouridæ* (by Jordan & Gilbert). Bull. U. S. Fish Comm., XXII, 577–630.

DAVID STARR JORDAN; EDWIN CHAPIN STARKS.

Osopsaron Jordan & Starks, 600; orthotype PTEROPSARON VERECUNDUM Jordan & Snyder.

Regania Jordan & Gilbert, 604; orthotype R. NIPPONICA J. & G.

Nezumia Jordan & Gilbert, 620; orthotype N. CONDYLURA J. & G.

Cleisthenes Jordan & Starks, 622; orthotype C. PINETORUM Jordan & Starks.

Alæops Jordan & Starks, 623; orthotype A. PLINTHUS Jordan & Starks.

Dexistes Jordan & Starks, 624; orthotype D. RIKUZENIUS Jordan & Starks.

Araias Jordan & Starks, 624; orthotype A. ARIOMMUS Jordan & Starks, the young of DEXISTES RIKUZENIUS Jordan & Starks. A synonym of DEXISTES Jordan & Starks.

Veræqua Jordan & Starks, 625; orthotype V. ACHNE Jordan & Starks.

1984. JORDAN & STARKS (1904). *A Review of the Scorpænoid Fishes of Japan.* Proc. U. S. Nat. Mus., XXVII, 91–175. (Jan. 22, 1904.)

DAVID STARR JORDAN; EDWIN CHAPIN STARKS.

Thysanichthys Jordan & Starks, 122; orthotype T. CROSSOTUS Jordan & Starks.

Sebastiscus Jordan & Starks, 124; orthotype SEBASTES MARMORATUS Cuv. & Val.

Lythrichthys Jordan & Starks, 140; orthotype L. EULABES Jordan & Starks.

Ebosia Jordan & Starks, 145; orthotype PTEROIS BLEEKERI Steind.

Decterias Jordan & Starks, 154; orthotype MINOUS PUSILLUS T. & S.

Erosa (Swainson) Jordan & Starks, 156; orthotype SYNANCEIA EROSA (Langsdorf) Cuv. & Val.

Inimicus Jordan & Starks, 158; orthotype PELOR JAPONICUM Cuv. & Val. A substitute for PELOR Cuv., preoccupied.

Ocosia Jordan & Starks, 162; orthotype O. VESPA Jordan & Starks.

Erisphex Jordan & Starks, 170; orthotype COCOTROPUS POTTII Steind.

1985. JORDAN & STARKS (1904). *A Review of the Cottidæ or Sculpins of the Waters of Japan.* Proc. U. S. Nat. Mus., XXVII, 231–335. (Jan. 28, 1904.)

DAVID STARR JORDAN; EDWIN CHAPIN STARKS.

Stlengis Jordan & Starks, 236; orthotype S. OSENSIS Jordan & Starks.

Schmidtia Jordan & Starks, 237; orthotype S. MISAKIA Jordan & Starks. Name preoccupied and changed to SCHMIDTINA, p. 961.

Daruma Jordan & Starks, 241; orthotype D. SAGAMIA Jordan & Starks.

Ricuzenius Jordan & Starks, 242; orthotype R. PINETORUM Jordan & Starks.

Rheopresbe Jordan & Starks, 270; orthotype R. FUJIYAMÆ Jordan & Starks.

Ainocottus Jordan & Starks, 283; orthotype A. ENSIGER Jordan & Starks.

Crossias Jordan & Starks, 296; orthotype C. ALLISI Jordan & Starks.

Cottiusculus (Schmidt) Jordan & Starks, 298; orthotype C. GONEZ Schmidt.

Elaphichthys Jordan & Starks, 301; orthotype CENTRIDERMICHTHYS ELONGATUS Steind.

Alcichthys Jordan & Starks, 301; orthotype CENTRIDERMICHTHYS ALCICORNIS Herzenstein.

Furcina Jordan & Starks, 303; orthotype F. ISHIKAWÆ Jordan & Starks.

Ocynectes Jordan & Starks, 306; orthotype O. MASCHALIS Jordan & Starks.

Bero Jordan & Starks, 317; orthotype CENTRIDERMICHTHYS ELEGANS Steind.

Vellitor Jordan & Starks, 318; orthotype PODABRUS CENTROPOMUS Rich. Substitute for PODABRUS Rich., preoccupied.

1986. JORDAN & STARKS (1904). *Schmidtina, a New Genus of Japanese Sculpins.* Proc. U. S. Nat. Mus. (June 27, 1904).

DAVID STARR JORDAN; EDWIN CHAPIN STARKS.

Schmidtina Jordan & Starks, 961; orthotype Schmidtia misakia Jordan & Starks. Substitute for Schmidtia Jordan & Starks, preoccupied.

1987. MEEK (1904). *The Fresh-Water Fishes of Mexico North of the Isthmus of Tehuantepec.* Field Mus. Zool., V, 1–252.

Seth Eugene Meek (1858–1912).

Cynodonichthys Meek, 101; orthotype C. tenuis Meek.
Paragambusia Meek, 133; orthotype Gambusia nicaraguensis Gthr.
Thorichthys Meek, 222; orthotype T. ellioti Meek.

1988. NIKOLSKII (1904). *Neue Fischarten aus Ostasien.* Ann. Mus. Zool. Acad. St. Petersburg, VIII.

Aleksandr Mikhailovich Nikolskii.

Ussuria Nikolskii, 362; orthotype U. leptocephala Nikolskii. Apparently a synonym of Misgurnus Lac.

1989. OGILBY (1904). *Studies in the Ichthyology of Queensland.* Proc. Roy. Soc. Queensland, XVIII, 27–30.

James Douglas Ogilby.

Daia Ogilby, 9; orthotype Centropogon indicus Day. Perhaps a synonym of Hypodytes Gistel (Paracentropogon Blkr.).
Notesthes Ogilby, 17; orthotype Centropogon robustus Gthr.
Liocranium Ogilby, 23; orthotype L. præpositum Ogilby.

1990. PELLÉGRIN (1904). *Contribution à l'étude anatomique, biologique et taxonomique des Poissons de la famille des Cichlidés.* Mem. Soc. Zool. France, XVI, 41–402.

Jacques Pellégrin.

Astatheros Pellégrin, 203; orthotype Heros heterodontus Vaillant & Pellégrin.
Astatoreochromis Pellégrin, 384; orthotype A. alluaudi Pellégrin.
Nanochromis Pellégrin, 273; orthotype Pseudoplesiops nudiceps Boulenger, Substitute for Pseudoplesiops Boulenger, preoccupied.
Lepidolamprologus Pellégrin, 295; orthotype Lamprologus elongatus Boulenger.
Astatotilapia Pellégrin, 299; orthotype Tilapia desfontainesi Lac.
Boulengerochromis Pellégrin, 304; orthotype Tilapia microlepis Boulenger.
Ophthalmotilapia Pellégrin, 345; orthotype Tilapia boops Boulenger.

1991. POPTA (1904). *Descriptions preliminaires des nouvelles espèces de Poissons recueillies au Borneo central par M. le Dr. A. W. Nieuwenhuis en 1898 et en 1900.* Notes Leyden Mus., XXIV, 179–202.

Canna M. L. Popta.

Neopangasius Popta, 180; orthotype N. nieuwenhuisii Popta.
Paracrossocheilus Popta, 200; orthotype P. bicornis Popta.

1992. REGAN (1904). *A Monograph of the Fishes of the Family Loricariidæ.* Trans. Zool. Soc. London, XVII, 191–350.

Charles Tate Regan.

Xenocara Regan, 251; orthotype Chætostomus latifrons Gthr.

1993. ROMANOVSKII (1904). *Notiz ueber den fossilen Fisch, Lyrolepis caucasicus.* Rom. Verh. Russ. Miner. Ges. St. Petersb.

Ghennadi Danilovich Romanovskii.

Lyrolepis Romanovskii, 1; orthotype L. caucasicus Romanovskii (fossil).

1994. RUTTER (1904). *Notes on Fishes of Gulf of California, with Description of a New Genus and Species.* Proc. Calif. Acad. Zool. (3), III, 251–253.

CLOUDSLEY M. RUTTER.

Pycnomma Rutter, 252; orthotype P. SEMISQUAMATA Rutter.

1995. SCHMIDT (1904). *Pisces marium orientalium imperii rossici.* St. Petersburg, 1904, 1–466.

PETR YULIEVICH SCHMIDT.

Cottiusculus Schmidt, 190; orthotype C. GONEZ Schmidt. First notice by Jordan & Starks, 1904.

Tilesina Schmidt, 134; orthotype T. GIBBOSA Schmidt.

Krusensterniella Schmidt, 198; orthotype K. NOTABILIS Schmidt.

Hadropareia Schmidt, 205; orthotype H. MIDDENDORFFI Schmidt.

Protopsetta Schmidt, 230; orthotype HIPPOGLOSSOIDES HERZENSTEINI Schmidt.

Acanthopsetta Schmidt, 237; orthotype A. NADESHNYI Schmidt.

1996. SIMIONESCU (1904). *Vorläufige Mitteilung über eine eligocäne Fischfauna aus den rumanischen Karpathen.* Verh. Geol. Reichsant, 1904, 147–149.

JOAN THEODOR SIMIONESCU.

Krambergeria Simionescu, 148; orthotype K. LANCEOLATA Simionescu (fossil).

1997. SMITH (1904). *A New Cottoid Fish from Bering Sea.* P. Soc. Wash., XVII, 163–164.

HUGH McCORMICK SMITH.

Thecopterus Smith, 163; orthotype T. ALEUTICUS Smith.

1998. SNYDER (1904). *A Catalogue of the Shore Fishes Collected by the Steamer "Albatross" about the Hawaiian Islands in 1902.* Bull. U. S. Fish Comm., XXII, 513–538.

JOHN OTTERBEIN SNYDER.

Veternio Snyder, 516; orthotype V. VERRENS Snyder.

Collybus Snyder, 525; orthotype C. DRACHME Snyder.

Merinthe Snyder, 535; orthotype SEBASTES MACROCEPHALUS Sauvage.

1999. STROMER (1904). *Nematognathi aus dem Fajum und dem Natronthale in Aegypten.* N. Jahrb. Min., 1904, I, 1–17.

ERNST STROMER VON REICHENBACH (1871–   ).

Fajumia Stromer, 3; orthotype F. SCHWEINFURTHI Stromer (fossil).

Socnopæa Stromer, 6; orthotype S. GRANDIS Stromer (fossil).

2000. VAILLANT (1904). *Quelques Reptiles, Batraciens et Poissons du Haut Tonquin.* Bull. Mus. Paris, 1904, 297–301.

LÉON VAILLANT.

Luciocyprinus Vaillant, 299; orthotype L. LANGSONI Vaillant.

2001. WAITE (1904). *Additions to the Fish Fauna of Lord Howe Island. No.4.* Rec. Austral. Mus., V, 135–186.

EDGAR RAVENSWOOD WAITE.

Xenogramma Waite, 157; orthotype X. CARINATUM Waite.

Allogobius Waite, 176; orthotype A. VIRIDIS Waite. A synonym of EVIOTA Jenkins.

Limnichthys Waite, 178; orthotype L. FASCIATUS Waite.

Lepadichthys Waite, 180; orthotype L. FRENATUS Waite.

Schizochirus Waite, 243; orthotype S. INSOLENS Waite.

1905.

2002. ANTIPA (1905). *Clupeinen des Westlichen Teiles des Schwarzen Meeres und der Donaumündungen.* Denkschr. Akad. Wiss. Wien, LXVIII.

GREGOR ANTIPA.

Sardina Antipa, 54; orthotype CLUPEA PILCHARDUS L. A synonym of SARDINIA Poey, the two names equally similar to the earlier SARDINIUS Von der Marck, a fossil sardine-like fish. For the genus typified by CLUPEA PILCHARDUS the name CLUPANODON Lac. has been used, but according to the present rules of nomenclature that name must stand for the Asiatic genus, KONOSIRUS.

2003. See Additions and Corrections, page 575.

2004. EASTMAN (1905). *Les types de poissons fossiles du Monte Bolca au Museum d'Histoire Naturelle de Paris.* Mem. Soc. Geol. Paris, XIII.

CHARLES ROCHESTER EASTMAN.

Eomyrus (Agassiz) Eastman, 15; logotype E. FORMOSISSIMUS Ag. (Ms.).

2005. EIGENMANN & WARD (1905). *The Gymnotidæ.* Proc. Wash. Ac., 1905, VII, 159–188.

CARL H. EIGENMANN; D. P. WARD.

Sternarchella Eigenmann & Ward, 163; orthotype STERNARCHUS SCHOTTI Steind.

Sternarchorhamphus Eigenmann & Ward, 165; logotype STERNARCHUS MULLERI Steind.

Sternarchogiton Eigenmann & Ward, 165; orthotype STERNARCHUS NATTERERI Steind.

2006. FORBES & RICHARDSON (1905). *On a New Shovelnose Sturgeon from the Mississippi River.* Bull. Illinois Lab., VII, 1905, 37–44.

STEPHEN ALFRED FORBES; ROBERT EARL RICHARDSON.

Parascaphirhynchus Forbes & Richardson, 37; orthotype P. ALBUS Forbes & Richardson.

2007. FOWLER (1905). *New, Rare, or Little Known Scombroids, I.* Proc. Acad. Nat. Sci. Phila., LVI, for 1904, 757–771. (Jan. 16, 1905.)

HENRY WEED FOWLER.

Sierra Fowler, 766; orthotype CYBIUM CAVALLA Cuv. A subgenus under SCOMBEROMORUS Lac.

Lepturacanthus Fowler, 770; orthotype TRICHIURUS SAVALA Cuv.

2008. FOWLER (1905). *New, Rare, or Little Known Scombroids, II.* Proc. Acad. Nat. Sci. Phila., LXII, 56–88. (Mar. 31, 1905.)

HENRY WEED FOWLER.

Rhaphiolepis Fowler, 59; orthotype CHORINEMUS TOL Cuv.

Vexillicaranx Fowler, 76; orthotype CARANX AFRICANUS Steind.

Elaphotoxon Fowler, 76; orthotype SCOMBER RUBER Bloch.

2009. FOWLER (1905). *Some Fishes from Borneo.* Proc. Acad. Nat. Sci. Phila., LVII, 455–523. (Aug. 14, 1905.)

HENRY WEED FOWLER.

Pristiopsis Fowler, 459; orthotype PRISTIS PERROTTETI Muller & Henle (lower caudal lobe developed).

Apodoglanis Fowler, 463; orthotype A. FURNESSI Fowler.

Vaillantella Fowler, 474; orthotype NEMACHEILUS EUEPIPTERUS Vaillant.

Homalopteroides Fowler, 476; orthotype Homaloptera wassinkii Blkr.
Labidorhamphus Fowler, 493; orthotype Hemirhamphus amblyurus Blkr.
Pampus (Bonaparte) Fowler, 499; orthotype Stromateus cinereus Bloch. Forgotten name revived; replaces Stromateoides Blkr.
Gigantogobius Fowler, 511; orthotype G. jordani Fowler.

2010. GILBERT (1905). *Aquatic Resources of the Hawaiian Islands. II. The Deep-Sea Fishes.* Bull. U. S. Fish Comm. for 1903, (Aug. 5, 1905.)
CHARLES HENRY GILBERT.

Metopomycter Gilbert, 585; orthotype M. denticulatus Gilbert.
Stemonidium Gilbert, 586; orthotype S. hypomelas Gilbert.
Nematoprora Gilbert, 587; orthotype N. polygonifera Gilbert.
Leptostomias Gilbert, 606; orthotype L. macronema Gilbert.
Lestidium Gilbert, 607; orthotype N. nudum Gilbert.
Hynnodus Gilbert, 617; orthotype H. atherinoides Gilbert.
Grammatonotus Gilbert, 618; orthotype G. laysanus Gilbert.
Stethopristes Gilbert, 622; orthotype S. eos Gilbert.
Cyttomimus Gilbert, 623; orthotype C. stelgis Gilbert.
Peloropsis Gilbert, 630; orthotype O. xenops Gilbert.
Plectrogenium Gilbert, 634; orthotype P. nanum Gilbert.
Snyderidia Gilbert, 634; orthotype S. canina Gilbert.
Bembradium Gilbert, 637; orthotype B. roseum Gilbert.
Chrionema Gilbert, 645; orthotype C. chryseres Gilbert.
Jordanicus Gilbert, 656; orthotype Fierasper umbratilis J. & E.
Tæniopsetta Gilbert, 680; orthotype T. radula Gilbert.
Samariscus Gilbert, 682; orthotype S. corallinus Gilbert.
Miopsaras Gilbert, 694; orthotype M. myops Gilbert.

2011. GILBERT & THOMPSON (1905). *Notes on the Fishes of Puget Sound.* Proc. U. S. Nat. Mus., XXVIII, 973-987. (Aug. 8, 1905.)
CHARLES HENRY GILBERT; JOSEPH C. THOMPSON.

Stelgidonotus Gilbert & Thompson, 977; orthotype S. latifrons Gilbert & Thompson.

2012. GILL (1905). *The Scorpænoid Fish, Neosebastes entaxis, as the Type of a Distinct Genus.* Proc. U. S. Nat. Mus. (February 15, 1905.)
THEODORE NICHOLAS GILL.

Sebastosemus Gill, 219; orthotype Neosebastes entaxis Jordan & Snyder.

2013. GILL (1905). *Notes on the Genera of Synanceine and Pelorine Fishes.* Proc. U. S. Nat. Mus., XXVIII, 221-225. (Feb. 23, 1905.)
THEODORE NICHOLAS GILL.

Simopias Gill, 224; orthotype Pelor filamentosus Cuv. & Val. A substitute for Pelor Cuv., based on a section of the genus distinct from the type of Inimicus Jordan & Starks.
Rhinopias Gill, 225; orthotype Scorpæna frondosa Gthr.

2014. GILL (1905). *On the Generic Characters of Prionotus Stearnsi.* Proc. U. S. Nat. Mus. (February 15, 1905.)
THEODORE NICHOLAS GILL.

Colotrigla Gill. 339; orthotype Prionotus stearnsi Jordan & Swain.
Fissala Gill, 342; orthotype Prionotus alatus G. & B.

2015. GILL (date not ascertained). *Mancopsetta.*

Mancopsetta Gill, —; orthotype LEPIDOPSETTA MACULATA Gthr.; substitute for LEPI-
DOPSETTA Gthr., preoccupied. (I am unable to place this genus.)

2016. GILL & SMITH (1905). *A New Family of Jugular Acanthopterygians
(Caristiidæ).* Proc. Biol. Soc. Wash., XVIII.

THEODORE GILL; HUGH MCCORMICK SMITH.

Caristius Gill & Smith, 249; orthotype C. JAPONICUS Gill & Smith.

2017. HAUG (1905). *Documents scientifiques de la mission saharienne Paleon-
tologique (Mission Foureau Lamy).* Paris, fasc. 3, 1905, 751–832.

EMILE HAUG.

Platyspondylus Haug, 751; orthotype P. FOUREAUI Haug (fossil).

2018. JORDAN & EVERMANN (1905). *The Aquatic Resources of the Hawaiian
Islands. I. The Shore Fishes.* Bull. U. S. Fish Comm. for 1903. (July 29,
1905.)

DAVID STARR JORDAN; BARTON WARREN EVERMANN.

Kelloggella (Jordan & Seale) Jordan & Evermann, 488; orthotype K. CARDINALIS
Jordan & Seale. More fully defined later in *Fishes of Samoa.*

Exallias Jordan & Evermann, 503; orthotype SALARIAS BREVIS Kner.

2019. JORDAN & SEALE (1905). *List of Fishes Collected by Dr. Bashford Dean
on the Island of Negros, Philippines.* Proc. U. S. Nat. Mus., XXVIII,
769–803.

DAVID STARR JORDAN; ALVIN SEALE.

Eleria Jordan & Seale, 774; orthotype E. PHILIPPINA Jordan & Seale.

Foa Jordan & Seale, 779; orthotype F. FO Jordan & Seale.

Drombus Jordan & Seale, 797; orthotype D. PALACKYI Jordan & Seale.

2020. JORDAN & STARKS (1905). *On a Collection of Fishes Made in Korea,
by Pierre Louis Jouy, with Descriptions of New Species.* Proc. U. S. Nat.
Mus., XXVIII, 193–211. (February 23, 1905.)

DAVID STARR JORDAN; EDWIN CHAPIN STARKS.

Longurio Jordan & Starks, 196; orthotype L. ATHYMIUS Jordan & Starks. A synonym of SAUROGOBIO Blkr.

Fusania Jordan & Starks, 198; orthotype F. ENSARCA Jordan & Starks. A synonym
of APHYOCYPRIS Gthr.

Coreius Jordan & Starks, 199; orthotype LABEO CETOPSIS Kner.

Larimichthys Jordan & Starks, 204; orthotype L. RATHBUNÆ Jordan & Starks.

2021. JORDAN & THOMPSON (1905). *The Fish Fauna of the Tortugas Archipelago.* Bull. Bur. Fish, XXIV, 231–266.

DAVID STARR JORDAN; JOSEPH C. THOMPSON.

Etelides Jordan & Thompson, 241; orthotype ETELIS AQUILIONARIS G. & B.

Execestides Jordan & Thompson, 253; orthotype E. EGREGIUS Jordan & Thompson.

2022. KRAMBERGER (1905). *Die obertriadische Fischfauna von Hallein in
Salzberg.* Bietr. Pal. Pesterr-Ung., XVIII, 193–224.

DRAGUTIN GORJANOVIC KRAMBERGER.

Spaniolepis Kramberger, 216; orthotype S. OVALIS Kramberger (fossil).

2023. LERICHE (1905). *Les poissons éocènes de la Belgique.* Mem. Mus. Bel-
giqüe Bruxelles, III, 1905, 228.

MAURICE LERICHE.

Cristigorina Leriche, 80; orthotype C. CRASSA Leriche (fossil).

Glyptorhynchus Leriche, 159; orthotype CŒLORHYNCHUS RECTUS Ag. (fossil). Substitute for CŒLORHYNCHUS Ag., preoccupied. A synonym of CYLINDRACANTHUS Leidy, 1856.

2024. LÖNNBERG (1905). *The Fishes of the Swedish South Pole Expedition.* Wissensch. Ergeb. Schwed. Sudpolar-Exp., 5, Lief. 6, 1906, 1–69.

AXEL JOHAN EINAR LÖNNBERG.

Artedidraco Lönnberg, 39; orthotype A. MIRUS Lönnberg.

2025. McGILCHRIST (1905). *Natural History Notes from the R. I. M. S. "Investigator" Capt. R. H. Heming, R. N. (retired), Commanding. Series III, No. 8. On a New Genus of Teleostean Fish Closely Allied to Chiasmodus.* Ann. Mag. Nat. Hist. (7), XV, 268–270.

A. C. McGILCHRIST.

Dysalotus McGilchrist, 268; orthotype D. ALCOCKI McGilchrist.

2026. PELLÉGRIN (1905). *Poisson nouveau de Mozambique.* Bull. Mus. Paris, XI, 145–146.

JACQUES PELLÉGRIN.

Xenopomichthys Pellégrin, 145; orthotype X. AURICULATUS Pellégrin.

2027. PELLÉGRIN (1905). *Mission Scientifique en Afrique Centrale.* Mem. Soc. Zool. de France.

JACQUES PELLÉGRIN.

Astatochromis Pellégrin, 183; orthotype A. ALLAUDI Pellégrin.

2028. POPTA (1905). *Suites des descriptions préliminaires des nouvelles espèces de poissons recueillis par M. le Dr. A. W. Nieuwenhuis en 1898 et en 1900.* Notes Leyden Mus., XXV, 171–187.

CANNA M. L. POPTA.

Neogastromyzon Popta, 180; orthotype N. NIEUWENHUISII Popta.

Parophiocephalus Popta, 183; orthotype P. UNIMACULATUS Popta. A synonym of BETTA Blkr.

2029. REGAN (1905). *A Synopsis of the Species of the Silurid Genera Parexostoma, Chimarrhichthys, and Exostoma.* Ann. Mag. Nat. Hist. (7), XV, 182.

CHARLES TATE REGAN.

Parexostoma Regan, 182; logotype EXOSTOMA STOLICZKÆ Day.

2030. REGAN (1905). *A Revision of the Fishes of the American Cichlid Genus Cichlasoma and of the Allied Genera.* Ann. Mag. Nat. Hist. (7), XV, 60–77; 225–243, 316–340.

CHARLES TATE REGAN.

Parapetenia Regan, 324; orthotype ACARA ADSPERSA Gthr.

2031. REGAN (1905). *A Revision of the Fishes of the South American Cichlid Genera Acara, Nannacara, Acaropsis, and Astronotus.* Ann. Mag. Nat. Hist. (7), XV, 329–427.

CHARLES TATE REGAN.

Nannacara Regan, 344; orthotype N. ANOMALA Regan.

2032. REGAN (1905). *(Gymnapogon).* Ann. Mag. Nat. Hist., XV.

CHARLES TATE REGAN.

Gymnapogon Regan, 19; orthotype G. JAPONICUS Regan.

2033. SAUTER (1905). *Notes from the Owston Collection.* Annot. Zool. Japan, V, 233–238.

HANS SAUTER.

Ijimaia Sauter, 233; orthotype I. DOFLEINI Sauter.

2034. STROMER (1905). *Die Fischreste des mittleren und oberen Eocens von Aegypten.* Beitr. Pal. Oesterr-Ung., XVIII, 37–58.

ERNST STROMER VON REICHENBACH.

Eopristis Stromer, 52; orthotype P. REINACHI Stromer.

2035. TOULA (1905). *Ueber einem dem Thunfische verwandten Raub-fish der Congerienschichten der Wiener Bucht.* . . . Jahrb. Geol. Reichsanst. Wien., LV.

FRANZ TOULA (1845–   ).

Pelamycybium Toula, 51; orthotype P. SINUS-VINDOBONENSIS Toula (fossil).

2036. WAITE (1905). *Notes on Fishes from Western Australia, III.* Rec. Austral. Mus., VI, 1905, 55–82.

EDGAR RAVENSWOOD WAITE.

Neatypus Waite, 64; orthotype N. OBLIQUUS Waite.
Bramichthys Waite, 72; orthotype B. WOODWARDI Waite.
Dipulus Waite, 77; orthotype D. CÆCUS Waite.

1906.

2037. ABEL (1906). *Fossile Flugfische.* Verh. Deutsche. Zool. Ges., XV, 47–48.

OTHENIO ABEL.

Dollopterus Abel, 47; orthotype DOLICHOPTERUS VOLITANS Compter (fossil).
Gigantopterus Abel, 48; orthotype G. TELLERI Abel (fossil).

2038. AMEGHINO (1906). *Les formations sedimentaires du crétacé supérieur et du Tertiaire de Patagonie, etc.* Ann. Mus. Nat. Buenos Aires, XV.

FLORENTINO AMEGHINO.

Pseudacrodus Ameghino, 176; orthotype P. PATAGONIENSIS Ameghino (fossil).
Carcharioides Ameghino, 183; orthotype C. TOTUSERRATUS Ameghino (fossil).

2039. BERG (1906). *Uebersicht der Marsipobranchii des russischen Reiches.* St. Petersburg, Bull. Ac. Ser. 5, XXIV, 169–183.

LEW SEMENOWITCH BERG.

Caspiomyzon Berg, 177; orthotype PETROMYZON WAGNERI Kessler.

2040. BERG (1906). *Uebersicht der Cataphracti (Fam. Cottidæ, Cottocomephoridæ und Comephoridæ) des Baikalsees.* Zool. Anz., XXX, 1906, 906–911.

LEW SEMENOWITCH BERG.

Asprocottus Berg, 907; orthotype A. HERZENSTEINI Berg.
Abyssocottus Berg, 908; orthotype A. KOROTNEFFI Berg.
Limnocottus Berg, 909; orthotype COTTUS GODLEWSKII Dybowski.

2041. BOULENGER (1906). *Fourth Contribution to the Ichthyology of Lake Tanganyika. Report on the Collection of Fishes Made by Dr. W. A. Cunnington during the Third Tanganyika Expedition, 1904–1905.* Trans. Zool. Soc. London, XVII, 1906, 537–576.

GEORGE ALBERT BOULENGER.

Dinotopterus Boulenger, 550; orthotype D. CUNNINGTONI Boulenger.

Phyllonemus Boulenger, 552; orthotype P. TYPUS Boulenger.
Haplotaxodon Boulenger, 566; orthotype H. MICROLEPIS Boulenger.
Cunningtonia Boulenger, 573; orthotype C. LONGIVENTRALIS Boulenger.

2042. BOULENGER (1906). *Descriptions of New Fishes Discovered by Mr. E Degen in Lake Victoria.* Ann. Mag. Nat. Hist., XVII, 433–452.
GEORGE ALBERT BOULENGER.
Platyæniodus Boulenger, 451; orthotype P. DEGENI Boulenger.

2043. BRAUER (1906). *Die Tiefsee-Fische. I. Syst. Theil. (Wissenschaftl. Ergebnisse der deutschen Tiefsee-Expedition).* Jena, 1906, 1–432.
AUGUST BRAUER.
Stylophthalmus Brauer, 70; orthotype S. PARADOXUS Brauer.
Oxyodon Brauer, 287; orthotype O. MACROPS Brauer.
Bassobythites Brauer, 307; orthotype B. BRUNSWIGI Brauer.

2044. EASTMAN (1906). *Structure and Relations of Mylostoma.* Bull. Mus. Comp. Zool., L, 1–29. (May, 1906.)
CHARLES ROCHESTER EASTMAN.
Dinomylostoma Eastman, 23; orthotype D. BEECHERI Eastman (fossil, teeth and body plates): (mentioned earlier, February, 1906, in a paper on ARTHRODIRES. Amer. Journ. Sci., XXI, 137). ARTHRODIRA, Devonian.

2045. EVERMANN & KENDALL (1906). *Notes on a Collection of Fishes from Argentina, South America, with Descriptions of Three New Species.* Proc. U. S. Nat. Mus., XXXI, 67–108. (July 25, 1906.)
BARTON WARREN EVERMANN; WILLIAM CONVERSE KENDALL.
Odontesthes Evermann & Kendall, 94; orthotype O. PERUGIÆ Evermann & Kendall.

2046. FOWLER (1905). *Some Cold-Blooded Vertebrates from the Florida Keys.* Proc. Acad. Nat. Sci. Phila., LVIII, 77–113. (May 29, 1906.)
HENRY WEED FOWLER.
Congrammus Fowler, 105; orthotype C. MOOREI Fowler.

2047. FOWLER (1906). *New, Rare, or Little Known Scombroids,. III.* Proc. Acad. Nat. Sci. Phila., LVIII, 114–122. (June 20, 1906.)
HENRY WEED FOWLER.
Glaucus (Klein) Fowler, 116; orthotype SCOMBER AMIA L. Revived as reprinted by Walbaum, *Artedi Piscium,* III, 585, 1792, to replace HYPODIS Raf., HYPACANTHA Raf., LICHIA Cuv., and PORTHMEUS Cuv. & Val. This choice of type antedates that of Jordan & Hubbs, 1915. SCOMBER GLAUCUS L., replacing CÆSIMORUS Lac., the law of tautonomy would apply here, and the names of Klein, however reprinted, are not yet declared eligible.
Pampanoa Fowler, 116; orthotype CHÆTODON GLAUCUS Bloch.
Pterorhombus Fowler, 118; orthotype FIATOLA FASCIATA Risso.
Priacanthopsis Fowler, 122; orthotype PEMPHERIS MULLERI Poey.

2048. FOWLER (1906). *Further Knowledge of Some Heterognathous Fishes.* Proc. Acad. Nat. Sci. Phila., LVIII, I, 293–351. (Sept. 25, 1906.)
HENRY WEED FOWLER.
The numerous genera indicated in this paper have been critically reviewed by Dr. Carl H. Eigenmann, Amer. Nat. Dec., 1907. From this paper I draw the comments on the different genera proposed.

Ophiocephalops Fowler, 293; orthotype ERYTHRINUS UNITÆNIATUS Ag.   A synonym of HOPLERYTHRINUS Gill.

Copeina Fowler, 294; orthotype PYRRHULINA ARGYROPS Fowler.   A substitute for HOLOTAXIS of Eigenmann, not of Cope.

Cyphocharax Fowler, 297; orthotype CURIMATUS SPILURUS Gthr.

Steindachnerina Fowler, 298; orthotype CURIMATUS TRACHYSTETHUS Cope.   Probably a synonym of CURIMATA Walbaum (CURIMATUS Cuv.).

Peltapleura Fowler, 300; orthotype SALMO CYPRINOIDES L.   A synonym of CURIMATA Walbaum.

Eigenmannina Fowler, 307; orthotype ANODUS MELANOPOGON Cope.

Chilomyzon Fowler, 309; orthotype PROCHILODUS STEINDACHNERI Fowler.   A synonym of PROCHILODUS.

Hemiodopsis Fowler, 318; orthotype HEMIODUS MICROLEPIS Kner.

Pithecocharax Fowler, 319; orthotype SALMO ANOSTOMUS L.   A substitute for ANOSTOMUS Cuv., regarded as preoccupied by ANASTOMUS Bonnaterre, 1788. But ANOSTOMUS dates from Gronow (1763), revived by Scopoli (1777).

Pœcilosomatops Fowler, 323; orthotype CHARACIDIUM ETHEOSTOMA Cope.   A synonym of NANOGNATHUS Boulenger, a section of CHARACIDIUM.

Garmanina Fowler, 326; orthotype RHYTODUS ARGENTEOFUSCUS Kner.

Abramites Fowler, 331; orthotype LEPORINUS HYPSELONOTUS Gthr.   A synonym of LEPORINUS Cuv. & Val.

2049. GARMAN (1906).   *New Plagiostomia.*   Bull. Mus. Comp. Zool., Harvard Coll., XLVI, 201–208.
SAMUEL GARMAN.

Parmaturus Garman, 201; orthotype P. PILOSUS Garman.

2050. GILCHRIST (1903).   *Descriptions of Fifteen New South African Fishes, with Notes on Other Species.*   Cape Town, Marine Investiga., IV, 1906, 143–171.
J. D. F. GILCHRIST.

Pseudocyttus Gilchrist, 152; orthotype P. MACULATUS Gilchrist.

Neostomias Gilchrist, 168; orthotype N. FILIFERUM Gilchrist.

2051. GRACIANOV (1906).   *Ueber eine besondere Gruppe der Rochen.*   Zool. Anz., XXX, 399–406.   (Dec., 1906.)
VALERIAN GRACIANOV (or GRATZIANOW).

Brachioptera Gratzianow, 402; orthotype B. RHINOCEROS Gratzianow.

Planerocephalus Gracianov, 403; orthotype P. ELLIOTI Gracianov.

2052. GRACIANOV (1906).   *Die Neunaugen des Russischen Reiches.*   Text in Russian.   Moskv. Dnevn. zool. otd. obsc. liub. jest., 3, 7–8, 1907, 18.
VALERIAN GRACIANOV.

Agnathomyzon Gracianov, 18; orthotype PETROMYZON WAGNERI Kessler.   A synonym of CASPIOMYZON Berg (September, 1906).

Haploglossa Gracianov, 18; orthotype A. CASPICUS Wagner = PETROMYZON WAGNERI Kessler.   A synonym of CASPIOMYZON Berg.

2053. JORDAN & SEALE (1906).   *Descriptions of Six New Species of Fishes from Japan.*   Proc. U. S. Nat. Mus., XXX, 143–148.
DAVID STARR JORDAN; ALVIN SEALE.

Sayonara Jordan & Seale, 145; orthotype S. SATSUMÆ Jordan & Seale.

2054. JORDAN & SEALE (1906). *The Fishes of Samoa. Descriptions of Species Found in the Archipelago, with a Provisional Check-List of the Fishes of Oceania.* Bull. Bur. Fish., XXV, 173–455. (Dec. 15, 1906.)

DAVID STARR JORDAN; ALVIN SEALE.

Anarchias Jordan & Seale, 204; orthotype A. ALLARDICEI Jordan & Seale.

Psettias (Jordan) Jordan & Seale, 237; orthotype PSETTUS SEBÆ Cuv. & Val.

Rooseveltia (Jordan & Evermann) Jordan & Seale, 265; orthotype SERRANUS BRIGHAMI Seale.

Oceanops Jordan & Seale, 277; orthotype LABRUS LATOVITTATA Lac.

Lo Seale, 360; orthotype AMPHACANTHUS VULPINUS Schlegel & Müller.

Abalistes Jordan & Seale, 364; orthotype BALISTES STELLARIS Bl. & Sch. Substitute for LEIURUS Sw., preoccupied.

Trimma Jordan & Seale, 391; orthotype T. CÆSIURA Jordan & Seale.

Vitreola Jordan & Seale, 393; orthotype V. SAGITTA Jordan & Seale.

Vaimosa Jordan & Seale, 395; orthotype V. FONTINALIS Jordan & Seale. A synonym of MUGILOGOBIUS Smitt, 1899.

Vailima Jordan & Seale, 398; orthotype V. STEVENSONI Jordan & Seale.

Exyrias Jordan & Seale, 405; orthotype GOBIUS PUNTANGOIDES Blkr.

Pselaphias Jordan & Seale, 406; orthotype GOBIUS OPHTHALMONEMUS Blkr.

Waitea Jordan & Seale, 407; orthotype GOBIUS MYSTACINUS Cuv. & Val.

Mars Jordan & Seale, 408; orthotype M. STRIGILLICEPS Jordan & Seale.

Kelloggella Jordan & Seale, 409; orthotype K. CARDINALIS Jordan & Seale. (First noticed by Jordan & Evermann, *Fishes of Hawaii*, 1905.)

Sauvagea Jordan & Seale, 420; orthotype LEPIDOBLENNIUS CALEDONICUS Sauvage. Substitute for LEPIDOBLENNIUS Sauvage, preoccupied.

2055. JORDAN & SNYDER (1906). *A Review of the Pœciliidæ or Killifishes of Japan.* Proc. U. S. Nat. Mus., XXXI, 287–290. (Sept. 10, 1906.)

DAVID STARR JORDAN; JOHN OTTERBEIN SNYDER.

Oryzias Jordan & Snyder, 289; orthotype PŒCILIA LATIPES T. & S.

2056. JORDAN & STARKS (1906). *Notes on a Collection of Fishes from Port Arthur, Manchuria, Obtained by James Francis Abbott.* Proc. U. S. Nat. Mus., XXXI, 515–526. (Sept. 10, 1906.)

DAVID STARR JORDAN; EDWIN CHAPIN STARKS.

Ranulina Jordan & Starks, 523; orthotype R. FIMBRIIDENS Jordan & Starks.

2057. JORDAN & STARKS (1906). *A Review of the Flounders and Soles of Japan.* Proc. U. S. Nat. Mus., XXXI, 161–246.

DAVID STARR JORDAN; EDWIN CHAPIN STARKS.

Cynopsetta (Schmidt) Jordan & Starks, 188; orthotype C. DUBIA Schmidt.

Limandella Jordan & Starks, 204; orthotype PLEURONECTES YOKOHAMÆ Gthr.

Amate Jordan & Starks, 228; orthotype ACHIRUS JAPONICUS T. & S.

2058. LERICHE (1908). *Note préliminaire sur les poissons nouveaux de l'Oligocène Belge.* Bull. Soc. Belge Geol., XXII.

MAURICE LERICHE.

Neocybium Leriche, 379; orthotype N. ROSTRATUM Leriche (fossil).

2059. OGILBY (1906). *(A New Genus of Sharks.)* Proc. Roy. Soc. Queensland, XX.

JAMES DOUGLAS OGILBY.

Brachælurus Ogilby, 27; orthotype CHILOSCYLLIUM MODESTUM Gthr.

2060. OSORIO (1906). *Description d'un poisson des profondeurs appartenant à un genre nouveau et trouvé sur les côtes du Portugal.* Lisboa, J. Sci., VII, 1906, 172–174.

BALTHAZAR OSORIO.

Lophocephalus Osorio, 173; orthotype L. ANTHRAX Osorio.

2061. PAPPENHEIM (1906). *Neue und ungenügend bekannte elektrische Fische (Mormyridæ) aus den deutsch-afrikanischen Schutzgebieten.* Sitzber. Ges. Naturf. Berlin, 1906, 260–264.

PAUL PAPPENHEIM.

Hippopotamyrus Pappenheim, 260; orthotype H. CASTOR Pappenheim.

2062. REGAN (1906). *A Revision of the South American Cichlid Genera Retroculus, Geophagus, Heterogramma, and Biotæcus.* Ann. Mag. Nat. Hist., XVII, 49–66.

CHARLES TATE REGAN.

Heterogramma Regan, 63; orthotype H. BORELLI Regan = H. RITENSE Haseman. Name preoccupied; replaced by APISTOGRAMMA Regan, 1913.

2063. REGAN (1906). *Notes on Some Loricariid Fishes, with Descriptions of Two New Species.* Ann. Mag. Nat. Hist. (7), XVII, 94–98.

CHARLES TATE REGAN.

Thysanocara Regan, 96; orthotype XENOCARA CIRRHOSA Regan.

2064. REGAN (1906). *Descriptions of New or Little Known Fishes from the Coast of Natal.* Ann. Natal Govt. Mus., I, pt. 1, 1–6.

CHARLES TATE REGAN.

Pliotrema Regan, 1; orthotype P. WARRENI Regan.

2065. SEALE (1906). *Fishes of the South Pacific.* Honolulu, H. I. Occ. Papers Bernice Bishop Mus., 4, 1906, 1–89.

ALVIN SEALE.

Deleastes Seale, 80; orthotype D. DÆCTOR Seale.

2066. SMITH & POPE (1906). *List of Fishes Collected in Japan in 1903, with Descriptions of New Genera and Species.* Proc. U. S. Nat. Mus., XXXI, 459–499. (Sept. 24, 1906.)

HUGH McCORMICK SMITH; THOMAS E. B. POPE.

Tosana Smith & Pope, 470; orthotype T. NIWÆ Smith & Pope.

Satsuma Smith & Pope, 472; orthotype S. MACROPS Smith & Pope. A synonym of MALAKICHTHYS Döderlein.

Lysodermus Smith & Pope, 483; orthotype L. SATSUMÆ Smith & Pope.

Lambdopsetta Smith & Pope, 496; orthotype L. KITAHARÆ Smith & Pope.

2067. SMITH & SEALE (1906). *Notes on a Collection of Fishes from the Island of Mindanao, Philippine Archipelago, with Descriptions of New Genera and Species.* Proc. Biol. Soc. Wash., XIX, 73–82.

HUGH McCORMICK SMITH; ALVIN SEALE.

Illana Smith & Seale, 73; orthotype I. CACABET Smith & Seale.

Caragobius Smith & Seale, 81; orthotype C. TYPHLOPS Smith & Seale.

2068. STEINDACHNER (1906). *Ueber eine neue Gattung und Art aus der Familie der Murænidæ, Nettastomops barbatula.* Anz. Ak. Wiss., 1906, 299–300.

FRANZ STEINDACHNER.

Nettastomops Steindachner, 299; orthotype N. BARBATULA Steind.

2069. STEINDACHNER (1906). *Zur Fischfauna der Samoa-Inseln.* Sb. Ak. Wiss. Wien., CXV, J-uly, 1906.
FRANZ STEINDACHNER.

Kræmeria Steindachner, 1409 (41); orthotype K. SAMOENSIS Steind. (TRICHONO-TIDÆ).

2070. WERNER (1906). *Beiträge zur Kenntnis der Fischfauna des Nils.* Ans Ak. Wiss. Wien., 1906, 325–327.
FRANZ WERNER.

Slatinia Werner, 326; orthotype S. MONGALLENSIS Werner.

1907.

2071. BERG (1907). *Die Cataphracti des Baikal-Sees. (Fam. Cottidæ, Cotto-comephoridæ und Comephoridæ.) Beiträge zur Osteologie und Systematik.* St. Petersberg und Berlin, 1907, 1–75.
LEW SEMENOWITCH BERG.

Cottinella Berg. 45; orthotype ABYSSOCOTTUS BOULENGERI Berg.

2072. BERG (1907). *Description of a New Cyprinoid Fish, Paraleucogobio nota-canthus, from North China.* Ann. Mag. Nat. Hist., XIX, 1907, 163–164.
LEW SEMENOWITCH BERG.

Paraleucogobio Berg, 163; orthotype P. NOTACANTHUS Berg.

2073. BERG (1907). *Verzeichniss der Fische von Russisch-Turkestan.* St. Peters-berg, Ann. Mus. Zool. Ac. Sc., X, 1905 (1907), 316–332.
LEW SEMENOWITCH BERG.

Aspiolucius Berg, 12; orthotype ASPIUS ESOCINUS Kessler.

2074. BERG (1907). *Description of a New Cyprinoid Fish, Acheilognathus sig-nifer, from Korea, with a Synopsis of All the Known Rhodeinæ.* Ann. Mag. Nat. Hist., XIX, 159–163.
LEW SEMENOWITCH BERG.

Pararhodeus Berg, 160; orthotype RHODEUS SYRIACUS Lortet.

2075. BERG (1907). *Beschreibungen einiger neuer Fische aus dem Stromgebiet des Amur.* St. Petersburg, Ann. Mus. Zool. Ac. Sc., XII, 1907, 418–423.
LEW SEMENOVITCH BERG.

Plagiognathops Berg, 419; orthotype PLAGIOGNATHUS JELSKII Dybowski = XENOCY-PRIS MICROLEPIS Blkr. A substitute for PLAGIOGNATHUS Berg, preoccupied in HEMIPTERA.

2076. EASTMAN (1907). *Devonic Fishes of the New York Formations.* Mem. N. Y. State Mus., X, 1907.
CHARLES ROCHESTER EASTMAN.

Protitanichthys Eastman, 144; orthotype P. FOSSATUS Eastman (fossil; front half of a cranium of an ARTHRODIRE). A snyonym of COCCOSTEUS Ag. (See Hussakof. Science, XXVIII, 311, 1909.)

2077. EIGENMANN (1907). *The Pœciliid Fishes of Rio Grande do Sul and the La Plata Basin.* Proc. U. S. Nat. Mus., XXXII, 1907, 425–433 (May 23, 1907). CARL H. EIGENMANN.

Acanthophacelus Eigenmann, 425; orthotype PŒCILIA RETICULATA Peters. A synonym of LEBISTES Philippi, 1861, according to Henn.

Ilyodon Eigenmann, 427; orthotype I. PARAGUAYENSE Eigenmann.

Phalloptychus Eigenmann, 430; orthotype GIRARDINUS JANUARIUS Hensel.

Phalloceros Eigenmann, 431; orthotype GIRARDINUS CAUDOMACULATUS Hensel.

2078. EIGENMANN (1907). *. Fowler's "Heterognathus Fishes," with a note on the Stethaprioninæ.* Amer. Naturalist, XLI, No. 492, 1907, 767–772. CARL H. EIGENMANN.

Phenacogaster Eigenmann, 769; orthotype ASTYANAX PECTINATUS Cope.

Astyacinus Eigenmann, 769; orthotype ASTYANAX MOORI Boulenger.

Fowlerina Eigenmann, 771; orthotype TETRAGONOPTERUS COMPRESSUS Gthr. Name preoccupied; replaced by EPHIPPICHARAX Fowler.

2079. EIGENMANN & BEAN (1907). *An Account of Amazon River Fishes Collected by J. B. Steere, with a Note on Pimelodus clarias.* Proc. U. S. Nat. Mus., XXXI, 659–668. (Jan. 16, 1907.) CARL H. EIGENMANN; BARTON APPLER BEAN.

Tænionema Eigenmann & Bean, 662; orthotype BRACHYPLATYSTOMA STEEREI Eigenmann & Bean.

Paracetopsis Eigenmann & Bean, 665; orthotype CETOPSIS OCCIDENTALIS.

2080. EIGENMANN & OGLE (1907). *An Annoted List of Characine Fishes in the United States Nat. Mus. and the Mus. of Ind. Univ., with Descriptions of New Species.* Proc. U. S. Nat. Mus., XXXIII, 1–36. (Sept. 10, 1907.) CARL H. EIGENMANN; FLETCHER OGLE.

Gilbertolus Eigenmann, 2; orthotype ANACYRTUS ALATUS Steind. Substitute for GILBERTELLA Eigenmann, preoccupied.

Evermannolus Eigenmann, 2; orthotype CYNOPOTAMUS BISERIALIS Garman. Substitute for EVERMANNELLA Eigenmann, preoccupied.

Phenacogrammus Eigenmann, 30; orthotype MICRALESTES INTERRUPTUS Boulenger.

2081. EIGENMANN, McATEE & WARD (1907). *On Further Collections of Fishes from Paraguay.* Ann. Carnegie Mus., 4, 1907, 110–157. CARL H. EIGENMANN; W. J. McATEE; D. P. WARD.

Homodiætus Eigenmann & Ward, 117; orthotype H. ANISTISTI Eigenmann & Ward.

Henonemus Eigenmann & Ward, 118; orthotype STEGOPHILUS INTERMEDIUS Eigenmann & Eigenmann.

Bryconamericus Eigenmann, 139; orthotype B. EXODON Eigenmann.

Deuterodon Eigenmann, 140; orthotype D. IGUAPE Eigenmann.

Chætobranchopsis Eigenmann & Ward, 144; orthotype C. AUSTRALIS Eigenmann & Ward.

2082. EVERMANN & SEALE (1907). *Fishes of the Philippine Islands.* Bull. Bur. of Fish., XXVI (1906), 1907, 49–110. BARTON WARREN EVERMANN; ALVIN SEALE.

Nesogrammus Evermann & Seale, 61; orthotype N. PIERSONI Evermann & Seale.

Hypomacrus Evermann & Seale, 101; orthotype H. ALBIENSIS Evermann & Seale.

2083. FOWLER (1907). *Further Knowledge of Some Heterognathous Fishes, II.*
Proc. Ac. Nat. Sci. Phila., LVIII, 431–483. (Jan. 7 and Jan. 16, 1907.)
HENRY WEED FOWLER.
Pellegrinina Fowler, 442; orthotype P. HETEROLEPIS Fowler.

Coscinoxyron Fowler, 450; orthotype CHALCINUS CULTER Cope.
Thoracocharax Fowler, 452; orthotype GASTROPELECUS STELLATUS Kner.
Cyrtocharax Fowler, 454; orthotype ANACYRTUS LIMÆSQUAMIS Cope. A synonym
of CYNOPOTAMUS.
Cynocharax Fowler, 457; orthotype ANACYRTUS AFFINIS Gthr.
Sphyrænocharax Fowler, 460; orthotype XIPHORHAMPHUS ABBREVIATUS Cope. A
synonym of ACESTRORHAMPHUS Eigenmann.
Belonocharax Fowler, 464; orthotype B. BEANI Fowler. A synonym of LUCIO-
CHARAX Steind., and of the imperfectly described CTENOLUCIUS Gill.
Waiteina Fowler, 473; orthotype MYLETES NIGRIPINNIS Cope. Probably a synonym
of COLOSSOMA Eigenmann.
Reganina Fowler, 475; orthotype MYLESTES BIDENS Ag. Probably a synonym of
COLOSSOMA.
Starksina Fowler, 476; orthotype MYLESTES HARNIARIUS Cope. Probably a synonym
of MYLOSOMA.
Sealeina Fowler, 478; orthotype MYLESTES LIPPINCOTTIANUS Cope.

2084. FOWLER (1907). *Some New and Little Known Percoid Fishes.* Proc.
Acad. Nat. Sci Phila., LVIII, 510–528. (Feb. 14, 1907.)
HENRY WEED FOWLER.
Boulengerina Fowler, 512; orthotype DULES MATO Lesson. Name preoccupied;
replaced by SAFOLE Jordan.
Astrapogon Fowler, 527; orthotype APOGONICHTHYS STELLATUS Cope.

2085. FOWLER (1907). *Notes on Serranidæ.* Proc. Acad. Nat. Sci. Phila., LIX,
249–269. (Aug. 16, 1907.)
HENRY WEED FOWLER.
Chrysoperca Fowler, 250; orthotype MORONE INTERRUPTA Gill.
Eudulus Fowler, 264; orthotype DULES AURIGA Cuv. Substitute for DULES Cuv.,
regarded as preoccupied by DULUS, a genus of birds.
Callidulus Fowler, 265; orthotype CENTROPRISTIS SUBLIGARIUS Cope.

2086. FOWLER (1907). *A Collection of Fishes from Victoria, Australia.* Proc.
Acad. Nat. Sci. Phila., LIX, 419–444.
HENRY WEED FOWLER.
Psychichthys Fowler, 419; orthotype HYDROLAGUS WAITEI Fowler.
Limiculina Fowler, 425; orthotype CENTRISCUS HUMEROSUS Rich. A synonym of
CENTRISCOPS Gill.
Macleayina Fowler, 426; orthotype HIPPOCAMPUS BLEEKERI Fowler.
Castelnauia Fowler, 426; orthotype SOLENOGNATHUS SPINOSISSIMUS Gthr.
Lesueurina Fowler, 440; orthotype L. PLATYCEPHALA Fowler. (Written also
LESUEURIELLA, a slip of proof-reading.)

2087. GRACIANOV (1907). *Versuch einer Uebersicht der Fische des Russischen
Reiches in systematicher und geographischer Hinsicht.* (Text in Russian.)
Trd. Otd. Ichtiol. Obsc. Akklimat. Moskva., 4, 1907, 1–567.
VALERIAN GRACIANOV.
Malacobatis Gracianov, 39; orthotype RAJA MUCOSA Pallas.

Hexagrammoides Gracianov, 289; orthotype H. NUDIGENIS Gracianov, which is probably HEXAGRAMMOS LAGOCEPHALUS (Pallas) or perhaps H. SUPERCILIOSUS (Pallas). A synonym of HEXAGRAMMOS Steller. It is very briefly described from a specimen from Bering Island. "Division of the dorsal fin not complete; in the first dorsal about 20 rays." The name "NUDIGENIS" indicates scaleless cheeks, but H. STELLERI, with the cheeks more fully naked, has more than 20 dorsal spines.

Taurulus Gracianov, 296; orthotype COTTUS BUBALIS Euphrasen.

Trigrammus Gracianov, 418; orthotype ERNOGRAMMUS STROSHI Schmidt.

Pseudophidium Gracianov, 420; orthotype OPHIDIUM GIGANTEUM Kittlitz.

2088. GRACIANOV (1907). *Uebersicht de Süsswassercottiden des russischen Reiches.* Zool. Anz., XXXI, 1907, 654–660.

VALERIAN GRACIANOV.

Cephalocottus Gracianov, 659; orthotype COTTUS AMBLYSTOMOPSIS Schmidt.

Mesocottus Gracianov, 660; orthotype COTTUS HAITEI Dybowski.

2089. HAY (1907). *A New Genus and Species of Fossil Shark Related to Edestus Leidy.* Science (new ser.), XXVI, 1907, 22–24.

OLIVER PERRY HAY.

Lissoprion Hay, 22; orthotype L. FERRIERI Hay (fossil).

2090. IHERING (1907). *Diverses especias novas de peixes nemathognathes de Brazil, notas preliminares.* Revista Mus. Paulista, n. s. I, 13–39.

RUDOLF VON IHERING.

Rhamdioglanis Ihering, 16; orthotype R. FRENATUS Ihering. A synonym of IMPARFINIS Eigenmann & Norris.

Aspidoras Ihering, 30; orthotype A. ROCHEI Ihering.

2090A. IHERING (1907). *Notas Preliminares, fasciculo I.* Revista Mus. Paulista, I. 13–39.

RUDOLF VON IHERING.

Rhamdioglanis Ihering, 16; orthotype R. FRENATUS Ihering.

2091. JAEKEL (1907). *Ueber Pholidosteus die Mundung und die Körperform der Placodermen.* Sitzber. Ges. Naturf. Freunde, 1907, 170–186.

OTTO JAEKEL.

Pholidosteus Jaekel, 171; orthotype P. FRIEDELII Jaekel.

2092. JORDAN (1907). *A Review of the Fishes of the Family Histiopteridæ Found in the Waters of Japan, with a Note on Tephritis Günther.* Proc. U. S. Nat. Mus., XXXII, 235–239. (Mar. 12, 1907.)

DAVID STARR JORDAN.

Gilchristia Jordan, 236; orthotype HISTIOPTERUS RICHARDSONI Gilchrist. A synonym of PSEUDOPENTACEROS Blkr.

Quadrarius Jordan, 236; orthotype PENTACEROS DECACANTHUS Gthr.

Quinquarius Jordan, 236; orthotype PENTACEROS JAPONICUS Döderlein. To replace PENTACEROS Cuv. & Val., regarded as preoccupied, a question not yet decided.

Zanclistius Jordan, 236; orthotype HISTIOPTERUS ELEVATUS Ramsay & Ogilby.

Evistias Jordan, 237; orthotype HISTIOPTERUS ACUTIROSTRIS T. & S. (not EVISTIUS Gill).

Velifracta Jordan, 239; orthotype PLEURONECTES SINENSIS Lac. Substitute for TEPHRITIS Gthr., preoccupied in flies (Fabricius, 1799). A synonym of the earlier substitute, TEPHRINECTES Gthr.

2093. JORDAN (1907). *The Fossil Fishes of California, with Supplementary Notes on Other Species of Extinct Fishes.* Univ. Cal. Publ. Geol., V, 1907, 95–144.

DAVID STARR JORDAN.

Xenesthes Jordan, 120; orthotype X. VELOX Jordan (fossil).

Etringus Jordan, 121; orthotype E. SCINTILLANS Jordan (fossil).

Rogenio Jordan, 128; orthotype R. SOLITUDINIS Jordan (fossil).

Merriamella Jordan, 131; orthotype M. DORYSSA Jordan (a fossil stickleback). A synonym of GASTEROSTEUS L.

Knightia Jordan, 131; orthotype CLUPEA HUMILIS Leidy (fossil); the name HUMILIS preoccupied; replaced by KNIGHTIA EOCÆNA Jordan.

Eobrycon Jordan, 140; orthotype TETRAGONOPTERUS AVUS Boulenger (fossil).

2094. JORDAN & HERRE (1907). *A Review of the Cirrhitoid Fishes of Japan* Proc. U. S. Nat. Mus., XXXIII, 157–167. (Oct. 23, 1907.)

DAVID STARR JORDAN; ALBERT CHRISTIAN HERRE (1868–   ).

Isobuna Jordan, 158; orthotype PARACIRRHITES JAPONICA Steind. A substitute for PARACIRRHITES Steind., preoccupied.

2095. JORDAN & SEALE (1907). *Fishes of the Islands of Luzon and Panay.* Bull. Bur. Fish., XXIV (1906), 1907, 1–48.

DAVID STARR JORDAN; ALVIN SEALE.

Gennadius Jordan & Seale, 37; orthotype SEBASTES STOLICZKÆ Day. A synonym of CENTROGENYS Rich., a Serranoid fish which resembles a Scorpænoid.

Elates Jordan & Seale, 39; orthotype E. THOMPSONI Jordan & Seale.

Creisson Jordan & Seale, 43; orthotype C. VALIDUS Jordan & Seale.

2096. JORDAN & STARKS (1907). *Note on Otohime, a New Genus of Gurnards.* Proc. U. S. Nat. Mus., XXXII, 131–133.

DAVID STARR JORDAN; EDWIN CHAPIN STARKS.

Otohime Jordan & Starks, 131; orthotype TRIGLA HEMISTICTA T. & S. A near relative of PTERYGOTRIGLA Waite; perhaps a synonym.

2097. MATTHEW (1907). *A New Genus and a New Species of Silurian Fish.* Proc. and Trans. Roy. Soc. Canada, Series I, Sec. 4, pp. 7–9.

GEORGE FREDERICK MATTHEW.

Ctenopleuron Matthew, 7; orthotype C. NEREPISENSE Matthew (fossil OSTRACOPHORE, New Brunswick).

2098. MAWSON AND WOODWARD (1907). *On the Cretaceous Formation of Bahia (Brazil) and on Vertebrate Fossils Collected Therein.* Quart. Journ. Geol. Soc., 63, 1907, 128–138.

JOSEPH MAWSON; ARTHUR SMITH WOODWARD.

Mawsonia Woodward, 134; orthotype M. GIGAS Woodw. (fossil).

2099. MEEK (1907). *Synopsis of the Fishes of the Great Lakes of Nicaragua.* Field Columb. Mus. Pub. Zool., Ser. 7, 1907, 97–132.

SETH EUGENE MEEK.

Erythrichthys Meek, 114; orthotype HEROS CITRINELLUS Gthr. Name twice preoccupied.

2100. OGILBY (1907). *Symbranchiate and Apodal Fishes New to Australia.* Proc. Roy. Soc. Queensland, XX, 1906, 1–15.

JAMES DOUGLAS OGILBY.

Rhabdura Ogilby, 13; orthotype MURÆNA MACRURA Blkr.· A synonym of EVEN-CHELYS J. & E.

2101. OGILBY (1907). *Some New Pediculate Fishes.* Proc. Roy. Soc. Queensland, XX, 1906, 17–25.
JAMES DOUGLAS OGILBY.

Rhycherus Ogilby, 17; orthotype R. WILDII Ogilby.

Tathicarpus Ogilby, 19; logotype T. BUTLERI Ogilby.

Æschynichthys Ogilby, 25; orthotype DICERATIAS BISPINOSUS Gthr. Substitute for DICERATIAS Gthr., regarded as preoccupied by a genus (DICERATIA) of Mollusks.

2102. PELLÉGRIN (1907). *Sur un Poisson acanthopterygien éocène (Parapygæus polyacanthus n. g. n. sp.).* Bull. Soc. Philom. Paris, ser. 9, 171–179.
JACQUES PELLÉGRIN.

Parapygæus Pellégrin, 172; orthotype P. POLYACANTHUS Pellégrin (fossil).

2103. PELLÉGRIN (1907). *Siluride nouveau du Fouta-Djalon.* Bull. Museum Paris, 1907, 25–27.
JACQUES PELLÉGRIN.

Paramphilius Pellégrin, 23; orthotype P. TRICHOMYCTEROIDES Pellégrin.

2104. REGAN (1907). *Pisces.* Biol. Centr. Amer., 1907, 33–168.
CHARLES TATE REGAN.

Xenatherina Regan, 64; orthotype MENIDIA LISA Meek.

Pogonopoma Regan, 205; orthotype PLECOSTOMUS WERTHHEIMERI Steind.

2105. REGAN (1907). *Descriptions of Two New Characinid Fishes from Argentina.* Ann. Mag. Nat. Hist., 1907 (7), XIX, 261–2.
CHARLES TATE .REGAN.

Pogonocharax Regan, 261; orthotype P. REHI Regan.

Phoxinopsis Regan, 262; orthotype P. TYPICUS Regan.

2106. REGAN (1907). *Descriptions of Two New Characinid Fishes from South America.* Ann. Mag. Nat. Hist., XX, 1907, 402–403.
CHARLES TATE REGAN.

Mimagoniates Regan, 402; orthotype M. BARBERI Regan.

Ctenocharax Regan, 403; orthotype C. BOGOTENSIS Regan. Probably a synonym of GRUNDULUS BOGOTENSIS Humboldt.

2107. RIBEIRO (1907). *Una novidade ichthyologica; Kosmos, Rio de Janeiro, IV.*
ALIPIO DE MIRANDA RIBEIRO.

Typhlobagrus Ribeiro, 21; orthotype T. KRONEI Ribeiro.

2108. SAUVAGE (1907). *Sur les poissons de la famille des Cichlidæ trouvés dans le terrain de Guelma.* Comptes Rendus, Acad. Sci., 145, 1907, 360–361.
HENRI ÉMILE SAUVAGE.

Palæochromis Sauvage, 361; logotype P. DARESTI Sauvage (fossil).

2109. WAITE (1907). *The Generic Name Crepidogaster.* Rec. Austr. Mus., VI, 1907.
EDGAR RAVENSWOOD WAITE.

Aspasmagaster Waite, 315; orthotype CREPIDOGASTER TASMANIENSIS Gthr. Substitute for CREPIDOGASTER Gthr., preoccupied.

2110. WEBER (1907): *Süsswasserfische von Neu-Guinea; ein Beitrag zur Frage nach dem früheren Zusammenhang von Neu-Guinea und Australien. In Nova Guinea, V.* Zoologie, Leiden, 1907, 529.

MAX WEBER.

Glossolepis Weber, 241; orthotype G. INCISUS Weber.

2111. WOODWARD (1907). *The Fossil Fishes of the English Chalk. Pt. III.* Monogr. Palæont. Soc., 61, 1907, 97–128.

ARTHUR SMITH WOODWARD.

Dinelops Woodward, 121; orthotype D. ORNATUS Woodw. (fossil).

1908.

2112. BARBOUR (1908). *Notes on Rhinomuræna.* Proc. Biol. Soc. Washington, XXI, 39–41.

THOMAS BARBOUR (1884–   ).

Rhinechidna Barbour, 41; orthotype RHINOMURÆNA ERITIMA Jordan & Seale.

2113. BERG (1908). *Vorläufige Bemerkungen über die europäischasiatischen Salmonien, insbesondere die Gattung Thymallus.* Ann. Mus. Zool. St. Petersb., XII, 1908, 500–514.

LEW SEMENOVICH BERG.

Salmothymus Berg, 502; orthotype SALMO OBTUSIROSTRIS Heckel.
Thymalloides Berg, 503; logotype SALMO ARCTICUS Pallas.

2114. BOULENGER (1908). *Diagnoses of New Fishes Discovered by Capt. E. L. Rhoades in Lake Nyassa.* Ann. Mag. Nat. Hist., II, 1908, 238–243.

GEORGE ALBERT BOULENGER.

Chilotilapia Boulenger, 243; orthotype C. RHOADESII Boulenger.

2115. DOLLO (1908). *Notolepis coatsi. Poisson pélagique nouveau recueilli par l'Expédition Antarctique Nationale Écossaise.* Edinburgh, Proc. Roy. Soc., XXVIII, 58–65.

LOUIS DOLLO.

Notolepis Dollo, 58; orthotype N. COATSI Dollo.

2116. EASTMAN (1908). *Devonian Fishes of Iowa. (With Special Description of the Auditory Organ and Other Soft Parts of Rhadinichthys deani Eastman . . . G. H. Parker.)* Iowa Geol. Surv. Rept., XVIII (1907), 1908, 29–360.

CHARLES ROCHESTER EASTMAN.

Palæophichthys Eastman, 253; orthotype P. PARVULUS Eastman (fossil). (CROSSOPTERYGIAN from Illinois coal measures.)

2117. EIGENMANN (1908). *Preliminary Descriptions of New Genera and Species of Tetragonopterid Characins. (Zoological Results of the Thayer Brazilian Expedition.)* Bull. Mus. Comp. Zool. Harvard Coll., LII, 1908, 91–106.

CARL H. EIGENMANN.

Gymnocorymbus Eigenmann, 93; orthotype G. THAYERI Eigenmann.
Ctenobrycon Eigenmann, 94; orthotype TETRAGONOPTERUS HAUXWELLIANUS Cope.
Thayeria Eigenmann, 94; orthotype T. OBLIQUA Eigenmann.

Psellogrammus Eigenmann, 96; orthotype HEMIGRAMMUS KENNEDYI Eigenmann.
Poptella Eigenmann, 99; orthotype TETRAGONOPTERUS LONGIPINNIS Eigenmann.
Pristella Eigenmann, 99; orthotype HOLOPRISTES RIDDEII Meek.
Hyphessobrycon Eigenmann, 100; orthotype HEMIGRAMMUS COMPRESSUS Meek.
Brycochandus Eigenmann, 106; orthotype B. DURBINI Eigenmann.

2118. FOWLER (1908). *Notes on Lancelets and Lampreys.* Proc. Acad. Nat.
    Sci. Phila., LIX, 1908, 461–466. (Jan. 27, 1908.)
                    HENRY WEED FOWLER.
Oceanomyzon Fowler, 461; orthotype O. WILSONI Fowler.

2119. GARMAN (1908). *New Plagiostomia and Chismopnea.* Bull. Mus. Comp.
    Zool. Harvard Coll., LI, 249–256.
                    SAMUEL GARMAN.
Aëtomylæus Garman, 252; orthotype MYLIOBATIS MACULATUS Gray.

2120. JORDAN & BRANNER (1908). *The Cretaceous Fishes of Ceará, Brazil.*
    Smithson. Inst. Misc. Collect., LII, 1908, 1–29.
        DAVID STARR JORDAN; JOHN CASPER BRANNER (1850–   ).
Tharrhias Jordan & Branner, 13; orthotype T. ARARIPIS Jordan & Branner (fossil).
Enneles Jordan & Branner, 23; orthotype E. AUDAX Jordan & Branner (fossil).
Cearana Jordan & Branner, 27; orthotype C. ROCHÆ Jordan & Branner (fossil).

2121. JORDAN & RICHARDSON (1908). *A Review of the Flat-Heads, Gur-
    nards, and Other Mail-Cheeked Fishes of the Waters of Japan.* Proc. U. S.
    Nat. Mus., XXXIII, 629–670.
        DAVID STARR JORDAN; ROBERT EARL RICHARDSON (1878–   ).
Rogadius Jordan & Richardson, 630; orthotype PLATYCEPHALUS ASPER Cuv. & Val.
Bambradon Jordan & Richardson, 643; orthotype BEMBRAS LÆVIS Nyström.
Ebisinus Jordan & Richardson, 665; orthotype DACTYLOPTERUS CHIROPHTHALMUS
    Blkr.
Dactyloptena Jordan & Richardson, 665; orthotype DACTYLOPTERUS ORIENTALIS Cuv.
    & Val.
Daicocus Jordan & Richardson, 667; orthotype DACTYLOPTERUS PETERSENI Nyström.

2122. JORDAN & RICHARDSON (1908). *Fishes from the Islands of the Philip-
    pine Archipelago.* Bull. Bur. Fish, XXVII (1907), 1908, 233–287.
        DAVID STARR JORDAN; ROBERT EARL RICHARDSON.
Aparrius Jordan & Richardson, 278; orthotype A. ACUTIPINNIS Jordan & Richardson.

2123. JORDAN & SNYDER (1908). *Descriptions of Three New Species of Caran-
    goid Fishes from Formosa.* Mem. Carneg. Mus., IV, 37–40.
        DAVID STARR JORDAN; JOHN OTTERBEIN SNYDER.
Ulua Jordan & Snyder, 39; orthotype U. RICHARDSONI Jordan & Snyder.

2124. JORDAN & STARKS (1908). *On a Collection of Fishes from Fiji, with
    Notes on Certain Hawaiian Fishes.* Proc. U. S. Nat. Mus., XXXIV. (Sept.
    14, 1908.)
        DAVID STARR JORDAN; EDWIN CHAPIN STARKS.
Rastrelliger Jordan & Starks, 607; orthotype SCOMBER BRACHYSOMUS Blkr.

2125. OGILBY (1908). *New or Little Known Fishes in the Queensland Museum.*
    Ann. Queensland Mus., IX, Pt. 1, 1908, 1–41.
                    JAMES DOUGLAS OGILBY.

Cirriscyllium Ogilby, 4; orthotype CHILOSCYLLIUM MODESTUM Gthr. A synonym (due to oversight) of BRACHÆLURUS Ogilby.

Nemapteryx Ogilby, 10; orthotype ARIUS STIRLINGI Ogilby.

Anyperistius Ogilby, 11; orthotype A. PERUGIÆ Ogilby = EUMEDA ELONGATA Perugia, not of Castelnau, which is the young of NEOSILURUS HYRTLI Steind. A synonym of TANDANUS Mitchell.

Jenynsella Ogilby, 15; orthotype J. WEATHERILLI Ogilby. A synonym of RETROPINNA Gill.

Squalomugil Ogilby, 28; orthotype MUGIL NASUTUS De Vis. A synonym of RHINOMUGIL Gill.

2126. OGILBY (1908). *Revision of the Batrachoididæ of Queensland.* Ann. Queensland Mus., IX, pt. 2, 43–57. (These genera of Toad-fishes had been indicated, not described, by Ogilby in the Ann. Rept. Amateur Fishermen's Association, Queensland, 1907.)

JAMES DOUGLAS OGILBY.

Batrachomœus Ogilby, 46; orthotype B. BROADBENTI Ogilby. Perhaps a synonym of PSEUDOBATRACHUS Castelnau, imperfectly described.

Halobatrachus Ogilby, 46; orthotype BATRACHUS DIDACTYLUS Bl. & Schn.

Coryzichthys Ogilby, 51; orthotype BATRACHUS DIEMENSIS Ogilby.

2127. OGILBY (1908).. *Descriptions of New Queensland Fishes.* Proc. Roy. Soc. Queensland, XXI, 1908, 87–98.

JAMES DOUGLAS OGILBY.

Eurycaulus Ogilby, 91; orthotype BELONE PLATYURA Cuv. & Val. (gill rakers present; tail keeled). Name preoccupied; replaced by TROPIDOCAULUS Ogilby.

Stenocaulus Ogilby, 91; orthotype BELONE KREFFTI Gthr.

Pseudomycterus Ogilby, 94; orthotype M. MACCULLOCHI Ogilby. A synonym or section of SCIÆNA L.

2128. OGILBY (1908). *On New Genera and Species of Fishes.* Proc. Roy. Soc. Queensland, XXI, 1–26.

JAMES DOUGLAS OGILBY.

Ptenonotus Ogilby, 13; orthotype EXOCŒTUS CIRRIGER Peters.

Merogymnus Ogilby, 18; orthotype M. EXIMIUS Ogilby.

2129. PASCOE (1908). *Marine Fossils in the Yenangyaung Oil-Field, Upper Burma.* Records Geol. Surv. India, XXXVI, 135–142.

E. H. PASCOE.

Twingonia Pascoe, 138, pl. 18; (no specific name proposed) (fossil, Miocene). Described as a foraminifer, but later shown by E. W. Vredenburg to be a fish otolith or ear bone (*fide* Hussakof).

2130. PELLÉGRIN (1908). *Characinides Americains nouveaux de la Collection du Museum d'Histoire Naturelle.* Bull. Mus. d'Hist. Nat. Paris, VII, 342–347.

JACQUES PELLÉGRIN.

Anostomoides Pellégrin, 346; orthotype A. ANALIS Pellégrin.

2131. REGAN (1908). *Report on the Marine Fishes Collected by Mr. J. Stanley Gardiner in the Indian Ocean.* Trans. Linn. Soc. London, XII, 1908, 217–225.

CHARLES TATE REGAN.

Borostomias Regan, 217; orthotype B. BRAUERI Regan.

Xenanthias Regan, 223; orthotype X. GARDINERI Regan.

Pogonoscorpius Regan, 236; orthotype P. SEYCHELLENSIS Regan.
Psammichthys Regan, 246; orthotype P. NUDUS Regan.
Sladenia Regan, 250; orthotype S. GARDINERI Regan.

2132. REGAN (1908). *Descriptions of New Fishes from Lake Candidius, Formosa, Collected by Dr. A. Moltrecht.* Ann. Mag. Nat. Hist., II, 1908, 358–360.
CHARLES TATE REGAN.
Pararasbora Regan, 360; orthotype P. MOLTRECHTI Regan.

2133. REGAN (1908). *A Synopsis of the Fishes of the Subfamily Salanginæ.* Ann. Mag. Nat Hist. (8), II, 444–446.
CHARLES TATE REGAN.
Hemisalanx Regan, 444; orthotype H. PROGNATHUS Regan.
Parasalanx Regan, 444; logotype P. GRACILLIMUS Regan.

2134. REGAN (1908). *A New Generic Name for An Orectolobid Shark.* Ann. Mag. Nat. Hist., II, 1908, 454–455.
CHARLES TATE REGAN.
Heteroscyllium Regan, 455; orthotype BRACHÆLURUS COLCLOUGHI Ogilby.

2135. REGAN (1908). *A Collection of Fresh-Water Fishes Made by Mr. C. F. Underwood in Costa Rica.* Ann. Mag. Nat. Hist,. II, 1908, 455–464.
CHARLES TATE REGAN.
Petalosoma Regan, 458; orthotype P. CULTRATUM Regan.
Xenorhychichthys Regan, 461; orthotype JOTURUS STIPES J. & G.
Tomocichla Regan, 463; orthotype T. UNDERWOODI Regan.

2136. REGAN (1908). *A Revision of the Sharks of the Family Orectolobidæ.* Proc. Zool. Soc. London, 1908, 347–364.
CHARLES TATE REGAN.
Eucrossorhinus Regan, 357; orthotype CROSSORHINUS DASYPOGON Blkr.

2137. RIBEIRO (1908). *Peixes da Ribeiro, resultados de excursao do Senhor Ricardo Krone . . .* Kosmos, Rio de Janeiro, V, 1–5.
ALIPIO DE MIRANDA RIBEIRO.
Cœlurichthys Ribeiro, 2; orthotype C. IPORANGÆ Ribeiro.

2137A. RIBEIRO (1908). *Peixes, da Ribeiro, resultados de excursão do Sr. Ricardo Krone, membro correspondente do Museu Nacional do Rio de Janeiro,* V, 1–5.
ALIPIO DE MIRANDA RIBEIRO.
Kronichthys Ribeiro, 1; orthotype K. SUBTERES Ribeiro; a synonym of HEMIPSILICHTHYS Eigenmann.

2138. SEALE & BEAN (1908). *On a Collection of Fishes from the Philippine Islands, Made by Maj. Edgar A. Mearns, Surgeon, U. S. Army, with Descriptions of Seven New Species.* Proc. U. S. Nat. Mus., XXXII, 1908, 229–248.
ALVIN SEALE; BARTON APPLER BEAN.
Mearnsella Seale & Bean, 229; orthotype M. ALESTES Seale & Bean.

2139. SNYDER (1908). *Descriptions of Eighteen New Species and Two New Genera of Fishes from Japan and the Riu Kiu Islands.* Proc. U. S. Nat. Mus., XXXV, 93–111. (Oct. 30, 1908).
JOHN OTTERBEIN SNYDER.

Doryptena Snyder, 102; orthotype D. OKINAWÆ Snyder.
Xenisthmus Snyder, 105; orthotype X. PRORIGER Snyder.

2140. STARKS (1908). *On a Communication Between the Air-Bladder and the Ear in Certain Spiny-Rayed Fishes.* Science, XXVIII, 1908, 613–614.

EDWIN CHAPIN STARKS.

Adioryx Starks, 614; orthotype HOLOCENTRUM SUBORBITALE Gill.

2141. STEINDACHNER (1908). *Ueber eine im Rio Juraguà bei Joinville, im Staate S. Catharina, Brasilien . . . eine neue Characinengattung.* Anz. Ak. Wiss. Wien., LXV.

FRANZ STEINDACHNER.

Zungaropsis Steindachner, 7; orthotype Z. MULTIMACULATUS Steind.
Joinvillea Steindachner, 30; orthotype J. ROSÆ Steind. A synonym of DEUTERODON Eigenmann.

2142. TANAKA (1908). *Notes on Some Rare Fishes of Japan, with Descriptions of Two New Genera and Six New Species.* Journ. Coll. Sci. Tokyo, XXIII, Art. 13, 1–24.

SHIGEHO TANAKA.

Tetronarcine Tanaka, 2; orthotype T. TOKIONIS Tanaka.
Gymnosimenchelys Tanaka, 2; orthotype G. LEPTOSOMUS Tanaka.
Paraceratias Tanaka, 18; orthotype CERATIAS MITSUKURII Tanaka.
Owstonia Tanaka, 46; orthotype O. TOTOMIENSIS Tanaka.

2143. WOODWARD (1908). *The Fossil Fishes of the Hawkesbury Series at St. Petersburg.* Mem. Geol. Surv. New South Wales, Palæontology, X, pp. vi, 31.

ARTHUR SMITH WOODWARD.

Elpisopholis Woodward, 17; orthotype E. DUNSTANI Woodw. (fossil) (PALÆONISCIDÆ).

1909.

2144. ANNANDALE (1909). *Report on the Fishes Taken by the Bengal Fisheries Steamer "Golden Crown." Part I, Batoidei.* Mem. Ind. Mus. Calcutta, 2, no. 1, 1909, 1–58.

NELSON ANNANDALE.

Bengalichthys Annandale, 47; orthotype B. IMPENNIS Annandale.

2145. BEAN & WEED (1909). *Description of a New Skate (Dactylobatus armatus) from Deep Water off the Southern Atlantic Coast of the United States.* Proc. U. S. Nat. Mus., XXXVI, 1909, 459–561. (May 27, 1909.)

BARTON APPLER BEAN; ALFRED C. WEED.

Dactylobatus Bean & Weed, 459; orthotype D. ARMATUS Bean & Weed.

2146. BOULENGER (1909). *Descriptions of New Fresh-Water Fishes Discovered by Mr. G. L. Bates in South Cameroon.* Ann. Mag. Nat. Hist., IV, 1909, 186–188.

GEORGE ALBERT BOULENGER.

Champsoborus Boulenger, 187; orthotype C. PELLEGRINI Boulenger.

2147. BROOM (1909). *The Fossil Fishes of the Upper Karoo Beds of South Africa.* Ann. S. Afr. Mus., VII, 251–268.

ROBERT BROOM.

Helichthys Broom, 254; orthotype H. BROWNI Broom (fossil). (PALÆONISCID.)

Hydropessum, 266; orthotype H. TANNEMEYERI Broom (fossil). (Imperfect Fish, GANOID.)

2148. CLIGNY (1909). *Sur un nouveau genre de Zéides.* Comptes Rendus. Acad. Sci. Paris, 148, 1909, 873–874.

ADOLPHE CLIGNY.

Parazenopsis Cligny, 874; orthotype P. ARGENTEUS Cligny.

2149. COCKERELL (1909). *The Nomenclature of the American Fishes Usually Called Leuciscus and Rutilus.* Proc. Biol. Soc. Wash., XXII, 1909, 215–217.

THEODORE DRU ALISON COCKERELL.

Temeculina Cockerell, 216; orthotype LEUCISCUS ORCUTTI Eigenmann & Eigenmann.

Margariscus Cockerell, 217; orthotype CLINOSTOMUS MARGARITA Cope.

2150. COCKERELL & ALLISTON (1909). *The Scales of Some American Cypinidæ.* Proc. Biol. Soc. Wash., XXII, 1909, 157–163.

THEODORE DRU ALISON COCKERELL; EDITH M. ALLISTON.

Macrhybopsis Cockerell & Alliston, 162; orthotype GOBIO GELIDUS Grd.

2151. COCKERELL & CALLAWAY (1909). *Observations on the Fishes of the Genus Notropis.* Proc. Biol. Soc. Wash., XXII, 1909, 189–196.

THEODORE DRU ALISON COCKERELL; OTIS CALLAWAY.

Coccogenia Cockerell & Callaway, 190; orthotype HYPSILEPIS COCCOGENIS Cope. A synonym of COCCOTIS Jordan.

2152. DOLLO (1909). *Cynomacrurus piriei, poisson abyssal nouveau recueilli par l'Expédition Antarctique Nationale Écossaise.* Proc. Roy. Soc. Edinb.. XXIX. 1909, 316–326.

LOUIS DOLLO.

Cynomacrurus Dollo, 316; orthotype C. PIRIEI Dollo.

2153. DURBIN (1909). *A New Genus and Twelve New Species of Tetragonopterid Characins. (Reports on the Expedition to British Guiana of the Indiana University and the Carnegie Museum, 1908, No. 2.)* Ann. Carneg. Mus.. VI. 1909, 55–72.

MARION LEE DURBIN.

Dermatocheir Durbin, 55; orthotype D. CATABLEPTA Durbin.

2154. EIGENMANN (1909). *Some New Genera and Species of Fishes from British Guiana. (Reports on the Expedition to British Guiana of the Indiana University and the Carnegie Museum, 1908, No. 1.)* Ann. Carnegie Mus., VI, 1909, 4–54.

CARL H. EIGENMANN.

Corymbophanes Eigenmann, 5; orthotype C. ANDERSONI Eigenmann.

Lithogenes Eigenmann, 6; orthotype L. VILLOSUS Eigenmann.

Pterodiscus Eigenmann, 12; orthotype P. LEVIS Eigenmann.

Carnegiella Eigenmann, 13; orthotype GASTEROPELECUS STRIGATUS Gthr.

Holobrycon Eigenmann, 33; orthotype BRYCON PESU M. & T.

Triurobrycon Eigenmann, 33; orthotype BRYCON LUNDII Lütken.

Pœcilocharax Eigenmann, 34; orthotype P. bovalii Eigenmann.
Pœcilobrycon Eigenmann, 43; orthotype P. harrisoni Eigenmann.
Archicheir Eigenmann, 46; orthotype A. minutus Eigenmann.
Acanthophacelus Eigenmann, 52; orthotype A. melanzonus Eigenmann.
Tomeurus Eigenmann, 53; orthotype T. gracilis Eigenmann.
Microcharax Eigenmann, 55; orthotype Nannostomus lateralis Boulenger.

2155. EIGENMANN (1909). *The Fresh-Water Fishes of Patagonia, and An Ex-amination of the Archiplata-Archihelenis Theory.* Princeton Univ., 3, (Zoology), 1909, 225–374.

Carl H. Eigenmann.

Hatcheria Eigenmann, 250; orthotype H. patagoniensis Eigenmann.

2156. GILCHRIST & THOMPSON (1909). *Descriptions of Fishes from the Coast of Natal, Pt. 2.* Ann. S. African Mus., VI, 1909, 213–276.

J. D. F. Gilchrist; Wardlaw Thompson.

Chæropsodes Gilchrist & Thompson, 260; orthotype C. pictus Gilchrist & Thompson.

2157. HUSSAKOF (1909). *The Systematic Relations of Certain American Ar-throdires.* Bull. Amer. Mus. Nat. Hist., XXVI, 263–272.

Louis Hussakof.

Brachygnathus Hussakof, 263; orthotype Dinichthys minor Newberry (fossil). Based on mandible and dorsomedian. Preoccupied; name replaced by Hussakofia Cossman.
Dinognathus Hussakof, 268; orthotype D. ferox Hussakof (fossil). Dental element (fossil: Upper Devonic, Ohio).

2158. JAEKEL (1909). *Fischreste aus den Mamfe-Schiefern.* (In *Beiträge zur Geologie vom Kammerun, bearb., V. C. Guillemain.*) Abh. Geol. Landes Anst. (N. F.), H. 62, 1909 (1910), 392–398.

Otto Jaekel.

Proportheus Jaekel, 396; orthotype P. kameruni Jaekel (fossil).

2159. JORDAN & RICHARDSON (1909). *A Catalogue of the Fishes of the Island of Formosa or Taiwan, Based on the Collection of Dr. Hans Sauter.* Mem. Carneg. Mus., IV, No. 4, 1909, 159–204.

David Starr Jordan; Robert Earl Richardson.

Candidia Jordan & Richardson, 169; orthotype Opsariichthys barbatus Regan.

2160. PELLÉGRIN (1909). *Characinidés du Brésil, rapportés par M. Jobert.* Bull. Museum Paris, 1909, 147–153.

Jacques Pellégrin.

Curimatella Pellégrin, 150; orthotype Curimatus alburnus M. & T.
Jobertina Pellégrin, 151; orthotype Charicidium interruptum Pellégrin.
2161. PELLÉGRIN (1909). *Characinidés américains nouveaux de la collection du Museum d'histoire naturelle.* Bull. Mus. Paris, 1908, 342–347.

Jacques Pellégrin.

Anostomoides Pellégrin, 346; orthotype A. atrianalis Pellégrin.

2162. SEALE (1909). *New Species of Philippine Fishes.* Philip. Journ. Sci., IV, 1909, 49–543.

Alvin Seale.

Biat Seale, 532; orthotype B. luzonica Seale.
Macgregorella Seale, 533; orthotype M. moroana Seale.

2163. SNYDER (1909). *Descriptions of New Genera and Species of Fishes from Japan and the Riu Kiu Islands.* Proc. U. S. Nat. Mus., XXXVI, 1909, 597–610. (June 18, 1909.)

JOHN OTTERBEIN SNYDER.

Expedio Snyder, 606; orthotype E. PARVULUS Snyder.
Inu Snyder, 607; orthotype I. KOMA Snyder.

2164. STECHE (1909). *Die Leuchtorgane von . . . zwei Oberflachen-fischen aus dem Malaischen Archipel.* Zeit. Wiss. Zool. XCIII.

OTTO STECHE.

Protoblepharon Steche, about 400; orthotype P. PALPEBRATUS.

2165. TANAKA (1909). *Descriptions of One New Genus and Ten New Species of Japanese Fishes.* Journ. Coll. Sci. Tokyo, XXVII, 1–27.

SHIGEHO TANAKA.

Anteliochimæra Tanaka, 1; orthotype A. CHÆTIRHAMPHUS Tanaka.

2166. TOULD (1909). *Eine jungstiäre Fauna von Gatun am Panama-kanal.* Wien. Jahrb. Geol. Rechs. Anst., LVIII, 1908 (1909), 673–760.

FRANZ TOULD.

Scaroides Toul, 687; orthotype S. GATUNENSIS Tould.

2167. WAITE (1909). *Scientific Results of the New Zealand Government Trawling Expedition.* Rec. Canterbury Mus., I. (July, 1909.)

EDGAR RAVENSWOOD WAITE.

Typhlonarke Waite, 16; orthotype ASTRAPE AYSONI Hamilton.
Arhynchobatis Waite, 20; orthotype A. ASPERRIMUS Waite.

2168. WEBER (1909). *Diagnosen neuer Fische der Siboga-Expedition.* Leiden; Notes Mus. Jentink, XXXI, 1909, 143–169.

MAX WEBER.

Rhabdamia Weber, 165; orthotype R. CLUPEIFORMIS Weber.
Siphamia Weber, 168; orthotype S. TUBIFER Weber.

———

1910.

2169. BOULENGER (1910). *On a Large Collection of Fishes Made by Dr. W. J. Ansorge in the Quanza and Bengo Rivers, Angola.* Ann. Mag. Nat. Hist., VI, 1910, 537–561.

GEORGE ALBERT BOULENGER.

Xenopomatichthys Boulenger, 542; orthotype X. ANSORGII Boulenger.
Nematogobius Boulenger, 560; orthotype N. ANSORGII Boulenger.

2170. CHAPMAN (1910). *A Study of the Batesford Limestone.* Proc. Roy. Soc. Vict. Melbourne, XX, 307.

FREDERICK REVANS CHAPMAN.

Metaceratodus Chapman, 25; orthotype CERATODUS WOLLASTONI Chapman (fossil).

2171. COCKERELL (1910). *The Scales of the African Cyprinid Fishes, with a Discussion of Related Asiatic and European Species.* Proc. Biol. Soc. Wash., XXIII, 141–152.

THEODORE DRU ALISON COCKERELL.

Acapoëta Cockerell, 149; orthotype VARICORHINUS TANGANICÆ Boulenger.

2172. COSSMAN (1910). *Rectifications de Nomenclature.* Revue Critique Paléozoologie, XIV, 74.

MAURICE COSSMAN.

Hussakofia Cossman, 74; orthotype DINICHTHYS MINOR Newberry (fossil). (ARTHRODIRE; Devonic, Ohio.) Substitute for BRACHYGNATHUS Hussakof, twice preoccupied.

2173. EIGENMANN (1910). *Catalogue and Bibliography of the Fresh-Water Fishes of Temperate and South Temperate America. (Rept. Princeton Univ. Expedition to Patagonia, 1896, 1899, III, Pt. 2)* Zoology, 375–511.

CARL H. EIGENMANN.

Megalonema Eigenmann, 383; orthotype M. PLATYCEPHALUM Eigenmann.

Hollandichthys Eigenmann, about 400; orthotype H. MULTIFASCIATUS Eigenmann & Norris.

2174. FOWLER (1910). *Notes on Batoid Fishes.* Proc. Acad. Nat. Sci. Phila., LXII, 1910, 468–475. (Aug. 17, 1910.)

HENRY WEED FOWLER.

Discotrygon Fowler, 468; orthotype DISCOBATIS MARGINIPINNIS Maclay & Macleay = RAIA LYMMA Forskål. A synonym of TÆNIURA M. & H. Name a substitute for DISCOBATIS Maclay & Macleay, regarded as preoccupied by DISCOBATUS.

Eunarce Fowler, 472; orthotype TORPEDO NARKE Risso.

Pteroplatytrygon Fowler, 474; orthotype TRYGON VIOLACEUM Bon.

2175. FOWLER (1910). *Descriptions of Four New Cyprinoids (Rhodeinæ).* Proc. Acad. Nat. Sci. Phila., LXII, 1910, 476–486. (Aug. 17, 1910.)

HENRY WEED FOWLER.

Rhodeops Fowler, 479; orthotype ACHEILOGNATHUS BREVIANALIS Fowler.

Hemigrammocypris Fowler, 483; orthotype H. RASBORELLA Fowler.

2176. FRAAS (1910). *Chimäridenreste aus dem oberen Lias von Holzmaden.* Stuttgart, Jahreshefte Ver. Natk., LXVI, 1910, 55–63.

OSCAR FRIEDRICH FRAAS.

Acanthorhina Fraas, 55; orthotype A. JAEKELI Fraas (fossil) (not ACANTHORHINUS Blainv.).

2177. FRANZ (1910). *Die Japanischen Knochfische der Sammlungen Haberer und Döflein. (Beiträge zur Naturgeschichte Ostasiens.)* Munchen Abh. Ak. Wiss. Math.-Phys. Kl. Suppl.-Bd., IV, 1910 (1911), Abh. 1, 1–135.

VIKTOR FRANZ.

Parabarbus Franz, 8; orthotype P. HABILIS Franz.

Cryptophthalmus Franz, 15; orthotype C. ROBUSTUS Franz. Name preoccupied; replaced by UNAGIUS Jordan, 1919.

Osteochromis Franz, 43; orthotype O. LARVATUS Franz.

Plagiopsetta Franz, 64; orthotype P. GLOSSA Franz.

Trypauchenophrys Franz, 68; orthotype T. ANOTUS Franz.

2178. GILBERT (1910) *Evesthes jordani, a Primitive Flounder from the Miocene of California.* Cal. Univ. Publ. Bull. Dept. Geol., V, 1910, 405-411.

JAMES ZACCHEUS GILBERT.

Evesthes J. Z. Gilbert, 407; orthotype E. JORDANI Gilbert (fossil).

2179. HAY (1910). *On Edestus and Related Genera.* Proc. U. S. Nat. Mus., XXXVII, 43–61.

OLIVER PERRY HAY.

Toxoprion Hay, 56; orthotype EDESTEUS LECONTEI Dean (fossil). (ELASMOBRANCH from Nevada.)

2180. HOLT & BYRNE (1910). *Preliminary Diagnosis of a New Stomiatoid Fish from Southwest of Ireland.* Ann. Mag. Nat. Hist., VI.

E. W. L. HOLT; L. W. BYRNE.

Grammatistomias Holt & Byrne, 295; orthotype G. FLAGELLIBARBA Holt & Byrne.

2181. JORDAN (1910). *Description of a Collection of Fossil Fishes from the Bituminous Shales at Riacho Doce, State of Alagoas, Brazil.* Ann. Carneg. Mus., VII, 1910, 23–34.

DAVID STARR JORDAN.

Ellipes Jordan, 24; orthotype E. BRANNERI Jordan (fossil). Name preoccupied; replaced by ELLIMMA Jordan, 1913.

Dastilbe Jordan, 29; orthotype D. CRANDALLI Jordan (fossil).

2182. JORDAN & RICHARDSON (1910). *A Review of the Serranidæ or Sea Bass of Japan.* Proc. U. S. Nat. Mus., XXXVII, 1910, 421–474.

DAVID STARR JORDAN; ROBERT EARL RICHARDSON.

Zalanthias Jordan & Richardson, 460; orthotype PSEUDANTHIAS KELLOGGI Jordan & Richardson.

Sacura Jordan & Richardson, 468; orthotype ANTHIAS MARGARITACEUS Hilgendorf.

2183. OGILBY (1910). *On New or Insufficiently Described Fishes.* Proc. Roy. Soc. Queensland, XXIII, 1910, 1–55.

JAMES DOUGLAS OGILBY.

Xystodus Ogilby, 5; orthotype X. BANFIELDI Ogilby.

Chilias Ogilby, 40; orthotype PERCA STRICTICEPS De Vis. A synonym of PARAPERCIS Blkr:

Cyneichthys Ogilby, 55; orthotype BLENNECHIS ANOLIUS Cuv. & Val.

Runulops Ogilby, 55; orthotype BLENNECHIS FASCIATUS Jenyns ("two South American species described by Jenyns," B. FASCIATUS and B. ORNATUS from Chile).

2184. OGILBY (1910). *On Some New Fishes from the Queensland Coast. Endeavour Series, I.* Proc. Roy. Soc. Queensland, XXIII, 1910, 85–139. (On account of a misunderstanding which does not concern the present work, this paper was suppressed by the Royal Society of Queensland, only a few separates being sent out. The new species have been elsewhere described by Mc-Culloch.)

JAMES DOUGLAS OGILBY.

Hyalorhynchus Ogilby, 118; orthotype H. PELLUCIDUS Ogilby = ELATES THOMPSONI Jordan & Richardson. A synonym of ELATES Jordan & Richardson.

2185. PAVLENKO (1910). *Fishes of Peter the Great Bay.* (Text in Russian.) Trd. Obsc. Jest, Kazani, 42, 1910, 2, 1–5.

M. N. PAVLENKO.

Agonocottus Pavlenko, 23; orthotype A. CATAPHRACTUS Pavlenko.

Askoldia Pavlenko, 50; orthotype A. VARIEGATA Pavlenko.

2186. PREOBRAZHENSKII (1910). *Ueber einige Vertreter der Familie der Psammosteidæ Ag.* S. B. Naturf. Ges. Univ. Jurjew. (Dorpat) IX, 21–36. (Text in Russian.)

J. A. PREOBRAZHENSKII.

Pycnosteus Preobrazhenskii, 34; orthotype P. PALÆFORMIS Preobrazhenskii (fossil). An imperfect dorsal plate of a Devonian OSTRACOPHORE. Probably a synonym of PSAMMOSTEUS Ag.

Dyptychosteus Preobrazhenskii, 35; orthotype D. TESSELLATUS Preobrazhenskii (fossil) (not PSAMMOSTEUS TESSELLATUS Traquair). Probably same as PSAMMOSTEUS PARADOXUS. Dorsal plate of a Devonian OSTRACOPHORE. A synonym of PSAMMOSTEUS Ag.

2187. SAUVAGE (1910). *Bull. Carte Geol. Algérie, 1910.*

HENRI ÉMILE SAUVAGE.

Palæochromis Sauvage, 308; orthotype P. DORESTII Sauvage.

2188. STOLLEY (1910). *Ueber mesozoischen Fischotolithen aus Norddeutschland.* Jahresbericht des Niedersächsischen Geol. Vereins, III, 246–257. (Reprinted with separate pagination in Hannover, 1912.)

ERNST STOLLEY.

Archæotolithus Stolley, 248; orthotype OTOLITHUS BORNHOLMIENSIS Malling & Grönwall, 1909 (fossil). Otolith from Lias of Denmark.

2189. WAITE (1911). *Additions to the Fish Fauna of New Zealand.* Proc. New Zealand Inst., 1910, 25–26.

EDGAR RAVENSWOOD WAITE.

Maccullochia Waite, 25; orthotype HISTIOPTERUS LABIOSUS Gthr. A substitute for RICHARDSONIA Castelnau, preoccupied. A synonym of PARISTIOPTERUS Blkr., 1876.

2190. WAITE (1910). *Notes on New Zealand Fishes.* Trans. N. Zeal. Inst., XLII, 1910, 384–391.

EDGAR RAVENSWOOD WAITE.

Triareus Waite, 387; orthotype MAUROLICUS AUSTRALIS Hector.

2191. WEYMOUTH (1910). *Notes on a Collection of Fishes from Cameron, Louisiana.* Proc. U. S. Nat. Mus., XXXVIII.

FRANK WALTER WEYMOUTH.

Leptocerdale Weymouth, 142; orthotype L. LONGIPINNIS Weymouth.

2192. WOODWARD (1910). *On Some Permo-Carboniferous Fishes from Madagascar.* Ann. Mag. Nat. Hist., V, 1910, 1–6.

ARTHUR SMITH WOODWARD.

Ecrinesomus Woodward, 2; orthotype E. DIXONI Woodw. (fossil).

2193. WOODWARD (1910). *On a Fossil Sole and a Fossil Eel from the Eocene of Egypt.* Geol. Mag. London, VII, 1910, 402–405.

ARTHUR SMITH WOODWARD.

Mylomyrus Woodward, 404; orthotype M. FRANGENS Woodw. (fossil).

1911.

2194. BOULENGER (1911). *Descriptions of New African Cyprinodont Fishes.*
Ann. Mag. Nat. Hist. (8), VIII, 260–268.

GEORGE ALBERT BOULENGER.

Mohanga Boulenger, 261; orthotype HAPLOCHILUS TANGANICANUS Boulenger. Same
as LAMPRICHTHYS Regan, a little later in the same year.

2195. BOULENGER (1911). *Catalogue of the Fresh-Water Fishes of Africa in the
British Museum.* London, 1911, 1–529.

GEORGE ALBERT BOULENGER.

Parauchenoglanis Boulenger, 364; orthotype PIMELODUS GUTTATUS Lönnberg. The
name GUTTATUS is preoccupied in PIMELODUS.

2196. BOULENGER (1911). *On a Third Collection of Fishes Made by Dr. E.
Bayon in Uganda, 1909–1910.* Genova Annal. Museo Civico. Ser. 3a, V, 1911,
64–78.

GEORGE ALBERT BOULENGER.

Bayonia Boulenger, 70; orthotype B. XENODONTA Boulenger.

2197. DEECKE (1911). *Paleontologische Betrachtungen.* IV. Ueber Fische.
Neues. Jahrb. Mineral., II.

JOHANNES ERNEST WILHELM DEECKE (1862–   ).

Perleidus Deecke, 70; orthotype P. ALTOLEPIS Deecke (fossil).

2198. EIGENMANN (1911). *New Characins in the Collection of the Carnegie
Museum.* Ann. Carnegie Mus., VIII, 1911, 164–181.

CARL H. EIGENMANN.

Probolodus Eigenmann, 164; orthotype P. HETEROSTOMUS Eigenmann.
Psalidodon Eigenmann, 165; orthotype P. GYMNODONTUS Eigenmann.
Spintherobolus Eigenmann. 167; orthotype S. PAPILLIFERUS Eigenmann.
Glandulocauda Eigenmann, 168; logotype G. MELANOGENYS Eigenmann.
Hysteronotus Eigenmann, 171; orthotype H. MEGALOSTOMUS Eigenmann.
Vesicatrus Eigenmann, 174; orthotype V. TEGATUS Eigenmann.

2199. EIGENMANN (1911). *Descriptions of Two New Tetragonopterid Fishes in
the British Museum.* Ann. Mag. Nat. Hist. (8), VII, 1911, 215–216.

CARL H. EIGENMANN.

Nematobrycon Eigenmann, 215; orthotype N. PALMERI Eigenmann.
Knodus Eigenmann, 216; orthotype K. MERIDÆ Eigenmann.

2200. ELLIS (1911). *On the Species of Hasemania, Hyphessobrycon, and Hemigrammus Collected by J. D. Haseman for the Carnegie Museum.* Ann. Carnegie Mus., VIII, 1911, 148–164.

MARIAN LEE ELLIS.

Hasemania Ellis, 148; orthotype H. MAXILLARIS Ellis.

2201. FOWLER (1911). *A New Albuloid Fish from Santo Domingo.* Proc. Acad.
Nat. Sci. Phila., LXII, 1911, 651–654 (Jan. 27, 1911).

HENRY WEED FOWLER.

Dixonina Fowler, 651; orthotype D. NEMOPTERA Fowler.

2202. FOWLER (1911). *Notes on Clupeoid Fishes.* Proc. Acad. Nat. Sci. Phila.,
LXIII, 1911, 204–211. (Apr. 6, 1911.)

HENRY WEED FOWLER.

Heringia Fowler, 207; orthotype CLUPEA AMAZONICA Steind.

Gudusia Fowler, 207; orthotype CLUPANODON CHAPRA Ham.

Anchoviella Fowler, 211; orthotype ENGRAULIS PERFASCIATUS Poey. A synonym of STOLEPHORUS Lac., as restricted by Blkr.

2203. FOWLER (1911). *Some Fishes from Venezuela.* Proc. Acad. Nat. Sci. Phila., LXIII, 1911, 419–437. (July 27, 1911.)

HENRY WEED FOWLER.

Apodastyanax Fowler, 422; orthotype A. STEWARDSONI Fowler.

2204. FOWLER (1911). *New Fresh-Water Fishes from Western Equador.* Proc. Acad. Nat. Sci. Phila., LXVIII, 1911, 493–520.

HENRY WEED FOWLER.

Rhoadsia Fowler, 497; orthotype R. ALTIPINNA Fowler.

2205. HASEMAN (1911). *Descriptions of Some New Species of Fishes, and Miscellaneous Notes on Others Obtained during the Expedition of the Carnegie Museum to Central South America.* Ann. Carneg. Mus., VII, 315–328.

JOHN D. HASEMAN.

Cephalosilurus Haseman, 317; orthotype C. FOWLERI Haseman.

Acestridium Haseman, 319; orthotype A. DISCUS Haseman.

Platysilurus Haseman, 320; orthotype P. BARBATUS Haseman.

2206. HASEMAN (1911). *Some New Species of Fishes from the Rio Iguassu.* Ann. Carnegie Mus., VII, 374–387.

JOHN D. HASEMAN.

Rhamdiopsis Haseman, 375; orthotype R. MOREIRAI Haseman.

2207. JORDAN & EVERMANN (1911). *A Review of the Salmonoid Fishes of the Great Lakes, with Notes on the White-Fishes of Other Regions.* Bull. U. S. Bureau of Fisheries for 1909 (1911), 1–41.

DAVID STARR JORDAN; BARTON WARREN EVERMANN.

Cisco Jordan & Evermann, 3; orthotype ARGYROSOMUS NIGRIPINNIS Gill.

Thrissomimus (Gill) Jordan & Evermann, 4; orthotype COREGONUS ARTEDI Le Sueur. Substitute for ARGYROSOMUS Ag., preoccupied.

2207A. JORDAN & THOMPSON (1911). *A Review of the Sciænoid Fishes of Japan.* Proc. U. S. Nat. Mus. for 1911 (Jan. 30), 241–261.

DAVID STARR JORDAN; WILLIAM FRANCIS THOMPSON.

Nibea Jordan & Thompson, 246; orthotype SCIÆNA MITSUKURII Jordan & Snyder.

Othonias Jordan & Thompson, 246; orthotype SCIÆNA MANCHURICA Jordan & Thompson.

2208. JORDAN & THOMPSON (1911). *A Review of the Families Lobotidæ and Lutianidæ Found in the Waters of Japan.* Proc. U. S. Nat. Mus., XXXIX, 435–471. (Jan. 30, 1911.)

DAVID STARR JORDAN; WILLIAM FRANCIS THOMPSON (1888–   ).

Etelinus Jordan & Thompson, 465; orthotype ETELIS MARSHI Jenkins.

2209. KENDALL & GOLDSBOROUGH (1911). *The Shore Fishes. (Reports on the Scientific Results of the Expedition to the Tropical Pacific in Charge of Alexander Agassiz, by the U. S. Fish Comm. Steamer "Albatross," from August, 1899, to March, 1900, Commander Jefferson F. Moser, U. S. N., Commanding, 13.)* Mem. Mus. Comp. Zool. Harvard Coll., XXVI, 1911, 239–344.

WILLIAM CONVERSE KENDALL; EDMUND LEE GOLDSBOROUGH.

Paragobioides Kendall & Goldsborough, 324; orthotype P. GRANDOCULIS Kendall & Goldsborough.

2210. KREYENBERG (1911). *Eine neue Cobitinen-Gattung aus China.* Zool. Anz., XXXVIII, 417–419.

M. KREYENBERG.

Gobiobotia Kreyenberg, 417; orthotype G. PAPPENHEIMI Kreyenberg.

2211. McCULLOCH (1911). *Report on the Fishes Obtained by the F. I. S. "Endeavour" on the Coast of New South Wales, Victoria, South Australia, and Tasmania. Part I. Zoological Results of the Fishing Experiments carried Out by the F. I. S. "Endeavour," 1909–1910,* I, 1911, 1–87.

ALLEN RIVERSTON McCULLOCH.

Austroberyx McCulloch, 39; orthotype BERYX AFFINIS Gthr. A synonym of TRACHICHTHODES Gilchrist.

2212. OGILBY (1911). *Description of New or Insufficiently Described Fishes from Queensland Waters.* Ann. Queensland Mus., X, 1911, 36–58.

JAMES DOUGLAS OGILBY.

Apistops Ogilby, 54; orthotype APISTUS CALOUNDRA De Vis.

Paratrigla Ogilby, 56; orthotype TRIGLA PLEURACANTHICA Ogilby.

2213. REGAN (1911). *The Anatomy and Classification of the Teleostean Fishes of the Orders Berycomorphi and Xenoberyces.* Ann. Mag. Nat. Hist. (8), VII, 1–9.

CHARLES TATE REGAN.

Caproberyx Regan, 8; orthotype BERYCOPSIS MAJOR Woodw. (fossil).

2214. REGAN (1911). *The Anatomy and Classification of the Teleostean Fishes of the Order Iniomi.* Ann. Mag. Nat. Hist. (8), VII, 120–133.

CHARLES TATE REGAN.

Hemipterois Regan, 126; orthotype BATHYPTEROIS GUENTHERI Alcock.

Bathysauropsis Regan, 126; orthotype CHLOROPHTHALMUS GRACILIS Gthr.

Parasudis Regan, 127; orthotype CHLOROPHTHALMUS TRUCULENTUS G. & B.

2215. REGAN (1911). *A Synopsis of the Marsipobranchs of the Order Hyperoartii.* Ann. Mag. Nat. Hist. (8), VII, 193–204.

CHARLES TATE REGAN.

Eudontomyzon Regan, 200; orthotype E. DANFORDI Regan.

2216. REGAN (1911). *The Classification of the Teleostean Fishes of the Order Synentognathi.* Ann. Mag. Nat. Hist. (8), VII, 327–335.

CHARLES TATE REGAN.

Lamprichthys Regan, 325; orthotype HAPLOCHILUS TANGANICANUS Boulenger. A synonym of MOHANGA Boulenger of a little earlier date.

Xenentodon Regan, 332; orthotype BELONE CANCILA Buchanan.

Xenopœcilus Regan, 374; orthotype HAPLOCHILUS SARASINORUM Popta.

2217. REGAN (1911). *The Classification of the Teleostean Fishes of the Order Ostariophysi. I. Cyprinoidea.* Ann. Mag. Nat. Hist. (8), VIII, 1911, 13–32.

CHARLES TATE REGAN.

Lepturichthys Regan, 31; orthotype HOMALOPTERA FIMBRIATA Gthr.

Hemimyzon Regan, 32; orthotype HOMALOPTERA FORMOSANA Boulenger.

2218. ROULE (1911). *(Certaines larves tiluriennes.)*

LOUIS ROULE (1861–   ).

Tiluropsis Roule, 329; orthotype a larval eel.

Tilurella Roule, 329; orthotype larval eel, not named in the Zoological Record.

2219. SNYDER (1911). *Descriptions of New Genera and Species of Fishes from Japan and the Riu Kiu Islands.* Proc. U. S. Nat. Mus., XL, 1911, 525–549. (May 26, 1911.)

JOHN OTTERBEIN SNYDER.

Jordanidia Snyder, 527; orthotype J. RAPTORIA Snyder.

Catalufa Snyder, 528; orthotype C. UMBRA Snyder.

Draculo Snyder, 545; orthotype D. MIRABILIS Snyder.

2220. STARKS & MANN (1911). *New and Rare Fishes from Southern California.* Univ. Calif. Pub. Zool., VIII, 9–19.

EDWIN CHAPIN STARKS; WILLIAM M. MANN.

Rusulus Starks & Mann, 14; orthotype R. SABURRÆ Starks & Mann.

Orthopnias Starks & Mann, 16; orthotype O. TRIACIS Starks & Mann.

2221. WAITE (1911). *Additions to the Fish Fauna of New Zealand, No. 2.* Proc. New Zealand Inst., 1910, 49–51. (June 24, 1911.)

EDGAR RAVENSWOOD WAITE.

Rexea Waite, 49; orthotype R. FURCIFERA Waite = GEMPYLUS SOLANDRI Cuv. & Val. A synonym of JORDANIDIA Snyder of slightly earlier date (May 26, 1911).

Pelotretis Waite, 50; orthotype P. FLAVILATUS Waite.

2222. ZUGMAYER (1911). *Poissons provenant des campagnes du yacht Princesse Alice.* Rés. Camp. Sci. Monaco, Fasc. 35, 1911, 1–159.

ERICH ZUGMAYER (1879–   ).

Asquamiceps Zugmayer, 10; orthotype A. VELARIS Zugmayer.

Nematostomias Zugmayer, 76; orthotype N. GLADIATOR Zugmayer.

Trichostomias Zugmayer, 78; orthotype T. VAILLANTI Zugmayer.

Poromitrella Zugmayer, 100; orthotype P. NIGRICEPS Zugmayer.

Platyberyx Zugmayer, 101; orthotype P. OPALESCENS Zugmayer.

Scopeloberyx Zugmayer, 103; orthotype S. OPERCULARIS Zugmayer.

Pachycara Zugmayer, 129; orthotype P. PLAGIOPHTHALMUS Zugmayer.

Barathrites Zugmayer, 132; orthotype B. IRIS Zugmayer.

Anotopterus Zugmayer, 138; orthotype A. PHARAO Zugmayer.

Benthalbella Zugmayer, 140; orthotype B. INFANS Zugmayer.

1912.

2223. BARBOUR (1912). *Two Preoccupied Names.* Proc. Biol. Soc. Washington. XXV, 1912, 121–126.

THOMAS BARBOUR.

Calliblennius Barbour, 187; orthotype ZACALLES BRYOPE Jordan & Snyder. Substitute for ZACALLES Jordan & Snyder, preoccupied.

2224. BEAN (1912). *Descriptions of New Fishes of Bermuda.* Proc. Biol. Soc. Washington, XXV, 1912, 121–126.

TARLETON HOFFMAN BEAN.

Eucrotus Bean, 123; orthotype E. VENTRALIS Bean.

Parasphyrænops Bean, 124; orthotype P. ATRIMANUS Bean.

2225. BEAUFORT (1912). *On Some New Gobiidæ from Ceram and Waigeu.* Zool. Anz., XXXIX, 1912, 136–143.

LIEVEN FERDINAND DE BEAUFORT.

Schismatogobius Beaufort, 139; orthotype S. BRUYNISI Beaufort.
Stiphodon (Weber), Beaufort, 143; orthotype SICYDIUM ELEGANS Steind.

2226. BERG (1912). *Faune de la Russie et des pays limitrophes. Poissons (Marsipobranchii et Pisces). Vol. III.* Pt. 1, 1–336, St. Petersberg.

LEW SEMENOWITCH BERG.

Pararutilus Berg, 43; orthotype LEUCISCUS FRISIO Nordman.
Acanthorutilus Berg, 42, 80; orthotype OREOLEUCISCUS DSAPCHYNENSIS Warpachowsky.
Hemiscaphirhynchus Berg, 309; orthotype PSEUDOSCAPHIRHYNCHUS KAUFMANNI Bogdanov.

2227. BOULENGER (1912). *Poissons recueillis dans la Région du Bas Congo par M. le Dr. W. J. Ansorge.* Ann. Mus. Congo, Bruxelles Zool., ser. 1, 2 fasc. 3, 1–25.

GEORGE ALBERT BOULENGER.

Ansorgia Boulenger, 17; orthotype A. VITTATA Boulenger.

2228. BURKE (1912). *A New Genus and Six New Species of the Family Cyclogasteridæ.* Proc. U. S. Nat. Mus., XLIII, 1912, 567–574.

CHARLES VICTOR BURKE.

Polypera Burke, 567; orthotype P. GREENI Burke.

2229. DUNCKER (1912). *Die Gattungen der Syngnathidæ.* Mett. Nat. Hist. Museum Hamburg, XXIX.

GEORG DUNCKER.

Acanthognathus Duncker, 228; orthotype SYNGNATHUS DACTYLOPHORUS Blkr.
Micrognathus Duncker, 235; orthotype SYNGNATHUS BREVIROSTRIS Rüppell.

2230. EIGENMANN (1912). *The Fresh-Water Fishes of British Guiana, including a Study of the Ecological Grouping of Species, and the Relation of the Fauna of the Plateau to that of the Lowlands.* Mem. Carnegie Mus., V, 1–578.

CARL H. EIGENMANN.

Chamaigenes Eigenmann, 120; orthotype ASPREDO FILAMENTOSUS Cuv. & Val.
Agmus Eigenmann, 128; orthotype A. LYRIFORMIS Eigenmann.
Megalonema Eigenmann, 150; orthotype M. PLATYCEPHALUM Eigenmann.
Microglanis Eigenmann, 155; orthotype M. PŒCILUS Eigenmann.
Brachyglanis Eigenmann, 156; orthotype B. FRENATA Eigenmann.
Leptoglanis Eigenmann, 158; orthotype L. ESSEQUIBENSIS Eigenmann.
Myoglanis Eigenmann, 159; orthotype M. PATAROËNSIS Eigenmann.
Chasmocranus Eigenmann, 160; orthotype C. LONGIOR Eigenmann.
Tympanopleura Eigenmann, 203; orthotype T. PIPERATA Eigenmann.
Ochmacanthus Eigenmann, 213; orthotype O. FLABELLIFER Eigenmann.
Lithoxus Eigenmann, 242; orthotype L. LITHOIDES Eigenmann.
Bivibranchia Eigenmann, 258; orthotype B. PROCTRACTILA Eigenmann.
Tylobranchus Eigenmann, 271; orthotype T. MACULOSUS Eigenmann.
Aphyodite Eigenmann, 314; orthotype A. GRAMMICA Eigenmann.

Acanthocharax Eigenmann, 405; orthotype A. MICROLEPIS Eigenmann.
Heterocharax Eigenmann, 406; orthotype H. MACROLEPIS Eigenmann.
Gymnorhamphichthys Eigenmann, 436; orthotype G. HYPOSOMUS Eigenmann.
Porotergus Eigenmann, 441; orthotype P. GYMNOTUS Eigenmann.
Rhinosardinia Eigenmann, 445; orthotype R. SERRATA Eigenmann = CLUPEA AMA-
ZONICA Steind. A synonym of HERINGIA Fowler, 1911.
Acarichthys Eigenmann, 500; orthotype ACARA HECKELII M. & T.
Soleonasus Eigenmann, 528; orthotype S. FINIS Eigenmann.

2231. EIGENMANN (1912). *Some Results of An Ichthyological Reconnaissance
of Colombia, South America.* Indiana Univ. Studies, 1912, No. 8, 1–27.
CARL H. EIGENMANN.
Xiliphius Eigenmann, 10; orthotype X. MAGDALENÆ Eigenmann.
Parastremma Eigenmann, 20; orthotype P. SADINA Eigenmann.
Genycharax Eigenmann, 22; orthotype G. TARPON Eigenmann.
Gephyrocharax Eigenmann, 24; orthotype G. CHOCOËNSIS Eigenmann.

2232. FOWLER (1912). *Descriptions of Nine New Eels, with Notes on Other
Species.* Proc. Acad. Nat. Sci. Phila., LXIV, 1912, 8–33.
HENRY WEED FOWLER.
Microconger Fowler, 9; orthotype LEPTOCEPHALUS CAUDALIS Fowler.
Ahynnodontophis Fowler, 25; orthotype GYMNOTHORAX STIGMANOTUS Fowler.

2233. GARMAN (1912). *Some Chinese Vertebrates. Pisces.* Mem. Comp. Zool.
Harvard Coll., XL, 111–123.
SAMUEL GARMAN.
Ageneiogarra Garman, 114; orthotype GARRA IMBERBA Garman.
Myloleuciscus Garman, 116; orthotype M. ATRIPINNIS Garman.
Coripareius Garman, 120; orthotype C. CETOPSIS Garman.

2234. GILBERT & BURKE (1912). *Fishes from the Bering Sea and Kamchatka.*
Bull. Bur. Fish., XXX (1910), 1912, 31–96.
CHARLES HENRY GILBERT; CHARLES VICTOR BURKE (1882–   ).
Archaulus Gilbert & Burke, 36; orthotype ARCHAULUS BISERIATUS Gilbert & Burke.
Thyriscus Gilbert & Burke, 43; orthotype T. ANOPLUS Gilbert & Burke.
Eurymen Gilbert & Burke, 64; orthotype E. GYRINUS Gilbert & Burke.
Elassodiscus Gilbert & Burke, 81; orthotype E. TREMEBUNDUS Gilbert & Burke.
Nectoliparis Gilbert & Burke, 82; orthotype N. PELAGICUS Gilbert & Burke.
Acantholiparis Gilbert & Burke, 83; orthotype A. OPERCULARIS Gilbert & Burke.
Gymnoclinus Gilbert & Burke, 86; orthotype G. CRISTULATUS Gilbert & Burke.
Alectridium Gilbert & Burke, 88; orthotype A. AURANTIACUM Gilbert & Burke.

2235. GILBERT & BURKE (1912). *New Cyclogasterid Fishes from Japan.* Proc.
U. S. Nat. Mus., XLII, 1912, 351–380.
CHARLES HENRY GILBERT; CHARLES VICTOR BURKE.
Ateleobrachium Gilbert & Burke, 97; orthotype A. PTEROTUM Gilbert & Burke.

2236. JORDAN (1912). *Note on the Generic Name Safole, replacing Boulengerina
for a Genus of Kuhliid Fishes.* Proc. U. S. Nat. Mus., XLII, 1912, 655. (Au-
gust 29, 1912.)
DAVID STARR JORDAN.
Safole Jordan, 655; orthotype DULES TÆNIURUS Cuv. & Val. Substitute for BOU-
LENGERINA Fowler, preoccupied.

2237. JORDAN & THOMPSON (1912). *A Review of the Sparidæ and Related Families of Perch-like Fishes Found in the Waters of Japan.* Proc. U. S. Nat. Mus., XLI. (Jan. 22, 1912.)

DAVID STARR JORDAN; WILLIAM FRANCIS THOMPSON.

Lethrinichthys Jordan & Thompson, 558; orthotype LETHRINUS NEMATACANTHUS Blkr.

Taius Jordan & Thompson, 570; orthotype CHRYSOPHRYS TUMIFRONS T. & S.

Evynnis Jordan & Thompson, 573; orthotype SPARUS CARDINALIS Lac.

2238. McCULLOCH (1912). *Notes on Some Australian Atherinidæ.* Proc. Roy. Soc. Queensland, XXIV, 1912, 47–53.

ALLAN RIVERSTON MCCULLOCH.

Craterocephalus McCulloch, 48; orthotype C. FLUVIATILIS McCulloch.

2239. McCULLOCH (1912). *Notes on Some Western Australian Fishes.* Rec. W. Austral. Mus., Perth 1, 1912, 78–97.

ALLAN RIVERSTON MCCULLOCH.

Mucogobius McCulloch, 93; orthotype GOBIUS MUCOSUS Gthr. A synonym of CALLOGOBIUS Blkr.

2240. MEEK (1912). *New Species of Fishes from Costa Rica.* Pub. Field Col. Mus., X.

SETH EUGENE MEEK.

Alfaro Meek, 72; orthotype A. ACUTIVENTRALIS Meek. A substitute for PETALOSOMA Regan, preoccupied.

2241. PELLÉGRIN (1912). *Poissons des côtes de l'Angola; Mission de M. Gruvel.* Bull. Soc. Zool. France, XXXVII.

JACQUES PELLÉGRIN.

Diagrammella Pellégrin, 295; orthotype DIAGRAMMA MACROPS Pellégrin.

2242. RADCLIFFE (1912). *Descriptions of Fifteen New Fishes of the Family Cheilodipteridæ, from the Philippine Islands and Contiguous Waters.* Proc. U. S. Nat. Mus., XLI, 431–446. (Jan. 31, 1912.)

LEWIS RADCLIFFE.

Amioides Smith & Radcliffe, 440; orthotype AMIA GROSSIDENS Smith & Radcliffe.

Neamia Smith & Radcliffe, 441; orthotype N. OCTOSPINA Smith & Radcliffe.

2243. RADCLIFFE (1912). *New Pediculate Fishes from the Philippine Islands and Contiguous Waters.* Proc. U. S. Nat. Mus., XLII, 199–214. (April 30, 1912.)

LEWIS RADCLIFFE.

Dermatias Radcliffe, 206; orthotype D. PLATYNOGASTER Radcliffe.

2244. RADCLIFFE (1912). *Description of a New Family, Two New Genera and Twenty-nine New Species of Anacanthine Fishes from the Philippine Islands and Contiguous Waters.* Proc. U. S. Nat. Mus., XLIII, 105–140. (Sept. 27, 1912.)

LEWIS RADCLIFFE.

Parateleopus Smith & Radcliffe, 139; orthotype P. MICROSTOMUS Smith & Radcliffe.

Macrouroides Smith & Radcliffe, 139; orthotype M. INFLATICEPS Smith & Radcliffe.

2245. REGAN (1912). *New Fishes from Aldabra and Assumption, Collected by Mr. J. C. F. Fryer.* Trans. Linn. Soc. London, XV, 1912, 301–302.
CHARLES TATE REGAN.
Xenoconger Regan, 301; orthotype X. FRYERI Regan.
Parioglossus Regan, 302; orthotype P. TÆNIATUS Regan.

2246. REGAN (1912). *The Classification of the Blennioid Fishes.* Ann. Mag. Nat. Hist. (8), X, 265–280.
CHARLES TATE REGAN.
Acanthoplesiops Regan, 266; orthotype ACANTHOCLINUS INDICUS Day.

2247. REGAN (1912). *The Classification of the Teleostean Fishes of the Order Pediculati.* Ann. Mag. Nat. Hist. (8), IX, 1912, 277–289.
CHARLES TATE REGAN.
Haplophryne Regan, 289; orthotype ACERATIAS MOLLIS Brauer.

2248. REGAN (1912). *The Anatomy and Classification of the Symbranchoid Eels.* Ann. Mag. Nat. Hist. (8), IX, 387–390.
CHARLES TATE REGAN.
Macrotrema Regan, 390; orthotype SYMBRANCHUS CALIGANS Cantor.

2249. REGAN (1912). *Description of Two New Eels from West Africa, Belonging to a New Genus and Family.* Ann. Mag. Nat. Hist., X.
CHARLES TATE REGAN.
Heterenchelys Regan, 323; orthotype H. MICROPHTHALMUS Regan.

2250. REGAN (1912). *A Revision of the Pœciliid Fishes of the Genera Rivulus, Pterolebias and Cynolebias.* Ann. Mag. Nat. Hist. (8), X, 1912, 494–508.
CHARLES TATE REGAN.
Petalurichthys Regan, 494; orthotype PETALOSOMA CULTRATUS Regan. Substitute for PETALOSOMA Regan, preoccupied in COLEOPTERA; a synonym of ALFARO Meek.

2251. REGAN (1912). *Sexual Differences in the Pœciliid Fishes of the Genus Cynolebias.* Ann. Mag. Nat. Hist. (8), X, 641–652.
CHARLES TATE REGAN.
Cynopœcilus Regan, 642; orthotype CYNOLEBIAS MELANOTÆNIA Regan.

2252. RIBEIRO (1912). *Fauna Brasiliense. Peixes 4 (A).* Arch. Mus. Rio de Janeiro, XVI, 1912, 1–504.
ALIPIO DE MIRANDA RIBEIRO.
Parasturisoma Ribeiro, 109; orthotype LORICARIA BREVIROSTRIS Eigenmann.
Mormyrostoma Ribeiro, 192; orthotype SILURUS CARINATUS L.
Tatia Ribeiro, 360; orthotype CENTROMOCHLUS INTERMEDIUS Steind.

2253. RIBEIRO (1912). *Loricariidæ, Callichthyidæ, Doradidæ e Trichomycteridæ. Commissaõ de Linhas Telegraphicas Estrategicas de Matto-Grosso ao Amazonas.* Annexo No. 5, Rio Janeiro, 1912, 1–31.
ALIPIO DE MIRANDA RIBEIRO.
Gyrinurus Ribeiro, 27; orthotype G. BATRACHOSTOMA Ribeiro (Sept., 1912). A synonym of OCHMACANTHUS Eigenmann, June, 1912.
Paravandellia Ribeiro, 28; orthotype P. OXYPTERA Ribeiro.

2254. SMITH (1912). *Description of a New Notidanoid Shark from the Philippine Islands, Representing a New Family.* Proc. U. S. Nat. Mus., XLI, 489–491. (February 8, 1912.)

HUGH McCORMICK SMITH.

Pentanchus Smith, 489; orthotype P. PROFUNDICOLUS Smith.

2255. SMITH (1912). *The Squaloid Sharks of the Philippine Archipelago, with Descriptions of New Genera and Species.* Proc. U. S. Nat. Mus., XLI, 677–685. (February 9, 1912.)

HUGH McCORMICK SMITH.

Nasisqualus Smith, 681; orthotype N. PROFUNDORUM Smith.
Squaliolus Smith, 684; orthotype S. LATICAUDUS Smith.

2256. SMITH & RADCLIFFE (1912). *Description of a New Family of Pediculate Fishes from Celebes.* Proc. U. S. Nat. Mus., XLII, 579–582. (Aug. 30, 1912.)

HUGH McCORMICK SMITH; LEWIS RADCLIFFE.

Thaumatichthys Smith & Radcliffe, 579; orthotype T. PAGIDOSTOMUS Smith & Radcliffe.

2257. SNYDER (1912). *The Fishes of the Streams Tributary to Monterey Bay.* Bull. U. S. Bur. Fish., XXXII, 47–72.

JOHN OTTERBEIN SNYDER.

Hesperoleucus Snyder, 63; orthotype POGONICHTHYS SYMMETRICUS Grd.

2258. TANAKA (1912). *Figures and Descriptions of Fishes of Japan.* Tokyo, V–X, 71–186.

SHIGEHO TANAKA.

Cyclolumpus Tanaka, 86; orthotype C. ASPERRIMUS Tanaka.
Heteroscymnus Tanaka, 102; orthotype H. LONGUS Tanaka.
Cirrhigaleus Tanaka, 151; orthotype C. BARBIFER Tanaka.
Calliscyllium Tanaka, 171; orthotype C. VENUSTUM Tanaka.

2259. ZUGMAYER (1912). *On a New Genus of Cyprinoid Fishes from High Asia.* Ann. Mag. Nat. Hist. (8), IX, 1912, 682.

ERICH ZUGMAYER.

Aspiopsis Zugmayer, 682; orthotype A. MERZBACHERI Zugmayer.

1913.

2260. BERG (1913). *A Review of the Clupeoid Fishes of the Caspian Sea,* etc. Ann. Mag. Nat. Hist., XI, 472–480.

LEO SEMENOWITCH BERG.

Clupeonella Berg, 472; orthotype C. CASPIA Berg. Not CLUPEONELLA Kessler; replaced by CASPIALOSA Berg.

2261. BROOM (1913). *On Some Fishes from Lower and Middle Karroo, South Africa.* Ann. S. Afr. Mus. Cape Town, XII, 1913, 1–5.

ROBERT BROOM.

Caruichthys Broom, 4; orthotype C. ORNATUS Broom (fossil).

2262. BROOM (1913). *On Some Fossil Fishes from the Diamond-Bearing Pipes of Kimberley.* Trans. Roy. Soc., S. Afr. III, 1913, 399–402.

ROBERT BROOM.

Disichthys Broom, 400; orthotype D. KIMBERLEYENSIS Broom (fossil). (PALÆO-NISCID.)

Pelichthys Broom, 401; orthotype P. KIMBERLEYENSIS Broom (fossil). (PALÆO-NISCID, S. Africa.)

2263. CHAPMAN (1913). *Note on the Occurrence of the Cainozoic Shark, Carcharoides, in Victoria.* Victoria Naturalist, XXX, 142–193.

FREDERICK REVANS CHAPMAN.

Carcharoides Chapman, 142; orthotype a fossil shark.

2264. CHAUDHURI (1913). *Zoological Results of the Abor Expedition, 1911–12.* XVIII. Fish Rec. Ind. Mus. Calcutta, VIII, 1913, 243–257.

BONAWARI LAL CHAUDHURI.

Aborichthys Chaudhuri, 245; orthotype A. KEMPI Chaudhuri.

2265. COCKERELL (1913). *Observations on Fish Scales.* Bull. Bureau of Fisheries, XXII (for 1912), Oct. 13, 1913.

THEODORE DRU ALLISON COCKERELL.

Myloleucops Cockerell, 136 (foot note); type LEUCISCUS ÆTHIOPS Basilewsky. A substitute for MYLOLEUCUS Gthr., preoccupied.

2266. EIGENMANN (1913). *Some Results from an Ichthyological Reconnaissance of Colombia, South America.* Pt. II, Indiana Univ. Studies, 1913, No. 18, 1–32.

CARL H. EIGENMANN.

Pterobrycon Eigenmann, 3; orthotype P. LANDONI Eigenmann.

Argopleura Eigenmann, 10; orthotype BRYCONAMERICUS MAGDALENÆ Eigenmann.

Zygogaster Eigenmann, 22; orthotype Z. FILIFER Eigenmann.

Microgenys Eigenmann, 22; orthotype M. MINUTUS Eigenmann.

2267. ELLIS (1913). *The Gymnotid Eels of Tropical America.* Mem. Carneg. Mus. VI, 1913, 109–195.

MAX MAPES ELLIS.

Orthosternarchus Ellis, 144; orthotype STERNARCHUS TEMANDUA Boulenger.

Odontosternarchus Ellis, 155; orthotype STERNARCHUS SACHSI Peters.

2268. ELLIS (1913). *The Plated Nematognaths.* Mem. Carneg. Mus., VIII, 1913, 384–413.

MARIAN LEE ELLIS.

Cascadura Ellis, 387; orthotype C. MACULOCEPHALA Ellis.

2269. FOWLER (1913). *Curimatus spilurus Cope, a wrongly identified Characin.* Proc. Acad. Nat. Sci. Phila., LXV, 1913, 673–675.

HENRY WEED FOWLER.

Xyrocharax Fowler, 673; orthotype CURIMATUS STIGMATURUS Fowler.

2270. FOWLER (1913). *Notes on Catostomoid Fishes.* Proc. Acad. Nat. Sci. Phila., LXV, 1913, 45–60.

HENRY WEED FOWLER.

Megastomatobus Fowler, 45; orthotype SCLEROGNATHUS CYPRINELLA Cuv. & Val. A synonym of SCLEROGNATHUS as restricted by Günther & Jordan. The first species named under SCLEROGNATHUS, however, was a CARPIODES.

Notolepidomyzon Fowler, 47; orthotype PANTOSTEUS ARIZONÆ (Gilbert).
Scartomyzon Fowler, 59; orthotype PTYCHOSTOMUS CERVINUS Cope.
Pithecomyzon Fowler, 54; orthotype CHASMISTES CUJUS Cope.

2271. FOWLER (1913). *Fishes from the Madeira River, Brazil.* Proc. Acad. Nat. Sci. Phila., LXV., 1913, 517–579.

HENRY WEED FOWLER.

Prionobrama Fowler, 553; orthotype P. MADEIRÆ Fowler.
Gnathocharax Fowler, 561; orthotype G. STEINDACHNERI Fowler.
Tyttocharax Fowler, 564; orthotype T. MADEIRÆ Fowler.

2272. FOWLER (1913). *Fowlerina Eigenmann, a Preoccupied Generic Name.* Science, XXXVIII, 1913, 51.

HENRY WEED FOWLER.

Ephippicharax Fowler, 51; orthotype TETRAGONOPTERUS COMPRESSUS Gthr. A substitute for FOWLERINA Eigenmann, preoccupied.

2273. GARMAN (1913). *The Plagiostomia (Sharks, Skates, and Rays).* Mem. Mus. Comp. Zool. Harvard Coll., XXXVI, 1913, 1–528.

SAMUEL GARMAN.

Vulpecula (Valmont) Garman, 31; orthotype VULPECULA MARINA (Valmont) Garman = SQUALUS VULPINUS Bonnaterre = SQUALUS VULPES Gmelin. Same as ALOPIAS Raf. and of earlier date, but Valmont's names are of doubtful validity.
Catulus (Valmont) Garman, 71; orthotype CATULUS VULGARIS Valmont = SQUALUS CATULUS L. Name preoccupied by CATULUS Kniphof, 1759, a genus of HEMIPTERA. Same as SCYLLIORHINUS Blainv.
Apristurus Garman, 96; orthotype SCYLLIORHINUS INDICUS Brauer.
Atelomycterus Garman, 100; orthotype SCYLLIUM MARMORATUM Bennett.
Haploblepharus Garman, 101; orthotype SCYLLIUM EDWARDSII (Cuv.) Voigt.
Zanobatus Garman, 291; orthotype PLATYRHINA SCHŒNLEINI M. & H.
Urobatis Garman, 401; orthotype LEIOBATUS SLOANI Blainv.
Pteromylæus Garman, 437; logotype MYLIOBATIS ASPERRIMUS J. & E.

2274. HOLT & BYRNE (1913). *Sixth Report on the Fishes of the Irish Atlantic Slope. The Families Stomiatidæ, Sternoptychidæ, and Salmonidæ.* Dublin Fish Ireland Sci. Invest., 1913, 1–27.

ERNEST W. L. HOLT; L. W. BYRNE.

Lamprotoxus Holt & Bryne, 2; orthotype L. FLAGELLIBARBA Holt & Byrne.

2275. JORDAN (1913). *Ellimma, a Genus of Fossil Herrings.* Proc. Biol. Soc. Washington, XXVI, 1913, 79.

DAVID STARR JORDAN.

Ellimma Jordan, 79; orthotype ELLIPES BRANNERI Jordan (fossil). Substitute for ELLIPES Jordan, preoccupied.

2276. JORDAN & METZ (1913). *A Catalogue of the Fishes Known from the Waters of Korea.* Mem. Carnegie Mus., VI, 1913, 1–65.

DAVID STARR JORDAN; CHARLES WILLIAM METZ (1889–   ).

Zunasia Jordan & Metz, 7; orthotype PRISTIGASTER CHINENSIS, Basilewsky, the young of ILISHA ELONGATA Gray. A synonym of ILISHA Gray.

2277. JORDAN & THOMPSON (1913). *Notes on a Collection of Fishes from the Island of Shikoku in Japan, with a Description of a New Species, Gnathypops iyonis.* Proc. U. S. Nat. Mus., XLVI, 65–72.

DAVID STARR JORDAN; WILLIAM FRANCIS THOMPSON.

Onigocia Jordan & Thompson, 70; orthotype PLATYCEPHALUS MACROLEPIS Blkr.

Inegocia Jordan & Thompson, 70; orthotype PLATYCEPHALUS JAPONICUS Krusenstern.

2278. LAHILLE (1913). *Nota sobre siete peces de las costas argentinas.* An. Mus. Nac. Buenos Aires, XXIV, 1913, 1–24.

FERNANDO LAHILLE.

Besnardia Lahille, 1; orthotype B. GYRINOPS Lahille.

2279. LANGER (1913). *Beiträge zur Morphologie der viviparen Cyprinodontiden.* Morph. Jahrb. Leipzig, XLVII, 1913, 193–307.

W. F. LANGER.

Gulapinnus Langer, 207; orthotype CNESTERODON DECEMMACULATUS Garman.

2279A. LERICHE (1913). *Les poissons paléocènes de Landana (Congo).* Ann. du Musée du Congo Belge. Geol. Paleont. Mineral., ser. 3, I, 69–80.

MAURICE LERICHE.

Hypolophites Leriche; orthotype H. MAYOMBENSIS Leriche (fossil shark).

2280. OGILBY (1913). *Edible Fishes of Queensland: I. Pempheridæ; II. Gadopsiform Percoids.* Mem. Queensland Mus. (Dec. 10, 1913).

JAMES DOUGLAS OGILBY.

Liopempheris Ogilby, 61; orthotype PEMPHERIS MULTIRADIATA Klunzinger; near CATALUFA Snyder, 1911; perhaps a synonym.

2281. PELLÉGRIN (1913). *Poisson des côtes de Mauritanie. Mission de M. Gruvel.* Paris Bull. Soc. Zool., XXXVIII, 1913, 116–118.

JACQUES PELLÉGRIN.

Panturichthys Pellégrin, 118; orthotype P. MAURITANICUS Pellégrin.

2282. PELLÉGRIN (1913). *Poissons nouveaux de Guinée Française recueillis par M. Pobéguin.* Bull. Soc. Zool. Paris, XXXVIII, 1913, 236–241.

JACQUES PELLÉGRIN.

Diagrammella Pellégrin, 295; orthotype DIAGRAMMA MACROPS Pellégrin.

2283. PELLÉGRIN (1913). *Sur un nouveau genre de Centrarchidés du Gabon.* Comptes Rendus Acad. Roy. 26, 1913, 1488–1489.

JACQUES PELLÉGRIN.

Parakuhlia Pellégrin, 1489; orthotype P. BOULENGERI Pellégrin.

2284. RADCLIFFE (1913). *Descriptions of Seven New Genera and Thirty-one New Species of Fishes of the Families Brotulidæ and Carapidæ from the Philippine Islands and the Dutch East Indies.* Proc. U. S. Nat. Mus., XLIV, 135–176.

LEWIS RADCLIFFE.

Homostolus Radcliffe, 147; orthotype H. ACER Radcliffe.

Enchelybrotula Radcliffe, 154; orthotype E. PAUCIDENS Radcliffe.

Mastigopterus Radcliffe, 159; orthotype M. IMPERATOR Radcliffe.

Hypopleuron Radcliffe, 164; orthotype H. CANINUM Radcliffe.

Luciobrotula Radcliffe, 170; orthotype L. BARTSCHI Radcliffe.

Xenobythites Radcliffe, 173; orthotype X. ARMIGER Radcliffe.

Pyramodon Radcliffe, 175; orthotype P. VENTRALIS Radcliffe.

2285. REGAN (1913). *The Antarctic Fishes of the Scottish National Antarctic Expedition.* Edinburgh Trans. Roy. Soc. 49, 1913, 229–292.

CHARLES TATE REGAN.

Eugnathosaurus Regan, 234; orthotype E. VORAX Regan.

Ophthalmolycus Regan, 243; orthotype LYCODES MACROPS Gthr.

Austrolycichthys Regan, 244; orthotype LYCODES BRACHYCEPHALUS Pappenheim.

Austrolycus Regan, 245; orthotype A. DEPRESSICEPS Regan.

Crossolycus Regan, 247; orthotype LYCODES FIMBRIATUS Steind.

Pagetopsis Regan, 286; orthotype CHAMPSOCEPHALUS MACROPTERUS Boulenger.

Chænocephalus Regan, 287; orthotype C. ACERATUS Regan.

2286. REGAN (1913). *The Classification of the Percoid Fishes.* Ann. Mag. Nat. Hist., XII, 1913, 111–115.

CHARLES TATE REGAN.

Centrodraco Regan, 145; orthotype DRACONETTA ACANTHOPOMA Regan.

2287. REGAN (1913). *Fishes from the River Ucayali, Peru, Collected by W. Mounsey.* Ann. Mag. Nat. Hist., XII, 1913, 281–283.

CHARLES TATE REGAN.

Apistogramma Regan, 282; orthotype HETEROGRAMMA BORELLI Regan. A substitute for HETEROGRAMMA Regan, preoccupied.

2288. REGAN (1913). *The Fishes of the San Juan River, Colombia.* Ann. Mag. Nat. Hist., XII, 1913, 462–473.

CHARLES TATE REGAN.

Xenurocharax Regan, 463; orthotype X. SPURRELLI Regan.

Nannorhamdia Regan, 467; orthotype N. SPURRELLI Regan.

2289. REGAN (1913). *Phallostethus dunckeri, a Remarkable New Cyprinodont Fish from Johore.* Ann. Mag. Nat. Hist. XII, 1913, 548–555.

CHARLES TATE REGAN.

Phallostethus Regan, 548; orthotype P. DUNCKERI Regan.

2290. REGAN (1913). *A Collection of Fishes Made by Professor Francisco Fuentes at Easter Island.* Proc. Zool. Soc. London, 1913, 368–374.

CHARLES TATE REGAN.

Girellops Regan, 369; orthotype GIRELLA NEBULOSA Kendall & Radcliffe.

2291. REGAN (1913). *A Revision of the Cyprinodont Fishes of the Sub-Family Pœciliinæ.* Proc. Zool. Soc. London. 1913, 977–1018.

CHARLES TATE REGAN.

Priapichthys Regan, 991; orthotype GAMBUSIA ANNECTENS Regan.

Priapella Regan, 992; orthotype GAMBUSIA BONITA Meek.

Pseudopœcilia Regan, 995; orthotype PŒCILIA FESTÆ Boulenger.

Pœcilopsis Regan, 997; orthotype P. ISTHMENSIS Regan.

Brachyrhaphis Regan, 997; orthotype GAMBUSIA RHABDOPHORA Regan.

Leptorhaphis Regan, 998; orthotype GAMBUSIA INFANS Woolman.

Pamphoria Regan, 1003; orthotype CNESTERODON SCALPRIDENS Garman.

Pamphorichthys Regan, 1003; orthotype HETERANDRIA MINOR Garman.

2292. RIBEIRO (1913). *Fauna Brasiliense. Peixes.* V (part). *Serranidæ* (39 pp.), *Hæmulidæ* (30 pp.), *and Sciænidæ* (46 pp.). Arch. Mus. Rio de Janeiro, XVII, 1913.

ALIPIO DE MIRANDA RIBEIRO.

Symphysoglyphus Ribeiro, 43; orthotype OTOLITHUS BAIRDI Steind.

2293. ROULE (1913). *Deuxième Expédition Antarctique Française.* Poissons, 1913, 1–25.

LOUIS ROULE.

Dolloidraco Roule, 15; orthotype D. LONGIDORSALIS Roule.

2294. ROULE (1913). *Étude sur les formes larvaires Tiluriennes de poissons apodes.* Ann. Inst. Ocean. Monaco, VI, f. 2, 1913, 1–23.

LOUIS ROULE.

Grimaldichthys Roule, 2; orthotype G. PROFUNDISSIMUS Roule.

2295. SMITH (1913). *The Hemiscylliid Sharks of the Philippine Archipelago, with Description of a New Genus from the China Sea. (Scientific Results of the Philippine Cruise of the Fisheries Steamer "Albatross," 1907–1910. No. 28.)* Proc. U. S. Nat. Mus., XLV, 567–569.

HUGH MCCORMICK SMITH.

Cirrhoscyllium Smith, 568; orthotype C. EXPOLITUM Smith.

2296–2297. SMITH (1913). *Description of a New Carcharioid Shark from the Sulu Archipelago.* Proc. U. S. Nat. Mus., XLV, 1913, 599–601.

HUGH MCCORMICK SMITH.

Eridacnis Smith, 599; orthotype E. RADCLIFFEI Smith.

2298. STARKS (1913). *The Fishes of the Stanford Expedition to Brazil.* Stanford Univ. Publ., University Series, 1913, 1–77.

EDWIN CHAPIN STARKS.

Platypogon Starks, 29; orthotype P. CÆRULIROSTRIS Starks.

2299. WEBER (1913). *Résultats de l'expédition scientifique néerlandaise à la Nouvelle-Guinée.* Nova Guinea, IX, 513–613.

MAX WEBER.

Oloplotosus Weber, 521; orthotype O. MARIÆ Weber.
Porochilus Weber, 523; orthotype P. OBBESI Weber.
Doiichthys Weber, 532; orthotype D. NOVÆ-GUINEÆ Weber.
Tetranesodon Weber, 545; orthotype T. CONORHYNCHUS Weber.

2300. WEBER (1913). *Siboga-Expeditie. Die Fische der Siboga-Expedition.* Leiden, 1–710.

MAX WEBER.

Ceromitus Weber, 54; orthotype C. FLAGELLIFER Weber.
Promacheon Weber, 84; orthotype P. SIBOGÆ Weber.
Apterygocampus Weber, 116; orthotype A. EPINNULATUS Weber.
Odontonema Weber, 148; orthotype O. KERBERTI Weber.
Leiogaster Weber, 179; orthotype L. MELANOPUS Weber.
Pteranthias Weber, 209; orthotype P. LONGIMANUS Weber.
Sphenanthias Weber, 210; orthotype S. SIBOGÆ Weber.
Nematochromis Weber, 265; orthotype N. ANNÆ Weber.
Cheiloprion Weber, 342; orthotype POMACENTRUS LABIATUS Day.
Cyttula Weber, 411; orthotype C. MACROPS Weber.
Lepidoblepharon Weber, 421; orthotype L. OPHTHALMOLEPIS Weber.
Liopteryx Weber, 423; orthotype BRACHYPLEURA XANTHOSTICTA Alcock.
Pleurosicya Weber, 456; orthotype P. BOLDINGHI Weber.
Tydemania Weber, 570; orthotype T. NAVIGATORIS Weber.

2301. WEBER (1913). *Neue Beiträge zur Kenntniss der Süsswasserfische von Celebes.** Amster. Bijdr. Dierk., 1913, 197–213.

MAX WEBER.

Adrianichthys Weber, 204; orthotype A. KRUYTI Weber.

2302. ZUGMAYER (1913). *Diagnoses des Stomiatidés provenant des campagnes du Yacht "Hirondelle II," 1911–1912.* Bull. Institut. Océanographique. Monaco, 1913.

ERICH ZUGMAYER.

Aristostomias Zugmayer, 1; orthotype A. GRIMALDII Zugmayer.

1914.

2303. BERG (1914). *Faune de la Russie et des Pays limitropes.* Bull. Acad. Imp. Sci. Petrograd. *Poissons, Marsipobranchii et Pisces.* Tome III, Ostariophysi, Livraison 2, 1914.

LEO SEMENOVICH BERG.

Chilogobio Berg, 488; orthotype C. SOLDATOVI Berg.
Pseudogobiops Berg, 500; orthotype GOBIO RIVULARIS Basilewsky.
Leucalburnus Berg, 5, 521; orthotype L. SATURNINI Berg.

2304. BOGOLIUBOV (1914). *Étude sur le Paracymatodus Traut.* Ann Géol. Univérs. Novo Aleksandrija, XVI, 197–199.

H. H. BOGOLIUBOV.

Paracymatodus Bogoliubov, 197; orthotype CYMATODUS RECLINATUS Trautschold.

2305. BOULENGER (1914). *Mission Stappers au Tanganika-Moero. Diagnoses de poissons nouveaux. 1. Acanthopterygiens, Opisthomes, Cyprinodontes.* Rev. Zool. Afr. Bruxelles, III, 1914, 442–447.

GEORGE ALBERT BOULENGER.

Luciolates Boulenger, 443; orthotype L. STAPPERSII Boulenger.
Stappersia Boulenger, 445; orthotype S. SINGULARIS Boulenger.

2306. BOULENGER (1914). *Cichlidæ.* Wissensch. Ergebniss. Deutsch. Zentral. Afrika-Exped., 1907–1908, V, Zool. III, 1914, 253–259.

GEORGE ALBERT BOULENGER.

Schubotzia Boulenger, 258; orthotype S. EDUARDINA Boulenger.

2307. BRANSON (1914). *The Devonian Fishes of Missouri.* Bull. Univ. Missouri. Science, II, 59–74.

EDWIN BAYER BRANSON.

Eoörodus Branson, 68; orthotype E. TYPUS Branson (fossil: shark-tooth).

2308. EASTMAN (1914). *Catalog of the Fossil Fishes in the Carnegie Museum.* Pt. II. Supplement to the *Catalog of Fishes from the Upper Eocene of Monte Bolca.* Mem. Carneg. Mus., VI, No. 5. 315–348.

CHARLES ROCHESTER EASTMAN.

Eobothus Eastman, 328; orthotype RHOMBUS MINIMUS Ag. (fossil flounder).
Eolabroides Eastman, 336; orthotype CRENILABRUS SZANOCHÆ Zigno (fossil labroid).
Gillidia Eastman, 345; orthotype TOXOTES ANTIQUUS Ag. (fossil).

---

* Reise von E. C. Abendanon.

2309. EASTMAN (1914). *Catalogue of the Fossil Fishes in the Carnegie Museum, IV.* Descriptive catalogue of the fossil fishes from the Lithographic Stone of Solenhofen, Bavaria. Mem. Carneg. Mus., VI, 1914, 389–449.

CHARLES ROCHESTER EASTMAN.

Parathrissops Eastman, 417; orthotype P. FURCATUS Eastman (fossil LEPTOLEPIDÆ, Jurassic, Solenhofen).

2310. EIGENMANN (1914). *Some Results from Studies of South American Fishes.* Indiana Univ. Bull., XII, 1914, 20–48.

CARL H. EIGENMANN.

Bleptonema Eigenmann, 44; orthotype B. PARAGUAYENSIS Eigenmann. A synonym of PRIONOBRAMA Eigenmann.

Parecbasis Eigenmann, 45; orthotype P. CYCLOLEPIS Eigenmann.

2311. EIGENMANN, HENN, AND WILSON. (1914). *New Fishes from Western Colombia, Equador and Peru.* Indiana Univ. Studies, XIX, 1914, 1–15.

CARL H. EIGENMANN; ARTHUR WILBUR HENN; CHARLES WILSON.

Landonia Eigenmann & Henn, 1; orthotype L. LATIDENS Eigenmann & Henn.

Phanagoniates Eigenmann & Henn, 2; orthotype P. WILSONI Eigenmann & Henn. Misprinted elsewhere as PHENAGONIATES.

Microbrycon Eigenmann, Henn & Wilson, 3; orthotype M. MINUTUS Eigenmann Henn & Wilson.

Ceratobranchia Eigenmann, Henn & Wilson, 4; orthotype C. OBTUSIROSTRIS Eigenmann, Henn & Wilson.

2312. FACCIOLÀ (1914). *Su di un nuovo tipo del Nettastomidi.* Roma, Bull. Soc. Zool. Ital. (3), III, 1914, 39–47.

LUIGI FACCIOLÀ.

Nettastomella Facciolà, 47; orthotype N. PHYSONEMA Facciolà.

2313. FOWLER (1914). *Fishes from the Rupununi River, British Guiana.* Proc. Acad. Nat. Sci. Phila., LXVI, 229–284.

HENRY WEED FOWLER.

Myocharax Fowler, 239; orthotype LEPORINUS DESMOTES Fowler.

Xiphocharax Fowler, 251; orthotype X. OGILVIEI Fowler.

Cobitiglanis Fowler, 268; orthotype OCHMACANTHUS TAXISTIGMA Fowler. A synonym of HENONEMUS Eigenmann & Ward.

Stoniella Fowler, 271; orthotype S. LEOPARDUS Fowler.

2314. JORDAN & THOMPSON (1914). *Record of the Fishes Obtained in Japan in 1911.* Mem. Carneg. Mus. VI, 1914, 205–313.

DAVID STARR JORDAN; WILLIAM FRANCIS THOMPSON.

Tanakia Jordan & Thompson, 230; orthotype RHODEUS ORYZÆ Jordan & Thompson.

Ectenias Jordan & Thompson, 241; orthotype E. BRUNNEUS Jordan & Thompson.

Icticus Jordan & Thompson, 242; orthotype I. ISCHANUS Jordan & Thompson.

Franzia Jordan & Thompson, 251; orthotype ANTHIAS NOBILIS Franz.

Calymmichthys Jordan & Thompson, 296; orthotype C. XENICUS Jordan & Thompson.

Spectrunculus Jordan & Thompson, 301; orthotype S. RADCLIFFEI Jordan & Thompson.

Tarphops Jordan & Thompson, 307; orthotype PSEUDORHOMBUS OLIGOLEPIS Blkr.

2315. McCULLOCH (1914). *Notes on Some Australian Pipe-Fishes.* Australian Zoologist, I, pt. 1, 1914, 29–31.

ALLAN RIVERSTON McCULLOCH.

Histiogamphelus McCulloch, 30; orthotype H. BRIGGSI McCulloch.

2316. McCULLOCH (1914). *Report on Some Fishes Obtained by the F. I. S. "Endeavour" on the Coast of Queensland, New South Wales, Victoria, Tasmania, South and South Western Australia.* Part II. Biol. Res. "Endeavour," II, Pt. 3, 77–165.

ALLAN RIVERSTON McCULLOCH.

Vinculum McCulloch, 110; orthotype CHÆTODON SEXFASCIATUS Rich.
Allocyttus McCulloch, 114; orthotype CYTTOSOMA VERRUCOSUM Gilchrist.

2317. MEEK (1914). *An Annoted List of Fishes Known To Occur in the Fresh Waters of Costa Rica.* Field Museum, Nat. Hist. Pub. Zool., Ser. X, 1914, 101–134.

SETH EUGENE MEEK.

Carlia Meek, 108; orthotype CHEIRODON EIGENMANNI Meek.

2318. OGILBY (1914). *(Ptenonotus.)* Ann. Queensland Museum, XXII.

JAMES DOUGLAS OGILBY.

Ptenonotus Ogilby, 13; orthotype EXOCŒTUS CIRRIGER Peters.

2319. PAPPENHEIM (1914). *Die Tiefseefische.* Südpol-Expedition. 1901–1903. XV, Zoologie, VI, 1914, 163–200.

KARL PAPPENHEIM.

Gnathostomias Pappenheim, 172; orthotype G. LONGIFILIS Pappenheim.
Neoceratias Pappenheim, 198; orthotype N. SPINIFER Pappenheim.

2320. REGAN (1914). *Diagnoses of New Marine Fishes Collected by the British Antarctic ("Terra Nova") Expedition.* Ann. Mag. Nat. Hist. XIII, 1914, 11–17.

CHARLES TATE REGAN.

Prinodraco Regan, 10; orthotype P. EVANSI Regan.
Pogonophryne Regan, 13; orthotype P. SCOTTI Regan.
Chænodraco Regan, 14; orthotype C. WILSONI Regan.
Serranops Regan, 15; orthotype S. MACULICAUDA Regan.
Cynophidium Regan, 16; orthotype C. PUNCTATUM Regan.
Lepidoperca Regan, 17; orthotype L. INORNATA Regan.

2321. REGAN (1914). *Note on Aristeus goldiei Macleay, and on Some Other Fishes from New Guinea.* Proc. Zool. Soc. London, 1914, 339–340.

CHARLES TATE REGAN.

Rhombosoma Regan, 339; orthotype NEMATOCENTRIS NOVÆ-GUINEÆ Weber = ARISTEUS GOLDIEI Macleay.

2322. REGAN (1914). *Report on the Fresh-Water Fishes Collected by the British Ornithologists' Union Expedition in Dutch New Guinea.* Trans. Zool. Soc. London, 1914, XX, 275–286.

CHARLES TATE REGAN.

Rhadinocentrus Regan, 280; orthotype R. ORNATUS Regan.
Anisocentrus Regan, 281; orthotype NEMATOCENTRIS RUBROSTRIATUS Weber.
Chilatherina Regan, 282; orthotype RHOMBATRACTUS FASCIATUS Weber.

2323. REGAN (1914). *Two New Cyprinid Fishes from Waziristan, Collected by Major G. E. Bruce.* Ann. Mag. Nat. Hist. XIII, 1914, 261–263.

CHARLES TATE REGAN.

Schizocypris Regan, 262; orthotype S. BRUCEI Regan.

2324. REGAN (1914). *Descriptions of Two New Cyprinodont Fishes from Mexico, Presented to the British Museum by Herr A. Rachow.* Ann. Mag. Nat. Hist., XIV, 1914, 65–67.

CHARLES TATE REGAN.

Heterophallus Regan, 66; orthotype H. RACHOVII Regan.

2325. REGAN (1914). *A Synopsis of the Fishes of the Family Macrorhamphosidæ.* Ann. Mag. Nat. Hist., XIII, 1914, 17–21.

CHARLES TATE REGAN.

Notopogon Regan, 14; logotype N. SCHOTELI Regan.
Scolopacichthys Regan, 21; orthotype CENTRISCUS ARMATUS Sauvage.

2326. REGAN (1914). *British Antarctic ("Terra Nova") Expedition.* Zoology, I, Fishes, 1914, 1–54.

CHARLES TATE REGAN.

Histiodraco Regan, 9; orthotype DOLLOIDRACO VELIFER Regan.
Pyramodon Regan, 20; orthotype CYNOPHIDIUM PUNCTATUM Regan.

2327. ZUGMAYER (1914.) *Diagnoses de quelques poissons nouveaux provenant des campagnes du yacht Hirondelle II (1911–1913).* Bull. Inst. Ocean. Monaco. 1914, No. 2888, 1–4.

ERICH ZUGMAYER.

Cetostoma Zugmayer, 3; orthotype C. REGANI Zugmayer.
Zoarcites Zugmayer, 3; orthotype Z. PARDALIS Zugmayer.

1915.

2328. BAMBER (1915). *Reports on the Marine Biology of the Sudanese Red Sea, from Collections Made by Cyril Crossland, etc.* XXII. *The Fishes.* Journ. Linn. Soc. London, XXXI, 1915, 477–485.

R. C. BAMBER.

Neenchelys Bamber, 479; orthotype N. MICROTRETUS Bamber.

2329. BERG (1915). *Compte-Rendu préliminaire sur les harénges collectionnés dans la mer Caspienne par l'Expédition Caspienne de l'Année, 1913.* Mat. Pozn. Russk. Rybolav, IV, No. 6, pp. 1–8.

LEW SEMENOVICH BERG.

Caspialosa Berg, 12; orthotype CLUPEONELLA CASPIA Berg. A substitute for CLUPEONELLA Berg; not of Kessler.

2330. BERG (1915). *Faune de la Russie.* Poissons, III, Pt. 2, 337–704.

LEW SEMENOWITCH BERG.

Chilogobio Berg, 490; orthotype C. CZERSKII Berg.

2331. EIGENMANN (1915). *The Cheilodontinæ, a Subfamily of Minute Characid Fishes of South America.* Mem. Carneg. Mus., VII, No. 1, 1–99.

CARL H. EIGENMANN.

Leptobrycon Eigenmann, 46; orthotype L. JUTUARANÆ Eigenmann.
Macropsobrycon Eigenmann, 48; orthotype M. URUGUAYANÆ Eigenmann.
Megalamphodus Eigenmann, 49; orthotype M. MEGALOPTERUS Eigenmann.
Microschemobrycon Eigenmann, 56; orthotype M. GUAPORENSIS Eigenmann.
Oligobrycon Eigenmann, 57; orthotype O. MICROSTOMUS Eigenmann.

Aphyocheirodon Eigenmann, 58; orthotype A. HEMIGRAMMUS Eigenmann.
Compsura Eigenmann, 60; orthotype C. HETERURA Eigenmann.
Mixobrycon Eigenmann, 62; orthotype CHEIRODON RIBEIROI Eigenmann.

2332. EIGENMANN (1915). *The Serrasalminæ and Mylinæ.* Ann. Carneg. Mus., IX, 1915, 226–272.

CARL H. EIGENMANN.

Gastropristis Eigenmann, 238; orthotype SERRASALMO TERNETZI Steind.
Rooseveltiella Eigenmann, 240; orthotype SERRASALMO NATTERERI Kner.
Pristobrycon Eigenmann, 245; orthotype SERRASALMO CALMONI Steind.

2333. FOWLER (1915). *Notes on Nematognathous Fishes.* Proc. Acad. Nat. Sci. Phila., LXVII, 223–243.

HENRY WEED FOWLER.

Cataphractops Fowler, 231; orthotype CALLICHTHYS MELAMPTERUS Cope.
Diapaletoplites Fowler, 237; orthotype HYPOPTOPOMA GULARE Cope.

2334. GILBERT (1915). *Fishes Collected by the United States Steamer "Albatross" in Southern California in 1904.* Proc. U. S. Nat. Mus., XLVIII, 305–380.

CHARLES HENRY GILBERT.

Xenognathus Gilbert, 311; orthotype X. PROFUNDORUM Gilbert.
Zastomias Gilbert, 322; orthotype Z. SCINTILLANS Gilbert.
Asterotheca Gilbert, 343; orthotype XENOCHIRUS PENTACANTHUS Gilbert.
Lipariscus Gilbert, 358; orthotype L. NANUS Gilbert.
Lycogramma Gilbert, 364; orthotype MAYNEA BRUNNEA Bean.
Monoceratias Gilbert, 379; orthotype M. ACANTHIAS Gilbert.

2335. HUBBS (1915). *Flounders and Soles from Japan Collected by the United States Bureau of Fisheries Steamer "Albatross" in 1906.* Proc. U. S. Nat. Mus., XLVIII, 449–496.

CARL L. HUBBS (1894–   ).

Citharoides Hubbs, 453; orthotype C. MACROLEPIDOTUS Hubbs.
Psettina Hubbs, 456; orthotype ENGYPROSOPON IIJIMÆ Jordan & Starks.
Læoptichthys Hubbs, 460; orthotype L. FRAGILIS Hubbs.
Gareus Hubbs, 486; orthotype PLEURONECTES OBSCURUS Herzenstein.

2336. McCULLOCH (1915). *Notes on and Descriptions of Australian Fishes.* Proc. Linn. Soc. New S. Wales, XL, 259–277.

ALLAN RIVERSTON McCULLOCH.

Lovettia McCulloch, 259; orthotype HAPLOCHITON SEALII Johnston.

2337. McCULLOCH (1915). *Notes and Illustrations of Queensland Fishes.* Mem. Queensland Mus., III, 1915, 47–56.

ALLAN RIVERSTON McCULLOCH.

Belonepterygion McCulloch, 51; orthotype ACANTHOCLINUS FASCIOLATUS Ogilby.

2338. McCULLOCH (1915). *Report on Some Fishes Obtained by the F. I. S. "Endeavour" on the Coast of Queensland, New South Wales, Victoria, Tasmania, South and South Western Australia, Part III.* Biol. Res. "Endeavour" 3, Pt. III, 97–170.

ALLAN RIVERSTON McCULLOCH.

Acanthopegasus McCulloch, 106; orthotype PEGASUS LANCIFER Kaup.

2339. McCULLOCH & WAITE (1915). *A Revision of the Genus Aracana and Its Allies.* Trans. Roy. Soc. Austral., XXXIX, 1915, 477–493.

ALLAN RIVERSTON McCULLOCH; EDGAR RAVENSWOOD WAITE.

Caprichthys McCulloch & Waite, 482; orthotype C. GYMNURA McCulloch & Waite.

2340. OGILBY (1915). *On Some New or Little Known Australian Fishes.* Mem. Queensland Mus., III, 1915, 117–129.

JAMES DOUGLAS OGILBY.

Reganichthys Ogilby, 123; orthotype R. MAGNIFICUS Ogilby. A synonym of GLAU-COSOMA T. & S.

Rhizoprion Ogilby, 132; orthotype CARCHARIAS CRENIDENS Klunzinger.

2341. PELLÉGRIN (1915). *Les poissons du bassin de l'Ogoué.* Comptes Rend. Ass. France. Adv. Science, XLIII, 500–505.

JACQUES PELLÉGRIN.

Hemistichodus Pellégrin, 500; orthotype H. VAILLANTI Pellégrin.

2342. RIBEIRO (1915). *Fauna Brasiliense, Peixes (Eleutherobranchios aspirophoros).* Archivos do Museo Nacional do Rio Janeiro, XVII, 1915. (In this large volume each family has a separate pagination.)

ALIPIO DE MIRANDA RIBEIRO.

Kronia Ribeiro (ATHERINIDÆ), 9; orthotype K. IGUAPENSIS Ribeiro.

Pseudothyrina Ribeiro (ATHERINIDÆ), 11; orthotype P. JHERINGI Ribeiro.

Toledia Ribeiro (STROMATEIDÆ), 5; orthotype T. MACROPHTHALMA Ribeiro.

Davidia Ribeiro (BALISTIDÆ), 9; orthotype ALUTERA PUNCTATA Ag. Not DAVIDIUS Selys, 1878, in NEUROPTERA, being named for a different person, David Starr Jordan.

Pseudomulloides Ribeiro (MULLIDÆ), 4; orthotype P. CARMINEUS Ribeiro.

Marcgravichthys Ribeiro (BATRACHOIDIDÆ), 3; orthotype BATRACHUS CRYPTOCEN-TRUS Cuv. & Val. Substitute for MARCGRAVIA Jordan, 1886, for the inadequate reason of the use of the name in botany. But both are synonyms of AMPHICH-THYS Sw., 1839.

Parablennius Ribeiro (BLENNIIDÆ), 3; orthotype BLENNIUS PILICORNIS Cuv. & Val.

2343–2344. ROULE (1915). *Sur un nouveau genre de poissons apodes et sur quelques particularités de la biologie de ces êtres.* C. R. Acad. Sci., 160, 1915, 283–284.

LOUIS ROULE.

Pseudophichthys Roule, 283; orthotype P. LATIDORSALIS Roule.

2345. SILVESTER (1915). *Fishes New to the Fauna of Porto Rico.* Washington, Carnegie Inst. Year Book, 14, 1915, 214–217.

CHARLES F. SILVESTER.

Mayerina Silvester, 214; orthotype M. MAYERI Silvester.

2346. SOLDATOV & PAVLENKO (1915). *Two New Genera of Cottidæ from Tartar Strait and Okhotsk Sea.* Ann. Mus. Zool. Ac. Sci. Petrograd, XX, 1915.

V. K. SOLDATOV; M. N. PAVLENKO.

Taurocottus Soldatov & Pavlenko, 149; orthotype T. BERGI Soldatov & Pavlenko.

Trichocottus Soldatov & Pavlenko, 151; orthotype T. BRASHNIKOVI Soldatov & Pavlenko.

2347. STEINDACHNER (1915). *Beiträge zur kenntniss der Flussfische Sudamenkas.* Denks. Kais. Ak. Wiss. Wien, V, 1915.

FRANZ STEINDACHNER.

Æquidens Steindachner, 34; orthotype TETRAGONOPTERUS FASSLII Steind. Name preoccupied by ÆQUIDENS Eigenmann & Bray.

Pseudepterus Steindachner, 82; orthotype AUCHENIPTERUS HASEMANI Steind.

2348. TANAKA (1915). *Ten New Species of Japanese Fishes.* (Text in Japanese). Tokyo Zool. Mag. XXVII, 1915, 556–568.

SHIGEHO TANAKA.

Kanekonia Tanaka, 556; orthotype K. FLORIDA Tanaka. (In Japanese, Nov. 15, 1915; English version Nov. 28, 1918).

Asterorhombus Tanaka, 567; orthotype A. STELLIFER Tanaka.

Scidorhombus Tanaka, 567; orthotype S. PALLIDUS Tanaka.

Lubricogobius Tanaka, 567; orthotype L. EXIGUUS Tanaka.

Henicichthys Tanaka, 568; orthotype H. FORAMINOSUS Tanaka.

Leptoscolopsis Tanaka, XIX; orthotype L. NAGASAKIENSIS Tanaka.

2349. WOODWARD (1915). *The Fossil Fishes of the English Wealden and Purbeck Formations,* I, 1–48. London Mon. Palaeont. Soc.

ARTHUR SMITH WOODWARD.

Hylæobatis Woodward, 19; orthotype H. PROBLEMATICA Woodw. (fossil) (MYLIO-BATIDÆ).

1916.

2350. BERG (1916). *Les poissons des eaux douccs de la Russie.* Moscow. (In Russian.) 1–563.

LEW SEMENOVICH BERG.

Leucalburnus Berg, 292; orthotype PHOXINUS SATUNINI Berg.

Acanthalburnus Berg, 299; orthotype ALBURNUS PUNCTULATUS Kessler.

Capoëtobrama Berg, 316; orthotype ACANTHOBRAMA KUSCHAKEWITSCHI Kessler.

2351. BOGOLIUBOV (1916). (Russian Title.) *Sur l'ichthyodorulite du Poly-rhizus concavus Trd. de Miatchkovo.* Gouv. de Moscou Ann. Geol. Miner. Novo Alexandrija, XVI.

N. N. BOGOLIUBOV.

Paracymatodus Bogoliubov, 197; orthotype CYMATODUS RECLINATUS Trautschold (fossil) (PETALODONTIDÆ). Substitute for CYMATODUS Trautschold (not of Newberry), preoccupied.

2352. BOULENGER (1916). *Catalogue of the Fresh-water Fishes of East Africa.*

GEORGE ALBERT BOULENGER.

Otoperca Boulenger, 130; orthotype LARIMUS AURITUS Cuv. & Val. A synonym of BRACHYDEUTERUS Gill.

Lobochilotes Boulenger, 280; orthotype TILAPIA LABIATA Boulenger.

Liauchenoglanis Boulenger, 314; orthotype L. MACULATUS Boulenger.

Champsochromis Boulenger, 433; orthotype C. CÆRULEA Boulenger.

2353. EIGENMANN (1916). *New and Rare Fishes from South American Rivers.* Ann. Carnegie Mus., X.

CARL H. EIGENMANN.

Cetopsorhamdia Eigenmann, 83; orthotype C. NASUS Eigenmann.

2354. EIGENMANN (1916). *On Apareiodon, a New Genus of Characid Fishes.* Ann. Carnegie Museum, X.

CARL H. EIGENMANN.

Apareiodon Eigenmann, 71; orthotype PARODON PIRACICABÆ Eigenmann.

2355. FOWLER (1916). *Notes on Fishes of the Orders Haplomi and Microcyprini.* Proc. Ac. Nat. Sci. Phila., LXVIII.
HENRY WEED FOWLER.

Galasaccus Fowler, 417; orthotype HYDRARGYRA SIMILIS Baird & Grd. A substitute for HYDRARGYRA as commonly used, HYDRARGYRA SWAMPINA Lac. being a true FUNDULUS.

Chriopeops Fowler, 425; orthotype LUCANIA GOODEI Jordan.

Oxzygonectes Fowler, 425; orthotype HAPLOCHILUS DOWI Gthr.

2356. GILBERT & HUBBS (1916). *Report on the Japanese Macruroid Fishes Collected by the United States Fisheries Steamer "Albatross" in 1901, with a Synopsis of the Genera.* Proc. U. S. Nat. Mus., LI, 135–214.
CHARLES HENRY GILBERT; CARL LEAVITT HUBBS.

Squalogadus Gilbert & Hubbs, 156; orthotype S. MODIFICATUS Gilbert & Hubbs.

2357. HENN (1916). *On Various South American Pœciliid Fishes.* Ann. Carnegie Mus., X, pp. 93–142.
ARTHUR WILBUR HENN.

Diphyacantha Henn, 113; orthotype D. CHOCOËNSIS Henn.

Neoheterandria Henn, 117; orthotype N. ELEGANS Henn.

Phalloptychus Henn, 121; orthotype P. EIGENMANNI Henn.

Phallatorhynus Henn, 126; orthotype P. FASCIOLATUS Henn.

2358. HUBBS (1916). *Notes on the Marine Fishes of California.* Univ. of California Pub. in Zoology, XVI, 1916, 153–169.
CARL LEAVITT HUBBS.

Lestidiops Hubbs, 154; orthotype L. SPHYRÆNOPSIS Hubbs.

2359. JORDAN (1916). *The Nomenclature of American Fishes as Affected by the Opinions of the International Commission on Zoological Nomenclature.* Copeia (April 12, 1916).
DAVID STARR JORDAN.

Sufflamen Jordan, 27; orthotype BALISTES CAPISTRATUS Shaw. Substitute for PACHYNATHUS Sw., preoccupied in Spiders as PACHYGNATHUS.

2360. McCULLOCH (1916). *Ichythyological Items.* Mem. Queensland Mus., V (July 10, 1916).
ALLAN RIVERSTON McCULLOCH.

Phyllichthys McCulloch, 67; orthotype P. PUNCTATUS McCulloch.

2361. MEEK & HILDEBRAND (1916). *The Fishes of the Fresh Waters of Panama.* Field Museum of Natural History, No. 191; X, Dec. 28, 1916.
SETH EUGENE MEEK; SAMUEL F. HILDEBRAND.

Leptoancistrus Meek & Hildebrand, 254; orthotype ACANTHICUS CANENSIS Meek & Hildebrand.

Pseudocheirodon Meek & Hildebrand, 275; orthotype P. AFFINIS Meek & Hildebrand.

Leptophilypnus Meek & Hildebrand, 361; orthotype L. FLUVIATILIS Meek & Hildebrand.

Microeleotris Meek & Hildebrand, 362; orthotype M. AFFINIS Meek and Hildebrand.

Hemieleotris Meek & Hildebrand, 364; orthotype Eleotris latifasciatus Meek & Hildebrand.

2362. OGILBY (1916). *Check List of the Cephalochordates, Selachians and Fishes of Queensland*, Part 1. Mem. Queensl. Mus., 1916.

James Douglas Ogilby.

Dasybatus (Klein; Walbaum) Ogilby, 87; orthotype Rajapastinaca L. Same as Dasyatis Raf. The names reproduced by Walbaum from Klein are rejected by the International Commission.

Cestracion (Walbaum after Klein) Ogilby, 81; orthotype Squalus zygæna L. The generic names quoted by Walbaum in 1792 from Klein are rejected by the International Commission.

2363. OGILBY (1916). *(Læpichthys.)* Mem. Queensland Mus., V.

James Douglas Ogilby.

Læphichthys Ogilby, 173; orthotype Acanthurus rostratus Gthr.

2364. REGAN (1916). *A Revision of the Clupeoid Fishes of the Genera Pomolobus. Brevoortia and Dorosoma and Their Allies.* Ann. Mag. Nat. Hist., 1916, 297 et seq.

Charles Tate Regan.

Paralosa Regan, 167; orthotype Alosa kanagurta Blkr. Name said to be preoccupied (Paralosa Blkr.), replaced by Hilsa Regan, 1916. But I do not find a Paralosa in Bleeker's writings.

Ethmalosa Regan, 302; orthotype Alausa dorsalis Cuv. & Val.

Hilsa Regan, 303; logotype Alosa kanagurta Blkr. Substitute for Paralosa Regan, said to be preoccupied.

Nematalosa Regan, 312; orthotype Clupea nasus Bloch.

Gonialosa Regan, 315; orthotype Chatoëssus modestus Day.

2365. REGAN (1916). *The Morphology of the Cyprinodont fishes of the subfamily Phallostethinæ,* etc. Proc. Zool. Soc. London.

Charles Tate Regan.

Neostethus Regan, 14; logotype N. lankesteri Regan.

2366. RIBEIRO (1916). *Fauna Brasiliensis, Peixes, V.* Archiv. Museu Nacional, XVII. Mullidæ.

Alipio de Miranda Ribeiro.

Pseudomulloides Ribeiro, 3; orthotype P. carmineus Ribeiro.

2367. ROULE (1916). *Notice préliminaire sur quelques espèces nouvelles provenant des croisières de S. A. S. le Prince de Monaco.* Bull. Inst. Oceanographique Monaco.

Louis Roule.

Belonopterois Roule, 13; orthotype B. viridensis Roule.

Barathrites Roule, 17; orthotype B. abyssorum Roule.

Echinomacrurus Roule, 22; orthotype E. mollis Roule.

Bathysolea Roule, 28; orthotype B. albida Roule.

2368. TANAKA (1916). *Nihon san gyorui no san shinshu (Three New Species of Japanese Fish).* Dobuts. Z. Tokyo, XXVII.

Shigeho Tanaka.

Paragyrops Tanaka, 141; orthotype P. editá Tanaka.

Pluviopsetta Tanaka, 141; orthotype P. taniguchi Tanaka.

2369. TANAKA (1916). *Nihon san gyorui no shi shinshu (Two New Species of Japanese Fishes).* Dobuts. Z. Tokyo, XXVIII.

SHIGEHO TANAKA.

Brevigobio Tanaka, 102; orthotype B. KAWABATÆ Tanaka.

2370. THOMPSON (1916). *Fishes Collected by the United States Bureau of Fisheries Steamer "Albatross" during 1888, between Montevideo, Uruguay, and Tome, Chile, on the Voyage Through the Straits of Magellan.* Proc. U. S. Nat. Mus. L. (May 20, 1916).

WILLIAM FRANCIS THOMPSON.

Ethmidium Thompson, 458; orthotype CLUPEA NOTACANTHOIDES Steind. = ALAUSA MACULATA Cuv. & Val.

2371. WAITE (1916). *Fishes, Australasian Antarctic Expedition.* Sci. Rep.

EDGAR RAVENSWOOD WAITE.

Aconichthys Waite, 30; orthotype A. HARRISSONI Waite.

Cygnodraco Waite, 32; orthotype C. MAWSONI Waite.

Dacodraco Waite, 35; type D. HUNTERI Waite.

Notosudis Waite, 56; orthotype N. HAMILTONI Waite.

Aurion Waite, 63; orthotype A. EFFULGENS Waite.

2372. WATSON & DAY (1916). *Notes on Some Palæozoic Fishes.* Mem. Lit. Phil. Doc. Manchester, LX.

DAVID MEREDITH SEARES WATSON; HENRY DAY.

Pentlandia Watson & Day, 34; orthotype DIPTERUS MACROPTERUS Traquair (fossil).

2373. WEBER & BEAUFORT (1916). *The Fishes of the Indo-Australian Archipelago, III, Ostariophysi, Apodes, Synbranchi.*

MAX WEBER; LIEVEN FERDINAND DE BEAUFORT.

Brachydanio Weber & Beaufort, 85; orthotype DANIO ALBOLINEATA Blyth.

Lissochilus Weber & Beaufort, 167; logotype L. SUMATRANUS Weber & Beaufort.

Hemerorhinus Weber & Beaufort, 281; orthotype SPHAGEBRANCHUS HEYNINGI Weber.

Cheiloprion Weber, 342; orthotype POMACENTRUS LABIATUS Day.

2374. WOODWARD (1916). *The Fossil Fishes of the English Wealden and Purbeck Formations, II.* Paleontological Society, LXIX, October, 1916.

ARTHUR SMITH WOODWARD.

Hylæobatis Woodward, 19; orthotype H. PROBLEMATICA Woodw. (fossil).

1917.

2375. EASTMAN (1917). *Fossil Fishes in the United States National Museum.* Proc. U. S. Nat. Mus., 1917. LII.

CHARLES ROCHESTER EASTMAN.

Parafundulus Eastman, 291; orthotype P. NEVADENSIS Eastman (Pleistocene of Nevada fossil).

2376. EIGENMANN (1917). *The American Characidæ.* Mem. Mus. Comp. Zool., XLIII, pt. 1, 1917.

CARL H. EIGENMANN.

Entomolepis Eigenmann, 63; orthotype TETRAGONOPTERUS STEINDACHNERI Eigenmann.

2377. EIGENMANN (1917). *New and Rare Species of South American Siluridæ in the Carnegie Museum.* Ann. Carneg. Mus., XI, 1917.

CARL H. EIGENMANN.

Cheirocerus Eigenmann, 398; orthotype C. EQUES Eigenmann.
Entomocorus Eigenmann, 403; orthotype E. BENJAMINI Eigenmann.

2378. EVERMANN & RADCLIFFE (1917). *Fishes of the West Coast of Peru and the Titicaca Basin.* Bull. U. S. Nat. Mus., XCV, 1917.

BARTON WARREN EVERMANN; LEWIS RADCLIFFE.

Epelytes Evermann & Radcliffe, 71; orthotype E. PUNCTATUS Evermann & Radcliffe.
Xenoscarus Evermann & Radcliffe, 129; orthotype X. DENTICULATUS Evermann & Radcliffe.

2379. JORDAN (1917). *Notes on Glossamia and Related Genera of Cardinal Fishes.* Copeia, May 24, 1917, pp. 46, 47.

DAVID STARR JORDAN.

Nectamia Jordan, 46; orthotype APOGON FUSCUS Quoy & Gaimard.
Xystramia Jordan, 46; orthotype GLOSSAMIA PANDIONIS G. & B.
Zoramia Jordan, 46; orthotype APOGON GRAEFFEI Gthr.

2380. JORDAN (1917). *The Genera of Fishes, etc.* Stanford Univ. Publ., University Series, June 1, 1917.

DAVID STARR JORDAN.

Thrissocles Jordan, 151; orthotype CLUPEA SETIROSTRIS Broussonet. Substitute for THRYSSA Cuv., preoccupied by THRISSA Raf.

2381. JORDAN (1917). *Changes in names of American Fishes.* Copeia, Oct. 4, 1917.

DAVID STARR JORDAN.

Stenesthes Jordan, 87; orthotype SPARUS CHRYSOPS L. = S. ARGYROPS L. Replaces STENOTOMUS Gill, regarded as preoccupied by STENOTOMA.
Xantocles Jordan, 88 (Oct. 4, 1917); orthotype ZANIOLEPIS FRENATUS Eigenmann.
Thoburnia Jordan & Snyder, 88 (Oct. 4, 1917); orthotype CATOSTOMUS RHOTHŒCUS Thoburn.

2382. JORDAN & HUBBS (1917). *Notes on a Collection of Fishes from Port Said, Egypt.* Ann. Carneg. Mus., XI, 461–468.

DAVID STARR JORDAN; CARL LEAVITT HUBBS.

Glaucus (Klein) Jordan & Hubbs, 463; orthotype SCOMBER GLAUCUS L. Same as CÆSIOMORUS Lac. and HYPODIS Raf.
Dacymba Jordan & Hubbs, 464; orthotype PRISTIPOMA BENNETTI Lowe.

2383. NICHOLS & GRISCOM (1917). *Fresh-Water Fishes of the Congo Basin, obtained by the American Congo Expedition, 1909–1915.* Bull. Am. Mus. Nat. Hist., XXXVII.

JOHN TREADWELL NICHOLS; LUDLOW GRISCOM.

Microstomatichthyoborus Nichols & Griscom, 685; orthotype M. BASHFORDDEANI Nichols & Griscom. It is to be hoped that no one will attempt to break this record as to length of generic name.
Amarginops Nichols & Griscom, 713; orthotype A. PLATUS Nichols & Griscom.
Acanthocleithron Nichols & Griscom, 720; orthotype A. CHAPINI Nichols & Griscom.
Gnathobagrus Nichols & Griscom, 711; orthotype G. DEPRESSUS Nichols & Griscom.

2384. REGAN (1917). *A Revision of the Clupeoid Fishes of the Genus Pellonula and of Related Genera in the Rivers of Africa.* Ann. Mag. Nat. Hist., 1917, 198–207.

CHARLES TATE REGAN.

Pœcilothrissa Regan, 201; orthotype P. CONGICA Regan.

Potamothrissa Regan, 203; orthotype PELLONULA OBTUSIROSTRIS Boulenger.

Cynothrissa Regan, 203; orthotype C. MENTO Regan.

Stolothrissa Regan, 206; orthotype S. TANGANICÆ Regan.

Limnothrissa Regan, 207; orthotype PELLONULA MICRODON Boulenger.

2385. SEALE (1917). *New Species of Apodal Fishes.* Bull. Mus. Comp. Zool., LXI.

ALVIN SEALE.

Garmannichthys Seale, 80; orthotype G. DENTATUS Seale.

2386. SNYDER (1917). *The Fishes of the Lahontan System of Nevada and North-eastern California.* Bull. Bur. Fisheries Washington, XXXV (Sept. 28, 1917).

JOHN OTTERBEIN SNYDER.

Leucidius Snyder, 64; orthotype L. PECTINIFER Snyder.

2387–2388. TANAKA (1917). *Figures and Descriptions of the Fishes of Japan, including Riukiu Islands, Bonin Islands, Formosa, Kurile Islands, and Southern Sakhalin.* (Issued in many parts, 1915 to 1918.)

SHIGEHO TANAKA.

Rosanthias Tanaka, 198; orthotype R. AMŒNUS Tanaka. (In Japanese, July 15, 1917; English version. p. 503, Sept. 28, 1918.)

Cyprinocirrhites Tanaka, 368; orthotype C. UI Tanaka (Japanese, Sept. 15, 1917; English version, p. 507, Sept. 28, 1918).

1918.

2389. ANNANDALE (1918). *Fish and Fisheries of the Inlé Lake.* Rec. Ind. Mus., XIV, 33–64.

NELSON ANNANDALE.

Chaudhuria Annandale, 33–64; orthotype C. CAUDATA Annandale. (A MASTACEMBELUS without spines (*fide* Regan, note on CHAUDHURIA. Ann. Mag. Nat. Hist., 198, Feb., 1919.)

2390. EIGENMANN (1918). *(South American Fishes).* Proc. Am. Philos. Soc., LVI, Jan., 1918.

CARL H. EIGENMANN.

Scleronema Eigenmann, 691; orthotype S. OPERCULATUM Eigenmann.

Branchoica Eigenmann, 702; orthotype B. BERTONII Eigenmann.

2391. EIGENMANN (1918). *The Pygidiidæ, a Family of South American Cat-Fishes.* Ann. Carneg. Mus., VII, Sept., 1918, 259–369.

CARL H. EIGENMANN.

Urinophilus Eigenmann, 359; logotype (indicated in Science, 1920, 441) VANDELLIA SANGUINEA Eigenmann.

2392. HUBBS (1918). *A Revision of the Viviparous Perches.* Proc. Biol. Soc. Washington, May 16, 1918.

CARL LEAVITT HUBBS.

Tocichthys Hubbs, 12; orthotype HYPERPROSOPON AGASSIZI Gill.
Amphigonopterus Hubbs, 13; orthotype ABEONA AURORA J. & G.

2393. HUBBS (1918). *Colpichthys, Thyrinops, and Austromenidia, New Genera of Atherinoid Fishes from the New World.* Proc. Ac. Nat. Sci. Phila. for 1917 (1918).
CARL LEAVITT HUBBS.
Colpichthys Hubbs, 305; orthotype ATHERINOPS REGIS Jenkins & Evermann.
Thyrinops Hubbs, 306; orthotype ATHERINICHTHYS PACHYLEPIS Gthr.
Austromenidia Hubbs, 307; orthotype BASILICHTHYS REGILLUS Abbott. Replaces BASILICHTHYS of Abbott, not of Girard.

2394. HUBBS (1918). *Supplementary Notes on Flounders from Japan, with Remarks on the Species of Hippoglossoides.* Ann. Zool Japonensis, IX, pt. 4, 1918, pp. 369–376.
CARL LEAVITT HUBBS.
Tanakius Hubbs, 370; orthotype MICROSTOMUS KITAHARÆ Jordan & Starks. (Not TANAKIA Jordan & Metz.)

2395. McCULLOCH & WAITE (1918). *Records of the South Australian Museum.*
ALLAN RIVERSTON McCULLOCH; EDGAR RAVENSWOOD WAITE.
Helcogramma McCulloch & Waite, 51; orthotype H. DECURRENS McCulloch & Waite.
Trianectes McCulloch & Waite, 53; orthotype T. BUCEPHALUS McCulloch & Waite.
Echinophryne McCulloch & Waite, 86; orthotype E. CRASSISPINA McCulloch & Waite.
Trichophryne McCulloch & Waite, 68; orthotype ANTENNARIUS MITCHELLI Morton.

2396. OGILBY (1918). Ann. Queensland Mus., VI.
JAMES DOUGLAS OGILBY.
Acanthagonia Ogilby, 45; orthotype A. POWERI Ogilby.

2397. RIBEIRO (1918). *Historia Natural. Zoologia: Cichlidæ.* Comm. Linhas Ielegr. de Matto Grosso ao Amazonas.
ALIPIO DE MIRANDA RIBEIRO.
Nannacara Ribeiro, 14; orthotype ACARA DORSIGERA Heckel.
Pseudopercis Ribeiro, 184; orthotype P. NUMIDA Ribeiro.

2398. RIBEIRO (1918). *Dezesetes especies de peixes Brasileiros.* Museu Paulista.
ALIPIO DE MIRANDA RIBEIRO.
Pleurophysus Ribeiro, 8; orthotype P. HYDROSTATICUS Ribeiro.
Tannayia Ribeiro, 14; orthotype T. MARGINATA Ribeiro.
Ceratocheilus Ribeiro, 16; orthotype C. OSTEOMYSTAX Ribeiro.

2399. RIBEIRO (1918). Commerc. Geogr. Geol. Soc. Saõ Paulo.
ALIPIO DE MIRANDA RIBEIRO.
Typhlobagrus Ribeiro, 3; orthotype T. KRONEI Ribeiro.

2400. RIBEIRO (1918). *Lista dos peixes Brasileiros do Museu Paulista.*
ALIPIO DE MIRANDA RIBEIRO.
Paragonus Ribeiro, 3; orthotype P. SERTORII Ribeiro. Name twice preoccupied: Gill, Guichenot; replaced by RIBEIROA Jordan, new name, given in honor of the excellent Brazilian naturalist who made this form known.

Gymnogeophagus Ribeiro, 6; orthotype G. CYANOPTERUS Ribeiro.
Decapogon Ribeiro; orthotype D. UROSTRIATUM Ribeiro.

2401. RIBEIRO (1918). *Dous generos e tres especies novas de peixes Brasileiros,*
. . . Museu Paulista.
ALIPIO DE MIRANDA RIBEIRO.
Homodietus (Ihering) Ribeiro, 23; orthotype H. PAULENSIS Rudolph. Same as
PSEUDOSTEGOPHILUS Ihering, on page 22 of the same paper.
Scleromystax Ribeiro; orthotype S. KRONEI Ribeiro, not of Ogilby.

2402. TANAKA (1918). *Nihon San Gyorui—No Juisshu Shin (Eleven New Spe-
cies of Japanese Fishes).* Journ. Zool. Studies, Imperial University, Tokyo,
XXXIX, No. 13. (In Japanese only.)
SHIGEHO TANAKA.
Vegetichthys Tanaka, 7; orthotype V. TUMIDUS Tanaka. (LUTIANIDÆ.)
Sebastella Tanaka, 10; orthotype S. LITTORALIS Tanaka (SCORPÆNIDÆ).

2403. TANAKA (1918). *Nihon San Gyorui—No Roku Shin (Six New Species
of Japanese Fishes).* Journ Zool. Studies, Imperial University, Tokyo, XXIX,
No. 345. (In Japanese only.)
SHIGEHO TANAKA.
Mustelichthys Tanaka, 301; orthotype M. UI Tanaka (SERRANIDÆ).

2404. TANAKA (1918). *Figures and Descriptions of the Fishes of Japan, includ-
ing Riukiu Islands, Bonin Islands, Formosa, Kurile Islands, Korea and
Southern Sakhalin.* (In English as well as in Japanese.)
SHIGEHO TANAKA.
Selenanthias Tanaka, 516; orthotype S. ANALIS Tanaka (Dec. 28, 1918).
Leptanthias Tanaka, 525; orthotype L. KASHIWÆ Tanaka (Dec. 28, 1918).

2405. WOODWARD (1918). *The Fossil Fishes of the English Wealden and Pur-
beck Formations, Part 2.* Palæontological Society, February, 1918.
ARTHUR SMITH WOODWARD.
Eomesodon Woodward, 56; orthotype MESODON BARNESI Woodw. (fossil).
Enchelyolepis Woodward, 80; orthotype MACROSEMIUS ANDREWSI Woodw. (Fossil).

2406. WOODWARD (1918). *On Two New Elasmobranchs . . . from the
Upper Jurassic Lithographic Stone of Bavaria.* Proc. Zool. Soc. London, 231–
235, June 11, 1918.
ARTHUR SMITH WOODWARD.
Protospinax Woodward, 233; orthotype P. ANNECTANS Woodw.

1919.

2407. COCKERELL (1919). *Some American Cretaceous Fish Scales, with Notes
on the Classification and Distribution of Cretaceous Fishes.* Bull. U. S.
Geol. Surv. for 1918.
THEODORE DRU ALLISON COCKERELL.
Helmintholepis Cockerell, 176; orthotype H. VERMICULATUS Cockerell (fossil
ELOPIDÆ).
Chicolepis Cockerell, 176; orthotype C. PUNCTATUS Cockerell (fossil PTEROTHRISSI-
DÆ).

Leucichthyops Cockerell, 180; orthotype L. VAGANS Cockerell (fossil SALMONIDÆ).

Erythrinolepis Cockerell, 182; orthotype E. MOWRIENSIS Cockerell (fossil ERYTHRI-NOLEPIDÆ).

Halecodon Cockerell, 183; orthotype H. DENTICULATUS Cockerell (fossil ENCHODON-TIDÆ).

Gonorhynchops Cockerell, 183; orthotype G. WOODWARDI Cockerell (fossil GONO-RHYNCHIDÆ: English Chalk).

Centrarchites Cockerell, 188; orthotype C. COLORADENSIS Cockerell (fossil EUCEN-TRARCHIDÆ: basal Eocene).

Benthosphyræna Cockerell, 172; orthotype ALEPOCEPHALUS MACROPTERUS Vaillant. This name, noted in passing by Cockerell, was found by him on specimens in the United States National Museum. It was doubtless a manuscript name of Goode & Bean, never published by them, or possibly of Gill. As the genus seems to be valid, the name is accepted by Cockerell and may date from his reference to it, which reads as follows: "BENTHOSPHYRÆNA MACROPTERA, also referred to the ALEPOCEPHALIDÆ, has small round scales wholly unlike those of ALEPOCEPHALUS." That BENTHOSPHYRÆNA was an unpublished name was not known to Cockerell when he examined the scales of its type species.

2408. EIGENMANN (1919). *Trogloglanis pattersoni, a New Blind Fish from San Antonio, Texas.* Proc. Am. Philos. Soc., LVIII, 397–400.

CARL H. EIGENMANN.

Trogloglanis Eigenmann, 397; orthotype T. PATTERSONI Eigenmann (a blind cave representative of the genus RABIDA).

2409. EIGENMANN & EIGENMANN (1919). *Steindachneridion.* Science, November, 1919.

CARL H. EIGENMANN; ROSA SMITH EIGENMANN.

Steindachneridion Eigenmann & Eigenmann; logotype STEINDACHNERIA AMBLY-URA Eigenmann & Eigenmann (SILURIDÆ). Substitute for STEINDACHNERIA Eigenmann & Eigenmann, 1888, preoccupied in MACROURIDÆ.

2410. HUSSAKOF & BRYANT (1919). *Catalog of the Fossil Fishes in the Museum of the Buffalo Society of Natural Sciences.* Bull. Buffalo Soc. Nat. Sci., Vol. XII.

LOUIS HUSSAKOF; WILLIAM L. BRYANT.

Perissognathus Hussakof & Bryant, 81; orthotype P. ADUNCUS (based on a mandible) (fossil ARTHRODIRA, Upper Devonic, New York).

Machærognathus Hussakof & Bryant, 83; orthotype M. WOODWARDI (fossil mandible ARTHRODIRA, Upper Devonic, New York).

Copanognathus Hussakof & Bryant, 84; orthotype C. CRASSUS (fossil mandible ARTHRODIRA, Upper Devonic, New York).

Deinodus Hussakof & Bryant, 123; orthotype D. BENNETTI (fossil dental plate PTYCTODONTIDÆ, Upper Devonic, New York).

Acmoniodus Hussakof & Bryant, 151; orthotype A. CLARKEI (fossil dental plate ELASMOBRANCHI, Upper Devonic, New York).

Atopacanthus Hussakof & Bryant, 157; orthotype A. DENTATUS (fossil spine ICHTHYODORULITES, Upper Devonic, New York).

2411. JORDAN (1919). *On Certain Genera of Atherine Fishes.* Proc. U. S. Nat. Mus., LV, 309–311.

DAVID STARR JORDAN.

Hubbesia Jordan, 310; orthotype MENIDIA GILBERTI Jordan & Bollman.

2412. JORDAN (1919). *Note on Gistel's Genera of Fishes.* Proc. Ac. Nat. Sci. Phila., April 10, 1919, pp. 335–340.

DAVID STARR JORDAN.

Urocles Jordan, 338, 343; orthotype M. LEPIDOTUS Ag. (fossil). Substitute for MEGALURUS Ag. and SYNERGUS Gistel, both names being preoccupied.

2413. JORDAN (1919). *New Genera of Fishes.* Proc. Ac. Nat. Sci. Phila., April 10, 1919, pp. 341–344.

DAVID STARR JORDAN.

Aoria Jordan, 341; orthotype BAGRUS LAMARII Cuv. & Val. Substitute for MACRONES Duméril, preoccupied (Newman, 1841, a genus of Beetles).

Azurella Jordan, 341; orthotype POMACENTRUS BAIRDI Gill.

Cotylichthys Jordan, 341; orthotype COTYLIS FIMBRIATA M. & T. Same as COTYLIS Gthr., not of Müller & Troschel, whose original type is a species of GOBIESOX Lac.

Extrarius Jordan, 341; orthotype HYBOPSIS TETRANEMUS Gilbert.

Eperlanio Jordan, 341; orthotype OSMERUS ALBATROSSIS J. & G.

Erythrocles Jordan, 341; orthotype ERYTHRICHTHYS SCHLEGELI Blkr. Substitute for ERYTHRICHTHYS T. & S., preoccupied by ERYTHRICHTHYS Bon.

Nautopædium Jordan, 342; orthotype PORICHTHYS PLECTRODON J. & G. = BATRACHUS POROSISSIMUS Cuv. & Val.

Irillion* Jordan, 342; orthotype COREGONUS OREGONIUS Jordan & Snyder.

Oshimia Jordan, 342; orthotype MICRACANTHUS MARCHEI Sauvage. Name a substitute for MICRACANTHUS Sauvage, 1878, preoccupied by MICROCANTHUS Sw. Named for Masamitsu Oshima, of the Bureau of Scientific Research of Formosa.

Rheocles Jordan & Hubbs, 343; orthotype ELEOTRIS SIKORÆ Sauvage.

Syletor Jordan, 343; orthotype PISODONOPHIS CRUENTIFER G. & B.

Unagius Jordan, 343; orthotype CRYPTOPHTHALMUS ROBUSTUS Franz. Substitute for CRYPTOPHTHALMUS Franz, preoccupied.

Verater Jordan, 343; orthotype OLIGOPUS ATER Risso. Substitute for PTERIDIUM Filippi & Verany, preoccupied by PTERIDIUM Scopoli.

Pnictes Jordan, 343; orthotype ACHIROPSIS ASPHYXIATUS Jordan & Goss. This genus differs from all other soles in having the gill opening on the right side completely closed.

Errex Jordan, 343; orthotype GLYPTOCEPHALUS ZACHIRUS Lockington.

Ainia Jordan, 344; orthotype CALLOPTERUS AGASSIZI Thiollière (fossil). Name a substitute for CALLOPTERUS, preoccupied in beetles; name from the locality Ain, in France.

Raiamas Jordan; orthotype CYPRINUS BOLA Ham. "Rajah Mas" of the anglers in India. Replaces BOLA Gthr., preoccupied.

Tarsistes Jordan, 344; orthotype RHYNCHOBATIS Philippi, 1858. No specific name. Generic name preoccupied as RHYNCHOBATUS. Species named TARSISTES PHILIPPII Jordan.

Vigil Jordan, 344; orthotype PLEUROLEPIS PELLUCIDUS Ag. A substitute for PLEUROLEPIS Ag., preoccupied in fossil fishes.

Orqueta Jordan, 344; orthotype MICROPTERYX (MICROPUS) POLYCENTRUS Kner. A substitute for MICROPUS Kner, four times preoccupied.

---

* "The merry wild Irillion rejoicing from fields of snow" (Dunsaney).

2414. JORDAN (1919). *Fossil Fishes of Southern California, I. Fossil Fishes of the Soledad Deposits* . . . pp. 1–12. Leland Stanford Junior University Publications, University Series, September 18, 1919.

DAVID STARR JORDAN.

Ganolytes Jordan, 6; orthotype G. CAMEO Jordan (fossil).

Rogenites Jordan, 8; orthotype ROGENIO BOWERSI Jordan (fossil).

Rhomurus Jordan, 9; orthotype R. FULCRATUS Jordan (fossil).

Auxides Jordan, 10; orthotype THYNNUS PROPTERYGIUS Ag. (fossil).

Bulbiceps Jordan, 12; orthotype B. RANINUS Jordan (fossil).

Eoperca Jordan, 12; orthotype MIOPLOSUS MULTIDENTATUS Cope (fossil). An Eocene ally of PERCA, differing from MIOPLOSUS in the presence of 12 dorsal spines and in the stronger and more numerous retrorse serrations of the preopercle.

2415. JORDAN (1919). *Description of a New Fossil Fish from Japan.* Proc. Cal. Ac. Sci, IX, 271–272. October 22, 1919.

DAVID STARR JORDAN.

Iquius Jordan, 271; orthotype I. NIPPONICUS Jordan (fossil CLUPEID).

2416. JORDAN (1919). *The Genera of Fishes, Part III.* Leland Stanford Junior University Publications, University Series, October 25, 1919.

DAVID STARR JORDAN.

Laides Jordan, 293; orthotype PANGASIUS HEXANEMA Blkr. Substitute for LAIS Blkr., 1860, preoccupied by LAIS Gistel, VIII, 1849, a tunicate, this being a substitute name for CYNTHIA, preoccupied.

2417. JORDAN (1919). *On Elephenor, a New Genus of Fishes from Japan.* Ann. Carneg. Museum, XII, December 16, 1919, pp. 329–343.

DAVID STARR JORDAN.

Elephenor Jordan, 328; orthotype PTERACLIS MACROPUS Bellotti.

2418. JORDAN & GILBERT (J. Z.) (1919). *Fossil Fishes of Southern California, II. Fossil Fishes of the (Miocene) Monterey Formations of Southern California.* Leland Stanford Junior University Publications, University Series, September 18, 1919, pp. 13–60.

DAVID STARR JORDAN; JAMES ZACCHEUS GILBERT.

Xyne Jordan & Gilbert (J. Z.), 25; orthotype X. GREX J. & G. (fossil CLUPEIDÆ).

Ellimmichthys Jordan, 27; orthotype DIPLOMYSTES LONGICOSTATUS Cope (fossil CLUPEID, Eocene, Brazil).

Alisea Jordan & Gilbert (J. Z), 26; orthotype A. GRANDIS J. & G. (fossil CLUPEID).

Smithites Jordan & Gilbert (J. Z.), 30; orthotype S. ELEGANS J. & G. (fossil DUSSUMIERIID).

Quæsita Jordan & Gilbert (J. Z.), 30; orthotype ·Q. QUISQUILIA J. & G. (fossil DUSSUMIERIID).

Azalois Jordan & Gilbert (J. Z.), 32; orthotype A. ANGELENSIS J. & G. (fossil DUSSUMIERIID).

Lygisma Jordan & Gilbert (J. Z.), 33; orthotype L. TENAX J. & G. (fossil DUSSUMIERIID).

Forfex Jordan, 36; orthotype F. HYPURALIS Jordan (fossil SCOMBERESOCID).

Zanteclites Jordan & Gilbert (J. Z.), 39; orthotype Z. HUBBSI J. & G. (fossil ATHERINID).

Eritima Jordan & Gilbert (J. Z.), 40; orthotype E. EVIDES J. & G. (fossil APOGONID).

Tunita Jordan & Gilbert (J. Z.), 42; orthotype T. OCTAVIA J. & G. (fossil mackerel).

Ozymandias Jordan, 43; orthotype O. GILBERTI Jordan (fossil of unknown family, probably SCOMBRID).

Lophar Jordan & Gilbert (J. Z.), 44; orthotype L. MIOCÆNUS J. & G. (fossil POMATOMID).

Eclipes Jordan & Gilbert (J. Z.), 47; orthotype E. VETERNUS J. & G. (fossil MERLU-CIID). Same as MERRIAMINA Jordan & J. Z. Gilbert.

Lompoquia Jordan & Gilbert (J. Z.), 49; orthotype L. RETROPES J. & G. (fossil SCIÆNOID).

Xyrinius Jordan & Gilbert (J. Z.), 50; orthotype X. HOUSHI J. & G. (fossil CLUPEID, mistaken for a LABROID).

Sebastavus Jordan & Gilbert (J. Z.), 50; orthotype S. VERTEBRALIS J. & G. (fossil SCORPÆNOID).

Rhomarchus Jordan & Gilbert (J. Z.), 51; orthotype R. ENSIGER J. & G. (fossil SCORPÆNOID).

Eoscorpius Jordan & Gilbert (J. Z.), 53; orthotype E. PRIMÆVUS J. & G. (fossil of uncertain relation, probably COTTOID).

Hayia Jordan & Gilbert (J. Z.), 55; orthotype H. DAULICA J. & G. (fossil of uncertain relations, probably COTTOID).

Merriamina Jordan & Gilbert (J. Z.), 56; orthotype M. ECTENES J. & G. (fossil MERLUCIID). A synonym of ECLIPES Jordan & J. Z. Gilbert.

Diatomœca Jordan & Gilbert (J. Z.), 58; orthotype D. ZATIMA Jordan & Gilbert (fossil closely allied to LAMPRIS).

Emmachære Jordan & Gilbert (J. Z.), 59; orthotype E. RHACHITES J. & G. (fossil SERRANID).

2419. JORDAN & GILBERT (J. Z.), (1919). *Fossil Fishes of Southern California, III. Fossil Fishes of the Pliocene Formations* . . . Leland Stanford Junior University Publ., University Series, Sept. 18, 1919, pp. 61–64.

DAVID STARR JORDAN; JAMES ZACCHEUS GILBERT.

Ectasis Jordan & Gilbert (J. Z.), 62; orthotype E. PRORIGER J. & G. (fossil, perhaps ELOPID).

Arnoldina Jordan & Gilbert (J. Z.), 63; orthotype A. INIISTIA J. & G. (fossil GADOID).

2420. JORDAN & HUBBS (1919). *Studies in Ichthyology: A Monographic Review of the Family of Atherinidæ or Silversides.* Leland Stanford Junior University Publications, Dec. 18, 1919.

DAVID STARR JORDAN; CARL LEAVITT HUBBS.

Archomenidia Jordan & Hubbs, 54; type ATHERINICHTHYS SALLEI Regan.

2421. OGILBY (1919). *(Tropidocaulus.)*

JAMES DOUGLAS OGILBY.

Tropidocaulus Ogilby; orthotype BELONE PLATYURA Cuv. & Val. Substitute for EURYCAULUS Ogilby, preoccupied.

2422. NICHOLS (1919). *Um novo genero do Cascudos do familia Loricariidæ (with English translation).* Revista do Mus. Paulista, XI.

JOHN TREADWELL NICHOLS.

Pseudotocinclus Nichols, 9; orthotype P. INTERMEDIUS Nichols.

2423. OSHIMA (1919). *Contributions to the Study of the Fresh-Water Fishes of the Island of Formosa.* Ann. Carneg. Museum, Pittsburgh, XII, Dec. 16, 1919, pp. 169–328.

MASAMITSU OSHIMA (1884–   ).

Formosania Oshima, 194; orthotype F. GILBERTI Oshima.
Acrossochilus Oshima, 206; orthotype GYMNOSTOMUS FORMOSANUS Regan.
Scaphesthes Oshima, 208; orthotype S. TAMUSIUENSIS Oshima.
Spinibarbus Oshima, 217; orthotype S. HOLLANDI Oshima.
Phoxiscus Oshima, 225; orthotype P. KIKUCHII Oshima.
Aristichthys Oshima, 246; orthotype LEUCISCUS NOBILIS (Gray), Richardson.
Cultriculus Oshima, 252; orthotype CULTER LEUCISCULUS Kner, not Richardson; = HEMICULTER KNERI Kreyenberg.

2424. TANAKA (1919). *Figures and Descriptions of the Fishes of Japan, including Riukiu Islands, Bonin Islands, Formosa, Kurile Islands, Korea and Southern Sakhaliu.* XXIX.

SHIGEHO TANAKA.

Selenanthias Tanaka, 516; orthotype S. ANALIS Tanaka.
Leptanthias Tanaka, 525; orthotype L. KASHIWÆ Tanaka.

1920.

2425. COCKERELL (1920). *(Characilepis).* Proc. U. S. Nat. Mus., 1920 (in press).

THEODORE DRU ALLISON COCKERELL.

Characilepis Cockerell; orthotype C. TRIPARTITUS Cockerell (fossil scales, Miocene of Peru).

2426. GILBERT & HUBBS (1920). *The Macrouroid Fishes of the Philippine Islands and the East Indies.* Bull. U. S. Bureau of Fisheries, 1919 (in press; not yet published).

CHARLES. HENRY GILBERT; CARL LEAVITT HUBBS.

Quincuncia Gilbert & Hubbs; orthotype CŒLORHYNCHUS ARGENTATUS Smith & Radcliffe.
Oxygadus Gilbert & Hubbs; orthotype MACRURUS PARALLELUS Gthr.
Hymenogadus Gilbert & Hubbs; orthotype HYMENOCEPHALUS GRACILIS Hubbs.
Papyrocephalus Gilbert & Hubbs; orthotype HYMENOCEPHALUS ATERRIMUS Gilbert.
Ventrifossa Gilbert & Hubbs; orthotype CORYPHÆNOIDES GARMANI J. & G. (Japan).
Atherodon Gilbert & Hubbs; orthotype OPTONURUS ATHERODON Gilbert & Cramer.
Lucigadella Gilbert & Hubbs; orthotype MACROURUS NIGROMARGINATUS Smith & Radcliffe.
Lucigadus Gilbert & Hubbs; orthotype MACROURUS LUCIFER Smith & Radcliffe.
Hyostomus Gilbert & Hubbs; orthotype MACRURUS HYOSTOMUS Smith & Radcliffe.

2427. JORDAN (1920). *New Genera of Fossil Fishes from Brazil.* Proc. Acad. Nat. Sci. Phila. for 1919 (March 11, 1920), pp. 208–210.

DAVID STARR JORDAN.

Ennelichthys Jordan, 208; orthotype E. DERBYI Jordan (fossil).
Brannerion Jordan, 209; orthotype CALAMOPLEURUS VESTITUS Jordan & Branner (fossil).
Vinctifer Jordan, 210; orthotype BELONOSTOMUS COMPTONI Ag. (fossil).

2428. JORDAN (1920). *The Genera of Fishes, part IV*.

DAVID STARR JORDAN.

Greeleya Jordan, 493; orthotype EXIMIA RUBELLIO Greeley. Substitute for EXIMIA Greeley, preoccupied.

Ribeiroa Jordan, 564; orthotype PARAGONUS SERTORII Ribeiro; substitute for PARA-GONUS Ribeiro, preoccupied.

Quisque Jordan, 571; orthotype Q. GILBERTI Jordan (a small herring from the Miocene of Elmodena, California, with very strong serræ on the belly, and the back also armed; body compressed; vertebræ, 32).

Ganoëssus Jordan, 571; orthotype GANOLYTES CLEPSYDRA J. & G. (fossil). A large herring, with enameled scales and hour-glass shaped vertebræ, differing from GANOLYTES in the much thinner scales. Vertebræ, 42.

Scomberessus Jordan, 571; orthotype SCOMBERESOX ACUTILLUS J. & G. (fossil). Differs from the living genus SCOMBERESOX in the much larger dorsal, of 16 rays.

Aristoscion Jordan, 571; orthotype A. EMPREPES Jordan. (A robust fossil SCIÆNOID, from the Miocene at Lompoc, allied to CYNOSCION, but the anal rays more numerous, 1.15 in number, the dorsal rays about XII — 1.20. and the caudal fin very large. Vertebræ 14 + 10 or 11.)

Beltion Jordan, 571; orthotype B. PERONIDES Jordan (fossil). (BELONIDÆ, distinguished from BELONE by needle-like jaws, moderate, even teeth, and larger scales; Miocene, Lompoc, California.)

Sebastoëssus Jordan, 571; orthotype S. APOSTATES Jordan (fossil). SCORPÆNIDÆ, near SEBASTODES, but the interneurals all slender and not dilated; fin spines moderate, the second anal spine not enlarged; vertebræ, 24; Miocene, Lompoc, California.

Sebastinus Jordan, 571; orthotype RIXATOR INEZIÆ J. & G. (fossil). Miocene SCORPÆNIDÆ; allied to SEBASTODES Gill, but with undilated interneural spines, and the second anal spine very elongate.

2429. JORDAN & GILBERT (1920). *The Fossil Fishes of the Diatom Beds of Lompoc, California*. Leland Stanford Junior University Publications, February, 1920.

DAVID STARR JORDAN; JAMES ZACCHEUS GILBERT.

Turio Jordan & Gilbert, 15; orthotype T. WILBURI J. & G. (fossil SCOMBRID).

Thyrsocles Jordan & Gilbert, 14, 19; orthotype THYRSITES KRIEGERI J. & G. (fossil SCOMBRID).

Thyrsion Jordan, 17; orthotype T. VELOX Jordan (fossil SCOMBRID). A synonym of THYRSOCLES J. & G.

Zaphleges Jordan, 23; orthotype Z. LONGURIO Jordan (fossil SCOMBRID).

Ocystias Jordan, 18; orthotype O. SAGITTA Jordan (fossil SCOMBRID).

Zelosis Jordan & Gilbert, 24; orthotype CLUPEA HADLEYI J. & G. (fossil, perhaps ALEPOCEPHALID).

Aræosteus Jordan & Gilbert, 25; orthotype A. ROTHI J. & G. (fossil, apparently ZAPRORID).

Rhythmias Jordan & Gilbert, 27; orthotype R. STARRII J. & G. (fossil SPARID).

Plectrites Jordan, 28; orthotype P. CLASSENI Jordan (fossil SPARID).

Lompochites Jordan, 29; orthotype L HOPKINSI Jordan (fossil, apparently CARANGOID).

Rixator Jordan & Gilbert, 31; orthotype R. PORTEOUSI J. & G. (fossil SCORPÆNID).

Zororhombus Jordan, 39; orthotype Z. VELIGER Jordan (fossil PLEURONECTID).

Atkinsonella Jordan, 41; orthotype A. STRIGILIS Jordan (fossil of uncertain relationship, apparently SPARID).

2431. OSHIMA (1920). *Notes on Fresh-Water Fishes of Formosa, with Descriptions of New Genera and Species.* Proc. Ac. Nat. Sci. Phila., 1920 (July 6).

MASAMITSU OSHIMA.

Lissocheilichthys Oshima, 124; orthotype L. MATSUDAI Oshima.
Scaphiodontella Oshima, 125; orthotype S. ALTICORPUS Oshima.
Leucisculus Oshima, 128; orthotype L. FUSCUS Oshima.
Rasborinus Oshima, 130; orthotype R. TAKAKAI Oshima.

2432. REGAN (1920). *The Classification of the Fishes of the Family Cichlidæ, I. The Tanganyika Genera.* Ann. Mag. Nat. Hist., V, January 1920.

CHARLES TATE REGAN.

Tylochromis Regan, 37; orthotype PELMATOCHROMIS JENTINKI Steind.
Heterotilapia Regan, 38; orthotype TILAPIA BUETTIKOFERI Boulenger.
Neotilapia Regan, 38; orthotype CHROMIS TANGANICÆ Gthr.
Otopharynx Regan, 38; orthotype TILAPIA AUROMARGINATA Boulenger.
Limnotilapia Regan, 39; orthotype TILAPIA DARDENNII Boulenger.
Cyathopharynx Regan, 42; orthotype TILAPIA GRANDOCULIS Boulenger.
Limnochromis Regan, 43; orthotype PELMATOCHROMIS AURITUS Boulenger.
Cyphotilapia Regan, 43; orthotype PELMATOCHROMIS FRONTOSUS Boulenger.
Lipochromis Regan, 45; orthotype PELMATOCHROMIS OBESUS Boulenger.
Neochromis Regan, 45; orthotype TILAPIA SIMOTES Boulenger.
Cnestrostoma Regan, 45; orthotype PARATILAPIA POLYODON Boulenger.
Mylochromis Regan, 45; orthotype TILAPIA LATERISTRIGA Gthr.
Sargochromis Regan, 45; orthotype PARATILAPIA CODRINGTONI Boulenger.
Labrochromis Regan, 45; orthotype TILAPIA PALLIDA Boulenger.
Serranochromis Regan, 45; orthotype CHROMYS THUNBERGI Castelnau.
Clinodon Regan, 45; orthotype HEMITILAPIA BAYONI Boulenger.
Callochromis Regan, 46; orthotype PELMATOCHROMIS MACROPS Boulenger.
Leptochromis Regan, 47; orthotype PARATILAPIA CALLIURA Boulenger.
Aulonocranus Regan, 47; orthotype PARATILAPIA DEWINDTI Boulenger.
Hemibates Regan, 49; orthotype PARATILAPIA STENOSOMA Boulenger.

2433. NICHOLS (1920). *A Contribution to the Ichthyology of Bermuda.* Proc. Biol. Soc. Wash., XXXIII, July 24, 1920, 59–64.

JOHN TREADWELL NICHOLS.

Bermudichthys Nichols, 62; orthotype B. SUBFURCATUS Nichols.

ADDITIONS AND CORRECTIONS

1817.

91. CUVIER (1817). *Règne Animal.* Ed. 1.

Myliobatis (Duméril) Cuvier, 137; logotype RAJA AQUILA L. As originally proposed MYLIOBATIS Duméril contained the same species as AËTOBATUS of Blainville (1916). A different type was assigned to the two by Müller & Henle (1838), who adopted both names, assigning an unwonted type to Blainville's genus. In strictness, MYLIOBATIS should be regarded as a synonym of AËTOBATUS and with RAJA AQUILA as the type of both, while RAJA NARINARI should stand as STOASODON Cantor.

1819.

103. RAFINESQUE (1819). *Prodrome,* etc.

Lepomis. The names LEPIDOPOMUS and LEPIOPOMUS have been produced by puristic emendations of this word. Other revisions of Rafinesque's "telescoped" Greek terms were proposed in 1878, as MYXOSTOMA for MOXOSTOMA, HOPLADELUS for OPLADELUS, ICHTHÆLURUS for ICTALURUS, ICHTHYOBUS for ICTIOBUS, PELODICHTHYS for PILODICTIS, etc.

1829.

136B. BANCROFT (1829). *Sea Devil,* etc.

Manta Bancroft, 444; orthotype M. AMERICANA Bancroft = RAIA BIROSTRIS Walbaum.

1831.

146A. CUVIER & VALENCIENNES (1831). *Histoire Naturelle des Poissons, VIII.*

Pelamys Cuvier & Valenciennes, 149; orthotype SCOMBER SARDA L. A synonym of SARDA Cuv.

Scorpis Cuvier & Valenciennes, 503; orthotype S. GEORGIANUS Cuv. & Val.

1837.

217A. MÜLLER & HENLE (1837). *Ueber die Gattungen der Plagiostomen.* Arch. Naturg., III.

JOHANNES MÜLLER; FREDERICH GUSTAV JAKOB HENLE.

Mr. A. R. McCulloch calls my attention to this paper, overlooked by me, in which the names GALEOCERDO, RHYNCHOBATUS, PRISTIOPHORUS, STEGOSTOMA appear.

The name TRYGONORRHINA was first used by Müller & Henle (Mag. Nat. Hist. 2, II, 90) and spelled with the double "r" as above.

Galeocerdo Müller & Henle, 308; orthotype G. ARCTICUS M. & H., not G. TIGRINUS as later given. See p. 192, Genera of Fishes, II.

Page 190, No. 217. The name GYMNURA appears not in the paper quoted, the date of which should be 1838, but in 217A, p. 400.

1838.

217A. MÜLLER & HENLE (1838). Mr. McCulloch informs me that a second article by Müller & Henle, dated Dec. 12, 1837, printed January, 1838, bearing the same title as No. 217 of our series, contains three additional genera:

Pristiurus Müller & Henle, 34; orthotype GALEUS MELASTOMUS Raf. This antedates Bonaparte's use of the same term (1841).

Leptocharias Müller & Henle, 36; orthotype TRIÆNODON SMITHI Gray.

Carcharodon Müller & Henle, 37; orthotype C. RONDELETI M. & H. = SQUALUS CARCHARIAS L. According to McCulloch, the name CARCHARODON is not used by Smith. The earliest reference to CARCHARODON is apparently that of Müller & Henle (Mag. Nat. Hist. 2, II, 1838, 37, January). Those authors saw Smith's specimens (*vide* p. 33) and perhaps heard his paper read at the Zoological Society in September, 1837. They therefore quoted the name as of Smith. But Smith's paper was not published until February 13, 1838, he having read Müller & Henle's paper "about a month previously, in which CARCHARODON appeared, and he evidently withdrew it with names he refers to in his paper" (McCulloch, *in lit.,* June 28, 1920).

1844.

316. McCLELLAND (1844). *Apodal Fishes.*

Strophidon McClelland, 187; logotype MURÆNA SATHETE Ham.

1846.

349A. COCCO (1846). *Phanerobranchus.* Georn. Gab. Lett. Messina Ann., V, VIII, 63–64.

ANASTASIO COCCO.

Phanerobranchus Cocco, 63; orthotype P. KROHNII Cocco, an undetermined larva; name preoccupied.

1849.

417. SMITH (1849). *Zoölogy of South Africa.*

Tilapia Smith, pl. 5; haplotype T. SPARRMANNI Smith. (Not CHROMIS NILOTICA as stated by error on page 244.) The last-named species is referred by Regan to SAROTHERODON.

1854.

No. 495, p. 256, should read:

495. BRANDT & LOWE (1854). *Description d'un nouveau genre de poisson de la famille des Murénoides.* Mém. Sav. Étrang. St. Petersburg, VII, 169–176.

JOHANN FRIEDRICH BRANDT; RICHARD THOMAS LOWE.

Belonopsis Brandt & Lowe, 174, etc.

1855.

533. COSTA ("1855") should be omitted and transferred to 850 (1864), the proper title being as follows:

533A. COSTA (1855). *(Cyrtorhynchus.)* Cos. Rendic. dell Accad. Pontaniana, 1855.

ORONZIO GABRIELE COSTA.

Cyrtorhynchus Costa; orthotype C. LEOPOLDI Costa, said to be a MYCTOPHID, of uncertain relations; figured by Costa, *Fauna del regno di Napoli.*

1858.

607–612. COSTA (1858). *(Uraleptus.)* This reference should be canceled, according to Mr. Henn. The genus URALEPTUS Costa should appear under No. 350, *Fauna del regno di Napoli* (1846, p. 30). GADELLA Lowe (1843) is still earlier.

612. COSTA (1858). *Luspia Casotti, nuovo genere di pesci fossili della calcarea tenera leccese, Napoli.*

ORONZIO GABRIELE COSTA.

Luspia Costa; orthotype L. CASOTTI Costa (fossil). An ally of SERRANUS Cuv.

1861.

740. GÜNTHER (1861). *Catalogue of Fishes, III.*

Cyttus Günther, 396; orthotype Capros australis Rich.

1862.

782. GILL (1862). The genera Lepidopsetta and Hypsopsetta date from this paper, page 330, not from 1864.

788. GÜNTHER (1862). *Catalogue, IV.*

Atypichthys, 510; orthotype Atypus strigatus Gthr.; substitute for Atypus Gthr., preoccupied.

792. Estheria Jones (1862) should be omitted, as Estheria middendorfi is not a fish but an Entomostracan, according to A. W. Henn. The name is twice preoccupied.

1863.

830A. JOHNSON (1863). *Description of Five New Species of Fish Obtained at Madeira.* Proc. Zool. Soc. London, 36–44.

James Yates Johnson

Neoscopelus Johnson, 44; orthotype N. macrolepidotus Johnson.

1864.

850. COSTA (1864). *Paleontologia del regno di Napoli, III.* Atti. Accad. Pontan. Napoli, VIII.

Oronzio Gabriele Costa.

Cæus Costa, 65; orthotype C. leopoldi Costa (fossil: indeterminate Elopidæ). Misprinted Cacus by authors.

1865.

892. COSTA (1865), page 336. The name Heterolepis Costa occurs in the volume quoted, but in a different paper, *Nuove osservazioni i terreni ad ittioliti delle provincie Napolitane,* etc.

1872.

1060. CASTELNAU (1872). *Ichthyology of Australia.*

Cæsioperca Castelnau, 49; orthotype Serranus rasor Rich.

Pseudaphritis Castelnau, 92; orthotype P. bassi Castelnau.

1878.

1234. CLARKE (1878). Read:

Ægeonichthys Clarke, 245; orthotype A. appelii Clarke.

1243. JORDAN (1878). *Manual of Vertebrates, 2d ed.*

Cristivomer Gill & Jordan, 80; type Salmo namaycush Walbaum. Here first defined; in No. 1244, the word occurs on page 89 as name only.

1880.

1310. KLUNZINGER (1880). *Müllersche Sammlung.*

Colpognathus Klunzinger, 339; orthotype Plectropoma dentex Cuv. & Val.

1884.

1423A. GIGLIOLI & ISSEL (1884). *Pelagos, Saggi sulta vita e sui prodotti del mare, Genova.*

Enrico Hillyer Giglioli; Raffaele Issel.

Hymenocephalus Giglioli, 228; orthotype H. italicus Giglioli.

1888.

1543. VAILLANT, p. 444.

Neostoma and **Eustomias** should be omitted. Both genera are of Filhol, 1884.

1894.

1689. BEAN, p. 463. Omit Cyclopsetta, a genus of Gill, of 1888.

1895.

1720. JORDAN (1895), p. 469. Add:

Garmannia Jordan, 495; orthotype Gobius paradoxus Gthr.

1896.

Omit 1755, p. 476; duplicated.

1898

1780. BOULENGER (1898). *Lake Tanganyika,* etc. Add:

Ectodus Boulenger, 21; logotype E. descampsi Boulenger.

Clariallabes Boulenger (1900); orthotype C. melas Boulenger. This generic
name, omitted in the Zoological Record, I have failed to locate.

1900.

1848, etc. The name of Jacques Pellegrin should be spelled without the acute accent.

1901.

1883. TRAQUAIR (1901). *Notes,* etc.

Cœlacanthopsis Traquair, 113; orthotype C. curta Traquair (fossil).

1903.

1938. GILCHRIST. Cyttosoma should be omitted (see No. 1977).

1904.

1975. FOWLER.

Boulengerina, inserted here by error, should be omitted. Its type is Dules mato
Lesson. That of Safole (No. 2084) is Dules tæniurus.

Omit 1993, p. 510; duplicated.

1905.

2003. BLAKE (1905). *Monograph of the Fauna of the Cornbrash, Pisces.* Pal.
Soc. Mon., LIX, 1–100.

J. F. Blake.

Macromesodon Blake, 32; orthotype Gyrodus macropterus Ag. Substitute for
Mesodon Wagner, preoccupied. A synonym of Typodus Quenstedt.

2018. JORDAN & EVERMANN (1905). *Aquatic Resources,* etc. Add:

Jenkinsiella Jordan & Evermann, 83; orthotype Microdonophis macgregori
Evermann.

1913.

Omit 2282, p. 549; duplicated.

1915.

2330. BERG (1915) should be omitted. Chilogobio is based on both C. czerskii
and C. soldatovi.

1916.

2367. ROULE.

Barathrites should be omitted. The genus is of Zugmayer (see No. 2222).

1918.

2401. IHERING.

Homodietus Ihering should be canceled. The genus Homodiætus is of Eigenmann & Ward (see page 522).

A Classification of Fishes

INCLUDING FAMILIES AND GENERA
AS FAR AS KNOWN

C

INTRODUCTORY NOTE

The present paper is a continuation and conclusion to a series of four memoirs issued from 1917 to 1920 under the title of "The Genera of Fishes." In it I have tried to place each genus and subgenus, living and fossil, thus far named or described, whether valid or not, in its proper family and to arrange the families themselves in as natural a sequence as is possible with a linear series which embraces highly divergent lines of evolution. I make no effort to distinguish between valid names, preoccupied names, and synonyms. The multitude of new forms recorded in the last fifty years precludes any such attempt. It will be observed that the number of families I have recognized is, in several groups, greater than accepted by any previous writer. This is due to the fact that analysis must precede synthesis, and it is better to lay a certain stress on aberrant forms than to include them uncritically in expanded groups, the definition of which is impaired or denied by their presence. Investigators who prefer to be more "conservative" can treat these groups as seems best.

In each family I have arranged the genera in the order of time. In most large groups our knowledge is too imperfect to make a detailed natural classification practicable in a compilation like this. In the list, moreover, are many nominal genera, especially among the fossils, in which an arrangement even by families is largely guesswork. As to sequence of orders and families, there is room for wide apparent divergence of opinion. Following recent custom I place the simplest or most ancient forms first, closing with highly specialized types of recent date.

The names of fossil genera are throughout printed in SMALL CAPITALS; those of doubtful eligibility—either because not binomial or else given without type, definition, or explanation—are in *Italics*. In my personal judgment these stillborn names should not enter the system unless later revived in regular form. Names attached to known species (typonyms) I have, however, admitted without question, and I make, of course, no distinction here between genera and subgenera. In each case I add the date of the genus with the page in "The Genera of Fishes," in which it is more fully treated.

In preparing this paper I have been under special obligation to Allan R. McCulloch of the Australian Museum; to my former student, Carl L. Hubbs of the University of Michigan; and to Theodore D. A. Cockerell of the University of Colorado. Each of these has brought errors or omissions to my attention, and Mr. Hubbs has suggested good reasons for certain important changes in sequence.

I may again emphasize the distinction between the ancient and modern periods in animal taxonomy. Beginning with 1758, the date of Linnæus' "Systema Naturæ," for sixty years a genus was usually regarded merely as a convenient pigeon-hole into which species are thrown. With Cuvier's "Règne Animal" (1817) came the conception of arrangement "according to organization" or comparative anatomy. With Darwin's "Origin of Species," forty years later (1859), zoologists began, though slowly at first, to realize that a classification must be more than an inventory; that its basis must be genetic, and the problems involved in a natural grouping are vastly more complex than even great morphologists like Cuvier and Agassiz had realized. To paraphrase a saying of Elliott Coues, genera and species are but larger and smaller twigs of a tree which we try to arrange as nearly as possible in accordance with nature's ramifications. This view applied to taxonomy involves the extensive subdivision of accepted genera, many of which were unnatural in the highest degree, a subdivision in many cases doubtless going too far, but our material must be first analyzed before we can group its parts in natural synthesis. No system of naming can progress beyond the knowledge on which it rests.

I may repeat a warning as old as science itself : that we must not expect a degree of accuracy which the subject in question does not permit.

DAVID STARR JORDAN.

STANFORD UNIVERSITY,
JANUARY 15, 1923

A CLASSIFICATION OF FISHES

Including Families and Genera as Far as Known

Class LEPTOCARDII

(*Myelozoa*)

Order AMPHIOXI

(*Cirrostomi*)

The Lancelets (or *Leptocardii*) are without doubt simplest in organization of all fish-like chordates or vertebrates.

Apparently their lowly structure is primitive but the geological history of the group is unknown and likely to remain so, as these creatures possess no hard parts. Whether the lampreys on the one hand or the lower and not fish-like acranite *Chordata* (*Tunicates* and *Enteropneustans*) are derived from *Leptocardians* can not be known.

Family 1. AMPHIOXIDIDÆ

Amphioxides [1] Gill, 1895, 467.

[1] This genus of three species, based on the absence of buccal cirri, the unsymmetrical mouth, and some other anatomical details, is now regarded as probably a larval form of *Asymmetron* rather than as a distinct genus.

Family 2. BRANCHIOSTOMIDÆ (*Amphioxidæ*)
(Lancelets)

Branchiostoma Costa, 1834, 180.   Dolichorhynchus [2] Willey, 1901, 495.
Amphioxus Yarrell, 1836, 186.

[2] Misprinted *Dolichorhamphus* by Hubbs, *Occ. Papers, Univ. Mich.*, 105, p. 13, 1922.

Family 3. EPIGONICHTHYIDÆ

Epigonichthys Peters, 1877, 391.   Paramphioxus [4] Haeckel, 1893, 461.
Asymmetron [3] Andrews, 1893.   Heteropleuron [4] Kirkaldy, 1895, 469.

[3] Studies Johns Hopkins Lab. V, 237; orthotype *Asymmetron lucayanum* Andrews (overlooked in "Genera of Fishes").

[4] *Paramphioxus* and *Heteropleuron* are synonyms of *Epigonichthys*.

Class MARSIPOBRANCHII

(Myzontes; Cyclostomi; Dermopteri)

The Lampreys and Hag-fishes stand next the Lancelets at the foot of the series. They have no hard parts, teeth excepted, and these have never been found fossil.

The name of CONODONTES has been given to a variety of tooth-like structures found in Devonian rocks in Russia and about Lake Erie. But the chemical composition of these differs from that of lamprey-teeth, and it is more likely that they are appendages of Crustaceans or teeth of snails (or worms) rather than of fishes. Most of the generic names given by Pander in 1856, enumerated on page 409 of the "Genera of Fishes," belong to the Palæozoic group of *Conodontes;* Pander thus classified the Conodonts:

(*a*) Simple structures, slender and sharp, DREPANODUS, ACODUS, DISTACO-DUS, MACHAIRODUS, PALTODUS, SCOLOPODUS, OISTODUS, ACONTIODUS.

(*b*) Complex structures, with several cusps. PRIONIODUS, BELODUS, LON-CHODUS, CTENOGNATHUS, CORDYLODUS, GNATHODUS, PRIONOGNATHUS, POLYGNATHUS, CENTRODUS (not of Giebel).

Nothing is known of the geological history of Lampreys, nor can any one say whether Sharks or Ostracophores were in the line of descent from them. The two orders of this class diverge widely from each other, and perhaps, as Lankester has contended, each should be regarded as a distinct class.

Order HYPEROTRETA

Family 4. MYXINIDÆ

(Hag-fishes)

Myxine Linnæus, 1758, 406.   Murænoblenna Lacépède, 1803, 68.
Gastrobranchus Bloch, 1797, 53.   Anopsus Rafinesque, 1815, 92.

Family 5. EPTATRETIDÆ

Eptatretus[5] (Duméril) Cloquet, 1819, 106.   Heptatrema Duméril, 1856, 268.
Homea[6] Fleming, 1822.   Polistotrema Gill, 1881, 416.
Bdellostoma Müller, 1835, 182.   Paramyxine Dean, 1904, 506.
Hexabranchus Schultze, 1835, 184.

---

[5] This name is often corrected to *Heptatretus.*
[6] Philos. Zool. II, 374; type *H. banksi* Fleming; omitted in "Genera of Fishes").

Order HYPEROARTIA

Family 6. PETROMYZONIDÆ
(*Lampreys*)

Petromyzon Linnæus, 1758, 11.
Ammocœtus[7] Duméril, 1808, 76.
Lampreda Rafinesque, 1815, 93.
Pricus Rafinesque, 1815, 93.
Lampetra Gray, 1851, 249.
Ichthyomyzon Girard, 1859, 291.
Scolecosoma Girard, 1859, 291.
Entosphenus Gill, 1862, 317.
Bathymyzon Gill, 1883, 424.
Caspiomyzon Berg, 1906, 516.

Agnathomyzon Gracianov, 1906, 518.
Haploglossa Gracianov, 1906, 518.
Oceanomyzon Fowler, 1908, 528.
Eudontomyzon Regan, 1911, 540.
Reighardina[7a] Creaser (Charles W.) and Hubbs, 1922.
Tetrapleurodon[7b] C. & H., 1922.
Lethenteron[7c] C. & H., 1922.
Okkelbergia[7d] C. & H., 1922.

[7] Usually written *Ammocœtes*.

[7a] "A Revision of the Holarctic Lampreys," *Occ. Pap. Mus. Zool. Michigan,* **120**: 2; orthotype *Ichthyomyzon fossor* Reighard & Cummins = *Petromyzon unicolor* Dekay.

[7b] *Loc. cit.,* 3; orthotype *Lampetra spadicea* Bean.

[7c] *Loc. cit.,* orthotype *Lampetra wilderi* Gage = *Petromyzon appendix* De Kay.

[7d] *Loc. cit.,* 8; orthotype *Petromyzon lamotteni* Le Sueur = *Ammocœtes æpyptera* Abbott.

Family 7. GEOTRIIDÆ[8]
(*Mordaciidæ*)

Geotria Gray, 1851, 249.
Caragola Gray, 1851, 249.
Velasia Gray, 1851, 249.
Mordacia Gray, 1851, 249.
Thysanochilus Philippi, 1857, 276.

Chilopterus Philippi, 1858, 283.
Exomegas Burmeister, 1868, 348.
Yarra Castelnau, 1872, 363.
Neomordacia Castelnau, 1872, 363.
Macrophthalmia Plate, 1897, 478.

[8] The name *Mordacia* may not be tenable, as its synonym *Caragola* has line priority.

Class OSTRACOPHORI

(Aspidoganoidei; Ostracodermi, name preoccupied; Protocephali)

In regarding the ostracophores as a distinct class I do not pretend to settle any of the vexing questions as to the origin and relations of the group. It may be that the four orders do not belong together, that the *Antiarcha* are sharks or else *Arthrodires,* that the group itself is a mailed variant of the Lampreys, that it is ancestral to the sharks, that it is derived from primitive sharks or that it is a variant of primitive crustaceans. In the latter case the ancestry of fishes and through them of other vertebrates, should be sought not in worm-like forms, but in marine creatures with hard shells, related to *Limulus* and remotely to existing spiders. In general, however, the outer parts are more often modified by natural selection than internal structures, hence of less value in determining relations. The outside of an animal tells where it has been, the inside what it really is.

In Hæckel's words, there are three ancestral documents in the study of evolution, morphology, embryology, and palæontology. From the first two no decisive answer is given to the question of the origin of the ostracophores. Paleontology shows that several of the great divisions of fishes had appeared in the Devonian. The oldest known fragment of a fish (*Astraspis*) is from the Ordovician in the Trenton horizon at Cañon City, Colorado. With this, however, are other fragments doubtfully referred to a chimæra (*Dictyorhabdus*) and a crossopterygian (*Eriptychius*). If these are correctly identified the differentiation of fishes had proceeded far when the Palæozoic era began.

Most naturalists have traced the hypothetic ancestry of vertebrates through lancelets or enteropneustans back to soft-bodied, worm-like forms which have left no traces in the rocks. Dr. Patten, however, regards *Pteraspis* as descended from primitive arachnids with bony carapace. From *Pteraspis* he derives the other ostracophores, and from these the sharks and other vertebrates appearing progressively later in time. This ingenious theory is developed in the *American Naturalist,* 1904. Patten regards the lateral fold in some sharks as the result of a fusion of the fringing appendages on the sides of the body. This is evident in his restoration of *Cephalaspis.*

Order HETEROSTRACI
(*Aspidorhini*)

Family 8. THELODONTIDÆ (*Cœlolepidæ*)

Thelodus is identical with *Cœlolepis* and should give the name to this family.

THELODUS Agassiz, 1839, 194.
CŒLOLEPIS Pander, 1856, 273.
NOSTOLEPIS Pander, 1856, 274.
PACHYLEPIS Pander, 1856, 274.
THELOLEPIS Pander, 1856, 274.
GOMPHODUS Pander, 1856, 274.
SCHEDIOSTEUS Pander, 1856, 273.

DASYLEPIS, Pander, 1856, 273.
DICTYOLEPIS Pander, 1856, 273.
ONISCOLEPIS Pander, 1856, 273.
MELITTOMALEPIS Pander, 1856, 273.
PRIONACANTHUS Pander, 1856, 273.
SPIRODUS Kade, 1858, 281.
LANARKIA Traquair, 1899, 488.

Family 9. PSAMMOSTEIDÆ

Psammolepis Agassiz, (1844) 1845, 217.
Placosteus Agassiz, (1844) 1845, 217.
PSAMMOSTEUS Agassiz, 1845, 225.
LOPHOSTEUS Pander, 1856, 274.
ACANTHASPIS Newberry, 1875, 379.

ACANTHOLEPIS Newberry, 1875, 379.
GANOSTEUS Rohon, 1902, 500.
PYCNOSTEUS Preobrazhenskii, 1910, 537.
DYPTYCHOSTEUS Preobrazhenskii, 1910, 537.

Family 10. ASTRASPIDÆ

ASTRASPIS[10] Walcott, 1892, 459.

[10] See Eastman, *Proc. U. S. Nat. Mus.*, **52**: 237, 1917.

Family 11. DREPANASPIDÆ

DREPANASPIS Schlüter, 1887, 439.

Family 12. CERASPIDÆ

CERASPIS Schlüter, 1887, 439.

Family 13. PTERASPIDÆ

SPHAGODUS[11] Agassiz, 1839, 194.
CHIASTOLEPIS[12] Eichwald, 1844, 220.
PTERASPIS Kner, 1847, 232.
ARCHÆOTEUTHIS Roemer, 1855, 264.
PALÆOTEUTHIS Roemer, 1855, 266.
TOLYPELEPIS Pander, 1856, 273.
SCAPHASPIS Lankester, 1865, 338.

CYATHASPIS[13] Lankester, 1865.
KALLOSTRAKON Lankester, 1870, 358.
HOLASPIS[14] Lankester, 1873, 370.
PALÆASPIS Claypole, 1884, 427, 430.
DIPLASPIS Matthew, 1887, 438.
TOLYPASPIS Schmidt, 1893, 462.
PHOLIDOSTEUS Jaekel, 1907, 524.

[11] A very doubtful genus, the type said to be part of the mouth of an Eurypterid Crustacean, as is also a tooth called AULACODUS.

[12] A very doubtful fragment of armor.

[13] *Rep. Brit. Assoc. for 1864*, p. 58. Omitted in "Genera of Fishes": type *Pteraspis banksi* Huxley & Salter.

[14] Preoccupied by *Holaspis* Gray, 1863, a reptile.

Order OSTEOSTRACI

Family 14. CEPHALASPIDÆ

CEPHALASPIS Agassiz, 1835, 181.
ZENASPIS Lankester, 1850, 246.
STEGANODICTYUM M'Coy, 1851, 249.

EUCEPHALASPIS Lankester, 1870, 358.
HEMICYLASPIS Lankester, 1870, 358.

Family 15. ATELEASPIDÆ

ATELEASPIS Traquair, 1899, 488.

Family 16. THYESTIDÆ (*Menaspidæ*)

SCLERODUS Agassiz, 1839, 194.
PLECTRODUS Agassiz, 1839, 194.
MENASPIS Ewald 1848, 233.
THYESTES Eichwald, 1854, 256.

AUCHENASPIS Egerton, 1857, 275.
DIDYMASPIS Lankester, 1867, 346.
EUKERASPIS Lankester, 1870, 358.
CEPHALOPTERUS Powrie, 1870, 359.

Family 17. ODONTODONTIDÆ (*Tremataspidæ*)
The name *Odontotodus* has priority over *Tremataspis*.

STIGMOLEPIS Pander, 1856, 273.
ODONTOTODUS Pander, 1856, 274.
COSCINODUS[14a] Pander, 1856.
MONOPLEURODUS[14a] Pander, 1856, 273.

STROSISPHERUS[14a] Pander, 1856, 274.
TREMATASPIS[15] Schmidt, 1866, 343.
ANCISTRODUS[14a] Rohon, 1893, 462.

[14a] The relationship of these forms is apparently quite unknown.

[15] The genus *Mycterops* Cope (1886) should be erased. According to Woodward it is based on a shield of an Eurypterid Crustacean.

Family 18. EUPHANEROPIDÆ

EUPHANEROPS Woodward, 1900, 491.

Order ANTIARCHA
(*Placodermi* part; *Placoganoidei*; *Azygostei* part)

It is not yet certain that this order should be retained among the ostracophores. Hay and others place it with the arthrodires in a subclass *Azygostei* or a superorder *Placodermi*.

Family 19. ASTROLEPIDÆ (*Bothrolepidæ*; *Pterichthyidæ*)

PLEIOPTERUS Agassiz, 1835, 181.
ASTROLEPIS[16] Eichwald, 1840, 206.
BOTHRIOLEPIS Eichwald, 1840, 206.
PTERICHTHYS Hugh Miller, 1841, 209.
PLACOTHORAX Agassiz, 1844, 218.
PAMPHRACTUS Agassiz, 1844, 218.
HOMOTHORAX Agassiz, 1845, 225.

ODONTACANTHUS Agassiz, 1845, 225.
NARCODES Agassiz, 1845, 225.
ACTINOLEPIS Agassiz, 1845, 225.
PTERICHTHYODES Bleeker, 1859, 289.
MICROBRACHIUM Traquair, 1888, 443.
Glyptosteus Agassiz, (1844) 1891, 217.
MILLERICHTHYS S. A. Miller, 1892, 458.

[16] Usually written *Asterolepis*.

Order ANASPIDA

Family 20. LASANIIDÆ

LASANIUS Traquair, 1899, 488.

Family 21. BIRKENIIDÆ

BIRKENIA Traquair, 1899, 488.   CTENOPLEURON Matthew, 1907, 525.

Order CYCLIAE

Family 22. PALÆOSPONDYLIDÆ

I here give the rank of a separate order to the singular little fish-like organism of the Scottish Devonian age known as *Palæospondylus gunni*. In doing so, I do not assume to decide what its relationships may be, whether a minute but specialized primitive lamprey, an offshoot from the arthrodires or the ostracophores or whatever else may be possible. Perhaps, as suggested by Huxley, it may be a "baby *Coccosteus*."

PALÆOSPONDYLUS Traquair, 1890, 453.

Class ARTHRODIRA

Order STEGOPHTHALAMI

Family 23. MACROPETALICHTHYIDÆ

MACROPETALICHTHYS Norwood and Owen, 1846, 229.

Family 24. ASTEROSTEIDÆ

ASTEROSTEUS Newberry, 1875, 379.

Order TEMNOTHORACI

Family 25. CHELONICHTHYIDÆ

ICHTHYOSAUROIDES Kutorga, 1837, 189.   HETEROSTEUS Asmuss, 1856, 267.
HOMOSTEUS[17] Asmuss, 1856, 267.   CHELONICHTHYS[16a] Agassiz, (1842) 1891,
                                           217.

[16a] Agassiz quotes this name from "Murchison's Report, 1842.
[17] Name misprinted *Homostius*.

Order ARTHROTHORACI

Family 26. COCCOSTEIDÆ

COCCOSTEUS Agassiz, 1844, 218.   LOPHOSTRACON Lankester, 1884, 429.
CHELYOPHORUS Agassiz, 1845, 225.   TITANICHTHYS[19] Dames, 1887, 437.
SIPHONODUS, Fischer von Waldheim,   GIGANTICHTHYS Dames, 1888, 440.
  1852, 250.   GLYPTASPIS Newberry, 1889, 447.
PHYSICHTHYS Meyer, 1855, 266.   TRACHOSTEUS Newberry, 1889, 447.
AGASSICHTHYS Newberry, 1857, 276.   PHLYCTÆNIUS Traquair, 1890, 452.
ASPIDOPHORUS Newberry, 1870, 359.   PHLYCTÆNASPIS Traquair, 1890, 453.
ASPIDICHTHYS Newberry, 1873, 371.   ASTEROPLAX Woodward, 1891, 456.
LIOGNATHUS Newberry, 1873, 371.   PONERICHTHYS Miller, 1892, 458.
BRACHYDIRUS[18] Koenen, 1880, 404.   LISPOGNATHUS Miller, 1892, 458.
ANOMALICHTHYS Koenen, 1883, 424.

---

[18] The name *Leptodus* was based on a Brachiopod shell and should be erased.
[19] According to Regan, as quoted by Cockerell, *Gigantichthys* (*Titanichthys* Dames, not Newberry) is based on a Reptilian tooth.

Family 27. DINICHTHYIDÆ

DINICHTHYS[20] Newberry, 1868, 371.   HUSSAKOFIA Cossman, 1910, 535.
GORGONICHTHYS Claypole, 1892, 457.   COPANOGNATHUS Hussakof and Bryant,
BRONTICHTHYS Claypole, 1894, 464.   1919, 566.
PLATYASPIS Koenen, 1895, 469.   MACHÆROGNATHUS Hussakof and Bry-
STENOGNATHUS Newberry, 1898, 479, 483.   ant, 1919, 566.
PHLYCTÆNIUM Zittel, 1900, 491.   PERISSOGNATHUS Hussakof and Bryant,
PACHYOSTEUS Jaekel, 1903, 503.   1919, 566.
BRACHYGNATHUS Hussakof, 1909, 533.

---

[20] *Proc. Amer. Asso. Adv. Sci. for 1867*, **17**: 148 (1868).

Family 28. TITANICHTHYIDÆ

TITANICHTHYS Newberry, 1885, 433.   PROTITANICHTHYS Eastman, 1907, 521.
HOLOPETALICHTHYS Koenen, 1895, 469.

Family 29. MYLOSTOMIDÆ

TYPODUS[21] Meyer, 1847, 232.   DINOMYLOSTOMA Eastman, 1906, 517.
MYLOSTOMA Newberry, 1883, 425.   DINOGNATHUS Hussakof, 1909, 533.

[21] Misprinted *Tylodus* in "Genera of Fishes," p. 232.

Family 30. SELENOSTEIDÆ

DIPLOGNATHUS Newberry, 1878, 397.   STENOSTEUS Dean, 1901, 492.
SELENOSTEUS Dean, 1901, 492.

Family 31. PHYLLOLEPIDÆ

PHYLLOLEPIS Agassiz, 1844, 219.   PENTAGONOLEPIS Lohest, 1890, 452.
HOLONEMA Newberry, 1889, 447.

Class ELASMOBRANCHII

(Selachii; Plagiostomata; Chondropterygii; Antacea)

Subclass SELACHII
(Sharks)

Whatever may have been the origin of sharks, whether from lamprey-like or other soft bodied forms, or through degeneration from more specialized types, I have little doubt that the most primitive form is the order *Pleuropterygii*. It seems to me, as to Dr. Dean, that the pectoral fin of *Cladoselache,* apparently derived or derivable from a lateral fold, comes nearest the primitive condition. It is conceivable that the robust pectoral spine of *Acanthoëssus* is homologous with the fin of *Cladoselache,* the anterior fin-rays being gathered together to be united in a thick spine.

If, however, we regard the pectoral limb as developed from gill-structures, the conception of origin would be changed. The jointed limb of *Orthacanthus* would be the nearest approach to the primitive pectoral limb, and from the ancestors or allies of the *Ichthyotomi* all other sharks and through them all the higher fishes as well as the higher vertebrates would be descended. Whether the pectoral limb or archipterygium of the *Crossopterygii* or the *Dipneusti* is homologous with that of the *Ichthyotomi,* or separately derived is not yet known.

The weight of authority, at present, favors the theory that the paired fins are derived from a fold of skin, and that of the known forms *Cladoselache* of the Devonian is the most primitive. We therefore adopt a sequence in accord with this theory, recognizing at the same time that weight of authority is unimportant when decisive evidence is wanting.

Leaving aside certain ichthyodorulites or detached fin-spines with other fragments of uncertain character the earliest known sharks (*Acanthoëssidæ*) appear in the lower Devonian, while *Orthacanthus* and certain *Heterodontids* occur in the Carboniferous. In the Triassic all the early forms of sharks, *Orodontidæ* excepted, disappear, and all or nearly all the living forms are descended from this Heterodontid type. To this statement the *Notidani* with *Chlamydoselachus* may form an exception, for their peculiar traits seem to be primitive, though no specimens are found before the Jurassic.

The chimæras represent a very early divergence from the sharks, being apparently nearly as old geologically as the sharks. If *Dictyorhabdus* from the Ordovician is really a chimæra, which Dr. Dean doubts, these fishes are older than any known sharks.

Order PLEUROPTERYGII
(*Cladoselachea*)
Family 32. CLADOSELACHIDÆ

CLADOSELACHE[22] Dean, 1894.

[22] "Contributions to the Morphology of *Cladoselache*," *Journ. Morphol.*, 9 : 87; type *Cladodus fyleri* Newberry. (Omitted in "Genera of Fishes.")

Order ACANTHODEI

This order does not differ widely from the *Pleuropterygii*. If we assume that its fin spines are simply massed rays, as Dr. Dean asserts, the essential structure of the two groups is identical.

Family 33. ACANTHOËSSIDÆ (*Acanthodidæ*)

The name *Acanthoëssus* has priority over *Acanthodes*. Those genera marked *Ich* are known only from *Ichthyodorulites,* fin spines of more or less uncertain relations.

ACANTHOËSSUS Agassiz, 1832, 141.
ACANTHODES Agassiz, 1833, 145, 177.
CHEIRACANTHUS Agassiz, 1835, 181.
HAPLACANTHUS Agassiz, 1845, 225.
HOMACANTHUS Agassiz, 1845, 225, Ich.
BYSSACANTHUS Agassiz, 1845, 225, Ich.
HOLACANTHODES Beyrich, 1848, 233.
PELONECTES Gistel, 1848, 234.
MACHÆRACANTHUS Newberry, 1857, 276 Ich.

MACHÆRIUS Rouault, 1858, 284, Ich.
ACANTHODOPSIS Hancock and Atthey, 1868, 352.
RHADINACANTHUS Traquair, 1888, 443.
MESACANTHUS Traquair, 1888, 443.
HETERACANTHUS[23] Newberry, 1889, 447 Ich.
TRAQUAIRIA Fritsch, 1893, 460.
PROTACANTHUS Fritsch, 1893, 460.

[23] *Ganiodus* Newberry should be erased. It is a misprint for *Goniodus* Agassiz, 1836.

Family 34. ISCHNACANTHIDÆ (*Ictinocephalidæ*)

ISCHNACANTHUS[24] Powrie, 1864.   *Ictinocephalus* Page, (1859) 1864, 334.

[24] *Quart. Journ. Geol.,* 419, 1864; type *Diplacanthus gracilis* Egerton. (Omitted in "Genera of Fishes.")

Family 35. DIPLACANTHIDÆ

DIPLACANTHUS Agassiz, 1842, 210, 225.
PAREXUS Agassiz, 1845, 225.
CLIMATIUS Agassiz, 1845, 225.

BRACHYACANTHUS Egerton, 1860, 295.
EUTHACANTHUS Powrie, 1864, 334.
MARSDENIUS Wellburn, 1902, 501.

Order ICHTHYOTOMI

Family 36. XENACANTHIDÆ (*Pleuracanthidæ; Orthacanthidæ*)

The name *Pleuracanthus* is preoccupied (*Coleoptera* Gray, 1832), as is also *Diplodus* Agassiz. The name next oldest, *Orthacanthus,* seems to belong to a different genus, while *Xenacanthus* appears to be an exact synonym of *Pleuracanthus*.

PLEURACANTHUS Agassiz, 1837, 186.
DIPLODUS Agassiz, 1843, 213.
ORTHACANTHUS Agassiz, 1843, 213.
ACANTHOPLEURUS Agassiz, 1844, 217.
XENACANTHUS Beyrich, 1848, 233.
TRIODUS Jordan, 1849, 243.
COMPSACANTHUS Newberry, 1856, 273.
DITTODUS Owen, 1867, 346.
OCHLODUS Owen, 1867, 347.
AGANODUS Owen, 1867, 347

PTERNODUS Owen, 1867, 347.
THRINACODUS St. John and Worthen, 1875, 380.
PHRICACANTHUS Davis, 1879, 399.
LOPHACANTHUS Stock, 1880, 406.
ANODONTACANTHUS Davis, 1881, 415.
DIDYMODUS Cope, 1883, 422.
DIACRANODUS[25] Garman, 1885, 432.
DOLIODUS Traquair, 1892, 459.

[25] Misprinted *Dicranodus* in "Genera of Fishes."

Family 37. CLADODONTIDÆ

CLADODUS Agassiz, 1843, 213.
CHILODUS Giebel, 1847, 231.
PHŒBODUS St. John and Worthen, 1875, 380.
LAMBDODUS St. John and Worthen, 1875, 380.
BATHYCHEILODUS St. John and Worthen, 1875, 380.

HYBOCLADODUS[26] St. John and Worthen, 1875, 380.
PTERNODUS Garman, 1885, 432.
DICENTRODUS Traquair, 1888, 443.
STYPTOBASIS Cope, 1891, 454.
PROTODUS Woodward, 1892, 459.
MONOCLADODUS Claypole, 1893, 460.
SYMMORIUM Cope, 1893, 460.

[26] The insertion of *Hybocladodus* Hay, 1899, in the "Genera of Fishes" is an error.

Family 38. CHONDRENCHELYIDÆ
CHONDRENCHELYS Traquair, 1888, 443.

Family 39. GYRACANTHIDÆ
GYRACANTHUS Agassiz, 1837, 186.   MITRODUS Owen, 1867, 347.

Order POLYSPONDYLI

We may here insert the hypothetical order *Polyspondyli*, regarded by Hasse as a primitive type known from fin-spines or Ichthydorulites only, and these all of uncertain relationship, some of them perhaps *Heterodontid*.

Onchus occurs in the upper Silurian.

Family 40. ONCHIDÆ
ONCHUS Agassiz, 1837, 186.
NAULAS Agassiz, 1845, 225.

LEPTOCHELES M'Coy, 1853, 408.
ARCHÆACANTHUS Kade, 1858, 281.

Order CESTRACIONTES
Suborder *JANASSIDES*

I here arrange provisionally certain families of sharks, some or all with flattened bodies like the rays, but occurring in Palæozoic time they are probably not genetically connected with the flattened skates and rays of recent times.   Except the genus *Janassa,* the species are mostly known from teeth only, these mostly large and fitted for grinding.

Family 41. JANASSIDÆ (*Petalodontidæ*)

JANASSA Münster, 1832, 176.
CTENOPTYCHIUS Agassiz, 1838, 190.
PETALODUS[27] Owen, 1840, 207.
DICTEA Münster, 1840, 207.
BYZENOS Münster, 1843, 216.
CARCHAROPSIS (Agassiz) Davis, (1843) 1883, 214, 432.
CLIMAXODUS M'Coy, 1848, 239.
GLOSSODUS M'Coy, 1848, 239.
POLYRHIZODUS M'Coy, 1848, 239.
PLATACANTHUS[28] M'Coy, 1848, 408.
DICRENODUS Romanovskii, 1853, 254.
PRISTICLADODUS M'Coy, 1855, 266.
SICARIUS Leidy, 1856, 272.
XYSTRACANTHUS Leidy, 1859, 292.
CTENOPETALUS (Agassiz) Morris & Roberts (1862) 1881, 320.
HARPACODUS (Agassiz) M. & R., (1862) 1881, 320, 415.
PETALORHYNCHUS (Agassiz) M. & R., (1862) 1881, 320.
DACTYLODUS Newberry and Worthen, 1866, 342.
ANTLIODUS N. & W., 1866, 349.
AGELEODUS Owen, 1867, 347.
CYMATODUS N. & W., 1870, 359.
PELTODUS N. & W., 1870, 359.

PETALODOPSIS Barkas, 1874, 372.
LISGODUS St. John & Worthen, 1875, 380.
TANAODUS St. J. & W., 1875, 380.
HARPACODUS (Agassiz) St. J. & W., 1875, 380.
CHOLODUS St. J. & W., 1875, 380.
CALOPODUS St. J. & W., 1875, 380.
FISSODUS St. J. & W., 1875, 380.
BATACANTHUS St. J. & W., 1875, 381, 425.
ANTACANTHUS[29] Devalque, 1877.
STRIGILINA Cope, 1877, 388.
SERRATODUS Koninck, 1878, 396.
STICHÆACANTHUS Koninck, 1878, 396.
CYMATODUS Trautschold, 1879, 401.
GLYPHANODUS Davis, 1883, 422.
PHODERACANTHUS Davis, 1883, 423.
PETALODOPSIS Davis, 1883, 423.
PNIGEACANTHUS St. John & Worthen, 1883, 425.
THORACODUS Cope, 1883, 422.
GOMPHACANTHUS Davis, 1884, 427.
AGANACANTHUS Traquair, 1884, 430.
CALLOPRISTODUS Traquair, 1888, 443.
MESOLOPHODUS Woodward, 1889, 449.
STETHACANTHUS Newberry, 1889, 447.
GAMPHACANTHUS Miller, 1892, 458.
PARACYMATODUS Bogoliubov, 1914, 552.

[27] Perhaps a synonym of *Ctenoptychius.*
[28] Later amended to *Platyacanthus.*
[29] *Ann. Soc. Geol. Belge,* 5 : 60; type *A. insignis* Devalque (omitted in "Genera of Fishes").

Family 42. PSAMMODONTIDÆ

PSAMMODUS[30] Agassiz, 1838.
APODOMETES Gistel, 1848, 237.
PALÆOBATES Meyer, 1849, 243.
CHARACODUS (Agassiz) Morris & Roberts, 1862, 320, 423.
MYLAX (Agassiz) M. & R., 1862, 320.
PLEUROGOMPHUS (Agassiz) M. & R., (1862) 1883, 320.
MESOGOMPHUS (Agassiz) M. & R., (1862) 1883, 320.
LOBODUS (Agassiz) M. & R., (1862) 1883, 320.

COPODUS (Agassiz) M. & R., (1862), 1883, 320, 422.
SOLENODUS, Trautschold, 1874, 377.
ARCHÆOBATIS Newberry, 1878, 397.
DIMYLEUS (Agassiz) Davis, 1883, 423.
MYLACODUS Davis, 1883, 423.
PINACODUS (Agassiz), Davis, 1883, 423.
HOMALODUS Davis, 1883, 423.
RHYMODUS (Agassiz) Davis, 1883, 423.
ASTRABODUS Davis, 1884, 427.
MAZODUS Newberry, 1889, 447.
LAGARODUS Jaekel, 1898, 480.

[30] *Poissons Fossiles,* 3 : 110; type *P. rugosus* Agassiz (omitted in "Genera of Fishes").

Family 43. **PERIPRISTIDÆ** (*Pristodontidæ*)

The name *Peripristis* is earlier than *Pristodus* and belongs to the same genus.

PERIPRISTIS[31] St. John, 1870, 359.

HOPLODUS[32] Etheridge, 1881.

DIODONTOPSODUS Davis, 1881, 415.

PRISTODUS (Agassiz) Davis, 1883, 424.

[31] Identical with *Pristodus,* not with *Ctenoptychius* as stated in "Genera of Fishes."

[32] R. Etheridge, *Jr. Geol. Mag.* (2) **2**: 244, 1825; type *Petalorhynchus benniei* Etheridge (omitted in "Genera of Fishes").

Family 44. **COCHLIODONTIDÆ**

This family, known only from the teeth, fitted for crushing shells, may be left provisionally with the *Janassides,* although it has much in common with the *Cestraciontes* or *Heterodontidæ*. The species belong mostly to the Carboniferous age.

COCHLIODUS Agassiz, 1838, 191.

CHOMATODUS Agassiz, 1838, 190.

HELODUS Agassiz, 1838, 190.

PŒCILODUS M'Coy, 1855, 266.

DICHELODUS Giebel, 1857, 409.

PSEPHODUS (Agassiz) Morris & Roberts, (1862) 1883, 320.

STREBLODUS (Agassiz) M. & R., (1862) 1883, 320.

TOMODUS (Agassiz) M. & R., (1862) 1883, 422.

XYSTRODUS (Agassiz) M. & R., (1862) 1883, 320.

DELTOPTYCHIUS (Agassiz) M. & R., (1862) 1883, 320.

DELTODUS (Agassiz) M. & R., (1862) 1883, 319.

LOPHODUS Romanowsky, 1864, 334.

TRIGONODUS Newberry & Worthen, 1866, 342.

ASPIDODUS N. & W., 1866, 342.

SANDALODUS N. & W., 1866, 342.

PLEURODUS Hancock & Atthey, 1872, 366, 213.

VENUSTODUS St. John & Worthen, 1875, 380.

PERIPLECTRODUS St. J. & W., 1875, 380.

PLATYODUS Newberry, 1875, 379.

HELODOPSIS Waagen, 1879, 401.

CRANODUS Trautschold, 1879, 401.

OXYTOMODUS Trautschold, 1880, 406.

VATICINODUS[33] St. J. & W., 1883, 405, 425.

STENOPTERODUS[34] St. J. & W., 1883, 405, 425.

CHITONODUS[35] St. J. & W., 1883, 405, 425.

DELTODOPSIS[36] St. J. & W., 1883, 405, 425.

ORTHOPLEURODUS[37] St. J. & W., 1880, 405, 425.

TÆNIODUS St. J. & W., 1883, 425.

RHAMPHODUS Davis, 1883, 422.

DIPLACODUS Davis, 1884, 427.

CYRTONODUS Davis, 1884, 427.

PLEUROPLAX Woodward, 1889, 449.

XENODUS Miller, 1892, 458.

ICANODUS Miller, 1892, 458.

PLATYXYSTRODUS Hay, 1899, 486.

[33] Logotype *V. vetustus* St. J. & W. The date of 1880 given in "Genera of Fishes" is an error.

[34] Logotype *S. planus* St. J. & W., as fixed by Hay.

[35] Logotype *C. springeri* St. J. & W.

[36] Orthotype *D. augustus* St. J. & W.

[37] Orthotype *O. carbonarius* St. J. & W.

Family 45. TAMIOBATIDÆ

This family is known from a single species, from the Devonian of Kentucky, one of the oldest sharks known. It has the depressed form of a ray, probably a matter of analogy only.

TAMIOBATIS Eastman, 1897, 476.

Suborder *PROSARTHRI*

Family 46. ORODONTIDÆ

The *Orodontidæ*, allies or ancestors of the *Heterodontidæ*, are the only kinds of sharks living in the Triassic, all the earlier types being then extinct.

ORODUS Agassiz, 1838, 190.

CAMPODUS Koninck, 1844, 220.

ASTEROPTYCHIUS Agassiz, 1848, 239.

ADIAPNEUSTES Gistel, 1848, 234.

PETRODUS M'Coy, 1848, 239.

OCTINASPIS Trautschold, 1874, 377.

MESODMODUS St. John & Worthen, 1875, 380.

ACONDYLACANTHUS St. J. & W., 1875, 380.

STEMMATODUS St. J. & W., 1875, 380.

LEIODUS St. J. & W., 1875, 380.

DESMIODUS St. J. & W., 1875, 380.

STEMMATIAS Hay, 1899, 486.

EOORODUS Branson, 1914, 552.

Family 47. HETERODONTIDÆ (*Cestraciontidæ, Hybodontidæ*)
(Bull-head Sharks)

To this family, characterized among other features by strong dorsal spines, many of the Ichthyodorulites are supposed to belong. Those genera of very uncertain relations are marked *Ich.*

Cestracion Klein, 1775, 40.

Heterodontus Blainville, 1816, 95.

Cestracion Cuvier, 1817, 97.

Centracion Gray, 1831, 138.

TRISTYCHIUS[38] Agassiz, 1837, 186.

PTYCHACANTHUS Agassiz, 1837, 186.

SPHENACANTHUS Agassiz, 1837, 186.

ASTERACANTHUS Agassiz, 1837, 186.

LEIACANTHUS Agassiz, 1837, 186.

HYBODUS[39] Agassiz, 1837, 186.

ACRODUS Agassiz, 1837, 189, 191.

CTENACANTHUS Agassiz, 1837, 186, Ich.

NEMACANTHUS Agassiz, 1837, 186, Ich.

STROPHODUS Agassiz, 1838, 191.

Oroptychus[40] Agassiz, 1842.

MERISTODON Agassiz, 1843, 214.

SPHENONCHUS Agassiz, 1843, 213.

WODNIKA[41] Münster, 1843, 216.

THECTODUS Meyer & Plieninger, 1844, 221.

CENTRODUS Giebel, 1847, 231.

CENTRODUS M'Coy, 1848, 243, 408.

PHYSONEMUS M'Coy, 1848, 239, Ich.

STYRACODUS Germar, 1849, 242.

SELACHIDEA Quenstedt, 1852, 251.

PYCNACANTHUS Fischer von Waldheim, 1852, 250, Ich.

DESMACANTHUS Quenstedt, 1858, 283, Ich.

PSILACANTHUS Quenstedt, 1858, 283, Ich.

XYSTRODUS Plieninger, 1860, 298.

LOPHODUS Symonds, 1861, 309.

Gyropleurodus Gill, 1862, 317.

Tropidodus Gill, 1862, 317.

---

[38] *Tristychius* Portlock should be omitted as an error for *Tristychius* Agassiz.

[39] Stensiö (1921) regards *Hybodus reticulatus* Agassiz as the type of this genus.

[40] A manuscript name, apparently never defined, quoted in the *Nomenclator Zoologicus from Misc. Coll.* as one of the "*Cestraciontes.*"

[41] The so-called *Wodnika ocoyæ* Jordan, "Fossil Fishes of Southern California," 1919, 69, from California Miocene proves to be not a tooth but a concretion.

NEMACANTHUS Mackie, 1863, 327.
CURTODUS Sauvage, 1867, 347.
LEPRACANTHUS Owen, 1869, 355, Ich.
LOPHODUS Newberry, 1870, 359.
PALÆOSPINAX Egerton, 1872, 365.
DREPANEPHORUS Egerton, 1872, 365.
AGASSIZODUS St. John & Worthen, 1875, 380.
AMACANTHUS St. J. & W., 1875, 381, Ich.
BYTHIACANTHUS St. J. & W., 1875, 381.
GLYMMATACANTHUS[42] St. J. & W., 1875, 381, 425, Ich.
GEISACANTHUS St. J. & W., 1875, 381, Ich.
HOPLONCHUS Davis, 1876, 385, Ich.
HYBODOPSIS Barkas, 1878, 392.
TOMODUS[43] Trautschold, 1879.
CHIASTODUS Trautschold, 1879, 401.
ARPAGODUS Trautschold, 1879, 401.
THAUMATACANTHUS Waagen, 1879, 401, Ich.
OXYTOMODUS[44] Trautschold, 1880.

BDELLODUS Quenstedt, 1882, 421.
MERISTODON Sauvage, 1883, 425.
DICLITODUS Davis, 1883, 422.
EUNEMACANTHUS St. J. & W., 1883, 425, Ich.
ACONDYLACANTHUS St. J. & W., 1883, 425, Ich.
LISPACANTHUS Davis, 1883, 422, Ich.
CHALAZACANTHUS Davis, 1883, 422, Ich.
ECHINODUS Davis, 1884, 427.
SYNECHODUS Woodward, 1888, 444.
ACRODONCHUS Fraas, 1889, 446.
POLYACRODUS Jaekel, 1889, 447.
HYBODONCHUS Fraas, 1889, 446.
FAYOLIA Renault & Zeiller, 1890, 452.
RHABDIODUS Rohon, 1893, 462.
EUPHYACANTHUS Traquair, 1894, 465.
ORTHYBODUS Jaekel, 1898, 480.
PARHYBODUS Jaekel, 1898, 480.
PSEUDACRODUS Ameghino, 1906, 516.

[42] Misprinted in "Genera of Fishes."

[43] *Nouv. Mem. Soc. Imp. Nat. Moscow,* **14** : 55; type *T. argutus* Trautschold (omitted in "Genera of Fishes").

[44] *Bull. Soc. Imp. Nat. Moscow* **2** : 140; substitute for *Tomodus*, preoccupied; type *Tomodus argutus* Trautschold. (Omitted in "Genera of Fishes.")

Family 48. EDESTIDÆ

A peculiar type of sharks, known only by singular bony structures now supposed fused symphyseal whorls of teeth of cestraciont sharks.

EDESTUS Leidy, 1856, 272.
CYNOPODIUS Traquair, 1881, 417.
EUCTENIUS Traquair, 1881, 417.
SPIRAXIS Newberry, 1885, 434, Ich.
PROSPIRAXIS Williams, 1887, 439.
PROTOPIRATA Trautschold, 1888, 443.

SPHENOPHORUS Newberry, 1889, 447.
ŒSTOPHORUS Miller, 1892, 458.
HELICOPRION Karpinskii, 1899, 487.
CAMPYLOPRION Eastman, 1902, 498.
LISSOPRION Hay, 1907, 524.
TOXOPRION Hay, 1910, 536.

Series *ICHTHYODORULITES*
(Shark-spines)

We place here a number of fin spines belonging perhaps to *Heterodontidæ* or to *Chimæridæ,* their relationship not yet determined.

PRISTACANTHUS Agassiz, 1837, 186, Ich.
Cricacanthus[45] Agassiz, 1843, 213, Ich.
COSMACANTHUS Agassiz, 1845, 225, Ich.
DIMERACANTHUS Keyserling, 1846, 229, Ich.
Gyropristis[46] (Agassiz) King, (1843) 1850, 213, Ich.
RYTIDOLEPIS Pander, 1856, 273, Ich.

STENACANTHUS Leidy, 1856, 273, Ich.
MARRACANTHUS St. John & Worthen, 1875, 381, Ich.
COMPSACANTHUS Davis, 1883, 422, Ich.
BRACHIACANTHUS Fritsch, 1889, 446, Ich.
TUBULACANTHUS Fritsch, 1889, 446, Ich.
PLATYACANTHUS Fritsch, 1889, 446, Ich.
CALLOGNATHUS Newberry, 1889, 447, Ich.

PHLYCTÆNACANTHUS Eastman, 1898, 479, Ich.

EUCENTRURUS Traquair, 1901, 495, Ich.

EUCTENODOPSIS Wellburn, 1901, 495.

ECZEMATOLEPIS Miller, 1892, 458, Ich.

ATOPACANTHUS Hussakof & Bryant, 1919, 566, Ich.

---

[45] Name only; the species, *C. jonesi* Agassiz, misprinted in "Genera of Fishes," 213.

[46] Name only; used in 1850 by W. King: "Permian Fossils," *Pal. Soc. 1850,* 222.

Order SELACHOPHIDICHTHYOIDEI
(*Pternodonta*)

This order and the next seem to find place somewhere between the *Cladoselachidæ* and the *Heterodontidæ,* although so far as geological evidence both groups long preceded them in time.

Family 49. CHLAMYDOSELACHIDÆ
(Frilled Sharks)

Chlamydoselachus Garman, 1884, 428.

Order NOTIDANI
(*Diplospondyli; Protoselachii*)

Suborder OPISTHARTHRI

Family 50. HEXANCHIDÆ
(Cow Sharks)

Hexanchus Rafinesque, 1810, 78.

Monopterhinus Blainville, 1816, 95.

Notidanus Cuvier, 1817, 97.

Holodus[48] Agassiz, 1842.

Heptranchias[47] Rafinesque, 1810, 78.

Notorhynchus Ayres, 1855, 262.

NOTIDANION[48a] Jordan & Hannibal, 1923.

---

[47] Sometimes perverted to *Heptanchus,* the original name a condensation of *Heptabranchus,* seven-gilled.

[48] Apparently a manuscript name never defined, quoted from "Misc. Coll." in the *Nomenclator Zoologicus,* Appendix, p. 3, as equivalent to *Notidanus.*

[48a] *Notidanion* Jordan & Hannibal, new genus; orthotype *Notidanus primigenius* Agassiz, *Poisson Fossiles,* plate 27, fig. 9: "Lateral teeth with the first cusp very large, the others few and large, 4 to 6 in number; anterior serrations very coarse."

Order EUSELACHII
Suborder GALEI
(*Scyllioidei*)

Family 51. HEMISCYLLIIDÆ

The sharks of this family are ovoviviparous, those of the other scyllioid genera have large oblong quadrate leathery egg-cases, with filaments at the corners. Some of the genera enumerated among the *Scylliorhinidæ* may prove to belong here.

Hemiscyllium A. Smith, 1837, 190.

Chiloscyllium Müller & Henle, 1838, 192.

Cirrhoscyllium H. M. Smith, 1913, 551.

Family 52. SCYLLIORHINIDÆ
(Cat-Sharks)

Catulus Valmont, 1768, 27.
Scylliorhinus Blainville, 1816, 95.
Scyllium Cuvier, 1817, 97.
Catulus Smith, 1837, 190.
Poroderma Smith, 1837, 190.
Pristiurus Müller & Henle, 1838, 573.
Pristidurus Bonaparte, 1839, 194.
Pristiurus Bonaparte, 1841, 208.
THYELLINA Agassiz, 1843, 214.
SCYLLIODUS Agassiz, 1843, 214.
PALÆOSCYLLIUM Wagner, 1857, 276.
ORTHODON Coquand, 1860, 295.
Cephaloscyllium Gill, 1861, 306.

Halælurus Gill, 1861, 306.
PARASCYLLIUM Gill, 1861, 306.
PALÆOSCYLLIUM von der Marck, 1863, 298.
MESITEIA Kramberger, 1885, 433.
CANTIOSCYLLIUM Woodward, 1889, 449.
Proscyllium Hilgendorf, 1904, 508.
Parmaturus Garman, 1906, 518.
Haploblepharus Garman, 1913, 548.
Apristurus Garman, 1913, 548.
Atelomycterus Garman, 1913, 548.
Catulus (Valmont) Garman, 1913, 548.

Family 53. PENTANCHIDÆ
First dorsal wanting.

Caninoa[49] Nardo, 1844, 221.
Caninotus Nardo, 1844, 222.

Thalassoklephtes Gistel, 1848, 235.
Pentanchus[50] Smith, 1912, 546.

---

[49] This genus, which has received three names, is perhaps mythical.

[50] This genus, distinguished by the absence of the first dorsal, is regarded by Regan as a mutilated scylliorhinoid; this is an error as a second (undescribed) species, of this family, locally known as *Kagurazame,* occurs in Sagami Bay.

Family 54. ORECTOLOBIDÆ

Orectolobus Bonaparte, 1837, 187.
Stegostoma Müller & Henle, 1837, 192, 573.
Crossorhinus M. & H., 1838, 192.
Synchismus Gill, 1861, 306.

Brachælurus Ogilby, 1906, 519.
Heteroscyllium Regan, 1908, 530.
Cirriscyllium Ogilby, 1908, 529.
Eucrossorhinus Regan, 1908, 530.
TOMODON[51] Leidy, 1865, 338.

---

[51] *Tomodon* = *Diplotomodon* Leidy, 1868, "Genera of Fishes," 352, is regarded by Dr. Hay as the tooth of a Mosasaur reptile. The name is preoccupied.

Family 55. GINGLYSTOMIDÆ
This family is scarcely distinct from the *Orectolobidæ.*

Nebrius Rüppell, 1835, 184.
Ginglymostoma Müller & Henle, 1837, 190.

PLICODUS Winkler, 1874, 377.
ACRODOBATIS Leidy, 1877, 391.
Nebrodes[52] Garman, 1913.

---

[52] *Plagiostomia,* 56; type *Nebrius concolor* Rüppell substituted for *Nebrius,* regarded as preoccupied. (Omitted in "Genera of Fishes.")

Family 56. PSEUDOTRIAKIDÆ
Pseudotriakis[53] Capello, 1868, 349.

---

[53] Also written *Pseudotriacis.*

Family 57. RHINEODONTIDÆ

Rhincodon[54] Smith, (1849), 174, 244.   Micristodus Gill, 1865, 337.

[54] Probably a misprint for *Rhineodon* or *Rhinodon.*

Series *LAMNOIDEI*
Family 58. CARCHARIIDÆ (*Odontaspidæ*)
(Sand Sharks)

Carcharias[55] Rafinesque, 1810, 77.   Oxytes Giebel, 1847, 231.
Odontaspis Agassiz, 1836, 184.   Eugomphodus Gill, 1864, 331.
Triglochis Müller & Henle, 1838, 192.   Xiphodolamia[55a] Leidy, 1877, 301.

[55] The status of the name *Carcharias* as determined by the International Commission of Zoological Nomenclature makes it identical with *Odontaspis,* although the obvious intent of Rafinesque was to apply it to the group called *Carcharhinus.*

[55a] This genus, having the teeth with bifid roots, the branches closely appressed, can not belong to the *Hexanchidæ,* in which family the teeth all have solid roots.

Family 59. MITSUKURINIDÆ
(Goblin Sharks)

Rhinognathus Davis, 1885, 431.   Mitsukurina Jordan, 1898, 480.
Scapanorhynchus Woodward, 1889, 449.

Family 60. LAMNIDÆ
(Mackerel Sharks; Man-Eaters)

Isurus Rafinesque, 1810, 78.   Otodus Agassiz, 1843, 214.
Lamna Cuvier, 1817, 97.   Exoles Gistel, 1848, 236.
Lamia Risso, 1826, 119.   Plectrostoma Gistel, 1848, 237.
Selanonius Fleming, 1828, 122.   Isuropsis Gill, 1861, 306.
Oxyrhina Agassiz, 1838, 190, 218.   Xenodolamia[56a] Leidy, 1877.
Corax Agassiz, 1843, 214.   Orthacodus Woodward, 1889, 449.
Sphenodus Agassiz, 1843, 214.   Pseudocorax Priem, 1897, 478.
Carcharodon Müller & Henle, 1838, 190, 192, 574.   Leptostyrax Williston, 1900, 491.
   Carcharocles[56] Jordan & Hannibal, 1923.

[56] New genus; orthotype *Carcharodon auriculatus* (Blainville). Similar to *Carcharodon,* but with a strong denticle on each side of the base of the tooth. Teeth narrower and more erect than in *Carcharodon,* their edges finely serrated.

[56a] *Journ Acad Nat. Sci. Phila.,* 8: 251; logotype *X. pravus* Leidy. (Omitted in "Genera of Fishes.")

Family 61. ALOPIIDÆ

Vulpecula Valmont, 1768, 27.   Alopecias Müller & Henle, 1838, 192.
Alopias Rafinesque, 1810, 78.   Vulpecula (Valmont) Garman, 1913, 548.

Family 62. CETORHINIDÆ
(Basking Sharks)

Tetroras[57] Rafinesque, 1810, 77.   Polyprosopus Couch, 1862, 313.
Cetorhinus Blainville, 1816, 95.   Hannovera Van Beneden, 1871, 360.
Selache Cuvier, 1817, 97.

[57] A name given to a doubtful or perhaps wholly mythical account apparently of the Basking Shark (*Cetorhinus*).

Series *GALEOIDEI*

Family 63. **GALEIDÆ** (*Galeorhinidæ; Carcharhinidæ*)
(Gray Sharks)

Galeus Valmont, 1768, 27.
Mustelus Valmont, 1768, 27.
Galeus Klein, 1775, 37.
Cynocephalus Klein, 1777, 42, 407.
Mustelus Linck, 1790, 49.
Aodon[58] Lacépède, 1798, 55.
Carcharias[59] Rafinesque, 1810, 77.
Galeus Rafinesque, 1810, 78.
Mustelus Leach, 1812, 84.
Mustellus Fischer, 1813, 85.
Carcharhinus Blainville, 1816, 95.
Galeorhinus Blainville, 1816, 95.
Carcharias Cuvier, 1817, 97.
Galeus Cuvier, 1817, 97.
Mustelus Cuvier, 1817, 97.
Leptocarias Smith, 1837, 190.
Scoliodon Müller & Henle, 1837, 190.
Galeocerdo M. & H., 1837, 192, 573.
Leptocharias M. & H., 1837, 193, 573.
Physodon[60] M. & H., 1838.
Aprion M. & H., 1838, 192.
Hypoprion M. & H., 1838, 192.
Triakis M. & H., 1838, 192.
Triænodon M. & H., 1838, 192.
Loxodon M. & H., 1838, 192.
Thalassorhinus M. & H., 1838, 192.
Prionodon M. & H., 1838, 192.
HEMIPRISTIS Agassiz, 1843, 214.
GLYPHIS Agassiz, 1843, 402.
Myrmillo Gistel, 1848, 237.
Prionace Cantor, 1849, 241.
Hemigaleus Bleeker, 1852, 250.

GALEODES Heckel, 1854, 259.
MICRODUS Emmons, 1857, 275.
PROTOGALEUS Molin, 1860, 298.
Platypodon Gill, 1861, 305, 331.
Hypoprionodon Gill, 1861, 306.
Eulamia Gill, 1861, 306.
Isogomphodon Gill, 1861, 306.
Lamiopsis Gill, 1861, 306.
Isoplagiodon Gill, 1861, 306.
Aprionodon Gill, 1861, 303, 306.
Cynocephalus (Klein) Gill, 1861, 306.
Boreogaleus Gill, 1861, 306.
Chænogaleus Gill, 1861, 306.
Rhinotriacis Gill, 1862, 317.
Eugaleus Gill, 1864, 330.
Pleuracromylon Gill, 1864, 330.
ALOPIOPSIS[61] Lioy, 1865, 388.
Leptocarcharias (Smith) Günther, 1870, 357.
Dirrhizodon Klunzinger, 1870, 358.
PSEUDOGALEUS Jaekel, 1894, 465.
Scylliogaleus Boulenger, 1902, 497.
Cynias[62] Gill, 1903.
HUNSRUCKIA Traquair, 1903, 506.
GEMUNDINA Traquair, 1903, 506.
CARCHARIOIDES Ameghino, 1906, 516.
Calliscyllium Tanaka, 1912, 546.
CARCHAROIDES Chapman, 1913, 547.
Eridacnis Smith, 1913, 551.
Rhizoprion Ogilby, 1915, 557.
GYRACE[63] Jordan & Hannibal, 1922.

---

[58] Perhaps imaginary.

[59] As "rigidly construed" the name *Carcharias* of Rafinesque must be excluded from this family.

[60] *Plagiostomen*, 30; orthotype *Carcharias mulleri* Müller & Henle. (Omitted in "Genera of Fishes.")

[61] Misprinted *Hopiopsis* by Zittel.

[62] *Proc. U. S. Nat. Mus.*, 960, 1903; orthotype *Squalus canis* Mitchill. (Omitted in "Genera of Fishes.")

[63] *Gyrace* new genus; orthotype *Scymnus occidentalis* Agassiz, California Miocene, allied to *Galeorhinus* but with the longer teeth peculiarly twisted, and coarsely serrated at base.

Family 64. SPHYRNIDÆ
(Hammer-head Sharks)

This family is hardly separable from the *Galeidæ,* differing only in the form of the head.

Cestracion Klein, 1775, 40.
Sphyrna Rafinesque, 1810, 82.
Sphyrnias Rafinesque, 1815, 92.
Cestrorhinus Blainville, 1816, 95.
Zygæna Cuvier, 1817, 97.
Sphyrichthys Thienemann, 1828, 173.
Platysqualus Swainson, 1839, 204.

Sphyra Van der Hoeven, 1849, 243.
Cestracion (Klein) Gill, 1861, 306, 303.
Eusphyra Gill, 1861, 306.
Reniceps Gill, 1861, 306.
Cestracion (Walbaum after Klein) Ogilby, 1916, 560.

Family 65. PRISTIOPHORIDÆ
(Saw Sharks)

Pristiophorus Müller & Henle, 1837, 192, 573.

Family 66. PLIOTREMIDÆ

Pliotrema Regan, 1906, 520.

Order TECTOSPONDYLI
Suborder SQUALOIDEI
Family 67. SQUALIDÆ
(Dog-fishes)

These sharks which, like the rays, lack the anal fin have also simplified or degenerate vertebræ in which the hard material is arranged in rings without radiation. In the most primitive family, *Squalidæ,* spines are present in each of the two dorsal fins, a character which may indicate descent from allies of the *Heterodontidæ,* in which group (*Cestraciontes*) such spines also occur.

Squalus Linnæus, 1758, 11.
Dalatias Rafinesque, 1810, 77.
Cerictius[64] Rafinesque, 1810, 78.
Etmopterus Rafinesque, 1810, 78.
Acanthorhinus Blainville, 1816, 95.
Spinax Cuvier, 1817, 97.
Acanthias Risso, 1826, 119.
Centroscyllium Müller & Henle, 1838, 193.
Centrophorus M. & H., 1838, 192.
Acanthidium Lowe, 1839, 195.
Spinax Bonaparte, 1845, 225.
Lepidorhinus Bonaparte, 1845, 225.
Acanthias Bonaparte, 1845, 225.
Carcharias Gistel, 1848, 234.

CENTROPTERUS Costa, 1861, 301.
Entoxychirus Gill, 1862, 317.
Scymnodon Bocage & Capello, 1864, 329.
Centroscymnus B. & C., 1864, 329.
Machephilus Johnson, 1867, 346.
CENTROPHOROIDES Davis, 1885, 431.
Paracentroscyllium Alcock, 1889, 444.
Deania Jordan & Snyder, 1902, 499.
Zameus Jordan & Fowler, 1903, 504.
Nasisqualus Smith, 1912, 546.
Centroselachus[65] Garman, 1913.
PROTOSPINAX Woodward, 1918, 565.
Atractophorus Gilchrist, 1922. (See index.)

---

[64] Mythical.

[65] *Plagiostomia,* 306; orthotype *Centrophorus crepidater* Bocage and Capello. (Omitted in "Genera of Fishes.")

Family 68. OXYNOTIDÆ

Oxynotus Rafinesque, 1810, 82.   Centrina Cuvier, 1817, 97.

Family 69. SCYMNORHINIDÆ (*Dalatiidæ*[66]; *Echinorhinidæ*)

Echinorhinus Blainville, 1816, 95.
Scymnus Cuvier, 1817, 97.
Goniodus Agassiz, 1836, 184.
Scymnorhinus Bonaparte, 1845, 225.
Borborodes Gistel, 1848, 237.

Euprotomicrus[67] Gill, 1864, 331.
Isistius Gill, 1864, 331.
Leius Kner, 1865, 338.
Cirrhigaleus Tanaka, 1912, 546.
Squaliolus Smith, 1912, 546.

[66]The earliest logotype of *Dalatias* Rafinesque was one of the *Squalidæ*, possibly *Centrophorus granulosus*.

[67] This genus and the next two are very aberrant, and each might stand as type of a distinct family. The same is also true of *Echinorhinus*.

Family 70. SOMNIOSIDÆ
(Sleeper Sharks)

Somniosus LeSueur, 1818, 107.
Læmargus Müller & Henle, 1838, 192.
Leiodon Wood, 1847, 232.

Rhinoscymnus Gill, 1864, 331.
Heteroscymnus Tanaka, 1912, 546.

Suborder *SQUATINOIDEI*
Family 71. SQUATINIDÆ
(Angel Sharks)

Rhina Klein, 1775, 39.
Squatina Duméril, 1906, 75.
Rhina Rafinesque, 1810, 78.
Squalraia De la Pylaie, 1835, 183.
THAUMAS Münster, 1842, 212.

PHORCYNUS Thiollière, 1854, 261.
Rhina (Klein) Gill, 1861, 306.
SCALDIA Le Hon, 1871, 361.
TRIGONODUS Winkler, 1878, 398.

Order BATOIDEI
(*Hypotremata*)
Suborder *SARCURA*
Family 72. PRISTIDÆ
(Saw-fishes)

Pristis Klein, 1775, 41.
Pristis Linck, 1790, 49.
Pristis Latham, 1794, 52.
Pristibatis Serville, 1829, 134.
MYRIOSTEON Gray, 1864, 332.
PROPRISTIS Dames, 1883, 422.

AMBLYPRISTIS Dames, 1888, 440.
SCLERORHYNCHUS Woodward, 1889, 449.
Pristiopsis Fowler, 1905, 512.
EOPRISTIS Stromer, 1905, 516.
ONYCHOPRISTIS[68] Stromer, 1918.

[68] *Abhandl. Bayer. Akad.*, **28** : 17 ; orthotype *Gigantichthys numidus* Haug.

Family 73. RHINIDÆ (*Rhamphobatidæ*; *Rhynchobatidæ*)

Rhina Bloch & Schneider, 1801, 59.
Rhynchobatus[69] Müller & Henle, 1837, 192, 573.

Rhamphobatis Gill, 1861, 306.

[69] Sometimes spelled *Rhynchobatis*. The above is the original form.

Family 74. RHINOBATIDÆ
(Guitar-fishes)

Rhinobatos Klein, 1775, 40.
Rhinobatus Linck, 1790, 49.
Rhinobatus Bloch & Schneider, 1801, 59.
Leiobatus Rafinesque, 1810, 78.
Squatinoraja Nardo, 1824, 116.
AELLOPOS Münster, 1836, 185.
Syrrhina Müller & Henle, 1838, 192.
Trygonorrhina[70] M. & H., 1838, 192.

EURYARTHRA Agassiz, 1843, 214.
CYCLARTHRUS Agassiz, 1843, 214.
Glaucostegus Bonaparte, 1845, 225.
Demiurga Gistel, 1848, 237.
SPATHOBATIS Thiollière, 1849, 244, 261.
Rhynchobatis Philippi, 1857, 276.
Zapteryx Jordan & Gilbert, 1880, 404.
Tarsistes Jordan, 1919, 567.

[70] This name appears first in *Ann. Mag. Nat. Hist.* (2), **2**: 90, 1838, spelled with "rr."

Family 75. ASTERODERMIDÆ

ASTERODERMUS[71] Agassiz, 1843, 214.

BELEMNOBATIS Thiollière, 1854, 261.

[71] *Asterodermus* Agassiz, 1843, is perhaps preoccupied by *Astrodermus* Bonelli, 1829.

Family 76. PLATYRHINIDÆ *(Discobatidæ; Analithidæ)*

Platyrhina Müller & Henle, 1838, 193.
Analithis Gistel, 1848, 237.
Platyrhinoidis Garman, 1880, 402.

Discobatus Garman, 1880, 402.
Zanobatus Garman, 1913, 548.
Arhynchobatis Waite, 1909, 534.

Family 77. RAJIDÆ
(Rays; Skates)

Raja[72] Linnæus, 1758, 11.
Leiobatus Klein, 1775, 38.
Dipturus Rafinesque, 1810, 78.
Cephaleutherus Rafinesque, 1810, 83.
Platopterus Rafinesque, 1815, 92.
Lævirajæ[73] Nardo, 1827, 121.
Propterygia Otto, 1824, 117.
Rajabatis De La Pylaie, 1835, 183.
Batis Bonaparte, 1837, 187.
Læviraja Bonaparte, 1837, 187.
Dasybatus Bonaparte, 1837, 187.
Sympterygia Müller & Henle, 1838, 193.

Uraptera M. & H., 1838, 193.
Hieroptera Fleming, 1841, 209.
ACTINOBATIS Agassiz, 1843, 214.
Psammobatis Günther, 1870, 357.
ONCOBATIS Leidy, 1871, 362.
Leucoraja Malm, 1877, 391.
Amblyraja Malm, 1877, 391.
Malacorhina Garman, 1877, 389.
ACANTHOBATIS Larrazet, 1886, 436.
PLATYSPONDYLUS Haug, 1905, 514.
Malacobatis Gracianov, 1907, 523.
Dactylobatus Bean & Weed, 1909, 531.

[72] Also spelled *Raia* by authors.
[73] Name given in plural form only.

Family 78. ARTHROPTERIDÆ

ARTHROPTERUS Agassiz, 1843, 214.

Suborder *NARCACIONTES*
Family 79. TORPEDINIDÆ *(Narcaciontidæ; Narcobatidæ)*
(Torpedos; Numb-fishes)

Torpedo[74] Forskål, 1775, 32.
Narcacion Klein, 1775, 39.

Torpedo Duméril, 1806, 75.
Torpedo Rafinesque, 1810, 82.

Narcobatus Blainville, 1816, 95.
Narke Kaup, 1826, 121.
Temera Gray, 1831, 138.
Narcine Henle, 1834, 180.
Astrape Müller & Henle, 1838, 193.
Syrraxis (Jourdan) Bonaparte, 1841, 208.
Narcopterus Agassiz, 1843, 214.
Discopyge Tochudi, 1846, 230.
Hypnos Duméril, 1852, 250.
Narcacion (Klein) Bleeker, 1866, 340.
Gonionarce Gill, 1861, 307.

Cyclonarce Gill, 1861, 307.
Tetranarce[75] Gill, 1861, 307.
Fimbriotorpedo Fritsch, 1886, 435.
Gymnotorpedo Fritsch, 1886, 435.
Benthobatis Alcock, 1898, 478.
Hypnarce Waite, 1902, 501.
Tetronarcine Tanaka, 1908, 531.
Bengalichthys Annandale, 1909, 531.
Typhlonarke Waite, 1909, 534.
Eunarce Fowler, 1910, 535.
Heteronarce[75a] Regan, 1921.

[74] In part; the name is used in an oblique case only, and the Electric Ray is confused with the Electric Cat-fish.

[75] Misprinted *Tetronarce*.

[75a] Orthotype *H. garmani* Regan.

Suborder *MASTICURA*
(*Centrobates*)

Family 80. DASYATIDÆ (*Trygonidæ; Dasybatidæ*)
(Sting-Rays)

Dasybatus Klein, 1775, 39.
Dasyatis Rafinesque, 1810, 78.
Uroxis Rafinesque, 1810, 83.
Trygonobatus Blainville, 1816, 94.
Leiobatus Blainville, 1816, 95.
Trygon (Adanson) Cuvier, 1817, 98.
Trygon Isidore Geoffroy St. Hilaire, 1826, 120.
Pastinacæ[76] Nardo, 1827, 121.
Pastinachus Rüppell, 1828, 122.
Anacanthus (Ehrenberg) Cuvier, 1829, 131.
Tæniura Müller & Henle, 1837, 189.
Himantura M. & H., 1837, 190.
Gymnura M. & H., 1837, 190.
Urogymnus M. & H., 1837, 190.
Hemitrygon M. & H., 1837, 190.
Pteroplatea M. & H., 1838, 193.
Trygonoptera M. & H., 1838, 193.
Urolophus M. & H., 1838, 193.
Hyplophus M. & H., 1838, 193.

Pastinaca Dekay, 1842, 210.
Cyclobatis Egerton, 1844, 220.
Rhachinotus Cantor, 1849, 241.
Pastinaca Gronow, 1854, 258.
Pastinaca Duméril, 1856, 268.
Alexandrinus Molin, 1861, 308.
Urotrygon Gill, 1863, 324.
Neotrygon Castelnau, 1873, 368.
Gryphodobatis Leidy, 1877, 391.
Heliobatis Marsh, 1877, 391.
Xiphotrygon Cope, 1879, 399.
Dasybatus (Klein) Garman, 1885, 432.
Discobatis Miklukho-Maclay & Macleay, 1886, 436.
Discotrygon Fowler, 1910, 535.
Pteroplatytrygon Fowler, 1910, 535.
Hypolophites Leriche, 1913, 549.
Amphotistius[77] Garman, 1913.
Urobatis Garman, 1913, 548.
Dasybatus (Klein, Walbaum) Ogilby, 1916, 560.

[76] Name given in plural form only.

[77] *Plagiostomia*, 8; type *Trygon sabina* Le Sueur. (Omitted in "Genera of Fishes.")

Family 81. POTAMOTRYGONIDÆ (*Elipesuridæ*)

Elipesurus Schomburgk, 1842, 212.
Paratrygon Duméril, 1865, 337.
Dynatobatis Larrazet, 1886, 436.

Disceus Garman, 1877, 389.
Potamotrygon Garman, 1877, 389.

Family 82. BRACHIOPTERIDÆ

Brachioptera Gracianov, 1906, 518. Phanerocephalus[78] Gracianov, 1906, 518.

[78] Misprinted *Planerocephalus*.

Family 83. PTYCHODONTIDÆ

Ptychodus Agassiz, 1839, 194. Hemiptychodus Jaekel, 1894, 465.
Sporetodus Cope, 1874, 375.

Family 84. MYLIOBATIDÆ (*Aëtobatidæ*)
(Eagle Rays)

Aëtobatus[79] Blainville, 1816, 95. Goniobatis Le Hon, 1871, 361.
Myliobatis (Duméril) Cuvier, 1817, 573. Mesobatis Leidy, 1877, 391.
Aëtoplatea Müller & Henle, 1838, 193. Bates Probst, 1877, 392.
Ptychopleurus Agassiz, 1838, 190. Apocopodon Cope, 1886, 434.
Stoasodon Cantor, 1849, 241. Promyliobatis Jaekel, 1894, 465.
Aulodus Dixon, 1850, 245. Aëtomylæus Garman, 1908, 528.
Goniobatis Agassiz, 1858, 277. Pteromylæus Garman, 1913, 548.
Holorhinus Gill, 1862, 317. Hylæobatis Woodward, 1915, 558.
Plinthicus Cope, 1869, 354.

[79] Also written *Aëtobatis*.

Family 85. RHINOPTERIDÆ

Rhinoptera (Kuhl) Cuvier, 1829, 173. Mylorhina Gill, 1865, 337.
Zygobatis Agassiz, 1836, 184, 214. Micromesus Gill, 1865, 337.

Family 86. MOBULIDÆ (*Mantidæ; Cephalopteridæ*)
(Devil Rays)

Cephalopterus Risso, 1810, 77. Ceratoptera Müller & Henle, 1838, 193.
Mobula Rafinesque, 1810, 82. Pterocephala Swainson, 1839, 204.
Apterurus Rafinesque, 1810, 83. Brachioptilon Newman, 1849, 244.
Dicerobatus Blainville, 1816, 95. Diabolichthys Holmes, 1856, 270.
Manta Bancroft, 1829, 174, 573. Ceratobatis Boulenger, 1897, 476.

Subclass *HOLOCEPHALI*
Order CHIMAEROIDEI
(*Chismopnea*) (Chimæras)

Family 87. PTYCTODONTIDÆ

Aulacosteus Eichwald, 1846 (1860), 295, 229. Rhynchodus Newberry, 1873, 371.
Ramphodus Jaekel, 1903, 504.
Ptyctodus Pander, 1858, 283. Acmoniodus Hussakof & Bryant, 1919, 566.
Rinodus Newberry & Worthen, 1866, 342. Deinodus H. & B., 1919, 566.

Family 88. SQUALORAIIDÆ

Squaloraia Riley, 1826, 173. Chalcodus Zittel, 1887, 439.
Spinacorhinus Agassiz, 1836, 184.

Family 89. MYRIACANTHIDÆ

CTENACANTHUS[80] Agassiz, 1836, 186.
MYRIACANTHUS Agassiz, 1837, 186.
PHYSONEMUS[80] M'Coy, 1848, 239.
XYSTRACANTHUS[80] Leidy, 1859, 292.
DREPANACANTHUS[80] Newberry &
Worthen, 1866, 342.
PROGNATHODUS Egerton, 1872, 365.

BATACANTHUS[80] St. John & Worthen,
1883, 425.
ACONDYLACANTHUS St. J. & W., 1883,
425.
METOPACANTHUS Zittel, 1887, 439.
CHIMÆROPSIS Zittel, 1887, 439.

[80] Probably all these represent frontal claspers of a chimæroid; see Eastman,
Proc. U. S. Nat. Mus., **52** : 262, 1917.

Family 90. CHIMÆRIDÆ
(Chimæras or Elephant-fishes)

Chimæra Linnæus, 1758, 11.
LEPTACANTHUS Agassiz, 1837, 186.
ORACANTHUS Agassiz, 1837, 186.
EDAPHODON Buckland, 1838, 191.
PASSALODON Buckland, 1838, 191.
Ameibodon Buckland, 1838, 191.
ISCHYODUS Egerton, 1843, 214.
ELASMODUS Egerton, 1843, 214.
GANODUS Agassiz, 1843, 214.
PSITTACODON Agassiz, 1843, 214.
ERISMACANTHUS[81] M'Coy, 1848, 239.
DIPRIACANTHUS[81] M'Coy, 1848, 408.
MYLOGNATHUS Leidy, 1856, 272.
CHIMÆRACANTHUS Quenstedt, 1858, 283.
Hydrolagus Gill, 1862, 317.
AULUXACANTHUS Sauvage, 1867, 347.
LEPTOMYLUS Cope, 1869, 354.
DIPRISTIS Gervais, 1869, 504.
SPHAGEPŒA Cope, 1870, 356.
DIPRISTIS Marsh, 1870, 359.
EUMYLODUS Leidy, 1873, 370.
LECRACANTHUS[81] St. John & Worthen,
1875, 381.
GAMPSACANTHUS[81] St. J. & W., 1875, 381.

ISOTÆNIA Cope, 1875, 379.
DIPHRISSA Cope, 1875, 379.
BRYACTINUS Cope, 1875, 379.
HEDRONCHUS Cope, 1876, 385.
MYLEDAPHUS Cope, 1876, 385.
ELASMOGNATHUS Newton, 1878, 397.
CLADACANTHUS (Agassiz) Davis, (1842)
1883, 213, 422.
HARPACANTHUS Traquair, 1886, 436.
ELASMODECTES Newton, 1888, 444.
STETHACANTHUS Newberry, 1889, 447.
VAILLANTOÖNIA[82] Meunier, 1891, 455.
Palæchimæra[83] Walcott, 1891, 456.
ELASMODECTES Newton, 1891, 456.
PALÆOMYLUS Woodward, 1891, 456.
DICTYORHABDUS[83] Walcott, 1892, 459.
PACHYMYLUS Woodward, 1892, 460.
BRACHYMYLUS Woodward, 1892, 460.
AMYLODON Storms, 1894, 466.
BELEMNACANTHUS Eastman, 1898, 479.
Bathyalopex Collett, 1901, 492.
Psychichthys Fowler, 1907, 523.
ACANTHORHINA Fraas, 1910, 535.

[81] These four nominal genera are regarded by Eastman as identical and based on
frontal claspers of male chimæroids.

[82] Fossil egg-case of a chimæroid.

[83] It seems by no means certain that *Dictyorhabdus* (*Palæchimæra*) is really a fish
relic. Occurring in the Ordovician, it is far older than any known *Chimæra*. It may
be a *Graptolite*. Cockerell informs me that he "has good material showing surface
sculpture and is entirely convinced that it is not a vertebrate."

Note on ICHTHYODORULITES: The fin-spines named below may have belonged to species of *Chimæridæ* or of some related family.

CYLINDRACANTHUS Leidy, 1856, 273.

RABDACANTHUS Pander, 1856, 274.

BYSSACANTHUS Salter, 1861, 309.

LISTRACANTHUS Newberry & Worthen, 1870, 359.

CYRTACANTHUS Newberry, 1873, 371.

OSTRACACANTHUS Davis, 1879. 399.

EUCTENIUS Traquair, 1881, 417.

EUACANTHUS Trautschold, 1883, 426.

GNATHACANTHUS Davis, 1883, 422.

APATEACANTHUS Woodward, 1891, 456.

GLYPTORHYNCHUS Leriche, 1905, 515.

Family 91. RHINOCHIMERIDÆ

Harriotta Goode & Bean, 1894, 464.

Rhinochimæra Garman, 1901, 493.

Anteliochimæra Tanaka, 1909, 534.

Family 92. CALLORHYNCHIDÆ

Callorhynchus Gronow, 1763, 18.

Callorhynchus (Gronow) Cuv., 1817, 98.

Class PISCES
(True Fishes)

Subclass CROSSOPTERYGII

The fishes of this group possess a jointed pectoral limb which may be homologous with that of the primitive sharks of the order *Ichthyotomi*. It is not certain, however, that this is not a case of analogy, in which case it does not indicate genetic relationship. The class is certainly very old in geologic time, standing at the base of the series of true fishes, while at the same time holding an ancestral relation to the amphibia and the higher vertebrates. A few species are still extant.

Order RHIPIDISTIA
(*Taxistia*)

Family 93. HOLOPTYCHIDÆ

HOLOPTYCHUS[84] Agassiz, 1839, 194.
DENDRODUS Owen, 1841, 209.
PLATYGNATHUS Agassiz, 1844, 218.
GLYPTOLEPIS Agassiz, 1844, 218.
SCLEROLEPIS Eichwald, 1844, 220.

LAMNODUS Agassiz, 1845, 225.
APEDODUS Leidy, 1858, 283.
ERIPTYCHIUS[85] Walcott, 1892, 459.
MAWSONIA Woodward, 1907, 525.

---

[84] Usually spelled *Holoptychius*.
[85] Scales of the Ordovician age, doubtful as to interpretation.

Family 94. MEGALICHTHYIDÆ[86] (*Rhizodontidæ*)

MEGALICHTHYS Agassiz & Hilbert, 1836, 185.
RHIZODUS Owen, 1840, 207.
CRICODUS Agassiz, 1844, 217.
SAURIPTERIS[87] Hall, 1843, 214.
COLONODUS M'Coy, 1848, 239.
GYROPTYCHIUS M'Coy, 1848, 239.
RABDIOLEPIS Emmons, 1857, 275.
POLYPLACODUS Pander, 1860, 298.
TRISTICHOPTERUS Egerton, 1861, 301.
STREPSODUS Young, 1866, 344.
RHIZODOPSIS Young, 1866, 344.
DENDROPTYCHIUS (Huxley) Young, 1866, 344.

GASTRODUS Owen, 1867, 347.
CHARACODUS Owen, 1867, 347
GANOLODUS Owen, 1867, 347.
ARCHICHTHYS Hancock & Atthey, 1870, 358.
LABYRINTHODONTOSAURUS Barkas, 1873, 366.
ORTHOGNATHUS Barkas, 1873, 366.
SIGMODUS[87a] Waagen, 1879.
EUSTHENOPTERON[88] Whiteaves, 1881, 418.
CŒLOSTEUS[89] Newberry, 1887.
DICTYONOSTEUS[90] Stromer, 1918.

---

[86] The name *Megalichthys* has priority over its synonym *Rhizodus*.
[87] Usually written *Sauripterus*.
[87a] *Loc. cit.*, 10; orthotype *S. dubius* Waagen. (Omitted in "Genera of Fishes.")
[88] Misprinted *Eusthenopleuron* in "Genera of Fishes."
[89] *Trans. N. Y. Ac. Sci.*, **6**: 137; type *C. ferox* Newberry. (Omitted in "Genera of Fishes."
[90] *Bull. Geol. Soc. Upsala,* **16**: 116; orthotype *D. arcticus* Stromer.

Family 95. OSTEOLEPIDÆ (*Rhombodipteridæ*)

OSTEOLEPIS Valenciennes, 1829, 134.
DIPLOPTERUS Agassiz, 1835, 181.
MEGALICHTHYS Agassiz, 1844, 217.
GLYPTOPOMUS Agassiz, 1844, 218.
TRIPTERUS M'Coy, 1848, 239.
PARABATRACHUS Owen, 1853, 254.
DIPLOPTERAX M'Coy, 1855, 265.
TRIPLOPTERUS, M'Coy, 1855, 266.
GYROLEPIS Kade, 1858, 281.
GLYPTOLÆMUS (Huxley) Anderson, 1859, 287.

SPOROLEPIS Romanowsky, 1864, 334.
PLINTHOLEPIS Romanowsky, 1864, 334.
RHOMBOPTYCHUS (Huxley) Young, 1866, 344.
ECTOSTEORHACHIS Cope, 1880, 402, 422.
THURSIUS Traquair, 1888, 443.
GLYPTOGNATHUS Woodward, 1891, 456.
POROLEPIS Woodward, 1891, 456.
GYROPELTUS Rohon, 1893, 462.
PALÆOSTEUS Rohon, 1893, 462.

Family 96. ONYCHODONTIDÆ

ONYCHODUS Newberry, 1857, 276.

Order ACTINISTIA
(*Cœlacanthini*)

Family 97. CŒLACANTHIDÆ

UNDINA Münster, 1834, 180.
MACROPOMA Agassiz, 1835, 181, 280.
LIBYS Münster,. 1842, 212.
CŒLACANTHUS Agassiz, 1844, 218.
HOPLOPYGUS Agassiz, 1844, 217.
HEPTANEMA Bellotti, 1857, 274.
KOKKODERMA[91] Quenstedt, 1858, 283.
GRAPHIURUS Kner, 1866, 341.
LOPHOPRIONOLEPIS Steinla, 1868, 350.
PEPLORHINA Cope, 1873, 368.
CONCHIOPSIS Cope, 1873, 368.

DIPLURUS Newberry, 1878, 396.
RHABDODERMA Reis, 1888, 443.
CŒLACANTHOPSIS Traquair, 1901, 576.
LEIODERMA[92] Stensiö, 1918.
WIMANIA[93] Stensiö, 1921.
SASSENIA[94] Stensiö, 1921.
AXELIA[95] Stensiö, 1921.
MYLACANTHUS[96] Stensiö, 1921.
SCLERACANTHUS[97] Stensiö, 1921.
DIPLOCERCIDES[98] (Jaekel) Stensiö, 1922.

[91] Later corrected to *Coccoderma*.
[92] Erik A. Son Stensiö, "Zur Kenntniss des Devons und des Kulms an der Klaas, Billenbag Spitzbergen," *Geol. Inst. Bull. Upsala*, **16**; orthotype *Leioderma sinuosa* Stensiö; name preoccupied, replaced by *Wimania* Stensiö, 1921.
[93] "Triassic Fishes from Spitzbergen," p. 51; orthotype *Leioderma sinuosa* Stensiö.
[94] *Loc. cit.*, p. 84; orthotype *Sassenia tuberculata* Stensiö.
[95] *Loc. cit.*, p. 89; orthotype *Axelia robusta* Stensiö.
[96] *Loc. cit.*, p. 107; orthotype *Mylacanthus lobatus* Stensiö.
[97] *Loc. cit.*, p. 111; orthotype *Scleracanthus asper* Stensiö.
[98] Stensiö, "Ueber Zwei Cœlacanthiden aus dem Oberdevon von Wildungen," *Palæontologisches Zeitschrift*, **4**: 168; orthotype *Holoptychius kayseri* von Koenen.

Family 98. TARRASIIDÆ

TARRASIUS Traquair, 1881, 418.
CONCHOPOMA Kner, 1868, 352.

PALÆOPHICHTHYS Eastman, 1908, 527.

Order CLADISTIA
Family 99. POLYPTERIDÆ
Polypterus St. Hilaire, 1798, 54.

Family 100. CALAMOICHTHYIDÆ
Herpetoichthys (Smith, J. A.) Günther,   Erpetoichthys Smith, 1865, 339.
    1865, 338.   Calamoichthys Smith, 1866, 343.

Subclass DIPNEUSTA
·(Dipnoi; name preoccupied)

Order CTENODIPTERINI
(*Ctenodipneusta*)

Family 101. DIPTERIDÆ

DIPTERUS Sedgwick & Murchison, 1828,   PALÆDAPHUS Van Beneden & De Ko-
    125.   ninck, 1864, 329.
CATOPTERUS Agassiz, 1833, 177.   GOMPHOLEPIS Barrande, 1872, 362.
POLYPHRACTUS Agassiz, 1844, 218.   GANORHYNCHUS Traquair, 1873, 371.
STAGONOLEPIS Agassiz, 1845, 225.   HELIODUS Newberry, 1875, 379.
OSTEOPLAX M'Coy, 1848, 239.   DIPNOITES Fritsch, 1888, 441.
CONCHODUS M'Coy, 1848, 239.   TYLODUS Rohon, 1893, 462.
CHEIRODUS Pander, 1858, 283.   CHELOMODUS Rohon, 1893, 462.
ARCHÆONECTES[99] Von Meyer, 1859.   PENTLANDIA Watson & Day, 1916, 561.

[99] *Palæontographica,* 7 : 12; type *A. pertusus* Von Meyer. (Omitted in "Genera of Fishes.")

Family 102. PHANEROPLEURIDÆ
PHANEROPLEURON (Huxley) Anderson,   SCAUMENACIA Traquair, 1893, 463.
    1859, 287.

Order SIRENOIDEI
Family 103. CTENODONTIDÆ

CTENODUS Agassiz, 1838, 191.   PTYONODUS Cope, 1877, 388.
RHADAMISTA Gistel, 1848, 235.   MEGAPLEURON Gaudry, 1883, 423.
CAMPYLOPLEURON Huxley, 1867, 346.   GNATHORHIZA[100] Cope, 1883, 422.
SAGENODUS Owen, 1867, 347.   HEMICTENODUS Jaekel, 1890, 452.

[100] A genus of uncertain relationship.

Family 104. URONEMIDÆ
URONEMUS Agassiz, 1844, 217.   GANOPRISTODUS Traquair, 1881, 417.

Family 105. CERATODONTIDÆ
(Lung-fishes)

CERATODUS Agassiz, 1838, 191.   GOSFORDIA Woodward, 1890, 453.
SCROPHA Gistel, 1848, 235.   Epiceratodus Teller, 1891, 456.
Neoceratodus[101] Castelnau, 1876, 384.   SYNTHETODUS Eastman, 1898, 476.
Ompax[101] Castelnau, 1879, 399.   METACERATODUS Chapman, 1910, 534.

[101] Perhaps mythical, in part at least.

Family 106. LEPIDOSIRENIDÆ
(Mud Sirens)

Lepidosiren Fitzinger, 1837, 189.   **Amphibichthys** Hogg, 1841, 209.

Family 107. PROTOPTERIDÆ

Protopterus Owen, 1839, 196.   **Rhinocryptis** Peters, 1844, 222.
Protomelus Hogg, 1841, 209.

Subclass *ACTINOPTERI*
Superorder *GANOIDEI*
Order CHONDROSTEI
(Lysopteri; Podopterygia; Lyomeri; Heterocerci; Glaniostomi)

Family 108. PALÆONISCIDÆ

PALÆONISCUM[102] Blainville, 1818, 108.

PALÆOTHRISSUM Blainville, 1818, 108.

ACROLEPIS Agassiz, 1833, 177.

GYROLEPIS Agassiz, 1833, 177.

AMBLYPTERUS Agassiz, 1833, 177.

PYGOPTERUS Agassiz, 1833, 177.

CHEIROLEPIS Agassiz, 1835, 181.

COCCOLEPIS Agassiz, 1844, 218.

MICROLEPIS Eichwald, 1844, 220.

ISODUS M'Coy, 1848, 239.

ELONICHTHYS Giebel, 1848, 233.

UROSTHENES Dana, 1848, 233.

TRACHELACANTHUS Fischer von Waldheim, 1850, 246.

PROPALÆONISCUS Pomel, 1853, 254.

OXYGNATHUS Egerton, 1855, 264.

MECOLEPIS Newberry, 1856, 273.

RHABDOLEPIS Troschel, 1857, 276.

EURYLEPIS Newberry, 1857, 276.

UROLEPIS Bellotti, 1857, 274.

COSMOLEPIS Egerton, 1858, 280.

CENTROLEPIS Egerton, 1858, 280.

THRISSONOTUS (Agassiz), (1844) 1858, 217.

MYRIOLEPIS Egerton, 1864, 330.

CYCLOPTYCHIUS Young, 1866, 344.

GANACRODUS Owen, 1867, 347.

PLATYSIAGUM Egerton, 1872, 365.

LEIOLEPIS Goldenburg, 1873, 369.

NEMATOPTYCHIUS Traquair, 1875, 382.

HYPTERUS Owen, 1876, 386.

SPHÆROLEPIS Fritsch, 1877, 389.

ACENTROPHORUS Traquair, 1877, 392.

GONATODUS Traquair, 1877, 392.

COSMOPTYCHIUS Traquair, 1877, 392.

MICROCONODUS Traquair, 1877, 392.

RHADINICHTHYS Traquair, 1877, 392.

HOLURUS Traquair, 1881, 418.

CRYPHIOLEPIS Traquair, 1881, 418.

CANOBIUS Traquair, 1881, 418.

EUPALÆONISCUS Rzehak, 1881, 417.

ANAGLYPHUS Rzehak, 1881, 417.

ACTINOPHORUS Newberry, 1889, 447.

ATHERSTONIA Woodward, 1889, 449.

APATEOLEPIS Woodward, 1890, 453.

DRYDENIUS Traquair, 1890, 453.

MESOPOMA Traquair, 1890, 453.

PALÆONISCINOTUS Rohon, 1890, 452.

ÆDUA Sauvage, 1891, 456.

ARCHÆONISCUS Sauvage, 1891, 456.

TEGEOLEPIS Miller, 1892, 458.

HAPLOLEPIS Miller, 1892, 458.

TRISSOLEPIS Fritsch, 1893, 460.

GANOLEPIS Woodward, 1893, 463.

PYRITOCEPHALUS Fritsch, 1894, 464.

SCELEPTOPHORUS Fritsch, 1894, 464.

MICROCONODUS Traquair, 1895, 470.

FARNELLIA Traquair, 1898, 484.

PSILICHTHYS[103] Hall, 1900.

SCHIZOSPONDYLUS Fritsch & Bayer, 1902, 498.

DIPHYODUS Lambe, 1902, 500.

---

[102] Written by Agassiz *Palæoniscus.*

[103] Thomas Sergeant Hall, "A New Genus and a New Species of Fish from the Mesozoic Rocks of Victoria," *Mem. Royal Soc. Victoria*, New Series, **12**: 121. Triassic, Melbourne.

ELPISOPHOLIS Woodward, 1908, 531.
HELICHTHYS Broom, 1909, 532.
HYDROPESSUM Broom, 1909, 532.
PELICHTHYS Broom, 1913, 547.
DISICHTHYS Broom, 1913, 547.

CARUICHTHYS Broom, 1913, 546.
BIRGERIA[104] Stensiö, 1921.
BOREOSOMUS[105] Stensiö, 1921.
GLAUCOLEPIS[106] Stensiö, 1921.
ACRORHABDUS[107] Stensiö, 1921.

[104] "Triassic Fishes, Spitzbergen," p. 151, 1921; orthotype *Saurichthys mougeoti* Agassiz.
[105] *Loc. cit.*, p. 211; orthotype *Acrolepis arcticus* Woodward.
[106] *Loc. cit.*, p. 200; orthotype *Glaucolepis gyrolepidoides* Stensiö.
[107] *Loc. cit.*, p. 220; orthotype *Acrorhabdus berteli* Stensiö.

Family 109. PLATYSOMIDÆ (*Scrotidæ*)

EURYNOTUS Agassiz, 1835, 181.
PLATYSOMUS Agassiz, 1835, 181.
GLOBULODUS Münster, 1842, 212.
CHEIRODUS M'Coy, 1848, 239, 408.
STROTES Gistel, 1848, 237.
NOTACMON Gistel, 1848, 236.
PLECTROLEPIS (Agassiz) Egerton, (1835) 1850, 181, 218, 246.
TURSEODUS Leidy, 1858, 283.
MESOLEPIS Young, 1866, 344.
EURYSOMUS[108] Young, 1866, 344.

AMPHICENTRUM Young, 1866, 344.
WARDICHTHYS Traquair, 1875, 382.
BENEDENIUS Traquair, 1878, 396.
PHANEROSTEON Traquair, 1881, 418.
CHEIRODOPSIS Traquair, 1881, 418.
HEMICLADODUS Davis, 1884, 427.
BENEDENICHTHYS Traquair, 1890, 453.
UROPTERYX (Agassiz) Woodward, 1891, 457.
PODODUS Agassiz, 1901, 217.
ECRINESOMUS Woodward, 1910, 537.

[108] Misprinted *Euryosomus* in "Genera of Fishes."

Family 110. DORYPTERIDÆ

DORYPTERUS Münster, 1842, 212.

Family 111. DICTYOPYGIDÆ (*Catopteridæ*)

CATOPTERUS Redfield, 1837, 190.
DICTYOPYGE Egerton, 1847, 231.

REDFIELDIUS Hay, 1899, 486.
PERLEIDUS[109] de Alessandri, 1910.

[109] Gulio de Alessandri, "Studii sui pesci triasici della Lombardia," *Mem. Soc. Ital. Sci. Nat. Mus. Civico Stor. Nat. Milano,* **7** : 49; orthotype *Semionotus altolepis* Deecke.

Family 112. ASTERODONTIDÆ (*Colobodontidæ*)

ASTERODON Münster, 1841, 209.
COLOBODUS Agassiz, 1844, 217.
THOLODUS Meyer, 1846, 229, 249.
OMPHALODUS Meyer, 1847, 322.
CENCHRODUS Meyer, 1847, 407.
HEMILOPAS Meyer, 1847, 408.
NEPHROTUS Meyer, 1851, 249.

CHARITODON[109a] Schmid, 1861.
EUPLEURODUS Gürich, 1884, 428.
DACTYLOLEPIS Kunisch, 1885, 433.
CRENILEPIS Dames, 1888, 440.
DOLLOPTERUS Abel, 1906, 516.
MERIDENSIA[110] Andersson, 1916.
PARALEPIDOTUS[111] Stolley, 1920.

[109a] *Loc. cit.*, 30; orthotype *C. glabridens* Schmid. (Omitted in "Genera of Fishes.")
[110] Erik Andersson, "Ueber Einige Triasfische aus der Cava Trefontane, Tessin," *Bull. Geol. Inst. Upsala,* **15** : 25; orthotype *Pholidophorus meridensis* de Alessandri.
[111] Ernst Stolley, "Beitrage zur Kenntnis des deutschen Muschelkalken," *Palæontographica,* **63** : 41; orthotype *Colobodus ornatus* Agassiz.

Family 113. BELONORHYNCHIDÆ

SAURICHTHYS Agassiz, 1834, 179.
STYLORHYNCHUS[112] Martín, 1873.
ICHTHYORHYNCHUS Bellotti, 1857, 274.
BELONORHYNCHUS Bronn, 1858, 279.

GIFFONUS Costa, 1862, 312.
BROWNEICHTHYS Woodward, 1889, 449.
SAURORHYNCHUS[113] (Münster) Reis, 1891, 207.

[112] Carl Martin, "Ein Beitrage zur Kenntniss fossiler Euganoiden," *Zeitsch. Deutsche Geol. Gesellsch.*, 25 : 725; type *S. tenuirostris* Martin. (Omitted in "Genera of Fishes.")

[113] Name only in 1840, defined by Reis in 1891; *Teleosaurus*, p. 346, is a genus of reptiles.

Family 114. DIPHYODONTIDÆ

DIPHYODUS Lambe, 1902, 500.

STENOPROPOMA Hay, 1903, 503.

Family 115. CHONDROSTEIDÆ

CHONDROSTEUS (Agassiz) Egerton, (1844) 1858, 217, 280.

GYROSTEUS (Agassiz) Egerton, (1844), 1858, 217, 280.

Order GLANIOSTOMI

Family 116. ACIPENSERIDÆ
(Sturgeons)

Acipenser Linnæus, 1758, 11.
Ichthyocolla Geoffroy, 1767, 24.
Sterleta Güldenstädt, 1772, 168.
Sturio Rafinesque, 1810, 82.
Dinectus[114] Rafinesque, 1818, 108.
Sterletus Rafinesque, 1820, 113.
Sterletus Brandt & Ratzeburg, 1833, 178.
Helops B. & R., 1833, 178.
Huso B. & R., 1833, 177.
Scaphirhynchus Müller, 1835, 182.
Sturio Müller, 1835, 182.
Antaceus Heckel & Fitzinger, 1836, 185.

Scaphirhynchus Heckel, 1836, 185.
Lioniscus H. & F., 1836, 185.
Sterledus Bonaparte, 1845, 225.
Ellops Gistel, 1848, 236.
Scaphirhynchops Gill, 1863, 323.
Schipa Brandt, 1869, 354.
Kessleria Boghdanov, 1882, 418.
Pseudoscaphirhynchus Nikolskii, 1900, 490.
Parascaphirhynchus Forbes & Richardson, 1905, 512.
Hemiscaphirhynchus Berg, 1912, 542.

Order SELACHOSTOMI

Family 117. POLYODONTIDÆ
(Paddle-fishes)

Polyodon Lacépède, 1797, 54.
Polyodon Bloch & Schneider, 1801, 59.
Spatularia Shaw, 1804, 74.
Platirostra Le Sueur, 1818, 107.

Proceros[115] Rafinesque, 1820, 113.
CROSSOPHOLIS[115a] Cope, 1866, 421, 434.
Psephurus Günther, 1873, 369.
PHOLIDURUS Woodward, 1889, 449.

[114] A mythical sturgeon.
[115] Mythical.
[115a] Misprinted *Crossophilus* in "Genera of Fishes," p. 434.

Order PYCNODONTI
Family 118. PYCNODONTIDÆ

PALÆOBALISTUM Blainville, 1818, 108.
PYCNODUS Agassiz, 1833, 177.
GYRODUS Agassiz, 1833, 177.
MICRODON Agassiz, 1833, 177.
SCAPHODUS Agassiz, 1844, 217, 453.
GYRONCHUS Agassiz, 1844, 217.
ACROTEMNUS Agassiz, 1844, 217.
PERIODUS Agassiz, 1844, 217.
PROSCINETES Gistel, 1848, 237.
COCCODUS Pictet, 1850, 247.
MESODON Wagner, 1851, 250.
ANCISTRODON Derby, 1852, 251.
GLOSSODUS Costa, 1853, 253.
CŒLODUS Heckel, 1856, 270.
STEMMATODUS Heckel, 1856, 270, 409.
ANOMIOPHTHALMUS Costa, 1856, 268.

HADRODUS[117] Leidy, 1858, 283.
TYPODUS[118] Quenstedt, 1858, 283.
MESTURUS Wagner, 1862, 321.
PHACODUS Cope, 1869, 354.
CROMMYODUS Cope 1870, 356.
ELLIPSODUS Cornuel, 1877, 389.
URANOPLOSUS Sauvage, 1879, 401.
COSMODUS Sauvage, 1879, 401.
ATHRODON Sauvage, 1880, 405.
XENOPHOLIS Davis, 1885, 431.
ANOMŒODUS Forir, 1887, 437.
GRYPODON Hay, 1899, 486.
POLYPSEPHIS Hay, 1899, 486.
MACROMESODON Blake, 1905, 576.
EOMESODON Woodward, 1918, 565.

[117] Perhaps the tooth of a reptile.
[118] This can not replace the preoccupied name *Mesodon*, as *Typodus* Meyer, 1847, is earlier.

Family 119. SEMIONOTIDÆ (*Lepidotidæ; Stylodontidæ; Dapediidæ; Sphærodontidæ*)

DAPEDIUM[119] Leach, 1822, 115.
TETRAGONOLEPIS Brown, 1830, 136, 177.
SEMIONOTUS Agassiz, 1832, 140.
LEPIDOTES[120] Agassiz, 1832, 141.
LEPIDOSAURUS Von Meyer, 1832, 141.
SPHŒRODUS Agassiz, 1833, 177.
AMBLYURUS Agassiz, 1836, 184.
SARGODON Plieninger, 1847, 232.
ISCHYPTERUS Egerton, 1847, 231.
PHOLIDOTUS Williamson, 1849, 244.
PLEUROLEPIS Quenstedt, 1852, 251.
PRIONOPLEURUS[121] Fischer de Waldheim, 1852.
SERROLEPIS Quenstedt, 1852, 251.
ÆCHMODUS Egerton, 1854, 256.
DIPTERONOTUS Egerton, 1854, 256.
HETEROSTICHUS Wagner, 1860, 300.
HOMŒOLEPIS Wagner, 1860, 300.

UROCOMUS Costa, 1862, 312.
SCROBODUS Von der Marck, 1863, 327.
HETEROSTROPHUS Wagner, 1863, 329.
PLESIODUS Wagner, 1863, 329.
CLEITHROLEPIS Egerton, 1864, 330.
EULEPIDOTUS Egerton, 1868, 349.
OMALOPLEURUS Costa, 1873, 368.
GYROLEPIDOTUS Rohon, 1889, 448.
ARCHÆOSEMIONOTUS Deecke, 1889, 445.
PRISTISOMUS Woodward, 1890, 453.
DOLICHOPTERUS Compter, 1891, 454.
LEPIDOPTERUS Pohlig, 1892, 459.
PROLEPIDOTUS Michael, 1893, 462.
ÆTHEOLEPIS Woodward, 1895, 470.
APHNELEPIS Woodward, 1895, 470.
PARAIKICHTHYS Ameghino, 1899, 484.
EOSEMIONOTUS[122] Stolley, 1920.

[119] Written *Dapedius* by Agassiz.
[120] Afterwards written *Lepidotus*.
[121] *Bull. Sci. Imp. Nat. Moscow*, 25 : 171; type *P. bronni* Fischer. (Omitted in "Genera of Fishes.")
[122] *Op. cit.*, 68; orthotype *Allolepidotus vogelii* Fritsch.

Order HOLOSTEI
Suborder *GINGLYMODI*
(*Lepidostei*)

Family 120. **ASPIDORHYNCHIDÆ**

ASPIDORHYNCHUS Agassiz, 1833, 177.
BELONOSTOMUS Agassiz, 1834, 180.
DICHELOSPONDYLUS Costa, 1856, 268.
OPHIRACHIS Costa, 1856, 268.

PLATYCERHYNCHUS Costa, 1864, 330.
HEMIRHYNCHUS Kner, 1867, 346.
APATEOPHOLIS Woodward, 1890, 453.
VINCTIFER Jordan, 1920, 570.

Family 121. **LEPISOSTEIDÆ**
(Gar-Pikes)

Acus Catesby, 1771, 31.
Psalisostomus Klein, 1781, 44.
Lepisosteus[123] Lacépède, 1803, 66.
Sarchirus Rafinesque, 1818, 109.
Litholepis[124] Rafinesque, 1818, 109.

Cylindrosteus Rafinesque, 1820, 112.
Atractosteus Rafinesque, 1820, 112.
PNEUMATOSTEUS Cope, 1870, 356.
CLASTES Cope, 1873, 368.

---

[123] Usually corrected to *Lepidosteus*.
[124] Mythical Gar-Pike.

Order HALECOMORPHI
(*Cycloganoidei; Protospondyli; Amioidei*)

Family 122. **PACHYCORMIDÆ**
(*Pelecopteridæ; Erisichthidæ; Protosphyrænidæ*)

SAUROPSIS Agassiz, 1832, 140.
PACHYCORMUS Agassiz, 1833, 177.
SAUROSTOMUS Agassiz, 1833, 177.
PROTOSPHYRÆNA Leidy, 1856, 273.
PACHYLEPIS Quenstedt, 1858, 283.
LYCODUS Quenstedt, 1858, 283.
HYPSOCORMUS Wagner, 1860, 300.
EUTHYNOTUS Wagner, 1860, 300.
CYCLOSPONDYLUS Quenstedt, 1867, 347.
ERISICHTHE Cope, 1872, 364.
CEPHENOPLOSUS Sauvage, 1874, 377.

PSEUDOTHRISSOPS Sauvage, 1875, 381.
HETEROTHRISSOPS Sauvage, 1875, 381.
PELECOPTERUS Cope, 1875, 379.
AGASSIZIA Vetter, 1881, 418.
DIPLOLEPIS Vetter, 1881, 418.
LEEDSIA Woodward, 1889, 448.
LEEDSICHTHYS Woodward, 1889, 449.
PARATHRISSOPS Sauvage, 1891, 456.
PROSAUROPSIS Sauvage, 1893, 462.
PROTOSAUROPSIS Sauvage, 1894, 465.

Family 123. **MACROSEMIIDÆ**

NOTAGOGUS Agassiz, 1833, 177.
MACROSEMIUS Agassiz, 1834, 179.
PROPTERUS Agassiz, 1834, 179.
OPHIOPSIS Agassiz, 1834, 179.
DISTICHOLEPIS Thiollière, 1850, 247.
PETALOPTERYX Pictet, 1850, 247.
BLENNIOMŒUS Costa, 1850, 245.
RHYNCHONCODES Costa, 1850, 245.

CALIGNATHUS Costa, 1853, 253.
HISTIONOTUS Egerton, 1854, 256.
LEGNONOTUS Egerton, 1854, 256.
ORTHURUS Kner, 1866, 341.
APHANEPYGUS Bassani, 1879, 398.
EUSEMIUS Vetter, 1881, 418.
MACREPISTIUS Cope, 1894, 464.
ENCHELYOLEPIS Woodward, 1918, 565.

Family 124. **FURIDÆ** (*Eugnathidæ; Isopholidæ*)

The name *Eugnathus* is preoccupied. Of the substitute names, *Furo* and *Isopholis,* the former has priority.

URÆUS Agassiz, 1832, 140.
PTYCHOLEPIS Agassiz, 1832, 140.
CATURUS Agassiz, 1834, 179.
EUGNATHUS Agassiz, 1839, 194.
AMBLYSEMIUS Agassiz, 1844, 217.
Conodus Agassiz, 1844, 217.
FURO Gistel, 1848, 236.
STROBILODUS Wagner, 1851, 250.
LOPHIOSTOMUS[125] Egerton, 1852, 250.
ENDACTIS Egerton, 1858, 280.
MACROPOMA (Agassiz) Egerton, (1844) 1858, 217.
CALLOPTERUS Thiollière, 1858, 284.
BRACHYICHTHYS Winkler, 1862, 320.
EURYCORMUS Wagner, 1863, 329.
DITAXIODUS Owen, 1866, 342.
THLATTODUS Owen, 1866, 342.

EURYPOMA Huxley, 1866, 341.
ISOCOLUM Egerton, 1868, 349.
OSTEORHACHIS Egerton, 1868, 349.
HETEROLEPIDOTUS Egerton, 1872, 365.
HARPACTES Egerton, 1876, 385.
HARPACTIRA Egerton, 1876, 385.
LISSOLEPIS Davis, 1884, 227.
ISOPHOLIS Zittel, 1887, 439.
CRENILEPIS Dames, 1888, 440.
NEORHOMBOLEPIS Woodward, 1888, 444.
ALLOLEPIDOTUS Deecke, 1889, 445.
OTOMITLA Felix, 1891, 454.
ASTHENOCORMUS Woodward, 1895, 470.
SPANIOLEPIS Kramberger, 1905, 514.
DOLLOPTERUS Abel, 1906, 516.
AINIA Jordan, 1919, 567.

---

[125] Misprinted *Lophistomus* in "Genera of Fishes."

Family 125. ARCHÆOMÆNIDÆ

ARCHÆOMÆNE Woodward, 1895, 470.

Family 126. LIODESMIDÆ

LIODESMUS Wagner, 1863, 377, 329.   LOPHIURUS Vetter, 1881, 418.

Family 127. AMIIDÆ (*Amiatidæ*)
(Bow-fins)

Amia Linnæus, 1766, 23.
Amiatus Rafinesque, 1815, 91.
MEGALURUS Agassiz, 1833, 177.
CYCLURUS Agassiz, 1844, 218.
NOTÆUS Agassiz, 1844, 218.
SYNERGUS Gistel, 1848, 236.

AMIOPSIS Kner, 1863, 327.
PAPPICHTHYS Cope, 1873, 368.
PROTAMIA Leidy, 1873, 370.
HYPAMIA Leidy, 1873, 370.
UROCLES Jordan, 1919, 567.

Superorder TELEOSTEI
(The Bony Fishes)

Order ISOSPONDYLI
(*Physostomi; Malacopterygii*)
(The Soft-rayed Fishes)

Suborder PHOLIDOPHOROIDEI

Family 128. PHOLIDOPHORIDÆ

PHOLIDOPHORUS Agassiz, 1832, 140.   CERAMURUS (Egerton) Brodie, 1845, 226.
MICROPS Agassiz, 1833, 177.   PERIERGUS Gistel, 1848, 237.

NOTHOSOMUS[126] (Agassiz) Egerton,
(1844) 1858, 217, 280.
THORACOPTERUS Bronn, 1858, 279.
PLEUROPHOLIS Egerton, 1858, 280.
PHOLIDOPLEURUS Bronn, 1858, 279.
MEGALOPTERUS Kner, 1866, 341.

PELTOPLEURUS Kner, 1866, 341.
PTERYGOPTERUS Kner, 1867, 346.
PROHALECITES Deecke, 1889, 445.
BALEIICHTHYS Rohon, 1890, 452.
GIGANTOPTERUS Abel, 1906, 516.

[126] Misprinted *Nothosoma* in "Genera of Fishes." *Nothosomus* Bassani, 434, "Genera of Fishes," should be canceled.

Family 129. OLIGOPLEURIDÆ

OLIGOPLEURUS Thiollière, 1850, 247.
IONOSCOPUS[127] Costa, 1853, 253.
Holochondrus[127a] Thiollière, 1854.
MACRORHIPIS Wagner, 1863, 329.
ŒNOSCOPUS Costa, 1864, 330.

OPSIGONUS[128] (Kramberger) Bassani,
1882, 418.
SPATHIURUS Davis, 1887, 431.
AMPHILAPHURUS Davis, 1887, 431.
ATTAKEOPSIS Thiollière, 1858, 284, 366.

[127] Altered to *Œnoscopus* by authors.
[127a] Name undefined.
[128] *Opsigonus,* 1895, "Genera of Fishes," p. 469, should be canceled.

Family 130. LEPTOLEPIDÆ

Ichthyolithus Germar, 1826, 120.
LEPTOLEPIS Agassiz, 1832, 141.
THRISSOPS Agassiz, 1833, 177.
ASCALABOS Münster, 1839, 196.
ÆTHALION Münster, 1842, 212.
MEGASTOMA Costa, 1845, 226.
OXYGONIUS (Agassiz) Brodie, 1845, 226.
THARSIS Giebel, 1848, 233.
LYCOPTERA Müller, 1848, 239.

SARGINITES ·Costa, 1850, 245.
EURYSTETHUS Sauvage, 1877, 392.
CTENOLEPIS (Agassiz) Woodward,
(1844) 1895, 217, 470.
VIDALIA Sauvage, 1903, 506.
THARRHIAS Jordan & Branner, 1908, 528.
CEARANA J. & B., 1908, 528.
PARATHRISSOPS Eastman, 1914, 553.
PACHYTHRISSOPS[128a] Woodward, 1918.

[128a] *Monogr. Pal. Soc.,* 1918, 128, orthotype *P.* lævis, Woodward.

Suborder *ELOPOIDEA*

Family 131. ELOPIDÆ
(Ten-pounders)

This family, of which but one genus is now living, shows much variation in dentition. *Spaniodon, Enneles,* and other genera differing widely from the usual type. It should perhaps be further subdivided. With the *Megalopidæ* and *Pachyrhizodontidæ,* the *Elopidæ* agrees with *Amia* in the presence of the gular plate, wanting in all other groups of fishes.

Elops Linnæus, 1766, 23.
Mugilomorus Lacépède, 1803, 67.
Trichonotus Rafinesque, 1815, 90.
RHACOLEPIS[129] Agassiz, 1841, 208.
CALAMOPLEURUS Agassiz, 1841, 208.

AULOLEPIS Geinitz, 1849, 408.
SAUROPSIDIUM Costa, 1850, 245.
SPANIODON Pictet, 1850, 247.
HISTIALOSA Gervais, 1855, 409.
ELOPOPSIS Heckel, 1856, 270.

[129] Misprinted *Phacolepis* in the original paper.

RHABDOLEPIS Von Der Marck, 1863, 327.
CÆUS[130] Costa, 1864, 264.
HYPTIUS Costa, 1864, 330.
HOLCOLEPIS Von Der Marck, 1865, 339, 352.
DYPTEROLEPIS (Steinla) Geinitz, 1868, 349.
MICROPETALOLEPIS (Steinla) Geinitz, 1868, 349.
PERIGRAMMATOLEPIS (Steinla) Geinitz, 1868, 349.
ACROGRAMMATOLEPIS (Steinla) Geinitz, 1868, 350.
LEPTOGRAMMATOLEPIS (Steinla) Geinitz, 1868, 349.
ANÆDOPOGON Cope, 1871, 360.
THRISSOPATER Günther, 1872, 365.
THRISSOPTEROIDES Von der Marck, 1873, 370.

OSMEROLEPIS Fritsch, 1875, 379.
HEMIELOPOPSIS Bassani, 1879, 398.
LEWISIA Davis, 1885, 431.
PROTELOPS Laube, 1885, 433.
HYPSOSPONDYLUS Kramberger, 1885, 433.
LYROLEPIS Romanovsky, 1886, 436.
ESOCELOPS Woodward, 1901, 495.
ELOPIDES (Ag.) Woodward, 1901, 495, 219.
NOTELOPS Woodward, 1901, 495.
EURYGNATHUS (Agassiz) Woodward, (1845) 1901, 224.
PARELOPS Fritsch & Bayer, 1902, 498.
DINELOPS Woodward, 1907, 527.
ENNELES Jordan & Branner, 1908, 528.
HELMINTHOLEPIS Cockerell, 1919, 565.
ECTASIS Jordan & Gilbert, 1919, 569.
ENNELICHTHYS Jordan, 1920, 570.
BRANNERION Jordan, 1920, 570.

[130] Misprinted *Cacus.*

Family 132. RAPHIOSAURIDÆ *(Pachyrhizodontidæ)*

MEGALODON Agassiz, 1835, 181.
RAPHIOSAURUS Owen, 1842, 212.
PACHYRHIZODUS (Agassiz) Dixon, 1850, 246.
CONOSAURUS Gibbes, 1850, 246.
ONCHOSAURUS[131] Gervais, 1852, 408.

ISCHYRHIZA[131] Leidy, 1856, 272.
CONOSAUROPS Leidy, 1868, 352.
ACRODONTOSAURUS Mason, 1869, 355.
ANOGMIUS Cope, 1871, 360.
CYCLOTOMODON Cope, 1876, 385.
ORICARDINUS, Cope, 1877, 388.

[131] It is likely that both *Onchosaurus* and *Ischyrhiza* are reptilian teeth.

Family 133. MEGALOPIDÆ
(Tarpons; Grande-Écailles)

Amia Browne, 1789, 46.
Megalops Lacépède, 1803, 66.
Oculeus (Commerson) Lacépède, 1803, 70.

Brisbania Castelnau, 1877, 388.
Tarpon Jordan & Evermann, 1896, 473.

Family 134. GANOLYTIDÆ

ETRINGUS Jordan, 1907, 525.
GANOLYTES Jordan, 1919, 568.

GANOËSSUS Jordan, 1920, 571.

Suborder *ALBULOIDEI*
Family 135. ALBULIDÆ
(Lady-fishes)

Albula Gronow, 1763, 20.
Albula Scopoli, 1777, 42.
Albula Bloch & Schneider, 1801, 59.
Butyrinus Lacépède, 1803, 65.
Glossodus Cuvier, 1815, 93.
PISODUS Agassiz, 1845, 227.

Conorhynchus (Nozemann) Gill, 1861, 303.
PETALOLEPIS (Steinla) Geinitz, 1868, 350.
APSOPELIX Cope, 1871, 360.
CYCLOIDES[132] Winkler, 1878, 398.
Dixonina Fowler, 1911, 538.

[132] Very doubtful, perhaps not a fish scale.

Family 136. **THRYPTODONTIDÆ** (*Plethodontidæ*)[133]

PLETHODUS Dixon, 1850, 245.
ANOGMIUS Cope, 1877, 388.

THRYPTODUS Loomis, 1900, 490.
PSEUDOTHRYPTODUS Loomis, 1900, 490.

[133] The name *Plethodontidæ* is already used for a family of Salamanders.

Suborder *CLUPEOIDEI*

Family 137. **CHANIDÆ**

Chanos Lacépède, 1803, 67.
Lutodeira Rüppell, 1828, 122, 184.
Scoliostomus Rüppell, 1828, 122.

Ptycholepis Gray, 1842, 211.
PROCHANOS Bassani, 1879, 398.
CHANOIDES Woodward, 1901, 495.

Family 138. **ANCYLOSTYLIDÆ**

A single genus, apparently related to the *Chanidæ*.

Ancylostylos Kramberger, 1895, 469.

Family 139. **CHIROCENTRIDÆ**

It is not certain that all of the genera named below belong to one family, or to the same family as the single living genus, *Chirocentrus*.

MONOPTEROS Volta, 1796, 169.
Chirocentrus Cuvier, 1817, 99.
SAUROCEPHALUS Harlan, 1824, 118.
SAURODON Hays, 1830, 136.
PLATINX Agassiz, 1835, 181.
Cœlogaster Agassiz, 1835, 181.
THRISSOPTERUS Heckel, 1856, 270.
ANDREIOPLEURA Costa, 1865, 336.
HETEROLEPIS Costa, 1865, 336.
TELEPHOLIS Marck & Schlüter, 1865, 339.
CREDEMNOLEPIS Steinla, 1868, 349.

POLYPTEROLEPIS Steinla, 1868, 350.
OÖLEPIS Steinla, 1868, 350.
AGRAMMATOLEPIS Steinla, 1868, 350.
CROMMIOLEPIS Steinla, 1868, 350.
CYCLOLEPIS Geinitz, 1868, 349.
HEMICYCLUS Geinitz, 1868, 350.
PTYCHOLEPIS Steinla, 1868, 350.
HELIOLEPIS Steinla, 1868, 350.
MIXOGRAMMATOLEPIS Steinla, 1868, 350.
CHIROMYSTUS Cope, 1886, 434.
EUBIODECTES Hay, 1903, 503.

Family 140. **ERYTHRINOLEPIDÆ**

ERYTHRINOLEPIS Cockerell, 1919, 566.

Family 141. **PTEROTHRISSIDÆ** (*Bathythrissidæ*)

KYMATOPETALEPIS (Steinla) Geinitz, 1868, 349.
Pterothrissus Hilgendorf, 1877, 390.

Bathythrissa Günther, 1877, 389.
SYNTEGMODUS Loomis, 1900, 490.
CHICOLEPIS Cockerell, 1919, 565.

Family 142. **ICHTHYODECTIDÆ**

HYPSODON Agassiz, 1887, 186, 219.
CHIROCENTRITES Heckel, 1849, 242.
TOMOGNATHUS Dixon, 1850, 246.
SPATHODACTYLUS Pictet & Loriol, 1859, 292.
ANDREIOPLEURA Costa, 1865, 336.

ICHTHYODECTES Cope, 1870, 356.
XIPHACTINUS Leidy, 1870, 358.
PRYMNETES Cope, 1871, 360.
PORTHEUS Cope, 1872, 364.
GILLICUS Hay, 1898, 480.
PROPORTHEUS Jackel, 1909, 533.

Family 143. SAURODONTIDÆ

Saurocephalus Harlan, 1824, 118.   Daptinus Cope, 1872, 364.
Saurodon Hays, 1830, 136.

Family 144. GONORHYNCHIDÆ

It is doubtful whether the extinct forms of this group (*Notogoneus, Charitostomus*) really belong to the same family as the living *Gonorhynchus*.

Gonorhynchus Gronow, 1763, 19.   **Rhynchana** Richardson, 1844, 407.
Gonorhynchus Scopoli, 1777, 42.

Family 145. NOTOGONEIDÆ

Anormurus Blainville, 1818, 171.   Notogoneus[134] Cope, 1885, 430.
Sphenolepis Agassiz, 1844, 218.   Protocatostomus Whitfield, 1890, 453.
Solenognathus Pictet, 1866, 342.   Gonorhynchops Cockerell, 1919, 566.
Charitosomus Marck, 1885, 433.

[134] Cockerell observes, in lit., May 13, 1922: "As to *Notogoneus*, I don't think it could be a characin. If not a gonorhynchid, then the type of a new family, *Notogoneidæ*. I suppose the *Gonorhynchid* type of scales, originating during the Mesozoic, may have persisted in several distinct branches of the original stem, of which our modern *Gonorhynchidæ* constitute one only. One of the other branches would be represented by *Notogoneus*. I think the resemblance is too close for convergence from entirely different stems."

The view expressed above implies that the Cretaceous *Gonorhynchops* and *Charitosomus* are probably not true *Gonorhynchidæ*, in the modern sense, but represent a family ancestral to the modern one. Whether such a family could be defined is another matter.

Family 146. HIODONTIDÆ
(Moon-fishes)

Hiodon[135] Le Sueur, 1818, 107.   **Clodalus** Rafinesque, 1820, 111.
Glossodon Rafinesque, 1818, 108.   **Glossodon** Heckel, 1842, 211.
Amphiodon Rafinesque, 1819, 110.   **Elattonistius** Gill & Jordan, 1877, 390.

[135] Later more correctly written *Hyodon*.

Family 147. CLUPEIDÆ
(Herrings)

Clupea Linnæus, 1758, 15.   *Trichis* (Plumier) Lacépède, 1803, 73.
Harengus Geoffroy, 1767, 24.   **Pristigaster** Cuvier, 1817, 99.
Harengus Catesby, 1771, 31.   **Pomolobus** Rafinesque, 1820, 111.
Harengus Klein, 1775, 38.   **Alosa**[136] Cuvier, 829, 130.
Alosa Linck, 1790, 49.   **Corica** Hamilton, 1822, 115.
Odontognathus Lacépède, 1800, 56.   **Raconda** Gray, 1831, 138.
Pomatias Bloch & Schneider, 1801, 60. **Apterygia** Gray, 1833, 179.
Gnathobolus B. & S., 1801, 60.   **Apterogasterus** De La Pylaie, 1835, 184.
Opisotomus (Commerson) Lacépède,   **Platygaster** Swainson, 1839, 203.
   1803, 70.   **Ilisha** Gray, 1845, 227.

[136] Also written *Alausa*.

Spratella Cuvier & Valenciennes, 1847, 231.
Kowala Cuv. & Val., 1847, 231.
Clupeonia Cuv. & Val., 1847, 231.
Rogenia Cuv. & Val., 1847, 231.
Sardinella Cuv. & Val., 1847, 231.
Harengula Cuv. & Val., 1847, 231.
Alausa Cuv. & Val., 1847, 231.
Meletta Cuv. & Val., 1847, 231.
Pellona Cuv. & Val., 1847, 231.
Amblygaster Bleeker, 1849, 240.
HISTIURUS Costa, 1850, 245.
Clupeoides Bleeker, 1851, 247.
Dascillus Gronow, 1854, 259.
Clupeichthys Bleeker, 1855, 263.
Sardinia Poey, 1860, 299.
ALOSINA Wagner, 1860, 300.
Opisthopterus Gill, 1861, 303.
Alausella Gill, 1861, 303.
Opisthonema Gill, 1861, 303.
Brevoortia Gill, 1861, 303.
SCOMBROCLUPEA Kner, 1863, 327.
UROPTERINA Lioy, 1865, 338.
PTERICEPHALINA Lioy, 1865, 339.
PSEUDOBERYX Pictet & Humbert, 1866, 342.
Pellonula Günther, 1868, 351.
Chirocentrodon Günther, 1868, 351.
Paralosa Bleeker, 1872, 362.
SYLLÆMUS Cope, 1875, 379.
Clupeonella Kessler, 1877, 391.
DIPLOMYSTUS Cope, 1877, 388.
CLUPEOPS Sauvage, 1880, 402.
HACQUETIA[137] Szajnocha, 1886, 436.
Hyperlophus Ogilby, 1892, 459.
Lile Jordan & Evermann, 1896, 473.

Potamalosa Ogilby, 1896, 475.
Omochetus Ogilby, 1897, 477.
Odaxothrissa Boulenger, 1899, 485.
LEPTICHTHYS Stewart, 1899, 487.
HISTIOTHRISSA Woodward, 1901, 495.
HALECOPSIS (Agassiz) Woodward, (1844) 1901, 217, 218, 496.
Microthrissa Boulenger, 1902, 496.
LYROLEPIS Romanovskii, 1904, 510.
COPEICHTHYS Dollo, 1904, 507.
Sardina Antipa, 1905, 512.
KNIGHTIA Jordan, 1907, 525.
ELLIPES Jordan, 1910, 536.
DASTILBE Jordan, 1910, 536.
Heringia Fowler, 1911, 539.
Gudusia Fowler, 1911, 539.
Rhinosardinia Eigenmann, 1912, 543.
Clupeonella Berg, 1913, 546.
ELLIMMA Jordan, 1913, 548.
Zunasia Jordan & Metz, 1913, 548.
Caspialosa Berg, 1915, 555.
Paralosa Regan, 1916, 560.
Ethmalosa Regan, 1916, 560.
Hilsa Regan, 1916, 560.
Ethmidium Thompson, 1916, 561.
Pœcilothrissa Regan, 1917, 563.
Potamothrissa Regan, 1917, 563.
Cynothrissa Regan, 1917, 563.
Stolothrissa Regan, 1917, 563.
Limnothrissa Regan, 1917, 563.
IQUIUS Jordan, 1919, 568.
XYNE Jordan & Gilbert, 1919, 568.
ELLIMMICHTHYS Jordan, 1919, 568.
ALISEA J. & G., 1919, 568.
XYRINIUS J. & G., 1919, 569.
QUISQUE Jordan, 1920, 571.

[137] Not identifiable.

Family 148. DUSSUMIERIIDÆ
(Round Herrings)

Stolephorus[138] Lacépède, 1803, 67.
Dussumieria Cuvier & Valenciennes, 1847, 231.
Spratelloides Bleeker, 1851, 248.
Etrumeus Bleeker, 1853, 253.
Leptogaster Bleeker, 1872, 362.

Perkinsia Eigenmann, Rosa S., 1891, 454.
Jenkinsia Jordan & Evermann, 1896, 473.
RHOMURUS Jordan, 1919, 568.
SMITHITES Jordan & Gilbert, 1919, 568.
QUÆSITA J. & G., 1919, 568.
LYGISMA J. & G., 1919, 568.

[138] This application of this name is still uncertain. It originally included a species of Dussumieriidæ and one of Engraulidæ.

Family 149. DOROSOMIDÆ
(Gizzard Shads)

Clupanodon[139] Lacépède, 1803, 68.
Thrissa Rafinesque, 1815, 90.
Dorosoma Rafinesque, 1820, 111.
Gonostoma Van Hasselt, 1823, 172.
Chatoëssus Cuvier, 1829, 130.
Anodontostoma Bleeker, 1849, 241.

Signalosa Evermann & Kendall, 1898, 480.
Konosirus Jordan & Snyder, 1900, 490.
Nematalosa Regan, 1916, 560.
Gonialosa Regan, 1916, 560.

[139] The proper application of this name has been questioned. Apparently it should replace *Konosirus* among the *Dorosomidæ*.

Family 150. ENGRAULIDÆ
(Anchovies)

Menidia Browne, 1789, 46.
Stolephorus[140] Lacépède, 1803, 67.
Encrasicholus (Commerson) Lacépède, 1803, 71.
Mystus Lacépède, 1803, 67.
Engraulis Cuvier, 1817, 98.
Thrissa[141] Cuvier, 1817, 98.
Encrasicholus Fleming, 1828, 122.
Coilia Gray, 1831, 138.
Trichosoma Swainson, 1839, 203.
Setipinna Swainson, 1839, 203.
Chætomus McClelland, 1844, 220.
Leptonurus Bleeker, 1849, 241.

Stethochætus Gronow, 1854, 259.
Osteoglossum Basilewski, 1855, 262.
Telara Günther, 1868, 351.
Cetengraulis Günther, 1868, 351.
Pterengraulis Günther, 1868, 351.
Heterothrissa Günther, 1868, 351.
Lycengraulis Günther, 1868, 351.
Lycothrissa Günther, 1868, 351.
Anchovia Jordan & Evermann, 1895, 468, 473.
Anchoviella Fowler, 1911, 539.
Thrissocles Jordan, 1917, 562.

[140] This genus is based on two species, belonging to different families. The final disposition of the name awaits decision.
[141] Later written *Thryssa* by Cuvier.

Family 151. ALEPOCEPHALIDÆ

Triurus Lacépède, 1800, 56.
Pomatias Bloch & Schneider, 1801, 60.
Alepocephalus Risso, 1820, 113.
Esunculus Kaup, 1856, 272.
Xenodermichthys Günther, 1878, 395.
Bathytroctes Günther, 1878, 395.
Platytroctes Günther, 1878, 395.
Aleposomus[142] Gill, 1884.
Leptoderma Vaillant, 1886, 436, 444.
Anomalopterus Vaillant, 1886, 436, 444.
Aulostomatomorpha Alcock, 1890, 449.
Tauredophidium Alcock, 1890, 449.

Narcetes Alcock, 1890, 449.
Conocara Goode & Bean, 1895, 467.
Talismania G. & B., 1895, 467.
Mitchillina Jordan & Evermann, 1896, 473.
Ericara Gill & Townsend, 1897, 477.
Leptochilichthys[143] Garman, 1899.
Dolichopteryx Brauer, 1901, 492.
Asquamiceps Zugmayer, 1911, 541.
Xenognathus Gilbert, 1915, 556.
Benthosphyraena Cockerell, 1919, 566.
Rouleina[144] Jordan, 1923.

[142] *Amer. Nat.*, **18**: 433; type *Aleposomus copei* Gill; regarded as a synonym of *Xenodermichthys*. (Omitted in "Genera of Fishes.")
[143] "Deep Sea Fish," 284; orthotype *L. agassizii*. (Omitted in "Genera of Fishes.")
[144] New genus; type *Aleposomus güntheri* Alcock, same as *Aleposomus* Roule, not of Gill, which is a synonym of *Xenodermichthys* Günther.

Family 152. MACRISTIIDÆ

Macristium Regan, 1903, 505.

Family 153. PHRACTOLÆMIDÆ

Phractolæmus[145] Boulenger, 1901.

---

[145] Boulenger, *Proc. Zool. Soc. Lond.*, 3, 1901; type *P. ansorgi* Boulenger. (Omitted in "Genera of Fishes.")

Family 154. KNERIIDÆ

Kneria Steindachner, 1866, 343.   **Xenopomatichthys** Boulenger, 1910, 534.

Family 155. CROMERIIDÆ

Cromeria Boulenger, 1901, 492.

Suborder *OSTEOGLOSSOIDEA*

Family 156. OSTEOGLOSSIDÆ

This family and the next two should perhaps be merged into one.

Osteoglossum (Vandelli) Cuvier, 1829, 131.

Ischnosoma Spix, 1829, 132.

CLADOCYCLUS[146] Agassiz, 1841, 208.

Scleropages Günther, 1864, 333.

PHAREODUS[147] Leidy 1873, 370.

DAPEDOGLOSSUS Cope, 1877, 388.

Platops Owen, 1901, 260.

BRYCHÆTUS (Agassiz) Woodward, 1901, 224, 495.

Pomaphractus (Agassiz) Woodward, (1845) 1901, 224.

---

[146] The original species of this genus from Brazil belongs certainly to or near the *Osteoglossidæ,* and is distinguished by the great development of intermuscular bones, attached to the occiput forming a brush three times as long as the head and extending backward along the sides, looking in the fossil example like the wing of a flying fish. Cockerell suggests that it may be an ancestral characin.

[147] This name seems to me tenable; the description of the genus is included in that of the species, as with *Protamia* and *Hypamia* of the same paper.

Family 157. ARAPAIMIDÆ

Sudis Cuvier, 1817, 99.   **Vastres** Cuvier & Valenciennes, 1846, 228.

Arapaima Müller, 1843, 216, 229.

Family 158. CLUPISUDIDÆ (*Heterotidæ*)

Clupisudis Swainson, 1839, 202.   **Helicobranchus** Hyrtl, 1854, 260.

Heterotis (Ehrenberg) Müller, 1843, 131, 216, 229.

Family 159. PANTODONTIDÆ

Pantodon Peters, 1877, 391.

Suborder *NOTOPTEROIDEI*

Family 160. NOTOPTERIDÆ

Notopterus Lacépède, 1800, 56.   **Xenomystus** Günther, 1868, 351.

Glanis Gronow, 1854, 259.

Suborder SCYPHOPHORI

Family 161. **MORMYRIDÆ**

Mormyrus Linnæus, 1758, 15.
Scrophicephalus Swainson, 1839, 198.
Mormyrops Müller, 1843, 216.
Petrocephalus Marcusen, 1854, 334, 260.
Phagrus Marcusen, 1854, 334.
Marcusenius Gill, 1862, 314.
Mormyrodes Gill, 1862, 314.
Hyperopisus Gill, 1862, 314.
Isichthys Gill, 1862, 317.

Gnathonemus Gill, 1862, 317.
Solenomormyrus Bleeker, 1874, 372.
Campylomormyrus Bleeker, 1874, 372.
Oxymormyrus Bleeker, 1874, 372.
Stomatorhinus Boulenger, 1898, 478.
Myomyrus Boulenger, 1898, 479.
Genyomyrus Boulenger, 1898, 479.
Hippopotamyrus Pappenheim, 1906, 520.

Family 162. **GYMNARCHIDÆ**

Gymnarchus Cuvier, 1829, 131.

Suborder SALMONOIDEI

Family 163. **SALMONIDÆ**

(Salmon; Trout)

The next three families are usually and perhaps properly united with *Salmonidæ*.

Salmo Linnæus, 1758, 14.
Truttæ[148] Linnæus, 1758, 14.
Trutta Geoffroy, 1767, 24.
Trutta Klein, 1775, 37.
Salvelini[149] Nilsson, 1832, 140.
Salvelinus Richardson, 1836, 186.
Baione Dekay, 1842, 210.
THAUMATURUS Reuss, 1844, 222.
Fario Cuvier & Valenciennes, 1848, 233.

Salar Cuv. & Val., 1848, 233.
Umbla Rapp, 1854, 261.
Oncorhynchus Suckley, 1860, 299.
Hypsifario Gill, 1862, 317.
Hucho Günther, 1866, 340.
RHABDOFARIO Cope, 1870, 356.
Cristivomer Gill & Jordan, 1878, 395.
Epitomynis Schulze, 1889, 448.
Phyllogephyra Boulenger, 1898, 479.

---

[148] This name with *Osmeri, Coregoni,* and *Characini* were given in plural form only by Linnæus.

[149] Given originally in plural form only, latinized by Richardson, 1836.

Family 164. **PLECOGLOSSIDÆ**

(Ayu)

Plecoglossus Temminck & Schlegel, 1846, 230.

Family 165. **COREGONIDÆ**

(White-fishes; Ciscoes)

Coregoni[150] Linnæus, 1758, 14.
Tripteronotus Lacépède, 1803, 65.
Coregonus Lac., 1803, 66.
Stenodus Richardson, 1836, 185.
Argyrosomus Agassiz, 1850, 245.
Brachymystax Günther, 1866, 340.
Luciotrutta Günther, 1866, 340.

Leucichthys Dybowski, 1874, 375.
Allosomus[151] Jordan, 1878.
Prosopium Milner, 1878, 395.
Cisco Jordan & Evermann, 1911, 539.
Thrissomimus Gill, 1911, 539.
Irillion Jordan, 1919, 567.
LEUCICHTHYOPS Cockerell, 1919, 566.

---

[150] Given in plural form only, as *Coregoni.*

[151] *Man. Vert.,* Ed. 2, 361; type *Coregonus tullibee* Richardson. (Omitted in "Genera of Fishes.")

Family 166. THYMALLIDÆ
(Grayling)

Thymallus Linck, 1790, 49.
Thymallus Cuvier, 1829, 130.
Choregon Minding, 1832, 140.
Orthocolus Gistel, 1848, 237.

CYCLOLEPIS Geinitz, 1868, 349.
PROTOTHYMALLUS Laube, 1901, 495.
Thymalloides Berg, 1908, 527.
Salmothymus Berg, 1908, 527.

Family 167. OSMERIDÆ
(Smelt)
A group scarcely distinct as a family from the *Argentinidæ*.

Osmeri[152] Linnæus, 1758, 14.
Osmerus Lacépède, 1803, 66.
Mallotus Cuvier, 1829, 130.
Eperlanus Gaimard, 1849, 249.
Eperlanus Basilewski, 1855, 262.
Thaleichthys Girard, 1859, 290.

Mesopus[153] Gill, 1862, 313.
Hypomesus Gill, 1862, 313.
Spirinchus Jordan & Evermann, 1896, 473.
Therobromus Lucas, 1898, 482.
Eperlanio Jordan, 1919, 567.

[152] In plural form only, as *Osmeri.*
[153] A name left in type through failure to correct proof sheets.

Family 168. ARGENTINIDÆ

Argentina Linnæus, 1758, 15.
Silus Reinhardt, 1833, 179.
Acantholepis Kröyer, 1846, 229.

Glossanodon Guichenot, 1866, 341.
Leuroglossus Gilbert, 1890, 451.

Family 169. MICROSTOMIDÆ

Microstoma Cuvier, 1817, 99.
Halaphya Günther, 1889, 447.

Nansenia Jordan & Evermann, 1896, 473.

Family 170. BATHYLAGIDÆ

Bathylagus Günther, 1878, 395.

Bathymacrops Gilchrist, 1922. (See Index.)

Family 171. RETROPINNIDÆ
(New Zealand White-bait)

Retropinna Gill, 1862, 313.

Richardsonia Steindachner, 1866, 343.

Family 172. SALANGIDÆ
(Chinese White-bait)

Albula Osbeck, 1762, 17.
Salanx Cuvier, 1817, 99.
Leucosoma Gray, 1831, 139.

Salangichthys Bleeker, 1860, 293.
Parasalanx Regan, 1908, 530.
Hemisalanx Regan, 1908, 530.

Family 173. GALAXIIDÆ

Galaxias Cuvier, 1817, 99.
Mesites Jenyns, 1842, 212.

Neochanna Günther, 1867, 345.
Austrocobitis Ogilby, 1899, 487.

Family 174. APLOCHITONIDÆ

Aplochiton[154] Jenyns, 1842, 212.
Farionella Cuvier & Valenciennes, 1849, 242.

Prototroctes Günther, 1864, 333.
Jenynsella Ogilby, 1908, 529.
Lovettia McCulloch, 1915, 556.

[154] Later spelled *Haplochiton.*

Suborder *ENCHODONTOIDEI*

Family 175. ENCHODONTIDÆ

Some of the genera named below have been referred, perhaps justly, to the *Dercetidæ,* a family which may be indeed closely related to the *Enchodontidæ.*

HALEC Agassiz, 1834, 180.
ENCHODUS Agassiz, 1835, 181.
ISODUS Heckel, 1849, 243.
SAURORHAMPHUS Heckel, 1850, 246.
POMOGNATHUS Dixon, 1850, 245.
EURYPHOLIS Pictet, 1850, 247.
PHASGANODUS Leidy, 1858, 283.
ISCHYROCEPHALUS von der Marck, 1858, 283.
PALÆOLYCUS von der Marck, 1863, 327.
PLINTHOPHORUS Günther, 1864, 333.

ARCHÆOGADUS Marck & Schlüter, 1865, 339.
TETHEODUS Cope, 1874, 375.
TRICHIURIDES Winkler, 1878, 398.
SOLENODON Kramberger, 1881, 416.
HOLCODON Kramberger, 1885, 433.
PHYLACTOCEPHALUS Davis, 1885, 431.
PANTOPHOLIS Davis, 1885, 431.
EURYGNATHUS Davis, 1885, 431.
DIPNOLEPIS Fritsch, 1893, 461.
HALECODON Cockerell, 1919, 566.

Suborder *STOMIATIOIDEI*

We may apply this term to a group of isospondylous fishes, which for the most part retain the adipose fin and which have the side of the upper jaw formed by the maxillaries, differing in this regard strongly from the *Iniomi,* with which they have been usually associated. Most members of both groups are bathypelagic fishes, degenerate in many regards. The division into families as here accepted must be regarded as provisional.

Family 176. STOMIATIDÆ

Stomias Cuvier, 1817, 99.
Echiostoma Lowe, 1843, 215.
Stomianodon Bleeker, 1849, 241.
Stomiasunculus Kaup, 1860, 297.
Lucifer Döderlein, 1882, 418.
Bathyophilus Giglioli, 1883, 423.
Hyperchoristus Gill, 1883, 424.
Eustomias Filhol, 1884, 427.
Pachystomias Günther, 1887, 438.
Opostomias Günther, 1887, 438.
Photonectes Günther, 1887, 438.
Eustomias Vaillant, 1888, 444.
Photostomias Collett, 1889, 444.
Thaumastomias Alcock, 1890, 449.

Grammatostomias[155] Goode & Bean, 1895, 468.
Dactylostomias Garman, 1899, 486.
Macrostomias Brauer, 1902, 497.
Melanostomias Brauer, 1902, 497.
Leptostomias Gilbert, 1905, 513.
Neostomias Gilchrist, 1906, 518.
Trichostomias Zugmayer, 1911, 541.
Nematostomias Zugmayer, 1911, 541.
Benthalbella Zugmayer, 1911, 541.
Lamprotoxus Holt & Bryne, 1913, 548.
Aristostomias Zugmayer, 1913, 552.
Gnathostomias Pappenheim, 1914, 554.
Zastomias Gilbert, 1915, 556.

---

[155] *Grammatostomias* Holt & Byrne, p. 536, should be canceled. *G. flagellibarba* Holt & Byrne is the type of *Lamprotoxus,* p. 548.

Family 177. TOMOGNATHIDÆ

TOMOGNATHUS Dixon, 1850, 246.

Family 178. ASTRONESTHIDÆ

Astronesthes Richardson, 1844, 222.
Phænodon Lowe, 1850, 247.

Bathylychnus Brauer, 1902, 497.
Borostomias Regan, 1908, 529.

Family 179. STYLOPHTHALMIDÆ

Stylophthalmus Brauer, 1906, 497, 517.

Stylophthalmoides Sanzo, 1922. (See Index.)

Family 180. CHAULIODONTIDÆ

Vipera Catesby, 1771, 32.
Chauliodus Bloch & Schneider, 1801, 59.

Leptodes Swainson, 1839, 203.

Family 181. GONOSTOMIDÆ

Gonostoma Rafinesque, 1810, 83.
Phosichthys[156] Hutton & Hector, 1872, 366.
Cyclothone Goode & Bean, 1882, 419.
Sigmops Gill, 1883, 424.
Neostoma[156a] Filhol, 1884, 427, 444.

Yarrella G. & B., 1895, 467.
Bonapartia G. & B., 1895, 467.
Manducus G. & B., 1895, 468.
Zaphotias G. & B., 1898, 482.
Triplophos Brauer, 1902, 497.
Azalois Jordan & Gilbert, 1919, 568.

[156] Later written *Photichthys*.
[156a] *Neostoma* Vaillant, p. 444, should be canceled.

Family 182. IDIACANTHIDÆ

Idiacanthus Peters, 1876, 386.

Bathyophis Günther, 1878, 394.

Family 183. MAUROLICIDÆ

Maurolicus Cocco, 1838, 191.
Ichthyococcus Bonaparte, 1840, 141, 206
Coccia Günther, 1864, 333.
Diplophos Günther, 1873, 369.
Vinciguerria Jordan & Evermann, 1895, 468.

Valenciennellus J. & E., 1895, 468.
Zalarges Jordan & Williams, 1895, 469.
Argyripnus Gilbert & Cramer, 1897, 477.
Lychnopoles Garman, 1899, 486.
Triareus Waite, 1910, 537.

Family 184. OPISTHOPROCTIDÆ

Opisthoproctus Vaillant, 1888, 443.

Winteria Brauer, 1901, 492.

Family 185. STERNOPTYCHIDÆ

Sternoptix[157] Hermann, 1781, 44.
Argyropelecus Cocco, 1829, 133.
Pleurothyris[158] Lowe, 1843, 215.

Polyipnus Günther, 1887, 438.
Sternoptychides Ogilby, 1889, 448.
Margrethia[158a] Jespersen & Täning, 1919.

[157] Usually written *Sternoptyx*.
[158] Misprinted *Pleurothysis* by Günther.
[158a] P. Jespersen and A. V. Täning. *Nat. Medd.,* 70 : 222; orthotype *M. obtusirostris* J. & T.: Cadiz Bay.

Family 186. ANOTOPTERIDÆ

Anotopterus Zugmayer, 1911, 541.

Order LYOPOMI

Family 187. HALOSAURIDÆ

ECHIDNOCEPHALUS von der Marck, 1858, 283.

Halosaurus Johnson, 1863, 327.

ENCHELURUS von der Marck, 1863, 328.

Halosaurichthys Alcock, 1889, 444.

Aldrovandia Goode & Bean, 1895, 468.

Halosauropsis Collett, 1896, 471.

Family 188. LIPOGENYIDÆ

Lipogenys Goode & Bean, 1894, 465.

Order HETEROMI

Family 189. PRONOTACANTHIDÆ

PRONOTACANTHUS Woodward, 1900, 491.

Family 190. NOTACANTHIDÆ
(Spiny-eels)

Notacanthus Bloch, 1787, 44.

Campylodon Fabricius, 1793, 53.

Acanthonotus Bloch, 1797, 53.

Polyacanthonotus Bleeker, 1874, 372.

Paradoxichthys Giglioli, 1882, 419.

Teratichthys Giglioli, 1882, 419.

Gigliolia Goode & Bean, 1894, 465.

Macdonaldia G. & B., 1894, 465.

Suborder *HOPLOPLEURI*
(*Dercetiformes*)

The relations of this extinct group are still obscure. Some have regarded them as allies of the *Enchodonts* among the *Isospondyli*. Others have placed them near the sticklebacks (*Gasterosteidæ*). Cockerell, who has the latest guess, places them with the *Heteromi*.

Family 191. DERCETIDÆ

DERCETIS Agassiz, 1834, 180.

PRIONOLEPIS Egerton, 1850, 245.

PALIMPHEMUS Heckel, 1862, 319.

PELARGORHYNCHUS von der Marck, 1858, 283.

LEPTOTRACHELUS von der Marck, 1863, 328.

ASPIDOPLEURUS Picket & Humbert, 1866, 342.

TRIÆNASPIS Cope, 1878, 394.

LEPTECODON Williston, 1899, 488.

APATEOPHOLIS Woodward, 1890, 453.

Family 192. STRATODONTIDÆ

CIMOLICHTHYS Leidy, 1856, 273.

STRATODUS Cope, 1872, 364.

EMPO Cope, 1872, 364.

Family 193. RHINELLIDÆ

RHINELLUS Agassiz, 1844, 217.

ICHTHYOTRINGA Cope, 1878, 394.

Order SYMBRANCHIA

Eel-like fishes of tropical rivers and swamps, having distinct maxillary and premaxillary, the latter forming the border of the jaw; apparently not related to the true eels.

Suborder *ICHTHYOCEPHALI*

Family 194. FLUTIDÆ (*Monopteridæ*)

Les Monoptères Lacépède, 1800, 56.
Fluta Bloch & Schneider, 1801, 60.
Monopterus (Lacépède) Duméril 1806, 56.

Ophicardia McClelland, 1844, 220.
Apterigia Basilewski, 1855, 262.

Suborder *HOLOSTOMI*

Family 195. SYNBRANCHIDÆ

Synbranchus[159] Bloch, 1795, 53.
Typhlobranchus Bloch & Schneider, 1801, 60.
Unibranchapertura Lacépède, 1803, 68.

Ophisternon McClelland, 1844, 220.
Tetrabranchus Bleeker, 1851, 248.
Unipertura Kaup, 1856, 268.
Macrotrema Regan, 1912, 545.

[159] Usually corrected to *Symbranchus*.

Family 196. AMPHIPNOIDÆ

Amphipnous Müller, 1839, 196.
Ophichthys Swainson, 1839, 205.

Pneumabranchus McClelland, 1844, 220.

Suborder *ALABIFORMES*

Family 197. ALABIDÆ

Alabes (Cuvier) Oken, 1817, 101.

Cheilobranchus Richardson, 1844, 223.

Order OPISTHOMI

Eel-shaped fishes, with dorsal spines and with the shoulder girdle attached behind the occiput as in the *Heteromi*.

Family 198. CHAUDHURIIDÆ

Chaudhuria Annandale, 1918, 563.

Family 199. MASTACEMBELIDÆ

Mastacembelus Gronow, 1763, 21.
Mastacembelus Scopoli, 1777, 42.
Macrognathus Lacépède, 1800, 56.

Rhynchobdella Bloch & Schneider, 1801, 59.
Pararhychobdella Bleeker, 1874, 372.

Order APODES

Suborder *CARENCHELI*

As these eel-shaped fishes have a distinct maxillary and premaxillary, it is very doubtful whether their place is among or even near the eels.

Family 200. DERICHTHYIDÆ

Derichthys Gill, 1884, 428.

Suborder *ARCHENCHELI*

Eel-like forms with well developed ventral fins. It is doubtful whether these fishes are really to be placed among the eels.

Family 201. **ANGUILLAVIDÆ**
ANGUILLAVUS Hay, 1903, 503.

Suborder ENCHELYCEPHALI
Series ANGUILLOIDEI
Body with rudimentary scales; eggs minute.

Family 202. **ANGUILLIDÆ**
(True Eels)

Anguilla Shaw, 1803, 73.   **Tribranchus** Peters, 1844, 407.

Family 203. **SIMENCHELYIDÆ**

Simenchelys[160] (Gill) Goode & Bean,   **Conchognathus** Collett, 1889, 444.
1879.   **Gymnosimenchelys** Tanaka, 1908, 531.

———
[160] *Bull. Essex Inst.*, 27; orthotype *Simenchelys parasilicus* Gill. (Omitted in
"Genera of Fishes.")

Family 204. **SYNAPHOBRANCHIDÆ**

Synaphobranchus Johnson, 1862, 319.   **Nettophichthys** Holt, 1891, 455.
Histiobranchus Gill, 1883, 424.

Family 205. **ILYOPHIDÆ**
Ilyophis Gilbert, 1891, 454.

Series CONGROIDEI
Body scaleless; eggs large.

Family 206. **DYSOMMIDÆ**
Dysomma Alcock, 1889, 444.   **Dysommopsis** Alcock, 1891, 454.

Family 207. **XENOCONGRIDÆ**
Xenoconger Regan, 1912, 545.

Family 208. **MYROCONGRIDÆ**
Myroconger Günther, 1870, 357.

Family 209. **MURÆNESOCIDÆ**

Murænesox McClelland, 1844, 220.   **Brachyconger** Bleeker, 1865, 335.
Cynoponticus Costa, 1846, 228.   **Congresox** Gill, 1890, 451.

Family 210. **SAUROMURÆNESOCIDÆ**
Sauromurænesox Alcock, 1889, 444.

Family 211. **CONGRIDÆ**[161] (*Leptocephalidæ*)
(Conger Eels)

Leptocephalus Gronow, 1763, 22.   **Conger** (Cuvier) Oken, 1817, 101.
Conger Klein, 1775, 37.   **Ariosoma** Swainson, 1838, 193, 198.
Leptocephalus Scopoli, 1777, 41 (22).   **Ophisoma** Swainson, 1839, 205.
Oxyurus Rafinesque, 1810, 79.

Congrus Richardson, 1844, 223.
Helmichthys Costa, 1854, 256.
Leptocephalichthys Bleeker, 1856, 267.
Hyoprorus Kölliker, 1853, 254.
Uroconger Kaup, 1856, 272.
Congermuræna Kaup, 1856, 272.
Congerodon Kaup, 1856, 271, 237.
Neoconger Girard, 1859, 291.
Hoplunnis Kaup, 1860, 297.
Gnathophis Kaup, 1860, 297.
Isognatha (Dekay) Gill, 1861, 303.
Diaphanichthys Peters, 1864, 334.
Oxyconger Bleeker, 1865, 335.
Pœcilogconger Günther, 1871, 361.
Coloconger Alcock, 1889, 444.
Promyllantor Alcock, 1890, 449.

Xenomystax Gilbert, 1891, 454.
Euleptocephalus Strömman, 1895, 470.
Todarus Grassi & Calandruccio, 1896, 471.
Scolenchelys Ogilby, 1897, 477.
Bathycongrus Ogilby, 1898, 483.
Congrellus Ogilby, 1898, 483.
Atopichthys Garman, 1899, 486.
Congrosoma Garman, 1899, 486.
Veternio Snyder, 1904, 511.
Metopomycter Gilbert, 1905, 513.
Microconger Fowler, 1912, 543.
Mayerina Silvester, 1915, 557.
Pseudophichthys Roule, 1915, 557.
Rhechias[162] Jordan, 1921.
Brachyconger[162a] Regan, 1922.
Endeconger[162b] Jordan, 1923.

[161] Usually called *Congridæ*; *Leptocephalus*, an earlier name than *Conger*, is the larval form of *Conger*, but the name has been long used for larval eels in general.

[162] *Proc. U. S. Nat. Mus.*, **59** : 644; orthotype *R. armiger* Jordan.

[162a] Two new fishes from New Britain and Japan: *Ann. Mag. Nat. Hist.* X. 217; orthotype *B. platyrhynchus*, Regan; not of Bleeker, 1865.

[162b] New name for *Brachyconger*, preoccupied.

Family 212. NETTASTOMIDÆ
(Duck-bill Eels)

Nettastoma Rafinesque, 1810, 81.
Chlopsis Rafinesque, 1810, 82.
Saurenchelys Peters, 1864, 334.

Nettastomops Steindachner, 1906, 521.
Nettastomella Facciolà, 1914, 353.
Venefica Jordan & Davis, 1892, 458.

Family 213. HETEROCONGRIDÆ

Heteroconger Bleeker, 1868, 348.

Tæniconger[163] Herre, 1923.

[163] Herre, "Preliminary Account of Philippine Eels" (in press) ; body ribbon-like, the depth eighty-six times in length; pectorals present.

Family 214. ENCHELIIDÆ

Enchelion Hay, 1903, 503.

Family 215. URENCHELYIDÆ

Urenchelys Woodward, 1900, 491.

Family 216. ECHELIDÆ (*Myridæ*)

Echelus Rafinesque, 1810, 81.
Enchelyopus Agassiz, 1844, 181, 218.
Myrophis Lütken, 1851, 249.
Chilorhinus Lütken, 1851, 249.
Murænichthys Bleeker, 1853, 252.
Myrus Kaup, 1856, 271.
Paranguilla Bleeker, 1865, 335.
Paramyrus Günther, 1870, 357.
Holopterura Cope, 1871, 361.
Bathymyrus Alcock, 1890, 450.

Ahlia Jordan & Davis, 1892, 458.
Verma Jordan & Evermann, 1896, 473.
Myropterura Ogilby, 1897, 477.
Eomyrus Storms, 1898, 484.
Rhynchorhinus (Agassiz) Woodward, 1901, 219, 496.
Eomyrus (Agassiz) Eastman, 1905, 512.
Mylomyrus Woodward, 1910, 537.
Garmannichthys Seale, 1917, 563.

Family 217. NEMICHTHYIDÆ
(Thread Eels)

Nemichthys Richardson, 1848, 240.
Leptorhynchus Lowe, 1852, 251.
Belonopsis Brandt, 1854, 256.
Labichthys Gill & Ryder, 1883, 424.
Spinivomer G. & R., 1883, 424.
Serrivomer G. & R., 1883, 424.

Gavialiceps Alcock, 1889, 444.
Avocettina Jordan & Davis, 1892, 458.
Nematoprora Gilbert, 1905, 513.
Stemonidium Gilbert, 1905, 513.
Ceromitus Weber, 1913, 551.

Family 218. CYEMIDÆ

Cyema Günther, 1878, 395.

Family 219. OPHICHTHYIDÆ
(Snake Eels)

Ophichthus[164] Ahl, 1787, 45.
Cœcula Vahl, 1794, 52.
Sphagebranchus Bloch, 1795, 53.
Ophisurus Lacépède, 1800, 56.
Cæcilia Lacépède, 1800, 56.
Colubrina[165] Lacépède, 1803, 65.
Apterichthys Duméril, 1806, 75.
Oxystomus Rafinesque, 1810, 83.
Dalophis Rafinesque, 1810, 81.
Cogrus Rafinesque, 1810, 81.
Pterurus Rafinesque, 1810, 82.
Helmictis Rafinesque, 1810, 83.
Typhlotes Fischer, 1813, 83.
Branderius Rafinesque, 1815, 92.
Leptognathus Swainson, 1839, 205.
Ophisternon McClelland, 1844, 220.
Ptyobranchus McClelland, 1844, 220.
Ophithorax McClelland, 1844, 220.
Leptorhynchus[165a] Smith, 1847, 244.
Anepistomon Gistel, 1848, 236.
Stethopterus Bleeker, 1853, 252.
Leiuranus Bleeker, 1853, 253.
Leiurus Kaup, 1856, 268.
Scytallurus Kaup, 1856, 268.
Centrurophis Kaup, 1856, 271.
Pœcilocephalus Kaup, 1856, 271.

Microdonophis Kaup, 1856, 271.
Cœcilophis Kaup, 1856, 271.
Herpetoichthys Kaup, 1856, 271.
Elapsopsis Kaup, 1856, 271.
Brachysomophis Kaup, 1856, 271.
Mystriophis Kaup, 1856, 271.
Murænopsis[165b] Kaup, 1856, 271.
Echiopsis Kaup, 1856, 271, 297.
Scytalophis Kaup, 1856, 271.
Leptorhinophis Kaup, 1856, 271.
Lamnostoma Kaup, 1856, 271.
Pisodonophis[166] Kaup, 1856, 271.
Ophisurapus Kaup, 1856, 271.
Cirrhimuræna Kaup, 1856, 271.
Callechelys Kaup, 1856, 271.
Anguisurus Kaup, 1856, 271.
Myrichthys Girard, 1859, 290.
Crotalopsis[167] Kaup, 1860, 297.
Cryptopterus Kaup, 1860, 297.
Achirophichthys Bleeker, 1865, 335.
Uranichthys Poey, 1867, 347.
Macrodonophis Poey, 1867, 347.
Letharchus Goode & Bean, 1882, 419.
Bascanichthys Jordan & Davis, 1892, 456.
Quassiremus J. & D., 1892, 458.
Scytalichthys J. & D., 1892, 458.

---

[164] Usually written *Ophichthys*.

[165] Probably fictitious; eel-shaped, long-nosed, with snake-like scales on head; abdominal ventrals, no dorsal; tail forked.

[165a] *Leptorhynchus* must be prior to its substitute *Anepistomon*.

[165b] *Murænopsis* as used by Kaup is a simple misprint for *Murænophis*.

[166] Written also *Pisoodontophis*.

[167] Misprinted *Crotalopis* in "Genera of Fishes."

Xyrias Jordan & Snyder, 1901, 494.
Chlevastes J. & S., 1901, 494.
Jenkinsiella Jordan & Evermann, 1905, 576.

Hemerorhinus Weber & Beaufort, 1916, 561.
Omochelys[168] Fowler, 1918.
Syletor Jordan, 1919, 567.
Acanthenchelys[168a] Regan, 1922.

---

[168] *Proc. Acad. Nat. Sci. Phila.*, 8, 1818; orthotype *Pisodonophis cruentifer* Goode & Bean. Same as *Syletor* Jordan, 1920; preoccupied in beetles (*Syletor* Tchitcherine, 1899).

[168a] "A new Eel from Tobago." *Ann. Mag. Nat. Hist.* X. 296; orthotype *A. spinicauda* Regan.

Family 220. NEENCHELYIDÆ

Neenchelys Bamber, 1915, 555.

Family 221. MORINGUIDÆ

Rataboura Gray, 1831, 138.
Moringua Gray, 1831, 138.
Pterurus Swainson, 1839, 205.
Pachyurus Swainson, 1839, 205.
Ptyobranchus McClelland, 1844, 220.
Aphthalmichthys Kaup, 1856, 272.

Pseudomoringua Bleeker, 1865, 335.
Stilbiscus Jordan & Bollman, 1888, 442.
Gordiichthys Jordan & Davis, 1892, 458.
Cryptophthalmus Franz, 1910, 535.
Unagius Jordan, 1919, 567.

Suborder *COLOCEPHALI*

Family 222. HETERENCHELYIDÆ

Heterenchelys Regan, 1912, 545.

Panturichthys Pellegrin, 1913, 549.

Family 223. MURÆNIDÆ
(Morays)

Muræna Linnæus, 1758, 11.
Echidna Forster, 1777, 42.
Gymnothorax Bloch, 1795, 53.
Murænophis Cuvier, 1798, 54.
Murænophis Lacépède, 1803, 68.
Gymnomuræna Lacépède, 1803, 68.
Megaderus Rafinesque, 1815, 92.
Gymnopsis Rafinesque, 1815, 92.
Haliophis Rüppell, 1828, 122.
Ichthyophis Lesson, 1828, 123.
Uropterygius Rüppell, 1835, 184.
Strophidon McClelland, 1844, 220, 574.
Lycodontis McClelland, 1844, 220.
Thærodontis McClelland, 1844, 220.
Channo-Muræna Richardson, 1844, 223.
Molarii[169] Richardson, 1844, 223.
Ichthyapus Brisout de Barneville, 1847, 231.

Enchelynassa Kaup, 1855, 265, 271.
Enchelycotte Kaup, 1856, 268.
Eurymyctera Kaup, 1856, 271.
Siderea Kaup, 1856, 271.
Enchelycore Kaup, 1856, 271.
Thyrsoidea Kaup, 1856, 271.
Limamuræna Kaup, 1856, 271.
Polyuranodon Kaup, 1856, 272.
Pœcilophis Kaup, 1856, 272.
Murænoblenna Kaup, 1856, 272.
Neomuræna Girard, 1859, 291.
Priodonophis Kaup, 1860, 297.
Tæniophis Kaup, 1860, 297.
Pseudomuræna Johnson, 1862, 319.
Pseudechidna Bleeker, 1863, 322.
Pythonichthys Poey, 1867, 347.
Rhinomuræna Garman, 1888, 441.
Rabula Jordan & Davis, 1892, 458.

---

[169] In plural form only.

Scutica Jordan & Evermann, 1896, 473.
Æmasia Jordan & Snyder, 1901, 494.
Scuticaria J. & S., 1901, 494.
Evenchelys Jordan & Evermann, 1902, 498.

Anarchias Jordan & Seale, 1906, 519.
Rhabdura Ogilby, 1907, 526.
Rhinechidna Barbour, 1908, 527.
Ahynnodontophis Fowler, 1912, 543.
DEPRANDUS[170] Jordan & J. Z. Gilbert, 1921.

[170] "Fish Fauna California Tertiary," 252; orthotype *D. lestes* Jordan & Gilbert.

Suborder *LYOMERI*

Family 224. SACCOPHARYNGIDÆ
(Gulpers)

Saccopharynx Mitchill, 1824, 117.   Ophiognathus Harwood, 1827, 121.

Family 225. EURYPHARYNGIDÆ
(Swallowers)

According to Dr. Roule, the nominal genera of this group are all identical with *Eurypharynx*.

Eurypharynx Vaillant, 1882, 421.
Gastrostomus Gill & Ryder, 1883, 424.

Megalopharynx Brauer, 1900, 489.
Macropharynx Brauer, 1902, 497.

Series *OSTARIOPHYSI*
(*Plectospondyli*)

Order HETEROGNATHI

Family 226. CHARACINIDÆ (*Characidæ*)
(Characins)

Some of the genera named below are perhaps referable to other families.

Characinus[171] Linnæus, 1758, 14.
Charax Gronow, 1763, 21.
Erythrinus Gronow, 1763, 21.
Charax (Gronow) Scopoli, 1777, 42.
Erythrinus (Gronow) Scopoli, 1777, 42.
Serrasalmus[172] Lacépède, 1803, 66.
Characinus Lacépède, 1803, 66.
Tetragonopterus Cuvier, 1815, 93.
Myletes Cuvier, 1815, 93.
Chalceus Cuvier, 1817, 96.
Hydrocyon Cuvier, 1817, 96.
Characinus Cuvier, 1817, 98.
Piabucus[173] (Cuvier) Oken, 1817, 98.
Erythrinus Agassiz, 1829, 173.
Salminus Agassiz, 1829, 132.

Xiphorhynchus Agassiz, 1829, 132.
Schizodon Agassiz, 1829, 132.
Cynodon[174] Spix, 1829, 132.
Rhaphiodon Agassiz, 1829, 132.
Erythrichthys Bonaparte, 1833, 175.
Macrodon Müller, 1843, 216.
Myleus Müller & Troschel, 1844, 221.
Alestes M. & T., 1844, 221.
Catoprion M. & T., 1844, 221.
Pygocentrus M. & T., 1844, 221.
Pygopristis M. & T., 1844, 221.
Epicyrtus M. & T., 1844, 221.
Xiphorhamphus M. & T., 1844, 221.
Hydrolycus M. & T., 1844, 221.
Brycon M. & T., 1844, 221.

[171] In plural form only, as *Characini*.
[172] Written *Serrasalmo* by Cuvier.
[173] Written *Piabuca* by authors.
[174] The dates of birth and death of Spix are incorrectly given in "Genera of Fishes," p. 173; read 1781–1826.

Exodon M. & T., 1844, 221.
Cnidon M. & T., 1845, 227.
Agoniates M. & T., 1845, 227.
Grundulus Cuvier & Valenciennes, 1846, 228.
Lebiasina Cuv. & Val., 1846, 229.
Pyrrhulina Cuv. & Val., 1846, 229.
Hydropardus Reinhardt, 1849, 244.
Cynopotamus Cuvier & Valenciennes, 1849, 242.
Chalcinus Cuv. & Val., 1849, 242.
Mylesinus Cuv. & Val., 1849, 242.
Tometes Cuv. & Val., 1849, 242.
Piabucina Cuv. & Val., 1849, 242.
Brycinus Cuv. & Val., 1849, 242.
Astyanax Baird & Girard, 1854, 255.
Cheirodon[175] Girard, 1854, 258.
Stevardia Gill, 1858, 281.
Corynopoma Gill, 1858, 281.
Nematopoma Gill, 1858, 281.
Pœcilurichthys Gill, 1858, 281.
Hemigrammus Gill, 1858, 281.
Bryconops Kner, 1859, 292.
Pseudochalceus[176] Kner, 1863, 327.
Chalcinopsis[177] Kner, 1863, 337.
Crenuchus Günther, 1863, 326.
Brachyalestes Günther, 1864, 333.
Hetererythrinus Günther, 1864, 332.
Hemibrycon Günther, 1864, 333.
Scissor Günther, 1864, 333.
Creatochanes Günther, 1864, 333.
Creagrutus Günther, 1864, 333.
Anacyrtus Günther, 1864, 333.
Rœstes Günther, 1864, 333.
Rœboides Günther, 1864, 333.
Hystricodon Günther, 1864, 333.
Sarcodaces Günther, 1864, 333.
Oligosarcus Günther, 1864, 333.
Psalidostoma[178] Kner, 1864, 338.
Piabina Reinhardt, 1866, 343.
Aphyocharax Günther, 1868, 351.
Megalobrycon Günther, 1869, 504.
Holotaxis Cope, 1870, 356.
Odontostilbe Cope, 1870, 356.

Stethaprion Cope, 1870, 356.
Plethodectes Cope, 1870, 356.
Proportheus Cope, 1872, 363.
Triportheus Cope, 1872, 363.
Iguanodectes Cope, 1872, 363.
Bryconæthiops Günther, 1873, 369.
Paragoniates Steindachner, 1876, 387.
Lutkenia Steindachner, 1876, 387.
Metynnis Cope, 1878, 394.
Bramocharax Gill, 1877, 389.
Leptogoniates Boulenger, 1887, 437.
Henochilus Garman, 1890, 451.
Pseudocorynopoma Perugia, 1891, 455.
Chalcinopelecus Holmberg, 1891, 455.
Bergia Steindachner, 1891, 456.
Brachychalcinus Boulenger, 1892, 457.
Asiphonichthys Cope, 1894, 464.
Diapoma Cope, 1894, 464.
Petersius Hilgendorf, 1894, 465.
Myloplus Gill, 1895, 467.
Hoplerythrinus Gill, 1895, 467.
Micralestes Boulenger, 1899, 485.
Neoborus Boulenger, 1899, 485.
Catabasis Eigenmann & Norris, 1900, 489.
Orthomyleus Eigenmann, 1903, 502.
Colossoma Eigenmann, 1903, 502.
Mylossoma Eigenmann, 1903, 502.
Boulengerella Eigenmann, 1903, 502.
Acnodon Eigenmann, 1903, 502.
Myleocollops Eigenmann, 1903, 502.
Gilbertella Eigenmann, 1903, 502.
Evermannella Eigenmann, 1903, 502.
Acestrocephalus Eigenmann, 1903, 502.
Holoshesthes[179] Eigenmann, 1903, 502.
Mœnkhausia Eigenmann, 1903, 502.
Markiana Eigenmann, 1903, 502.
Othonophanes Eigenmann, 1903, 502.
Holoprion Eigenmann, 1903, 502.
Holopristis Eigenmann, 1903, 502.
Stichonodon Eigenmann, 1903, 502.
Bryconodon Eigenmann, 1903, 502.
Acestrorhynchus Eigenmann, 1903, 502.
Piaractus Eigenmann, 1903, 502.

---

[175] Often corrected to *Chirodon*.
[176] This name first appears in *Sitzb. Acad. Wiss. München*, 226, 1863.
[177] Kner, *Sitzb. Acad. Wiss. München*, 225, 1863.
[178] First published in 1864.
[179] Later written, more correctly, *Holesthes*.

Acestrorhamphus Eigenmann, 1903, 502.
Gymnocharacinus Steindachner, 1903, 506.
Hoplias Gill, 1903, 503.
Eucynopotamus Fowler, 1904, 507.
Copeina Fowler, 1906, 518.
Ophiocephalops Fowler, 1906, 518.
Cyphocharax Fowler, 1906, 518.
Steindachnerina Fowler, 1906, 518.
Peltapleura Fowler, 1906, 518.
Phoxinopsis Regan, 1907, 526.
Mimagoniates Regan, 1907, 526.
Ctenocharax Regan, 1907, 526.
EOBRYCON Jordan, 1907, 525.
Pellegrinina Fowler, 1907, 523.
Coscinoxyron Fowler, 1907, 523.
Cynocharax Fowler, 1907, 523.
Cyrtocharax Fowler, 1907, 523.
Sphyrænocharax Fowler, 1907, 523.
Belonocharax Fowler, 1907, 523.
Waiteina Fowler, 1907, 523.
Reganina Fowler, 1907, 523.
Starksina Fowler, 1907, 523.
Sealeina Fowler, 1907, 523.
Bryconamericus Eigenmann, 1907, 522.
Gilbertolus Eigenmann, 1907, 522.
Evermannolus Eigenmann, 1907, 522.
Phenacogrammus Eigenmann, 1907, 522.
Phenacogaster Eigenmann, 1907, 522.
Thayeria Eigenmann, 1907, 527.
Deuterodon Eigenmann, 1907, 522.
Astyacinus Eigenmann, 1907, 522.
Fowlerina Eigenmann, 1907, 522.
Joinvillea Steindachner, 1908, 531.
Pristella Eigenmann, 1908, 528.
Poptella Eigenmann, 1908, 528.
Psellogrammus Eigenmann, 1908, 528.
Thayeria Eigenmann, 1908, 527.
Ctenobrycon Eigenmann, 1908, 527.
Cœlurichthys Ribeiro, 1908, 530.
Gymnocorymbus Eigenmann, 1908, 527.
Brycochandus Eigenmann, 1908, 528.
Hyphessobrycon Eigenmann, 1908, 528.
Champsoborus Boulenger, 1909, 531.
Dermatocheir Durbin, 1909, 532.
Triurobrycon Eigenmann, 1909, 532.
Holobrycon Eigenmann, 1909, 532.
Pterodiscus Eigenmann, 1909, 532.
Pœcilocharax Eigenmann, 1909, 533.

Hollandichthys Eigenmann, 1910, 535.
Rhoadsia Fowler, 1911, 539.
Probolodus Eigenmann, 1911, 538.
Psalidodon Eigenmann, 1911, 538.
Spintherobolus Eigenmann, 1911, 538.
Glandulocauda Eigenmann, 1911, 538.
Vesicatrus Eigenmann, 1911, 538.
Hysteronotus Eigenmann, 1911, 538.
Nematobrycon Eigenmann, 1911, 538.
Hasemania Ellis, 1911, 538.
Apodastyanax Fowler, 1911, 539.
Knodus Eigenmann, 1911, 538.
Genycharax Eigenmann, 1912, 543.
Gephyrocharax Eigenmann, 1912, 543.
Acanthocharax Eigenmann, 1912, 543.
Heterocharax Eigenmann, 1912, 543.
Aphyodite Eigenmann, 1912, 542.
Tylobranchus Eigenmann, 1912, 542.
Bivibranchia Eigenmann, 1912, 542.
Parastremma Eigenmann, 1912, 543.
Tyttocharax Fowler, 1913, 548.
Apistogramma Regan, 1913, 550.
Xenurocharax Regan, 1913, 550.
Ephippicharax Fowler, 1913, 548.
Gnathocharax Fowler, 1913, 548.
Pterobrycon Eigenmann, 1913, 547.
Prionobrama Fowler, 1913, 548.
Argopleura Eigenmann, 1913, 547.
Microgenys Eigenmann, 1913, 547.
Zygogaster Eigenmann, 1913, 547.
Ceratobranchia Eigenmann, 1914, 553.
Phanagoniates Eigenmann & Henn, 1914, 553.
Landonia E. & H., 1914, 553.
Parecbasis Eigenmann, 1914, 553.
Bleptonema Eigenmann, 1914, 553.
Microbrycon Eigenmann, 1914, 553.
Xiphocharax Fowler, 1914, 553.
Carlia Meek, 1914, 554.
Gastropristis Eigenmann, 1915, 556.
Rooseveltiella Eigenmann, 1915, 556.
Pristobrycon Eigenmann, 1915, 556.
Mixobrycon Eigenmann, 1915, 556.
Compsura Eigenmann, 1915, 556.
Aphyocheirodon Eigenmann, 1915, 556.
Oligobrycon Eigenmann, 1915, 555.
Microschemobrycon Eigenmann, 1915, 555.
Megalamphodus Eigenmann, 1915, 555.

Macropsobrycon Eigenmann, 1915, 555.
Leptobrycon Eigenmann, 1915, 555.
Æquidens Steindachner, 1915, 558.
Pseudocheirodon Meek & Hildebrand, 1916, 559.

Entomolepis Eigenmann, 1917, 561.
Bertoniolus[180] Fowler, 1918.
CHARACILEPIS Cockerell, 1920, 570.
Phenacobrycon[181] Eigenmann, 1920.

[180] "A New Characin from Paraguay," *Proc. Acad. Nat. Sci. Phila.*, 141, 1918; orthotype *B. paraguayensis* Fowler.

[181] Eigenmann, *Indiana University Studies*, **45**: 18; orthotype *Bryconamericus henni* Eigenmann.

Family 227. GASTEROPELECIDÆ

Gasteropelecus Gronow, 1763, 22.
Gasteropelecus Pallas, 1769, 22.

Thoracocharax Fowler, 1907, 523.
Carnegiella Eigenmann, 1909, 532.

Family 228. XIPHOSTOMIDÆ

Xiphostoma Agassiz, 1829, 132.
Ctenolucius Gill, 1861, 302.

Luciocharax Steindachner, 1876, 387.

Family 229. ANOSTOMIDÆ (*Curimatidæ: Anodontidæ*)

Anostomos Gronow, 1763, 20.
Anostomus (Gronow) Scopoli, 1777, 42.
Curimata Walbaum, 1792, 50.
Anostomus (Gronow) Cuvier, 1817, 98.
Curimatus (Cuvier) Oken, 1817, 98.
Leporinus Spix, 1829, 132.
Prochilodus Agassiz, 1829, 132.
Pacu Spix, 1829, 132.
Anodus Spix, 1829, 132.
Mormyrhynchus Swainson, 1839, 203.
Chilodus Müller & Troschel, 1844, 221.
Histiodromus Gistel, 1848, 234.
Microdus Kner, 1859, 292.
Rhytiodus Kner, 1859, 292.
Cænotropus Günther, 1864, 333.
Læmolyta Cope, 1872, 363.
Leporellus Lütken, 1874, 376.

Curimatopsis Steindachner, 1876, 387.
Elopomorphus Gill, 1878, 394.
Potamorhina Cope, 1878, 394.
Psectrogaster Eigenmann & Eigenmann, 1889, 445.
Curimatella[181a] E. & E., 1889, 446.
Semitapicis E. & E., 1889, 446.
Schizodontopsis Garman, 1890, 451.
Lahiliella Eigenmann, 1903, 502.
Pithecocharax Fowler, 1906, 518.
Chilomyzon Fowler, 1906, 518.
Eigenmannina Fowler, 1906, 518.
Garmanina Fowler, 1906, 518.
Abramites Fowler, 1906, 518.
Peltapleura Fowler, 1906, 518.
Anostomoides Pellegrin, 1908, 529, 533.
Xyrocharax Fowler, 1913, 547.

[181a] *Curimatella* Pellégrin is introduced by error on page 533.

Family 230. HEMIODONTIDÆ

Hemiodus Müller, 1843, 216, 243.
Parodon Cuvier & Valenciennes, 1849, 242.
Centrophorus Kner, 1859, 292.
Characidium Reinhardt, 1866, 343.
Saccodon[182] Kner, 1863, 327.
Nannostomus Günther, 1872, 365.
Chorimycterus Cope, 1894, 464.
Nanognathus Boulenger, 1895, 466.

Anisistia Eigenmann, 1903, 502.
Pœcilosomatops Fowler, 1906, 518.
Hemiodopsis Fowler, 1906, 518.
Jobertina Pellegrin, 1909, 533.
Pœcilobrycon Eigenmann, 1909, 533.
Archicheir Eigenmann, 1909, 533.
Microcharax Eigenmann, 1909, 533.
Apareiodon Eigenmann, 1916, 559.

[182] Kner, *Sitzb. Acad. Wiss. München*, 225, 1863; orthotype *Saccodon wagneri* Kner.

Family 231. **CITHARINIDÆ**

Citharinus[183] (Cuvier) Oken, 1817, 98, 243.

Distichodus Müller & Troschel, 1845, 227, 242.

Ichthyborus[184] Günther, 1864, 333.

Phago Günther, 1865, 337.

Xenocharax Günther, 1867, 345.

Nannocharax Günther, 1867, 345.

Nannæthiops Günther, 1871, 361.

Monostichodus Vaillant, 1900, 491.

Hemistichodus Pellegrin, 1900, 491, 557.

Neolebias Steindachner, 1894, 465.

Eugnathichthys Boulenger, 1898, 479.

Paraphago Boulenger, 1898, 485.

Neoborus Boulenger, 1898, 485.

Mesoborus Pellegrin, 1900, 490.

Citharidium Boulenger, 1902, 497.

Microstomatichthyoborus Nichols & Griscom, 1917, 562.

Paradistichodus[185] Pellegrin, 1922.

[183] The original spelling, *Cytharinus,* may be a lapse for *Citharinus,* as Cuvier wrote "les Citharines" and Valenciennes *"Citharinus."*

[184] Not *Ichthyoborus* Kaup, 1845, a genus of birds.

[185] "Poissons de l'Oubranchi-Chari," *Bull. Zool. Soc. France,* **47** : 70; orthotype *Nannocharax dimidatus* Pellegrin.

Order GYMNONOTI

Family 232. **RHAMPHICHTHYIDÆ**

Rhamphichthys Müller & Troschel, 1849, 243.

Altona Kaup, 1856, 268.

Hypopomus Gill, 1864, 330.

Brachyrhamphichthys Günther, 1870, 357.

Steatogenys Boulenger, 1898, 479.

Gymnorhamphichthys Eigenmann, 1912, 543.

Family 233. **APTERONOTIDÆ** (*Sternarchidæ*)

Apteronotus Lacépède, 1800, 56.

Sternarchus Bloch & Schneider, 1801, 59.

Carapus[186] Rafinesque, 1810, 82.

Sternarchorhynchus Castlenau, 1855, 264.

Memarchus Kaup, 1856, 268.

Rhamphosternarchus Günther, 1870, 357.

Sternarchorhamphus Eigenmann & Ward, 1905, 512.

Sternarchella E. & W., 1905, 512.

Sternarchogiton E. & W., 1905, 512.

Porotergus Eigenmann, 1912, 543.

Orthosternarchus Ellis, 1913, 547.

Odontosternarchus Ellis, 1913, 547.

[186] The final disposition of this generic name is open to question. The decision of the International Commission removes it from this group to replace *Fierasfer* Cuvier.

Family 234. **GYMNOTIDÆ**
(Carapos)

Gymnotus Linnæus, 1758, 12.

Carapus Cuvier, 1817, 101, 243.

Carapo Oken, 1817, 101.

Sternopygus Müller & Troschel, 1849, 243.

Giton Kaup, 1856, 268.

Gymnotes Gill, 1864, 330.

Cryptops Eigenmann, 1894, 464.

Eigenmannia Jordan & Evermann, 1896, 473.

Family 235. **ELECTROPHORIDÆ**
(Electric Eels)

Gymnotus Linnæus, 1766 (not of 1758).   Electrophorus Gill, 1864, 330.

Order EVENTOGNATHI

Family 236. CATOSTOMIDÆ

(Suckers; Buffalo Fishes)

Catostomus Le Sueur, 1817, 96.
Hypentelium Rafinesque, 1818, 109.
Cycleptus Rafinesque, 1819, 110.
Ictiobus[187] Rafinesque, 1820, 112.
Carpiodes Rafinesque, 1820, 112.
Moxostoma[188] Rafinesque, 1820, 112.
Decactylus Rafinesque, 1820, 113.
Teretulus Rafinesque, 1820, 113.
Eurystomus[189] Rafinesque, 1820, 113.
Stomocatus Bonaparte, 1840, 206.
Sclerognathus Cuvier & Valenciennes, 1844, 219.
Hylomyzon Agassiz, 1855, 261.
Ptychostomus Agassiz, 1855, 261.
Bubalichthys Agassiz, 1855, 261.
Acomus Girard, 1856, 269.
Minomus Girard, 1856, 269.

Placopharynx Cope, 1870, 356.
Amyzon Cope, 1872, 364.
Pantosteus Cope, 1876, 384.
Erimyzon Jordan, 1876, 386.
Lagochila Jordan & Brayton, 1877, 391.
Myxocyprinus Gill, 1878, 394.
Minytrema Jordan, 1878, 395.
Chasmistes Jordan, 1878, 396.
Quassilabia Jordan & Brayton, 1878, 395.
Lipomyzon Cope, 1881, 415.
Xyrauchen Kirsch, 1888, 443.
Deltistes Seale, 1896, 475.
Megastomatobus Fowler, 1913, 547.
Pithecomyzon Fowler, 1913, 548.
Scartomyzon Fowler, 1913, 548.
Notolepidomyzon Fowler, 1913, 548.
Thoburnia Jordan & Snyder, 1917, 562.

---

[187] Spelled *Ichthyobus* by authors.
[188] Name corrected to *Myxostoma*.
[189] Mythical.

Family 237. CYPRINIDÆ

(Carps, Dace, Minnows, Chubs, etc.)

Cyprinus Linnæus, 1758, 15.
Leuciscus Klein, 1775, 37.
Brama Klein, 1775, 37.
Mystus Klein, 1775, 38.
Labeo Cuvier, 1817, 99.
Tinca Cuvier, 1817, 99.
Gobio Cuvier, 1817, 99.
Abramis Cuvier, 1817, 99.
Barbus Cuvier, 1817, 99.
Leuciscus Cuvier, 1817, 99.
Cirrhinus[190] (Cuvier) Oken, 1817, 99.
Exoglossum Rafinesque, 1818, 109.
Maxillingua Rafinesque, 1818, 109.
Notropis Rafinesque, 1818, 108.
Notemigonus Rafinesque, 1819, 110.
Hemiplus Rafinesque, 1820, 110.
Minnilus Rafinesque, 1820, 111.
Dobula Rafinesque, 1820, 112.
Phoxinus Rafinesque, 1820, 112.
Alburnus Rafinesque, 1820, 112.
Luxilus Rafinesque, 1820, 112.

Chrosomus Rafinesque, 1820, 112.
Semotilus Rafinesque, 1820, 112.
Rutilus Rafinesque, 1820, 112.
Pimephales Rafinesque, 1820, 112.
Plargyrus Rafinesque, 1820, 112.
Garra Hamilton, 1822, 115.
Morulius Hamilton, 1822, 115.
Cabdio Hamilton, 1822, 115.
Puntius Hamilton, 1822, 115.
Barilius Hamilton, 1822, 115.
Bangana Hamilton, 1822, 115.
Chela Hamilton, 1822, 115.
Danio Hamilton, 1822, 115.
Diplocheilus Van Hasselt, 1823, 172.
Crossocheilus Van Hasselt, 1823, 116.
Lobocheilus Van Hasselt, 1823, 116.
Hampala Van Hasselt, 1823, 116.
Nandina Gray, 1831, 138.
Carassius Nilsson, 1832, 176.
Bengala Gray, 1833, 139, 179.
Tor Gray, 1833, 139, 179.

---

[190] Spelled *Cirrhina* by authors.

Phoxinus Agassiz, 1835, 180.
Rhodeus Agassiz, 1835, 180.
Chondrostoma Agassiz, 1835, 180.
Aspius Agassiz, 1835, 180.
Pelecus Agassiz, 1835, 180.
Varicorhinus Rüppell, 1836, 186.
Labeobarbus Rüppell, 1836, 186.
Squalius Bonaparte, 1837, 142, 187.
Telestes Bonaparte, 1837, 142, 187.
Scardinius Bonaparte, 1837, 142, 187.
Cyprinopsis Bonaparte, 1837, 142, 187.
Schizothorax Heckel, 1838, 192.
Platycara McClelland, 1839, 195.
Gonorhynchus McClelland, 1839, 195.
Oreinus McClelland, 1839, 195.
Systomus McClelland, 1839, 195.
Perilampus McClelland, 1839, 195.
Opsarius McClelland, 1839, 195.
Esomus Swainson, 1839, 202.
Chedrus Swainson, 1839, 202.
Salmophasia Swainson, 1839, 202.
Salmostoma Swainson, 1839, 198.
Phoxinellus Heckel, 1840, 211, 288.
Aulopyge Heckel, 1841, 209.
Rohtee Sykes, 1841, 210.
Racoma McClelland, 1842, 212.
Stilbe Dekay, 1842, 210.
Capoëta Cuvier & Valenciennes, 1842, 210.
Rohita Cuv. & Val., 1842, 210.
Nuria Cuv. & Val., 1842, 210.
Dangila Cuv. & Val., 1842, 210.
Tylognathus Heckel, 1842, 211.
Discognathus Heckel, 1842, 211.
Luciobarbus Heckel, 1842, 211.
Leucosomus Heckel, 1842, 211.
Argyreus Heckel, 1842, 211
Pachystomus Heckel, 1842, 211.
Osteobrama Heckel, 1842, 211.
Acanthobrama Heckel, 1842, 211.
Ballerus Heckel, 1842, 211.
Chondrorhynchus Heckel, 1842, 211.
Chondrochilus Heckel, 1842, 211.
Gymnostomus Heckel, 1842, 211.
Scaphiodon Heckel, 1842, 211.
Dillonia Heckel, 1842, 211.
Cyprinion Heckel, 1842, 211.

Gibelion Heckel, 1842, 211.
SORICIDENS Münster, 1842, 212.
CAPITODUS Münster, 1842, 212.
Aspidoparia Heckel, 1843, 215.
Schizopyge Heckel, 1843, 215.
Cyrene Heckel, 1843, 215.
Mola Heckel, 1843, 215.
Leucos Heckel, 1843, 215.
Catla Cuvier & Valenciennes, 1844, 219.
Cenisophius Bonaparte, 1845, 226.
Gardonus Bonaparte, 1845, 226.
Pigus Bonaparte, 1845, 226.
Cephalus Bonaparte, 1845, 226.
Microlepis Bonaparte, 1845, 226.
Hegerius Bonaparte, 1845, 226.
Rhinichthys Agassiz, 1848, 232.
Englottogaster Gistel, 1848, 236.
Cheilobarbus Smith, 1849, 244.
Pseudobarbus Smith, 1849, 244.
Abrostomus Smith, 1849, 244.
Hypsolepis[191] Baird, 1850, 245, 255.
Gila Baird & Girard, 1853, 252.
Ceratichthys[192] B. & G., 1853, 252, 270.
Pogonichthys Girard, 1854, 257.
Hybopsis Agassiz, 1854, 254.
Cochlognathus Baird & Girard, 1854, 255.
Lavinia Girard, 1854, 257.
TARSICHTHYS Troschel, 1854, 261.
Cheilonemus Baird, 1855, 266.
Culter Basilewski, 1855, 262.
Nasus Basilewski, 1855, 262.
Cephalus Basilewski, 1855, 262.
Leptocephalus Basilewski, 1855, 262.
Luciosoma Bleeker, 1855, 264.
Epalzeorhynchos Bleeker, 1855, 264.
Schismatorhynchos Bleeker, 1855, 264.
Mylopharodon Ayres, 1855, 262.
Acrocheilus Agassiz, 1855, 261.
Campostoma Agassiz, 1855, 261.
Hyborhynchus Agassiz, 1855, 261.
Hybognathus Agassiz, 1855, 261.
Ptychocheilus Agassiz, 1855, 262.
Mylocheilus Agassiz, 1855, 262.
Opsaridium Peters, 1855, 409.
Agosia Girard, 1856, 269.
Nocomis Girard, 1856, 269.

[191] Later written *Hypsilepis*.
[192] Misprinted *Ceraticthys*.

Cliola Girard, 1856, 269.
Alburnops Girard, 1856, 270.
Alburnellus Girard, 1856, 269.
Richardsonius Girard, 1856, 270.
Codoma Girard, 1856, 270.
Dionda Girard, 1856, 269.
Orthodon Girard, 1856, 269.
Algoma Girard, 1856, 269.
Algansea Girard, 1856, 269.
Clinostomus Girard, 1856, 270.
Hudsonius Girard, 1856, 270.
Tiaroga Girard, 1856, 270.
Tigoma Girard, 1856, 270.
Siboma Girard, 1856, 270.
Cyprinella Girard, 1856, 270.
Moniana Girard, 1856, 270.
Cheonda Girard, 1856, 270.
Blicca Heckel, 1858, 281.
Carpio Heckel, 1858, 281.
Lepidocephalus Bleeker, 1858, 279.
Idus Heckel, 1858, 281.
Leucaspis Heckel & Kner, 1858, 281.
Balantiocheilos Bleeker, 1859, 287.
Lissorhynchus Bleeker, 1859, 287.
Acanthobrama Bleeker, 1859, 287.
Morara Bleeker, 1859, 287.
Macrochirichthys Bleeker, 1859, 288.
Laubuca Bleeker, 1859, 288.
Pseudophoxinus[192a] Bleeker, 1859, 288.
Hemibarbus Bleeker, 1859, 288.
Hypselobarbus Bleeker, 1859, 288.
Rohteichthys Bleeker, 1859, 288.
Barbodes Bleeker, 1859, 288.
Anematichthys Bleeker, 1859, 288.
Siaja Bleeker, 1859, 288.
Hemiculter Bleeker, 1859, 288.
Pseudoculter Bleeker, 1859, 288.
Gonoproktopterys Bleeker, 1859, 288.
Pseudogobio Bleeker, 1859, 287.
Smilogaster Bleeker, 1859, 287.
Opistocheilus Bleeker, 1859, 287.
Semiplotus Bleeker, 1859, 287.
Rohitichthys Bleeker, 1859, 287.
Barbichthys Bleeker, 1859, 287.
Discognathichthys Bleeker, 1859, 287.
Diplocheilichthys Bleeker, 1859, 287.
Chanodichthys Bleeker, 1859, 288.
Leptobarbus Bleeker, 1859, 288.

Trinematichthys Bleeker, 1859, 288.
Hypophthalmichthys Bleeker, 1859, 288.
Barbichthys Bleeker, 1859, 287.
Chrysophekadion Bleeker, 1859, 287.
Amblypharyngodon Bleeker, 1859, 288.
Devario Bleeker, 1859, 288.
Tambra Bleeker, 1859, 287.
Mrigala Bleeker, 1859, 287.
Albulichthys Bleeker, 1859, 287.
Acheilognathus Bleeker, 1859, 287.
Amblyrhynchichthys Bleeker, 1859, 287.
Thynnichthys Bleeker, 1859, 288.
Elopichthys Bleeker, 1859, 288.
Shacra Bleeker, 1859, 288.
Bendilisis Bleeker, 1859, 288.
Pseudorasbora Bleeker, 1859, 288.
Rasbora Bleeker, 1859, 288.
Gnathopogon Bleeker, 1859, 288.
Rasborichthys Bleeker, 1859, 288.
Alburnoides Jeitteles, 1861, 308.
Bungia Keyserling, 1861, 308.
Owsianka Dybowski, 1862, 313.
Brama (Klein) Bleeker, 1862, 312.
Bliccopsis Siebold, 1863, 308.
Leptobarbus Bleeker, 1863, 321.
Acra Bleeker, 1863, 321.
Rohitodes Bleeker, 1863, 321.
Paraschizothorax Bleeker, 1863, 322.
Carpionichthys Bleeker, 1863, 322.
Abramidopsis Siebold, 1863, 328.
Isocephalus Heckel, 1863, 321.
Crossocheilichthys Bleeker, 1863, 321.
Gobionichthys Bleeker, 1863, 321.
Cirrhinichthys Bleeker, 1863, 321.
Opsariichthys Bleeker, 1863, 322.
Paraphoxinus Bleeker, 1863, 322.
Trachybrama Heckel, 1863, 322.
Paracheilognathus Bleeker, 1863, 322.
Pseudoperilampus Bleeker, 1863, 322.
Platygobio Gill, 1863, 323.
Peletrophus Günther, 1864, 333.
Scardiniopsis Jäckel, 1864, 334.
Pseudolaubuca Bleeker, 1865, 335.
Acanthobrama Bleeker, 1865, 335.
Paracanthobrama Bleeker, 1865, 335.
Parabramis Bleeker, 1865, 335.
Paralaubuca Bleeker, 1865, 335.

[192a] *Pseudophoxinus* "Heckel," page 211, "Genera of Fishes," should be canceled.

Paradanio (Bleeker) Day, 1865, 337.
Brachygramma Day, 1865, 337.
Stilbius Gill, 1865, 337.
Ericymba Cope, 1865, 336.
Cyprinopsis Blanchard, 1866, 339.
Pachychilon Steindachner, 1866, 343.
Ctenopharyngodon Steindachner, 1866, 343.
Ptychobarbus Steindachner, 1866, 343.
Diptychus Steindachner, 1866, 343.
Schizopygopsis Steindachner, 1866, 343.
Phenacobius Cope, 1867, 344.
Photogenis Cope, 1867, 344.
Enteromius Cope, 1867, 344.
Graodus Günther, 1868, 351.
Megarasbora Günther, 1868, 351.
Aphyocypris Günther, 1868, 351.
Pteropsarion Günther, 1868, 351.
Mystacoleucus Günther, 1868, 351.
Opsaridium (Peters) Günther, 1868, 351.
Bola Günther, 1868, 351.
Schacra Günther, 1868, 351.
Ochetobius Günther, 1868, 351.
Squaliobarbus Günther, 1868, 351.
Eustira Günther, 1868, 351.
Securicula Günther, 1868, 351.
Cachius Günther, 1868, 351.
Oreonectes Günther, 1868, 351.
Gymnocypris Günther, 1868, 351.
Barynotus Günther, 1868, 351.
Xenocypris Günther, 1868, 351.
Osteochilus Günther, 1868, 351.
Mayoa Day, 1869, 354.
Gobiobarbus Dybowski, 1869, 354.
Micraspius Dybowski, 1869, 354.
Pseudaspius Dybowski, 1869, 354.
Ladislavia Dybowski, 1869, 354.
Sarcocheilichthys Bleeker, 1869, 353.
Acrocheilichthys Bleeker, 1869, 353.
Abramocephalus Steindachner, 1869, 355.
Pseudobrama Bleeker, 1870, 355.
Luciobrama Bleeker, 1870, 355.
Rhinogobio Bleeker, 1870, 355.
Acanthorhodeus Bleeker, 1870, 355.
Saurogobio Bleeker, 1870, 355.

Hemitremia Cope, 1870, 356.
Anchybopsis Cope, 1870, 356.
Mylocyprinus Leidy, 1870, 359.
Lithichthys Cope, 1870, 356.
Diastichus Cope, 1870, 356.
Oligobelus Cope, 1870, 356.
Protoporus Cope, 1872, 363.
Myloleucus Cope, 1872, 363.
Sarcidium Cope, 1872, 363.
Coliscus Cope, 1872, 363.
Apocope Cope, 1872, 363.
Neocarassius[192b] Castelnau, 1872, 363.
Gobiosoma Dybowski, 1872, 364.
Onychodon Dybowski, 1872, 364.
Barbodon Dybokski, 1872, 364.
Plagiognathus Dybowski, 1872, 364.
Squalidus Dybowski, 1872, 364.
Megalobrama Dybowski, 1872, 364.
Orfus Fitzinger, 1873, 369.
Zopa Fitzinger, 1873, 368.
Rubellus Fitzinger, 1873, 369.
Vimba Fitzinger, 1873, 368.
Habrolepis Fitzinger, 1873, 369.
Cephalopsis Fitzinger, 1873, 369.
Bathystoma Fitzinger, 1873, 369.
Machærochilus Fitzinger, 1873, 369.
Toxabramis Günther, 1873, 369.
Mylöleucus Günther, 1873, 369.
Sinibarbus Sauvage, 1874, 376.
Eritrema Cope, 1876, 384.
Pristiodon Dybowski, 1877, 389.
Episema Cope & Jordan, 1877, 390.
Hemitrichas Peters, 1877, 392.
Aturius Dubalen, 1878, 394.
Cosmochilus Sauvage, 1878, 397.
Hydrophlox Jordan, 1878, 395.
Erogala Jordan & Brayton, 1878, 395.
Couesius Jordan, 1878, 395.
Chriope Jordan, 1878, 396.
Symmetrurus Jordan, 1878, 396.
Zophendum Jordan, 1878, 396.
Agenigobio Sauvage, 1878, 397.
Heteroleuciscus Sauvage, 1874, 376.
Megagobio Kessler, 1876, 386.
Lythrurus Jordan, 1876, 386.
Erinemus Jordan, 1876, 386.

[192b] This nominal genus is no doubt based on an escaped gold fish. There are no Cyprinidæ in Australia nor in South America.

HEXAPSEPHUS Günther, 1876, 385.
BRACHYSPONDYLUS[192c] von der Marck, 1876.
Aspiorhynchus Kessler, 1879, 400.
ENOPLOPHTHALMUS Sauvage, 1880, 405.
PARALEUCISCUS Sauvage, 1880, 406.
Probarbus Sauvage, 1880, 405.
Opsopœodus Hay, 1880, 403.
Semilabeo Peters, 1880, 405.
Mylopharyngodon Peters, 1880, 404.
Distœchodon Peters, 1880, 404.
Paratylognathus Sauvage, 1880, 405.
Stypodon Garman, 1881, 415.
Parachela Steindachner, 1881, 417.
Tirodon Hay, 1882, 419.
Coccotis Jordan, 1882, 419.
Erimystax Jordan, 1882, 419.
Trycherodon Forbes, 1882, 420.
Pachychilon Steindachner, 1882, 421.
Cosmochilus Sauvage, 1882, 421.
Gymnognathus Sauvage, 1884, 429.
Siphateles Cope, 1883, 421.
Tribolodon Sauvage, 1883, 425.
Hypargyrus Forbes, 1884, 428.
Rhodeoides Thominot, 1884, 430.
Oxygeneum Forbes, 1885, 431.
Luxilinus Jordan, 1885, 432.
Hemiculterella Varpakhovskii, 1887, 439.
Miniellus Jordan, 1888, 442.
Platypharodon Herzenstein, 1888, 442.
Chuanchia Herzenstein, 1888, 442.
Acanthonotus (Tickell) Day, 1888, 440.
Metallites Schulze, 1889, 448.
Liparus Schulze, 1889, 448.
Rhynchocypris Günther, 1889, 446.
Parapelecus Günther, 1889, 447.
Scaphiodonichthys Vinciguerra, 1889, 448.
Scombrocypris[193] Günther, 1889, 446.
Matsya Day, 1889, 445.
Spirlingus Fatio, 1890, 450.
Eucirrhichthys Perugia, 1892, 459.
Gymnodiptychus Herzenstein, 1892, 458.
Acanthogobio Herzenstein, 1892, 458.
Pungtungia Herzenstein, 1892, 458.

Oxybarbus Vaillant, 1893, 463.
APHELICHTHYS Cope, 1893, 460.
Engraulicypris Günther, 1893, 461.
Nematabramis Boulenger, 1894, 464.
Neobola Vinciguerra, 1895, 470.
Evarra Woolman, 1895, 470.
Leucogobio Günther, 1896, 471.
Onychostoma Günther, 1896, 471.
Azteca Jordan & Evermann, 1896, 473.
Orcella J. & E., 1896, 473.
Iotichthys J. & E., 1896, 473.
Opsopœa J. & E., 1896, 473.
Yuriria J. & E., 1896, 473.
Apiorhynchus Nikolskii, 1897, 477.
Apiostoma Nikolskii, 1897, 477.
Catlocarpio Boulenger, 1898, 478.
Aztecula Jordan & Evermann, 1898, 482.
Chelæthiops Boulenger, 1899, 485.
Xystrosus Jordan & Snyder, 1899, 487.
Falcula J. & S., 1899, 487.
Xenendum J. & S., 1899, 487.
Oreoleuciscus Varpachovskii, 1899, 488.
Orcula Jordan & Evermann, 1900, 489.
Ischikauia J. & S., 1900, 490.
Otakia J. & S., 1900, 490.
Leptocypris Boulenger, 1900, 488.
Gyrinochilus Vaillant, 1902, 500.
Culticula Abbott, 1901, 491.
Pterocapoëta Günther, 1902, 498.
Zacco Jordan & Evermann, 1902, 498.
Paracrossochilus Popta, 1902, 500.
Zezera Jordan & Fowler, 1903, 505.
Biwia J. & F., 1903, 505.
Abbottina J. & F., 1903, 505.
Falcularius Jordan & Snyder, 1903, 504.
Paracrossocheilus Popta, 1904, 510.
Paranotropis Fowler, 1904, 507.
Luciocyprinus Vaillant, 1904, 511.
Longurio[194] Jordan & Starks, 1905, 514.
Fusania J. & S., 1905, 514.
Coreius J. & S., 1905, 514.
Neogastromyzon Popta, 1905, 515.
Xenopomichthys Pellegrin, 1905, 515.
Pararhodeus Berg, 1907, 521.
Plagiognathops Berg, 1907, 521.

[192c] *Paleontogr.*, **22**: 411; orthotype *B. saropterix* Marck. (Omitted in "Genera of Fishes.")

[193] Duplicated under the misprint *Scomberopsis* in "Genera of Fishes," page 447.

[194] The name *Longurio*, a synonym of *Saurogobio*, is preoccupied in *Diptera*.

Aspiolucius Berg, 1907, 521.
Paraleucogobio Berg, 1907, 521.
Pogonocharax[195] Regan, 1907, 526.
Pararasbora Regan, 1908, 530.
Mearnsella Seale & Bean, 1908, 530.
Coccogenia Cockerell & Callaway, 1909, 532.
Candidia Jordan & Richardson, 1909, 533.
Margariscus Cockerell, 1909, 532.
Macrhybopsis Cockerell & Allison, 1909, 532.
Temeculina Cockerell, 1909, 532.
Acapoëta Cockerell, 1910, 535.
Rhodeops Fowler, 1910, 535.
Hemigrammocypris Fowler, 1910, 535.
Parabarbus Franz, 1910, 535.
Aspiopsis Zugmayer, 1912, 546.
Hesperoleucus Snyder, 1912, 546.
Ageneiogarra Garman, 1912, 543.
Myloleuciscus Garman, 1912, 543.
Acanthorutilus Berg, 1912, 542.
Coripareius Garman, 1912, 543.
Pararutilus Berg, 1912, 542.
Myloleucops Cockerell, 1913, 547.
Leucalburnus Berg, 1914, 552.
Pseudogobiops Berg, 1914, 552.
Chilogobio Berg, 1914, 552.
Schizocypris Regan, 1914, 554.
Tanakia Jordan & Thompson, 1914, 553.

Chilogobio Berg, 1915, 555.
Lissochilus Weber & Beaufort, 1916, 561.
Brevigobio Tanaka, 1916, 561.
Brachydanio Weber & Beaufort, 1916, 561.
Capoëtobrama Berg, 1916, 558.
Acanthalburnus Berg, 1916, 558.
Leucalburnus Berg, 1916, 558.
Leucidius Snyder, 1917, 563.
Sambwa[196] Annandale, 1918.
Microrasbora[197] Annandale, 1918.
Georgichthys[198] Nichols, 1918.
Spinibarbus Oshima, 1919, 570.
Phoxiscus Oshima, 1919, 570.
Scaphesthes Oshima, 1919, 570.
Aristichthys Oshima, 1919, 570.
Cultriculus Oshima, 1919, 570.
Formosania Oshima 1919, 570.
Acrossochilus Oshima, 1919, 570.
Raiamas Jordan, 1919, 567.
Extrarius Jordan, 1919, 567.
Lissocheilichthys Oshima, 1920, 572.
Leucisculus Oshima, 1920, 572.
Rasborinus Oshima, 1920, 572.
Scaphiodontella Oshima, 1920, 572.
Exoglossops[199] Fowler & Bean, 1920.
ALISODON[200] Hay, 1921.

[195] This generic name was based, according to Dr. J. T. Nichols, on an aquarium fish from India wrongly reported from Argentina, a cyprinoid, not a characim. It is a synonym of *Esomus* Swainson, 1839, and of *Nuria* Valenciennes, 1842.

[196] *Records Ind. Mus.*, **14**: 48; orthotype *S. resplendens* Annandale.

[197] *Records Ind. Mus.*, **14**: 50; logotype *M. rubescens* Annandale.

[198] Nichols, *Proc. Biol. Soc. Wash.*, **31**: 17; orthotype *Georgichthys scaphignathus* Nichols.

[199] "Fishes from Soochow, China," *Proc. U. S. Nat. Mus.*, **58**: pl. 311; type *E. geei* Fowler & Bean.

[200] *Proc. U. S. Nat. Mus.*, **58**: 132; orthotype *A. mirus* Hay (fossil); Pleistocene, Bulverde, Texas.

Family 238. MEDIDÆ

Meda Girard, 1856, 269.
Plagopterus Cope, 1874, 375.

Lepidomeda Cope, 1874, 375.

Family 239. COBITIDÆ
(Loaches)

Cobitis Linnæus, 1758, 14.
Barbatula Linck, 1790, 49.
Misgurnus Lacépède, 1803, 65.

Nemacheilus Van Hasselt, 1823, 116.
Acanthopsis Van Hasselt, 1823, 116.
Oxygaster Van Hasselt, 1823, 116.

Acanthophthalmus Van Hasselt, 1823, 116.

Botia Gray, 1831, 138.

Pterozygus De la Pylaie, 1835, 183.

Acanthopsis Agassiz, 1835, 180.

Somileptes Swainson, 1839, 204.

Acoura[201] Swainson, 1839, 204.

Canthophrys Swainson, 1839, 204.

Diacantha Swainson, 1839, 204.

Psilorhynchus McClelland, 1839, 195.

Schistura McClelland, 1839, 195.

Hymenphysa[202] McClelland, 1839, 195.

Aperioptus Richardson, 1848, 240.

Cobitichthys Bleeker, 1858, 279.

Syncrossus Blyth, 1860, 294.

Prostheacanthus Blyth, 1860, 294.

Pangio Blyth, 1860, 294.

Apua Blyth, 1860, 294.

Acanthocobitis Peters, 1861, 308.

Paracobitis Bleeker, 1863, 322.

Lepidocephalichthys Bleeker, 1863, 322.

Platacanthus Day, 1865, 367.

Octonema[202a] Martens, 1868, 352.

Leptobotia Bleeker, 1870, 355.

Oreias Sauvage, 1874, 376.

Diplophysa Kessler, 1874, 376.

Pseudodon Kessler, 1874, 376.

Paramisgurnus Sauvage, 1878, 397.

Crossostoma Sauvage, 1878, 397.

Octonema Herzenstein, 1887, 438.

Lefua Herzenstein, 1888, 442.

Modigliania Perugia, 1893, 462.

Elxis Jordan & Fowler, 1903, 505.

Orthrias J. & F., 1903, 505.

Ussuria Nikolskii, 1904, 510.

Vaillantella Fowler, 1905, 512.

Gobiobotia Kreyenberg, 1911, 540.

Aborichthys Chaudhuri, 1913, 547.

---

[201] Spelled also *Acourus* by Swainson.

[202] Usually corrected to *Hymenophysa*.

[202a] This name is apparently not preoccupied and is not identical with *Lefua*.

Family 240. HOMALOPTERIDÆ

Homaloptera Van Hasselt, 1823, 116.

Balitora Gray, 1833, 178.

Helgia Vinciguerra, 1889, 448.

Glaniopsis Boulenger, 1899, 485.

Ellopostoma Vaillant, 1902, 500.

Parhomaloptera Vaillant, 1902, 500.

Gyrinocheilus Vaillant, 1902, 500.

Homalopteroides Fowler, 1905, 513.

Lepturichthys Regan, 1911, 540.

Hemimyzon Regan, 1911, 540.

Blavania Hora, 1922. (See Index.)

Family 241. ADIPOSIIDÆ

The genus *Adiposia,* with a well-developed adipose fin, approaches the *Nematognathi,* and may stand as a separate family pending investigation. Adiposia[203] Annandale & Hora, 1921.

---

[203] "The Fish of Seistan," *Records of the Indian Museum,* **18**: Part IV, pp. 157, 187; type *Nemachilus macmahoni* Chaudhuri.

Order NEMATOGNATHII

Family 242. DIPLOMYSTIDÆ

Diplomystes Duméril, 1856, 269.

Diplomystes (Duméril) Bleeker, 1862, 311.

Family 243. ARIIDÆ (*Tachysuridæ*)
(Sea Cat-fishes)

The name *Arius* may be provisionally retained in place of *Tachysuras,* known from a drawing only, perhaps intended for some fresh-water form.

Tachysurus Lacépède, 1803, 66.
Bagre (Cuvier) Oken, 1817, 100.
Glanis Agassiz, 1829, 131.
Breviceps Swainson, 1838, 194.
Felichthys Swainson, 1839, 203.
Arius Cuvier & Valenciennes, 1840, 206.
Galeichthys Cuv. & Val., 1840, 206.
Batrachocephalus Bleeker, 1847, 233.
Osteogeneiosus Bleeker, 1847, 230.
Ketengus Bleeker, 1847, 230.
Ariodes Müller & Troschel, 1849, 243.
Mystus Gronow, 1854, 259.
Ailurichthys[204] Baird & Girard, 1854, 255.
Genidens Castelnau, 1855, 264.
Hexanematichthys Bleeker, 1858, 278.
Selenaspis Bleeker, 1858, 278.
Cephalocassis Bleeker, 1858, 278.
Netuma Bleeker, 1858, 278.
Guiritinga Bleeker, 1858, 278.
Osteogeneiosus Bleeker, 1858, 278.
Batrachocephalus Bleeker, 1858, 278.
Ketengus Bleeker, 1858, 278.

Hemipimelodus Bleeker, 1858, 278.
Ariopsis Gill, 1861, 303.
Parancistrus Bleeker, 1862, 311.
Hemiarius Bleeker, 1862, 311.
Pseudarius Bleeker, 1862, 311.
Catastoma (Kuhl & Van Hasselt) Bleeker, 1862, 312.
Leptarius Gill, 1863, 324.
Notarius Gill, 1863, 324.
Paradiplomystes Bleeker, 1863, 312.
Rhineastes Cope, 1872, 364.
Astephus Cope, 1873, 368.
Neoarius Castelnau, 1877, 388.
Cathorops Jordan & Gilbert, 1882, 420.
Nedystoma Ogilby, 1898, 483.
Cinetodus Ogilby, 1898, 483.
Aspistor Jordan & Evermann, 1898, 482.
Pachyula Ogilby, 1898, 483.
Ancharius Steindachner, 1880, 406.
Nemapteryx Ogilby, 1908, 529.
Doiichthys Weber, 1913, 551.
Tetranesodon Weber, 1913, 551.

---

[204] Often written Aelurichthys.

Family 244. DORADIDÆ

Cataphractus Catesby, 1771, 32.
Ageneiosus Lacépède, 1803, 66.
Doras Lacépède, 1803, 65.
Ceratorhynchus Agassiz, 1829, 131.
Centrochir Agassiz, 1829, 132.
Auchenipterus[205] Cuvier & Valenciennes, 1840.
Trachelyopterus Cuv. & Val., 1840, 206.
Euanemus Müller, 1843, 216.
Oxydoras Kner, 1855, 265.
Pseudodoras Bleeker, 1858, 278.
Hemidoras Bleeker, 1858, 278.
Davalla Bleeker, 1858, 278.
Trachycorystes Bleeker, 1858, 279.
Asterophysus Kner, 1858, 282.
Centromochlus Kner, 1858, 282.
Platydoras Bleeker, 1862, 311.
Lithodoras Bleeker, 1862, 311.
Rhinodoras Bleeker, 1862, 311.
Astrodoras Bleeker, 1862, 311.
Pterodoras Bleeker, 1862, 311.
Acanthodoras Bleeker, 1862, 311.
Amblydoras Bleeker, 1862, 311.
Parauchenipterus Bleeker, 1862, 311.

Trachycorystes Bleeker, 1862, 311.
Pseudauchenipterus Bleeker, 1862, 311.
Auchenipterichthys Bleeker, 1862, 311.
Tetranematichthys Bleeker, 1862, 312.
Pseudageneiosus Bleeker, 1862, 312.
Trachelyopterichthys Bleeker, 1862, 311.
Dianema Cope, 1872, 363.
Physopyxis Cope, 1872, 363.
Zathorax Cope, 1872, 363.
Glanidium Lütken, 1874, 376.
Wertheimeria Steindachner, 1876, 386.
Epapterus Cope, 1878, 393.
Pseudepapterus Steindachner, 1893. (See Index.)
Leptodoras Boulenger, 1898, 479.
Mormyrostoma Ribeiro, 1912, 545.
Parasturisoma Ribeiro, 1912, 545.
Tatia Ribeiro, 1912, 545.
Tympanopleura Eigenmann, 1912, 542.
Pseudepterus Steindachner, 1915, 558.
Entomocorus Eigenmann, 1917, 562.
Ceratocheilus Ribeiro, 1918, 564.
Tannayia Ribeiro, 1918, 564.

---

[205] Histoire naturelle des Poissons, 15: 207; orthotype Hypophthalmus nuchalis Spix; later made type of Euanemus. (Omitted in "Genera of Fishes.")

Family 245. PLOTOSIDÆ

Plotosus Lacépède, 1803, 66.
Tandanus Mitchell, 1839, 196.
Deportator Gistel, 1848, 237.
Paraplotosus[206] Bleeker, 1862.
Chœroplotosus Kner, 1864, 334.
Copidoglanis[205a] Günther, 1864, 332.
Cnidoglanis Günther, 1864, 332.
Gastromyzon Günther, 1874, 375.
Neoplotosus Castelnau, 1875, 378.
Neosilurus Castelnau, 1877, 388.
Eumeda Castelnau, 1878, 393.

Cainosilurus Macleay, 1881, 416.
Endorrhis Ogilby, 1898, 483.
Euristhmus Ogilby, 1899, 487.
Anyperistius Ogilby, 1908, 529.
Porochilus Weber, 1913, 551.
Lambertia Perugia, 1894, 465.
Ostophycephalus[207] Ogilby, 1899, 487.
Oloplotosus Weber, 1913, 551.
Anodontoglanis Rendahl, 1922 (See Index A).

[205a] Misprinted Copiodoglanis.
[206] "Systema Silurorum Revisum," 100; orthotype Plotosus albilabris Cuvier & Valenciennes. (Omitted in "Genera of Fishes.")
[207] Based on a starveling example of Cnidoglanis, according to Waite.

Family 246. SILURIDÆ
(Sheat-fishes)

Silurus Linnæus, 1758, 14.
Ompok Lacépède, 1803, 65.
Callichrous Hamilton, 1822, 114.
Wallago Bleeker, 1851, 247.
Glanis Agassiz, 1856, 267.
Belodontichthys Bleeker, 1858, 279.
Pseudosilurus Bleeker, 1858, 279.
Kryptopterus[208] Bleeker, 1858, 279.
Kryptopterichthys Bleeker, 1858, 279.
Micronema Bleeker, 1858, 279.

Phalacronotus Bleeker, 1858, 279.
Hemisilurus Bleeker, 1858, 279.
Silurichthys Bleeker, 1858, 278.
Silurodes Bleeker, 1860, 293, 312.
Pterocryptis Peters, 1861, 308.
Parasilurus Bleeker, 1862, 310.
Silurodon Kner, 1864, 334, 367.
Neosilurus Steindachner, 1867, 348.
Diastatomycter Vaillant, 1891, 456.
Apodoglanis Fowler, 1905, 512.

[208] Written Cryptopterus by purists.

Family 247. AMEIURIDÆ
(Horned Pout; Channel Cats)

Bagre Catesby, 1771, 31.
Noturus Rafinesque, 1818, 109.
Pilodictis[209] Rafinesque, 1819, 110.
Ictalurus[210] Rafinesque, 1820, 113.
Elliops Rafinesque, 1820, 113.
Leptops Rafinesque, 1820, 112.
Opladelus[211] Rafinesque, 1820, 112.
Ameiurus[212] Rafinesque, 1820, 112.
Ilictis Rafinesque, 1820, 112.

Synechoglanis Gill, 1858, 280.
Schilbeodes Bleeker, 1858, 279.
Gronias Cope, 1864, 329.
Rabidus Jordan & Evermann, 1896, 473.
Haustor J. & E., 1896, 473.
Villarius Rutter, 1896, 475.
Istlarius Jordan & Snyder, 1899, 486.
Trogloglanis Eigenmann, 1919, 566.

[209] Mythical, the name corrected to Pelodichthys.
[210] Corrected to Ichthælurus.
[211] Name corrected to Hopladelus.
[212] Also spelled Amiurus.

Family 248. BAGRIDÆ

Mystus Gronow, 1763, 21.
Mystus (Gronow) Scopoli, 1777, 42.
Porcus Isidore Geoffroy St. Hilaire,
  1818, 107.
BUCKLANDIUM König, 1825, 118.
Bagrus Cuvier & Valenciennes, 1839,
  194.
Olyra McClelland, 1842, 212.
GLYPTOCEPHALUS Agassiz, 1844, 217.
Bagroides Bleeker, 1851, 248.
Macrones Duméril, 1856, 269.
Clarotes Kner, 1855, 265.
Pseudobagrus Bleeker, 1858, 278.
Melanodactylus Bleeker, 1858, 278.
Chrysichthys Bleeker, 1858, 278.
Octonematichthys Bleeker, 1858, 278.
Rita Bleeker, 1858, 278.
Auchenaspis Bleeker, 1858, 278.
Rama Bleeker, 1858, 279.
Leiocassis Bleeker, 1858, 278.
Bagrichthys Bleeker, 1860, 293, 312.
Pseudobagrus Bleeker, 1860, 293.
Hemibagrus Bleeker, 1862, 312.
Pseudobagrichthys Bleeker, 1862, 312.
Aspidobagrus Bleeker, 1862, 312.
Hypselobagrus Bleeker, 1862, 313.

Branchiosteus Gill, 1862, 318.
Heterobagrus Bleeker, 1864, 329.
Auchenoglanis Günther, 1864, 334.
Rhinobagrus Bleeker, 1865, 335.
Pelteobagrus Bleeker, 1865, 335.
Gogrius Day, 1867, 344.
Liobagrus Hilgendorf, 1878, 395.
Cranoglanis Peters, 1880, 405.
Neobagrus Bellotti, 1892, 457.
Chrysobagrus Boulenger, 1899, 485.
Gephyroglanis Boulenger, 1899, 485.
Notoglanidium Günther, 1902, 498.
Leptoglanis Boulenger, 1902, 496.
Fluvidraco Jordan & Fowler, 1903, 505.
FAJUMIA[213] Stromer, 1904, 511.
SOCNOPÆA Stromer, 1904, 511.
Phyllonemus Boulenger, 1906, 517.
Parauchenoglanis Boulenger, 1911, 538.
Liauchenoglanis Boulenger, 1916, 558.
Amarginops Nichols & Griscom, 1917,
  562.
Gnathobagrus Nichols & Griscom, 1917,
  562.
Aoria Jordan, 1919, 567.
Macronoides[214] Hora, 1921.
Laguvia[215] Hora, 1921.

[213] Stromer compares this genus to *Platystoma*, a Brazilian pimelodid. But it probably belongs with *Socnopæa* to the *Bagridæ*.
[214] Sunder Lal Hora, "Fish and Fisheries of Manipur," *Rec. Ind. Mus.*, **22**: Part III, 179; logotype *Batasio affinis* Blyth.
[215] "Fishes from the Eastern Himalayas," *Rec. Ind. Mus.*, **22**: Part V, 740; logotype *Pimelodus asperus* McClelland.

Family 249. AMBLYCIPITIDÆ

Amblyceps Blyth, 1858, 279.
Acrochordonichthys Bleeker, 1858, 279.

Akysis Bleeker, 1860, 293.
Sosia Vaillant, 1902, 500.

Family 250. SISORIDÆ

Sisor Hamilton, 1822, 115.
Glyptosternon[216] McClelland, 1842, 212.
Erethistes Müller & Troschel, 1849, 243.
Bagarius Bleeker, 1853, 253.
Gagata Bleeker, 1858, 279.
Hara Blyth, 1860, 294.
Exostoma Blyth, 1860, 294.
Glyptothorax Blyth, 1860, 294.
Pseudecheneis Blyth, 1860, 294.

Batasio Blyth, 1860, 294.
Aglyptosternon Bleeker, 1862, 312.
Euclyptosternum Günther, 1864, 332.
Callomystax Günther, 1864, 332.
Chiloglanis Peters, 1868, 353.
Nangra Day, 1876, 385.
Breitensteinia Steindachner, 1881, 417.
Parexostoma Regan, 1905, 515.

[216] Spelled later *Glyptosternum*.

Family 251. AMPHILIIDÆ

Amphilius Günther, 1864, 332.
Chimarrichthys[217] Sauvage, 1874.
Doumea Sauvage, 1879, 401.
Anoplopterus Pfeffer, 1889, 448.
Peltura Perugia, 1892, 459.
Chimarrhoglanis Vaillant, 1897, 478.

Andersonia Boulenger, 1900, 488.
Phractura Boulenger, 1900, 488.
Belonoglanis Boulenger, 1902, 496.
Paraphractura Boulenger, 1902, 496.
Trachyglanis Boulenger, 1902, 496.
Paramphilius Pellegrin, 1907, 526.

---

[217] Sauvage, *Rev. Zool.* (3) **2**: 332; type *C. davidi* Sauvage. (Omitted in "Genera of Fishes.")

Family 252. LEPIDOGLANIDÆ

Said to resemble *Chimarrhichthys* but to have cycloid scales like a cyprinoid, a character hard to explain.

Lepidoglanis Vaillant, 1889, 448.

Family 253. CHACIDÆ

Chaca Gray, 1831, 138.

Family 254. SCHLBEIDÆ

Schilbe (Cuvier) Oken, 1817, 100.
Ailia Gray, 1831, 138.
Acanthonotus Gray, 1831, 138.
Silonia Swainson, 1839, 203.
Clupisoma Swainson, 1839, 204.
Pachypterus Swainson, 1839, 204.
Pusichthys Swainson, 1839, 204.
Silundia Cuvier & Valenciennes, 1840, 206.
Eutropius Müller & Troschel, 1849, 243.
Schilbeichthys Bleeker, 1858, 279.
Siluranodon Bleeker, 1858, 279.
Helicophagus Bleeker, 1858, 277.

Lais Bleeker, 1860, 293.
Eutropiichthys Bleeker, 1862, 310.
Pseudeutropius Bleeker, 1862, 310.
Gymnallabes Günther, 1867, 345.
Ailichthys Day, 1871, 361.
Anopleutropius Vaillant, 1893, 463.
Parailia Boulenger, 1899, 485.
Physailia Boulenger, 1901, 492.
Pseudolais Vaillant, 1902, 500.
Ansorgia Boulenger, 1912, 542.
Laides Jordan, 1919, 568.
Pareutropius[218] Regan, 1920.

---

[218] "Three New Fishes from the Tanganyika Territory," *Ann. Mag. Nat. Hist.*, July, 1920, 103; type *P. micristius* Regan.

Family 255. CLARIIDÆ

Clarias Gronow, 1763, 20.
Clarias (Gronow) Scopoli, 1777, 42.
Macropteronotus Lacépède, 1803, 65.
Heterobranchus Isidore Geoffroy St. Hilaire, 1818, 107.
Heteropneustes Müller, 1839, 196.
Saccobranchus Cuvier & Valenciennes, 1840, 206.

Phagorus McClelland, 1844, 220.
Cossyphus McClelland, 1844, 220.
Channalabes Günther, 1873, 369.
Clariallabes Boulenger, 1898, 576.
Allabenchelys Boulenger, 1902, 497.
Dinotopterus Boulenger, 1906, 516.

Family 256. PANGASIIDÆ

Pangasius Cuvier & Valenciennes, 1840, 206.

Pseudopangasius Bleeker, 1862, 310.
Neopangasius Popta, 1904, 510.

Family 257. **MOCHOKIDÆ**

Synodontis Cuvier, 1817, 100.
Mochokus[219] Joannis, 1835, 182.
Hemisynodontis Bleeker, 1862, 311.
Pseudosynodontis Bleeker, 1862, 311.
Brachysynodontis Bleeker, 1862, 311.
Leiosynodontis Bleeker, 1862, 311.
Atopochilus Sauvage, 1879, 401.

Oxyglanis Vinciguerra, 1898, 484.
Euchilichthys Boulenger, 1900, 488.
Microsynodontis Boulenger, 1903, 501.
Slatinia Werner, 1906, 521.
Acanthocleithron Nichols & Griscom, 1917, 562.

[219] Spelled *Mochocus* by Günther.

Family 258. **MALAPTERURIDÆ** *(Torpedinidæ)*
(Electric Cat-Fishes)

Torpedo[220] Forskål, 1771, 32.
Malapterurus Lacépède, 1803, 65.

Torpedo (Forskål) Gill, 1890.

[220] The species examined by Forskål is an electric cat-fish; the synonym is mixed with the electric ray.

Family 259. **HELOGENIDÆ**

Helogenes Günther, 1863, 326.

Family 260. **HYPOPHTHALMIDÆ**

Hypophthalmus Spix, 1829, 131.
Notophthalmus Hyrtl, 1859, 292.
Davalla Bleeker, 1858, 278.

Pseudohypophthalmus Bleeker, 1862, 312.

Family 261. **PIMELODIDÆ**

Pimelodus Lacépède, 1803, 65.
Phractocephalus Agassiz, 1829, 131.
Platystoma Agassiz, 1829, 131.
Sorubim Spix, 1829, 132.
Pirarara Spix, 1829, 173.
Pteronotus Swainson, 1839, 204.
Calophysus Müller, 1843, 216.
Abron Gistel, 1848, 237.
Sciades Müller & Troschel, 1849, 243.
Conostome[221] Duméril, 1856, 269.
Conorhynchos Bleeker, 1858, 278.
Zungaro Bleeker, 1858, 278.
Rhamdia Bleeker, 1858, 278.
Pinirampus Bleeker, 1858, 278.
Sciadeichthys Bleeker, 1858, 278.
Heptapterus Bleeker, 1858, 278.
Piratinga Bleeker, 1858, 279.
Platynematichthys Bleeker, 1858, 279.
Piramutana Bleeker, 1858, 279.
Batrachoglanis Gill, 1858, 281.
Pimelonotus Gill, 1858, 281.

Pimelotropis Gill, 1859, 291.
Brachyplatystoma Bleeker, 1862, 312.
Hemisorubim Bleeker, 1862, 312.
Pseudoplatystoma Bleeker, 1862, 312.
Hemiplatystoma Bleeker, 1862, 312.
Sorubimichthys Bleeker, 1862, 312.
Platystomatichthys Bleeker, 1862, 312.
Leiarius Bleeker, 1862, 312.
Pseudariodes Bleeker, 1862, 312.
Malacobagrus Bleeker, 1862, 312.
Parabagrus Bleeker, 1862, 312.
Pseudorhamdia Bleeker, 1862, 310.
Pseudopimelodus Bleeker, 1863, 322.
Notoglanis Günther, 1864, 332.
Bagropsis Lütken, 1874, 376.
Pimelodina Steindachner, 1876, 387.
Lophiosilurus Steindachner, 1876, 387.
Nannoglanis Boulenger, 1887, 437.
Rhamdella Eigenmann & Eigenmann, 1888, 440.
Luciopimelodus E. & E., 1888, 440.

[221] In French form only.

Pimelodella E. & E., 1888, 440.
Duopalatinus E. & E., 1888, 440.
Sciadeoides E. & E., 1888, 440.
Steindachneria E. & E., 1888, 440.
Nemuroglanis E. & E., 1889, 445.
Acentronichthys E. & E., 1889, 445.
Phrenatobius Goeldi, 1898, 480.
Paulicea Ihering, 1898, 480.
Perugia Eigenmann & Norris, 1900, 489.
Iheringichthys E. & N., 1900, 489.
Bergiella E. &. N., 1900, 489.
Gœldiella E. & N., 1900, 489.
Imparfinis E. & N., 1900, 489.
Bergiaria E. & N., 1901, 493.
Typhlobagrus Ribeiro, 1907, 526.
Rhamdioglanis Ihering, 1907, 524.
Tænionema Eigenmann & Bean, 1907, 522.
Zungaropsis Steindachner, 1908, 531.
Megalonema Eigenmann, 1910, 535, 542.
Cephalosilurus Haseman, 1911, 539.

Platysilurus Haseman, 1911, 539.
Rhamdiopsis Haseman, 1911, 539.
Microglanis Eigenmann, 1912, 542.
Brachyglanis Eigenmann, 1912, 542.
Leptoglanis Eigenmann, 1912, 542.
Myoglanis Eigenmann, 1912, 542.
Chasmocranus Eigenmann, 1912, 542.
Nannorhamdia Regan, 1913, 550.
Cetopsorhamdia Eigenmann, 1916, 558.
Cheirocerus Eigenmann, 1917, 562.
Leptorhamdia Eigenmann, 1918.
Typhlobagrus Ribeiro, 1918, 564.
Steindachneridion Eigenmann & Eigen-
    mann, 1919, 566.
Laimumena Sauvage, 1884, 430.
Platypogon Starks, 1913, 551.
Breviglanis[222] Eigenmann, 1909.
Chasmocephalus[223] Eigenmann, 1909.

[222] "Fresh Water Fish S. A.," 384; type *B. frenata* Eigenmann.
[223] "Catal. and Bibliogr. S. Am. Fresh Water Fishes," 1909, 384; type *Ch. longior*, Eigenmann. (Omitted in "Genera of Fishes.")

Family 262. BUNOCEPHALIDÆ

Bunocephalus Kner, 1855, 265.
Bunocephalichthys Bleeker, 1858, 279.

Dysichthys Cope, 1874, 374.
Xiliphius Eigenmann, 1912, 543.

Family 263. ASPREDINIDÆ

Aspredo Gronow, 1763, 20.
Aspredo (Gronow) Scopoli, 1777, 42.
Platystacus Klein, 1779, 168.
Platystacus (Klein) Bloch, 1794, 53.
Platistus Rafinesque, 1815, 171.

Cotylephorus Swainson, 1839, 204.
Aspredinichthys Bleeker, 1858, 279.
Agmus Eigenmann, 1912, 542.
Chamaigenes Eigenmann, 1912, 542.

Family 264. PYGIDIIDÆ

Eremophilus Humboldt, 1806, 76.
Thrycomycterus[224] Humboldt & Valen-
    ciennes, 1811, 170.
Pygidium Meyen, 1835, 182.
Trichomycterus Valenciennes, 1846, 230.
Vandellia Cuvier & Valenciennes, 1846,
    407.
Nematogenys Girard, 1854, 258.
Pareiodon[225] Kner, 1855, 265.
Stegophilus Reinhardt, 1858, 284.

Astemomycterus Guichenot, 1860, 296.
Trachypoma Giebel, 1870, 357.
Pariolius Cope, 1872, 363.
Tridens Eigenmann & Eigenmann, 1889,
    446.
Pseudostegophilus E. & E., 1889, 446.
Miuroglanis E. & E., 1889, 446.
Acanthopoma Lütken, 1891, 455.
Henonemus Eigenmann & Ward, 1907,
    522.

[224] Later spelled *Trichomycterus.*
[225] Later spelled *Pariodon.*

Homodiætus E. & W., 1907, 522.
Hatcheria Eigenmann, 1909, 533.
Ochmacanthus Eigenmann, 1912, 542.
Paravandellia Ribeiro, 1912, 545.
Gyrinurus Ribeiro, 1912, 545.

Cobitiglanis Fowler, 1914, 553.
Pleurophysus Ribeiro, 1918, 564.
Urinophilus Eigenmann, 1918, 563.
Branchoica Eigenmann, 1918, 563.
Scleronema Eigenmann, 1918, 563.

Family 265. CETOPSIDÆ

Cetopsis Agassiz, 1829, 131.
Hemicetopsis Bleeker, 1862, 311.
Pseudocetopsis Bleeker, 1862, 311.
Paracetopsis (Guichenot) Bleeker, 1862, 312.

Paracetopsis Eigenmann & Bean, 1907, 522.
Cetopsogiton[226] E. & B., 1910.

---

[226] *Rept. Princeton Univ. Exp. Patagonia*, 111–398, 1910; type *Cetopsis occidentals* Steindachner. (Omitted in "Zoological Record.")   Same as *Paracetopsis* Eigenmann & Bean, preoccupied.

Family 266. ARGIDÆ

Astroblepus Humboldt, 1806, 76.
Cyclopium Swainson, 1839, 204.
Arges Cuvier & Valenciennes, 1840, 206.

Brontes Cuv. & Val., 1840, 206.
Strephon Gistel, 1848, 235.
Stygogenes Günther, 1864, 332.

Family 267. CALLICHTHYIDÆ

Callichthys Gronow, 1763, 21.
Callichthys (Gronow) Scopoli, 1777, 42.
Cataphractus Bloch, 1793, 51.
Corydoras Lacépède, 1803, 66.
Cordorinus Rafinesque, 1815, 91.
Hoplisoma Swainson, 1839, 203.
Hoplosternum Gill, 1858, 281.
Scleromystax Günther, 1864, 332.

Brochis Cope, 1872, 363.
Chænothorax Cope, 1878, 394.
Decapogon[226a] Eigenmann & Eigenmann, 1888, 440.
Osteogaster Cope, 1894, 464.
Aspidoras Ihering 1907, 524.
Cascadura Ellis, 1913, 547.
Cataphractops Fowler, 1915, 556.

[226a] *Decapogon* Ribeiro, p. 565, "Genera of Fishes," should be canceled.

Family 268. LORICARIIDÆ

Loricaria Linnæus, 1758, 14.
Plecostomus Gronow, 1763, 21.
Hypostomus Lacépède, 1803, 66.
Acanthicus Spix, 1829, 131.
Rhinelepis Spix, 1829, 131.
Sturisoma Swainson, 1839, 203.
Ancistrus Kner, 1854, 260.
Acestra Kner, 1854, 260.
Hemiodon Kner, 1854, 260.
Acanthodemus Castelnau, 1855, 264.
Pterygoplichthys Gill, 1858, 281.
Hemiancistrus Bleeker, 1862, 311.
Pseudancistrus Bleeker, 1862, 311.
Plecostomus (Gronow) Bleeker, 1862.
Pseudorinelepis Bleeker, 1862, 311.
Hemiloricaria Bleeker, 1862, 311.
Pseudacanthicus Bleeker, 1862, 311.

Parahemiodon Bleeker, 1862, 311.
Pseudoloricaria Bleeker, 1862, 311.
Loricariichthys Bleeker, 1862, 311.
Oxyloricaria Bleeker, 1862, 311.
Rineloricaria Bleeker, 1862, 311.
Pseudohemiodon Bleeker, 1862, 311.
Hemiodontichthys Bleeker, 1862, 311.
Cochliodon Heckel, 1864, 332.
Hypoptopoma Günther, 1868, 351.
Liposarcus Günther, 1864, 332.
Otocinclus Cope, 1872, 363.
Xenomystus Lütken, 1873, 370.
Harttia Steindachner, 1876, 387.
Hassar Eigenmann & Eigenmann, 1888, 440.
Neoplecostomus E. & E., 1888, 440.
Farlowella E. & E., 1889, 445.

Oxyropsis E. & E., 1889, 445.

Hisonotus E. & E., 1889, 445.

Parotocinclus E. & E., 1889, 445.

Panaque E. & E., 1889, 446.

Delturus E. & E., 1889, 446.

Hemipsilichthys E. & E., 1889, 446.

Microlepidogaster E. & E., 1889, 445.

Aristommata Holmberg, 1893, 462.

Xenocara Regan, 1904, 510.

Reganella[227] Eigenmann, 1905.

Thysanocara[228] Regan, 1906, 520.

Pogonopoma Regan, 1907, 526.

Kronichthys Ribeiro, 1908, 530.

Corymbophanes Eigenmann, 1909, 532.

Lithogenes Eigenmann, 1909, 532.

Canthopomus[229] Eigenmann, 1909.

Acestridium Haseman, 1911, 539.

Lithoxus Eigenmann, 1912, 542.

Stoniella Fowler, 1914, 553.

Diapaletoplites Fowler, 1915, 556.

Leptoancistrus Meek & Hildebrand, 1916, 559.

Pareiorhaphis[230] Ribeiro, 1918.

Pseudotocinclus Nichols, 1919, 569.

Cheirododus[231] Eigenmann, 1921.

[227] Eigenmann, *Science,* **21**: 793, May 18, 1905; orthotype *Hemiodon depressus* Kner; substitute for *Hemiodon;* preoccupied in mollusks.

[228] Type *Ancistrus cirrhosa* Kner.

[229] Cat. Fresh Water Fishes S. A. 407; type *Rhinelepis genibarbis* Cuv. & Val. (Omitted in "Genera of Fishes.")

[230] *Rev. Soc. Sci. Rio de Janeiro,* **2**: 106; orthotype *Hemipsilichthys calmoni* Ribeiro.

[231] Orthotype *Plecostomus hondæ* Regan.

Order INIOMI

The fishes of this order, mostly inhabitants of depths of the sea and thus more or less degenerate structurally, bear a strong external resemblance to the stomiatoid *Isospondyli,* with which they have been usually associated and with which they agree in possessing an adipose dorsal fin, as in the *Salmonidæ, Characinidæ, Siluridæ,* and *Percopsidæ.* The persistence of this primitive character does not necessarily indicate close relationship among the fishes possessing it. In the *Iniomi,* the premaxillary is well developed, excluding the maxillary from the margin of the upper jaw.

Family 269. CHEIROTHRICIDÆ

Megapus Marck & Schlüter, 1865, 339.

Cheirothrix[232] Pictet & Humbert, 1866, 342.

Exocœtoides Davis, 1885, 431.

Megistopus Landois, 1894, 465.

[232] Spelled *Chirothrix* by authors.

Family 270. AULOPIDÆ

Aulopus Cuvier, 1817, 98.

Sardinioides von der Marck, 1858, 283.

Scopelosaurus Bleeker, 1860, 294.

Family 271. SYNODONTIDÆ
(Lizard Fishes)

Synodus Gronow, 1763, 20.

Saurus Catesby, 1771, 29.

Synodus (Gronow) Scopoli, 1777, 42.

Soarus[233]Linck, 1790, 49.

[233] A misprint for *Saurus,* the genus undescribed and unidentifiable.

Tirus Rafinesque, 1810, 80.
Saurus Cuvier, 1817, 98.
Alpismaris Risso, 1826, 120.
Laurida Swainson, 1839, 203.
Triurus Swainson, 1839, 203.
Saurida Cuvier & Valenciennes, 1849, 242.
Sauridichthys Bleeker, 1858, 278.

Trachinocephalus Gill, 1861, 303.
Bathysaurus Günther, 1878, 395.
Pelopsia Facciolà, 1883, 423.
Bathylaco Goode & Bean, 1895, 467.
Goodella Ogilby, 1897, 477.
Xystodus Ogilby, 1910, 536.

Family 272. HARPODONTIDÆ
(Bombay Ducks)

Harpodon Le Sueur, 1825, 172.

Family 273. CHLOROPHTHALMIDÆ

Chlorophthalmus Bonaparte, 1840, 142, 206.

Hyphalonedrus[234] Goode, 1880.
Bathysauropsis Regan, 1911, 540.

[234] *Proc. U. S. Nat. Mus.*, 111: 183; type *H. chalybeius* Goode. (Omitted in "Genera of Fishes.")

Family 274. PARALEPIDÆ (*Sudidæ*)

Sudis Rafinesque, 1810, 80.
Paralepis Cuvier, 1817, 104.
Prymnothonus Richardson, 1844, 223.
Arctozenus Gill, 1864, 330.
ANAPTERUS Sauvage, 1873, 371.
Symproptopterus Cocco, 1885, 430.
Lestidium Gilbert, 1905, 513.

Notolepis Dollo, 1908, 527.
Parasudis Regan, 1911, 540.
Lestidiops Hubbs, 1916, 559.
Notosudis Waite, 1916, 561.
TROSSULUS[235] Jordan, 1921.
LESTICHTHYS[236] Jordan, 1921.

[235] "Fish Fauna Cal. Tertiary," 250; orthotype *Trossulus exoletus* Jordan.
[236] *Loc. cit.*, 250; orthotype *Lestichthys porteousi* Jordan.

Family 275. EVERMANELLIDÆ (*Odontostomidæ; Scopelarchidæ*)
The members of the small group are not all well known, and may prove incongruous.

Odontostomus Cocco, 1838, 191.
Neosudis Castelnau, 1873, 368.
Scopelarchus Alcock, 1896, 470.

Evermanella Fowler, 1901, 493.
Dissomma Brauer, 1902, 497.
Promacheon Weber, 1913, 551.

Family 276. BATHYPTEROIDÆ

Bathypterois Günther, 1878, 395.
Synapteretmus[237] Goode & Bean, 1895.

Hemipterois Regan, 1911, 540.
Belonopterois Roule, 1916, 560.

[237] "Oceanic Ichthyology," 64; orthotype *Bathypterois quadrifilis* Günther. (Omitted in "Genera of Fishes.")

Family 277. BENTHOSAURIDÆ

Benthosaurus Goode & Bean, 1886, 435.

Family 278. **MYCTOPHIDÆ** (*Scopelidæ*)
(Lantern Fishes)

Of the numerous fossil genera mentioned below, several may belong to related families, rather than to *Myctophidæ*.

Myctophum Rafinesque, 1810, 82.
Scopelus Cuvier, 1817, 98.
Macrostoma Risso, 1826, 120.
Nyctophus Cocco, 1829, 133.
Alysia Lowe, 1839, 195.
Lampanyctus Bonaparte, 1840, 142, 206.
RHINELLUS Agassiz, 1844, 217.
OSMEROIDES Agassiz, 1844, 219.
HOLOSTEUS Agassiz, 1844, 219.
ACROGNATHUS Agassiz, 1844, 219.
Phanerobranchus Cocco, 1846, 574.
Cyrtorhynchus Costa, 1855, 574.
RHAMPHORNIMIA Costa, 1857, 275.
SARDINIUS von der Marck, 1858, 283.
PALIMPHEMUS Kner, 1862, 319.
Neoscopelus Johnson, 1863, 575.
LEPTOSOMUS von der Marck, 1863, 328.
MICROCŒLIA von der Marck, 1863, 328.
TACHYNECTES von der Marck, 1863, 328.
Dasyscopelus Günther, 1864, 333.
Notoscopelus Günther, 1864, 333.
BRACHYSPONDYLUS Marck & Schlüter, 1865, 339.
DERMATOPTYCHUS M. & S., 1865, 339.
DACTYLOPOGON M. & S., 1865, 339.
OPISTHOPTERYX[238] Pictet & Humbert, 1866, 342.

HEMISAURIDA Kner, 1867, 346.
TYDEUS Sauvage, 1870, 359.
PARASCOPELUS Sauvage, 1873, 371.
Gymnoscopelus Günther, 1873, 369.
Neoscopelus Castelnau, 1875, 378.
ICHTHYOTRINGA Cope, 1878, 394.
SCOPELOIDES Wettstein, 1886, 436.
Nannobrachium Günther, 1887, 438.
OMIODON Bassani, 1888, 440.
Scopelengys[239] Alcock, 1890, 449.
Stenobrachius Eigenmann & Eigenmann, 1890, 450.
Catablemella E. & E., 1890, 450.
Diaphus E. & E., 1891, 454.
Tarletonbeania E. & E., 1891, 454.
Rhinoscopelus Lütken, 1892, 458.
Benthosema Goode & Bean, 1895, 467.
Collettia G. & B., 1895, 467.
Lampadena G. & B., 1895, 467.
Aëthoprora G. & B., 1895, 467.
Electrona[240] G. & B., 1895, 467.
NEMATONOTUS Woodward, 1899, 488.
Centrobranchus Fowler, 1904, 507.
Scopelopsis[241] Brauer, 1906.
Protoblepharon Steche, 1909, 534.
Nyctimaster[242] Jordan, 1922.

[238] Misprinted *Opistopteryx*.

[239] Misprinted *Scopelogenys* in *Zoological Record* and in "Genera of Fishes."

[240] "Oceanic Ichthyology," 91; orthotype *Scopelus rissoi* Cocco. (Omitted in "Genera of Fishes.")

[241] "Die Tiefseéfische," 146; type *S. multipunctatus* Brauer. (Omitted in "Genera of Fishes.")

[242] *Proc. U. S. Nat. Mus.*, 645, 1921; orthotype *Lampanyctus jordani* Gilbert.

Family 279. **IPNOPIDÆ**

Ipnops Günther, 1878, 395.

Family 280. **OMOSUDIDÆ**

Omosudis Günther, 1887, 438.

Family 281. **MALACOSTEIDÆ**

Malacosteus Ayres, 1849, 240.
Bathylychnus Brauer, 1902, 497.

Family 282. **PLAGYODONTIDÆ** (*Alepisauridæ*)
(Hand-saw Fishes)

Plagyodus[243] (Steller) Pallas, 1811, 84.   **Plagyodus** (Steller) Günther, 1867, 345.
Alepisaurus Lowe, 1833, 179.   **Apateodus** Woodward, 1901, 496.
Istieus Agassiz, 1844, 219.   **Eugnathosaurus** Regan, 1913, 550.
Caulopus Gill, 1862, 314.

---

[243] This name has been questioned as appearing in an oblique case only, as "*Plagyodontem.*"

Suborder **CETUNCULI**
(New suborder)
The two families which follow are quite distinct from the other *Iniomi*, doubtless forming a separate suborder among the deep-sea fishes.

Family 283. **RONDELETIIDÆ**
Rondeletia Goode & Bean, 1894, 465.

Family 284. **CETOMIMIDÆ**

Cetomimus Goode & Bean, 1894, 465.   **Pelecinomimus** Gilchrist, 1922. (See
Cetostoma Zugmayer, 1914, 555.   Index.)

Suborder **CHONDRICHTHYES**
A curious group of very uncertain relationships.

Family 285. **ATELEOPODIDÆ**

Ateleopus Temminck & Schlegel, 1846,   **Ijimaia** Sauter, 1905, 516.
230.   **Parateleopus** Smith & Radcliffe, 1912,
Podateles Boulenger, 1902, 497.   544.

Order **XENOMI**

Family 286. **DALLIIDÆ**
(Alaska Black-fish)

Dallia Bean, 1879, 398.

Order **HAPLOMI**

The *Haplomi* are nearly related to the *Isospondyli*, although lacking the mesocoracoids characteristic of the latter order and of the *Ostariophysi*. Externally they have much in common with the *Microcyprini* or *Cyprinodontes*, with which most writers have united them. The maxillary, however, continues, as in the *Isospondyli*, to form the side of the upper jaw, but in the *Haplomi*, as in all the families which follow in this linear series, it ceases to bear teeth, and is gradually relegated to a subordinate position. It would appear therefore that the *Haplomi* and *Cyprinodontes* are actually rather closely related, though definable as separate orders or suborders. The *Xenomi* are doubtless near the *Haplomi*, but are quite as distinct from the latter as the *Cyprinodontes*.

The following remarks of Jordan & Evermann ("Fish. N. M. Am.," I, 623) are still pertinent: "While our knowledge of the osteology and embryology of most of the families of fishes is very incomplete, it is evident that the relationships of the groups can not be shown in any linear series, or by any conceivable arrangement of orders and suborders. The living teleost fishes have sprung from many lines of descent, their relationships are extremely diverse, and their differences are of every possible degree of value. The ordinary schemes have magnified the importance of a few common characters, at the same time neglecting other differences of equal importance. No system of arrangement which throws these fishes into large groups can ever be definite or permanent."

Family 287. CROSSOGNATHIDÆ

CROSSOGNATHUS Pictet & Loriol, 1859, 292.

APSOPELIX Cope, 1871, 360.
PELYCORAPIS[244] Cope, 1874, 375.

[244] Misprinted *Pelecorapis.*

Family 288. ESOCIDÆ (*Luciidæ*)

(Pikes)

Esox[245] Linnæus, 1758, 14.
Lucius Geoffroy, 1767, 24.
Lucius Klein, 1775, 40.
Lucius Rafinesque, 1810, 80.

Picorellus[246] Rafinesque, 1820, 112.
Mascalongus Jordan, 1878, 396.
Kenoza Jordan & Evermann, 1896, 473.

[245] This use of *Esox* follows the decision of the International Commission of Zoological Nomenclature. The name has been also employed in place of *Belone.*
[246] A mythical pickerel.

Family 289. UMBRIDÆ
(Mud Minnows)

Umbra (Kramer) Gronow, 1763, 20.
Umbra (Gronow) Scopoli, 1777, 42.

Melanura Agassiz, 1853, 252.

Order CYPRINODONTES
(*Microcyprini*)

The *Microcyprini* or *Cyprinodontes* have been usually placed in one or two families. It seems to us that more should be recognized. These fishes bear considerable resemblance to the *Haplomi*, with which group they have usually been united. But the structure of the mouth is quite different, the premaxillary forming the whole side of the upper jaw, the toothless maxillary lying above it as a sort of secondary bone, as in all the percoid fishes. The posterior insertion of the dorsal fin is characteristic of both groups as well as of most of the *Synentognathi*, far from which they should not be placed in a natural system.

Hitherto the viviparous and ovipafous Cyprinodonts have been placed in a single family, but the differences within the group are extensive and if the species were not all of small size they would certainly be assigned to two or more family groups. As here understood, the oviparous species, with the anal fin not modified in the male, constitute the family of *Cyprinodontidæ.*

Family 290. **CYPRINODONTIDÆ**
(Killifishes)

The *Fundulinæ,* with pointed teeth, and the *Cyprinodontinæ,* with the teeth incisor-like, form a natural family widely diffused in the North Temperate Zone.

Fundulus Lacépède, 1803, 65.
Hydrargira[247] Lacépède, 1803, 67.
Cyprinodon Lacépède, 1803, 68.
Prinodon Rafinesque, 1815, 90.
Lebia[248] (Cuvier) Oken, 1817, 99.
Aphanius Nardo, 1827, 121.
Aplocheilus[249] McClelland, 1839, 195.
Panchax Cuvier & Valenciennes, 1846, 407.
Encrotes Gistel, 1848, 236.
Zygonectes Agassiz, 1853, 252.
Tellia Gervais, 1853, 253.
Fundulichthys Bleeker, 1859, 288.
Adinia Girard, 1859, 290.
Lucania Girard, 1859, 290.
Micromugil Gulia, 1861, 307.
Rivulus Poey, 1860, 299.
Aplocheilichthys Bleeker, 1861, 321.
Lebistes[250] de Filippi, 1862.
Micristius Gill, 1865, 337.
Lycocyprinus Peters, 1868, 353.
Nothobranchius Peters, 1868, 353.

Problebias Sauvage, 1874, 376.
Pachystetus (Aymard) Sauvage, 1874, 376.
Cynolebias Steindachner, 1876, 387.
Xenisma Jordan, 1877, 390.
Jordanella Goode & Bean, 1879, 399.
Borborys (Broussonet) G. & B., 1885, 432.
Gephyrura Cope, 1891, 454.
Proballostomus Cope, 1891, 454.
Neolebias Steindachner, 1894, 465.
Pterolebias Garman, 1895, 467.
Fontinus Jordan & Evermann, 1896, 473.
Pachylebias Woodward, 1901, 496.
Procatopus Boulenger, 1904, 506.
Oryzias Jordan & Snyder, 1906, 519.
Mohanga Boulenger, 1911, 538.
Lamprichthys Regan, 1911, 540.
Galasaccus Fowler, 1916, 559.
Chriopeops Fowler, 1916, 559.
Oxyzygonectes Fowler, 1916, 559.
Parafundulus Eastman, 1917, 561.

---

[247] Usually corrected to *Hydrargyra.*
[248] Later spelled *Lebias.*
[249] Later corrected to *Haplochilus.*
[250] *Archiv. Zool. Anat.,* etc., 1: 69; orthotype *L. pœciloides* de Filippi. (Omitted in "Genera of Fishes.")

Family 291. **ORESTIIDÆ**

This little group comprises allies of *Fundulus,* which have the lower pharyngeals enlarged and provided with molar teeth. The ventral fins are wanting. The family, probably once more widely diffused, is now confined to Lake Titicaca and to the desert streams about Death Valley, Nevada.

Orestias Cuvier & Valenciennes, 1846, 228.   Empetrichthys Gilbert, 1893, 461.

Family 292. CHARACODONTIDÆ

This family and the next include those viviparous Cyprinodonts in which the first rays of the male are somewhat modified but not transformed into an intromittent organ.

Characodon Günther, 1866, 340.   **Zoogoneticus** Meek, 1902, 500.

Family 293. GOODEIDÆ

Girardinichthys Bleeker, 1860, 294.   **Xenendum** Jordan & Snyder, 1899, 487.
Limnurgus Günther, 1866, 340.   **Skiffia** Meek, 1902, 500.
Goodea Jordan, 1879. 400.   **Chapalichthys** Meek, 1902, 500.

Family 294. PŒCILIDÆ
(Top Minnows)

This family includes provisionally those viviparous Cyprinodonts having the anal fin modified into an intromittent organ. All the *Pœciliidæ* are confined to warmer parts of America. It is possible that the family should be further subdivided. The *Anablepinæ* with "four eyes" (the eye crossed by a partition) and the *Gambusiinæ* with teeth as in the *Fundulinæ,* may well be detached on account of differences in dentition. In the *Pœciliinæ* the teeth are slender and movable, while in the *Gambusiinæ* they are conic and fixed as in the *Fundulinæ.* The name *Pœciliidæ* is probably sufficiently distinct from *Pœcilidæ,* a family of beetles.

Pœcilia Bloch & Schneider, 1801, 59.
Mollienesia Le Sueur, 1821, 114.
Alazon Gistel, 1848, 237.
Xiphophorus Heckel, 1848, 238.
Heterandria Agassiz, 1853, 251.
Pœcilops Pomel, 1853, 254.
Limia Poey, 1854, 261.
Gambusia Poey, 1854, 260.
Girardinus[251] Poey, 1854, 261.
Belonesox Kner, 1859, 292, 298.
Pseudoxiphophorus Bleeker, 1859, 287.
Hemixiphophorus Bleeker, 1859, 288.
Trifarcius Poey, 1860, 299.
Limia Poey, 1861, 308.
Pœcilodes Steindachner, 1863, 328.
Platypœcilus Günther, 1866, 341.
Acropœcilia Hilgendorf, 1889, 447.
Glaridodon Garman, 1895, 467.
Cnestrodon Garman, 1895, 467.
Gambusinus Jordan & Evermann, 1896, 473.
Glaridichthys Garman, 1896, 471.
Toxus Eigenmann, 1903, 501.
Cynodonichthys Meek, 1904, 510.

Paragambusia Meek, 1904, 510.
Acanthophacelus Eigenmann, 1907, 522, 533.
Ilyodon Eigenmann, 1907, 522.
Phalloptychus Eigenmann, 1907, 522, 559.
Phalloceros Eigenmann, 1907, 522.
Petalosoma[252] Regan, 1908, 530.
Tomeurus Eigenmann, 1909, 533.
Alfaro Meek, 1912, 544.
Cynopœcilus Regan, 1912, 545.
Petalurichthys Regan, 1912, 545.
Gulapinnus Langer, 1913, 549.
Priapichthys Regan, 1913, 550.
Pseudopœcilia Regan, 1913, 550.
Priapella Regan, 1913, 550.
Pœcilopsis Regan, 1913, 550.
Brachyrhaphis Regan, 1913, 550.
Leptorhaphis Regan, 1913, 550.
Pamphoria Regan, 1913, 550.
Pamphorichthys Regan, 1913, 550.
Heterophallus Regan, 1914, 555.
Diphyacantha Henn, 1916, 559.
Phallatorhynus Henn, 1916, 559.
Neoheterandria Henn, 1916, 559.

---

[251] According to Garman this genus is distinct from *Heterandria*.
[252] Misprinted *Palosoma* in Dean's "Bibliography of Fishes."

Family 295. FITZROYIIDÆ

Fitzroyia Günther, 1866, 340.   Jenynsia Günther, 1866, 341.

Family 296. ANABLEPIDÆ
(Quatro Ojos)

Anableps Gronow, 1763, 20.   Anableps (Gronow) Bloch & Schneider,
Anableps (Gronow) Scopoli, 1777, 42.   1801, 59.

Family 297. PHALLOSTETHIDÆ

Phallostethus Regan, 1913, 550.   Neostethus Regan, 1916, 560.

Family 298. AMBLYOPSIDÆ
(Cave Blind Fishes)

Amblyopsis Dekay, 1842, 210.   Typhlichthys Girard, 1859, 290.
Chologaster Agassiz, 1853, 251.   Troglichthys Eigenmann, 1899, 485.

Family 299. ADRIANICHTHYIDÆ

Adrianichthys Weber, 1913, 552.   Xenopœcilus Regan, 1911, 540.

Order SYNENTOGNATHI
Family 300. FORFICIDÆ

FORFEX Jordan, 1919, 568.   ZELOTES[253] Jordan, 1921.

----

[253] "Fossil Fish. S. Cal.," 36; orthotype *Z. alhambræ* Jordan.

Family 300A. ROGENIIDÆ

ROGENIO Jordan, 1907, 525.

Family 301. XENESTHIDÆ

The genus *Xenesthes* is based on a long and strong premaxillary bone, seven inches in length, armed anteriorly with a row of large teeth. The toothless maxillary is apparently united with this bone. As the *Ichthyodectidæ* have the premaxillary short, forming the front only of the edge of the upper jaw, *Xenesthes* can not be allied to them. It is perhaps near the *Belonidæ,* although the type specimen shows none of the small teeth characteristic of the latter, and the snout could not have been attenuate. It bears also a certain resemblance to *Plagyodus,* but in the latter the bones are soft and smooth, while in *Xenesthes* they are rigid and rough.
XENESTHES Jordan, 1907, 524.

Family 302. BELONIDÆ
(Hound Fishes; Needle Fishes)

Mastacembelus Klein, 1775, 40.   Tylosurus Cocco, 1833, 178.
Raphistoma[254] Rafinesque, 1815, 91.   Ramphistoma Swainson, 1839, 203.
Belone Cuvier, 1817, 99.   Potamorrhaphis Günther, 1866, 340.
Strongylura[254a] Van Hasselt, 1824.   *Athlennes* Jordan & Fordice, 1886, 435.

----

[254] Written *Ramphistoma* by authors.
[254a] *Bull. Sci. Nat. Ferussac,* **2**: 374; orthotype *Strongylura caudimaculata* Van Hasselt= *Belone strongylura* Van Hasselt, 1823. (Omitted in "Genera of Fishes.")

Ablennes J. & F., 1886, 435.
Eurycaulus Ogilby, 1908, 529.
Stenocaulus Ogilby, 1908, 529.
Xenentodon Regan, 1911, 540.

Petalichthys[255] Regan, 1911.
Platybelone[255a] Fowler, 1919.
Tropidocaulus Ogilby, 1919, 569.
BELTION Jordan, 1920, 571.

[255] *Ann. Mag. Nat. Hist.*, 332; orthotype *P. capensis* Regan. (Omitted in "Genera of Fishes.")

[255a] *Proc. Acad. Nat. Sci. Phila.*, **71**: 2; orthotype *Belone platyura* Bennett (same as *Tropidocaulus* of slightly later date).

Family 303.   SCOMBERESOCIDÆ
(Sauries)

Scomberesox[256] Lacépède, 1803, 67.
Sayris Rafinesque, 1810, 81.
Grammiconotus Achille Costa, 1862, 312.

Cololabis Gill, 1896, 473.
SCOMBERESSUS Jordan, 1920, 571.

[256] Often written *Scombresox*.

Family 304.   HEMIRAMPHIDÆ
(Half-beaks)

Hemi-Ramphus[257] Cuvier, 1817, 99.
Dermogenys[258] Van Hasselt, 1823, 172.
COBITOPSIS Pomel, 1853, 254.
Hyporhamphus Gill, 1859, 289.
Euleptorhamphus Gill, 1859, 289, 326.
Zenarchopterus Gill, 1863, 326.
Oxyporhamphus Gill, 1863, 326.
Hemirhamphodon Bleeker, 1866, 340.
Arrhamphus Günther, 1866, 340.
HEMILAMPRONITES Geinitz, 1868, 350.

Chriodorus Goode & Bean, 1882, 419.
Hemiexocœtus Fowler, 1901, 493.
Labidorhamphus Fowler, 1905, 513.
Eulepidorhamphus Fowler, 1919.
ROGENITES Jordan, 1919, 568.
ZELOSIS Jordan & Gilbert, 1920, 571.
Rhamphodermogenys[259] Fowler & Bean, 1922.
Nomarhamphus Weber, 1922 (See Index A).

[257] Usually spelled *Hemirhamphus*.
[258] Spelled *Dermatogenys* by Bleeker.
[258a] *Loc. cit.*, 7; orthotype *Hemirhamphus sajori* Schlegel.
[259] Fowler & Bean, "Fishes of Formosa and the Philippine Islands," *Proc. U. S. Nat. Mus.*, **152**: 17; orthotype *Dermogenys bakeri* Fowler & Bean.

Family 305.   EXOCŒTIDÆ
(Flying Fishes)

Exocœtus Linnæus, 1758, 15.
Hirundo Catesby, 1771, 30.
Pterichthus (Commerson) Lacépède, 1803, 71.
Cypselurus[260] Swainson, 1839, 203.
Cheilopogon Lowe, 1840, 207.
Ptenichthys Müller, 1843, 216.

Halocypselus Weinland, 1858, 284.
Parexocœtus Bleeker, 1866, 340.
Fodiator Jordan & Meek, 1885, 433.
Exonautes Jordan & Evermann, 1896, 472.
Evolantia Heller & Snodgrass, 1903, 503.
Ptenonotus Ogilby, 1908, 529, 554.

[260] Spelled *Cypsilurus* by misprint, as decided by the International Commission of Nomenclature.

Order ANACANTHINI

This large group, usually associated with the blennies and brotulids on the one hand and the flounders on the other, is quite distinct from both. Its origin and relationship are obscure. The absence of fin spines and the frequent presence of more than five ventral rays indicate its descent from some early type in which the ventral rays have not been reduced to I, 5, the normal number in percomorphous fishes, and no spines have developed in the fins. The anterior position of the ventrals separate these fishes from the forms with a duct to the air bladder, *Physostomi, Malacopteri, Ostariophysi, Isospondyli,* etc. Regan suggests their possible derivation from allies of *Aulopidæ* or other primitive *Iniomi.*

Family 306. MACROUROIDIDÆ

Macrouroides Smith & Radcliffe, 1912, 544.

Squalogadus Gilbert & Hubbs, 1916, 559.

Family 307. MACROURIDÆ (*Coryphænoididæ; Macruridæ*)
(Grenadiers; Rat-tails)

Coryphænoides Gunner, 1761, 16.
Macrourus[261] Bloch, 1786, 44.
Trachyrincus[262] Giorna, 1805, 74, 407.
Cœlorhynchus Giorna, 1805, 74.
Lepidoleprus Risso, 1810, 77.
Oxycephas Rafinesque, 1810, 79.
Krohnius Cocco, 1844, 219.
Coccolus (Kaup) Bonaparte, 1844, 219, 226, 284.
Lepturus Gronow, 1854, 259.
Malacocephalus Günther, 1862, 318.
Macruronus Günther, 1872, 365.
Macruroplus Bleeker, 1874, 372.
Paramacrurus Bleeker, 1874, 372.
Oxymacrurus Bleeker, 1874, 372.
Bathygadus[263] Günther, 1878.
Lepidorhynchus Bleeker, 1879, 398.
Chalinura[264] Goode & Bean, 1882, 419.
Hymenocephalus Giglioli & Issel, 1884, 575.
Trachonurus Günther, 1887, 437.
Lionurus Günther, 1887, 437.
Mystaconurus Günther, 1887, 437.

Optonurus Günther, 1887, 437.
Nematonurus Günther, 1887, 437.
Cetonurus Günther, 1887, 437.
Lyconus Günther, 1887, 437.
Steindachneria Goode & Bean, 1888, 441.
Abyssicola Goode & Bean, 1895, 468.
Moseleya Goode and Bean, 1895, 468.
Steindachnerella Eigenmann, 1897, 476.
Cœlocephalus Gilbert & Cramer, 1897, 477.
Albatrossia Jordan & Gilbert, 1898, 482.
Bogoslovius J. & E., 1898, 482.
Matæocephalus Berg, 1899, 484.
Dolloa Jordan, 1900, 489.
Gadomus Regan, 1903, 505.
Melanobranchus Regan, 1903, 505.
Nezumia Jordan & Gilbert, 1904, 509.
Regania Jordan & Gilbert, 1904, 509.
Cynomacrurus Dollo, 1909, 532.
Ateleobrachium[265] Gilbert & Burke, 1912, 543.
Echinomacrurus Roule, 1916, 560.
Lyconodes Gilchrist, 1922. (See Index.)

---

[261] Usually corrected as *Macrurus.*
[262] Usually corrected to *Trachyrhynchus.*
[263] *Ann. Mag. Nat. Hist.,* 23; orthotype *Bathygadus cottoides* Günther. (Omitted in "Genera of Fishes.")
[264] Written *Chalinurus* by Günther.
[265] *Bull. U. S. Bur. Fish.,* **30** : 94, 1912. (Erroneously given in "Genera of Fishes.")

Hyomacrurus Gilbert & Hubbs, 1920.
(See Index.)
Hyostomus Gilbert & Hubbs, 1920, 570.
Atherodon G. & H., 1920, 570.
Ventrifossa G. & H., 1920, 570.
Papyrocephalus G. & H., 1920, 570.

Quincuncia G. & H., 1920, 570.
Hymenogadus G. & H., 1920, 570.
Oxygadus G. & H., 1920, 570.
Lucigadus G. & H., 1920, 570.
Lucigadella G. & H., 1920, 570.

Family 308. BREGMACEROTIDÆ

Bregmaceros (Cantor) Thompson, 1840, 208.
Calloptilum Richardson, 1844, 222.

Asthenurus Tickell, 1865, 339.
Auchenoceros Günther, 1889, 447.

Family 309. GAIDROPSARIDÆ
(Rocklings)

Enchelyopus Bloch & Schneider, 1801, 57.
Gaidropsarus Rafinesque, 1810, 81.
Dropsarus Rafinesque, 1815, 88.
Onos Risso, 1826, 119.
Motella Cuvier, 1829, 131.

Ciliata Couch, 1832, 176.
Couchia Thompson, 1856, 274.
Molvella Kaup, 1858, 281.
Rhinonemus Gill, 1863, 325.

Family 310. GADIDÆ
(Cod-fishes)

Gadus Linnæus, 1758, 12.
Callarias Klein, 1777, 42.
Phycis Röse, 1793, 51.
Phycis Bloch & Schneider, 1801, 58.
Phycis Rafinesque, 1810, 79.
Strinsia Rafinesque, 1810, 81.
Eleginus Fischer, 1813, 85.
Trisopterus Rafinesque, 1814, 87.
Morrhua (Cuvier) Oken, 1817, 100.
Merlangus (Cuvier) Oken, 1817, 100.
Raniceps (Cuvier) Oken, 1817, 100.
Lota (Cuvier) Oken, 1817, 100.
Mustela (Cuvier) Oken, 1817, 100.
Brosme[266] (Cuvier) Oken, 1817, 100.
Lotta Risso, 1826, 119.
Morua Risso, 1826, 119.
Mora Risso, 1826, 119.
Molva Fleming, 1828, 122.
Brosmus Stark, 1828, 123.
Tilesia Swainson, 1839, 203.
Cephus Swainson, 1839, 203.
Lepidion Swainson, 1839, 203.
Gadella Lowe, 1843, 215.
Nemopteryx Agassiz, 1844, 219.
Asellus Valenciennes, 1844, 223.
Merlinus Agassiz, 1845, 224.

Pollachius (Nilsson) Bonaparte, 1845, 226.
Gastronemus Bonaparte, 1845, 226.
Tinimogaster Bonaparte, 1845, 226.
Uraleptus Costa, 1846, 574.
Cerdo Gistel, 1848, 235.
Gadiculus Guichenot, 1850, 246.
Lotella Kaup, 1858, 281.
Physiculus Kaup, 1858, 281.
Uraleptus Costa, 1858, 279.
Pteridium Filippi & Verany, 1859, 289.
Palæobrosmius Rath, 1859, 292.
Palæogadus Rath, 1859, 292.
Hypsiptera Günther, 1860, 297.
Algoa Castelnau, 1861, 301.
Melanogrammus Gill, 1862, 316.
Brachygadus Gill, 1862, 316.
Læmonema Günther, 1862, 318.
Pseudophycis Günther, 1862, 318.
Halargyreus Günther, 1862, 318.
Boreogadus Günther, 1862, 318.
Haloporphyrus Günther, 1862, 318.
Urophycis Gill, 1863, 325.
Micromesistius Gill, 1863, 325.
Brachygadus Gill, 1863, 325.
Leptogadus Gill, 1863, 325.

---

[266] Later written *Brosmius* by Cuvier.

Odontogadus Gill, 1863, 325.
Microgadus Gill, 1863, 325.
Æglefinus Malm, 1877, 391.
Antimora[267] Günther, 1878.
Melanonus Günther, 1878, 394.
Pleurogadus Bean, 1885, 433.
Salilota Günther, 1887, 437.
Brosmiculus Vaillant, 1888, 444.
Sympodoichthys Facciolà, 1888, 441.
Austrophycis Ogilby, 1897, 477.
Emphycus Jordan & Evermann, 1898, 481.
Theragra Lucas, 1898, 481.
Microlepidium Garman, 1899, 486.

Leptophycis Garman, 1899, 486.
Rhinocephalus (Agassiz) Woodward, 1901, 223.
Tripterophycis Boulenger, 1902, 497.
Melanonosoma Gilchrist, 1902, 498, 501.
Læmonemodus Gilchrist, 1903, 503.
Gargilius[268] (Jensen) Schmidt, 1906.
Arnoldina Jordan & Gilbert, 1919, 569.
Eclipes J. &. G., 1919, 569.
Merriamina J. & G., 1919, 569.
Verater Jordan, 1919, 567.
Arnoldites[269] Jordan, 1923.

[267] Ann. Mag. Nat. Hist., 2: 18; orthotype A. rostrata Günther. (Omitted in "Genera of Fishes.")

[268] Gargilius (Jensen, Adolf Severien) Schmidt (Ernest Johannes), "Contrib. Life Hist. Eel," 177, 1906; no description and no specific name given; a small deep-water gadoid. (See Holt & Byrne, Fish. Ireland Invest., 5: 58.)

[269] A new generic name to replace Arnoldina, preoccupied in Unionidæ (Hannibal, 1917).

Family 311. MERLUCCIIDÆ
(Hakes)

Merluccius[270] Rafinesque, 1810, 79.
Onus Rafinesque, 1810, 81.
Merlangus Rafinesque, 1810, 83.
Stomodon Mitchill, 1814, 86.
Hydronus Minding, 1832, 176.

Merlus Guichenot, 1848, 238.
Merlucius Gronow, 1854, 259.
Homalopomus Girard, 1856, 269.
Epicopus Günther, 1860, 296.

[270] Often spelled Merlucius.

Family 312. MURÆNOLEPIDÆ
Murænolepis Günther, 1880, 403.

Family 313. ERETMOPHORIDÆ
Hypsirhynchus Facciolà, 1884, 427, 468.   Eretmophorus Giglioli, 1889, 446.

Order SALMOPERCÆ

Adipose fin present; dorsal fin with spines.

The Salmopercæ and Xenarchi are detached fragments of a lost fresh water fauna of the Cretaceous or Eocene. They are closely related to each other, but neither has much affinity with other known fishes. The several orders which follow are often spoken of as *transitional* between soft-rayed and spiny-rayed fishes. The word is misleading. More accurately they are diverging offshoots from transitional forms now extinct; no linear series can express their varied relations, their line of evolution in some cases leading distinctly downward, that is, toward less than ancestral specialization. Some groups as the *Percesoces* that seem to lead upward toward the *Percoidei* may really be declining downward from perch-like ancestry.

In general the forward movement is marked by the development of spines in the fins, the thoracic attachment of the ventral fins, the reduction in the number of ventral rays to six, the first being normally spinous, the loss of the mesocoracoid, the relegation of the maxillary to a subordinate position in the upper jaw, the roughening of the scales, the obliteration of the duct to the air bladder, with a number of minor characters. These transitions do not appear uniformly, and they are sometimes disguised by special traits of degeneration among the varied groups naturally regarded as Acanthopterygians or spiny-rayed. But nowhere in the vast range of variety among fishes is it possible to arrange a linear series which shall express the actual fact. The most one can do is keep up a sort of marching abreast from the *Ganoids,* through the *Isospondyli* and transitional types to the specialized *Acanthopteri* and their peculiar offshoots. In this group and the next, which is closely related, the pneumatic duct is obsolete; dorsal and anal each with a few spines; ventral rays more than five.

Family 314. PERCOPSIDÆ
(Sand Rollers; Trout Perch)

Adipose fin present; ventrals subabdominal, without spine.

Percopsis Agassiz, 1848, 232.   **Columbia** Eigenmann, 1892, 457.
Salmoperca Thompson, 1851, 249.

Order XENARCHI

Adipose fin wanting; ventrals subthoracic, with a spine.

Family 315. APHREDODERIDÆ
(Pirate Perch)

Aphredoderus[271] Cuvier & Valenciennes,   **Sternotremia** Nelson, 1876, 386.
1833, 178.   **Asternotremia** Nelson, 1877, 390.
Erismatopterus Cope, 1870, 356.   Amphiplaga Cope, 1877, 388.
Trichophanes Cope, 1872, 364.

---

[271] Corrected to *Aphododerus* and *Aphrodedirus* by authors.

Family 316. ASINEOPIDÆ

Asineops Cope, 1872, 356.

Order ALLOTRIOGNATHI
Suborder *HISTICHTHYES*
Family 317. VELIFERIDÆ

Velifer Temminck & Schlegel, 1850, 247.

Family 318. LOPHOTIDÆ

Lophotes Giorna, 1805, 74.
Leptopus Rafinesque, 1814, 86.

Podoleptus Rafinesque, 1815, 92.
Eumecichthys[272] Regan, 1907.

[272] *Proc. Zool. Soc. Lond.*, 638, 1907; orthotype *Lophotes fiskii* Günther. (Omitted in "Genera of Fishes.")

Suborder *TAENIOSOMI*
Family 319. REGALECIDÆ
(Oar Fishes; Sea Serpents)

Regalecus Brünnich, 1771, 28.
Regalecus Ascanius, 1788, 45.
Gymnetrus Bloch, 1788, 45.
Cephalepis Rafinesque, 1810, 82.

Xypterus Rafinesque, 1810, 82.
Epidesmus Ranzani, 1818, 107.
Xiphichthys Swainson, 1839, 201.

Family 320. TRACHIPTERIDÆ
(King-of-the-Herring)

Trachipterus[273] Gouan, 1770, 28.
Gymnogaster Brünnich, 1788, 168.
Bogmarus Bloch & Schneider, 1801, 60.
Argyctius Rafinesque, 1810, 80.

Nemochirus Rafinesque, 1815, 86, 407.
Nemotherus Costa, 1834, 180.
Vogmarus Reid, 1849, 244.

[273] Usually corrected to *Trachypterus.*

Suborder *ATELAXIA*
Family 321. STYLEPHORIDÆ

Stylephorus Shaw, 1791, 50.

Family 322. GIGANTURIDÆ

Gigantura Brauer, 1901, 492.

Order SELENICHTHYES

Ventrals many-rayed; basal bones of pectoral set horizontally; humeral arch and pelvic bones enormously enlarged; body very deep, compressed.

Family 323. LAMPRIDÆ
(Opah; Moon-fish)

Lampris Retzius, 1799, 54.
Chrysotosus Lacépède, 1803, 65.
Echemythes Gistel, 1848, 235.

DIATOMŒCA Jordan & J. Z. Gilbert, 1919, 569.

Family 324. SEMIOPHORIDÆ

SEMIOPHORUS Agassiz, 1838, 191.

Order HETEROSOMATA
(Flounders; Flat Fishes)

Both eyes on the same side of the body.

The flounders and soles, having no spines and the ventral fins thoracic with an increased number of rays, should not be placed far from the percomorphous series. Boulenger suggests that the group may have close affinities with the *Amphistiidæ* and *Zeidæ,* constituting with these the order *Zeorhombi.* This view of the case is plausible, but it is far from established.

Family 325. PSETTODIDÆ

Psettodes Bennett, 1830, 174. Sphagomorus Cope, 1867, 344.

Family 326. BOTHIDE

Solea Catesby, 1771, 31.
Psetta Klein, 1775, 38.
Europus Klein, 1775, 38.
Rhombus Klein, 1775, 38.
Citharus Röse, 1793, 52.
Bothus Rafinesque, 1810, 79.
Scophthalmus Rafinesque, 1810, 82.
Rhombus Cuvier, 1817, 100.
Rhomboides Goldfuss, 1820, 171.
Zeugopterus Gottsche, 1835, 182.
Citharus Reinhardt, 1838, 193.
Psetta Swainson, 1839, 203.
Platophrys Swainson, 1839, 203.
Syacium Ranzani, 1840, 207.
Peloria Cocco, 1844, 219.
Passer Valenciennes, 1855, 266.
Rhomboidichthys Bleeker, 1856, 267.
Lophopsetta Gill, 1861, 303, 331.
Citharus Bleeker, 1862, 310.
Hemirhombus Bleeker, 1862, 310.
Citharichthys Bleeker, 1862, 310.
Arnoglossus Bleeker, 1862, 310.
Orthopsetta Gill, 1862, 317.
Engyprosopon Günther, 1862, 319.
Phrynorhombus Günther, 1862, 319.
Lepidorhombus Günther, 1862, 319.
Metoponops Gill, 1864, 331.
Bascanius[274] Schiödte, 1868, 353.
Nematops Günther, 1880, 403.
Lophonectes Günther, 1880, 403.
Monolene Goode, 1880, 403.
Thyris Goode, 1880, 403.

Anticitharus Günther, 1880, 403.
Læops Günther, 1880, 403.
Thysanopsetta Günther, 1880, 403.
Lepidopsetta Günther, 1880, 403.
Etropus Jordan & Gilbert, 1881, 416.
Lophorhombus Macleay, 1882, 421.
Delothyris Goode, 1883, 424.
Charybdia Facciolà, 1885, 431.
Aramaca Jordan & Goss, 1885, 433.
Cyclopsetta Gill, 1888, 441.
Azevia Jordan & Goss, 1888, 442.
Eucitharus Gill, 1888, 441.
Trichopsetta Gill, 1888, 441.
Engyophrys Jordan & Bollman, 1889, 447.
Psettylis Alcock, 1890, 450.
Scianectes Alcock, 1890, 450.
Caulopsetta Gill, 1893, 461.
Chascanopsetta Alcock, 1894, 463.
Embassichthys Jordan & Evermann, 1896, 475.
Boopsetta Alcock, 1896, 470.
Pelecanichthys Gilbert & Cramer, 1897, 477.
Perissias Jordan & Evermann, 1898, 481.
Apsetta Kyle, 1900, 490.
Scæops[275] Jordan and Starks, 1904.
Tæniopsetta Gilbert, 1905, 513.
Mancopsetta Gill, 1905, 514.
Lambdopsetta Smith & Pope, 1906, 520.
Trachypterophrys[276] Franz, 1910.
Plagiopsetta Franz, 1910, 535.
Laiopteryx[277] Weber, 1913, 551.

---

[274] An undetermined larva.

[275] *Scæops* Jordan & Starks, *loc. cit.,* 623; orthotype *Rhombus grandisquama* Temminck & Schlegel.

[276] *Op. cit.,* 60; orthotype *P. raptator* Franz; a synonym of *Chascanopsetta* Alcock, according to McCulloch. (Omitted in "Genera of Fishes.")

[277] Misprinted *Liopteryx* in "Genera of Fishes."

Pseudocitharichthys[278] Weber, 1913. Asterorhombus Tanaka, 1915, 558.
Lepidoblepharon Weber, 1913, 551. Scidorhombus Tanaka, 1915, 558.
Eobothus Eastman, 1914, 552. Zororhombus Jordan, 1920, 571.
Læoptichthys Hubbs, 1915, 556. Paracitharus[279] Regan, 1920.
Citharoides Hubbs, 1915, 556. Crossorhombus[280] Regan, 1920.
Psettina Hubbs, 1915, 556. Platotichthys[281] Nichols, 1921.

[278] *Op. cit.* 413; orthotype *P. aureus* Weber. (Omitted in "Genera of Fishes.")
[279] "A Review of the Flat Fishes (*Heterosomata*) of Natal," *Ann. Durban Museum,* **2**: 209, March 25, 1920; type *Arnoglossus macrolepis* Günther.
[280] Regan, *op. cit.,* 211; type *Platophrys dimorphus* Günther.
[281] "List of Turks Island Fishes," *Bull. Amer. Mus. Nat. Hist.,* **44**: 21; orthotype *P. chartes* Nichols.

Family 327. PARALICHTHYIDÆ
(Bastard Halibuts)

To the *Bothidæ* Regan refers all sinistral or "left-handed" flounders. The genera named below, however, seem to me quite as closely related to the *Hippoglossidæ* as to *Bothus* and *Platophrys.* They may be allowed to stand as a separate transitional subfamily or family pending further investigation.

Paralichthys[282] Girard, 1859. Notosema Goode & Bean, 1882, 419.
Chænopsetta Gill, 1861, 303. Xystreurys Jordan & Gilbert, 1890, 451.
Uropsetta Gill, 1862, 317. Lioglossina Gilbert, 1890, 451.
Pseudorhombus Bleeker, 1862, 310. Verecundum Jordan, 1890, 452.
Tephritis Günther, 1862, 318. Gastropsetta Goode & Bean, 1894, 463.
Tephrinectes Günther, 1862, 319. Ramularia Jordan & Evermann, 1898, 481.
Ancylopsetta Gill, 1864, 331. Rhombiscus Jordan & Snyder, 1900, 490.
Neorhombus Castelnau, 1875, 378. Evesthes J. Z. Gilbert, 1905, 536.
Hippoglossina Steindachner, 1876, 387. Velifracta Jordan, 1907, 524.
Teratorhombus Macleay, 1881, 416. Tarphops Jordan & Thompson, 1914, 553.

[282] Omitted from No. 657, "Genera of Fishes"; orthotype *Paralichthys maculosus* Girard.

Family 328. HIPPOGLOSSIDÆ
(Halibuts)

Hippoglossus Cuvier, 1817, 100. Verasper Jordan & Gilbert, 1898, 481.
Hippoglossoides Gottsche, 1835, 182. Cynopsetta Schmidt, 1903, 506, 519.
Reinhardtius Gill, 1861, 302. Xystrias[283] Jordan & Starks, 1904.
Drepanopsetta Gill, 1861, 303. Veræqua J. & S., 1904, 509.
Platysomatichthys Bleeker, 1862, 310. Dexistes J. & S., 1904, 509.
Pomatopsetta Gill, 1864, 331. Araias J. & S., 1904, 509.
Atheresthes Jordan & Gilbert, 1880, 404. Cleisthenes J. & S., 1904, 509.
Eopsetta Jordan & Goss, 1885, 433. Protopsetta Schmidt, 1904, 511.
Lyopsetta Jordan & Goss, 1885, 433. Acanthopsetta Schmidt, 1904, 511.

[283] *Loc. cit.,* 623; orthotype *Hippoglossus grigorjewi* Herzenstein. (Omitted in "Genera of Fishes.")

Family 329. PLEURONECTIDÆ

Pleuronectes Linnæus, 1758, 13.
Passer Klein, 1775, 38.
Platessa Cuvier, 1817, 100.
Microstomus Gottsche, 1835, 182.
Limanda Gottsche, 1835, 182.
Glyptocephalus Gottsche, 1835, 182.
Cynicoglossus Bonaparte, 1837, 187.
Cynoglossa Bonaparte, 1845, 226.
Platichthys Girard, 1854, 257.
Psettichthys Girard, 1854, 257.
Parophrys Girard, 1854, 257.
Pleuronichthys Girard, 1854, 257.
Hypsopsetta Gill, 1862, 575, 317.
Lepidopsetta Gill, 1862, 575, 317.
Myzopsetta Gill, 1861, 308, 331.
Clidoderma Bleeker, 1862, 310.

Pseudopleuronectes Bleeker, 1862, 310.
Heteroprosopon Bleeker, 1862, 310.
Brachyprosopon Bleeker, 1862, 310.
Liopsetta Gill, 1864, 331.
Euchalarodus Gill, 1864, 331.
Pœcilopsetta Günther, 1880, 403.
Flesus Moreau, 1881, 416.
Isopsetta Lockington, 1882, 420.
Inopsetta Jordan & Goss, 1885, 433.
Kareius Jordan & Snyder, 1900, 490.
Alæops Jordan & Starks, 1904, 509.
Limandella Jordan & Starks, 1906, 519.
Gareus Hubbs, 1915, 556.
Pluviopsetta Tanaka, 1916, 560.
Tanakius Hubbs, 1918, 564.
Errex Jordan, 1919, 567.

Family 230. SAMARIDÆ (*Paralichthodidæ*)

Samaris Gray, 1831, 139.
Brachypleura Günther, 1862, 319.

Paralichthodes Gilchrist, 1902, 498.
Samariscus Gilbert, 1905, 513.

Family 331. RHOMBOSOLEIDÆ

Psammodiscus Günther, 1862, 319.
Rhombosolea Günther, 1862, 319.
Ammotretis Günther, 1862, 319.

Peltorhamphus Günther, 1862, 319.
Oncopterus Steindachner, 1875, 381.

Family 332. ACHIRIDÆ

Eyes and color dextral. Ventral of the right side continuous along the ridge of the abdomen and joined to the anal; caudal free from dorsal and anal.

Achirus Lacépède, 1803, 65.
Trinectes Rafinesque, 1832, 142.
Apionichthys Kaup, 1858, 282.
Soleotalpa Günther, 1862, 319.
Gymnachirus Kaup, 1858, 282.

Achiropsis Steindachner, 1876, 387.
Baiostoma Bean, 1882, 419.
Amate Jordan & Starks, 1906, 519.
Soleonasus Eigenmann, 1912, 543.
Pnictes Jordan, 1919, 567.

Family 333. SOLEIDÆ
(Soles)

Eyes and color dextral; caudal free from dorsal and anal, as are the ventral fins also.

Solea Klein, 1775, 40.
Solea Quensel, 1806, 74.
Solea Rafinesque, 1810, 81.
Odontolepis Fischer, 1813, 85.
Monochirus Rafinesque, 1814, 87.
Solea Cuvier, 1817, 100.
Monochirus (Cuvier) Oken, 1817, 100.

Leptosoma Nardo, 1827, 121.
Monochir Cuvier, 1829, 131.
Monochirus Bonaparte, 1837, 141, 187.
Microchirus Bonaparte, 1837, 187.
Spanius Gistel, 1848, 236.
Monochirus Kaup, 1858, 282.
Aseraggodes Kaup, 1858, 282.

Heteromycteris Kaup, 1858, 282.
Soleichthys Bleeker, 1860, 294.
Pegusa Günther, 1862, 319.
Microbuglossus Günther, 1862, 319.
Pardachirus Günther, 1862, 319.
Buglossus Günther, 1862, 319.
Liachirus Günther, 1862, 319.

Bowenia Haast, 1873, 369.
Quenselia Jordan, 1888, 442.
Pelotretis Waite, 1911, 541.
Phyllichthys McCulloch, 1916, 559.
Bathysolea Roule, 1916, 560.
Eusolea[284] Roule, 1919.

[284] New name, perhaps not intended as a generic name distinct from *Solea*. *Solea capellonis* Steindachner the only species mentioned.

Family 334. SYNAPTURIDÆ

Eyes dextral; ventrals distinct; dorsal and anal united with caudal.

Brachirus[285] Swainson, 1839, 203.
Synaptura Cantor, 1849, 241.
Achiroides Bleeker, 1851, 247.
Æsopia Kaup, 1858, 282.
Euryglossa Kaup, 1858, 282.

Eurypleura Kaup, 1858, 282.
Anisochirus Günther, 1862, 319.
Zebrias Jordan & Snyder, 1900, 490.
Austroglossus[286] Regan, 1920.

[285] Not *Brachyrus* Swainson of page 71. The name *Brachirus* is first used by Swainson, 1839, II, page 71, for an ally of *Pterois,* with a brief diagnosis, no species being mentioned. On page 180 the same diagnosis is given without reference to species, the name being changed to *Dendrochirus.* On page 264 the definition is expanded, the word is spelled *Brachyrus* and two species, *zebra* and *brachypterus,* are mentioned. The first of these has been taken by the first reviser, Swain (1882), as type. On page 303 Swainson again uses the name *Brachirus,* this time for a genus of Soles, mentioning five species, *plagiusa, orientalis, zebra, commersoni (anus) jerreus,* and *pan.*

Of these, *plagiusa* is a *Symphurus,* not related to the others and not agreeing with the generic diagnosis. In 1849, recognizing that the name *Brachirus,* which he changes to *Brachyurus,* was preoccupied, Cantor offered *Synaptura* as a substitute, at the same time redescribing *commersonianus* and *zebra.* Bleeker next described *commersonianus* as the sole species of *Synaptura,* referring the others to *Brachirus,* a plan inadmissable, as *Synaptura* was a substitute name for *Brachirus,* and thus subject to the same limitations. Kaup (1858) makes *orientalis* type of a new genus (*Euryglossa*) which Günther (1862) recognizes as a subgenus, leaving *zebra, commersoniana* and *pan* in *Synaptura.* *Zebra* is the type of *Zebrias* Jordan & Snyder (1900); the two other species are congeneric, *commersonianus,* after Bleeker, presumably serving as Günther's type. Swain, however, definitely named *orientalis* as type as the first species mentioned under *Brachirus* which corresponds to the author's definition. It is, however, more convenient as well as according better to the rules to regard *commersonianus* as type of *Synaptura.*

The name *Brachirus* must go with its first use, for *Pterois zebra,* thus replacing *Dendrochirus.*

[286] *Op. cit.,* 217; orthotype *Synaptura pectoralis* Kaup; right pectoral longer than head; dorsal rays in increased number.

Family 335. CYNOGLOSSIDÆ

(Tongue-fishes)

Eyes and color sinistral; dorsal and anal joined to the caudal; pectorals wanting; ventrals, if present, free from anal.

Plagusia Browne, 1789, 46.
Symphurus Rafinesque, 1810, 81.
Plagiusa Rafinesque, 1815, 88.
Plagusia (Browne) Cuvier, 1817. 101.
Cynoglossus Hamilton, 1822, 114.
Plagusia Swainson, 1839, 203.
Bibronia Cocco, 1844, 219.
Plagusia Bonaparte, 1845, 187, 226.
Euporista Gistel, 1848, 237.
Eupnœa Gistel, 1848, 238.
Cantoria Kaup, 1858, 282.
Grammichthys Kaup, 1858, 282.
Aphoristia Kaup, 1858, 282.

Icania Kaup, 1858, 282.
Trulla Kaup, 1858, 282.
Arelia Kaup, 1858, 282.
Glossichthys Gill, 1861, 303.
Ammopleurops Günther, 1862, 319, 278.
Paraplagusia Bleeker, 1865, 336.
Rhinoplagusia Bleeker, 1870, 355.
Acedia Jordan, 1888, 442.
Usinosita[287] Jordan & Snyder, 1900, 490.
Areliscus Jordan & Snyder, 1900, 490.
Cynoglossoides Von Bonde, 1922. (See Index.)

[287] Misprinted *Usinostia.*

Superorder *ACANTHOPTERYGII*
Order ZEOIDEI
Family 336. ZEIDÆ

(John Dories)

A family of uncertain relations, having perhaps some affinities with the flounders, with the increased number of ventral rays characteristic of the berycoid fishes, and the adnate post-temporal of *Capros* and the chætodonts.

Zeus Linnæus, 1758, 13.
Oreosoma Cuvier, 1829, 127.
Cyttus Günther, 1861, 575.
Cyttopsis Gill, 1862, 314.
Zenopsis Gill, 1862, 314.
Cyttoides Wettstein, 1886, 436.
Zenion Jordan & Evermann, 1896, 473.
Rhombocyttus Gill, 1893, 461.
Capromimus Gill, 1893, 461.

Cyttosoma Gilchrist, 1903, 503, 504.
Zen Jordan, 1903, 504.
Stethopristes Gilbert, 1905, 513.
Cyttomimus Gilbert, 1905, 513.
Pseudocyttus Gilchrist, 1906, 518.
Neocyttus[288] Gilchrist, 1906.
Parazenopsis Cligny, 1909, 532.
Cyttula Weber, 1913, 551.
Allocyttus McCulloch, 1914, 554.

[288] *Op. cit.,* 4 : 153; orthotype *N. rhomboidalis* Gilchrist. (Omitted in "Genera of Fishes.")

Family 337. GRAMMICOLEPIDÆ

Grammicolepis Poey, 1873, 371.
Vesposus[289] Jordan, 1921.

Xenolepidichthys Gilchrist, 1922. (See Index.)

[289] *Proc. U. S. Nat. Mus.,* **70** : 649; orthotype *Vesposus egregius* Jordan.

Order XENOBERYCES
Family 338. STEPHANOBERYCIDÆ

Stephanoberyx Gill, 1883, 424.
Acanthochænus Gill, 1884, 428.

Malacosarcus Günther, 1887, 437.

Family 339. MELAMPHAIDÆ

A small group of deep-sea fishes not altogether homogeneous.

Metopias Lowe, 1843, 215.
AULOLEPIS Geinitz, 1849, 408.
Anoplogaster Günther, 1859, 291.
Melamphaës Günther, 1864, 333.
Poromitra Goode & Bean, 1881, 419.
Caulolepis Gill, 1883, 424.

Plectromus Gill, 1883, 424.
Scopelogadus Vaillant, 1888, 444.
Lophocephalus Osorio, 1906, 520.
Scopeloberyx Zugmayer, 1911, 541.
Poromitrella Zugmayer, 1911, 541.

Order BERYCOIDEI

Family 340. POLYMIXIIDÆ

Polymixia Lowe, 1838, 192.
Nemobrama Valenciennes, 1844, 223.
PYCNOSTERINX[290] Heckel, 1849, 243.

IMOGASTER Costa, 1857, 275.
PLATYCORMUS von der Marck, 1858, 283.
Dinemus Poey, 1860, 298.

[290] Misprinted in "Genera of Fishes." This genus and the two which follow are probably berycoid, but their relations are uncertain.

Family 341. BERYCOPSIDÆ

BERYCOPSIS Dixon, 1850, 245.

STENOSTOMA Dixon, 1850, 245.

Family 342. BERYCIDÆ

Beryx Cuvier, 1829, 127.
PRISTIGENYS Agassiz, 1839, 191, 194.
Centroberyx Gill, 1862, 315.
MACROLEPIS von der Marck, 1863, 327, 346.
HETEROLEPIS Costa, 1865, 336.
HEMIGONOLEPIS Steinla, 1868, 350.
PRIONOLEPIS Steinla 1868, 350.
GONIOLEPIS Steinla, 1868, 350.

HEMICYCLOLEPIS Steinla, 1868, 350.
SPINACITES Fritsch, 1893, 461.
ELECTROLEPIS Fritsch, 1893, 461.
LOBOPTERUS Kramberger, 1895, 469.
Trachichthodes Gilchrist, 1903, 503.
BRADYURUS Gill, 1904, 508.
Austroberyx McCulloch, 1911, 540.

Family 343. HOPLOPTERYGIDÆ

HOPLOPTERYX Agassiz, 1838, 191.

Family 344. DIRETMIDÆ

Diretmus Johnson, 1863, 326.

Discus[291] Campbell, 1879, 399.

[291] A synonym of *Diretmus*.

Family 345. MONOCENTRIDÆ
(Pine-cone Fishes)

Monocentris Bloch & Schneider, 1801, 58.
Lepisacanthus Lacépède, 1802, 62.
Ericius Tilesius, 1809, 76.

Lepicantha Rafinesque, 1815, 89.
Cleidopus De Vis, 1883, 423.

Family 346. ANOMALOPIDÆ

Heterophthalmus Bleeker, 1856, 267.
Anomalops Kner, 1868, 352.

Photoblepharon Weber, 1902, 501.

Family 347. TRACHICHTHYIDÆ

Trachichthys Shaw, 1799, 55.
Hoplostethus Cuvier & Valenciennes, 1829, 133.
Acrogaster Agassiz, 1838, 191.
Sphenocephalus Agassiz, 1838, 191.

Aipichthys Steindachner, 1859, 293.
Acanthophoria Kramberger, 1895, 469.
Paratrachichthys Waite, 1899, 488.
Gephyroberyx[292] Boulenger, 1902, 497.
Leiogaster Weber, 1913, 551.

[292] Misprinted in "Genera of Fishes."

Family 348. HOLOCENTRIDÆ
(Soldier Fishes)

Holocentrus[293] Gronow, 1763, 19.
Farer Forskål, 1775, 34.
Holocentrus (Gronow) Scopoli, 1777, 42.
Erythrinus[294] (Plumier) Lacépède, 1803, 72.
Corniger Agassiz, 1829, 132.
Myripristis Cuvier, 1829, 127.
Ostichthys (Langsdorff) Cuvier & Valenciennes, 1829, 174.
Rhynchichthys Cuv. & Val., 1831, 137.
Homonotus Dixon, 1850, 245.
Plectrypops Gill, 1862, 315.
Rhinoberyx Gill, 1862, 315.
Rhamphoberyx Gill, 1863, 324.

Neomyripristis Castelnau, 1873, 367.
Holotrachys Günther, 1873, 369.
Neoniphon Castelnau, 1875, 378.
Harpage De Vis, 1884, 427.
Ostichthys (Langsdorff) Jordan & Evermann, 1896, 473.
Flammeo J. & E., 1898, 482.
Howella Ogilby, 1898, 483.
Trachichthyoides Woodward, 1902, 501.
Beanea Steindachner, 1902, 500.
Sargocentron Fowler, 1904, 507.
Adioryx Starks, 1908, 531.
Caproberyx Regan, 1911, 540.

[293] Often written *Holocentrum*.
[294] Also written *Eritrinus* by Plumier.

Family 349. DINOPTERYGIDÆ

Dinopteryx Woodward, 1901, 496.

Family 350. CTENOTHRISSIDÆ

Aulolepis Agassiz, 1844, 218.

Ctenothrissa Woodward, 1899, 488.

Family 351. CARISTIIDÆ

Caristius Gill & Smith, 1905, 514.

Platyberyx Zugmayer, 1911, 541.

Order THORACOSTEI
Suborder HEMIBRANCHII
(*Phthinobranchii*)
Family 352. GASTEROSTEIDÆ
(Sticklebacks)

Gasterosteus Linnæus, 1758, 13.
Gasteracanthus Pallas, 1811, 84.
Spinachia Cuvier, 1817, 105.

Leiurus Swainson, 1839, 200.
Polycanthus Swainson, 1839, 200.
Apeltes Dekay, 1842, 210.

174 [674]   THORACOSTEI

Pungitius[295] Coste, 1846.
Pygosteus (Brevoort) Gill, 1861, 302.
Gastræa Sauvage, 1874, 377.

Gasterostea Sauvage, 1874, 377.
Eucalia Jordan, 1876, 386.
MERRIAMELLA Jordan, 1907, 525.

[295] Coste, P., "Nidification des épinoches et des épinochettes," *Mém. Savants Étrangers,* Paris, 1846, **10**: 574–588. Quoted by Gill, "Smithsonian Report for 1905," 505, 1907; type *Gasterosteus pungitius* Linnæus; replaces *Pygosteus* Brevoort. I owe this note to William Converse Kendall, not having seen Coste's paper.

Family 353. AULORHYNCHIDÆ

Aulorhynchus Gill, 1861, 305.
Aulichthys Brevoort, 1862, 315.

Auliscops Peters, 1866, 342.

Family 354. PROTOSYNGNATHIDÆ

PROTOSYNGNATHUS Von der Marck, 1876, 386.

Suborder *LOPHOBRANCHII*
Family 355. SOLENOSTOMIDÆ

Solenostomus[296] Lacépède, 1803, 67.
SOLENORHYNCHUS Heckel, 1854, 259.

Solenichthus[297] Bleeker, 1865, 336.

[296] Written *Solenostoma* by Rafinesque.
[297] Later written *Solenostomatichthys* by Bleeker.

Family 356. SYNGNATHIDÆ (*Hippocampidæ*)
(Pipe Fishes; Sea Horses)

The pipe fishes, elongate, with caudal fin, grade almost imperceptibly into the more specialized sea horses (*Hippocampus*), in which the caudal is wanting and the tail prehensile.

Syngnathus Linnæus, 1758, 16.
Acus P. L. S. Müller, 1767, 24.
Acus Valmont, 1791, 50.
Typhle Rafinesque, 1810, 78.
Siphostoma Rafinesque, 1810, 79.
Hippocampus Rafinesque, 1810, 79.
Nerophis Rafinesque, 1810, 82.
Hippocampus Leach, 1814, 85.
Tiphlinus Rafinesque, 1815, 91.
Hippocampus Cuvier, 1817, 98.
Scyphius Risso, 1826, 119.
Nematosoma Eichwald, 1831, 176.
Acestra Jardine, 1831, 137.
CALAMOSTOMA Agassiz, 1833, 177.
Acus Swainson, 1839, 205.
Solegnathus[298]Swainson, 1839, 205.

Phyllopteryx Swainson, 1839, 205.
Hippichthys Bleeker, 1849, 241.
Syngnathoides Bleeker, 1851, 248.
Microphis Kaup, 1853, 254.
Hemimarsupium Kaup, 1853, 254.
Doryichthys Kaup, 1853, 254.
Chœroichthys Kaup, 1853, 254.
Stigmatopora[299] Kaup, 1853, 254.
Doryrhamphus Kaup, 1853, 254.
Leptoichthys Kaup, 1853, 254.
Leptonotus Kaup, 1853, 253.
Ichthyocampus Kaup, 1853, 253.
Corythoichthys[300] Kaup, 1853, 253.
Trachyrhampus Kaup, 1853, 253.
Halicampus Kaup, 1853, 253.
Gasterotokeus Heckel, 1853, 253.

[298] Later spelled *Solenognathus.*
[299] Spelled *Stigmatophora* by error.
[300] Spelled *Corythroichthys* by error.

Acentronura Kaup, 1853, 253.
Cœlonotus Peters, 1855, 266.
Siphonostoma[300a] Kaup, 1856.
Hemithylacus Kaup, 1856, 272.
Solenognathus Bleeker, 1856, 268.
Haliichthys Gray, 1859.
Dermatostethus Gill, 1862, 317.
Pseudosyngnathus Kner, 1863, 327, 338.
Belonichthys Peters, 1868, 353.
Protocampus Günther, 1870, 357.
Entelurus Duméril, 1870, 357.
Nannocampus Günther, 1870, 357.
Urocampus Günther, 1870, 357.
Hymenolomus Duméril, 1870, 357.

Atelurus Duméril, 1870, 357.
Typhlus (Bibron) Bleeker, 1870, 357.
Osphyolax Cope, 1875, 379.
Penetropteryx Lunel, 1881, 416.
Acmonotus Philippi, 1896, 475.
Phycodurus Gill, 1896, 471.
Yozia Jordan & Snyder, 1901, 495.
Castelnauina Fowler, 1907, 523.
Macleayina Fowler, 1907, 523.
Acanthognathus Duncker, 1912, 542.
Micrognathus Duncker, 1912, 542.
Apterygocampus Weber, 1913, 551.
Histiogamphelus McCulloch, 1914, 553.
Lissocampus[301] Waite & Hale, 1921.

[300a] *Op. cit.*, 46; orthotype *Syngnathus typhle* Linnæus. A synonym of *Typhle*. (Omitted in "Genera of Fishes.")

[301] *Records South Australian Museum*, 1 : 306, Jan. 29, 1921; orthotype *L. caudalis* Waite & Hale.

Order HYPOSTOMIDES
Family 357. PEGASIDÆ
(Sea Moths)

Pegasus Linnæus, 1758, 16.
Cataphractus Gronow, 1763, 21.
Eurypegasus Bleeker, 1863, 322.
Parapegasus Bleeker, 1870, 357.

Leptopegasus Bleeker, 1873, 367.
Zalises Jordan & Snyder, 1901, 494.
Acanthopegasus McCulloch, 1915, 556.

Order AULOSTOMI
Family 358. AULOSTOMIDÆ
(Trumpet Fishes)

Aulostomus[302] Lacépède, 1803, 66.

Polypterichthys Bleeker, 1853, 252.

[302] Often written *Aulostoma*.

Family 359. FISTULARIIDÆ
(Cornet Fishes)

Fistularia Linnæus, 1758, 14.
Solenostomus Gronow, 1763, 21.
Petimbuabo Catesby, 1771, 31.
Solenostomus Klein, 1778, 43.
Solenostomus Browne, 1789, 46.

Aulus (Commerson) Lacépède, 1803, 70.
Cannorhynchus Cantor, 1849, 241.
Flagellaria[303] Gronow, 1854.
Solenostomus (Gronow) Gill, 1861, 302.
Protaulopsis Woodward, 1901, 496.

[303] Type *F. fistularis* Gronow, *F. tabacaria* Linnæus. (Omitted in "Genera of Fishes.")

Family 360. UROSPHENIDÆ
Urosphen Agassiz, 1844, 181, 218.

Family 361. RHAMPHOSIDÆ
Rhamphosus Agassiz, 1844, 218.

Family 362. CENTRISCIDÆ (*Amphisilidæ*)

Centriscus Linnæus, 1758, 16.
Amphisilen Klein, 1775, 38.
Amphisile Cuvier, 1817, 106.

Acentrachme Gill, 1862, 315.
AULORHAMPHUS Zigno, 1890, 453.
Æoliscus Jordan & Starks, 1902, 499.

Family 363. MACRORHAMPHOSIDÆ

Macrorhamphosus Lacépède, 1803, 66.
Macrognathus Gronow, 1854, 259.
Orthichthys Gill, 1862, 315.
Centriscops Gill, 1862, 315.

Limiculina Fowler, 1907, 523.
Notopogon Regan, 1914, 555.
Scolopacichthys Regan, 1914, 555.

Order LABYRINTHICI

Family 364. LUCIOCEPHALIDÆ

Diplopterus[304] Gray, 1831, 138.

Luciocephalus Bleeker, 1851, 247.

[304] This name is preoccupied; the genus is identical with *Luciocephalus*.

Family 365. OPHICEPHALIDÆ

Channa Gronow, 1763, 22.
Channa (Gronow) Scopoli, 1777, 42.

Ophicephalus[305] Bloch, 1794, 53.
Philypnoides Bleeker, 1849, 241.

[305] Usually spelled *Ophiocephalus*.

Family 366. HELOSTOMIDÆ

Helostoma (Kuhl) Cuvier, 1829, 129.
Spirobranchus Cuvier, 1829, 129.

Sandelia Castelnau, 1861, 301.

Family 367. POLYACANTHIDÆ

Polyacanthus (Kuhl) Cuvier, 1829, 129, 173.

Family 368. OSPHRONEMIDÆ

Osphronemus[306] Lacépède, 1801, 61.
Trichogaster Bloch & Schneider, 1801, 58.
Trichopodus Lacépède, 1802, 61.
Macropodus[307] Lacépède, 1802, 63.
Platypopus Lacépède, 1804, 169.
Trichopus Shaw, 1804, 73.
Colisa Cuvier & Valenciennes, 1831, 137.
Ctenops McClelland, 1845, 227.
Pedites Gistel, 1848, 236.

Lithulcus Gistel, 1848, 237.
Betta Bleeker, 1850, 245.
Trichopsis Canestrini, 1860, 295.
Sphærichthys Canestrini, 1860, 295.
Parosphromenus Bleeker, 1879, 398.
Pseudosphromenus Bleeker, 1879, 398.
Micracanthus Sauvage, 1879, 400.
Parophiocephalus[308] Popta, 1905, 515.
Oshimia Jordan, 1919, 567.

[306] Written *Osphromenus* by authors.
[307] Written *Macropus* by authors.
[308] A synonym of *Betta*.

Family 369. ANABANTIDÆ
(Climbing Perch)

Anabas Cuvier, 1817, 106.
Coius Hamilton, 1822, 114.

Ctenopoma Peters, 1846, 229.

Order PERCOMORPHI
Suborder PERCESOCES
Family 370. NANNATHERINIDÆ

Nannatherina[309]Regan, 1906.

[309] *Ann. Mag. Nat. Hist.*, **17**: 451; orthotype *N. ralstoni* Regan. (Omitted in "Genera of Fishes.")

Family 371. BEDOTIIDÆ

Bedotia Regan, 1903, 506.

Family 372. MELANOTÆNIIDÆ

A group of river fishes of the Australian region, closely related to the *Atherinidæ* but showing certain traits which set them off as a distinct subfamily. The number of nominal genera had been inordinately multiplied.

Melanotænia Gill, 1862, 316.
Pseudomugil Kner, 1864, 334.
Strabo Kner & Steindachner, 1866, 341.
Nematocentris Peters, 1870, 356.
Atherinosoma Castelnau, 1872, 363.
Zantecla Castelnau, 1873, 367.
Neoatherina Castelnau, 1875, 378.
Aida Castelnau, 1875, 378.
Aristeus Castelnau, 1879, 393, 399.

Rhombatractus[310] Gill, 1894, 464.
Telmatherina Boulenger, 1897, 476.
Glossolepis Weber, 1907, 527.
Rhombosoma Regan, 1914, 554.
Rhadinocentrus Regan, 1914, 554.
Anisocentrus Regan, 1914, 554.
Centratherina[311] Regan, 1914.
Chilatherina Regan, 1914, 554.

[310] The date 1879 on page 399, "Genera of Fishes," is an error.

[311] Regan, *loc. cit.;* orthotype *Rhombatractus crassispinosus* Weber. (Omitted in "Genera of Fishes.")

Family 373. ATHERINIDÆ
(Silversides; Hardy Heads; Pescados del Rey; Peixerey)

Atherina Linnæus, 1758, 15.
Aphia[311a] Risso, 1826, 119.
Hepsetia Bonaparte, 1837, 187.
Menidia Bonaparte, 1837, 187.
Membras Bonaparte, 1837, 142, 187.
Chirostoma Swainson, 1839, 200.
Argyrea Dekay, 1842, 210.
RHAMPHOGNATHUS Agassiz, 1844, 219.
MESOGASTER Agassiz, 1844, 219.
Atherinoides Bleeker, 1853, 253.
Atherinichthys Bleeker, 1853, 253.
Heterognathus Girard, 1854, 258.
Basilichthys Girard, 1854, 258.
Atherinopsis Girard, 1854, 257.
Labidesthes Cope, 1870, 356.

Protistius Cope, 1874, 375.
Atherinops Steindachner, 1875, 381.
Atherinella Steindachner, 1875, 381.
Gastropterus Cope, 1878, 394.
Leuresthes Jordan & Gilbert, 1880, 403.
Thyrina Jordan & Culver, 1895, 468.
Tropidostethus Ogilby, 1895, 469.
Lethostole Jordan & Evermann, 1896, 473.
Eurystole J. & E., 1896, 473.
Kirtlandia J. & E., 1896, 473.
Eslopsarum J. & E., 1896, 472.
Tæniomembras Ogilby, 1898, 483.
Pisciregia Abbott, 1899, 484.
Iso Jordan & Starks, 1901, 495.
Atherion J. & S., 1901, 485.

[311a] These minute translucent fishes were regarded by Cuvier as the young of *Atherina*. But Risso describes their egg-laying, and they may be really a distinct form.

Melaniris Meek, 1902, 500.
Ischnomembras Fowler, 1903, 503.
Atherinomorus Fowler, 1903, 503.
Phoxargyrea Fowler, 1903, 503.
Odontesthes Evermann & Kendall, 1906, 517.
Xenatherina Regan, 1907, 526.
Craterocephalus McCulloch, 1912, 544.
Pseudothyrina Ribeiro, 1915, 557.
Kronia Ribeiro, 1915, 557.

Colpichthys Hubbs, 1918, 564.
Austromenidia Hubbs, 1918, 564.
Thyrinops Hubbs, 1918, 564.
Hubbesia Jordan, 1919, 566.
Rheocles Jordan & Hubbs, 1919, 567.
ZANTECLITES Jordan & J. Z. Gilbert, 1919, 568.
Archomenidia Jordan & Hubbs, 1919, 569.
Melanorhinus[312] Metzelaar, 1920.

[312] Metzelaar (Jan), "Over Tropisch Atlantisch Visschen, Amsterdam," 38; type *Melanorhinus boeki* Metzelaar.

Family 374. MUGILIDÆ
(Mullets)

Mugil Linnæus, 1758, 15.
Albula Catesby, 1771, 30.
Cestreus Klein, 1777, 43.
Chelon Röse, 1793, 52.
Cephalus (Plumier) Lacépède, 1803, 73.
Agonostomus Bennett, 1830, 174.
Nestis Cuvier & Valenciennes, 1836, 185.
Dajaus Cuv. & Val., 1836, 185.
Cestræus Cuv. & Val., 1836, 185.
Arnion Gistel, 1848, 237.
Joturus Poey, 1860, 299.
Myxus Günther, 1861, 307.

Rhinomugil Gill, 1863, 324.
Chænomugil Gill, 1863, 324.
Gonostomyxus Macdonald, 1869, 355.
Neomyxus Steindachner, 1878, 397.
Querimana Jordan & Gilbert, 1882, 420.
Æschrichthys Macleay, 1883, 424.
Liza Jordan & Swain, 1884, 428.
Trachystoma Ogilby, 1887, 438.
Neomugil Vaillant, 1894, 466.
Œdalechilus Fowler, 1903, 502.
Squalomugil Ogilby, 1908, 529.
Xenorhynchichthys Regan, 1908, 530.

Family 375. SPHYRÆNIDÆ
(Barracudas)

Umbla Catesby, 1771, 29.
Sphyræna Klein, 1778, 43.
Sphyræna Röse, 1793, 52.
Sphyræna Bloch & Schneider, 1801, 58.

Acus (Plumier) Lacépède, 1803, 73.
Sphærina[313] Swainson, 1839.
Agriosphyræna Fowler, 1903, 502.

[313] A misprint for *Sphyræna*.

Suborder *RHEGNOPTERI*
Family 376. POLYNEMIDÆ
(Thread-fins)

Polynemus Linnæus, 1758, 15.
Trichidion Klein, 1775, 40.
Polydactylus Lacépède, 1803, 67.
Galeoides Günther, 1860, 297.
Pentanemus Günther, 1860, 296.

Trichidion (Klein) Gill, 1861, 302.
Polistonemus Gill, 1861, 305.
Eleutheronema Bleeker, 1862, 310.
Eleutherochir Bleeker, 1879, 398.

Series *SCOMBRIFORMES*

Family 377. SCOMBRIDÆ

(Mackerels)

Scomber Linnæus, 1758, 14.
Pelamys Klein, 1775, 37.
Scomberomorus Lacépède, 1802, 62.
Macrorhynchus (Lacépède) Duméril, 1806, 56, 75.
Polipturus Rafinesque, 1815, 89.
Cybium Cuvier, 1829, 129.
Apolectus Bennett, 1831, 173, 175, 284.
Apodontis Bennett, 1831, 173, 284.
Xiphopterus Agassiz, 1835, 181.
Dictyodus Owen, 1839, 196.
Sphyrænodus (Owen) Agassiz, 1841, 209, 219.
Palimphyes Agassiz, 1844, 218, 231.
Isurus (Egerton) Agassiz, 1844, 189, 218.
Gasteroschisma Richardson, 1845, 227.
Acropoma Owen, 1854, 260.
Eucynodus Owen, 1854, 260.
Cordylus Gronow, 1854, 259.
Acanthocybium Gill, 1862, 314.
Lepidocybium Gill, 1862, 314.
Stereodus Owen, 1865, 339.
Scomberodon Van Beneden, 1871, 360.
Chriomitra Lockington, 1879, 400.
Megalolepis Kramberger, 1879, 400.
Pneumatophorus Jordan & Gilbert, 1882, 420.

Amphodon Storms, 1886, 436.
Lepidothynnus Günther, 1889, 447.
Cœlopoma (Agassiz) Woodward, 1901, 224.
Scombrinus (Agassiz) Woodward, 1901, 224, 496.
Rhonchus (Agassiz) Woodward, 1901, 224, 496.
Isurichthys Woodward, 1901, 496.
Eocœlopoma Woodward, 1901, 496.
Scombramphodon Woodward, 1901, 496.
Chenogaster Lahille, 1903, 505.
Krambergeria Simionescu, 1904, 511.
Sierra Fowler, 1905, 512.
Pelamycybium Toula, 1905, 516.
Neocybium Leriche, 1906, 519.
Nesogrammus Evermann & Seale, 1907, 522.
Rastrelliger Jordan & Starks, 1908, 528.
Auxides Jordan, 1919, 568.
Tunita J. & G., 1919, 569.
Turio J. & G., 1920, 571.
Thyrsocles J. & G., 1920, 571.
Thyrsion Jordan, 1920, 571.
Zaphleges Jordan, 1920, 571.
Ocystias Jordan, 1920, 571.
Xestias[314] Jordan, 1921.

[314] "Fish Fauna, Cal. Tertiary," 270; type *X. iratus* Jordan.

Family 378. THUNNIDÆ

(Tunnies; Albacores)

"The bonitos and tunnies are quite different from other teleostomous fish in having dark red muscles deeply situated on both sides of the vertebral column."—(K. Kishinouye, *Proc. Pan-Pacific Scientific Conference*, 1: 229 1921.)

Thynnus Browne, 1789, 48.
Thynnus Cuvier, 1817, 105.
Orcynus Cuvier, 1817, 105.
Sarda Cuvier, 1829, 129.
Auxis Cuvier, 1829, 129.
Pelamys Cuvier & Valenciennes, 1831, 573.

Palamita Bonaparte, 1831, 175.
Thunnus South, 1845, 227.
Creotroctes Gistel, 1848, 237.
Orycnus Gill, 1862, 315.
Gymnosarda Gill, 1862, 315.
Orcynopsis[315] Gill, 1862, 314.
Grammatorcynus[316] Gill, 1862, 314.

[315] Misprinted *Orycnopsis*.
[316] Misprinted *Grammatorycnus*.

Orycnus Gill 1862, 315.
Thynnichthys Giglioli, 1880, 402.
Pelamichthys Giglioli, 1880, 402.
Euthynnus (Lütken) Jordan & Gilbert, 1882, 420.
Germo Jordan, 1888, 442.
Albacora Jordan, 1888, 442.

Cœlocephalus (Agassiz) Woodward, 1901, 219.
Phalacrus (Ag.) Woodward, 1901, 224.
Eothynnus Woodward, 1901, 496.
Cariniceps (Owen) Woodward, (1854) 1901, 260, 496.
Ozymandias Jordan, 1919, 569.

Family 379. GEMPYLIDÆ
(Snake Mackerels; Escolares)

Acinaces Bory de St. Vincent, 1804, 170.
Macrorhynchus (Lacépède) Duméril, 1806, 56.
Thyrsites Cuvier, 1829, 129.
Gempylus Cuvier, 1829, 129.
Lemnisoma Lesson, 1830, 140.
Ruvettus Cocco, 1833, 178.
Acanthoderma Cantraine, 1837, 188.
Aplurus Lowe, 1841, 209.
Prometheus Lowe, 1841, 209.
Epinnula Poey, 1854, 260.
Dicrotus Gumber, 1860, 297.
Thyrsitocephalus Rath, 1859, 292.
Thyrsitops Gill, 1862, 314.

Nesiarchus Johnson, 1862, 319.
Nealotus Johnson, 1865, 338.
Acanthonotus Sauvage, 1870, 359.
Hemithyrsites Sauvage, 1873, 371.
Gyrionemus Vaillant, 1888, 443.
Bathysoma Davis, 1890, 450.
Promethichthys Gill, 1893, 461.
Escolar Jordan & Evermann, 1896, 468.
Bipinnula Jordan & Evermann, 1896, 474.
Machærope Ogilby, 1898, 483.
Xenogramma Waite, 1904, 511.
Jordanidia Snyder, 1911, 541.
Rexia Waite, 1911, 541.

Family 380. TRICHIURIDÆ (*Lepidopidæ*)
(Hair-tails)

Trichiurus Linnæus, 1758, 12.
Gymnogaster Gronow, 1767, 22.
Lepidopus Gonau, 1770, 28.
Enchelyopus Klein, 1775, 57.
Vandellia Shaw, 1803, 73.
Scarcina Rafinesque, 1810, 79.
Xiphotheca[217] Montagu, 1812, 84.
Anenchelum Blainville, 1818, 107.
Aphanopus Lowe, 1839, 195.

Zypothyca Swainson, 1839, 200.
Lepidopides Heckel, 1850, 246.
Lepturus (Artedi) Gill, 1861, 302.
Enchelyopus (Klein) Bleeker, 1862, 310.
Eupleurogrammus Gill, 1862, 314.
Evoxymetopon Poey, 1863, 325.
Trichiurichthys Sauvage, 1873, 371.
Benthodesmus Goode & Bean, 1882, 419.
Lepturacanthus Fowler, 1905, 512.

[217] Written *Ziphotheca* by Swainson.

Series *XIPHIIFORMES*
Family 381. PALÆORHYNCHIDÆ

Palæorhynchum Blainville, 1818, 108.
Hemirhynchus Agassiz, 1844, 219.

Homalorhynchus Van Beneden, 1873, 367.

Family 382. BLOCHIIDÆ

Blochius Volta, 1796, 54.
Cœlorhynchus Agassiz, 1844, 219.

Gliscus Gistel, 1848, 235.

Family 383. ISTIOPHORIDÆ
(Sail Fishes; Spear Fishes)

Istiophorus[318] Lacépède, 1802, 62.
Makaira[319] Lacépède, 1803, 65.
Notistium Herrmann, 1804, 74.
Tetrapturus Rafinesque, 1810, 80.
Skeponopodus Nardo, 1833, 179.

Zanclurus Swainson, 1839, 200.
ENCHEIZIPHIUS Rütimeyer, 1857, 276.
EMBALORHYNCHUS Marsh, 1870, 359.
BRACHYRHYNCHUS Van Beneden, 1871, 360.

[318] Written *Histiophorus* by authors.
[319] Also spelled *Machæra* and *Macaria* by authors.

Family 384. XIPHIIDÆ
(Sword Fishes)

Xiphias Linnæus, 1758, 12.
ACESTRUS (Agassiz) Woodward, 1901, 221, 496.

Family 385. XIPHIORHYNCHIDÆ

XIPHIORHYNCHUS Van Beneden, 1871, 360.

OMMATOLAMPES Fischer von Waldheim, 1850, 246.

Series *LUVARIFORMES*

Family 386. LUVARIDÆ

Luvarus Rafinesque, 1810, 79.
Ausonia Risso, 1826, 120.

Scrofaria Gistel, 1848, 235.

Series *CORYPHAENIFORMES*

Family 387. CORYPHÆNIDÆ
(*Dolphins; Dorados*)

Coryphæna Linnæus, 1758, 12.
Hippurus Klein, 1779, 43.
Caranxomorus Lacépède, 1802, 61.
Coryphus (Commerson) Lacépède, 1802, 70.

Lepimphis Rafinesque, 1810, 79.
Lampugus Cuvier & Valenciennes, 1833, 178.
Sarda Gronow, 1854, 259.

Series *BRAMIFORMES*

Family 388. BRAMIDÆ
(Sea-breams)

The fishes of this family have some traits in common with the *Selenichthyes,* from which they may be derived.

Brama Bloch & Schneider, 1801, 58.
Lepodus Rafinesque, 1810, 80.
Taractes Lowe, 1843, 215.
Tylometopon (Van Bemmelen) Bleeker, 1876, 383.

Argo Döderlein, 1883, 426.
Collybus Snyder, 1904, 511.
Eumegistus[320] Jordan & Jordan, 1923.

[320] Jordan (David Starr) and Jordan (Eric Knight), "Fishes of Hawaii," *Mem. Carnegie Museum,* 35; orthotype *Eumegistus illustris* Jordan & Jordan.

Family 389. **STEINEGERIIDÆ**

Steinegeria Jordan & Evermann, 1886,
435.

Family 390. **DIANIDÆ**

Diana Risso, 1826, 119.   **Astrodermus** (Bonelli) Cuvier, 1829, 129.
Proctostegus Nardo, 1827, 121.

Family 391. **PTERACLIDÆ**

Pteraclis Gronow, 1763, 22.   **Oligopodes** (Risso) Cuvier, 1817, 105.
Pteraclis Gronow, 1772, 22.   **Oligopus**[320a] Risso, 1826, 105.
Pteridium Scopoli, 1777, 41.   **Pterycombus** Fries, 1837, 189.
Oligopodus Lacépède, 1800, 57.   **Centropholis** Hilgendorf, 1878, 395.
Pteraclidus Rafinesque, 1815, 88.   **Bentenia** Jordan & Snyder, 1901, 494.

[320a] Type *Oligopus niger* Risso.

Family 392.. **ELEPHENORIDÆ**

Elephenor Jordan, 1919, 568.

Series *STROMATEIFORMES*

Family 393. **STROMATEIDÆ**

Teeth in the gullet; scales small, mostly silvery; first dorsal more or less
rudimentary.

Stromateus Linnæus, 1758, 12.   **Poronotus** Gill, 1861, 302.
Rhombus Lacépède, 1800, 56.   **Psenopsis** Gill, 1862, 314.
Lepterus Rafinesque, 1810, 80.   Aspidolepis Geinitz, 1868, 349.
Fiatola Cuvier, 1817, 106.   **Palometa** Jordan & Evermann, 1896, 474.
Seserinus (Cuvier) Oken, 1817, 106.   **Pterorhombus** Fowler, 1906, 517.
Fiatola Risso, 1826, 119.   **Eucrotus** B. A. Bean, 1912, 541.
Peprilus Cuvier, 1829, 129.   **Toledia** Ribeiro, 1915, 557.

Family 394. **PAMPIDÆ**

The fishes of this group are closely similar to the *Stromateidæ* in character and appearance, differing, however, in the narrowly restricted gill
openings.

Pampus Bonaparte, 1837, 187.   *Leptolepis* (Van Hasselt) Guichenot,
Stromateoides Bleeker, 1851, 248.   1867, 345.
Chondroplites Gill, 1862, 314.   **Pampus** (Bonaparte) Fowler, 1905, 513.

Family 395. **CENTROLOPHIDÆ**

This group or family is closely allied to the *Stromateidæ,* differing in
the continuous dorsal, larger scales, and much more elongate form.

Centrolophus Lacépède, 1803, 64.   **Leirus**[321] Lowe, 1833, 179.
Acentrolophus Nardo, 1827, 121.   **Mupus** Cocco, 1833, 178.
Schedophilus Cocco, 1829, 133.   **Gymnocephalus** Cocco, 1838, 191.

[321] Usually written *Lirus.*

Pompilus Lowe, 1839, 195.
Palinurus[322] Dekay, 1842, 288.
Crius Valenciennes, 1844, 223.
Palinurichthys Bleeker, 1859, 288.
Palinurichthys Gill, 1860, 295.

Pammelas Günther, 1860, 297.
Hoplocoryphis Gill, 1862, 314.
Icichthys Jordan & Gilbert, 1880, 404.
Ectenias Jordan & Thompson, 1914, 553.

[322] "N. Y. Fauna Fishes," 118; orthotype *P. perciformis* Dekay; name preoccupied. (Omitted in "Genera of Fishes.")

Family 396. NOMEIDÆ (*Psenidæ*)

This group is closely allied to the *Stromateidæ,* having, like the latter, teeth in the gullet. It differs in the presence of a well-developed first dorsal fin.

Nomeus Cuvier, 1817, 105.
Psenes Cuvier & Valenciennes, 1833, 142.
Cubiceps Lowe, 1843, 215.
Seriolella Guichenot, 1848, 238.
Atimostoma Smith, 1849, 244.
Carangodes Heckel, 1856, 270.
Hyperoglyphe Günther, 1859, 291.

Navarchus Filippi & Verany, 1859, 291.
Neptomenus Günther, 1860, 297.
Trachelocirrus Doumet, 1863, 323.
Bathyseriola Alcock, 1890, 449.
Ariomma[323] Jordan & Snyder, 1904, 508.
Icticus[324] Jordan & Thompson, 1914, 553.

[323] Probably not distinguishable from *Cubiceps.*

[324] *Icticus,* like *Icichthys, Schedophilus,* and *Icosteus,* is pelagic, the body soft and limp like a wet rag. *Icticus* has, however, a distinct spinous dorsal, and *Icosteus* is covered with tough naked skin beset on the lateral line and fin rays with prickles.

Family 397. TETRAGONURIDÆ

Allied to the *Stromateidæ,* according to Regan, the species having likewise teeth in the gullet. The external characters are, however, very different.

Tetragonurus Risso, 1810, 77.

Ctenodax Macleay, 1886, 436.

Series *ICOSTEIFORMES*
Family 398. ICOSTEIDÆ
(Rag Fishes)

This family and the next seem related to the *Centrolophidæ,* but the bones are very feeble and the gullet teeth seem to be wanting.

Icosteus[325] Lockington, 1880, 404.

Schedophilopsis Steindachner, 1881, 417.

[325] In *Icosteus* the ventral rays, I, 4; in *Acrotus* the fin is entirely wanting.

Family 399. ACROTIDÆ

Acrotus Bean, 1887, 437.

Series *CARANGIFORMES*

Vertebræ strong, 24 to 26 in number.

Family 400. **APOLECTIDÆ**

This group, with the general aspect of *Stromateus,* has the vertebræ 10 + 15 as in *Caranx.* The species lack the gullet teeth characteristic of the *Stromateidæ.* The dorsal and anal are without spines, in the adult at least, and the keeled tail is without scutes. The soft dorsal is inserted farther forward than in *Caranx,* in front of middle of body.

Apolectus Cuvier & Valenciennes, 1831, 174, 284.   **Parastromateus** Bleeker, 1865, 336.

Family 401. **CARANGIDÆ**
(Cavallas)

These fishes, having much in common with the scombroid fishes, have nevertheless the vertebræ 10 + 14 = 24, as in typical percoids, toward which they form a gradual transition.

Glaucus Klein, 1775, 38.

Saurus Browne, 1789, 48.

Rhomboida Browne, 1789, 48.

Trachurus (Plumier) Lacépède, 1802, 72.

Trachinotus[326] Lacépède, 1802, 61.

Pelamis (Plumier) Lacépède, 1802, 72.

Cæsiomorus Lacépède, 1802, 61.

Caranx Lacépède, 1802, 60.

Scomberoides Lacépède, 1802, 60.

Centronotus Lacépède, 1802, 62.

Acanthinion Lacépède, 1803, 64.

Gallus Lacépède, 1803, 64.

Argyreiosus Lacépède, 1803, 64.

Selene Lacépède, 1803, 64.

Guaperva (Plumier) Lacépède, 1803, 72.

Hypacantha[327] Rafinesque, 1810, 80.

Trachurus Rafinesque, 1810, 79.

Tricropterus Rafinesque, 1810, 79.

Hypodis Rafinesque, 1810, 79.

Naucrates Rafinesque, 1810, 80.

Alectis Rafinesque, 1815, 88.

Baillonus Rafinesque, 1815, 89.

Orcynus Rafinesque, 1815, 89.

Seriola Cuvier, 1817, 105.

Citula Cuvier, 1817, 104.

Atropus (Cuvier) Oken, 1817, 105.

Lichia Cuvier, 1817, 105.

Blepharis Cuvier, 1817, 105, 129.

Vomer Cuvier, 1817, 105.

TERATICHTHYS König, 1825, 172.

Micropteryx Agassiz, 1829, 132.

Scyris Cuvier, 1829, 129.

Olistus Cuvier, 1829, 129.

Chorinemus Cuvier & Valenciennes, 1832, 137.

Pompilus Minding, 1832, 140.

Gallichthys Cuvier & Valenciennes, 1833, 178.

Porthmeus Cuv. & Val., 1833, 178.

Nauclerus Cuv. & Val., 1833, 178.

Hynnis Cuv. & Val., 1833, 178.

ACANTHONEMUS Agassiz, 1834, 180.

CARANGOPSIS Agassiz, 1835, 181, 218.

Platysomus Swainson, 1839, 201.

Zonichthys Swainson, 1839, 200.

Alepes Swainson, 1839, 200.

Elagatis Bennett, 1840, 206.

Paropsis[327a] Jenyns, 1842.

DUCTOR Agassiz, 1844, 218.

PLEIONEMUS Agassiz, 1844, 218.

ARCHÆUS Agassiz, 1844, 218.

Xystophorus Richardson, 1844, 223.

Carangus[328] Agassiz, 1845.

Selenia[329] Bonaparte, 1845, 226.

[326] Often corrected to *Trachynotus.*

[327] Misprinted *Hypacantus;* corrected in appendix to *Hypacantha.*

[327a] "Voyage 'Beagle'," 65; orthotype *P. signata* Jenyns. (Omitted in "Genera of Fishes.")

[328] This is the first use of *Carangus* as a generic name. No explanation is given, merely a reference to Cuvier & Valenciennes. The presumable type is *Scomber carangus* Bloch.

[329] The name *Selenia* is preoccupied; *Uraspis,* with the bucklers hooked forward is a distinct genus.

Carangoides Bleeker, 1851, 248.
Selar[330] Bleeker, 1851, 248.
Decapterus Bleeker, 1851, 248.
Gnathanodon Bleeker, 1851, 247.
Leioglossus Bleeker, 1851, 248.
Megalaspis Bleeker, 1851, 248.
Selaroides Bleeker, 1851, 248.
Leptaspis Bleeker, 1852, 250.
Carangichthys Bleeker, 1852, 250.
Seriolichthys Bleeker, 1854, 255.
Thynnus Gronow, 1854, 259.
Trachurus Gronow, 1854, 259.
VOMEROPSIS, Heckel, 1854, 259.
Uraspis Bleeker, 1855, 264.
Bothrolæmus Holbrook, 1855, 265.
AIPICHTHYS Steindachner, 1859, 298.
Carangus[331] Girard, 1859, 291.
Doliodon Girard, 1859, 291.
Chloroscombrus Girard, 1859, 291.
ARCHÆOIDES Rath, 1859, 292.
Decaptus Poey, 1861, 309.
Blepharichthys Gill, 1861, 302.
Hemicaranx[332] Bleeker, 1862.
Carangops Gill, 1862, 316.
Trachurops Gill, 1862, 315.
Gymnepignathus Gill, 1862, 316.
Eustomatodus Gill, 1862, 316.
Evepigymnus Gill, 1862, 316.
Paratractus Gill, 1862, 317.
Halatractus Gill, 1862, 317.

Naucratopsis Gill, 1862, 317.
Irex Valenciennes, 1862, 320.
Glaucus (Klein) Bleeker, 1863, 321.
Oligoplites Gill, 1863, 324.
Micropus Kner, 1868, 352.
PSEUDOVOMER Sauvage, 1870, 359.
DESMICHTHYS Sauvage, 1877, 392.
ACANTHONEMOPSIS Bosniaski, 1878, 393.
PAREQUULA Sauvage, 1880, 406.
Lepidomegas Thominot, 1880, 406.
Hypocaranx Klunzinger, 1884, 429.
Parona Berg, 1895, 466.
Zalocys Jordan & McGregor, 1898, 482.
Campogramma Regan, 1903, 505.
Rastrum Fowler, 1904, 507.
Rhaphiolepis Fowler, 1905, 512.
Vexillicaranx Fowler, 1905, 512.
Elaphotoxon Fowler, 1905, 512.
Eleria Jordan & Seale, 1905, 514.
Glaucus (Klein) Fowler, 1906, 517.
Pampanoa Fowler, 1906, 517.
Ulua Jordan & Snyder, 1908, 528.
Glaucus (Klein) Jordan & Hubbs, 1917, 562.
Orqueta Jordan, 1919, 567.
LOMPOCHITES Jordan, 1920, 571.
Atule[333] Jordan, 1923.
Longirostrum[334] Wakiya, 1923.
Seriolina[334a] Wakiya, 1923.

---

[330] Logotype, by first reviser, *Caranx boops*, hence equivalent to *Trachurops* Gill, which name *Selar* must replace.

[331] "*Carangus* Griffith," page 180, "Genera of Fishes," should be suppressed; *Carangues* is merely quoted as a French name.

[332] *Versl. Kong. Akad. Wet.*, **14**: 134; orthotype *H. marginatus* Bleeker. (Omitted in "Genera of Fishes.")

[333] "Fishes of Hawaii," *Carnegie Mus.*, in press; orthotype *Caranx affinis* Rüppell.

[334] Yoshiro Wakiya, "Carangoid Fishes of Japan," Ms., 1923; type *Caranx platessa* Cuvier & Valenciennes.

[334a] *Loc. cit.;* orthotype *Seriola intermedia* Temminck & Schlege (gill rakers obsolete).

Family 402. NEMATISTIIDÆ
(Peacock Fishes)

Nematistius Gill, 1862, 316.   Seriolophus Guichenot, 1867, 345.

Family 403. MENIDÆ
Vertebræ $10 + 15 = 25$.

Mene Lacépède, 1803, 68.   GASTERONEMUS Agassiz, 1833, 177.
Meneus Rafinesque, 1815, 90.   GASTERACANTHUS[335] Agassiz, 1833.

---

[335] A manuscript name, printed but not adopted by Agassiz, being preoccupied; replaced by *Gastronemus. Poissons Fossiles,* **5**: 20; orthotype *G. rhomboidalis* Agassiz ms.

Family 404. POMATOMIDÆ
(Blue Fishes)

Saltatrix Catesby, 1771, 30.
Pomatomus Lacépède, 1803, 63.
Gonenion Rafinesque, 1810, 80.
Lopharis Rafinesque, 1810, 80.
Temnodon Cuvier, 1817, 106.

Sypterus Eichwald, 1841, 208.
Chromis Gronow, 1854, 259.
Sparactodon Rochebrune, 1880, 405.
LOPHAR Jordan & Gilbert, 1919, 569.

Family 405. RACHYCENTRIDÆ
(Sergeant Fishes)

Rachycentron[336] Kaup, 1826, 121.
Spinax (Commerson), Cuvier & Valenciennes, 1831, 175.

Elacate Cuvier & Valenciennes, 1831, 129.
Meladerma Swainson, 1839, 200.

[336] Name later corrected to *Rachycentrum*.

Family 406. LACTARIIDÆ

Lactarius Cuvier & Valenciennes, 1833, 178.

Platylepes Swainson, 1839, 200.

Family 407. LEIOGNATHIDÆ [337]

Leiognathus Lacépède, 1803, 64.
Halex Commerson, 1803, 71.
Equula Cuvier, 1817, 105.
Gazza Rüppell, 1835, 184.
Argyrlepes Swainson, 1839, 200.

Secutor Gistel, 1848, 235.
Deveximentum Fowler, 1904, 507.
Eubleekeria Fowler, 1904, 507.
Equulites Fowler, 1904, 507.
Aurigequula[338] Fowler, 1918.

[337] The resemblance of *Leiognathus* to *Gerres* seems to be superficial, not indicating any special affinity.
[338] *Proc. Acad. Nat. Sci. Phila.*, 70 : 17; orthotype *Leiognathus fasciatus* Fowler.

Family 408. BATHYCLUPEIDÆ

A deep-water group of uncertain affinities.

Bathyclupea Alcock, 1891, 454.

Series *KURTIFORMES*

Family 409. KURTIDÆ

A single genus of uncertain relationships.

Kurtus[339] Bloch, 1786, 44.

[339] Corrected to *Cyrtus* and *Curtus* by authors.

Family 410. PERCIDÆ
(Perch)

Perca Linnæus, 1758, 13.
Cernua Schäfer, 1761, 16.
Schraitzer Schäfer, 1761, 16.
Asperulus Schäfer, 1761, 16.
Asper Schäfer, 1761, 16.

Acerina Güldenstädt, 1774, 32.
Percis Klein, 1775, 39.
Asperulus Klein, 1775, 40.
Gymnocephalus Bloch, 1793, 51.
Cephimmus Rafinesque, 1815, 89.

Zingel (Cuvier) Oken, 1817, 104.
Sander (Cuvier) Oken, 1817, 104.
Acerina Cuvier, 1817, 104, 125.
Pogostoma[340] Rafinesque, 1818, 108.
Aplocentrus[340] Rafinesque, 1819, 110.
Leucops[340] Rafinesque, 1819, 110.
Pomacampsis[340] Rafinesque, 1820, 111.
Stizostedion[341] Rafinesque, 1820, 111.
Sandat Cloquet, 1827, 122.
Cernua Fleming, 1828, 123.
Sandrus Stark, 1828, 123.
Cingla Stark, 1828, 123.
Lucioperca Cuvier & Valenciennes, 1828, 124.
Aspro Cuv. & Val., 1828, 124.
Sandat Bory de St. Vincent, 1828, 124.
Acerina Cuvier, 1820, 129.
Schilus Krynicki, 1832, 176.
Podocys Agassiz, 1838, 191.
Percarina Nordmann, 1840, 207.

Pachygaster Giebel, 1847, 231.
Peroptera Gistel, 1848, 237.
Gremilla Gistel, 1848, 234.
Sandroserrus Gervais, 1852, 251.
Asperulus (Klein) Gill, 1861, 303.
Leptoperca Gill, 1861, 306.
Cynoperca Gill & Jordan, 1877, 390.
Mimoperca Gill & Jordan, 1877, 390.
Mioplosus Cope, 1877, 388.
Epitrachys Schulze, 1889, 448.
Podocephalus (Agassiz) Woodward, (1845) 1901, 224.
Brachygnathus (Agassiz) Woodward, (1845) 1901, 224.
Cœloperca (Agassiz) Woodward, (1845) 1901, 224.
Percostoma (Agassiz) Woodward, (1845) 1901, 224.
Cristigorina Leriche, 1905, 515.
Eoperca Jordan, 1919, 568.

[340] All these genera of Rafinesque appear to be mythical, an opinion derived from drawings in his private notebook preserved in the Smithsonian Institution.

[341] Name corrected to *Stizostethium*.

Family 411. ETHEOSTOMIDÆ
(Darters; Crawl-a-bottoms)

These dwarf or rather concentrated perches are peculiar to the waters of the eastern United States. They differ from the *Percidæ* in having six branchiostegals instead of seven, the head unarmed and the air-bladder obsolete or nearly so. Anal spines two, rarely one.

Etheostoma Rafinesque, 1819, 110.
Diplesion Rafinesque, 1820, 111.
Percina Haldeman, 1842, 211.
Pileoma Dekay, 1842, 210.
Boleosoma Dekay, 1842, 210.
Pœcilosoma Agassiz, 1850, 245.
Hadropterus Aggasiz, 1854, 254.
Pœcilichthys Agassiz, 1854, 254.
Catonotus Agassiz, 1854, 254.
Hyostoma Agassiz, 1854, 254.
Oligocephalus Girard, 1859, 290.
Arlina Girard, 1859, 290.
Estrella Girard, 1859, 290.
Boleichthys Girard, 1859, 290.
Alvordius Girard, 1859, 290.
Alvarius Girard, 1859, 290.

Pleurolepis (Agassiz) Putnam, 1863, 328.
Microperca Putnam, 1863, 328.
Hololepis[341a] (Agassiz) Putnam, 1863.
Nothonotus (Agassiz) Putnam, 1863, 328.
Cottogaster Putnam, 1863, 328.
Asproperca Heckel, 1860, 295.
Hypohomus Cope, 1870, 355.
Astatichthys Vaillant, 1873, 371.
Plesioperca Vaillant, 1873, 371.
Imostoma Jordan, 1877, 390.
Ammocrypta Jordan, 1877, 390.
Nanostoma (Putnam) Jordan, 1877, 390.
Ericosma Jordan & Copeland, 1877, 390.
Rheocrypta Jordan, 1877, 390.
Vaillantia Jordan, 1878, 395.
Ioa Jordan & Brayton, 1878, 395.

[341a] Omitted in "Genera of Fishes"; type *Boleosoma barratti* Holbrook (near *Boleichthys* or identical with it).

Ulocentra Jordan, 1878, 395.
Serraria Gilbert, 1884, 428.
Crystallaria Jordan & Gilbert, 1885, 432.
Rhothœca Jordan, 1885, 432.
Copelandellus Jordan & Evermann, 1896, 474.
Psychromaster J. & E., 1896, 474.

Claricola J. & E., 1896, 474.
Nivicola Jordan, 1896, 474.
Rafinesquiellus Jordan & Evermann, 1896, 474.
Torrentaria J. & E., 1896, 474.
Swainia Jordan & Evermann, 1896, 474.
Vigil Jordan, 1919, 567.

Family 412. PERCICHTHYIDÆ

This family and the next are confined to rivers of the Chilean region. They are intermediate between *Percidæ* and *Moronidæ*.

Percichthys Girard, 1854, 258.
Deuteropterus Gill, 1861, 304.

Percosoma Gill, 1861, 304.

Family 413. PERCILIIDÆ

Percilia Girard, 1854, 258.

Family 414. APOGONIDÆ (*Amiidæ; Chilodipteridæ*)
(Cardinal Fishes)

Anal spines two, rarely one; vent normal in position.

Amia Gronow, 1763, 20.
Apogon Lacépède, 1802, 62.
Cheilodipterus[342] Lacépède, 1802, 63.
Aspro (Commerson) Lacépède, 1803, 70.
Ostorhinchus Lacépède, 1803, 63.
Dipterodon Lacépède, 1803, 63.
Epigonus Rafinesque, 1810, 83.
Clodipterus Rafinesque, 1815, 89.
Macrolepis Rafinesque, 1815, 89.
Pomatomus (Risso) Cuvier & Valenciennes, 1828, 124.
Apogonoides Bleeker, 1849, 240.
Microichthys Rüppell, 1852, 251.
Amia Gronow, 1854, 259.
Apogonichthys Bleeker, 1854, 256.
Monoprion Poey, 1860, 298.
Glossamia Gill, 1863, 323.
Lepidamia Gill, 1863, 323.
Archamia Gill, 1863, 323.
Paramia Bleeker, 1863, 322.
Pseudamia Bleeker, 1865, 336.
Mionorus Krefft, 1867, 346.
Pristiapogon Klunzinger, 1870, 358.
Dinolestes[343] Klunzinger, 1872, 366.

Lanioperca[343] Günther, 1872, 365.
Neosphyræna[343] Castelnau, 1872, 362.
Vincentia Castelnau, 1872, 363.
Percamia Bleeker, 1876, 382.
Telescops Bleeker, 1876, 382.
Gulliveria Castelnau, 1878, 393.
Pomatomichthys Giglioli, 1880, 402.
Monosira Poey, 1881, 417.
Melanostoma Steindachner & Döderlein, 1883, 426.
Synagrops Günther, 1887, 437.
Telescopias Jordan & Snyder, 1901, 495.
Fowleria Jordan & Evermann, 1903, 504.
Galeagra Heller & Snodgrass, 1903, 503.
Foa Jordan & Seale, 1905, 514.
Hynnodus Gilbert, 1905, 513.
Gymnapogon Regan, 1905, 515.
Astrapogon Fowler, 1907, 523.
Siphamia Weber, 1909, 534.
Rhabdamia Weber, 1909, 534.
Neamia Smith & Radcliffe, 1912, 544.
Amioides Smith & Radcliffe, 1912, 544.
Nectamia Jordan, 1917, 562.
Xystramia Jordan, 1917, 562.

---

[342] Written *Chilodipterus* by authors.

[343] These three generic names refer to the same fish, *Dinolestes muelleri*, of Australia.

Zoramia Jordan, 1917, 562.
ERITIMA Jordan & Gilbert, 1919, 569.
Adenapogon³⁴⁴ McCulloch, 1921.
Brephamia³⁴⁵ Jordan, 1922.

Scombrolabrax³⁴⁷ Roule, 1922.
Maccullochina³⁴⁶ Jordan, 1923.
Scepterias³⁴⁸ Jordan & Jordan, 1923.

³⁴⁴ *Records Austral. Mus.,* **13**: 132; orthotype *Apogon roseigaster* Ramsay & Ogilby.
³⁴⁵ "Fishes of Hawaii," *Mem. Carnegie Museum,* 43; orthotype *Amia parvula* Smith & Radcliffe.
³⁴⁶ *Loc. cit.* 44; orthotype *Scombrops serratospinosa* Smith & Radcliffe.
³⁴⁷ Roule, *Bull. Sci. Oceanogr.,* No. 408, March 23, 1922; type *S. heterolepis* Roule; a deep-sea fish from Madeira. Spinous dorsal deeply divided; anal spine single; scales cycloid, very irregular as to size.
³⁴⁸ *Op. cit.* 44; orthotype *S. fragilis* Jordan & Jordan, from Hawaii.

Family 415. SCOMBROPIDÆ

Allied to the *Apogonidæ,* but with three or four anal spines. The number in the *Apogonidæ,* as in the *Percidæ* and *Etheostomatidæ,* is but two or rarely one.

Scombrops Temminck & Schlegel, 1845, 227.
Sphyrænops Poey, 1860, 299.
Latebrus Poey, 1860, 298.
Amiichthys Poey, 1886, 435.

Hypoclydonia Goode & Bean, 1895, 468.
Apogonops Ogilby, 1896, 475.
Parasphyrænops Bean, 1912, 542.
Neoscombrops Gilchrist, 1922. (See Index.)

Family 416. ACROPOMATIDÆ

Three anal spines; vent anterior.

Acropoma Temminck & Schlegel, 1843, 216.
Brephostoma Alcock, 1889, 444.

Parascombrops³⁴⁹ Alcock, 1890, 450.
Oxyodon Brauer, 1906, 517.

³⁴⁹ A synonym of *Acropoma.*

Family 417. AMBASSIDÆ

Chanda Hamilton, 1822, 114.
Ambassis (Commerson) Cuvier & Valenciennes, 1828, 124.
Priopis (Kuhl & Van Hasselt) Cuvier & Valenciennes, 1830, 174.
Hamiltonia Swainson, 1839, 200.

Bogoda Bleeker, 1853, 253, 275.
Parambassis Bleeker, 1874, 374.
Pseudambassis Bleeker, 1876, 383.
Acanthoperca Castelnau, 1878, 393.
Pseudoambassis Castelnau, 1878, 393.
Tetracentrum Macleay, 1883, 425.

Family 418. CENTRARCHIDÆ (*Eucentrarchidæ; Micropteridæ*)
(Black Bass; Sun Fishes; Pumpkin-seeds)

The name *Centrarchus* of Cuvier was framed for those sun fishes which had more than three anal spines. The first species mentioned (*æneus*) is an *Ambloplites,* and this, it appears, was the only one actually seen by the author. But he also includes, without question, the *Labrus irideus* and *Labrus macropterus* of Lacépède, *irideus* being formally indicated as type of *Centrarchus* by Agassiz in 1854, by Holbrook in 1860, and by Jordan &

Gilbert in 1882. There seems no reason for changing this arrangement as with Culver *irideus* was not a *"species inquirenda."*

Micropterus Lacépède, 1803, 63.
Pomoxis Rafinesque, 1818, 109.
Calliurus Rafinesque, 1819,110.
Lepomis[350] Rafinesque, 1819, 110.
Pomotis Rafinesque, 1819, 110.
Apomotis Rafinesque, 1819, 110.
Aplesion Rafinesque, 1820, 111.
Ichthelis[351]Rafinesque, 1820.
Dioplites Rafinesque, 1820, 111.
Nemocampsis Rafinesque, 1820, 111.
Ambloplites Rafinesque, 1820, 111.
Aplites Rafinesque, 1820, 111.
Telipomis Rafinesque, 1820, 111.
Huro Cuvier & Valenciennes, 1828, 124.
Centrarchus Cuvier, 1829, 126.
Pomotis Cuvier, 1829, 126.
Gristes[352] Cuvier, 1829, 126.
Bryttus Cuvier & Valenciennes, 1831, 137.

Archoplites Gill, 1861, 305.
Mesogonistius Gill, 1864, 332.
Chænobryttus Gill, 1864, 332.
Hyperistius Gill, 1864, 332.
Eucentrarchus Gill, 1864, 332.
Acantharchus Gill, 1864, 332.
Enneacanthus Gill, 1864, 332.
Hemioplites Cope, 1867, 344.
Glossoplites[353] Jordan, 1876.
Helioperca Jordan, 1877, 390.
Xystroplites Jordan, 1877, 388.
Erichæta Jordan, 1877, 391.
Eupomotis Gill & Jordan, 1877, 389.
Xenotis Jordan, 1877, 390.
Copelandia Jordan, 1877, 390.
PLIOPLARCHUS Cope, 1883, 422.
OLIGOPLARCHUS Cope, 1891, 454.
CENTRARCHITES Cockerell, 1919, 566.

[350] Later corrected to *Lepiopomus,* an inacceptable alteration.

[351] *Ichthyologia Ohiensis,* 27; type *Labrus auritus* Linnæus; same as *Lepomis.* (Omitted in "Genera of Fishes.")

[352] Later written *Grystes.*

[353] "Manual Vertebrates," Ed. 1, 223; type *Calliurus melanops* Girard; same as *Chænobryttus* Gill. (Omitted in "Genera of Fishes.")

Family 419. ELASSOMIDÆ
(Dwarf Sun Fishes)

Elassoma Jordan, 1877, 390.

Family 420. KUHLIIDÆ
(Mountain Bass; Sesele)

Kuhlia Gill, 1861, 304.
Moronopsis Gill, 1863, 324.
Nannoperca Günther, 1862, 307.
Paradules Bleeker, 1863, 322.
Paradules Klunzinger, 1872, 366.
Edelia Castelnau, 1873, 368.
Microperca Castelnau, 1873, 368.

Platysoma[354] (Liénard) Scudder, (1832) 1882.
Herops De Vis, 1885, 431.
Boulengerina Fowler, 1904, 507, 523.
Safole Jordan, 1912, 543.
Parakuhlia Pellegrin, 1913, 549.

[354] *Platysoma* Liénard (as *"Platysome"*), *Proc. Zool Soc. Lond.,* 1832, II, 112; type *Dules caudavittatus* Cuvier & Valenciennes; latinized by Scudder 1882; name preoccupied; same as *Safole* Jordan.

Family 421. CENTROPOMIDÆ
(Robalos)

Macrocephalus Browne, 1789, 47.
Centropomus Lacépède, 1803, 63.
Oxylabrax Bleeker, 1876, 382.

Macrocephalus (Browne) Bleeker, 1876, 383.
Platycephalus Ribeiro, 1902, 500.

Family 422. LATIDÆ

This group and the next two are usually united with the *Serranidæ* (including *Epinephelidæ*), forming thus a vast and divergent family, which for our purpose may be conveniently separated into some of its component parts. These are usually and perhaps justly regarded as subfamilies. Boulenger places *Lates* with the *Centropomidæ* and Regan adds the *Ambassidæ* also.

Lates Cuvier & Valenciennes, 1828, 124.
Psammoperca Richardson, 1844, 223.
Cnidon Müller & Troschel, 1849, 243.
Hypopterus Gill, 1861, 304.
Pseudolates Alleyne & Macleay, 1877, 387.

Paralates Sauvage, 1883, 425.
Platylates Storms, 1887, 439.
Pseudolates Priem, 1898, 483.
Prolates Priem, 1899, 487.
Luciolates Boulenger, 1914, 552.

Family 423. MORONIDÆ
(White Bass)

Maxillary without supplemental bone. The fossil forms referred to this group and the preceding can not be placed with accuracy.

Labrax Klein, 1775, 39.
Morone Mitchill, 1814, 86.
Roccus Mitchill, 1814, 86.
Lepibema Rafinesque, 1820, 111.
Labrax (Klein) Cuvier, 1829.
Smerdis Agassiz, 1833, 177.
Cyclopoma Agassiz, 1833, 177.

Dapalis Gistel, 1848, 237.
Platacanthus Fischer von Waldheim, 1850, 246.
Dicentrarchus Gill, 1860, 296.
Percalates Ramsay & Ogilby, 1887, 439.
Chrysoperca Fowler, 1907, 523.

Family 424. OLIGORIDÆ

Maxillary with a supplemental bone, as in *Epinephelus*. This group is very close to the *Moronidæ*, both being usually merged in the *Serranidæ*.

Perca-Labrax Temminck & Schlegel, 1842, 213.
Lateobrax Bleeker, 1857, 274.
Percalabrax (Schlegel) Günther, 1859, 291.

Oligorus Günther, 1859, 291.
Odontolabrax Bleeker, 1873, 367.
Homodemus De Vis, 1885, 431.
Malakichthys Döderlein, 1883, 426.
Satsuma Smith & Pope, 1906, 520.

Family 425. NIPHONIDÆ

Niphon[355] Cuvier & Valenciennes, 1828.

[355] Cuvier & Valenciennes, *Hist. Poss.*, **2**: 131; orthotype *Niphon spinosus* Cuvier & Valenciennes. (Omitted in "Genera of Fishes.")

Family 426. EPINEPHELIDÆ
(Sea Bass; Groupers; Garrupas)

The great group of genera commonly called percoid (although centering rather around *Epinephelus, Perca* being a somewhat aberrant form) is here provisionally divided into several families obviously of very unequal value. Several of them usually and very properly, no doubt, given the rank of subfamily. Even after the removal of various outlying genera, the *Epinephelidæ* are very much diversified and further subdivision is natural. Whatever rank the divisions may receive, the presence of supplemental maxillary by which character 'the *Epinephelidæ* are set off from the *Serranidæ* is a matter of importance, perhaps adequate to define a separate family, although in some groups, as the *Centrarchidæ*, it disappears by degrees.

Cugupuguacu Catesby, 1771, 30.
Daba Forskål, 1775, 34.
Louti Forskål, 1775, 34.
Bodianus[356] Block, 1790, 49.
Epinephelus Bloch, 1793, 51.
Grammistes Bloch & Schneider, 1801, 58.
Cephalopholis B. & S., 1801, 59.
Alphestes B. & S., 1801, 59.
Chrysomelanus (Plumier) Lacépède, 1803, 72.
Notognidion Rafinesque, 1810, 80.
Polyprion (Cuvier) Oken, 1817, 104, 117.
Plectropomus[357] (Cuv.) Oken, 1817, 103.
Rypticus[358] Cuvier, 1829, 125.
Macquaria Cuvier & Valenciennes, 1830, 135.
Merou Bonaparte, 1831, 175.
Cerna Bonaparte, 1837, 141, 187.
Uriphaëton Swainson, 1839, 198.
Variola Swainson, 1839, 198.
Cynichthys Swainson, 1839, 198.
Cromileptes Swainson, 1839, 198.
Centrogenys Richardson, 1842, 212.
Aulacocephalus Temminck & Schlegel, 1842, 213.
Myriodon Brisout de Barneville, 1847, 230.
Paschalestes Gistel, 1848, 236.
Serranichthys Bleeker, 1855, 264.
Smecticus Valenciennes, 1855, 266.
Stereolepis Ayres, 1858, 277.
Trachypoma Günther, 1859, 291.
Pogonoperca Günther, 1859, 291.
Anyperodon Günther, 1859, 291.
Promicropterus Gill, 1861, 302.
Liopropoma Gill, 1861, 304.
Dermatolepis Gill, 1861, 304.
Prospinus Poey, 1861, 308.
Siniperca Gill, 1862, 313.

Lioperca Gill, 1862, 315.
Schistorus Gill, 1862, 315.
Gonioplectrus Gill, 1862, 315.
Plectroplites Gill, 1862, 315.
Acanthistius Gill, 1862, 315.
Labroperca Gill, 1862, 315.
Mycteroperca Gill, 1862, 315.
Hyporthodus Gill, 1862, 304.
Chorististium Gill, 1862, 313.
Petrometopon Gill, 1865, 337.
Enneacentrus Gill, 1865, 337.
Trisotropis Gill, 1865, 337.
Plectroperca Peters, 1865, 339.
Priacanthichthys Day, 1868, 349.
Promicrops Poey, 1868, 353.
Eleutheractis Cope, 1871, 361.
Ctenolates Günther, 1871, 361.
Menephorus Poey, 1871, 362.
Murrayia Castelnau, 1872, 362.
Riverina Castelnau, 1872, 362.
Actenolepis Dybowski, 1872, 364.
Bostockia Castelnau, 1873, 368.
Odontolabrax Bleeker, 1873, 367.
Hectoria Castelnau, 1873, 368.
Pikea Steindachner, 1874, 377.
Phaëthonichthys Bleeker, 1876, 382.
Hemilutjanus Bleeker, 1876, 382.
Homalogrystes Alleynes & Macleay, 1877, 387.
Itaiara Vaillant & Bocourt, 1878, 397.
Megaperca Hilgendorf, 1878, 395.
Doderleinia Steindachner, 1883, 426.
Acanthocephalus Döderlein, 1883, 426.
Labracopsis Steindachner & Döderlein, 1883, 426.
Garrupa Jordan & Eigenmann, 1890, 452.
Dinoperca Boulenger, 1895, 466.
Pseudalphestes Boulenger, 1895, 466.

---

[356] *Bodianus* belongs here by first restriction, not by tautonomy, as *Bodianus bodianus* is a labroid fish.
[357] Later written *Plectropoma*.
[358] Later corrected to *Rhypticus*.

Pomodon Boulenger, 1895, 466.
Xystroperca Jordan & Evermann, 1896, 474.
Archoperca J. & E., 1896, 474.
Enneistus J. & E., 1896, 474.
Coreoperca Herzenstein, 1896, 472.
Centristhmus Garman, 1899, 486.
Epinephelides Ogilby, 1899, 487.
Bryttosus Jordan & Snyder, 1900, 490.
Eteliscus J. & S., 1900, 490.

Corusculus J. & E., 1901, 494.
Gilbertella Waite, 1902, 501.
Æthaloperca Fowler, 1904, 507.
Gennadius Jordan & Seale, 1907, 525.
Epelytes Evermann & Radcliffe, 1917, 562.
Mustelichthys Tanaka, 1918, 565.
EMMACHÆRE Jordan & J. Z. Gilbert, 1919, 569.
Rhabdosebastes[359] Fowler & Bean, 1922.

[359] Fowler & Bean, "Fishes from Formosa and the Philippine Islands," *Proc. U. S. Nat. Mus.,* **62**: 60; orthotype *Sebastes stoliczæ* Day; same as *Gennadius* Jordan & Seale, both being synonyms of *Centrogenys*. This highly aberrant form with the lower pharyngeals united, bears a deceptive resemblance to species of *Sebastes*.

Family 427. SERRANIDÆ
(Sea Bass)

Supplemental maxillary wanting.

Anthias Bloch, 1792, 50.
Hepatus Röse, 1793, 52.
Aylopon Rafinesque, 1810, 80.
Serranus Cuvier, 1817, 103.
Dules Cuvier, 1828, 126.
Centropristes[360] Cuvier, 1829, 126.
Dulichthys Bonaparte, 1831, 175.
ACANUS Agassiz, 1838, 191.
Callanthias Lowe, 1839, 195.
Prionodes Jenyns, 1842, 211.
Caprodon Temminck & Schlegel, 1843, 216.
Diplectrum Holbrook, 1855, 265.
Paralabrax Girard, 1856, 269.
LUSPIA Costa, 1858, 574.
Atractoperca Gill, 1861, 305.
Triloburus Gill, 1861, 302.
Brachyrhinus Gill, 1862, 315.
Hypoplectrus Gill, 1862, 315.
Hypoplectrodes Gill, 1862, 315.
Mentiperca Gill, 1862, 314.
Haliperca Gill, 1862, 315.
Gonioperca Gill, 1862, 315.
Pronotogrammus Gill, 1863, 323.
Paranthias Guichenot, 1868, 352.
Holanthias Günther, 1869, 351.
Paracentropristes Klunzinger, 1870, 358.
Pseudoserranus Klunzinger, 1870, 358.
Cæsioperca Castelnau, 1872, 375.

Pseudanthias Bleeker, 1873, 367.
Plectranthias Bleeker, 1873, 367.
Dactylanthias Bleeker, 1873, 367.
Hemianthias Steindachner, 1874, 377.
Neoanthias Castelnau, 1878, 393.
Symphysanodon Bleeker, 1878, 393.
Cratinus Steindachner, 1878, 397.
Colpognathus Klunzinger, 1880, 575.
Bathyanthias Günther, 1880, 403.
PROPERCA Sauvage, 1880, 405.
Creolus Jordan & Gilbert, 1882, 420.
Hyposerranus Klunzinger, 1884, 429.
Neoniphon De Vis, 1885, 431.
Gilbertia[361] Jordan & Eigenmann, 1890, 452.
Serranellus J. & E., 1890, 452.
Chelidoperca Boulenger, 1895, 466.
Ocyanthias Jordan & Evermann, 1896, 474.
Anagramma Ogilby, 1899, 487.
Grammatonotus Gilbert, 1905, 513.
Sayonara Jordan & Seale, 1906, 518.
Eudulus Fowler, 1907, 523.
Callidulus Fowler, 1907, 523.
Xenanthias Regan, 1908, 529.
Zalanthias Jordan & Richardson, 1910, 536.
Sacura Jordan & Richardson, 1910, 536.

[360] Later written *Centropristis*.
[361] This genus is probably distinct from *Hypoplectrodes*.

Sphenanthias Weber, 1913, 551.
Pteranthias Weber, 1913, 551.
Lepidoperca Regan, 1914, 554.
Serranops Regan, 1914, 554.
Franzia Jordan & Thompson, 1914, 554.
Rosanthias Tanaka, 1917, 563.

Selenanthias Tanaka, 1918, 565.
Leptanthias Tanaka, 1918, 565.
Rhyacanthias[362] Jordan, 1921.
Percanthias[363] Tanaka, 1922.
Parosphenanthias Gilchrist, 1922. (See Index.)

[362] Jordan, *U. S. Nat. Mus.,* **59**: 547; orthotype *R. carlsmithi* Jordan.
[363] Tanaka, *Fishes Japan,* **32**: 591; orthotype *Callanthias japonicus* Franz.

Family 428. PLESIOPIDÆ

This family and the next three are closely related, and also allied to the *Anthiinæ* among the *Serranidæ.*

Plesiops (Cuvier) Oken, 1817, 102.
Pharopteryx Rüppell, 1828, 122, 251.
Trachinops[364] Günther, 1861.
Ruppelia Castelnau, 1873, 367.
Bleeckeria Castelnau, 1873, 368.

Paraplesiops Bleeker, 1875, 378, 383.
Cirrhiptera (Kuhl &,Van Hasselt) Bleeker, 1876, 383.
Tosana Smith & Pope, 1906, 520.
Belonepterygion McCulloch, 1915, 556.

[364] Günther, *Cat. Fish,* **3**: 366; type *Trachinops tæniatus* Günther. (Omitted in "Genera of Fishes.")

Family 429. ACANTHOCLINIDÆ
Acanthoclinus Jenyns, 1842, 212.   Acanthoplesiops Regan, 1912, 545.

Family 430. PSEUDOPLESIOPIDÆ
Pseudoplesiops Bleeker, 1858, 278.

Family 431. PSEUDOCHROMIDÆ

Pseudochromis Rüppell, 1835, 184.
Labristoma Swainson, 1839, 200.
Cichlops Müller & Troschel, 1858, 278.
Gramma Poey, 1868, 353.
Dampieria Castelnau, 1875, 378.
Pseudogramma Bleeker, 1875, 378.

Leptochromis Bleeker, 1876, 382.
Labracinus (Schlegel) Bleeker, 1876, 383.
Stigmatonotus Peters, 1876, 386.
Onar De Vis, 1885, 431.
Nesiotes De Vis, 1884, 427.
Nematochromis Weber, 1913, 551.

Family 432. RHEGMATIDÆ
Rhegma Gilbert, 1900, 490.

Family 433. PRIACANTHIDÆ
(Big-eyes)

Abuhamrur Forskål, 1771, 34.
Priacanthus Cuvier, 1817, 104.
Boops Gronow, 1854, 258.

Bogota Blyth, 1860, 294, 383.
Pseudopriacanthus Bleeker, 1869, 353.

Family 434. PEMPHERIDÆ
(Catalufas)

A group of uncertain relationships, usually placed near the scombroid fishes.

Pempheris Cuvier, 1829, 128.
Parapriacanthus Steindachner, 1870, 360.
Pempherichthys Klunzinger, 1871, 361.
Leptobrama Steindachner, 1878, 397.

Neopempheris Macleay, 1881, 416.
Priacanthopsis Fowler, 1906, 517.
Catalufa Snyder, 1911, 541.
Liopempheris Ogilby, 1913, 549.

Family 435. DIPLOPRIONIDÆ

Diploprion (Kuhl & Van Hasselt) Cuvier
& Valenciennes, 1828, 124.

Family 436. LOBOTIDÆ
(Flashers)

Coius Hamilton, 1822, 172.
Lobotes Cuvier, 1829, 128.

Datnioides Bleeker, 1852, 252.
Verrugato[364a] Jordan, 1923.

[364a] *Verrugato* Jordan, 1924; new generic name; type *Lobotes pacificus* Gilbert; distinguished by the feebleness of head serrations.

Family 437. GLAUCOSOMIDÆ

Glaucosoma Temminck & Schlegel, 1843, 216.

Breviperca[365] Castelnau, 1873, 378.
Reganichthys Ogilby, 1915, 557.

[365] Young of *Glaucosoma*.

Family 438. ARRIPIDÆ

Arripis Jenyns, 1842, 211.
Homodon Brisout de Barneville, 1847, 230.

Family 439. XENICHTHYIDÆ

These genera are placed by Regan among the *Pomadasidæ*.

Xenichthys Gill, 1863, 324.
Xenistius Jordan & Gilbert, 1882, 420.

Xenocys Jordan & Bollman, 1889, 447.

Family 440. HOPLOPAGRIDÆ

Hoplopagrus Gill, 1861, 304

Family 441. LUTIANIDÆ
(Snappers)

To this family Regan refers the allies of *Cæsio,* together with *Aphareus, Propoma,* and *Hoplopagrus.*

Salpa Catesby, 1771, 30.
Turdus Catesby, 1771, 30.
Naqua Forskål, 1771, 34.
Anthea Catesby, 1771, 31.
Hobar Forskål, 1775, 34.
Lutianus[366] Bloch, 1790, 49.
Sciænus (Commerson) Lacépède, 1802, 69.
Sarda (Plumier) Lacépède, 1802, 72.
Pagrus (Plumier) Lacépède, 1803, 72.
Sargus Plumier, 1803, 72.
Diacope Cuvier, 1815, 94.
Mesoprion Cuvier & Valenciennes, 1828, 124.

Etelis Cuv. & Val., 1828, 124.
Aprion Cuv. & Val., 1830, 135.
Apsilus Cuv. & Val., 1830, 135.
Elastoma Swainson, 1839, 198.
Chætopterus Temminck & Schlegel, 1844, 223.
Genyoroge Cantor, 1849, 241.
Pristipomoides Bleeker, 1852, 250.
Macrops Duméril, 1856, 269.
Neomænis Girard, 1859, 291.
Macolor Bleeker, 1860, 294.
Ocyurus Gill, 1862, 315.
Proamblys Gill, 1862, 315.

[366] Also spelled *Lutjanus.*

Rhomboplites Gill, 1862, 315.
Platyinius[367] Gill, 1862, 315.
Hypoplites Gill, 1862, 315.
Evoplites Gill, 1862, 315.
Tropidinius (Gill) Poey, 1868, 353.
Neomesoprion Castelnau, 1875, 378.
Rabirubia Jordan & Fesler, 1893, 462.
Raizero Jordan & Fesler, 1893, 462.
Bowersia[367] Jordan & Evermann, 1903, 504.
Bennettia Fowler, 1904, 507.

Parkia Fowler, 1904, 507.
Etelides Jordan & J. C. Thompson, 1905, 514.
Rooseveltia Jordan & Evermann, 1906, 519.
Ulaula[368] Jordan & W. F. Thompson, 1911.
Etelinus Jordan & W. F. Thompson, 1911, 539.
Vegetichthys Tanaka, 1918, 565.
Rhomboplitoides[369] Fowler, 1918.

[367] *Platyinus* and *Bowersia* are apparently synonymous with *Pristipomoides*.
[368] Jordan & W. F. Thompson, *Proc. U. S. Nat. Mus.,* **39**: 459, 1911; type *Bowersia ulaula* Jordan & Evermann, *Chætopterus sieboldi* Bleeker; replaces *Chætopterus* Temminck & Schlegel, preoccupied.
[369] Fowler, *Proc. Ac. Nat. Sci. Phila.,* **70**: 33; orthotype *R. megalops* Fowler. (Omitted in "Genera of Fishes.")

Family 442. VERILIDÆ

Verilus Poey, 1860, 298.

Family 443. APHAREIDÆ

Aphareus Cuvier & Valenciennes, 1830, 135.

Family 444. POMADASIDÆ (*Hæmulidæ; Pristipomidæ*)
(Grunts)

Ghanan Forskål, 1775, 34.
Gaterin Forskål, 1775, 34.
Cæsio Lacépède, 1802, 61.
Plectorhinchus[370] Lacépède, 1802, 61.
Pomadasys Lacépède, 1803, 64.
Cheloniger (Plumier) Lacépède, 1803, 73.
Diagramma Cuvier, 1815, 94.
Scolopsis[371] Cuvier, 1815, 94.
Pristipomus[372] (Cuvier) Oken, 1817, 103.
Diabasis Desmarest, 1823, 116.
Anomalodon Bowdich, 1825, 119.
Hæmulon[373] Cuvier, 1829, 128.
Conodon Cuvier & Valenciennes, 1830, 135.
Lycogenis (Kuhl & Van Hasselt) Cuv. & Val., 1830, 174.
Hapalogenys Richardson, 1844, 222.
Pinjalo Bleeker, 1845, 225.
Heterodon Bleeker, 1845, 225.
Heterognathodon Bleeker, 1848, 233.

Hyperoglyphe Günther, 1859, 291.
Odontonectes Günther, 1859, 291.
Orthopristis Girard, 1859, 291.
Apostata (Heckel) Canestrini, 1860, 294.
Anisotremus Gill, 1860, 301.
Orthostœchus Gill, 1862, 316.
Microlepidotus Gill, 1862, 316.
Genytremus Gill, 1862, 316.
Genyatremus Gill, 1862, 316.
Pristocantharus Gill, 1862, 316.
Brachydeuterus Gill, 1862, 313.
Anarmostus Scudder, 1863, 328.
Bathystoma Scudder, 1863, 328.
Brachygenys (Scudder) Poey, 1868, 353.
Hæmulopsis Steindachner, 1869, 355.
Parapristipoma Bleeker, 1873, 367.
Paracæsio Bleeker, 1875, 378.
Paraconodon Bleeker, 1876, 382.
Gymnocæsio Bleeker, 1876, 382.
Liocæsio Bleeker, 1876, 382.

[370] Written *Plectorhynchus* by authors.
[371] Later called *Scolopsides*.
[372] Later written by Cuvier *Pristipoma*.
[373] Also written incorrectly *Hæmylum*.

Pterocæsio Bleeker, 1876, 382.
Pseudopristipoma Sauvage, 1880, 405.
Propoma Günther, 1880, 403.
Lythrulon Jordan & Swain, 1884, 428.
Eurumetopos Morton, 1887, 438.
Isaciella Jordan & Fesler, 1893, 462.
Isacia Jordan & Fesler, 1893, 462.
Rhonciscus Jordan & Evermann, 1896, 472.
Rhencus J. & E., 1896, 472.

Evapristis J. & E., 1896, 472.
Parascolopsis Boulenger, 1901, 492.
Mylacrodon Regan, 1903, 505.
Euelatichthys Fowler, 1904, 508.
Spilotichthys Fowler, 1904, 508.
Diagrammella Pellegrin, 1912, 544, 549.
Leptoscolopsis Tanaka, 1915, 558.
Otoperca Boulenger, 1916, 558.
Dacymba Jordan & Hubbs, 1917, 562.

Family 445. THERAPONIDÆ

Djabub Forskål, 1775, 34.
Therapon[374] Cuvier, 1817, 104.
Coius Hamilton, 1822, 172.
Helotes Cuvier, 1829, 127.
Pelates Cuvier, 1829, 126.
Pterapon Gray, 1833, 139, 179.
Mesopristes[375] Bleeker, 1845, 225, 382.

Datnioides (Brisout de Barneville)
  Canestrini, 1860, 294.
Pseudohelotes Guimaraës, 1882, 419.
Autisthes De Vis, 1885, 431.
Hephæstus De Vis, 1885, 431.
Plagiogeneion Forbes, 1890, 450.
Eutherapon Fowler, 1904, 508.

[374] Misprinted *Terapon.*
[375] Replaces *Datnia* Cuvier & Valenciennes, which is not the same as *Datnia* Cuvier, the latter a synonym of *Sparus* Linnæus.

Family 446. BANJOSIDÆ

Anoplus Temminck & Schlegel, 1842, 213.   Banjos Bleeker, 1876, 382.

Family 447. LETHRINIDÆ

Schour Forskål, 1775, 34.
Lethrinus Cuvier, 1829, 128.
Pentapus[376] Cuvier & Valenciennes, 1830. 258.
Monotaxis Bennett, 1830, 136.
Leiopsis Bennett, 1830, 174.

Sphœrodon Rüppell, 1830, 184.
Mænioides[377] Richardson, 1843.
Maina Gistel, 1848, 236.
Lethrinella Fowler, 1904, 508.
Lethrinichthys Jordan & Thompson, 1912, 544.

[376] *Pentapus* is transferred (with its synonyms, *Leiopsis* and *Mænoides*) from the *Pomadasidæ* to this group by Regan.
[377] Richardson, "Icones Piscium," 8; logotype *M. aurofrenatus Richardson* = *Sparus vittatus* Cuvier & Valenciennes, a synonym of *Pentapus*. (Omitted in "Genera of Fishes.") (McCulloch.)

Family 448. NEOLETHRINIDÆ

Palate and whole roof of mouth said to be covered with small molar teeth.

Neolethrinus Castelnau, 1875, 378.

Family 449. SPARIDÆ
(Pargos)

The genera related to *Sparus* are very differently distributed by Regan from a study of the skeletons. The present arrangement, mainly according dentition, approximates more nearly the earlie rarrangement of Günther and is at least not unnatural. Regan adds to this group all the Mediterranean genera (*Boops, Scatharus, Oblada,* etc.) here placed in *Girellidæ.* He also adds *Dentex* and its relatives, leaving *Nemipterus* (and its synonyms) with *Scolopsis* and *Heterognathodon* to form a separate family of *Nemipteridæ.*

Sparus Linnæus, 1758, 13.

Cynædus Gronow, 1763, 19.

Aurata Catesby, 1771, 30.

Sargus Klein, 1775, 39.

Cynædus (Gronow) Scopoli, 1777, 42.

Mylio (Commerson) Lacépède, 1802, 70.

Diplodus Rafinesque, 1810, 82.

Sargus Cuvier, 1817, 102.

Pagrus Cuvier, 1817, 103.

Aurata (Cuvier) Oken, 1817, 103.

Labeo Bowdich, 1825, 118.

Aurata Risso, 1826, 120.

Charax Risso, 1826, 173.

Daurada Stark, 1828, 123.

Pagellus Cuvier, 1828, 128.

Datnia[378] Cuvier, 1829, 126.

Chrysophris[379] Cuvier, 1829, 128.

Boridia[380] Cuvier & Valenciennes, 1830, 135.

Sparnodus Agassiz, 1838, 191.

Chrysoblephus Swainson, 1839, 199.

Sphærodus Agassiz, 1839, 194.

Argyrops Swainson, 1839, 199.

Lithognathus Swainson, 1839, 199.

Calamus Swainson, 1839, 199.

Sciænurus Agassiz, 1845, 223.

Radamas Münster, 1846, 229.

Asima Giebel, 1848, 233, 242.

Cæso Gistel, 1848, 235.

Denius Gistel, 1848, 237.

Eudynama Gistel, 1848, 238.

Trigonodon Sismonda, 1849, 244.

Lagodon Holbrook, 1855, 265.

Dichelodus Giebel, 1857, 409.

Pagrichthys Bleeker, 1860, 293.

Lobodus Costa, 1866, 340.

Symphorus Günther, 1872, 365.

Grammateus Poey, 1874, 376.

Sparoides Probst, 1874, 376.

Puntazzo Bleeker, 1876, 383.

Sparosoma Sauvage, 1883, 425.

Pseudosphærodon Noetling, 1885, 434.

Pagrosomus Gill, 1893, 461.

Sparosomus Gill, 1893, 461.

Otrynter Jordan & Evermann, 1896, 472.

Salema J. & E., 1896, 472.

Calopomus (Agassiz) Woodward, (1845) 1901, 224.

Taius Jordan & Thompson, 1912, 544.

Evynnis[381] J. & T., 1912, 544.

Parargyrops[382] Tanaka, 1916, 560.

Stenesthes Jordan, 1917, 562.

Rhythmias Jordan & Gilbert, 1920, 571.

Plectrites Jordan, 1920, 571.

Atkinsonella Jordan, 1920, 572.

---

[378] The original type species of this genus is said to be a *Sparus.*

[379] Also written *Chrysophrys.*

[380] This little genus is placed by Regan with the *Pomadasidæ.*

[381] This genus, otherwise close to *Pagrosomus,* differs from all other *Sparoid* fishes in having a few stout conical teeth on the vomer.

[382] Misprinted *Paragyrops* in "Genera of Fishes."

Family 450. DENTICIDÆ (*Nemipteridæ*)

Regan refers *Dentex* to the *Sparidæ,* leaving most of the other genera to form the *Nemipteridæ.*

Synagris Klein, 1775, 38.

Dentex Cuvier, 1815, 94.

Nemipterus Swainson, 1839, 199.

Synagris Günther, 1859, 291.

Sparopsis[383] Kner, 1868, 352.

Gymnocranius Klunzinger, 1870, 358.

---

[383] *Sparopsis* and *Anemura* are apparently synonymous with *Synagris* as restricted.

Polysteganus Klunzinger, 1870, 358.
Gnathodentex Bleeker, 1873, 367.
Paradentex Bleeker, 1876, 382.
Synagris (Klein) Bleeker, 1876, 383.

Ctenodentex Storms, 1896, 475, 484.
Anemura[383] Fowler, 1904, 507.
Odontoglyphis Fowler, 1904, 507.
Euthyopteroma Fowler, 1904, 507.

Family 451. GIRELLIDÆ

Vegetable-feeders with incisor teeth, mostly movable, and no molars.

Coracinus Gronow, 1763, 19.
Boops Cuvier, 1815, 94.
Cantharus Cuvier, 1817, 103.
Dipterodon Cuvier, 1829, 128.
Oblada Cuvier, 1829, 128.
Crenidens Cuvier & Valenciennes, 1830, 135.
Scatharus Cuv. & Val., 1830, 135.
Box Cuv. & Val., 1830, 135.
Sarpa Bonaparte, 1831, 175.
Girella Gray, 1833, 179.
Exocallus De la Pylaie, 1835, 183.
Melanichthys Temminck & Schlegel, 1844, 223.
Spondyliosoma Cantor, 1849, 241.

Camarina Ayres, 1854, 255.
Doydixodon Valenciennes, 1855, 266.
Gymnocrotaphus Günther, 1859, 291.
Proteracanthus Günther, 1859, 291.
Tephræops Günther, 1859, 291.
Pachymetopon Günther, 1859, 291.
Incisidens Gill, 1862, 316.
Melambaphes Günther, 1863, 326.
Glyphodes Guichenot, 1864, 332.
Tripterodon Playfair, 1866, 343.
Girellichthys Klunzinger, 1872, 366.
Neotephræops Castelnau, 1872, 362.
Aplodon Duméril, 1883, 426.
Dichistius Gill, 1888, 441.
Girellops Regan, 1913, 550.

Family 452. KYPHOSÏDÆ
(Pilot Fishes)

Tahhmel Forskål, 1775, 34.
Kyphosus Lacépède, 1802, 61.
Pimelepterus Lacépède, 1803, 63.
Xyster Lacépède, 1803, 68.
Dorsuarius Lacépède, 1803, 68.

Xysterus Rafinesque, 1815, 90.
Seleima Bowdich, 1825, 119.
Opisthistius Gill, 1862, 316.
Sectator Jordan & Fesler, 1893, 462.

Family 453. HISTIOPTERIDÆ
(Boar Fishes)

Pentaceros Cuvier, 1829, 125.
Histiopterus[384] Temminck & Schlegel, 1843.
Richardsonia Castelnau, 1872, 362.
Pseudopentaceros Bleeker, 1876, 382.
Paristiopterus Bleeker, 1876, 382.
Pentaceropsis Steindachner & Döderlein, 1883, 426.

Prosoplismus Waite, 1903, 506.
Zanclistius Jordan, 1907, 524.
Evistias Jordan, 1907, 524.
Quadrarius Jordan, 1907, 524.
Quinquarius Jordan, 1907, 524.
Gilchristia Jordan, 1907, 524.
Maccullochia Waite, 1910, 537.

---

[384] "Fauna Japonica," 86; orthotype *Histiopterus typus* Temminck & Schlegel. (Omitted in "Genera of Fishes.")

Family 454. ENOPLOSIDÆ

Enoplosus Lacépède, 1803, 64.

Family 455. INERMIIDÆ

This small family has no visible relation to the *Emmelichthyidæ,* to which group the genus *Inermia* has been referred.   The narrow maxillary is devoid of scales and slips under the preorbital. The dorsal fins are separate, the first of many spines.

Dipterygonotus Bleeker, 1849, 240.   Inermia Poey, 1861, 299.

Family 456. MÆNIDÆ (Merolepidæ)

Mænas Klein, 1775, 39.
Spicara Rafinesque, 1810, 80.
Centracanthus[385] Rafinesque, 1810, 80.
Merolepis Rafinesque, 1810, 81.

Smaris Cuvier, 1815, 93.
Mæna Cuvier, 1829, 128.
Mænas (Klein) Bleeker, 1876, 382.

[385] Later spelled Centracantha, originally misspelled Centracantus.

Family 457. EMMELICHTHYIDÆ (Erythrichthyidæ)

This family differs widely from the Mænidæ in the large, scaly uncovered maxillary. The dorsal fin is divided into two.

Emmelichthys Richardson, 1844, 223.
Erythrichthys Temminck & Schlegel, 1845, 227.

Boxaodon Guichenot, 1848, 238.
Plagiogeneion Forbes, 1890, 450.
Erythrocles Jordan, 1919, 567.

Family 458. GERRIDÆ (Xystæmidæ)
(Mojarras; Silver Perch)

Mormyrus Catesby, 1771, 29.
Gerres (Cuvier) Quoy & Gaimard, 1824, 117.
Diapterus Ranzani, 1840, 208.
Podager Gistel, 1848, 236.
Catochænum Cantor, 1849, 241.
Pentaprion Bleeker, 1850, 245.
Eucinostomus Baird, 1857, 274.
Synistius Gill, 1862, 315.

Clara Gill, 1862, 314.
Moharra Poey, 1875, 379.
Gerreomorpha Alleyne & Macleay, 1877, 387.
Parequula Steindachner, 1879, 401.
Chthalmopteryx Ogilby, 1887, 438.
Xystæma Jordan & Evermann, 1895, 468.
Ulæma J. & E., 1895, 468.
Pertica Fowler, 1904, 508.

Family 459. OPLEGNATHIDÆ

Oplegnathus Richardson, 1840, 208.
Scaradon Temminck & Schlegel, 1844, 223.

Scarostoma Kner, 1867, 346.

Family 460. MULLIDÆ
(Surmullets)

Mullus Linnæus, 1758, 14.
Upeneus[386] Cuvier, 1829, 127.
Mulloides Bleeker, 1849, 240.
Upeneoides Bleeker, 1849, 240.
Megalepis Bianconi, 1855, 262.
Upeneichthys Bleeker, 1855, 263.

Acanthomullus Valenciennes, 1861, 309.
Pseudupeneus Bleeker, 1862, 310.
Parupeneus Bleeker, 1863, 322.
Mullhypeneus Poey, 1867, 347.
Brachymullus Bleeker, 1876, 383.
Pseudomulloides Ribeiro, 1915, 557, 560.

[386] Name corrected to Hypeneus by authors.

Family 461. SCIÆNIDÆ

(Croakers; Roncadores)

Vertebræ of the usual number, 10 + 14.

Sciæna Linnæus, 1758, 13.

Alburnus Catesby, 1771, 30.

Cromis Browne, 1789, 47.

Johnius Bloch, 1793, 51.

Lonchiurus[386a] Bloch, 1793, 51.

Pogonias Lacépède, 1802, 61.

Leiostomus[386b] Lacépède, 1803, 64.

Pogonathus Lacépède, 1803, 65.

Chromis (Plumier) Lacépède, 1803, 73.

Coracinus Pallas, 1811, 84.

Equietus Rafinesque, 1815, 89.

Stellifer (Cuvier) Oken, 1817, 104.

Umbrina Cuvier, 1817, 104.

Aplodinotus[386c] Rafinesque, 1819, 110.

Amblodon Rafinesque, 1819, 110.

Bola Hamilton, 1822, 114.

Stelliferus Stark, 1828, 123.

Pachyurus Agassiz, 1829, 132.

Corvina Cuvier, 1829, 128.

Larimus Cuvier & Valenciennes, 1830, 134.

Micropogon Cuv. & Val.,1830, 135.

Lepipterus Cuv. & Val., 1830, 135.

Micropogonias Bonaparte, 1831, 175.

Argyrosomus De la Pylaie, 1835, 184.

Cheilotrema Tschudi, 1846, 230.

Apeches Gistel, 1848, 236.

Melantha Gistel, 1848, 238.

Attilus Gistel, 1848, 238.

Homoprion Holbrook, 1855, 265.

Sciænoides Blyth, 1860, 294.

Collichthys Günther, 1860, 296.

Rhinoscion Gill, 1861, 304.

Plagioscion Gill, 1861, 304.

Anomiolepis Gill, 1861, 304.

Menticirrhus Gill, 1861, 304.

Pachypops Gill, 1861, 304.

Genyonemus Gill, 1861, 304.

Haploidonotus (Rafinesque) Gill, 1861, 304.

Bairdiella Gill, 1861, 302.

Pseudotolithus Bleeker, 1861, 321.

Odontoscion Gill, 1862, 313.

Cirrhimens Gill, 1862, 313.

Amblyscion Gill, 1863, 324.

Ophioscion Gill, 1863, 324.

Diplolepis Steindachner, 1863, 328.

Hemisciæna Bleeker, 1863, 321.

Pseudosciæna Bleeker, 1863, 321.

Sciænops Gill, 1863, 323.

Polycirrhus Bocourt, 1869, 354.

Paralonchurus Bocourt, 1869, 354.

Larimodon (Kaup) Bleeker, 1876, 383.

Eutychelithus Jordan, 1876, 386.

Pareques Gill, 1876, 385.

Roncador Jordan & Gilbert, 1880, 403.

Corvula Jordan & Eigenmann, 1889, 447.

Callaus Jordan & Eigenmann, 1889, 447.

Polyclemus Berg, 1895, 466.

Asperina Ostroumoff, 1896, 473.

Elattarchus Jordan & Evermann, 1896, 472.

Zonoscion J. & E., 1896, 472.

Zaclemus Gilbert, 1896, 473.

Nector Jordan & Evermann, 1898, 480.

Zestis J. & E., 1898, 481.

Zestidium J. & E., 1898, 481.

Stellicarens J. & E., 1898, 481.

Sigmurus Gilbert, 1898, 481.

Cilus Delfin, 1900, 489.

Larimichthys Jordan & Starks, 1905, 514.

Pseudomycterus Ogilby, 1908, 529.

Twingonia Pascoe, 1908, 529.

Othonias Jordan & Thompson, 1911, 539.

Nibea J. & T., 1911, 539.

[386a] Often written *Lonchurus*.

[386b] Often written *Liostomus*.

[386c] Corrected to *Haploidonotus* by Gill.

Family 462. OTOLITHIDÆ

(Weak Fishes; Queen Fishes)

Vertebræ 14 + 10, not 10 + 14 as in *Sciænidæ* and the types allied to *Sciæna, Serranus, Sparus,* and *Lutianus.*

Otolithes[387] (Cuvier) Oken, 1817, 104.
Ancylodon (Cuvier) Oken, 1817, 104.
Macrodon Schinz, 1822, 115.
Nebris Cuvier & Valenciennes, 1830, 134.
Nomalus Gistel, 1848, 234.
Seriphus Ayres, 1854, 255.
Cestreus Gronow, 1854, 258.
Cynoscion Gill, 1861, 304.
Atractoscion Gill, 1862, 313.
Archoscion Gill, 1862, 313.
Apseudobranchus Gill, 1862, 313.

Isopisthus Gill, 1862, 313.
Paraplesichthys (Kaup) Bleeker, 1876, 383.
Sagenichthys Berg, 1895, 466.
Buccone Jordan & Evermann, 1896, 472.
ARCHÆOTOLITHUS Stolley, 1910, 537.
Symphysoglyphus Ribeiro, 1913, 551.
LOMPOQUIA Jordan & Gilbert, 1919, 569.
ARISTOSCION Jordan, 1920, 571.
IOSCION[388] Jordan, 1921.

[387] Later written *Otolithus* by Cuvier.
[388] Jordan, "Fishes of the California Tertiary," 283; orthotype *Ioscion morgani* Jordan.

Family 463. SILLAGINIDÆ

Sillago Cuvier, 1817, 101.
Sillaginodes Gill, 1861, 306.
Sillaginopsis Gill, 1861, 306.

Neosillago Castelnau, 1875, 378.
Iosillago Macleay, 1878, 396.

Family 464. NANDIDÆ

This group may require further subdivision.

Nandus Cuvier & Valenciennes, 1831, 137.
Bedula Gray, 1833, 179.
Pomanotis Guichenot, 1847, 232.
Acharnes Müller & Troschel, 1848, 239.

Pristolepis Jerdon, 1848, 239, 246.
Catopra Bleeker, 1851, 248.
Badis Bleeker, 1853, 253.
Paranandus Day, 1865, 336.

Family 465. POLYCENTRIDÆ

Monocirrhus Heckel, 1840, 207.
Polycentrus Müller & Troschel, 1848, 239.

Polycentropsis Boulenger; 1901, 492.

Family 466. MALACANTHIDÆ

This family and the next four bear strong resemblance to the trachiniform fishes, differing mainly in the thoracic insertion of the ventrals.

Malacanthus Cuvier, 1829, 130

Oceanops Jordan & Seale, 1906, 519.

Family 467. BRANCHIOSTEGIDÆ (*Latilidæ*)

Coryphænoides Lacépède, 1802, 62.
Branchiostegus Rafinesque, 1815, 90.
Latilus Cuvier & Valenciennes, 1830, 135.
Caulolatilus Gill, 1862, 316.

Dekaya Cooper, 1863, 323.
Prolatilus Gill, 1865, 337.
Lopholatilus Goode & Beane, 1879, 400.
Hoplolatilus Günther, 1887, 438.

Family 468. LABRACOGLOSSIDÆ

Platystethus Günther, 1860, 296.
Labracoglossa[389] Peters, 1866, 342.
Cypselichthys Steindachner & Döderlein, 1883, 426.

Evistius Gill, 1893, 461.
Bathystethus Gill, 1893, 461.

[389] This genus is without teeth in the gullet and has no apparent affinity with the *Stromateidæ*.

Family 469. CEPOLIDÆ

Cepola Linnæus, 1764, 22.
Tænia Röse, 1793, 52.

Acanthocepola Bleeker, 1874, 372.

Family 470. HENICICHTHYIDÆ

Henicichthys[390] Tanaka, 1915, 558.

[390] Description in Japanese. The following is the translation as furnished me by Mr. Tanaka: *Hemichthys foraminosus (Henichthyidæ* new family) proposed in Japanese in *Zool. Mag., 27*: No. 325, Nov. 15, 1915, p. 568. Head 2 4-5 in length (exclusive of caudal), depth 5 3-8. Eye lateral; mouth large, somewhat oblique; maxillary extending behind posterior border of eye; teeth in jaws acute, arranged in a row; vomer and palatines toothed. Two dorsals somewhat close together. Dorsal, VI, 11; anal, 11; pectoral 13; ventral I, 5. Ventral fins close together, beneath pectoral; caudal nearly truncated. Scaleless; body and head with many pores which are arranged longitudinally and transversely. Color pale, plain. Length (exclusive of caudal) 5-5 cm. Nagasaki. The species seems to be a trachinoid fish and also to be included in a new family.

Series TRICHODONTIFORMES

Family 471. TRICHODONTIDÆ

The trachinoid fishes (*Jugulares*) follow in a natural sequence, a fact not to be shown in a linear series.

Trichodon (Steller) Cuvier, 1829, 127.
Chætichthys Gistel, 1848, 238.

Arctoscopus Jordan & Evermann, 1896, 474.

Suborder CIRRHITOIDEI

Family 472. CIRRHITIDÆ

The fishes of this group bear some resemblance to the *Cataphracti*, which may be an indication of common origin.

Cirrhitus[390a] Lacépède, 1803, 65.
Cirrhitichthys[390b] Bleeker, 1856, 274.
Oxycirrhites Bleeker, 1857, 274.
Amblycirrhitus Gill, 1862, 313.
Cirrhitopsis Gill, 1862, 313.
Lacepedia Castelnau, 1873, 367.
Neocirrhites Castelnau, 1873, 368.
Paracirrhites Bleeker, 1875, 378.

Zeodrius Castelnau, 1878, 393.
Paracirrhites Steindachner & Döderlein, 1883, 426.
Psilocranium Macleay, 1884, 429.
Dactylophora De Vis, 1884, 427.
Cirrhitoidea Jenkins, 1903, 504.
Isobuna Jordan, 1907, 525.
Cyprinocirrhites Tanaka, 1917, 563.

[390a] Written *Cirrhites* by authors.
[390b] This name appears first in *Ned. Tydschr. Dierk., 10*: 474, 1856.

Family 473. CHIRONEMIDÆ

Chironemus Cuvier, 1828, 126.
Threpterius Richardson, 1850, 247.

Sciænoides (Solander) Richardson, 1850, 247.

Family 474. CHEILODACTYLIDÆ

Cheilodactylus Lacépède, 1803, 65.
Nemadactylus[391] Richardson, 1839, 197.
Pteronemus Van der Hoeven, 1849, 243, 383.
Trichopterus Gronow, 1854, 259.

Chirodactylus Gill, 1862, 313.
Acantholatris Gill, 1862, 313.
Dactylopagrus Gill, 1862, 313.
Dactylosparus Gill, 1862, 314.

---

[391] Spelled also *Nematodactylus.*

Family 475. APLODACTYLIDÆ

Aplodactylus[392] Cuvier & Valenciennes, 1831, 137.
Crinodus Gill, 1862, 313.

Dactylosargus Gill, 1862, 313.
Goniistius Gill, 1862, 314.
Parhaplodactylus Thominot, 1883, 426.

---

[392] Often written *Haplodactylus.*

Family 476. LATRIDÆ

Latris Richardson, 1839, 197.
Latridopsis Gill, 1862, 314.

Mendosoma Guichenot, 1848, 238.

Series *GADOPSIFORMES*

Family 477. GADOPSIDÆ

Gadopsis Richardson, 1844, 223.

Series *CAPRIFORMES*

A small group of uncertain relations, bearing a strong resemblance to the *Zeidæ* in form and having, like the *Zeidæ* and the *Chætodontidæ,* the post-temporal solidly joined to the skull. But the ventral rays are always I, 5, as in the *Chætodonts,* and the gill membranes free from the isthmus, as in *Zeus.* A comparative study of the skeletons of these transitional groups, *Zeus, Capros, Toxotes, Scorpis, Ephippus, Platax, Drepane,* and *Amphistium,* is much to be desired.

Family 478. CAPROIDÆ

Capros Lacépède, 1803, 65.
Caprophonus Müller & Troschel, 1849, 243.

Metoponichthys Kramberger, 1882, 420.
Proantigonia Kramberger, 1882, 420.

Family 479. ANTIGONIIDÆ

Antigonia Lowe, 1843, 215.
Hypsinotus Temminck & Schlegel, 1844, 223.

Acanthagonia Ogilby, 1918, 564.

Family 480. AMPHISTIIDÆ

A group of uncertain relations, apparently allied to the *Ephippidæ,* regarded by Boulenger as an ally of *Zeus* and perhaps ancestral to the

Heterosomata. As Regan finds the ventrals I, 5, we have separated it from the zeoid fishes.

AMPHISTIUM Agassiz, 1835, 181, 218.   MACROSTOMA[393] Agassiz, 1838, 191.

[393] A genus of doubtful affinity, resembling a chætodont, the name preoccupied.

Series *EPHIPPIFORMES*

This group has many characters in common with the perciform fishes, especially in having the post-temporal not solidly united to the skull. It agrees, however, with the *Squamipennes* in many regards, notably the restriction of the gill openings to the sides; the compressed body and scaly fins mark a transition from the *Scorpidæ* and related forms to the *Chætodontidæ*.

Family 481. SCORPIDÆ
(Half-moon Fishes)

Scorpis Cuvier & Valenciennes, 1831, 573.
Atypus Günther, 1860, 296.
Atypichthys Günther, 1862, 575.
Helotosoma Kaup, 1863, 327.
Parapsettus Steindachner, 1875, 381.
Parascorpis Bleeker, 1875, 377.
Cæsiosoma Kaup, 1876, 383.

Agenor Castelnau, 1878, 393.
Diodyxodon Thominot, 1881, 417.
Tilodon Thominot, 1881, 417.
Neptotichthys Hutton, 1890, 451.
Medialuna Jordan & Fesler, 1893, 462.
Atyposoma Boulenger, 1899, 485.
Neatypus Waite, 1905, 516.

Family 482. MONODACTYLIDÆ

A group of uncertain affinity, seeming close to the *Xyphosidæ* on the one hand and to the chætodontoid forms on the other. I place them provisionally with the *Ephippidæ*, but it may be that the genera will require to be rearranged. From the allies of *Chætodon*, the *Scorpidæ, Monodactylidæ, Platacidæ, Drepanidæ,* and *Ephippidæ* differ in not having the post-temporal coössified with the skull.

Monodactylus Lacépède, 1802, 61.
Centropodus Lacépède, 1802, 62.
Psettus (Commerson) Lacépède, 1802, 70.
Acanthopodus Lacépède, 1803, 64.

Psettus (Commerson) Cuvier, 1829, 128.
Schuettea Steindachner, 1866, 343.
Psettias Jordan, 1906, 519.

Family 483. PLATACIDÆ
Platax Cuvier, 1817, 105.

Family 484. EPHIPPIDÆ

Chætodipterus Lacépède, 1803, 64.
Ephippus Cuvier, 1817, 105.
Ilarches Cantor, 1849, 241.

Parephippus Gill, 1861, 305.
EPHIPPITES Von Ammon, 1901, 492.
Bramichthys Waite, 1905, 516.

Family 485. DREPANIDÆ

Drepane Cuvier & Valenciennes, 1831, 136, 175.
Drepanichthys[394] Bonaparte, 1831.

Enixe Gistel, 1848, 235.
Harpochirus Cantor, 1849, 241.
Cryptosmilia Cope, 1867, 344.

[394] Bonaparte, "Saggio," etc. A substitute for *Drepane*, as are also *Enixe* and *Harpochirus, Drepane* being regarded as preoccupied by the earlier names, *Drepana, Drepania, Drepanus, Drepanis,* and *Drepanius. Drepanichthys* is quoted by Agassiz, "Nomenclator," 23, 1843. (Overlooked in "Genera of Fishes.")

Suborder SQUAMIPENNES
(Epelasmia)

Post-temporal firmly attached to the skull, the interspace between the three forks filled in by bone; gill openings restricted to the sides; soft fins largely scaly.

Series TOXOTIFORMES
Family 486. TOXOTIDÆ

This family, of uncertain relationship, agrees with the *Chætodontidæ*, according to Günther in the union of the post-temporal with the skull.

Toxotes Cuvier, 1817, 105.   Amblytoxotes Bleeker, 1876, 383.
Trompe Gistel, 1848, 237.

Series CHAETODONTIFORMES
Family 487. SCATOPHAGIDÆ

In this family the post-temporal is solidly united to the skull as in *Chætodontidæ*.

Scatophagus Cuvier & Valenciennes, 1831, 136.   Prenes Gistel, 1848, 237.
Cacodoxus Cantor, 1849, 241.

Family 488. CHÆTODONTIDÆ
(Butterfly Fishes; Coral Fishes)

Chætodon Linnæus, 1758, 13.
Acarauna Catesby, 1771, 32.
Tetragonoptrus Klein, 1775, 40.
Holacanthus Lacépède, 1803, 64.
Pomacanthus Lacépède, 1803, 64.
Chelmon[395] Cuvier, 1817, 105.
Heniochus Cuvier, 1817, 105.
Taurichthys Cuvier, 1829, 128.
Rabdophorus Swainson, 1839, 198.
Microcanthus[396] Swainson, 1839, 198.
Genicanthus Swainson, 1839, 198.
Megaprotodon Guichenot, 1848, 238.
Diphreutes Cantor, 1849, 241.
Centropyge Kaup, 1860, 298.
Linophora Kaup, 1860, 297.
Eteira Kaup, 1860, 297.
Citharœdus Kaup, 1860, 297.
Coradion Kaup, 1860, 297.
Therapaina Kaup, 1860, 298.
Sarothrodus Gill, 1861, 304, 316.
Prognathodes Gill, 1862, 316.

Pomacanthodes Gill, 1862, 316.
Tholichthys Günther, 1868, 351.
Tetragonoptrus (Klein) Bleeker, 1868, 348, 383.
Neochætodon Castelnau, 1873, 368.
Parachætodon Bleeker, 1874, 372.
Oxychætodon Bleeker, 1876, 383.
Chætodontops Bleeker, 1876, 383.
Lepidochætodon Bleeker, 1876, 383.
Acanthochætodon Bleeker, 1876, 383.
Chætodontoplus Bleeker, 1876, 383.
Gonochætodon Bleeker, 1876, 383.
Hemitaurichthys Bleeker, 1876, 383.
Chelmonops Bleeker, 1876, 383.
Hemichætodon Bleeker, 1876, 384.
Anisochætodon Klunzinger, 1884, 429.
Angelichthys Jordan & Evermann, 1896, 473.
Forcipiger Jordan & McGregor, 1898, 481, 482.
Osteochromis Franz, 1910, 535.

---
[395] Later written *Chelmo*.
[396] A misprint for *Micracanthus*.

Vinculum McCulloch, 1914, 554.   Tifia[398] Jordan, 1923.
Loa[397] Jordan, 1921.   Xiphypops[399] Jordan, 1923.

[397] Jordan, *Proc. U. S. Nat. Mus.*, 653, 1921; type *Loa excelsa* Jordan.
[398] "Fishes of Hawaii," *Mem. Carnegie Museum,* 1923, 60; orthotype *Chætodon corallicola* Snyder.
[399] *Loc. cit.,* 64; orthotype *Holacanthus fisheri* Snyder.

Series *ZANCLIFORMES*
Family 489. ZANCLIDÆ

Zanclus (Commerson) Lacépède, 1803, 70.   Gonopterus Gronow, 1854, 258.
Zanclus (Commerson) Cuvier & Valen-   Gnathocentrum Guichenot, 1866, 341.
ciennes, 1831, 136.

Series *ACANTHURIFORMES*
Family 490. ACANTHURIDÆ (*Teuthididæ; Hepatidæ*)

Hepatus[400] Gronow, 1763, 20.
Teuthis[401] Linnæus, 1766, 23.
Rhombotides Klein, 1775, 38.
Acanthurus Forskål, 1775, 33.
Harpurus Forster, 1778, 43.
Teuthis Browne, 1789, 48.
Monoceros Bloch & Schneider, 1801, 58.
Naso Lacépède, 1802, 61.
Naseus (Commerson) Lacépède, 1802, 69.
Aspisurus Lacépède, 1803, 64.
Prionurus Lacépède, 1804, 169.
Nasonus Rafinesque, 1815, 88.
CHIRURGUS Blainville, 1818, 108.
Priodon Quoy & Gaimard, 1824, 118.
Axinurus Cuvier, 1829, 129.
Priodontichthys Bonaparte, 1831, 175.
Scopas Bonaparte, 1831, 175.
Ctenodon Bonaparte, 1833, 175.

Keris[402] Cuvier & Valenciennes, 1835, 181.
Zebrasoma Swainson, 1839, 201.
Callicanthus Swainson, 1839, 201.
Ctenodon Swainson, 1839, 201.
Acronurus Gronow, 1854, 259.
Acronurus (Gronow) Günther, 1861, 307.
Rhombotides (Klein) Bleeker, 1863, 322.
Scopas[403] (Bonaparte) Kner, 1865.
CALAMOSTOMA Steindachner, 1863, 328.
Ctenodon (Bonaparte) Klunzinger, 1884, 429.
Ctenochætus Gill, 1884, 428.
Colocopus Gill, 1884, 428.
APOSTASIS Kramberger, 1891, 455.
Xesurus Jordan & Evermann, 1896, 473.
Acanthocaulus Waite, 1900, 491.
Hepatus (Gronow) Jordan & Seale, 1906.
Læphichthys Ogilby, 1916, 560.

[400] If Gronow's names (1763) are accepted *Hepatus* would replace *Acanthurus.* The original *Teuthis* of Linnæus, 1768, was exactly equivalent to the non-binomial *Hepatus* of Gronow. The generic distinction between the two species, *hepatus* and *javus,* was recognized by Forskål, 1775, who named the type of the former *Acanthurus,* the other *Siganus.* Cuvier & Valenciennes used the names *Acanthurus* and *Amphacanthus.* Neither of these writers adopted *Teuthis* as a generic name. The first restriction of *Teuthis* was that of Cantor, who used it for the *Siganus* group, although by good rights he should have taken it instead of *Acanthurus.* For the present we follow Cantor and Günther.
[401] This name, commonly used as a synonym of *Siganus,* has been often misspelled as *Teuthys, Theutis,* etc.
[402] Corrected to *Ceris.*
[403] "Novara Fische," 212; type *Acanthurus scopas* Cuvier & Valenciennes. (Omitted in "Genera of Fishes.")

Family 491. PYGÆIDÆ

Pygæus Agassiz, 1838, 191.   Parapygæus Pellegrin, 1907, 526.

NOTE.—The *Plectognathi* follow naturally as next in this descent, a fact which can not be shown in a linear series.

Suborder *AMPHACANTHI*

Family 492. TEUTHIDÆ (*Siganidæ*)

Teuthis[404] Linnæus, 1766, 23 (in part).   Buro Lacépède, 1803, 67.
Siganus Forskål, 1775, 32.   Buronus Rafinesque, 1815, 90.
Centrogaster Houttuyn, 1782, 44.   Archæoteuthis[405] Wettstein, 1886, 436.
Amphacanthus Bloch and Schneider,   Lo Seale, 1906, 519.
1801, 59.

[404] The proper application of the name *Teuthis* is not yet definitely settled. Its first restriction, that of Cantor, 1849, makes it identical with *Siganus,* but the first use of the name (pre-Linnæus) was for a species of *Acanthurus.*

[405] Misspelled *Archæteuthis* in "Genera of Fishes."

Order CATAPHRACTI

(*Loricati; Pareioplitæ; Sclerogeni*)

Series *SCORPAENIFORMES*

Family 493. SCORPÆNIDÆ

(Scorpion Fishes; Rock Cod)

Scorpæna Linnæus, 1758, 13.   Pteroleptus Swainson, 1839, 201.
Pseudopterus Klein, 1775, 40.   Pteropterus Swainson, 1839, 201.
Scorpius (Plumier) Lacépède, 1802, 72.   Brachyrus[408] Swainson, 1839, 201.
Tænianotus Lacépède, 1803, 63.   Platypterus Swainson, 1839, 201.
Panotus Rafinesque, 1815, 89.   Pterichthys Swainson, 1839, 201.
Pterois (Cuvier) Oken, 1817, 104.   Gymnapistes Swainson, 1839, 201.
Ampheristus König, 1825, 118.   Goniognathus Agassiz, 1844, 218.
Apistes[406] Cuvier, 1829, 127.   Hypodytes Gistel, 1848, 234.
Sebastes Cuvier, 1829, 127.   Corythobatus Cantor, 1849, 241.
Scorpænopsis Heckel, 1837, 189.   Prosopodasys Cantor, 1849, 241.
Pteropterus Swainson, 1839, 198.   Scorpænichthys Girard, 1854, 257.
Trichosomus Swainson, 1839, 197.   Pteroidichthys Bleeker, 1856, 267.
Trichophasia Swainson, 1839, 197.   Scorpænichthys Bleeker, 1856, 267.
Brachirus Swainson, 1839, 201.   Ctenopoma Heckel, 1856, 270.
Dendrochirus[407] Swainson, 1839, 198.   Scorpænodes Bleeker, 1857, 275.
Macrochyrus Swainson, 1839, 201.   Polemius Kaup, 1858, 282.

[406] Spelled also *Apistus.*

[407] A synonym of *Brachirus,* used for the same group on page 71, *loc. cit., Dentrochirus* occurring on page 180, and *Brachyrus* on page 264.

[408] Spelled *Brachirus* by Swainson on page 71, *loc. cit., Brachyrus* on page 264; replaced by *Dendrochirus* on page 180, while the name *Brachirus* is on page 303, taken for a genus of the soles, *Synaptura* Cantor, of which genus the proper type is *Synaptura commersoniana.*

Cocotropus Kaup, 1858, 282.
SCORPÆNOPTERUS Steindachner, 1859, 293.
Tetraroge Günther, 1860, 296.
Glyptauchen Günther, 1860, 296.
Pentaroge Günther, 1860, 296.
Centropogon Günther, 1860, 296.
Pontinus Poey, 1860, 299.
Sebastodes Gill, 1861, 305.
Sebastichthys Gill, 1862, 317.
Sebastopsis Gill, 1862, 316.
Setarches Johnson, 1863, 319.
Sebastoplus Gill, 1863, 324.
Sebastosomus Gill, 1864, 330.
Sebastomus Gill, 1864, 330.
Neosebastes Guichenot, 1867, 345.
Sebastopsis Sauvage, 1873, 371.
Pseudopterus (Klein) Bleeker, 1876, 384.
Paracentropogon Bleeker, 1876, 384.
Parascorpæna Bleeker, 1876, 384.
Parapterois Bleeker, 1876, 384.
Pteroidichthys Bleeker, 1876, 384.
Amblyapistus Bleeker, 1876, 384.
Cottapistus Bleeker, 1876, 384.
Pseudosebastes Sauvage, 1878, 397.
Eusebastes Sauvage, 1878, 397.
Sebastapistes Gill, 1878, 397.
Lioscorpius Günther, 1880, 403.
Sebastolobus Gill, 1881, 415.
Auctospina Eigenmann & Beeson, 1893, 460.
Primospina E. & B., 1893, 460.
Acutomentum E. & B., 1893, 460.
Pteropodus E. & B., 1893, 460.
Helicolenus Goode & Bean, 1895, 468.

Hispanicus Jordan & Evermann, 1896, 474.
Eosebastes J. & E., 1896, 474.
Rosicola J. &. E., 1896, 474.
Emmelas J. & E., 1898, 481.
SCORPÆNOIDES Priem, 1899, 487.
Ectreposebastes Garman, 1899, 486.
Iracundus Jordan & Evermann, 1903, 504.
Merinthe Snyder, 1904, 511.
Erisphex Jordan & Starks, 1904, 509.
Decterias J. & S., 1904, 509.
Ebosia J. & S., 1904, 509.
Lythrichthys J. & S., 1904, 509.
Sebastiscus J. & S., 1904, 509.
Thysanichthys J. & S., 1904, 509.
Daia Ogilby, 1904, 510.
Notesthes Ogilby, 1904, 510.
Liocranium Ogilby, 1904, 510.
Plectrogenium Gilbert, 1905, 513.
Sebastosemus Gill, 1905, 513.
Hypomacrus Evermann & Seale, 1907, 522.
Pogonoscorpius Regan, 1908, 530.
Vespicula[409] Jordan & Richardson, 1910.
Apistops Ogilby, 1911, 540.
Kanekonia Tanaka, 1915, 558.
Sebastella Tanaka, 1918, 565.
SEBASTAVUS Jordan & Gilbert, 1919, 569.
RHOMARCHUS Jordan & Gilbert, 1919, 569.
RIXATOR Jordan & Gilbert, 1920, 571.
SEBASTOËSSUS Jordan, 1920, 571.
SEBASTINUS Jordan, 1920, 571.
Metzelaaria[410] Jordan, 1923.

[409] Check List, "Fishes Philippines," 52; orthotype *Prosopodasys gogorzæ* Jordan & Seale. (Omitted in "Genera of Fishes.")

[410] New genus; type *Scorpæna tridecimspinosa* Metzelaar from Surinam; allied to *Auctospina* and *Pterodus,* with thirteen dorsal spines, but with nasal and maxillary tentacles.

Family 494. CONGIOPODIDÆ (*Agriopidæ*)

Congiopus[411] Perry, 1811, 170.
Agriopus Cuvier, 1829, 127.
Cephalinus Gronow, 1854, 259.

Zanclorhynchus Günther, 1880, 403.
Snyderina Jordan & Starks, 1901, 495.
Ocosia Jordan & Starks, 1904, 509.

[411] Written *Congiopodus* by some writers.

Family 495. APLOACTIDÆ

Aploactis Temminck & Schlegel, 1843, 216.
Sthenopus Richardson, 1848, 240.

Trichopleura Kaup, 1858, 282.
Paraploactis Bleeker, 1865, 335, 384.
Aploactisoma Castelnau, 1872, 363.

Family 496. SYNANCEJIDÆ

Synanceja[412] Bloch & Schneider, 1803, 58.
Minous Cuvier & Valenciennes, 1829, 133.
Pelor Cuvier, 1829, 127.
Spurco (Commerson) Cuvier & Valenciennes, 1829, 174.
Synanchia[413] Swainson, 1839, 201.
Erosa Swainson, 1839, 197.
Bufichthys Swainson, 1839, 201.
Trachicephalus Swainson, 1839, 201.
Synancidium Müller, 1843, 216.
Choridactylus[414] Richardson, 1848, 240.
Glyptauchen[415] Günther, 1860.
Polycaulus Günther, 1860, 296.
Uranoblepus Gill, 1861, 302.
Synanceichthys[416] Bleeker, 1863.

Leptosynanceia Bleeker, 1874, 374.
Synanchia (Swainson) Bleeker, 1874, 374.
Pseudosynanceia Day, 1875, 379.
Emmydrichthys Jordan & Rutter, 1896, 472.
Choridactylodes Gilchrist, 1902, 498.
Inimicus Jordan & Starks, 1904, 509.
Erosa (Swainson) Jordan & Starks, 1904, 509.
Simopias Gill, 1905, 513.
Rhinopias Gill, 1905, 513.
Peloropsis Gilbert, 1905, 513.
Deleastes Seale, 1906, 520.
Lysodermus Smith & Pope, 1906, 520.

[412] Usually written *Synanceia;* spelled by Swainson *Synanchia.*
[413] A variant of *Synanceja.*
[414] Name later written *Chorismodactylus.*
[415] "Cat. Fishes," 2: 121; orthotype *Apistes panduratus* Richardson. (Omitted in "Genera of Fishes.")
[416] "Ternate," 11: 324; type *Synanceja verrucosa* Bloch & Schneider. (Omitted in "Genera of Fishes.")

Family 497. CARACANTHIDÆ

Micropus Gray, 1831, 139.
Caracanthus Kröyer, 1845, 227.
Amphiprionichthys Bleeker, 1855, 263.

Crossoderma Guichenot, 1869, 354.
Trachycephalus De Vis, 1884, 427.

Family 498. PATÆCIDÆ

Patæcus Richardson, 1844, 222.

Neopatæcus Steindachner, 1884, 430.

Family 498a. GNATHANACANTHIDÆ
(Velvet Fishes)

Gnathanacanthus Bleeker, 1855, 263.
Holoxenus Günther, 1876, 385.

Beridia Castelnau, 1877, 388.

Series *HEXAGRAMMIFORMES*
Family 499. ANOPLOPOMIDÆ
(Skil Fishes; Sable Fishes)

Anoplopoma Ayres, 1858, 277.
Scombrocottus Peters, 1872, 366.

Eoscorpius Jordan & J. Z. Gilbert, 1919, 569.

Family 500. ERILEPIDÆ
(Fat-Priests)

Myriolepis Lockington, 1880, 404.
Erilepis Gill, 1893, 461.

Ebisus Jordan & Snyder, 1901, 494.

Family 501. HEXAGRAMMIDÆ
(Greenlings)

Hexagrammos[416a] (Steller) Tilesius, 1809, 76.
Labrax Pallas, 1810, 77.
Lebius (Steller) Pallas, 1811, 84.
Chirus (Steller) Pallas, 1811, 84.
Chiropsis Girard, 1857, 275, 410.
Agrammus Günther, 1860, 296.

Grammatopleurus Gill, 1861, 305.
Pleurogrammus Gill, 1861, 305.
Acantholebius Gill, 1861, 305.
Octogrammus Bleeker, 1874, 372.
Hexagrammoides Gracianov, 1907, 524.
ACHRESTOGRAMMUS[417] Jordan, 1921.
ZEMIAGRAMMUS[418] Jordan, 1921.

[416a] Usually corrected to Hexagrammus.
[417] "Fossil Fishes Cal. Tertiary," 289; orthotype Hexagrammos achrestus Jordan & Gilbert.
[418] Loc. cit., 289; orthotype Zemiagrammus isistius Jordan.

Family 502. OPHIODONTIDÆ
(Cultus Cods)

Ophiodon Girard, 1854, 257.　　　　Oplopoma Girard, 1856, 269.

Family 503. OXYLEBIIDÆ

Oxylebius Gill, 1862, 316.

Family 504. ZANIOLEPIDÆ

Zaniolepis Girard, 1857, 275.　　　Xantocles Jordan, 1917, 562.

Series PLATYCEPHALIFORMES
Family 505. HOPLICHTHYIDÆ

Hoplichthys[419] Cuvier & Valenciennes, 1829, 133.

[419] Also spelled Oplichthys.

Family 506. PLATYCEPHALIDÆ
(Flat-heads)

Platycephalus Bloch, 1795, 53.
Calliomorus Lacépède, 1800, 56.
Centranodon Lacépède, 1803, 66.
Bembras Cuvier & Valenciennes, 1829, 133.
Amora[420] Gray, 1833, 179.
Neoplatycephalus Casťelnau, 1872, 362.
Parabembras Bleeker, 1874, 372.
Thysanophrys Ogilby, 1898, 483.

Insidiator Jordan & Snyder, 1900, 490.
Grammoplites Fowler, 1904, 508.
Bembradium Gilbert, 1905, 513.
Elates Jordan & Seale, 1907, 525.
Rogadius Jordan & Richardson, 1908, 528.
Bambradon J. & R., 1908, 528.
Hyalorhynchus Qgilby, 1910, 536.
Inegocia Jordan & Thompson, 1913, 549.
Onigocia J. & T., 1913, 549.

[420] Name corrected to Anaora on a plate examined.

Series COTTIFORMES
Family 507. JORDANIIDÆ

Ventral rays I, 5; head narrow; scales present, often reduced each to a few radiating spinules.

"The scales of *Jordania zonope* treated with hot caustic potash are nothing but rows of strong ctenoid spines placed as they would be in true scales. In the dorsal region the rows are curved as they would be were they margins of ctenoid scales. In *Lepidocottus brevis* (Agassiz) the ctenoid elements are as in *Jordania*, but complete scales are present with circuli and basal radii as usual. It must be supposed that *Jordania* came from such an ancestor and represents the survival of certain elements of scale structure without the scales." (Cockerell, *Science*, Dec. 19, 1919, 568.)

LEPIDOCOTTUS Sauvage, 1875, 380.
Paricelinus Eigenmann & Eigenmann, 1889, 445.

Jordania Starks, 1895, 469.
Alcidea Jordan & Evermann, 1898, 481.
EOCOTTUS Woodward, 1901, 496.

Family 508. ICELIDÆ

Hemilepidotus Cuvier, 1829, 127.
Triglops Reinhardt, 1832, 175.
Temnistia Richardson, 1836, 136.
Enophrys Swainson, 1839, 202.
Icelus Kröyer, 1845, 226.
Aspicottus Girard, 1854, 257.
Clypeocottus Ayres, 1854, 257.
Calycilepidotus Ayres, 1855, 282.
Artedius Girard, 1856, 269.
Ceratocottus Gill, 1859, 290.
Melletes Bean, 1879, 398.
PARAPERCA Sauvage, 1875, 380.
Chitonotus Lockington, 1881, 416.
Prionistius Bean, 1883, 421.
Icelinus Jordan, 1885, 433.
Artediellus Jordan, 1885, 433.
Radulinus Gilbert, 1890, 451.
Ruscarius Jordan & Starks, 1895, 469.
Astrolytes J. & S., 1895, 469.
Rastrinus Jordan & Evermann, 1896, 474.

Axyrias Starks, 1896, 475.
Tarandichthys Jordan & Evermann, 1896, 472.
Sternias[421] J. & E., 1898.
Stelgistrum Jordan & Gilbert, 1898, 482.
Archistes J. & G., 1898, 482.
Stlengis Jordan & Starks, 1904, 509.
Schmidtia J. & S., 1904, 509.
Daruma J. & S., 1904, 509.
Ricuzenius J. &. S., 1904, 509.
Schmidtina J. & S., 1904, 510.
Stelgidonotus Jordan & Thompson, 1905, 513.
Taurulus Gracianov, 1907, 524.
Agonocottus Pavlenko, 1910, 537.
Orthopnias Starks & Mann, 1911, 541.
Archaulus Gilbert & Burke, 1912, 543.
Thyriscus Gilbert & Burke, 1912, 543.
Pterygiocottus[422] Bean & Weed, 1920.

---

[421] "Fishes of North and Middle America," II, 1926; orthotype *Triglops xenostethus* Gilbert. (Omitted in "Genera of Fishes.")

[422] "Notes on a Collection of Fishes from Vancouver Island, British Columbia," *Trans. Roy. Soc. Canada*, 13: 73; orthotype *P. macouni* Bean & Weed; an ally of *Artedius*.

Family 509. BLEPSIIDÆ

Skin covered with velvety prickles, form compressed.

Blepsias Cuvier, 1829, 127.
Peropus Lay & Bennett, 1839, 125.
Nautichthys Girard, 1859, 290.

Histiocottus Gill, 1888, 441.
Nautiscus Jordan & Evermann, 1898, 481.

Family 510. SCORPÆNICHTHYIDÆ

(Cabezones)

Ventral rays I, 5; head broad; skin smooth.

Scorpænichthys Girard, 1854, 257.

Family 511. **COTTIDÆ**
(Sculpins)

Ventral rays I, 2, to I,'4; skin naked or with imbedded plates or prickles; general form more or less depressed.

Even after the elimination of several aberrant genera, *Jordania, Scorpænichthys, Ascelichthys, Synchirus, Psychrolutes, Blepsias, Abyssocottus, Hemitripterus, Ereunias, Cottocomephorus, Comephorus, Rhamphocottus,* etc., the *Cottidæ* are still extremely diversified and may, after study of the skeletons, require further subdivision into families or subfamilies. For purpose of laying stress on distinctions, we further separate certain more or less natural groups, the form centering about *Icelus* being apparently derived from the ancestors of *Jordania,* while *Cottus, Myoxocephalus* and their varied associates may have sprung from ancestors of *Scorpænichthys.* Both *Jordania* and *Scorpænichthys* have normal ventrals (I, 5), but little else in common. Most of the allies of *Icelus* and *Hemilepidotus* have developed scales, while none of the *Cottidæ* as here restricted have true scales other than prickles or imbedded dermal plates. But in a general treatment of the group, such a minute subdivision of families may be found superfluous.

Cottus Linnæus, 1758, 12.
Uranoscopus Gronow, 1763, 19.
Myoxocephalus (Steller) Tilesius, 1811, 83.
Pegedictis[423] Rafinesque, 1820, 113.
Trachydermus Heckel, 1837, 189.
Gymnocanthus[424] Swainson, 1839, 202.
Uranidea Dekay, 1842, 210.
Centridermichthys Richardson, 1844, 222.
Phobetor Kröyer, 1845, 226.
Scorpichthes Bonaparte, 1845, 226.
Podabrus Richardson, 1848, 240.
Aphobus Gistel, 1848, 237.
Pseudoblennius Temminck & Schlegel, 1850, 247.
Pseudoclinus T. & S., 1850, 247.
Cottopsis Girard, 1851, 249.

Acanthocottus Girard, 1851, 249.
Leptocottus Girard, 1854, 257.
Triglopsis Girard, 1854, 257.
Leiocottus Girard, 1856, 269.
Oligocottus Girard, 1856, 269.
Hoplocottus Kaup, 1858, 282.
Boreocottus Gill, 1859, 290.
Porocottus Gill, 1859, 290.
Ptyonotus Günther, 1860, 296.
Potamocottus Gill, 1861, 307.
Oncocottus Gill, 1861, 302, 313.
Clinocottus Gill, 1861, 305.
Blennicottus Gill, 1861, 305.
Megalocottus Gill, 1861, 305.
Bunocottus[425] Kner, 1868, 352.
Cottunculus Collett, 1875, 379.
Elaphocottus Sauvage, 1878, 397.
Tauridea Jordan & Rice, 1878, 395.

---

[423] Description confused with that of *Catonotus.*

[424] Corrected usually to *Gymnacanthus.*

[425] This genus, apparently cottoid, is described by Kner as lacking ventral fins, although these structures are plainly shown in his excellent figure. If the description is correct, the genus should stand as a separate family. The fish, *Bunocottus apus,* is said to be from deep water, off Cape Horn. But no other cottoid is known south of the Tropic of Cancer. If Kner's account is correct *Bunocottus* may be an aberrant scorpænoid.

Malacocottus Bean, 1890, 450.
Dasycottus Bean, 1890, 450.
Argyrocottus Herzenstein, 1892, 458.
Zesticelus Jordan & Evermann, 1896, 474.
Oxycottus J. & E., 1898, 481.
Sigmistes Rutter, 1898, 482.
Dialarchus Greeley, 1901, 493.
Rusciculus Greeley, 1901, 493.
Eximia Greeley, 1901, 493.
Thecopterus H. M. Smith, 1904, 511.
Cottiusculus Schmidt, 1904, 509, 511.
Rheopresbe Jordan & Starks, 1904, 509.
Ainocottus J. & S., 1904, 509.
Crossias J. & S., 1904, 509.

Elaphichthys J. & S., 1904, 509.
Alcichthys J. & S., 1904, 509.
Furcina J. & S., 1904, 509.
Ocynectes J. & S., 1904, 509.
Bero J. & S., 1904, 509.
Vellitor J. & S., 1904, 509.
Cephalocottus Gracianov, 1907, 524.
Mesocottus Gracianov, 1907, 524.
Rusulus Starks & Mann, 1911, 541.
Trichocottus Soldatov & Pavlenko, 1915, 557.
Taurocottus S. & P., 1915, 557.
HAYIA Jordan & Gilbert, 1919, 569.
Greeleya Jordan, 1920, 571.

Family 512. ABYSSOCOTTIDÆ

Degenerate sculpins found in the depths of Lake Baikal.

Procottus Gracianov, 1902, 498.
Batrachocottus Berg, 1903, 501.
Abyssocottus Berg, 1906, 516.

Asprocottus Berg, 1906, 516.
Linnocottus Berg, 1906, 516.
Cottinella Berg, 1907, 521.

Family 513. ASCELICHTHYIDÆ

Ventral fins wanting; head broad; skins smooth.

Ascelichthys Jordan & Gilbert, 1880, 404.

Family 514. PSYCHROLUTIDÆ

Psychrolutes Günther, 1861, 307.
Gilbertina Jordan & Starks, 1895, 469.

Gilbertidia Berg, 1899, 484.
Eurymen Gilbert & Burke, 1912, 543.

Family 515. NEOPHRYNICHTHYIDÆ

Until the New Zealand genus *Neophrynichthys* can be critically examined, it seems best to regard it as constituting a distinct family. As no other cottoid forms have been found south of the Tropic of Cancer, I associate with it provisionally a genus from Buenos Aires.

Neophrynichthys Günther, 1876, 385.
Besnardia Lahille, 1913, 549.

Family 516. SYNCHIRIDÆ

Synchirus Bean, 1889, 444.

Family 517. EREUNIIDÆ

Ereunias Jordan & Snyder, 1901, 493.

Family 518. RHAMPHOCOTTIDÆ

Rhamphocottus Günther, 1874, 375.

Family 519. HEMITRIPTERIDÆ
(Sea Ravens)

Hemitripterus Cuvier, 1829, 117.
Ulca Jordan & Evermann, 1896, 472.

Family 520. COTTOCOMEPHORIDÆ

Cottocomephorus Pellegrin, 1900, 491.   Baicalocottus Berg, 1903, 501.

Family 521. COMEPHORIDÆ
(Baical Fishes)

Comephorus Lacépède, 1800, 56.   Elæorhoüs Pallas, 1811, 83.

Family 522. AGONIDÆ
(Sea Poachers; Alligator Fishes)

Percis Scopoli, 1777, 41.
Cataphractus Klein, 1777, 43.
Agonus[426] Bloch & Schneider, 1801.
Aspidophorus Lacépède, 1802, 62.
Phalangistes Pallas, 1811, 83.
Cataphractus (Klein) Fleming, 1828, 123.
Hippocephalus Swainson, 1839, 202.
Brachyopsis Gill, 1861, 304, 305.
Agonopsis Gill, 1861, 305.
Leptagonus Gill, 1861, 305.
Hypsagonus Gill, 1861, 305.
Podothecus Gill, 1861, 304.
Paragonus Gill, 1862, 305.
Agonomalus Guichenot, 1866, 341.
Paragonus Guichenot, 1869, 355.
Siphagonus Steindachner, 1876, 387.
Odontopyxis Lockington, 1879, 400.

Bothragonus (Gill) Jordan & Gilbert, 1882, 420.
Bathyagonus Gilbert, 1890, 451.
Xenochirus Gilbert, 1890, 451.
Cheiragonus Herzenstein, 1890, 451.
Xystes Jordan & Starks, 1895, 469.
Averruncus J. & S., 1895, 469.
Stelgis Cramer, 1895, 469.
Pallasina Cramer, 1895, 469.
Stellerina Cramer, 1896, 474.
Sarritor Cramer, 1896, 474.
Occa Jordan & Evermann, 1898, 481.
Draciscus Jordan & Snyder, 1901, 493.
Xeneretmus Gilbert, 1903, 504.
Tilesina Schmidt, 1904, 511.
Asterotheca Gilbert, 1915, 556.
Paragonus Ribeiro, 1918, 564.
Ribeiroa Jordan, 1920, 571.

---

[426] "Systema Ichthyologia," 104; type *Cottus cataphractus* Linnæus. (Omitted in "Genera of Fishes.")

Family 523. ASPIDOPHOROIDIDÆ

Aspidophoroides Lacépède, 1802, 62.
Canthirhynchus Swainson, 1839, 202.

Anoplagonus Gill, 1861, 305.
Ulcina[427] Cramer, 1896.

---

[427] In Jordan & Evermann's "Check List," 1896, 449; type *Aspidophoroides olriki* Lütken. (Omitted in "Genera of Fishes.")

Series *CYCLOPTERIFORMES*
Family 524. CYCLOPTERIDÆ
(Lump Fishes)

Cyclopterus Linnæus, 1758, 12.
Oncotion Klein, 1777, 42.
Lumpus Rafinesque, 1815, 90.
Lumpus Cuvier, 1817, 171.

Eumicrotremus Gill, 1862, 317, 331.
Cyclopteroides Garman, 1892, 457.
Lethotremus Gilbert, 1895, 469.
Cyclolumpus Tanaka, 1912, 546.

Family 525. LIPAROPIDÆ

Aptocyclus De la Pylaie, 1835, 183.
Cyclopterichthys Steindachner, 1881, 417.

Liparops Garman, 1892, 457.

Family 526.   LIPARIDÆ   (*Cyclogasteridæ*)
(Sea Snails)

Cyclogaster Gronow, 1763, 18.
Liparis Scopoli, 1777, 41.
Liparis Röse, 1793, 52.
Liparius Rafinesque, 1815, 90.
Massaria Gistel, 1848, 236.
Careproctus Kröyer, 1861, 308.
Neoliparis Steindachner, 1875, 381.
Paraliparis Collett, 1878, 393.
Amitra[428] Goode, 1880, 402.
Monomitra Goode, 1883, 424.
Gymnolycodes Vaillant, 1888, 444.
Enantioliparis Vaillant, 1888, 443.
Bathyphasma Gilbert, 1895, 467.
Gyrinichthys Gilbert, 1895, 467.
Rhinoliparis Gilbert, 1895, 467.
Hilgendorfia Goode & Bean, 1895, 468.
Amitrichthys[428] Jordan & Evermann, 1896, 474.
Allurus J. & E., 1896, 474.

Allochir J. & E., 1896, 474.
Caremitra J. & E., 1896, 474.
Lyoliparis J. & E., 1896, 474.
Crystallichthys Jordan & Gilbert, 1898, 482.
Prognurus J. & G., 1898, 482.
Allinectes Jordan & Evermann, 1898, 482.
Crystallias Jordan & Snyder, 1902, 499.
Trismegistus J. & S., 1904, 508.
Acantholiparis Gilbert & Burke, 1912, 543.
Elassodiscus G. & B., 1912, 543.
Polypera Burke, 1912, 542.
Nectoliparis[429] Gilbert & Burke, 1912, 543.
Lipariscus Gilbert, 1915, 556.
Cyclogaster (Gronow) Burke, 1912.

---

[428] A synonym of *Paraliparis* proper, according to Sigurd Johnsen.

[429] According to Sigurd Johnsen of Bergen *Nectoliparis* is a larval *Paraliparis.* If so, it would seem to be an ally of *P. holomelas,* having a wide gill-slit, and may be generically distinct from *Paraliparis,* the type of which has a narrow slit as in the group called *Amitrichthys.*

Series *TRIGLIFORMES*
Family 527.   PERISTEDIIDÆ

Peristedion[430] Lacépède, 1802, 62.
Octonus Rafinesque, 1810, 82.
Peristethus Kaup, 1858, 282.
Vulsiculus Jordan & Evermann, 1896, 475.

Gargariscus[431] Smith, 1917.
Heminodus[432] Smith, 1917.
Scalicus[433] Jordan, 1922.

---

[430] Written *Peristethus* by some authors.

[431] "New Genera of Deep Sea Gurnards (*Peristediidæ*) from the Philippine Islands," *Proc. Biol Soc. Wash.,* 145; type *Gargariscus semidentatus* Smith. (Overlooked in "Genera of Fishes.")

[432] Smith, *op. cit.,* 146; type *Heminodus philippinus* Smith.

[433] New genus: type *Peristedion amiscus* Jordan & Starks; distinguished by the shovel-shaped snout without horns at the angles.   Japan.

Family 528. TRIGLIDÆ

Trigla Linnæus, 1758, 14.
Corystion Klein, 1775, 41.
Prionotus Lacépède, 1802, 62.
Lepidotrigla Günther, 1860, 296.
Hoplonotus Guichenot, 1866, 341.
Ornichthys Swainson, 1839, 201.
Microtrigla Kaup, 1873, 369.
Chelidonichthys Kaup, 1873, 370.
Palænichthys Kaup, 1873, 370.
Lyrichthys Kaup, 1873, 370.
Trigloides Van Beneden, 1871, 360.

Chriolax Jordan & Gilbert, 1878, 396.
Podopteryx Sauvage, 1880, 406.
Bellator Jordan & Evermann, 1896, 475.
Merulinus J. & E., 1898, 481.
Gurnardus J. & E., 1898, 481.
Pterygotrigla Waite, 1899, 488.
Colotrigla Gill, 1905, 513.
Fissala Gill, 1905, 513.
Otohime Jordan & Starks, 1907, 525.
Paratrigla Ogilby, 1911, 540.
Exolissus[434] Jordan, 1923.

[434] New genus: type *Prionotus alepis* Alcock, *Nat. Hist. Notes,* etc., 1889, 303. This genera differs from *Prionotus* in the absence of scales.

Series *DACTYLOPTERIFORMES*

Family 529. DACTYLOPTERIDÆ *(Cephalacanthidæ)*

Cephalacanthus Lacépède, 1803, 62.
Dactylopterus Lacépède, 1803 62.
Cephacandia Rafinesque, 1815, 89.
Gonocephalus Gronow, 1854, 258.

Dactyloptena Jordan & Richardson, 1908, 528.
Daicocus J. & R., 1908, 528.
Ebisinus J. & R., 1908, 528.

Order HOLCONOTI

Family 530. EMBIOTOCIDÆ

Ditrema Temminck & Schlegel, 1844, 223.
Embiotoca Agassiz, 1853, 252.
Rhacochilus Agassiz, 1854 (May), 255.
Holconotus Agassiz, 1854 (May), 255.
Amphistichus Agassiz, 1854 (May), 255.
Cymatogaster Gibbons, 1854 (May 18), 257.
Hyperprosopon Gibbons, 1854 (May 18), 257.
Micrometrus Gibbons, 1854 (May 30), 257.
Pachylabrus Gibbons, 1854 (June 21), 257.
Mytilophagus Gibbons, 1854 (July), 257.
Cymatogaster Gibbons, 1854 (June 21; not of May 18), 257.
Damalichthys Girard, 1854 (August), 258.
Abeona Girard, 1854, 258.

Ennichthys Girard, 1854, 258.
Phanerodon Girard, 1854, 258.
Metrogaster (Agassiz) Alex. Agassiz, 1861, 300.
Bramopsis (Agassiz), A. Agassiz, 1861, 300.
Tæniotoca A. Agassiz, 1861, 300.
Hypsurus A. Agassiz, 1861, 300.
Brachyistius Gill, 1862, 316.
Hypocritichthys Gill, 1862, 316.
Mænichthys (Kaup) Bleeker, 1876, 383.
Sema Jordan, 1878, 396.
Neoditrema Steindachner & Döderlein, 1883, 426.
Zalembius Jordan & Evermann, 1896, 473.
Tocichthys Hubbs, 1918, 563.
Amphigonopterus Hubbs, 1918, 563.

Family 531. HYSTEROCARPIDÆ

Hysterocarpus Gibbons, 1854, 257.
Sargosomus (Agassiz) A. Agassiz, 1861, 300.

Dacentrus Jordan, 1878, 396.

Order CHROMIDES

Family 532. POMACENTRIDÆ

Abudefduf Forskål, 1775, 33.
Prochilus Klein, 1775, 39.
Amphiprion Bloch & Schneider, 1801, 59.
Glyphisodon[435] Lacépède, 1803, 64.
Pomacentrus Lacépède, 1803, 64.
Chromis Cuvier, 1815, 93.
Premnas Cuvier, 1817, 106.
Heliases[436] Cuvier, 1829, 128.
Dascyllus[437] Cuvier, 1829, 128.
Pristotis Rüppell, 1835, 184.
Chrysiptera Swainson, 1839, 199.
ODONTEUS Agassiz, 1839, 194.
Stegastes Jenyns, 1842, 211.
Pirene Gistel, 1848, 235.
Tetradrachmum Cantor, 1849, 241.
Furcaria Poey, 1860, 299.
Hypsypops Gill, 1861, 305.
Lepidozygus Günther, 1862, 318.
Microspathodon Günther, 1862, 318.
Parma Günther, 1862, 318.
Euschistodus Gill, 1862, 315.
Acanthochromis Gill, 1863, 324.
Dischistodus Gill, 1863, 324.
Ayresia Cooper, 1863, 323.
Pomataprion Gill, 1863, 324.

Onychognathus Troschel, 1866, 343.
Jerdonia Day, 1870, 356.
Paraglyphidodon Bleeker, 1876, 384.
Heptadecacanthus Alleyne & Macleay, 1877, 387.
Parapomacentrus Bleeker, 1877, 387.
Glyphidodontops Bleeker, 1877, 387.
Brachypomacentrus Bleeker, 1877, 387.
Eupomacentrus Bleeker, 1877, 387.
Amblypomacentrus Bleeker, 1877, 387.
Daya[438] Bleeker, 1877.
Hemiglyphidodon Bleeker, 1879, 398, 402.
Hermosilla Jenkins & Evermann, 1888, 442.
Nexilarius[439] Gilbert, 1896, 475.
Azurina Jordan & McGregor, 1898, 481.
Nexilosus Heller & Snodgrass, 1903, 503.
Actinicola Fowler, 1904, 508.
Cheiloprion Weber, 1913, 551, 561.
Hoplochromis[440] Fowler, 1918.
Ctenoglyphiodon[441] Fowler, 1918.
Azurella Jordan, 1919, 567.
Actinochromis[442].
Centrochromis[443] Norman, 1922.

[435] Often written *Glyphidodon.*
[436] Also written *Heliastes.*
[437] Also written *Dascillus.*
[438] Bleeker, *Verh. Holl. Mij. Harlem,* 71; orthotype *Pomacentrus jerdoni* Day. (Omitted in "Genera of Fishes.")
[439] Misprinted *Nexilaris.*
[440] *Proc. Acad. Nat. Sci. Phila.,* **70** : 66; orthotype *Heliastes cœruleus* Cuv. & Val.
[441] *Proc. Acad. Nat. Sci. Phila.,* 1918, 59; type *Abudefduf melanopselion* Fowler.
[442] Orthotype *A. lividus.* (Further details lacking.)
[443] J. R. Norman, "Fishes from Tobago," *Ann. Mag Nat. Hist.,* **9** : 533; orthotype *Glyphidodon rudis* Poey.

Family 533. PRISCACARIDÆ

This extinct group of the American Eocene differs from the *Cichlidæ* in the possession of vomerine teeth.

PRISCACARA Cope, 1877, 388.   COCKERELLITES[444] Jordan, 1923.

[444] New genus, distinguished from *Priscacara* by the weak ventral spine and the long soft dorsal of 13 or 14 rays; orthotype *Priscacara liops* Cope. Named for Theodore Dru Alison Cockerell, of the University of Colorado.

Family 534. CICHLIDÆ

This vast group covers a wide variety of forms, river fishes of both tropics, and doubtless requires subdivision.

Cichla Bloch & Schneider, 1801, 59.

Chromis Cuvier, 1817, 102.

Etroplus Cuvier & Valenciennes, 1830, 135.

Cichlaurus Swainson, 1839, 198.

Cichlasoma Swainson, 1839, 200.

Astronotus Swainson, 1839, 200.

Chætolabrus Swainson, 1839, 199.

Microgaster Swainson, 1839, 199.

Tilapia[445] Andrew Smith, 1840, 244.

Uaru Heckel, 1840, 207.

Symphysodon Heckel, 1840, 207.

Pterophyllum Heckel, 1840, 207.

Acara Heckel, 1840, 207.

Heros Heckel, 1840, 207.

Chætobranchus Heckel, 1840, 207.

Geophagus Heckel, 1840, 207.

Crenicichla Heckel, 1840, 207.

Batrachops Heckel, 1840, 207.

Chætostoma[446] Heckel, 1846, 230.

Sarotherodon Rüppell, 1852, 251.

Coptodon Gervais, 1853, 253.

Herichthys Baird & Girard, 1854, 255.

Plataxoides Castelnau, 1855, 264.

Hemichromis Peters, 1857, 276.

Parachromis Agassiz, 1858, 277.

Hypsophrys Agassiz, 1858, 277.

Amphilophus Agassiz, 1858, 277.

Baiodon Agassiz, 1858, 277.

Chromichthys Duméril, 1858, 280.

Haligenes Günther, 1859, 292.

Hoplarchus Kaup, 1860, 297.

Satanoperca Günther, 1862, 318.

Hygrogonus Günther, 1862, 318.

Mesops Günther, 1862, 318.

Mesonauta Günther, 1862, 318.

Petenia Günther, 1862, 318.

Chromis (Cuvier) Günther, 1862, 318.

Theraps Günther, 1862, 318.

Pseudetroplus (Bleeker) Günther, 1862, 318.

Nandopsis Gill, 1862, 315.

Melanogenes Bleeker, 1863, 321.

Neetroplus Günther, 1866, 341.

Paretroplus Bleeker, 1868, 348.

Paratilapia Bleeker, 1868, 348.

Acaropsis Steindachner, 1875, 381.

Crenicara Steindachner, 1875, 381.

Dicrossus (Agassiz) Steindachner, 1875, 381.

Saraca Steindachner, 1875, 381.

Paracara Bleeker, 1878, 392.

Ptychochromis Steindachner, 1880, 406.

Hoplotilapia Steindachner, 1888, 443.

Haplochromis Hilgendorf, 1888, 442.

Oreochromis Günther, 1889, 446.

Lamprologus Schilthuis, 1890, 452.

Ctenochromis[447] Pfeffer, 1893.

Retroculus Eigenmann & Bray, 1894, 464.

Æquidens Eigenmann & Bray, 1894, 464.

Pelmatochromis Steindachner, 1894, 465.

Docimodus Boulenger, 1896, 471.

Corematodus Boulenger, 1896, 471.

Boggiania Perugia, 1897, 478.

Chromidotilapia Boulenger, 1898, 479.

Telmatochromis Boulenger, 1898, 479.

Julidochromis Boulenger, 1898, 479.

Eretmodus Boulenger, 1898, 479.

Bathybates Boulenger, 1898, 479.

Tropheus Boulenger, 1898, 479.

Petrochromis Boulenger, 1898, 479.

Simochromis Boulenger, 1898, 479.

Perissodus Boulenger, 1898, 479.

Plecodus Boulenger, 1898, 479.

Ectodus Boulenger, 1898, 576.

Xenochromis Boulenger, 1899, 485.

Xenotilapia Boulenger, 1899, 485.

Trematocara Boulenger, 1899, 485.

Steatocranus Boulenger, 1899, 485.

Grammatotria Boulenger, 1899, 485.

Pseudoplesiops Boulenger, 1899, 485.

Spathodus Boulenger, 1900, 488.

Asprotilapia Boulenger, 1901, 492.

Gephyrochromis Boulenger, 1901, 492.

---

[445] Type *Tilapia sparrmanni* Smith.

[446] Written *Chætostomus* by Kner, 1854.

[447] Type *Ctenochromis pectoralis,* a synonym of *Haplochromis.* (Omitted in "Genera of Fishes" and in *Zoological Record.*)

Hemitilapia Boulenger, 1902, 497.
Cyrtocara Boulenger, 1902, 497.
Chilochromis Boulenger, 1902, 497.
Biotæcus Eigenmann & Kennedy, 1903, 502.
Biotodoma Eigenmann & Kennedy, 1903, 502.
Ophthalmotilapia Pellegrin, 1904, 510.
Boulengerochromis Pellegrin, 1904, 510.
Astatotilapia Pellegrin, 1904, 510.
Nannochromis Pellegrin, 1904, 510.
Astatoreochromis Pellegrin, 1904, 510.
Lepidolamprologus Pellegrin, 1904, 510.
Astatheros Pellegrin, 1904, 510.
Thorichthys Meek, 1904, 510.
Nannacara Regan, 1905, 515.
Parapetenia Regan, 1905, 515.
Astatochromis Pellegrin, 1905, 515.
Platytæniodus Boulenger, 1906, 517.
Haplotaxodon Boulenger, 1906, 517.
Cunningtonia Boulenger, 1906, 517.
Heterogramma Regan, 1906, 520.
Enantiopus Boulenger, 1906. (See Index.)
Chætobranchopsis Eigenmann & Ward, 1907, 522.
Erythrichthys Meek, 1907, 525.
PALÆOCHROMIS Sauvage, 1907, 526, 537.
Tomocichla Regan, 1908, 530.
Chilotilapia Boulenger, 1908, 527.
Bayonia Boulenger, 1911, 538.
Acarichthys Eigenmann, 1912, 543.
Stappersia Boulenger, 1914, 552.
Schubotzia Boulenger, 1914, 552.

Champsochromis[448] Boulenger, 1916, 558.
Lobochilotes Boulenger, 1916, 558.
Gymnogeophagus Ribeiro, 1918, 565.
Nannacara Ribeiro, 1918, 564.
Pseudopercis Ribeiro, 1918, 564.
Otopharynx Regan, 1920, 572.
Limnotilapia Regan, 1920, 572.
Cyathopharynx Regan, 1920, 572.
Limnochromis Regan, 1920, 572.
Cyphotilapia Regan, 1920, 572.
Lipochromis Regan, 1920, 572.
Neochromis Regan, 1920, 572.
Cnestrostoma Regan, 1920, 572.
Mylochromis Regan, 1920, 572.
Sargochromis Regan, 1920, 572.
Labrochromis Regan, 1920, 572.
Serranochromis Regan, 1920, 572.
Clinodon Regan, 1920, 572.
Callochromis Regan, 1920, 572.
Leptochromis Regan, 1920, 572.
Aulonocranus Regan, 1920, 572.
Hemibates Regan, 1920, 572.
Neotilapia Regan, 1920, 572.
Heterotilapia Regan, 1920, 572.
Tylochromis Regan, 1920, 572.
Pseudotropheus[449] Regan, 1921.
Cynotilapia[450] Regan, 1921.
Lethrinops[451] Regan, 1921.
Rhamphochromis[452] Regan, 1921.
Aulonocara[453] Regan, 1921.
Heterochromis[453a] Regan, 1922.
Parachromis[453b] Regan, 1922.
Macropleurodus Regan, 1922. (See Index.)

[448] A synonym of *Haplochromis*.

[449] "Cichloid Fishes of Lake Nyassa," *Proc. Zool. Soc. Lond.*, 681; orthotype *Chromis williamsi* Günther.

[450] *Loc. cit.*, 684; orthotype *C. afra* Günther.

[451] *Loc. cit.*, 718; orthotype *Chromis lethrinus* Günther.

[452] *Loc. cit.;* orthotype *Hemichromis longiceps* Günther.

[453] *Loc. cit.;* orthotype *A. nyassæ* Regan.

[453a] "Classification of the Fishes of the Family *Cichlidæ* II"; African and Syrian Genera, *Ann. Mag. Nat. Hist.* X., September, 1922, 252; orthotype *Paratilapia multidens* Pellegrin.

[453b] *Loc. cit.*, 251; orthotype *Hemichromis sacer* Günther.

Order PHARYNGOGNATHI

Family 535. PHARYNGOPILIDÆ (*Phyllodontidæ*)

A small extinct family of Labroid fishes, with the upper pharyngeals united into one, the median teeth enlarged.

PHYLLODUS Agassiz, 1844, 217.
LABRODON Gervais, 1857, 275.
NUMMOPALATUS Rouault, 1858, 284.
DIAPHYODUS Schaffhäutl, 1863, 328.

PHARYNGOPILUS Cocchi, 1864, 329.
EGERTONIA Cocchi, 1864, 329.
PARAPHYLLODUS[454] Sauvage, 1875.

[454] *Bull. Geol. Soc. France*, **3**: 615; name given to species of *Phyllodus* with the median teeth small. (Omitted in "Genera of Fishes.")

Family 536. LABRIDÆ

This family may be naturally separated into the tropical forms, having the vertebræ in normal number ($10 + 15 = 25$), and those with the number increased ($10 + 17 = 27$ to $20 + 21 = 41$). Those with the increased numbers of vertebræ inhabit the colder seas.

Labrus Linnæus, 1758, 13.
Scarus Gronow, 1763, 19.
Suillus Catesby, 1771, 30.
Cicla Klein, 1775, 39.
Helops Browne, 1789, 47.
Bodianus[455] Bloch, 1790, 49.
Cicla Röse, 1793, 52.
Hiatula Lacépède, 1800, 57.
Harpe Lacépède, 1803, 63.
Symphodus Rafinesque, 1810, 79.
Tautoga Mitchill, 1814, 86.
Corycus[456] Cuvier, 1815, 93.
Crenilabrus Cuvier, 1815, 93.
Diastodon Bowdick, 1825, 119.
Lachnolaimus[457] Cuvier, 1829, 130.
Clepticus Cuvier, 1829, 130.
Ctenolabrus Cuvier & Valenciennes, 1839, 194.
Acantholabrus C. & V., 1839, 194.
Cossyphus C. & V., 1839, 194.
Malapterus[458] C. & V., 1839.
Cynædus Swainson, 1839, 200.
Chœrodon Bleeker, 1847, 230.

Ronchifex Gistel, 1848, 235.
PLATYLÆMUS Dixon, 1850, 245.
Chœrops Rüppell, 1852, 251.
Pteragogus Peters, 1855, 266.
Xiphocheilus Bleeker, 1856, 268.
Cossyphodes Bleeker, 1860, 293.
Pseudodax[459] Bleeker, 1861.
Hypsigenys Günther, 1861, 308.
Semicossyphus Günther, 1861, 308.
Tautogolabrus Günther, 1862, 318.
Trochocopus Günther, 1862, 318.
Decodon Günther, 1862, 318.
Lepidaplois Gill, 1862, 314, 325.
Euhypsocara Gill, 1863, 325.
Gymnopropoma Gill, 1863, 325.
Achœrodus Gill, 1863, 325.
Pimelometopon Gill, 1864, 330.
Heterochœrops Steindachner, 1866, 343.
PHACODUS Cope, 1869, 354.
CROMMYODUS Cope, 1870, 356.
PROTAUTOGA Leidy, 1873, 370.
Torresia Castelnau, 1875, 378.
Platychœrops Klunzinger, 1880, 404.

[455] By tautonomy, not by first restriction.

[456] Later written *Coricus*.

[457] Corrected later to *Lachnolœmus*.

[458] *Poissons*, **12**: 355; type *M. reticulatus* Cuvier & Valenciennes; spelled *Malacopterus* by Günther. (Omitted in "Genera of Fishes.")

[459] Bleeker, "Scaroidei," 2; orthotype *Odax moluccensis* Cuvier & Valenciennes. (Omitted in "Genera of Fishes.")

STYLODUS Sauvage, 1882, 421.
PSEUDOSPHÆRODON Noetling, 1885, 434.
Graus Philippi, 1887, 438.
Lappanella Jordan, 1890, 452.
PROTAUTOGA di Alessandri, 1896, 471.
Teleogramma Boulenger, 1899, 453.
Verreo Jordan & Snyder, 1902, 499.

Verriculus Jordan & Evermann, 1903, 504.
Chœropsodes Gilchrist & Thompson, 1909, 533.
GILLIDIA Eastman, 1914, 553.
EOLABROIDES Eastman, 1914, 552.

Family 537. CORIDÆ

Platiglossus Klein, 1775, 40.
Gomphosus Lacépède, 1802, 61.
Coris Lacépède, 1802, 61.
Hemipteronotus Lacépède, 1802, 61.
Acarauna Sewastianof, 1802, 68.
Elops (Commerson) Lacépède, 1802, 70.
Hologymnosus Lacépède, 1802, 63.
Cheilinus Lacépède, 1802, 63.
Cheilio Lacépède, 1803, 63.
Tautoga Mitchill, 1814, 86.
Micropodus Rafinesque, 1815, 89.
Xyrichthys[460] Cuvier, 1815, 93.
Epibulus Cuvier, 1815, 93.
Aygula Rafinesque, 1815, 90.
Julis Cuvier, 1815, 94.
Novacula Cuvier, 1817, 102.
Anampses (Cuvier) Quoy & Gaimard, 1824, 117.
Amorphocephalus Bowdich, 1825, 119.
Elops (Commerson) Bonaparte, 1831, 175.
Halichœres Rüppell, 1835, 184.
Thalassoma Swainson, 1839, 199.
Crassilabrus Swainson, 1839, 199.
Urichthys Swainson, 1839, 199.
Ichthycallus Swainson, 1839, 200.
Eupemis Swainson, 1839, 200.
Chlorichthys Swainson, 1839, 200.
Thalliurus Swainson, 1839, 200.
Cirrhilabrus Temminck & Schlegel, 1845, 227.
Neanis Gistel, 1848, 236.
Cheilinoides Bleeker, 1851, 248.
Labroides Bleeker, 1851, 248.

Labrichthys Bleeker, 1854, 256.
Duymæria Bleeker, 1856, 267.
Xiphocheilus[461] Bleeker, 1856, 268.
Labrastrum Guichenot, 1860, 296.
Stethojulis Günther, 1861, 308.
Hemigymnus Günther, 1861, 308.
Doratonotus Günther, 1862, 318.
Cymolutes Günther, 1862, 318.
Ophthalmolepis[462] Bleeker, 1862, 300.
Macropharyngodon Bleeker, 1862, 309.
Platyglossus (Klein) Bleeker, 1862, 309.
Hemicoris Bleeker, 1862, 309.
Pseudojulis Bleeker, 1862, 309.
Pseudocheilinus Bleeker, 1862, 309.
Pseudocoris Bleeker, 1862, 309.
Guntheria Bleeker, 1862, 309.
Pseudolabrus Bleeker, 1862, 309.
Hemitautoga Bleeker, 1862, 309.
Diproctacanthus Bleeker, 1862, 309.
Novaculichthys Bleeker, 1862, 309.
Leptojulis Bleeker, 1862, 309.
Oxycheilinus[463] Gill, 1862, 315.
Malacocentrus Gill, 1862, 314.
Chœrojulis Gill, 1862, 314.
Iniistius Gill, 1862, 314.
Dimalacocentrus Gill, 1863, 325.
Oxyjulis Gill, 1863, 325.
Cheiliopsis Steindachner, 1863, 328.
Thysanocheilus Kner, 1865, 338.
Parajulis Bleeker, 1865, 335.
Cheilolabrus Alleyne & Macleay, 1877, 387.
Austrolabrus Steindachner, 1884, 430.

---

[460] Spelled also *Xirichthys.*

[461] Corrected to *Xiphochilus.*

[462] The date of this paper (No. 712) is 1862, and the genera which follow are included in it as well as in No. 754. The type of *Hemicoris* is *Halichœres variegatus* Rüppell; that of *Guntheria, G. cæruleovittatus* Bleeker; that of *Hemitautoga, Labrus centiquadrus* Lacépède; and of *Novaculichthys, Labrus tæniourus* Lacépède.

[463] Orthotype *Cheilinus arenatus.*

Julichthys De Vis, 1885, 431.
Eupetrichthys Ramsey & Ogilby, 1887, 439.
Xyrula Jordan, 1890, 452.
Pictilabrus Gill, 1891, 455.
Julidio Jordan & Evermann, 1896, 473.
Emmeekia J. & E., 1896, 473.

Iridio J. &. E., 1896, 473.
Ampheces Jordan & Snyder, 1902, 499.
Octocynodon Fowler, 1904, 508.
Chœropsodes Gilchrist & Thompson, 1909, 533.
Bermudichthys Nichols, 1920, 572.
Hinalea[464] Jordan & Jordan, 1923.

[464] "Fishes of Hawaii," *Mem. Carnegie Mus.* 69; orthotype *Julis axillaris* Quoy and Gaimard; differs from *Stethojulis* in lacking posterior canines.

Family 538. NEOLABRIDÆ

Allied to the *Coridæ*, but with the weak dorsal spines only 3 or 4 in number.

Neolabrus Steindachner, 1875, 381.

Family 539. SPARISOMIDÆ (*Scarichthyidæ*)

This group is well distinguished from the *Scaridæ,* by its very different pharyngeal dentition as well as by numerous other characters of fins and teeth.

Leptoscarus Swainson, 1839, 199.
Sparisoma Swainson, 1839, 199.
Scarus (Forskål) Bleeker, 1849, 240.
Scarichthys Bleeker, 1859, 287.
Taurinichthys Cocchi, 1864, 329.
Cryptotomus Cope, 1871, 361.

Heteroscarus Castelnau, 1872, 363.
Calotomus Gilbert, 1890, 451.
Scaroides Tould, 1899, 534.
Scaridea Jenkins, 1901, 493.
Xenoscarus Evermann & Radcliffe, 1917, 562.

Family 540. SCARIDÆ (*Callyodontidæ*)

Callyodon Gronow, 1763, 19.
Psittacus Catesby, 1771, 31.
Novacula Catesby, 1771, 31.
Scarus[465] Forskål, 1775, 33.
Callyodon[466] (Gronow) Scopoli, 1777, 42.
Callyodon (Gronow) Bloch, 1788, 45.
Mormyra Browne, 1789, 46.
Calliodon Bloch & Schneider, 1801, 59.
Odax (Commerson) Lacépède, 1802, 70.

Aper (Plumier) Lacépède, 1803, 72.
Calliodon Cuvier, 1829, 130.
Hemistoma Swainson, 1839, 199.
Petronason Swainson, 1839, 199.
Erychthys Swainson, 1839, 199.
Chlorurus Swainson, 1839, 199.
Amphiscarus Swainson, 1839, 199.
Euscarus Jordan & Evermann, 1896, 473.
Loro J. & E., 1896, 473.

[465] I use this familiar name pending final decision as to the retention of Gronow's non-binomial post-Linnæan generic names.
[466] Also often written *Calliodon*.

Family 541. ODACIDÆ

Odax Cuvier, 1829, 130.
Olisthops[467] Richardson, 1850, 247.
Coridodax Günther, 1862, 318.

Heteroscarus Castelnau, 1872, 363.
Neoodax Castelnau, 1875, 378.

[467] Spelled *Olistherops* by Günther.

Family 542. SIPHONOGNATHIDÆ

Siphonognathus Richardson, 1857, 276.

Order GOBIOIDEA

Family 543. RHYACICHTHYIDÆ (*Platypteridæ*)

Platyptera (Kuhl & Van Hasselt)   Rhyacichthys Boulenger, 1901, 492.
Cuvier, 1829, 130, 189.

Family 544. ELEOTRIDÆ
(Sleepers)

Eleotris Gronow, 1763, 20.
Pelmatia Browne, 1789, 47.
Gobiomorus Lacépède, 1800, 57.
Gobiomoroides Lacépède, 1800, 57.
Eleotris Bloch & Schneider, 1801, 58.
Bostrychus Lacépède, 1802, 61.
Bostrychoides Lacépède, 1802, 61.
Asellus (Plumier) Lacépède, 1802, 71.
Bostrichthys Duméril, 1806, 75.
Bostrictis Rafinesque, 1815, 88.
Psilus Fischer, 1813, 85.
Psiloides Fischer, 1813, 85.
Pterops Rafinesque, 1815, 88.
Ictiopogon Rafinesque, 1815, 92.
Epiphthalmus Rafinesque, 1815, 90.
Prochilus Cuvier, 1817, 104.
Asterropteryx Rüppell, 1828, 122.
Philypnus Cuvier & Valenciennes, 1837, 188.
Cestreus McClelland, 1842, 212.
Culius Bleeker, 1856, 267, 372.
Butis Bleeker, 1856, 267, 372.
Belobranchus Bleeker, 1856, 268, 372.
Eleotriodes Bleeker, 1857, 275, 373.
Lembus Günther, 1859, 291.
Erotelis Poey, 1860, 299.
Dormitator Gill, 1861, 302, 316.
Mogurnda Gill, 1863, 326.
Gobiomorphus Gill, 1863, 326.
Ophiocara Gill, 1863, 326.
Hypseleotris Gill, 1863, 326.
Odonteleotris Gill, 1863, 326.

Calleleotris Gill, 1863, 326.
Ptereleotris Gill, 1863, 326.
Valenciennea Bleeker, 1868, 348.
Valenciennesia Bleeker, 1874, 373.
Amblyeleotris Bleeker, 1874, 372.
Eleotrioides Bleeker, 1874, 372.
Guavina Bleeker, 1874, 372.
Oxyeleotris Bleeker, 1874, 372.
Gymneleotris Bleeker, 1874, 372.
Gymnobutis Bleeker, 1874, 372.
Prionobutis Bleeker, 1874, 372.
Odontobutis Bleeker, 1874, 372.
Priolepis (Ehrenberg) Bleeker, 1874, 373.
Brachyeleotris Bleeker, 1874, 373.
Hetereleotris Bleeker, 1874, 373.
Pogoneleotris Bleeker, 1875, 377.
Giuris Sauvage, 1879, 401.
Ioglossus Bean, 1882, 419, 420.
Alexurus Jordan, 1895, 468.
Krefftius Ogilby, 1897, 478.
Carassiops Ogilby, 1897, 477.
Mulgoa Ogilby, 1897, 478.
Ophiorrhinus Ogilby, 1897, 478.
Caulichthys Ogilby, 1898, 483.
Vireosa Jordan & Snyder, 1901, 494.
Eviota Jenkins, 1903, 504.
Parioglossus[468] Regan, 1912, 545.
Microeleotris Meek & Hildebrand, 1916, 559.
Hemieleotris M. & H., 1916, 560.
Leptophilypnus M. & H., 1916, 559.

---

[468] Misprinted *Pariglossus* by authors.

Family 545. GOBIIDÆ
(Gobies)

Gobius Linnæus, 1758, 12.
Gobio Klein, 1779, 43.
Benthophilus Eichwald, 1831, 176.
Oplopomus (Ehrenberg) Cuvier & Valenciennes, 1837, 188.
Cryptocentrus (Ehrenberg) Cuvier & Valenciennes, 1837, 188, 374.

Apocryptes Cuvier & Valenciennes, 1837, 188.
Hexacanthus Nordmann, 1838, 193.
Ruppellia Swainson, 1839, 202.
Gobileptes Swainson, 1839, 198.
Brachyochirus Nardo, 1841, 209.
Chæturichthys Richardson, 1844, 222.

Gobiodon (Kuhl & Van Hasselt) Bleeker, 1856, 267.
Oxyurichthys Bleeker, 1858, 279.
Lepidogobius Gill, 1858, 280.
Tridentiger Gill, 1858, 280.
Triænophorus Gill, 1858, 280.
Triænophorichthys Gill, 1858, 280.
Chænogobius Gill, 1858, 280.
Ctenogobius Gill, 1858, 281.
Gobionellus Girard, 1859, 291.
Gobiosoma Girard, 1859, 291.
Euctenogobius Gill, 1859, 291.
Evorthodus Gill, 1859, 289.
Acanthogobius Gill, 1859, 290.
Glossogobius Gill, 1859, 290.
Rhinogobius Gill, 1859, 290.
Luciogobius Gill, 1859, 290.
Smaragdus Poey, 1860, 299.
Chonophorus Poey, 1860, 299.
Awaous Steindachner, 1860, 299.
Cyclogobius Steindachner, 1860, 299.
Oplopomus Steindachner, 1860, 299.
Gobiopsis Steindachner, 1860, 299.
Oxyurichthys Bleeker, 1860, 294.
Latrunculus Günther, 1861, 307.
Lentipes Günther, 1861, 307.
Lophogobius Gill, 1862, 316.
Eucyclogobius Gill, 1862, 316.
Gillichthys Cooper, 1863, 323.
Boreogobius Gill, 1863, 326.
Crystallogobius Gill, 1863, 326.
Gymnogobius Gill, 1863, 326.
Ophiogobius Gill, 1863, 326.
Synechogobius Gill, 1863, 326.
Pterogobius Gill, 1863, 326.
Coryphopterus Gill, 1863, 325.
Deltentosteus Gill, 1863, 325.
Pomatoschistus Gill, 1863, 325.
Cotylopus Guichenot, 1864, 332.
Gillia Günther, 1864, 334.
Orthostomus Kner, 1868, 352.
Gobiichthys Klunzinger, 1871, 361.
Paragobiodon Bleeker, 1873, 367.
Lophiogobius Günther, 1873, 369.
Ellerya Castelnau, 1873, 367.
Alepidogobius Bleeker, 1874, 373.
Gobiopterus Bleeker, 1874, 373.
Leptogobius Bleeker, 1874, 373.
Triænopogon Bleeker, 1874, 373.

Sicydiops Bleeker, 1874, 373.
Microsicydium Bleeker, 1874, 373.
Brachygobius Bleeker, 1874, 373.
Callogobius Bleeker, 1874, 373.
Platygobius Bleeker, 1874, 373.
Mesogobius Bleeker, 1874, 373.
Stenogobius Bleeker, 1874, 373.
Hemigobius Bleeker, 1874, 373.
Oligolepis Bleeker, 1874, 373.
Gnatholepis Bleeker, 1874, 373.
Hypogymnogobius Bleeker, 1874, 373.
Actinogobius Bleeker, 1874, 373.
Cephalogobius Bleeker, 1874, 373.
Centrogobius Bleeker, 1874, 373.
Acentrogobius Bleeker, 1874, 373.
Porogobius Bleeker, 1874, 373.
Pseudogobiodon Bleeker, 1874, 373.
Oxymetopon Bleeker, 1874, 373.
Paragobiodon Bleeker, 1874, 373.
Amblygobius Bleeker, 1874, 373.
Zonogobius Bleeker, 1874, 374.
Odontogobius Bleeker, 1874, 374.
Stigmatogobius Bleeker, 1874, 374.
Amblychæturichthys Bleeker, 1874, 374.
Parachæturichthys Bleeker, 1874, 374.
Apocryptodon Bleeker, 1874, 374.
Parapocrytes Bleeker, 1874, 374.
Gobileptes (Swainson) Bleeker, 1874, 374.
Pseudapocryptes Bleeker, 1874, 374.
Orthostomus Kner, 1874, 373.
Gobiopus Gill, 1874, 375.
Microgobius Poey, 1875, 379.
Latrunculodes Collett, 1875, 379.
Leptogobius Bleeker, 1875, 377.
Apocryptichthys Day, 1876, 385.
Paroxyurichthys Bleeker, 1876, 384.
Lebetus Winther, 1877, 392.
Perccottus Dybowski, 1877, 389.
Bathygobius[469] Bleeker, 1878, 393.
Typhlogobius Steindachner, 1880, 406.
Leucopsarion Hilgendorf, 1880, 403.
Othonops Rosa Smith (Eigenmann), 1881, 417.
Saccostoma Sauvage, 1882, 421.
Leme De Vis, 1884, 427.
Clevelandia Eigenmann & Eigenmann, 1888, 440.
Barbulifer E. &. E., 1888, 440.

---

[469] An unfortunate name for a very common shallow-water form.

Bollmannia Jordan, 1889, 447.
Chriolepis Gilbert, 1891, 455.
Salarigobius Pfeffer, 1893, 462.
Garmannia Jordan, 1895, 576.
Quietula Jordan & Evermann, 1895, 469.
Aboma Jordan & Starks, 1895, 468.
Evermannia Jordan, 1895, 468.
Stiphodon[470] Weber, 1895, 470.
Sicya Jordan & Evermann, 1896, 474.
Zalypnus J. & E., 1896, 474.
Lythrypnus J. & E., 1896, 474.
Ilypnus J. & E., 1896, 474.
Sicyosus J. & E., 1898, 482.
Enypnias J. & E., 1898, 481.
Trichopharynx Ogilby, 1898, 483.
Austrogobio Ogilby, 1898, 483.
Oreogobius Boulenger, 1899, 485.
Mapo Smitt, 1899, 487.
Lebistes Smitt, 1899, 487.
Caffrogobius Smitt, 1899, 487.
Mugilogobius Smitt, 1899, 487.
Proterorhinus Smitt, 1899, 487.
Eichwaldia Smitt, 1899, 487.
Trifissus Jordan & Snyder, 1900, 490.
Chasmias J. & S., 1901, 494.
Eutæniichthys J. & S., 1901, 494.
Clariger J. & S., 1901, 494.
Astrabe J. & S., 1901, 494.
Ainosus J. & S., 1901, 494.
Sagamia J. & S., 1901, 494.
Suruga J. & S., 1901, 494.
Chloëa J. & S., 1901, 494.
Hazeus J. & S., 1901, 494.
Chasmichthys J. & S., 1901, 494, 504.
Mistichthys H. M. Smith, 1902, 500.
Chlamydes Jenkins, 1903, 504.
Vitraria Jordan & Evermann, 1903, 504.
Quisquilius J. & E., 1903, 504.

Elacatinus Jordan, 1904, 508.
Pycnomma Rutter, 1904, 511.
Allogobius Waite, 1904, 511.
Gigantogobius Fowler, 1905, 513.
Kelloggella Jordan & Seale, 1905, 514, 519.
Drombus J. & S., 1905, 514.
Waitea J. & S., 1906, 519.
Pselaphias J. & S., 1906, 519.
Exyrias J. & S., 1906, 519.
Mars J. & S., 1906, 519.
Vailima J. & S., 1906, 519.
Vaimosa J. & S., 1906, 519.
Vitreola J. & S., 1906, 519.
Trimma J. & S., 1906, 519.
Caragobius Smith & Seale, 1906, 520.
Illana S. & S., 1906, 520.
Ranulina Jordan & Starks, 1906, 519.
Creisson Jordan & Seale, 1907, 525.
Xenisthmus Snyder, 1908, 531.
Doryptena Snyder, 1908, 531.
Aparrius Jordan & Richardson, 1908, 528.
Macgregorella Seale, 1909, 533.
Biat Seale, 1909, 533.
Expedio Snyder, 1909, 534.
Inu Snyder, 1909, 534.
Nematogobius Boulenger, 1910, 534.
Mucogobius McCulloch, 1912, 544.
Schismatogobius Beaufort, 1912, 542.
Pleurosicya Weber, 1913, 551.
Lubricogobius Tanaka, 1915, 558.
Evermannichthys[471] Metzelaar, 1920.
Xenogobius[472] Metzelaar, 1920.
Micropercops[473] Fowler & Gee, 1920.
Ulcigobius[474] Fowler, 1920.
Aprolepis[475] Hubbs, 1921.
Radcliffella[476] Hubbs, 1921.

---

[470] The reference on page 542 in "Genera of Fishes" should be canceled.

[471] *Over Tropisch Atlantisch Visschen,* 139; type *Evermannichthys spongicola* Metzelaar.

[472] *Op. cit.,* 140; type *Xenogobius weberi* Metzelaar.

[473] *Proc. U. S. Nat. Mus.,* **58**: 318; type *Micropercops dabryi* Fowler & Bean.

[474] *Proc. Acad. Nat. Sci. Phila.,* 1918, 69: type *Drombus maculipinnis* Fowler.

[475] *Occ. Papers Univ. Michigan,* **99**: 1; orthotype *Aprolepis barbaræ* Hubbs.

[476] *Occ. Papers Univ. Mich.,* **99**: 2; orthotype *Garmannia spongicola* Radcliffe.

Family 546. PERIOPHTHALMIDÆ

Apocryptes[477] Osbeck, 1762, 17.
Periophthalmus Bloch & Schneider,
1801, 58.
Scartelaos Swainson, 1839, 202.
Apocryptes (Osbeck) Cuvier & Valenciennes, 1837, 188.

Boleophthalmus C. & V., 1837, 188.
Euchoristopus Gill, 1863, 326.
Boleops Gill, 1863, 326.
Periophthalmodon Bleeker, 1874, 374.

[477] A binomial name, but first issued in 1757, before the "Systema Naturæ" of Linnæus; reprinted in 1762. The name was revived by Cuvier & Valenciennes, its designated type being that of Osbeck, identical with *Bolcophthalmus,* described on a subsequent page of the "Histoire Naturelle des Poissons." The name *Apocryptes* was later transferred to a different group.

Family 547. GOBIOIDIDÆ

Gobioides Lacépède, 1800, 57.
Tænioides Lacépède, 1800, 57.
Plecopodus Rafinesque, 1815, 90.
Gymnurus Rafinesque, 1815, 88.
Amblyopus Cuvier & Valenciennes, 1837, 188.
Psilosomus Swainson, 1839, 198.

Ognichodes Swainson, 1839, 202.
Tyntlastes Günther, 1862, 319.
Brachyamblyopus Bleeker, 1874, 374.
Odontamblyopus Bleeker, 1874, 374.
Cayennia Sauvage, 1880, 405.
Paragobioides Kendall & Goldsborough, 1911, 540.

Family 548. TRYPAUCHENIDÆ

Trypauchen[478] Cuvier & Valenciennes, 1837.
Trypauchenichthys Bleeker, 1860, 293, 374.

Ctenotrypauchen Steindachner, 1867, 348.
Trypauchenopsis Volz, 1903, 506.
Trypauchenophrys Franz, 1910, 535.

[478] "Histoire Naturelle des Poissons," XII, 162; type *Gobius vagina* Bloch & Schneider. (Omitted in "Genera of Fishes.")

Family 549. DOLIICHTHYIDÆ

Doliichthys Sauvage, 1874, 376.

Family 550. PSAMMICHTHYIDÆ

Psammichthys Regan, 1908, 530.

Order DISCOCEPHALI

Family 551. OPISTHOMYZONIDÆ

Opisthomyzon Cope, 1889, 445.

Family 552. ECHENEIDÆ
(Remoras; Shark-pilots)

Echeneis Linnæus, 1758, 12.
Remora Forster, 1771, 28.
Remora Catesby, 1771, 31.
Phtheirichthys Gill, 1862, 316.
Remilegia Gill, 1862, 316.
Remora Gill, 1862, 316.

Rhombochirus Gill, 1863, 324.
Remoropsis Gill, 1863, 324.
Leptecheneis Gill, 1864, 330.
Remorina Jordan & Evermann, 1896, 475.

Order JUGULARES
(*Deripia*)

Series *TRACHINIFORMES*
Family 553. TRACHINIDÆ

This group forms a natural sequence from *Trichodontidæ, Champso-dontidæ,* etc.

Trachinus Linnæus, 1758, 12.
Corystion Rafinesque, 1810, 79.
Pseudotrachinus Bleeker, 1862, 309.

Echiichthys Bleeker, 1862, 309.
PSEUDELEGINUS Sauvage, 1873, 371.
TRACHINOPSIS Sauvage, 1875, 380.

Family 554. CALLIPTERYGIDÆ

CALLIPTERYX Agassiz, 1838, 191.

Series *NOTOTHENIIFORMES*
Family 555. NOTOTHENIIDÆ

Eleginus Cuvier & Valenciennes, 1830, 135.
Aphritis Cuvier & Valenciennes, 1831, 137.
Notothenia Richardson, 1844, 222.
Macronotothen Gill, 1861, 306.
Eleginops Gill, 1861, 306, 455.
Pseudaphritis Castelnau, 1872, 575.

Phricus Berg, 1895, 466.
Dissostichus Smitt, 1898, 483.
Racovitzia Dollo, 1900, 489.
Artedidraco Lönnberg, 1905, 515.
Dolloidraco Roule, 1913, 551.
Pogonophryne Regan, 1914, 554.
Histiodraco Regan, 1914, 555.

Family 556. BATHYDRACONIDÆ

Bathydraco Günther, 1878, 394.
Gerlachea Dollo, 1900, 489.
Prinodraco Regan, 1914, 554.

Cygnodraco Waite, 1916, 561.
Aconichthys Waite, 1916, 561.

Family 557. CHANNICHTHYIDÆ

Channichthys Richardson, 1844, 222.
Pagetodes Richardson, 1844, 222.
Chænichthys Richardson, 1844, 222.
Champsocephalus Gill, 1861, 306.
Ponerodon Alcock, 1890, 449.

Cryodraco Dollo, 1900, 489.
Chænocephalus Regan, 1913, 550.
Pagetopsis Regan, 1913, 550.
Chænodraco Regan, 1914, 554.
Dacodraco Waite, 1916, 561.

Family 558. BOVICTIDÆ

Bovictus[479] Cuvier & Valenciennes, 1831, 137.

Cottoperca Steindachner, 1875, 381.
Aurion Waite, 1916, 561.

---

[479] Later corrected to *Bovichthys.*

Family 559. HARPAGIFERIDÆ

Harpagifer Richardson, 1844, 222.

Sclerocottus Fischer (J. G.), 1885, 431.

Series *CALLIONYMIFORMES*

Family 560. **DRACONETTIDÆ**

Draconetta Jordan & Fowler, 1903, 505.   **Centrodraco** Regan, 1913, 550.

Family 561–562. **CALLIONYMIDÆ**

(Dragonets)

Callionymus Linnæus, 1758, 12.
Synchiropus Gill, 1859, 290.
Dactylopus Gill, 1859, 290.
Vulsus Günther, 1861, 307.

Calliurichthys Jordan & Fowler, 1903, 505.
Draculo Snyder, 1911, 541.
Calymmichthys Jordan & Thompson, 1914, 553.

Series *PERCOPHIDIFORMES*

Family 563. **PERCOPHIDIDÆ**

Percophis Quoy & Gaimard, 1824, 118.

Family 564. **MUGILOIDIDÆ** (*Pinguipedidæ*)

Mugiloides Lacépède, 1803, 67.
Myxonum Rafinesque, 1815, 90.

Pinguipes Cuvier, 1829, 127.
Pseudopercis Ribeiro, 1903, 506.

Family 565. **PARAPERCIDÆ**

Percis Bloch & Schneider, 1801, 58.
Parapercis Bleeker, 1863, 322.
Parapercis Steindachner, 1884, 430.
Neopercis Steindachner, 1885, 434.
Bathypercis Alcock, 1893, 460.

Macrias Gill & Townsend, 1901, 493.
Pleuragramma Boulenger, 1902, 496.
Osurus Jordan & Evermann, 1903, 506.
Chilias Ogilby, 1910, 536.

Family 566. **PTEROPSARIDÆ** (*Bembropidæ*)

Bembrops Steindachner, 1876, 387.
Hypsicometes Goode, 1880, 403.
Bathypercis Alcock, 1880, 403.
Chrionema Gilbert, 1905, 513.

Acanthaphritis Günther, 1880, 403.
Pteropsaron[480] Jordan & Snyder, 1902, 499.
Osopsaron Jordan & Starks, 1904, 509.

[480] A synonym of *Acanthaphritis* according to Regan, but the two genera differ in several respects.

Family 567. **HEMEROCŒTIDÆ**

Hemerocœtes Cuvier & Valenciennes, 1837, 189.

Family 568. **CHIMARRHICHTHYIDÆ**

Chimarrhichthys Haast, 1874, 375.

Family 569. **CREEDIIDÆ**

Creedia Ogilby, 1898, 483.

Family 570. **LIMNICHTHYIDÆ**

Limnichthys Waite, 1904, 511.   **Schizochirus** Waite, 1904, 511.

Family 571. TRICHONOTIDÆ

Trichonotus Bloch & Schneider, 1801, 58.   Lesueurina Fowler, 1907, 523.
Tæniolabrus Steindachner, 1867, 348.   Squamicreedia[481] Rendahl, 1921.
Kræmeria Steindachner, 1906, 521.

---

[481] "Results of Dr. E. Mjöberg's Swedish Scientific Expeditions to Australia,1910-13," XVIII; Hjalmar Rendahl, "Fische," 20, 1921; orthotype *Squamicreedia obtusa* Rendahl.

Family 572. OXUDERCIDÆ

Oxuderces Valenciennes, 1842, 213.

Series *AMMODYTIFORMES*
Family 573. AMMODYTIDÆ
(Sand-lances)

Ammodytes Linnæus, 1758, 12.   Hyperoplus Günther, 1862, 318.
Argyrotænia Gill, 1861.

Family 574. BLEEKERIIDÆ

Bleekeria Günther, 1862, 318.   Embolichthys Jordan, 1903, 504.
Rhynchias Gill, 1898, 482.

Family 575. HYPOPTYCHIDÆ

Hypoptychus Steindachner, 1880, 406.

Series *BATHYMASTERIFORMES*
Family 576. BATHYMASTERIDÆ
(Ronquils)

Bathymaster Cope, 1873, 368.   Rathbunella Jordan & Evermann, 1896,
Ronquilus Jordan & Starks, 1895, 469.   474.

Family 577. ZAPRORIDÆ

A group of uncertain relationships, having no ventral fins; the skeleton not studied. It may belong near *Bathymaster,* but it is perhaps an ally of *Centrolophus,* possibly of *Cryptacanthodes.*

Zaprora Jordan, 1896, 472.   Aræosteus Jordan & Gilbert, 1920, 571.

Series *URANOSCOPIFORMES*
Family 578. CHIASMODONTIDÆ
(Black Swallowers)

Chiasmodon[482] Johnson, 1863, 327.   Dysalotus McGilchrist, 1905, 515.
Pseudoseopelus Lütken, 1892, 458.   Odontonema Weber, 1913, 551.

---

[482] Spelled *Chiasmodus* by Günther.

Family 579. OPISTHOGNATHIDÆ
(Jaw Fishes)

Opisthognathus (Cuvier) Oken, 1817, 101.   Lonchopisthus Gill, 1862, 316.
          Stalix Jordan & Snyder, 1902, 499.
Gnathypops Gill, 1862, 316.   Merogymnus Ogilby, 1908, 529.

Family 580. OWSTONIIDÆ

Owstonia Tanaka, 1908, 531.

Family 581. CHAMPSODONTIDÆ

Champsodon Günther, 1867, 345.   Centropercis Ogilby, 1895, 469.

Family 582. URANOSCOPIDÆ
(Star-gazers)

Uranoscopus Linnæus, 1758, 12.
Ichthyscopus Swainson, 1839, 202.
Astroscopus Brevoort, 1860, 295.
Kathetostoma[483] Günther, 1860, 296.
Agnus Günther, 1860, 296.
Upsilonphorus Gill, 1861, 305.
Nematagnus Gill, 1861, 305.

Genyagnus Gill, 1861, 305.
Gnathagnus Gill, 1861, 305.
Synnema Haast, 1873, 369.
Ariscopus Jordan & Snyder, 1902, 499.
Execestides Jordan & Thompson, 1905, 514.

[483] Later written *Cathetostoma*.

Family 583. LEPTOSCOPIDÆ

Leptoscopus Gill, 1859, 290.   Crapatalus Günther, 1861, 308.

Family 584. DACTYLOSCOPIDÆ

Dactyloscopus Gill, 1859, 290.
Crapatalus Günther, 1861, 208.
Myxodagnus Gill, 1861, 305.
Dactylagnus Gill, 1862, 318.

Gillellus Gilbert, 1890, 451.
Esloscopus Jordan & Evermann, 1896, 474.

Series *BLENNIIFORMES*
Family 585. CLINIDÆ

Clinus Cuvier, 1817, 101.
Tripterygion[484] Risso, 1826, 119.
Myxodes Cuvier, 1829, 129.
Cirrhibarbus Cuvier, 1829, 129.
Enneapterygius Rüppell, 1835, 184.
Cristiceps Cuvier & Valenciennes, 1836, 185.
PTERYGOCEPHALUS Agassiz, 1839, 194.
Clinitrachus Swainson, 1839, 197.
Blennophis Swainson, 1839, 202.
Labrisomus Swainson, 1839, 202.
Ophisomus Swainson, 1839, 202.
Lepisoma Dekay, 1842, 210.
Heterostichus Girard, 1854, 258.
Neoclinus[485] Girard, 1858.
Pterognathus Girard, 1859, 290.

Entomacrodus Gill, 1859, 291.
Malacoctenus Gill, 1860, 295.
Gobioclinus Gill, 1860, 295.
Blennioclinus Gill, 1860, 295.
Auchenionchus Gill, 1860, 296.
Calliclinus Gill, 1860, 296.
Ophthalmolophus Gill, 1860, 296.
Auchenopterus Günther, 1861, 307.
Cremnobates Günther, 1861, 308.
Gibbonsia Cooper, 1863, 323.
Lepidoblennius Steindachner, 1867, 348.
Lepidoblennius Sauvage, 1874, 376.
Blakea Steindachner, 1876, 387.
Acanthoclinus Mocquard, 1885, 433.
Petraites Ogilby, 1885, 434.
Paraclinus Mocquard, 1888, 443.

[484] Also corrected to *Tripterygium*.
[485] Girard, *U. S. Pac. R. R. Surv.*, 10 : 114; orthotype *Neoclinus blanchardi* Girard. (Omitted in "Genera of Fishes.")

Dialommus Gilbert, 1890, 451.
Cryptotrema Gilbert, 1890, 451.
Cologrammus Gill, 1893, 461.
Notoclinus Gill, 1893, 461.
Ericentrus Gill, 1893, 461.
Scleropteryx De Vis, 1894, 465.
Enneanectes Jordan & Gilbert, 1895, 468.
Exerpes Jordan & Evermann, 1896, 472.
Mnierpes J. & E., 1896, 474.
Starksia Jordan & Evermann, 1896, 472.
Emmnion Jordan, 1897, 477.
Pterognathus (Girard) Jordan & Evermann, 1898, 481.
Corallicola Jordan & Evermann, 1898, 481
Gillias Evermann & Marsh, 1899, 485.

Auchenistius Evermann & Marsh, 1899, 485.
Coralliozetus E. & M., 1899, 485.
Zacalles Jordan & Snyder, 1902, 499.
Ericteis Jordan, 1904, 508.
Acteis Jordan, 1904, 508.
Sauvagea Jordan & Seale, 1906, 519.
Congrammus Fowler, 1906, 517.
Calliblennius Barbour, 1912, 541.
Helcogramma McCulloch & Waite, 1918, 564.
Trianectes M. & W., 1918, 564.
Histioclinus[486] Metzelaar, 1921.
Tekla Nichols, 1922. (See Index.)

---

[486] *Trop. Atl. Visch.*, 157; orthotype *Histioclinus veliger* Metzelaar.

Family 586. NOTOGRAPTIDÆ

Notograptus Günther, 1867, 345.
Sticharium Günther, 1867, 345.

Blanchardia Castelnau, 1875, 378.

Family 587. PERONEDYIDÆ

Body eel-shaped, without pectoral fins, scales rudimentary; three lateral lines; dorsal mostly of spines.

Peronedys Steindachner, 1884, 430.

Eucentronotus Ogilby, 1898, 483.

Family 588. OPHIOCLINIDÆ

A group of slender, eel-shaped blennies, with rudimentary scales, the fin rays mainly spinous, here provisionally associated, their actual relationship undetermined.

Ophioclinus Castelnau, 1872, 363.
Neoblennius Castelnau, 1875, 378.

Neogunellus Castelnau, 1875, 378.
Stenophus Castelnau, 1875, 378.

Family 589. BLENNIIDÆ

Blennius Linnæus, 1758, 12.
Salaria Forskål, 1775, 33.
Blennus Klein, 1779, 44.
Alticus (Commerson) Lacépède, 1800, 69.
Salarias Cuvier, 1817, 101.
Pholis Cuvier, 1817, 101.
Petroscirtes Rüppell, 1828, 173.
Aspidontus Quoy & Gaimard, 1834, 180.
Ichthyocoris Bonaparte, 1836, 142, 206.
Chasmodes Cuvier & Valenciennes, 1836, 185.
Alticus[487] (Commerson) Cuv. & Val., 1836.

Blennechis Cuv. & Val., 1836, 185.
Omobranchus (Ehrenberg) Cuv. & Val., 1836, 185.
Blennitrachus Swainson, 1839, 198.
Rupiscartes Swainson, 1839, 202.
Erpichthys Swainson, 1839, 202.
Cirripectes Swainson, 1839, 202.
Blennophis[488] Valenciennes, 1844.
Adonis Gronow, 1854, 258.
Entomacrodus Gill, 1859, 289.
Ophioblennius Gill, 1860, 296.
Hypleurochilus Gill, 1861, 302.

---

[487] "Histoire Naturelle Poissons," XI, 337; the name mentioned but not accepted.

[488] Orthotype *Blennophis webbi* Valenciennes; preoccupied; replaced by *Ophioblennius* Gill.

Hypsoblennius Gill, 1861, 302.
Salarichthys[489] Guichenot, 1867, 346.
Enchelyurus Peters, 1868, 353.
Alticus (Commerson) Bleeker, 1869, 353.
Isesthes Jordan & Gilbert, 1882, 420.
Scartella Jordan, 1886, 435.
Lipophrys Gill, 1896, 471.
Scartes Jordan & Evermann, 1896. 475.
Macrurrhynchus Ogilby, 1896, 475.
Homesthes Gilbert, 1898, 481.

Blenniolus Jordan & Evermann, 1898, 481.
Scartichthys Jordan & Evermann, 1898, 481.
Oncolepis Bassani, 1898, 478.
Brannerella Gilbert, 1900, 489.
Graviceps Fowler, 1903, 502.
Exallias Jordan & Evermann, 1905, 514.
Cyneichthys Ogilby, 1910, 536.
Parablennius Ribeiro, 1915, 557.

[489] Written also *Salariichthys.*

Family 590. EMBLEMARIIDÆ

This family and the three which follow are entirely provisional, designed to receive aberrant genera which by differences in dentition confuse the definition of *Blenniidæ*.

Emblemaria Jordan & Gilbert, 1882, 420.   Acanthemblemaria[490] Metzelaar, 1920.

[490] "Trop. Atl. Visschen," 159; orthotype *A. spinosa* Metzelaar.

Family 591. RUNULIDÆ

A group of naked blennies with the mouth small, transverse, and inferior, the members perhaps not related one to another.

Andamia Blyth, 1859, 291.   Runula Jordan & Bollman, 1889, 447.
Heteroclinus Castelnau, 1872, 363.   Runulops Ogilby, 1910, 536.

Family 592. ATOPOCLINIDÆ

Teeth solid, trenchant.

Atopoclinus Vaillant, 1894, 466.

Family 593. CHÆNOPSIDÆ

A provisional group of elongate, naked blennies of the tropics, perhaps unrelated among themselves, and resembling on one hand the *Pholidæ* and on the other the *Clinidæ*.

Pholidichthys Bleeker, 1856, 267.   Psednoblennius Jenkins & Evermann, 1888, 442.
Gunnellichthys Bleeker, 1858, 277.   Lucioblennius Gilbert, 1890, 451.
Chænopsis Gill, 1863, 323.
Stathmonotus Bean, 1885, 430.

Family 594. CEBIDICHTHYIDÆ

The three genera here associated differ from the allied *Pholidæ* in having the posterior half of the dorsal fin of soft rays.

Cebidichthys[491] Ayres, 1855, 262.   Zoarchias Jordan & Snyder, 1902, 499.
Neozoarces Steindachner, 1880, 406.

[491] Often misspelled *Cebedichthys.*

Family 595. PHOLIDÆ

A varied group of subarctic blennies which may require further sub-division.

Pholis Gronow, 1763, 19.

Pholis (Gronow) Scopoli, 1777, 42.

Pholis Röse, 1793, 52.

Pholidus Rafinesque, 1815, 88.

Murænoides Lacépède, 1800, 56.

Centronotus Bloch & Schneider, 1801, 58.

Dactyleptus Rafinesque, 1815, 88.

Gunnellus Cuvier & Valenciennes, 1836, 185.

Ophisomus Swainson, 1839, 202.

Chirolophis Swainson, 1839, 202.

Carelophus Kröyer, 1845, 226.

Dictyosoma Temminck & Schlegel, 1845, 227.

Apodichthys Girard, 1854, 258.

Blenniops Nilsson, 1855, 266.

Anoplarchus Gill, 1861, 305.

Asternopteryx (Rüppell) Günther, 1861, 307.

Urocentrus Kner, 1868, 352.

Opisthocentrus Kner, 1868, 352.

Gunnellops Bleeker, 1874, 372.

Plectobranchus Gilbert, 1890, 451.

Blenniophidium Boulenger, 1892, 457.

Plagiogrammus Bean, 1893, 460.

Xererpes Jordan & Gilbert, 1895, 469.

Bryostemma Jordan & Starks, 1895, 469.

Rhodymenichthys Jordan & Evermann, 1896, 475.

Ulvicola Gilbert & Starks, 1897, 477.

Alectrias Jordan & Evermann, 1898, 482.

Pholidapus Bean & Bean, 1896, 471.

Enedrias Jordan & Evermann, 1898, 481.

Eulophias Smith, 1902, 500.

Azuma Jordan & Snyder, 1902, 499.

Bryolophus J. & S., 1902, 499.

Abryois J. & S., 1902, 499.

Trigrammus Gracianov, 1907, 524.

Askoldia Pavlenko, 1910, 537.

Gymnoclinus Gilbert & Burke, 1912, 543.

Alectridium Gilbert & Burke, 1912, 543.

Family 596. XIPHISTERIDÆ (*Xiphidiontidæ*)

Xiphidion[492] Girard, 1857, 290.

Xiphister Jordan, 1879, 400.

Xiphistes Jordan & Starks, 1895, 469.

Phytichthys[493] Hubbs, 1923.

---

[492] Preoccupied by *Xiphidion* Seville, a genus of *Orthoptera,* the word later corrected to *Xiphidium. Xiphidion* was first used by Girard, *U. S. Pac. R. R. Surv.,* **6**: 20, 1857.

[493] Hubbs, ms.: new genus; orthotype *Xiphister chirus* Jordan & Gilbert; substitute for *Xiphistes,* preoccupied in insects.

Family 597. STICHÆIDÆ

Stichæus[494] Reinhardt, 1837.

Eumesogrammus Gill, 1864, 331.

Stichæopsis Steindachner & Kner, 1870, 360.

Notogrammus[495] Bean, 1881.

Dinogunnellus Herzenstein, 1890, 451.

Ulvarius Jordan & Evermann, 1896, 475.

Ernogrammus J. & E., 1898, 481.

Ozorthe J. & E., 1898, 481.

Trigrammus Gracianov, 1907, 524.

---

[494] *Dansk. Vid. Nat. Afh.,* 109; orthotype *Stichæ punctatus* Reinhardt. (Omitted in "Genera of Fishes.")

[495] Bean, *Proc. U. S. Nat. Mus.,* **4**: 147; orthotype *N. rothrocki* Bean. A synonym of *Stichæus.* (Omitted in "Genera of Fishes.")

Family 598.   LUMPENIDÆ

Lumpenus[496] Reinhardt, 1837.
Ctenodon Nilsson, 1855, 266.
Leptogunnellus Ayres, 1855, 262.
Leptoblennius Gill, 1860, 295.

Leptoclinus Gill, 1861, 302.
Centroblennius Gill, 1861, 302.
Anisarchus Gill, 1864, 331.
Poroclinus Bean, 1890, 450.

[496] *Dansk. Vid. Selsk. Nat.*, **6**: 110; orthotype *Blennius lumpenus* Fabricius, not of Linnæus. (Omitted in "Genera of Fishes.")

Family 599. PTILICHTHYIDÆ

Ptilichthys Bean, 1881, 415.

Family 600. CRYPTACANTHODIDÆ

Cryptacanthodes Storer, 1839, 197.
Delolepis Bean, 1881, 415.

Lyconectes Gilbert, 1895, 467.

Family 601. ANARHICHADIDÆ

Anarchichas[497] Linnæus, 1758, 12.
Lathargus[498] Klein, 1775, 38.
PALANARRHICHAS LeHon, 1871, 361.

Lycichthys Gill, 1877, 389.
LAPARUS (Agassiz) Woodward, (1845) 1901, 224.

[497] Often spelled *Anarrhichas*.
[498] Misprinted *Latargus*.

Family 602. ANARRHICHTHYIDÆ

Anarrhichthys Ayres, 1855, 262.

Family 603. XIPHASIIDÆ

Xiphasia[499] Swainson, 1839, 201.
Nemophis Kaup, 1858, 281.

Xiphogadus Günther, 1862, 318.
Plagiotremus Gill, 1863, 323.

[499] Also spelled *Ziphasia* by Swainson.

Family 604. XENOCEPHALIDÆ

Xenocephalus Kaup, 1858, 281.

Series *ZOARCIFORMES*

Family 605. CONGROGADIDÆ

Haliophis Rüppell, 1828, 122.
Machærium Richardson, 1843, 216.
Congrogadus Günther, 1862, 318.

Halidesmus Günther, 1871, 361.
Blennodesmus Günther, 1871, 361.
Hierichthys Jordan & Fowler, 1902, 499.

Family 606. CERDALIDÆ (*Microdesmidæ*)

Microdesmus Günther, 1864, 333.
Cerdale Jordan & Gilbert, 1881, 416.

Leptocerdale Weymouth, 1910, 537.

Family 607. SCYTALINIDÆ

Scytalina Jordan & Gilbert, 1880, 404.

Scytaliscus Jordan & Gilbert, 1883, 424.

Family 608. ZOARCIDÆ (*Lycodidæ*)

Enchelyopus Gronow, 1763, 19.
Zoarces[500] Cuvier, 1829, 129.
Lycodes Reinhardt, 1838, 193.
Gymnelis Reinhardt, 1838, 193.
Phycocœtes Jenyns, 1842, 212.
Iluocœtes Jenyns, 1842, 212.
Uronectes Günther, 1862, 318.
Macrozoarces Gill, 1863, 325.
Maynea Cunningham, 1871, 361.
Paralycodes Bleeker, 1874, 372.
Lycodalepis Bleeker, 1874, 372.
Lycodopsis Collett, 1879, 399.
Leurynnis Lockington, 1879, 400.
Hypolycodes Hector, 1880, 403.
Melanostigma Günther, 1881, 416.
Bassozetus Gill, 1883, 424.
Lycocara Gill, 1884, 428.
Lycenchelys Gill, 1884, 428.
Gymnelichthys Fischer, 1885, 431.
Lycodophis Vaillant, 1888, 444.
Lyconema Gilbert, 1895, 467.

Nemalycodes Herzenstein, 1896, 472.
Furcella Jordan & Evermann, 1896, 475.
Aprodon Gilbert, 1890, 451.
Platea Steindachner, 1898, 484.
Lycias Jordan & Evermann, 1898, 481.
Embryx J. & E., 1898, 481.
Furcimanus J. & E., 1898, 482.
Selachophidium Gilchrist, 1903, 503.
Petrotyx Heller & Snodgrass, 1903, 503.
Eutyx Heller & Snodgrass, 1903, 503.
Krusensterniella Schmidt, 1904, 511.
Hadropareia Schmidt, 1904, 511.
Pachycara Zugmayer, 1911, 541.
Mastigopterus Radcliffe, 1913, 549.
Ophthalmolycus Regan, 1913, 550.
Austrolycus Regan, 1913, 550.
Austrolycichthys Regan, 1913, 550.
Crossolycus Regan, 1913, 550.
Zoarcites Zugmeyer, 1914, 555.
Lycogramma Gilbert, 1915, 556.

[500] Also spelled *Zoarchus*.

Family 609. LYCODAPODIDÆ

Lycodapus Gilbert, 1890, 451.

Snyderidia Gilbert, 1905, 513.

Family 610. DEREPODICHTHYIDÆ

Derepodichthys Gilbert, 1895, 467.

Series *BROTULIFORMES*

Family 611. BROTULIDÆ

Brotula[501] Cuvier, 1829.
Bythites Reinhardt, 1837, 190.
Tilurus[502] Kölliker, 1854, 254.
Dinematichthys Bleeker, 1855, 263.
Gadopsis Filippi, 1855, 265.
Brotulophis Kaup, 1858, 281.
Brotella Kaup, 1858, 281.
Hoplophycis Kaup, 1858, 281.
Sirembo Bleeker, 1858, 277.
Halias Ayres, 1859, 293.
Lucifuga Poey, 1860, 298.
Brosmophycis Gill, 1861, 305.

Nematobrotula Gill, 1863, 325.
Stygicola Gill, 1863, 325.
Hoplobrotula Gill, 1863, 325.
Acanthonus Günther, 1878, 394.
Bathynectes Günther, 1878, 394.
Aphyonus Günther, 1878, 394.
Typhlonus Günther, 1878, 394.
Dicrolene Goode & Bean, 1882, 419.
Lycodonus Goode & Bean, 1882, 419.
Poromitra Goode & Bean, 1882, 419.
Barathrodemus Goode & Bean, 1882, 419.

[501] *Régne Animal;* type *Enchelyopus barbatus* Bloch and Schneider. (Omitted in "Genera of Fishes.")

[502] This genus and its allies, *Tilurella, Tiluropis,* and *Grimaldichthys,* are larval forms, supposed to be of brotulids.

Bellottia Giglioli, 1883, 423.
Porogadus Goode & Bean, 1885, 432.
Bathyonus Goode & Bean, 1885, 432.
Neobythites Goode & Bean, 1885, 432.
Barathronus Goode & Bean, 1886, 435.
Myxocephalus Steindachner & Döderlein, 1887, 439.
Catætyx Günther, 1887, 437.
Pteroidonus Günther, 1887, 437.
Mixonus Günther, 1887, 437.
Diplacanthopoma Günther, 1887, 437.
Nematonus Günther, 1887, 437.
Alexeterion Vaillant, 1888, 444.
Pycnocraspedum Alcock, 1889, 444.
Paradicrolene Alcock, 1889, 444.
Saccogaster Alcock, 1889, 444.
Glyptophidium Alcock, 1889, 444.
Dermatorus Alcock, 1890, 449.
Monomitopus Alcock, 1890, 449.
Paradicrolene Alcock, 1890, 449.
Bothrocara Bean, 1890, 450.
Lamprogrammus Alcock, 1891, 454.
Hephthocara Alcock, 1892, 457.
Benthocometes Goode & Bean, 1895, 468.
Dicromita G. & B., 1895, 468.
Celema G. & B., 1895, 468.
Penopus G. & B., 1895, 468.
Mœbia G. & B., 1895, 468.
Alcockia G. & B., 1895, 468.
Dermatopsis Ogilby, 1896, 475.

Monothrix Ogilby, 1897, 477.
Ogilbia Jordan & Evermann, 1898, 480.
Diancistrus Ogilby, 1898, 483.
Holcomycteronus Garman, 1899, 486.
Pseudonus Garman, 1899, 486.
Leucicorus Garman, 1899, 486.
Sciadonus Garman, 1899, 486.
Eretmichthys Garman, 1899, 486.
Bothrocaropsis Garman, 1899, 486.
Watasea Jordan & Snyder, 1901, 494.
Dipulus Waite, 1905, 516.
Bassobythites Brauer, 1906, 517.
Barathrites Zumayer, 1911, 541.
Leucochlamys[503] Zugmayer, 1911.
Parabrotula[504] Zugmayer, 1911.
Tilurella Roule, 1911, 541.
Tiluropsis Roule, 1911, 541.
Xenobythites Radcliffe, 1913, 549.
Pyramodon Radcliffe, 1913, 549.
Hypopleuron Radcliffe, 1913, 549.
Mastigopterus Radcliffe, 1913, 549.
Homostolus Radcliffe, 1913, 549.
Enchelybrotula Radcliffe, 1913, 549.
Hypopleuron Radcliffe, 1913, 549.
Luciobrotula Radcliffe, 1913, 549.
Grimaldichthys Roule, 1913, 551.
Pyramodon Regan, 1914, 555.
Spectrunculus Jordan & Thompson, 1914, 553.
Cynophidium Regan, 1914, 554.

---

[503] *Rés. Camp. Sci. Monaco,* 131; type *Leucochlamys cryptophthalmus* Zugmayer. (Omitted in "Genera of Fishes.")

[504] *Loc. cit.,* 129; orthotype *Parabrotula plagiosphelmus* Zugmayer. (Omitted in "Genera of Fishes.")

Series *OPHIDIIFORMES*
Family 612. **RHODICHTHYIDÆ**

Rhodichthys Collett, 1878, 393.

Family 613. **OPHIDIIDÆ**

Ophidion[505] Linnæus, 1758, 12.
Xiphiurus Smith, 1849, 244.
Cepolophis Kaup, 1856, 272.
Genypterus Philippi, 1857, 276.
Leptophidium Gill, 1863, 324.

Otophidium Gill, 1885, 433.
Lepophidium Gill, 1895, 467.
Chilara Jordan & Evermann, 1896, 475.
Rissola J. & E., 1896, 475.
Pseudophidium Gracianov, 1907, 524.

---

[505] Often spelled *Ophidium.*

238 [738]   XENOPTERYGII

Series CARAPIFORMES
Family 614. CARAPIDÆ (*Fierasferidæ*)
(Pearl Fishes)

Carapus[506] Rafinesque, 1810.
Fierasfer (Cuvier) Oken, 1817, 93, 101.
Echiodon Thompson, 1837, 190.
Encheliophis Müller, 1843, 215.
Diaphasia Lowe, 1843, 215.
Oxybeles Richardson, 1844, 223.
Porobronchus Kaup, 1860, 297.

Helminthodes Gill, 1864, 331.
Helminthostoma Cocco, 1870, 357.
Vexillifer Gasco, 1870, 357.
Lefroyia Jones, 1874, 375.
Rhizoiketicus Vaillant, 1893, 463.
Jordanicus Gilbert, 1905, 513.

[506] A genus based on a definition borrowed from Lacépède and including two unrelated forms; one *Fierasfer*; one *Gymnotus;* no type named. By a decision of the International Commission of Zoological Nomenclature, *Carapus* is held to replace *Fierasfer,* as the species first referred to it by a subsequent author was *Ohpidium imberbe* Linnæus.

Suborder HAPLODOCI
Family 615. BATRACHOIDIDÆ

Batrachoides Lacépède, 1800, 57.
Batrachus Bloch & Schneider, 1801, 57.
Batrictius Rafinesque, 1815, 88.
Opsanus Rafinesque, 1818, 108.
Amphichthys Swainson, 1839, 202.
Porichthys Girard, 1854, 258.
Thalassophryne Günther, 1861, 307.
Halophryne Gill, 1863, 324.
Pseudobatrachus Castelnau, 1875, 378.

Phrynotitan Gill, 1885, 432.
Marcgravia Jordan, 1886, 435.
Thalassothia Berg, 1895, 466.
Dæctor Jordan & Evermann, 1898, 481.
Coryzichthys Ogilby, 1908, 529.
Halobatrachus Ogilby, 1908, 529.
Batrachomœus Ogilby, 1908, 529.
Marcgravichthys Ribeiro, 1915, 557.
Nautopædium Jordan, 1919, 567.

Order XENOPTERYGII *(Xenopteri)*
Family 616. GOBIESOCIDÆ

Lepadogaster Gouan, 1770, 28.
Gobiesox Lacépède, 1800, 57.
Piescephalus Rafinesque, 1810, 81.
Megaphalus Rafinesque, 1815, 89.
Gouania Nardo, 1832, 176.
Rupisuga Swainson, 1839, 206.
Cotylis Müller, 1843, 215.
Sicyases Müller, 1843, 216.
Trachelochismus de Barneville, 1846, 228.
Chorisochismus de Barneville, 1846, 228.
Sicyogaster Brisout de Barneville, 1846, 228.
Tomicodon Brisout de Barneville, 1846, 228.
Apepton Gistel, 1848, 236.

Leptopterygius Troschel, 1860, 300.
Crepidogaster Günther, 1861, 307.
Diplocrepis Günther, 1861, 307.
Caularchus Gill, 1862, 317.
Mirbelia Canestrini, 1864, 329.
Rimicola Jordan & Evermann, 1896, 472.
Arbaciosa J. & E., 1896, 472.
Bryssetæres J. & E., 1896, 472.
Caulistius J. & E., 1896, 475.
Bryssophilus J. & E., 1898, 481.
Aspasma Jordan & Fowler, 1902, 498.
Lepadichthys Waite, 1904, 511.
Aspasmagaster Waite, 1907, 526.
Bulbiceps Jordan, 1919, 568.
Cotylichthys Jordan, 1919, 567.

Order PLECTOGNATHI

Suborder *SCLERODERMI*

This group is closely related to the *Squamipinnes,* especially to the *Acanthuridæ,* from ancestors of which it must have been derived.

Family 617. SPINACANTHIDÆ

Spinacanthus Agassiz, 1844, 219.

Protobalistum (Massalongo) Zigno, 1887, 439.

Family 618. TRIACANTHIDÆ

Triacanthus (Cuvier) Oken, 1817, 98.
Acanthopleurus Agassiz, 1844, 217.
Triacanthodes Bleeker, 1858, 277.

Hollardia Poey, 1861, 308.
Halimochirus Alcock, 1899, 484.
Tydemania Weber, 1913, 551.

Family 619. BALISTIDÆ

Balistes Linnæus, 1758, 15.
Capriscus Klein, 1777, 43.
Capriscus Röse, 1793, 52.
Capriscus Rafinesque, 1810, 82.
Balistapus Tilesius, 1820, 113.
Xenodon Rüppell, 1835, 184.
Pachynathus[507] Swainson, 1839, 199.
Zenodon[508] Swainson, 1839, 204.
Rhinecanthus Swainson, 1839, 204.
Chalisoma Swainson, 1839, 204.
Canthidermis Swainson, 1839, 204.
Melichthys[509] Swainson, 1839, 204.
Capriscus Swainson, 1839, 204.
Leiurus Swainson, 1839, 205.

Ancistrodon[510] (Debey) Roemer, 1843.
Acanthoderma Agassiz, 1844, 217.
Agoreion Gistel, 1848, 237.
Odonus Gistel, 1848, 238.
Erythrodon Rüpell, 1852, 251.
Pyrodon Kaup, 1855, 265.
Xanthichthys Kaup, 1856, 274.
Pseudobalistes Bleeker, 1866, 340.
Parabalistes Bleeker, 1866, 340.
Ankistrodus Koninck, 1870, 358.
Protacanthodes Gill, 1888, 441.
Grypodon[511] Hay, 1899.
Abalistes Jordan & Seale, 1906, 519.
Sufflamen Jordan, 1916, 559.

---

[507] Regarded by the International Commission as a misprint for *Pachygnathus,* a name already in use.

[508] Perhaps a misprint for *Xenodon.*

[509] Written later *Melanichthys.*

[510] "Texas," 420; orthotype *Ancistrodon texanus* Roemer. (Omitted in "Genera of Fishes.")

[511] Substitute for *Ancistrodon;* preoccupied in snakes.

Family 620. MONACANTHIDÆ

Unicornis Catesby, 1771, 31.
Monoceros (Plumier) Lacépède, 1800, 71.
Monacanthus (Cuvier) Oken, 1817, 98.
Alutera[512] (Cuvier) Oken, 1817, 98.
Amanses Gray, 1833, 139, 179.
Cantherines Swainson, 1839, 205.
Chætodermis Swainson, 1839, 205.

Trichoderma Swainson, 1839, 205.
Stephanolepis Gill, 1861, 304.
Ceratacanthus Gill, 1861, 303.
Pseudaluterius[513] Bleeker, 1865, 336.
Pseudomonacanthus Bleeker, 1866, 340.
Acanthaluteres Bleeker, 1866, 340.
Brachaluteres Bleeker, 1866, 340.

---

[513] Variously written *Aleuteres, Alutarius, Aleisterius,* etc.

Pseudaluteres[513] Bleeker, 1866, 340.
Paraluteres Bleeker, 1866, 340.
Oxymonacanthus Bleeker, 1866, 340.
Paramonacanthus Bleeker, 1866, 340.
Liomonacanthus Bleeker, 1866, 340.

Paramonacanthus Steindachner, 1867, 348.
Osbeckia Jordan & Evermann, 1896, 474.
Rudarius Jordan & Fowler, 1902, 498.
Davidia Ribeiro, 1915, 557.

---

[513] Same as *Pseudalutarius*.

Family 621. PSILOCEPHALIDÆ

Anacanthus Gray, 1833.
Psilocephalus Swainson, 1839, 205.

Pogonognathus Bleeker, 1849, 240.

Suborder OSTRACODERMI
Family 622. OSTRACIIDÆ

Ostracion Linnæus, 1758, 15.
Acarana Gray, 1833, 139, 179, 192.
Lactophrys Swainson, 1839, 204.
Rhinesomus Swainson, 1839, 204.
Tetrosomus Swainson, 1839, 204.
Platycanthus Swainson, 1839, 204.
Anoplocapros Kaup, 1855, 265.
Kentrocapros Kaup, 1855, 265.

Cibotion (Klein) Kaup, 1855, 265.
Capropygia Gray, 1855, 265.
Acanthostracion Bleeker, 1866, 339.
Chapinus Jordan & Evermann, 1896, 474.
Lactoria Jordan & Fowler, 1902, 498.
Caprichthys McCulloch & Waite, 1915, 557.

Suborder GYMNODONTES
Family 623. TRIODONTIDÆ

Triodon Cuvier, 1829, 131.

Family 624. TETRAODONTIDÆ (*Chonerhinidæ*)

Tetraodon[514] Linnæus, 1758, 15.
Orbis Catesby, 1771, 31.
Crayracion Klein, 1777, 42.
Ovoides[515] Lacépède, 1798.
Sphœroides[515] Lacépède, 1798.
Ovoides Cuvier, 1800, 55.
Orbis (Plumier) Lacépède, 1800, 71.
Ovum Bloch & Schneider, 1801, 60.
Spheroides (Lacépède) Duméril, 1806, 55, 75.
Ovoides (Lacépède) Duméril, 1806, 75.
Orbis Fischer, 1813, 85.
Oonidus Rafinesque, 1815, 91.
Orbidus Rafinesque, 1815, 91.

Physogaster Müller, 1839, 196.
Arothron Müller, 1839, 196.
Cheilichthys Müller, 1839, 196.
Chelonodon Müller, 1839, 196.
Leiodon Swainson, 1839, 198.
Lagocephalus Swainson, 1839, 205.
Cirrhisomus Swainson, 1839, 205.
Leisomus Swainson, 1839, 205.
Gastrophysus Müller, 1843, 216.
Anchisomus Kaub, 1854, 261.
Holacanthus Gronow, 1854, 258.
Chonerhinos[516] Bleeker, 1854, 256.
Dichotomycter Bibron, 1855, 263.
Epipedorhynchus Bibron, 1855, 263.

---

[514] Often written *Tetrodon*.

[515] In the *Allgemeine Literar Zeitung*, Berlin, No. 287, 1798, is an unsigned review of Lacépède's Volumes I and II, Latin names are supplied page 674 to "les Ovoides," and page 675 to "les Spheroides." *Sphœroides* of this reviewer, 1798, replaces *Spheroides* Duméril, 1806. For this reference I am indebted to Mr. Remington Kellogg.

[516] Usually corrected to *Chonerhinus*.

Geneion Bibron, 1855, 263.
Catophorhynchus Bibron, 1855, 263.
Batrachops Bibron, 1855, 263.
Monotreta[517] Bibron, 1855, 263.
Ephippion Bibron, 1855, 263.
Aphanacanthus Bibron, 1855, 263.
Xenopterus Bibron, 1855, 263.
Amblyrhynchotus Bibron, 1855, 263.
Stenometopus Bibron, 1855, 263.
Dilobomycter Bibron, 1855, 263.
Promecocophalus Bibron, 1855, 262.

Apsicephalus Hollard, 1857, 275.
Brachycephalus Hollard, 1857, 275.
Pleuranacanthus (Bibron) Bleeker, 1865, 336.
Uranostoma (Bibron) Bleeker, 1865, 336.
Crayracion (Klein) Bleeker, 1865, 339.
Hemiconiatus Günther, 1870, 357.
Liosaccus Günther, 1870, 357.
Guentheridia Gilbert & Starks, 1904, 508.

[517] Also spelled *Monotretus.*

Family 625. CANTHIGASTERIDÆ (*Tropidichthyidæ*)
(Sharp-nosed Puffers)

Canthigaster Swainson, 1839, 198.
Psilonotus Swainson, 1839, 205.
Prilonotus[518] (Kaup) Müller, 1854, 261.
Tropidichthys Bleeker, 1854, 256.

Rhynchotus Bibron, 1855, 263.
Anosmius Peters, 1855, 266.
Eumycterias Jenkins, 1901, 493.

[518] Perhaps a slip for *Psilonotus.*

Family 626. DIODONTIDÆ

Diodon Linnæus, 1758, 15.
Orbis Müller (P. L. S.),1767, 24.
Chilomycterus Bibron, 1846, 228.
MEGALURITES Costa, 1850, 245.
ENNEODON Heckel, 1854, 259.
Cyclichthys Kaup, 1855, 265.
Cyanichthys Kaup, 1855, 265.
Dicotylichthys Kaup, 1855, 265.
HEPTADIODON Bronn, 1855, 264.
Atopomycterus[519] (Verraux) Bleeker, 1865, 336.

Paradiodon Bleeker, 1865, 336.
Trichodiodon Bleeker, 1866, 340.
Trichocyclus Günther, 1870, 357.
GYMNODUS Delfortrie, 1871, 361.
PROGYMNODON Dames, 1883, 422.
Lyosphæra Evermann & Kendall, 1898, 480.
Euchilomycterus Waite, 1900, 491.
Allomycterus[520] McCulloch, 1921.

[519] Orthotype *A. diversispinis* Verraux; incorrectly stated in "Genera of Fishes."
[520] *Rec. Austr. Mus.,* 13 : 141; orthotype *Diodon jaculiferus* Cuvier.

Family 627. MOLIDÆ (*Orthagoriscidæ*)

Mola Kœlreuter, 1770, 28.
Mola Linck, 1790, 49.
Mola Cuvier, 1798, 54.
Orthagoriscus Bloch & Schneider, 1801, 60.
Cephalus Shaw, 1804, 73.
Orthragus Rafinesque, 1810, 78.
Diplanchias Rafinesque, 1810, 78.
Pedalion Swainson, 1838, 193.
Ozodura Ranzani, 1839, 196.

Tympanomium Ranzani, 1839, 196.
Trematopsis Ranzani, 1839, 196.
Pallasia Nardo, 1839, 196.
Ranzania Nardo, 1839, 196.
Molacanthus Swainson, 1839, 205.
Acanthosoma Dekay, 1842, 210.
Centaurus Kaup, 1855, 265.
Masturus Gill, 1884, 428.
Chelonopsis[521] Jordan, 1900.

[521] "Guide to the Study of Fishes," II, 425. This name, never intended as generic, is a printer's error for *Mola chelonopsis,* a fossil *Mola.*

Order PEDICULATI

Family 628. LOPHIIDÆ

Lophius Linnæus, 1758, 11.

Batrachus Klein, 1775, 40.

Lophidius Rafinesque, 1815, 92.

Lophiopsides Guichenot, 1867, 345.

Lophiomus Gill, 1882, 419.

Lophiodes Goode & Bean, 1895, 468.

Chirolophius Regan, 1903, 505.

Sladenia Regan, 1908, 530.

Family 629. ANTENNARIIDÆ

Antennarius (Commerson) Lacépède, 1798, 69.

Histrio[522] Fischer, 1813, 84.

Antennarius (Commerson) Cuvier, 1817, 104.

Chironectes Cuvier, 1817, 96.

Batrachopus[523] Goldfuss, 1820, 114, 172.

Capellaria Gistel, 1848, 235.

Pterophryne Gill, 1863, 324.

Saccarius Günther, 1861, 307.

Histiophryne Gill, 1863, 324.

Pterophrynoides[524] Gill, 1878.

Tetrabrachium Günther, 1880, 403.

Histiocephalus Zigno, 1890, 453.

Histionotopterus Eastman, 1904, 507.

Trianectes McCulloch, 1918, 564.

Echinophryne McCulloch, 1918, 564.

Trichophryne McCulloch, 1918, 564.

[522] This name may be questioned on account of an error in the published diagnosis, which transposes the generic character of *Lophius* and *Histrio*.

[523] Misprinted *Batrachops* on page 114.

[524] *Proc. U. S. Nat. Mus.*, 1 : 215; orthotype *Lophius histrio* Gill; a substitute for *Pterophryne*, regarded as preoccupied by *Pterophrynus*, a synonym of *Histrio* if the latter is eligible. (Omitted in "Genera of Fishes.")

Family 630. BRACHIONICHTHYIDÆ

Brachionichthys Bleeker, 1855, 263.

Sympterichthys Gill, 1878, 394.

Family 631. CHAUNACIDÆ

Chaunax Lowe, 1846, 229.

Family 632. OGCOCEPHALIDÆ (*Oncocephalidæ; Malthidæ*)

Ogcocephalus[525] Fischer, 1813, 85.

Malthe[526] Cuvier, 1817, 105.

Halieutæa Cuvier & Valenciennes, 1837, 189.

Astrocanthus Swainson, 1839, 205.

Halieutichthys Poey, 1863, 324.

Dibranchus Peters, 1875, 379.

Halieutella Goode & Bean, 1885, 432.

Allector Heller & Snodgrass, 1903, 503.

Cœlophrys Brauer, 1902, 497.

Brephostoma Alcock, 1889, 444.

Halieutopsis Garman, 1899, 486.

Dibranchichthys Garman, 1899, 486.

Dibranchopsis Garman, 1899, 486.

Halicmetus Alcock, 1891, 454.

Malthopsis Alcock, 1891, 454.

Zalieutes[527] Jordan & Evermann, 1896.

Æschynichthys Ogilby, 1907, 526.

Tathicarpus Ogilby, 1907, 526.

Rhycherus Ogilby, 1907, 526.

Dermatias Radcliffe, 1912, 544.

[525] Sometimes corrected to *Oncocephalus*.

[526] Afterwards written *Malthæa*.

[527] Jordan & Evermann, "Check List," 511; type *Malthe elater* Jordan & Gilbert. (Omitted in "Genera of Fishes.")

Family 633. CERATIIDÆ

Ceratias Kröyer, 1845, 227.
Mancalias Gill, 1878, 394.
Typhlopsaras Gill, 1883, 423.
Cryptopsaras Gill, 1883, 423.
Miopsaras Gilbert, 1905, 513.

Paraceratias Tanaka, 1908, 531.
Neoceratias Pappenheim, 1914, 554.
Monoceratias Gilbert, 1915, 556.
Thaumatichthys Smith & Radcliffe, 1912, 546.

Family 634. HIMANTOLOPHIDÆ

Himantolophus Reinhardt, 1838, 193.
Oneirodes[528] Lütken, 1871, 362.
Corynolophus[529] Gill, 1878.
Ægæonichthys Clarke, 1878, 394, 575.

Diceratias Günther, 1887, 437.
Linophryne Collett, 1886, 434.
Paroneirodes Alcock, 1890, 449.
Dolopichthys Garman, 1899, 486.

[528] Misprinted *Oneiroides* in "Genera of Fishes."
[529] *Proc. U. S. Nat. Mus.*, 219; type *Himantolophus reinhardti* Lütken. (Omitted in "Genera of Fishes.")

Family 635. MELANOCETIDÆ

Melanocetus Günther, 1864, 333.   Lyocetus Günther, 1887, 437.

Family 636. CAULOPHRYNIDÆ

Caulophryne Goode & Bean, 1895, 468.

Family 637. ACERATIIDÆ

Aceratias Brauer, 1902, 497.   Haplophryne Regan, 1912, 545.

Family 638. GIGACTINIDÆ

Gigantactis Brauer, 1902, 497.

NOTE.—In the Report of the Fisheries and Marine Biological Survey of South Africa, received as these pages go through the press, are two important papers, the first on "The *Heterosomata* (Flat Fishes)" by Cecil von Bonde, the other "Deep Sea Fishes, Part I," by J. D. F. Gilchrist. The following new genera are proposed:

Cynoglossoides von Bonde, 1822, 23; orthotype *Cynoglossus attenuatus* Gilchrist (*Cynoglossidæ*).
Atractophorus Gilchrist, 48; orthotype *A. annectens* Gilchrist (*Squalidæ*).
Bathymacrops Gilchrist, 53; orthotype *B. macrolepis* G. (*Bathylagidæ*).
Pelecinomimus Gilchrist, 56; orthotype *P. picklei* G. (*Cetomimidæ*).
Lyconodes Gilchrist, 59; orthotype *L. argenteus* G. (*Macrouridæ*).
Neoscombrops Gilchrist, 67; orthotype *N. annectens* G. (*Scombropidæ*).
Parasphenanthias Gilchrist, 69; orthotype *P. weberi* G. (*Serranidæ*).
Xenolepidichthys Gilchrist, 73; orthotype *X. dalgleishi* G. (*Grammicolepidæ*).

(Note from page iii of Index A, original publication.)

Index

A. NAMES OF GENERA

All initial numerals in lightface type refer to the page numbers in *The Genera of Fishes*. Numbers in boldface type (**596**) refer to family numbers in *A Classification of Fishes*. Names of genera included in *Classification* but not placed in a numbered family are referred to by the adjunct page number: (p. [597]). All italic numbers (*532*) refer to footnotes relating to genera listed under that family number in *Classification*. Obvious misspellings have been corrected by the compilers of the index and are shown thus: Branchoica [Branchioica]. Where changes would cause a significant shift in position, a cross-reference is made in the new position: [Typhlinus] Tiphlinus. *In all cases, the* bracketed *name or number is the correct, current one.* Some genera placed in the wrong families by Jordan have been correctly placed: [b. in **000**]. The numbered footnotes herein are taken verbatim from the original indexes of *Classification*.

748 INDEX

Alexurus, 468, **544**
Alfaro, 544, **294**
Algansea, 269, **237**
Algoa, 301, **310**
Algoma, 269, **237**
Alisea, 568, **147**
Alisodon, **237**, *237*
Allabenchelys, 497, **255**
Allector, 503, **632**
Allinectes, 482, **526**
Allochir, 474, **526**
Allocyttus, 554, **336**
Allogobius, 511, **545**
Allolepidotus, 445, *119*, **124**
Allomycterus, **626**, *626*
Allosomus, **165**, *165*
Allurus, 474, **526**
Alopecias, 192, **61**
Alopias, 27, 78, **61**
Alopiopsis, 338, **63**
Alosa, 49, 130, **147**
Alosina, 300, **147**
Alphestes, 51, 59, **426**
Alpismaris, 210, **271**
Alticus, 69, 353, **589**, *589*
Altona, 268, **231**
Alutarius, *620*
Alutera, 98, **620**
Alvarius, 290, **411**
Alvordius, 290, **411**
Alysia, 195, **278**
Amacanthus, 381, **47**
Amanses, 139, 179, **620**
Amarginops, 562, **248**
Amate, 519, **332**
Ambassis, 124, **417**
Amblodon, 110, **461**
Ambloplites, 111, **418**
Amblyapistus, 384, **493**
Amblyceps, 279, **249**
Amblychaeturichthys, 374, **545**
Amblycirrhitus, 313, **472**
Amblydoras, 311, **244**
Amblyeleotris, 372, **544**
Amblygaster, 240, **147**
Amblygobius, 373, **545**
Amblyopsis, 210, **298**
Amblyopus, 188, **547**
Amblypharyngodon, 288, **237**
Amblypomacentrus, 387, **532**
Amblypristis, 440, **72**
Amblypterus, 144, 177, **108**
Amblyraja, 391, **77**

Amblyrhynchichthys, 288, **237**
Amblyrhynchotus, 263, **624**
Amblyscion, 324, **461**
Amblysemius, 217, **124**
Amblytoxotes, 383, **486**
Amblyurus, 184, **119**
Ameibodon, 191, **90**
Ameiurus, 31, 112, **247**
Amia, 20, 23, 41, 46, 62, 259, **127, 133, 414**, *414*
Amiatus, 23, 91, **127**
Amiichthys, 435, **415**
Amioides, 544, **414**
Amiopsis, 327, **127**
Amitra, 402, **526**
Amitrichthys, 474, **526**, *526*
Amiurus, *247*
Ammocoetes, 97, *6*
Ammocoetus, 76, 170, **6**
Ammocrypta, 390, **411**
Ammodytes, 12, **573**
Ammopleurops, 278, 319, **335**
Ammotretis, 319, **331**
Amora, 179, **506**
Amorphocephalus, 119, **537**
Amphacanthus, 59, *490*, **492**
Ampheces, 499, **537**
Ampheristus, 118, **493**
Amphibichthys, 209, **106**
Amphicentrum, 344, **109**
Amphichthys, 202, **615**
Amphigonopterus, 563, **530**
Amphilaphurus, 431, **129**
Amphilius, 332, **251**
Amphilophus, 277, **534**
Amphiodon, 110, **146**
Amphioxides, 467, **1**
Amphioxus, 186, **2**
Amphiplaga, 388, **315**
Amphipnous, 196, **196**
Amphiprion, 39, 59, **532**
Amphiprionichthys, 263, **497**
Amphiscarus, 199, **540**
Amphisile, 38, 106, **362**
Amphisilen, 38, **362**
Amphistichus, 255, **530**
Amphistium, 181, 218, **480**
Amphodon, 436, **377**
Amphotistius, **80**, *80*
Amylodon, 466, **90**
Amyzon, 364, **236**
Anabas, 106, **369**

Anableps, 20, 42, 59, **296**
Anacanthus, 131, 138, **80, 621**
Anacyrtus, 14, 333, **226**
Anaedopogon, 360, **131**
Anaglyphus, 417, **108**
Anagramma, 487, **427**
Analithis, 237, **76**
Anampses, 117, 130, **537**
Anaora, *506*
Anapterus, 371, **274**
Anarchias, 519, **223**
Anarchichas, 12, **601**
Anarhichas, 38
Anarmostus, 328, **444**
Anarrhichas, 12
Anarrhichthys, 262, **602**
Anarrichas, *601*
Ancharius, 406, **243**
Anchisomus, 261, **624**
Anchovia, 468, 473, **150**
Anchoviella, 46, 67, 71, 539, **150**
Anchybopsis, 356, **237**
Ancistrodon, 251, **118, 619**, *619*
Ancistrodus, 462, **17**
Ancistrus, 260, **268**, *268*
Ancylodon, 104, **462**
Ancylopsetta, 331, **327**
Ancylostylos, 469, **138**
Andamia, 291, **591**
Andersonia, 488, **251**
Andreiopleura, 336, **139, 142**
Anema, 296
Anematichthys, 288, **237**
Anemura, 507, **450**, *450*
Anenchelum, 107, **380**
Anepistomon, 236, **219**, *219*
Angelichthys, 473, **488**
Anguilla, 73, **202**
Anguillavus, 503, **201**
Anguisurus, 271, **219**
Anisarchus, 331, **598**
Anisitsia, 502, **230**
Anisocentrus, 554, **372**
Anisochaetodon, 429, **488**
Anisochirus, 319, **334**
Anisotremus, 302, **444**
Ankistrodus, 358, **619**
Anodontacanthus, 415, **36**
Anodontoglanis[1], **245**
Anodontostoma, 241, **149**
Anodus, 132, **229**

---

[1] *Anodontoglanis* Rendahl, "Fish," N.W. Australia." *Zool. Mus. Kristiania*, 1922, 168; orthotype *A. dahli* Rendahl.

Bembras, 133, **506**
Bembrops, 387, **566**
Bendilisis, 288, **237**
Benedenichthys, 453, **109**
Benedenius, 396, **109**
Bengala, 139, 179, **237**
Bengalichthys, 531, **79**
Bennettia, 507, **441**
Bentenia, 494, **391**
Benthalbella, 541, **176**
Benthobatis, 478, **79**
Benthocometes, 468, **611**
Benthodesmus, 419, **380**
Benthophilus, 176, **545**
Benthosaurus, 435, **277**
Benthosema, 467, **278**
Benthosphyraena, 566, **151**
Bergia, 456, **226**
Bergiaria, 493, **261**
Bergiella, 489, **261**
Beridia, 388, **498a**
Bermudichthys, 572, **537**
Bero, 509, **511**
Bertoniolus, **226**, *226*
Berycopsis, 245, **341**
Beryx, 127, **342**
Besnardia, 549, **515**
Betta, 245, **368**, *368*
[Bhavania] see Blavania
Biat, 533, **545**
Bibronia, 219, **335**
Biotaecus, 502, **534**
Biotodoma, 502, **534**
Bipinnula, 474, **379**
Birgeria, **108**, *108*
Birkenia, 488, **21**
Bivibranchia, 542, **226**
Biwia, 505, **237**
Blakea, 387, **585**
Blanchardia, 378, **586**
Blavania [Bhavania][2], **240**
Bleeckeria, 368, **428**
Bleekeria, 318, **574**
Blennechis, 185, **589**
Blennicottus, 305, **511**
Blennioclinus, 295, **585**
Blenniolus, 481, **589**
Blenniomoeus, 245, **123**
Blenniophidium, 457, **595**
Blenniops, 266, **595**
Blennitrachus, 198, **589**
Blennius, 12, 33, 39, 41, 44, **589**, *598*
Blennodesmus, 361, **605**
Blennophis, 202, **585, 589**, *589*
Blennus, 44, **589**

Blepharichthys, 302, 410, **401**
Blepharis, 105, 129, **401**
Blepsias, 127, **509**
Bleptonema, 553, **226**
Blicca, 281, **237**
Bliccopsis, 328, **237**
Blochius, 54, **382**
Bodianus, 47, 49, 63, 72, 171, **426**, *426*, **536**
Boggiania, 478, **534**
Bogmarus, 60, **320**
Bogoda, 253, 275, **417**
Bogoslovius, 482, **307**
Bogota, 294, 383, **433**
Bola, 114, 351, **237, 461**
Bolcophthalmus, *546*
Boleichthys, 290, **411**, *411*
Boleophthalmus, 17, 188, **546**
Boleops, 326, **546**
Boleosoma, 210, **411**, *411*
Bollmannia, 447, **545**
Bonapartia, 467, **181**
Boops, 94, 102, 258, **433, 451**
Boopsetta, 470, **326**
Boopsidea, 301
Borborodes, 237, **69**
Borborys, 432, **290**
Boreocottus, 290, **511**
Boreogadus, 318, **310**
Boreogaleus, 306, **63**
Boreogobius, 326, **545**
Boreosomus, **108**, *108*
Boridia, 135, **449**
Borostomias, 529, **178**
Bostockia, 368, **426**
Bostrichthys, 61, 75, **544**
Bostrichus, 61
Bostrictis, 88, **544**
Bostrychoides, 61, **544**
Bostrychus, 61, **544**
Bothragonus, 420, **522**
Bothriolepis, 206, **19**
Bothrocara, 450, **611**
Bothrocaropsis, 486, **611**
Bothrolaemus, 265, **401**
Bothrosteus, 224
Bothus, 38, 79, **326**
Botia, 138, **239**
Boulengerella, 502, **226** [b. in **228**]
Boulengerina, 507, 523, 576, **420**
Boulengerochromis, 510, **534**

Bovichthys, *558*
Bovictus, 137, **558**
Bowenia, 369, **333**
Bowersia, 504, **441**, *441*
Box, 135, **451**
Boxaodon, 238, **457**
Brachaelurus, 519, **54**
Brachaluteres, 340, **620**
Brachiacanthus, 446 (p. [596])
Brachionichthys, 263, **630**
Brachioptera, 518, **82**
Brachioptilon, 244, **86**
Brachirus, 201, 203, **334**, *334*, **493**, *493*
Brachyacanthus, 295, **35**
Brachyalestes, 333, **226**
Brachyamblyopus, 374, **547**
Brachycephalus, 275, **624**
Brachychalcinus, 457, **226**
Brachyconger, 335, **209, 211**, *211*
Brachydanio, 561, **237**
Brachydeuterus, 313, **444**
Brachydirus, 404, 424, **26**
Brachyeleotris, 373, **544**
Brachygadus, 316, 325, **310**
Brachygenys, 353, **444**
Brachyglanis, 542, **261**
Brachygnathus, 224, 533, **410**
Brachygobius, 373, **545**
Brachygramma, 337, **237**
Brachyichthys, 321, **124**
Brachyistius, 316, **530**
Brachymesistius, 325
Brachymullus, 383, **460**
Brachymylus, 460, **90**
Brachymystax, 340, **165**
Brachyochirus, 209, **545**
Brachyopsis, 304, 305, **522**
Brachyplatystoma, 312, **261**
Brachypleura, 319, **330**
Brachypomacentrus, 387, **532**
Brachyprosopon, 310, **329**
Brachyrhamphichthys, 357, **232**
Brachyrhaphis, 550, **294**
Brachyrhinus, 315, **427**
Brachyrhynchus, 360, **383**
Brachyrus, 201, *334*, **493**
Brachysomophis, 271, **219**
Brachyspondylus, 339, **237**, *237*, **275**
Brachysynodontis, 311, **257**
Brachyurus, *334*, *493*

[2] *Bhavania* Hora, *Rec. Ind. Mus.*, 1920, 202; orthotype *B. annandalei* Hora.

[3] *Coius* Hamilton, "Genera of Fishes," 1822, 172.

Cryptops, 464, **234**
Cryptopsaras, 423, **633**
Cryptopterus, 297, **219**, *246*
Cryptosmilia, 344, **485**
Cryptotomus, 19, 361, **539**
Cryptotrema, 451, **585**
Crystallaria, 432, **411**
Crystallias, 499, **526**
Crystallichthys, 482, **526**
Crystallogobius⁴, 326, **545**
Ctenacanthus, 186, **47, 89**
Ctenobrycon, 527, **226**
Ctenochaetus, 428, **490**
Ctenocharax, 526, **226**
Ctenochromis, **534**, *534*
Ctenodax, 436, **397**
Ctenodentex, 476, 484, **450**
Ctenodon, 175, 201, 266, 429, **490, 598**
Ctenodus, 191, **103**
Ctenoglyphiodon [Ctenoglyphidodon], **532**, *532*
Ctenognathus, 409
Ctenogobius, 281, **545**
Ctenolabrus, 194, **536**
Ctenolates, 361, **426**
Ctenolepis, 217, 470, **130**
Ctenolucius, 302, **226**
Ctenopetalus, 320, 415, **41**
Ctenopharyngodon, 343, **237**
Ctenopleuron, 525, **21**
Ctenopoma, 229, 266, 270, 280, **369, 493**
Ctenops, 227, **368**
Ctenoptychius, 190, **41**, *41*, **43**
Ctenothrissa, 488, **350**
Ctenotrypauchen, 348, **548**
Cubiceps, 215, **396**, *396*
Cugupuguacu, 30, **426**
Culius, 267, 372, **544**
Culter, 262, **237**
Culticula, 491, **237**
Cultriculus, 570, **237**
Cunningtonia, 517, **534**
Curimata, 50, **229**
Curimatella, 446, 533, **229**, *229*
Curimatopsis, 387, **229**
Curimatus, 50, 98, **229**
Curtodus, 347, **47**
Curtus, *409*
Cyanichthys, 265, **626**
Cyathaspis, **13**, *13*
Cyathopharynx, 572, **534**

Cybium, 129, **377**
Cyclarthrus, 214, **74**
Cycleptus, 110, **236**
Cyclichthys, 265, **626**
Cyclobatis, 220, **80**
Cyclocheilichthys, 288
Cyclogaster, 18, 19, 41, 52, **526**
Cyclogobius, 299, **545**
Cycloides, 398, **135**
Cyclolepis, 349, **139, 166**
Cyclolumpus, 546, **524**
Cyclonarce, 307, **79**
Cyclopium, 204, **266**
Cyclopoma, 145, 177, **423**
Cyclopsetta, 441, 463, **326**
Cyclopterichthys, 417, **525**
Cyclopteroides, 457, **524**
Cyclopterus, 12, 42, **524**
Cycloptychius, 344, **108**
Cyclospondylus, 347, **122**
Cyclothone, 419, **181**
Cyclotomodon, 385, **132**
Cyclurus, 218, **127**
Cyema, 395, **218**
Cygnodraco, 561, **556**
Cylindracanthus, 273, **107**
Cylindrosteus, 112, **121**
Cymatodus, 359, 401, **41**
Cymatogaster, 257, **530**
Cymolutes, 318, **537**
Cynaedus, 19, 42, 200, **449, 536**
Cyneichthys, 536, **589**
Cynias, **63**, *63*
Cynichthys, 198, **426**
Cynicoglossus, 187, **329**
Cynocephalus, 27, 42, 306, 407, **63**
Cynocharax, 523, **226**
Cynodon, 132, **226**
Cynodonichthys, 510, **294** [b. in **290**]
Cynoglossa, 226, **329**
Cynoglossoides, **335** (p. [7743])
Cynoglossus, 114, **335**
Cynolebias, 387, **290**
Cynomacrurus, 532, **307**
Cynoperca, 390, **410**
Cynophidium, 554, **611**
Cynopodius, 417, **48**
Cynopoecilus, 545, **294** [b. in **290**]
Cynoponticus, 228, **209**

Cynopotamus, 242, **226**
Cynopsetta, 506, 519, **328**
Cynoscion, 304, **462**
Cynothrissa, 563, **147**
Cynotilapia, **534**, *534*
Cyphocharax, 518, **226**
Cyphomalepis, 273
Cyphotilapia, 572, **534**
Cyprinella, 270, **237**
Cyprinion, 211, **237**
Cyprinocirrhites, 563, **472**
Cyprinodon, 68, **290**
Cyprinopsis, 187, 339, **237**
Cyprinus, 15, 36, **237**
Cypselichthys, 426, **468**
Cypselurus, 203, **305**
Cypsilurus, *305*
Cyrene, 215, **237**
Cyrtacanthus, 371, **107**
Cyrtocara, 497, **534**
Cyrtocharax, 523, **226**
Cyrtonodus, 427, **44**
Cyrtorhynchus, 574, **278**
Cyrtus, 44, *409*
Cytharinus, *231*
Cyttoides, 436, **336**
Cyttomimus, 513, **336**
Cyttopsis, 314, **336**
Cyttosoma, 503, 504, 508, 576, **336**
Cyttula, 551, **336**
Cyttus, 575, **336**

Daba, 33, 34, 51, **426**
Dacentrus, 396, **531**
Dacodraco, 561, **557**
Dactylagnus, 318, **584**
Dactylanthias, 367, **427**
Dactyleptus, 88, **595**
Dactylobatus, 531, **77**
Dactylodus, 342, **41**
Dactylolepis, 433, **112**
Dactylopagrus, 313, **474**
Dactylophora, 427, **472**
Dactylopogon, 339, **278**
Dactyloptena, 528, **529**
Dactylopterus, 62, **529**
Dactylopus, 290, **561–562**
Dactylosargus, 313, **475**
Dactyloscopus, 290, **584**
Dactylosparus, 314, **474**
Dactylostomias, 486, **176**
Dacymba, 562, **144**
Daector, 481, **615**
Dagysa, 30

---

⁴ *Crystallogobius (Latrunculodes)* and *Boreogobius (Latrunculus,* preoccupied) are minute, transparent, annual gobies. *Aphia,* with forked caudal fin, is an atherine, not a goby.

[5] *Enantiopus* Boulenger, *loc. cit.*, 1906; orthotype *E. melanogenys* Boul.

[6] "Genera of Fishes," p. 570; the name *Hyostomus* in proof-sheets changed to *Hyomacrurus* in publication. *Hyomacrurus* has priority.

Leme, 427, **545**
Lemnisoma, 140, **379**
Lentipes, 307, **545**
Lepadichthys, 511, **616**
Lepadogaster, 28, **616**
Lepibema, 111, **423**
Lepicantha, 89, **345**
Lepidamia, 323, **414**
Lepidaplois, 314, 325, **536**
Lepidion, 203, **310**
Lepidoblennius, 348, 376, **585**
Lepidoblepharon, 551, **326**
Lepidocephalichthys, 322, **239**
Lepidocephalus, 279, **237**
Lepidochaetodon, 383, **488**
Lepidocottus, 380, **507**
Lepidocybium, 314, **377**
Lepidoglanis, 448, **252** [b. in **240**]
Lepidogobius, 280, **545**
Lepidolamprologus, 510, **534**
Lepidoleprus, 77, **307**
Lepidomeda, 375, **238**
Lepidomegas, 406, **401**
Lepidoperca, 554, **427**
Lepidopides, 246, **380**
Lepidopomus, 573
Lepidopsetta, 317, 331, 403, 575, **326, 329**
Lepidopterus, 459, **119**
Lepidopus, 28, 73, **380**
Lepidorhinus, 225, **67**
Lepidorhombus, 319, **326**
Lepidorhynchus, 398, **307**
Lepidosaurus, 141, **119**
Lepidosiren, 189, **106**
Lepidosoma, 407
Lepidosteus, 66, *121*
Lepidotes, 141, **119**
Lepidothynnus, 447, **377**
Lepidotrigla, 296, **528**
Lepidotus, *119*
Lepidozygus, 318, **532**
Lepimphis, 79, 90, **387**
Lepiopomus, 573, *418*
Lepipterus, 89, 135, **461**
Lepisacanthus, 62, **345**
Lepisoma, 210, **585**
Lepisosteus, 31, 44, 66, **121**
Lepodus, 37, 80, **388**
Lepomis, 110, 573, **418**
Lepomus, 89
Lepophidium, 467, **613**
Leporellus, 376, 410, **229**
Leporinus, 132, **229**
Lepracanthus, 355, **47**

Leptacanthus, 186, **90**
[Leptagoniates] Leptogoniates
Leptagonus, 305, **522**
Leptanthias, 565, **427**
Leptarius, 324, **243**
Leptaspis, 250, **401**
Leptecheneis, 330, **552**
Leptecodon, 488, **191**
Lepterus, 80, **393**
Leptichthys, 487, **147**
Leptoancistrus, 559, **268**
Leptobarbus, 288, 321, **237**
Leptoblennius, 295, 302, **598**
Leptobotia, 355, **239**
Leptobrama, 397, **434**
Leptobrycon, 555, **226**
Leptocarcharias, 357, **63**
Leptocarias, 190, **63**
Leptocephalichthys, 267, **211**
Leptocephalus, 22, 41, 262, **211**, *211*, **237**
Leptocerdale, 537, **606**
Leptocharias, 193, 573, **63**
Leptocheles, 408, **40**
Leptochilichthys, **151**, *151*
Leptochromis, 382, 572, **431, 534**
Leptoclinus, 302, **598**
Leptocottus, 257, **511**
Leptocypris, 488, **237**
Leptoderma, 436, 444, **151**
Leptodes, 203, **180**
Leptodoras, 479, **244**
Leptodus, 424, *26*
Leptogadus, 325, **310**
Leptogaster, 362, **148**
Leptoglanis, 496, 542, **248, 261**
Leptognathus, 205, **219**
Leptogobius, 373, 377, **545**
Leptogoniates [Leptagoniates], 437, **226**
Leptogrammatolepis, 349, 350, **131**
Leptogunnellus, 262, **598**
Leptoichthys, 254, **356**
Leptojulis, 309, **537**
Leptolepis, 141, 345, **130, 394**
Leptomylus, 354, **90**
Leptonotus, 253, **356**
Leptonurus, 241, **150**
Leptopegasus, 367, **357**
Leptoperca, 16, 306, **410**
Leptophidium, 324, **613**
Leptophilypnus, 559, **544**
Leptophycis, 486, **310**

Leptops, 112, **247**
Leptopterygius, 300, **616**
Leptopus, 86, **318**
Leptorhamdia, 550, **261**
Leptorhaphis, 550, **294**
Leptorhinophis, 271, **219**
Leptorhynchus, 244, 251, **217, 219**, *219*
Leptoscarus, 199, **539**
Leptoscolopsis, 558, **444**
Leptoscopus, 290, **583**
Leptosoma, 121, **333**
Leptosomus, 328, **278**
Leptostomias, 513, **176**
Leptostyrax, 491, **60**
Leptosynanceia, 374, **496**
Leptotrachelus, 328, **191**
Lepturacanthus, 512, **380**
Lepturichthys, 540, **240**
Lepturus, 259, 302, **307, 380**
Les Monoptères, 56, **194**
Lestichthys, 274, *274*
Lestidiops, 559, **274**
Lestidium, 513, **274**
Lesueuriella, 523
Lesueurina, 523, **571**
Letharchus, 419, **219**
Lethenteron, **6**
Lethostole, 473, **373**
Lethotremus, 469, **524**
Lethrinella, 508, **447**
Lethrinichthys, 544, **447**
Lethrinops, **534**, *534*
Lethrinus, 34, 128, **447**
Leucalburnus, 552, 558, **237**
Leucaspis [Leucaspius], 281, **237**
Leucichthyops, 566, **165**
Leucichthys, 375, **165**
Leucicorus, 486, **611**
Leucidius, 563, **237**
Leucisculus, 572, **237**
Leuciscus, 37, 99, **237**
Leucochlamys, **611**, *611*
Leucogobio, 471, **237**
Leucops, 110, **410**
Leucopsarion, 403, **545**
Leucoraja, 391, **77**
Leucos, 215, **237**
Leucosoma, 139, 211, **172, 237**
Leuresthes, 403, **373**
Leuroglossus, 457, **168**
Leurynnis, 400, **608**
Lewisia, 431, **131**
Liachirus, 319, **333**
Liauchenoglanis, 558, **248**
Libys, 212, **97**

Lysodermus, 520, **496**
Lythrichthys, 509, **493**
Lythrulon, 428, **444**
Lythrurus, 386, **237**
Lythrypnus, 474, **545**

Macaria, *383*
Maccullochia, 537, **453**
Maccullochina, **414**, *414*
Macdonaldia, 465, **190**
Macgregorella, 533, **545**
Machaera, *383*
Machaeracanthus, 276, **33**
Machaerium, 216, **605**
Machaerius, 284, **33**
Machaerochilus, 369, **237**
Machaerognathus, 566, **27**
Machaerope, 483, **379**
Machairodus, 409
Machephilus, 346, **67**
Macleayina, 523, **356**
Macolor, 294, **441**
Macquaria, 135, **426**
Macrepistius, 464, **123**
Macrhybopsis, 532, **237**
Macrias, 493, **565**
Macristium, 505, **151**
Macrocephalus, 47, 383, **421**
Macrochirichthys, 288, **237**
Macrochyrus, 201, **493**
Macrodon, 115, 216, **226**, **462**
Macrodonophis, 347, **219**
Macrognathus, 56, 59, 259, **199, 363**
Macrolepis, 89, 327, 346, **342, 414**
Macromesodon, 576, **118**
Macrones, 269, **248**
Macronoides, **248**, *248*
Macronotothen, 306, **555**
Macropetalichthys, 229, **23**
Macropharyngodon, 309, **537**
Macropharynx, 497, **225**
Macrophthalmia, 478, **7**
Macropleurodus[7], **534**
Macropodus, 63, **368**
Macropoma, 181, 217, 280, **97, 124**
Macrops, 269, *368*, **441**
Macropsobrycon, 555, **226**
Macropteronotus, 65, **255**
Macrorhamphosus, 66, **363**
Macrorhipis, 329, **129**

Macrorhynchus, 56, 75, **377, 379**
Macrosemius, 179, **123**
Macrostoma, 120, 191, **278, 480**
Macrostomias, 497, **176**
Macrotrema, 545, **195**
Macrouroides, 544, **306**
Macrourus, 44, **307**
Macrozoarces, 325, **608**
Macruronus, 365, **307**
Macruroplus, 372, **307**
Macrurrhynchus, 475, **589**
Macrurus, *307*
Macullochia, 537, **453**
Maena, 39, 128, **456**
Maenas, 39, 382, **456**
Maenichthys, 383, **530**
Maenioides, **447**, *447*
Maenoides, *447*
Maina, 236, **447**
Makaira, 65, **383**
Malacanthus, 130, **466**
Malacobagrus, 312, **261**
Malacobatis, 523, **77**
Malacocentrus, 314, **537**
Malacocephalus, 318, **307**
Malacocottus, 450, **511**
Malacoctenus, 295, **585**
Malacopterus, *536*
Malacorhina, 389, **77**
Malacosarcus, 437, **338**
Malacosteus, 240, **281**
Malakichthys, 426, **424**
Malapterurus, 65, **258**
Malapterus, **536**, *536*
Mallotus, 130, **167**
Malthaea, *632*
Malthe, 105, 130, **632**, *632*
Malthopsis, 454, **632**
Mancalias, 394, **633**
Mancopsetta, 514, **326**
Manducus, 468, **181**
Manta, 174, 573, **86**
Mapo, 487, **545**
Marcgravia, 435, **615**
Marcgravichthys, 557, **615**
Marcusenius, 314, **161**
Margariscus, 532, **237**
Margrethia, **185**, *185*
Markiana, 502, **226**
Marracanthus, 381
  (p. [596])
Mars, 519, **545**
Marsdenius, 501, **35**

Mascalongus, 396, **288**
Massaria, 236, **526**
Mastacembelus, 21, 40, 42, **199, 302**
Mastigopterus, 549, **608, 611**
Masturus, 428, **627**
Mataeocephalus, 484, **307**
Matsya, 445, **237**
Maturacus, 90
Maurolicus, 191, **183**
Mawsonia, 525, **93**
Maxillingua, 109, **237**
Mayerina, 557, **211**
Maynea, 361, **608**
Mayoa, 354, **237**
Mazodus, 447, **42**
Mearnsella, 530, **237**
Mecolepis, 273, **108**
Meda, 269, **238**
Medialuna, 462, **481**
Megabatus, 93
Megaderus, 92, **223**
Megagobio, 386, **237**
Megalamphodus, 555, **226**
Megalaspis, 248, **401**
Megalepis, 262, **460**
Megalichthys, 185, 217, **94**, *94*, **95**
Megalobrama, 364, **237**
Megalobrycon, 504, **226**
Megalocottus, 305, **511**
Megalodon, 181, **132**
Megalolepis, 400, **377**
Megalonema, 342, 535, **261**
Megalopharynx, 489, **225**
Megalops, 66, 70, 90, **133**
Megalopterus, 341, **128**
Megalurites, 245, **626**
Megalurus, 144, 177, **127**
Megaperca, 395, **426**
Megaphalus, 89, **616**
Megapleuron, 432, 459, **103**
Megaprotodon, 238, **488**
Megapus, 339, **269**
Megarasbora, 351, **237**
Megastoma, 226, **130**
Megastomatobus, 547, **236**
Megistopus, 465, **269**
Meladerma, 200, **405**
Melambaphes, 326, **451**
Melamphaës, 333, **339**
Melananosoma, 498
Melanichthys, 223, **451**, *619*
Melanictis, 92

[7] *Mucropleurodus* [*Macropleurodus*] Regan, "Cichlid Fishes of Lake Victoria," *Proc. Zool. Soc. Lond.*, 1922, 189; orthotype *Paratilapia bicolor* Boulenger.

[8] *Nomorhamphus* Weber & de Beaufort, "Indo-Austr. Fishes IV," 1922, 141; orthotype *N. celebensis* W. & de B.

[9] *Denks. Akad. Wien*, 1893, 83; orthotype *Auchenipterus hasemani* Steind.

[10] *Rhyacanthias* is apparently a synonym of *Symphysansdon*.

[11] Sanzo, *Mem. Acad. Roma*, 1920; orthotype *S. loviancae Sanzo* (larvae of some Stomiatid).

Taenioconger, **213**, *213*
Taeniodus, 425, **44**
Taenioides, 57, **547**
Taeniolabrus, 348, **571**
Taeniomembras, 483, **373**
Taenionema, 522, **261**
Taeniophis, 297, **223**
Taeniopsetta, 513, **326**
Taeniotoca, 300, **530**
Taeniura, 189, 193, **80**
Tahhmel, 34, **452**
Taius, 544, **449**
Talismania, 467, **151**
Tambra, 288, **237**
Tamiobatis, 476, **45**
Tanakia, 553, **237**
Tanakius, 564, **329**
Tanaodus, 380, **41**
Tandanus, 196, **245**
Tangus, 91
Tannayia [Taunayia], 564, **244**
Taractes, 215, **388**
Tarandichthys, 472, **508**
Tarletonbeania, 454, **278**
Tarphops, 553, **327**
Tarpon, 473, **133**
Tarrasius, 418, **98**
Tarsichthys, 261, **237**
Tarsistes, 567, **74**
Tasica, 88
Tathicarpus, 526, **632**
Tatia, 545, **244**
[Taunayia] Tannayia
Taunis, 88
Tauredophidium, 449, **151**
Taurichthys, 128, **488**
Tauridea, 395, **511**
Taurinichthys, 329, **539**
Taurocottus, 557, **511**
Taurulus, 524, **508**
Tautoga, 57, 86, **536**, **537**
Tautogolabrus, 318, **536**
Tegeolepis, 458, **108**
Tekla[12], **585**
Telara, 351, **150**
Teleogramma, 485, **536** [b. in **534**]
Teleosaurus, 346, *113*
Telepholis, 339, **139**
Telescopias, 495, **414**
Telescops, 382, **414**
Telestes, 141, 187, **237**
Telipomis, 111, **418**
Tellia, 253, **290**
Telmatherina, 476, **372**
Telmatochromis, 479, **534**

Temeculina, 532, **237**
Temera, 138, **79**
Temnistia, 186, **508**
Temnodon, 106, **404**
Tephraeops, 291, **451**
Tephrinectes, 319, **327**
Tephritis, 318, **327**
Terapon, 104, *445*
Teratichthys, 172, 419, **190**, **401**
Teratorhombus, 416, **327**
Teretulus, 113, **236**
Tetheodus, 375, **175**
Tetrabrachium, 403, **629**
Tetrabranchus, 248, **195**
Tetracentrum, 425, **417**
Tetrades, 335
Tetradrachmum, 241, **532**
Tetragonolepis, 136, 177, **119**
Tetragonopterus, 93, 96, **226**
Tetragonoptrus, 13, 40, 348, 383, **488**
Tetragonurus, 77, **397**
Tetranarce, 307, **79**
Tetranematichthys, 312, **244**
Tetranesodon, 551, **243**
Tetraodon, 15, 31, 42, 55, 71, 167, **624**
Tetrapleurodon, **6**
Tetrapturus, 80, **383**
Tetraroge, 296, **493**
Tetrodon, *624*
Tetronarce, *79*
Tetronarcine, 531, **79**
Tetroras, 77, **62**
Tetrosomus, 204, **622**
Teuthis, 20, 23, 32, 38, 41, 43, 48, 59, 64, **490**, *490*, **492**, *492*
Teuthys, *490*
Thaerodontis, 220, **223**
Thalassoklephtes, 235, **53**
Thalassoma, 199, **537**
Thalassophryne, 307, **615**
Thalassorhinus, 192, **63**
Thalassothia, 466, **615**
Thaleichthys, 290, **167**
Thalliurus, 200, **537**
Tharrhias, 528, **130**
Tharsis, 233, **130**
Thaumas, 212, **71**
Thaumastomias, 449, **176**
Thaumatacanthus, 401, **47**
Thaumatichthys, 546, **633**
Thaumaturus, 222, **163**
Thayeria, 527, **226**

Thecopterus, 511, **511**
Thectodus, 221, **47**
Thelodus, 194, 309, **8**
Thelolepis, 274
Theragra, 481, **310**
Therapaina, 298, 383, **488**
Therapon, 34, 104, **445**
Theraps, 318, **534**
Therobromus, 482, **167**
Theutis, 23, *490*
Theutys, 23
Thlattodus, 342, **124**
Thoburnia, 562, **236**
Tholichthys, 351, **488**
Tholodus, 229, 249, **112**
Thoracocharax, 523, **227**
Thoracodus, 422, **41**
Thoracopterus, 279, **128**
Thorichthys, 510, **534**
Threpterius, 247, **473**
Thrinacodus, 380, **36**
Thrissa, 68, 90, 98, **149**, **150**
Thrissocles, 98, 130, 151, 562, **150**
Thrissomimus, 539, **165**
Thrissonotus, 217, **108**
Thrissopater, 365, **131**
Thrissops, 144, 177, **130**
Thrissopteroides, 370, 484, **131**
Thrissopterus, 270, **139**
Thrycomycterus, 170, **264**
Thryptodus, 490, **136**
Thryssa, *150*
Thunnus, 227, **378**
Thursius, 443, **95**
Thyellina, 214, **52**
Thyestes, 256, **16**
Thymalloides, 527, **166**
Thymallus, 49, 130, **166**
Thynnichthys, 288, 402, **237**, **378**
Thynnus, 48, 105, 259, **378**, **401**
Thyrina, 468, **373**
Thyrinops, 564, **373**
Thyris, 403, **326**
Thyriscus, 543, **508**
Thyrsion, 571, **377**
Thyrsites, 129, **379**
Thyrsitocephalus, 292, **379**
Thyrsitops, 314, **379**
Thyrsocles, 571, **377**
Thyrsoidea, 271, **223**
Thysanichthys, 509, **493**
Thysanocara, 520, **268**
Thysanocheilus, 338, **537**

[12] *Tekla* Nichols, "Copeia," 1922, 69; orthotype *Crmnobates* fasciatur Steindachner.

Tropidichthys, 256, **625**
Tropidinius, 353, **441**
Tropidocaulus, 569, **302**, *302*
Tropidodus, 317, **47**
Tropidostethus, 469, **373**
Trossulus, **274**, *274*
Trulla, 282, **335**
Trutta, 14, 24, 37, **163**
Truttae, 14, 37, **163**
Trycherodon, 420, **237**
Trygon, 39, 98, 120, 172, **80**, *80*
Trygonobatus, 94, **80**
Trygonoptera, 193, **80**
Trygonorrhina, 192, **74**, *74*
Trypauchen, **548**, *548*
Trypauchenichthys, 293, 374, **548**
Trypauchenophrys, 535, **548**
Trypauchenopsis, 506, **548**
Tubulacanthus, 446
  (p. [596])
Tunita, 569, **377**
Turdus, 30, 39, **441**
Turio, 571, **377**
Turseodus, 283, **109**
Twingonia, 529, **461**
Tydemania, 551, **617**
Tydeus, 359, **278**
Tylobronchus, 542, **226**
Tylochromis, 572, **534**
Tylodus, 232, 462, **29**, *29*, **101**
Tylognathus, 211, **237**
Tylometopon, 367, 383, **388**
Tylosurus, 143, 178, **302**
Tympanomium, 196, **627**
Tympanopleura, 542, **244**
Tyntlastes, 319, **547**
Typhle, 78, 170, **356**, *356*
Typhlichthys, 290, **298**
[Typhlinus] Tiphlinus, 91
Typhlobagrus, 526, 564, **261**
Typhlobranchus, 60, **195**
Typhlogobius, 406, **545**
Typhlonarke, 534, **79**
Typhlonus, 394, **611**
Typhlopsaras, 423, **633**
Typhlotes, 85, **219**
Typhlus, 357, **356**
Typodus, 283, **29**, **118**, *118*
Tyttocharax, 548, **226**

Uaru, 207, **534**
Ulaema, 468, **458**
Ulaula, **441**, *441*
Ulca, 472, **519**
Ulcigobius, **545**, *545*

Ulcina, 215, **523**, *523*
Ulocentra, 395, **411**
Ulua, 528, **401**
Ulvarius, 475, **597**
Ulvicola, 477, **595**
Umbla, 29, 30, 43, 58, 261, **163**, **375**
Umbra, 20, 42, **289**
Umbrina, 13, 73, 104, **461**
Unagius, 567, **221**
Undina, 180, **97**
Unibranchapertura, 68, **195**
Unicornis, 30, 31, **620**
Unipertura, 268, **195**
Upeneichthys, 263, **460**
Upeneoides, 240, **460**
Upeneus, 127, **460**
Upsilonophorus, 305, **582**
Uraeus, 140, **124**
Uraleptus, 279, 574, **310**
Uranichthys, 347, **219**
Uranidea, 210, **511**
Uranoblepus, 302, **496**
Uranoplosus, 401, **118**
Uranoscopus, 12, 19, **511**, **582**
Uranostoma, 336, **624**
Uraptera, 193, **77**
Uraspis, 264, **401**, *401*
Urenchelys, 491, **215**
Urichthys, 199, **537**
Urinophilus, 563, **264**
Uriphaëton, 198, **426**
Urobatis, 548, **80**
Urocampus, 357, **356**
Urocentrus, 352, **595**
Urocles, 567, **127**
Urocomus, 312, **119**
Uroconger, 272, **211**
Urogymnus, 190, **80**
Urolepis, 274, **108**
Urolophus, 193, **80**
Uronectes, 318, **608**
Uronemus, 217, **104**
Urophycis, 325, **310**
Uropsetta, 317, **327**
Uropterina, 338, **147**
Uropterygius, 68, 184, **223**
Uropteryx, 457, **109**
Urosphen, 181, 218, **360**
Urosthenes, 233, **108**
Urotrygon, 324, **80**
Uroxis, 83, **80**
Usinosita, 490, **335**
Usinostia, *335*
Ussuria, 510, **239**

Vailima, 519, **545**

Vaillantella, 512, **239**
Vaillantia, 395, **411**
Vaillantoönia, 455, **90**
Vaimosa, 519, **545**
Valenciennea, 348, **544**
Valenciennellus, 468, **183**
Valenciennesia, 348, 373, **544**
Vandellia, 73, 407, **264**, **380**
Varicorhinus, 186, **237**
Variola, 34, 198, **426**
Vastres, 228, **157**
Vaticinodus, 405, 425, **44**
Vegetichthys, 565, **441**
Velasia, 249, **7**
Velifer, 247, **317**
Velifracta, 524, **327**
Vellitor, 509, **511**
Venefica, 458, **212**
Ventrifossa, 570, **307**
Venustodus, 380, **44**
Veraequa, 509, **328**
Verasper, 481, **328**
Verater, 567, **310**
Verecundum, 452, **327**
Verilus, 298, **442**
Verma, 473, **216**
Verreo, 499, **536**
Verriculus, 504, **536**
Verrugato, 436, *436*
Vesicatrus, 538, **226**
Vespicula, **493**, *493*
Vesposus, **337**, *337*
Veternio, 511, **211**
Vexillicaranx, 512, **401**
Vexillifer, 357, **614**
Vidalia, 506, **130**
Vigil, 567, **411**
Villarius, 476, **247**
Vimba, 368, **237**
Vincentia, 363, **414**
Vinciguerria, 468, **183**
Vinctifer, 570, **120**
Vinculum, 554, **488**
Vipera, 32, **180**
Vireosa, 494, **544**
Vitraria, 504, **545**
Vitreola, 519, **545**
Vogmarus, 244, **320**
Vomer, 48, 105, **401**
Vomeropsis, 259, **401**
Vulpecula, 27, 548, **61**
Vulsiculus, 475, **527**
Vulsus, 307, **561–62**

Waitea, 519, **545**
Waiteina, 523, **226**
Wallago, 247, 312, **246**
Wardichthys, 382, **109**

B. NAMES OF FAMILIES AND HIGHER GROUPS

The page numbers given herein are in brackets to indicate that they are the *adjunct* page numbers appearing inside the original page numbers of *Classification*.

Abyssocottidae, [714]
Acanthoclinidae, [694]
Acanthoëssidae, [591]
Acanthodei, [591]
Acanthodidae, [591]
Acanthopterygii, [671]
Acanthuridae, [707]
Acanthuriformes, [707]
Aceratiidae, [743]
Achiridae, [669]
Acipenseridae, [613]
Acropomatidae, [689]
Acrotidae, [683]
Actinistia, [609]
Actinopteri, [611]
Adiposiidae, [645]
Adrianichthyidae, [660]
Aëtobatidae, [605]
Agonidae, [715]
Agriopidae, [709]
Alabidae, [629]
Alabiformes, [629]
Alepisauridae, [656]
Alepocephalidae, [622]
Albulidae, [618]
Albuloidei, [618]
Allotriognathi, [665]
Alopiidae, [599]
Ambassidae, [689]
Amblycipitidae, [648]
Amblyopsidae, [660]
Ameiuridae, [647]
Amiatidae, [616]
Amiidae, [616]
Amiidae, [688]
Amioidei, [615]
Ammodytidae, [730]
Ammodytiformes, [730]
Amphacanthi, [708]
Amphiliidae, [649]
Amphioxi, [581]
Amphioxidae, [581]
Amphioxididae, [581]

Amphipnoidae, [629]
Amphisilidae, [676]
Amphistiidae, [704]
Anabantidae, [676]
Anablepidae, [660]
Analithidae, [603]
Anarhichadidae, [735]
Anarrhichthyidae, [735]
Anaspida, [587]
Ancylostylidae, [619]
Anguillavidae, [630]
Anguillidae, [630]
Anguilloidei, [630]
Anodontidae, [637]
Anomalopidae, [672]
Anoplopomidae, [710]
Anostomidae, [637]
Anotopteridae, [627]
Antacea, [590]
Antennariidae, [742]
Antiarcha, [586]
Antigoniidae, [704]
Aphareidae, [696]
Aphredoderidae, [665]
Aploactidae, [709]
Aplochitonidae, [625]
Aplodactylidae, [704]
Apodes, [629]
Apogonidae, [688]
Apolectidae, [684]
Apteronotidae, [638]
Arapaimidae, [623]
Archaeomaenidae, [616]
Archencheli, [629]
Argentinidae, [625]
Argidae, [652]
Ariidae, [645]
Arripidae[13], [695]
Arthrodira, [588]
Arthropteridae, [603]
Arthrothoraci, [588]
Ascelichthyidae, [714]
Asineopidae, [665]

Aspredinidae, [651]
Asterodermidae, [603]
Asterodontidae, [612]
Asterosteidae, [588]
Aspidoganoidei, [584]
Aspidophoroididae, [715]
Aspidorhynchidae, [615]
Aspidorhini, [585]
Astraspidae, [585]
Astrolepidae, [586]
Astronesthidae, [627]
Atelaxia, [666]
Ateleaspidae, [586]
Ateleopodidae, [656]
Atherinidae, [677]
Atopoclinidae, [733]
Aulopidae, [653]
Aulorhynchidae, [674]
Aulostomi, [675]
Aulostomidae, [675]
Azygostei, [586]

Bagridae, [648]
Balistidae, [739]
Banjosidae, [697]
Bathyclupeidae, [686]
Bathylagidae, [625]
Bathymasteriformes, [730]
Bathydraconidae, [728]
Bathymasteridae, [730]
Bathypteroidae, [654]
Bathythrissidae, [619]
Batoidei, [602]
Batrachoididae, [738]
Bedotiidae, [677]
Belonidae, [660]
Belonorhynchidae, [613]
Bembropidae, [729]
Benthosauridae, [654]
Berycidae, [672]
Berycoidei, [672]
Berycopsidae, [672]
Birkeniidae, [587]

[13] In words of this type classical authority prefers the shorter form to the less euphonious *Arripididæ*, and the like. (Dr. H. R. Fairclough.)

Orectolobidae, [598]
Orestiidae, [658]
Orodontidae, [595]
Orthacanthidae, [591]
Osmeridae, [625]
Osphronemidae, [676]
Ostariophysi, [634]
Osteoglossidae, [623]
Osteoglossoidea, [623]
Osteolepidae, [609]
Osteostraci, [586]
Ostraciidae, [740]
Ostracodermi, [740]
Ostracodermi, [584]
Ostracophori, [584]
Otolithidae, [702]
Oxudercidae, [730]
Oxylebiidae, [711]
Oxynotidae, [602]
Owstoniidae, [731]

Pachycormidae, [615]
Pachyrhizodontidae, [618]
Palaeoniscidae, [611]
Palaeorhynchidae, [680]
Palaeospondylidae, [587]
Pampidae, [682]
Pangasiidae, [649]
Pantodontidae, [623]
Paralepidae, [654]
Paralichthodidae, [669]
Paralichthyidae, [668]
Parapercidae, [729]
Pareioplitae, [708]
Pataecidae, [710]
Pediculati, [742]
Pegasidae, [675]
Pelecopteridae, [615]
Pempheridae, [694]
Pentanchidae, [598]
Percesoces, [677]
Percichthyidae, [688]
Percidae, [686]
Perciliidae, [688]
Percomorphi, [677]
Percophididae, [729]
Percopsidae, [665]
Periophthalmidae, [727]
Peripristidae, [594]
Peristediidae, [716]
Peronedyidae, [732]
Petalodontidae, [593]
Petromyzonidae, [583]
Phallostethidae, [660]
Phaneropleuridae, [610]
Pharyngognathi, [721]
Pharyngopilidae, [721]
Pholidae, [734]

Pholidophoridae, [616]
Pholidophoroidei, [616]
Phractolaemidae, [623]
Phthinobranchii, [673]
Phyllodontidae, [721]
Phyllolepidae, [589]
Physostomi, [616]
Pimelodidae, [650]
Pinguipedidae, [729]
Pisces, [608]
Placodermi, [586]
Placoganoidei, [586]
Plagiostomata, [590]
Plagyodontidae, [656]
Platacidae, [705]
Platycephalidae, [711]
Platycephaliformes, [711]
Platypteridae, [724]
Platyrhinidae, [603]
Platysomidae, [612]
Plecoglossidae, [624]
Plectognathi, [739]
Plectospondyli, [634]
Plesiopidae, [694]
Plethodontidae, [619]
Pleuracanthidae, [591]
Pleuronectidae, [669]
Pleuropterygii, [591]
Pliotremidae, [601]
Plotosidae, [647]
Podopterygia, [611]
Poecilidae, [659]
Polyacanthidae, [676]
Polycentridae, [702]
Polymixiidae, [672]
Polynemidae, [678]
Polyodontidae, [613]
Polypteridae, [610]
Polyspondyli, [592]
Pomacentridae, [718]
Pomadasidae, [696]
Pomatomidae, [686]
Potamotrygonidae, [604]
Priacanthidae, [694]
Priscacaridae, [718]
Pristiophoridae, [601]
Pristidae, [602]
Pristipomidae, [696]
Pristodontidae, [594]
Pronotacanthidae, [628]
Prosarthri, [595]
Protocephali, [584]
Protopteridae, [611]
Protoselachii, [597]
Protosphyraenidae, [615]
Protospondyli, [615]
Protosyngnathidae, [674]
Psammichthyidae, [727]

Psammodontidae, [593]
Psammosteidae, [585]
Psenidae, [683]
Psettodidae, [667]
Pseudochromidae, [694]
Pseudoplesiopidae, [694]
Pseudotriakidae, [598]
Psilocephalidae, [740]
Psychrolutidae, [714]
Pteraclidae, [682]
Pteraspidae, [585]
Pterichthyidae, [586]
Pternodonta, [597]
Pteropsaridae, [729]
Pterothrissidae, [619]
Ptilichthyidae, [735]
Ptychodontidae, [605]
Ptyctodontidae, [605]
Pycnodonti, [614]
Pycnodontidae, [614]
Pygaeidae, [708]
Pygidiidae, [651]

Rachycentridae, [686]
Rajidae, [603]
Raphiosauridae, [618]
Regalecidae, [666]
Retropinnidae, [625]
Rhamphichthyidae, [638]
Rhamphobatidae, [602]
Rhamphocottidae, [714]
Rhamphosidae, [675]
Rhegmatidae, [694]
Rhegnopteri, [678]
Rhinellidae, [628]
Rhineodontidae, [599]
Rhinidae, [602]
Rhinobatidae, [603]
Rhinochimeridae, [607]
Rhinopteridae, [605]
Rhipidistia, [608]
Rhizodontidae, [608]
Rhodichthyidae, [737]
Rhombodipteridae, [609]
Rhombosoleidae, [669]
Rhyacichthyidae, [724]
Rhynchobatidae, [602]
Rogeniidae, [660]
Rondeletiidae, [656]
Runulidae, [733]

Saccopharyngidae, [634]
Salangidae, [625]
Salmonidae, [624]
Salmonoidei, [624]
Salmopercae, [664]
Samaridae, [669]
Sarcura, [602]